Make Reform a Reality— with
HRW MATHEMATICS!

INTEGRATING

- **MATHEMATICS**
- **TECHNOLOGY**
- **EXPLORATIONS**
- **APPLICATIONS**
- **ASSESSMENT**

EXPLORE • COMMUNICATE • APPLY

The Sign of Progress– SATISFIED TEACHERS!

> **I liked it! This material makes the best possible use of technology. The graphics calculator saves students from spending time on tedious calculations and allows them a way to check their work.**
>
> Linda Craine
> Abilene, TX

> **I am impressed by the connection between mathematics and the real world. The examples are interesting and informative, and they motivate students to apply concepts.**
>
> Audrey Beres
> Bridgeport, CT

> **Your concrete use of algebra tiles is sound and well presented.**
>
> Robert E. Bauman
> Appleton, WI

> **The text captures the nature of modern-day mathematics. The calculator approach is wonderful and well-supported with written exercises.**
>
> Kenneth Dupuis
> Sterling Heights, MI

> **I really like the way you integrate material, especially cooperative-learning activities!**
>
> Dianne Hershey
> Jonesburg, GA

> **The real-life examples are meaningful to the students—so much, in fact, that many have even taken it upon themselves to discuss their work with their parents.**
>
> Melanie Gasperec
> Olympia Fields, IL

Making Progress– The Authors of
HRW MATHEMATICS!

WADE ELLIS, JR.

"Integration can cultivate an appreciation for the relevance of mathematics—provided you meaningfully unify material with an intuitive, common-sense approach. Otherwise, diversity fosters confusion and becomes a liability."

Professor Ellis has co-authored numerous books and articles on how to integrate technology realistically and meaningfully into the mathematics curriculum. He was a key contributor to the landmark study, ***Everybody Counts: A Report to the Nation on the Future of Mathematics Education***.

KATHLEEN A. HOLLOWELL

"Mathematics classrooms should become laboratories of learning where excited students collect data, look for patterns, make and test conjectures, and explain their reasoning."

Dr. Hollowell's keen understanding of what takes place in the mathematics classroom recently helped her win a major NSF research grant to enhance mathematics teaching methods. She is particularly well-versed in the special challenge of motivating students and making the classroom a more dynamic place to learn.

JAMES E. SCHULTZ

"Technology has the capability of opening new worlds of mathematics to more students than ever."

Dr. Schultz is a co-author of the ***NCTM Curriculum and Evaluation Standards for Mathematics*** and ***A Core Curriculum: Making Mathematics Count for Everyone.*** He is especially well regarded for his inventive and skillful integration of mathematics and technology. Jim's dynamic vision of classroom reform recently earned him the prestigious Morton Chair at Ohio University.

Reform You Can Relate To—

"The sign of progress? Get real!"

"When I teach advanced algebra, my students rarely have problems understanding basic concepts and skills, but sometimes I wonder if I'm taking them as far as they can go. Fortunately, my students are making more progress than ever with the real-world approach they get in *HRW Advanced Algebra*.

"The first lesson opens with an introduction to what I think of as a mandatory tool for making maximum progress— the graphics calculator. And it's immediately put to work with some great skill-building applications, including processing and representing collected data."

HRW ADVANCED ALGEBRA!

"Every text should function this well!"

"It just makes so much sense to cover this ground first. I've seen too many students become bored early on because a book asks them to take on equations without providing anything with which to equate or relate. But HRW always establishes a clear relationship between relevant data and equations that students encounter every day."

"The same holds true for linear functions (Chapters 2-4) and nonlinear functions (Chapters 5-9), which are strongly emphasized in **HRW Advanced Algebra**. My students first discover the properties of functions and then use them as models to solve real-world problems."

"Let's be discrete— this is advanced stuff!"

"For me, being an advanced algebra teacher means a lot more than presenting challenging material—it means teaching contemporary concepts and skills that can advance the success of my students and our society as a whole. I think it's great that HRW devotes three chapters (11-13) to discrete math topics, and my students sincerely enjoy seeing how information is communicated in today's society. The applications are terrific, and they run the gamut from checking e-mail messages to undertaking quality control for a business."

"It ought to be mandatory!"

"The graphics calculator is one of those technological advances that you immediately warm-up to. It's intuitive and engaging, and it really helps my students understand concepts in much greater depth. So it doesn't surprise me that a graphics calculator is required in **HRW Advanced Algebra**."

"It really makes a difference. For example, my students have always found chapters on matrices to be tedious, boring, and insignificant. But the graphics calculator allows them to explore some really motivating applications, like figuring the number of days you'll spend in each country on a visit to South America."

KNOCK DOWN THE WALLS!

"No sooner do I open a new chapter in *HRW Advanced Algebra* than its applications take my students beyond the walls of the classroom— to the places where math really matters.

"Every chapter begins with a credible connection to the real world, including a lively *portfolio activity.* HRW really helps me to knock down the walls that some of my students build around themselves when it comes to learning math."

Look for Some

"Turn snoring to exploring with Exploratory Lessons!"

"If you really want to get your students off to a good start, give them a job to do. That's exactly what HRW does with its *Exploratory Lessons*, which present concepts using a discovery approach.

"For example, I have my students pretend they're going on a long hiking trip, and if they plan to travel the full route, they'll need to explore the relationship between distance, rate, and time."

"Set an example with Expository Lessons"

"Another lesson format, *Expository Lessons*, provides step-by-step examples in a relevant, applied, or hands-on context. Students begin with applications or activities and stay involved in what they're doing. Sometimes they go on 'mini explorations' when a discovery approach is particularly useful."

Sure Signs of Progress!

"But first ask Why?"

"**HRW Advanced Algebra** begins each lesson from a student's point of view, asking *"Why should I learn this?"* This spirit of inquiry is kept alive throughout HRW's concept development with a series of strategically positioned questions.

"My students are prompted to reflect upon and clarify their understandings as they move from explorations and examples to *Try This* practice, integrated *Critical Thinking,* and open-ended assessment questions." And how can you miss them—they're all highlighted in yellow?"

"It's great exercise!"

"**HRW Advanced Algebra** pumps up math comprehension in four sessions of the best, no-nonsense math workout I've ever come across. In *Communicate,* my students discuss, explain, or write about math to exercise one of the most powerful and underdeveloped problem-solving muscles—the logic of language.

"In the next session, they break out in a healthy sweat with some robust *Practice and Apply* problems and applications. And when it's time to wind down, they can *Look Back* and review what they've learned, and *Look Beyond* to prepare for future workouts."

"Expect to see some healthy changes."

"When it's time to take a deep breath and measure individual or group progress, my students stretch their minds with portfolio activities, long-term *Chapter Projects,* and *Eyewitness Math* activities. And they can further examine their progress with *Chapter Reviews* and *Chapter* and *Cumulative Assessments.*"

SO WHAT'S THE NEXT PROGRESSION?

Plug Into Math with "Explore!"

PUT SOME ELECTRICITY IN THE AIR!

"My classroom really comes to life whenever we plug into technology. With HRW, technology is more than a computational toy—it's one of the most serious instructional advances ever to hit mathematics education.

"HRW Advanced Algebra requires a graphics calculator and features some highly practical software. Both are seamlessly integrated into the text at the right place and the right time with all the help you need.

"Content is always covered in a reasonable and balanced progression, so my students feel as comfortable as I do. The whole approach really motivates my students to *explore*, *communicate*, and *apply* technology. As a result, they've come away with a much deeper understanding of challenging concepts."

Reduced from actual size

▶ **APPLICATION**

Graphics Calculator

You can use the functions for braking distance and reaction distance found in a state driving manual to graph the stopping distance function.

First enter the function $R(x)$ for reaction distance, $Y_1 = \frac{11x}{10}$, and the function $B(x)$ for braking distance, $Y_2 = \frac{x^2}{19}$. For the third function enter the sum of Y_1 and Y_2, $Y_3 = Y_1 + Y_2$.

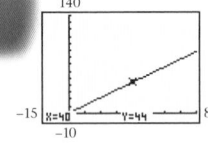
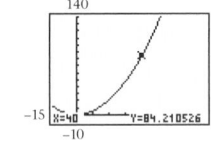
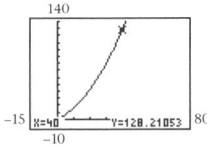

$Y_1 = R(x) = \frac{11x}{10}$ $Y_2 = B(x) = \frac{x^2}{19}$ $Y_3 = S(x) = R(x) + B(x) = \frac{11x}{10} + \frac{x^2}{19}$

You can use the trace feature to find the reaction distance, braking distance, and total stopping distance for any rate of speed.

For example, by tracing Y_3, you can find that when a car is traveling 40 miles per hour, the total stopping distance is approximately 128.2 feet. ❖

CRITICAL *Thinking*

How does the real-world domain of Y_3 compare with the real-world domains of Y_1 and Y_2?

LESSON 2.6   Operations With Functions   **99**

"I had fun coming up with a bunch of scenarios for hitting the brakes in an auto race—something I never would have had time to do without the graphics calculator."

"HRW's use of technology allows me to go far beyond the surface of a math problem. Right now my class is exploring operations with functions by using the trace features on a graphics calculator to complete a table for measuring stopping distance."

TECHNOLOGY!

"Communicate!"

Reduced from actual size

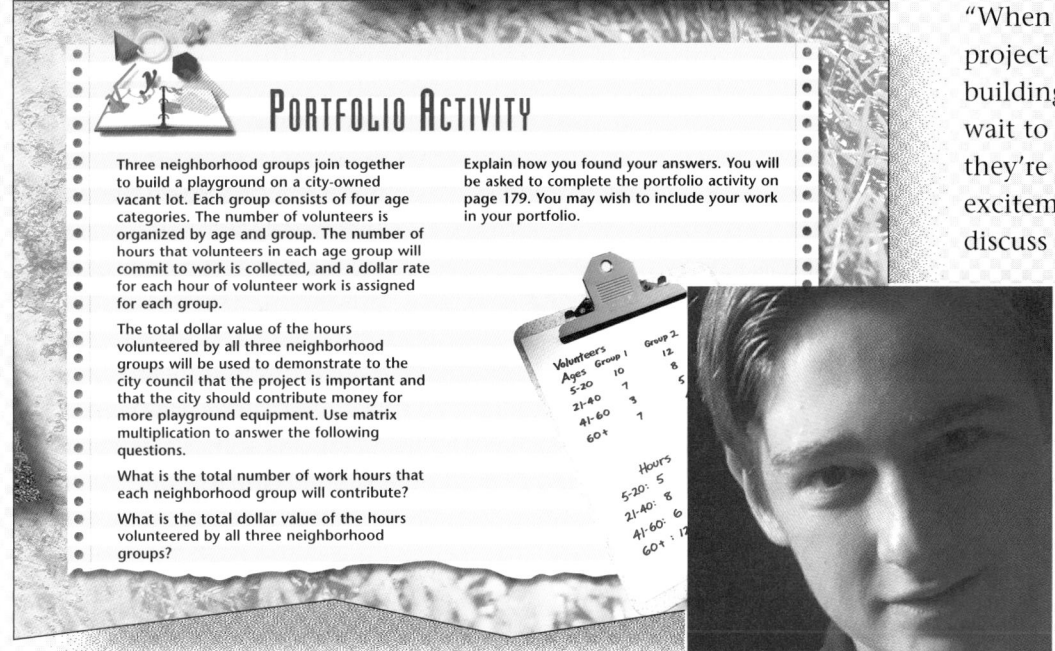

PORTFOLIO ACTIVITY

Three neighborhood groups join together to build a playground on a city-owned vacant lot. Each group consists of four age categories. The number of volunteers is organized by age and group. The number of hours that volunteers in each age group will commit to work is collected, and a dollar rate for each hour of volunteer work is assigned for each group.

The total dollar value of the hours volunteered by all three neighborhood groups will be used to demonstrate to the city council that the project is important and that the city should contribute money for more playground equipment. Use matrix multiplication to answer the following questions.

What is the total number of work hours that each neighborhood group will contribute?

What is the total dollar value of the hours volunteered by all three neighborhood groups?

Explain how you found your answers. You will be asked to complete the portfolio activity on page 179. You may wish to include your work in your portfolio.

"When my students complete a project such as making plans for building a playground, they can't wait to tell their friends about what they're doing. HRW harnesses that excitement when it asks students to discuss what they've discovered."

"Graphics calculators are cool—I can see the results of matrix operations so much easier and faster."

"Apply!"

"Technology is incredibly useful, as my class discovered when they planned a talent show. I had my students use graphics calculators to estimate the profit they could make by adjusting the cost of tickets."

"I told my group not to raise the ticket prices any higher, since the graphics calculator clearly showed how ticket price, attendance, and revenue were variables that determined profit."

Reduced from actual size

Exploration 2 *Revenue Function*

Graphics Calculator

Use your graphics calculator for Steps 4–6 of this exploration.

The student council plans to run a talent show to raise money. Last year, tickets sold for $5 each and 300 people attended. This year, the student council wants to make an even bigger profit than last year. They estimate that for each dollar increase in ticket price, attendance will drop by 20 people, and for each dollar decrease in ticket price, attendance will increase by 20 people.

Let x be the increase or decrease in the ticket price, in dollars.

1 Complete the table below.

2 Write the ticket price, $t(x)$, as a function of x. What type of function is t?

3 Write the attendance, $a(x)$, as a function of x. What type of function is a?

Look for exciting signs

"Don't miss the action!"

"It seems strange to me that math books tend to be so stationary because I've always thought of mathematics as an active and practical discipline— something you do.

"That's probably why I'm so comfortable with **HRW Advanced Algebra.** Students learn by doing with thought-provoking materials and innovative explorations. This approach really helps my students to get interested in, and take responsibility for, their own work."

"Explore!"

"When I want my students to make discoveries, HRW gives them something worth exploring, like using a graphics calculator to explore the transformation of trigonometric functions.

"In this case, my students explored how vibration relates to pitch with stringed instruments. They did a terrific job when I challenged them to raise and lower the notes at different intervals. Now they know what people mean when they say that music is really mathematical."

Reduced from actual size

Music Since sound is generated by vibrations and travels in waves, it can be modeled with circular functions. When an object vibrates twice as fast, it produces a sound with a pitch one octave higher than before. Amplitude is associated with loudness. Suppose a sound can be modeled with the function $g(\theta) = 15 \sin(200\theta)$.

24. Write a function for a sound with the same pitch as the sound modeled by g but 3 times as loud.

25. Write a function for a sound with the same volume as the sound modeled by g but one octave higher.

26. Write a function for a sound with the same volume as the sound modeled by g but two octaves higher.

"I had no trouble shifting the notes because everything seemed so logical. Handling functions, and for that matter the whole idea of playing an instrument, is a lot less mysterious to me now."

# of life in your class!

"Communicate!"

"HRW has some out-of-this-world challenges that always manage to sustain real-world interest. For example, my class really took off when I asked them to discuss the moon's orbit and describe steps for finding the equation of an ellipse."

Reduced from actual size

Practice & Apply

Write an equation for each of the following ellipses.

5. Vertices at $(4, 0)$ and $(-4, 0)$; co-vertices at $(0, 2)$ and $(0, -2)$

6. Vertices at $(7, 0)$ and $(-7, 0)$; co-vertices at $(0, 5)$ and $(0, -5)$

7. Vertices at $(6, 0)$ and $(-6, 0)$; co-vertices at $(0, 3)$ and $(0, -3)$

8. Vertices at $\left(0, \frac{11}{2}\right)$ and $\left(0, -\frac{11}{2}\right)$; co-vertices at $\left(\frac{5}{3}, 0\right)$ and $\left(-\frac{5}{3}, 0\right)$

Determine the vertices and co-vertices for each ellipse.

9. $x^2 - 6x + 2y^2 + 8y + 1 = 0$

10. $5x^2 + 20x + 4y^2 + 24y + 36 = 0$

11. $2x^2 + 6y^2 - 6x - 12y = \frac{3}{2}$

12. $16x^2 + 4y^2 - 96x + 8y + 84 = 0$

Astronomy The Moon orbits Earth in an elliptical path with Earth at one focus. The major axis of the orbit is 774,000 kilometers, and the minor axis is 773,000 kilometers.

13. Using $(0, 0)$ as the center of the ellipse, write an equation for the orbit of the Moon.

14. How far from Earth is the Moon when the Moon comes closest to Earth?

15. How far from Earth is the Moon when the Moon is farthest from Earth?

"The graphics calculator really helped my group to see the elliptical path of the moon as it orbits the earth."

"Apply!"

"You won't find busy work in ***HRW Advanced Algebra***. Students always understand the purpose and relevance of what they're doing, and at the same time, they get to use their imaginations. I think that's the secret to success in motivating my students.

"My class recently used the graphics calculator to find out if it really takes 4.34 years for light to travel from Alpha Centauri to Earth."

4.34 light-years, or 2.55×10^{13} miles?

Reduced from actual size

APPLICATION

The speed of light is 1.86×10^5 miles per second. How can you find out if it really takes 4.34 years for light to travel from Alpha Centauri to Earth?

Recall that distance equals rate multiplied by time, or $d = r \cdot t$. Thus, $t = \frac{d}{r}$.

So substitute 2.55×10^{13} miles for distance, d, and substitute 1.86×10^5 miles per second for rate, r, in $t = \frac{d}{r}$ to find the time, t, in seconds.

$$t = \frac{d}{r} = \frac{2.55 \times 10^{13}}{1.86 \times 10^5} = \frac{2.55}{1.86} \times \frac{10^{13}}{10^5}$$
$$= 1.37 \times 10^{(13-5)}, \text{ or } 1.37 \times 10^8 \text{ seconds}$$

Convert 1.37×10^8 seconds to years.
$$1.37 \times 10^8 \text{ sec} \times \frac{1 \text{ hr}}{3600 \text{ sec}} \times \frac{1 \text{ day}}{24 \text{ hr}} \times \frac{1 \text{ yr}}{365 \text{ days}} = 4.34 \text{ years}$$

Thus, it takes 4.34 years for light to travel 2.55×10^{13} miles, which is the distance from Alpha Centauri to Earth. ❖

Does it really take 4.34 years for light to travel from Alpha Centauri to Earth?

A summary of the rules of exponents is given below.

RULES OF EXPONENTS

Let a represent any real number, where $a \neq 0$, and let m and n be integers, where $n \neq 0$.

1. $a^0 = 1$ **2.** $a^m \cdot a^n = a^{(m+n)}$ **3.** $a^{-n} = \frac{1}{a^n}$

4. $\frac{a^m}{a^n} = a^{(m-n)}$ **5.** $(a^m)^n = a^{m \cdot n}$ **6.** $a^{\frac{m}{n}}$

"It's incredible to think how big the universe is. I mean, when you consider how fast light moves, it's amazing to calculate that it would take more than four years for light to reach anywhere!"

"Students climb out of the zip lock™ bag!"

"Sometimes it seems as if algebra lives in two worlds. In the everyday world, it touches everything from personal finances to fun and games. But in textbooks, algebraic concepts are often isolated from one another, not to mention from the real world itself."

*"**HRW Advanced Algebra** helps my students bring those worlds together with seamlessly connected concepts, activities, applications, technology, disciplines and cultures. My students make smooth, intuitive transitions as they explore, communicate, and apply mathematics."*

"Explore!"

"I like to take my students to the places where math happens. With HRW, that means looking at urban population growth patterns and projecting census figures in the year 2000."

"This challenge certainly got my students to see the relevance of exponential functions. In fact, we always get so much more out of class when concepts are connected to real-world challenges."

"The numbers I came up with really showed me how important exponential functions are in preparing for the future."

Reduced from actual size

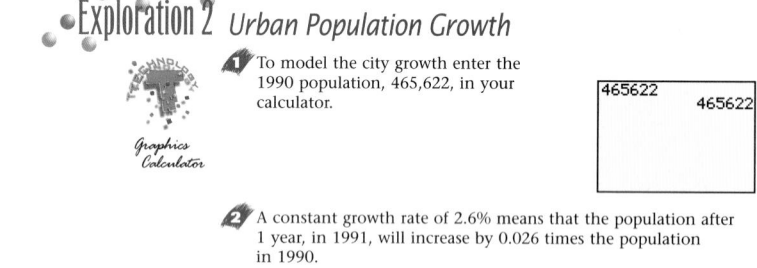

Exploration 2 Urban Population Growth

Graphics Calculator

1 To model the city growth enter the 1990 population, 465,622, in your calculator.

```
465622
          465622
```

2 A constant growth rate of 2.6% means that the population after 1 year, in 1991, will increase by 0.026 times the population in 1990.

Let *x* represent the 1990 population. The 1991 population will be

$$x + 0.026x = x(1 + 0.026) = 1.026x.$$

Multiply the 1990 population by 1.026.

`[ × ] [ 1 ] [ . ] [ 0 ] [ 2 ] [ 6 ] [ ENTER ]`

This gives the projected 1991 population.

```
465622
          465622
Ans*1.026
          477728.172
```

3 Continue by pressing `[ENTER]` to find the projected population for each successive year until the year 2000. Record each result in a table.

Round your population values to the nearest whole number, since it is not realistic to count a fraction of a person.

1990	1991	1992	199...

SEAMLESS INTEGRATION!

"Communicate!"

Reduced from actual size

EXAMPLE 1

Investment Suppose the parents in a family want to have $10,000 available for each of their children by the time each child reaches his or her 18th birthday. At each child's 3rd birthday, the parents deposit $3000 in a fund for college education. The fund earns 8.5% interest compounded annually. Will $10,000 be available for college expenses on the child's 18th birthday?

Solution▸

Make a table or use a spreadsheet to show the growth of the college fund. The interest rate is assumed to be fixed, so each year the multiplier is 1.085.

		C6	=3000*(1.085)^A6	
				Investment
	A	**B**	**C**	**D**
1	Years	Child's Age	Balance ($)	
2	*x*		*y*	
3	0	3	$3000.00	
4	1	4	$3255.00	
5	2	5	$3531.68	
6	3	6	$3831.87	
7	4	7	$4157.58	
8				
9	n	n + 3	3000(1.085)^n	

LESSON 7.2   The Exponential Function

"Sometimes the most important connection students make is with one another, but you've got to have an interesting and relevant context to make that a rewarding experience. HRW comes up with connections you can build on—like managing money and saving for college—something students really would and should talk about.

"My class uses spreadsheet software to chart compound interest for a college tuition fund. They compare the fund's exponential growth to escalating college expenses."

"The payoff on the fund was unbelievable! In fact, my business class discussed the same topic, and I was able to show them what a huge difference compounding interest makes."

"Apply!"

"Integration works best when you have applications that stand at the intersection of concepts, disciplines, technology, or cultures, but most importantly, student interest. HRW does exactly that when, for example, my students apply their knowledge of probability as they develop strategies for pitching a baseball."

"I was pretty confident about which section of the target was easiest to hit and where I should aim. And sure enough, I turned out to be a pretty good pitcher—at least on paper!"

Reduced from actual size

Sports A baseball pitcher is practicing a pitch by throwing at the yellow target shown. The black square has 4-inch sides. The white circle has a 7-inch radius. Assuming there is an equally likely chance of hitting any part of the yellow target (but not missing it), find the probability that a pitch will hit the following part of the target.

27. The black square

28. Anywhere in the circle

29. In the circle but not in the black square

30. Not in the black square

31. Nowhere in the circle

Buffon's Problem Get a toothpick and measure its length. Call its length *l*. Draw parallel lines at distance *d* apart, where $l < d$.

32. By randomly dropping the toothpick on the paper 20 times, find the probability that the toothpick will intersect one of the lines.

33. This problem was first considered and solved by the French naturalist Comte de Buffon. His solution is the value $\frac{2l}{\pi d}$. How does your approximation compare with this value?

34. If you made more trials, do you think your value would approach Buffon's value? Explain.

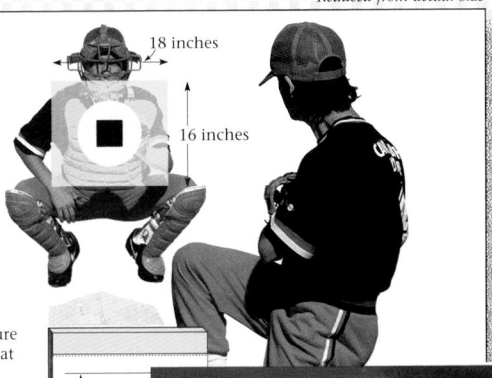

18 inches

16 inches

d

d

d

l

Integrated Instruction

Reduced from actual size

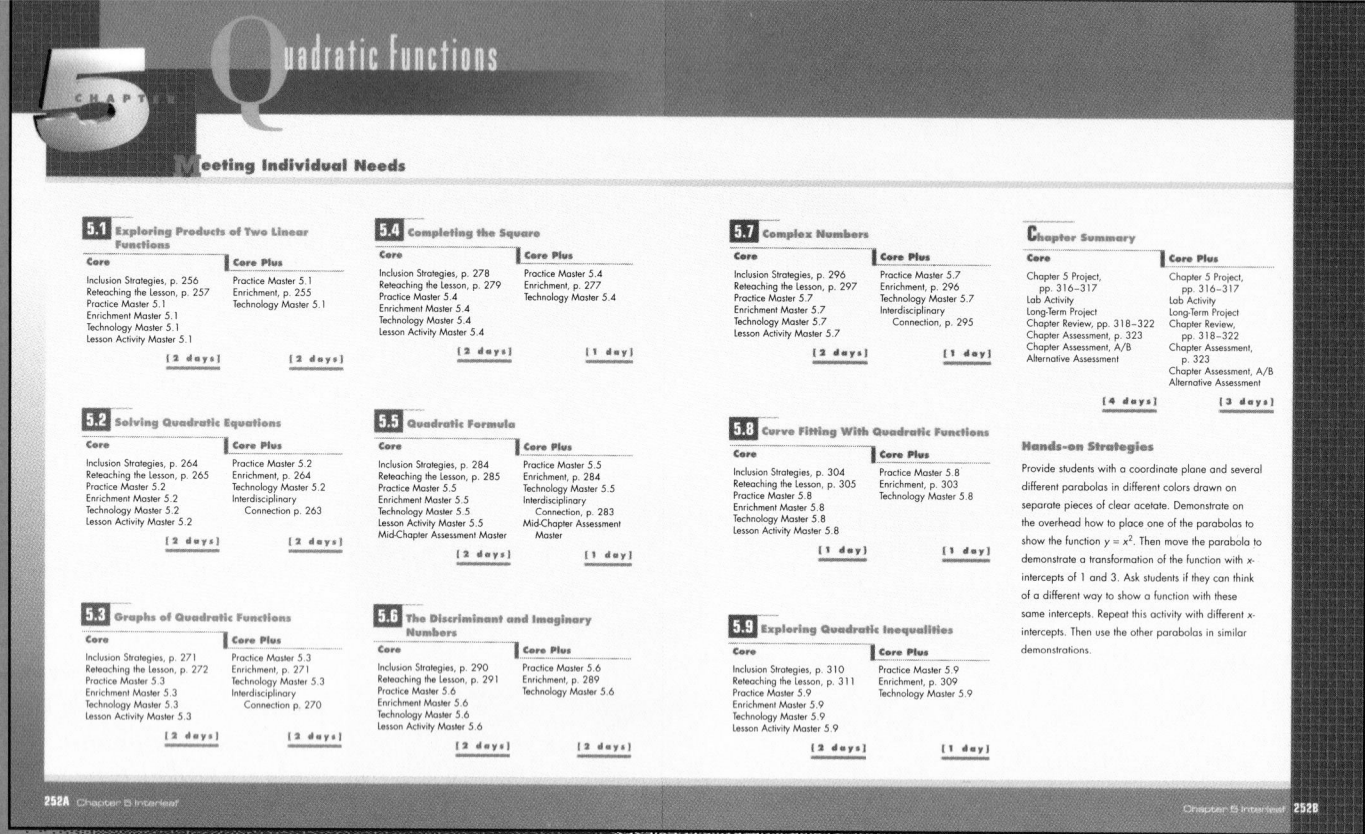

"Get full instructional support!"

"As a teacher, nothing is more important than the transfer of knowledge between me and my students. *HRW Advanced Algebra* gives teachers a full stream of instructional options and suggestions, so I can maintain a seamless relationship between me, the text, and my students."

"HRW really supports instruction today rather than someone's vision of math instruction in the year 2010. I get useful information, like the *Alternative Teaching Strategies* you'll find in every lesson. Plus, I really appreciate HRW's easy-to-follow layout and organization."

"Before..."

"Before you begin a chapter, HRW's opening *Side Columns* on the *Chapter Openers* help you prepare to introduce new material by providing *Background Information, Chapter Objectives*, a list of available *Resources*, and more."

"Then there's a series of *Interleaves*, which I find particularly useful. I read them before starting a chapter, then refer back to them whenever necessary. Interleaves include everything from a *Planning Guide* to *Reading, Visual*, and *Hands-on Strategies* for helping individual students."

"You also get a quick-look reference to the chapter's *Cooperative Learning Activities, Cultural Connections*, and *Portfolio Assessment*, as well as a full page dedicated to *Technology* instruction."

begins with You!

Reduced from actual size

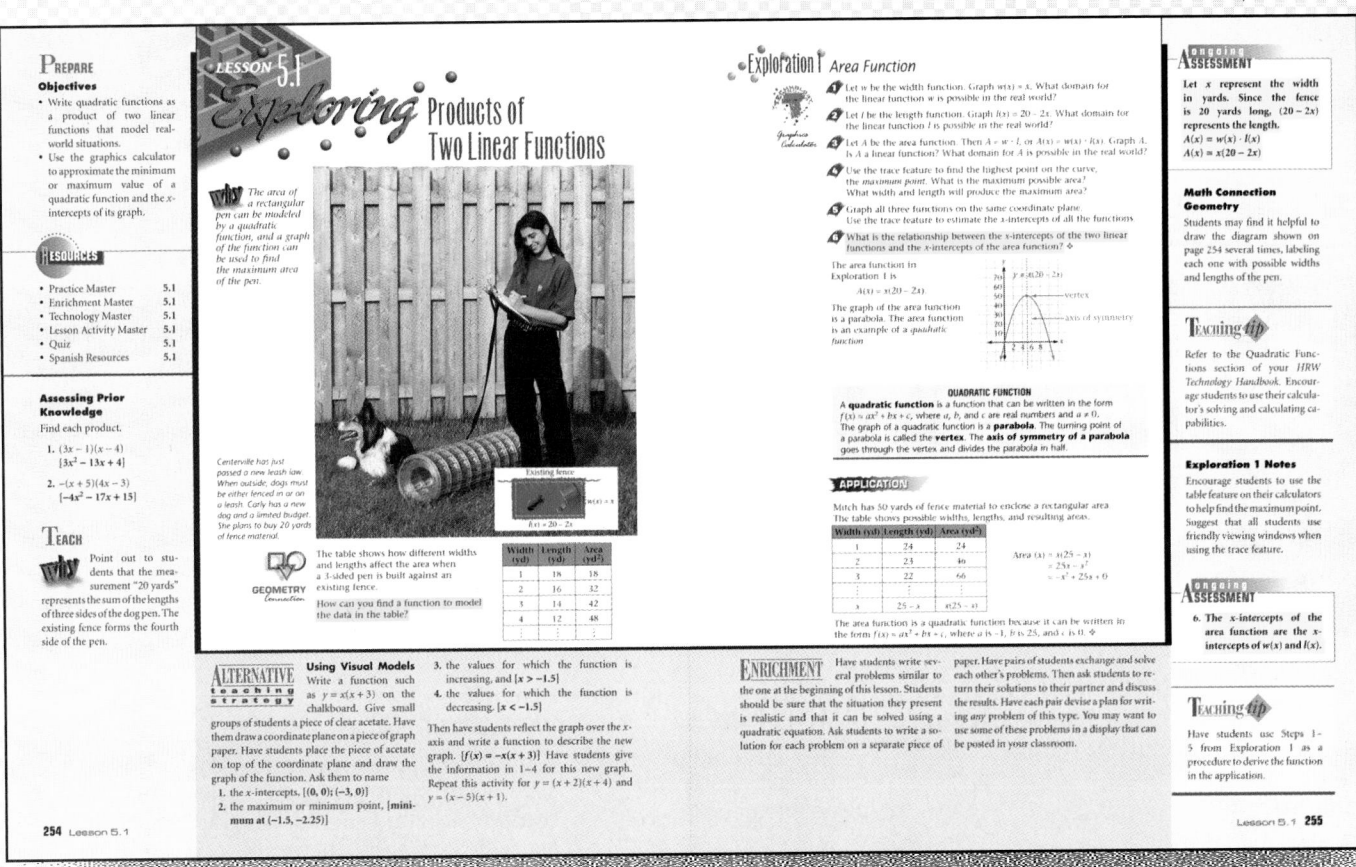

"During..."

"As you move through a lesson, you'll notice side column support is designed to lead you through content in a timely and logical progression.

"Key sections include *PREPARE*, with *Objectives*, *Resources*, and *Prior Knowledge Assessment*, and *TEACH*, which features some substantial teaching strategies, notes, and tips, from *Critical Thinking* to *Alternative Teaching Strategies* to *Interdisciplinary Connections*."

"And After Your Lessons!"

"*HRW Advanced Algebra* gives you plenty of ways to measure the progress of your students with *Ongoing* and *Alternative Assessment*, and *Practice* and *Technology Masters*.

"You'll also get lots of practical tips for the *Chapter Project*, which applies skills presented throughout the chapter. And finally, you'll find some great ideas in *Eyewitness Math*, a fun cooperative-learning activity that springs from today's headlines and stories."

Reform That's Based on What You Need!

HRW MATHEMATICS...
This *is* Progress

TECHNOLOGY HANDBOOK

Teacher's guide for using technology with ***HRW Mathematics.***

TEACHING RESOURCES

Support for each lesson in four convenient booklets.

TEACHING TRANSPARENCIES

Over 100 full-color visuals with suggested lesson plans for their use.

PRACTICE WORKBOOK

A full page of practice for each lesson helps students comprehend and review learned skills and concepts.

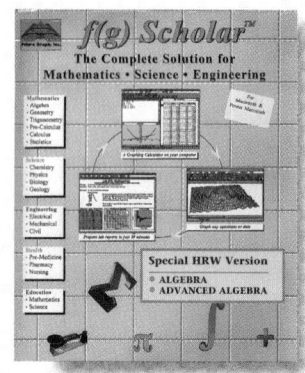

***f(g) Scholar*™**

A dynamic mathematics visualization tool, integrating calculator, spreadsheet, and graphics capabilities.

TEST GENERATOR

A variety of assessment types delivered on user-friendly software. (Two versions available– Macintosh® and IBM®PC and compatibles.)

TEST GENERATOR: ASSESSMENT ITEM LISTING

Printout of all items included on the Test Generator.

SOLUTION KEY

Worked-out solutions for all *Exercises & Problems.*

SPANISH RESOURCES

Spanish translation of objectives, main ideas of lessons, and terminology.

LAB ACTIVITIES AND LONG-TERM PROJECTS

Hands-on activities and projects to be used before, during, or after each chapter.

EXPLORE • COMMUNICATE • APPLY

HRW
ADVANCED
ALGEBRA

TEACHER'S EDITION

explore

communicate

APPLY

Integrating

Mathematics
Technology
Explorations
Applications
Assessment

$g(\theta)=15\sin(200\theta)$

$h(t)=-16t^2+27t+240$

HOLT, RINEHART AND WINSTON
Harcourt Brace & Company
Austin • New York • Orlando • Atlanta • San Francisco • Boston • Dallas • Toronto • London

A U T H O R S

Wade Ellis, Jr.

Professor Ellis has gained tremendous recognition for his reform-minded and visionary math publications. He has made invaluable contributions to teacher inservice training through a continual stream of hands-on workshops, practical tutorials, instructional videotapes, and a host of other insightful presentations, many focusing on how technology should be implemented in the classroom. He has been a member of the National Research Council's Mathematical Sciences Education Board, the MAA Committee on the Mathematical Education of teachers, and is a former Visiting Professor of Mathematics at West Point.

James E. Schultz

Dr. Schultz is one of the math education community's most renowned mathematics educators and authors. He is especially well regarded for his inventive and skillful integration of mathematics and technology. He helped establish standards for mathematics instruction as a co-author of the NCTM "Curriculum and Evaluation Standards for Mathematics" and "A Core Curriculum, Making Mathematics Count for Everyone." Following over 25 years of successful experience teaching at the high school and college levels, his dynamic vision recently earned him the prestigious Robert L. Morton Mathematics Education Professorship at Ohio University.

Kathleen A. Hollowell

Dr. Hollowell is widely respected for her keen understanding of what takes place in the mathematics classroom. Her impressive credentials feature extensive experience as a high school mathematics and computer science teacher, making her particularly well-versed in the special challenges associated with integrating math and technology. She currently serves as Associate Director of the Secondary Mathematics Inservice Program, Department of Mathematical Sciences, University of Delaware and is a past-president of the Delaware Council of Teachers of Mathematics.

Printed in the United States of America

1 2 3 4 5 6 7 063 00 99 98 97 96 95

ISBN: 0-03-097774-6

CONTRIBUTING AUTHORS

Martin Cohen   A professor of Mathematics Education at the University of Pittsburgh since 1976, Dr. Cohen is the author of many successful secondary mathematics textbooks and articles. He is very active in the NCTM and is a frequent guest speaker at local and national mathematics conferences.

Martin Engelbrecht   A mathematics teacher at Culver Academies, Culver, Indiana, Mr. Engelbrecht also teaches statistics at Purdue University, North Central. An innovative teacher and writer, he integrates applied mathematics with technology to make mathematics accessible to all students.

PROGRAM CONCEPTUALIZER

Larry Hatfield   Dr. Hatfield is Department Head and Professor of Mathematics Education at the University of Georgia. He is a recipient of the Josiah T. Meigs Award for Excellence in Teaching, his university's highest recognition for teaching. He has served at the National Science Foundation and is Director of the NSF-funded Project LITMUS.

●●

Editorial Director of Math
Richard Monnard

Executive Editor
Gary Standafer

Senior Editor
Ronald Huenerfauth

Project Editors
Kelli R. Flanagan
Nell Wackwitz Eichler

Design and Photo
Pun Nio
Diane Motz
Lori Male
Julie Ray
Rhonda Holcomb
Robin Bouvette
Candace Moore
Ophelia Wong
Sam Dudgeon
Victoria Smith
Katie Kwun

Editorial Staff
Steve Oelenberger
Michael Funderburk
Andrew Roberts
Joel Riemer
Janet Harrington
Editorial Permissions

Jane Gallion
Desktop Systems Operator

Jill Lawson
Department Secretary

Production and Manufacturing
Gene Rumann
Amber Martin
Pronk&Associates
Jenine Street
Shirley Cantrell

CONTENT CONSULTANT

Deborah Hadd   An award-winning mathematics teacher in Okaloosa County, Florida, Ms. Hadd has over 20 years experience at the middle school and high school levels. She is a teacher at Baker School in Baker, Florida, and is very busy giving workshops on such topics as interdisciplinary instruction, cooperative learning, and the use of manipulatives.

MULTICULTURAL CONSULTANT

Beatrice Lumpkin   A former high school teacher and associate professor of mathematics at Malcolm X College in Chicago, Professor Lumpkin is a consultant for many public schools for the enrichment of mathematics education through its multicultural connections. She served as a principal teacher-writer for the *Chicago Public Schools Algebra Framework* and has served as a contributing author to many other mathematics and science publications that include multicultural curriculum.

REVIEWERS

Sandra J. Beasley
Redlands High School
Redlands, California

Francis Beaudry
Windsor Locks High School
Windsor Locks, Connecticut

Audrey Beres
Bassick High School
Bridgeport, Connecticut

Dave Brittain
Rufus King High School
Milwaukee, Wisconsin

Kenneth Dupuis
Adlai Stevenson High School
Sterling Heights, Michigan

Carol A. Fiedler
Milford High School
Milford, Massachusetts

Karen M. Lesko
Pacific High School
San Bernardino, California

Betty Mayberry
Gallatin High School
Gallatin, Tennessee

Patricia McMann
Roosevelt High School
Wyandotte, Michigan

Barbara J. Schewe
Preble High School
Green Bay, Wisconsin

TABLE OF CONTENTS

CHAPTER **1** **From Data to Equations** 2

CHAPTER **2** **Operations With Numbers and Functions** 60

 **MATH** *Connections*

Geometry 33, 43, 50, 52, 58, 59, 72, 82, 87, 88
Statistics 24, 25, 28, 88, 111
Transformations 67, 103

APPLICATIONS

Science
Anatomy 28, 59
Astronomy 32, 33
Biology 96
Ecology 8, 10
Electronics 34
Meteorology 12, 20
Physics 21, 31, 33, 34, 69, 97, 101, 108
Physiology 88
Projectile Motion 81
Science 66

Social Studies
Current Events 66
Demographics 13, 96

Government 35
History 44
National Debt 22

Language Arts
Communicate 12, 20, 27, 33, 42, 51, 65, 72, 80, 87, 95, 100
Eyewitness Math 74

Business and Economics
Accounting 103
Business 10, 13, 81, 91, 95, 109, 111
Economics 14, 43, 87
Investment 73, 89
Marketing 27
Small Business 52, 59

Life Skills
Construction 21
Consumer Awareness 47, 97
Consumer Economics 42, 53, 82, 95
Depreciation 21
Family Savings 58
Time Management 84

Sports and Leisure
Entertainment 13
Hobby 29
Leisure 43
Recreation 90
Sports 28

MATH *Connections*
Coordinate Geometry
  117, 118, 140, 160,
  188, 209, 211, 214,
  215
Geometry 119, 141,
  187, 188, 202, 208
Maximum/Minimum
  223, 230, 234
Probability 197
Statistics 217
Transformations 119,
  140, 145, 216, 249

APPLICATIONS

Science
Animal Nutrition 178
Medicine 131
Motion 126
Physics 161, 188
Physiology 160
Projectile Motion 153
Temperature 124, 133

Social Studies
Community Planning 232
Geography 168, 197
Human Development 151
Human Resources 196
Political Science 146
Sociology 169

Language Arts
Communicate 117, 125, 132,
139, 144, 151, 168, 176, 185,
195, 201, 207, 214, 224, 231, 239
Eyewitness Math 218

Business and Economics
Banking 187
Business 160, 180, 226, 228,
235, 248
Cost Analysis 134
Economics 127
Fund Raising 248
Inventory Control 170
Investment 201, 208
Manufacturing 196, 226, 227,
231, 241
Payroll 166
Sales Tax 146
Small Business 169, 186, 240

Life Skills
Architecture 216
Consumer Economics 128, 145,
161
Farming 232
Landscaping 216
Nutrition 174, 183, 186, 223, 239

Postage 144
Postal Service 185
Telephone Rates 144, 145
Time Management 132, 224

Sports and Leisure
Band Expenses 178
Entertainment 207, 225
Sports 152, 159, 160,
171, 177, 249
Tourism 232
Travel 140

Other
Animation 215
Aviation 147, 152
Communications 233, 240
Data Processing 164
Magic Addition Squares 209

MATH *Connections*

Coordinate Geometry
  266, 268, 323
Geometry 254, 286,
  323, 330, 358, 364,
  371
Maximum/Minimum
  257, 259, 287, 311,
  323, 327, 330, 354,
  355, 359, 364, 371
Statistics 301
Transformations 260,
  275, 300, 314

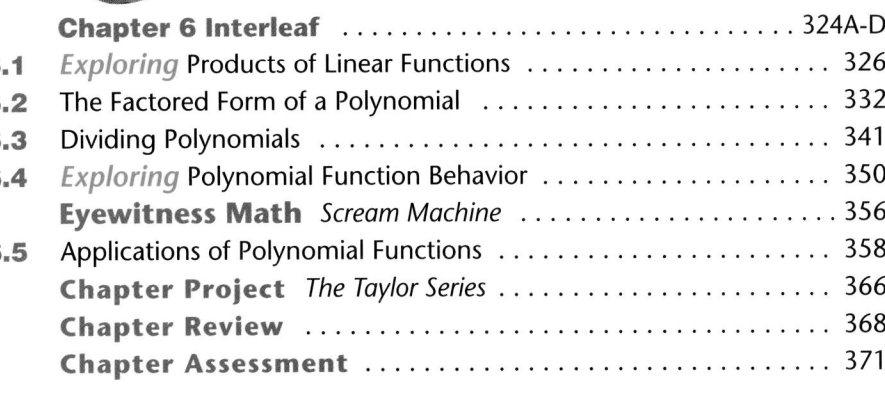

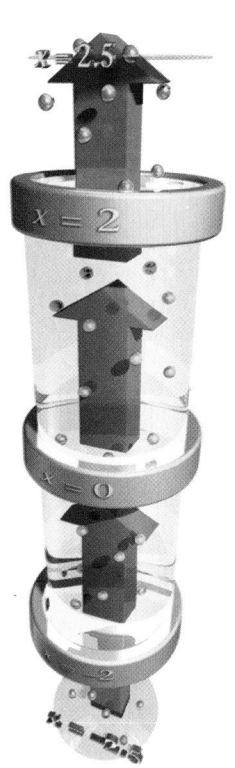

APPLICATIONS

Science
Astronomy 304, 306, 315
Ecology 267, 301
Electronics 281
Free Fall 314
Physics 272, 314, 323, 332, 340, 354
Radioactive Decay 365
Thermodynamics 339

Social Studies
Geography 348

Language Arts
Communicate 259, 267, 273, 280, 285, 292, 299, 306, 312, 329, 339, 347, 353, 363
Eyewitness Math 356

Business and Economics
Banking 359, 365
Business 286, 293, 300, 323, 350
Business Management 301, 307, 322
Economics 355, 364
Sales 313, 322

Small Business 259, 260, 261, 310, 312, 314

Life Skills
Investment 370, 371

Sports and Leisure
Recreation 274
Sports 275, 313

Other
Rescue 273
Transportation 314

MATH *Connections*

Coordinate Geometry
   438, 442, 445
Geometry   432, 434,
   436, 449, 458, 459,
   460, 467, 470, 481
Statistics   445, 467
Transformations   388,
   405, 412, 455, 458,
   463

APPLICATIONS

 **MATH** *Connections*

Geometry 499, 543, 576, 577
Maximum/Minimum 512, 515, 526
Statistics 496, 521
Transformations 503, 504, 508, 515, 537, 545, 549

APPLICATIONS

Science
Acoustics 559
Anatomy 559
Astronomy 556, 557, 558, 590, 591
Chemistry 500, 506, 508
Electronics 497, 520, 532
Free Fall 584
Geology 550
Marine Biology 567, 591
Marine Research 564
Medicine 501
Optics 500, 513
Physics 499, 500, 515, 532
Radio 550
Radio Transmission 547
Satellite 541

Language Arts
Communicate 498, 507, 514, 520, 526, 542, 549, 557, 566, 576, 583
Eyewitness Math 522

Business and Economics
Business 507, 516
Business Management 494
Civil Engineering 543
Economics 501

Life Skills
Architecture 558
Construction 499
Health Care 521
Metal Working 543
Time Management 527

Sports and Leisure
Cycling 499, 532
Music 533
Sports 524, 532, 533, 568, 584, 590
Travel 496, 527

Other
Navigation 567
Pyrotechnics 542
Transportation 549, 559

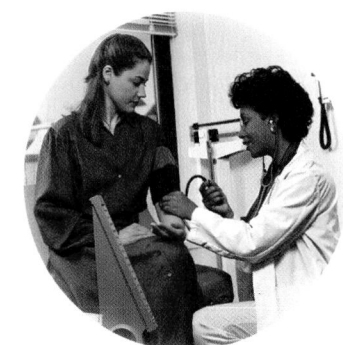

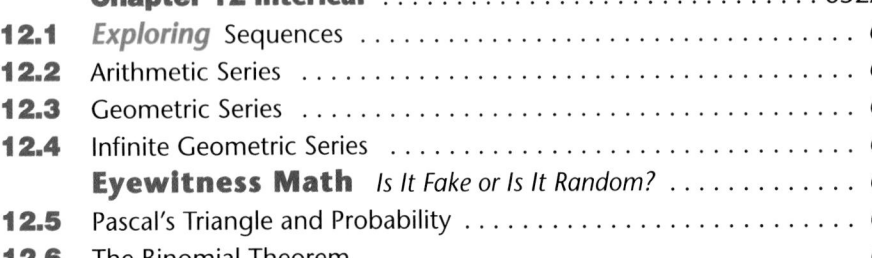

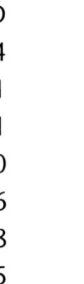

MATH *Connections*

Coordinate Geometry 644, 645, 679
Geometry 600, 626, 642, 646, 651, 657, 678, 679, 683, 685, 699
Maximum/Minimum 646
Statistics 669, 674, 677
Transformations 694

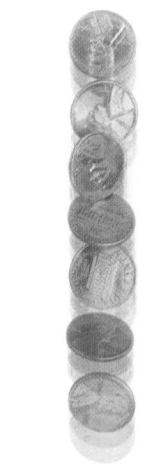

APPLICATIONS

MATH *Connections*

Coordinate Geometry
    780, 782, 785, 789
Geometry 770, 771,
    776, 809
Maximum/Minimum
    745, 752, 782, 795,
    801, 807
Probability 808, 809
Transformations 729,
    801, 809

A P P L I C A T I O N S

Science
Acoustics 801
Astronomy 800
Biology 751
Ecology 735, 736, 737
Electricity 801
Genetics 766
Geology 809
Medicine 704
Physics 752, 783
Science 750
Seismology 796

Social Studies
Demographics 727
Geography 801
Radio Broadcasting 714
Sociology 744

Language Arts
Communicate 718, 726, 734,
743, 750, 759, 774, 781, 787,
793, 799
Eyewitness Math 738

Business and Economics
Business 720, 800
Construction 806
Grocery Sales 760
Investment 728
Marketing 727
Mortgages 761
Quality Control 760
Road Construction 806

Life Skills
Automotive Maintenance 756
Automotive Mechanics 760

Employment 724, 767
Health 767
Health Care 728, 729
Home Improvement 784
Nutrition 744
Time Management 758

Sports and Leisure
Entertainment 721, 767
Recreation 806
Sports 743, 745, 766, 776,
795, 806, 807, 809

Other
Academics 759
Education 719, 720, 751, 767
Navigation 777, 782
Surveying 772, 775
Technology 727, 751, 752

TECHNOLOGY

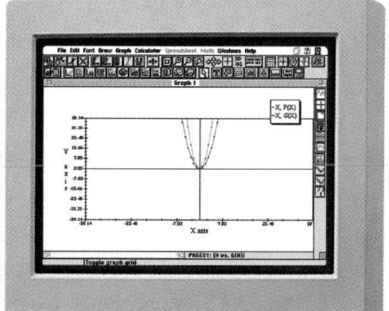

Interactive Learning

Students learn best if they are active participants in the learning process. Technology allows today's mathematics classroom to become a laboratory where students explore and experiment with mathematical concepts rather than just memorize isolated concepts. Students make generalizations and reach conclusions about mathematical concepts and relationships and apply them to real-life situations. *HRW Advanced Algebra* encourages instruction that utilizes technology with numerous explorations and examples. Where appropriate, students explore, work examples, solve exercises, make and test conjectures, and confirm mathematical ideas for themselves.

Role of the Teacher

Technology is changing the role of the teacher in the mathematics classroom. The National Council of Teachers of Mathematics' *Curriculum and Evaluation Standards for School Mathematics* describes "the emergence of a new classroom dynamic in which teachers and students become natural partners in developing mathematical ideas and solving mathematical problems."[1]

As a facilitator of learning, the teacher becomes a guide, leading students to their own mathematical discoveries and generalizations. Mathematics is no longer a static subject or a group of abstract symbols. Mathematics becomes a dynamic field of related concepts that can be explored and experimented with in the same way that science concepts can be. The tedious arithmetic calculations that discouraged, if not prohibited, students from making statistical analysis of data are no longer a barrier between students and their understanding and application of real-life problems.

Technology allows students to quickly create visual representations of functions. Students concentrate on the effect of parameter changes on a function, rather than on calculating and plotting points. Students can more effectively explore and learn about the more complex algebraic concepts that require comparisons.

Changes in the Classroom

Teachers are finding that technology encourages cooperative learning. Cooperative-learning groups allow students to compare results, brainstorm, and reach conclusions based on group results. As in real life where scientists or financial analysts often consult each other, students in the technology-oriented classroom learn to communicate and consult with each other. Students learn that such consultations are not "cheating," but rather are a method of sharing information that will be used to solve a problem.

Another change in the classroom will be the role of the change in the teacher from a dispenser of facts and formulas to a guide along the exploration trail. Teachers will become monitors who keep students headed in the right directions. As students gain confidence in their ability, they may ask more questions and even higher-level questions. Teachers no longer always need to have the "right" answer—teachers can suggest further exploration or research. In this way, teachers model real-life situations in which experts with the right answer are not always available or in which there is no right answer.

Real-World Applications

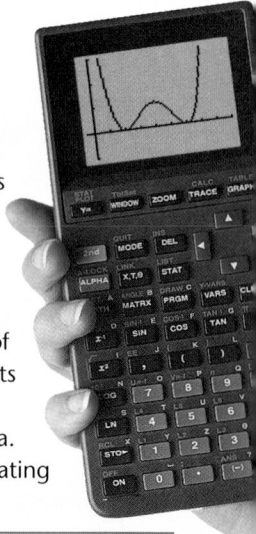

HRW Advanced Algebra integrates technology throughout the text. This use of technology allows students to explore real-world applications that would not be practical to explore without the use of such technology. Students are able to study realistic rather than contrived data. For instance, when calculating

[1]*National Council of Teachers of Mathematics. Curriculum and Evaluation Standards for School Mathematics. Reston, VA: The Council, 1989. (p. 128)*

by hand, most students study only simple interest. The use of technology allows students to explore compound interest, which is the way that interest is usually calculated in the financial world today.

Real data about a student's local town or neighborhood can be used to find the average price or rental cost of a home. Students can study data that are more closely related to their own lives, such as the cost of a 4-year college degree based on current costs and projections for the time students will actually be in college. The same can be done for projecting the cost of new cars when they graduate from high school.

C6		=3000*(1.085)^A6
	Investment	
	A	**B**	**C**
1	Years	Child's Age	Balance ($)
2	x		y
3	0	3	$3000.00
4	1	4	$3255.00
5	2	5	$3531.68
6	3	6	$3831.87
7	4	7	$4157.58
8			
9	n	n + 3	3000(1.085)^n

Mathematics is no longer just a study of symbols and equations. Mathematics becomes a useful tool to help students in their own lives.

Exploration Lessons

Exploratory lessons occur throughout *HRW Advanced Algebra* . Many of these exploratory lessons utilize the power of technology as a tool of exploration. For

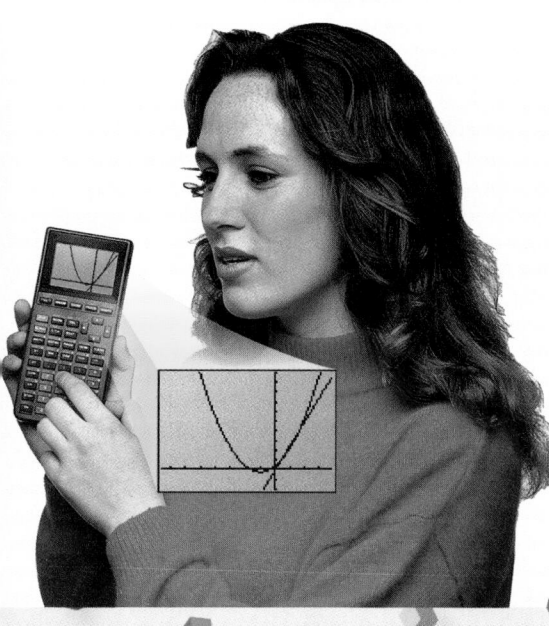

example, in Lesson 5.1 students use a graphics calculator to explore profit as a function of revenue. This is one of many graphics calculator explorations which simulate real-life situations.

The use of technology in simulations allows students to concentrate on the concept being studied rather than tedious calculations.

Expository Lessons

The power of technology is often used to help students analyze functions that would be difficult or impossible to study without the use of technology. In Lesson 6.4, students investigate the effect of multiple zeros on the graph of a quartic polynomial function.

In lesson 5.1, a graphics calculator helps students study the maximum or minimum values of a parabola. In lesson 5.9, calculators are used to find the intersection of a line and a parabola.

Graphics Calculators

Graphics calculators perform many specialized and complex mathematical actions, including graphing. These calculators use certain keys to access full-screen menus that show additional operations. The scope of capabilities is wide, including traditional arithmetic calculations, matrix operations, statistics mode, function graphing, parametric graphing, polar graphing, sequence graphing, and more. Graphics calculators are capable of running user-defined programs. When using the graphing capabilities of the calculators, the use of a range-setting window adds greatly to the usefullness of the technology.

The trace and zoom functions allow students to find the coordinates of points on the curve and enlarge sections of a graph.

Spreadsheets

A computer spreadsheet program is another valuable technological tool used in the mathematics classroom. By utilizing a spreadsheet, students can enter, store, and sort the type of real-world data that would have been impractical in yesterday's classroom. Spreadsheets facilitate complex mathematics operations and data analysis. Students learn important mathematical concepts by entering formulas to generate simple or complex lists of data to solve problems. The spreadsheet shown is used to determine the balance in a savings account after x years.

f(g) Scholar™

An effective, powerful yet easy to use tool in the classroom is *f(g) Scholar™*. This tool performs like a traditional graphics calculator, but it is used in conjunction with a personal computer. The advantages of *f(g) Scholar™* are a larger screen for use with groups when individual calculators are not available and the ability to print out parameters, equations, and results of computations or graphs. This powerful tool has outstanding spreadsheet capabilities. The screen display offers the option of displaying the data from the calculator, the graph, and the spreadsheet all at once.

Explorations

Facilitator of Learning

An increasing number of educators are realizing that more learning takes place when students construct knowledge for themselves. This approach necessitates changes in both the form of instructional materials and the vehicles used to deliver instruction.

Instead of presenting students with rules, theorems, principles, and worked out examples this approach calls for students to be given questions to investigate, problems to explore, and conjectures to verify or disprove. A teacher is no longer the source of all information. Instead, the teacher acts as a facilitator, presenting the questions to explore and pointing to areas that need further discussion and clarification.

An activity-oriented approach gives teachers several advantages. Perhaps the most important of these is the ability to provide for learning styles other than verbal and visual. Students who need hands-on tactile experiences—the kinesthetic learners—are no longer penalized by materials and instructional approaches that emphasize primarily listening and recording information.

Students who need to work in collaboration with others to understand algebra concepts—the interpersonal learners—now have the opportunities that occur during small-group instruction. In fact, all students benefit from the increased written and verbal communication that occurs when active learning is implemented in the classroom.

So, what specific form should constructivist materials take in algebra? Although many variations are possible, materials suitable for active learning usually involve identification of patterns, making and testing conjectures, and exploring alternate approaches to problems. Students should be encouraged to try any approach that occurs as they work on a problem. Concrete models, graphic representations, tabular methods such as those facilitated by spreadsheets—all of these have an appropriate place in today's algebra classroom.

In many classroom situations, a key teacher decision will be how much constructivism to use, and when to use it. Some topics and situations will continue to require direct instruction; others will open opportunities for both students and teachers to experience the richer knowledge formation that results from students' active and personal involvement in their own learning.

New Tools

Rapid and continuing changes in technology provide teachers and students in the algebra classroom with more choices for strategies and procedures than in past years. Graphics calculators, spreadsheets, and computer programs are excellent tools for use in student explorations. These tools allow students to experiment with and explore many concepts that would otherwise be too tedious and time consuming.

Another technology tool for the math classroom is the hand-held computer. Although more expensive that the typical graphics calculator, they are only a fraction of the cost of a desk-top computer. In addition to numerous graphics capabilities, this device contains geometry graphics software, a powerful computer algebra system and advanced programming software.

Communicating In *HRW Advanced Algebra*

The exploratory approach in *HRW Advanced Algebra* provides many opportunities for students to communicate. While in a traditional classroom setting or while in cooperative learning groups, students communicate orally as they complete Explorations, answer Critical Thinking questions, and explain results of simulations and projects. Students communicate in writing as they complete written assignments, keep a journal, and build a portfolio of their work. Students communicate graphically as they make drawings to illustrate a concept or as they display data in a graph.

Active Learning in *HRW Advanced Algebra*

In the past, many algebra textbooks included little, if any, material appropriate for active learning situations. Thus, teachers who realized a need for this type of approach had to search out or create special materials.

In marked contrast, *HRW Advanced Algebra* has been structured to include the frequent use of activity-style lessons and activities. In lessons with the word "Exploring," the entire instructional focus is centered around exploration and discovery. Most are appropriate for small-group work; many allow for use of alternate technologies; all give students a chance to discover new ideas on their own.

As students begin a set of exercises at the end of a lesson, they may notice that they do not start with a set of simple practice or drill problems. Instead, each exercise set starts with *Communicate* questions. Here, students can talk or write about key ideas as a check on what they have learned from the lesson. Students may be asked to compare and contrast two different concepts, explain how they would solve a particular problem, or discuss why a particular answer is not reasonable for a given problem.

Students for whom active learning is a new approach may at first be confused by the way in which material is presented. They may have come to expect that they are simply to repeat whatever the teacher or the textbook has explained. But, in *HRW Advanced Algebra*, the textbook doesn't "give it all away." Students must explore and think a bit to get at the central concepts. With a few experiences, students will come to enjoy doing algebra in their

own way, rather than being tied to someone else's thinking processes.

Real-World Applications

Teachers who implement activity-style learning can expect to see gradual improvement in long-term recall of information, clearer understanding of connections among concepts, more creative approaches to problem solving,

and more sophisticated ways of using and communicating mathematical ideas.

In addition, teachers will be preparing students for future real-world problem–solving situations. Few, if any, of the problems and decisions that face people in their personal and professional lives come in tidy, textbook packaging. Real problems are vaguely defined and messy, with missing or too much information, and require difficult compromises to be made. Students who have learned to tackle algebraic problems by actively exploring approaches, collaborating with others in group or team approaches, and comparing results to look for alternative strategies will be equipped to transfer these methodologies to personal and professional challenges later in their lives.

Activity-style instruction that makes frequent use of explorations and investigations gives students more than just a knowledge of algebraic concepts and skills. Careful and consistent implementation of these techniques will prepare all students to face and solve future challenges.

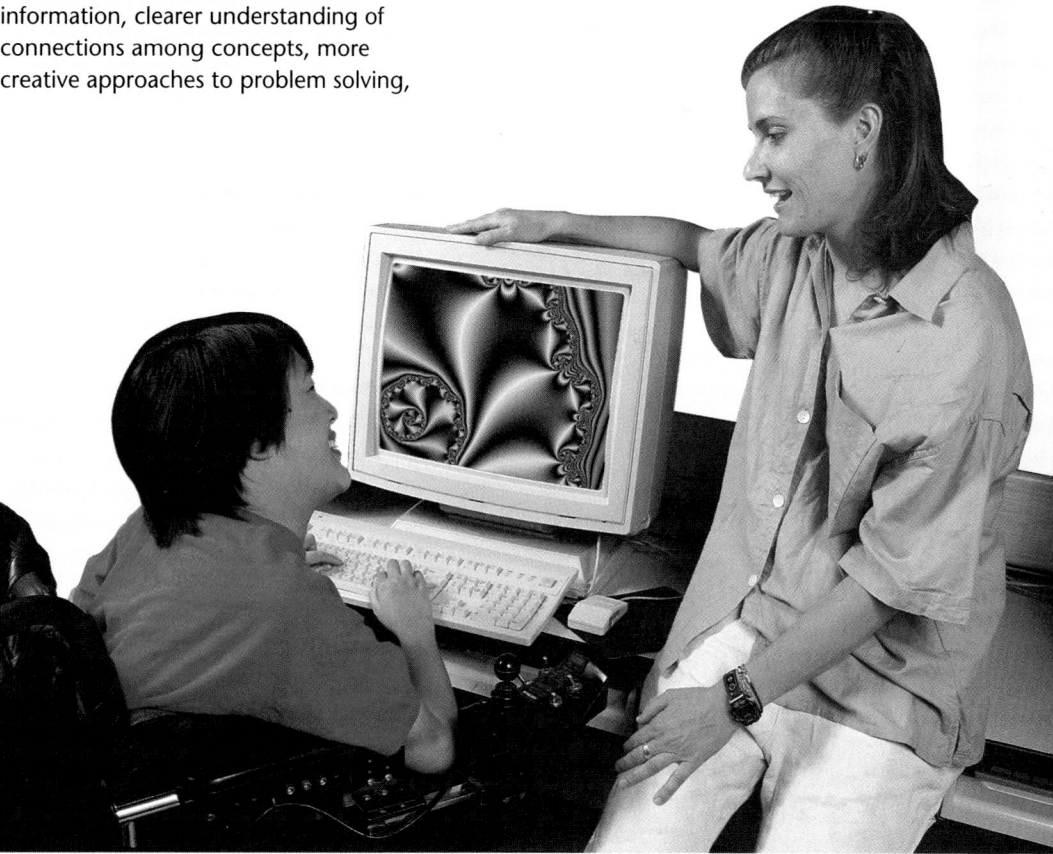

ASSESSMENT

Assessment Goals

An essential aspect of any learning environment, such as an algebra classroom, is the process of assessing or evaluating what students have learned. Informally this has been done using paper-and-pencil tests given by the teacher on a regular basis to measure students' performance against the material being studied. Formal evaluations using standardized tests are generally conducted over a period of years to establish performance records for both individuals and groups of students within a school or school district. Both types of tests are very good at measuring the ability of a student to use a particular mathematical skill or to recall a specific fact. They fall short, however, in evaluating other key goals of learning mathematics, such as being able to solve problems, to reason critically, to understand concepts, and to communicate mathematically, both verbally and in writing. Other techniques, usually referred to as alternative assessment, are needed to evaluate students' performance on these *process* goals of instruction.

The goals of an alternative assessment program are to provide a means of evaluating students' progress in non-skill areas of learning mathematics. Thus, the design and structure of alternative assessment techniques must be quite different from the skill-oriented, paper-and-pencil tests of the past.

Types of Alternative Assessment

In the world outside of school, a person's work is evaluated by what that person can do, that is, by the results the person achieves, and not by taking a test. For example, a musician may demonstrate skill by making music, a pilot by flying an airplane, a writer by writing a book, and a surgeon by performing an operation. Students learn to think mathematically and to solve problems on a continuous basis over a long period of time as they study mathematics at many grade levels. Students, too, can demonstrate what they have learned and understand by collecting a representative sample of their best work in a **portfolio**. A portfolio should illustrate achievements in problem solving, critical thinking, writing mathematics, mathematical applications or connections, and any other activity that demonstrates an understanding of both concepts and skills.

Specific examples of the kinds of work that students can include in their portfolios are solutions to nonroutine problems, graphs, tables or charts, computer printouts, group reports or reports of individual research, simulations, and examples of artwork or models. Each entry should be dated and should be chosen to show the students' growth in mathematical competence and maturity.

A portfolio is just one way for students to demonstrate their performance on a mathematical task. Performance assessment can also be achieved in other ways, such as by asking students questions and evaluating their answers, by observing their work in cooperative learning groups, by having students give verbal presentations, by working on extended projects and investigations, and by keeping journals.

Peer assessment and self-evaluation are also valuable methods of assessing students' performance. Students should be able to critique their classmates' work and their own work against standards set by the teacher. In order to evaluate their own work, students need to know the teacher's goals of instruction and the criteria that have been established (scoring rubrics) for evaluating performance against the goals. Students can help to design their own self-assessment forms that they then fill out on a regular basis and give to the teacher. They can also help to construct test items that are incorporated into tests given to their classmates. This work is ideally done in small groups of four students. The teacher can then choose items from each group to construct the test for the entire class. Another alternative testing technique is to have students work on *take-home* tests that pose more open-ended and non routine questions and problems. Students can devote more time to such tests and, in so doing, demonstrate their understanding of concepts and skills and their ability to do mathematics independently.

Scoring

The use of alternative assessment techniques implies the need to have a set of standards against which students' work is judged. Numerical grades are no longer sufficient because growth in understanding and problem solving cannot be measured by a single number or letter grade. Instead, scoring rubrics or criteria can be devised that allow the teacher more flexibility to recognize and comment upon all aspects of a student's work, pointing out both strengths and weaknesses that need to be corrected.

A scoring rubric can be created for each major instructional goal, such as being able to solve problems or communicate mathematically. A rubric generally consists of four or five short descriptive paragraphs that can be used to evaluate a piece of work. For example, if a five-point paragraph scale is used, a rating of 5 may denote that the student has completed all aspects of the assignment and has a comprehensive understanding of problems. The content of paragraph 5 specifies the details of what constitutes the rating of highly satisfactory. On the other hand, a rating of 1 designates essentially an unsatisfactory performance, and paragraph 1 would detail what is unsatisfactory. The other three paragraphs provide an opportunity for the teacher to recognize significant accomplishments by

the student and also aspects of the work that need improvement. Thus, scoring rubrics are a far more realistic and educationally substantive way to evaluate a student's performance than a single grade, which is usually determined by an answer being right or wrong.

The guide pictured, the Kentucky Mathematics Portfolio, was developed by the Kentucky Department of Education for use by school districts throughout that state. This guide is illustrated to show an example of an excellent and effective holistic scoring guide currently in use by math teachers who are practicing performance assessment in their classrooms. The scorer uses the Workspace/Annotations section of the guide to gather evidence about a student's mathematical ability. The top of the guide is then used to assign a single performance rating, based on an overall view of the full contents of the student's portfolio. The lower right-hand corner of the Holistic Scoring Guide lists the Types and Tools for the Breadth of Entries that are appropriate for a student to place in their portfolio.

Portfolio Holistic Scoring Guide

An individual portfolio is likely to be characterized by some, but not all, of the descriptors for a particular level. Therefore, the overall score should be the level at which the appropriate descriptors for a portfolio are clustered.

		NOVICE	APPRENTICE	PROFICIENT	DISTINGUISHED
PROBLEM SOLVING	Understanding Strategies	• Indicates a basic understanding of problems and uses strategies	• Indicates an understanding of problems and selects appropriate strategies	• Indicates a broad understanding of problems with alternate strategies	• Indicates a comprehensive understanding of problems with efficient, sophisticated strategies
	Execution/Extensions	• Implements strategies with minor mathematical errors in the solution without observations or extensions	• Accurately implements strategies with solutions, with limited observations or extensions	• Accurately and efficiently implements and analyzes strategies with correct solutions with extensions	• Accurately and efficiently implements and evaluates sophisticated strategies with correct solutions and includes analysis, justifications and extensions
REASONING		• Uses mathematical reasoning	• Uses appropriate mathematical reasoning	• Uses perceptive mathematical reasoning	• Uses perceptive, creative, and complex mathematical reasoning
MATHEMATICAL COMMUNICATION	Language	• Uses appropriate mathematical language some of the time	• Uses appropriate mathematical language	• Uses precise and appropriate mathematical language most of the time	• Uses sophisticated, precise, and appropriate mathematical language throughout
	Representations	• Uses few mathematical representations	• Uses a variety of mathematical representations accurately and appropriately	• Uses a wide variety of mathematical representations accurately and appropriately; uses multiple representations within some entries	• Uses a wide variety of mathematical representations accurately and appropriately uses multiple representations within entries and states their connections
UNDERSTANDING/CONNECTING CORE CONCEPTS		• Indicates a basic understanding of core concepts	• Indicates an understanding of core concepts with limited connections	• Indicates a broad understanding of some core concepts with connections	• Indicates a comprehensive understanding of core concepts with connections throughout
TYPES AND TOOLS		• Includes few types; uses few tools	• Includes a variety of types; uses tools appropriately	• Includes a wide variety of types; uses a wide variety of tools appropriately	• Includes all types; uses a wide variety of tools appropriately and insightfully

PERFORMANCE DESCRIPTORS

PROBLEM SOLVING
• Understands the features of a problem (understands the question, restates the problem in own words)
• Explores (draws a diagram, constructs a model and/or chart, records data, looks for patterns)
• Selects an appropriate strategy (guesses and checks, makes an exhaustive list, solves a simpler but similar problem, works backward, estimates a solution)
• Solves (implements a strategy with an accurate solution)
• Reviews, revises, and extends (verifies, explores, analyzes, evaluates strategies/solutions; formulates a rule)

REASONING
• Observes data, records and recognizes patterns, makes mathematical conjectures (inductive reasoning)
• Validates mathematical conjectures through logical arguments or counter-examples; constructs valid arguments (deductive reasoning)

MATHEMATICAL COMMUNICATION
• Provides quality explanations and expresses concepts, ideas, and reflections clearly
• Uses appropriate mathematical notation and terminology
• Provides various mathematical representations (models, graphs, charts, diagrams, words, pictures, numerals, symbols, equations)

UNDERSTANDING/CONNECTING CORE CONCEPTS
• Demonstrates an understanding of core concepts
• Recognizes, makes, or applies the connections among the mathematical core concepts to other disciplines, and to the real world

WORKSPACE/ANNOTATIONS

PORTFOLIO CONTENTS
• Table of Contents
• Letter to Reviewer
• 5-7 Best Entries

BREADTH OF ENTRIES
TYPES
• Investigations/Discovery
• Applications
• Non-routine Problems
• Projects
• Interdisciplinary
• Writing

TOOLS
• Calculators
• Computer and Other Technology
• Models Manipulatives
• Measurement Instruments
• Others

GROUP ENTRY

Place an X on each continuum to indicate the degree of understanding demonstrated for each core concept

DEGREE OF UNDERSTANDING OF CORE CONCEPTS

Basic

NUMBER	
MATHEMATICAL PROCEDURES	
SPACE & DIMENSIONALITY	
MEASUREMENT	
CHANGE	
MATHEMATICAL STRUCTURE	
DATA: STATISTICS AND PROBABILITY	

Within the interleaf pages that precede each chapter of this Annotated Teachers Edition is a list of seven activity domains that correspond to the Breadth of Entries. Each of these activity domains is correlated to specific examples of activities in the pupil book that are appropriate for portfolio development.

Assessment and HRW Advanced Algebra

HRW Advanced Algebra provides many opportunities to employ alternative assessment techniques to evaluate students' performance. These opportunities are an integral part of the textbook itself and can be found in the Explorations, Try This, the interactive questions (which are highlighted), Critical Thinking questions, the Chapter Projects, , and in the exercise and problem sets.

Throughout the textbook, students are asked to explain their work; describe what they are doing; compare and contrast different approaches; analyze a problem; make sketches, graphs, tables, and other models; hypothesize, conjecture, and look for counterexamples; and make and prove generalizations.

All of these activities, including the more traditional responses to routine problems, provide the teacher with a wealth of assessment opportunities to see how well students are progressing in their understanding and knowledge of algebra. The assessment task can be aligned with the major process goals of instruction and scoring rubrics established for each goal. For example, a teacher may decide to organize his or her assessment tasks in the following general areas of doing mathematics.

- Problem solving
- Reasoning
- Communication
- Connections

Within each of these areas, specific goals can be written and shared with students. In this way, the assessment process becomes an integral part, not only of evaluating students' progress, but also of the instructional process itself. The results of assessment can and should be used to modify the instructional approach to enhance learning for all students.

In addition to the many opportunities for performance assessment found in *HRW Advanced Algebra*, a variety of assessment types are integrated into the chapter-end material. The Chapter Review, the Chapter Assessment, and the Cumulative Assessment include both traditional and

alternative assessment. All Chapter Tests include both multiple choice and open-ended type questions. The Cumulative Reviews are formatted in the style of college preparatory exams. In additional to multiple-choice and free response questions, each Cumulative Assessment contains quantitative comparison questions which emphasize concepts of equalities, inequalities, and estimation. Other types of college entrance exam questions found in the Cumulative Review are student-produced response questions with gridded solutions. These Cumulative Assessments expose students to the new types of assessment that they will encounter when they take the latest form of college entrance examinations.

Eyewitness Math

Special two-page features called Eyewitness Math appear in almost every other chapter. These feature pages provide students with opportunities to read about current developments in mathematics and to solve real-life problems by working together in cooperative groups. Students' performance on Eyewitness Math can be assessed through group reports in writing or orally.

Quantitative Comparison

For items 1–4 write
A if the quantity in Column A is greater than the quantity in Column B;
B if the quantity in Column B is greater than the quantity in Column A;
C if the two quantities are equal; or
D if the relationship cannot be determined from the given information.

	Column A	Column B	Answers
1.	$\boxed{x}$ $\lvert 2x + 1 \rvert + 2 = 7$ $\boxed{y}$		Ⓐ Ⓑ Ⓒ Ⓓ [Lesson 3.4]
2.	$\boxed{\lceil 5.001 \rceil - \lceil 3.125 \rceil}$	$\boxed{\lceil \sqrt{3} \rceil - \lfloor \sqrt{3} \rfloor}$	Ⓐ Ⓑ Ⓒ Ⓓ [Lesson 3.5]
3.	x, given that $\begin{bmatrix} -1 & 2 & -1 \end{bmatrix} \begin{bmatrix} 0 \\ -2 \\ 3 \end{bmatrix} = \begin{bmatrix} x \end{bmatrix}$	$\begin{bmatrix} -5 & 0 \end{bmatrix} \begin{bmatrix} 0 \\ 2 \end{bmatrix} = \begin{bmatrix} x \end{bmatrix}$	Ⓐ Ⓑ Ⓒ Ⓓ [Lessons 4.2]
4.	$\boxed{x}$ $\begin{cases} 2x + 3y = 8 \\ -3x + y = -1 \end{cases}$ $\boxed{y}$		Ⓐ Ⓑ Ⓒ Ⓓ [Lessons 4.4, 4.6]

A Message from the Authors

It is certainly a challenge in our rapidly changing world for textbooks to capture the essence of what students need and teachers can provide. Frequent visits to schools confirm that mathematics programs often continue to focus on skills — many of which are diminishing in importance — even though there is an increasing need for students to be able to understand and apply concepts to solve problems in real-world settings using appropriate technology. To make matters worse, limited school budgets make it difficult to implement desired changes while teachers are faced with significant challenges that compete for their time and energy.

The authors are dedicated to the idea that mathematics programs should help all students gain mathematical power in a technological society. Based on careful examination of current recommendations and school mathematics programs, the authors have developed a program that strikes a balance in maintaining the strengths of former approaches while moving to mathematics content and methods of learning that are up-to-date and relevant to the present and future lives of students. In education, as in many other areas, even well-intended change should not be so rapid that students, teachers, and parents cannot cope with it. This program makes carefully chosen strides in the most vital areas while staying within the comfort zones of students, teachers, and parents.

Our textbooks reflect a vision of mathematics instruction that includes three components:
- active students
- solving interesting and relevant problems
- using appropriate technology

For example, in these books students use readily available technology. Thus, important problems that are traditionally attempted by 40 percent of high school juniors using advanced techniques are now accessible to almost 100 percent of high school freshmen using simple, inexpensive technology. This earlier study of important topics makes mathematics more interesting and more relevant.

This program successfully solves a long-standing dilemma: textbooks don't feature the use of technology because schools don't have the equipment. And schools don't have the equipment because it's not used in the textbooks. Thus, we have seen too many textbooks that do not reflect the needs of the students. In this series, technology is highly profiled but only a limited amount is required. Courses that are strongly rooted in mathematical content can be enhanced by including the technology as it becomes available.

We wish you well in pursuing this timely, balanced approach!

CHAPTER 1

From Data to Equations

Meeting Individual Needs

1.1 Tables and Graphs of Linear Equations

Core	Core Plus
Inclusion Strategies, p. 10 Reteaching the Lesson, p. 11 Practice Master 1.1 Enrichment Master 1.1 Technology Master 1.1 Lesson Activity Master 1.1	Practice Master 1.1 Enrichment, p. 10 Technology Master 1.1
[1 day]	**[1 day]**

1.4 Direct Variation and Proportion

Core	Core Plus
Inclusion Strategies, p. 31 Reteaching the Lesson, p. 32 Practice Master 1.4 Enrichment Master 1.4 Technology Master 1.4 Lesson Activity Master 1.4	Practice Master 1.4 Enrichment, p. 31 Technology Master 1.4
[1 day]	**[1 day]**

1.2 Exploring Slopes and Intercepts

Core	Core Plus
Inclusion Strategies, p. 17 Reteaching the Lesson, p. 18 Practice Master 1.2 Enrichment Master 1.2 Technology Master 1.2 Lesson Activity Master 1.2	Practice Master 1.2 Enrichment, p. 16 Technology Master 1.2
[1 day]	**[1 day]**

1.5 Solving Equations

Core	Core Plus
Inclusion Strategies, p. 38 Reteaching the Lesson, p. 41 Practice Master 1.5 Enrichment Master 1.5 Technology Master 1.5 Lesson Activity Master 1.5	Practice Master 1.5 Enrichment, p. 38 Technology Master 1.5 Interdisciplinary Connection, p. 37
[2 days]	**[1 day]**

1.3 Scatter Plots and Correlation

Core	Core Plus

Inclusion Strategies, p. 25
Reteaching the Lesson, p. 26
Practice Master 1.3
Enrichment Master 1.3
Technology Master 1.3
Lesson Activity Master 1.3
Mid-Chapter Assessment Master
	Practice Master 1.3
	Enrichment, p. 25
	Technology Master 1.3
	Interdisciplinary
	    Connection, p. 24
	Mid-Chapter Assessment
	    Master

[1 day]	**[1 day]**

1.6 Solving Inequalities

Core	Core Plus
Inclusion Strategies, p. 47 Reteaching the Lesson, p. 48 Practice Master 1.6 Enrichment Master 1.6 Technology Master 1.6 Lesson Activity Master 1.6	Practice Master 1.6 Enrichment, p. 47 Technology Master 1.6 Interdisciplinary Connection, p. 46
[2 days]	**[1 day]**

Chapter Summary

Core	Core Plus
Chapter 1 Project, pp. 54–55	Chapter 1 Project, pp. 54–55
Lab Activity	Lab Activity
Long-Term Project	Long-Term Project
Chapter Review, pp. 56–58	Chapter Review, pp. 56–58
Chapter Assessment, p. 59	Chapter Assessment, p. 59
Chapter Assessment Masters, A/B	Chapter Assessment Masters, A/B
Alternative Assessment	Alternative Assessment
[3 days]	**[3 days]**

Hands-on Strategies

Have students perform a science experiment for an interdisciplinary connection. Data collected in the experiment should be used in a scatter plot. Have students write an equation, graph the line of best fit, and determine the correlation. Documentation of the experiment along with pictures, and/or a demonstration can be compiled and used in a portfolio. An activity like this is most relevant when it is coordinated with the students' science classes. They might be allowed to use data from an experiment in these classes for their scatter plots.

Reading Strategies

To introduce Chapter 1, have students skim lessons and answer the following questions in a journal.

- What do you think the chapter (or the lesson) is about?
- What do you hope to learn from this chapter?

After or during the reading of the lesson, have students make a list of words, phrases, equations, etc. that they think are important to the lesson or are confusing to the student. They should also list any questions that they have about the lesson and try to respond to the items on their list. After instruction, students should respond to their lists.

Visual Strategies

Students should preview each lesson by looking at the graphs and pictures. Have them compare and contrast the graphs in each successive lesson, noting what new ideas or concepts have been introduced. Have them respond in their journals to the following questions:

- How does the information about graphs from previous lessons help me to understand the graphs in today's lessons?
- How are the graphs in today's lesson the same and/or different from those in previous lessons?
- What are some real world applications for using graphs like those in the lessons?

Cooperative Learning

GROUP ACTIVITIES	
Egyptian measurements	Portfolio Activity
Slopes and intercepts	Lesson 1.2, Exploration 3
Direct variation and proportion	Lesson 1.3, Examples 1 and 2
Solving equations	Lesson 1.5, Cultural Connection
Scatter plots	Chapter 1 Project

You may wish to have students work with partners for some of the above activities. Additional suggestions for cooperative group activities are noted in the teacher's notes in each lesson.

Multicultural

The cultural connections in this chapter include references to Africa.

CULTURAL CONNECTIONS	
Africa: Ancient units of measure	Lesson 1.3, Exercises 27–30
Africa: Egyptian method of solving equations	Lesson 1.5, Exercise 31

Portfolio Assessment

Below are portfolio activities for the chapter. They are listed under seven activity domains that are appropriate for portfolio development.

1. **Investigation/Exploration** In Explorations 1, 2, and 3 in Lesson 1.2, students investigate properties of graphs of linear functions.

2. **Applications** Recommended are any of the following: Ecology (Lesson 1.1), Business (Lesson 1.1), Entertainment (Lesson 1.1), Economics (Lessons 1.1, 1.5), Depreciation (Lesson 1.2), Construction (Lesson 1.2), and Government (Lesson 1.4).

3. **Non-Routine Problems** The Portfolio Activity on page 41 asks students to graph data using ancient Egyptian units of measure.

4. **Project** In the Chapter Project students investigate the relationship between arm length and body height. They compile data from measurements of the students in class and make a scatter plot.

5. **Interdisciplinary Topics** Students may choose from the following: Demographics (Lesson 1.1), Ecology (Lesson 1.1), Meteorology (Lesson 1.2), Physics (Lesson 1.2), Marketing (Lesson 1.3), Anatomy (Lesson 1.3), Astronomy (Lesson 1.4), and Electronics (Lesson 1.4).

6. **Writing** The Alternative Assessment activities on p. 43 and p. 52 ask students to write summaries and descriptions of how to solve equations and inequalities. In addition, any of the Communication questions at the beginning of the exercise sets are appropriate.

7. **Tools** Chapter 1 uses graphics calculators or graph paper. The technology masters are recommended to measure the individual student's proficiency in using graphics calculators.

Technology

Technology provides important tools for manipulating and visualizing algebraic concepts. Several lessons in Chapter 1 can be significantly enhanced by using appropriate technology. Scientific calculators make arithmetic calculations more efficient and let students focus on the underlying algebraic concepts. Graphics calculators provide a visual model of the functions, and students save the time needed to create hand drawn graphs. For more information on the use of technology, refer to the *HRW Technology Handbook*.

Graphics Calculator

Graphing Linear Equations Several lessons in Chapter 1 and throughout the book suggest the use of a graphics calculator to model equations. Follow the steps to graph a linear equation:

1. Set the ranges for the graphing window by pressing the **WINDOW** key (TI-82) or the **RANGE** key (Casio, TI-81) and using the arrow and number keys. When finished press **ENTER** or **EXE**.

2. Enter the equation by pressing the **Y=** key(TI) or the **GRAPH** key (Casio). Then use the number and variable keys to enter the functions to be graphed.

3. Press **GRAPH** (TI) or **EXE**(Casio) to draw the graph.

Graphing Linear Inequalities Lesson 1.6 suggests the use of a graphics calculator to graph linear inequalities. Follow the steps to create a graph. The instructions listed below apply to the Casio. Instructions for using other calculators can be found in the *HRW Technology Handbook*.

1. Press **MODE** and then press **SHIFT**. Press ÷ to choose inequalities.

2. Press **GRAPH**. When **Y>**, **Y<**, **Y≥**, and **Y≤** appear across the bottom of the screen corresponding to the function keys (**F1, F2, F3, F4**), press the function key matching the appropriate inequality type.

3. Enter the inequality using the **X,Ø,T** key along with the number and operation keys.

4. Press **EXE** to show the graph. If your graph does not fit in the window, use the **Range** key to reset the window range.

Integrated Software

f(g) Scholar™ is an integrated computer-based mathematics productivity tool that combines calculator, spreadsheet, and graphics capabilities to provide a dynamic and interactive environment for explorations in mathematics. It is appropriate to use *f(g) Scholar*™ for any lesson needing a spreadsheet, calculator, graphics calculator, or any combination of the three.

From Data to Equations

From Data to Equations

ABOUT THE CHAPTER

Background Information

Collecting and organizing data are important skills in mathematics and in other areas. In this chapter students will look at relationships between sets of data and their graphs.

CHAPTER RESOURCES

- Practice Masters
- Enrichment Masters
- Technology Masters
- Lesson Activity Masters
- Lab Activity Masters
- Long-Term Project Masters
- Assessments Masters
  Chapter Assessments, A/B
  Mid-Chapter Assessment
  Alternative Assessment, A/B
- Teaching Transparencies
- Spanish Resources

CHAPTER OBJECTIVES

- Identify linear equations and linear relationships between variables in a table.
- Write and graph an equation describing a real-world linear relationship.
- Write an equation in slope-intercept form, given two points on the line or the slope and a point on the line.
- Graph a scatter plot and identify the data correlation.
- Use a graphics calculator to find the correlation coefficient and to make predictions using the line of best fit.
- Solve problems involving direct variation.

LESSONS

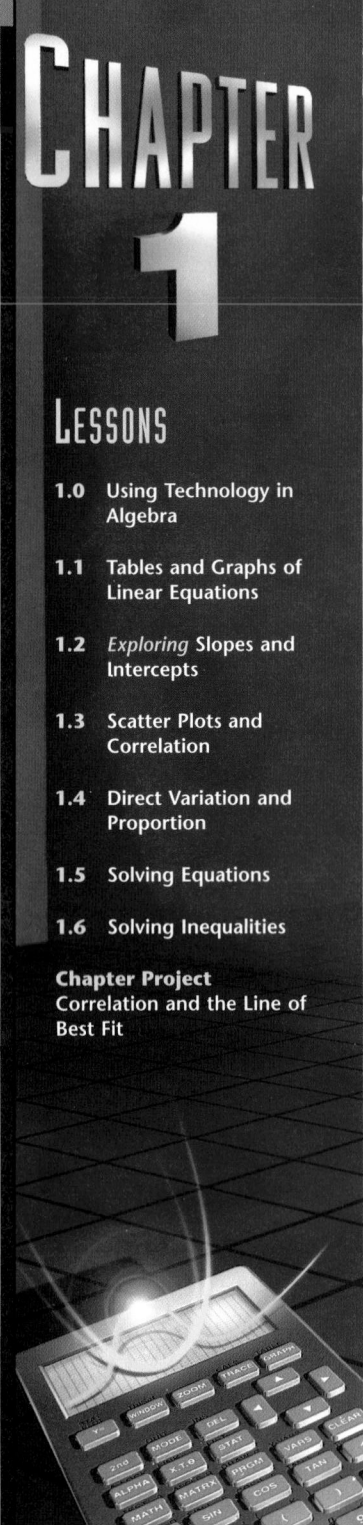

Collecting, organizing, and representing data are important to contemporary society. A major focus of this chapter will be investigating, analyzing, and describing data that can be gathered from the world around us. Often this data can be used to influence public opinion about critical issues.

The collected data is organized in a table.

Students in a high school biology class collect samples from the rivers and streams to measure the level of contaminants.

ABOUT THE PHOTOS

The picture shows students collecting data. These data are converted into mathematical models as a way to study the results of an experiment. The data are then plotted for a visual representation which can help people understand and draw conclusions based on the results.

Contaminants (parts per liter)

Distance From City Limits (miles)

A graph of the dots represents the decreasing levels of contaminants in the river as the distance from the city limits increases.

REPRESENTING

INFLUENCING

The students use the results of their studies to convince voters of the harmful effects of pollution on natural aquatic habitats.

PORTFOLIO ACTIVITY

The collection and investigation of data, commonly referred to as the study of statistics, dates back to the days of the first pharaohs of Egypt.

Since the Nile River provided water for many African farms, the pharaohs kept a close watch on the height of the annual Nile flood. Readings of the water level during the flood were taken from "Nilometers" that were carved in the rocks. Good floods meant good crops, and the pharaoh knew that good crops yielded high tax collections. Everyone wanted floods high enough to water the field, but not so high that they would cause damage.

The table on page 29 gives realistic data for flood height, crop value, and tax collection. The exercises that follow the table may be used for a portfolio activity.

A Nilometer is an instrument for measuring the rise of the Nile during flood time. It consists of a graduated pillar with a water chamber.

ABOUT THE CHAPTER PROJECT

The Chapter 1 Project on pages 54 and 55 involves collecting data in class relating to various kinds of student information. Students are then asked to plot the data on scatter plots and discuss the correlation between the variables. This activity is best done in groups of four or five.

- Solve problems by writing and solving linear equations.
- Solve and graph linear inequalities.

The portfolio represents a profile of the student's work throughout the course in algebra. It should be the best that student can do, but it can also show developmental progress. Portfolio Activities can provide a start, but more should be included. The reviewer should be able to see whether the student is willing to take risks, persevere to complete difficult problems, try various approaches, and continue beyond the answer with *What-if* questions or additional research.

PORTFOLIO ACTIVITY

In ancient Egypt, data relating to flood height, crop value, and tax collection were collected in measurements called cubits and hekats. Students are asked to make a scatter plot using these ancient units of measure.

Students can work together to complete Exercises 27–30 on page 29. Have them think of modern ways of showing similar information. For Exercise 30, have them brainstorm to think of different conditions that would make the crop value zero in the sixth year. Have each cooperative group do research reports on the economy of ancient Egypt. Have them make scatter plots using other units of measure that were present in the economy.

Objectives

Lesson 1.0 is a unique introductory lesson. It does not follow the same format as the other lessons in the text. This lesson does not contain instructional examples. It does, however, contain a set of exercises to be worked out by students in small groups.

One of the main objectives of this lesson is to introduce students to the many ways that technology is used in the real world. Another objective is to give students the opportunity to work in a cooperative group to explore mathematical concepts from real-world settings using their graphics calculators.

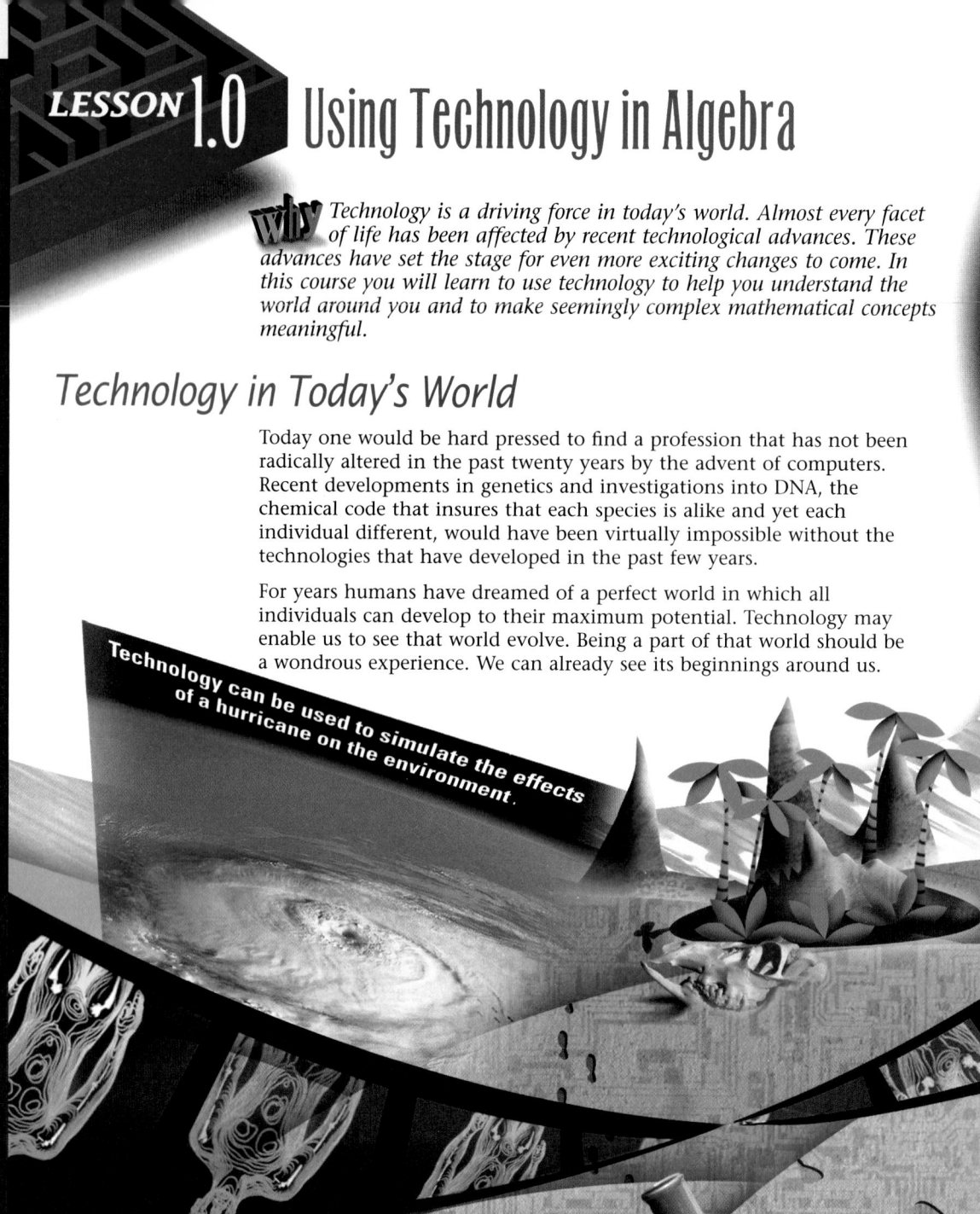

LESSON 1.0 Using Technology in Algebra

why *Technology is a driving force in today's world. Almost every facet of life has been affected by recent technological advances. These advances have set the stage for even more exciting changes to come. In this course you will learn to use technology to help you understand the world around you and to make seemingly complex mathematical concepts meaningful.*

Technology in Today's World

Today one would be hard pressed to find a profession that has not been radically altered in the past twenty years by the advent of computers. Recent developments in genetics and investigations into DNA, the chemical code that insures that each species is alike and yet each individual different, would have been virtually impossible without the technologies that have developed in the past few years.

For years humans have dreamed of a perfect world in which all individuals can develop to their maximum potential. Technology may enable us to see that world evolve. Being a part of that world should be a wondrous experience. We can already see its beginnings around us.

Technology can be used to simulate the effects of a hurricane on the environment.

Technology is used in almost every business.

Technology is used to perform quality control in medical sciences.

Technology is used to create special effects in motion pictures.

The most often used expression heard from students taking algebra is, "Why do I need to learn this?" This lesson can be used to give your students an overview of how they will study algebra this year. Point out that each lesson will begin with a *Why* statement, and this statement usually reflects the utility of the mathematics presented in the lesson.

As students proceed, discuss how the concepts learned in algebra are necessary or useful in any career that the student might choose.

Technology in Advanced Algebra

Often algebraic procedures involve difficult computational techniques that are hard to remember, even with a thorough understanding of the concepts involved. Today, technology in the form of graphics calculators, graphing software, and spreadsheets can make calculations easier and concepts more visual.

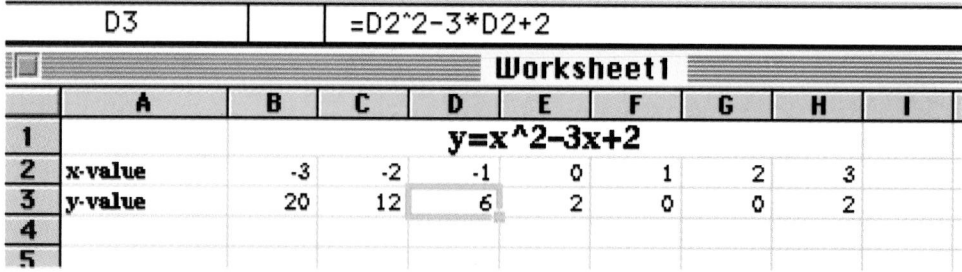

| D3 | | =D2^2-3*D2+2 |

Worksheet1

	A	B	C	D	E	F	G	H	I
1				y=x^2-3x+2					
2	x-value	-3	-2	-1	0	1	2	3	
3	y-value	20	12	6	2	0	0	2	
4									
5									

Graphics calculators can provide graphs of the most complicated functions, reduce computation with matrices to the simple task of inputting data, and make statistical analysis of data an intellectual exercise rather than an exercise in endless computation.

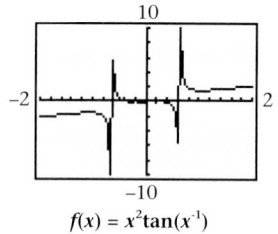

$f(x) = x^2\tan(x^{-1})$

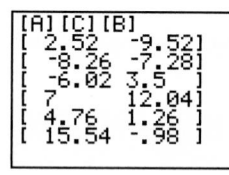

Matrix multiplication

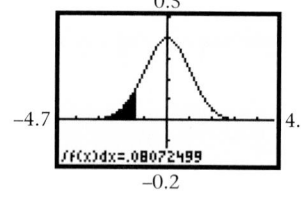

Areas under the normal curve

Graphic and scientific calculators have made working with logarithmic and trigonometric tables obsolete. Now with the push of a button the data is available to whatever precision is required.

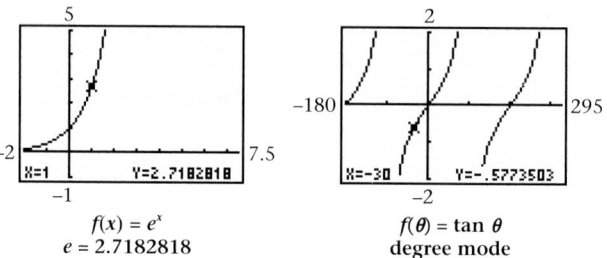

$f(x) = e^x$
$e = 2.7182818$

$f(\theta) = \tan\theta$
degree mode

You will find that the use of this technology will enable you to concentrate on the concepts involved in the mathematics you are studying. You will be able to focus on the application of the mathematics, rather than on tedious computational procedures. The technology will free you to investigate, to conjecture, and to solve problems. When you use technology effectively, mathematics becomes even more useful and exciting.

You Can Begin Now...

Use the owner's manual to your graphics calculator to discover how to perform each of the following operations.

- Enter and graph a function.
- Trace a function you have already graphed.
- Enter data in two lists in order to graph a scatter plot.
- Find and graph the line of best fit for a scatter plot.

1 The function $y = 34x^2 - 4206x + 132,539$ can be used to estimate the cost of a popular sports car for the years 1953 through the present. In the formula, x represents the years since 1900. For example, for the year 1970, x is 70. Graph the function, and use the trace feature to approximate the cost for 1960 and 1980. Predict the cost for the year 2000.

2 Graph the functions $y = \sqrt{25 - x^2}$ and $y = -2x + 10$, and use the trace feature to find the points of intersection.

3 A student athlete is interested in the relationship between the distance from the goal and her ability to make a basket. At each distance, she makes 20 attempts at the goal. Graph a scatter plot to represent this data.

Distance from the hoop (ft)	10	12	14	16	18	20	22	24	26
Baskets made by student	16	15	12	13	11	10	6	5	2

4 Find and graph the line of best fit for the scatter plot in Exercise 3. Trace the line of best fit, and predict the number of baskets the student athlete will make at 6, 8, and 15 feet from the goal. Do these values seem reasonable?

1. Cost for 1960: $2579; 1980: $13,659; 2000: $51,939

2. (3, 4) and (5, 0)

3 & 4.

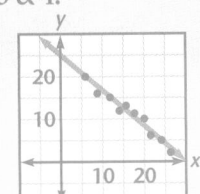

6 feet: ≈ 20 baskets; 8 feet: ≈ 18 baskets; 15 feet: 12–13 baskets. Answers may vary. A sample answer: The values at 8 and 15 feet seem reasonable. However, the value at 6 feet seems highly unlikely.

Objectives

• Identify linear equations and linear relationships between variables in a table.
• Write and graph an equation describing a real-world linear relationship.

RESOURCES

• Practice Master   **1.1**
• Enrichment Master   **1.1**
• Technology Master   **1.1**
• Lesson Activity Master   **1.1**
• Quiz   **1.1**
• Spanish Resources   **1.1**

Assessing Prior Knowledge

1. Plot the ordered pairs (2, 10), (−1, 4), and (5, 16).

2. Connect the points.

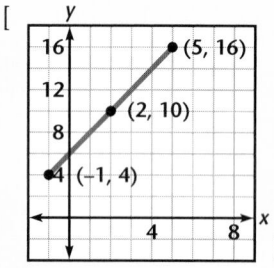

3. Estimate the y-value when the x-value is zero. [**6**]

TEACH

Discuss how tables of data make reading information easier. Explain how the table represents the relationship between the x- and y-values. Students should see that an increase in gallons of gas burned results in an increase in CO_2 emissions.

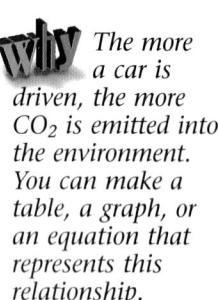

LESSON **1.1**

Tables and Graphs of Linear Equations

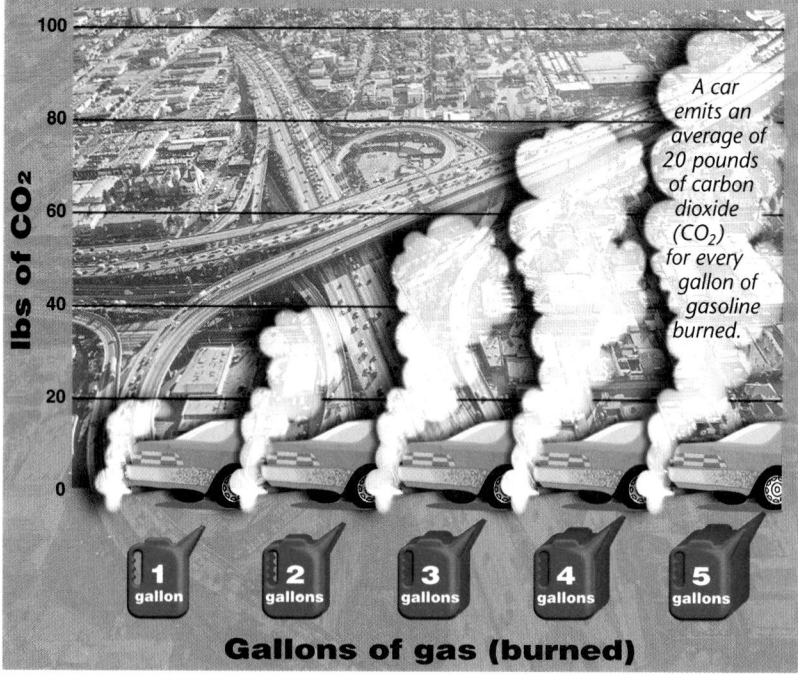

A car emits an average of 20 pounds of carbon dioxide (CO_2) for every gallon of gasoline burned.

The more a car is driven, the more CO_2 is emitted into the environment. You can make a table, a graph, or an equation that represents this relationship.

Ecology For 1 gallon of gasoline burned, 20 pounds of CO_2 are emitted; for 2 gallons of gasoline burned, 40 pounds of CO_2 are emitted; for 3 gallons of gasoline burned, 60 pounds of CO_2 are emitted; and so on.

CO_2 Table

Gallons of gas	x	0	1	2	3	4	5
Pounds of CO_2	y	0	20	40	60	80	100

Recognizing Linear Relationships

Exploration Looking for Patterns

Look for patterns in the CO_2 table.

1 Predict the next x- and y-values.

2 What pattern did you use to predict the next numbers?

3 Graph the points represented by the table. Can you draw one straight line through all of the points?

ALTERNATIVE **teaching strategy**

Using Visual Models

Have students graph each of the following relationships and then use several values of their own to extend each table. Ask students which of the four relationships are nonlinear. Have them explain their answers.

Feet	3	6	15
Yards	1	2	5

[Linear]

Cost ($)	10	30	50	100
Tax ($)	0.6	1.8	3.0	6.0

[Linear]

Look for patterns in each of the following tables.

Table 1		Table 2		Table 3		Table 4		Table 5		Table 6	
x	y	x	y	x	y	x	y	x	y	x	y
−2	4	−1	−1	8	2	1	1	10	−2	−3	0
−1	9	0	−3	5	6	2	4	8	3	4	1
0	14	1	−9	2	10	3	9	6	8	11	3
1	19	2	−27	−1	14	4	16	4	13	18	6

4 For each table, find the differences in consecutive x-values and in consecutive y-values. Which tables have a constant difference between consecutive x-values? Which tables also have a constant difference between consecutive y-values?

5 For each table, graph the points represented. Which tables represent points that can all fit on a straight line? Are these tables the same tables you listed in Step 4?

6 How can you determine if variables in a table represent points that will all fit on one line? ❖

When a constant difference in the consecutive x-values results in a constant difference in consecutive y-values, the variables are said to be **linearly related**.

EXAMPLE 1

A Are the variables in the table at the right linearly related?

x	2	5	8	11	14	17
y	5	11	17	23	29	35

B What is the y-value that corresponds to an x-value of 20?

Solution▸

A The difference between consecutive x-values, 3, is constant.
The difference between consecutive y-values, 6, is constant.
Therefore, the variables are linearly related.

B The difference in x-values is 3, so 20 is the next x-value in the table.
The corresponding y-value is the next y-value in the table: $35 + 6$, or 41. ❖

Hours	2	3	4	6
Miles	80	110	150	150

[Nonlinear. The difference between consecutive y-values (miles) is not constant.]

Gallons	1	2	3
Quarts	4	8	12

[Linear]

Throughout the text, answers to exploration steps can be found in Additional Answers beginning on page 842.

Exploration Notes

Students will have an oportunity to discover that, for all linear equations, a constant difference between consecutive x-values results in a constant difference in consecutive y-values.

Throughout the Pupil's Edition, text highlighted in yellow is designated to be used for ongoing assessment. The answers are provided in the Teacher's Edition side copy.

ongoing ASSESSMENT

6. If the differences between consecutive x-values and consecutive y-values are constant, then the points will all fit on one line.

Alternate Example 1

A. Are the variables in the table below linearly related?

x	1	4	7	10	13	16
y	2	11	20	29	38	47

B. What is the y-value that corresponds to an x-value of 19?

[A. Yes; B. 56]

TEACHING tip

Throughout the textbook, information necessary to work the examples may not appear in the written part of the example. Remind students to look for pertinent information in the art, photos, and captions.

EXAMPLE 2

Are the variables in $y = x^2$ linearly related?

Solution

Make a table of values for the equation using x-values that have a constant difference.

x	2	4	6	8
y	4	16	36	64

The difference in the x-values is a constant 2, but the difference in y-values is not constant. The variables, x and y, are not linearly related. ❖

Writing Equations

Ecology Examine the CO_2 table again. Can you write an equation to represent the linear relationship between gallons of gasoline and pounds of CO_2? The amount of CO_2 emitted, y, depends on x, the amount of gasoline that is burned. Since y is always 20 times x, you can write $y = 20x$. This is an example of a **linear equation**.

The graph of $y = 20x$ is a line.

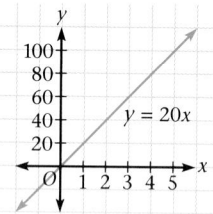

EXAMPLE 3

Business

Ⓐ Make a table of values for 1, 2, 3, 4, and 5 calls.

Ⓑ Write an equation to represent the total monthly cost for the Catch-A-Call answering service.

CATCH-a-CALL
Answering Service
$8.00/month *
*Plus $2.00 per call received

Solution

Ⓐ Substitute the values of c in the equation to find t.

☎ calls, c	1	2	3	4	5
$ total cost, t	10	12	14	16	18

Ⓑ The initial monthly charge is $8. Let c represent the number of calls and t represent the total monthly cost, in dollars. Since $2 is charged for each call received, multiply c by 2. The total monthly cost, t, is given by adding the cost for each call, $2c$, to $8. Therefore, $t = 2c + 8$. ❖

CRITICAL Thinking

In Example 3, is it possible to be charged for $1\frac{1}{2}$ calls or for 2.7 calls? What kind of numbers are possible for values of c?

Graphs of Linear Equations

EXAMPLE 4

Use a spreadsheet to sketch the graph of $y = -3x + 1$.

Solution

Spreadsheet

Make a table of values using the x-values -1, 0, 1, and 2. Enter these values in one row of your spreadsheet. In the second row, enter the appropriate formula. For the x-value -1, the formula is $-3*B7+1$. Extend this formula for all x-values. Graph the resulting ordered pairs.

The equation $y = -3x + 1$ is a linear equation because its graph is a line. ❖

EXAMPLE 5

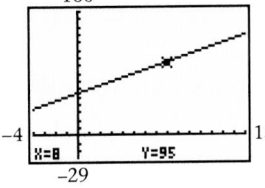
Graphics Calculator

A 110-gallon tank is exactly half full when Darius begins to fill it. The water flows at a rate of 5 gallons per minute.

Ⓐ Write an equation to find the volume, v, in gallons of water per time, t, in minutes.

Ⓑ Graph the equation, and approximate the volume to the nearest gallon after 8 minutes.

Solution

Ⓐ The 110-gallon tank is half full, so the volume begins with 55 gallons. Water is added at a rate of 5 gallons per minute, or $5t$. Therefore, $v = 55 + 5t$.

Ⓑ Graph $y = 55 + 5x$. Use the trace feature to find the value of y when x is 8.

Since y is 95 when x is 8, the volume after 8 minutes is 95 gallons. ❖

Try This Use your graphics calculator to estimate, to the nearest tenth of a minute, how long it will take to fill the tank in Example 5.

A graph that slopes upward from left to right is steadily *increasing*. A graph that slopes downward from left to right is steadily *decreasing*.

Look at the graphs in Examples 4 and 5. Which is increasing, and which is decreasing?

RETEACHING the lesson

Using Tables Have students write several tables of their own data, some that are linearly related and some that are not. Remind students that linearly related data have constant differences in both the x- and y-values, whereas data that are not linearly related do not. Then have students exchange data. Ask each student to determine whether the data in each table they received are linearly related. Then have students graph the data to verify their answers.

Alternate Example 4

Sketch the graph of $y = 2x - 1$.

[Make a table of values.

x	-1	0	1
y	-3	-1	1

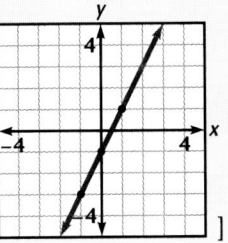
]

Alternate Example 5

A 200-gallon tank is exactly half full when Darla begins to fill it. The water flows at a rate of 10 gallons per minute.

A. Write an equation to find the volume, v, in gallons of water per time, t, in minutes. [$v = 100 + 10t$]

B. Graph the equation, and find the volume in gallons after 8 minutes.

[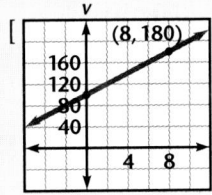

180 gallons]

Try This questions provide an opportunity for ongoing assessment. The answers are provided in the Teacher's Edition side copy.

Aongoing ASSESSMENT

Try This

11.0 minutes

Aongoing ASSESSMENT

The graph in Example 4 is decreasing, while the graph in Example 5 is increasing.

ASSESS

Selected Answers

Odd-numbered Exercises 7–43

Communicate Exercises provide an opportunity for students to discuss their discoveries, conjectures, and the lesson concepts. Throughout the text, the answers to all Communicate Exercises can be found in Additional Answers beginning on page 842.

Assignment Guide

Core 1–8, 10–14, 16–25, 27–45

Core Plus 1–5, 7–11, 13–23, 25–45

Technology

Graphics calculators are needed for Exercises 11, 19, and 45.

Error Analysis

Watch for students who look only at the x- or y-values for a constant difference. Encourage students always to look at both the x- and y-values in every data table they examine.

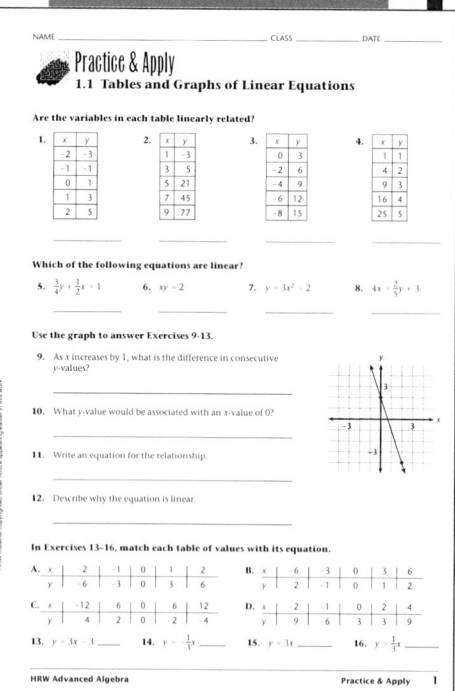

EXERCISES & PROBLEMS

Communicate

1. How can you tell if an equation is a linear equation?

2. Suppose your state has a sales tax of 7%. Let p represent the price of an item and t represent the tax on the item. Make a table showing the tax on items with prices $6, $8, $10, $12, and $14. How can you tell if the variables are linearly related?

3. How can you write an equation that represents the relationship in Exercise 2?

4. Describe a table of values for variables that are linearly related and whose graph is steadily decreasing.

5. Describe a table of values for variables that are linearly related and whose graph is steadily increasing.

Practice & Apply

Explain whether the variables in each table are linearly related.

6.

x	y
2	10
4	16
6	22
8	28
10	34
12	40

linear

7.

x	y
12	−3
9	0
6	3
3	6
0	9
−3	12

linear

8.

x	y
1	1
2	2
3	4
4	8
5	16
6	32

not linear

9.

x	y
−2	−5
−3	−1
−4	3
−5	7
−6	11
−7	15

linear

Meteorology At 6:00 A.M., the temperature was 67° Fahrenheit. During a cold front the temperature began to drop at a steady rate of 4°F per hour.

10. Write a linear equation relating the temperature in degrees, t, to the time in hours, h, as the cold front passed through. $t = 67 - 4h$

11. Use your graphics calculator to estimate, to the nearest 15 minutes, how long it would take to reach freezing if the drop in temperature continued at the same rate. 8 h, 45 min

6. linear. The difference in the x-values is a constant 2 and the difference in the y-values is a constant 6.

7. linear. The difference in the x-values is a constant −3 and the difference in the y-values is a constant 3.

8. not linear. The differnce in the x-values is a constant 1 and but the difference in the y-values is not constant.

9. linear. The differnce in the x-values is a constant −1 and the difference in the y-values is a constant 4.

Which of the following equations are linear equations?

12. $y = -x$ **13.** $2x - y = 12$ **14.** $y = -x^2 + 1$ **15.** $x^2 + y^2 = 4$

   linear   linear   not linear   not linear

Demographics A city community college plans to increase its enrollment capacity to keep up with population growth. The college has an enrollment of 2200 students and plans to increase the enrollment by 70 students each year.

16. Make a table of values. Let y represent the number of years, and let t represent the total number of students.

17. What will the enrollment capacity be 3 years from now? 2410

18. Write a linear equation that could be used to find the total enrollment capacity after y number of years. $t = 2200 + 70y$

19 Graph the equation you wrote for Exercise 18 on your graphics calculator. What range of values did you use for each variable to view the graph? Explain why you used those values.

20. Make a table of values for the linear equation $y = -4x - 3$. What are the constant differences in the x-values and the y-values?

Business An attorney charges clients an initial fee of $250, plus $150 per hour for each hour she works on a case.

21. Write a linear equation relating hours, h, and total cost, c. $c = 250 + 150h$

22. Sketch a graph of the equation.

23. How much would the attorney charge for a case that required a total of 52 hours? $8050

Explain whether the variables in each table are linearly related.

24.

x	y
-3	-11
-1	-7
1	-3
3	1
5	5

linear

25.

x	y
-5	9
-4	8
-3	6
-2	3
-1	-1

not linear

26.

x	y
2	-1
-1	11
-4	23
-7	35
-10	47

linear

Entertainment A group of friends go to an amusement park. With a green admission ticket, each ride costs $0.25, but with a red admission ticket, each ride costs $0.75.

27. Write two linear equations that describe the two options.

28. Graph the two equations on the same coordinate plane. Do the two lines intersect? yes

29. Which is the better option? Explain. What factors are included in your decision?

16. Sample table:

y	0	1	2	3
t	2200	2270	2340	2410

19. Sample answer: Values of 0 to 20 for x and values of 2000 to 4000 for y will display data for the enrollment over 20 years.

20. Sample table:

x	-2	-1	0	1	2
y	5	1	-3	-7	-11

The constant difference in consecutive x-values is 1.

The constant difference in consecutive y-values is -4.

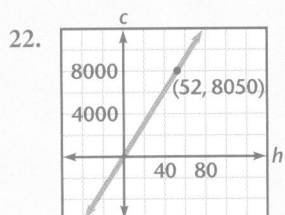
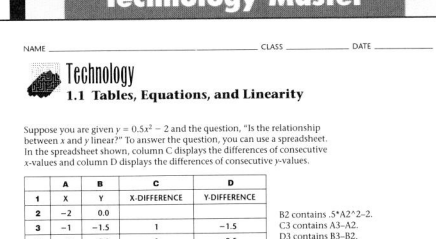

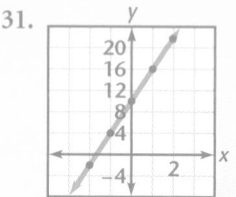
For Exercises 30–32, use the linear equation $y = 6x + 10$.

30. Make a table of x- and y-values using x-values of –2, –1, 0, 1, and 2.

31. Sketch the graph of the equation.

32. Is the graph increasing or decreasing?
increasing

Economics The table gives the supply and demand for different prices of a video game at a toy warehouse.

33. Graph the (*price, supply*) line and the (*price, demand*) line on the same coordinate plane. What are the coordinates of the point where the two lines intersect? (37.5, 325)

34. Explain how the point of intersection of these two lines can be used in marketing.

Price	Supply	Demand
$20.00	150	500
$30.00	250	400
$50.00	450	200

Look Back

35. $-12 - 4 \div [6 - (-2)] =$ __ −12.5   **36.** $(8h)(0.5k) =$ __ 4hk

37. Find the Greatest Common Factor (GCF) of $13r$ and $26r^2$. 13r

38. Let x be –4. Evaluate $x^2 - 4x + 1$. 33

39. Let x be 8, and let y be 15. Evaluate $15x - 8y$. 0

40. Is 12 a solution of $6x - 14 = 58$? yes

41. Is 28 a solution of $5y + 44 = 96$? no

Determine whether each of the following is true or false.

42. $-8 > -4$ false   **43.** $0 > -6$ true   **44.** $-3 < 14$ true

Look Beyond

45 Use your graphics calculator to graph the equations $y = 2x + 3$, $y = 2x - 4$, and $y = 2x$ on the same coordinate plane. How are their graphs alike or different?

33.

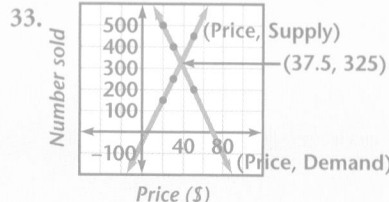

Price ($)

sold. If the price is set higher, the demand will be lower than the supply and there will be unsold product in the warehouse. If the price is set lower, the demand will be greater than the supply, and customers must wait for video games.

34. At the point of intersection, supply and demand are equal. By choosing that price for their merchandise, manufacturers can ensure that all merchandise in stock will be

45. The graphs are all linear. The lines are all steadily increasing at the same rate; they are parallel to each other. The three line have different x- and y-intercepts.

Exploring

Slopes and Intercepts

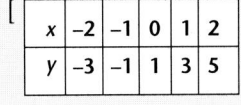 Traveling on a mountain road, Reneé notices a sign warning of the upcoming steep grade. Will she need to put her truck in a lower gear? The steepness of a road offers information that allows you to plan ahead and make predictions.

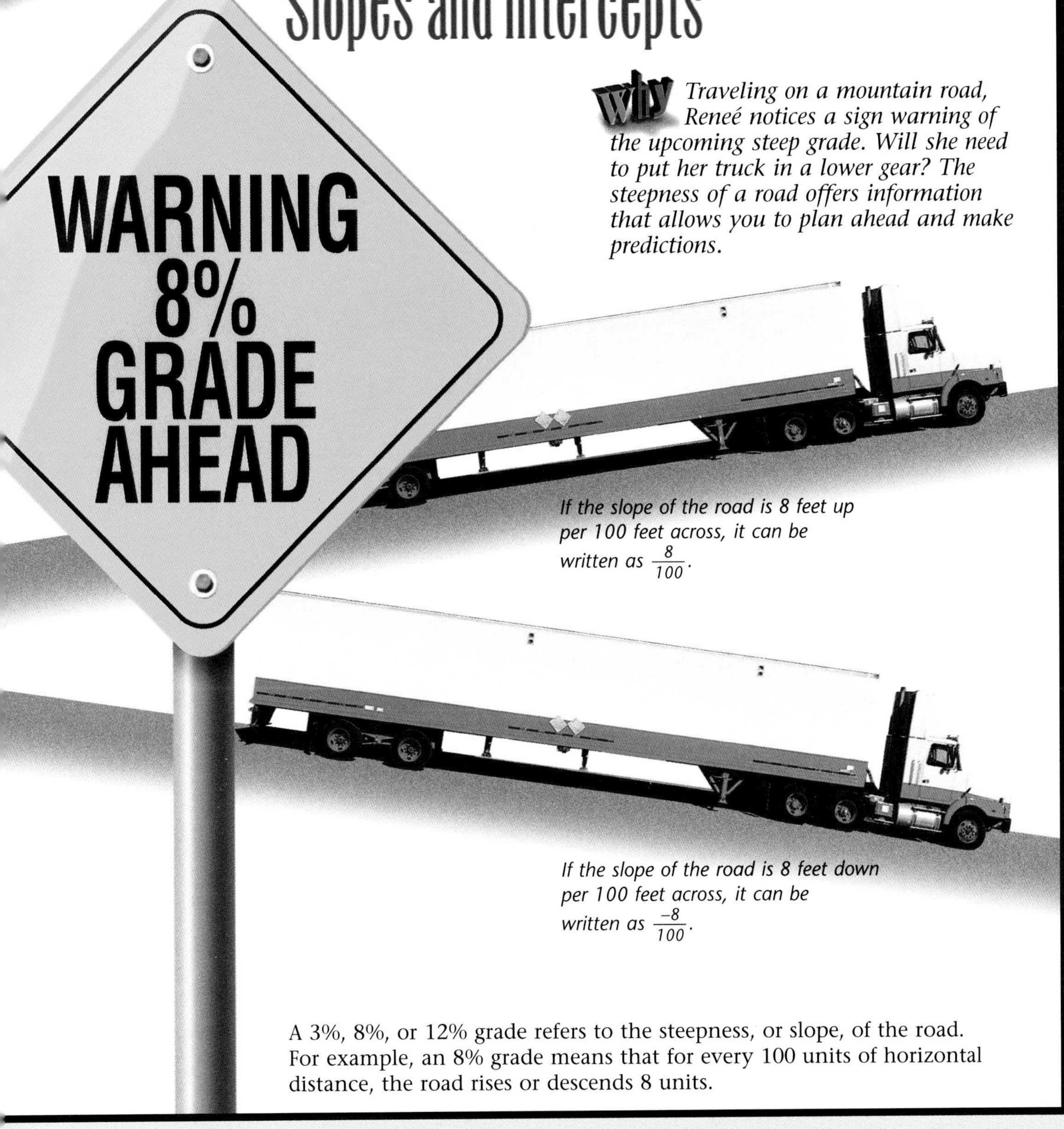

WARNING 8% GRADE AHEAD

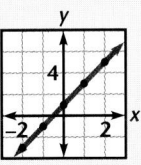

If the slope of the road is 8 feet up per 100 feet across, it can be written as $\frac{8}{100}$.

If the slope of the road is 8 feet down per 100 feet across, it can be written as $\frac{-8}{100}$.

A 3%, 8%, or 12% grade refers to the steepness, or slope, of the road. For example, an 8% grade means that for every 100 units of horizontal distance, the road rises or descends 8 units.

 **ALTERNATIVE teaching strategy**

Cooperative Learning Have students work in groups of two or three to graph lines with slopes of 0.5, 1, 2, −0.5, −1, and −2. Encourage each group to discuss the meaning of slope, keeping in mind that some students may not recall the exact meaning of slope. After each group has completed the assignment, ask for a description of what happens to the graph of a line as slope increases and decreases. Ask them the meaning of positive and negative slope.

PREPARE

Objectives

• Write an equation in slope-intercept form, given two points on the line or the slope and a point on the line.

RESOURCES

• Practice Master   1.2
• Enrichment Master   1.2
• Technology Master   1.2
• Lesson Activity Master   1.2
• Quiz   1.2
• Spanish Resources   1.2

Assessing Prior Knowledge

1. Graph $y = 2x + 1$ by making a table of values.

[
x	−2	−1	0	1	2
y	−3	−1	1	3	5

]

2. What is slope? [**Slope is rise over run, or y-change divided by x-change.**]

3. Will the slope change anywhere on a line? [**No**]

TEACH

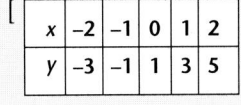 Discuss why slope is relevant in real life. Have students describe situations in which it is necessary to know slope.

The explorations can be completed in small groups. Students should find completing Exploration 3 in groups especially beneficial because they can divide the many tasks.

CRITICAL *Thinking*

$y = (2/3)x$ is better because 2/3 is actually 0.666..., a non-terminating decimal. If you enter $y = 0.67x$, the calculator will not be using all the decimal places that it can calculate.

The answers to exploration steps can be found in Additional Answers beginning on page 842.

Exploration 1 Notes

Be sure students understand that no matter what pair of points is chosen on a line, the slope is always the same. Step 10 in this exploration helps students realize that if a linear equation has been solved for y, then the coefficient of x is the slope. Before beginning this exploration, it may be helpful to assure students understand the meaning of the work *coefficient*.

TEACHING *tip*

Explain how to find slope from the graphed line using rise over run. Emphasize that the rise means the same as $y_2 - y_1$ and that the run means the same as $x_2 - x_1$.

Slope of a Line

In the following explorations, you will be using your graphics calculator to graph lines. When an equation has a fraction, such as $y = \frac{1}{2}x$, you can enter the equation using the equivalent decimal number: $y = 0.5x$. If the fraction in the equation has a *repeating* decimal-number equivalent, such as $y = \frac{1}{3}x$, you can enter the equation with the fraction in parentheses: $y = (1/3)x$.

 **CRITICAL** *Thinking*   Is it better to enter the equation $y = \frac{2}{3}x$ into your graphics calculator as $y = (2/3)x$ or $y = 0.67x$? Explain.

•Exploration 1 *Finding the Quotient of Differences*

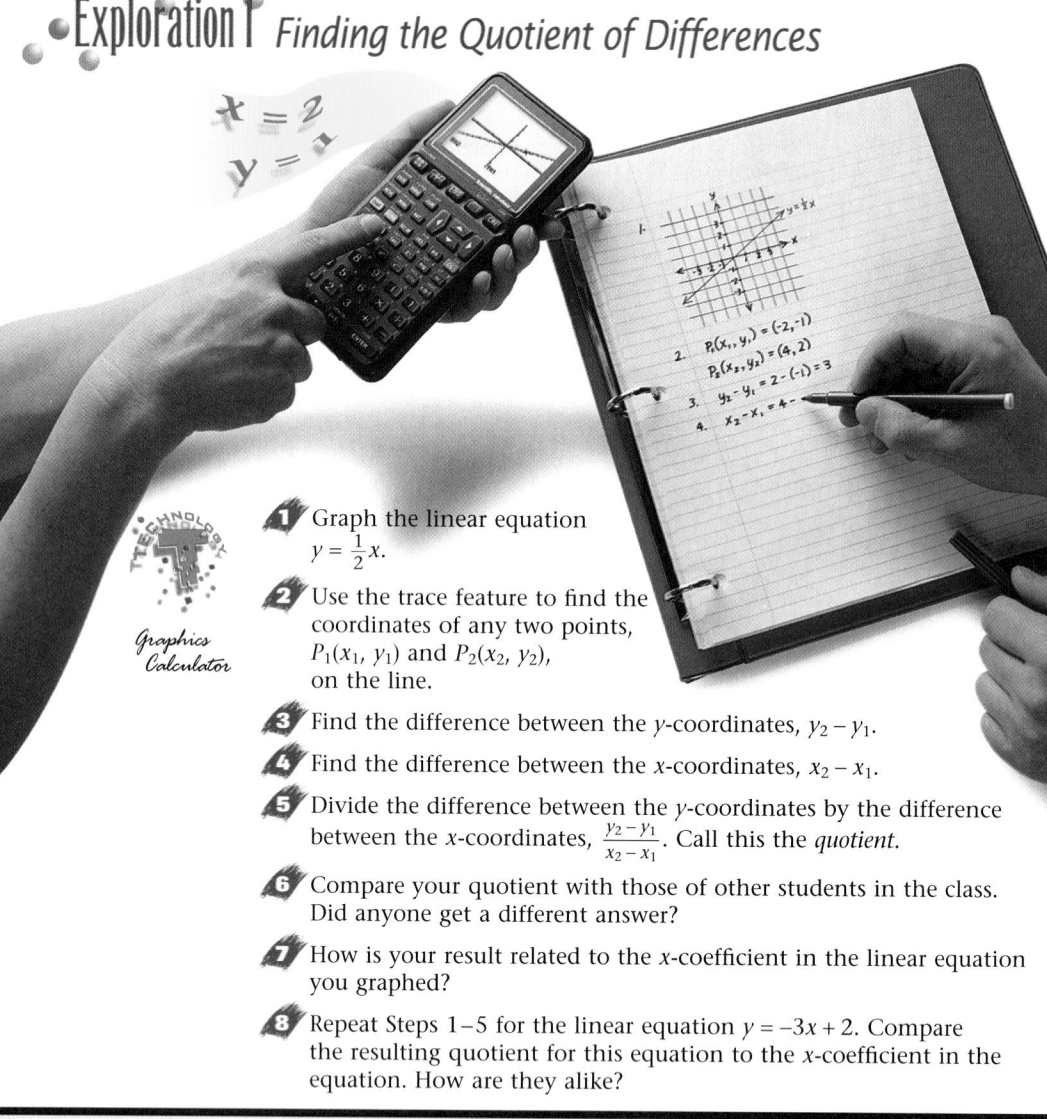

Graphics Calculator

1 Graph the linear equation $y = \frac{1}{2}x$.

2 Use the trace feature to find the coordinates of any two points, $P_1(x_1, y_1)$ and $P_2(x_2, y_2)$, on the line.

3 Find the difference between the y-coordinates, $y_2 - y_1$.

4 Find the difference between the x-coordinates, $x_2 - x_1$.

5 Divide the difference between the y-coordinates by the difference between the x-coordinates, $\frac{y_2 - y_1}{x_2 - x_1}$. Call this the *quotient*.

6 Compare your quotient with those of other students in the class. Did anyone get a different answer?

7 How is your result related to the x-coefficient in the linear equation you graphed?

8 Repeat Steps 1–5 for the linear equation $y = -3x + 2$. Compare the resulting quotient for this equation to the x-coefficient in the equation. How are they alike?

ENRICHMENT   A family takes a car trip. They travel for 3 hours at an average rate of 40 miles per hour. They rest for 1 hour and 30 minutes. They travel for 2 hours at 50 miles per hour. Graph the situation. Give the slope for each part of the graph. How many miles did the family travel?

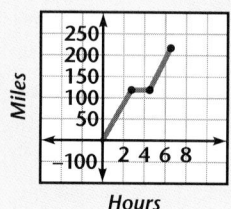

[Slopes: $m = 40$; $m = 0$, $m = 50$; 220 miles]

9 Predict the quotient that you will get for the equation $y = 5x + \frac{3}{4}$. Repeat Steps 1–5 to find the quotient. Was your prediction correct?

10 Describe the relationship between the *x*-coefficient in the linear equations in this exploration and the steepness of their graphs. ❖

The quotient you found in Exploration 1 is called the *slope*.

SLOPE OF A LINE

The **slope of a line** is the ratio of the change in vertical direction to the change in horizontal direction.

$$\text{slope} = \frac{\text{change in } y}{\text{change in } x} = \frac{y_2 - y_1}{x_2 - x_1}$$

The slope is used to describe the steepness of an incline or a decline.

Exploration 2 *Horizontal and Vertical Lines*

Graphics Calculator

Use your graphics calculator for Steps 1–4 of this exploration.

1 Graph the following equations on the same coordinate plane.
$$y = 3 \qquad y = 5 \qquad y = -2 \qquad y = -3$$

2 How are all of the graphs alike?

3 Find the slope of each line. Are any of the slopes the same?

4 Predict the slope of the line $y = 4$. Graph the equation, and find the slope. Was your prediction correct?

5 Make a generalization about the slopes of horizontal lines.

The graph of $x = 3$ is a vertical line.

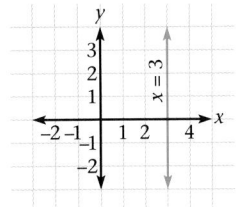

6 Use graph paper to graph the following equations on the same coordinate plane.
$$x = 4 \qquad x = -5 \qquad x = -1 \qquad x = -2$$

7 How are all of the graphs alike?

8 What can you say about the slopes of these lines?

9 Predict the slope of the line $x = -4$. Graph the equation. Was your prediction correct?

10 Make a generalization about the slopes of vertical lines. ❖

CRITICAL *Thinking*

Is it possible to use your graphics calculator to complete Step 6 in Exploration 2? If so, explain how. If not, explain why.

INCLUSION strategies

Using Cognitive Strategies Inability to calculate slope does not indicate that the student does not understand the concept of slope. It may be necessary for some students to review operations with positive and negative integers.

ongoing **ASSESSMENT**

10. The larger the absolute value of the *x*-coefficient, the steeper the graph.

Exploration 2 Notes

Students should conclude that all horizontal lines have an equation $y =$ (any number) and a slope of 0. They also should conclude that all vertical lines have an equation $x =$ (any number) and an undefined slope.

TEACHING *tip*

$y = 8$ means that whatever the *x*-value, the *y*-value will always be 8. The numerator of the slope formula will always be $8 - 8$, which is 0.

ongoing **ASSESSMENT**

5. Horizontal lines have a slope of 0.

10. Vertical lines have an undefined slope.

TEACHING *tip*

Refer to your *HRW Technology Handbook* or the calculator manual for instructions on graphing vertical lines.

CRITICAL *Thinking*

Answers will depend on the graphics calculator being used. On some calculators, vertical lines can be graphed using the draw feature. However, vertical lines cannot be graphed as functions.

Lesson 1.2 **17**

Aongoing
SSESSMENT

5. Graphs of linear equations with equal slopes are parallel.

10. The constant term in the equation is the *y*-coordinate of the point where the graph crosses the *y*-axis.

Exploration 3 Equal Slopes and y-Intercepts

Graphics Calculator

1 Graph the following linear equations on the same coordinate plane.
$$y = \tfrac{1}{4}x + 1 \qquad y = -\tfrac{1}{4}x + 1 \qquad y = \tfrac{1}{4}x - 1 \qquad y = -\tfrac{1}{4}x - 1$$

2 Repeat the steps from Exploration 1 for these equations, and record your results in the following table. Use the completed table for Steps 3–7.

Equation	$y_2 - y_1$	$x_2 - x_1$	$\dfrac{y_2 - y_1}{x_2 - x_1}$
$y = \tfrac{1}{4}x + 1$			
$y = -\tfrac{1}{4}x + 1$			
$y = \tfrac{1}{4}x - 1$			
$y = -\tfrac{1}{4}x - 1$			

3 Find two linear equations that have equal slopes. How are their graphs alike? How are their graphs different?

4 Are there two other linear equations that have equal slopes? How are their graphs alike?

5 What can you say about the graphs of linear equations that have equal slopes?

6 Which two lines cross the *y*-axis at (0, 1)? What parts of these two equations are alike?

7 Which two lines cross the *y*-axis at (0, –1)? What parts of these two equations are alike?

8 Graph the two equations $y = 3x + 2$ and $y = x + 2$ on the same coordinate plane. Where do they cross the *y*-axis? How do the crossing points compare with their equations?

9 Predict where each of the following equations will cross the *y*-axis. Graph each equation to see if your predictions were correct.
$$y = \tfrac{1}{3}x + 4 \qquad y = -\tfrac{1}{3}x - 1 \qquad y = \tfrac{3}{4}x + 8 \qquad y = x - 5$$

10 Explain the connection between the equations in this exploration and the points where their graphs intersect the *y*-axis. ❖

The point where the graph of a line crosses the *y*-axis is called the **y-intercept**.

RETEACHING
the lesson

English Language Development Ask students to graph the line $y = 2x + 1$. Have them write a verbal description of the location of that line and share their description with the class. Then give pairs of students the equation $y = 4x + 3$ on a piece of paper, and ask one student in each pair to describe to their partner how to graph the equation. The second student should graph the line. After students have completed this activity, explain the definition of slope and *y*-intercept.

The type of equations you graphed in Explorations 1, 2, and 3 are called linear equations in *slope-intercept form*.

SLOPE-INTERCEPT FORM
A linear equation is in **slope-intercept form** when it is in the form $y = mx + b$, where m represents the slope and b represents the y-intercept.

APPLICATION

At noon a pump starts emptying an oil storage tank at a constant rate.

Two points on the line modeling this relationship are (3, 1800) and (7, 600).

The slope is $m = \frac{1800 - 600}{3 - 7} = \frac{1200}{-4} = -300$.

To write an equation for this line in slope-intercept form, find the y-intercept, b.

Substitute the slope, m, and the x- and y-values from either one of the points in the slope-intercept form.

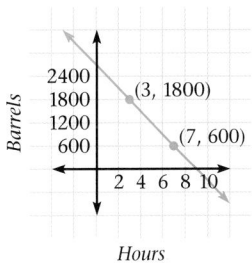

Barrels

Hours

Substitute x- and y-values.

$$y = mx + b$$
$$1800 = -300(3) + b$$
$$1800 = -900 + b$$
$$2700 = b \qquad \text{The y-intercept, } b, \text{ is } 2700.$$

Therefore, the equation of the line is $y = -300x + 2700$.

How many barrels of oil will remain at 8 P.M.? Substitute 8 for x (the hours pumping), and solve for y (the number of barrels remaining).

$$y = -300x + 2700$$
$$y = -300(8) + 2700$$
$$y = -2400 + 2700$$
$$y = 300$$

At 8 P.M. there will be 300 barrels remaining. ❖

TEACHING tip

Remind students that when something occurs at a constant rate, a linear equation should be used to represent the situation. Explain that an equation can be used to make predictions if one of the variables is known. Make sure students understand how to find the y-intercept, given the slope and a point.

Applications appear in the Exploration lessons. They provide a practical example of the concepts presented in an exploration. The Application also reviews the concepts from the explorations. This can help students that have difficulty identifying or comprehending the key concepts.

ASSESS

Selected Answers

Odd-numbered Exercises 7–47

The answers to Communicate Exercises can be found in Additional Answers beginning on page 842.

Assignment Guide

Core 1–11, 14–15, 17–20, 23–49

Core Plus 1–5, 8–13, 15–16, 19–49

Technology

A graphics calculator should be used to complete Exercises 30 and 49.

Error Analysis

Watch for students who try to calculate slope using the difference of the x-values divided by the difference of the y-values. Emphasize the relationship between rise/run and the formula for slope.

Practice Master

NAME ______________ CLASS ________ DATE ________

Practice & Apply

1.2 Exploring Slopes and Intercepts

Name the slope m and the y-intercept b of each line.

1. $y = 0.2x - 3$ ______ 2. $y = -4x + 7$ ______ 3. $y = -x$ ______
4. $y = \frac{1}{2}x - 6$ ______ 5. $y = 5$ ______ 6. $y = -\frac{3}{4}x + 8$ ______

Write a linear equation with the indicated slope m and y-intercept b.

7. $m = 3; b = 4$ ______ 8. $m = -2; b = 0$ ______
9. $m = -\frac{1}{4}; b = 5$ ______ 10. $m = 0; b = -6$ ______

For Exercises 11–14, (a) find the slope of a line passing through the indicated points, and (b) write the equation of the line in slope-intercept form.

11. (2, 3) and (5, 9) ______
12. (−1, 4) and (3, −8) ______
13. $\left(\frac{1}{3}, -2\right)$ and $\left(2, -\frac{1}{3}\right)$ ______
14. (−3, −5) and (−1, −2) ______
15. Which equation has a 0 slope? ______
    A. $y = 4x - 4$ B. $x = 4$ C. $y = 4$ D. $y = -4x + 4$
16. Which equation does not have the same slope as $3x + y = 6$? ______
    A. $-3x - 6 = -y$ B. $6x + 2y = 12$ C. $x + \frac{1}{3}y = 6$ D. $-6 + y = -3x$
17. To find the slope of the line through (−1, 9) and (2, 3), Jane found the slope as $\frac{9-3}{-1-2}$. Paul found the slope as $\frac{3-9}{2-(-1)}$. Compare the two answers.

______

The linear equation $y = 9.97 + 1.05x$ models the extent of air and water pollution where x is the amount of industrialization of a geographical region and y is the amount of air and water pollution.

18. Use your own graph paper to graph the equation.
19. Suppose the geographical region has an industrialization value of 80. What would the pollution value be? ______

2 Practice & Apply HRW Advanced Algebra

EXERCISES & PROBLEMS

Communicate

1. Suppose (−1, 7), (0, 4), and (2, −2) are all points on one line. Is the slope of the line the same no matter which two points you use to find the slope? Why or why not?
2. Describe the graph of a line that has a negative slope. Describe the graph of a line that has a positive slope.
3. Describe the slope of a horizontal line. Describe the slope of a vertical line.
4. How can you write the equation of a line with a slope of −3 and a y-intercept of 7?
5. Suppose the points (3, −2) and (4, 5) are on one line. How can you write the equation of the line?

Numbers in blue represent negative degrees or values.

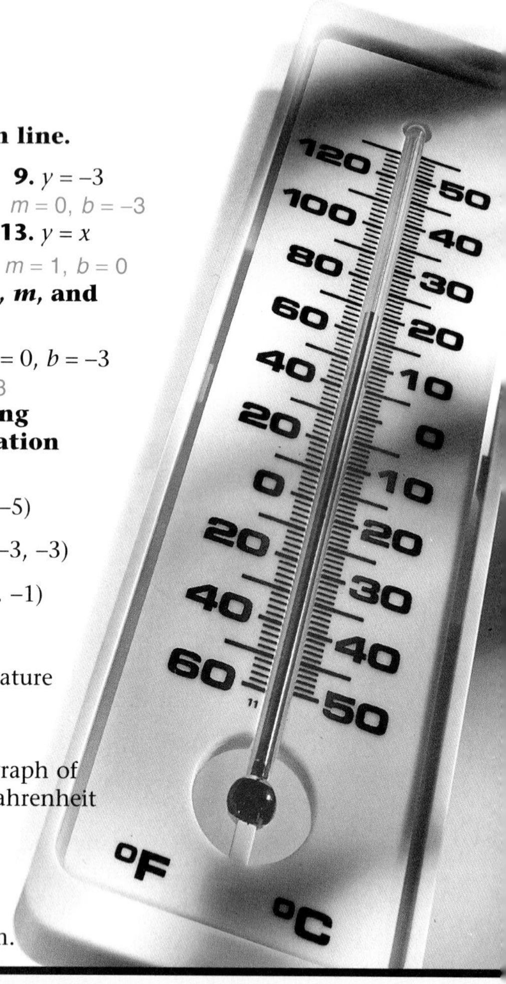

Practice & Apply

Name the slope, m, and the y-intercept, b, of each line.

6. $y = 0.6x - 4$ $m = 0.6$, $b = -4$
7. $y = -2x + 1$ $m = -2$, $b = 1$
8. $y = -\frac{1}{2}x - 4$ $m = -\frac{1}{2}$, $b = -4$
9. $y = -3$ $m = 0$, $b = -3$
10. $x = 2$
11. $y = \frac{1}{3}x - 7$ $m = \frac{1}{3}$, $b = -7$
12. $y = -x + 6$ $m = -1$, $b = 6$
13. $y = x$ $m = 1$, $b = 0$

Write a linear equation with the indicated slope, m, and y-intercept, b.

14. $m = 2$, $b = 0.75$ $y = 2x + 0.75$
15. $m = -5$, $b = 0$ $y = -5x$
16. $m = 0$, $b = -3$ $y = -3$

For Exercises 17–22, find the slope of a line passing through the indicated points, and write the equation of the line in slope-intercept form.

17. (0, 0) and (3, 30) 10, $y = 10x$
18. (1, −3) and (3, −5)
19. $\left(\frac{1}{2}, -3\right)$ and $\left(3, -\frac{1}{2}\right)$ 1, $y = x - 3\frac{1}{2}$
20. (−10, −4) and (−3, −3)
21. (−6, −6) and (−3, 1) $\frac{7}{3}$, $y = \frac{7}{3}x + 8$
22. (−2, 8) and (−2, −1)

Meteorology The thermometer shows temperatures in degrees Fahrenheit and in degrees Celsius. Room temperature is 20˚C, or 68˚F.

23. Use the thermometer to name two other Fahrenheit temperatures and their Celsius equivalents. Make a graph of the points with Celsius on the horizontal axis and Fahrenheit on the vertical axis. Draw a line through the points.
24. Identify the slope of the line. 1.8 or $\frac{9}{5}$
25. Find the y-intercept of the line. 32
26. Write the equation of the line in slope-intercept form.

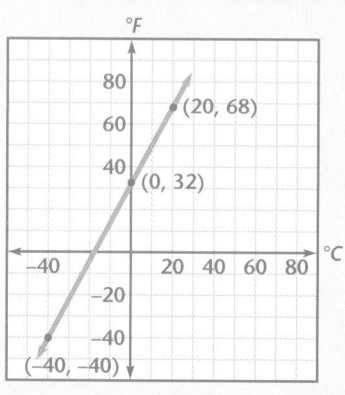

10. m is undefined, no y-intercept

18. $m = -1$, $y = -x - 2$

20. $m = \frac{1}{7}$, $y = \frac{1}{7}x - \frac{18}{7}$

22. m is undefined, $x = -2$

23. Answers may vary. Two equivalent temperatures in the form (C, F) are (−40, −40) and (0, 32). See graph at right.

26. $F = 1.8C + 32$, or $F = \frac{9}{5}C + 32$

Depreciation Abdul buys a computer for $3600. Its value decreases constantly by $600 per year.

27. What is the slope of the line that models this depreciation? −600

28. Find the *y*-intercept of the line. 3600

29. Write the linear equation, in slope-intercept form, of the line. $y = -600x + 3600$

30 Graph the equation you wrote for Exercise 29 on your graphics calculator. Find the value of Abdul's computer after 4.5 years.
$900

Physics There is a linear relationship between the weight hanging on a spring and the length that it stretches.

31. Find the slope of the line that models the stretch of the spring depending on weight.

32. Find the *y*-intercept of the line that $m = \frac{3}{16}$ models the stretch of the spring depending on weight. 0

33. In slope-intercept form, write the equation that models the stretch of the spring depending on weight. $s = \frac{3}{16}w$

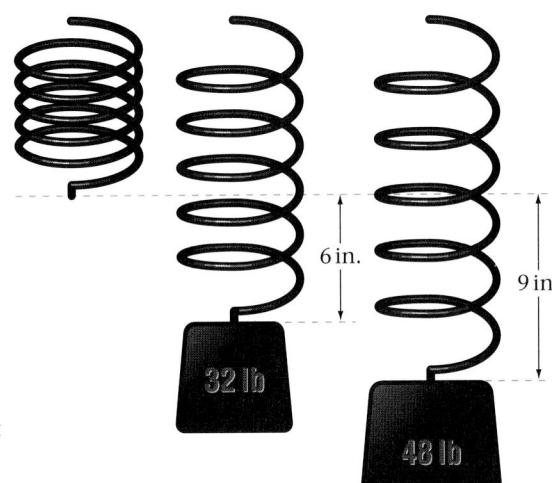

6 in.

9 in.

The more weight that is hanging on a spring, the more the spring stretches.

Construction Slopes are used in the planning and construction of the roof of a building.

34. Find the pitch of a roof if the rise is 12 feet and the span is 30 feet. $\frac{4}{5}$ or 0.8

35. Find the pitch of a roof if the rise is 18 feet and the span is 60 feet. $\frac{3}{5}$ or 0.6

36. Find the pitch of a roof if the rise is 4 feet and the span is 50 feet. $\frac{4}{25}$ or 0.16

A carpenter must know the pitch (slope) of a roof to cut rafters of the proper length.

$$pitch = \frac{rise\ of\ a\ roof}{\frac{1}{2} \times span\ of\ a\ roof}$$

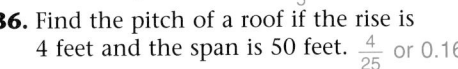

rise

span

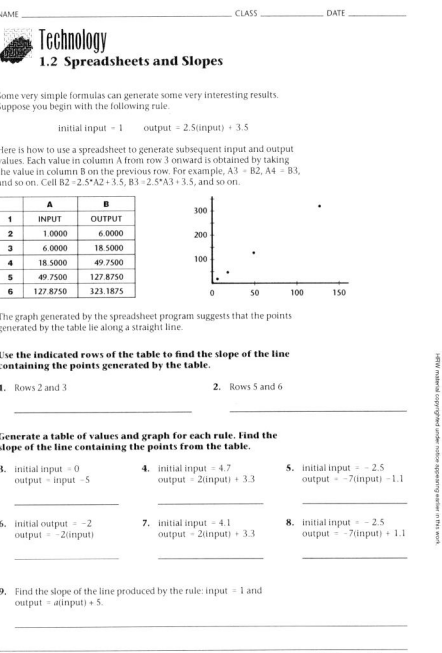

Technology Master

NAME _____ CLASS _____ DATE _____

Technology
1.2 Spreadsheets and Slopes

Some very simple formulas can generate some very interesting results. Suppose you begin with the following rule.

initial input = 1   output = 2.5(input) + 3.5

Here is how to use a spreadsheet to generate subsequent input and output values. Each value in column A from row 3 onward is obtained by taking the value in column B on the previous row. For example, A3 = B2, A4 = B3, and so on. Cell B2 = 2.5*A2 + 3.5, B3 = 2.5*A3 + 3.5, and so on.

	A	B
1	INPUT	OUTPUT
2	1.0000	6.0000
3	6.0000	18.5000
4	18.5000	49.7500
5	49.7500	127.8750
6	127.8750	323.1875

The graph generated by the spreadsheet program suggests that the points generated by the table lie along a straight line.

Use the indicated rows of the table to find the slope of the line containing the points generated by the table.

1. Rows 2 and 3

2. Rows 5 and 6

Generate a table of values and graph for each rule. Find the slope of the line containing the points from the table.

3. initial input = 0
output = input − 5

4. initial input = 4.7
output = 2(input) + 3.3

5. initial input = −2.5
output = −7(input) − 1.1

6. initial output = −2
output = −2(input)

7. initial input = 4.1
output = 2(input) + 3.3

8. initial input = −2.5
output = −7(input) + 1.1

9. Find the slope of the line produced by the rule: input = 1 and output = a(input) + 5.

14   Technology   HRW Advanced Algebra

National Debt The table gives the public debt (in billions of dollars) of the United States for some years between 1880 and 1990.

37. How much did the public debt increase between the years 1880 to 1990? $3231.3 billion

38. In what decade did the public debt increase the most? What is the slope of the graph for that decade? 1980s, $232.36 billion / year

39. In what decade did the public debt decrease the most? What is the slope of the graph for that decade? 1920s, – $0.81 billion / year

YEAR	DEBT
1880	2.0
1890	1.1
1900	1.2
1910	1.1
1920	24.2
1930	16.1
1940	43.0
1950	256.1
1960	284.1
1970	313.8
1980	909.7
1990	3233.3

Look Back

40. Find the value of V in the equation $V = lwh$ when l is 2, w is 3, and h is 5. 30

For Exercises 41–44, let x be 6.15, y be 2.05, and z be 0.0123. Evaluate each expression.

41. xy 12.6075   **42.** $x + y + z$ 8.2123 **43.** $3.0495 - y$ **44.** $x \div z$ 500
0.9995

45. Use the formula $D = rt$ to find the distance, D, when the rate, r, is 50 and the time, t, is 4. 200

46. Use the formula $C = \pi d$ to find the circumference, C, when the diameter, d, is 8. (Use 3.14 for π.) 25.12

47. Use the formula $I = prt$ to find the interest, I, when the principal, p, is 1000, the interest rate, r, is 0.08, and the time, t, is 2. $160

48. Use the formula $P = 4s$ to find the perimeter, P, when the length of a side, s, is 16. 64

Look Beyond

Look Beyond

Exercise 49 foreshadows systems of equations that will be solved using traditional paper-and-pencil computations or a graphics calculator.

49 Use your graphics calculator to graph $y = 2x - 3$ and $y = x + 5$ on the same coordinate plane. Find the coordinates of the point of intersection. (8, 13)

LESSON 1.3 Scatter Plots and Correlation

 *Have you ever wondered if studying longer results in getting a better grade on a test? One way to find out how two events are related is to gather data and make a scatter plot.*

Time-for-Study Table	
h	s
1	60
1	55
2	68
2	70
2	100
3	65
3	70
3	80
3	74
3	68
4	70
4	80
4	88
5	92
5	75
6	85
6	94
6	100
6	98
7	95

Time-for-Study Graph

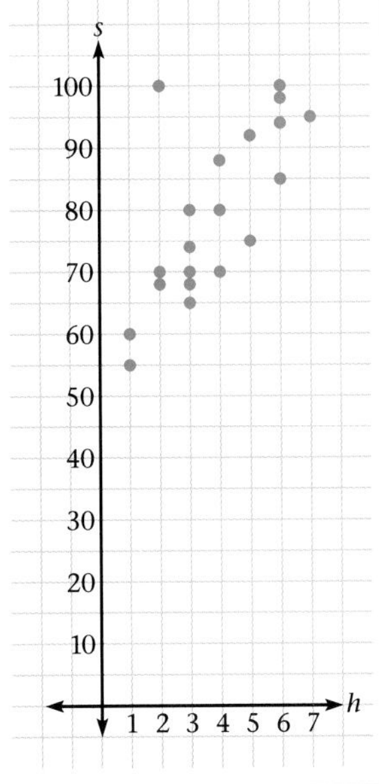

Brian surveyed 20 classmates. The data from his survey is shown in the table. Hours studied is represented by h, and test scores are represented by s.

The relationship between the hours studied and resulting test scores can be viewed by graphing the ordered pairs (h, s). This graph is called a **scatter plot**. What relationship between h and s can you see in this graph?

ALTERNATIVE teaching strategy

Using Cognitive Strategies Have students brainstorm to think of general statements that describe a relationship between two different variables. Get students started by offering the following examples: the distance a student lives from school and the time it takes to get to school; years of schooling and earnings in later life; age and height; the number of students present in a classs and the number of empty chairs. Have students work in small groups to gather data pertaining to one of the ideas. Have each group draw a scatter plot of the data and analyze it.

PREPARE

Objectives
- Graph a scatter plot and identify the data correlation.
- Use a graphics calculator to find the correlation coefficient and to make predictions using the line of best fit.

RESOURCES
- Practice Master — 1.3
- Enrichment Master — 1.3
- Technology Master — 1.3
- Lesson Activity Master — 1.3
- Quiz — 1.3
- Spanish Resources — 1.3

Assessing Prior Knowledge

1. Describe the graph of the equation $y = 2x + 4$. Is the slope positive or negative? [**The graph is a line with a slope of 2 and a y-intercept of 4; positive**]

2. From looking at a graph, how can you tell if the slope is positive or negative? [**A line with positive slope rises from left to right; a line with negative slope falls from left to right.**]

TEACH

 It is important to study relationships between variables to see how different variables affect each other. Have students make a list of variables that are related, such as exercise and longevity.

Use Transparency 1

CRITICAL Thinking

Similar to positive and negative slopes, scatter plots showing a positive correlation show points that slope upward to the right, and scatter plots showing a negative correlation show points that slope downward to the right. Scatter plots showing zero correlation, however, show no pattern, which differs from horizontal lines that have zero slope.

Alternate Example 1

The table shows the number of juniors and seniors taking Advanced Algebra at Western High School for each of the past 8 years.

Yr.	1	2	3	4	5	6	7	8
Jr.	33	29	37	45	43	43	51	61
Sr.	30	25	28	26	37	35	38	42

A. Graph the data for juniors and seniors on two separate scatter plots.

[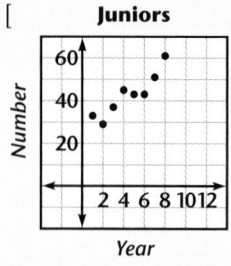

Juniors

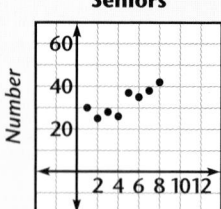

Seniors

]

B. Identify the correlation for each scatter plot. [**The points in both scatter plots rise upward to the right. Thus, the relationship has a positive correlation**]

C. Identify any outliers. [**There are no obvious outlines.**]

Correlation and Line of Best Fit

STATISTICS Connection

Notice that the points in the Time-for-Study graph tend to slope upward to the right. This indicates that the test scores increase as the hours spent studying increase. This relationship is called a *positive correlation*. Sometimes there is a *negative correlation*, and sometimes there is no correlation, or a *zero correlation*.

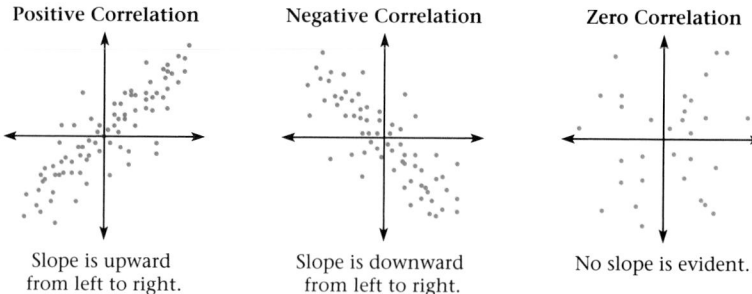

Positive Correlation	Negative Correlation	Zero Correlation
Slope is upward from left to right.	Slope is downward from left to right.	No slope is evident.

CRITICAL Thinking

Compare correlation and slope. How are they alike? How are they different?

Sometimes points on a scatter plot appear to stand out from the rest of the points. These points are called **outliers** and are sometimes omitted when the data is used by statisticians. In the Time-for-Study graph, the point (2, 100) is considered an outlier.

EXAMPLE 1

Gordon and Lydia want to find a relationship between making a basket and the distance from the hoop. They each tried 20 shots at distances ranging from 10 feet to 26 feet. The table shows their data.

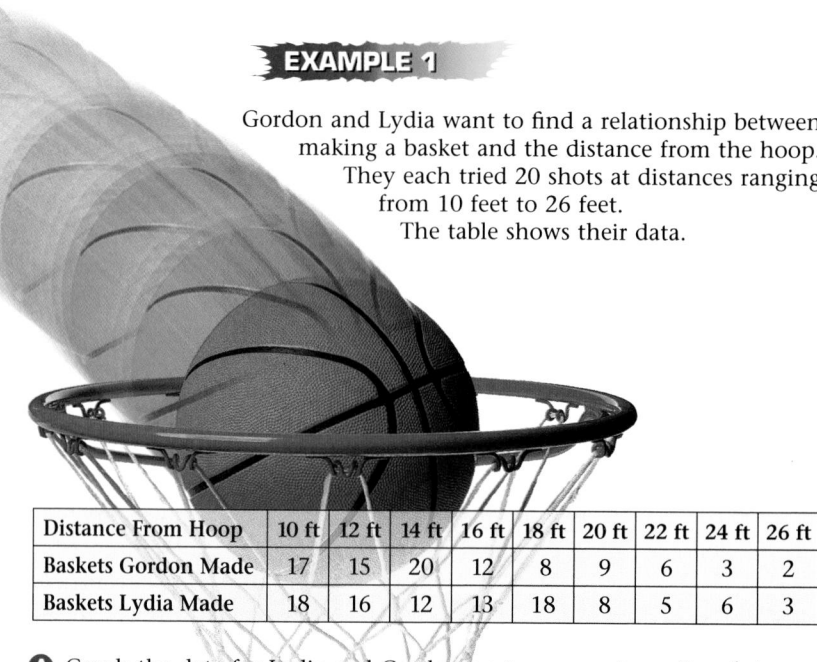

Distance From Hoop	10 ft	12 ft	14 ft	16 ft	18 ft	20 ft	22 ft	24 ft	26 ft
Baskets Gordon Made	17	15	20	12	8	9	6	3	2
Baskets Lydia Made	18	16	12	13	18	8	5	6	3

Ⓐ Graph the data for Lydia and Gordon on two separate scatter plots.

Ⓑ Identify the correlation for each scatter plot.

Ⓒ Identify any outliers.

interdisciplinary CONNECTION **Medical Research** Correlation is used continually in research that employs statistical methods. For example, studies have considered whether there is a correlation between smoking cigarettes and getting different heart and lung diseases. Have students make predictions about whether these studies have a positive or negative correlation. Then have students research the actual findings and compare them with their predictions.

Solution

A

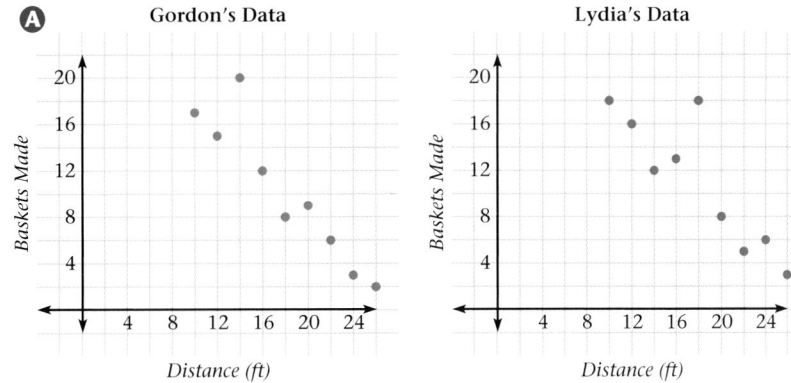

Gordon's Data

Lydia's Data

B The points in both scatter plots slope downward to the right. Thus, the relationship between distance from the hoop and baskets made has a negative correlation for both Gordon and Lydia.

C The point (14, 20) on Gordon's graph may be considered an outlier. The point (18, 18) on Lydia's graph may be considered an outlier. ❖

What are some examples of outliers in data from the real world?

•Exploration *Fitting a Line*

1 Graph the following ordered pairs in a scatter plot.

(−10, −5)	(−6, −8)	(−5, −3)	(−3, −8)	(−2, −1)
(−2, −4)	(−2, 6)	(−1, 3)	(3, 1)	(4, −2)
(5, 5)	(6, 9)	(7, 3)	(8, 8)	(10, 4)

2 Use a straightedge to draw the line that you think fits best through all of the data points.

3 Compare your line with other students' lines. Did any two students draw the same line through the same points?

4 Explain how you would define a *line of best fit*. ❖

Correlation Coefficient

STATISTICS
Connection

The **line of best fit** is the straight line that fits closest to all of the data points in a scatter plot.

The line of best fit may go through some or none of the points. A *correlation coefficient* measures the mutual relationship, or the correlation, between the two sets of data.

The **correlation coefficient**, represented by the variable r, describes how closely points in a scatter plot cluster around the line of best fit.

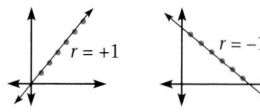

Emphasize that a negative correlation means that as one variable
increases, the other decreases. A
positive correlation means that
as one of the variables increases,
the other one also increases.

Use Transparency ▶ 3

Alternate Example 2

Enter the data in Alternate
Example 1 for the juniors into
your graphics calculator.
A. Find the correlation coefficient, r, to the nearest tenth.
[$r \approx 0.9$]
B. Plot the line of best fit,
and use it to predict the
number of juniors who will
take Advanced Algebra in
the ninth year. [**Using the
line of best fit, there should
be about 60 juniors in
Advanced Algebra in the
ninth year.**]

CRITICAL
Thinking

The closer the correlation coefficient is to 1 or −1, the more
accurate the prediction will be.

When all the points fit on one line,
$r = +1$ or $r = -1$.

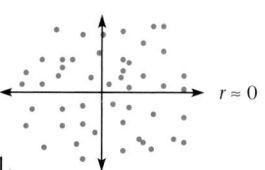

When the points are random and no
specific line can be considered a
best fit, $r \approx 0$.

The value of r can be any value from
−1 to +1. The closer the points are to
the line of best fit, the closer r is to −1 or +1.

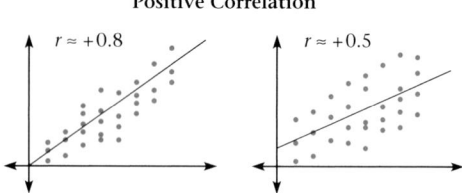

Negative Correlation

$r \approx -0.8$ $r \approx -0.5$

Positive Correlation

$r \approx +0.8$ $r \approx +0.5$

EXAMPLE 2

*Graphics
Calculator*

Enter the data in the Time-for-Study table at the beginning of the lesson
into your graphics calculator.

Ⓐ Find the correlation coefficient, r, to the nearest tenth.

Ⓑ Plot the line of best fit on the same coordinate plane as your scatter
plot for this data set. Use the line of best fit to predict the test score
for a student who studies half an hour.

Solution▶

Ⓐ With the data stored in your
calculator, select the statistics
menu on your calculator, and use
the *linear regression* command.
You can see that $r \approx +0.7$.

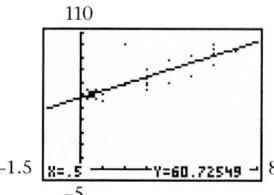

Ⓑ The line of best fit is shown with
the scatter plot. Place the cursor
on the point on the line with
an x-coordinate of 0.5, and read
the y-coordinate. According to
the line of best fit, a student who
studies half an hour will make
approximately 61 on the exam. ❖

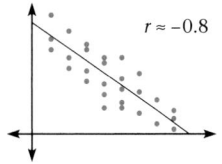 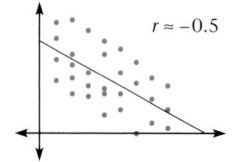

CRITICAL
Thinking

Explain how the correlation coefficient, r, is related to the accuracy of any
prediction you make using the line of best fit.

 **RETEACHING
the
lesson**

Independent Learning
Have students gather data
relating to one of the
following categories: lati-
tude north and mean annual temperature; age
and weight during the first two years of life; U.S.

population during the past 100 years; field goals
made and the distance of the kick for your football team in the past three years. In each case
students should gather the information, organize it, prepare tables, plot points, determine a
line of best fit, find the correlation coefficient,
and make a prediction.

EXERCISES & PROBLEMS

Communicate

1. Give examples of when high positive correlations might occur and when high negative correlations might occur.

Describe the scatter plots that have the following correlation coefficients, *r*.

2. $r = 0.02$ **3.** $r = -0.61$ **4.** $r = 0.96$

Name the type of correlation you would expect for each pair of variables below: positive, negative, or zero. Explain why.

5. Average hours spent watching TV per night and grade point average

6. Speed of a car and distance traveled to a complete stop after brakes are applied

Practice & Apply

For Exercises 7–10, use the following table.

x	1	2	3	4	5	6	7	8	9	10
y	–6	–3	–4	0	–6	3	5	4	12	7

7. Draw a scatter plot to represent this data.

8. What is the correlation between the two variables *x* and *y*: positive, negative, or zero? positive

9. Are there any outliers? If so, name them. (9, 12),(5, –6)

10. Find the correlation coefficient, *r*, to the nearest tenth. 0.9

Marketing Fifteen people of various ages were polled and asked to estimate the number of cassette tapes they had bought in the previous year. The following table contains the collected data.

Age	18	20	20	22	24	25	25	26	28	30	31	32	33	35	45
Cassettes	12	15	18	12	10	8	6	6	4	4	2	2	3	6	1

11. Let *x* represent age, and let *y* represent the number of cassettes purchased. Draw a scatter plot to display the data.

12. What type of correlation best describes this data: positive, negative, or zero? negative

13. Find the correlation coefficient, *r*, to the nearest tenth for the age and the number of cassettes purchased. –0.8

Assess

Selected Answers
Odd-numbered Exercises 7–25, 31–39.

The answers to Communicate Exercises can be found in Additional Answers beiginning on page 842.

Assignment Guide
Core 1–41

Core Plus 1–42

Technology
Graphics calculators are needed for Exercises 10, 13, 15–22, and 24–28.

7.
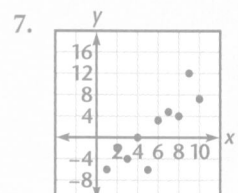

11.
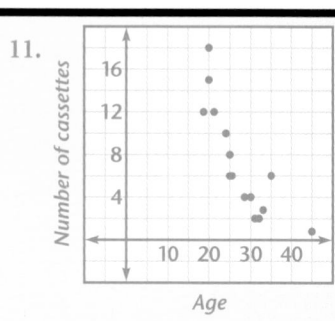

Anatomy The following tables give sample heights and shoe sizes of some adults. **Complete Exercises 14–22.**

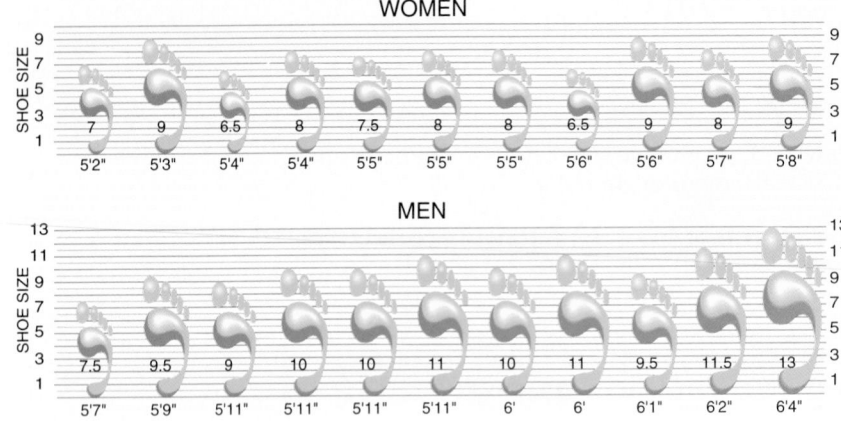

WOMEN

SHOE SIZE										
7	9	6.5	8	7.5	8	8	6.5	9	8	9
5'2"	5'3"	5'4"	5'4"	5'5"	5'5"	5'5"	5'6"	5'6"	5'7"	5'8"

MEN

SHOE SIZE										
7.5	9.5	9	10	10	11	10	11	9.5	11.5	13
5'7"	5'9"	5'11"	5'11"	5'11"	5'11"	6'	6'	6'1"	6'2"	6'4"

14. How will you enter the height 5'5" into your graphics calculator? 65

15. Enter the data for women, and display the data in a scatter plot. What is the equation of the line of best fit? $y = 0.17x - 2.97$

16. Enter the data for men, and display the data in a scatter plot. What is the equation of the line of best fit? $y = 0.52x - 26.91$

17. Find the correlation coefficient, r, for women to the nearest tenth. 0.3

18. Find the correlation coefficient, r, for men to the nearest tenth. 0.9

Use the line of best fit for this data to predict the following.

19. The shoe size of a woman 5'1" tall $7\frac{1}{2}$

20. The shoe size of a man 6'2" tall $11\frac{1}{2}$

21. The height of a man with size 12 shoe 6'3"

22. The height of a woman with size 8 shoe $5'4\frac{1}{2}"$

23. **Statistics** Suppose research indicates a strong negative correlation between the number of pages in a book and the popularity of the book. Does this mean that in order to write a bestseller, an author should make it as short as possible? Explain.

Sports The table gives the average speed of each winner of the Indy 500 from 1963 to 1980. Display the data in a scatter plot.

24. What is the equation of the slope of the line of best fit to the nearest tenth? $m \approx 0.5$

25. Find the correlation coefficient, r, to the nearest tenth. Explain how the value of r describes the data. 0.4

26. Use the line of best fit to predict the winning speed, to the nearest tenth of a mile, in the year 2000. 167.2 mph

Year	Speed
1963	143.1
1964	147.4
1965	151.4
1966	144.3
1967	151.2
1968	152.9
1969	156.9
1970	155.7
1971	157.7
1972	163.5
1973	159.0
1974	158.6
1975	149.2
1976	148.7
1977	161.3
1978	161.4
1979	158.9
1980	142.9

23. Answers may vary. One possible answer: It is usually best for the author to keep the book as short as possible, because many people will not buy a book they think they will never finish. However, there may be other contributing factors, such as the cost of the book. A bestseller that is lengthy may be considered an outlier.

25. $r \approx 0.4$. There is a weak positive correlation between the year and the average winning speed.

Cultural Connection: Africa Refer to the introduction in the portfolio activity on page 3. The following table gives realistic data for flood height (in cubits), the crop value (in hekats), and the tax collection (in hekats) for 10 consecutive years.

Portfolio Activity Use the information in the table to complete Exercises 27–30.

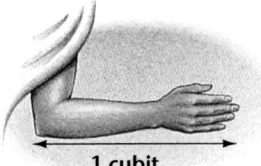

1 cubit

A cubit is an ancient unit of measure for length. The cubit originated from the length from the end of the middle finger to the elbow, approximately 20 inches.

Year	1	2	3	4	5	6	7	8	9	10
Flood Height (cubits)	12	13	13.4	14	16	8	10	15	14.2	14.6
Crop Value (hekats)	450	500	530	550	600	0	200	580	570	530
Tax Collection (hekats)	45	50	50	55	60	0	20	58	57	53

27 Make a scatter plot of flood height and crop value. Find the correlation coefficient, *r*, to the nearest thousandth. 0.962

28 Make a scatter plot of crop value and tax collection. Find the correlation coefficient, *r*, to the nearest thousandth. 0.999

29. Express the tax collection as an equation with crop value as the independent variable. $t = 0.099v + 0.011$

30. Why do you think that there was no crop value or tax collection in the sixth year?

A hekat is an ancient dry measure derived from a large bowl that holds about 16 cups of grain. Tax was collected in grain rather than money.

Look Back

Write a decimal to the nearest hundredth for each fraction.

31. $\frac{1}{3}$ **32.** $-\frac{3}{5}$ **33.** $\frac{17}{4}$ **34.** $-\frac{23}{80}$ **35.** $\frac{5}{3}$ **36.** $\frac{7}{20}$

0.33   −0.60   4.25   −0.29   1.67   0.35

Find the value of the variable that makes the equation true.

37. $648 = 24x$ 27   **38.** $p - 21 = 37$ 58   **39.** $19 = y - 47$ 66   **40.** $21 = \frac{1}{3}x$ 63

The Indianapolis 500 auto race is held each year on Memorial Day.

Look Beyond

Hobby Twelve screws are needed to build a "Wobble-not" stool.

41. If 25 stools are to be made, how many screws are needed? Explain how you found your answer. 300

42. Write a linear equation to represent this situation. Identify the slope and *y*-intercept. $y = 12x$, $m = 12$, $b = 0$

30. The flood height in the sixth year was unusually low. There was probably insufficient water for plants to grow, resulting in a lost crop, and thus no taxes.

41. Let *n* represent the number of stools. Since each stool requires 12 screws, *n* stools will require 12*n* screws. To find the number of screws for 25 stools, substitute 25 for *n*. So, $12n = 12(25)$, or 300 screws.

PREPARE

Objectives

- Solve problems involving direct variation.

RESOURCES

- Practice Master 1.4
- Enrichment Master 1.4
- Technology Master 1.4
- Lesson Activity Master 1.4
- Quiz 1.4
- Spanish Resources 1.4

Assessing Prior Knowledge

1. Using the equation $y = 4x$, complete the following ordered pairs: $(2, ?)$, $(3, ?)$, and $(4, ?)$. **[8, 12, 16]**
2. How are the x-values and y-values related in each ordered pair? [**The y-value is 4 times the x-value.**]
3. Solve for x. $\frac{1}{2}x = \frac{6}{8}$ **[1.5]**

TEACH

why Discuss the idea of a constant rate. Have students provide real-life examples of related data that increase or decrease at a constant rate. Be sure students understand how predictions can be made if change occurs at a constant rate.

LESSON 1.4 Direct Variation and Proportion

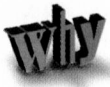

why *Many events in nature are related to each other by a constant multiple. The distance you are from lightning in a storm is a constant multiple of the time between seeing lightning and hearing the thunder.*

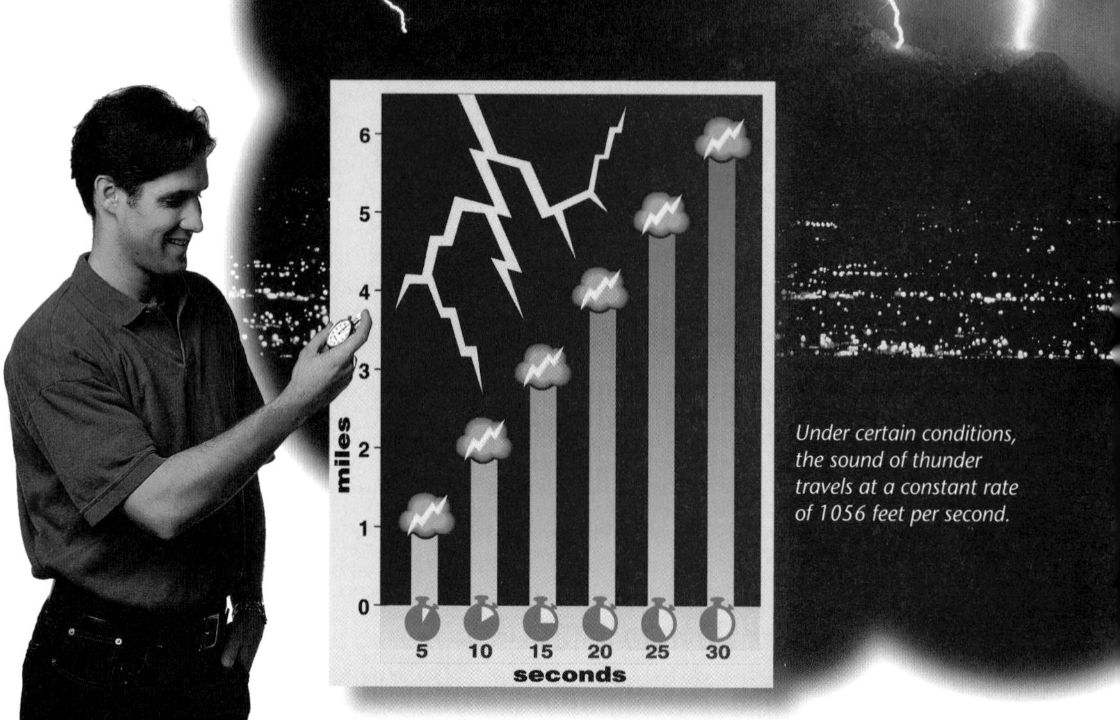

Under certain conditions, the sound of thunder travels at a constant rate of 1056 feet per second.

If you count 5 seconds between the time you see a bolt of lightning and the time you hear the thunder, the lightning is 5280 feet, or 1 mile, away. If you count 10 seconds, the lightning is 2 miles away. If you count 15 seconds, the lightning is 3 miles away, and so on.

ALTERNATIVE teaching strategy

Using Visual Models Direct variation may be introduced using circular geometric figures. Each figure shows that the circumference of a circle varies directly as its diameter. The constant is π. The circumference of a circle is π times the diameter. Explain that the constant need not be an integer.

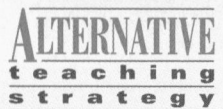

Direct Variation

In a *direct variation*, one variable is a constant multiple of the other. This constant multiple is often denoted by the letter k, after the German word *konstante*, meaning constant.

DIRECT VARIATION

In a **direct variation**, y varies directly as x. Thus, $y = kx$, where k is the **constant of variation** and $k \neq 0$.

EXAMPLE 1

Physics The distance, d, that a bolt of lightning is from you varies directly as the number of seconds, s, that you count after hearing the thunder, or $d = ks$. Suppose that you count 10 seconds when the lightning is 2 miles away. Find the constant of variation, k, and write the appropriate equation of direct variation.

Solution

Substitute 2 for d and 10 for s in the equation $d = ks$.

$$d = ks$$
$$2 = k \cdot 10$$
$$\frac{1}{5} = k$$

Thus, k is $\frac{1}{5}$, or 0.2.

To write the equation, substitute 0.2 for k in the equation $d = ks$. Thus, $d = 0.2s$ is the equation of direct variation. ❖

Try This Suppose y varies directly as x, and y is −42 when x is 6. Find the constant of variation, k, and write the appropriate equation of direct variation.

Proportion

Notice that these ordered pairs form a proportion.

(1, 6) and (2, 12) → $\frac{6}{1} = \frac{12}{2}$

In fact, any two ordered pairs that satisfy the direct variation $y = 6x$ form a proportion.

(1, 6) and (3, 18) → $\frac{6}{1} = \frac{18}{3}$ (−1, −6) and (5, 30) → $\frac{-6}{-1} = \frac{30}{5}$

(4, 24) and (−8, −48) → $\frac{24}{4} = \frac{-48}{-8}$ (−7, −42) and (−9, −54) → $\frac{-42}{-7} = \frac{-54}{-9}$

PROPORTION

In a direct variation, any two ordered pairs (x_1, y_1) and (x_2, y_2) that satisfy a direct variation equation are **proportional**. $\frac{y_1}{x_1} = \frac{y_2}{x_2}$

In the proportion $\frac{y_1}{x_1} = \frac{y_2}{x_2}$, notice that $y_1x_2 = x_1y_2$. This is commonly called *cross multiplication*. Also if $y_1x_2 = x_1y_2$, then $y_1 = \frac{x_1y_2}{x_2}$.

Alternate Example 1

The distance, d, traveled at a constant rate, k, varies directly with the time, t, in hours. Suppose it takes 2.5 hours to travel 75 miles at a constant rate. Find this constant and write an equation of direct variation.
[$k = 30$ mph, $d = 30t$]

Cooperative Learning

After students have read Example 1, have them work in small groups to list events that are related by direct variation. Have several groups read their lists for class discussion.

TEACHING tip

Make sure students understand that direct variation means that the y-value is equal to a number, called the constant, multiplied by the x-value. Tell students that if they know both the x- and y-values, they can always find the constant, k, by substituting the ordered pair (x, y) into $y = kx$.

ongoing ASSESSMENT

Try This

$k = -7$; $y = -7x$

ENRICHMENT Ask students whether the following situation represents direct variation and if so to write an equation: Potatoes cost k cents per pound. There is a 6% tax added to the cost. Write an equation for the cost of p pounds including tax.
[$c = 1.06kp$; $1.06k$ is the constant of variation]

INCLUSION strategies **Using Visual Models** Graphs could be used to demonstrate to students that the constant, k, is the number that is multiplied by the x-value and that this number is the same at all places on the graph. Show, using rise/run, that k is also the slope of the line.

Alternate Example 2

Wages for workers at a particular store are paid by the hour. A person working 18 hours earned $114.30. How many hours must this person work to earn $127.00? [20]

ACHING *tip*

If students struggle with using proportions, have them find the direct variation equations and substitute the given information.

Aongoing
ASSESSMENT

Try This

150 pounds

Aongoing
ASSESSMENT

Answers may vary. Four possible examples are miles driven and gallons of gas used, number of items purchased and sales tax paid, height and normal body weight, or the more rain received, the fewer times a lawn is watered.

CRITICAL
Thinking

If $y_1 = kx_1$, then $k = \frac{y_1}{x_1}$.
If $y_2 = kx_2$, then $k = \frac{y_2}{x_2}$.
Thus, $\frac{y_1}{x_1} = \frac{y_2}{x_2}$.

EXAMPLE 2

Astronomy The weight of an object on Mars y, varies directly as its weight on Earth, x. How much would a person weighing 120 pounds on Earth weigh on Mars?

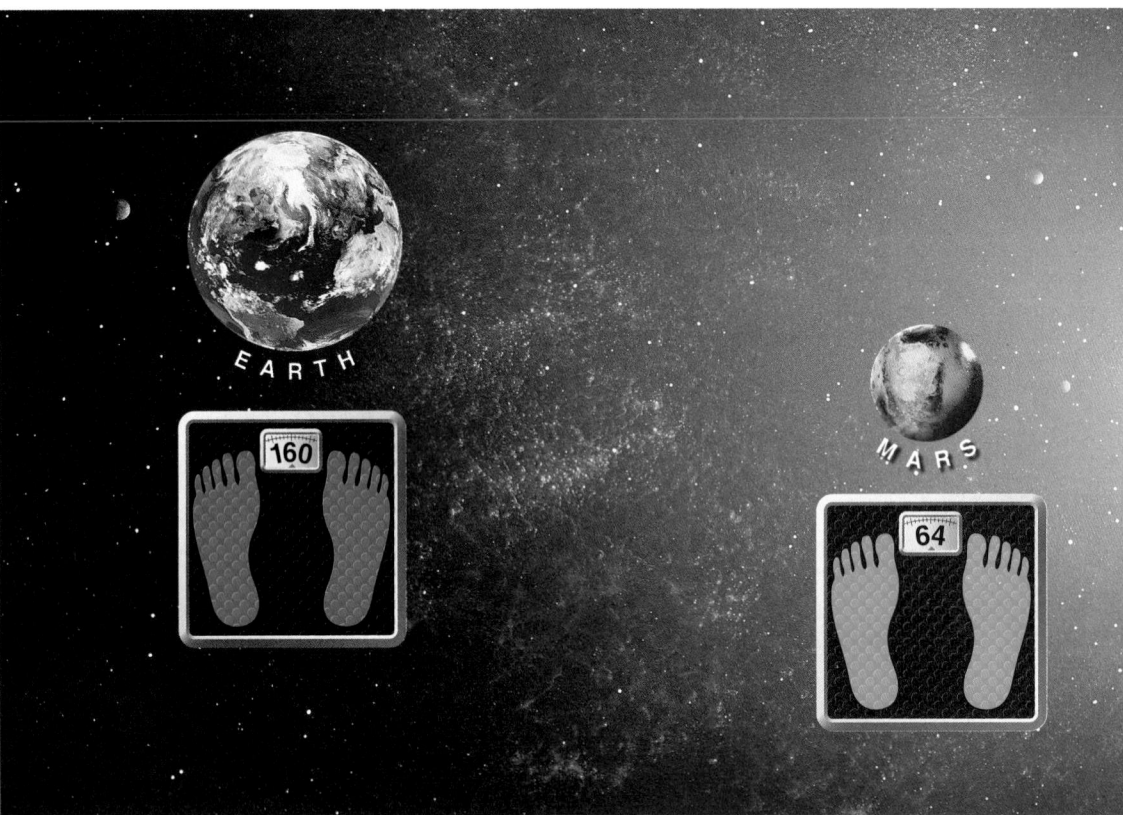

Solution

Since y varies directly as x, $\frac{y_1}{x_1} = \frac{y_2}{x_2}$. Write a proportion.

$$\begin{array}{c} \text{Weight on Mars} \rightarrow \\ \text{Weight on Earth} \rightarrow \end{array} \frac{y}{120} = \frac{64}{160}$$

Cross-multiply.
$$160y = 7680$$
$$y = 48$$

Therefore, a person weighing 120 pounds on Earth would weigh 48 pounds on Mars. ❖

Try This How much would a person who weighs 60 pounds on Mars weigh on Earth?

What are some examples of direct variation in your daily life?

 **CRITICAL** *Thinking*

Prove that any two ordered pairs (x_1, y_1) and (x_2, y_2) that satisfy a direct variation, $y = kx$, are proportional. In other words, show that if $y_1 = kx_1$ and $y_2 = kx_2$, where k is constant, then $\frac{y_1}{x_1} = \frac{y_2}{x_2}$.

 **RETEACHING the lesson**

Using Algorithms This lesson can be taught by emphasizing that direct variation equations are all linear equations where the y-intercept is zero and the slope is the constant. Give students the two points (3, 21) and (6, 42), and have them find the linear equation. Explain that the equation they should get, $y = 7x$, is a direct variation equation with a constant of 7. Point out that the two ordered pairs are proportional in different ways: $\frac{3}{21} = \frac{6}{42}$, $\frac{21}{3} = \frac{42}{6}$, $\frac{3}{6} = \frac{21}{42}$, or $\frac{6}{3} = \frac{42}{21}$.

EXERCISES & PROBLEMS

Communicate

1. Suppose that y varies directly as x, and y is 18 when x is 9. How do you write the equation of direct variation?

2. Are all linear equations direct variations? Are all direct variations linear equations? Explain your answers.

3. **Geometry** What is the constant of variation in the formula for circumference?

Explain how to write an equation that describes each direct variation.

4. p varies directly as q

5. a is directly proportional to b

$C = 2\pi r$

Practice & Apply

For each of the following values of x and y, y varies directly as x. Find the constant of variation, and write the equation.

6. y is 2 when x is 1 $2, y = 2x$

7. y is –16 when x is 2 $-8, y = -8x$

8. y is $\frac{4}{5}$ when x is $\frac{1}{5}$ $4, y = 4x$

9. y is –2 when x is 9 $-\frac{2}{9}, y = -\frac{2}{9}x$

10. y is 5 when x is –0.1 $-50, y = -50x$ **11.** y is 1.8 when x is 30 $0.06, y = 0.06x$

12. **Physics** How far away has lightning struck if 16 seconds pass between seeing the lightning and hearing the thunder?   3.2 mi

13. **Astronomy** Refer to Example 2. How much would a person weighing 135 pounds on Earth weigh on Mars?   54 lb

Write an equation of direct variation that relates the two variables.

14. **Geometry** The perimeter, p, of a square varies directly as the length, l, of a side. $p = 4l$

15. The distance, d, that a car travels at 60 miles per hour is directly proportional to the time in hours, t, traveled. $d = 60t$

For Exercises 16–18, a varies directly as b. Find the value for the indicated variable.

16. If a is 2.8 when b is 7, find a when b is –4. $a = -1.6$

17. If a is 6.3 when b is 70, find b when a is 5.4. $b = 60$

18. If a is –5 when b is 2.5, find b when a is 6. $b = -3$

ASSESS

Selected Answers

Odd-numbered Exercises 7–47

The answers to Communicate Exercises can be found in Additional Answers beginning on page 842.

Assignment guide

Core 1–10, 12–30, 32–33, 35–47, 49

Core Plus 1–5, 7–25, 27–31, 33–49

Technology

Calculators are helpful in this lesson. Graphics calculators are needed for Exercises 48 and 49.

Error Analysis

Be sure students understand how to set up proportions. Demonstrate all of the possibilities. Remind students that proportions can be solved by cross-multiplying or by multiplying both sides by the common denominator. Explain that an equation such as $2x = \frac{3}{5}$ can be treated as a proportion if $2x$ is written as $\frac{2x}{1}$.

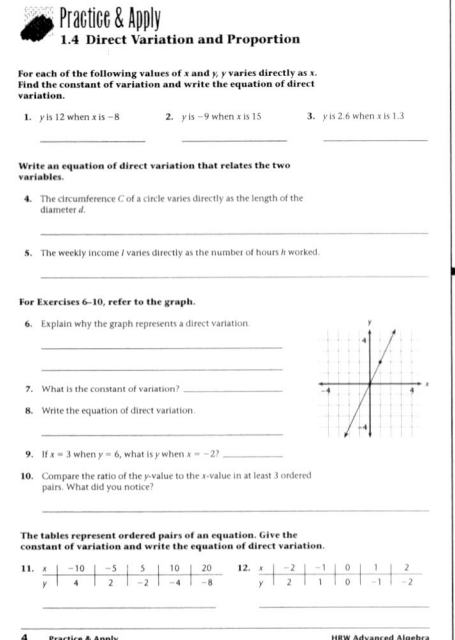
Physics The increase of pressure in your ears varies directly with the depth below the surface. The increase of pressure varies with depth at *different* rates in fresh water and salt water.

OCEAN
At 80 feet below the surface in salt water, the increase of pressure is about 35.6 pounds per square inch.

LAKE
At 80 feet below the surface in fresh water, the increase of pressure is 34.64 pounds per square inch.

19. What is the constant of variation for the increase of pressure in salt water? in fresh water? 0.445. 0.433

20. Write equations of direct variation for the increase of pressure in salt water and in fresh water. $p = 0.445d$, $p = 0.433d$

21. What is the pressure at 100 feet below the surface in salt water? in fresh water? 44.5 psi, 43.3 psi

Electronics Ohm's law states that the voltage, *V*, in an electric circuit varies directly with the electric current, *I*, measured in amperes. The constant of variation is the amount of resistance, *R*, in ohms: $V = IR$.

22. An iron is carrying a current of 5.5 amperes and is plugged into a 110-volt electrical outlet. Find the electric resistance of the iron. 20 ohm

23. Find the current carried by a heater with 11 ohms of resistance that is plugged into a 110-volt outlet. 10 amp

24. Find the current carried, to the nearest hundredth, by a night light with 300 ohms of resistance that is plugged into a 110-volt outlet. 0.37 amp

25. Find the current carried, to the nearest hundredth, by a 120-watt light bulb with 385 ohms of resistance that is plugged into a 110-volt outlet. 0.29 amp

For each of the following values of *x* and *y*, *y* varies directly as *x*.
Find the constant of variation, and write the equation of direct
variation.

26. *y* is 24 when *x* is 8   **27.** *y* is 12 when *x* is $\frac{1}{4}$ **28.** *y* is $-\frac{5}{8}$ when *x* is -1 $\frac{5}{8}$, $y = \frac{5}{8}x$
     3, $y = 3x$ 48, $y = 48x$

29. *y* is 4 when *x* is 0.2   **30.** *y* is 0.6 when *x* is -3 **31.** *y* is -1.2 when *x* is 4
     20, $y = 20x$ -0.2, $y = -0.2x$ -0.3, $y = -0.3x$

Each of the following tables represents ordered pairs of an equation.
Tell whether the equation is a direct variation. If so, give the
constant of variation, and write the equation of direct variation.

32.

x	*y*
2	-14
3	-21
4	-28
5	-35
6	-42

yes, -7, $y = -7x$

33.

x	*y*
5	0.10
6	0.12
7	0.14
8	0.16
9	0.18

yes, 0.02, $y = 0.02x$

34.

x	*y*
-2	1
-1	0.5
0	0
1	-0.5
2	-1

yes, -0.5, $y = -0.5x$

35. If *p* varies directly as *q*, and *q* varies directly as *r*, will *p* vary directly
as *r*? Explain your answer.   yes

36. Prove that if *x* varies directly as *y*, then *y* varies directly as *x*.

37. Government   In order to determine how people felt
about a school bond issue, a public opinion poll was taken.
In a sample of 300 voters, 240 favored the school bond measure.
If the number of people who favored the school bond measure
is directly proportional to the number of people who vote,
how many are likely to vote *for* the school bond
measure out of 75,000 voters?   60,000

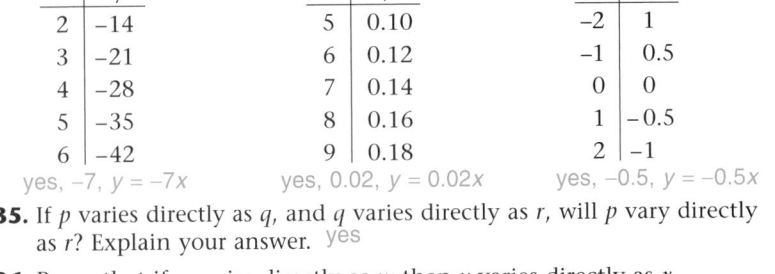

Look Back

Write the prime factorization for each number.

38. 261 $3^2 \cdot 29$ **39.** 860 $2^2 \cdot 5 \cdot 43$ **40.** 315 $5 \cdot 3^2 \cdot 7$

41. 180 $2^2 \cdot 3^2 \cdot 5$ **42.** 154 $2 \cdot 7 \cdot 11$ **43.** 490 $2 \cdot 5 \cdot 7^2$

Simplify each expression. $4x + 18$

44. $6x(x - 2)$ $6x^2 - 12x$ **45.** $6(3 - 8x) + 52x$ **46.** $2x(3 + 7x) - 4x$ $14x^2 + 2x$

47. Find the slope of the equation $3x + 12y = -9$. [**Lesson 1.2**] $-\frac{1}{4}$

Look Beyond

48 An equation of the form $xy = k$, where $x \neq 0$ and *k* is a constant such
that $k \neq 0$, is called an inverse variation. Choose a positive value for *k*
and use your graphics calculator to graph the equation. Describe the graph.

49 Use your graphics calculator to graph the equations $2x - y = -7$ and
$y = -1$ on the same coordinate plane. Find the coordinates of the point
of intersection. $(-4, -1)$

35. Yes. If *p* varies directly as *q*, then $p = kq$ for
some constant *k*. If *q* varies directly as *r*,
then $q = cr$ for some constant *c*. Substitute
cr for *q* into $p = kq$ to obtain $p = kcr$. Since
k and *c* are constants, *kc* is also a constant.
So, *p* varies directly as *r*.

36. If *x* varies directly as *y*, then $x = ky$ for
some constant *k*. Therefore, $y = \frac{1}{k}x$. Since
k is a constant, $\frac{1}{k}$ is also a constant. Hence,
y varies directly as *x*.

48. Answers may vary depending on the choice
of *k*, but they should show a graph made
up of two curves. One part of the curve is
in the first quadrant and the other part is
in the third quadrant. The *x*- and *y*-axes
are asymptotes for the curves.

Look Beyond

Students may be surprised that
the graph in Exercise 48 is a
curve. If so, have them substitute
values to see how *y* varies with
x. Exercise 49 reviews systems of
linear equations. Have students
who recall how to solve this
type of system describe the
substitution method to the class.

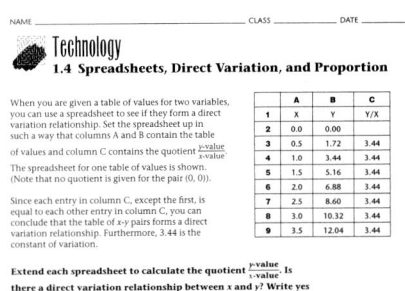

Technology Master

PREPARE

Objectives

- Solve problems by writing and solving linear equations.

RESOURCES

- Practice Master 1.5
- Enrichment Master 1.5
- Technology Master 1.5
- Lesson Activity Master 1.5
- Quiz 1.5
- Spanish Resources 1.5

Assessing Prior Knowledge

1. Solve for x. $12 = 5x + 2$
   [$x = 2$]
2. Use the Distributive Property to simplify $5(9x - 6)$.
   [$45x - 30$]

TEACH

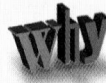

why Be sure students understand why equations can be compared to balanced scales. Before formally introducing the properties of equality, explain to students that given $a = b$, statements such as $a + c = b + c$ or $ac = bc$ hold true.

LESSON 1.5 Solving Equations

why *A mathematical equation is like a balanced scale. To keep both sides equal, any operation that is performed must be performed on both sides.*

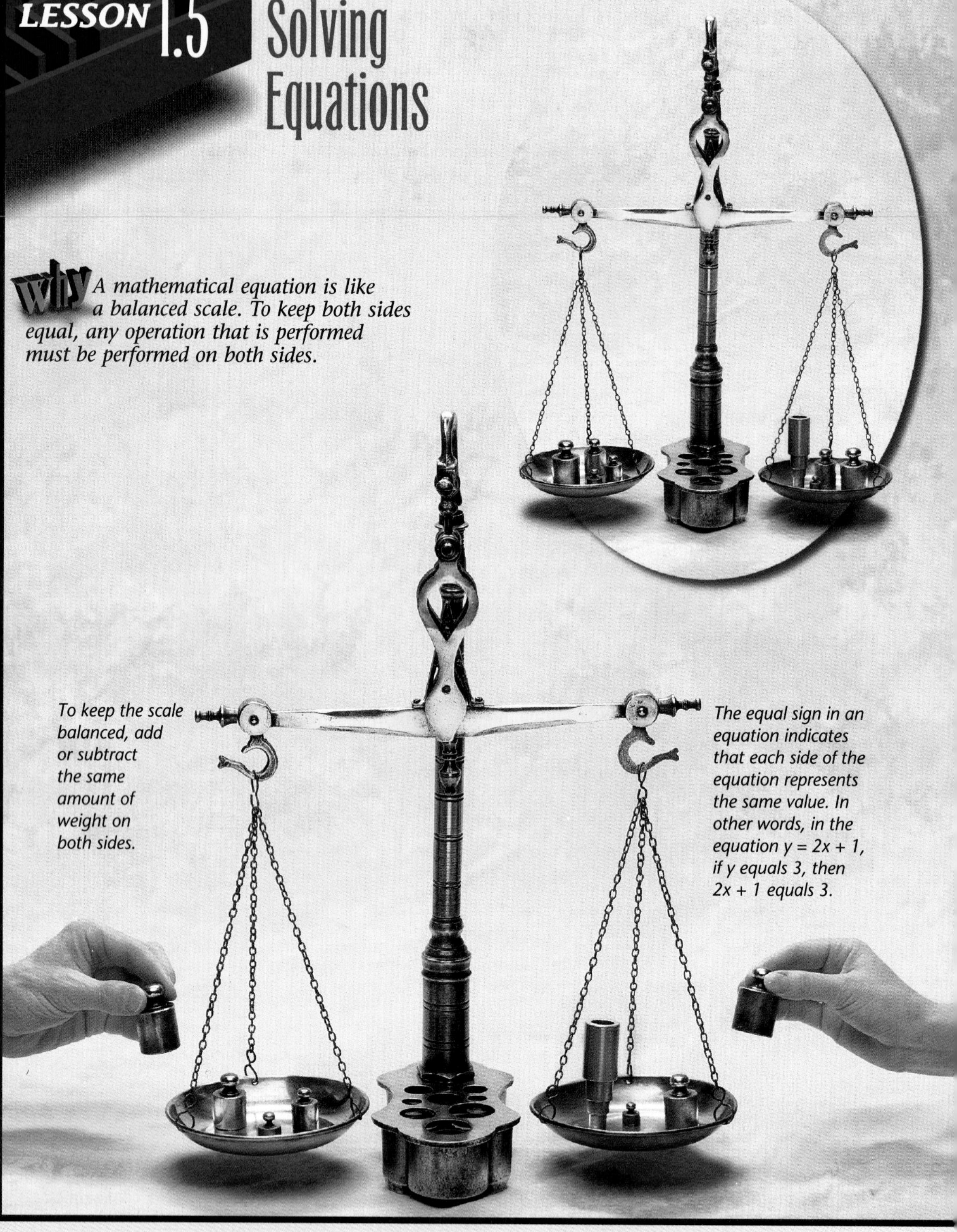

To keep the scale balanced, add or subtract the same amount of weight on both sides.

The equal sign in an equation indicates that each side of the equation represents the same value. In other words, in the equation $y = 2x + 1$, if y equals 3, then $2x + 1$ equals 3.

ALTERNATIVE teaching strategy

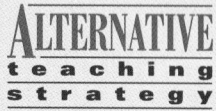

Cooperative Learning Present the following problem. Suppose a car leaves Center City and travels at an average speed of 40 miles per hour. An hour later another car leaves and travels along the same road at an average speed of 50 miles per hour. If the time for the first car is x hours, then the time for the second car is $(x - 1)$ hours because it will always be on the road an hour less than the first car. The equations are $y = 40x$ and $y = 50(x - 1)$. Have students find the number of hours when the miles traveled by both cars is the same.

Solving Equations by Graphing

How can you solve the equation $3 = 2x + 1$ using a graphics calculator?

The equation $3 = 2x + 1$ cannot be entered into the graphics calculator.

Let each side of the equation be equal to the same value, y. Then $y = 3$, and $y = 2x + 1$.

Look at the graphs of $y = 2x + 1$ and $y = 3$. Finding the intersection of these two lines is the same as solving $3 = 2x + 1$ for x.

The intersection of the lines is the point $(1, 3)$. The solution to $3 = 2x + 1$ is the x-coordinate of the intersection. Therefore, the solution to $3 = 2x + 1$ is 1.

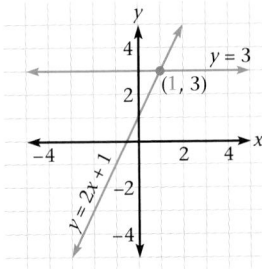

EXAMPLE 1

Solve by graphing. **A** $-1 = 2x + 1$ **B** $-x + 1 = \frac{1}{2}x + 4$

Solution

A Graph the equations $y = -1$ and $y = 2x + 1$ on the same coordinate plane.

The intersection of the lines is the point $(-1, -1)$. The x-coordinate of this point is -1.

Thus, the solution to $-1 = 2x + 1$ is -1.

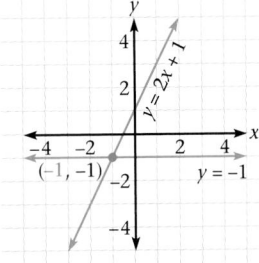

B Graph the equations $y = -x + 1$ and $y = \frac{1}{2}x + 4$ on the same coordinate plane.

The intersection of the lines is the point $(-2, 3)$. The x-coordinate of this point is -2.

Thus, the solution to $-x + 1 = \frac{1}{2}x + 4$ is -2. ❖

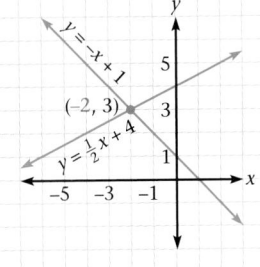

CRITICAL *Thinking*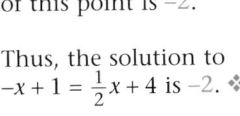

In Example 1, the solutions given are the x-coordinates of the points of intersection. Why is the y-coordinate not given in the solution?

interdisciplinary  **CONNECTION** **Business** Systems of equations are used frequently in business. When solving a system of equations in which one equation represents the formula for cost and the other the formula for revenue, the point of intersection is the break-even point.

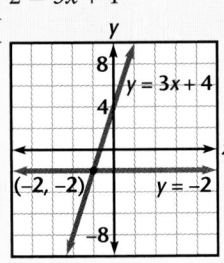
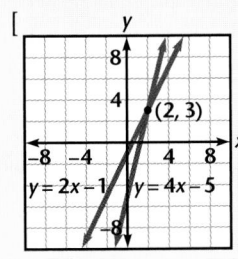

Alternate Example 2

The equation relating Celsius and Fahrenheit temperatures is $F = \frac{9}{5}C + 32$. Find the number of degrees Celsius equal to 122°F. [**50°C**]

TEACHING *tip*

Explain that substitution can be used in any system of two equations with two unknowns, by isolating one variable and then substituting in the other equation. Answers should always be checked by substituting in the original equations.

EXAMPLE 2

Graphics Calculator

The equation relating Celsius and Fahrenheit temperatures is $F = \frac{9}{5}C + 32$. Find the number of degrees Celsius equal to 86°F.

Solution

Solve $86 = \frac{9}{5}C + 32$. Graph $F = \frac{9}{5}C + 32$ and $F = 86$ on the same coordinate plane.

Move the trace cursor to the intersection, and read the coordinates.

The intersection of the lines shows that C is 30 when F is 86. ❖

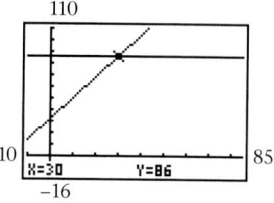

Equality and Equivalence

Equations can also be solved by using algebraic techniques. Review the Properties of Equality and follow the steps to solve the problem in Example 2.

PROPERTIES OF EQUALITY

Let a, b, and c represent any real numbers.

Addition Property of Equality
If $a = b$, then $a + c = b + c$.

Subtraction Property of Equality
If $a = b$, then $a - c = b - c$.

Multiplication Property of Equality
If $a = b$, then $ac = bc$.

Division Property of Equality
If $a = b$, then $\frac{a}{c} = \frac{b}{c}$ (for $c \neq 0$).

**ENRICHMENT** Explain how to solve systems of equations such as $2x + 4y = 8$ and $-2x + 3y = 6$ by elimination. [**(0, 2)**]

**INCLUSION** **strategies** **Hands-On Strategies** This lesson could be taught by using graph paper instead of graphics calculators. Elimination, adding like terms of the equations, can also be used to solve systems of equations.

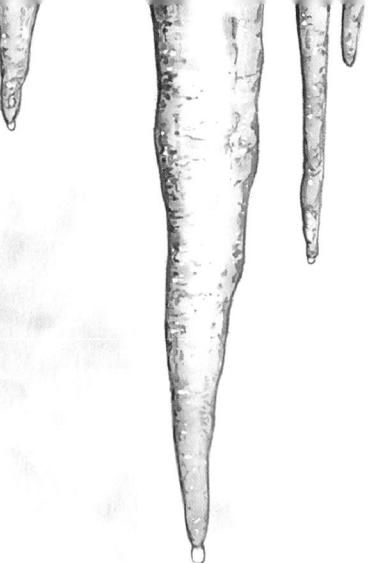

Substitute 86 for F in the equation.

$$F = \frac{9}{5}C + 32$$

$$86 = \frac{9}{5}C + 32$$

$$86 - 32 = \frac{9}{5}C + 32 - 32 \qquad \text{Subtraction Property of Equality}$$

$$54 = \frac{9}{5}C$$

$$5(54) = 5\left(\frac{9}{5}C\right) \qquad \text{Multiplication Property of Equality}$$

$$270 = 9C$$

$$\frac{270}{9} = \frac{9C}{9} \qquad \text{Division Property of Equality}$$

$$30 = C$$

Thus, $86°$ Fahrenheit is equivalent to $30°$ Celsius.

To solve some equations, other properties such as the Distributive Property are needed.

DISTRIBUTIVE PROPERTY

For any real numbers a, b, and c,
$$a(b + c) = ab + ac \text{ and } (b + c)a = ba + ca.$$

Equivalent means to have the same value. When you solve an equation, the equation resulting from each step is an *equivalent equation*. **Equivalent equations** have the same solutions.

SUBSTITUTION PROPERTY

If two expressions are equivalent, you can replace one with the other.

What are some examples of equivalent equations?

TEACHING *tip*

Have students share equivalent equations with the class. Discuss the different equations to give students additional insight into the meaning of the term equivalency. Then have students examine this alternative statement of the substitution property and determine whether or not it applies to their examples: For real numbers a, b, and c, if $a = b$ and $b = c$, then $a = c$.

Aon g o i n g **SSESSMENT**

Answers may vary. The two equations $3x = 12$ and $x = 4$ are equivalent, as are the equations $x - 5 = 0$ and $x = 5$.

Alternate Example 3

Given that $y = 4 - 2x$, solve $3x + 5y = 6$ for x. Then find y. $[x = 2, y = 0]$

Aongoing ASSESSMENT

Try This

$y = 2, x = 5$

Cooperative Learning

Have pairs of students complete the *Try This* exercise. One student in each pair should solve by graphing, and the other by substitution. Then ask partners to compare answers for consistency.

Alternate Example 4

Given the equation $V = \frac{1}{3}b^2h$, solve for h in terms of V and b. $[h = \frac{3V}{b^2}]$

EXAMPLE 3

Given that $y = 2x - 5$, solve $3x + 2y = 11$ for x. Then find y.

Solution

Since y is equal to $2x - 5$, substitute $2x - 5$ for y in the equation $3x + 2y = 11$, and solve for x.

$$3x + 2y = 11$$
$$3x + 2(2x - 5) = 11 \qquad \text{Substitution Property}$$
$$3x + 4x - 10 = 11 \qquad \text{Distributive Property}$$
$$7x - 10 = 11$$
$$7x = 21$$
$$x = 3$$

Since x is 3, substitute 3 for x in either equation, and solve for y.

$$y = 2x - 5$$
$$y = 2(3) - 5 \qquad \text{Substitution Property}$$
$$y = 1$$

Check

Substitute 3 for x and 1 for y in the other equation to check.

$$3x + 2y = 11$$
$$3(3) + 2(1) \stackrel{?}{=} 11$$
$$9 + 2 \stackrel{?}{=} 11$$
$$11 = 11 \qquad \text{True}$$

Thus, x is 3 and y is 1. ❖

Try This Given that $x = 3y - 1$, use substitution to solve $5x - 2y = 21$ for y. Then find x.

EXAMPLE 4

Given the equation $V = \frac{1}{3}s^2h$, solve for h in terms of V and s.

Solution

$$V = \frac{1}{3}s^2h$$
$$3V = s^2h \qquad \text{Multiplication Property of Equality}$$
$$\frac{3V}{s^2} = h \qquad \text{Division Property of Equality} ❖$$

The Egyptian Method of Solving Equations

Cultural Connection: Africa The foundation for much of today's high school mathematics came from Egypt and Babylonia.

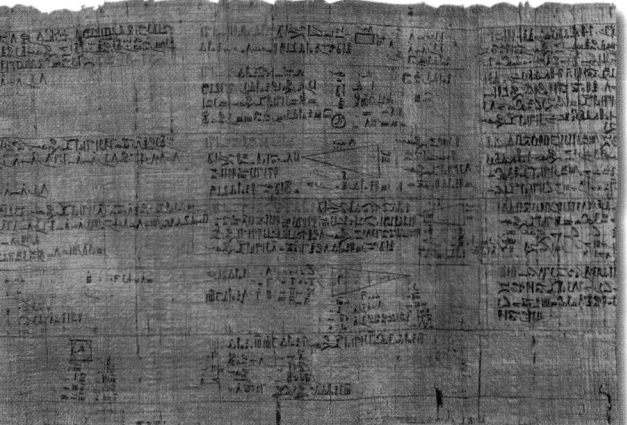

There are many equations in The Rhind Mathematical Papyrus, a book written by the scribe Ahmose almost 4000 years ago. Proportional thinking was the basis of the ancient Egyptian method of solving equations.

Ahmose set up an equation "machine" and substituted any convenient number, a "false" value, in this machine. For example, given that the sum of a quantity and its quarter is 15, then $x + \frac{1}{4}x = 15$.

First guess a number. This number is not expected to work, but the resulting error leads to the correct answer.

Guess that the quantity is 4.

$$x + \frac{1}{4}x = 15$$
$$(4) + \frac{1}{4}(4) \stackrel{?}{=} 15$$
$$5 \neq 15$$

The result, $5 \neq 15$, is used to obtain the *correction factor*. How many times must 5 be multiplied to get 15? The answer is 3, so the correction factor is 3.

Multiply the false value, 4, by the correction factor, 3. The product, 12, is the correct value for x.

Check this answer to see that this clever method really works!

$$x + \frac{1}{4}x = 15$$
$$(12) + \frac{1}{4}(12) \stackrel{?}{=} 15$$
$$12 + 3 = 15 \quad \text{True}$$

The Egyptian mathematicians always completed a problem by proving that the corrected value worked in the original equation.

THE EGYPTIAN METHOD

1. Guess a number, and try it.
2. Use the result to obtain the correction factor.
3. Multiply the false value by the correction factor to get the correct value.
4. Check the correct value in the equation.

Review the distributive and substitution properties. Have students work in pairs to solve systems of equations, such as the one in Example 3. Students should use substitution and then graph the equations on a graphics calculator.

Assess

Selected Answers

Odd-numbered Exercises 5–53

The answers to Communicate Exercises can be found in Additional Answers beginning on page 842.

Assignment Guide

Core 1–9, 11–12, 14–25, 31–35, 38–57

Core Plus 1–4, 6–10, 12–14, 20–31, 34–57

Technology

Graphics calculators are needed for Exercises 5–13, 15–30, 32–37 and 46–52.

Error Analysis

Watch for students who choose the *y*-coordinate as their answer when the equation is written only in terms of *x*. Encourage students to check their answers in the original equation. Emphasize that if the answer does not make the equation true, then it is not the correct answer.

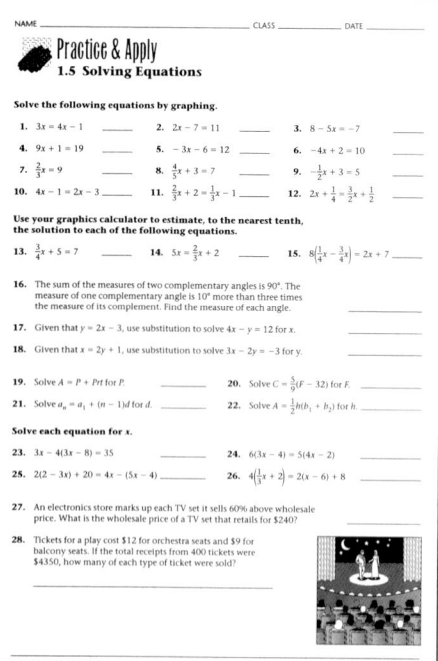

EXERCISES & PROBLEMS

Communicate

1. Explain how to solve $x - 2 = -5x + 30$ by graphing.

The sum of two numbers is 45. The greater number is 1 more than 3 times the smaller number.

2. Explain how you would organize this information to find the two numbers. Write an equation in terms of x to represent this relationship.

3. Describe the steps that you would take to solve the equation. What properties would you use and why?

4. When you solve for x, is that the complete answer? Why or why not?

Practice & Apply

Use your graphics calculator to solve the following equations.

5 $20 = 6x - 10$ 5

6 $5x - 3 = 12$ 3

7 $4 - 5x = 19$ –3

8 $5 = -2x - 3$ –4

9 $3x = x + 2$ 1

10 $x - 5 = -2x - 2$ 1

Use your graphics calculator to estimate, to the nearest tenth, the solution to each of the following equations.

11 $7x = -2x + 5$ 0.6

12 $\frac{2}{3}x - 9 = -\frac{1}{2}x + 4$ 11.1

13 $6\left(\frac{1}{3}x - \frac{4}{3}\right) = -x - 6$ 0.7

14. Consumer Economics Victor's receipt for his car repair work is shown. If the mechanic charged $40 per hour, how long did the repair take to complete?

5 hr

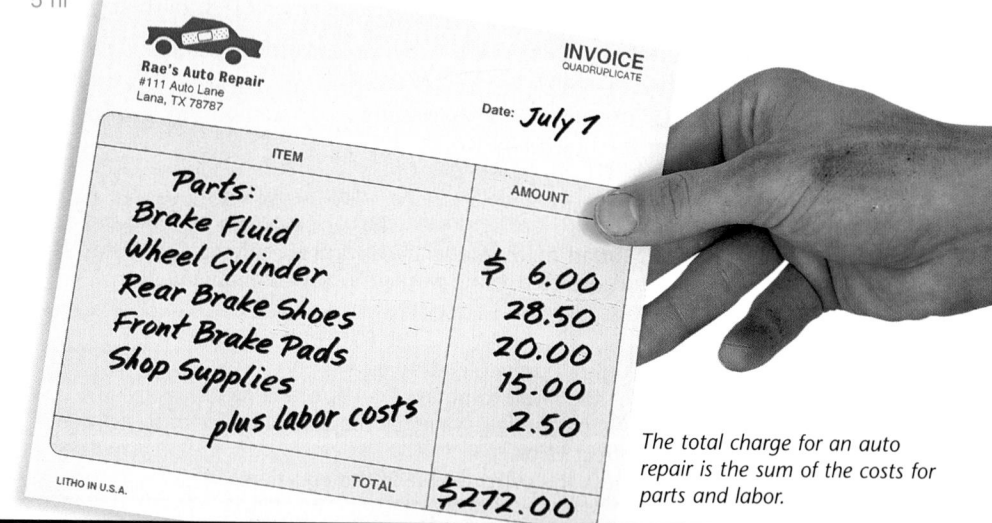

The total charge for an auto repair is the sum of the costs for parts and labor.

Use your graphics calculator to solve the following equations.

15. $1 = 2x - 5$ 3
16. $\frac{1}{5}x + 3 = 2$ −5
17. $-2x - 7 = 9$ −8
18. $4x + 8 = -16$ −6
19. $\frac{1}{4}x - \frac{5}{2} = -2$ 2
20. $\frac{1}{6}x + \frac{3}{2} = 2$ 3
21. $5x + 15 = 5$ −2
22. $0 = \frac{1}{2}x + 2$ −4
23. $2x - 1 = -5$ −2
24. $-2x - 3 = 1$ −2
25. $-5 = \frac{3}{2}x - 2$ −2
26. $\frac{1}{2}x + \frac{5}{2} = 2$ −1
27. $\frac{1}{3}x = -x + 4$ 3
28. $7 - 6x = 2x - 9$ 2
29. $3x - 8 = 2x + 2$ 10
30. $x - 5 = -\frac{3}{2}x + \frac{5}{2}$ 3

31. **Cultural Connection: Africa** Use the Egyptian method to solve the following. The sum of a quantity and its seventh is 20. What is the quantity? $17\frac{1}{2}$

Use your graphics calculator to estimate, to the nearest tenth, the solution to each equation.

32. $3x + 1 = \frac{1}{2}$ −0.2
33. $3x - 3 = 5$ 2.7
34. $-\frac{1}{3}x + 1 = \frac{3}{2}x - 1$ 1.1
35. $-\frac{3}{5}x + 12 = 4$ 13.3
36. $\frac{1}{4}x - 3 = 6x$ −0.5
37. $-2x + 5 = -\frac{1}{3}x - 6$ 6.6

38. **Geometry** The measure of one supplementary angle is 45° more than twice the measure of its supplement. Find the measure of each angle. 45°, 135°

180°

The sum of measures of two supplementary angles is 180°.

39. Given that $y = 4x + 7$, use substitution to solve $-2x - y = 19$ for x. $x = -\frac{13}{3}$
40. Given that $x = -y + 9$, use substitution to solve $3x - 5y = 59$ for x. 13
41. **Leisure** If 180 of those surveyed responded that they listen to CDs mostly on a portable CD player, how many people were surveyed? 900

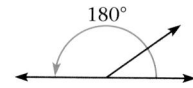

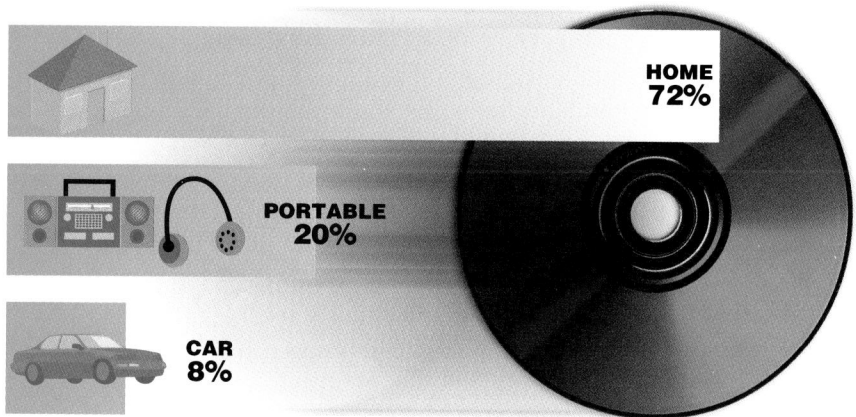

HOME
72%
PORTABLE
20%
CAR
8%

42. **Economics** Atietie wants to buy a used car that will cost $185.00 per month. If Atietie earns $5.35 per hour, what is the minimum number of hours that Atietie must work each month to pay for the car? 35 hr
43. Solve $Ax + By = C$ for y.
44. Solve $\frac{1}{2}bh = A$ for b.
45. Solve $p = 2l + 2w$ for w.

43. $y = \frac{C - Ax}{B}$
44. $b = \frac{2A}{h}$
45. $w = \frac{p - 2l}{2}$

alternative ASSESSMENT

Authentic Assessment

Have students describe in writing two methods that can be used to solve equations. Ask students to describe when it is best to use each method.

Technology Master

NAME ____________________ CLASS __________ DATE __________

Technology
1.5 Spreadsheets and Solving Equations

You can use a spreadsheet to help you solve an equation like $0.826x - 2.961 = 3.61x - 20.501$. Set up a spreadsheet so that column A contains a range of values for x, such as 0 to 10, column B contains the values of $0.826x - 2.961$, the left side of the equation, and column C contains the values of $3.61x - 20.501$, the right side of the equation. From the spreadsheet, you can observe that the values in columns B and C are closest when $x = 6$. Revise column A by using 5.6, 5.7, 5.8, 5.9, 6.0, 6.1, 6.2, 6.3, 6.4, 6.5 as shown. The values in columns B and C are closest when $x = 6.3$. Thus, the solution of $0.826x - 2.961 = 3.61x - 20.501$ is about 6.3, to the nearest tenth.

	A	B	C
1	X	LEFT	RIGHT
2	5.6	1.6646	−0.2850
3	5.7	1.7472	0.0760
4	5.8	1.8298	0.4370
5	5.9	1.9124	0.7980
6	6.0	1.9950	1.1590
7	6.1	2.0776	1.5200
8	6.2	2.1602	1.8810
9	6.3	2.2428	2.2420
10	6.4	2.3254	2.6030
11	6.5	2.4080	2.9640

Notice from the spreadsheet that the values in columns B and C are "far apart," then closer together, and then farther apart. This condition must be satisfied for the equation to have a solution between the least and greatest x-values selected.

Use a spreadsheet to approximate any solution in the given range to the nearest tenth. If no solution exists, state so.

1. $2.3x - 5 = 4x - 7.55$ for all x
2. $6x + 4 = 5x + 14.3$ for all x
3. $-2x + 6 = 3x + 14.5$ for all x
4. $1.3x - 4 = 2x - 0.5$ for all x
5. $2x - 1.9 = 3x - 5.9$ for $-4 \le x \le 0$
6. $6.2x + 9 = 4x + 22.2$ for $5 \le x \le 7$
7. $13.9x + 5 = 17.1x + 22.5$ for $-9 \le x \le 0$
8. $6.3x - 17 = -18x + 206.8$ for $0 \le x \le 20$
9. $33.6x + 17 = 93$ for $-9 \le x \le 0$
10. $100.37 = -19x - 206.8$ for all x

HRW Advanced Algebra Technology 17

 **Look Back**

Graph each linear equation, and find the slope. [Lesson 1.2]

46 $y = 2x - 6$ $m = 2$ **47** $3x + 4y = 9$ $m = -\frac{3}{4}$ **48** $y = 2$ $m = 0$

History The following table shows the number of African American elected officials from these southern states from 1941 to 1985.
[Lessons 1.2, 1.3]

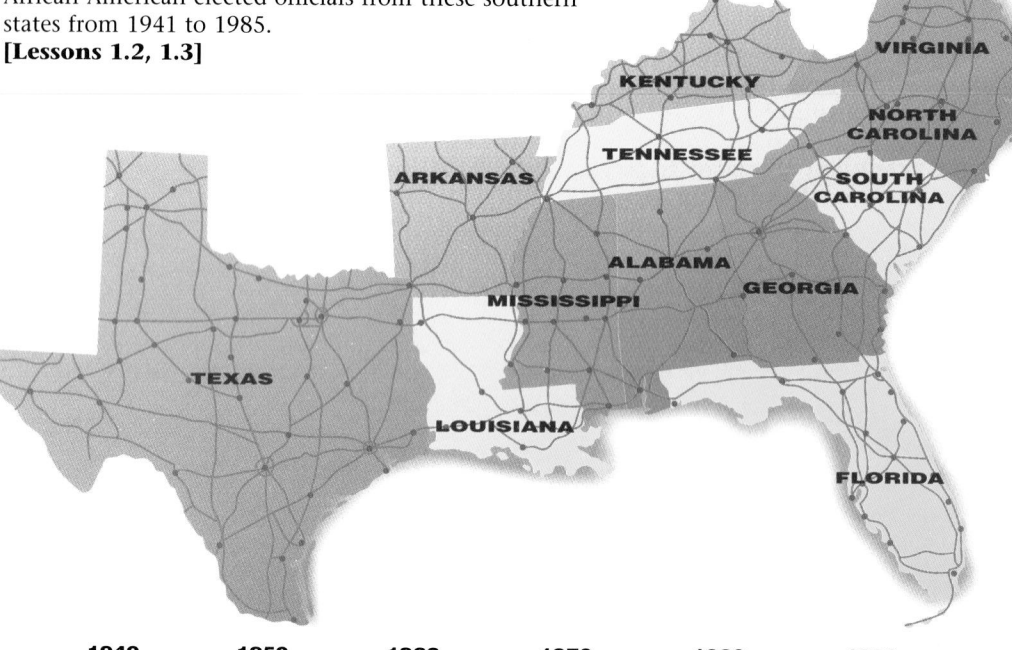

1940	1950		1960	1970			1980		1990
1941	1947	1951		1965	1970	1975	1980	1985	
2	6	15		87	103	1913	2981	3801	

49 Make a scatter plot for this information, with the years on the *x*-axis. What viewing window did you use to view this plot?

50 Use the scatter plot to find the slope of the line of best fit for the data from the years 1941–1965.   3.8

51 Use the scatter plot to find the slope of the line of best fit for the data from the years 1965–1985.   206

52. Explain how the slopes of the two lines are different. What happened in the 1960s that might explain the extreme change?

Look Beyond

Explain what each expression means.

53. $y > -5$ **54.** $-3 < x < 3$ **55.** $-1 \le y \le 1$ **56.** $x \le -3$

52. The slope of the line representing 1965–1985 is much steeper than the slope for 1941–1965. The rapid increase in the number of African Americans elected for public office can possibly be attributed to the civil rights legislation in the early 1960s.

53. *y* is greater than −5

54. *x* is greater than −3 and less than 3

55. *y* is greater than or equal to −1 and less than or equal to 1

56. *x* is less than or equal to −3

LESSON 1.6 Solving Inequalities

 *You can see from the tug-of-war game that one group is stronger than the other group. There are many times when two items or groups that are compared are not equal. A situation like this can be modeled by an inequality.*

In the last lesson you studied equal expressions written as an equation. In this lesson you will study expressions that are not equal, written as an *inequality*.

INEQUALITY

An **inequality** is a mathematical sentence that contains $>$, $<$, $\geq$, $\leq$, or $\neq$.

Properties of Inequalities

•Exploration 1 *Adding and Subtracting*

1 Write the inequality $-3 < 5$. Add -4 to each side of the inequality. Is the resulting inequality true or false?

2 Choose any two numbers. Write an inequality with $>$. Subtract a negative number of your choice from each side of the inequality. Is the resulting inequality true or false?

3 Describe what happens to an inequality when you add or subtract a negative number from both sides of the inequality. Do you get the same result when you add or subtract a positive number from both sides of inequality? Explain. ❖

 **Using Symbols** Have students write inequalities for the condition given in each statement below. Then review what the students have written, discussing the accuracy of the symbols used and the direction of the inequality symbols.

a. To qualify for the police force in Fairview County, a candidate must be at least 66 in. tall and not more than 78 in. tall.

[$66 \leq h \leq 78$]

b. To make the honor roll, a student must have a grade point average equal to or higher than 3.0. [$a \geq 3.0$]

c. Children under 12 pay half price. [$c < 12$]

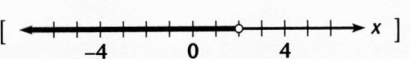

Exploration 2 Notes

Exploration 2 allows students to examine what happens to inequalities when both sides of the inequality are multiplied or divided by both positive and negative numbers.

Aongoing
SSESSMENT

3. Multiplying both sides of an inequality by a *negative* number produces an equivalent inequality *provided the order of the inequality is reversed.* Multiplying both sides of −3 < 5, by −2 gives 6 > −10, which is true. Multiplying both sides of an inequality by a positive number directly produces an equivalent inequality. For example, multiplying both sides of 5 > 2 by 2 gives 10 > 4, which is true.

When solving inequalities, if you add or subtract any real number on both sides of the inequality, the resulting inequality is equivalent.

The Addition and Subtraction Properties of Inequality are very much like the Addition and Subtraction Properties of Equality.

ADDITION AND SUBTRACTION PROPERTIES OF INEQUALITY

Let a, b, and c represent any real numbers.

Addition Property of Inequality

If $a < b$, then $a + c < b + c$.

If $a > b$, then $a + c > b + c$.

Subtraction Property of Inequality

If $a < b$, then $a - c < b - c$.

If $a > b$, then $a - c > b - c$.

Exploration 2 *Multiplying and Dividing*

1 Write the inequality −3 < 5. Multiply each side of the inequality by −2. Is the resulting inequality true or false?

2 Choose any two numbers. Write an inequality with >. Multiply each side of the inequality by a negative number of your choice. Is the resulting inequality true or false?

3 Describe what happens to an inequality when you multiply both sides by a negative number. Do you get the same result when you multiply both sides of an inequality by a positive number? Explain. ❖

When solving inequalities, if you multiply or divide both sides of the inequality by a *negative* number, you must reverse the direction of the inequality symbol.

MULTIPLICATION AND DIVISION PROPERTIES OF INEQUALITY

Let a, b, c, and d represent any real numbers, where $c > 0$ and $d < 0$.

Multiplication Property of Inequality

If $a < b$, then $ac < bc$ and $ad > bd$.

If $a > b$, then $ac > bc$ and $ad < bd$.

Division Property of Inequality

If $a < b$, then $\frac{a}{c} < \frac{b}{c}$ and $\frac{a}{d} > \frac{b}{d}$.

If $a > b$, then $\frac{a}{c} > \frac{b}{c}$ and $\frac{a}{d} < \frac{b}{d}$.

---

interdisciplinary
CONNECTION

Product Packaging

Suppose that for certain kinds of packaging, boxes must be between 15 and 20 inches in length, between 12 and 15 inches in width, and between 10 and 14 inches in height. Have students write inequalities to describe this situation.

[$15 \leq l \leq 20$, $12 \leq w \leq 15$, $10 \leq h \leq 14$]

Inequalities in One Variable

The number of solutions to an inequality, such as $x < 1$, is countless. The solution to an inequality in one variable is the set of all values that make the inequality true.

EXAMPLE 1

Consumer Awareness

David is interested in renting a car for one week. He compares Hire-a-Heap and Wagon Rentals. His choice of rental car agency should be based on how many miles he will drive that week. If the car is rented from Wagon Rentals, how many miles can the car be driven during the week and still cost less than if the car is rented from Hire-a-Heap?

Solution

The weekly cost for Hire-a-Heap is $210. Let m represent the number of miles David drives during the week. Then the weekly cost for Wagon Rentals is $120 + \$0.20m$.

$$\left\{ \begin{array}{c} \text{weekly cost for} \\ \text{Wagon Rentals} \end{array} \right\} < \left\{ \begin{array}{c} \text{weekly cost for} \\ \text{Hire-a-Heap} \end{array} \right\}$$
$$120 + 0.20m < 210$$

Solve the inequality for m.

$$120 + 0.20m < 210$$
$$0.20m < 90 \quad \text{Subtraction Property of Inequality}$$
$$m < 450 \quad \text{Division Property of Inequality}$$

Thus, if the car is driven less than 450 miles during the week, a car from Wagon Rentals will cost less than one from Hire-a-Heap. ❖

EXAMPLE 2

Solve $4 - 3p \geq 16 + p$ for p.

Solution

$$4 - 3p \geq 16 + p$$
$$4 - 4p \geq 16$$
$$-4p \geq 12$$

Reverse the inequality symbol. $\quad p \leq -3$

Thus, the solution is $p \leq -3$. ❖

Graph the inequality
$3t - 1 \geq -2t + 9$.
$[t \geq 2;$

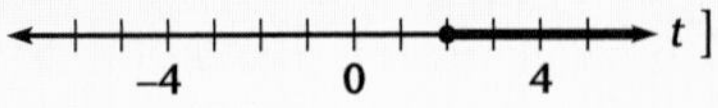

$t]$

$\mathbf{T}_{EACHING}$ *tip*

Remind students that when graphing inequalities on a number line, the circle is shaded in when $\leq$ and $\geq$ are in the statement, and an open circle is used for statements including $<$ and $>$.

Use Transparency ▶ **5**

Graphing Inequalities in One Variable

There are an infinite number of solutions for $p \leq -3$. Since it is impossible to list all the solutions, you can represent the solution set with a graph.

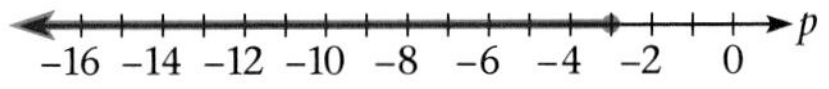

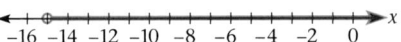

graph of $p \leq -3$

EXAMPLE 3

Graph the inequality $-2x - 5 < -x + 10$.

Solution▶

Solve the inequality for x.

$$-2x - 5 < -x + 10$$
$$-2x < -x + 15$$
$$-x < 15$$

Reverse the inequality symbol. $\quad x > -15$

Graph $x > -15$.

Thus, the solution set includes all numbers greater than -15, but does not include -15.

Check▶

Select a number from the solution set, say -5. Make sure that it works in the original inequality.

$$-2x - 5 < -x + 10$$
$$-2(-5) - 5 \stackrel{?}{<} -(-5) + 10$$
$$10 - 5 \stackrel{?}{<} 5 + 10$$
$$5 < 15 \qquad \text{True} \diamond$$

Linear Inequalities

The graph of a line separates the plane into three distinct regions: two *half-planes* and the line itself. The line is the boundary of the two half-planes and is called the **boundary line**. It divides the coordinate plane into two half-planes for a linear inequality.

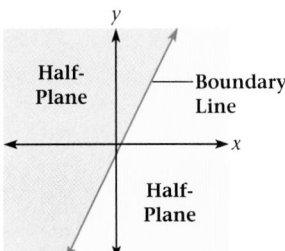

$\mathbf{R}_{ETEACHING}$ **the** **lesson**

Using Visual Models Review by asking students how to find the solutions for an equation by looking at its graph. Repeat the process when graphing inequalities. Students can more easily see the difference between equations and inequalities by using real-life examples of each. For instance, ask which of the following statements can be described with an equation and which can be described with an inequality: Mica will drive 250 miles today. [**equation**] Mica will drive at least 250 miles today. [**inequality**]

A **linear inequality** has a boundary line that can be expressed in the form $y = mx + b$. The solution to a linear inequality is the set of all ordered pairs that make the inequality true.

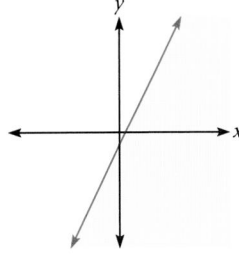

When the inequality is ≤ or ≥, the solution includes points on the boundary line, and the graph will have a solid boundary line.

When the inequality is < or >, the solution does not include points on the boundary line, and the graph will have a dashed boundary line.

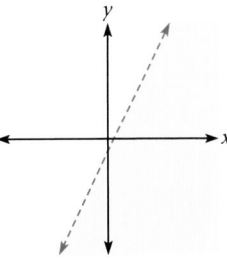

How many solutions does a linear inequality have? Explain.

EXAMPLE 4

Graph $2x + y < 4$.

Solution➤

The boundary line is $2x + y = 4$.
Graph the boundary line.

$$2x + y = 4$$
$$y = -2x + 4$$

Use a dashed line to show that the points on the line are *not* solutions to $2x + y < 4$.

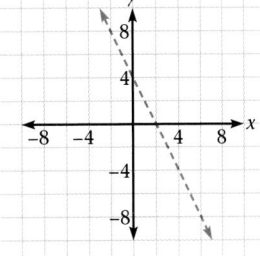

To determine which half-plane is the solution, select a point from each half-plane, and see which ordered pair makes the inequality true.

(0, 0)	(4, 0)
$y < -2x + 4$	$y < -2x + 4$
$0 \overset{?}{<} -2(0) + 4$	$0 \overset{?}{<} -2(4) + 4$
$0 < 4$ True	$0 < -4$ False

Since the coordinates of (0, 0) make the inequality true, shade the half-plane in which this point is located. ❖

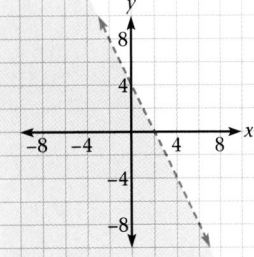

Try This Graph the inequality $-4x > 8 + 3y$.

Aongoing SSESSMENT

A linear inequality has an infinite number of solutions. The solution set contains all the points in the half-plane where an inequality is true.

Alternate Example 4

Graph $-x + 4 \geq 2y$.

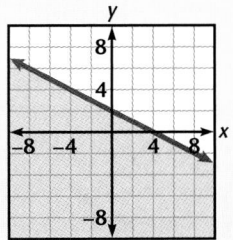

Aongoing SSESSMENT

Try This

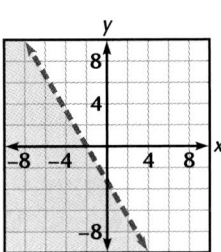

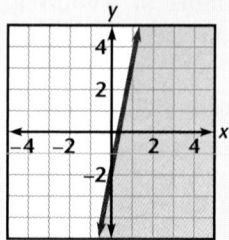
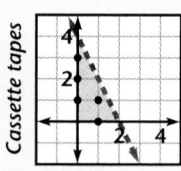
EXAMPLE 5

Write the inequality represented by the graph.

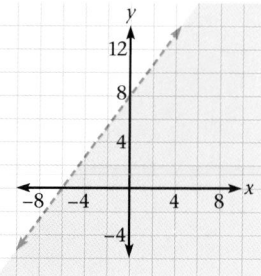

Solution▶

Write the equation of the boundary line. The line goes through the points $(-6, 0)$ and $(0, 8)$; therefore, the slope is $\frac{8-0}{0-(-6)} = \frac{8}{6} = \frac{4}{3}$. The y-intercept is 8.

So the equation for the boundary line is $y = \frac{4}{3}x + 8$.

Select a point from the shaded area, say $(0, 0)$, and substitute the values in the boundary line equation to see which inequality symbol is correct.

$$y = \frac{4}{3}x + 8$$
$$0 \stackrel{?}{=} \frac{4}{3}(0) + 8$$
$$0 < 8 \qquad \text{True}$$

The boundary line is dashed, so use the inequality symbol <, not ≤. Thus, the inequality $y < \frac{4}{3}x + 8$ describes the graph. ❖

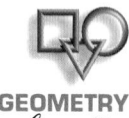

CRITICAL *Thinking*

How can an inequality, such as $x < 2$ or $y \ge -4$, be graphed on the coordinate plane?

Applications of Linear Inequalities

Some real-world situations can be modeled with linear inequalities.

EXAMPLE 6

GEOMETRY *Connection*

Gene raises keeshonds, and one of the female keeshonds is going to have pups soon. Gene wants to build a rectangular pen for the mother and her pups. If he has 60 feet of fencing, what are the possible dimensions for the pen?

Each cage has a 60 foot perimeter.

Solution▸

Write an inequality that describes the situation.

perimeter of rectangle ≤ 60

$2w + 2l \le 60$

Graph the boundary line, $2w + 2l = 60$.

Since $2w + 2l \le 60$ includes the points on the line, graph a solid line segment.

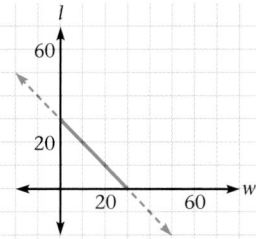

Since the possible widths and lengths for the pen cannot be negative or zero, the boundary line is dashed for values of w and l that are not positive.

To determine which half-plane is the solution, select a point from each half-plane and see which ordered pair makes the inequality true.

Point (1, 1):

$2w + 2l \le 60$

$2(1) + 2(1) \overset{?}{\le} 60$

$4 \le 60$ True

Point (30, 30):

$2w + 2l \le 60$

$2(30) + 2(30) \overset{?}{\le} 60$

$120 \le 60$ False

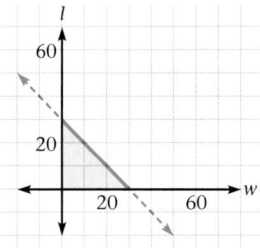

The solution is the set of all ordered pairs in the same half-plane as (1, 1), such that $w > 0$ and $l > 0$. ❖

What are some applications for inequalities that include solutions with only whole-number coordinates?

EXERCISES & PROBLEMS

Communicate

1. How is the method of solving an inequality different from solving an equation?

2. How can you denote that x is nonnegative with an inequality?

3. Describe how to graph $7x - 7 > 0$ on graph paper.

4. Is $x < y$ the same inequality as $-x < -y$? Explain.

CRITICAL
Thinking

Possibilities include any applications that deal with numbers of people, objects, events, or purchasing transactions involving actual items.

ASSESS

Selected Answers

Odd-numbered Exercises 5–47

The answers to Communicate Exercises can be found in Additional Answers beginning on page 842.

Assignment guide

Core 1–25, 31–48

Core Plus 1–14, 20–48

Technology

Graphics calculators are needed for Exercises 10, 11, and 45–47.

Error Analysis

Students may forget to change the direction of the inequality when both sides are multiplied or divided by a negative number. Some students do not understand that inequalities and their graphs are the solutions. They may think that there should be one specific number for the solution.

ALTERNATIVE ASSESSMENT

Performance Assessment

Have the students write a summary of inequalities and rules that apply to inequalities.

9.

```
<---+--+--+--◇--+--+--+---> x
     60   62   64
```

10. Solve for y to obtain $y > -\frac{1}{3}x + \frac{10}{3}$. For Y1, enter the expression $(-1/3)X + 10/3$. Define the viewing window similar to $X_{min} = -2$, $X_{max} = 15$, $Y_{min} = -5$, $Y_{max} = 5$. Use the test point $(0, 0)$ to find that the solution set is all the points above the line, not including the line.

The answers to Exercises 11, 15–30 can be found in Additional Answers beginning on page 842.

Practice Master

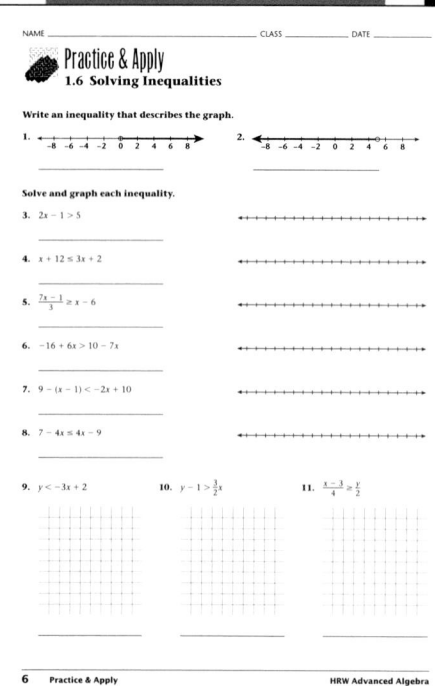

NAME _____ CLASS _____ DATE _____

Practice & Apply
1.6 Solving Inequalities

Write an inequality that describes the graph.

1.
2.

Solve and graph each inequality.

3. $2x - 1 > 5$

4. $x + 12 \le 3x + 2$

5. $\frac{7x-1}{3} \ge x - 6$

6. $-16 + 6x > 10 - 7x$

7. $9 - (x - 1) < -2x + 10$

8. $7 - 4x \le 4x - 9$

9. $y < -3x + 2$ 10. $y - 1 > \frac{3}{2}x$ 11. $\frac{x-3}{4} \ge \frac{y}{2}$

6   Practice & Apply   HRW Advanced Algebra

Practice & Apply

Write an inequality that describes each graph.

5.

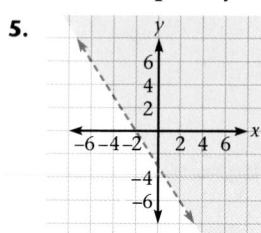

$y > -\frac{3}{2}x - 3$

6.

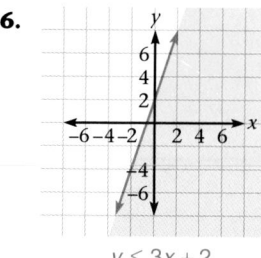

$y \le 3x + 2$

Small Business The revenue (money earned) from selling x units of a product is $R = 54x$. The cost of producing x units is $C = 40x + 868$. In order to obtain a profit, the revenue must be greater than the cost.

7. Write an inequality in one variable to describe this relationship. $54x > 40x + 868$

8. How many units of the product must be sold to earn a profit? more than 62

9. Graph the solution.

Use your graphics calculator to graph each inequality. Describe the steps you took to display the graph and what range of values you used for each variable.

10. $x + 3y > 10$

11. $2y + 5x \ge -8$

Geometry The perimeter of a rectangle with length x and width y cannot exceed 200 feet.

12. Write a linear inequality to describe this restriction. $2x + 2y \le 200$

13. Sketch a graph showing the possible combinations of lengths and widths.

14. Explain how this restriction could possibly occur in the real world.

Solve and graph each inequality. Use a number line to graph the inequalities with only one variable.

15. $3 - x \ge -5$

16. $2 + 5x \ge x - 3$

17. $4y - 12 > 7y - 15$

18. $3(4x - 5) < 8x + 3$

19. $\frac{x+3}{6} + \frac{x}{4} \le 1$

20. $-2 < \frac{x-3}{4}$

21. $\frac{3x-8}{5} \le \frac{x+4}{3}$

22. $\frac{2y+3}{-5} < 3 - y$

23. $y > 4 - 2x$

24. $3x - 7y < 21$

25. $-x - 5y \ge 15$

26. $4 \le -2(x - 3) - 7$

27. $-\frac{5}{4}x + 8 < -2y$

28. $\frac{6y+5}{3} \le 7x$

29. $\frac{x}{3} - \frac{y}{4} > -2$

30. $2x - 7y \ge 3(4x - 1) + 8$

31. **Geometry** Let a, b, and c be the lengths of the sides of any triangle. Write an inequality to explain why $a + b = c$ is not possible.

13.

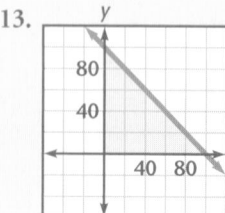

14. This restriction could occur if there were only enough money to buy 200 feet of material, or if only 200 feet of fencing were available.

31. Since the sum of the lengths of any two sides of a triangle is greater than the length of the third side, $a + b > c$. If $a + b = c$ is true, then by substitution, $c > c$, which cannot be true.

Consumer Economics Suppose you want to buy x bags of popcorn and y bags of peanuts, and you only have $2.00.

32. Write an inequality to describe the number of bags of popcorn and the number of bags of peanuts you could possibly buy. $0.6x + 0.75y \leq 2.00$

33. Solve the inequality you wrote for Exercise 32 for y.
$y \leq -0.8x + 2.67$
34. Describe the values of x and y that are possible for this situation.

35. Sketch the graph showing how many bags of each you could buy. How do the possible values for x and y affect the graph?

For Exercises 36 and 37, write an inequality that describes each graph.

36.

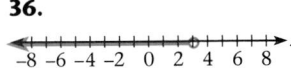

$x < 3$

37.

$x \geq -1$

 **Look Back**

Find the slope of the line defined by each linear equation. [Lesson 1.2]

38. $y + 2x = 3$ -2 **39.** $3x - y = 6$ 3 **40.** $x - 3y = -8$ $\frac{1}{3}$ **41.** $2x - 4y = 3(x - y) + 7$ -1

Write the equation of a line that passes through the given points.   [Lesson 1.2]

42. $(1, 2)$ and $(3, -1)$ **43.** $(5, -2)$ and $(-4, -9)$ **44.** $(8, -30)$ and $(-1, -6)$
$y = -\frac{3}{2}x + \frac{7}{2}$ $y = \frac{7}{9}x - \frac{53}{9}$ $y = -\frac{8}{3}x - \frac{26}{3}$

For Exercises 45–47, use your graphics calculator to draw a scatter plot for this data.   [Lesson 1.3]

x	1.0	1.3	1.5	1.6	1.8	1.9	2.0	2.2	2.3	2.5
y	58	47	50	39	40	35	41	31	34	36

45 What type of correlation does this data have: positive, negative, or zero? negative

46 Find the correlation coefficient, r. -0.87

47 Plot the line of best fit, and predict the value of y when x is 2.8. 26

Look Beyond

48. Graph $y > 2x - 1$ and $y \leq x$ on the same coordinate plane. Describe the intersection.

47.

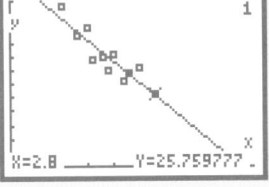

Therefore, when x is 2.8, $y \approx 26$.

48.

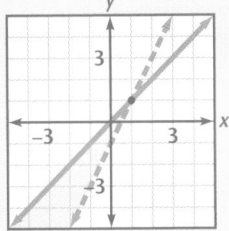

---

34. First, both x and y must be ≥ 0 and must be whole numbers. Since popcorn is 60 cents per bag, the maximum value for x is 3. Since peanuts are 75 cents per bag, the maximum value for y is 2. Thus, $(1, 0)$, $(2, 0)$, $(3, 0)$, $(1, 1)$, $(2, 1)$, $(0, 1)$, and $(0, 2)$ give a total cost $\leq \$2.00$. No bags of popcorn and peanuts, the point $(0, 0)$, is not included in the solution set.

35.

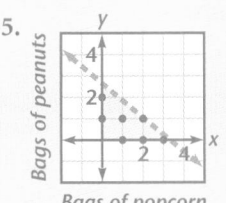

Bags of peanuts

Bags of popcorn

Only ordered pairs containing whole numbers within the triangular shaded region are possible values.

Look Beyond

Encourage students to determine which parts of the overlapping regions satisfy both inequalities, and to check the solution.

The intersection is a wedge-shaped region extending down and to the left from the point of intersection of the boundary lines, $(1, 1)$. This region includes its upper boundary line, $y = x$, but does not include its lower boundary line, $y = 2x - 1$, nor the intersection point $(1, 1)$.

FOCUS

Scientists, as well as professionals in other fields, often have to interpret the data that they have collected. Without a table or graph, relationships between data sets are hard to determine. Students will interpret the data they collect in this project by presenting it graphically as a scatter plot.

MOTIVATE

Have students find the relationship between the distance from the top of the forehead to the eyes and the entire length of the head by taking measurements and making a scatter plot. Ask students to think about other relationships that are present in the measurements of the human face.

Discuss the following questions.

- Do you think taller people have longer arm spans?
- Do you think taller people have bigger hand spans?
- How do you think the distance from the top of a person's head to the ceiling is related to his or her height?
- How do you think the value of someone's pocket change is related to his or her height?

Collect the following information from each student in the classroom, and record your data in the table.

Person	Height	Arm span	Hand span	Distance from head to ceiling	Coin value
1					
2					
3					
⋮					

Draw a scatter plot for each of the following.

1. Height and arm span

2. Height and hand span

3. Height and distance from top of one's head to the ceiling

4. Height and pocket-change value

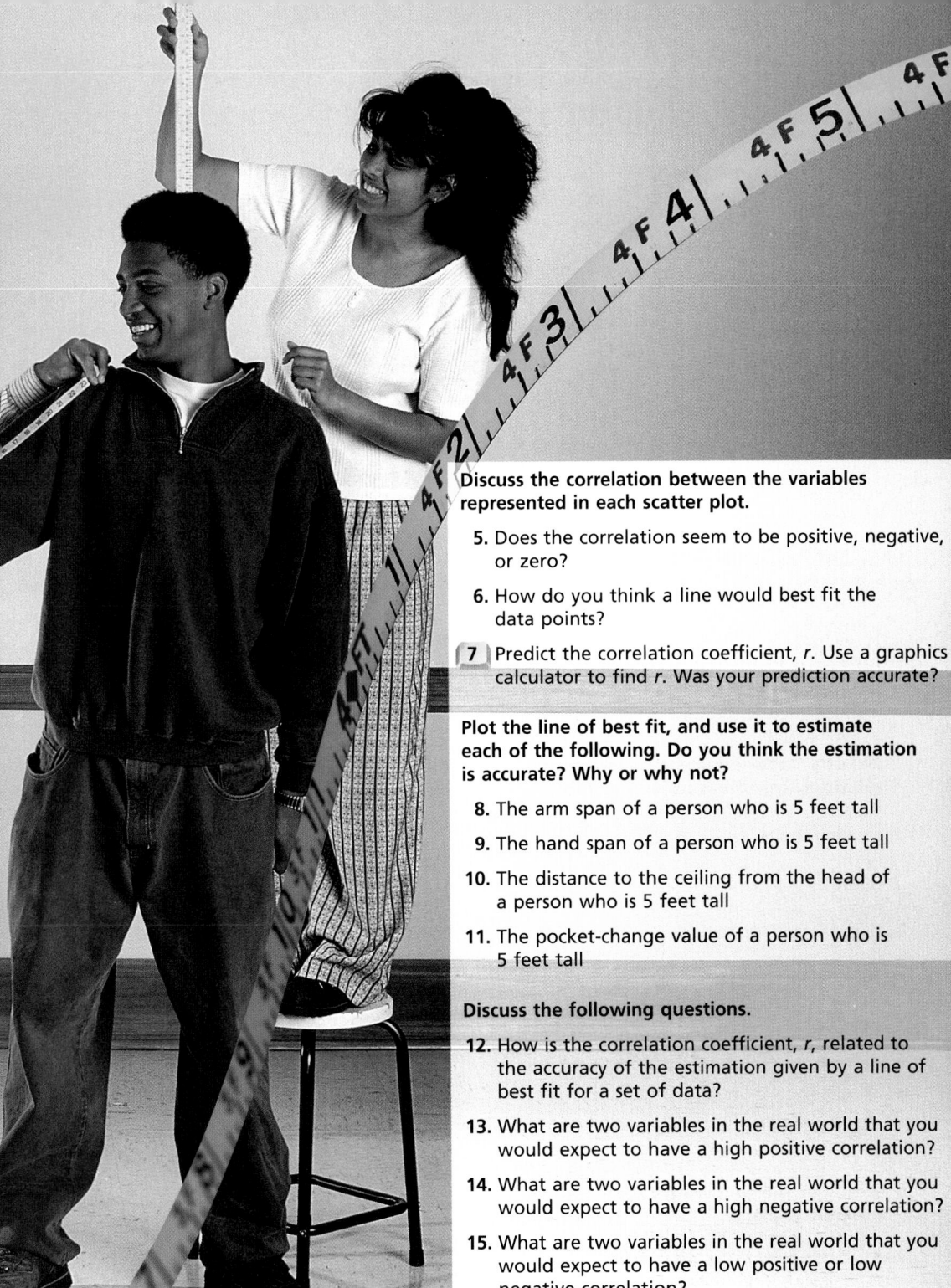

The measurements needed for this exercise can be taken in small groups. Because the larger the data set the greater the accuracy of the scatter plot, have groups share their data with the entire class. Students in small groups can then graph the class data.

Discuss

Discuss the proportions of other primates, such as gorillas. Students should notice that the arm lengths of many primates are proportionally longer than the arm lengths of humans. Have students find pictures that illustrate this in a biology textbook or another source.

Discuss the correlation between the variables represented in each scatter plot.

5. Does the correlation seem to be positive, negative, or zero?

6. How do you think a line would best fit the data points?

7. Predict the correlation coefficient, r. Use a graphics calculator to find r. Was your prediction accurate?

Plot the line of best fit, and use it to estimate each of the following. Do you think the estimation is accurate? Why or why not?

8. The arm span of a person who is 5 feet tall

9. The hand span of a person who is 5 feet tall

10. The distance to the ceiling from the head of a person who is 5 feet tall

11. The pocket-change value of a person who is 5 feet tall

Discuss the following questions.

12. How is the correlation coefficient, r, related to the accuracy of the estimation given by a line of best fit for a set of data?

13. What are two variables in the real world that you would expect to have a high positive correlation?

14. What are two variables in the real world that you would expect to have a high negative correlation?

15. What are two variables in the real world that you would expect to have a low positive or low negative correlation?

Chapter 1 Review

1. $y = -\frac{1}{3} + 5$ is linear. The equation is of the form $y = mx + b$.

2. $2y = x^2 - 9$ is not linear. The equation contains x^2 which means it is a quadratic function.

3. $x + 4y = 11$ is linear. It is in the standard form $Ax + By = C$.

4.
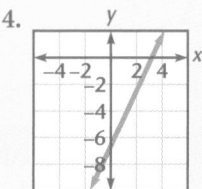

5. x-differences are constant, 2; y-differences are constant, 2. So this is a linear relation.

6. x-differences are not constant; y-differences are constant, 3. So this is not linear.

Vocabulary

addition property of equality	38	linear relationship	9
addition property of inequality	46	line of best fit	25
boundary line	48	multiplication property of equality	38
constant of variation	31	multiplication property of inequality	46
correlation coefficient	25	outlier	24
direct variation	31	proportion	31
distributive property	39	scatter plot	23
division property of equality	38	slope of a line	17
division property of inequality	46	slope-intercept form of a line	19
equivalent	39	substitution property	39
inequality	45	subtraction property of equality	38
linear equation	10	subtraction property of inequality	46
linear inequality	49	y-intercept	18

Key Skills & Exercises

Lesson 1.1

➤ **Key Skills**

Identify linear equations and linear relationships of variables in a table.

When a constant difference in consecutive x-values results in a constant difference in the consecutive y-values in a table, the variables are linearly related.

Write an equation describing a linear relationship, and graph it.

$y = 2x - 3$ is a linear equation.

x	y
1	−1
2	1
3	3

➤ **Exercises**

Determine which of the following equations are linear. Explain.

1. $y = -\frac{1}{3}x + 5$ linear   2. $2y = x^2 - 9$ not linear   3. $x + 4y = 11$ linear

4. Graph $2x - y = 7$.

Which of the following tables contain variables that are linearly related? Explain.

5.
x	−3	−5	−7	−9
y	−3	−1	1	3

linear

6.
x	2	4	8	16
y	−3	−6	−9	−12

not linear

Lesson 1.2

➤ Key Skills

Write an equation in slope-intercept form, given two points on the line or the slope and a point on the line.

The slope of a line that passes through the points $(4, -1)$ and $(-3, -2)$ is

$$m = \frac{y_2 - y_1}{x_2 - x_1} = \frac{-2 - (-1)}{-3 - 4} = \frac{-1}{-7} = \frac{1}{7}.$$

The slope-intercept form of a line is $y = mx + b$, where m is the slope and b is the y-intercept.

➤ Exercises

Find the slope of the line that contains the indicated points.

7. $(-4, 1)$ and $(-1, 3)$ $\frac{2}{3}$ **8.** $(5, -3)$ and $(-5, -3)$ 0 **9.** $(6, -1)$ and $(6, -5)$ undefined

Write the equation of each line described.

10. The line with slope $-\frac{3}{4}$ and containing the point $(-2, 7)$ $y = -\frac{3}{4}x + \frac{11}{2}$

11. The line containing the points $(-2, -3)$ and $(-1, 3)$ $y = 6x + 9$

Lesson 1.3

➤ Key Skills

Graph a scatter plot, and identify the data correlation.

Positive Correlation:
   Slope is upward from left to right.
Negative Correlation:
   Slope is downward from left to right.
Zero Correlation:
   No slope is evident.

Use a graphics calculator to find the correlation coefficient and to make predictions using the line of best fit.

Enter the data into a graphics calculator. Use the linear regression command to find the correlation coefficient, r, and to plot the line of best fit.

➤ Exercises

Use your graphics calculator to make scatter plot for the data.

x	0	1	1	2	3	3	3	4	4	5	5	5
y	12	11	13	2	10	7	9	8	6	7	3	5

12. Identify the correlation. negative   **13.** Identify any outliers. $(2, 2)$

14 Find the correlation coefficient, r. -0.65

15 Use the line of best fit to predict the value of y when x is 7. 2.5

Lesson 1.4

➤ Key Skills

Solve problems involving direct variation.

In a direct variation, y varies directly as x. Thus, $y = kx$, where k is the constant of variation and $k \neq 0$.
Two ordered pairs that satisfy a direct variation equation are proportional.

➤ Exercises

For Exercises 16–18, y varies directly as x, and y is –6 when x is 3.

16. Write the equation of direct variation. $y = -2x$

17. Find y when x is 8. -16 **18.** Find x when y is 3. $-\frac{3}{2}$

22. $x \le -5$

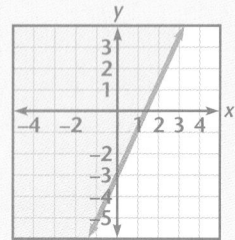

23. $y \ge 2x - 3$

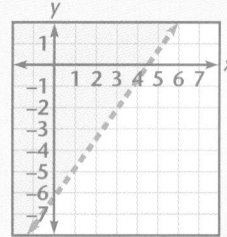

24. $y > \frac{4}{3}x - 6$

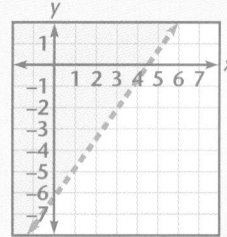

Lesson 1.5

➤ **Key Skills**

Solve problems by writing and solving linear equations.

To solve $2x - 6(-3x + 5) = 10$ for x by graphing, set each side of the equation equal to y. Graph both equations. The solution is the x-coordinate of the point of intersection.

➤ **Exercises**

Use your graphics calculator to approximate the solution for x in the following equations. Then solve for x using algebra.

19 $4x - 3 = 15 - 2x$ 3   **20** $8x - 10(x + 4) = x - 4$ **21** $3x - 7 + 10x = 8x + 3$ 2

-12

Lesson 1.6

➤ **Key Skills**

Solve and graph linear inequalities.

Solve $4x - 3y \le 9$, and graph the solution.

$$4x - 3y \le 9$$
$$-3y \le -4x + 9$$
$$y \ge \frac{4}{3}x - 3$$

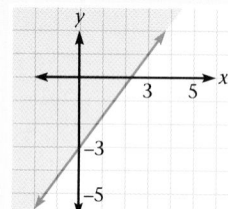

➤ **Exercises**

Solve and graph each inequality.

22. $-3x - 8 \ge 7$ $x \le -5$ **23.** $2x - y \le 3$ $y \ge 2x - 3$ **24.** $-4x + 3y > -18$ $y > \frac{4}{3}x - 6$

25. Write the inequality that is graphed.

$$y < -\frac{4}{3}x - 2$$

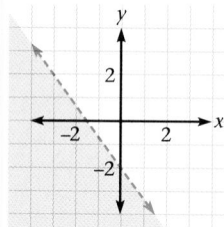

Applications

Geometry The length of the hypotenuse of a 30-60-90 triangle varies directly as the length of the side opposite the 30° angle. The length of the hypotenuse, h, is 45 when the side opposite the 30° angle, s, is 22.5.

26. Write the equation of direct variation. $h = 2s$

27. Find the length of the side opposite the 30° angle when the hypotenuse is 13 inches long.   6.5 in.

Family Savings The majority of American families save less than $\frac{1}{10}$ of their income.

28. Write a linear inequality that models this savings. $s < \frac{1}{10}i$

29. Graph the solution set.

29.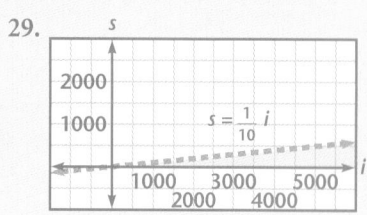

Chapter 1 Assessment

Anatomy The following table lists the heights and weights for 10 young male adults with medium frames. The heights are in inches, and the weights are in pounds.

Height	67	66	70	67	67	68	69	66	70	71
Weights	146	145	157	148	145	149	151	141	154	159

1 Graph a scatter plot for this data. Is the correlation positive, negative, or zero? positive

2 Find the correlation coefficient, r. 0.97

3 Plot the line of best fit on the same coordinate plane as the scatter plot. Predict the weight (to the nearest pound) of a young male who is 63 inches tall with medium frame. 134 lb

4. Are the variables in this table linearly related? Explain.

x	−2	4	10	16
y	1	5	10	15

not linear

For Items 5 and 6, determine whether or not each equation is linear. Explain why or why not.

5. $2x - \frac{2}{5}y = 7$ linear

6. $y = x^3 - 12$ not linear

7. Graph the equation $y = -5x - 12$.

8. Find the slope of the line containing the points (−3, 6) and (−5, 8). −1

9. Write the equation of the line containing the points (2, −4) and (3, −1). $y = 3x - 10$

For Items 10–12, q varies directly as p. Find the constant of variation, and write the equation.

10. q is 21 when p is 7 $3, q = 3p$

11. q is 1 when p is $\frac{1}{3}$ $3, q = 3p$

12. q is 6 when p is 18 $\frac{1}{3}, q = \frac{1}{3}p$

Solve each equation for x.

13. $-(2x + 3) = 6x - 11$ 1

14. $6x - 4(x - 7) = -2$ −15

Small Business A mobile computer repair service, Drive Doctors, charges $25 per hour, plus a base fee of $25 to go to the work site or home and to make the estimate of repairs. Another mobile computer repair service, Mighty Bytes, charges $30 per hour, and a base fee of $30 for the on-site estimate. On separate calls, the two repair services earned the same amount of money, but Mighty Bytes worked half an hour less than Drive Doctors.

$30 + 30x = 25 + 25(x + 0.5)$

15. Wri'e an equation to find the number of hours that each took for the repair.

16. Find the number of hours that each took for the repair. MB: 1.5 hr; DD: 2 hr

Solve and graph each inequality.

17. $x - 2(x - 9) \geq -2 + 6x$ $x \leq \frac{20}{7}$

18. $5x - 6(x + 9) < -3y$ $y < \frac{1}{3}x + 18$

Geometry Tim has 80 feet of fencing and wants to make a rectangular outdoor cage for his iguana.

19. Write a linear inequality to represent the possible perimeter of the base of the cage. $2w + 2l \leq 80$

20. Graph the solutions.

4. The x-values have a constant difference of 6, but consecutive y-values do not have a common difference. So the variables in the table are not linear.

5. $2x - \frac{2}{5}y = 7$ is linear. The equation may be expressed in the $y = mx + b$ form.

6. $y = x^3 - 12$ is not linear. The x-term is cubed.

7.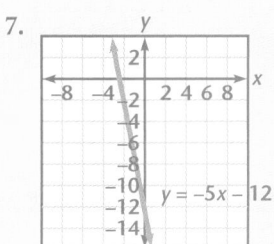

17. $x \leq \frac{20}{7}$

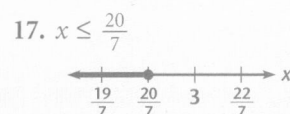

18. $y < \frac{1}{3}x + 18$

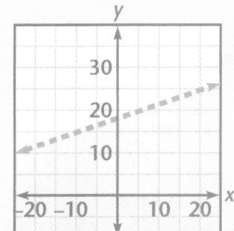

20.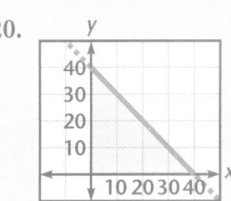

Operations With Numbers and Functions

Meeting Individual Needs

2.1 Exploring Numbers

Core	Core Plus
Inclusion Strategies, p. 63	Practice Master 2.1
Reteaching the Lesson, p. 64	Enrichment, p. 63
Practice Master 2.1	Technology Master 2.1
Enrichment Master 2.1	
Technology Master 2.1	
Lesson Activity Master 2.1	
[1 day]	**[1 day]**

2.4 Using Notation to Represent Functions

Core	Core Plus
Inclusion Strategies, p. 85	Practice Master 2.4
Reteaching the Lesson, p. 86	Enrichment, p. 84
Practice Master 2.4	Technology Master 2.4
Enrichment Master 2.4	
Technology Master 2.4	
Lesson Activity Master 2.4	
[2 days]	**[2 days]**

2.2 Exploring Properties of Exponents

Core	Core Plus
Inclusion Strategies, p. 70	Practice Master 2.2
Reteaching the Lesson, p. 71	Enrichment, p. 70
Practice Master 2.2	Technology Master 2.2
Enrichment Master 2.2	
Technology Master 2.2	
Lesson Activity Master 2.2	
[2 days]	**[2 days]**

2.5 Linear Functions

Core	Core Plus
Inclusion Strategies, p. 92	Practice Master 2.5
Reteaching the Lesson, p. 93	Enrichment, p. 92
Practice Master 2.5	Technology Master 2.5
Enrichment Master 2.5	
Technology Master 2.5	
Lesson Activity Master 2.5	
Interdisciplinary Connection, p. 91	
[2 days]	**[2 days]**

2.3 Definition of a Function

Core	Core Plus
Inclusion Strategies, p. 78	Practice Master 2.3
Reteaching the Lesson, p. 79	Enrichment, p. 78
Practice Master 2.3	Technology Master 2.3
Enrichment Master 2.3	Mid–Chapter Assessment
Technology Master 2.3	Master
Lesson Activity Master 2.3	
Interdisciplinary Connection, p. 77	
Mid–Chapter Assessment Master	
[1 day]	**[1 day]**

2.6 Exploring Operations with Functions

Core	Core Plus
Inclusion Strategies, p. 99	Practice Master 2.6
Reteaching the Lesson, p. 100	Enrichment, p. 99
Practice Master 2.6	Technology Master 2.6
Enrichment Master 2.6	
Technology Master 2.6	
Lesson Activity Master 2.6	
[1 day]	**[1 day]**

Chapter Summary

Core	Core Plus
Eyewitness Math, pp. 74–75	Eyewitness Math, pp. 74–75
Chapter 2 Project, pp. 104–105	Chapter 2 Project, pp. 104–105
Lab Activity	Lab Activity
Long-Term Project	Long-Term Project
Chapter Review, pp. 106–108	Chapter Review, pp. 106–108
Chapter Assessment, p. 109	Chapter Assessment, p. 109
Chapter Assessment, A/B	Chapter Assessment, A/B
Alternative Assessment	Alternative Assessment
Cumulative Assessment, pp. 110–111	Cumulative Assessment, pp. 110–111
[3 days]	**[3 days]**

Hands-on Strategies

In groups have students develop their own number systems, or change the operations within our own number system. They might invent numerals based on twelve digits instead of ten. They might also invent new operations. For instance, $2 @ 2 = 2 + 5 - 7 + 2$ and $2 @ 3 = 2 + 5 - 7 + 3$ and so forth. Encourage students to be as creative as possible, but emphasize that the numbers or operations they invent must be logical. When the groups are finished, they can teach their new mathematics to the class and give a short "math test" over their number systems or operations.

Reading Strategies

As students read each lesson, have them list unfamiliar words. Before instruction have them try to define the words using any prior knowledge they may have. Wherever a procedure is presented in a lesson (such as how to graph a linear function), have students write or verbalize step by step instructions for the procedure. Have students paraphrase the information to ensure that they comprehend the information.

Visual Strategies

In many lessons in this chapter there will be photos, tables, charts, and graphs. Have students first determine what information is being presented. For correlations, have them describe what information appears in each format and what variables are being correlated, if it applies. Have them also look at the pictures and determine how they contribute to the understanding of the lesson.

Cooperative Learning

GROUP ACTIVITIES	
Mayan Numerals	Portfolio Activity
Properties of real numbers	Lesson 2.1, Exploration 3
Fractional exponents	Lesson 2.2, Exploration 4
The method of false position	Eyewitness math
Linear functions	Lesson 2.5, Exercises 25–28
Operations with functions	Lesson 2.6, Cooperative Learning
Bouncing Ball Experiment	Chapter 2 Project

You may wish to have students work with partners for some of the above activities. Additional suggestions for cooperative group activities are noted in the teacher's notes in each lesson.

Multicultural

The cultural connections in this chapter include references to Asia, Africa, and North America.

CULTURAL CONNECTIONS	
Asia: Rational and irrational numbers	Lesson 2.1, Exercises 41–45
Asia: Mathematicians of the Islamic Empire	Lesson 2.2
Asia: Palestinian formula for the area of a circle	Lesson 2.4, Exercises 34–36
Africa: Linear functions	Lesson 2.5, Example 2
North America: Mayan numerals	Lesson 2.6, Exercises 23–25

Portfolio

Below are portfolio activities for the chapter. They are listed under seven activity domains that are appropriate for portfolio development.

1. **Investigation/Exploration** Students investigate and discover mathematical rules in the Portfolio Activity. Also appropriate is the Chapter Project, where students are asked to find the relationship between the distance a ball is dropped and the height of its rebound.

2. **Applications** Recommended to be included are the following: Investment (Lesson 2.2), Projectile Motion (Lesson 2.3), Consumer Awareness (Lesson 2.3), Consumer Economics (Lesson 2.3), Demographics (Lesson 2.5), Economics (Lesson 2.5).

3. **Non-Routine Problems** The Palestinian formula for area of a circle (Lesson 2.4) gives students a new look at something with which they are very familiar. Also recommended are the Chapter Project, Eye Witness Math, and the Portfolio Activity.

4. **Project** Two projects in this lesson are The Bouncing Ball Experiment and the activity in the Hands on Strategies.

Both use cooperative learning to lead students to the discovery of mathematical relationships.

5. **Interdisciplinary Topics** Recommended to be included are the following: Science (Lesson 2.1), Economics (Lesson 2.4), Physiology (Lesson 2.4), Biology (Lesson 2.5), and Physics (Lesson 2.5).

6. **Writing** The following activities use writing to solve problems or to aid in understanding: Enrichment p. 63 (Lesson 2.1), Reteaching the Lesson p. 79 (Lesson 2.3), Current Events exercises (Lesson 2.1). Also appropriate are the Communicate questions at the beginning of the exercise sets.

7. **Tools** The material in Chapter 2 makes use of calculators, graphics calculators, and/or graph paper. Manipulatives are needed for the Portfolio Activity. The Chapter Project requires the use of a ball and a ruler or tape measure

Technology

Technology provides important tools for manipulating and visualizing algebraic concepts. Some lessons in Chapter 2 can be significantly enhanced by using spreadsheets. Refer to the HRW Technology Handbook for additional information and detailed instructions for spreadsheets, calculators, and other technologies.

Spreadsheets

The sample problem below will help you model the set-up and the use of spreadsheets with students.

Sample Problem A town's population grows from 10,000 to 11,000 in one year. Calculate and graph the population growth for ten years if:
 a. the population grows by the same amount each year
 b. the population grows by the same rate each year

1. **Spreadsheet Set-up** Spreadsheets are very similar to matrices because cells are the spreadsheet's equivalent of matrix elements. You can enter text, numbers, or formulas in an individual cell and format each in a variety of ways. Cells addressed A1, B1 provide column labels. Label A1 as **Year**, B1 as **Fixed Amount**, and C1 as **Fixed Rate**. Column B: Entries will increase by a fixed amount.

	A	B	C
1	Year	Fixed Amt.	Fixed Rate
2	0	11,000	11,000
3	1	12,000	+C2*1.1
4	2	13,000	
5	3	14,000	
6	4	15,000	
7	5	16,000	
8	6	17,000	
9	7	18,000	
10	8	19,000	
11	9	20,000	

2. **Copying formulas** At the end of the first year, the new population is 11,000, so the fixed amount is 1000 added each year. Different software uses different symbols (= or +) to indicate a formula. For example, enter the symbol, +, and the formula B2+1000 in B3. Then copy the formula to B4 through B12.

For a fixed rate increase, each current year should be multiplied by 1.1. Thus after +, the formula becomes **C2*1.1**. Enter this formula in C3 and copy it to C4 through C12.

3. **Formatting columns B & C** Place cursor on cell B2, press **Range**, **Format**, Enter, and how many decimal places to include. If you want only whole dollars, specify **0** decimal places. Use the arrow keys to highlight the remainder the of range to format, and press **Enter**.

4. **Graphing Data** Pull up command menu and highlight **Graph**. The computer will give you several choices. First, select the type of graph (line graph) and press **Enter**. Then select data for the X-register. When the computer asks for a range, highlight A2 to A12. For the A register use B2 to B12, and for the B-register use C2 to C12. You can also specify titles and data labels, if desired. Then press View to see the graph.

Integrated Software

f(g) Scholar™ is an integrated computer-based mathematics productivity tool that combines calculator, spreadsheet, and graphics capabilities to provide a dynamic and interactive environment for explorations in mathematics. It is appropriate to use *f(g) Scholar*™ for any lesson needing a spreadsheet, calculator, graphics calculator, or any combination of the three.

2 Operations With Numbers and Functions

A BOUT THE CHAPTER

Background Information

In this chapter students will become familiar with properties of real numbers and functions. They will graph both linear and nonlinear functions.

CHAPTER RESOURCES

- Practice Masters
- Enrichment Masters
- Technology Masters
- Lesson Activity Masters
- Lab Activity Masters
- Long-Term Project Masters
- Assessment Masters
  Chapter Assessments A/B
  Mid-Chapter Assessment
  Alternative Assessments, A/B
- Teaching Transparencies
- Cumulative Assessment
- Spanish Resources

C HAPTER OBJECTIVES

- Compare and identify number systems.
- Identify properties of real numbers, and use these properties to perform operations with rational numbers.
- Perform operations, and evaluate expressions using the properties of exponents.
- Identify and compare relations and functions, and use the vertical-line test to identify functions from their graphs.
- Use functions to model real-world applications, and give appropriate domain and range restrictions for the situation.

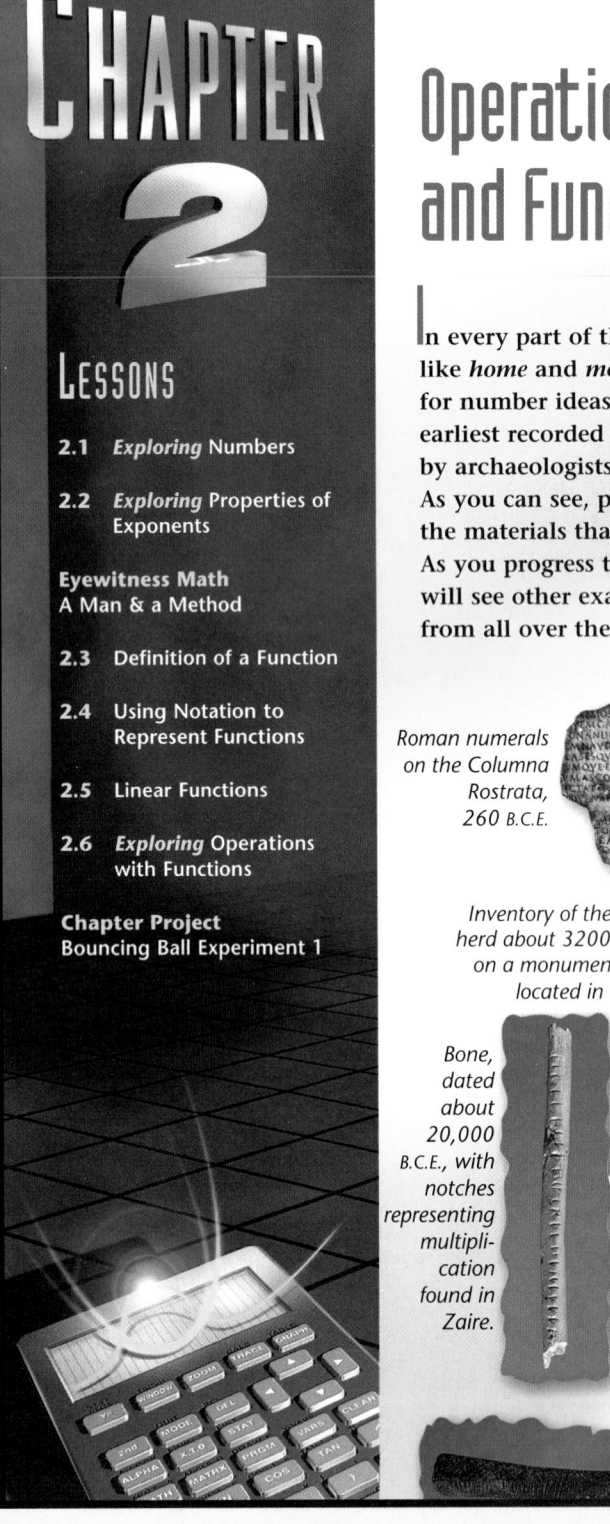

CHAPTER 2

LESSONS

Operations With Numbers and Functions

In every part of the world, as words were developed for ideas like *home* and *mother*, words were also being developed for number ideas like *one* and *many*. A few of the earliest recorded numbers that have been found by archaeologists are shown on these pages. As you can see, people wrote and recorded on the materials that were available to them. As you progress through this book, you will see other examples of early numbers from all over the world.

Roman numerals on the Columna Rostrata, 260 B.C.E.

Inventory of the king's cattle herd about 3200 B.C.E., found on a monument now located in Cairo.

Bone, dated about 20,000 B.C.E., with notches representing multiplication found in Zaire.

Fossilized Baboon bone with engraved tallies, dated about 35,000 B.C.E., found in southern Africa.

EASTE
EURO

ITALY

EGYP

CENTRAL
AFRICA

SOUTHER
AF

A BOUT THE PHOTOS

The way numbers work has very much to do with whether or not they can be easily added, subtracted, multiplied, or divided. For instance, the rules for adding and subtracting with Roman numerals are fairly straightforward. However, there are not simple rules for multiplying Roman numerals. The photos on pages 60 and 61 show the evidence we have that numbers existed and the form in which they existed in different locations all over the world.

Antler with tally marks, dated about 33,000 B.C.E., found in what is now the Czech Republic.

Counting rods made of bone, from the Shǎnxi province, fifth century B.C.E.

CHINA

- Use function notation to define, evaluate, and operate with functions.
- Use the slope formula to write and identify increasing and decreasing linear functions.
- Identify and use properties of functions to add, subtract, multiply, and divide functions.

PORTFOLIO ACTIVITY

In the Portfolio Activity exercises on page 102, students will perform operations using Maya numerals and develop rules for the operations. Have students work in groups to devise the rules for adding Maya numerals. Then have them compare Maya numerals with other ancient number systems as well as with our own decimal system. Have students invent their own number system. Be sure they include the rules for operations as well as the symbols. Then ask students to consider how our own number system would change if it were based on 12 digits instead of 10.

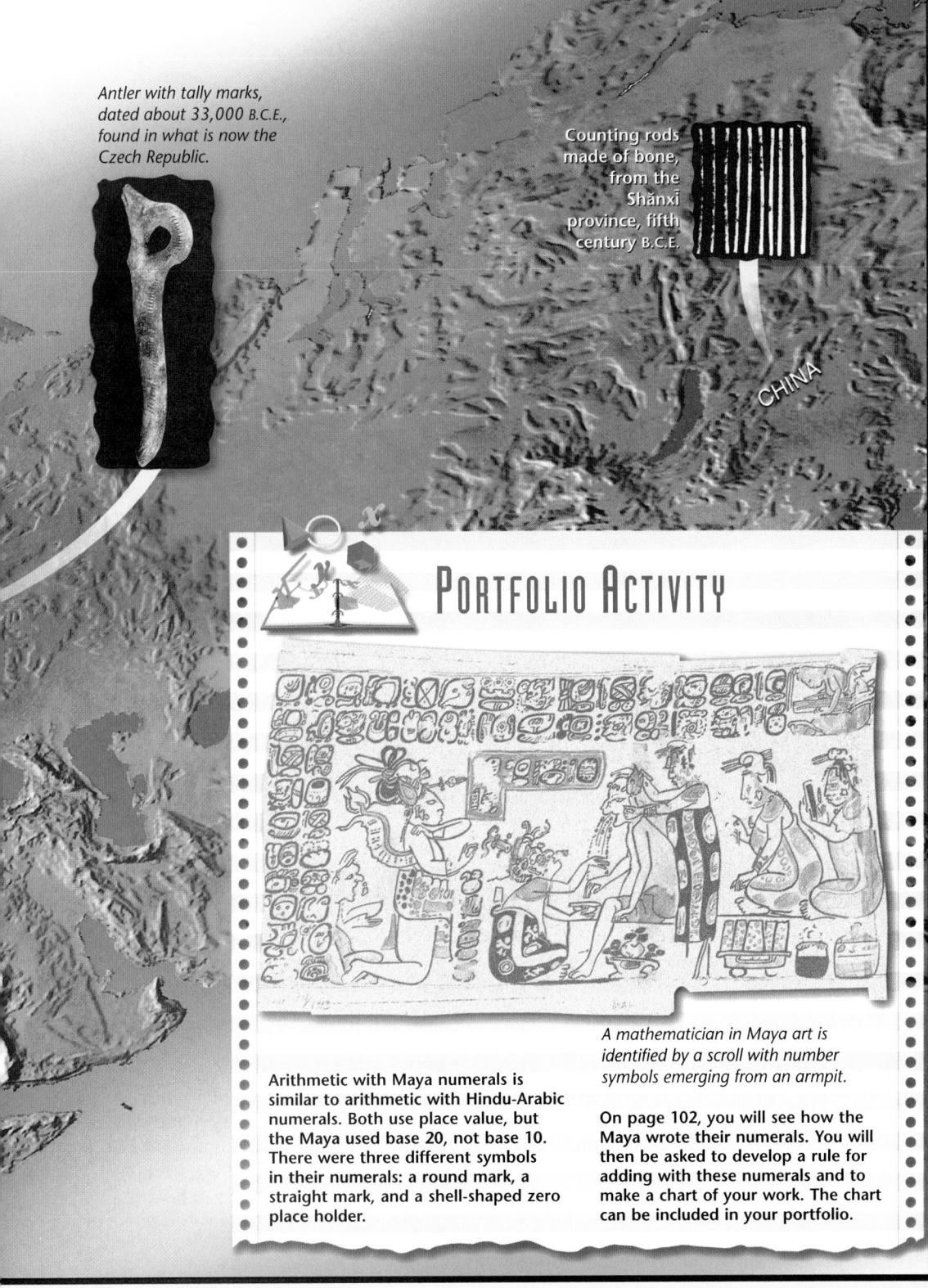

PORTFOLIO ACTIVITY

A mathematician in Maya art is identified by a scroll with number symbols emerging from an armpit.

Arithmetic with Maya numerals is similar to arithmetic with Hindu-Arabic numerals. Both use place value, but the Maya used base 20, not base 10. There were three different symbols in their numerals: a round mark, a straight mark, and a shell-shaped zero place holder.

On page 102, you will see how the Maya wrote their numerals. You will then be asked to develop a rule for adding with these numerals and to make a chart of your work. The chart can be included in your portfolio.

ABOUT THE CHAPTER PROJECT

In the Chapter Project on pages 104 and 105, pairs of students bounce a ball and record the height of each bounce in a table. Each pair will make a scatter plot to determine the correlation between the original height of the ball and the height of its rebound. The project closes with students comparing each other's work and analyzing their results as a class.

PREPARE

Objectives

- Compare and identify number systems.
- Identify properties of real numbers, and use these properties to perform operations with rational numbers.

RESOURCES

- Practice Master   2.1
- Enrichment Master   2.1
- Technology Master   2.1
- Lesson Activity Master   2.1
- Quiz   2.1
- Spanish Resources   2.1

Assessing Prior Knowledge

Evaluate each expression if $x = 0.2$, $y = -3$, and $z = \frac{1}{3}$.

1. $x - z \quad \left[-\frac{2}{15}\right]$

2. $y + z \quad \left[-2\frac{2}{3}\right]$

3. $\frac{1}{y+x} \quad \left[-\frac{5}{14}\right]$

4. $yz \quad [-1]$

TEACH

 Discuss the fact that numbers are present everywhere. Have students describe types of numbers that are present in their daily lives. Discuss why different types of numbers are needed.

Use Transparency ▸ 6

Exploring Numbers

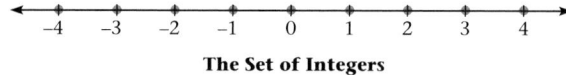 *Anna picked up a copy of the New York Times and found that every article she read contained numbers of some sort. How can Anna begin to understand the meaning and importance of all of these numbers?*

Number Systems

Mathematicians classify numbers into various sets depending on how they are used. For instance, the set of **whole numbers**, **W** = {0, 1, 2, 3, 4, . . .}, is used to express ideas such as the number of correct responses on an exam or the number of students in a classroom. When the whole numbers are combined with their opposites, the set of **integers**, **I**, is formed. For example, integers are useful in expressing a gain (+3) or a loss (–5).

```
  ◄──┼────┼────┼────┼────┼────┼────┼────┼────►
    -4   -3   -2   -1    0    1    2    3    4
```

The Set of Integers

In order to measure the dimensions of a classroom or to give partial credit on an exam, it is necessary to have numbers that fall *between* the integers. Numbers like $2\frac{1}{2}$, $\frac{78}{40}$, $0.\overline{333}$, and -5.60222759 are examples of numbers that are not integers. These numbers are all *rational* numbers. The set of **rational numbers**, **Q**, contains all the numbers that can be written as the quotient of two integers, where the denominator is not zero. For example, $\frac{2}{3}$ and $\frac{3}{2}$ are rational numbers, but $\frac{2}{0}$ is not.

ALTERNATIVE
teaching
strategy

Inviting Participation Provide each student with a piece of grid paper. Have each student create a large grid of 12 squares (3 squares wide by 4 squares high), and randomly write one property given on page 65 in each square. Since there are only 11 properties, have the students write **free** in the remaining square. Write an example of each of the properties on the chalkboard, such as $3(x + 2) = 3x + 6$, and ask each student to cross out the corresponding square. The first student correctly to cross out either a row or a column is the winner.

Does the set of rational numbers include the set of integers? Explain.

There are points on the number line that cannot be written as a quotient of two integers. Examples of these **irrational** (*not* rational) **numbers** are π and $\sqrt{2}$.

The set of **real numbers**, **R**, is the combination of all rational and all irrational numbers.

Use what you have learned about sets of numbers to explain this diagram.

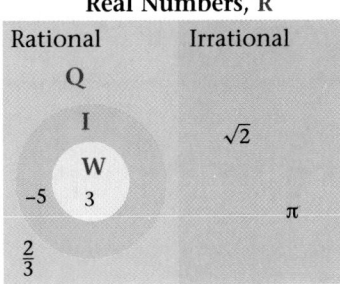

Real Numbers, R

Rational   Irrational

Q

I $\sqrt{2}$

W

−5   3

$\frac{2}{3}$ π

Operations With Rational Numbers

Exploration 1   Adding Rational Numbers

Let $a = 3$, $b = \frac{1}{2}$, $c = -5$, $d = \frac{3}{4}$, $e = -\frac{2}{3}$, and $f = -1$.

Substitute the given values for the variables in each equation. Which statements are true?

1. $a + c = c + a$
2. $b - d = d - b$
3. $d - (a - f) = (d - a) - f$
4. $b + (c + e) = (b + c) + e$
5. $c - (f - a) = c - f + a$
6. $-(d + c) = -d + c$
7. $-(b - e) = -b + e$
8. $b - (e + f) = b - e + f$
9. $-f + c = -(f + c)$
10. $e + d = d - (-e)$
11. Pick your own nonzero rational-number values for each variable. Substitute these values in the equations.
12. Which statements seem to be true for all rational numbers? ❖

Exploration 2   Multiplying Rational Numbers

Let $a = 3$, $b = \frac{1}{2}$, $c = -5$, $d = \frac{3}{4}$, $e = -\frac{2}{3}$, and $f = -1$.

Substitute the given values for the variables in each equation. Which statements are true?

1. $a \cdot c = c \cdot a$
2. $b \div d = d \div b$
3. $\frac{d}{a \div e} = \frac{d \div a}{e}$
4. $\frac{1}{b \div e} = \frac{e}{b}$
5. $\frac{1}{f} \cdot \frac{1}{c} = \frac{1}{f \cdot c}$
6. $\frac{c \div d}{b} = c \div (d \cdot b)$
7. $\frac{d}{c} = \frac{c}{\frac{1}{d}}$
8. $e \cdot d = \frac{d}{\frac{1}{e}}$

11 Pick your own rational-number values for each variable. Substitute these values in the equations.

12 Which statements seem to be true for all rational numbers? ❖

•Exploration 3 *Multiplication and Addition*

Let $a = 3$, $b = \dfrac{1}{2}$, $c = -5$, $d = \dfrac{3}{4}$, $e = -\dfrac{2}{3}$, and $f = -1$.

Substitute the given values for the variables in each equation. Which statements are true?

1 $\dfrac{1}{d+c} = \dfrac{1}{d} + \dfrac{1}{c}$

2 $a(e + f) = ae + af$

3 $-(a - e) = -a - e$

4 $(b - c)d = bd - cd$

5 $-e(b - a) = -eb - ea$

6 $-f(d + e) = -fd - fe$

7 $(d + a)(e - b) = de - db + ae - ab$

8 $(f - b)(a - c) = fa - ba + fc - bc$

9 Pick your own rational-number values for each variable. Substitute these values in the equations.

10 Which statements seem to be true for all rational numbers? ❖

The Real-Number System

There are many rules and properties for the set of rational numbers. You have reviewed some of the rules and properties in the explorations above. The properties that are true for the set of rational numbers are also true for the set of real numbers. The table of properties defines the *real number system*.

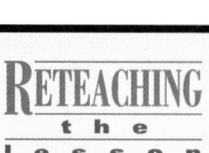

Properties for the Set of Real Numbers
with the Operations (+) and (·)

Let a, b, and c represent any real numbers.

Property	Addition	Multiplication
Closure	$a + b$ is a real number.	$a \cdot b$ is a real number.
Associative	$(a + b) + c = a + (b + c)$	$(a \cdot b) \cdot c = a \cdot (b \cdot c)$
Commutative	$a + b = b + a$	$a \cdot b = b \cdot a$
Identity	There is a number 0 such that $a + 0 = a = 0 + a$.	There is a number 1 such that $a \cdot 1 = a = 1 \cdot a$.
Inverse	For any number a, there is a number $-a$ such that $a + (-a) = 0 = (-a) + a$.	For any nonzero number a, there is a number $\frac{1}{a}$ such that $a \cdot \frac{1}{a} = 1 = \frac{1}{a} \cdot a$.
Distributive	$a \cdot (b + c) = a \cdot b + a \cdot c$	

EXERCISES & PROBLEMS

Communicate

1. Discuss two times in the past week when you added numbers. What type of numbers did you add?

2. Use a number line to explain what happens when you add two integers.

3. Step 1 of Exploration 1 and Step 1 of Exploration 2 both illustrate the commutative property. Explain this property. Why do you think the word *commutative* is appropriate for this property?

4. Step 4 of Exploration 1 and Step 9 of Exploration 2 both illustrate the associative property. Explain this property. Why do you think the word *associative* is appropriate for this property?

5. Describe one possible way numbers developed historically. Consider the needs of early people and the increasingly complex needs of modern people.

ASSESS

Selected Answers

Odd-numbered problems 7–55

Assignment Guide

Core 1–18, 21–34, 37–55

Core Plus 1–8, 13–24, 28–56

Technology

Graphics calculators are needed for Exercises 42 and 44, 46–50, and 56.

Error analysis

Watch for students who do not recall how to add, subtract, multiply, and divide fractions. Provide students with several worked-out examples for each operation, along with several exercises for them to complete on their own. Remind students that it is easier to multiply and divide fractions if those fractions are in reduced form.

alternative

ASSESSMENT

Performance Assessment

Have the students go through the exploration problems, and determine whether each statement demonstrates a property. If a statement does, which property is expressed? If it does not, explain why not.

Practice & Apply

Current Events Use your local newspaper for Exercises 6–8.

6. Name three articles that include numerical data. Answers may vary.

7. Summarize each article you listed in Exercise 6. Include the reason the article depends on numbers and the impact each article has on your life. Answers may vary.

8. What types of numbers were included in each article? What units, if any, were used with the numbers? Answers may vary.

Simplify.

9. $\frac{3x}{4} + \frac{2}{3}$ $\frac{9x+8}{12}$ **10.** $\frac{3}{6} + \frac{-2x}{3}$ $\frac{3-4x}{6}$ **11.** $\frac{2x}{7} + \frac{15x}{2}$ $\frac{109x}{14}$ **12.** $\frac{y}{3} - \frac{3y}{4}$ $-\frac{5y}{12}$

13. $\frac{x-1}{2} + \frac{x}{4}$ $\frac{3x-2}{4}$ **14.** $\frac{5y}{6} + \frac{3y}{5}$ $\frac{43y}{30}$ **15.** $\frac{3x}{4} + \frac{2(x-1)}{3}$ $\frac{17x-8}{12}$ **16.** $\frac{-3(1-x)}{4} + \frac{2x}{3}$ $\frac{17x-9}{12}$

17. $\frac{x+2}{3} + \frac{x}{6}$ $\frac{3x+4}{6}$ **18.** $\frac{1+2x}{3} - \frac{x}{2}$ $\frac{x+2}{6}$ **19.** $\frac{3}{2x} + \frac{4}{6x}$ $\frac{13}{6x}$ **20.** $\frac{3}{x} + \frac{5}{x^2}$ $\frac{3x+5}{x^2}$

Complete the investigation in Exercises 21–24.

21. Count the number of items in your home that display numbers. Answers may vary.

22. How many different types of numbers are represented? Answers may vary.

23. Name two examples of integers and two examples of rational numbers that you discovered. Answers may vary.

24. Explain the significance of the numbers you named in Exercise 23. Answers may vary.

Simplify.

25. $\frac{2x}{7} \cdot \frac{14}{2}$ $2x$

26. $\frac{3}{7x} \cdot \frac{4x}{28}$ $\frac{3}{49}$

27. $\frac{a}{4} \cdot \frac{12}{a^2}$ $\frac{3}{a}$

28. $\frac{ab}{6a} \cdot \frac{2b}{ab^2}$ $\frac{1}{3a}$

29. $\frac{2bc}{3a} \cdot \frac{ab}{4c}$ $\frac{b^2}{6}$

30. $\frac{24a}{3(b-1)} \cdot \frac{b-1}{3a^2}$ $\frac{8}{3a}$

31. $\frac{4(x+1)}{3x-3} \cdot \frac{x-1}{x+1}$ $\frac{4}{3}$

32. $\frac{x+2}{2x} \cdot \frac{4}{2x+4}$ $\frac{1}{x}$

33. $\frac{x-3}{4x+12} \div \frac{2x-6}{4}$ $\frac{1}{2x+6}$

34. $\frac{x+1}{2} \div \frac{5x+5}{6}$ $\frac{3}{5}$

35. $\frac{x^2+x}{3} \div x^2$ $\frac{x+1}{3x}$

36. $\frac{6x-27}{4x} \div (4x-18)$ $\frac{3}{8x}$

Science Use a scientific magazine for Exercises 37–40.

37. How many articles in the magazine include numbers? Which one of these articles interests you the most? Answers may vary.

38. How do numbers contribute to the article that interests you? Answers may vary.

39. Summarize the article. Include the reason you chose the article. Answers may vary.

40. Are numbers the main focus of the article? If so, explain the importance of understanding numbers in order to understand the article. If not, how could numbers play a larger role in the article? Answers may vary.

Cultural Connection: Asia

Ancient Babylonians used irrational numbers much as engineers and computer scientists do today. They found rational numbers that were approximations of irrational numbers. For example, the Babylonians knew that the diagonal of a square was $\sqrt{2}$ times the length of a side. For the value of $\sqrt{2}$, the Babylonians used 1.4142. They thought this value was good enough for their practical purposes. Use a calculator in the following exercises.

41. Use the "Pythagorean" Right-Triangle Theorem to see why the diagonal of a square is $\sqrt{2}$ times the length of a side.

42. Is the Babylonian value of 1.4142 equal in value to $\sqrt{2}$? Explain why or why not. HINT: Find $(1.4142)^2$.

43. Why do you think the Babylonians were satisfied with their value for $\sqrt{2}$?

44. Find $\sqrt{2}$ on your calculator. Write down the result. Enter this number in your calculator, and square it. Is the result equal to 2? Explain why or why not.

45. Supercomputers give $\sqrt{2}$ to many more decimal places than your calculator. Do you think this decimal expansion will ever terminate, giving $\sqrt{2}$ as an exact decimal value? Explain.

A Babylonian cuneiform tablet showing the calculation of areas

Look Back

Use your graphics calculator to graph the following linear equations. What viewing window shows both the x- and the y-intercepts? Find the slope of each line. [Lessons 1.1, 1.2]

46. $y = -3x$ $m = -3$ **47.** $y = 2x - 1$ $m = 2$ **48.** $y = x + 5$ $m = 1$ **49.** $y = \dfrac{3x - 1}{4}$ $m = \dfrac{3}{4}$ **50.** $y = \dfrac{-x}{2} + 1$ $m = -\dfrac{1}{2}$

Substitute 2 for x in each equation, and simplify.

51. $y = -3x$ $y = -6$ **52.** $y = 2x - 1$ $y = 3$ **53.** $y = x + 5$ $y = 7$ **54.** $y = \dfrac{3x - 1}{4}$ $y = \dfrac{5}{4}$ **55.** $y = \dfrac{-x}{2} + 1$ $y = 0$

Look Beyond

56. **Transformations** Use your graphics calculator to graph $y = x^{-1}$, $y = \dfrac{1}{x}$, $y = \dfrac{-1}{x}$, and $y = -x$ on the same coordinate plane. How are these graphs alike? How are they different? Make a table of values including x-values −2, −1, 0, 1, and 2 for each equation. Are the resulting y-values the same for any two equations?

41. Let a be the length of the side of a square and let c be the length of the diagonal of the square. Then

$$a^2 + a^2 = c^2$$
$$2a^2 = c^2$$
$$\text{or } c = \sqrt{2}a.$$

42. No. $(1.4142)^2 = 1.99996164$, not 2.

43. It is likely that the Babylonians used measurement tools that could not measure any physical object so precisely that they needed a more accurate approximation of $\sqrt{2}$ for their calculations.

44. Answers may vary depending upon the student's calculator but should be similar to 1.414213562. Assuming the answer to the first question is 1.414213562, the calculator displays $(1.414213562)^2 = 1.999999999$, not 2.

45. No. It can be shown that $\sqrt{2}$ is an irrational number and therefore cannot be written as the quotient of two integers. Any terminating decimal can be written as the quotient of two integers.

46. Window must include the origin; $m = -3$

47. Window must include $(0, -1)$ and $(0.5, 0)$; $m = 2$

48. Window must include $(0, 5)$ and $(-5, 0)$; $m = 1$

49. Window must include $(0, -0.25)$ and $\left(\dfrac{1}{3}, 0\right)$; $m = \dfrac{3}{4}$

50. Window must include $(0, 1)$ and $(2, 0)$; $m = -\dfrac{1}{2}$

The answer to Exercise 56 can be found in Additional Answers beginning on page 842.

Look Beyond

This problem allows students to explore new types of functions and compare transformations.

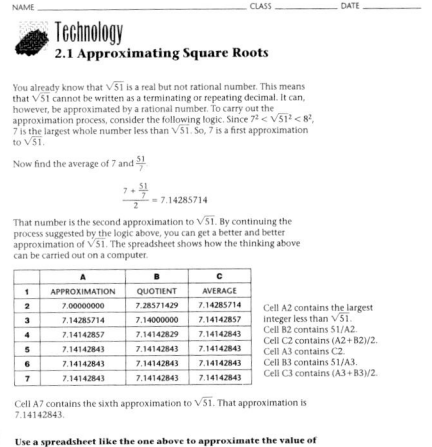

PREPARE

Objectives

- Perform operations, and evaluate expressions using the properties of exponents.

RESOURCES

- Practice Master 2.2
- Enrichment Master 2.2
- Technology Master 2.2
- Lesson Activity Master 2.2
- Quiz 2.2
- Spanish Resources 2.2

Assessing Prior Knowledge

Simplify.

1. 3^3 [27]
2. 7^2 [49]
3. 2^5 [32]
4. 4^3 [64]

TEACH

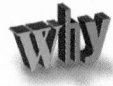

why Avogadro's number (6.02×10^{23}) represents the number of molecules in a mole, the fundamental unit for measuring the amount of a substance. Have students write this number without using scientific notation to illustrate the convenience of such notation.

LESSON 2.2 Exploring Properties of Exponents

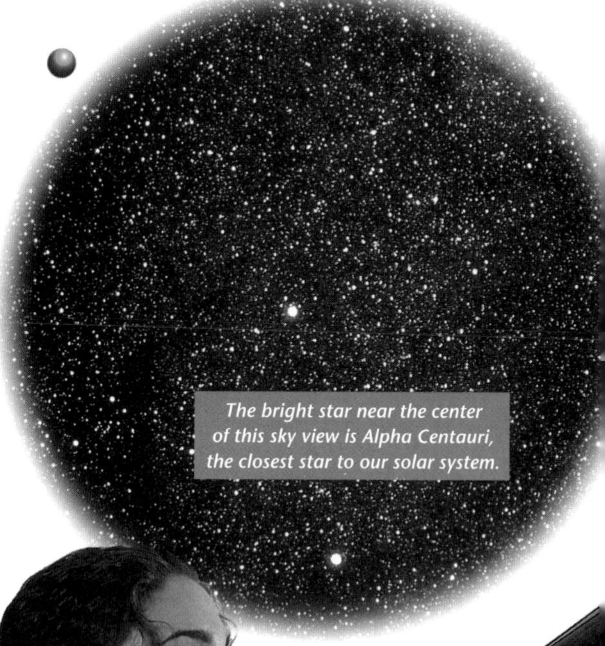

The bright star near the center of this sky view is Alpha Centauri, the closest star to our solar system.

why *You can use exponents to write 186,000 in scientific notation, 1.86×10^5. Numbers written in scientific notation are easy to read and to compare. However, performing simple operations with them requires an understanding of how exponents are added, subtracted, multiplied, and divided.*

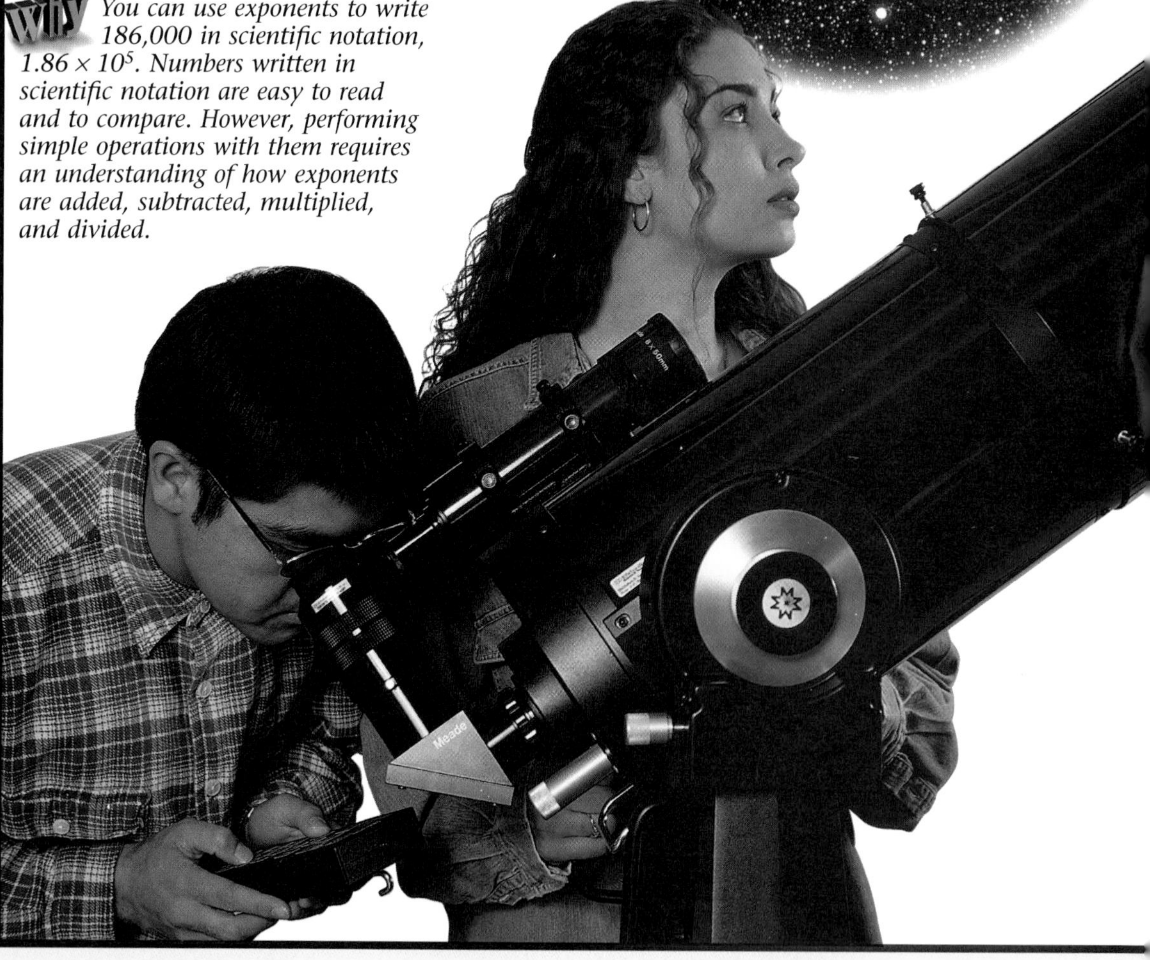

ALTERNATIVE teaching strategy

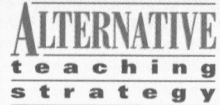

Using Visual Models Students understand the rules of exponents more clearly if they understand why the rules work. Present an example of each rule using numbers. Emphasize that students can create their own examples and determine the rules of exponents without having to memorize them. For example, to show that $(3)^2(3)^3 = 3^5$, write out what this means: $(3 \cdot 3)(3 \cdot 3 \cdot 3)$, which is equal to 3^5. Therefore, the rule states that exponents should be added when bases are the same.

Physics Light travels about 1.86×10^5 miles per second. The star Alpha Centauri is about 4.34 light-years away from Earth. That means that if Alpha Centauri were to explode and disappear today, we would continue to see the star in our sky for the next 4.34 years. One light-year, the distance light travels in one year, is approximately 5.88×10^{12} miles.

Recall the definition of an exponent. Let a represent any real number except 0, and let b represent any whole number except 0. Then

$$a^b \text{ means } \underbrace{a \cdot a \cdot a \cdot \ldots \cdot a}_{b \text{ factors of } a}, \quad a^{-b} \text{ means } \frac{1}{a^b}, \quad \text{and } a^{\frac{1}{b}} \text{ means } \sqrt[b]{a}.$$

Exploration 1 · Multiplication With Exponents

Graphics Calculator

1 Simplify the expression in each column. Then use your calculator to evaluate each expression.

Column 1	Column 2	Column 3
$5^2 \times 5^3$	$5^{(2+3)}$	$5^{(2 \times 3)}$
$7^4 \times 7^2$	$7^{(4+2)}$	$7^{(4 \times 2)}$
$3^4 \times 3^1$	$3^{(4+1)}$	$3^{(4 \times 1)}$

2 Compare the values in each column. Which two columns have the same values? What patterns do you notice between the entries in these two columns?

3 Use the pattern you discovered in Step 2 to complete the following statement: $8^{15} \cdot 8^9 = 8^?$. ❖

Exploration 2 · Division With Exponents

Graphics Calculator

1 Simplify the expression in each column. Then use your calculator to evaluate each expression.

Column 1	Column 2	Column 3
$\frac{8^8}{8^3}$	$8^{(8-3)}$	$8^{(8 \div 3)}$
$\frac{4^7}{4^4}$	$4^{(7-4)}$	$4^{(7 \div 4)}$
$\frac{3^5}{3^5}$	$3^{(5-5)}$	$3^{(5 \div 5)}$

2 Compare the values in each column. Which two columns have the same values? What patterns do you notice between the entries in these two columns?

3 Use the pattern you discovered in Step 2 to complete the following statement: $\frac{7^{212}}{7^{210}} = 7^?$. ❖

For a power raised to a power, use $(3^2)^3 = (3^2)(3^2)(3^2) = 3 \cdot 3 \cdot 3 \cdot 3 \cdot 3 \cdot 3$, or 3^6. This shows that when a power is raised to a power, the exponents are multiplied. To show how the division rule works, use $(3^3) \div (3^2) = (3 \cdot 3 \cdot 3) \div (3 \cdot 3)$, which is equal to 3^1. This shows that when powers with the same bases are divided, the bottom exponent is subtracted from the top. The rule stating that *any number raised to the power of zero equals one* can be shown using the division rule. $\frac{3^4}{3^4} = 3^{4-4}$, or 3^0. Since $\frac{3^4}{3^4} = 1$, $3^0 = 1$.

Cooperative Learning

Students can generalize rules about exponents by studying a list of examples in which each rule is applied. Have groups present the properties they discover and find additional examples of each property being applied.

Exploration 1 Notes

Students should see a pattern develop that shows the rule for multiplication of exponents: $(a^m)(a^n) = a^{m+n}$.

Aongoing ASSESSMENT

3. $8^{15} \cdot 8^9 = 8^{(15+9)} = 8^{24}$

TEACHING tip

Remind students that when multiplying or dividing exponential numbers with like bases, the exponent changes, not the base.

Exploration 2 Notes

As in Exploration 1, students should recognize a pattern in order to discover the division with exponents rule. $a^m \div a^n = a^{(m-n)}$

Aongoing ASSESSMENT

3. $\frac{7^{212}}{7^{210}} = 7^{(212-210)} = 7^2$

Exploration 3 — Negative Exponents

Graphics Calculator

1. Simplify the expression in each column. Then use your calculator to find the decimal equivalent of each fraction.

Column 1	Column 2	Column 3	Column 4
4^2	$(-4)^2$	$\frac{1}{4^2}$	$-\frac{1}{4^2}$
2^1	$(-2)^1$	$\frac{1}{2}$	$-\frac{1}{2}$
3^{-5}	$(-3)^5$	$\frac{1}{3^5}$	$-\frac{1}{3^5}$

2. Compare the values in each column. Which two columns have the same values? What patterns do you notice between the entries in these two columns?

3. Use the pattern you discovered in Step 2 to complete the following statement: $6^{-9} = \underline{?}$. ❖

CRITICAL Thinking

What can you conclude from Explorations 1, 2, and 3? Make a general rule about operations with exponents for each of the three explorations.

Mathematicians of the Islamic Empire

Cultural Connection: Asia Mathematicians of the Islamic Empire lived in different countries and practiced a variety of religions. Many were considered physicians, although they spent much of their time working on mathematics, physics, and chemistry.

Al-Samaw'al is an outstanding example of a mathematician from the Islamic Empire. He was born in Baghdad. His full name was al-Samaw'al Yahya ben Yahudi-al-Maghribi, which means al-Samaw'al, son of Yahya, a Jew from North Africa. At the age of 19, in 1130 C.E., al-Samaw'al wrote the *Shining Book of Calculations*, in which he gave, among other things, the rules for exponents. For example, he proved the following rules.

$$2^2 \cdot 2^3 \cdot 2^3 = 2^8 = 256, \quad 3^{-2} \cdot 3^{-3} = 3^{-5} = \frac{1}{243}, \quad \text{and} \quad \frac{3^3}{3^2} = 3$$

Exploration 4 Fractional Exponents

Graphics Calculator

1 Complete the following tables.

Table 1			Table 2	
x	x^2		x	$x^{\frac{1}{2}}$
2			4	
3			9	
4			16	
5			25	

Table 3			Table 4	
x	x^3		x	$x^{\frac{1}{3}}$
2			8	
3			27	
4			64	
5			125	

2 Explain how Table 2 is related to Table 1. Explain how Table 4 is related to Table 3. Describe how the terms $x^{\frac{1}{n}}$ and x^n are related for any integer n.

3 The expression $x^{\frac{1}{n}}$ can also be written $\sqrt[n]{x}$, read "nth root of x." Complete the following statement by writing an equivalent expression and evaluating: $64^{\frac{1}{5}} = \sqrt[?]{?} = ?.$ ❖

When n is 2, $\sqrt[n]{x}$ is the square root of x, written $\sqrt{x}$.

APPLICATION

The speed of light is 1.86×10^5 miles per second. How can you find out if it really takes 4.34 years for light to travel from Alpha Centauri to Earth?

Recall that distance equals rate multiplied by time, or $d = r \cdot t$. Thus, $t = \frac{d}{r}$.

So substitute 2.55×10^{13} miles for distance, d, and substitute 1.86×10^5 miles per second for rate, r, in $t = \frac{d}{r}$ to find the time, t, in seconds.

$$t = \frac{d}{r} = \frac{2.55 \times 10^{13}}{1.86 \times 10^5} = \frac{2.55}{1.86} \times \frac{10^{13}}{10^5}$$
$$= 1.37 \times 10^{(13-5)}, \text{ or } 1.37 \times 10^8 \text{ seconds}$$

Convert 1.37×10^8 seconds to years.
$$1.37 \times 10^8 \text{ sec} \times \frac{1 \text{ hr}}{3600 \text{ sec}} \times \frac{1 \text{ day}}{24 \text{ hr}} \times \frac{1 \text{ yr}}{365 \text{ days}} = 4.34 \text{ years}$$

Thus, it takes 4.34 years for light to travel 2.55×10^{13} miles, which is the distance from Alpha Centauri to Earth. ❖

Does it really take 4.34 years for light to travel from Alpha Centauri to Earth?

4.34 light-years, or 2.55×10^{13} miles?

A summary of the rules of exponents is given below.

RULES OF EXPONENTS

Let a represent any real number, where $a \neq 0$, and let m and n be integers, where $n \neq 0$.

1. $a^0 = 1$ **2.** $a^m \cdot a^n = a^{(m+n)}$ **3.** $a^{-n} = \frac{1}{a^n}$

4. $a^m \div a^n = a^{(m-n)}$ **5.** $(a^m)^n = a^{m \cdot n}$ **6.** $a^{\frac{m}{n}} = \sqrt[n]{a^m}$

RETEACHING **the lesson**

Using Symbols In groups, have students identify each rule for exponents from lists of examples in which that rule is applied (see Cooperative Learning on page 69). Have students identify the rules of exponents that are to be applied in each of the exercises.

Exploration 4 Notes

Encourage students to substitute a wide range of real numbers, besides 0 and 1, into the equations. Explain the special properties of the numbers 0 and 1 as both exponents and bases.

Aongoing **SSESSMENT**

2. $x^{\frac{1}{n}}$ is the nth root of x. x^n is the nth power of x. Also, $(x^n)^{\frac{1}{n}} = x$ and $(x^{\frac{1}{n}})^n = x$.

Assignment Guide

Core 1–20, 24–45

Core Plus 1–5, 12–46

Technology

Graphics calculators are needed for Exercises 30 and 35–40.

Error Analysis

Watch for students who multiply exponents when they should add them and divide exponents when they should subtract them. Provide students with ample opportunities to complete basic exercises that emphasize these properties.

EXERCISES & PROBLEMS

Communicate

For Exercises 1–2, let *x* represent any nonzero real number, and let *a* and *b* represent any integers.

1. Explain how to perform the multiplication $x^a x^b$.

2. Explain how to perform the division $\frac{x^a}{x^b}$.

3. What does a negative exponent mean? Give an example to illustrate your explanation.

4. What does an exponent of zero mean? Give an example to illustrate your explanation.

5. What does a fractional exponent mean? Give an example to illustrate your explanation.

Practice & Apply

Simplify. Use only positive exponents in your answer.

6. $(3^5)(3^{-2})$ 27

7. $\left(2^{\frac{1}{2}}\right)(2^3)$ $2^{\frac{7}{2}} \approx 11.3137...$

8. $\left((-1)^3\right)^{\frac{8}{3}}$ 1

9. $\left((2)^{11}\right)^{\frac{2}{11}}$ 4

10. $(5^3)\left(5^{1.4}\right)$ $5^{4.4} \approx 1189.7837...$

11. $\left(\frac{-7c^3}{10d^4}\right)^{-2}$ $\frac{100d^8}{49c^6}$

12. $\left(\frac{3a^{-1}}{5c^2}\right)^2$ $\frac{9}{25a^2c^4}$

13. $\left(\frac{-2mn^2p^{-2}}{3m^3n^{-1}p}\right)^3$ $\frac{-8n^9}{27m^6p^9}$

14. $\frac{x^{8a}y^{4b}}{x^{2a}y^b}$ $x^{6a}y^{3b}$

15. $\left(\frac{x^2y^4}{x^3y^{-3}}\right)^{-5}$ $\frac{x^5}{y^{35}}$

16. $(x^3)(x^{-2})$ x

17. $(2xy^2)(3x^3y)$ $6x^4y^3$

18. $(3ab^2c)(4bc)$ $12ab^3c^2$

19. $(4x^{-1}y^2)(2x^3y^{-3})$ $\frac{8x^2}{y}$

20. $5y^2 \cdot 2y^{-5}$ $\frac{10}{y^3}$

21. $(2x^{-4})^{-3}$ $\frac{1}{8}x^{12}$

22. $2(-3x^{-2})^4$ $\frac{162}{x^8}$

23. $(3x^{-8}y^2)(5x^3y^{-4})^{-1}$ $\frac{3y^6}{5x^{11}}$

Geometry A room has a square floor. The length of a side of the floor is *x* meters. **Complete Exercises 24–28.**

24. Use exponents to write the formula for the area of the floor. $y = x^2$

25. If *x* is 3 meters, find the area of the floor.   9 m²

26. Use rules of exponents to find the formula for the area of the floor if the length of a side, *x*, is doubled. $y = 4x^2$

27. If *x* is 3 meters, find the area of the floor described in Exercise 26.   36 m²

28. Draw both square floors from Exercises 24–27 on graph paper. Describe the relationship of the smaller area to the larger area. Include a ratio in your description.

29. Prove or disprove that $(2 + 3)^2 = 2^2 + 3^2$. As a result of your findings, write a general rule for the relationship between $(a + b)^2$ and $a^2 + b^2$.

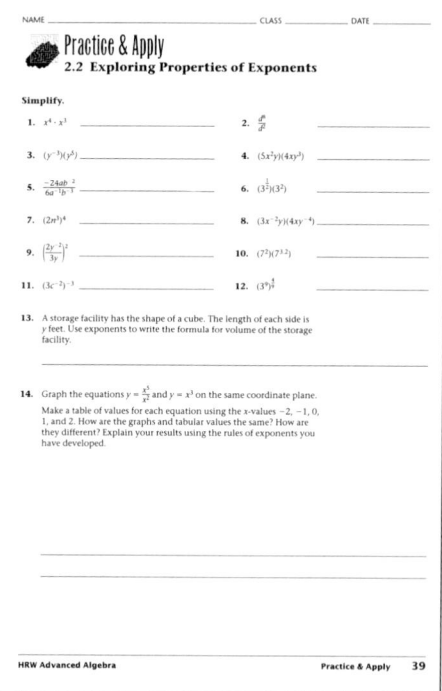

28.

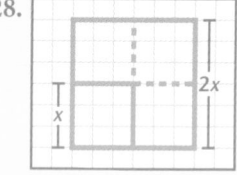

The ratio of the smaller area to the larger area is $\frac{x^2}{4x^2} = \frac{1}{4}$. So, doubling the length of the side of a square quadruples the area of the square.

29. $(2 + 3)^2 = 5^2 = 25$, but $2^2 + 3^2 = 4 + 9 = 13$. So, $(2 + 3)^2 \neq 2^2 + 3^2$. In general, for real numbers *a* and *b* and any integer *m*, $(a + b)^m \neq a^m + b^m$.

For Exercises 30–33, let $y_1 = (x+3)^2$, $y_2 = x^2 + 6x + 9$, and $y_3 = x^2 + 3^2$.

30 Use your graphics calculator to graph each equation. Make a table of values for each equation using the x-values $-2, -1, 0, 1,$ and 2.

31. How are the graphs and the table values *alike* for the three equations?

32. How are the graphs and the table values *different* for the three equations?

33. Write a general rule for expanding $(x+a)^2$, where a represents any real number.

Investment Jeanne has invested $300 at 6% interest compounded once per year. The formula for the amount, A, of money she will have in t years is $A = 300(1.06)^t$.

34. What will be the value of the original investment after 1 year? after 2 years? $318, $337.08

35 Use your graphics calculator to graph the formula $A = 300(1.06)^t$. Use the trace feature of your calculator to determine the amount of the investment after 6 years. $425.56

36 Use the trace feature to determine how many years must pass before the amount will be greater than $600. 12

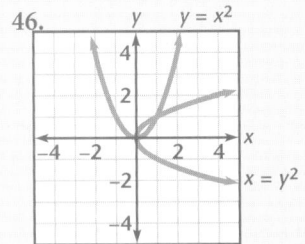

Look Back

Use your graphics calculator to graph each equation. Is the equation linear? Explain why or why not. Give the slope and the y-intercept for each *linear* equation. **[Lesson 1.2]**

37 $y = x + 1$ $m = 1, b = 1$ **38** $y = x^3$ not linear   **39** $y = \dfrac{4}{x+2}$ not linear   **40** $y = \dfrac{x+2}{4}$

41. Write the equation of a horizontal line. Write the equation of a vertical line. What is the slope of any horizontal line? Can you find the slope of a vertical line? Explain. **[Lesson 1.3]**

Let a, b, and c represent any real numbers. Name the property illustrated in each equation. **[Lesson 2.1]**

42. $ab = ba$ Comm Prop for Mult

43. $a + (b + c) = (a + b) + c$ Assoc Prop for Add

44. $a\left(\dfrac{1}{a}\right) = 1$, where $a \neq 0$ Inverse Prop for Mult

45. $a + b$ is real. Closure Prop for Add

Look Beyond

46. Use graph paper to graph $y = x^2$ and $x = y^2$. What are the values of y when x is 9 in each equation?

The answers to Exercises 30–33 can be found in Additional Answers beginning on page 842.

37. The equation is linear since its graph is a line. The slope is 1, and the y-intercept is $(0, 1)$.

38. The equation is not linear since its graph is not a line.

39. The equation is not linear since its graph is not a line.

40. The equation is linear since its graph is a line. The slope is $\frac{1}{4}$, and the y-intercept is $(0, \frac{1}{2})$.

41. Answers may vary but should be equivalent to the following. The line $y = b$ is a horizontal line. The line $x = c$ is a vertical line. Any horizontal line has slope 0. The slope of any vertical line is undefined since for the points (c, y_1) and (c, y_2) on $x = c$, $m = \dfrac{y_2 - y_1}{c - c} = \dfrac{y_2 - y_1}{0}$, which is undefined.

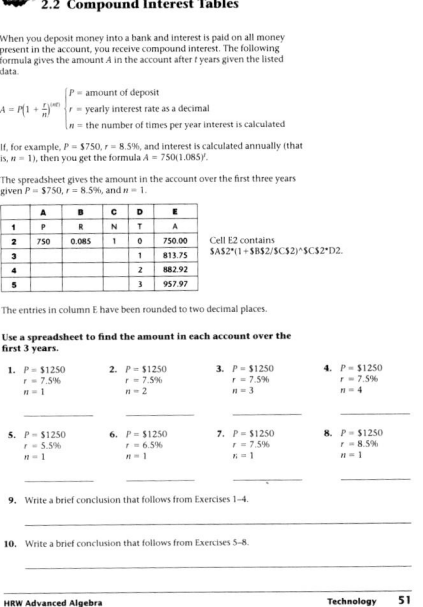

A MAN & A METHOD

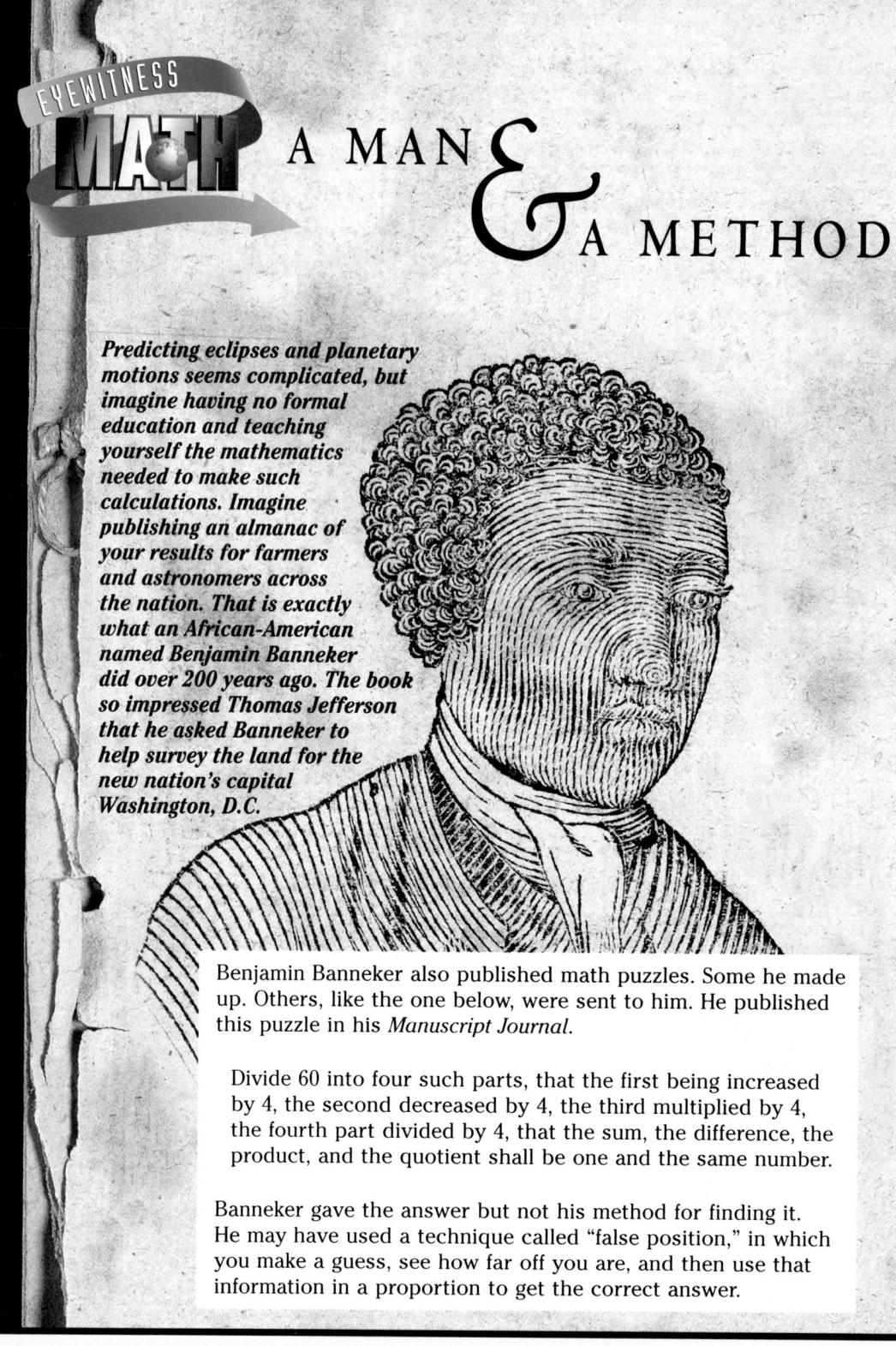

FOCUS

Students will learn about analyzing mathematical methods

MOTIVATE

The work of an 18th century African American astronomer, surveyor, and mathematical puzzler provides the basis for comparing algebraic and non-algebraic methods.

As you discuss the title page from Benjamin Banneker's almanac, help students acquire an appreciation for the marvelous accomplishments of this self-taught African American. Banneker was 57 years old when he borrowed some astronomy books and equipment and taught himself enough to publish his own almanac of astronomy.

The wood-carving of Banneker shown on page 74 is a self-portrait which appeared on the title page of his almanac.

As you discuss the puzzle, make sure students follow the unfamiliar style. You may want to discuss how our language has changed over the last 2 centuries.

Predicting eclipses and planetary motions seems complicated, but imagine having no formal education and teaching yourself the mathematics needed to make such calculations. Imagine publishing an almanac of your results for farmers and astronomers across the nation. That is exactly what an African-American named Benjamin Banneker did over 200 years ago. The book so impressed Thomas Jefferson that he asked Banneker to help survey the land for the new nation's capital Washington, D.C.

Benjamin Banneker also published math puzzles. Some he made up. Others, like the one below, were sent to him. He published this puzzle in his *Manuscript Journal*.

Divide 60 into four such parts, that the first being increased by 4, the second decreased by 4, the third multiplied by 4, the fourth part divided by 4, that the sum, the difference, the product, and the quotient shall be one and the same number.

Banneker gave the answer but not his method for finding it. He may have used a technique called "false position," in which you make a guess, see how far off you are, and then use that information in a proportion to get the correct answer.

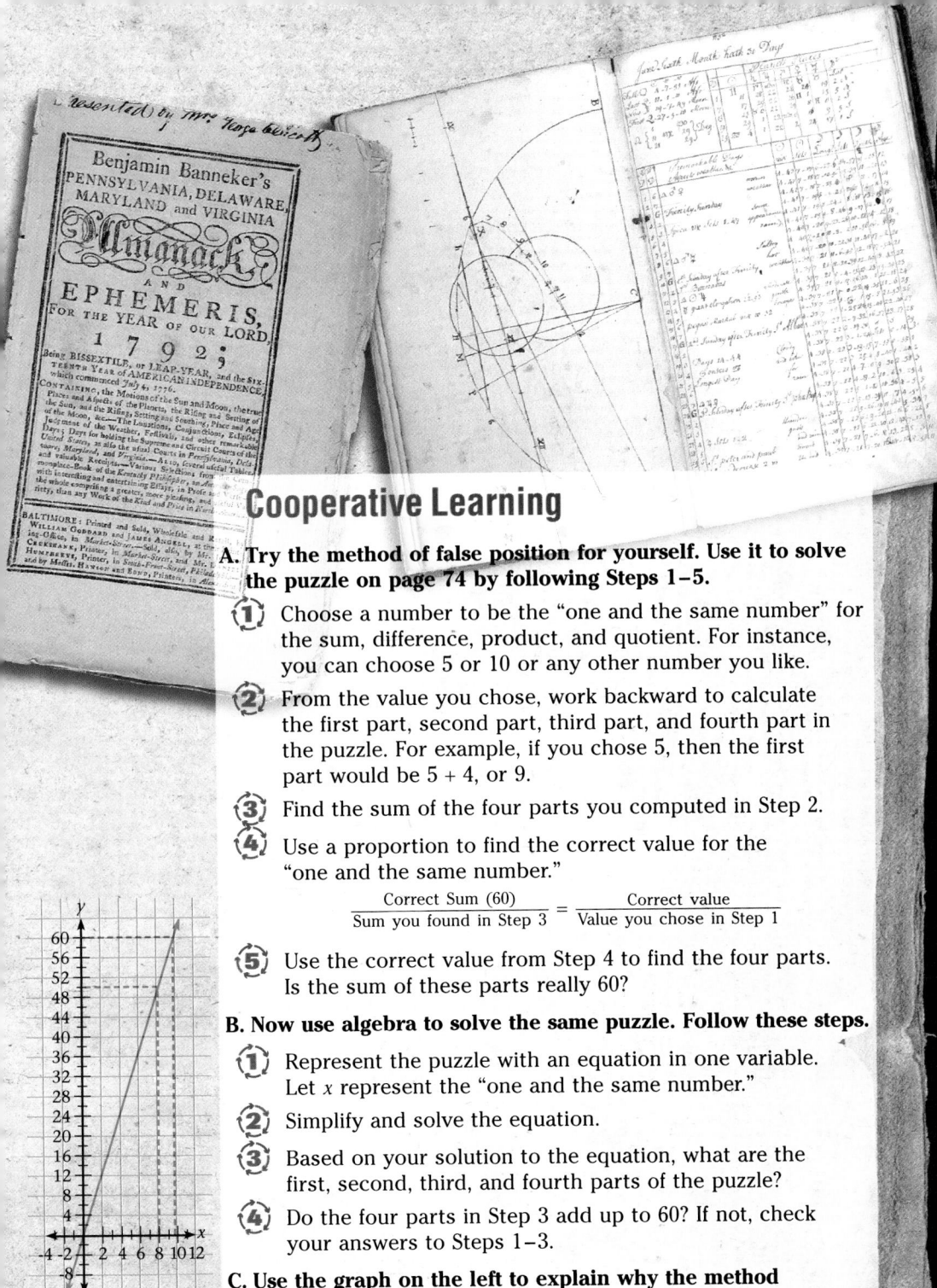

Cooperative Learning

A. Try the method of false position for yourself. Use it to solve the puzzle on page 74 by following Steps 1–5.

(1) Choose a number to be the "one and the same number" for the sum, difference, product, and quotient. For instance, you can choose 5 or 10 or any other number you like.

(2) From the value you chose, work backward to calculate the first part, second part, third part, and fourth part in the puzzle. For example, if you chose 5, then the first part would be 5 + 4, or 9.

(3) Find the sum of the four parts you computed in Step 2.

(4) Use a proportion to find the correct value for the "one and the same number."

$$\frac{\text{Correct Sum (60)}}{\text{Sum you found in Step 3}} = \frac{\text{Correct value}}{\text{Value you chose in Step 1}}$$

(5) Use the correct value from Step 4 to find the four parts. Is the sum of these parts really 60?

B. Now use algebra to solve the same puzzle. Follow these steps.

(1) Represent the puzzle with an equation in one variable. Let *x* represent the "one and the same number."

(2) Simplify and solve the equation.

(3) Based on your solution to the equation, what are the first, second, third, and fourth parts of the puzzle?

(4) Do the four parts in Step 3 add up to 60? If not, check your answers to Steps 1–3.

C. Use the graph on the left to explain why the method of false position works for this puzzle.

Cooperative Learning

Since students are probably not familiar with the method of false position, you might want to complete Activity 1 as a class. Choose at least 2 different values in Part A to help assure students that the method works no matter which "false" value is first chosen.

In Part B, emphasize the use of the inverse operation to work backward to calculate each of the four parts from the chosen value.

In Part C, be sure students see the connection between the values on the graph and the values in the proportion used in the false-position method.

Discuss

Have students answer the following questions.
- How do the values on the graph relate to the slope of the line?
- What triangles do you see in the figure? Are any congruent? Are any similar?

PREPARE

Objectives

- Identify and compare relations and functions, and use the vertical-line test to identify functions from their graphs.
- Use functions to model real-world applications, and give appropriate domain and range restrictions for the situation.

RESOURCES

- Practice Master   2.3
- Enrichment Master   2.3
- Technology Master   2.3
- Lesson Activity Master   2.3
- Quiz   2.3
- Spanish Resources   2.3

Assessing Prior Knowledge

1. Complete the table for $y = 2x^2 + 1$.

x	y
−3	[19]
−2	[9]
−1	[3]
0	[1]
1	[3]
2	[9]
3	[19]

2. Are there any x-values that correspond to the same y-values? If so, name them. [Yes, −3 and 3 correspond to 19; −2 and 2 correspond to 9; and −1 and 1 correspond to 3.]

LESSON 2.3 — Definition of a Function

Why *In science, formulas are used to solve problems. A physics formula can be used to calculate how far a free-falling object is from the ground at any time after being dropped.*

Don and Joni are learning to be parachuting instructors. Their teacher tells them to wait 6 seconds after leaving the plane to pull their rip cords. How far will Don and Joni be from the ground in 6 seconds?

The distance, d, from the parachutists to the ground can be measured in meters or in feet. The time, t, is measured in seconds. The acceleration due to gravity, represented by g, is 9.8 meters per second squared, or 32 feet per second squared. Since the height of the plane, $h_0 = 3600$ feet, is given in feet, use 32 feet per second squared for acceleration.

The free-fall formula, $d = h_0 - \frac{1}{2}gt^2$, can be simplified. Substitute 32 for g and 3600 for h_0.

$$d = 3600 - \frac{1}{2}(32)t^2 \quad d = 3600 - 16t^2$$

By substituting values for t, this equation can be used to generate a table, a set of ordered pairs, and a graph.

Skydivers who jumped out of a plane from a height of 3600 feet can free-fall for hundreds of feet before opening their chutes.

ALTERNATIVE teaching strategy

Technology   Use graphics calculators to demonstrate finding the approximate domain and range of functions. From examples, have students discover rules for what graphs of functions should look like.

Free-Fall Distance

	Table		Ordered Pairs	Graph
Time t (sec)	$d = 3600 - 16t^2$	Distance d (ft)	(t, d)	
0	$3600 - 16(0)^2$	3600	(0, 3600)	
1	$3600 - 16(1)^2$	3584	(1, 3584)	
2	$3600 - 16(2)^2$	3536	(2, 3536)	
3	$3600 - 16(3)^2$	3456	(3, 3456)	
4	$3600 - 16(4)^2$	3344	(4, 3344)	
5	$3600 - 16(5)^2$	3200	(5, 3200)	
6	$3600 - 16(6)^2$	3024	(6, 3024)	

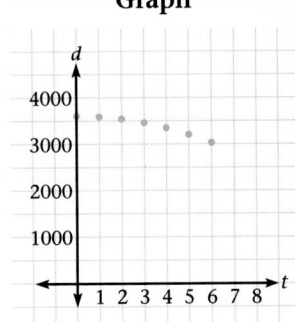

After 6 seconds, Joni and Don are 3024 feet from the ground.

Tables and Functions

Examine the Free-Fall Distance information. The table, the ordered pairs, and the graph are three different representations of the same *relationship* between time and distance. In general, a *relation* is any correspondence between two sets. When each value of the first variable is related to *one and only one* value of the second variable, the relation is called a *function*.

In the Free-Fall Distance chart, how many d-values are matched with each t-value? Does any t-value give more than one d-value? Explain.

Ordered pairs are often used in the definition for relation and function. Since all three representations are equivalent, ordered pairs, tables, and graphs can be used to describe relations and functions.

RELATION AND FUNCTION
A **relation** is any set of ordered pairs.
A **function** is a relation in which, for each first coordinate, there is exactly *one* corresponding second coordinate.

TEACH

Discuss the fact that formulas are used continually in the real world. Formulas allow people to determine information. By substituting information into equations, new information can be determined. Ask students to suggest some occupations that require the use of formulas.

ongoing ASSESSMENT

As the t-value increases, the d-value decreases, so only one d-value corresponds to each t-value, and each t-value corresponds to only one d-value.

interdisciplinary

CONNECTION

Science Sonar is often used to locate animals or objects in deep water. When a high frequency sound is bounced off an object, the time it takes for that sound to return indicates the location of the object. Have students work in small groups to design a graph and plot points that could simulate data taken by sonar. Ask students to answers the following questions.

- What measurements would be represented on the x-axis? [**time**]
- What measurements would be represented on the y-axis? [**depth**]
- What type of function is the graph of your data? [**Answers may vary. A sample response is *linear***]

Alternate Example 1

Does the table represent a function? Why or why not?

A.

x	y
2	2
4	3
6	4
8	5

B.

x	y
3	4
3	7
5	-4
6	3

[A. Yes. Each *x*-value is paired with exactly one *y*-value. B. No. The *x*-value 3 is paired with two different *y*-values, 4 and 7.]

C.

x	y
-2	7
-2	6
3	-2
3	-4

D.

x	y
1	-2
2	-2
3	3
4	-3

[C. No. The *x*-value -2 is paired with two *y*-values, 7 and 6; the *x*-value 3 is paired with two *y*-values, −2 and −4. D. Yes. Each *x*-value is paired with exactly one *y*-value.]

Alternate Example 2

Identify whether the graph represents a function. Why or why not?

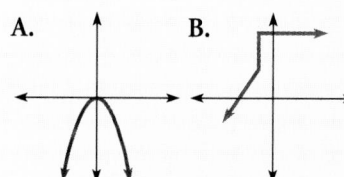

[A. Function; any vertical line intersects the graph in exactly one place. B. Not a function; the vertical line $x = -1$ intersects the graph in more than one place.]

Use Transparency ▶ 7

EXAMPLE 1

Does each table below represent a function? Why or why not?

A

x	y
1	3
2	4
3	5
4	7

B

x	y
0	3
2	-5
2	1
4	7

C

x	y
1	6
2	6
3	9
4	9

D

x	y
4	-2
4	2
9	-3
9	3

Solution▶

A Yes. Each *x*-value is paired with exactly one *y*-value.

B No. The *x*-value 2 is paired with two different *y*-values, −5 and 1.

C Yes. Although different *x*-values are paired with the same *y*-value (the *x*-values 1 and 2 with the *y*-value 6, and the *x*-values 3 and 4 with the *y*-value 9), there is still only *one* *y*-value for each individual *x*-value.

D No. The *x*-value 4 is paired with two *y*-values, −2 and 2; the *x*-value 9 is paired with two *y*-values, −3 and 3. ❖

Graphing and the Vertical-Line Test

Since a function has exactly one *y*-value for each *x*-value, a vertical line will cross the graph of a function in *no more than one point*.

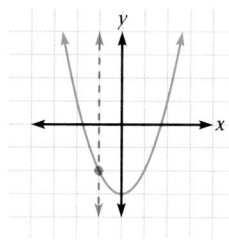

Function

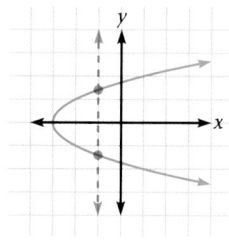

Not a Function

THE VERTICAL-LINE TEST

If a vertical line crosses the graph of a relation in **more than one point**, the relation is **not** a function.

EXAMPLE 2

Determine which relation is a function. Give a reason for your answer.

A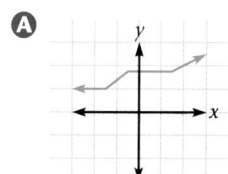

B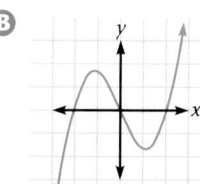

C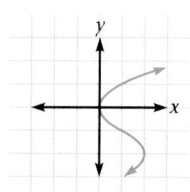

ENRICHMENT Have students find examples of functions in the real world. Ask them to bring in pictures or sketches of structures resembling parabolas, linear functions, and so on. Also have students look through newspapers and magazines for examples of the graphs of functions. Ask students to describe what these graphs show the reader.

INCLUSION strategies **Independent Learning** There are many different tools that may be used to teach graphing: graph paper, graphics calculators, computers, an overhead projector, or a chalkboard. Different students will respond differently to different graphing methods. It is best to use a variety of methods and allow students to choose the method that works best for them.

Solution

Ⓐ Function; any vertical line intersects the graph in exactly one place.

Ⓑ Function; any vertical line intersects the graph in exactly one place.

Ⓒ Not a function; the vertical line $x = 1$ intersects the graph in more than one place. ❖

Domain and Range

In the free-fall function $d = 3600 - 16t^2$, what values for the time, t, are possible? What corresponding values for the distance, d, are possible?

> ### DOMAIN AND RANGE
> The **domain** of a relation is the set of possible values for the first coordinates of the relation.
> The **range** of a relation is the set of possible values for the second coordinates of the relation.

Graphics Calculator

EXAMPLE 3

Use your graphics calculator to graph the free-fall function $d = 3600 - 16t^2$. Find the real-world domain and range.

Solution

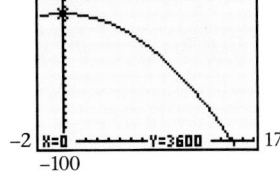

The distance, d, from Earth can only range from the original height of the plane, 3600 feet, to the ground, 0 feet. So the range is the set of all real numbers d such that $0 \le d \le 3600$.

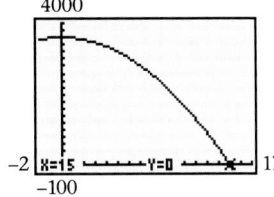

The domain consists of the values of time, t, from 0 seconds until the time the object hits the ground.
To find the value of t when the object hits the ground, solve for t when d is 0.

$$0 = 3600 - 16t^2$$
$$16t^2 = 3600$$
$$t^2 = \frac{3600}{16}$$
$$t^2 = 225$$
$$t = 15$$

The domain is the set of all real numbers, t, such that $0 \le t \le 15$. ❖

What does the point (15, 0) represent in Example 3?

Alternate Example 3
Use your graphics calculator to graph $h = 3136 - 16t^2$. Assume this function refers to the height of a free-falling object with respect to time. Find the real-world domain and range.

[**domain: all real numbers t such that $0 \le t \le 14$; range: all real numbers d such that $0 \le d \le 3136$**]

Cooperative Learning
Place students in small groups of three or four. Ask each group to write at least three equations and to graph each equation. Have groups exchange and graph each other's equations. Have them discuss the graphs.

Ongoing ASSESSMENT

The point (15, 0) represents the fact that after 15 seconds, a free-falling object will hit the ground.

TEACHING tip

The domain is all the possible numbers that the x-value can be. The range is all the possible numbers that the y-value can be.

Alternate Example 4

Use your graphics calculator to find the domain and range of $y = x^2 + 1$. [**domain: any real number; range: any real number greater than or equal to 1**]

Aongoing SSESSMENT

Try This

domain: any real number except 1; range: any real number except 0. Make certain students enter the function correctly as $y = 1/(x - 1)$.

CRITICAL Thinking

The domain and range of a function are restricted only by division by zero or the impossibility of taking the square root of numbers less than zero. *Real-world restrictions* on the domain and range are related to values that represent actual and possible events, such as lengths or speeds always being positive.

Practice Master

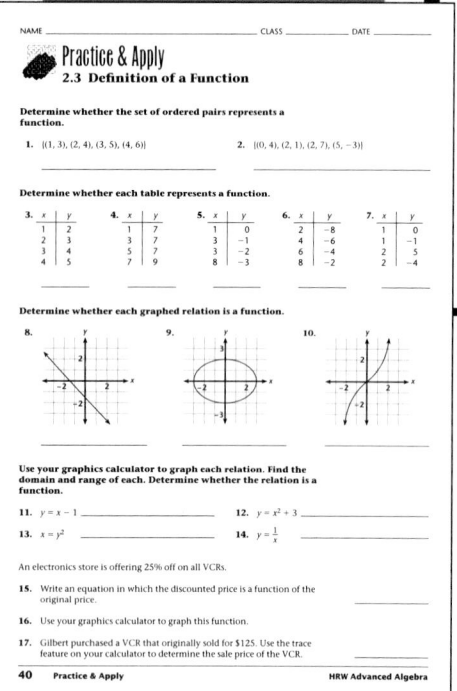

NAME _____ CLASS _____ DATE _____

Practice & Apply
2.3 Definition of a Function

Determine whether the set of ordered pairs represents a function.

1. {(1, 3), (2, 4), (3, 5), (4, 6)}   2. {(0, 4), (2, 1), (2, 7), (5, −3)}

Determine whether each table represents a function.

Determine whether each graphed relation is a function.

8.   9.   10.

Use your graphics calculator to graph each relation. Find the domain and range of each. Determine whether the relation is a function.

11. $y = x - 1$ _____   12. $y = x^2 + 3$ _____
13. $x = y^2$ _____   14. $y = \frac{1}{x}$ _____

An electronics store is offering 25% off on all VCRs.

15. Write an equation in which the discounted price is a function of the original price. _____

16. Use your graphics calculator to graph this function.

17. Gilbert purchased a VCR that originally sold for $125. Use the trace feature on your calculator to determine the sale price of the VCR. _____

40   Practice & Apply   HRW Advanced Algebra

EXAMPLE 4

Use your graphics calculator to find the domain and range of $y = x^2 - 1$.

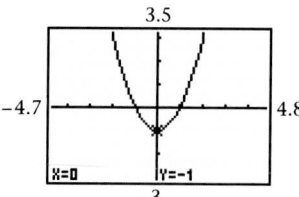
Graphics Calculator

Solution

Since x can be any real number, the domain is **R**, the set of real numbers.

The lowest point on the graph is $(0, -1)$. As x gets larger *or* smaller than 0, y continues to increase. Thus, the range is all real numbers, y, greater than or equal to -1. ❖

Try This Use your graphics calculator to find the domain and range of $y = \frac{1}{x - 1}$.

CRITICAL Thinking

What is the difference between the domain or range of a function and the *real-world* domain or range of that function?

EXERCISES & PROBLEMS

Communicate

1. What is the difference between a relation and a function?

2. Describe three ways to represent a function.

3. Explain how to find the domain and the range of a set of ordered pairs. Give one example.

4. Explain how to find the domain and the range of a function from its graph. Give one example.

5. Draw the graph of a relation that is not a function.

6. What is the vertical-line test? Is the relation shown at the right a function? Explain why or why not.

Practice & Apply

For Exercises 7–9, determine whether the set of ordered pairs represents a function.

7. {(0, 0), (1, 1), (2, 2)}   function

8. {(1, 2), (2, 2), (3, 2)}   function

9. {(1, −1), (1, −2), (1, −3)}   not a function

10. What is the range of the function $y = 5x$ when the domain is {3, 4, 5, 6}?   {15, 20, 25, 30}

The height of the rocket at any time, t is given by the function $h = 64t - 16t^2$.

Assignment Guide

Core 1–7, 9–17, 19–29, 32–47

Core Plus 1–6, 8–14, 16–23, 26–48

Technology

Graphics calculators are needed for problems 11–14, 24–31, 33–35, and 47–48.

Error Analysis

Watch for students who have difficulty determining the domain and range of functions. Encourage them to graph each function and use the graph to find its domain and range.

Projectile Motion For Exercises 11–14, use your graphics calculator to graph the function for the height of the toy rocket.

11 Use the trace feature to find the time required for the rocket to reach its maximum height. 2 sec

12 What is the maximum height of the rocket? 64 ft

13 How long will it take the rocket to land? 4 sec

14 What are the real-world domain and range of *h*? d: $0 \le t \le 4$
r: $0 \le h \le 64$

For Exercises 15–18, determine whether each table represents a function. Explain.

15.
x	y
3	9
–3	9
2	4
–2	4
function

16.
x	y
2	2
4	3
4	4
6	5
not a function

17.
x	y
1	0
3	–1
5	–2
7	–1
function

18.
x	y
1	9
2	9
3	9
4	9
function

For Exercises 19 and 20, determine whether each graphed relation is a function. Explain.

19.
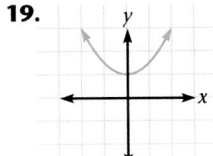
function

20.
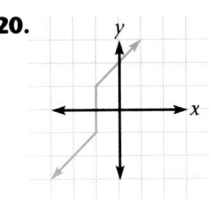
not a function

Business The Universal Rent-All Company rents carpet cleaners for $0.75 per hour, plus a base fee of $25.00.

21. Write a function relating cost and hours rented. $c = 0.75h + 25$

22. Use this function to determine the cost of renting the cleaner for one day (24 hours). $43

23. If the cost of renting the cleaner is $37.00, for how many hours was the cleaner rented? 16

15. function; For each *x*-value there is only one *y*-value.

16. not a function; When *x* is 4 there are two different *y*-values, 3 and 4.

17. function; For each *x*-value there is only one *y*-value.

18. function; For each *x*-value there is only one *y*-value.

19. function; Any vertical line intersects the graph in only one point.

20. not a function; The vertical line $x = -1$ intersects the graph in multiple points.

Technology Master

NAME _____ CLASS _____ DATE _____

Technology
2.3 Building Functions

You can build functions of your own by chaining together a set of operations. For example, the following set of steps will define a function that you can evaluate with a spreadsheet.

Step 1 Begin with a real number *x*.
Step 2 Multiply *x* by itself.
Step 3 To the result of step 2 add 4 times the number in step 1.
Step 4 From the result of step 3 subtract 5. Call the result *y*.

	STEP 1	STEP 2	STEP 3	STEP 4
	A	B	C	D
1				
2	–2	4	–4	–9
3	–1	1	–3	–8
4	0	0	0	–5
5	1	1	5	0
6	2	4	12	7

Column A contains the values of *x*.
Cell B2 contains A2*A2.
Cell C2 contains B2+4*A2.
Cell D2 contains C2–5.

Make a spreadsheet for the function defined by each set of steps. Use a reasonable domain for the function.

1. Step 1 Begin with a real number *x*.
   Step 2 Multiply *x* by itself.
   Step 3 From the result of step 2 subtract 3 times the number in step 1.
   Step 4 From the result of step 3 subtract 1. Call the result *y*.

2. Step 1 Begin with a real number *x*.
   Step 2 Multiply *x* by 3 times itself.
   Step 3 From the result of step 2 subtract 4 times the number in step 1.
   Step 4 From the result of step 3 subtract 4. Call the result *y*.

3. Step 1 Begin with a real number *x*.
   Step 2 Multiply *x* by twice itself.
   Step 3 To the result of step 2 add 3 times the number in step 1.
   Step 4 From the result of step 3 subtract 2. Call the result *y*.

4. Step 1 Begin with a real number *x*.
   Step 2 Multiply *x* by itself.
   Step 3 To the result of step 2 add the number in step 1.
   Step 4 From the result of step 3 subtract 1. Call the result *y*.

Plot the ordered pairs from each spreadsheet.

5. Exercise 1   6. Exercise 2   7. Exercise 3   8. Exercise 4

9. Use a spreadsheet to find the minimum value of *y* from Exercise 1.

52   Technology   HRW Advanced Algebra

For Exercises 24–31, use your graphics calculator to graph each function. Find the domain and range of each function.

[24] $y = 4x$ R, R   **[25]** $y = 2x - 5$ R, R   **[26]** $y = x^3$ R, R   **[27]** $y = x^2 + 2$ R, $y \geq 2$

[28] $y = \left(\dfrac{x}{3}\right)^2$ R, $y \geq 0$ **[29]** $y = \dfrac{x}{2}$ R, R   **[30]** $y = |x|$ R, $y \geq 0$ **[31]** $y = \dfrac{2}{x}$ $x \neq 0$, $y \neq 0$

32. Draw the graph of a function for which the domain is all real numbers, x, such that $-2 \leq x \leq 5$, and the range is all real numbers, y, such that $-3 \leq y \leq 3$.

Consumer Economics A clothing store is offering 30% off on all out-of-season clothing.

[33] Write an equation in which the original price is a function of the discounted price. Use your graphics calculator to graph this function. $y = \dfrac{10}{7}x$

[34] Jason spent $47.25 on sale items. Use the trace feature to estimate the original cost of the clothing. $67.50

[35] Helena purchased out-of-season items that originally cost $52. Use the trace feature on your calculator to determine the sale price of these items. $36.40

Geometry The cube shown has volume V.

36. Express the volume of the cube as a function of the length of its sides. $v = s^3$

37. If the volume of the cube is 27 m³, find the area of one face of the cube. 9 m²

Look Back

For Exercises 38–41, write the equation of the line that contains the given points.   [Lesson 1.2]

38. (1, 4) and (−3, 0)   **39.** (0, 2) and (−1, 1)   **40.** (2, 3) and (0, 0)   **41.** (−2, −5) and (5, −1)
$y = x + 3$ $y = x + 2$ $y = \dfrac{3}{2}x$ $y = \dfrac{4}{7}x - \dfrac{27}{7}$

Simplify.

42. $\dfrac{x}{7} + \dfrac{-x}{2}$ $\dfrac{-5x}{14}$ **43.** $\dfrac{x}{2} - \dfrac{5x}{10}$ 0 **44.** $\dfrac{2x}{3} + \dfrac{3x}{2}$ $\dfrac{13x}{6}$ **45.** $\dfrac{x+2}{3} + \dfrac{2x}{6}$ $\dfrac{2x+2}{3}$ **46.** $\dfrac{3}{4x} - \dfrac{1-x}{5x^2}$ $\dfrac{19x-4}{20x^2}$

Look Beyond

[47] Use your graphics calculator to graph $y = x^2 - 3x - 10$. Explain why this is a function. Find the domain and range of this function.

[48] Use your graphics calculator to graph $y = 2^x$. Explain why this is a function. Find the domain and range of this function.

32. Answers may vary. One possible answer is $f(x) = 0.5x^2 - 1.5x - 2$. Its graph is shown.

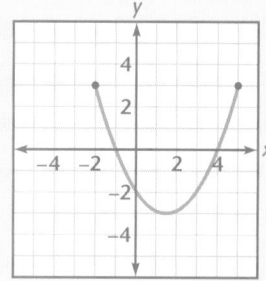

47. This is a function because for each x-value there is only one corresponding y-value. The domain is any real number. The range is any real number greater than or equal to -12.25.

48. This is a function because for each x-value there is only one corresponding y-value. The domain is any real number. The range is any real number greater than 0.

LESSON 2.4 Using Notation to Represent Functions

Why It is not always easy to use words to describe a relationship between quantities, such as distance and time. Function notation is descriptive. It enables complex relationships to be expressed simply.

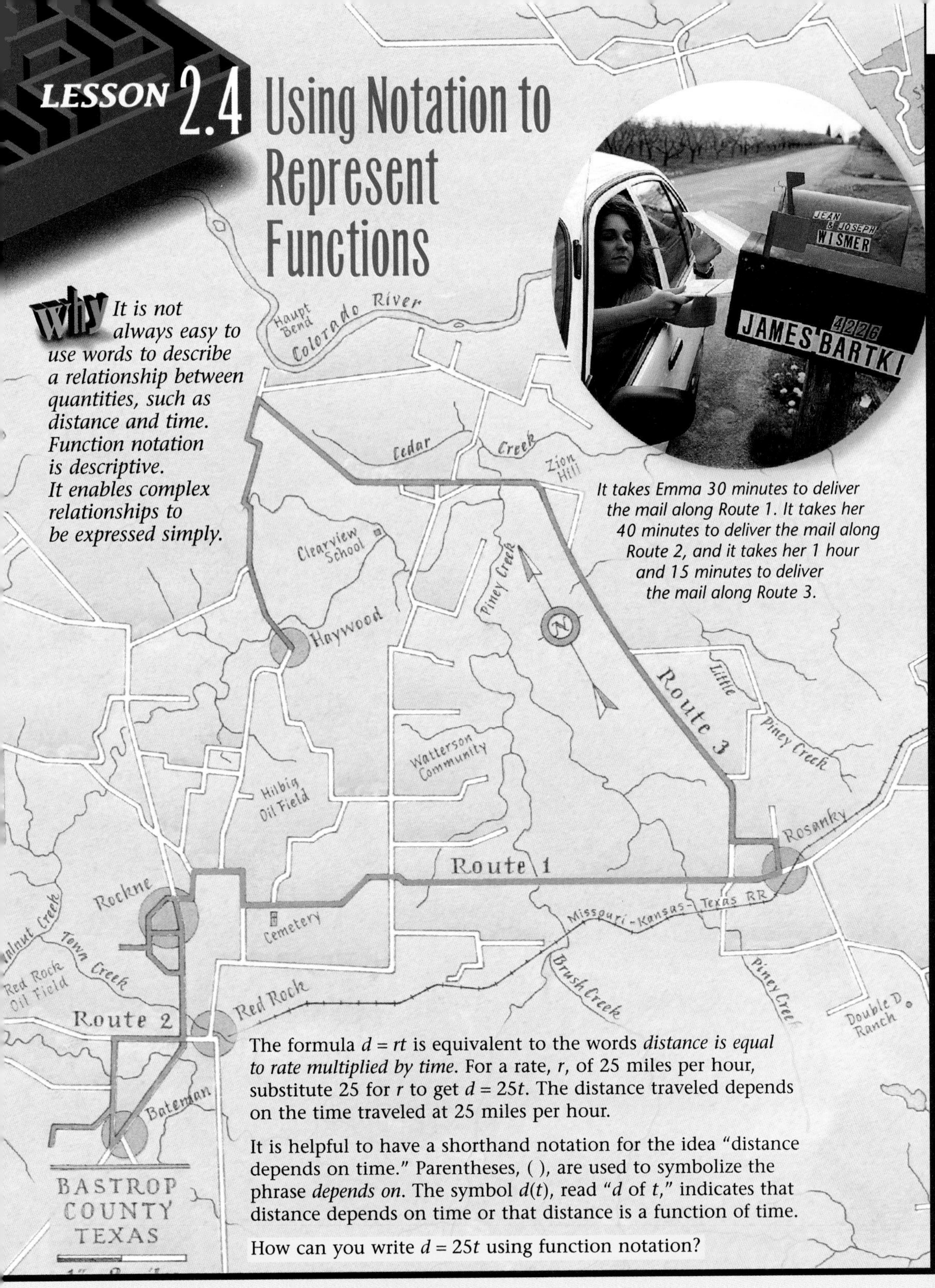

It takes Emma 30 minutes to deliver the mail along Route 1. It takes her 40 minutes to deliver the mail along Route 2, and it takes her 1 hour and 15 minutes to deliver the mail along Route 3.

The formula $d = rt$ is equivalent to the words *distance is equal to rate multiplied by time*. For a rate, r, of 25 miles per hour, substitute 25 for r to get $d = 25t$. The distance traveled depends on the time traveled at 25 miles per hour.

It is helpful to have a shorthand notation for the idea "distance depends on time." Parentheses, ( ), are used to symbolize the phrase *depends on*. The symbol $d(t)$, read "d of t," indicates that distance depends on time or that distance is a function of time.

How can you write $d = 25t$ using function notation?

ALTERNATIVE teaching strategy

Using Symbols Have students rewrite functions from earlier lessons using function notation. Have them use their graphics calculators to find the domain and range of each function.

PREPARE

Objectives

• Use function notation to define, evaluate, and operate with functions.

RESOURCES

• Practice Master   2.4
• Enrichment Master   2.4
• Technology Master   2.4
• Lesson Activity Master   2.4
• Quiz   2.4
• Spanish Resources   2.4

Assessing Prior Knowledge

1. Given $y = x - 3$, find y if $x = 6$. [$y = 3$]

2. Given $y = 4x + 5$, find y if $x = -2$. [$y = -3$]

TEACH

Why Using symbols to express an idea saves time and space; it also makes a long statement easier to read and understand. By using symbols rather than words, mathematical manipulations can be applied and problems can be solved more easily. Writing equations using function notation illustrates which variable is dependent and which is independent, making the relationship more easily understood.

ongoing ASSESSMENT

$d(t) = 25t$

Cooperative Learning

Have students work in pairs to write five functions. For each of these functions, have them find $f(3), f(x-2)$, and $f(x+h)$.

Alternate Example 1

Suppose Doug drives at an average speed of 20 miles per hour when he delivers mail through the Boise National Forest along State Highway 21 from Idaho City to Stanley, Idaho.

A. Write the function for his distance in terms of time. [$\boldsymbol{d(t) = 20t}$]

B. If it takes Doug about 3 hours 45 minutes to drive from Idaho City to Stanley, how long is this route? [**about 75 miles**]

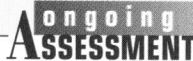

Ongoing Assessment

Try This

25 miles

Suppose Emma averages 25 miles per hour along Route 1, and it takes her 30 minutes to deliver the mail for Route 1. To find out how long Route 1 is, substitute 0.5 (half an hour) for t in $d(t) = 25t$.

$$\begin{aligned} d(t) &= 25t \\ d(0.5) &= 25(0.5) \\ &= 12.5 \end{aligned}$$

Thus, Route 1 is 12.5 miles long.

You can graph the function $y = 25x$ on your graphics calculator and use the trace function to verify that Route 1 is 12.5 miles long.

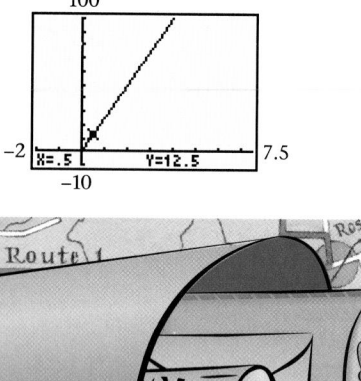

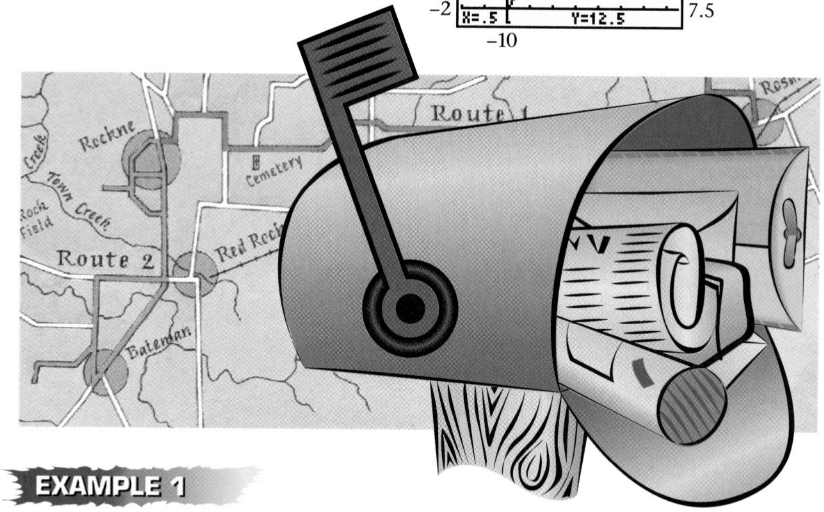

EXAMPLE 1

Time Management

Suppose Emma drives at an average speed of 15 miles per hour when she delivers mail along Route 2, from Bateman to Rockne.

A Write the function for her distance in terms of time.

B If it takes Emma 40 minutes to drive from Bateman to Rockne, how long is Route 2?

Solution

A The rate, r, is 15, so the function for the distance in terms of time is $d(t) = 15t$.

B Since 40 minutes is the same as $\frac{2}{3}$ hour, substitute $\frac{2}{3}$ for t in $d(t) = 15t$.

$$\begin{aligned} d(t) &= 15t \\ d\left(\tfrac{2}{3}\right) &= 15\left(\tfrac{2}{3}\right) = 10 \end{aligned}$$

Thus, Route 2 is 10 miles long. ❖

Try This Suppose Emma drives at an average speed of 20 miles per hour when she delivers the mail along Route 3, from Rosanky to the Haywood community. If it takes her 1 hour and 15 minutes, how long is Route 3?

ENRICHMENT Have students use the Palestinian formula for the area of a circle, given on page 86, to find the area of each circle to the nearest tenth whose diameter or radius is given below.

1. diameter = 3 [$\mathbf{7\frac{1}{14}}$]
2. diameter = 18 [$\mathbf{254\frac{4}{7}}$]
3. radius = 7 [**154**]

Usually a positive value is used for time. Is it possible that a negative value for *t* could have any meaning? If so, what?

Some letters, such as *f*, *g*, and *h*, are commonly used to represent functions. However, function notation is usually *defined* in terms of *x*, *f*(*x*), and *y*.

FUNCTION NOTATION: $y = f(x)$

The **independent** variable, *x*, is the quantity inside the parentheses, or the input.

The **dependent** variable, *f*(*x*), is the resulting expression, or the output.

What is the set of possible values for the independent variable, *x*, called? What is the set of corresponding values of the dependent variable, *f*(*x*), called?

Evaluating a Function

EXAMPLE 2

Let $f(x) = x^2 + 6$.

Ⓐ Identify the independent and the dependent variables.

Ⓑ Evaluate $f(-1)$, $f(a)$, and $f(t + 2)$.

Ⓒ Make a table of values to summarize your results from Part Ⓑ.

Solution▸

Ⓐ The independent variable is *x*, and the dependent variable is *f*(*x*).

Ⓑ $f(x) = x^2 + 6$, so

$$f(-1) = (-1)^2 + 6 \qquad f(a) = (a)^2 + 6$$
$$= 1 + 6 \qquad\qquad\quad = a^2 + 6$$
$$= 7$$

$$f(t + 2) = (t + 2)^2 + 6$$
$$= (t^2 + 4t + 4) + 6$$
$$= t^2 + 4t + 10$$

Ⓒ

x	−1	*a*	*t* + 2
f(*x*)	7	$a^2 + 6$	$t^2 + 4t + 10$

❖

Try This Let $g(x) = \dfrac{3}{x + 2}$.

Ⓐ Identify the independent and the dependent variables.

Ⓑ Evaluate $g(0)$, $g(-b)$, and $g(h + 3)$.

Ⓒ Make a table of values to summarize your results from Part Ⓑ.

Using Symbols Some students may become confused with function notation. Show students that *f*(*x*) means the same thing as *y*.

Yes. Answers should include the idea that if $t = 0$ represents the starting time of some event, then negative values for *t* represent the time before the event began. For example, if *t* represents the number of weeks after a baby's birth, then a negative value for *t* would represent the number of weeks before the birth.

Aongoing
SSESSMENT

domain; range

Alternate Example 2

Let $g(x) = x^2 - 4$.

A. Identify the independent and dependent variables. [**The independent variable is *x*, and the dependent variable is *g*(*x*).**]

B. Evaluate $g(8)$, $g(a)$, and $g(t - 1)$. [$g(8) = 60$, $g(a) = a^2 - 4$, **and** $g(t - 1) = t^2 - 2t - 3$]

C. Make a table of values to summarize your results from Part B.

x	8	*a*	*t* − 1
y	60	$a^2 - 4$	$t^2 - 2t - 3$

Aongoing
SSESSMENT

Try This

A. The independent variable is *x*, and the dependent variable is *g*(*x*).

B. $g(0) = \dfrac{3}{2}$;

$g(-b) = \dfrac{3}{-b + 2}$, $b \neq 2$;

$g(h + 3) = \dfrac{3}{h + 5}$, $h \neq -5$

C.

x	0	−*b*	*h* + 3
g(*x*)	$\dfrac{3}{2}$	$\dfrac{3}{-b + 2}$	$\dfrac{3}{h + 5}$

EXAMPLE 3

Graphics Calculator

Graph the function $f(x) = x^2 - 3x$, and describe its graph. Make a table of values, and find the domain and range of f.

Solution

The shape of the graph of f is called a *parabola*. Parabolas have either a minimum or a maximum point, called the *vertex*, or turning point.

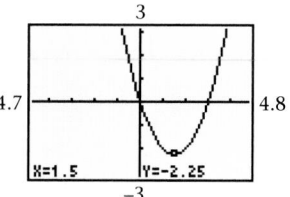

This parabola has a *minimum* vertex point at $(1.5, -2.25)$.

Use the trace feature to make a table of values.

x	-2	-1	0	1	1.5	2	3
$f(x)$	10	4	0	-2	-2.25	-2	0

Since x can be any real number, the domain of f is the set of all real numbers, **R**. Since the y-coordinate of the vertex is the lowest value of the function, the range of f is all real numbers, y, such that $y \geq -2.25$. ❖

Try This Use your graphics calculator to graph the function $f(x) = -x^2 - 2x + 3$, and describe its graph. Make a table of values, and find the domain and range of f.

Cultural Connection: Asia
The Palestinian work *Mishnat ha-Middot* gives the area of a circle as a function of the diameter. This work is thought to have been written by Rabbi Nehemiah in about 150 C.E. To calculate the area of a circle, *Mishnat ha-Middot* gave the method written on the scroll.

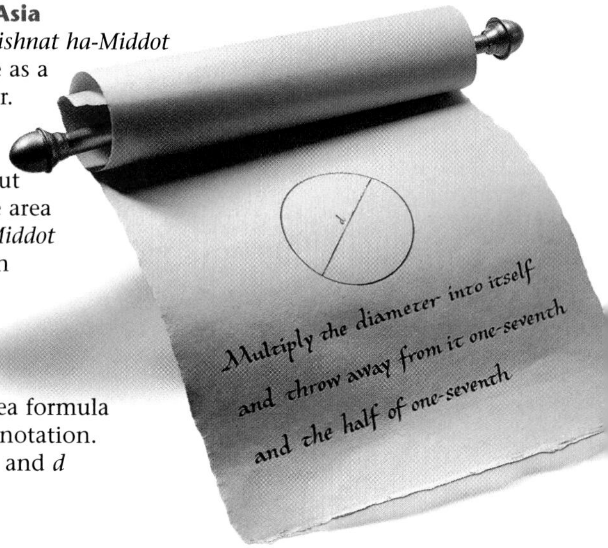

Multiply the diameter into itself and throw away from it one-seventh and the half of one-seventh

EXAMPLE 4

Write the Palestinian area formula using modern function notation. Let A represent the area and d represent the diameter.

Solution

$$A(d) = d^2 - \left(\frac{1}{7} \cdot d^2 \right) - \left(\frac{1}{2} \cdot \frac{1}{7} \cdot d^2 \right)$$
$$= d^2 - \frac{d^2}{7} - \frac{d^2}{14}$$
$$= \frac{14d^2}{14} - \frac{2d^2}{14} - \frac{d^2}{14}$$
$$= \frac{11d^2}{14} \quad ❖$$

For example, the diameter is 7 long. Its multiplication into itself is 7^2, or 49. A seventh of 49 and half of one-seventh of 49 is $\frac{1}{7} \cdot 49 + \frac{1}{2} \cdot \frac{1}{7} \cdot 49$, or $10\frac{1}{2}$. So the area is $49 - 10\frac{1}{2}$, or $38\frac{1}{2}$.

Exercises & Problems

ASSESS

Selected Answers
Odd-numbered problems 7–45

Assignment Guide
Core 1–18, 21–47

Core Plus 1–23, 26–47

Technology
Graphics calculators are needed for Exercises 6–8, 16–20, 26–29, 31–33, and 40–42.

Communicate

1. Explain what $f(x) = 3x$ means.

2. Explain what $a(r) = \pi r^2$ means.

3. In $g(x) = 4x$, which is the independent variable, and which is the dependent variable?

4. Draw the general shape of a parabola. Indicate the location of the vertex on your drawing.

5. The perimeter of a square is a function of the length of the sides of the square. Express this relationship using function notation.

Practice & Apply

6 Use your graphics calculator to graph the function $f(x) = 2x - 8$. Find the domain and range of f. R, R

7 Use your graphics calculator to graph the function $g(x) = x^2 + x$. Find the domain and range of g. R, $y \geq -0.25$

8 Explain how the coordinates of the vertex help you find the range of the function g in Exercise 7.

9. Economics The total cost of x items varies directly as the number of items sold. If each item costs 30 cents, express the cost using function notation. $c(x) = 0.30x$

In Exercises 10–13, let $f(x) = 3x - 7$ and $g(x) = x^2 - 8x + 1$.

10. Find $f(a^2)$. $3a^2 - 7$ **11.** Find $g(-3)$. 34

12. Find $f(h - b)$.   **13.** Find $g(a + b)$.
     $3h - 3b - 7$ $a^2 + 2ab + b^2 - 8a - 8b + 1$

Geometry The volume of a sphere is a function of its radius.

14. Express the volume of a sphere using function notation. $v(r) = \frac{4}{3}\pi r^3$

15. Find the volume of a sphere to the nearest whole number if the radius is 3. $v(3) \approx 113$ cubic units

r

$$V = \frac{4}{3}\pi r^3$$

For Exercises 16–18, let $f(x) = x^2 + x + 7$. Use your graphics calculator to find the indicated properties of f.   (−0.5, 6.75) $y \geq 6.75$

16 Domain of f R   **17** Vertex of f **18** Range of f

8. Since the vertex is the minimum point of this parabola, the range of g is greater than or equal to the y-coordinate of the vertex.

Physiology The percent concentration, c, of a particular medication in the bloodstream t hours after the medication has been injected is given by the function $c(t) = \dfrac{4.06t}{0.64t^2 + 8.1}$. Use your graphics calculator to graph this function.

19 Find the approximate time, to the nearest minute, that it will take for the medication to reach its maximum concentration. What is the maximum percent concentration? Between 3hr, 30min and 3hr, 36min; 0.89

20 Approximately how many hours after injection will the concentration of the medication be half of its maximum? Approx. 1hr and 13hr, 18min.

For Exercises 21–23, let $f(x) = 6x + 4$.

21. Find $f(a)$. $6a + 4$ **22.** Find $f(b)$. $6b + 4$ **23.** Find $\dfrac{f(a) - f(b)}{a - b}$. 6

Geometry The length of a rectangle is 3 more than its width.

24. Express the area of the rectangle as a function of its length using function notation. $a(l) = l(l - 3)$

25. Find the area if the length is $2x$. $4x^2 - 6x$ sq units

Statistics Investigate the relationship between the time of sunrise and the season of the year using the data provided. Graph a scatter plot with the day of the year on the horizontal axis and the time of sunrise on the vertical axis.

26 Find a linear function that best fits the relationship between sunrise and the day of the year from January through June.

27 What is the correlation coefficient for the line of best fit from Exercise 26?

28 Find a linear function that expresses the relationship between sunrise and the day of the year from July through December.

29 What is the correlation coefficient for the line of best fit from Exercise 28?

30. Describe the shape of the graph of the function that has a close correlation between the day of the year and the sunrise for the entire year.

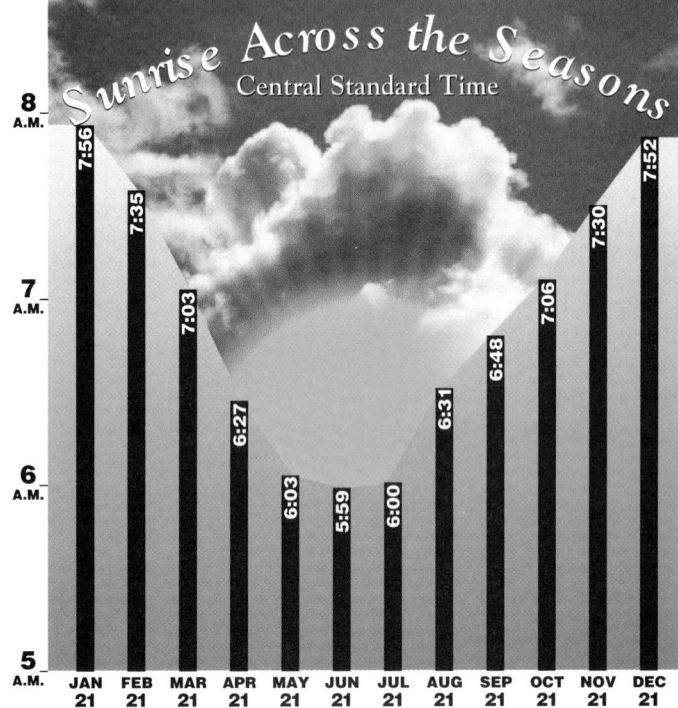

Sunrise Across the Seasons
Central Standard Time

JAN 21	FEB 21	MAR 21	APR 21	MAY 21	JUN 21	JUL 21	AUG 21	SEP 21	OCT 21	NOV 21	DEC 21
7:56	7:35	7:03	6:27	6:03	5:59	6:00	6:31	6:48	7:06	7:30	7:52

26. $f(x) = -0.014x + 8.20$

27. $r \approx -0.98$

28. $f(x) = 0.012x + 3.69$

29. $r \approx 0.997$

30. A function with a V-shaped graph and a minimum point close to (174, 5.7).

Investment The amount, *a*, of money accumulated in an investment is related to the principal amount of money invested, *p*, the annual interest rate, *r*, and the length of time in years, *t*, that the money has been invested. It is expressed by the function $a(t) = p(1 + r)^t$.

31 Let $r = 0.05$, $p = \$100$, and *t* be variable. Make a table of values using positive values, negative values, and fractional values for *t*. Describe how the value of *t* affects the value of *a*.

32 Use your graphics calculator to graph *a*. Use the trace feature to find the amount, *a*, after 3.6 years. $119.20

33 Use the trace feature to find the time, to the nearest month, necessary to double the principal amount. Between 14 yr and 14 yr 3 mos

Cultural Connection: Asia The ancient Palestinian formula for the area of a circle is $A(d) = \frac{11d^2}{14}$, where *d* is the diameter.

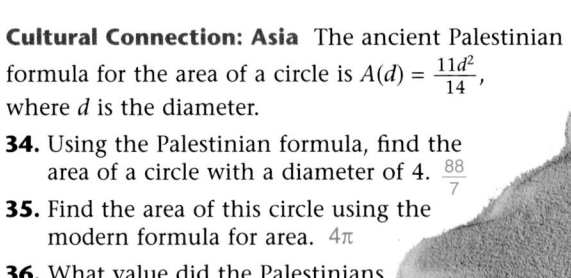

34. Using the Palestinian formula, find the area of a circle with a diameter of 4. $\frac{88}{7}$

35. Find the area of this circle using the modern formula for area. 4π

36. What value did the Palestinians use to approximate π? $\frac{22}{7}$

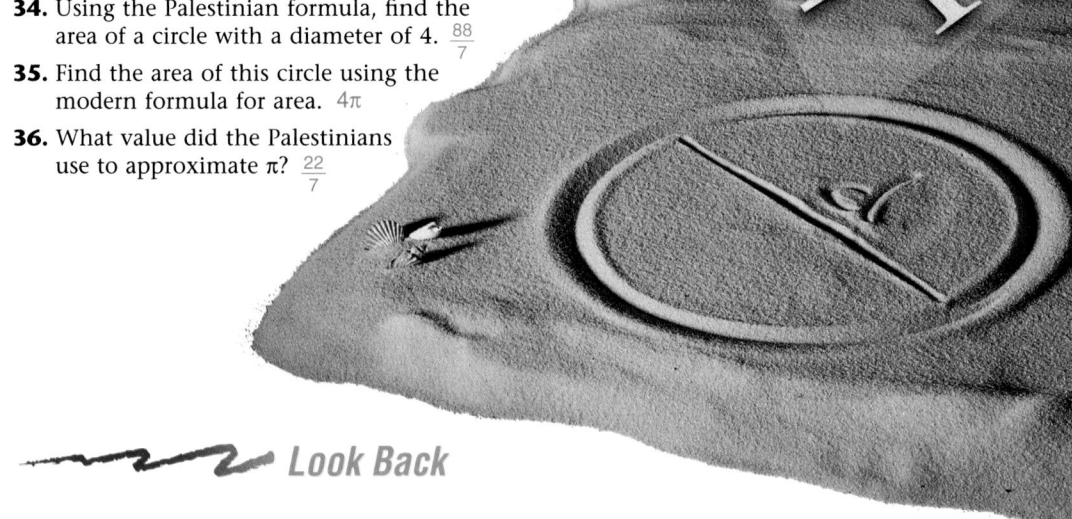

Look Back

For Exercises 37–39, *y* varies directly as *x*. Find the constant of variation, *k*, and write the equation of direct variation. **[Lesson 1.4]**

37. *y* is 3 when *x* is 6
0.5, $y = 0.5x$

38. *y* is 15 when *x* is 0.3
50, $y = 50x$

39. *y* is 7.5 when *x* is −25
−0.3, $y = -0.3x$

For Exercises 40–42, solve for *x* by graphing. **[Lesson 1.5]**

40 $3x - 2 = 5 - 4(1 - x)$ −3

41 $x - (3x - 2) = 7x - 2$
$\frac{4}{9} \approx 0.44$

42 $2x + 3 - 7(-x) = 0$ $-\frac{1}{3} \approx -0.33$

Simplify. **[Lesson 2.2]**

43. $(3^5)(3^{-2})$ 27

44. $\left(2^{\frac{1}{2}}\right)\left(2^3\right) 2^{\frac{7}{2}} \approx 11.31$

45. $\left((-1)^3\right)^{\frac{8}{3}}$ 1

46. $\left((2)^{11}\right)^{\frac{2}{11}}$ 4

Look Beyond

47. If $f(x) = 2x - 5$, find $f(-3x + 2)$. $-6x - 1$

31. Answers may vary. One example is shown.

t	−4	−2	0.5	3	3.5	20	30
a(t)	82.27	90.70	102.47	115.76	118.62	265.33	432.19

As *t* increases, *a(t)* increases.

Look Beyond
Exercise 47 gives students an advance look at the composition of functions, which is introduced in Chapter 3.

Technology Master

NAME _____ CLASS _____ DATE _____

Technology
2.4 Domains and Ranges

Sometimes you can easily determine the domain and range of a function. At other times, you may need to use a graphics calculator to get a picture of the function before determining the domain and range. Suppose that you are given the function

$$f(x) = \frac{1}{(x-1)(x+1)}$$

To determine the domain and range, graph the function on a graphics calculator as shown. From the graph, it appears that the function is defined for all real numbers except −1 and 1. Also from the graph, it appears that f(x) can be any real number greater than zero and less than or equal to −1. So, the range is all real numbers except −1 < f(x) ≤ 0.

Use a graphics calculator to find the domain and range of each function.

1. $f(x) = x$ 2. $f(x) = x^2$ 3. $f(x) = x^3$ 4. $f(x) = x^4$

5. $f(x) = x^5$ 6. $f(x) = x^6$ 7. $f(x) = x^7$ 8. $f(x) = x^8$

9. $f(x) = \frac{1}{(x-1)(x)(x+1)}$ 10. $f(x) = \frac{1}{(x-2)(x-1)(x)(x+1)(x+2)}$

11. Write a brief description of the domains and ranges of the functions in Exercises 1–8. Describe how the graphs are similar and how they are different.

12. Write a brief description of the domains and ranges of the functions in Exercises 9 and 10. Describe how the graphs are similar and how they are different.

HRW Advanced Algebra   **Technology** 53

PREPARE

Objectives

- Use the slope formula to write and identify increasing and decreasing linear functions.

- Practice Master   2.5
- Enrichment Master   2.5
- Technology Master   2.5
- Lesson Activity Master   2.5
- Quiz   2.5
- Spanish Resource   2.5

Assessing Prior Knowledge

Let $f(x) = -\frac{3}{4}x + 1$.

1. Find $f(4)$. **[−2]**

2. Find $f(-4)$. **[4]**

3. Find $f(2)$. **[−0.5]**

4. Find $f(0.5)$. **[0.625]**

TEACH

 This lesson relates functions and linear equations. Students may recall that the only linear equations that are not functions are of the form $x = a$.

ongoing ASSESSMENT

$y = mx + b$ where m is the slope and b is the y-intercept

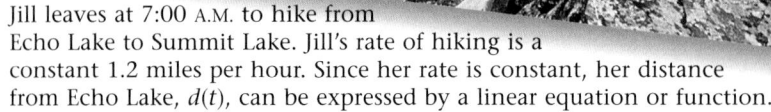

LESSON 2.5 Linear Functions

why *You have studied equations of lines and slopes of lines. Functions and function notation allow you to obtain more information from the equation of a line.*

Mount Evans is about a 1-hour drive from Denver, Colorado, in the Colorado Rockies. Echo Lake is 10.5 miles from Summit Lake. Notice that Summit Lake is at the base of Mount Evans.

Recreation Jill leaves at 7:00 A.M. to hike from Echo Lake to Summit Lake. Jill's rate of hiking is a constant 1.2 miles per hour. Since her rate is constant, her distance from Echo Lake, $d(t)$, can be expressed by a linear equation or function.

When Jill begins hiking from Echo Lake, t is 0 and the distance, $d(t)$, from Echo Lake is 0 miles. Therefore, $(t, d) = (0, 0)$ is on the graph of the line.

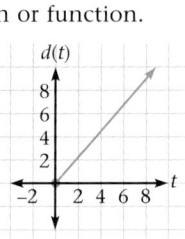

What is the slope-intercept form of any linear equation?

The y-intercept is 0 and the slope is 1.2, so the linear function for Jill's distance from Echo Lake is $d(t) = 1.2t + 0$, or $d(t) = 1.2t$, where $d(t)$ represents the distance from Echo Lake at any time, t.

To find the time it will take to reach Summit Lake, substitute 10.5 for $d(t)$, and find t.

$$d(t) = 1.2t$$
$$10.5 = 1.2t$$
$$t = 8.75$$

Thus, it will take Jill 8.75 hours to reach Summit Lake.

Since Jill began the hike at 7:00 A.M., she will arrive at approximately 3:45 P.M.

 **Using Visual Models**
Have students use graphics calculators to graph a series of equations. Have them determine the relationship between the value for m and the direction that the line slants. Using two points on a line and the value of m, guide them to discover the formula for slope.

Linear Functions and The Slope Formula

A **linear function** is a function of the form $f(x) = mx + b$, where m is the slope, or **rate of change**, and b is the y-intercept. A linear function with slope zero is a **constant function**, $f(x) = b$.

Recall the formula for the slope of a line.

	Point 1	Point 2	Slope formula
Equation: $y = mx + b$	$P_1(x_1, y_1)$	$P_2(x_2, y_2)$	$\frac{y_2 - y_1}{x_2 - x_1}$
Function: $f(x) = mx + b$	$P_1(x_1, f(x_1))$	$P_2(x_2, f(x_2))$	$\frac{f(x_2) - f(x_1)}{x_2 - x_1}$

SLOPE FORMULA FOR FUNCTIONS

Let f represent any linear function, and let $(x_1, f(x_1))$ and $(x_2, f(x_2))$ be two points on the graph of f.

The slope of the graph of f is $m = \frac{f(x_2) - f(x_1)}{x_2 - x_1}$.

EXAMPLE 1

Business Mrs. Davies, the plumber, charges a constant hourly fee in addition to a base charge. She charges a total of $165 for a 4-hour job and $255 for a 7-hour job.

A Write a function for Mrs. Davies' charges.

B How much will Mrs. Davies charge for a 5-hour job?

Solution

A Mrs. Davies charges a constant hourly rate, so the function for her charges is linear. Let $f(x)$ represent the total charge for x hours worked. Then $f(4) = 165$ represents a 4-hour job, and $f(7) = 255$ represents a 7-hour job.

Use the points $(4, f(4))$ and $(7, f(7))$ to find m.

$$m = \frac{f(7) - f(4)}{7 - 4} = \frac{255 - 165}{3} = 30$$

Substitute 30 for m in $f(x) = mx + b$.

Substitute the x-value and the y-value from either point in the linear function to find b.

$$f(x) = mx + b$$
$$f(x) = 30x + b$$
$$(165) = 30(4) + b$$
$$45 = b$$

Thus, $f(x) = 30x + 45$ is the linear function for Mrs. Davies' charges.

B For a 5-hour job, find $f(5)$. $f(5) = 30(5) + 45 = 195$

Thus, Mrs. Davies will charge $195 for a 5-hour job. ❖

In Example 1, what do the slope and the y-intercept represent?

interdisciplinary CONNECTION

Physiology The formula used for calculating a person's maximum heart rate per minute, $r = 220 - a$, is a linear function where a is the person's age. Have students calculate the maximum heart rate for people of various ages including their own heart rates. How do these values compare with the values given by the linear function? Explain why the number 220 is based on the average number of heartbeats per minute of infants at birth.

A. Use a graphics calculator to plot the points represented in the table below on the same coordinate plane. What range of values did you use?

x	y
9	6.75
2	1.5
4	3
1.8	1.35

[*x*-values from 0 to 10 and *y*-values from 0 to 7 will be sufficient.]

B. Write an equation of the linear function that is represented. Plot this line on the same coordinate plane as the data points. [*y* = 0.75*x*]

C. Is this function a direct variation? If so, state the constant of variation. [yes; 0.75]

Cooperative Learning

Have students work in small groups. Ask each student to develop five quiz questions pertaining to this lesson, along with solutions. Have students within each group answer each other's questions and discuss their answers.

TEACHING *tip*

Remind students that direct variation means that (0, 0) is on the graph and as one variable increases or decreases, the other variable does the same thing at the same rate. In other words, a direct variation function is a non-constant linear function that goes through (0, 0).

Functions of Direct Variation

Cultural Connection: Africa   The early civilizations of Africa resembled modern society in many ways. There were thousands of bookkeepers to calculate the wages for hundreds of thousands of government workers. An artifact from the Illahun Temple of ancient Egypt gave wages for several different temple employees as shown in this table.

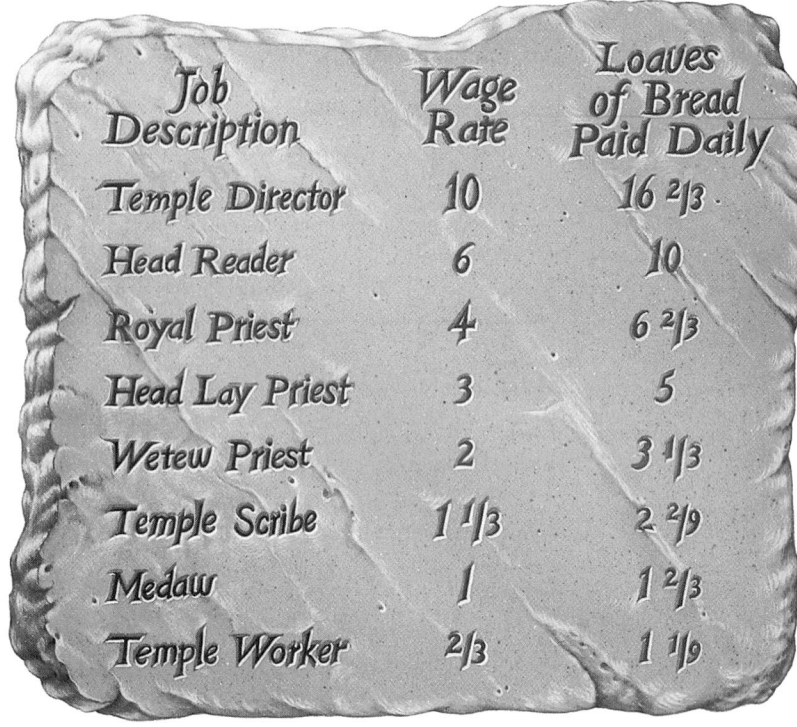

Job Description	Wage Rate	Loaves of Bread Paid Daily
Temple Director	10	16 2/3
Head Reader	6	10
Royal Priest	4	6 2/3
Head Lay Priest	3	5
Wetew Priest	2	3 1/3
Temple Scribe	1 1/3	2 2/9
Medaw	1	1 2/3
Temple Worker	2/3	1 1/9

EXAMPLE 2

Graphics Calculator

Ⓐ Plot the points represented in the table on the same coordinate plane. Place wage rate on the horizontal axis and loaves of bread on the vertical axis.

Ⓑ Write the equation of the linear function that is represented. Plot this line on the same coordinate plane as the data points.

Ⓒ Is this function a direct variation? If so, state the constant of variation.

Solution

Ⓐ

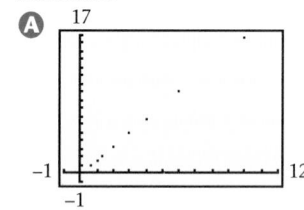

ENRICHMENT   Give students several linear equations like the ones below. Ask them to find the slope and the *y*-intercept, and determine whether the graph of each equation is increasing or decreasing.

1. $-2y + 6x = 8$
   [$m = 3, b = -4$, increasing]

2. $7x + 9y = 2$
   [$m = -\frac{7}{9}, b = \frac{2}{9}$, decreasing]

INCLUSION
strategies

Using Visual Models
Even though a linear function can be classified as increasing or decreasing without viewing a graph of the function, students may need to complete several examples in which they draw the graphs before they actually grasp this concept. Graphic calculators provide a quick and easy way for students to graph functions. Encourage students to use graphics calculators as necessary.

B Use any two points from the table to find the slope. Using $P_1 = (3, 5)$, $P_2 = (6, 10)$, the slope is $m = \frac{10-5}{6-3} = \frac{5}{3}$.

To find b, substitute $\frac{5}{3}$ for m in the linear function.

$$f(x) = \frac{5}{3}x + b$$

Substitute the x-value and the y-value from one of the given points into the linear function.

$$f(x) = \frac{5}{3}x + b$$

Using (6, 10):
$$10 = \frac{5}{3}(6) + b$$
$$10 = 10 + b$$
$$0 = b$$

Thus, the equation of the linear function represented in the table is $f(x) = \frac{5}{3}x + 0$, or $f(x) = \frac{5}{3}x$.

C Since the function is in the form $f(x) = kx$, the function f is a direct variation. Wages vary directly as a person's position in the temple. The constant of variation, k, for $f(x) = \frac{5}{3}x$ is $\frac{5}{3}$. ❖

Increasing and Decreasing Functions

The slope of a linear function falls into three categories. Either the slope is positive, $m > 0$; the slope is zero, $m = 0$; or the slope is negative, $m < 0$.

 **CRITICAL** *Thinking*

Can the slope of a linear function be undefined? Explain.

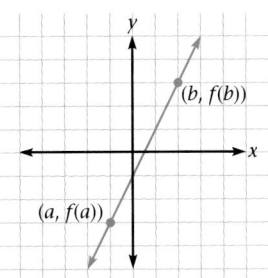

In $f(x) = 2x - 1$, $m > 0$.
The function f is increasing.

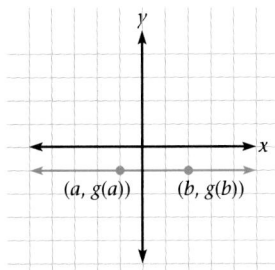

In $g(x) = -1$, $m = 0$.
The function g is neither increasing nor decreasing.

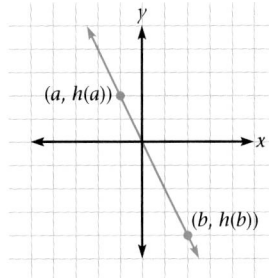

In $h(x) = -2x$, $m < 0$.
The function h is decreasing.

What type of function is neither increasing nor decreasing?

Using Visual Models The slope formula can be taught by presenting students with a graph and then using two points to find the slope by first determining the rise over the run. Next, write those two points as ordered pairs and show that the difference in the y-values is the rise and the difference in the x-values is the run. To introduce increasing and decreasing functions, graph several lines on the board and ask students to determine whether the lines are increasing or decreasing.

Let f be a linear function containing the points $(5, 11)$ and $(-3, -5)$.

A. Determine whether f is increasing, decreasing, or neither. [**increasing**]

B. Find the slope of f, and check your result from Part A. [$m = 2$. **Since** $m > 0$, f **is increasing.**]

C. Write the linear function for f. [$f(x) = 2x + 1$]

TEACHING tip

Ask the students why the slope of linear functions falls into only the three categories of positive slope, negative slope, and zero slope, rather than four categories, which includes no slope. After they have thought about this, remind them that vertical lines — lines that have no slope — are not functions because every x-value has the same y-value.

---

ongoing ASSESSMENT

Try This

A. Since $4 < 6$, let $x_1 = 4$ and $x_2 = 6$. So $f(x_1) = 3$ and $f(x_2) = 5$. Since $3 < 5$, or $f(x_1) < f(x_2)$, f is an increasing function.

B. The slope, m, is $\frac{5-3}{6-4} = 1$. Since $m > 0$, this confirms that f is increasing.

C. $f(x) = x - 1$

INCREASING AND DECREASING LINEAR FUNCTIONS

Let f be any linear function, and let $a < b$, where a and b are real numbers. For all a and b:

a function f is **increasing** if $f(a) < f(b)$.

a function f is **decreasing** if $f(a) > f(b)$.

a function f is **neither increasing nor decreasing** if $f(a) = f(b)$.

EXAMPLE 3

Let f be a linear function containing the points $(-1, 4)$ and $(2, 3)$.

Ⓐ Determine if f is increasing, decreasing, or neither.

Ⓑ Find the slope of f, and check your result from Part **Ⓐ**.

Ⓒ Write the linear function for f.

Solution

Ⓐ To use the definition of increasing and decreasing functions, a must be less than b. Since $-1 < 2$, let $a = -1$ and $b = 2$.

Then, $f(a) = 4$, $f(b) = 3$, and $f(a) > f(b)$.

By definition, since $a < b$ and $f(a) > f(b)$, f is decreasing.

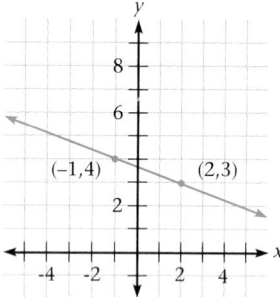

Ⓑ Let $(-1, 4)$ be (x_1, y_1) and let $(2, 3)$ be (x_2, y_2).
Then $m = \frac{f(x_2) - f(x_1)}{x_2 - x_1} = \frac{3-4}{2-(-1)} = -\frac{1}{3}$.
Since $m < 0$, f is decreasing.

Ⓒ Substitute $-\frac{1}{3}$ for m in $f(x) = mx + b$. Then substitute the coordinates from either point, and solve for b.

$$f(x) = mx + b$$
$$f(x) = -\frac{1}{3}x + b$$
$$3 = -\frac{1}{3}(2) + b$$
$$\frac{11}{3} = b$$

Thus, the linear function is $f(x) = -\frac{1}{3}x + \frac{11}{3}$. ❖

Try This Let f be a linear function containing the points $(6, 5)$ and $(4, 3)$.

Ⓐ Determine if f is increasing, decreasing, or neither.

Ⓑ Find the slope of f, and check your result.

Ⓒ Write the linear function for f.

EXERCISES & PROBLEMS

Communicate

1. What types of functions have a constant rate of change?

2. Explain how to identify a linear function.

3. Explain the slope formula for functions.

4. Which linear functions are constant functions?

5. When is a function a direct variation?

6. Explain how to write a linear function given two points on the graph of the line.

Practice & Apply

Find the slope of the linear function containing the given points, and write the function.

7. (3, 1); (2, 5) -4, $f(x) = -4x + 13$

8. (1, −1); (0, −2) 1, $f(x) = x - 2$

9. (0, 4); (−2, −3) $\frac{7}{2}$, $f(x) = \frac{7}{2}x + 4$

10. (0, −1); (1, −2) -1, $f(x) = -x - 1$

Write the linear function f using the given information.

11. $f(0) = -1$; f decreases by 3 when x increases by 1 $f(x) = -3x - 1$

12. $m = \frac{1}{3}$; $f(0) = -2$ $f(x) = \frac{1}{3}x - 2$

Write the linear function of each graph.

13.

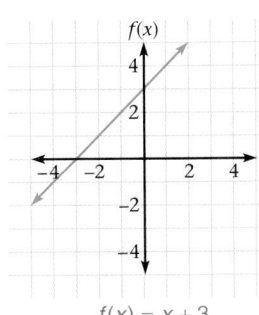

$f(x) = x + 3$

14.

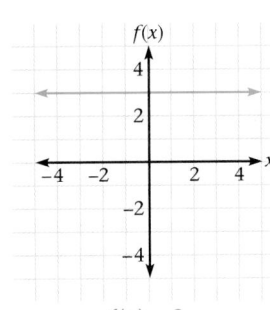

$f(x) = 3$

15.

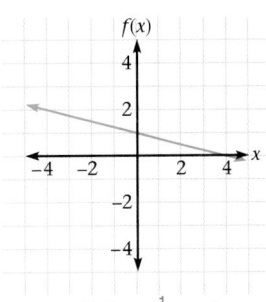

$f(x) = -\frac{1}{4}x + 1$

16. **Business** A salesperson receives a monthly base pay of $700, plus 12% of all sales. Write a linear function for the salesperson's earnings. Discuss the meaning of the slope and the y-intercept. $f(x) = 0.12x + 700$

17. **Consumer Economics** It is quite expensive to have your car towed away. One towing company charges an initial fee of $120, plus $80 for each day that your car has not been reclaimed after the first day. Write a linear function for the towing company's charges. $f(x) = 80x + 120$

16. $f(x) = 0.12x + 700$, where x represents the sales in dollars and $f(x)$ represents the salesperson's monthly earnings. The slope is 0.12, which represents the rate of commission received on sales. The y-intercept is 700, which represents the base salary of $700, the monthly base pay.

ASSESS

Selected Answers
Odd-numbered problems 7–47

Assignment Guide
Core 1–9, 11–20, 22–28, 31–50

Core Plus 1–6, 9–17, 19–51

Technology
Graphics calculators are needed for Exercises 18–21, 39, and 44.

Error Analysis
Watch for students who make mistakes finding the slope of lines when negative numbers are involved. Encourage students to write out each step used to find the slope rather than trying to complete the work mentally.

Use your graphics calculator to graph each function. Find all the x-intercepts for each function.

18 $f(x) = 2x - 5$
(2.5, 0)

19 $f(x) = -3x + 1$
(0.33..., 0)

20 $f(x) = \frac{x-2}{3}$
(2, 0)

21 $f(x) = x^2 + 2$
none

Demographics A particular school's growth in enrollment is approximately linear.

1970 — 2500 students
1990 — 4000 students

22. Find the slope of the linear function for the school's enrollment. 75

23. Estimate the number of students the school had in 1985. 3625

24. How many students does the school expect in 2000? 4750

25. Look up data for your own school. Construct a scatter plot to represent the data for the past 10 years. Answers may vary.

26. Determine if the enrollment increase or decrease over the past 10 years could be approximated by a linear function. Answers may vary.

27. Use your linear function to determine the number of students enrolled in your school in 1990. How accurate is your value? Answers may vary.

28. Use your linear function to predict the number of students that will be enrolled in your school in 10 years. Do you think your prediction will be accurate? Why or why not? Answers may vary.

Use the definition of increasing and decreasing functions to prove each statement for $f(x) = mx + b$.

29. The function f is increasing when $m > 0$.

30. The function f is decreasing when $m < 0$.

Biology The number of times a cricket chirps per minute is thought to be a function of the air temperature. The linear function for the number of cricket chirps is $f(C) = 7C - 30$, where C is the temperature in degrees Celsius.

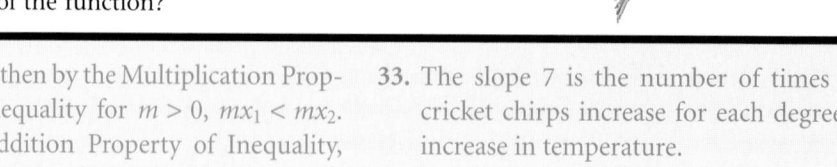

31. How many times per minute would a cricket chirp at 15°C? 75

32. How many more times per minute would a cricket chirp at 30°C than at 24°C? 42

33. How is the increase or decrease in temperature related to the slope of the function?

29. If $x_1 < x_2$, then by the Multiplication Property of Inequality for $m > 0$, $mx_1 < mx_2$. By the Addition Property of Inequality, $mx_1 + b < mx_2 + b$. So, $f(x_1) < f(x_2)$, and f is increasing.

30. If $x_1 < x_2$, then by the Multiplication Property of Inequality for $m < 0$, $mx_1 > mx_2$. By the Addition Property of Inequality, $mx_1 + b > mx + b$. So, $f(x_1) > f(x_2)$, and f is decreasing.

33. The slope 7 is the number of times the cricket chirps increase for each degree of increase in temperature.

Consumer Awareness Adie is on a business trip and needs a rental car for one day. The car rental agency charges $35.00, plus $0.20 for each mile driven.

34. Write a linear equation to describe the total charges. $f(x) = 0.20x + 35$

35. What will Adie pay for driving the rental car 85 miles in one day? $52

36. If the bill is $48.20, how many miles did Adie drive during the day? 66 mi

Physics In 1787 the French scientist Jacques Charles observed that the volume of a fixed amount of air is a linear function of the temperature of the air. Thus when t is temperature measured in degrees Celsius, $f(t)$ is volume measured in cubic centimeters.

37. Jacques Charles found that $f(30) = 505$ and $f(90) = 605$. Write the linear volume function. $f(t) = \frac{5}{3}t + 455$

38. What is the temperature when the volume is 0? What is the special name for this number? −273°C, absolute zero

39. Plot the volume function. What is the approximate temperature in degrees Celsius when the volume is 60 cubic centimeters? −237°C

Look Back

Simplify. [Lesson 2.3]

40. $\frac{7}{5} + \frac{-3}{8}$ $\frac{41}{40}$ **41.** $\frac{1}{2} + 2\frac{7}{8}$ $\frac{27}{8}$ **42.** $-\frac{3}{4} \times \frac{17}{34}$ $-\frac{3}{8}$ **43.** $\frac{6}{5} \div \frac{27}{75}$ $\frac{10}{3}$

44. What is the domain of $f(x) = \frac{x-1}{x+2}$? [Lesson 2.3] $x \neq -2$

Let $f(x) = 2x + 3$. Evaluate. [Lesson 2.4]

45. $f(a + b)$ **46.** $f(4 + w)$ **47.** $f(c - 2)$ **48.** $f(m^2)$
$2a + 2b + 3$ $2w + 11$ $2c - 1$ $2m^2 + 3$

Look Beyond

For Exercises 49 and 50, let $f(x) = x + 1$ and $g(x) = x - 1$.

49. Find $f(x) + g(x)$. $2x$

50. Find $h(3)$ if $h(x) = f(x) + g(x)$. 6

51. Give an example of a function where $f(x) = f(-x)$ for all x.

51. Answers may vary. Some examples are $f(x) = 5$, $f(x) = x^2$, and $f(x) = |x|$.

Jacques Charles lived from 1746 to 1823. The apparatus used today to verify Charles's law is exactly the same as the apparatus Charles used to determine it over 200 years ago.

Looking Beyond

These exercises preview adding functions, which will be presented in the next lesson. They also allow students to begin using function notation to solve problems.

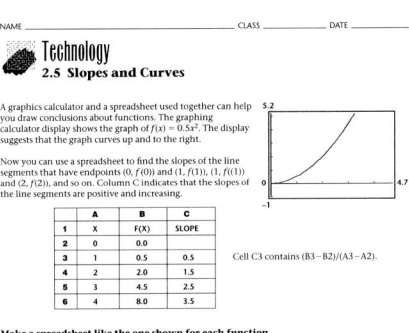

Exploring
Operations With Functions

PREPARE

Objectives

• Identify and use properties of functions to add, subtract, multiply, and divide functions.

RESOURCES

• Practice Master   2.6
• Enrichment Master   2.6
• Technology Master   2.6
• Lesson Activity Master   2.6
• Quiz   2.6
• Spanish Resources   2.6

Assessing Prior Knowledge

Simplify.

1. $(x - 2) + (x + 5)$
   $[2x + 3]$

2. $(3x + 4) + (-2x - 6)$
   $[x - 2]$

3. $(x + 2)(x - 3)$
   $[x^2 - x - 6]$

4. $(2x - 1)(x + 5)$
   $[2x^2 + 9x - 5]$

TEACH

Why This lesson explores adding and multiplying functions. Discuss several real-world situations in which functions might need to be added or multiplied.

Why *Reaction distance and braking distance are both functions of the driving speed of a car. By adding these two functions, you can find the total stopping distance. It is important to understand how to add and multiply functions.*

STARTING POINT	BEGINNING TO BRAKE	STOPPING POINT
0 seconds	1.1 seconds	2.7 second

REACTION DISTANCE | BRAKING DISTANCE
STOPPING DISTANCE

ALTERNATIVE teaching strategy

Using Algorithms Have students add and multiply expressions, such as those in the **Assessing Prior Knowledge** section. Set y equal to each sum or product, and graph each equation. Then, students ought to have a better grasp of the concept of performing operations with functions. Be sure to emphasize that an equation such as $y = 4x + 6$ is the same as the function $f(x) = 4x + 6$.

•Exploration 1 *Adding Functions*

Graphics Calculator

Let $Y_1 = X - 3$, $Y_2 = X + 1$, and $Y_3 = Y_1 + Y_2$. Graph all three functions on the same coordinate plane. Make certain that the viewing window includes both the *x*- and the *y*-intercepts of all three functions.

Use the trace feature to complete a table for each function.

x	Y_1	Y_2	Y_3
❶ −2			
❷ −1			
❸ 0			
❹ 1			
❺ 2			

❻ Determine the domain and the range of each function.

❼ How do the values of Y_3 compare with the values of Y_1 and Y_2 for each value of *x*? ❖

You can find the **sum or the difference of two functions** by adding or subtracting their dependent variables. In other words, the sum of two functions is $(f + g)(x) = f(x) + g(x)$, and the difference is $(f - g)(x) = f(x) - g(x)$.

APPLICATION

Graphics Calculator

You can use the functions for braking distance and reaction distance found in a state driving manual to graph the stopping distance function.

First enter the function $R(x)$ for reaction distance, $Y_1 = \frac{11x}{10}$, and the function $B(x)$ for braking distance, $Y_2 = \frac{x^2}{19}$. For the third function enter the sum of Y_1 and Y_2, $Y_3 = Y_1 + Y_2$.

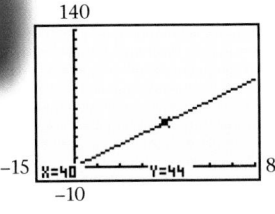

$Y_1 = R(x) = \frac{11x}{10}$

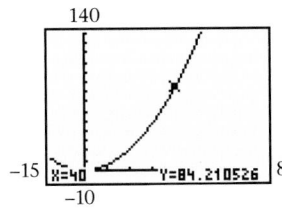

$Y_2 = B(x) = \frac{x^2}{19}$

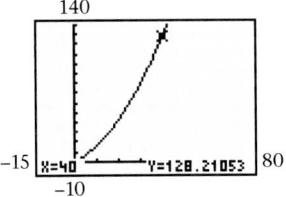

$Y_3 = S(x) = R(x) + B(x) = \frac{11x}{10} + \frac{x^2}{19}$

You can use the trace feature to find the reaction distance, braking distance, and total stopping distance for any rate of speed.

For example, by tracing Y_3, you can find that when a car is traveling 40 miles per hour, the total stopping distance is approximately 128.2 feet. ❖

CRITICAL *Thinking*

How does the real-world domain of Y_3 compare with the real-world domains of Y_1 and Y_2?

Use Transparency 9

Exploration 1 Notes

Students use graphics calculators to explore the graphs, domains, and ranges of the sum of two functions. Emphasize that the addition is done by the calculator and the results are shown graphically, instead of in the form of an equation. Encourage students to use additional *x*-values as necessary.

Aongoing **SSESSMENT**

7. **For each value of *x*, Y_3 is the sum of the values of Y_1 and Y_2.**

CRITICAL *Thinking*

The real-world domains of the reaction distance, Y_1, and the braking distance, Y_2, are positive values of speed. The real-world domain of the stopping distance, Y_3, is also positive values of speed.

ENRICHMENT Briefly explain composite functions. Show students that if $f(x) = 2x + 3$ and $g(x) = 3x$,

$$f(g(x)) = 2(3x) + 3 = 6x + 3.$$

Then have students use these same functions to find that

$$g(f(x)) = 3(2x + 3) = 6x + 9.$$

Provide students with several other examples, and then allow them to write several of their own.

INCLUSION **strategies**

Using Algorithms Some students may have trouble understanding the concept of adding or multiplying functions. If students practice adding and multiplying expressions like those presented in the **Assessing Prior Knowledge** section on page 98, it will enable them to better understand the operations as they apply to functions.

Students use graphics calculators to explore the graphs, domains, and ranges of the multiplication of two functions. Encourage students to use additional x-values as necessary.

Aongoing
SSESSMENT

> 7. For each value of x, Y_3 is the product of Y_1 and Y_2.

Cooperative Learning

Have students work in pairs. Give each pair two functions such as $Y_1 = 5X - 6$ and $Y_2 = -3X + 4$ and have them define $Y_3 = Y_1 + Y_2$. Have one student in each pair use a graphics calculator to determine the value for Y_3 when x is 2, and have the other student determine the same value by first finding Y_3 and then substituting 2 for x. Have partners compare answers and then repeat the process for multiplication.

Exploration 2 *Multiplying Functions*

Graphics Calculator

Let $Y_1 = X - 3$, $Y_2 = X + 1$, and $Y_3 = Y_1 Y_2$. Graph all three functions on the same coordinate plane. Make certain that the viewing window includes both the x- and the y-intercepts of all three functions.

Use the trace feature to complete the table.

	x	Y_1	Y_2	Y_3
1	–2			
2	–1			
3	0			
4	1			
5	2			

6 Determine the domain and the range of each function.

7 How do the values of Y_3 compare with the values of Y_2 and Y_1 for each individual value of x? ❖

You can find the **product or the quotient of two functions** by multiplying or dividing the two dependent variables. In other words, the product of two functions is $(f \cdot g)(x) = f(x) \cdot g(x)$, and the quotient of two functions is $(f \div g)(x) = f(x) \div g(x) = \frac{f(x)}{g(x)}$, where $g(x) \neq 0$.

Exercises & Problems

Communicate

1. Explain what happens when two functions are added. How do the domain and range of the sum function compare with the domain and range of the individual functions?

2. Is there a number a such that $(f + g)(a) \neq (g + f)(a)$? Explain.

3. Is there a number b such that $(f \cdot g)(b) \neq (g \cdot f)(b)$? Explain.

4. Explain what happens when two functions are multiplied. How do the domain and range of the product function compare with the domain and range of the individual functions?

5. Explain what special domain and range restrictions occur when two functions are divided.

RETEACHING the lesson

Technology This lesson can be taught without using graphics calculators by presenting students with two functions and then having them find the sum and product of the functions. In order for students to visualize the graphs, have them graph the functions on graph paper.

Practice & Apply

Let $f(x) = 2x^2 + x - 3$ and $g(x) = x + 1$. Use your graphics calculator to graph each function. Find the domain and range of each function.

6 f **7** g R; R

R; $y \geq -3.125$ R; $y \geq -2.5$

8 $f + g$ R; R

9 $f \cdot g$ R; R

10 $f \cdot \dfrac{1}{g}$ $x \neq -1$; R

11 $g \cdot \dfrac{1}{f}$ $x \neq -\dfrac{3}{2}$, $x \neq 1$; R

Use the given functions f and g to find the combinations $(f + g)(x)$ and $(f \cdot g)(x)$.

12. $f(x) = 2x$, $g(x) = x^2$ $x^2 + 2x; 2x^3$

13. $f(x) = \dfrac{1}{x}$, $g(x) = 3x$ $3x + \dfrac{1}{x}$; 3, $x \neq 0$

14. $f(x) = x + 2$, $g(x) = x$

$2x + 2$; $x^2 + 2x$

15. $f(x) = \dfrac{1}{x+1}$, $g(x) = x - 1$ $\dfrac{x^2}{x+1}$, $x \neq -1$, $\dfrac{x-1}{x+1}$, $x \neq -1$

16. $f(x) = x^2 - 4$, $g(x) = x + 2$

$x^2 + x - 2$; $x^3 + 2x^2 - 4x - 8$

17. $f(x) = x - 3$, $g(x) = x^2 - 9$ $x^2 + x - 12$; $x^3 - 3x^2 - 9x + 27$

18. $f(x) = \sqrt{x}$, $g(x) = \sqrt{x - 1}$

$\sqrt{x} + \sqrt{x - 1}$, $x \geq 1$; $\sqrt{x^2 - x}$, $x \geq 1$

19. $f(x) = \dfrac{1}{x-1}$, $g(x) = x^2 - 1$ $\dfrac{x^3 - x^2 - x + 2}{x - 1}$; $x + 1$, $x \neq 1$

Physics Refer to the application on page 99. Tyler is testing the functions for stopping distance and reaction distance. He assumes that the braking distance function is correct.

Tyler conducted three separate test runs with each car traveling at 20 miles per hour.

20. Find the average stopping distance for the car traveling at 20 miles per hour.

21. Determine the braking distance at 20 miles per hour. Then determine the reaction distance based on the average stopping distance found.
21.05 ft, 21.62 ft

22 Use your graphics calculator to graph the original functions R, B, and S on the same coordinate plane. How accurate are the R and S functions at 20 miles per hour? What factors could influence the reaction distance? What factors could influence the braking distance?

20. 42.67 feet

22. Reaction distances may vary depending on the driver's eyesight, skill, alertness, or preoccupation with something else, such as a child or music. Braking distances may vary depending on the type of road surface, weather conditions, and the condition of the brakes on the car.

ASSESS

Selected Answers
Odd-numbered problems 7–21 and 27–31.

Assignment Guide
Core 1–18, 20–33

Core Plus 1–11, 14–33

Technology
Graphing calculators are needed for Exercises 6–11, 22, and 26–28.

Cultural Connection: The Americas Refer to the Portfolio Activity in the Chapter Opener. The Maya wrote the numeral 1 as •, the numeral 5 as ▬, and the zero placeholder as ⬮. Since the Maya base is 20, the numerals from 1 through 19 do not use the zero.

| 2 | 4 | 6 | 9 | 10 | 11 | 15 | 19 |

The numbers greater than 19 were written using the zero in the units or 20s position when needed as a placeholder. For instance,

$$20 = \text{⬮} \times 20^0 + \text{•} \times 20^1$$

$$\text{and } 400 = \text{⬮} \times 20^0 + \text{⬮} \times 20^1 + \text{•} \times 20^2$$

These numerals were written in vertical columns, each row representing a power of 20. Several Maya numerals are shown.

400s (20^2)						
20s (20^1)						
Units (20^0)						
	20	21	40	85	201	503

Portfolio Activity Use pinto beans to represent the round marks, toothpicks for the long marks, and lima beans for the zero placeholder.

23. Make a chart showing the Maya numerals 17, 31, 48, and 93. Write each numeral using exponential notation, base 20.

24. Represent the sums 17 + 31, 31 + 48, and 17 + 93 using Maya numerals. Include these numerals on your chart.

25. Make a rule for addition using Maya numerals. Include an example of your rule on your chart.

23.

20s (20^1)		•	••	••••
Units (20^0)	•• ▬▬	• ▬	••• ▬	••• ▬
	17	31	48	93

24.

••	•••	▬
•••	••••	▬▬
17 + 31 = 48	31 + 48 = 79	17 + 93 = 110

25. Answers may vary but should be similar to the following two answers. When the sum has 5 dots, replace those 5 dots with a horizontal bar. When the sum has 4 horizontal bars, replace them with the 20s symbol.

 **Transformations** Let $f(x) = x^2 - 2x - 3$.

26 Graph f and $-f$ on the same coordinate plane. What are the points of intersection? Compare the graphs of f and $-f$.　　　(−1, 0), (3, 0)

27 Let $v(x) = f(x) + 3$. Compare the graphs of f and v. What happens when 3 is subtracted from f?

28 Is $f(x - 2) = f(x) - 2$? Use your graphics calculator to determine your answer.　　no

Look Back

Accounting Suppose the tax bracket on adjusted gross income—taxable income—for a single person is given by the table.　　**[Lesson 2.5]**

If your taxable income is:	but not over:	Then tax is:	of the amount over:
$0	$22, 100	15%	$0
$22,100	$53,500	$3,315 + 28%	$22,100
$53,500	$115,000	$12,107+ 31%	$53,500
$115,000	$250,000	$31,172 + 36%	$115,000
$250,000	—	$79,772 + 39.6%	$250,000

29. Write a linear function describing the situation for people in the 28% bracket. What is the domain of this function?　$t(x) = 0.28x - 2873$
$22,100 < x \leq 53,500$

30. Write the linear functions describing each remaining income bracket. Find the domain for each of these linear functions.

31. Graph each of the tax rates on the same coordinate plane. Do the endpoints of the graphs connect? Do any of the graphs intersect? Why or why not?

32. Together, these graphs form the graph of the entire tax-rate schedule for a single person. What is the tax for a single person with a taxable income of $14,050? of $25,700? of $53,500? of $300,000?
$2107.50, $4323.00, $12,107.00, $99,572.00

Look Beyond

33. Let $f(x) = 3x$ and $g(x) = \frac{x}{3}$. Evaluate $f(x)$ for $x = \{1, 2, 3, 4\}$. Now evaluate $g(x)$ for $x = \{3, 6, 9, 12\}$. What is the relationship between the ordered pairs of f and the ordered pairs of g? Explain.

26. (−1, 0) and (3, 0); the graphs of f and $-f$ are opposites (i. e., the graphs of f and $-f$ are reflections of each other with respect to the x-axis)

27. The graphs of f and v have the same shape but are located in different positions on the coordinate plane. The graph of v is the graph of f shifted up three units. If 3 is subtracted from f, the new graph is the graph of f shifted down three units.

30. 15% bracket: $t(x) = 0.15x$; domain: all real numbers such that $0 \leq x \leq 22,100$

31% bracket: $t(x) = 0.31x - 4478$; domain: all real numbers such that $53,500 < x \leq 115,000$

36% bracket: $t(x) = 0.36x - 10,228$; domain: all real numbers such that $115,000 < x \leq 250,000$

39.6% bracket: $t(x) = 0.396x - 19,228$; domain: all real numbers such that $x > 250,000$

The answer to Exercise 31 can be found in Additional Answers beginning on page 842.

33. $f(1) = 3, f(2) = 6, f(3) = 9, f(4) = 12; g(3) = 1, g(6) = 2, g(9) = 3, g(12) = 4;$ The x- and y-values of f and g are interchanged. The domain of f is the range of g and the domain of g is the range of f. This happens because, for each x-value in the domain of f, $g(f(x)) = g(3x) = \frac{(3x)}{3} = x$, and for each x-value in the domain of g, $f(g(x)) = f(\frac{x}{3}) = 3(\frac{x}{3}) = x$.

Look Beyond

This exercise allows students to examine inverse functions, which are presented in the next chapter.

Technology Master

FOCUS

Students investigate the correlation between different initial
heights of balls and how high
they bounce. Students will produce a graph of the data and determine the type of relationship
that exists.

MOTIVATE

In real life, studies like this are
done by professionals in fields
such as engineering and marketing to determine the effective-
ness of products or of materials
used in products. What types of
products contain materials that
might be studied in ways that
relate to the Bouncing Ball Experiment? [**One possible example is different types of rubber
and plastic (made into balls)
to determine their elasticity.
These data could then be used
to determine the best material
for the soles of athletic shoes.**]
How might these products be
studied to determine their effec-
tiveness? [**One possible example is the way different materials react to different types of
surfaces.**]

bouncing ball
EXPERIMENT 1

Work in groups of two to complete this project.
Each group should produce a graph of the data found.
This data will be used in several ways to develop a function
that describes the experimental results.

Materials needed for each group of students:

- meter sticks (or some other way of measuring in centimeters)
- a "bouncing" ball (tennis ball, racquetball, handball, rubber ball)
- an area without carpeting
- poster board and three colored pencils
- graphics calculator

The Experiment:

Each group will drop a ball from varying heights and record
how high the ball rebounds each time.

1. Begin with the ball at 250 centimeters. Record the rebound of
   the ball dropped from this height 10 times. Keep a record of all
   results in a table. See the sample table shown below.

2. Place the ball at 200 centimeters, and repeat Step 1.

3. Continue collecting data at 50-centimeter intervals until you reach
   50 centimeters.

Height	1	2	3	4	5	6	7	8	9	10	Average	Low	High
250 cm													
200 cm													
150 cm													
100 cm													
50 cm													

4. Construct a graph on the poster board using your data points.
   Mark the average point with a large ×. Mark the high and low values
   with a connected bar. This bar is commonly known as an *error bar*.

5 Rebound height should be a function of the initial height. Use your graphics calculator to enter the *high* and *low* data points. Make a scatter plot, and find the equation for the line of best fit. Plot this line on the same coordinate plane you used in Step 4 above, using some color other than black.

6 Use the *average* data points and your calculator to determine the equation for this different line of best fit. Draw the line of best fit for the average data on the same coordinate plane you used in Step 5. Make certain the two lines are drawn in different colors or in different ways.

What Do You Think?

- Which set of data points produced the strongest correlation? List several reasons why.

- Compare your group's results with the results of the other groups. How are the graphs produced by each group similar? How are the graphs different? Determine which group produced the strongest linear correlation.

- As a class, list several reasons why one group would produce different results from the rest of the class. Develop a set of standards that might improve the results of this experiment.

Chapter 2 Review

1. Answers may vary but should be equivalent to the given answers.
   real but not rational: $\sqrt{x}$, where $x \geq 0$; rational but not an integer: $\frac{a}{b}$, where a and b are integers such that $a \neq kb$, k is an integer, and $b \neq 0$; an integer that is not a whole number: $-x$, where x is an integer and $x \neq 0$.

2. Answers may vary but should be equivalent to the given answers. $\frac{a}{b} + \left(-\frac{a}{b}\right) = 0$, where a and b are integers and $b \neq 0$, illustrates the Inverse Property for Addition; $\frac{a}{b} \cdot \left(\frac{b}{a}\right) = 1$, where a and b are nonzero integers, illustrates the inverse property for multiplication.

Vocabulary

associative property	65	inverse property	65
closure property	65	irrational numbers	63
commutative property	65	linear function	91
constant function	91	product of functions	100
decreasing function	94	quotient of functions	100
dependent variable	85	range	79
difference of functions	99	rate of change	91
distributive property	65	rational numbers	62
domain	79	real numbers	63
function	77	relation	77
function notation	85	rules of exponents	71
identity property	65	slope formula for functions	91
increasing function	94	sum of functions	99
independent variable	85	vertical-line test	78
integers	62	whole numbers	62

Key Skills & Exercises

Lesson 2.1

> **Key Skills**

Compare and identify number systems.

All real numbers are either rational or irrational. The rational numbers include the integers, and the integers include the whole numbers.

Identify properties of real numbers, and use these properties to perform operations with rational numbers.

The real numbers under addition and multiplication follow the closure, associative, commutative, identity, inverse, and distributive properties.

> **Exercises**

1. Give a real number that is not a rational number, a rational number that is not an integer, and an integer that is not a whole number.

2. Give examples illustrating the inverse property for addition and for multiplication using rational numbers.

Simplify using the properties of real numbers.

3. $\frac{-3(x+2)}{4} + \frac{-x}{3}$ $\quad \frac{-13x - 18}{12}$

4. $\frac{bc}{3a} \cdot \frac{9a^2}{b^2c}$ $\quad \frac{3a}{b}$

5. $\frac{x-2}{5} \div \frac{x^2-4}{15}$ $\quad \frac{3}{x+2}$, where $x \neq \pm 2$

Lesson 2.2

➤ Key Skills

Perform operations, and evaluate expressions using the properties of exponents.

$$\left(\frac{(5^3)(5^{-2})}{(5^2)}\right)^2 = \left(\frac{5^{3-2}}{5^2}\right)^2 = \left(\frac{5^1}{5^2}\right)^2 = (5^{1-2})^2 = (5^{-1})^2 = 5^{-2} = \frac{1}{25}$$

➤ Exercises

6. Determine which of the following expressions are equivalent. all equivalent

   a. a^4 **b.** $a^{-7}a^{11}$ **c.** $\frac{a^{-6}}{a^{-10}}$ **d.** $\left(\frac{a^3}{a^5}\right)^{-2}$ **e.** $\left(\frac{16a^{-4}b^2}{2^4a^{-5}(b^{-1})^{-2}}\right)^4$

Lesson 2.3

➤ Key Skills

Identify and compare relations and functions, and use the vertical-line test to identify functions from their graphs.

A function can have only one second coordinate for each first coordinate. The relation (0, 2), (1, 4), (2, 4), (2, 5) is not a function, because the x-value 2 is paired with two y-values, 4 and 5.

Use functions to model real-world applications, and give appropriate domain and range restrictions for the situation.

Use your graphics calculator to graph $y = \frac{1}{x^2 - 1}$. Find the domain and range. Trace to see that the domain does not include x-values of 1 or −1. This is because when x is 1 or −1, the denominator of $\frac{1}{x^2 - 1}$ is 0.

➤ Exercises

Determine whether each relation graphed is a function. Explain your answer.

7.
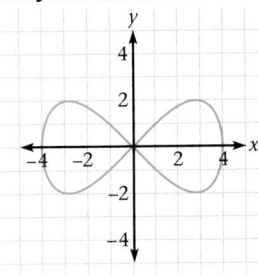
not a function

8.
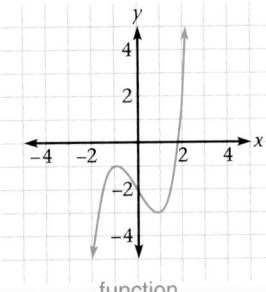
function

Find the domain and range of the functions shown.

9.
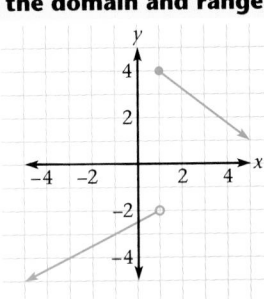
R; $y \leq 4$

10.
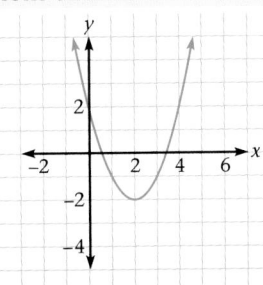
R; $y \geq -2$

7. not a function: many vertical lines intersect the graph at two points.

8. function: Any vertical line intersects the graph at only one point.

17. $(f+g)(x) = x^2 + x + 3$;
$(f \cdot g)(x) = x^3 + 3x^2$

18. $(f+g)(x) = x^3 - x^2$;
$(f \cdot g)(x) = -x^5 + 5x^3 + 5x^2 - 25$

19. $(f+g)(x) = 2x^2 + x - 6$;
$(f \cdot g)(x) = 2x^3 + 4x^2 - 8x - 16$

20. $(f+g)(x) = 9$;
$(f \cdot g)(x) = -x^4 + 3x^2 + 18$

Lesson 2.4

➤ Key Skills

Use function notation to define, evaluate, and operate with functions.

In the function notation $y = f(x)$, x is the independent variable, or input, and $f(x)$ is the dependent variable, or output.

➤ Exercises

For Exercises 11–13, let $g(x) = 2x^2 - 3x - 3$. Find each value.

11. $g(a)$ $2a^2 - 3a - 3$ **12.** $g(a+3)$ $2a^2 + 9a + 6$ **13.** $g(2a+6)$ $8a^2 + 42a + 51$

Lesson 2.5

➤ Key Skills

Use the slope formula to write and identify increasing and decreasing linear functions.

Find the slope of the linear function with $f(1) = 4$ and $f(3) = 8$.

$$m = \frac{f(x_2) - f(x_1)}{x_2 - x_1} = \frac{8 - 4}{3 - 1} = 2$$

Since the slope is greater than zero, the function is increasing.

➤ Exercises

Write the linear function containing the given points or having the given function values. Determine whether each is increasing or decreasing.

14. $(-2, -3)$ and $(0, 5)$ **15.** $(-2, 4)$ and $(-5, 7)$ **16.** $f(2) = 1$ and $f(-3) = -2$
$f(x) = 4x + 5$, increasing $f(x) = -x + 2$, decreasing $f(x) = \frac{3}{5}x - \frac{1}{5}$, increasing

Lesson 2.6

➤ Key Skills

Identify and use properties of functions to add, subtract, multiply, and divide functions.

For functions with real-number coefficients, the following properties are true.

$(f+g)x = f(x) + g(x)$ $(f \cdot g)x = f(x) \cdot g(x)$

$(f-g)x = f(x) - g(x)$ $(f \div g)x = f(x) \div g(x) = \frac{f(x)}{g(x)}$

➤ Exercises

For Exercises 17–20, find the combinations $(f+g)(x)$ and $(f \cdot g)(x)$.

17. $f(x) = x^2$, $g(x) = x + 3$ **18.** $f(x) = x^3 - 5$, $g(x) = -x^2 + 5$
19. $f(x) = 2x^2 - 8$, $g(x) = x + 2$ **20.** $f(x) = x^2 + 3$, $g(x) = -x^2 + 6$

21. Using f and g from Exercise 19, find $\frac{f(x)}{g(x)}$. What domain restriction is necessary?

$2x - 4$; $x \neq -2$

Applications

Physics A bungee jumper falling from a 256-foot platform is at a height of $h = 256 - 16t^2$ at any time, t, in seconds until the 144-foot cord starts slowing the fall. Use your graphics calculator to graph h.

22 Find the approximate length of time the jumper free-falls before the cord begins to stretch. 3 sec

23 If an object fell from 256 feet without a bungee cord to stop it, how long would it take to hit the ground? 4 sec

Chapter 2 Assessment

Simplify. Use positive exponents in your answer.

1. $(2x^5y)^{-1}(6x^3y^2)$ $\dfrac{3y}{x^2}$ **2.** $\left(\dfrac{3r^3st^{-3}}{2r^2s^{-2}t}\right)^2$ $\dfrac{9r^2s^6}{4t^8}$

3. $\dfrac{-2(x+1)}{5} + \dfrac{3x}{2}$ **4.** $\dfrac{8c}{c+1} \div \dfrac{2c^2}{5(c^2-1)}$

5. Find the equation of the linear function for which $f(-2) = 3$ and $f(4) = 0$. $f(x) = -\frac{1}{2}x + 2$

6. Does the table represent a function? Why or why not? no

7. Determine whether the linear function containing the points (1, –2) and (–3, 5.5) is increasing or decreasing. decreasing

8. What are the domain and range of the relation graphed at the right? Is it a function? Explain.

d: {–4, –2, 0, 1, 2, 3, 4}
r: {–3, –2, 1, 3, 4}
not a function

x	–4	–3	–2	–2	0
y	3	3	–1	6	3

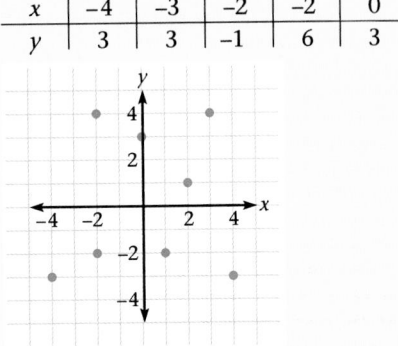

Business A chain of hardware stores sells an average of 3000 fire extinguishers per month at $20 each. From experience, they estimate that for a month-long sale, as long as the price doesn't drop below half price, each $1 reduction in price will increase sales by 500 units.

9. Write a function P in function notation that gives the sale price in terms of the number of dollars the price is reduced. $P(x) = 20 - x$

10. Write a function S in function notation that gives the predicted sales in terms of the number of dollars the price is reduced. $S(x) = 3000 + 500x$

11. The revenue from fire extinguisher sales is found by multiplying the number sold times the price. Write a function R in terms of P and S that gives the estimated revenue for a month-long sale on fire extinguishers. $R(x) = (S \cdot P)(x) = 60{,}000 + 7000x - 500x^2$

12. What are the independent and dependent variables in your revenue function?

13 Graph the revenue function on your graphics calculator, and trace to find $R(3)$. What does this represent? $76,500; revenue if price is $17

14 Trace to find the price reduction that will bring in the greatest revenue. What is the maximum predicted revenue? $7 discount; $84,500

15 What are the two discounts that will bring in an estimated $82,500 in revenue? $5 and $9

16 What are the real-world domain and range of the revenue function for the conditions described? $0 \le x \le 10$; $60{,}000 \le y \le 84{,}500$

Use the functions f and g to find the combinations $(f - g)(x)$ and $(f \div g)(x)$.

17. $f(x) = 12x$, $g(x) = 4x$ **18.** $f(x) = x^2 - 9$, $g(x) = x + 3$

19. $f(x) = \dfrac{4}{x^2 - 25}$, $g(x) = \dfrac{2}{3(x+5)}$ **20.** $f(x) = 9 - 5x$, $g(x) = x^3$

---

3. $\dfrac{11x - 4}{10}$

4. $\dfrac{20(c-1)}{c}$, $c \ne \pm 1, 0$

6. no; When $x = -2$ there are two y-values, –1 and 6.

8. domain:
{–4, – 2, 0, 1, 2, 3, 4};
range:
{–3, – 2, 1, 3, 4};
This is not a function because when $x = -2$ there are two y-values, 4 and –2.

12. The independent variable is x, the number of dollars of the price reduction. The dependent variable is $R(x)$, the estimated revenue for the month of the sale.

17. $(f - g)(x) = 8x$;
$(f \div g)(x) = 3, x \ne 0$

18. $(f - g)(x) = x^2 - x - 12$;
$(f \div g)(x) = x - 3, x \ne -3$

19. $(f - g)(x) = \dfrac{-2x + 22}{3(x^2 - 25)}$;
$(f \div g)(x) = \dfrac{6}{x - 5}$,
$x \ne -5$ or 5

20. $(f - g)(x) = -x^3 - 5x + 9$;
$(f \div g)(x) = \dfrac{9 - 5x}{x^3}$,
$x \ne 0$

COLLEGE ENTRANCE-EXAM PRACTICE

Multiple-Choice and Quantitative-Comparison Samples

The first half of the Cumulative Assessment contains two types of items found on standardized tests — multiple-choice questions and quantitative-comparison questions. Quantitative-comparison items emphasize the concepts of equality, inequality, and estimation.

Free-Response Grid Samples

The second half of the Cumulative Assessment is a free-response section. A portion of this part of the Cumulative Assessment consists of student-produced response items commonly found on college entrance exams. These questions require the use of machine-scored answer grids. You may wish to have students practice answering these items in preparation for standardized tests.

Sample answer-grid masters are available in the *Chapter Teaching Resources Booklets*.

Chapters 1–2 Cumulative Assessment

College Entrance Exam Practice

Quantitative Comparison For items 1–4 write

A if the quantity in Column A is greater than the quantity in Column B;

B if the quantity in Column B is greater than the quantity in Column A;

C if the two quantities are equal; or

D if the relationship cannot be determined from the given information.

	Column A	Column B	Answers
B 1.	$f(2) - f(4)$, where $f(x) = x^2 + 3$	$f(-5) + f(-6)$, where $f(x) = 9 + 2x$	Ⓐ Ⓑ Ⓒ Ⓓ [Lesson 2.4]
B 2.	$\frac{x^3}{x} - x^2$, where $x \neq 0$	$x^2 \cdot x^{-2}$	Ⓐ Ⓑ Ⓒ Ⓓ [Lesson 2.2]
B 3.	The value of y, where $y = 3(4x - 2) + 7$ and $x = -1$	The value of y, where $-12x + y = 1$ and $x = 1$	Ⓐ Ⓑ Ⓒ Ⓓ [Lesson 1.5]
B 4.	Slope of the line graphed		Ⓐ Ⓑ Ⓒ Ⓓ [Lesson 1.2]

5. What value should replace the "?" in the table so that
d x and y will be linearly related?   **[Lesson 1.1]**

x	3	5	7	9	11
y	2	?	–4	–7	–10

   a. 1   **b.** 4   **c.** –2   **d.** –1

6. Which one of the following is *not* a linear equation?   **[Lesson 1.1]**
c   **a.** $2x + 3y = 11$ **b.** $y = \frac{3 - 4x}{7}$ **c.** $y = \frac{7}{3 - 4x}$ **d.** $x = 3 - y$

7. Which correlation coefficient is most likely to apply to the
d   relationship between the price of a given type of headphones and how many are sold at that price?   **[Lesson 1.3]**
   a. $r = 0.85$ **b.** $r = 0.43$ **c.** $r = -0.03$ **d.** $r = -0.68$

8. If m varies directly as n, and m is –3 when n is 12, what is n when
a m is 1.5?   **[Lesson 1.4]**
   a. –6   **b.** 16.5   **c.** 6   **d.** –0.375

9. Which of the following statements is false? **[Lesson 2.1]**

d **a.** All integers are rational. **b.** No rational number is irrational.
  c. All irrational numbers are real. **d.** All integers are whole numbers.

10. Which of the following expressions is equivalent to $\frac{-(1-2x)}{2} + \frac{2-2x}{3}$?
  [Lesson 2.1]

b **a.** $\frac{1}{5}$ **b.** $\frac{2x+1}{6}$ **c.** $\frac{1}{6}$ **d.** $\frac{4x+1}{6}$

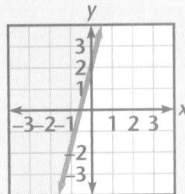

Statistics The following data represent the resale price of a certain computer model at monthly intervals.   **[Lesson 1.3]**

Months	1	2	3	4	5	6	7	8	9	10
Price ($)	3250	3150	3100	2850	2800	2700	2700	2450	2350	2300

11. Find the correlation coefficient. −0.99

12. Use the line of best fit to predict the price after 12 months. $2050

13. Solve and graph the inequality $2(1-2x) \geq -y + 4$.   **[Lesson 1.6]** $y \geq 4x + 2$

14. Write a relation that is not a function, and explain why it is not a function.   **[Lesson 2.3]**

Use the functions f and g to find the combinations $(f-g)(x)$ and $(f \div g)(x)$.   [Lesson 2.6]

15. $f(x) = 2x^2 - 2$, $g(x) = x - 1$
  $2x^2 - x - 1$; $2x + 2$, $x \neq 1$

16. $f(x) = \frac{1}{x-2}$, $g(x) = \frac{2}{x^2-4}$ $\frac{x}{x^2-4}$, $x \neq \pm 2$;
  $\frac{x+2}{2}$, $x \neq \pm 2$

17. Solve $2x - 5 = x - 1$ by graphing on your graphics calculator. Then give the coordinates of the point of intersection. 4; (4, 3)
  [Lesson 1.5]

Free-Response Grid The following questions may be answered using a free-response grid commonly used by standardized test services.

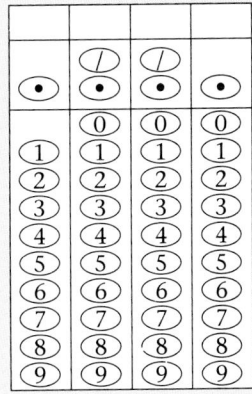

Business An electric utility company charges a fee of $6.50 per month, plus 9.25 cents per kilowatt-hour of electricity used. **[Lesson 2.5]**

18. What is the charge for using 1500 kilowatt-hours of electricity in a month? $145.25

19. If the monthly bill is $97.15, how much electricity was used? 980 kwatt-hr

20. What is the maximum number of kilowatt-hours of electricity that can be used in one month without costing more than $75? 740 kwatt-hr

13. $y \geq 4x + 2$

14. Answers may vary but should have at least one value in the domain for which there are two or more corresponding values in the range.

CHAPTER 3

Equations and Functions

Meeting Individual Needs

* The material in lessons designated as optional
is not prerequisite to any core material presented later in the text.

3.1 Exploring Symmetry

Core

Inclusion Strategies, p. 116
Reteaching the Lesson, p. 117
Practice Master 3.1
Enrichment Master 3.1
Technology Master 3.1
Lesson Activity Master 3.1

[1 day]

Core Plus

Practice Master 3.1
Enrichment, p. 116
Technology Master 3.1
Interdisciplinary
    Connection, p. 115

[1 day]

3.2 Inverse Functions

Core

Reteaching the Lesson, p. 123
Inclusion Strategies, p. 122
Practice Master 3.2
Enrichment Master 3.2
Technology Master 3.2
Lesson Activity Master 3.2

[1 day]

Core Plus

Practice Master 3.2
Enrichment, p. 122
Technology Master 3.2
Interdisciplinary
    Connection p. 121

[1 day]

3.3 Composition of Functions

Core

Reteaching the Lesson, p. 131
Inclusion Strategies, p. 130
Practice Master 3.3
Enrichment Master 3.3
Technology Master 3.3
Lesson Activity Master 3.3
Mid-Chapter Assessment Master

[2 days]

Core Plus

Practice Master 3.3
Enrichment, p. 130
Technology Master 3.3
Interdisciplinary
    Connection, p. 129
Mid-Chapter Assessment
    Master

[2 days]

3.4 The Absolute Value Function

Core

Reteaching the Lesson, p. 139
Inclusion Strategies, p. 138
Practice Master 3.4
Enrichment Master 3.4
Technology Master 3.4
Lesson Activity Master 3.4

[1 day]

Core Plus

Practice Master 3.4
Enrichment, p. 138
Technology Master 3.4
Interdisciplinary
    Connection, p. 137

[1 day]

3.5 Step Functions

Core

Reteaching the Lesson, p. 144
Inclusion Strategies, p. 143
Practice Master 3.5
Enrichment Master 3.5
Technology Master 3.5
Lesson Activity Master 3.5

[1 day]

Core Plus

Practice Master 3.5
Enrichment, p. 143
Technology Master 3.5

[1 day]

3.6 Parametric Equations

Core

Reteaching the Lesson, p. 150
Inclusion Strategies, p. 149
Practice Master 3.6
Enrichment Master 3.6
Technology Master 3.6
Lesson Activity Master 3.6

[Optional*]

Core Plus

Practice Master 3.6
Enrichment, p. 149
Technology Master 3.6
Interdisciplinary
    Connection, p. 148

[2 days]

Chapter Summary

Core	Core Plus
Chapter 3 Project, pp. 154–155	Chapter 3 Project, pp. 154–155
Lab Activity	Lab Activity
Long-Term Project	Long-Term Project
Chapter Review, pp. 156–160	Chapter Review, pp. 156–160
Chapter Assessment, p. 161	Chapter Assessment, p. 161
Chapter Assessment, A/B	Chapter Assessment, A/B
Alternative Assessment	Alternative Assessment
[3 days]	**[3 days]**

Hands-on Strategies

Students can learn many basic aspects of equations, functions, and their graphs by using graph paper and graphing by hand. For example, they should use graph paper and straight pins to complete the explorations in Lesson 3.1. Throughout the rest of the chapter, after students are proficient at graphing by hand, they should use a graphics calculator or software like *f(g) Scholar*™ to graph functions. Students may use a graphics calculator to duplicate the processes shown in the examples. They should also be encouraged to use available technology to complete the exercises whenever possible.

Reading Strategies

Be sure to continually evaluate students' ability to read and comprehend the material. Assign a partner to help any student who struggles with the reading. After students have read each lesson, have them respond to the following questions in their journals before they begin to complete the exercises provided.

- What is the main idea of this lesson?
- List and describe any words, phrases, equations, or graphs that you have not seen before. Why do you think they are important?

As you go through each lesson as a class, ask students to record their responses to the Explorations, Critical Thinking, and Ongoing Assessment questions in their journals.

Visual Strategies

Most of the work in this chapter involves graphs. Encourage students to note any graphs they find puzzling or unfamiliar. Students should be familiar with most of the graphs because they appear in previous mathematics courses. Ask students to describe how the graphs are used in algebra. You may want to have students complete this work in small groups, where they can discuss the graphs as they go through the chapter.

Cooperative Learning

GROUP ACTIVITIES	
Exploring functions	Lesson 3.0, Portfolio Activity
Symmetry	Lesson 3.1, Explorations 1, 2, 3
Inverse of a function	Lesson 3.2, Exploration
Functional symmetry	Chapter 3 Project

You may wish to have students work with partners for some of the above activities. Additional suggestions for cooperative group activities are noted in the teacher's notes in each lesson.

Multicultural

The cultural connections in this chapter include references to Africa and Asia.

CULTURAL CONNECTIONS	
Africa: Geometric transformations by the Nabdan people	Lesson 3.3, Exercises 42–49
Asia: Measuring mass in ancient Pakistan	Lesson 3.4, Exercises 35–37

Portfolio Assessment

Below are portfolio activities for the chapter. They are listed under seven activity domains that are appropriate for portfolio development.

1. **Investigation/Exploration** Four projects listed in the teacher's notes ask students to investigate a method or a relationship. They are the Alternative Teaching Strategy on page 114, Interdisciplinary Connection on page 116, Enrichment on page 122, and Enrichment on page 130. The Interdisciplinary Connection on page 148 asks students to explore important formulas from science.

2. **Applications** The following are recommended: temperature (Lesson 3.2), economics (Lessons 3.3 and 3.5), medicine (Lesson 3.3), telephone rates (Lesson 3.5) and projectile motion (Lesson 3.6).

3. **Non-Routine Problems** The Portfolio Activity on page 113 asks students to examine the constraints that affect the movement of a bicycle and the person riding it.

4. **Project** The Chapter Project on pages 154–155 asks students to create their own design by using several different functional symmetries. They must repeat the design several times and determine the functional

relationship between the original function and the image designs. Students then repeat the process to create another design.

5. **Interdisciplinary Topics** Students may choose from the following: coordinate geometry (Lesson 3.1), political science (Lesson 3.5), and from the teacher notes, Science on page 121 (Lesson 3.2), Biology on page 129 (Lesson 3.3), Psychology on page 137 (Lesson 3.4), and Physics on page 148 (Lesson 3.6).

6. **Writing** The interleaf material at the beginning of this chapter has information for journal writing. Also recommended are the Communicate questions at the beginning of the exercise sets.

7. **Tools** Chapter 3 uses graph paper and straight pins, and graphic calculators or computer software, such as *f(g)* ***Scholar*** ™. The technology masters are recommended to measure individual students' proficiency.

Technology

The advantage that graphing utilities provide in this chapter cannot be overstated. Without technology, such as graphic calculators and computer software such as *f(g) Scholar™*, students would spend excessive time and paper calculating and constructing graphs by hand. This would impose limitations on the number of graphs that students could construct to observe and explore concepts. This can easily obstruct a student's motivation and enthusiasm.

Graphics Calculator

Graphing Functions Graphics calculators are used extensively in this chapter to enhance the visualization of concepts and to expedite understanding of the many graphical representations of functions. Refer to your *HRW Technology Handbook* or to the calculator's manual for instructions for graphing functions on specific calculators. The steps given below illustrate how to display the graph of a function on the TI-82.

1. Adjust the viewing window by pressing the WINDOW key and choosing WINDOW.

2. Enter appropriate values for Xmin, Xmax, Xscl, Ymin, Ymax, and Yscl. Appropriate values generally include any window that shows all *x*- and *y*-intercepts, and any maximum or minimum values of the function. For example, to view the function $y = x^2 - 4$, you could use the following viewing window:

	min	max	scl
x	−4.7	4.7	1
y	−5.1	1.1	1

This will show both *x*-intercepts, 2 and −2, the *y*-intercept, −4, and the minimum point, (0, −4). This is also a square viewing window since −4.7 + 4.7 = 9.4 is a multiple of 94, and −5.1 + 1.1 = 6.2 is a multiple of 62.

3. Press the Y= key to enter the function. Then, press the X,T,θ key to enter the variable, x^2 to square the variable, and finally − 2 to subtract 2.

4. Press the GRAPH key to graph the function. Press TRACE to find specific points on the graph. In this example, the coordinates of the vertex are shown.

5. To see a table of values for each function, press **TABLE**, or 2nd GRAPH.

Integrated Software

f(g) Scholar™ is an integrated computer-based mathematics productivity tool that combines calculator, spreadsheet, and graphics capabilities to provide a dynamic and interactive environment for explorations in mathematics. Use *f(g) Scholar™* for any lesson needing a spreadsheet, calculator, graphics calculator, or any combination of the three.

Equations and Functions

Equations and Functions

Functions are a fundamental concept in algebra. They are used to express relationships among variables. For example, a function can describe the forward motion with respect to time of a bicycle racer who is speeding along the straightaway of a race course. There are functions that describe the relationship between one turn of the bicycle pedal and the distance covered in one turn. You are probably familiar with other relationships. In this chapter, you will study many functions and their properties. You will see why the concept of function is one of the most important in mathematics.

ABOUT THE CHAPTER

Background Information

Functions can be used to describe many relationships that occur in everyday life. As one part of the relationship changes, other parts change as well. As a pattern develops, a function can describe the changes taking place. Then predictions can be made about the relationship.

CHAPTER RESOURCES

- Practice Masters
- Enrichment Masters
- Technology Masters
- Lesson Activity Masters
- Lab Activity Masters
- Long-Term Project Masters
- Assessment Masters
  Chapter Assessments, A/B
  Mid-Chapter Assessment
  Alternative Assessments, A/B
- Teaching Transparencies
- Spanish Resources

CHAPTER OBJECTIVES

- Identify the image and pre-image points and the axis of symmetry of a set of ordered pairs.
- Determine the relationship of coordinates of points reflected over the y-axis, the x-axis, and the line $y = x$.
- Determine and define the inverse of a function, and use the horizontal-line test.
- Define the composition of functions, and describe the relationship between the dependent and independent variables.

ABOUT THE PHOTOS

The photographs of the bicycle rider should illustrate the relationships of the motions of a bike and the rider. These motions should indicate how the different parts of the bike and the rider can be functions of one another. This will also assist students in performing the Portfolio Activity using the reproduction of Leonardo da Vinci's bicycle. The da Vinci bicycle also illustrates how art, mechanics, and mathematics go hand-in-hand.

PORTFOLIO ACTIVITY

Use the drawing of a bicycle by Leonardo da Vinci, 1452–1519. Reproduce this drawing on a large piece of poster board. Consider and show all of the different ways the bicycle can move. You can include a person and the person's movements on your diagram. For each different way the bike can move, write down the constraints that control or affect that movement. Indicate on your diagram the direction of motion of each part of the bike. On page 146, you will be asked to complete this activity. You may wish to include your work in your portfolio.

- Identify special properties of composition, and use these properties to analyze functions.
- Develop a composition test to determine whether two functions are inverses.
- Define and graph translations, reflections, and scalar transformations of the absolute value function.
- Solve equations involving absolute value symbols by using graphing and algebraic methods.
- Define and graph the greatest-integer and the rounding-up functions and use them to model real-world applications.
- Define and graph a system of parametric equations, and use them to model real-world applications.
- Determine the linear function represented by a system of parametric equations.

PORTFOLIO ACTIVITY

Encourage students to discuss a bicycle and how it moves. Suggest that they begin their diagram by showing pairs of relationships, such as the relationship between the pedal gear and the wheels.

Have students work in small groups. If possible, have one student actually ride a bicycle while the other members of the group record, in writing and in their diagram, the movements they observe. Each group should then compare their individual observations and compile a final list and diagram. Have each group describe their work to the class.

ABOUT THE CHAPTER PROJECT

The Chapter 3 Project on pages 154 and 155 asks students to use all of the ideas presented in this chapter to produce a unique design. As students study their classmates' designs, they will gain a better understanding of the role symmetry plays in many types of functions.

Exploring Symmetry

PREPARE

Objectives

- Identify the image and pre-image points and the axis of symmetry of a set of ordered pairs.
- Determine the relationship of coordinates of points reflected over the y-axis, the x-axis, and the line $y = x$.

RESOURCES

- Practice Master   3.1
- Technology Master   3.1
- Enrichment Master   3.1
- Lesson Activity Master   3.1
- Quiz   3.1
- Spanish Resources   3.1

Assessing Prior Knowledge

1. Plot the following points.

   $A(2, 4)$ $B(-2, 4)$
   $C(4, 2)$ $D(-4, -2)$

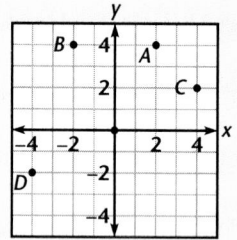

2. Which two of the above points, if any, are the same distance from the x-axis? the y-axis? [**A and B, C and D; A and B, C and D**]

TEACH

Discuss different ways in which symmetry occurs, such as in music, art, poetry, nature, architecture, and science. Symmetry is advantageous because you need to know only half of the form to know the shape of the whole thing.

why *"In an everyday sense, symmetry carries the meaning of balance, proportion, harmony, regularity of form. Beauty is sometimes linked to symmetry…"*

James R. Newman
Editor and Writer

We say a drawing is symmetric when an axis of symmetry exists. An axis of symmetry is simply a line over which the drawing mirrors itself.

How many lines of symmetry can you find in this Navajo rug design? How many lines of symmetry can you find in the design on the Zuni clay jar?

ALTERNATIVE teaching strategy

Technology Have students use a graphics calculator to draw points and their images with respect to some line or axis of symmetry. Refer to Explorations 1–3 for more information. Students will observe the relationship between the coordinates of each point and its image when the student must enter the data into the calculator. The visual display can provide helpful feedback for many students.

Exploration 1 · Symmetry and the y-Axis

You will need graph paper and a sharp pencil or a straight pin for this exploration.

1. Plot the points $A(1, -1)$, $B(2, 2)$, $C(3, 0)$, and $D(0, 4)$ on the same coordinate plane. Use a pin to poke small holes through the graph paper at the points A, B, C, and D.

2. Fold the graph paper along the y-axis. Make certain you fold the graph paper so that you can see the coordinate plane you drew. Use the pin to poke image points through *each* pin hole you made in Step 1.

3. Determine the coordinates of the new *image* points, A', B', C', and D'.

4. How are the coordinates of the points A', B', C', and D' similar to the coordinates of the original *pre-image* points A, B, C, and D? Which coordinates have the same sign? Which coordinates have opposite signs?

5. Explain what it means for the image points to be *symmetric* to the pre-image points with respect to the y-axis.

6. Let $E(a, b)$ be any point on the coordinate plane. What are the coordinates of the image of E with respect to the y-axis? ❖

Exploration 2 · Symmetry and the x-Axis

You will need graph paper and a sharp pencil or a straight pin for this exploration.

1. Plot the points $A(1, -1)$, $B(2, 2)$, $C(3, 0)$, and $D(0, 4)$ on the same coordinate plane. Use a pin to poke small holes through the graph paper at the points A, B, C, and D.

2. Fold the graph paper along the x-axis. Make certain you fold the graph paper so that you can see the coordinate plane you drew. Use the pin to poke image points through *each* pin hole you made in Step 1.

3. Determine the coordinates of the new *image* points, A', B', C', and D'.

4. How are the coordinates of the points A', B', C', and D' similar to the coordinates of the original *pre-image* points A, B, C, and D? Which coordinates have the same sign? Which coordinates have opposite signs?

5. Explain what it means for the image points to be *symmetric* to the pre-image points with respect to the x-axis.

6. Let $E(a, b)$ be any point on the coordinate plane. What are the coordinates of the image of E with respect to the x-axis? ❖

Cooperative Learning

Have students work in pairs to complete the explorations. Have them take turns plotting the points, poking the holes, and folding the paper to answer each of the questions.

Exploration 1 Notes

Be sure students label the axes and points carefully. Encourage them to fold the graph paper carefully along the y-axis. After completing Step 2, be sure students unfold their paper and *then* label the image points.

Aongoing ASSESSMENT

5. The line segment joining each pre-image point to its image point is perpendicular to the y-axis. These line segments have the same length on each side of the y-axis.

6. $E'(-a, b)$

Exploration 2 Notes

As with Exploration 1, be sure students label the axes and points carefully and fold the graph paper carefully. Notice the relationship between the image points in Exploration 2 and the image points in Exploration 1.

Aongoing ASSESSMENT

5. The line segment joining each pre-image point to its image point is perpendicular to the x-axis. These line segments have the same length on each side of the x-axis.

6. $E'(a, -b)$

Be sure students carefully draw the line $y = x$ and the points A, B, C, and D. These points are the same as those they plotted in Explorations 1 and 2. Have students compare the lines of symmetry and the relationships between the image points in this exploration with the two previous explorations.

Aongoing SSESSMENT

6. The line segment joining each pre-image point to its image point is perpendicular to the line $y = x$. These line segments have the same length on each side of the line $y = x$.

Teaching tip

Be sure students notice that the symmetries they have explored are reflections of the pre-image points over the given line $y = x$.

Use Transparency ▶ 10

Exploration 3 Symmetry With Respect to the Line y = x

You will need graph paper and a sharp pencil or a straight pin for this exploration.

1 Draw the line $y = x$ on your graph paper.

2 Plot the points $A(1, -1)$, $B(2, 2)$, $C(3, 0)$, and $D(0, 4)$ on the same coordinate plane. Use a pin to poke small holes through the graph paper at the points A, B, C, and D.

3 Fold the graph paper along the line $y = x$. Make certain you fold the graph paper so that you can see the coordinate plane you drew. Use the pin to poke image points through *each* pin hole you made in Step 2.

4 Determine the coordinates of the new *image* points, A', B', C', and D'.

5 How are the coordinates of the points A', B', C', and D' similar to the coordinates of the original *pre-image* points A, B, C, and D?

6 Explain what it means for the image points to be *symmetric* to the pre-image points with respect to the line $y = x$.

7 Draw an x- and y-axis on a new piece of graph paper. Plot the points $(2, 3)$, $(4, 6)$, $(1, 0)$, and $(-3, -4)$, and the line $y = x$.

8 Plot the coordinates of the points you believe will be symmetric to the original points with respect to the line $y = x$.

ENRICHMENT
There are many examples of symmetry in nature. Have students find several examples. Ask them to identify the lines of symmetry and explain why they think each symmetry exists. Encourage students to share their examples and ideas with the rest of the class.

INCLUSION strategies
Using Symbols Examples of notations that can be used to indicate the mapping of one point to another are given below.
Reflection of $A(1, -1)$ over the x-axis:
$$r_x(1, -1) = (1, 1)$$
Reflection of $A(1, -1)$ over the y-axis:
$$r_y(1, -1) = (-1, -1)$$
Reflection of $A(1, -1)$ over the line $y = x$:
$$r_{y=x}(1, -1) = (-1, 1)$$

9 Fold the paper along the line $y = x$. Make certain you fold the graph paper so that you can see the coordinate plane you drew. Use a pin to poke holes through *each* pre-image point, and check your guesses. Were you correct? Compare your work with that of a classmate. Were your results the same? Why or why not?

10 Let $E(a, b)$ be any point on the coordinate plane. What are the coordinates of the image of E with respect to the line $y = x$? ❖

EXTENSION

COORDINATE GEOMETRY
Connection

Graph a parallelogram with vertices $A(3, 0)$, $B(4, 1)$, $C(4, 3)$, and $D(3, 2)$. Draw the line segments $\overline{AB}$, $\overline{BC}$, $\overline{CD}$, and $\overline{DA}$. Reflect the parallelogram over the line $y = x$. The image parallelogram is found by determining the image of each vertex. The image vertices are $A'(0, 3)$, $B'(1, 4)$, $C'(3, 4)$, and $D'(2, 3)$. Draw the line segments $\overline{A'B'}$, $\overline{B'C'}$, $\overline{C'D'}$, and $\overline{D'A'}$.

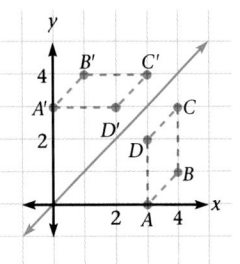

The image parallelogram $A'B'C'D'$ is a reflection of the pre-image parallelogram $ABCD$ over the line $y = x$. ❖

CRITICAL
Thinking

What can you say about the image of any point on $\overline{AB}$? What can you say about the image of the midpoint on $\overline{AB}$?

EXERCISES & PROBLEMS

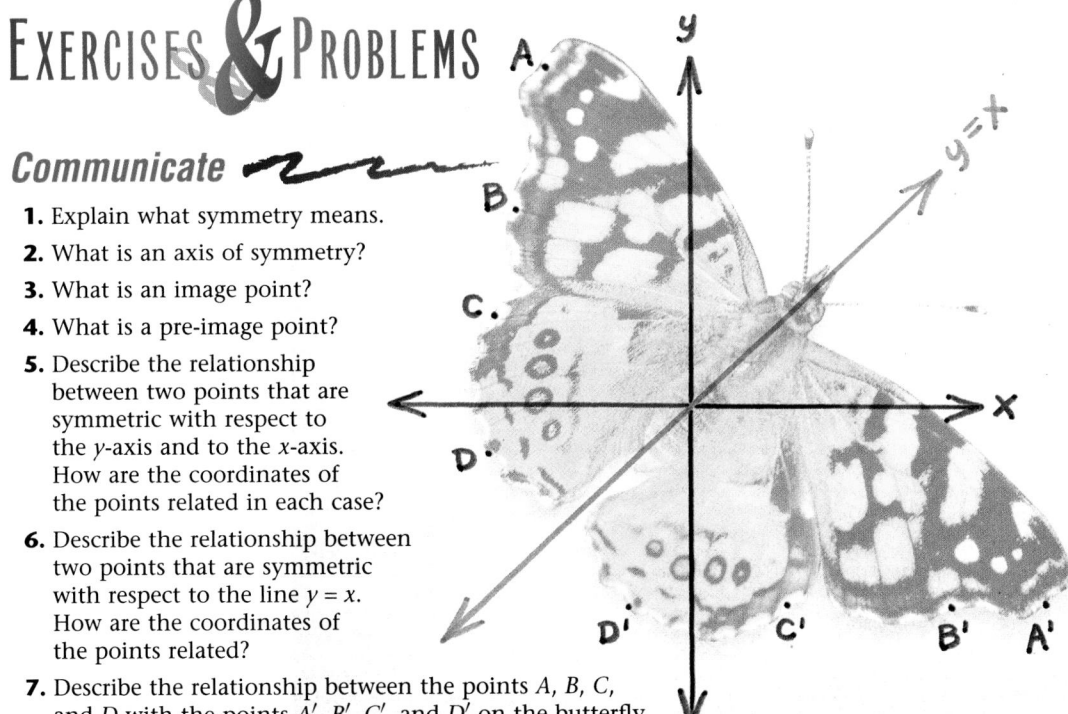

Communicate

1. Explain what symmetry means.
2. What is an axis of symmetry?
3. What is an image point?
4. What is a pre-image point?
5. Describe the relationship between two points that are symmetric with respect to the y-axis and to the x-axis. How are the coordinates of the points related in each case?
6. Describe the relationship between two points that are symmetric with respect to the line $y = x$. How are the coordinates of the points related?
7. Describe the relationship between the points A, B, C, and D with the points A', B', C', and D' on the butterfly.

ongoing
ASSESSMENT

10. $E(b, a)$

Math Connection
Coordinate Geometry

Some students may need to graph points A, B, C, and D, reflect them over the line $y = x$, and then draw the parallelograms. Have students connect the vertices of the parallelogram one side at a time, drawing corresponding sides one after the other. For example, after they draw $\overline{AB}$, have them draw $\overline{A'B'}$, and so on. Then provide students another set of vertices for another parallelogram and have them follow the procedure given in the text for reflecting the parallelogram over the line $y = x$.

CRITICAL
Thinking

The image of any point on $\overline{AB}$ will lie on $\overline{A'B'}$. The image of the midpoint of $\overline{AB}$ will be the midpoint of $\overline{A'B'}$.

ASSESS

Selected Answers

Odd numbered problems 9–37.

Assignment Guide

Core 1–10, 12–17, 19–24, 26–40

Core Plus 1–7, 1–14, 17–21, 24–40

Technology

A graphics calculator is needed for Exercises 36–38.

RETEACHING
the
lesson

Using Visual Models Have students plot the points $A(3, 2)$, $B(2, 3)$, $C(-2, 3)$, $D(-3, -2)$, and $E(3, -2)$ on the coordinate plane. Ask them which pairs of points, if any, have the x-axis as an axis of symmetry [**A and E**], the y-axis as an axis of symmetry [**B and C, and D and E**], the line $y = x$ as an axis of symmetry [**A and B, and C and E**]. Then have students generalize their answers using (a, b) in place of $(3, 2)$.

Error Analysis

Some students have difficulty identifying pre-image and image points. Encourage these students to draw a line segment from the image point to the pre-image point. If the points are truly symmetric, the line segment drawn will be perpendicular to the line of symmetry.

8. image points:
{(−2, 1), (−4, 2), (−6, 3), (−8, 4)}

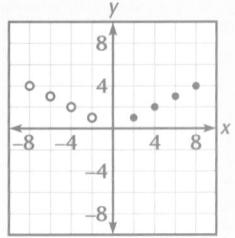

9. image points:
{(3, −6), (1, −2), (−1, 2), (−3, 6)}

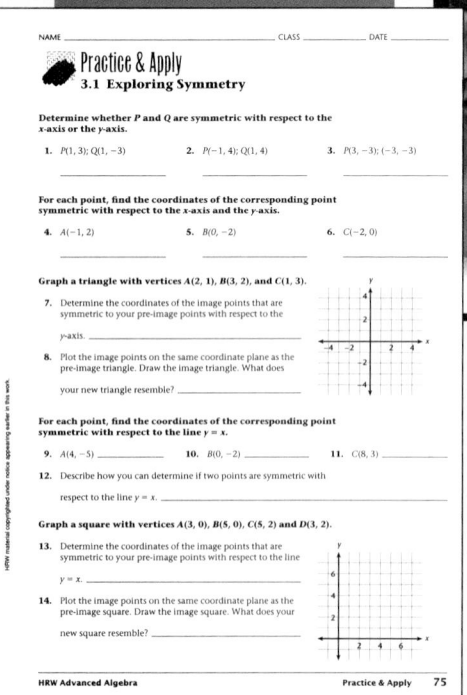

118 Lesson 3.1

Practice & Apply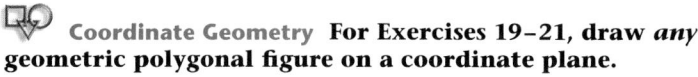

Plot each set of ordered pairs. Then plot the image points that are symmetric to the set with respect to the *y*-axis.

8. {(2, 1), (4, 2), (6, 3), (8, 4)}   **9.** {(−3, −6), (−1, −2), (1, 2), (3, 6)}

10. {(5, 2), (4, 3), (3, 4), (2, 5)}   **11.** {(9, −2), (3, −1), (1, 0), (3, 1), (9, 2)}

Coordinate Geometry For Exercises 12–14, draw a rectangle on a coordinate plane.

12. Identify the coordinates of the vertices (or corners) of your rectangle. Answers may vary.

13. Determine the coordinates of the image points that are symmetric to your pre-image points with respect to the *y*-axis. Answers may vary.

14. Plot the image points on the same coordinate plane as the pre-image rectangle. Draw the image rectangle. Answers may vary.

Plot each set of ordered pairs. Then plot the image points that are symmetric to the set with respect to the *x*-axis.

15. {(2, 1), (4, 2), (6, 3), (8, 4)}   **16.** {(−3, −6), (−1, −2), (1, 2), (3, 6)}

17. {(5, 2), (4, 3), (3, 4), (2, 5)}   **18.** {(9, −2), (3, −1), (1, 0), (3, 1), (9, 2)}

Coordinate Geometry For Exercises 19–21, draw *any* geometric polygonal figure on a coordinate plane.

19. Identify the coordinates of the vertices of your figure. Answers may vary.

20. Determine the coordinates of the image vertices that are symmetric to your pre-image vertices with respect to the *x*-axis. Answers may vary

21. Plot the image vertices on the same coordinate plane as your pre-image figure. Draw the image figure. Answers may vary.

10. image points:
{(−5, 2), (−4, 3), (−3, 4), (−2, 5)}

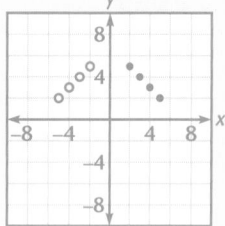

11. image points:
{(−9, −2), (−3, −1), (−1, 0), (−3, 1), (−9, 2)}

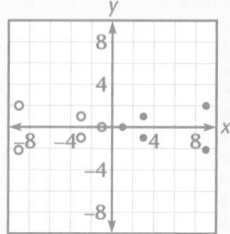

Plot each set of ordered pairs. Then plot the image points that are symmetric to the set with respect to the line $y = x$.

22. {(2, 1), (4, 2), (6, 3), (8, 4)}

23. {(−3, −6), (−1, −2), (1, 2), (3, 6)}

24. {(5, 2), (4, 3), (3, 4), (2, 5)}

25. {(9, −2), (3, −1), (1, 0), (3, 1), (9, 2)}

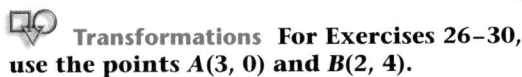 **Transformations For Exercises 26–30, use the points $A(3, 0)$ and $B(2, 4)$.**

26. Plot A and B and the line $y = x$ on a coordinate plane.

27. Determine the equation of the line that contains points A and B. Draw this line on the same coordinate plane. $y = -4x + 12$

28. Determine the coordinates of the image points that are symmetric to A and B with respect to the line $y = x$. $A'(0, 3)$ and $B'(4, 2)$

29. Draw the line that contains the image points. Determine the equation of the image line. $y = -\frac{1}{4}x + 3$

30. What is the relationship between graphs of the image line and the pre-image line? between their equations?

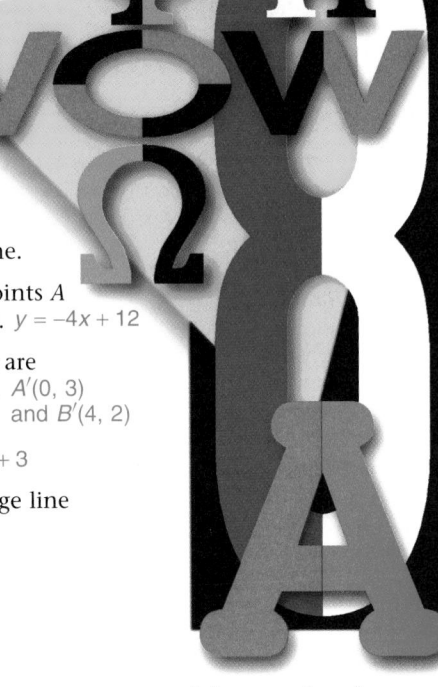

Look Back

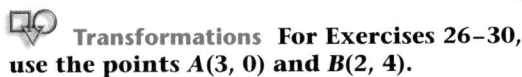 **Geometry The perimeter of a rectangle with length x and width y cannot exceed 100 meters.   [Lesson 1.6]**

31. Write a linear inequality to describe this situation. $2x + 2y \leq 100$

32. Graph the linear inequality you wrote for Exercise 31.

33. What values of x and y are realistic? $0 < x < 50, 0 < y < 50$

34. Use the x- and y-coordinates of three different solutions to find possible dimensions of the rectangle.

35. What possible values of x and y do you think will result in the largest area? Justify your response.

Use your graphics calculator to determine the domain and the range of each function.   [Lesson 2.3]

36. $g(x) = x + 1$ R; R   **37.** $h(x) = x^2 + 2$ R; $y \geq 2$ **38.** $f(x) = x^3$ R; R

Look Beyond

Solve for x in terms of y.

39. $y = \frac{2}{x}$ $x = \frac{2}{y}$ **40.** $y = \frac{x - 5}{3}$ $x = 3y + 5$

Numerical Palindromes

Start with any whole number. Add to it the number formed by reversing its digits. To this sum add the number formed by reversing the sum's digits. Continue this process. Eventually you will end up with a numerical palindrome.

$$1564$$
$$+ 4651$$
$$6215$$
$$+ 5126$$
$$11341$$
$$+14311$$

A numerical palindrome ⟶ 25652

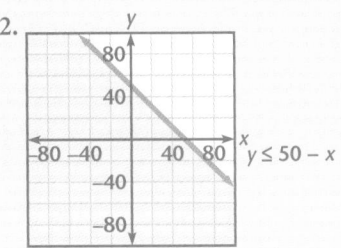
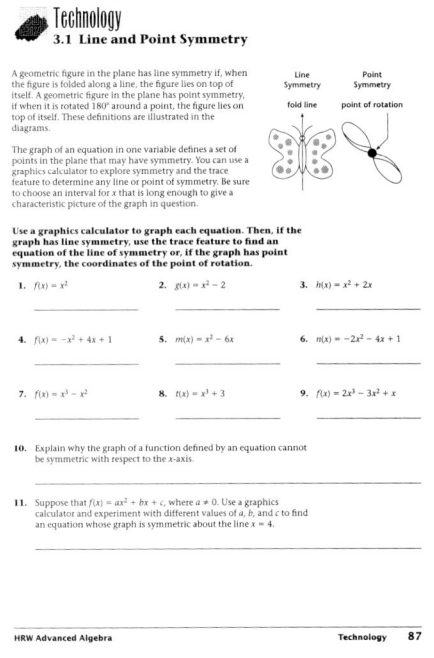

PREPARE

Objectives

- Determine the inverse of a function.
- Define the inverse of a function, and use the horizontal-line test to determine whether the inverse is a function.

RESOURCES

- Practice Master 3.2
- Enrichment Master 3.2
- Technology Master 3.2
- Lesson Activity Master 3.2
- Quiz 3.2
- Spanish Resources 3.2

Assessing Prior Knowledge

Plot the set of ordered pairs. Then plot the image points that are symmetric to the set with respect to the line $y = x$.

$\{(1, -3), (2, 2), (3, 2)\}$

[

]

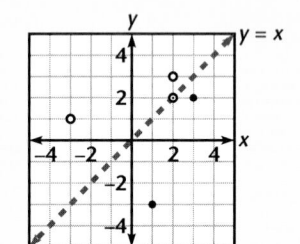

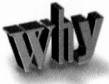

TEACH

why Review the notion of independent and dependent variables. Be sure students understand that when a function is evaluated, the value of the independent variable is substituted to find the value of the dependent variable.

LESSON 3.2 Inverse Functions

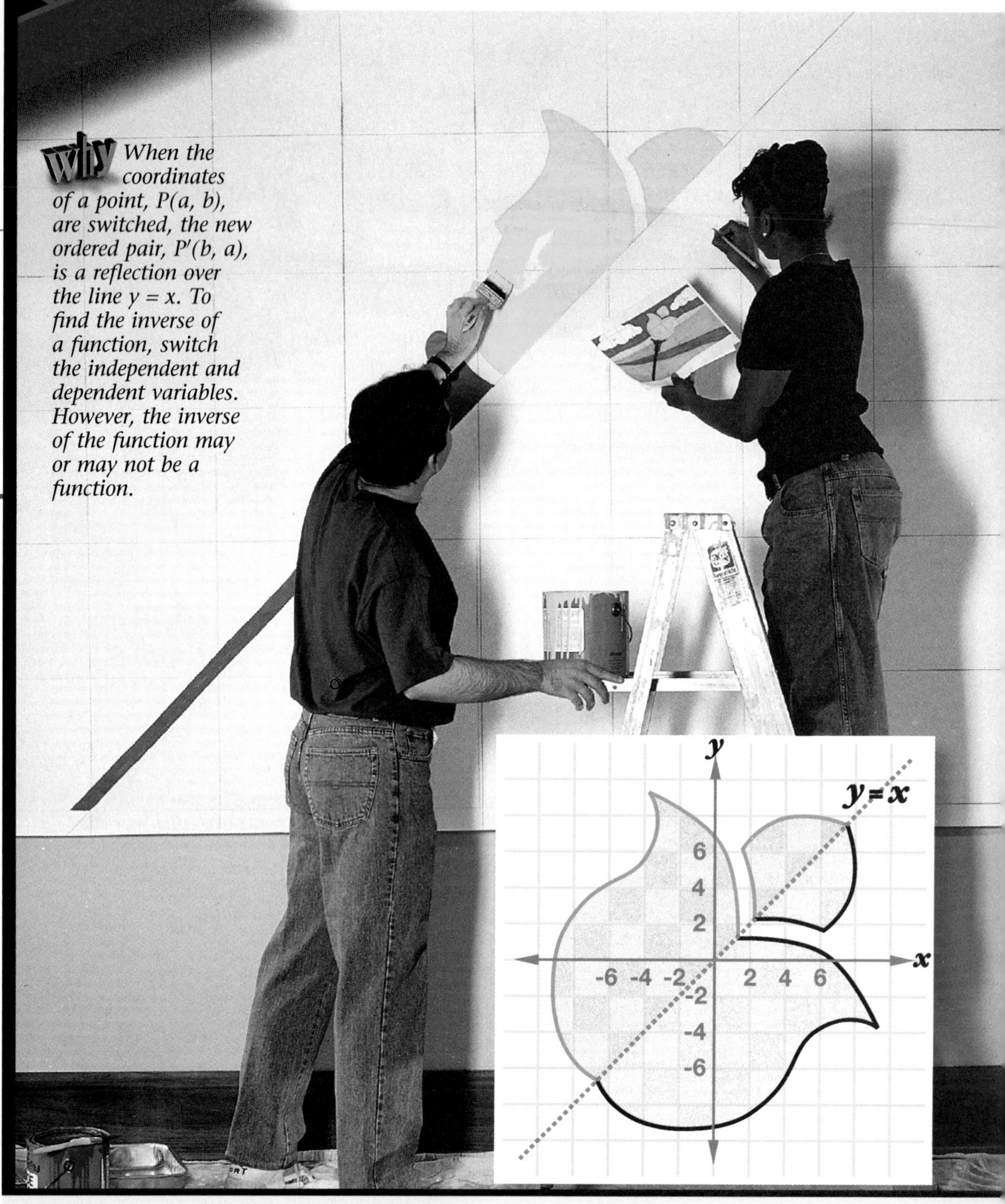

why *When the coordinates of a point, $P(a, b)$, are switched, the new ordered pair, $P'(b, a)$, is a reflection over the line $y = x$. To find the inverse of a function, switch the independent and dependent variables. However, the inverse of the function may or may not be a function.*

ALTERNATIVE teaching strategy

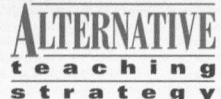

Hands-On Strategies Have students graph a function and the line $y = x$ on the same coordinate plane. Then ask them to put a piece of tracing paper over the graph and trace the graph. Have them flip the tracing of the graph over the line $y = x$ to find the graph of the inverse of the function. To determine whether the inverse is a function, have students use the horizontal-line test.

Exploration *The Inverse of a Function*

1. Plot the line $y = x$ on a coordinate plane.

2. Plot the function $f = \{(-2, 2), (-1, 1), (2, 2), (-3, -5)\}$ on the same coordinate plane.

3. Reflect the points in f over the axis of symmetry $y = x$. Find the coordinates of each of these image points, and plot them on the same coordinate plane. The relation defined by the set of image points is the *inverse* of f.

4. Use the definition of a function to determine whether the inverse of f is a function.

5. Find the domain and the range of f and its inverse. What do you notice about the domain of f and the range of its inverse? What do you notice about the range of f and the domain of its inverse? ❖

Reflecting a function with respect to the line $y = x$ simply reverses the order of the coordinates. Each ordered pair (a, b) in the function becomes (b, a) when reflected. This process of switching the x- and y-coordinates is called *finding the inverse* of a function.

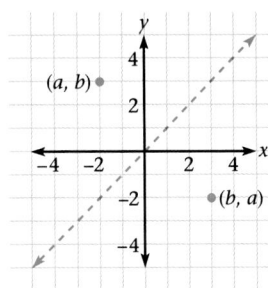

THE INVERSE OF A FUNCTION
The **inverse of a function** f can be obtained by reversing the order of the coordinates in each ordered pair of f.

A function, f, will have an **inverse function** when each second coordinate of f corresponds to exactly one first coordinate. To indicate that the inverse of a function is a function, use the notation f^{-1}, read *f inverse*.

The inverse of the cubic function, $f(x) = x^3$, is a reflection over the line $y = x$, and its inverse, $f^{-1}(x) = \sqrt[3]{x}$, is also a function, the cube root function.

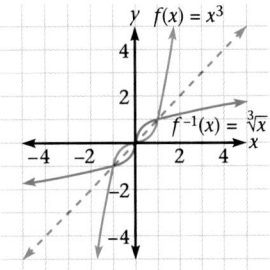

Science Look through physics and physical science books for formulas. Provide students with several examples of scientific formulas to explore. Ask students to determine whether each formula is a function. Then have them find the inverse of the formula and determine whether the inverse is a function.

Point out to students that the domain of a function for a specific interval can be determined by scanning the *x*-axis to see which values are included in the graph. An interval for the range can be determined in a similar manner by scanning the *y*-axis.

Alternate Example 1

Find the inverse of each function. Determine whether its inverse is a function.

A. $f = \{(0, 0), (1, 2), (4, -2), (3, 2)\}$ **[inverse: $\{(0, 0), (2, 1), (-2, 4), (2, 3)\}$; The inverse is not a function since the first coordinate, 2, is paired with two second coordinates, 1 and 3.]**

B. $g = \{(7, 5), (-2, 4), (3, -4), (-8, -5)\}$ **[inverse: $\{(5, 7), (4, -2), (-4, 3), (-5, -8)\}$; The inverse is a function since each first coordinate of g has exactly one second coordinate.]**

TEACHING *tip*

Review the vertical-line test for functions. Be sure students understand how this relates to the horizontal-line test.

EXAMPLE 1

Find the inverse of each function. Determine whether its inverse is a function.

A $f = \{(-3, -6), (-1, -2), (1, 2), (3, 6)\}$

B $g = \{(2, 1), (3, 2), (4, 2), (5, 1)\}$

Solution▸

A The inverse of f is $\{(-6, -3), (-2, -1), (2, 1), (6, 3)\}$. The inverse is a function since each first coordinate has exactly one second coordinate.

B The inverse of g is $\{(1, 2), (2, 3), (2, 4), (1, 5)\}$. The inverse is not a function, since the first coordinate 1 is paired with two second coordinates, 2 and 5, and the first coordinate 2 is paired with two second coordinates, 3 and 4. ❖

Graphing and the Horizontal-Line Test

How can you tell whether a function has an inverse function by examining its graph? You used the vertical-line test to determine that each *first* coordinate of a function is paired with exactly one *second* coordinate. You can use a *horizontal-line test* to determine that each *second* coordinate of a function is paired with exactly one *first* coordinate. When the graph of any function passes the horizontal-line test, its inverse function exists. A function is **invertible** when it has an inverse that is a function.

ENRICHMENT The inverse of a function switches the coordinates of each ordered pair. If the inverse is also a function, the original function is called *one-to-one*. Have students examine several functions to determine which are one-to-one. Have them note and explain any additional characteristics of one-to-one functions. Students should discover that one-to-one functions are strictly increasing or strictly decreasing. Functions that are not one-to-one will change from increasing to decreasing or vice versa.

INCLUSION **strategies** **Using Cognitive Strategies** Ask students if they can think of any graphs that are not functions but whose inverses are functions. If they need a hint, suggest that they think about the functions they have worked with, whose inverses are not functions.

HORIZONTAL-LINE TEST

If a horizontal line crosses the graph of a function in more than one point, the inverse of the function is *not* a function.

Graphics Calculator

EXAMPLE 2

Use the horizontal-line test to determine whether each of the following functions has an inverse function.

A $f(x) = 5x - 2$ **B** $g(x) = x^2 + 1$ **C** $h(x) = \dfrac{x}{x-1}$

Solution▶

Graph each function. Then use the horizontal-line test to see if its inverse function exists.

A
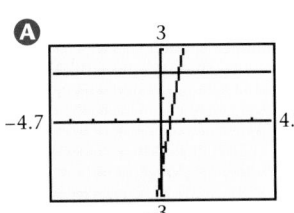

The inverse function f^{-1} exists because any horizontal line intersects the graph of f in exactly *one* place.

B
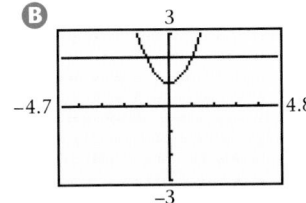

The inverse function does not exist because a horizontal line can cross the graph of g in 2 places.

C
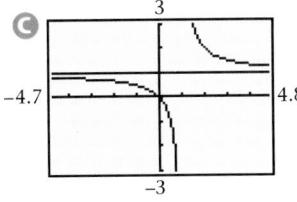

The inverse function h^{-1} exists because any horizontal line that *does* intersect the graph of h will cross in exactly *one* place. ❖

Try This Graph each function. Use the horizontal-line test to determine if its inverse function exists.

A $f(x) = x^2 - 2$

B $g(x) = x - 3$

C $h(x) = \dfrac{3}{x+1}$

RETEACHING
the lesson

Using Manipulatives
Have students use a pencil or a piece of spaghetti to represent a horizontal line. Have students move the pencil or piece of spaghetti up and down the graph until they have gathered enough information to determine whether the inverse is a function.

TEACHING *tip*

Encourage students to use several different horizontal lines for the test by choosing different equations. For example, students could use a calculator to graph the function as $Y1$ and let $Y2 = 5, Y3 = 3, Y4 = 1, Y5 = -1, Y6 = -3$, and so on.

Alternate Example 2

Use the horizontal-line test to determine whether each of the following functions has an inverse function.

A. $f(x) = 3x + 1$ [**The inverse function exists because any horizontal line intersects the graph of f in exactly one place.**]

B. $g(x) = x^2 - 3$ [**The inverse function does not exist because a horizontal line can cross the graph of g in 2 places.**]

C. $h(x) = \dfrac{x}{x+1}$ [**The inverse function exists because any horizontal line intersects the graph of h in exactly one place.**]

A *ongoing* **SSESSMENT**

Try This

A. The inverse function does not exist, since the line $y = 1$ intersects the graph in more than one place.

B. The inverse function exists.

C. The inverse function exists.

Alternate Example 3

Let $f(x) = -3 + 4x$.
A. Find the inverse of f.
    $[f^{-1}(x) = \frac{1}{4}x + \frac{3}{4}]$
B. Graph f and f^{-1} on the same coordinate plane with the line $y = x$. Use the trace feature to find $f(1)$ and $f^{-1}(1)$.
    $[f(1) = 1, f^{-1}(1) = 1]$
C. If (a, b) is a point on the graph of f, what is $f(a)$ and $f^{-1}(b)$?
    $[f(a) = b \text{ and } f^{-1}(b) = a)]$

CRITICAL *Thinking*

If the graphs of f and f^{-1} intersect at a point (c, d), then $c = d$. This is due to the fact that the point (c, d) lies on the line $y = x$.

Alternate Example 4

Find the inverse function of $K = C + 273$, where K represents kelvins, the temperature of a substance on the absolute temperature scale, and C represents degrees Celsius. Use the inverse function to find 4 K in degrees Celsius.
$[C = K - 273; 4 \text{ K is equivalent to } -269°C.]$

Finding the Inverse Function

To find the inverse function *when it exists*, switch the dependent and the independent variables of the function.

EXAMPLE 3

Graphics Calculator

Let $f(x) = 2x + 5$.

A Find the inverse of f.

B Graph f and f^{-1} on the same coordinate plane with the line $y = x$. Use the trace feature to find $f(-3)$ and $f^{-1}(-1)$.

C If (a, b) is a point on the graph of f, what is $f(a)$ and $f^{-1}(b)$?

Solution▸

A • Replace $f(x)$ by y. $y = 2x + 5$
   • Interchange x and y. $x = 2y + 5$
   • Solve for y in terms of x. $y = \frac{x-5}{2}$
   • Replace y with $f^{-1}(x)$. $f^{-1}(x) = \frac{x-5}{2}$

B
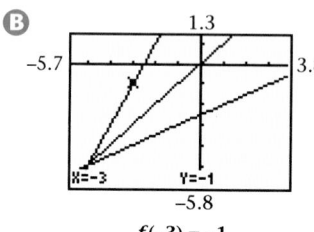

$f(-3) = -1$

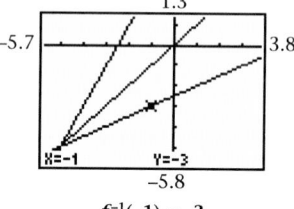

$f^{-1}(-1) = -3$

C Part **B** shows that if (a, b) is a point on the graph of f, then $f(a) = b$ and $f^{-1}(b) = a$. ❖

CRITICAL *Thinking*

Suppose f and f^{-1} intersect at $P(c, d)$. What can you say about c and d?

EXAMPLE 4

Temperature

Find the inverse function of $F = \frac{9}{5}C + 32$. Use the inverse function to find 77°F in degrees Celsius.

Solution▸

To find the inverse function, solve for C in terms of F.

$$\frac{9}{5}C + 32 = F$$
$$\frac{9}{5}C = F - 32$$
$$C = \frac{5}{9}(F - 32)$$

At 77°F, $C = \frac{5}{9}(77 - 32) = 25$. Thus, 77°F is equivalent to 25°C. ❖

EXERCISES & PROBLEMS

Communicate

1. Describe the steps you take to find the inverse of a function.
2. Explain why $f(x) = x^2$ has an inverse relation but not an inverse function.
3. Describe the horizontal-line test, and explain why it works.
4. Compare the terms *inverse function* and *inverse of a function*. How are they alike, and how are they different?
5. Explain how to determine from their graphs when two functions are inverses.

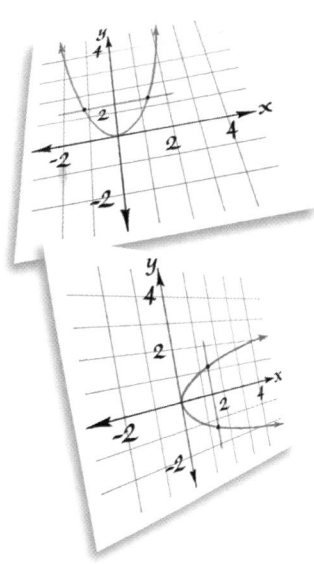

Practice & Apply

Find the inverse of each function. Plot each function and its inverse on the same coordinate plane. Determine which inverses are functions.

6. {(2, 1), (4, 2), (6, 3), (8, 4)}

7. {(−3, −6), (−1, 2), (1, 2), (3, 6)}

8. {(5, 2), (4, 3), (3, 4), (2, 5)}

9. {(9, −2), (4, −1), (1, 0), (3, 1), (7, 2)}

Use your graphics calculator to determine if f and g are inverse functions.

10. $f(x) = \frac{x+8}{3}$ and $g(x) = 3x - 8$ yes

11. $f(x) = 5x + 1$ and $g(x) = \frac{x-1}{5}$ yes

12. $f(x) = -\frac{1}{2}x + 3$ and $g(x) = -2x + 6$ yes

13. $f(x) = \frac{1}{x}$ and $g(x) = \frac{1}{x}$ yes

14. $f(x) = \frac{x-1}{4}$ and $g(x) = 4x - 1$ no

15. $f(x) = \frac{x-3}{2}$ and $g(x) = 2x + 3$ yes

Find the inverse of each function. Check your result using your graphics calculator.

16. $f(x) = 8x - 1$
$f^{-1}(x) = \frac{x+1}{8}$

17. $f(x) = x + \frac{3}{5}$
$f^{-1}(x) = x - \frac{3}{5}$

18. $f(x) = \frac{2x-3}{4}$
$f^{-1}(x) = \frac{4x+3}{2}$

19. $f(x) = \frac{x}{3} + 7$
$f^{-1}(x) = 3x - 21$

Determine whether each function has an inverse function. Draw a rough sketch of the inverse function if it exists.

20.

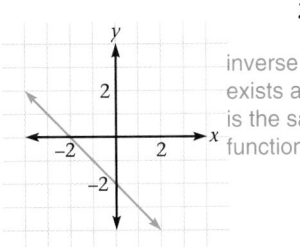

inverse exists and is the same function

21.

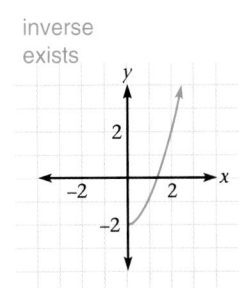

no inverse function

22.

inverse exists

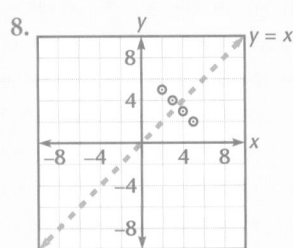

6.

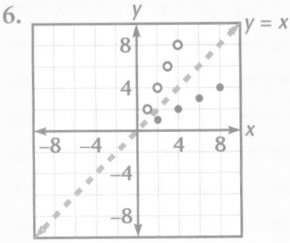

{(1, 2),(2, 4),(3, 6),(4, 8)}; function

7.

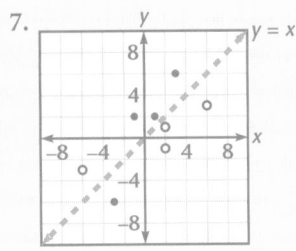

{(−6, −3),(2, −1),(2, 1),(6, 3)}; not a function

ASSESS

Selected Answers
Odd-numbered problems 7–59.

Assignment Guide
Core 1–8, 10–14, 16–18, 20–28, 30–61

Core Plus 1–5, 8–9, 12–15, 17–23, 26–61

Technology
A graphics calculator is needed for Exercises 10–19, 30, 31, 33, 60, and 61.

Error Analysis
Be sure students use the horizontal-line test not the vertical line test on the original relation to determine whether its inverse is a function.

8.

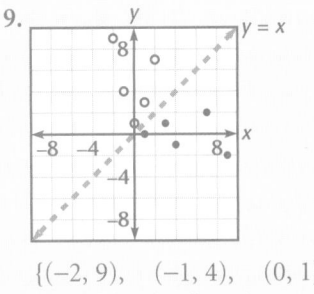

{(2, 5), (3, 4), (4, 3), (5, 2)}; function

9.

{(−2, 9), (−1, 4), (0, 1), (1, 3), (2, 7)}; function

The answers to Exercises 20 and 22 can be found in Additional Answers beginning on page 842.

Performance Assessment

Ask each student to write a function of his or her own. Then have students exchange functions and graph the function and its inverse on the same coordinate plane.

23. One possible graph is given. Since a horizontal line can be drawn that intersects the graph at more than one point, the function is not invertible.

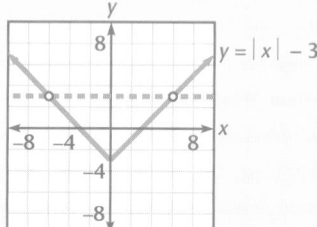

$y = |x| - 3$

NAME _____ CLASS _____ DATE _____

Practice & Apply
3.2 Inverse Functions

Find the inverse of each function. Identify which inverses are functions.

1. {(4, 3), (2, 2), (0, 1), (−2, 0)} _____

2. {(−3, −5), (−2, 1), (1, 3), (2, 6)} _____

3. {(6, −2), (5, −1), (3, 0), (5, −2)} _____

4. {(6, 8), (4, 6), (2, 3), (1, 5)} _____

Use your graphics calculator to determine if f and g are inverse functions.

5. $f(x) = \frac{2x+1}{3}$ and $g(x) = \frac{3x-1}{2}$ _____

6. $f(x) = 5x − 1$ and $g(x) = \frac{x+1}{5}$ _____

7. $f(x) = x + 1$ and $g(x) = \frac{1}{x+1}$ _____

8. $f(x) = \frac{x}{3} − 2$ and $g(x) = 3x + 2$ _____

Find the inverse of each function.

9. $f(x) = 7x + 4$ _____ 10. $f(x) = \frac{x}{5} − 6$ _____ 11. $f(x) = \frac{3x-1}{2}$ _____

In Exercises 12–15, $f(x) = 3x + 4$.

12. What are the slope and y-intercept of f? _____

13. Find the inverse of $f(x) = 3x + 4$. _____

14. What are the slope and the y-intercept of f^{-1}? _____

15. What is the relationship between the slope of f and the slope of f^{-1}? _____

---

23. Sketch a function whose inverse is not a function. Explain how to use the horizontal-line test to show that your function is not invertible.

For Exercises 24–29, find the inverse of f, and graph f and f^{-1} on the same coordinate plane.

24. $f(x) = 5 − x$ $f^{-1}(x) = 5 − x$

25. $f(x) = \frac{x-3}{8}$ $f^{-1}(x) = 8x + 3$

26. $f(x) = −2x + 1$ $f^{-1}(x) = \frac{1-x}{2}$

27. $f(x) = \frac{5-x}{7}$ $f^{-1}(x) = 5 − 7x$

28. $f(x) = \frac{2}{x} − 7$ $f^{-1}(x) = \frac{2}{x+7}$

29. $f(x) = \frac{1}{x}$ $f^{-1}(x) = \frac{1}{x}$

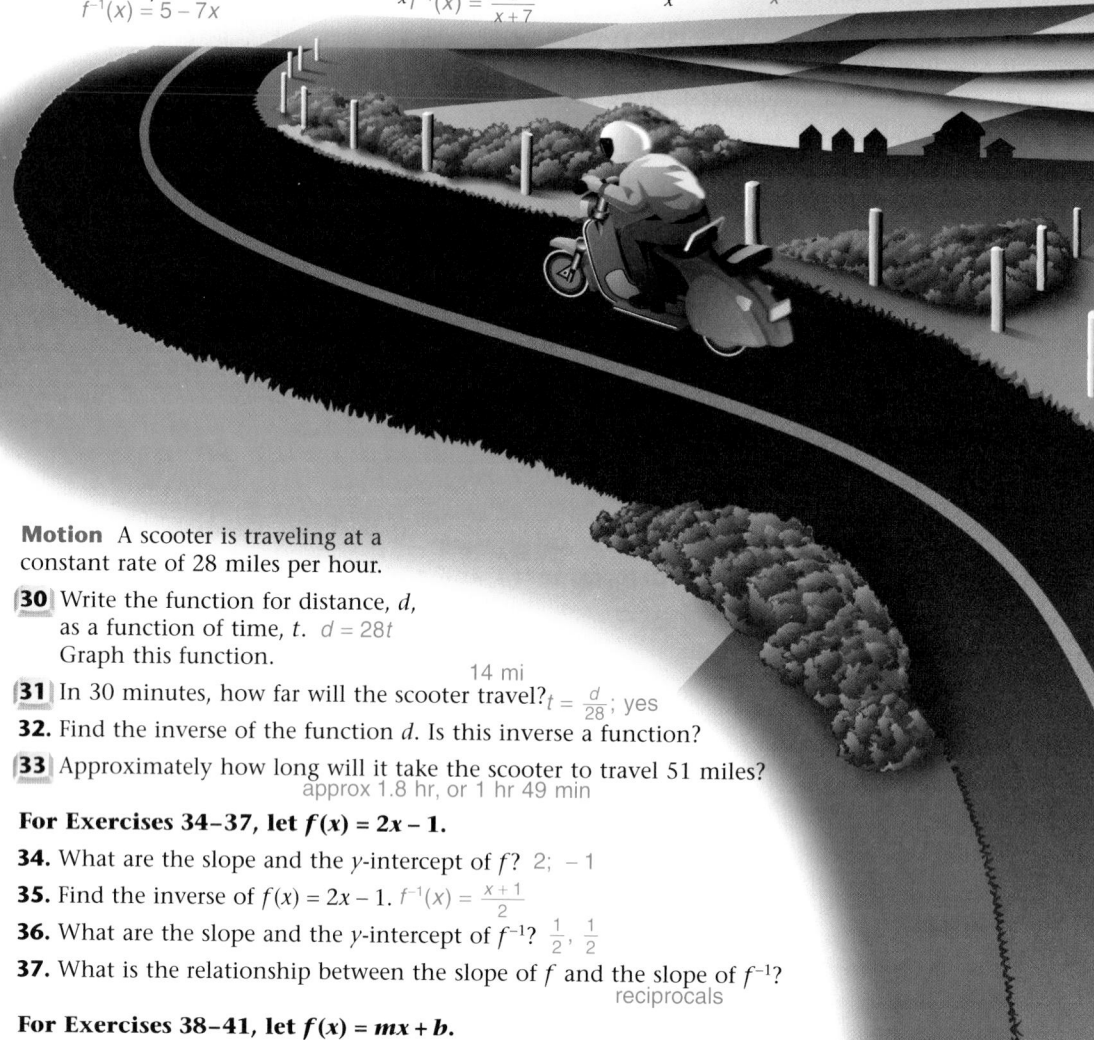

Motion A scooter is traveling at a constant rate of 28 miles per hour.

30 Write the function for distance, d, as a function of time, t. $d = 28t$
Graph this function.

31 In 30 minutes, how far will the scooter travel?   14 mi

32. Find the inverse of the function d. Is this inverse a function? $t = \frac{d}{28}$; yes

33 Approximately how long will it take the scooter to travel 51 miles?   approx 1.8 hr, or 1 hr 49 min

For Exercises 34–37, let $f(x) = 2x − 1$.

34. What are the slope and the y-intercept of f?   2; −1

35. Find the inverse of $f(x) = 2x − 1$. $f^{-1}(x) = \frac{x+1}{2}$

36. What are the slope and the y-intercept of f^{-1}? $\frac{1}{2}, \frac{1}{2}$

37. What is the relationship between the slope of f and the slope of f^{-1}?   reciprocals

For Exercises 38–41, let $f(x) = mx + b$.

38. What are the slope and y-intercept of f?   m, b

39. Find the inverse of f. $f^{-1}(x) = \frac{x-b}{m}$

40. What are the slope and y-intercept of f^{-1}? $\frac{1}{m}$; $-\frac{b}{m}$

41. What is the relationship between the slope of f and the slope of f^{-1}? Can the graph of a linear function and its inverse ever be perpendicular? Explain.

The answers for Exercises 24−29 can be found in Additional Answers beginning on page 842.

41. They are reciprocals. The graph of a linear function and its inverse can never be perpendicular since two lines are perpendicular only when their slopes are negative reciprocals of each other. The slope of any line perpendicular to f is $-\frac{1}{m}$.

For Exercises 42–46, let $f(x) = 3$.

42. Graph f.

43. Use the graph of f to construct the graph of its inverse.

44. Write the equation representing the inverse of f. $x = 3$

45. Is the inverse of f a function? Use the horizontal-line test to explain your answer. no

46. Does the constant function, $f(x) = c$, where c is any real number, have an inverse function? Why or why not? no

Economics Rick works at a factory where he is paid $9.50 per hour plus $0.75 per puzzle.

47. Write a function for Rick's daily 8-hour wage with respect to the number of puzzles produced. $f(x) = 9.5(8) + 0.75x$

48. How much will Rick earn in one 8-hour day if he produces an average of 12 puzzles per hour? $148

49. What is the inverse of Rick's daily 8-hour wage function?

50. What does each variable in the inverse function represent?

51. If Rick made $130.00 in one 8-hour day, how many puzzles did he average each hour of that day? 9

Look Back

Write the linear function described.
[Lessons 1.2, 2.5]

52. Slope 4 and y-intercept -2 $f(x) = 4x - 2$

53. Contains the point $(-1, -3)$ and b is 2 $f(x) = 5x + 2$

54. Contains the points $(0, 4)$ and $(-1, 2)$ $f(x) = 2x + 4$

55. Slope $\frac{1}{4}$ and contains the point $(1, 5)$ $f(x) = \frac{1}{4}x + \frac{19}{4}$

Simplify, and write with only positive exponents.
[Lesson 2.2]

56. $(5^2)^{-3}$ $\frac{1}{5^6}$ **57.** $(2^3)^2$ 2^6 **58.** $(ab^2)^{-1}(b^3)$ $\frac{b}{a}$ **59.** $\left(\frac{a^{-2}b^3}{a^4b^{-3}}\right)^4$ $\frac{b^{24}}{a^{24}}$

Look Beyond

Use your graphics calculator to graph $f(x) = |x|$, $(y = \boxed{\text{ABS}}\, x)$.

60 What are the domain and range of this function? R; $y \geq 0$

61 Does f have an inverse that is a function? Justify your response. no

---

49. $f^{-1}(x) = \frac{x - 9.5(8)}{0.75}$ assuming x is the number of puzzles produced each 8-hour day.

50. In the inverse function, x represents the wages per 8-hour shift, and $f^{-1}(x)$ represents the number of puzzles made per 8-hour shift.

61. By the horizontal-line test, the inverse of f is not a function, since any horizontal line $y = b$ where $b > 0$ intersects the graph of f at more than one point.

Look Beyond

After students graph $y = |x|$ on their graphics calculator, have them transfer the graph to paper and draw the inverse. Looking at the graph of the inverse may help students in answering Exercise 61.

42–43.

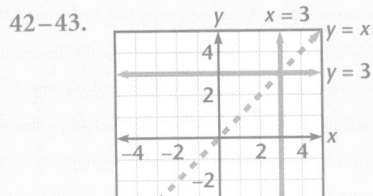

45. No. The inverse cannot be a function by the horizontal-line test, since the line $y = 3$ intersects f in more than one point.

46. No. By the horizontal-line test, the constant function $f(x) = c$ never has an inverse function, since the horizontal line $y = c$ always intersects f in more than one point.

Objectives

- Define the composition of functions, and describe the relationship between the dependent and independent variables.
- Identify special properties of composition, and use these properties to analyze functions.
- Develop a composition test to determine whether two functions are inverses.

RESOURCES

- Practice Master   3.3
- Enrichment Master   3.3
- Technology Master   3.3
- Lesson Activity Master   3.3
- Quiz   3.3
- Spanish Resources   3.3

Assessing Prior Knowledge

Evaluate each function for $x = -1$, $x = 2$, and $x = 0$.

1. $g(x) = x^2 - 2x$ $[g(-1) = 3,$ $g(2) = 0, g(0) = 0]$

2. $h(x) = \frac{x+1}{2}$ $[h(-1) = 0,$ $h(2) = 1.5, h(0) = 0.5]$

TEACH

 Many examples exist of events in which the outcome of one event affects the options available for a following event. Ask students to name several such events.

LESSON 3.3 Composition of Functions

 Life is composed of a series of events and decisions. The outcome of one event can affect the options available for following events. Situations like this can be modeled by a composition of functions.

TODAY ONLY

10% OFF

Every item in the store!

$100 REBATE

Consumer Economics   Electro City is offering a cash rebate of $100 and a 10% discount on the price of a new entertainment center. If you want to buy the entertainment center, which option would result in the lower price: taking the $100 rebate first and then the 10% discount, or taking the 10% discount first and then the $100 rebate?

ALTERNATIVE teaching strategy

Using Manipulatives Write functions on index cards as described below.

For the function $f(x) = 2x - 3$, write "$f(x)$" on one side of a card and "$2x - 3$" on the other side. For the function $g(x) = 4x + 3$, write "$g(x)$" on one side of a card and "$4x + 3$" on the other side. Write $f(\ )$ on the chalkboard and put the $g(x)$ card in the parentheses. Turn the $g(x)$ card over to show that $g(x)$ is replaced with $4x + 3$. Then have students put $4x + 3$ in place of x in $f(x)$ and simplify. Repeat the activity with $g(\ )$.

Let x represent the price of the entertainment center. There are two functions to consider, a discount function and a rebate function.

Let d represent the discount function. A 10% discount on the price is the same as paying 90% of the price.
$$d(x) = 0.9x$$

Let r represent the rebate function, $100 subtracted from the price.
$$r(x) = x - 100$$

When the discount function is used first:

$$\begin{aligned} r(d(x)) &= r(0.9x) \\ &= (0.9x) - 100 \\ &= 0.9x - 100 \end{aligned}$$

When the rebate function is used first:

$$\begin{aligned} d(r(x)) &= d(x - 100) \\ &= 0.9(x - 100) \\ &= 0.9x - 90 \end{aligned}$$

When you compare the functions, you can see that taking the discount *first* results in a lower final sales price.

The operation that combines functions by substituting one function *into* another function is called the *composition of functions*.

EXAMPLE 1
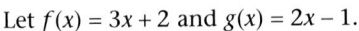

Graphics Calculator

Let $f(x) = 3x + 2$ and $g(x) = 2x - 1$.

Ⓐ Find $f(g(x))$. **Ⓑ** Graph $f(g(x))$. **Ⓒ** Find $f(g(1))$.

Solution▸

Ⓐ • Write the function f.
• Substitute $g(x)$ for every x in the function f.
• Replace each $g(x)$ with $2x - 1$.
• Simplify.

$f(x) = 3x + 2$

$\begin{aligned} f(g(x)) &= 3[g(x)] + 2 \\ &= 3(2x - 1) + 2 \\ &= 6x - 1 \end{aligned}$

Ⓑ Enter f as Y_1 and g as Y_2 in your graphics calculator. Let Y_3 represent $f(g(x))$.

$$f(g(x)) = 3[g(x)] + 2$$
$$Y_3 = 3Y_2 + 2$$

Graph Y_3.

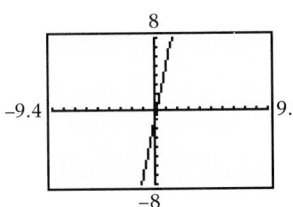

Ⓒ Evaluate $f(g(x))$ when x is 1.

$$\begin{aligned} f(g(x)) &= 6x - 1 \\ f(g(1)) &= 6(1) - 1 \\ &= 5 \end{aligned}$$

Check with your graphics calculator.

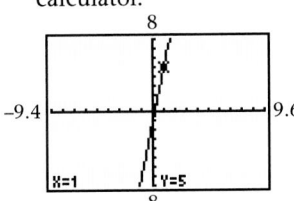

Thus, $f(g(1))$ is 5. ❖

Try This Let $f(x) = 2x$ and $g(x) = -4x + 1$.

Ⓐ Find $g(f(x))$. **Ⓑ** Graph $g(f(x))$. **Ⓒ** Find $g(f(-3))$.

---

interdisciplinary

CONNECTION

Biology In the food chain, barracuda eat bass and bass eat shrimp. Suppose the function $b(x) = 1000 + \sqrt{2x}$, where x is the number of bass, can be used to estimate the number of barracuda. Also suppose that the function $s(x) = 2500 + \sqrt{x}$, where x is the number of shrimp, can be used to estimate the number of bass. Ask students the following questions.

• What composition of functions can be used to describe the number of barracuda in terms of the number of shrimp?
$[b(s(x)) = 1000 + \sqrt{2(2500 + \sqrt{x})}]$

• Suppose there are 500,000 shrimp. How many barracuda are there?
[about 1080 barracuda]

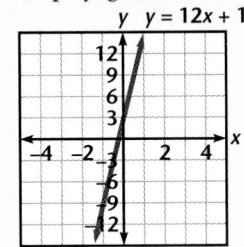 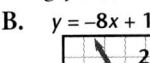 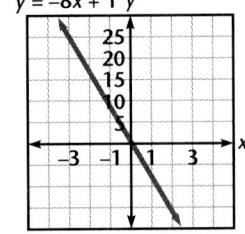

Students may want to make a copy and fill in the column under $g(x) = 2x - 1$ to show the substitutions. For example, for the -2 row, write $2(-2) - 1$. Then simplify this expression to get -5. The same process can be used for the column under $f(x) = 3x + 2$.

Exploration 1 Notes

You may want to have students complete several compositions of functions before they attempt to answer Step 4. Students can either create their own functions or use the functions you give them.

Ongoing ASSESSMENT

4. **The composition of functions is not commutative. For most functions, the result will be different depending on the order in which the functions are composed.**

Use Transparency ▶ 12

When finding the composition $f(g(x))$, the range values of g become the domain values of f.

x = input for g	$g(x) = 2x - 1$	$g(x)$ = input for f	$f(x) = 3x + 2$	$f(g(x))$
-2	$\longrightarrow$	-5	$\longrightarrow$	-13
-1	$\longrightarrow$	-3	$\longrightarrow$	-7
0	$\longrightarrow$	-1	$\longrightarrow$	-1
1	$\longrightarrow$	1	$\longrightarrow$	5
2	$\longrightarrow$	3	$\longrightarrow$	11

When f and g are both nonconstant linear functions, the domain and range of each function is the set of all real numbers, and $f(g(x))$ is always defined.

COMPOSITION OF FUNCTIONS

The **composition of functions** f and g, $(f \circ g)(x)$, is defined as $(f(g(x))$. The domain of f must include the range of g.

•Exploration 1 The Commutative Property

Let $f(x) = -x$, $g(x) = 0.5x^2$, and $h(x) = \frac{1-x}{2}$.

1 Find the compositions $f \circ h$ and $h \circ f$. Compare these composite functions. How are they the same? How are they different?

2 Find the compositions $g \circ h$ and $h \circ g$. Compare these composite functions. How are they the same? How are they different?

3 Create your own function. Call your function y. Find the compositions $y \circ h$ and $h \circ y$. Are these compositions equivalent?

4 Do you think the composition of functions is commutative? Why or why not? ❖

ENRICHMENT Have students explore the associativity of the composition of functions. For example, for functions f, g, and h, find $(f \circ g) \circ h$ and $f \circ (g \circ h)$. Ask students to compare these composite functions and state how they are the same, how they are different, and whether they are equivalent. Have students complete enough examples to determine whether the composition of functions is associative and why or why not. **[composition is associative]**

INCLUSION strategies **Using Visual Models** Allow students to use the concept of the function machine as presented at the top of page 130 until they understand the concept. Encourage students to draw their own machines or provide handouts containing the machines. Simulations of the function machine can be done using calculators. The resulting value from the first calculator becomes the input for a second calculator.

Exploration 2 *The Inverse Property*

Graphics Calculator

Let $f(x) = x + 3$, $h(x) = \frac{1-x}{2}$, and $I(x) = x$.

1 Find the inverse functions for f, h and I.

2 Graph f and f^{-1} on the same coordinate plane. What is the equation of the axis of symmetry of the two graphs?

Find $f \circ f^{-1}$ and $f^{-1} \circ f$. How do $f \circ f^{-1}$ and $f^{-1} \circ f$ compare with the equation of their axis of symmetry?

3 Graph h and h^{-1} on the same coordinate plane. What is the equation of the axis of symmetry of the two graphs?

Find $h \circ h^{-1}$ and $h^{-1} \circ h$. How do $h \circ h^{-1}$ and $h^{-1} \circ h$ compare with the equation of their axis of symmetry?

4 Graph I. How does this graph compare with the graphs of the axes of symmetry you found in Steps 2 and 3?

5 What do you think is the composition of any function and its inverse? ❖

CRITICAL Thinking Is the composition of inverse functions commutative? Explain.

EXAMPLE 2

Medicine The volume of a spherical cancer tumor is related to the diameter of the tumor, d, by the function $V(d) = \frac{\pi d^3}{6} + 1.5$, where d is in millimeters. The function that relates the diameter of the tumor to the time, t (in days), after the detection of the tumor is $d(t) = 0.0006t^2 + 0.02t + 0.005$. Use your graphics calculator to find the volume of a spherical cancer tumor 50 days after detection.

Solution▸

The volume, V, depends on diameter, d, which depends on time, t.

Therefore, graph $V(d(t))$.

First enter function d as Y_1.

Then graph V as $Y_2 = \frac{\pi (Y_1)^3}{6} + 1.5$.

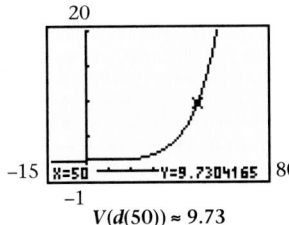
$V(d(50)) \approx 9.73$

Thus, 50 days after detection, the spherical cancer tumor has a volume of approximately 9.73 cubic millimeters. ❖

In the composite function $V(d(t))$, what is the independent variable, and what is the dependent variable?

CRITICAL Thinking Would the composite function $V(d(t))$ be defined if the range of $d(t)$ included negative numbers? Explain.

RETEACHING the lesson

Using Patterns Present several compositions with numbers instead of variables. For example, use $f(g(2))$ instead of $f(g(x))$. Remind students that when evaluating an algebraic expression, the first part to be evaluated is the part *inside* the parentheses. Apply this concept to the composition of functions by instructing students to start with the function inside the parentheses, evaluate it, and then substitute this value in the function outside of the parentheses.

Exploration 2 Notes

You may want to have students complete several compositions of a function with its inverse before they attempt to answer Step 5. Students can either create their own functions or use functions you give them.

Aongoing **SSESSMENT**

5. The composition of any function and its inverse is the identity function, $y = x$.

CRITICAL Thinking

Yes, the composition of inverse functions is commutative because the result is always the identity function, $y = x$.

Alternate Example 2

Use your graphics calculator to find the volume of a spherical cancer 100 days after detection. [approximately 270.09 cubic millimeters]

Aongoing **SSESSMENT**

The independent variable is $d(t)$ and the dependent variable is $V(d(t))$.

CRITICAL Thinking

Yes, the composition function $V(d(t))$ would be defined, but the values would not be realistic for negative values of the independent variable, $d(t)$, the diameter.

Assignment Guide

Core 1–15, 18–24, 26–59

Core Plus 1–9, 12–19, 22–60

Technology

A graphics calculator is needed for Exercises 8, 9, 19, 30–32, and 60.

Error Analysis

Watch for students who do not use the functions in the correct order. Remind students that $f(g(x))$ does not, in general, equal $g(f(x))$.

EXERCISES & PROBLEMS

Communicate

1. Explain the process of composing two functions.

For Exercises 2–5, let $f(x) = x^2$ and $g(x) = x + 2$.

2. What are the domain and range of f and g?

3. Find $f \circ g$ and $g \circ f$.

4. Find the domain and range of $f \circ g$ and $g \circ f$.

5. Explain why the domain of f must include the range of g in order to find $f \circ g$.

6. What results from the composition of any invertible function with its inverse function?

7. Explain whether or not composition is a commutative operation.

Practice & Apply

For Exercises 8–17, let $f(x) = \frac{-3x+2}{5}$ and $g(x) = 4x - 1$.

8. Graph f. Find the domain and range of f. R; R

9. Graph g. Find the domain and range of g. R; R

10. Find $f \circ g$. $\frac{-12x+5}{5}$ 11. Find $g \circ f$. $\frac{-12x+3}{5}$ 12. Find $f \circ f$. $\frac{9x+4}{25}$ 13. Find $g \circ g$. $16x - 5$

14. Find $(f \circ g)(4)$. $-\frac{43}{5}$ 15. Find $(g \circ g)(-1)$. -21 16. Find $(g \circ f)(0)$. $\frac{3}{5}$ 17. Find $(f \circ f)(1)$. $\frac{13}{25}$

Time Management Gene believes that the function $C(x) = 0.2x^2 + x + 850$ closely approximates the cost of a daily production run of x picture frames. She knows that the number of picture frames produced depends on the time, t, in hours since the beginning of a shift and is given by the function $x(t) = 90t$.

18. Write the function that gives cost as a function of time. $C(t) = 1620t^2 + 90t + 850$

19. Graph $C(t)$ on your graphics calculator. Find the approximate cost, C, of producing x picture frames 5 hours after the beginning of a shift. $41,800

For Exercises 20–25, determine whether the composition of the given pairs of functions is commutative.

20. $f(x) = \frac{x+3}{2}$ and $g(x) = 2x - 3$ yes   21. $f(x) = \frac{1}{2}x - 1$ and $g(x) = 2x + 2$ yes

22. $f(x) = -\frac{1}{2}x$ and $g(x) = -2x$ yes   23. $f(x) = -\frac{1}{x}$ and $g(x) = -\frac{1}{x}$ yes

24. $f(x) = \frac{x+1}{3}$ and $g(x) = 3x + 1$ no   25. $f(x) = 8x^3$ and $g(x) = \frac{\sqrt[3]{x}}{2}$ yes

---

NAME _____ CLASS _____ DATE _____

Practice & Apply
3.3 Composition of Functions

For Exercises 1–4, let $f(x) = x - 1$ and $g(x) = 3x + 2$.

1. Find the domain and range of f and g. _____

2. Find $f \circ g$ and $g \circ f$. _____

3. Is composition a commutative operation? _____

4. What are the domain and range of $f \circ g$ and $g \circ f$? _____

For Exercises 5–12, let $f(x) = 2x + 3$ and $g(x) = x - \frac{1}{2}$. Find each value.

5. $f \circ g$ _____   6. $g \circ f$ _____   7. $f \circ f$ _____   8. $g \circ g$ _____

9. $(f \circ g)(-3)$ _____   10. $(g \circ g)(2)$ _____

11. $(g \circ f)(0)$ _____   12. $(f \circ f)(1)$ _____

13. The circumference C of a circle and the diameter d are given by $C = f(d) = \pi d$ and $d = g(r) = 2r$. Find $(f \circ g)(r)$. Interpret the result. _____

For Exercises 14–22, let $f(x) = x^2 - 2$, $g(x) = -3x + 1$, and $h(x) = 2x$. Find each value.

14. $f \circ g$ _____   15. $g \circ f$ _____   16. $f \circ h$ _____   17. $h \circ g$ _____

18. $f \circ (g \circ h)$ _____   19. $(f \circ h)(2)$ _____

20. $(h \circ g)(0)$ _____   21. $(g \circ f)(-1)$ _____

Write $h(x)$ as the composition of two functions f and g for which $(f \circ g)(x) = h(x)$.

22. $h(x) = (x + 5)^2$ _____   23. $h(x) = 3x - 4$ _____

Let $f(x) = x - 1$ and $g(x) = 2x + 4$. Jim found $f(g(x))$ as $2x + 3$. Jim then evaluated $f(g(x))$ for $x = 2$. Megan evaluated $g(x)$ for $x = 2$ and then used the result to evaluate $f(x)$.

24. Find $f(g(2))$ using Jim's method. _____

25. Find $f(g(2))$ using Megan's method. _____

26. Explain the result. _____

Portfolio Assessment

Have students write an explanation of the composition of functions. Be sure they include information about the commutativity of compositions and composition of inverses. Ask students to include examples with their explanations.

Temperature The temperature in degrees Celsius is $\frac{5}{9}$ times the difference of the Fahrenheit temperature minus 32.

26. Write a function expressing kelvins in terms of degrees Fahrenheit. $K(F) = \frac{5}{9}(F - 32) + 273$

27. Are kelvins and degrees Fahrenheit linearly related? yes

28. Water freezes at 32°F. At what temperature does water freeze in kelvins? 273 kelvins

29. Absolute zero, defined as 0 kelvins, is the temperature at which all molecular motion vanishes and no heat energy may be extracted from an object. At what temperature is absolute zero when given in degrees Fahrenheit? −459.4°F

For Exercises 30–40, let $f(x) = 2x$, $g(x) = x^2 + 2$, and $h(x) = -4x + 3$.

30 Use your graphics calculator to find the domain and range of f. R, R

31 Use your graphics calculator to find the domain and range of g. R, $y \geq 2$

32 Use your graphics calculator to find the domain and range of h. R, R

33. Find $f \circ g$. $2x^2 + 4$ **34.** Find $g \circ f$. $4x^2 + 2$ **35.** Find $f \circ h$. $-8x + 6$ **36.** Find $h \circ g$. $-4x^2 - 5$

37. Find $f \circ f$. $4x$ **38.** Find $(h \circ f)(-2)$. 19 **39.** Find $(g \circ h)(1)$. 3 **40.** Find $(f \circ g)(0)$. 4

41. Let $h(x) = x^2 - 9$. Find two functions f and g for which $(f \circ g)(x) = h(x)$. Answers may vary. Sample: $f(x) = x - 9$; $g(x) = x^2$

42.

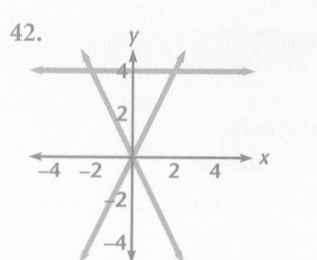

45.

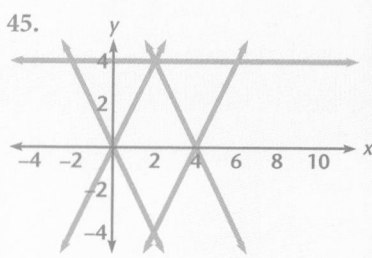

A pattern of alternating triangles is formed similar to the pattern shown in the book.

47.

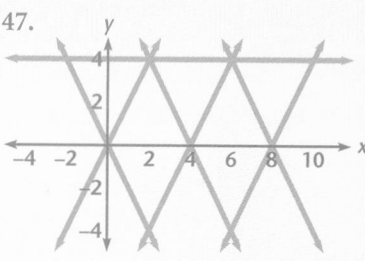

Look Beyond

This exercise provides students with experience graphing and finding the domain and range of an exponential function. Such functions are discussed in Chapter 7.

Cultural Connection: Africa The Nabdam people of West Africa, south of the Niger river, used simple geometric transformations to create strip patterns. A pattern of triangles can be formed in the strip between $y = 0$ and $y = 4$ by continued compositions of $f(x) = 2x$, $g(x) = -2x$, and $h(x) = x - 4$.

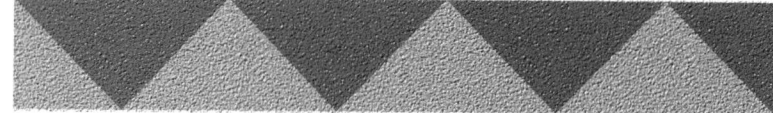

42. Graph f and g on the same coordinate plane. Plot the lines $y = 0$ and $y = 4$ on the same coordinate plane as f and g.

43. Find $f \circ h$. $2x - 8$ **44.** Find $g \circ h$. $-2x + 8$

45. Graph $f \circ h$ and $g \circ h$ on the same coordinate plane as f and g. Describe the pattern formed in the strip between $y = 0$ and $y = 4$.

46. What compositions will give you the next triangles in the strip? Find these compositions. $(f \circ h) \circ h = 2x - 16$; $(g \circ h) \circ h = -2x + 16$

47. Graph the composite functions you found in Exercise 46 on the same coordinate plane as f and g.

48. Describe the pattern of compositions that will complete this strip design. Graph them to check your answer. Explain.

49. Make up your own functions f and g and transformation function h. Between which values of y will your pattern be formed? Describe your strip design. Answers may vary.

Look Back

Cost Analysis The total revenue, R, is directly proportional to the number of cameras, x, sold. When 500 cameras are sold, the revenue is $3800. **[Lesson 1.4]**

50. Find the revenue when 600 cameras are sold. $4560

51. What is the constant of variation? $7\frac{3}{5}$, or 7.6

52. In this situation, what does the constant of variation measure? the revenue from selling one camera

Simplify. **[Lesson 2.2]**

53. $2^3 \cdot 2^5$ 2^8 **54.** $25^{-\frac{1}{2}}$ $\frac{1}{\sqrt{25}} = \frac{1}{5}$ **55.** $\left(3^2\right)^3$ 3^6

Let $f(x) = 3x - 2$ and $g(x) = 5x + 2$. **[Lesson 2.6]**

56. Find $(f + g)(x)$. $8x$ **57.** Find $(f \cdot g)(x)$. $15x^2 - 4x - 4$

58. Find $(f + g)(-1)$. -8 **59.** Find $(f \cdot g)(a)$. $15a^2 - 4a - 4$

Look Beyond

60 Let $f(x) = 2^x$. Use your graphics calculator to graph this function. What viewing window did you use? Determine the domain and range of f.

48. $(((f \circ h) \circ h) \circ h) = 2x - 24$, $(((g \circ h) \circ h) \circ h) = -2x + 24$

60. A viewing window that includes the points $(-4, 0)$ and $(2, 4)$ is sufficient. The domain is the set of all real numbers, and the range is the set of all positive real numbers.

The Absolute Value Function

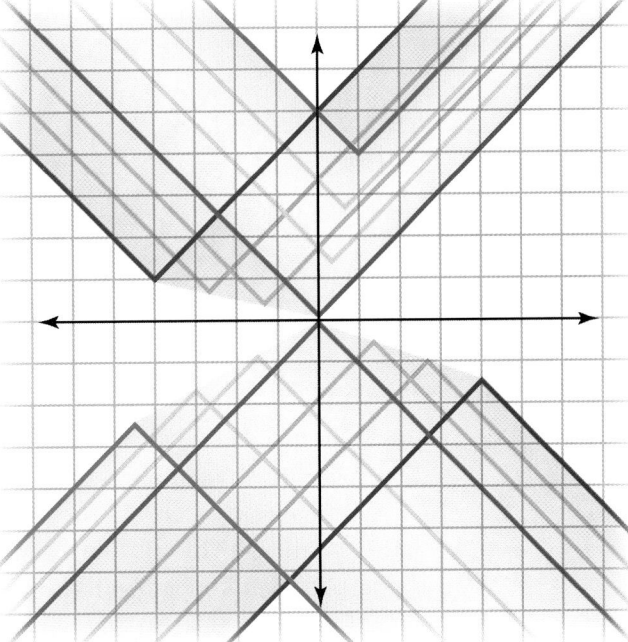

why *All linear functions, $f(x) = mx + b$, have graphs that are straight lines. Changing the values of m and b moves the graph on the coordinate plane, but the graph remains a straight line. There is another function that behaves like a straight line but has a shape like a ∨. You will see how you can make changes to the function that will move the ∨ on the coordinate plane but maintain the basic ∨ shape.*

On the following number line, the coordinate of *A* is –2 and the coordinate of *B* is 2. Although their coordinates are different, their distances from the origin are the same.

$$\begin{array}{c} A \quad\quad O \quad\quad B \\ \xleftarrow{\quad\underset{-4}{|}\;\;\underset{-3}{|}\;\;\underset{-2}{\bullet}\;\;\underset{-1}{|}\;\;\underset{0}{|}\;\;\underset{1}{|}\;\;\underset{2}{\bullet}\;\;\underset{3}{|}\;\;\underset{4}{|}\quad}\xrightarrow{}x \end{array}$$

The distance from the origin to a point *x* units from the origin is called the *absolute value* of *x*.

The **absolute value function** is defined as:

$$f(x) = |x| = \begin{cases} x, & \text{if } x \geq 0 \\ -x, & \text{if } x < 0 \end{cases}$$

Thus, $|2| = 2$ and $|-2| = 2$.

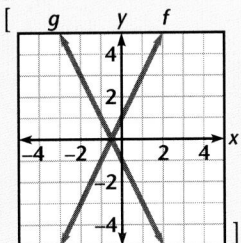

You can graph $f(x) = |x|$ with your graphics calculator. On some calculators, the absolute value feature is a function key. On other calculators, the absolute value feature is in the [MATH] menu. With either type of calculator, set $Y_1 = $ [ABS] (X). The graph of the absolute value function has a basic ∨ shape.

Graphics Calculator

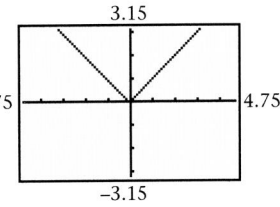

The domain of $f(x) = |x|$ is all real numbers *x*, and the range is all real numbers *y*, such that $y \geq 0$. The axis of symmetry is the *y*-axis.

Describe the inverse of the absolute value function, $f(x) = |x|$.

PREPARE

Objectives

- Define and graph translations, reflections, and scalar transformations of the absolute value function.
- Solve equations involving absolute value symbols by using graphing and algebraic methods.

RESOURCES

• Practice Master	3.4
• Enrichment Master	3.4
• Technology Master	3.4
• Lesson Activity Master	3.4
• Quiz	3.4
• Spanish Resources	3.4

Assessing Prior Knowledge

Graph on the same plane.

1. $f(x) = 2x + 1$

2. $g(x) = -2x - 1$

TEACH

why Have students graph simple linear functions. Ask students to find two linear functions that result in a ∨-shaped graph.

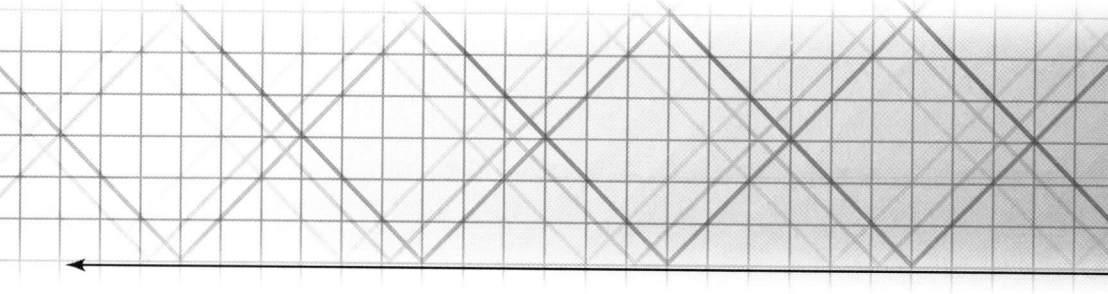

Translations

Absolute value functions can be shifted up or down and can be moved to the left or to the right. These sliding moves are called *translations*.

EXAMPLE 1

Graphics Calculator

Graph the function $g(x) = |x| + 1$.

Ⓐ Explain the relationship between the graph of g and the graph of $f(x) = |x|$.

Ⓑ Solve $|x| + 1 = 7$ by using the trace feature on your calculator.

Ⓒ Solve $|x| + 1 = 7$ by using the definition of absolute value.

Solution➤

Ⓐ The graph of g is the graph of f *translated* one unit up. In general, the graph of $g(x) = |x| + k$ is the graph of f translated k units up when k is positive, and translated k units down when k is negative. The axis of symmetry is $x = 0$, and the y-intercept is $(0, k)$.

Ⓑ
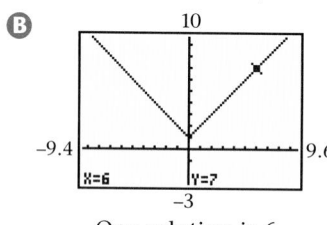
One solution is 6.

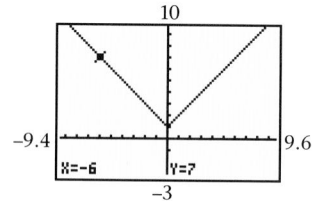
One solution is −6.

Ⓒ Isolate the absolute value symbol on one side of the equation.
Replace the absolute value symbol with ±( ).
Solve the two equations separately.

$$|x| + 1 = 7$$
$$|x| = 6$$
$$\pm(x) = 6$$
$$x = 6 \text{ or } -x = 6$$
$$x = -6$$

Check each value of x in the original equation.

$$|x| + 1 = 7 \qquad |x| + 1 = 7$$
$$|6| + 1 \stackrel{?}{=} 7 \qquad |-6| + 1 \stackrel{?}{=} 7 \qquad \text{Thus, the solutions are}$$
$$6 + 1 = 7 \text{ True} \qquad 6 + 1 = 7 \text{ True} \qquad 6 \text{ and } -6. \; ❖$$

Try This Solve $|2x| - 3 = 4$.

Hands-On Strategies Have students work in small groups. Have each member of each group graph the function $f(x) = |x|$ and the following transformations of the function f. Be sure each student records the function that produces each transformation.

1. translation
2. reflection
3. scalar transformation

Have members of each group compare their graphs and functions in each category and note the similarities and differences. Ask each group to summarize their work by making a general statement about the changes to a basic type of function that will produce each transformation.

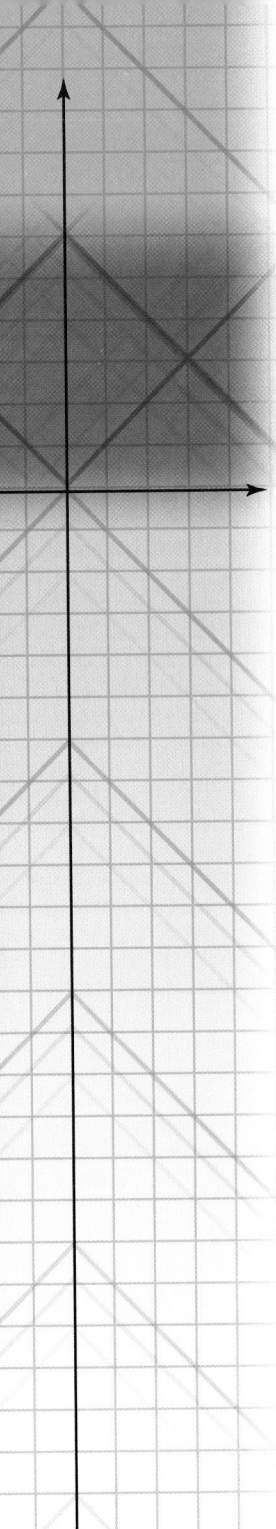

EXAMPLE 2

Graph $j(x) = |x + 1|$.

Ⓐ Explain the relationship between the graph of j and the graph of $f(x) = |x|$.

Ⓑ Solve $|x + 1| = 3$ by using the trace feature on your calculator.

Ⓒ Solve $|x + 1| = 3$ by using the definition of absolute value.

Solution

Ⓐ The graph of j is the graph of f *translated* one unit to the left. In general, the graph of $j(x) = |x - h|$ is the graph of f translated h units to the right when h is positive, and translated h units to the left when h is negative. The axis of symmetry is $x = h$.

Ⓑ
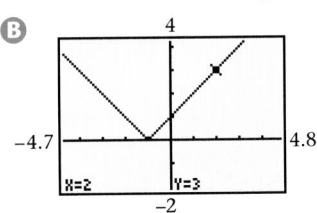
One solution is 2.

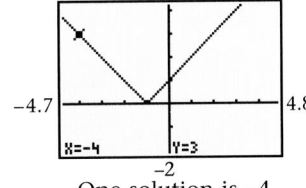
One solution is −4.

Ⓒ Replace the absolute value symbol with ±( ).

$$|x + 1| = 3$$
$$\pm(x + 1) = 3$$

Solve the two equations separately.

$$+(x + 1) = 3 \qquad -(x + 1) = 3$$
$$x + 1 = 3 \qquad -x - 1 = 3$$
$$x = 2 \qquad x = -4$$

Check

Check each value of x using the original equation.

$$|x + 1| = 3 \qquad\qquad |x + 1| = 3$$
$$|(2) + 1| \overset{?}{=} 3 \qquad\qquad |(-4) + 1| \overset{?}{=} 3$$
$$|3| = 3 \text{ True} \qquad\qquad |-3| = 3 \text{ True}$$

Therefore, the solutions are 2 and −4. ❖

Reflections

The ∨ can be *reflected* over a line parallel to the x-axis, making the shape appear inverted, ∧.

EXAMPLE 3

Describe the graph of $j(x) = -|x|$.

Solution

The graph of j is a reflection of the graph of $f(x) = |x|$ with respect to the x-axis. ❖

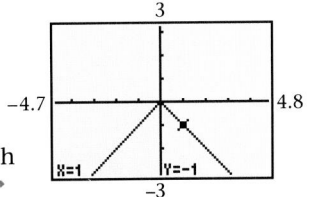

---

interdisciplinary

CONNECTION

Psychology Psychologists use absolute value functions to describe the margin of error of results of their studies. For example, the function $f(x) = |60 - x|$ can be used to describe the margin of error when people are asked to estimate the length of a minute. Have students find other examples of margins of error.

Graph $k(x) = -3|x|$ and $j(x) = -\frac{1}{3}|x|$.

A. Compare the graphs of k and j with the graph of $f(x) = |x|$. [**k is narrower than f and j is wider than f; both functions are reflections of f over the x-axis.**]

B. Describe the function $g(x) = a|x|$ for any negative real number a. Give the axis of symmetry, the domain, and the range of g. [**g multiplies every value of f by a. When a < −1, the angle between the rays is smaller than the angle between the rays in the graph of f. When −1 < a < 0, the angle between the rays is larger than the angle between the rays of the graph of f. The axis of symmetry in both cases is x = 0. The range of g is the set of nonpositive numbers.**]

Aongoing
SSESSMENT

For $a < 0$, the domain of k is the set of all real numbers and the range of k is the set of all negative real numbers and zero.

CRITICAL
Thinking

Given any values for x and y, $|x + y| \leq |x| + |y|$. If either x or y is negative, their sum prior to taking the absolute value will be less than the sum of the individual absolute value of x and of y. This is also known as the Triangle Inequality.

Scalar Transformations

The angle between the rays of both the ∨ and ∧ shapes can be made larger or smaller by applying a *scalar* transformation.

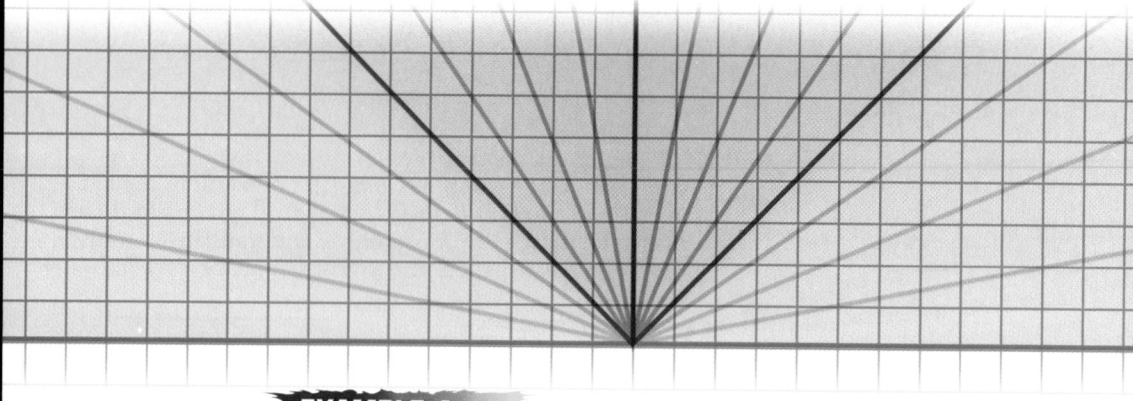

EXAMPLE 4

Graphics Calculator

Graph $g(x) = 2|x|$ and $h(x) = \frac{1}{2}|x|$.

Ⓐ Compare the graphs of g and h with the graph of $f(x) = |x|$.

Ⓑ Describe the function $k(x) = a|x|$ for any positive real number a. Give the axis of symmetry, the domain, and the range of k.

Solution▸

Ⓐ

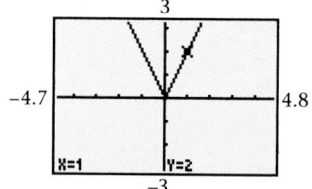

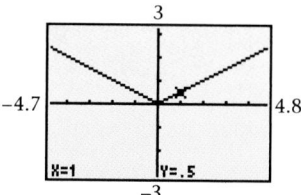

The graph of g is narrower than the graph of f.

The graph of h is wider than the graph of f.

Ⓑ The function k multiplies every value of f by a. When $a > 1$, as in function g, the angle between the rays is smaller than the angle between the rays in the graph of f.

When $0 < a < 1$, as in function h, the angle between the rays is larger than the angle between the rays of the graph of f.

The axis of symmetry in both cases is $x = 0$.

The domain and range of k are the same as the domain and range of f. ❖

In Example 4, Part **Ⓑ**, what would be the domain and range of k if $a < 0$?

CRITICAL
Thinking

Substitute positive and negative values for the variables in the expressions $|x + y|$ and $|x| + |y|$. Which symbol, $\leq$, $<$, $=$, $>$, or $\geq$, can always be placed between these two expressions? Explain.

ENRICHMENT Discuss whether an absolute value function can occur in Quadrants I and IV only or in Quadrants II and III only. Be sure students can explain their answers.

INCLUSION
strategies **Using Visual Models** Use an overhead projector, a transparency with a coordinate plane, and a clear piece of acetate. Draw the graph of the function $f(x) = |x|$ on the piece of acetate and demonstrate several different transformations of this function. Be sure to record the equations of the functions the graphs show. Then ask students to demonstrate transformations of their own on the overhead.

EXERCISES & PROBLEMS

Communicate

1. Explain the idea of the absolute value function.

2. Compare the function $g(x) = |x + c|$, where c is any real number, with the function $f(x) = |x|$. Describe the axis of symmetry, the domain, and the range of g.

3. Compare the function $h(x) = |x| + d$, where d is any real number, with the function $f(x) = |x|$. Describe the axis of symmetry, the domain, and the range of h.

4. Compare the function $k(x) = a|x|$, where a is any positive real number, with the function $f(x) = |x|$. Describe the axis of symmetry, the domain, and the range of k.

5. Compare the function $j(x) = -a|x|$, where a is any positive real number, with the function $f(x) = |x|$. Describe the axis of symmetry, the domain, and the range of j.

6. Do absolute value graphs always form rays that meet at a right angle? If not, give an example.

7. Describe how to solve an absolute value equation by graphing and by using the definition of absolute value.

Practice & Apply

For Exercises 8–13, write the function that is graphed.

8. $f(x) = |x + 3|$

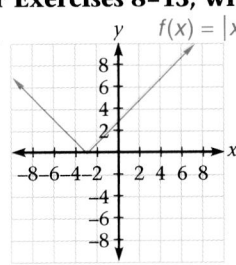

9. $f(x) = |x| - 3$

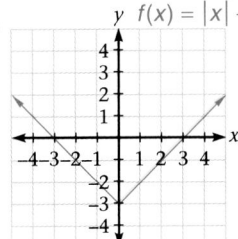

10. $f(x) = |x + 2| - 4$

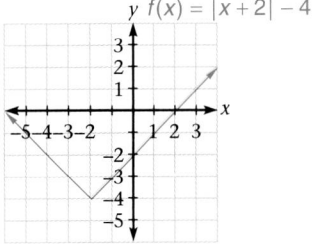

11. $f(x) = -|x + 2|$

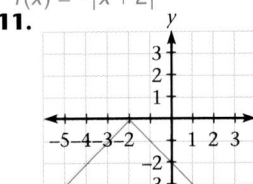

12. $f(x) = 4 - |x - 5|$

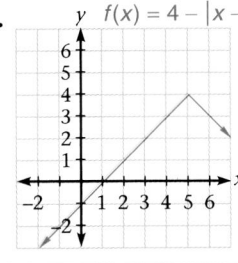

13. $f(x) = 2|x - 3|$

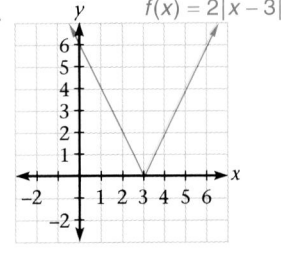

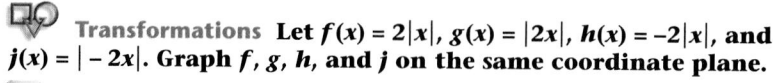

ALTERNATIVE ASSESSMENT

Performance Assessment

Have students write and graph a function for each transformation of $f(x) = |x|$.

1. translation up or down

2. translation right or left

3. reflection over the x-axis

4. scalar transformation

Error Analysis

In Exercises 22 and 23, students must use a square viewing window. For most calculators, the viewing window $-9.4 \le x \le 9.4$ and $0 \le y \le 12.4$ will give the best results.

22. The figure is a square.

30. The graphs of f, g, and j produce the same narrowing scalar transformation of $y = |x|$. The graph of h is a reflection through the x-axis of the graphs of f, g, and j.

Practice Master

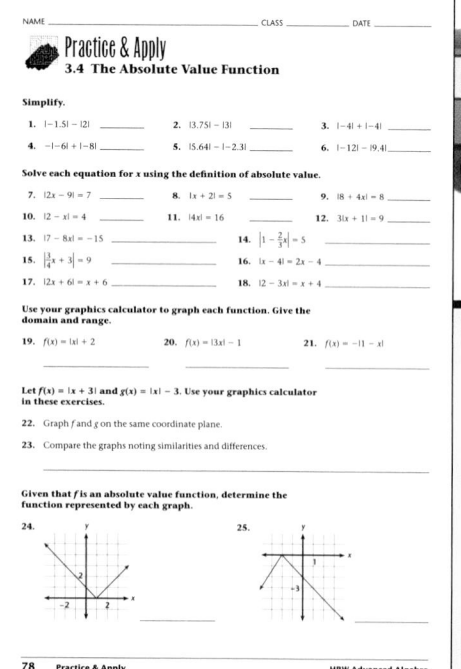

Simplify.

14. $|4| - |-5|$ -1 15. $|-7| - |2.2|$ 4.8 16. $|3| + |-4|$ 7 17. $-|-1| - |1|$ -2

18. $|-1| - |3|$ -2 19. $|3| + |-3|$ 6 20. $|6| - |-2.67|$ 3.33 21. $-|2.75| + |-2.75|$ 0

Coordinate Geometry Let $f(x) = |x|$ and $g(x) = 10 - |x|$. Graph f and g on your graphics calculator.

22. Describe the polygonal figure that is formed.

23. What is the area of this figure? 50 sq units

Solve each equation using the definition of absolute value. Use your graphics calculator to check.

24. $|x| - 5 = 3$ $8; -8$ 25. $|2x| - 1 = 8$ $\frac{9}{2}, -\frac{9}{2}$ 26. $|4 - 5x| = 0$ $\frac{4}{5}$

27. $|2x + 3| = 7$ $2; -5$ 28. $|1 - 3x| = -1$ no solution 29. $|2x - 3| + 2 = x + 5$ $0; 6$

Transformations Let $f(x) = 2|x|$, $g(x) = |2x|$, $h(x) = -2|x|$, and $j(x) = |-2x|$. Graph f, g, h, and j on the same coordinate plane.

30. Compare each graph with the graph of $y = |x|$. How are they alike and how are they different?

31. Make a rule about scalar transformations of the form $t(x) = |ax|$, where a represents *any* real number.

Travel An express train to St. Louis leaves Houston traveling at an average speed of 80 miles per hour. After about 3 hours it passes through Dallas and continues down the track. The distance between the train and Dallas after leaving the station in Houston is given by $d(t) = 80|3 - t|$. Use your graphics calculator to graph d.

32. After 2 hours, how far has the train traveled? 160 mi Where is the train at this time?

33. After 4 hours, how far has the train traveled? Where is the train at this time? 320 mi

34. Explain why this function contains absolute value symbols. Is there a way to write this function without absolute value symbols? Explain your answer using an example.

31. Given $t(x) = |ax|$, then $t(x) = |a||x|$. So, if $a \ge 0$, then $t(x) = a|x|$, and if $a < 0$, then $t(x) = -a|x|$.

32. 160 mi; 80 mi from Dallas, on the Houston side

33. 320 mi; 80 mi from Dallas, on the St. Louis side

34. The function uses absolute value symbols so that the distances from Dallas on the St. Louis side will not be given as negative. Avoid the absolute value symbols by separating the function into two functions: $d(t) = 80(3 - t)$ for $t \le 3$, and $d(t) = -80(3 - t)$ or $80(t - 3)$ for $t > 3$. Then $d(1) = 160$, $d(5) = 160$, and so on, and all of the distances from Dallas will be positive.

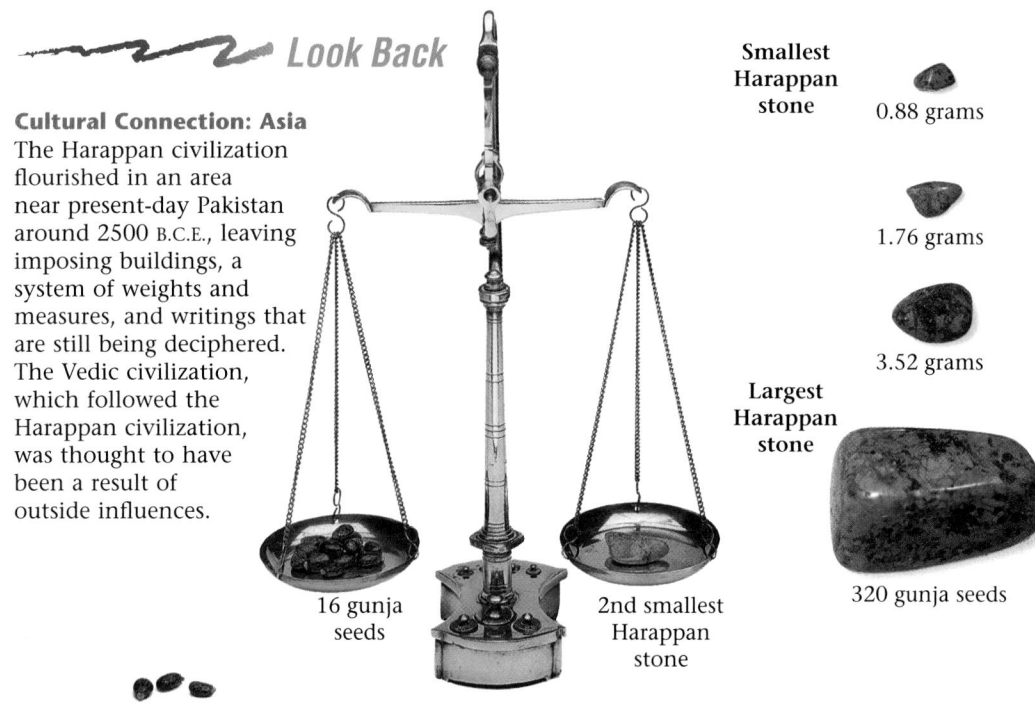

Look Back

Cultural Connection: Asia
The Harappan civilization flourished in an area near present-day Pakistan around 2500 B.C.E., leaving imposing buildings, a system of weights and measures, and writings that are still being deciphered. The Vedic civilization, which followed the Harappan civilization, was thought to have been a result of outside influences.

Smallest Harappan stone — 0.88 grams

1.76 grams

3.52 grams

Largest Harappan stone — 320 gunja seeds

16 gunja seeds

2nd smallest Harappan stone

A set of Harappan balancing stones, recently found by archaeologists, is compared with Vedic gunja seeds. The Vedic civilization used gunja seeds to weigh precious metals. The smallest Harappan stone has the same mass as 8 gunja seeds. The second-smallest Harappan stone has the same mass as 16 gunja seeds.   **[Lesson 2.5]**

35 Write a linear function that relates the mass of the Harappan stones to the number of gunja seeds. Use your graphics calculator to graph the function. $f(x) = 0.11x$

36 How many gunja seeds are equivalent to the third smallest Harappan stone? 32

37 The largest Harappan stone is equivalent to 320 gunja seeds. What is the mass of this stone? 35.2 g

This Harappan seal was found in the ruins of a house of the Kassite Period.

Graph the relation that is symmetric to the given function with respect to the line $y = x$.   [Lesson 3.1]

38. $f(x) = 2x - 1$ **39.** $f(x) = 3 - x$ **40.** $f(x) = x^2$

Look Beyond

   Geometry   Suppose a rectangle has a perimeter of 24 centimeters.

41. Make a table of possible lengths, widths, and areas.

42 Use your graphics calculator to graph a function for area in terms of length. Use the trace feature on your calculator to determine the maximum area of the rectangle. What length and width give this area?

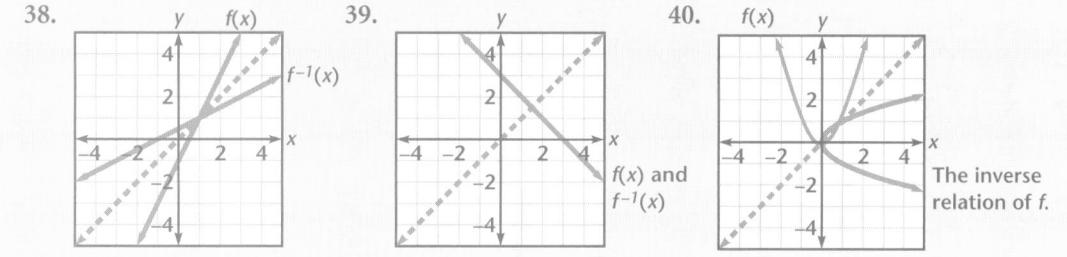

38.

39.

$f(x)$ and $f^{-1}(x)$

40.

The inverse relation of f.

---

41. $a(l) = l(12 - l)$,   where l represents the length and $a(l)$ represents the area

Width (cm)	Length (cm)	Area (cm²)
1	11	11
2	10	20
3	9	27
4	8	32
5	7	35
6	6	36
7	5	35

42. $A = l(12 - l)$; The maximum area is 36 cm². The length and width are both 6 cm.

Look Beyond

The graph of the function is part of a parabola. Only the part of the parabola in the first quadrant is considered in this problem since the dimensions of a rectangle cannot be negative. The maximum area corresponds to the vertex of the parabola.

- Define and graph the greatest-integer and the rounding-up functions.
- Define step functions, and use them to model real-world applications.

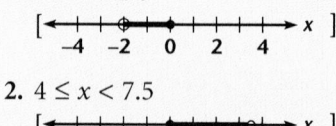

RESOURCES

- Practice Master   **3.5**
- Enrichment Master   **3.5**
- Technology Master   **3.5**
- Lesson Activity Master   **3.5**
- Quiz   **3.5**
- Spanish Resources   **3.5**

Assessing Prior Knowledge

Graph each inequality on a number line.

1. $-2 < x \le 0$

2. $4 \le x < 7.5$

TEACH

Ask students to think of everyday examples of step functions. Many examples are given in the lesson, but many other examples exist, such as the percent of federal income tax U.S. citizens pay. Encourage students to make a table for the example they choose.

Aongoing ASSESSMENT

Yes, the graph of sales tax represents a function. No, its inverse is not a function because the graph of sales tax does not pass the horizontal-line test.

142 Lesson 3.5

LESSON 3.5 Step Functions

Examine the table showing a 7% sales tax.

Amount spent x (¢)	Sales tax $t(x)$ (¢)
$0 \le x < 8$	0
$8 \le x < 22$	1
$22 \le x < 36$	2
$36 \le x < 50$	3
$50 \le x < 64$	4
$64 \le x < 78$	5
$78 \le x < 92$	6
$92 \le x < 106$	7

The graph of the table used to find sales tax looks like a flight of stairs. This function remains constant over a certain part of the domain and then changes to another constant value.

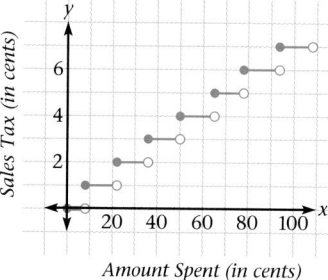

The graph shows the sales tax, y, for the amount of purchase, x, given in the table.

How much tax should Theresa charge on a purchase of 85¢? The x-value of 85 is in the interval $78 \le x \le 92$.

When x is 85, y is 6. So Theresa should add a 6¢ tax on an 85¢ purchase.

Does this graph represent a function? Is the inverse of the graph a function? Explain.

ALTERNATIVE teaching strategy

Using Visual Models

Use a transparency showing a coordinate plane on an overhead projector.

Place a clear piece of acetate on top of the coordinate plane, and draw the graph of a step function. Use the graph to illustrate examples of the greatest-integer and rounding-up functions by simply changing the scale on the axes.

Greatest-Integer Function

The sales-tax graph represents an entire class of functions called **step functions**. Step functions look like a series of steps. The **greatest-integer function**, denoted $f(x) = [x]$, is another example of a step function. The function converts a real number, x, into the largest integer that is *less than or equal to x*. Some examples are shown below.

$$[6.3] = 6 \qquad [5.7] = 5 \qquad [\pi] = 3 \qquad [0.48] = 0$$

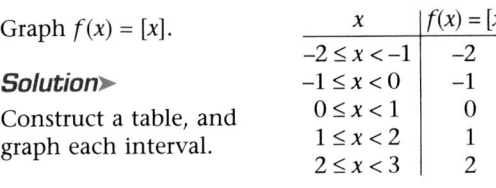**EXAMPLE 1**

Graph $f(x) = [x]$.

Solution

Construct a table, and graph each interval.

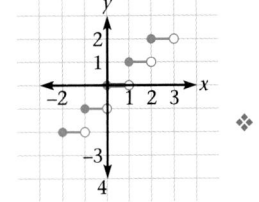

x	$f(x) = [x]$
$-2 \le x < -1$	-2
$-1 \le x < 0$	-1
$0 \le x < 1$	0
$1 \le x < 2$	1
$2 \le x < 3$	2

Graphics Calculator

To graph the greatest-integer function from Example 1 on your graphics calculator, set your calculator to disconnected (or dot) graphing mode. Then graph f using the integer INS function operation.

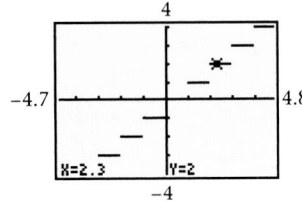

You can use the graph of $f(x) = [x]$ to evaluate $[2.3]$. Use the trace feature to find the x-value 2.3 on the graph. The corresponding y-value is 2. Thus, $[2.3] = 2$.

What is $[-1.2]$?

The Rounding-Up Function

Telephone charges for long-distance calls are based on the total number of minutes of the call. If a call lasts 2 minutes and 30 seconds, you are charged for a 3-minute call. Thus, a function describing telephone charges is a *rounding-up function*.

The **rounding-up function**, or least-integer function, denoted $f(x) = \lceil x \rceil$, converts a real number, x, into the smallest integer *greater than or equal to x*. Some examples are shown below.

$$\lceil 2.83 \rceil = 3 \qquad \lceil \pi \rceil = 4 \qquad \lceil 5.01 \rceil = 6 \qquad \lceil -1.25 \rceil = -1$$

CRITICAL *Thinking*

The greatest-integer and rounding-up functions are both increasing functions. Can you think of a real-life example of a *decreasing* step function?

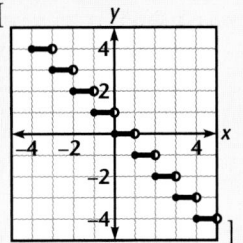

Alternate Example 2

What is the cost of a call from Austin to Denver that lasts 2 minutes and 59 seconds? [**$0.81**]

ASSESS

Selected Answers

Odd-numbered problems 7–45.

Assignment Guide

Core 1–12, 15–47

Core Plus 1–5, 10–17, 22–47

Technology

A graphics calculator is needed for Exercises 36, 37, 41–44, and 46.

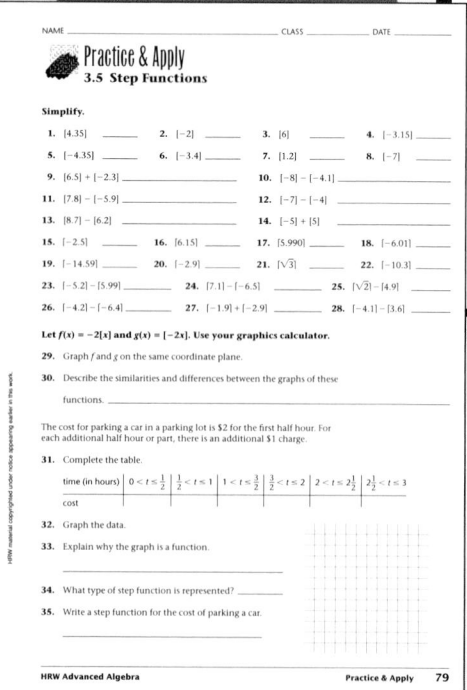

Practice Master

NAME _____ CLASS _____ DATE _____

Practice & Apply
3.5 Step Functions

Simplify.

1. [4.35] _____   2. [−2] _____   3. [6] _____   4. [−3.15] _____
5. [−4.35] _____   6. [−3.4] _____   7. [1.2] _____   8. [−7] _____
9. [6.5] + [−2.3] _____   10. [−8] − [−4.1] _____
11. [7.8] − [5.9] _____   12. [−7] − [−4] _____
13. [8.7] − [6.2] _____   14. [−5] + [5] _____
15. [−2.5] _____   16. [6.15] _____   17. [5.990] _____   18. [−6.01] _____
19. [−14.59] _____   20. [−2.9] _____   21. [√3] _____   22. [−10.3] _____
23. [−5.2] − [5.99] _____   24. [7.1] − [−6.5] _____   25. [√2] − [4.9] _____
26. [−4.2] − [−6.4] _____   27. [−1.9] + [−2.9] _____   28. [−4.1] − [3.6] _____

Let $f(x) = -2[x]$ **and** $g(x) = [-2x]$. **Use your graphics calculator.**

29. Graph f and g on the same coordinate plane.
30. Describe the similarities and differences between the graphs of these functions.

The cost for parking a car in a parking lot is $2 for the first half hour. For each additional half hour or part, there is an additional $1 charge.

31. Complete the table.

time (in hours)	$0 < t \leq \frac{1}{2}$	$\frac{1}{2} < t \leq 1$	$1 < t \leq \frac{3}{2}$	$\frac{3}{2} < t \leq 2$	$2 < t \leq 2\frac{1}{2}$	$2\frac{1}{2} < t \leq 3$
cost						

32. Graph the data.
33. Explain why the graph is a function.
34. What type of step function is represented? _____
35. Write a step function for the cost of parking a car.

HRW Advanced Algebra   Practice & Apply   79

EXAMPLE 2

Telephone Rates Suppose a daytime call from Austin, Texas to Denver, Colorado is charged as shown in the following table (excluding taxes). What would be the cost for a call that lasts 5 minutes and 1 second?

Length of call (minutes)	$0 < t \leq 1$	$1 < t \leq 2$	$2 < t \leq 3$
Cost of call (dollars)	$1(0.27) = 0.27$	$2(0.27) = 0.54$	$3(0.27) = 0.81$

Solution▸

Round up the value of t: 5 minutes and 1 second rounds up to 6 minutes. The cost of the call is $6(0.27)$, or $1.62. ❖

Try This What would be the cost for a call from Austin, Texas to Denver, Colorado that lasts 8 minutes and 30 seconds?

EXERCISES & PROBLEMS

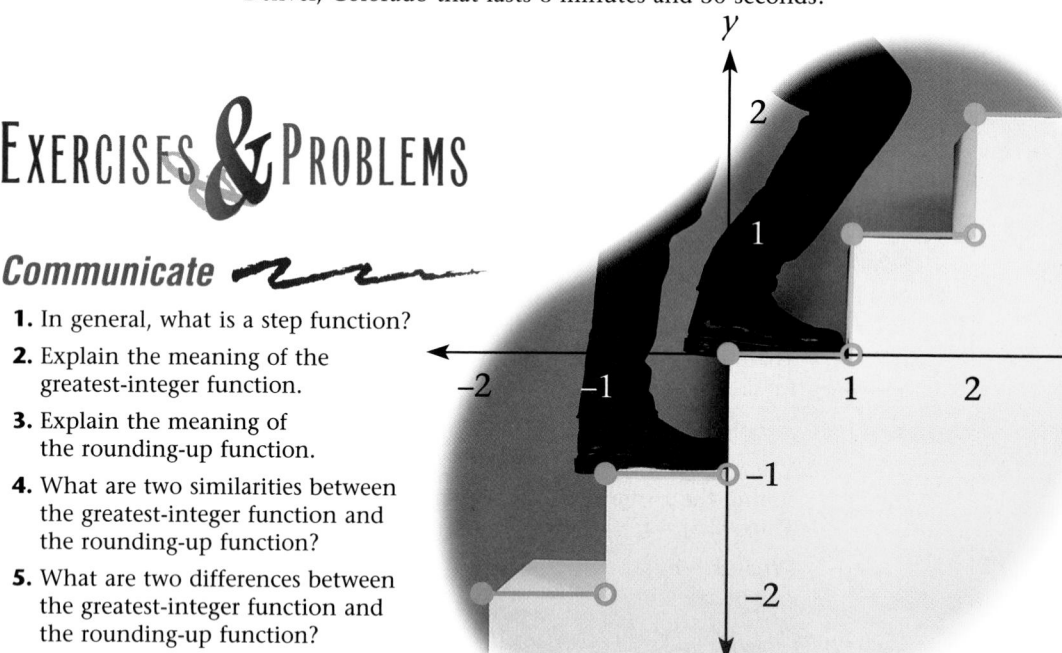

Communicate

1. In general, what is a step function?
2. Explain the meaning of the greatest-integer function.
3. Explain the meaning of the rounding-up function.
4. What are two similarities between the greatest-integer function and the rounding-up function?
5. What are two differences between the greatest-integer function and the rounding-up function?

Practice & Apply

Simplify.

6. [−1.5] −2
7. [−6.1] − [−3.2] −3
8. [5.1] + [−2.01] 2
9. [−2.5] − [1.999] −4
10. [−6.105] −7
11. [4.0] − [−7.001] 12
12. [−2.99] + [2.99] −1
13. [√3] − [1.7] 0

Postage In 1995, mailing a letter by first-class delivery in the United States cost 32¢ for the first ounce or fraction of an ounce, and 23¢ for each additional ounce or fraction of an ounce. Thus, the cost of mailing a first-class letter is a function of weight. **Complete Exercises 14–16.**

14. Construct a table relating the weight of a letter to the cost of mailing it.

RETEACHING the lesson

Inviting Participation Have students use their graphics calculator to explore the greatest-integer function. Ask students to find the value of the greatest integer for a list of numbers you provide. Discuss their results and the meaning of the greatest-integer function. Then present an example like the sales tax example on page 142. Discuss the rounding-up function and how it differs from the greatest-integer function. Present an example like the telephone rate example on page 144 to explore further the rounding-up function.

15. Is this step function for postage a greatest-integer function or a rounding-up function? rounding-up

16. In 1995, how much did it cost to mail a first-class letter weighing 3.2 ounces? $1.01

17. Graph the rounding-up function, $f(x) = \lceil x \rceil$, for all real numbers x such that $-3 \le x \le 3$.

Simplify.

18. $\lceil -1.5 \rceil$ -1 **19.** $\lceil -6.1 \rceil - \lceil -3.2 \rceil$ -3 **20.** $\lceil 5.1 \rceil + \lceil -2.01 \rceil$ 4 **21.** $\lceil -2.5 \rceil - \lceil 1.999 \rceil$ -4

22. $\lceil -6.105 \rceil$ -6 **23.** $\lceil 4.0 \rceil - \lceil -7.001 \rceil$ 11 **24.** $\lceil -2.99 \rceil + \lceil 2.99 \rceil$ 1 **25.** $\lceil \sqrt{3} \rceil - \lceil 1.7 \rceil$ 0

Transformations For Exercises 26–28, let $f(x) = [x]$, $g(x) = [x + 1.5]$, and $h(x) = [x] + 1.5$.

26. Graph f, g, and h on separate coordinate planes. Make certain that your axes include the x- and y-intercepts.

27. Describe the similarities and differences between the graphs of these functions.

28. Describe the type of transformation represented in each graph. Explain, in general terms, how the greatest-integer function is transformed.

Consumer Economics The Break-n-Fix Repair Store charges $45 for a service call that involves up to one hour of labor. For each additional half-hour of labor, or fraction thereof, there is a charge of $20.

29. Construct a table to represent the charges for all real numbers, t, such that $0 < t \le 10$, where t is time measured in hours.

30. Graph the data in your table.

31. Write a step function for the cost of the repair.
$C(t) = 45 + 20\lceil 2(t - 1) \rceil$
32. What type of step function is represented?
rounding-up
33. Determine the cost of a repair that takes 3.75 hours.
$165

Telephone Rates Suppose that the weekday rate for a long-distance telephone call from Chicago to New York City is $0.27 per minute, or fraction thereof, plus 9% tax on the total charge.

34. Construct a table to represent the total cost in terms of time (in minutes), t, such that $0 < t \le 10$.

35. Write a step function for the total cost of a weekday call from Chicago to New York City.
$c(t) = 0.27\lceil t \rceil (1.09)$
36 Graph the data in your table. What type of step function is represented? rounding-up

37 Determine the cost of a weekday call from Chicago to New York City that lasts 7.52 minutes. $2.35

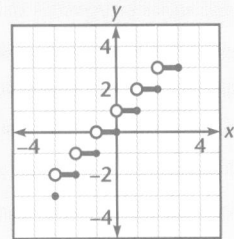

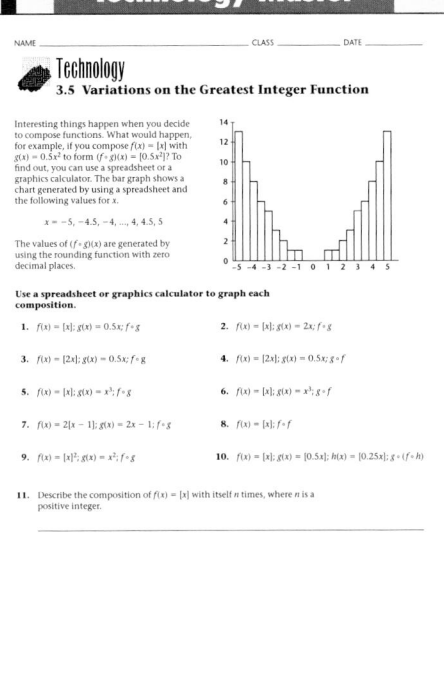

14.

Weight (oz)	Cost (¢)
$0 < t \le 1$	32
$1 < t \le 2$	55
$2 < t \le 3$	78
$3 < t \le 4$	101

Performance Assessment

Have students research the telephone rates between where they live and another city. Ask them to complete Exercises 34–36 for their data. Then have them determine the cost of a call that lasts 8 minutes 30 seconds.

Sales Tax The amount of sales tax you pay varies for different states and cities.

38. Find out what the sales tax is in your city. You may want to get a sales tax table. Answers may vary.

39. Draw the graph of the amount of tax as a function of the purchase price. Include purchase prices from $0 to $2.00.
Answers may vary.

40. **Portfolio Activity** Complete the Portfolio Activity described on page 113. Answers may vary.

Look Back

Political Science The following table shows the number of Hispanic elected officials in selected states for 1984 and 1992. **[Lesson 1.2]**

State	1984	1992
Illinois	25	42
New York	65	93
Colorado	175	204
Arizona	241	350
New Mexico	556	661
California	460	797
Texas	1427	2030

41. Plot the data from Texas. Let the years be on the x-axis. Find the slope of the line. 75.375

42. Plot the data from California. Let the years be on the x-axis. Find the slope of the line. 42.125

43. What does the slope of the line represent? How do the slopes for Texas and California compare?

44. Predict the number of Hispanic elected officials in Texas and in California in the year 2010. 3387; 1555

45. Would your response to Exercise 44 change if you were provided with more data? Explain. yes

Look Beyond

46. Use your graphics calculator to find the point of intersection of the lines $x - y = 5$ and $2x + 3y = 0$. (3, −2)

47. The **signum function**, sg, is defined as $sg(x) = \begin{cases} 1 \text{ if } x > 0 \\ 0 \text{ if } x = 0 \\ -1 \text{ if } x < 0 \end{cases}$.

Graph this function, and determine the domain and range of the signum function. R; {−1, 0, 1}

Pictured from the top: Congress Member Lucille Roybal-Allard, and Assembly Members Martha Escutia, Joe Baca and Grace Napolitano

Look Beyond

Exercise 46 previews using graphing to solve systems of equations, which is presented in Chapter 4. Exercise 47 gives students additional experience working with functions and their graphs.

43. The slope represents the average increase in the number of Hispanic elected officials from 1984 to 1992. The number of Hispanic elected officials is increasing faster in Texas than in California. The slope is greater for Texas than for California.

45. Yes. It is very likely that the prediction in Exercise 44 would be different if more data points were available. Trying to find a pattern based on two data points per state is really just guessing. With more data points, you could more confidently predict if the growth can best be modeled by a line or by some sort of curve, such as an exponential curve.

The answer to Exercise 47 can be found in Additional Answers beginning on page 842.

LESSON 3.6 Parametric Equations

PREPARE

Objectives

• Define and graph a system of parametric equations, and use them to model real-world applications.
• Determine the linear function represented by a system of parametric equations.

RESOURCES

• Practice Master   3.6
• Enrichment Master   3.6
• Technology Master   3.6
• Lesson Activity Master 3.6
• Quiz   3.6
• Spanish Resources   3.6

This lesson may be considered optional in the core course. See page 112A.

Assessing Prior Knowledge

Write an equation for each line described below.

1. slope −2, contains (−4, −3)
   $[y = -2x - 11]$

2. contains (−4, 2) and (6, 7)
   $[y = \frac{1}{2}x + 4]$

3. contains (5, 2) and (3, 2)
   $[y = 2]$

TEACH

Discuss an everyday situation in which two related quantities depend on a third quantity such as time. For example, both the surface area of an expanding balloon and its volume depend on the radius of the balloon.

When a plane is taking off, it has vertical motion as well as horizontal motion. Both motions are dependent on the same variable, time. Parametric equations allow you to examine two quantities, such as horizontal and vertical motion, in terms of a third quantity, time.

Rise of 46 ft

Run of 126 ft

Motion of airplane in 1 second during takeoff.

Aviation A plane takes off at a rate of 126 feet per second horizontally and 46 feet per second vertically. In 3 seconds, the run will be 126 · 3, or 378 feet, and the rise will be 46 · 3, or 138 feet.

t	x(t)	y(t)
0	0	0
1	126	46
2	252	92
3	378	138
4	504	184

The table gives the position of the plane in the horizontal (x) and in the vertical (y) directions each second after takeoff.

Both horizontal and vertical positions are functions of time, t.

$$\begin{cases} x(t) = 126t \\ y(t) = 46t \end{cases}$$

The variable t is a **parameter**, and $x(t)$ and $y(t)$ form a **system of parametric equations**.

Here, t is measured in seconds, and both x and y are measured in feet. To find out how far the plane will rise in half of a minute, or 30 seconds, you can evaluate $y(30)$.

$$y(t) = 46t$$
$$y(30) = 46(30), \text{ or } 1380$$

The plane will rise 1380 feet in half a minute.

ALTERNATIVE
t e a c h i n g
s t r a t e g y

Using Cognitive Strategies Begin the study of parametric equations by looking at the linear equation they describe. For example, the parametric equations presented in the opener, $x(t) = 126t$ and $y(t) = 46t$, describe the linear equation $y = \frac{46}{126}x$. Use the following alternative method to find this linear equation.

$x = 126t \rightarrow t = \frac{x}{126}$

$y = 46t \rightarrow t = \frac{y}{46}$

$\frac{x}{126} = \frac{y}{46}$

$46x = 126y$

$y = \frac{46}{126}x$

Reverse this process to show students how to get the parametric equations from the linear equation.

Alternate Example 1

Graph the parametric system using your graphics calculator.

$$\begin{cases} x(t) = t + 3 \\ y(t) = t - 3 \end{cases}$$

[Set viewing window for most calculators at:
Tmin = −4, Tmax = 4,
Tstep = 0.1, Xmin = −1.7,
Xmax = 7.7, Xscl = 1,
Ymin = −6.1, Ymax = 0.1,
Yscl = 1]

ongoing
ASSESSMENT

Try This

Any viewing window that includes $-1.7 < x \le 0$ and $-0.05 < y \le 2.5$.

ongoing
ASSESSMENT

There is exactly one *x*-value and one *y*-value for each *t*-value.

Graphing Parametric Equations

You can use the parametric mode of your graphing calculator to view the graph that represents the motion of the plane.

With your calculator in parametric mode, enter $x(t) = 126t$ and $y(t) = 46t$. Set the viewing window as follows:

Tmin = 0	Tmax = 60	Tstep = 0.1
Xmin = −200	Xmax = 4750	Xscl = 100
Ymin = −600	Ymax = 1575	Yscl = 100

Why were these range values chosen? Why are the Xmax and Ymax so large? Why does T range from 0 to 60? What happens when you change the Tstep?

Try different values for these variables to see the effect on the graph.

Graph the parametric equations. Use the trace feature to determine the horizontal and vertical displacement 30 seconds after takeoff.

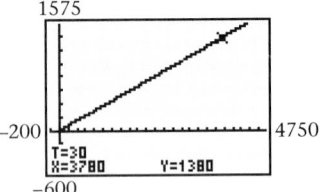

As you can see from the calculator screen shown, the three values displayed are the *t*-value and the corresponding horizontal, *x*, and vertical, *y*, displacements. Thus, 30 seconds after takeoff, the plane will have run 3780 feet and will have risen 1380 feet.

EXAMPLE 1

Graph the parametric system using your graphics calculator.

$$\begin{cases} x(t) = t + 2 \\ y(t) = t - 2 \end{cases}$$

Solution

Leave your calculator in parametric mode. Enter $x(t) = t + 2$ and $y(t) = t - 2$. Use the viewing window shown (or any similar viewing window).

Tmin = −3	Tmax = 3	Tstep = 0.1
Xmin = −1.5	Xmax = 8	Xscl = 1
Ymin = −5	Ymax = 1	Yscl = 1

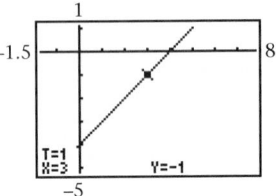

Try This Graph the parametric system using your graphics calculator. What viewing window shows both the *x*- and *y*-intercepts?

$$\begin{cases} x(t) = 2t - 1 \\ y(t) = 3t + 1 \end{cases}$$

How many *x*- and *y*-values correspond to each *t*-value?

interdisciplinary
CONNECTION
Physics Have students do research to find formulas used in physics that can be written as parametric equations. Ask students to state the advantages to writing the formulas in parametric form.

Combining Parametric Equations

Notice that the graph of the parametric system of equations in Example 1 is a straight line. Is there a single linear function, f, that is equivalent to these parametric equations?

EXAMPLE 2

Graphics Calculator

Find the linear function that is equivalent to the parametric system of equations given in Example 1.

Solution▶

The function, f, can be determined by two points since it is a linear function. Use the trace feature to find the coordinates of two points on the graph of f.

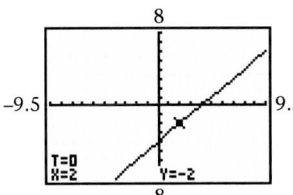 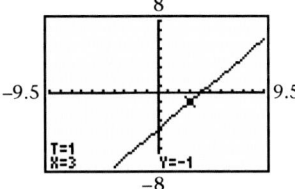

Using the points $(2, -2)$ and $(3, -1)$, you can find that the slope is 1 and the y-intercept is -4.

Thus, the linear function is $f(x) = x - 4$.

Check▶

Graph f with your calculator in function mode and use the trace feature to find the points $(2, -2)$ and $(3, -1)$.

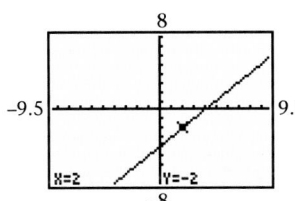 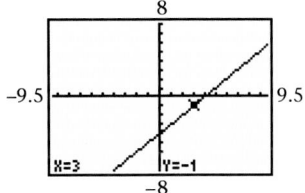

Thus, the parametric system $\begin{cases} x(t) = t + 2 \\ y(t) = t - 2 \end{cases}$ can be represented by the linear function $f(x) = x - 4$. ❖

Try This   What linear function is equivalent to $\begin{cases} x(t) = 2t - 1 \\ y(t) = 3t + 1 \end{cases}$?

What information did the parametric system provide that is missing from the single linear function?

CRITICAL Thinking

Are there other parametric systems equivalent to $f(x) = x - 4$? HINT: Examine the system shown. Can you devise a method for determining other equivalent parametric systems?

$\begin{cases} x(t) = t + 1 \\ y(t) = t - 3 \end{cases}$

TEACHING *tip*

Applications of Parametric Functions

Parametric graphing allows you to simulate a variety of different physical events.

Graphics Calculator

EXAMPLE 3

An outfielder is throwing a baseball to the catcher 53 meters away in order to stop a runner from scoring. The ball is released 2 meters above the ground at a 30° angle with the ground and with an initial velocity of 24 meters per second. The catcher wants the ball to land in his mitt, which is $\frac{1}{2}$ meter off the ground. The parametric equations describing the motion of the ball are $x(t) = 21t$ and $y(t) = 2 + 12t - 4.9t^2$.

Ⓐ Use your graphics calculator in parametric mode to graph the path of the ball. Find the time it takes the ball to reach its highest point.

Ⓑ Use the graph to determine how long it will take the ball to hit the ground if the catcher does not interfere. What are the x and y positions of the ball when it hits the ground?

Ⓒ Will the catcher be able to catch the ball? Explain.

Solution

Ⓐ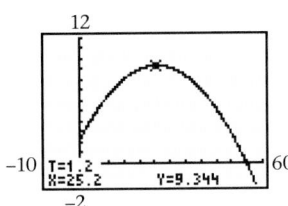

At approximately $t = 1.2$ seconds, the ball will reach its highest position, about 9.3 meters above the ground.

Ⓑ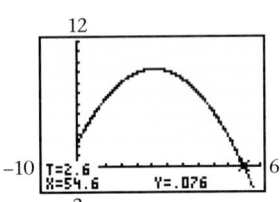

At about 2.6 seconds, the ball will hit the ground. Its horizontal distance is about 54.6 meters, and its vertical distance is approximately 0 meters.

Ⓒ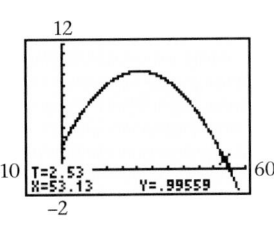

Yes. When the ball is about 53 meters away from the fielder, the ball will be about 1 meter off of the ground and directly in front of the catcher. The ball is certainly within range for the catcher to catch the ball and stop the runner from scoring. ❖

Exercises & Problems

Communicate

1. Identify two reasons why parametric representation of functions is useful.

2. Explain the meaning of the parametric variables.

3. Explain the relationship between the parametric representation of a function and the function itself.

4. Explain how to convert parametric equations of lines into the linear functions they represent.

5. Explain what information is lost when two parametric equations are combined into one linear function.

Practice & Apply

Use your graphics calculator to graph each parametric system. Then combine each system into one linear function. Graph the function to check your result.

6 $\begin{cases} x(t) = 2t \\ y(t) = t - 1 \end{cases}$ $y = \frac{1}{2}x - 1$

7 $\begin{cases} x(t) = t + 3 \\ y(t) = 3t - 1 \end{cases}$ $y = 3x - 10$

8 $\begin{cases} x(t) = t \\ y(t) = 3 - 2t \end{cases}$ $y = -2x + 3$

9 $\begin{cases} x(t) = t - 2 \\ y(t) = t + 7 \end{cases}$ $y = x + 9$

10 $\begin{cases} x(t) = 3t \\ y(t) = 1 - t \end{cases}$ $y = -\frac{1}{3}x + 1$

11 $\begin{cases} x(t) = 2t + 1 \\ y(t) = -t \end{cases}$ $y = -\frac{1}{2}x + \frac{1}{2}$

Human Development A newborn baby weighs 7 pounds and is 21 inches long. Each month, for the first 6 months, he grows an average of 1 inch and gains an average of 1.75 pounds.

12 Write a system of parametric equations describing the height and weight for t months ($0 \le t \le 6$). Use your graphics calculator to graph your system. $h(t) = t + 21$
$w(t) = 1.75t + 7$

13 How many months will it take for the baby to reach 15 pounds? How long is he at this time? ≈ 4.6 mo; ≈ 25.6 in.

14 How many months will it take for the baby to reach a length of 2 feet? How heavy is he at this time? 3 mos; 12.25 lbs

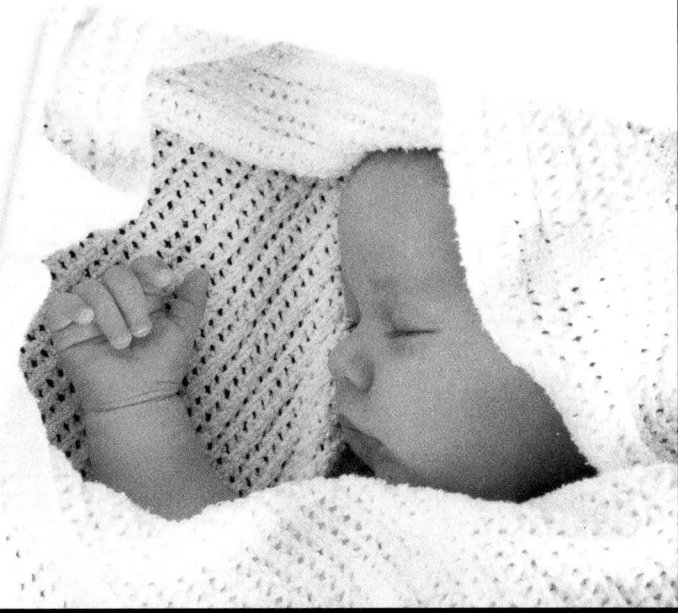

Selected Answers
Odd-numbered problems 7–29.

Assignment Guide
Core 1–10, 12–31
Core Plus 1–5, 9–31

Technology
A graphics calculator is needed for Exercises 6–14, 16 & 17, 19–26, and 31.

Error Analysis
Watch for students who choose a value for Tmax that is too large or too small. If a large value is chosen, a long time may elapse before the trace function can be used. If a small value is chosen, the graph will be incomplete. If either of these situations occurs, encourage students to adjust this value accordingly.

Performance Assessment

Have students write a problem that can be solved using parametric equations. Be sure students provide a solution to their problem. Ask students to exchange problems with several other classmates and solve each other's problems.

Aviation Thirty seconds after takeoff, a plane is moving at 120 feet per second horizontally and at 48 feet per second vertically. **Complete Exercises 15–17.**

15. Give the parametric representation of the flight path of the airplane. $x(t) = 120t$ $y(t) = 48t$

16 Use your graphics calculator to graph the path of the plane. What viewing window did you use? sample: $0 \leq x \leq 18{,}000$; $0 \leq y \leq 7500$; $0 \leq t \leq 150$

17 Use the trace feature to determine the position of the plane 2 minutes after takeoff. $x = 14{,}400$ ft; $y = 5760$ ft

18. Write a system of parametric equations for the line through $(-1, 3)$ and $(2, 9)$. sample: $x(t) = t$ $y(t) = 2t + 5$

Sports Frannie throws a ball from one end of a 200-meter-long field, from a height of 2.2 meters, at a 45° angle with the ground, and with an initial velocity of 20 meters per second in the horizontal direction. The following system of parametric equations describes the ball's path.

$$\begin{cases} x(t) = 20t \\ y(t) = 2.2 + 20t - 4.9t^2 \end{cases}$$

Use your graphics calculator to graph the path of the ball.

19 How high does the ball get? How long does it take for the ball to reach this height? approx 22.6 m; approx 2 sec

20 How far will the ball go before it hits the ground? How long will it take to hit the ground? approx 84 m; approx 4.2 sec

21 How far must Hilder run in order to catch the ball if he is originally 100 meters away from Frannie and directly in the path of the ball? approx 16 m

21. Depending on how high above the ground Hilder catches the ball, he must run approximately 16 meters to catch it.

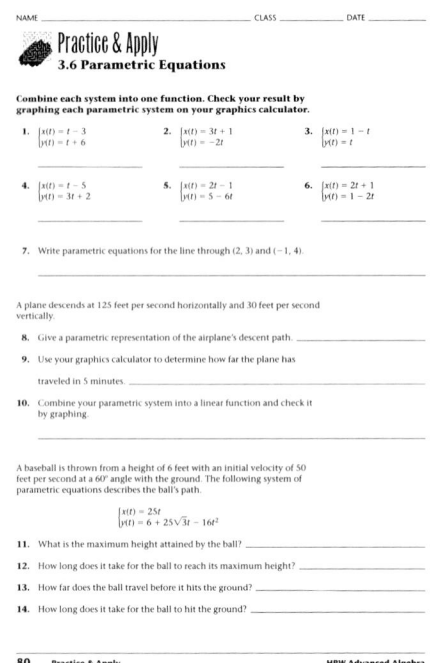

NAME _____ CLASS _____ DATE _____

Practice & Apply
3.6 Parametric Equations

Combine each system into one function. Check your result by graphing each parametric system on your graphics calculator.

1. $\begin{cases} x(t) = t - 3 \\ y(t) = t + 6 \end{cases}$ **2.** $\begin{cases} x(t) = 3t + 1 \\ y(t) = -2t \end{cases}$ **3.** $\begin{cases} x(t) = 1 - t \\ y(t) = t \end{cases}$

4. $\begin{cases} x(t) = t - 5 \\ y(t) = 3t + 2 \end{cases}$ **5.** $\begin{cases} x(t) = 2t - 1 \\ y(t) = 5 - 6t \end{cases}$ **6.** $\begin{cases} x(t) = 2t + 1 \\ y(t) = 1 - 2t \end{cases}$

7. Write parametric equations for the line through (2, 3) and (−1, 4).

A plane descends at 125 feet per second horizontally and 30 feet per second vertically.

8. Give a parametric representation of the airplane's descent path. _____

9. Use your graphics calculator to determine how far the plane has traveled in 5 minutes. _____

10. Combine your parametric system into a linear function and check it by graphing. _____

A baseball is thrown from a height of 6 feet with an initial velocity of 50 feet per second at a 60° angle with the ground. The following system of parametric equations describes the ball's path.

$$\begin{cases} x(t) = 25t \\ y(t) = 6 + 25\sqrt{3}t - 16t^2 \end{cases}$$

11. What is the maximum height attained by the ball? _____

12. How long does it take for the ball to reach its maximum height? _____

13. How far does the ball travel before it hits the ground? _____

14. How long does it take for the ball to hit the ground? _____

80   Practice & Apply   HRW Advanced Algebra

Projectile Motion The server in a volleyball game serves the ball at an angle of 35° with the ground and from a height of 2 meters. The server is 9 meters from the 2.2-meter-high net. The ball must not touch the net when served, but must land within the 9-meter region on the other side of the net. The system of parametric equations that describes the ball's path is

$$\begin{cases} x(t) = (0.8192)v_0 t \\ y(t) = 2 + (0.5736)v_0 t - 4.9t^2 \end{cases}$$

where v_0 is the initial velocity of the ball.

22 Determine an initial velocity, to the nearest whole number, in meters per second, that will enable the ball to be a good serve. Explain. 10–12 m/s

23 How far will the ball travel horizontally with this initial velocity before it hits the ground? How long will it take the ball to hit the ground? Answers may vary.

24 How high above the net will the ball travel? Answers may vary.

Look Back

Let $f(x) = 2x + 5$. Use your graphics calculator to graph f. **[Lessons 2.3, 3.2]**

25 Find the domain and the range of f. R; R

26 Determine whether f^{-1} exists. If so, graph f^{-1} on the same coordinate plane with f and describe the graph of each.

Let $f(x) = x^2 + 2$ and $g(x) = \dfrac{1}{x+1}$, where $x \neq -1$. **[Lesson 3.3]**

27. Find $(f \circ g)(x)$. **28.** Find $(g \circ f)(x)$. **29.** Find $(f \circ f)(x)$. **30.** Find $(g \circ g)(x)$.

$\dfrac{2x^2+4x+3}{x^2+2x+1}$, $x \neq -1$ $\dfrac{1}{x^2+3}$ $x^4 + 4x^2 + 6$ $\dfrac{x+1}{x+2}$, $x \neq -1, -2$

Look Beyond

31 Use your graphics calculator (in radian mode) to graph the parametric system

$$\begin{cases} x(t) = \cos t \\ y(t) = \sin t \end{cases}$$

where sin and cos are the names of functions on your calculator. Use the following viewing window.

Tmin = 0	Tmax = 10	Tstep = 0.1
Xmin = −4.7	Xmax = 4.8	Xscl = 1
Ymin = −3.1	Ymax = 3.2	Yscl = 1

Was the figure drawn in the clockwise or counterclockwise direction? What is the shape of the graph? counterclockwise; circle

26. $f^{-1}(x)$ exists and is $\dfrac{x-5}{2}$. The function and its inverse form lines that intersect at $(-5, -5)$ and are symmetric to $y = x$.

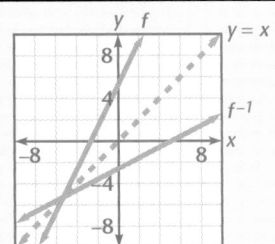

22. The serve will be good if the initial velocity is 10–12 meters per second.

23. When $v_0 = 10$, the ball travels ≈ 11.9 meters in ≈ 1.45 seconds.
When $v_0 = 11$, the ball travels ≈ 14 meters in ≈ 1.55 seconds.
When $v_0 = 12$, the ball travels ≈ 16.2 meters in ≈ 1.65 seconds.

24. When $v_0 = 10$, the ball is about 0.2 meters above the net.
When $v_0 = 11$, the ball is about 1.2 meters above the net.
When $v_0 = 12$, the ball is about 2 meters above the net.

Look Beyond

This exercise is an informal introduction to basic concepts of trigonometric and analytical functions that are developed fully in Chapters 8, 10, and 14.

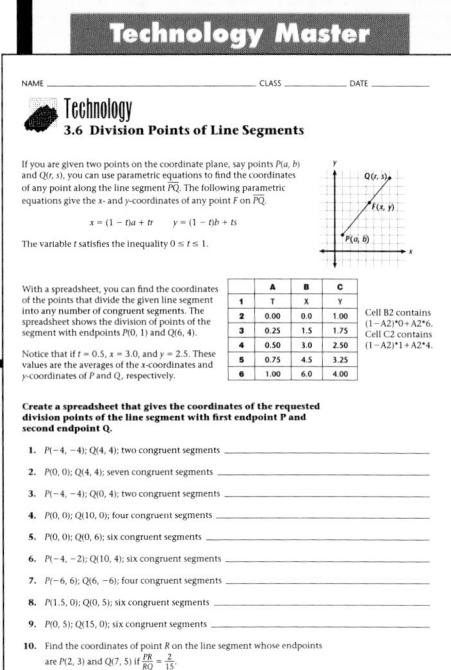

FOCUS

Symmetry is a part of many designs. Encourage students to find examples of several designs that involve symmetry and bring a photo or draw a picture to share with the class. Discuss the different designs and the symmetry found in them.

MOTIVATE

Invite a graphic designer to come and talk to your class about ways he or she uses symmetry in his or her designs. Ask students to take notes during the presentation, especially noting any ideas they can use while completing their projects.

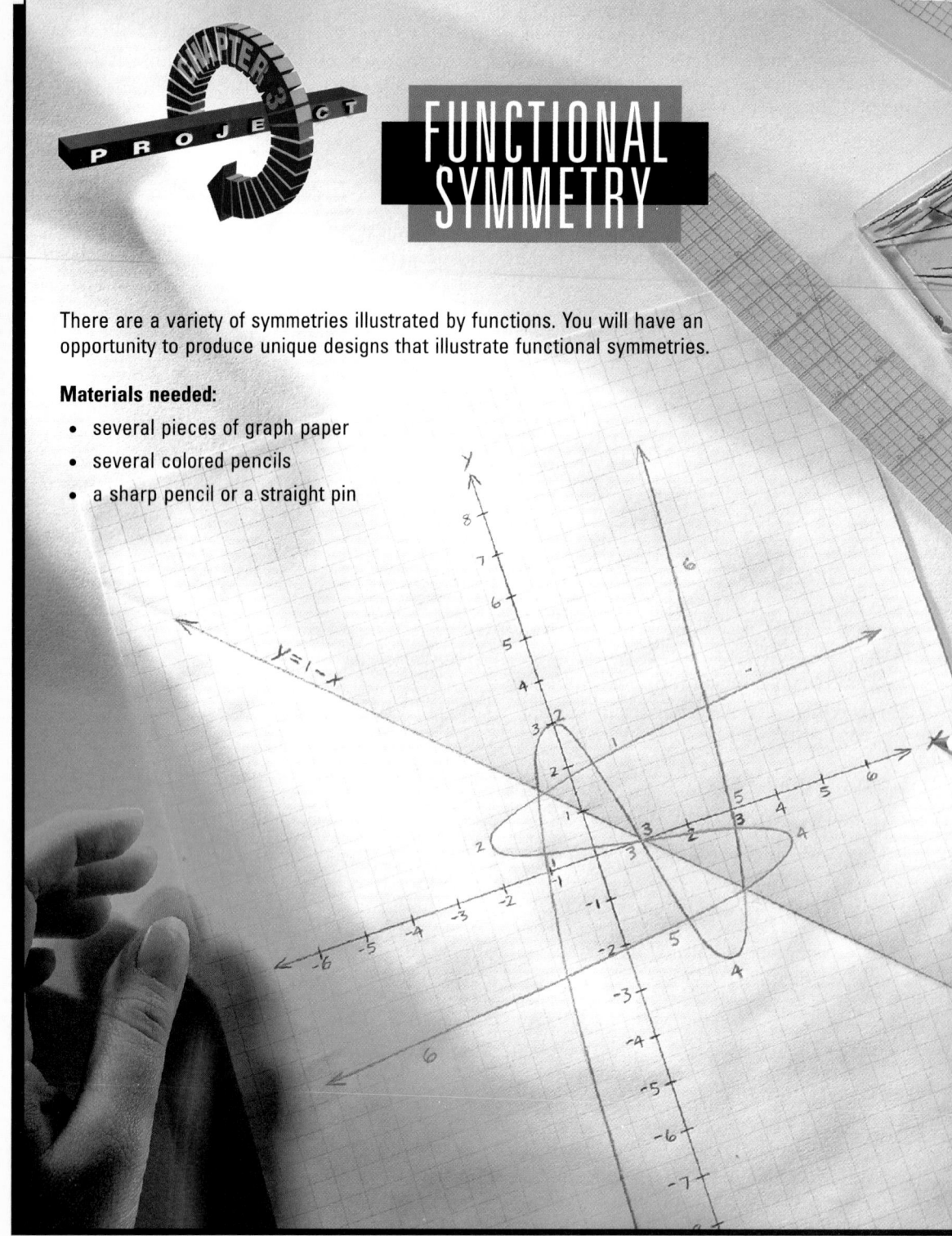

FUNCTIONAL SYMMETRY

There are a variety of symmetries illustrated by functions. You will have an opportunity to produce unique designs that illustrate functional symmetries.

Materials needed:

- several pieces of graph paper
- several colored pencils
- a sharp pencil or a straight pin

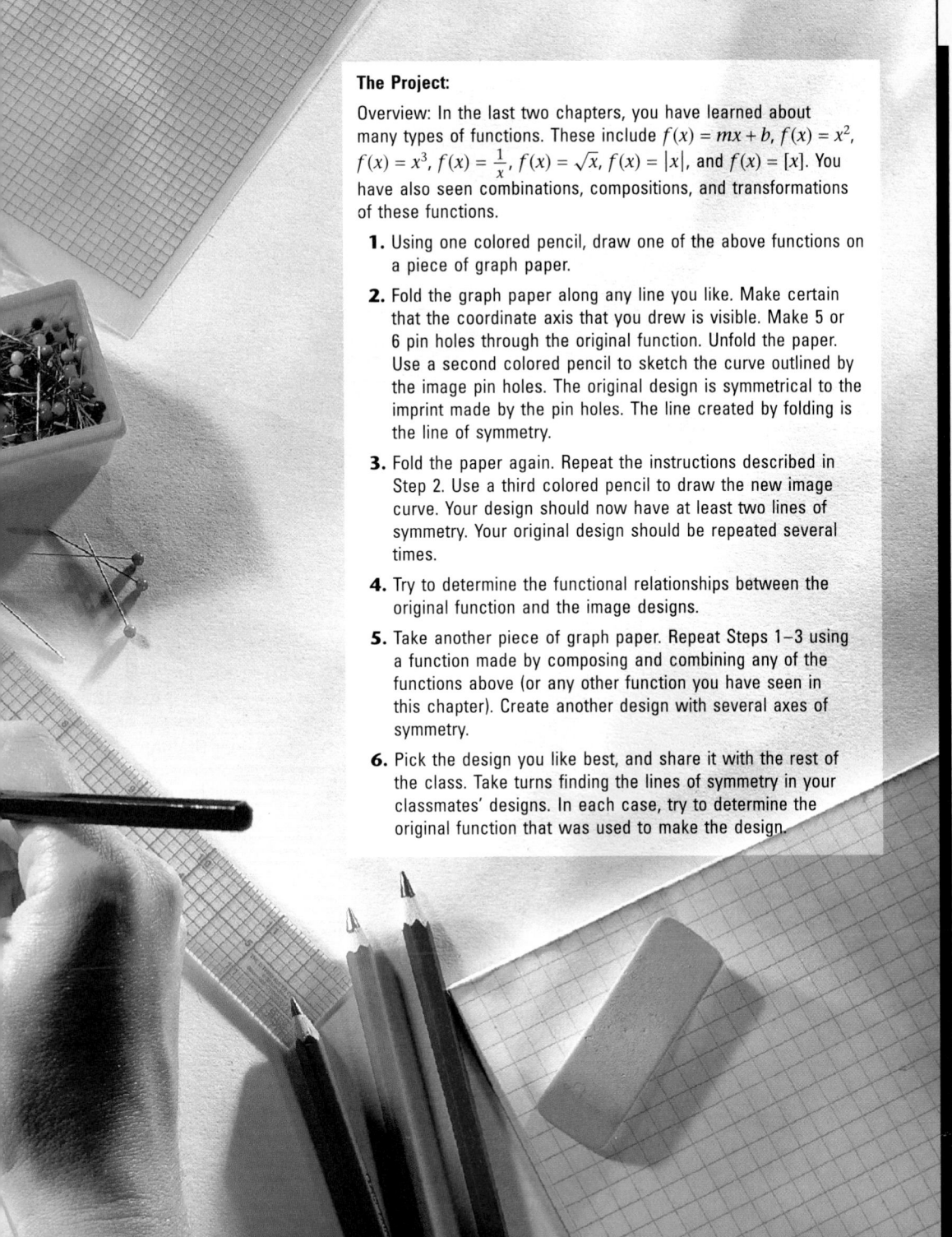

The Project:

Overview: In the last two chapters, you have learned about many types of functions. These include $f(x) = mx + b$, $f(x) = x^2$, $f(x) = x^3$, $f(x) = \frac{1}{x}$, $f(x) = \sqrt{x}$, $f(x) = |x|$, and $f(x) = [x]$. You have also seen combinations, compositions, and transformations of these functions.

1. Using one colored pencil, draw one of the above functions on a piece of graph paper.

2. Fold the graph paper along any line you like. Make certain that the coordinate axis that you drew is visible. Make 5 or 6 pin holes through the original function. Unfold the paper. Use a second colored pencil to sketch the curve outlined by the image pin holes. The original design is symmetrical to the imprint made by the pin holes. The line created by folding is the line of symmetry.

3. Fold the paper again. Repeat the instructions described in Step 2. Use a third colored pencil to draw the new image curve. Your design should now have at least two lines of symmetry. Your original design should be repeated several times.

4. Try to determine the functional relationships between the original function and the image designs.

5. Take another piece of graph paper. Repeat Steps 1–3 using a function made by composing and combining any of the functions above (or any other function you have seen in this chapter). Create another design with several axes of symmetry.

6. Pick the design you like best, and share it with the rest of the class. Take turns finding the lines of symmetry in your classmates' designs. In each case, try to determine the original function that was used to make the design.

Cooperative Learning

Have small groups of students discuss ideas for various ways functions can be used to make a design. Encourage experimentation by the group, either individually or as a whole. Be sure that all group members participate in making the final design.

Discuss

Have students look at the different designs made by each group and the different designs that students brought to class. As a class, discuss the importance of symmetry.

Chapter 3 Review

Answers (margin column)

1. x-axis:
   {(0, −1), (4, 0), (5, −5)};
   y-axis:
   {(0, 1), (−4, 0), (−5, 5)};
   y = x:
   {(1, 0), (0, 4), (5, 5)}

2. x-axis:
   {(−2, −3), (−4, 7), (6, −5),
   (1, 4)}
   y-axis:
   {(2, 3), (4, −7), (−6, 5),
   (−1, −4)};
   y = x:
   {(3, −2), (−7, −4), (5, 6),
   (−4, 1)}

Vocabulary

Key Skills & Exercises

Lesson 3.1

> **Key Skills**

Identify the image and pre-image points and the axis of symmetry of a set of ordered pairs.

The image of point $A(2, 4)$ reflected over the x-axis is $A'(2, -4)$. The image of point A reflected over the y-axis is $A''(-2, 4)$.

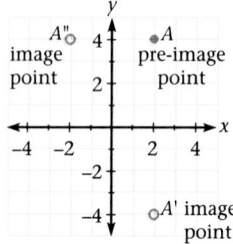

Determine the relationship of coordinates of points reflected over the y-axis, the x-axis, and the line y = x.

Reflecting point $B(1, 3)$ over the line $y = x$ switches the x- and y-coordinates of pre-image point B to form its image point $B'(3, 1)$.

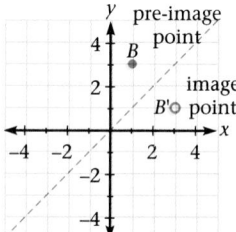

> **Exercises**

Give the image of each set with respect to reflections over the x-axis, the y-axis, and the line y = x.

1. {(0, 1), (4, 0), (5, 5)}

2. {(−2, 3), (−4, −7), (6, 5), (1, −4)}

Lesson 3.2

➤ Key Skills

Determine the inverse of a function.

Find the inverse of
$f = \{(-2, 4), (-3, -1), (2, 2), (3, 4)\}$.
Is the inverse a function?

Reverse the order of the coordinates in each ordered pair. The inverse of f is $\{(4, -2), (-1, -3), (2, 2), (4, 3)\}$. Because the x-coordinate, 4, of the inverse has two y-coordinates, -2 and 3, the inverse is *not* a function.

Define the inverse of a function, and use the horizontal-line test to determine whether the inverse is a function.

If a function, f, has an inverse function, any horizontal line that intersects the graph of f will do so in only one point. To find f^{-1}, replace $f(x)$ with y, interchange x and y, and solve for y. Then, $y = f^{-1}(x)$.

➤ Exercises

Find the inverse of each function. Is the inverse a function?

3. $\{(-3, -3), (4, 2), (2, 4), (5, 5)\}$
$\{(-3, -3),(2, 4),(4, 2),(5, 5)\}$; yes

4. $\{(-2, 7), (-5, -2), (0, 3), (1, -2)\}$
$\{(7, -2),(-2, -5),(3, 0),(-2, 1)\}$; no

Use your graphics calculator and the horizontal-line test to determine whether each function has an inverse function. Then find the inverse function, if it exists.

5. $f(x) = \frac{3x + 3}{2}$
yes; $f^{-1}(x) = \frac{2}{3}x - 1$

6. $f(x) = x^2 - x - 1$
no inverse function

7. $f(x) = 2\left(\frac{2x}{3} + 1\right)$
yes; $f^{-1}(x) = \frac{3}{4}x - \frac{3}{2}$

Lesson 3.3

➤ Key Skills

Define the composition of functions, and describe the relationship between the dependent and independent variables of functions that are composed.

Let $f(x) = 2x + 5$ and $g(x) = 3x^2$.

$$(f \circ g)(x) = f(g(x))$$
$$= 2(3x^2) + 5$$
$$= 6x^2 + 5$$

The range of g must be included in the domain of f. Since the range of g is all real numbers, y, such that $y \geq 0$, and the domain of f is all real numbers, this condition is met.

Identify special properties of composition, and use these properties to analyze functions.

Composition of functions is associative, but, in general, not commutative. Composition has an identity function, $I(x) = x$, such that $I \circ f = f = f \circ I$.

Develop a composition test to determine whether two functions are inverses.

For any invertible function f,
$(f \circ f^{-1})(x) = x = (f^{-1} \circ f)(x)$.

➤ Exercises

For Exercises 8–10, let $f(x) = 2x^2 - 1$ and $g(x) = 2x + 3$.
Find the following.

8. $f \circ g$ $8x^2 + 24x + 17$ **9.** $g \circ f$ $4x^2 + 1$ **10.** $(g \circ f)(1)$ 5

For Exercises 11–14, use composition to determine whether the pairs of functions are inverses.

11. $f(x) = x + 1$, $g(x) = -x - 1$ no

12. $f(x) = x - 2$, $g(x) = x + 2$ yes

13. $f(x) = 2x - 1$, $g(x) = \frac{x+1}{2}$ yes

14. $f(x) = 3x - 5$, $g(x) = \frac{x-5}{3}$ no

15. Show that the functions $f(x) = 2x^2$ and $g(x) = x + 5$ are not commutative under composition. $(f \circ g)(x) = 2x^2 + 20x + 50$ but $(g \circ f)(x) = 2x^2 + 5$

Lesson 3.4

➤ *Key Skills*

Define and graph translations, reflections, and scalar transformations of the absolute value function.

The graph of $f(x) = |x| + k$ is a vertical translation of the graph of $g(x) = |x|$ and has y-intercept $(0, k)$.

The graph of $f(x) = |x - h|$ is a horizontal translation of the graph of $g(x) = |x|$ and has axis of symmetry $x = h$.

The graph of $f(x) = a|x|$, where $a > 0$, is a scalar transformation of the graph of $g(x) = |x|$. It is narrower than the graph of g when $a > 1$ and wider when $0 < a < 1$.

The graph of $f(x) = -a|x|$, where $a > 0$, is a reflection of the graph of $g(x) = a|x|$ with respect to the x-axis.

Solve equations involving absolute value symbols by using graphing and algebraic methods.

Isolate the absolute value symbol on one side of the equation, replace the absolute value symbol with $\pm(\ )$, and solve the two equations separately. Be sure to check your answers.

➤ *Exercises*

Solve each equation for x.

16. $1 - |x| = -7$ $-8, 8$

17. $|x + 4| = 3$ $-1, -7$

18. $5 - |3 - 2x| = 1$ $-\frac{1}{2}, \frac{7}{2}$

Lesson 3.5

➤ *Key Skills*

Define and graph the greatest-integer and the rounding-up functions.

The greatest-integer function gives the largest integer that is less than or equal to a given value. The rounding-up function gives the smallest integer that is greater than or equal to a given value.

$$[3.01] = 3 \qquad \lceil 3.01 \rceil = 4$$
$$[-3.01] = -4 \qquad \lceil -3.01 \rceil = -3$$

Define step functions, and use them to model real-world applications.

The *step function* remains constant over a certain part of the domain and then changes to another constant value. Its graph looks like a flight of stairs.

➤ *Exercises*

Simplify.

19. $[-2.1] + \lceil 5.4 \rceil$ 2

20. $\lceil -2.1 \rceil + \lceil 5.4 \rceil$ 4

21. $[-7.99] - [-7.01]$ 0

Lesson 3.6

➤ *Key Skills*

Define and graph a system of parametric equations, and use them to model real-world applications.

A hiker heads 20° west of north, averaging 2 miles per hour. The parametric equations $x(t) = 1.9t$ and $y(t) = 0.7t$ give her distance to the north and west, respectively, from the starting point after t hours. Graph and trace to find her position after 4 hours.

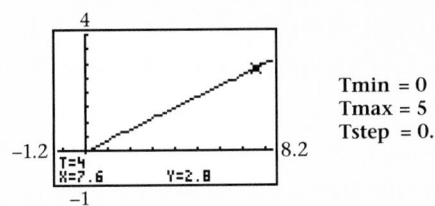

Tmin = 0
Tmax = 5
Tstep = 0.1

After 4 hours, she is $x = 7.6$ miles north and $y = 2.8$ miles west of her starting point.

Determine the linear function represented by a system of parametric equations.

To find a single linear function, f, represented by a parametric system $\begin{cases} x(t) = at + b \\ y(t) = ct + d \end{cases}$ find two points on the line. Use the slope formula and the slope-intercept form of a line to find a linear function for the graph.

26. It takes about 4.1 seconds to reach the wall. When the ball has traveled 400 feet horizontally, it is still almost 17 feet above the ground, or nearly 5 feet above the wall. It appears to be a home run.

27. The ball would hit the wall between 7 and 8 feet above the ground, and would not be a home run.

➤ *Exercises*

For Exercises 22–24, combine each parametric system into one linear function.

22. $\begin{cases} x(t) = 2t + 1 \\ y(t) = 2t - 3 \end{cases}$
    $f(x) = x - 4$

23. $\begin{cases} x(t) = t - 2 \\ y(t) = 4t + 1 \end{cases}$
    $f(x) = 4x + 9$

24. $\begin{cases} x(t) = 2t + 1 \\ y(t) = 8t + 13 \end{cases}$
    $f(x) = 4x + 9$

Applications

Sports A baseball player hits a pitch 3 feet above the ground with an initial velocity of 120 feet per second and at an angle of 35° above the horizontal. A 12-foot-high home-run wall is 400 feet away from the hitter in the direction of the ball's path. The parametric equations that describe the ball's path (disregarding air resistance) are given.

$$\begin{cases} x(t) = (0.82)120t \\ y(t) = 3 + (0.57)120t - 16t^2 \end{cases}$$

25. Graph the path of the ball on your calculator. What is the maximum height that the ball reaches? approx 76 ft

26. How long does it take the ball to reach the wall? Is it a home run? Explain. approx 4.1 sec; yes

27. What would happen if the ball were hit at 118 feet per second? (Replace 120 in the parametric equations with 118 to find the answer.)

28. $900

29. $65

30. $5125

34.

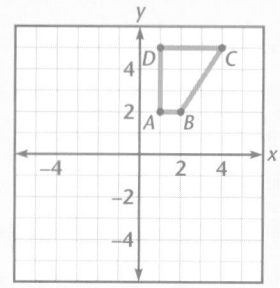

35.

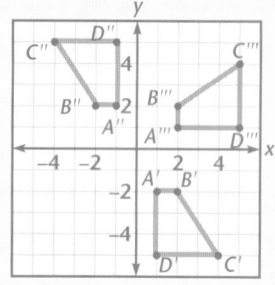

Business For a certain move involving at least 1000 pounds, a moving company charges $f(x) = 900 + 65\left[\frac{x-1000}{100}\right]$ dollars, where x represents the weight in pounds. Graph f on your graphics calculator.

28 What is the minimum charge?

29 With each 100-pound increase, how much does the charge increase?

30 What is the charge on a 7500-pound move?

Physiology Normal sea-level air pressure is about 30 units (in inches of mercury). The pressure below the ocean surface is approximated by $p(d) = 30 + \frac{9}{10}d$, where d is the depth in feet. The volume of nitrogen in liters (adjusted for sea level) that can dissolve in the body at a given pressure is roughly $V(p) = \frac{p}{30}$. If a diver surfaces too fast, the extra nitrogen can form bubbles inside body tissues, causing "the bends."

31. What are the independent and dependent variables of the functions p and V? *d, p(d); p, V(p)*

32. Write a function that gives the adjusted volume of nitrogen in the body as a function of depth. What are the independent and dependent variables? *V(p(d)) = 1 + 0.03d; d, V(p(d))*

33 Graph the function from Exercise 32 on your calculator. About how much nitrogen can dissolve in the body at sea level? About how much can dissolve at a depth of 300 feet? *1 liter; 10 liters*

Coordinate Geometry A quadrilateral has the vertices $A(1, 2)$, $B(2, 2)$, $C(4, 5)$, and $D(1, 5)$.

34. Plot the vertices on the same coordinate plane, and draw the line segments $\overline{AB}$, $\overline{BC}$, $\overline{CD}$, and $\overline{DA}$.

35. On the same coordinate plane, plot the images of the figure with respect to the x-axis, the y-axis, and the line $y = x$. Label the vertices of the respective images A', B', C', and D'; A'', B'', C'', and D''; and A''', B''', C''', and D'''.

Sports A golfer, trying to take a shortcut by going over a tree and a stream, hits a drive at an angle of 45° above the horizontal and with an initial horizontal velocity of 100 feet per second. The tree is 75 feet tall and stands 190 yards (570 feet) away, directly in the path of the ball. The far bank of the 10-yard-wide stream is 205 yards (615 feet) away. The parametric equations describing the position of the ball in feet, where t is in seconds, are $x(t) = 100t$ and $y(t) = 100t - 16t^2$. Graph the path of the drive on your graphics calculator.

36 Disregarding the tree, is the ball hit hard enough to clear the stream? How far does it travel (in yards) before landing? *Yes; $208\frac{1}{3}$ yd*

37 Will the shot clear the tree? If so, by how much? If not, how much too low is it? Round your answers to the nearest foot. *No; 25 ft*

Chapter 3 Assessment

Consumer Economics   A downtown parking garage charges $2.25 for each hour, or fraction thereof, up to a maximum of $22.50 per day.

1. What kind of step function does this charge schedule represent? rounding-up

2. Write a function that represents the cost of parking. What domain restriction is needed to ensure that this function is accurate? $c(t) = 2.25\lceil t \rceil$, $0 \le t \le 10$

3. What is the cost for parking 7 hours and 21 minutes? What is the cost for parking 16 hours? $18; $22.50

Plot each set of ordered pairs. Then plot the image points that are symmetric to each set with respect to the y-axis.

4. {(−2, 1), (−2, −1), (1, 4), (2, −5)}   **5.** {(0, 5), (2, −4), (−2, −5), (3, 4)}

6. What function is commutative with every other function under the operation of composition? What other function is commutative under composition with a given invertible function f?

7. What can you say about two points that are symmetric with respect to the line $y = x$?

Physics   The stretch of a spring is directly proportional to the load that it is supporting. For a particular spring, $f(x) = 2.5x$ gives the stretch in centimeters for a load, x, in kilograms.

8. Find the inverse function for f. $f^{-1}(x) = 0.4x$

9 Graph f and f^{-1} on your graphics calculator. If the spring stretches 9.5 centimeters, what is the load? 3.8 kg

10 As you trace along f, what do the x- and y-coordinates represent? As you trace along f^{-1}, what do the x- and y-coordinates represent?

For Items 11–13, let $f(x) = -2x - 1$ and $g(x) = 3x^2 + x + 3$.

11. Find $f \circ g$.   **12.** Find $g \circ f$.   **13.** Find $(f \circ g)(-2)$. −27

14 Write the linear function represented by the following system of parametric equations.

$$\begin{cases} x(t) = -4t + 3 \\ y(t) = -2t \end{cases} \quad f(x) = \tfrac{1}{2}x - \tfrac{3}{2}$$

15 Does $f(x) = x^4 - 3x^2 + x + 1$ have an inverse function? How can you tell? no; does not pass horizontal-line test

Find the inverse of each function. Check your results using composition of functions.

16. $f(x) = 2x - 3$ $f^{-1}(x) = \dfrac{x+3}{2}$ **17.** $f(x) = -\left(\dfrac{3x+1}{4}\right)$ $f^{-1}(x) = \dfrac{-4x-1}{3}$

18 For $f(x) = |x|$, write an equation that represents the translation of 3 units to the left and 5 units down. Check by graphing the translation on your graphics calculator. $f'(x) = |x + 3| - 5$

Solve each equation for x.

19. $5 + |x - 3| = 7$ 1, 5   **20.** $|3.2x + 0.2| - 4.1 = 5.2$ 2.84375, −2.96875

---

4. {(2, 1), (2, −1), (−1, 4), (−2, −5)}

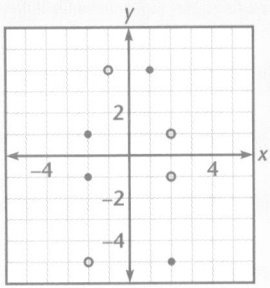

5. {(0, 5), (−2, −4), (2, −5), (−3, 4)}

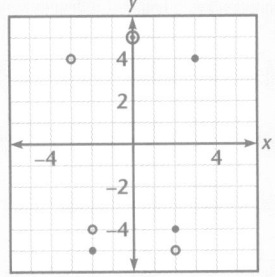

6. The identity function, $I(x) = x$; the inverse of f, f^{-1}

7. The x-coordinate of one is the y-coordinate of the other, and vice versa. They are inverses.

10. For f, the x-coordinate represents the load and the y-coordinate represents the amount of stretch for that load. For f^{-1}, the x-coordinate represents the stretch and the y-coordinate represents the load that corresponds to that amount of stretch.

11. $(f \circ g)(x) = -6x^2 - 2x - 7$

12. $(g \circ f)(x) = 12x^2 + 10x + 5$

14. $f(x) = \tfrac{1}{2}x - \tfrac{3}{2}$

17. $f^{-1}(x) = \dfrac{-4x-1}{3}$

CHAPTER 4
Matrices and Systems of Equations

Meeting Individual Needs

4.1 Using Matrices to Represent Data

Core

Inclusion Strategies, p. 166
Reteaching the Lesson, p. 167
Practice Master 4.1
Enrichment Master 4.1
Technology Master 4.1
Lesson Activity Master 4.1

[1 day]

Core Plus

Practice Master 4.1
Enrichment, p. 165
Technology Master 4.1

[1 day]

4.2 Matrix Multiplication

Core

Inclusion Strategies, p. 173
Reteaching the Lesson, p. 174
Practice Master 4.2
Enrichment Master 4.2
Technology Master 4.2
Lesson Activity Master 4.2

[2 days]

Core Plus

Practice Master 4.2
Enrichment, p. 173
Technology Master 4.2
Interdisciplinary
    Connection, p. 172

[1 day]

4.3 Systems of Two Linear Equations

Core

Inclusion Strategies, p. 182
Reteaching the Lesson, p. 183
Practice Master 4.3
Enrichment Master 4.3
Technology Master 4.3
Lesson Activity Master 4.3

[2 days]

Core Plus

Practice Master 4.3
Enrichment, p. 182
Technology Master 4.3
Interdisciplinary
    Connection, p. 181

[2 days]

4.4 Using Matrix Row Operations

Core

Inclusion Strategies, p. 190
Reteaching the Lesson, p. 191
Practice Master 4.4
Enrichment Master 4.4
Technology Master 4.4
Lesson Activity Master 4.4

[2 days]

Core Plus

Practice Master 4.4
Enrichment, p. 190
Technology Master 4.4

[2 days]

4.5 The Inverse of a Matrix

Core

Inclusion Strategies, p. 199
Reteaching the Lesson, p. 200
Practice Master 4.5
Enrichment Master 4.5
Technology Master 4.5
Lesson Activity Master 4.5
Mid-Chapter Assessment Master

[1 day]

Core Plus

Practice Master 4.5
Enrichment, p. 199
Technology Master 4.5
Mid-Chapter Assessment
    Master

[1 day]

4.6 Using Matrix Algebra

Core

Inclusion Strategies, p. 205
Reteaching the Lesson, p. 206
Practice Master 4.6
Enrichment Master 4.6
Technology Master 4.6
Lesson Activity Master 4.6

[2 days]

Core Plus

Practice Master 4.6
Enrichment, p. 204
Technology Master 4.6

[2 days]

4.7 Exploring Transformations Using Matrices

Core

Inclusion Strategies p. 212
Reteaching the Lesson p. 213
Practice Master 4.7
Enrichment Master 4.7
Technology Master 4.7
Lesson Activity Master 4.7

[1 day]

Core Plus

Practice Master 4.7
Enrichment, p. 212
Technology Master 4.7
Interdisciplinary
    Connection p. 211

[1 day]

4.8 Exploring Systems of Linear Inequalities

Core

Inclusion Strategies, p. 222
Reteaching the Lesson, p. 223
Practice Master 4.8
Enrichment Master 4.8
Technology Master 4.8
Lesson Activity Master 4.8

[1 day]

Core Plus

Practice Master 4.8
Enrichment, p. 222
Technology Master 4.8
Interdisciplinary
    Connection, p. 221

[1 day]

4.9 Introduction to Linear Programming

Core

Reteaching the Lesson, p. 230
Inclusion Strategies, p. 229
Practice Master 4.9
Enrichment Master 4.9
Technology Master 4.9
Lesson Activity Master 4.9

[Optional*]

Core Plus

Practice Master 4.9
Enrichment, p. 229
Technology Master 4.9

[2 days]

4.10 Linear Programming in Two Variables

Core

Reteaching the Lesson, p. 237
Inclusion Strategies, p. 236
Practice Master 4.10
Enrichment Master 4.10
Technology Master 4.10
Lesson Activity Master 4.10

[Optional*]

Core Plus

Practice Master 4.10
Enrichment, p. 236
Technology Master 4.10
Interdisciplinary
    Connection, p. 235

[1 day]

Chapter Summary

Core

Eyewitness Math,
    pp. 218–219
Chapter 4 Project,
    pp. 242–243
Lab Activity
Long-Term Project
Chapter Review, pp. 244–248
Chapter Assessment, p. 249
Chapter Assessment, A/B
Alternative Assessment
Cumulative Assessment,
    pp. 250–251

[4 days]

Core Plus

Chapter 4 Project,
    pp. 242–243
Lab Activity
Long-Term Project
Chapter Review,
    pp. 244–248
Chapter Assessment,
    p. 249
Chapter Assessment, A/B
Alternative Assessment
Cumulative Assessment,
    pp. 250–251

[3 days]

Cooperative Learning

GROUP ACTIVITIES	
Exploring matrix operations	Lesson 4.0, Portfolio Activity
Transformations	Lesson 4.7, Explorations 1–4
3-D or Not 3-D	Eyewitness Math
Linear inequalities	Lesson 4.8, Explorations 1–3
Flipped Out and Twisted	Chapter 4 Project

You may wish to have students work with partners for some of the above activities. Additional suggestions for cooperative group activities are noted in the teacher's notes in each lesson.

Multicultural

The cultural connections in this chapter include references to China.

CULTURAL CONNECTIONS	
China: The use of a counting board to solve systems of equations	Lesson 4.4

Portfolio Assessment

Below are portfolio activities for the chapter. They are listed under seven activity domains that are appropriate for portfolio development.

1. **Investigation/Exploration** Two projects listed in the teacher's notes ask students to investigate a method and/or a relationship. They are *Cooperative Learning* on page 199 (Lesson 4.5) and Why on page 210 (Lesson 4.7).

2. **Applications** Recommended are any of the following: data processing (Lesson 4.1), payroll (Lesson 4.1), nutrition (Lessons 4.2, 4.3, and 4.8), sports (Lesson 4.2), business (Lessons 4.3, 4.9, and 4.10), manufacturing (Lesson 4.10), investment (Lesson 4.5).

3. **Non-Routine Problems** Exercises 34–38 on page 209 in Lesson 4.6 give students an opportunity to explore Magic Addition Squares.

4. **Project** The Chapter Project on pages 242 and 243 asks students to create their own design using several different transformations. They use transformation matrices for rotations and reflections to determine the vertices of the transformations. The teacher's notes suggest that students complete the project by discussing the importance of symmetry in the designs they share with the class.

5. **Interdisciplinary Topics** Students may choose from geography (Lessons 4.1 and 4.4), physics (Lesson 4.3), geometry (Lesson 4.7), and Maximum/Minimum (Lessons 4.9 and 4.10).

6. **Writing** There are several Alternative Assessment Activities that suggest that students write about mathematics. Also recommended are the Communicate exercises at the beginning of the exercise sets.

7. **Tools** Chapter 4 uses graphic calculators or computer software, such as *f(g) Scholar*™. The technology masters are recommended to measure individual students' proficiency.

Technology

Graphing calculators and computer software such as *f(g) Scholar*™ provide students with an opportunity to explore matrices and their sums, products, and inverses without performing extensive pencil-and-paper calculations. The advantage of using technology is the reduction of some of the frustration that students often experience when working with the many computations associated with matrices. Technology allows students to focus on the use of matrix algebra and linear programming methods for problem solving. For more information on the use of technology, refer to the *HRW Technology Handbook*.

Graphics Calculator

When graphics calculators are used to multiply matrices, students can concentrate on analyzing and using the results, rather than on the many tedious and time-consuming calculations. General instructions for using a graphics calculator to enter, multiply, add, and invert matrices are given in your *HRW Technology Handbook* or in the calculator's manual.

Row Operations with Matrices Instructions are given below for performing row operations using the graphics calculator. The images shown and the specific key instructions used are from the TI-82. Given Matrix A below, add (-2) times row 1 to row 2. This results in a 0 in the second row, first column of the matrix.

1. Enter the matrix into the calculator. Then display the matrix on the screen.

```
[A]
[[1   -2  3   0 ]
 [2    1  7  -3 ]
 [-2   4 -1   4 ]]
```

2. Use the matrix math menu to perform the desired row operation.

```
*row+(-2,[A],1,
2)
[[1   -2  3   0 ]
 [0    5  1  -3 ]
 [-2   4 -1   4 ]]
```

3. Store the resulting matrix as matrix [B].

```
Ans→[B]
[[1   -2  3   0 ]
 [0    5  1  -3 ]
 [-2   4 -1   4 ]]
```

4. To get a "1" in the second row, second column of matrix B, multiply row 2 by 0.2.

```
*row(0.2,[B],2)
[[1   -2  3   0 ]
 [0    1  .2 -.6]
 [-2   4 -1   4 ]]
```

The graphics calculator will allow you to swap rows and add the elements of one row to the elements of another row. In fact, any row operation required in linear programming can be performed on the graphics calculator.

Integrated Software

f(g) Scholar™ is an integrated computer-based mathematics productivity tool that combines calculator, spreadsheet, and graphics capabilities to provide a dynamic and interactive environment for explorations in mathematics. It is appropriate to use *f(g) Scholar*™ for any lesson needing a spreadsheet, calculator, graphics calculator, or any combination of the three.

4 Matrices and Systems of Equations

ABOUT THE CHAPTER

Background Information

It is easy to read and use data that is well organized. Matrices provide a very easy-to-use method for organizing and manipulating data that might otherwise be difficult to work with. As the use of data becomes more important in our society, so will the use of matrices.

CHAPTER RESOURCES

- Practice Masters
- Enrichment Masters
- Technology Masters
- Lesson Activity Masters
- Lab Activity Masters
- Long-Term Project Masters
- Assessment Masters
  - Chapter Assessments, A/B
  - Mid-Chapter Assessment
  - Alternative Assessments, A/B
- Teaching Transparencies
- Cumulative Assessment
- Spanish Resources

CHAPTER OBJECTIVES

- Use matrices to store and represent data.
- Add and subtract matrices.
- Perform matrix multiplication.
- Identify each type of system of two linear equations.
- Solve a system of two linear equations using elimination by adding.
- Represent a system of linear equations with an augmented matrix.
- Solve a system of linear equations using the row reduction method and back substitution.

CHAPTER 4

Matrices and Systems of Equations

LESSONS

Matrices can be used to store information in a very compact and organized manner. Matrices are commonly used to solve systems of equations. Many small businesses, financial analysts, scientists, and teachers use matrices to store data. By performing operations on matrices, you can manipulate the data to determine other beneficial information.

ABOUT THE PHOTOS

Students are shown examples of neighborhood groups working together to reclaim a vacant lot and turn it into a community playground. These photographs relate to the Portfolio Activity on page 163.

- Find the inverse of a matrix.
- Use matrix algebra to solve a system of equations.
- Use matrices to represent and transform objects.
- Graph the solution to a system of linear inequalities.
- Graph a feasible region determined by constraints.
- Find the maximum and minimum values of the objective function determined by the feasible region.

PORTFOLIO ACTIVITY

Begin this activity by asking students to study the tables and to discuss how they are organized. Be sure the students see how the tables relate to matrices of the data. Encourage discussion of other ways to organize the data.

Have students work in small groups. Ask each group to work together to answer the questions. Be sure all students participate in the work. Encourage each group to put their final answers for the questions in a table with the rows and columns labeled. Ask each group to present their answers to the class, describing any strategies they used to get their final answers.

As an extension to this activity, have each group write their own problem that can be solved using matrix multiplication.

The answers to the Portfolio Activity can be found in the Teacher's Edition on page 179.

PORTFOLIO ACTIVITY

Three neighborhood groups join together to build a playground on a city-owned vacant lot. Each group consists of four age categories. The number of volunteers is organized by age and group. The number of hours that volunteers in each age group will commit to work is collected, and a dollar rate for each hour of volunteer work is assigned for each group.

The total dollar value of the hours volunteered by all three neighborhood groups will be used to demonstrate to the city council that the project is important and that the city should contribute money for more playground equipment. Use matrix multiplication to answer the following questions.

What is the total number of work hours that each neighborhood group will contribute?

What is the total dollar value of the hours volunteered by all three neighborhood groups?

Explain how you found your answers. You will be asked to complete the portfolio activity on page 179. You may wish to include your work in your portfolio.

Volunteers Ages	Group 1	Group 2	Group 3
5-20	10	12	9
21-40	7	8	11
41-60	3	5	4
60+	7	4	6

Hours		Hourly Dollar Value	
5-20: 5		Group 1	$12.50
21-40: 8		Group 2	$11.75
41-60: 6		Group 3	$13.00
60+: 12			

ABOUT THE CHAPTER PROJECT

The Chapter 4 Project on pages 242 and 243 asks students to use transformation matrices to produce a design. They will find many examples of such symmetric designs all around them. Students will be given an opportunity to create any design they wish, and then transform that design in as many ways as they wish. The answers to the Chapter Project may vary.

PREPARE

Objectives

- Use matrices to store and represent data.
- Add and subtract matrices.

RESOURCES

- Practice Master 4.1
- Enrichment Master 4.1
- Technology Master 4.1
- Lesson Activity Master 4.1
- Quiz 4.1
- Spanish Resources 4.1

Assessing Prior Knowledge

Complete the following table.

x	1	2	3
y = 2x + 6	[8]	[10]	[12]
y = 12x − 18	[−6]	[6]	[18]

TEACH

why Point out that storing data in a matrix is similar to organizing it in a table or spreadsheet. Create a list of additional personnel information that can be stored in a matrix. Be sure students understand that the matrix shown actually has 508 rows of information, only 4 of which are displayed. Ask students to give examples of information that might occur in row 100.

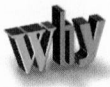

LESSON 4.1 Using Matrices to Represent Data

Danielle works in the phone-order department of a nationwide distributing company. One of her tasks is to process orders. She enters the data into the computer database where the information is organized and stored.

why Matrices and computer databases can be used to store information in an organized manner.

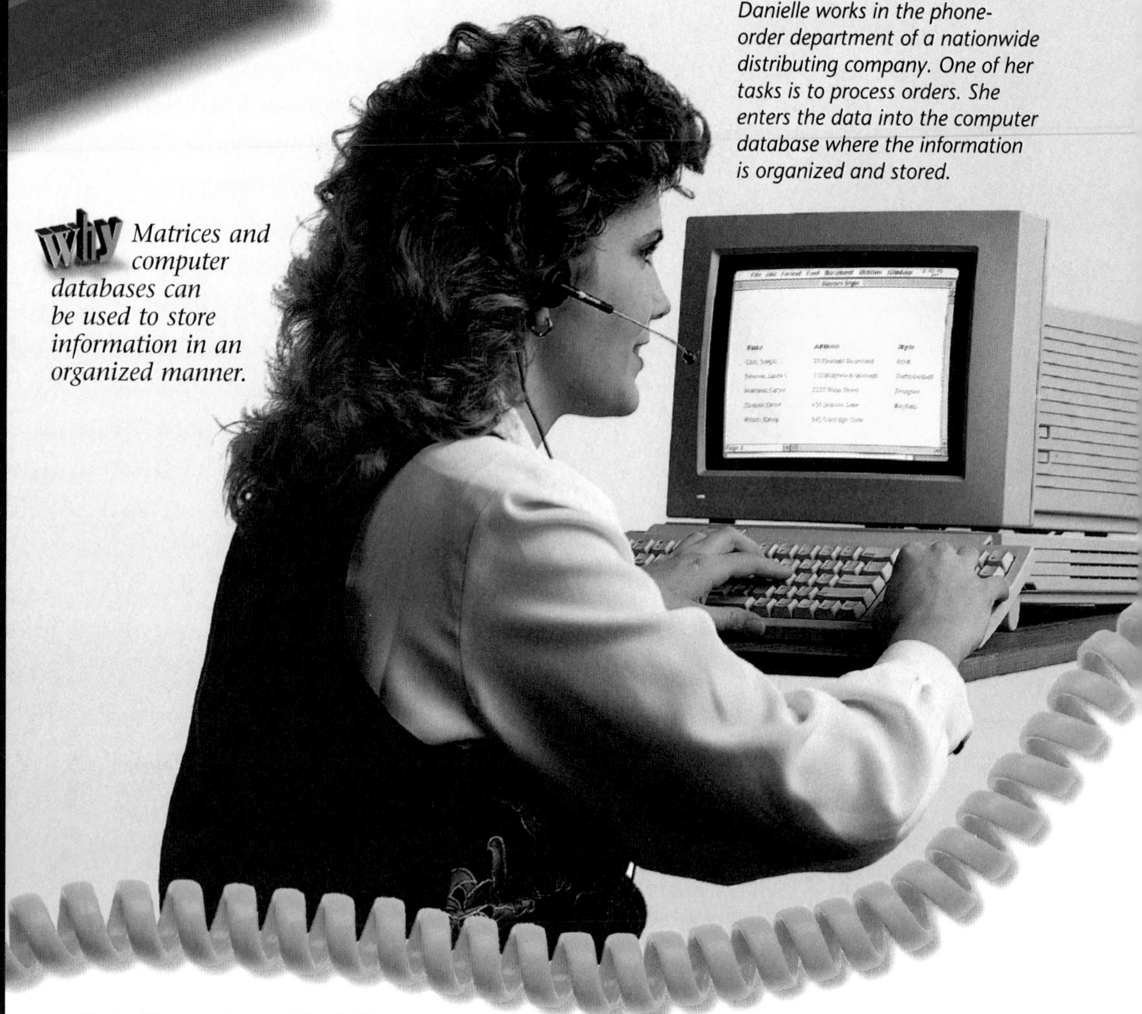

Data Processing The following matrix contains names and addresses.

	Last Name	First Name	Address	City	State	Zip
1	Abrams	Beth	123 42nd Ave.	New York	NY	01257
2	Andrews	Julian	181 Poe Terr.	Columbus	OH	43230
3	Ballesteros	Silvia	514 Main St.	Sonora	CA	95370
⋮	⋮	⋮	⋮	⋮	⋮	⋮
508	Zadhava	Vati	1282 Oak Ln.	Ames	IA	61630

The vertical dots in each column indicate that some information is omitted in the matrix. The numbers on the left side of the matrix are row labels, and the headings above the matrix are column labels. Labels are not part of the matrix. In this matrix, there are 508 rows and 6 columns.

ALTERNATIVE teaching strategy

Hands-On Provide students with several BINGO cards. Ask them to discuss the similarities and differences among the cards. Students should notice that all BINGO cards have the same dimensions. Some cards may have some of the same elements, but no two cards are exactly the same. No two have exactly the same elements in every location. Extend this activity by asking students to write the element given in a particular row and column that the teacher or one of the students calls out. Show students that BINGO cards are set up and read like matrices.

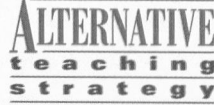

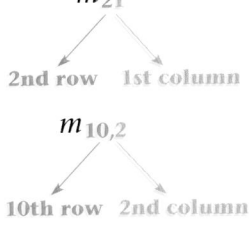

MATRIX

A **matrix** is any rectangular arrangement of elements in rows and columns enclosed by brackets.

Matrix Notation

We need to establish a convention that will allow us to locate any element in the matrix of names and addresses. Naming the matrix M, let the entry m_{ab} represent the data in row a, column b. When either a or b is greater than 9, use a comma to separate them: $m_{a,b}$. Using this notation, the matrix of names and addresses becomes

m_{21}

2nd row   1st column

$m_{10,2}$

10th row   2nd column

$$M = \begin{bmatrix} m_{11} & m_{12} & m_{13} & m_{14} & m_{15} & m_{16} \\ m_{21} & m_{22} & m_{23} & m_{24} & m_{25} & m_{26} \\ m_{31} & m_{32} & m_{33} & m_{34} & m_{35} & m_{36} \\ \vdots & \vdots & \vdots & \vdots & \vdots & \vdots \\ m_{508,1} & m_{508,2} & m_{508,3} & m_{508,4} & m_{508,5} & m_{508,6} \end{bmatrix}$$

where m_{34} is Sonora, m_{25} is OH, and $m_{508,3}$ is 1282 Oak Ln.

What is the name of the entry where "Beth" is stored?

EXAMPLE 1

Refer to matrix M, containing names and addresses. Find the data entered in each location.

A m_{11} **B** m_{23} **C** m_{32} **D** $m_{508,6}$

Solution

A The element located in row 1, column 1 is Abrams. Thus, m_{11} is Abrams.

B The element located in row 2, column 3 is 181 Poe Terr. Thus, m_{23} is 181 Poe Terr.

C The element located in row 3, column 2 is Silvia. Thus, m_{32} is Silvia.

D The element located in row 508, column 6 is 61630. Thus, $m_{508,6}$ is 61630. ❖

ENRICHMENT Ask students to enter and display matrices of various dimensions on their graphics calculator. Refer to the Matrices section of your *HRW Technology Handbook* for specific instructions on entering matrices for each type of graphics calculator. Encourage students to use negative numbers, decimals, and whole numbers. Ask students to determine how many matrices their calculators will store at one time and the maximum dimensions possible for these matrices.

Alternate Example 1

Mr. Montel uses the matrix below to organize his grade book. He named this matrix N.

	Name	Q_1	Q_2	T_1
1	T. Albertson	80	92	83
2	S. Ballard	79	75	72
3	V. Champeny	91	93	95
$\vdots$	$\vdots$	$\vdots$	$\vdots$	$\vdots$
20	R. Zelade	91	93	95

Refer to matrix N containing names and grades for quizzes and tests. Find the data entered in each location.

A. n_{24} [**72**]
B. n_{13} [**92**]
C. n_{31} [**V. Champeny**]
D. $n_{20,4}$ [**95**]

Use Transparency 14

Alternate Example 2

Refer to matrix N in Alternate Example 1.

A. What grade did S. Ballard receive on Quiz 2? Name this location. **[75; the information is given in row 2, column 3, n_{23}]**

B. Describe the data in $n_{20,2}$. **[The data shown in row 20, column 2 is 91. This is the score that R. Zelade received on Quiz 1.]**

Aongoing SSESSMENT

$R_{3×2}$ has 3 rows and 2 columns.

Alternate Example 3

What are the dimensions of matrix B?

$$B = \begin{bmatrix} 2 & -3 & 7 & 8 & -4 & 1 \\ 9 & 0 & -4 & -2 & 1 & 5 \end{bmatrix}$$

$[2 × 6]$

TEACHING tip

Be sure students see the consistency in naming an element of a matrix and in giving the dimensions of a matrix. In both cases, the number of rows is given first.

EXAMPLE 2

Payroll   Matrix P holds payroll information for a small company. The first column contains the social security number of each employee. The remaining columns, from left to right, contain the number of hours worked, hourly wage, gross pay, federal tax, state tax, and net pay for each employee.

$$P = \begin{bmatrix} 463\text{-}33\text{-}3835 & 36 & 23.78 & 856.08 & 128.41 & 34.24 & 693.43 \\ 461\text{-}53\text{-}8150 & 40 & 11.26 & 450.40 & 67.56 & 18.02 & 364.82 \\ 128\text{-}22\text{-}9241 & 40 & 8.52 & 340.80 & 51.12 & 13.63 & 276.05 \\ 288\text{-}44\text{-}8653 & 46 & 9.28 & 454.72 & 68.21 & 18.19 & 368.32 \\ 356\text{-}30\text{-}7402 & 43 & 6.25 & 278.12 & 41.72 & 11.12 & 225.28 \end{bmatrix}$$

(A) What is the hourly wage for employee 288-44-8653? Name this location.

(B) Write an expression, in matrix notation, for the total number of hours worked by each employee.

Solution

(A) Locate employee 288-44-8653 in row 4, column 1 and the hourly wage in row 4, column 3. The value given is 9.28. Therefore, employee 288-44-8653 makes \$9.28 per hour. The name of this location is p_{43}.

(B) The hours worked are in column 2. Thus, in matrix notation, the total number of hours worked by all employees is

$$p_{12} + p_{22} + p_{32} + p_{42} + p_{52}. \;❖$$

Matrix Dimensions

The dimensions of a matrix are given by the number of rows and the number of columns. Matrix P in Example 2 has 5 rows and 7 columns. Thus, the dimensions of matrix P are $5 × 7$, read 5 by 7. This is usually written $P_{5×7}$.

How many rows and columns does a matrix named $R_{3×2}$ have?

EXAMPLE 3

What are the dimensions of matrix A?

$$A = \begin{bmatrix} 4 & -3 & 7 & 1 \\ 2 & 0 & 5 & -2 \\ -6 & 1 & -4 & 3 \end{bmatrix}$$

Solution

Since matrix A has 3 rows and 4 columns, its dimensions are $3 × 4$. ❖

INCLUSION **strategies**

Using Visual Models Provide large grids of different sizes, labeling each cell with the name of the location of the cell. Make the cells large enough so that you can place small index cards on top of each cell without overlapping. Then write the actual elements of a matrix on index cards and place them in their appropriate cell on the grid. Students can then lift each card up to see the name of the location of the element underneath. Reverse this activity by asking students to place the cards at locations you specify.

Matrix Equality

When each pair of corresponding elements of two matrices are equal, the two matrices are said to be equal.

$$P = \begin{bmatrix} p_{11} & p_{12} & p_{13} \\ p_{21} & p_{22} & p_{23} \end{bmatrix} \quad Q = \begin{bmatrix} q_{11} & q_{12} & q_{13} \\ q_{21} & q_{22} & q_{23} \end{bmatrix}$$

If $P = Q$, then $p_{11} = q_{11}$, $p_{12} = q_{12}$, $p_{13} = q_{13}$, and so on.

You can represent linear equations with matrices.

EXAMPLE 4

Use matrices to represent the linear equations $\begin{cases} 5x - 2y = 3 \\ x + 4y = 7 \end{cases}$.

Solution▸

Store the left side of each equation in one matrix and the right side of each equation in another matrix.

$$\begin{cases} 5x - 2y = 3 \\ x + 4y = 7 \end{cases} \rightarrow \begin{bmatrix} 5x - 2y \\ x + 4y \end{bmatrix} = \begin{bmatrix} 3 \\ 7 \end{bmatrix} ❖$$

CRITICAL Thinking
What must be true about the dimensions of two matrices that are equal?

Matrix Addition

•Exploration• Operating With Matrices

Graphics Calculator

1 Enter the elements of matrix A into your graphics calculator. Display matrix A to check your entries.

$$A = \begin{bmatrix} 4 & -3 & 7 & 1 \\ 2 & 0 & 5 & -2 \\ -6 & 1 & -4 & 3 \end{bmatrix}$$

2 With matrix A stored in your calculator, enter matrix B. What are the dimensions of A? of B? What do matrices A and B have in common?

$$B = \begin{bmatrix} 8 & -4 & 2 & 3 \\ 0 & 5 & -6 & 0 \\ -2 & 7 & 1 & -3 \end{bmatrix}$$

3 Find $A + B$. Explain how your calculator performed this addition. Why must matrices A, B, and $A + B$ have the same dimensions?

4 Find $B - A$. Explain how your calculator performed this subtraction. Why must matrices A, B, and $B - A$ have the same dimensions?

5 Find $A - A$. Name the resulting matrix C.

6 Find $C + B$. What is a good name for matrix C?

7 Find $-1 \times B$. Describe the relationship between the resulting matrix and matrix B. ❖

RETEACHING the lesson

Inviting Participation Draw the following table on the chalkboard.

	Kitchen	Living Room	Own Bedroom	Other Bedroom	Bathroom
# of sq feet					
# of pictures on the wall					

Ask students to fill in the table with information from their home. Have students write the data in their table in a matrix. Then have students work in groups to find the sum of all of the matrices in their group. Ask them to use this sum to order the rooms from largest to smallest based upon square feet. Then ask them to use their sum to predict which room in a house is most likely to have the greatest number of pictures on the wall.

Alternate Example 4

Use matrices to represent the linear equations $\begin{cases} 3x + 27 = 15 \\ 5x - y = 12 \end{cases}$

$$\left[\begin{bmatrix} 3x + 2y \\ 5x - y \end{bmatrix} = \begin{bmatrix} 15 \\ 12 \end{bmatrix} \right]$$

CRITICAL Thinking

The dimensions must be the same.

Exploration Notes

Students should realize that two matrices cannot be added or subtracted unless they have exactly the same dimensions.

Ongoing ASSESSMENT

3. Matrices A and B must have the same dimensions so that corresponding elements can be added. The dimensions of the sum matrix, $A + B$, must be the same as A and B.

4. Matrices B and A must have the same dimensions so that corresponding elements can be subtracted. The difference matrix, $B - A$, must have the same dimensions as B and A.

6. Matrix C is the zero matrix, or the additive identity matrix.

7. Each element in the resulting matrix has the opposite sign of its corresponding element in matrix B.

ASSESS

Selected Answers
Odd-numbered Exercises 7–45

Assignment Guide

Core 1–8, 10–11, 13–36, 41–49

Core Plus 1–6, 9, 11–49

Technology

A graphics calculator could be used for Exercises 35 and 36. You may want to have students use pencil and paper first and then use the calculator to check their answers.

EXERCISES & PROBLEMS

Communicate

1. List several different businesses that might use matrices to store data. For each business, explain what data could be stored in the matrix.

2. Explain how to find the dimensions of a matrix.

3. How do you find the element m_{52} in a matrix named M?

4. Describe the location of the terms b_{21}, b_{22}, b_{23}, and b_{24} in a matrix named B.

5. Describe the location of the terms a_{11}, a_{22}, and a_{33} in a matrix named A.

6. Design a matrix to hold information about your friends or family. What information might be stored in the matrix?

Practice & Apply

Give the dimensions of each matrix.

7. $A = \begin{bmatrix} 5 & 7 & -3 & 0 \\ -2 & 1 & 8 & 11 \end{bmatrix}$
2×4

8. $B = \begin{bmatrix} \text{Akim} & \text{Maria} & \text{Hy} \\ \text{Bess} & \text{Jamay} & \text{Cary} \\ \text{Devon} & \text{Kelly} & \text{Isabel} \\ \text{Larry} & \text{Milan} & \text{Robbi} \end{bmatrix}$ 4×3

9. $C = \begin{bmatrix} 7 \\ 2 \\ 6 \end{bmatrix}$ 3×1

Refer to the matrices in Exercises 7–9. Give the contents of the following entries.

10. Entry a_{23} 8

11. Entry b_{12} Maria

12. Entry c_{31} 6

Geography Tracy and Ron both collect maps. Together they have a variety of maps from the 1960s to the 1990s. Matrix M shows the number of each type of map they have.

$M = \begin{array}{l} \\ \\ \text{Europe} \\ \text{Asia} \\ \text{North America} \\ \text{Africa} \end{array} \begin{array}{cccc} \textbf{'60s} & \textbf{'70s} & \textbf{'80s} & \textbf{'90s} \\ 3 & 1 & 4 & 2 \\ 5 & 3 & 6 & 3 \\ 2 & 7 & 9 & 5 \\ 8 & 5 & 4 & 6 \end{array}$

Continents

13. What are the dimensions of matrix M? 4×5

14. Describe the data in location m_{42}.

15. Describe the data in location m_{21}.

16. Write an expression, in matrix notation, for the total number of maps of Africa that Ron and Tracy have.
$m_{42} + m_{43} + m_{44} + m_{45}$

17. Write an expression, in matrix notation, for the total number of maps from the 1960s that Tracy and Ron have. $m_{12} + m_{22} + m_{32} + m_{42}$

14. $m_{42} = 8$; Tracy and Ron have 8 maps of Africa from the 1960s.

15. $m_{21} = $ Asia; Tracy and Ron's map collection includes maps of Asia.

For Exercises 18–20, give the dimensions of the matrix, and use matrix notation to describe the contents of any three entries.

18. $L = \begin{bmatrix} 2 & 7 & 6 & 4 & 1 \\ 5 & 11 & 0 & 6 & 3 \\ 14 & 8 & 17 & 9 & 6 \end{bmatrix}$
    3×5

19. $Z = \begin{bmatrix} 4 & 0 & 7 \end{bmatrix}$
    1×3

20. $N = \begin{bmatrix} \text{Blue} & 32 & \text{Long} & 175 \\ \text{Gray} & 36 & \text{Regular} & 215 \end{bmatrix}$
    2×4

Use a matrix to represent each of the following sets of linear equations.

21. $\begin{cases} 3x - 2y = -8 \\ -2x + 9y = 3 \end{cases}$ $\begin{bmatrix} 3x - 2y \\ -2x + 9y \end{bmatrix} = \begin{bmatrix} -8 \\ 3 \end{bmatrix}$

22. $\begin{cases} 5x - 6y = 0 \\ 7x + 3y = 0 \end{cases}$ $\begin{bmatrix} 5x - 6y \\ 7x + 3y \end{bmatrix} = \begin{bmatrix} 0 \\ 0 \end{bmatrix}$

23. $\begin{cases} -x + 7y = 10 \\ -3x - 8y = 1 \\ 5x - 2y = 4 \end{cases}$

24. $\begin{cases} -3x - 6y = 5 \\ 4x + 8y = 5 \\ -5x - y = 1 \end{cases}$

Sociology The following matrix shows the number of activities for three extracurricular groups during the fall semester of school.

	Aug.	Sept.	Oct.	Nov.	Dec.
Drama productions	0	1	2	1	2
Soccer games	1	4	3	3	0
Journalism publications	1	2	3	3	2

25. What are the dimensions of this matrix? 3×5

26. Name this matrix E. Write an expression, in matrix notation, for the total number of activities that occurred in September. $e_{12} + e_{22} + e_{32}$

27. Write an expression for the total number of drama productions in the fall. $e_{11} + e_{12} + e_{13} + e_{14} + e_{15}$

28. Write an expression, in matrix notation, for the month during which the most activities occurred. $e_{13} + e_{23} + e_{33}$

For Exercises 29 and 30, create and name the matrices with the given elements.

29. g_{12} is 4, g_{32} is 7, g_{42} is 8, g_{21} is 9, g_{11} is 3, g_{22} is 0, g_{31} is 5, and g_{41} is 2.

30. w_{11} is *plastic*, w_{22} is *6-top*, w_{32} is *2-top*, w_{13} is *picnic*, w_{21} is *wood*, w_{12} is *4-top*, w_{33} is *round*, w_{31} is *marble*, and w_{23} is *dinner*.

Small Business A local farmer sells fresh produce from her farm. One weekend she sold 27 squash, 31 tomatoes, 24 peppers, and 18 melons. On the same weekend, the produce section of a grocery store sold 48 squash, 72 tomatoes, 61 peppers, and 25 melons.

31. Create a 2×4 matrix to store this data. Name this matrix P.

32. Could you have created a matrix with different dimensions? Explain.

33. What is the location of the number of peppers that the local farmer sold?

34. Use matrix notation to describe the data stored in the second row, first column.

18. 3×5; Answers may vary. Examples are: l_{11} is 2, l_{24} is 6, and l_{33} is 17.

19. 1×3; z_{11} is 4, z_{12} is 0, and z_{13} is 7

20. 2×4; Answers may vary. Examples are: n_{14} is 175, n_{22} is 36, and n_{23} is Regular

23. $\begin{bmatrix} -x + 7y \\ -3x - 8y \\ 5x - 2y \end{bmatrix} = \begin{bmatrix} 10 \\ 1 \\ 4 \end{bmatrix}$

24. $\begin{bmatrix} -3x - 6y \\ 4x + 8y \\ -5x - y \end{bmatrix} = \begin{bmatrix} 5 \\ 5 \\ 1 \end{bmatrix}$

29. $G = \begin{bmatrix} 3 & 4 \\ 9 & 0 \\ 5 & 7 \\ 2 & 8 \end{bmatrix}$

The answers to Exercises 30–34 can be found in Additional Answers beginning on page 842.

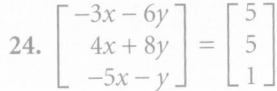

alternative ASSESSMENT

Portfolio Assessment

Have students complete the following project:

- Cut out information given in a matrix-like format from a magazine or newspaper. Put this information in a matrix with row and column labels.
- Comment on why you think the information was given in a matrix-like format.
- Determine the dimensions of your matrix.
- Use matrix notation to identify three locations in your matrix. Describe the data located in each of these locations.

For Exercises 35–37, let $A = \begin{bmatrix} 7 & 3 & -1 & 5 \\ -2 & 8 & 0 & -4 \end{bmatrix}$ and $B = \begin{bmatrix} 6 & 0 & 11 & -3 \\ -5 & 2 & -8 & 9 \end{bmatrix}$.

35. Find $A + B$. $\begin{bmatrix} 13 & 3 & 10 & 2 \\ -7 & 10 & -8 & 5 \end{bmatrix}$ **36.** Find $B - A$. $\begin{bmatrix} -1 & -3 & 12 & -8 \\ -3 & -6 & -8 & 13 \end{bmatrix}$

37. Is it true that $A + B = B + A$? Name this property for matrices.
yes, comm prop of matrix add

Inventory Control A record store manager wishes to organize information about the inventory of music. The store carries records, tapes, and compact discs of country, jazz, rock, blues, and classical music.

38. Give possible numbers for each type of music in each type of format.

39. Create a matrix to store this information. Name this matrix M.

40. Indicate what the number in location m_{23} refers to in your matrix.

Look Back

41. Write the equation of the line that contains the points $(4, 0)$ and $(-9, 11)$. **[Lesson 1.2]**

42. If y varies directly as x, and y is 49 when x is 14, find x when y is 63. **[Lesson 1.4]** 18

43. Describe what it means for two equations to be equivalent. Be aware that stating *two equations are equal* is incorrect. **[Lesson 1.5]**

44. Find the inverse of $f(x) = 2x - 1$. **[Lesson 3.2]** $f^{-1}(x) = \frac{x+1}{2}$

For Exercises 45 and 46, let $f(x) = \frac{1}{x+1}$ and $g(x) = 2x + 5$. **[Lesson 3.3]**

45. Find $(f \circ g)(x)$, and give the domain and range of this composite function. $\frac{1}{2x+6}$; $x \neq -3$; $y \neq 0$

46. Find $(g \circ f)(x)$, and give the domain and range of this composite function. $\frac{5x+7}{x+1}$; $x \neq -1$; $y \neq 5$

Look Beyond

Look Beyond

These exercises will help prepare students for their work with matrix multiplication in the next lesson. Be sure to emphasize the use of the distributive property. It may be helpful to have students draw diagrams to connect Exercises 48 and 49.

47. Explain how the Distributive Property is used to multiply $(x + 4)(x - 3)$.

48. It is a sign of good sportsmanship to shake hands with the opposing team after a game is over. Each member of one team shakes the hand of each member of the other team. Explain how this demonstrates the Distributive Property.

49. Think of two matrices, A and B, as opposing teams. Matrix A is a team of rows, and matrix B is a team of columns. Draw a diagram to illustrate how matrix A can shake hands with matrix B.

38. Answers may vary. Examples are records: 10 country, 15 jazz, 5 rock, 8 blues, 3 classical; tapes: 5 country, 9 jazz, 1 rock, 13 blues, 6 classical; compact discs: 2 country, 1 jazz, 9 rock, 5 blues, 10 classical

The answers to Exercises 39 and 49 can be found in Additional Answers beginning on page 842.

40. Answers may vary. An example is $m_{23} = 1$ rock tape.

41. $y = -\frac{11}{13}x + \frac{44}{13}$

43. When two equations are equivalent, they have the same solution(s).

47. Each term of one factor is multiplied by the other factor: $x(x - 3) + 4(x - 3)$. Then each term inside the parentheses is multiplied

by the factor outside the parentheses, and like terms are combined: $x^2 - 3x + 4x - 12$ or $x^2 + x - 12$.

48. The handshakes are member-by-member distributed from one team to each member of the other team; that is, each member of one team will shake hands with each member of the other team.

LESSON 4.2 Matrix Multiplication

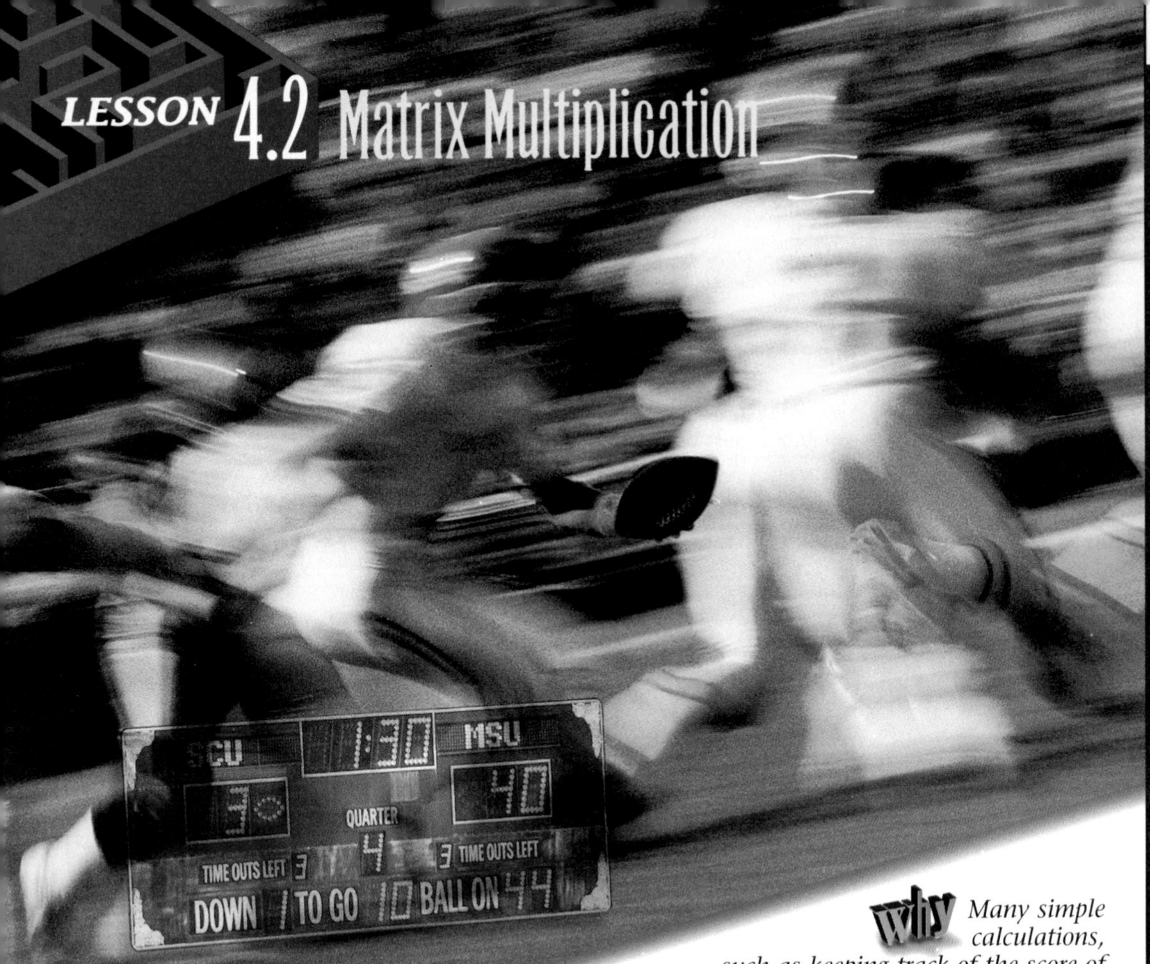

why *Many simple calculations, such as keeping track of the score of a football game, can be done using matrix multiplication. When you understand how to multiply matrices, you can perform more complicated calculations.*

Sports Suppose a football team scores 5 touchdowns, 4 extra points, and 2 field goals. A touchdown is worth 6 points, an extra point is worth 1 point, and a field goal is worth 3 points. To determine the final score, evaluate as follows.

(5 touchdowns)(6 points) + (4 extra points)(1 point) + (2 field goals)(3 points)

= 30 points + 4 points + 6 points

= 40 points total

You can perform the same operation using matrix multiplication.

$$\begin{array}{c} \text{Touchdowns} \quad \text{Extra points} \quad \text{Field goals} \\ \begin{bmatrix} 5 & 4 & 2 \end{bmatrix} \end{array} \begin{array}{c} \text{Point values} \\ \begin{bmatrix} 6 \\ 1 \\ 3 \end{bmatrix} \end{array} = [(5)(6) + (4)(1) + (2)(3)] = [40]$$

ALTERNATIVE teaching strategy

Using Visual Models Use an overhead projector and colored markers or highlighters to demonstrate matrix multiplication. Use one color to circle or highlight specific rows and columns that are to be multiplied. Then use the same color to circle or highlight the corresponding element in the product matrix. Use different colors for other rows, columns, and corresponding elements.

PREPARE

Objectives

- Perform matrix multiplication.

RESOURCES

- Practice Master 4.2
- Enrichment Master 4.2
- Technology Master 4.2
- Lesson Activity Master 4.2
- Quiz 4.2
- Spanish Resources 4.2

Assessing Prior Knowledge

Give the dimensions of each matrix.

1. $M = \begin{bmatrix} -5 & 0 & 7 & -11 \\ 0 & 3 & 7 & -19 \end{bmatrix}$
   [2 × 4]

2. $N = \begin{bmatrix} -15 & 7 \\ 10 & 9 \\ 6 & 6 \end{bmatrix}$
   [3 × 2]

3. $B = \begin{bmatrix} -7 & 0 & 13 \\ -2 & 4 & -2 \\ 12 & 1 & 10 \end{bmatrix}$
   [3 × 3]

4. $A = \begin{bmatrix} 6 & 16 & 5 & 9 & -3 \end{bmatrix}$
   [1 × 5]

TEACH

why Point out that matrix multiplication is used in many disciplines; football statistics are one example. Have students work together to create a list of other scores that can be computed using matrix multiplication.

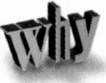

No. Two 4×3 matrices cannot be multiplied because their inner dimensions, 3 and 4, are not the same.

Alternate Example 1

Let $M = \begin{bmatrix} -2 & 1 & 8 \\ -9 & 3 & -4 \\ 15 & 1 & 12 \end{bmatrix}$ and

$N = \begin{bmatrix} -5 & 13 & -6 & -1 & 0 \\ -1 & 1 & 3 & 2 & 4 \\ 5 & 4 & 5 & 4 & 5 \end{bmatrix}$

A. Is it possible to find the product MN? Explain. [Yes. The inner dimensions of $M_{3\times3}$ and $N_{3\times5}$, 3 and 3, are equal.]

B. Is it possible to find the product NM? Explain. [No. The inner dimensions of $N_{3\times5}$ and $M_{3\times3}$, 5 and 3, are not equal.]

Alternate Example 2

Let $X = \begin{bmatrix} -5 & 0 & 0 \\ -2 & 1 & 10 \\ 16 & 6 & 11 \end{bmatrix}$ and $Y = [-2 \ -3 \ 8]$.

What are the dimensions of the product matrix YX? [3×1]

 Use Transparency 15

Inner Dimensions

Refer to the football example on page 171. Notice that a 1×3 matrix is multiplied by a 3×1 matrix. This multiplication is possible because the *inner dimensions*, 3 and 3, are the same.

$1 \times 3 \qquad 3 \times 1$

Inner dimensions

INNER DIMENSIONS

To perform matrix multiplication, the **inner dimensions** of the two matrices *must* be the same.

Can you multiply two 4×3 matrices? Explain your answer.

EXAMPLE 1

Let $A = \begin{bmatrix} 2 & -1 & 3 \\ 0 & 5 & 1 \end{bmatrix}$ and $B = \begin{bmatrix} 1 & 0 \\ 2 & -1 \end{bmatrix}$.

Ⓐ Is it possible to find the product BA? Explain.

Ⓑ Is it possible to find the product AB? Explain.

Solution

Ⓐ Yes. The inner dimensions of $B_{2\times2} \cdot A_{2\times3}$, 2 and 2, are equal.

Ⓑ No. The inner dimensions of $A_{2\times3} \cdot B_{2\times2}$, 3 and 2, are not equal. ❖

Outer Dimensions

Examine matrices A and B in Example 1. To find the dimensions of the product matrix BA, look at the *outer* dimensions of B and A.

$B_{2\times2} \cdot A_{2\times3} = BA_{2\times3}$

Outer dimensions

If two matrices are multiplied, their **outer dimensions** become the dimensions of the resulting product matrix. The outer dimensions of $B_{2\times2} \cdot A_{2\times3}$ are 2 and 3, so the dimensions of the product BA are 2×3.

EXAMPLE 2

Let $C = [2 \ 0 \ -1]$ and $D = \begin{bmatrix} 1 & 3 & 2 \\ 7 & -2 & 1 \\ 0 & 4 & 3 \end{bmatrix}$.

What are the dimensions of the product matrix CD?

Solution

The outer dimensions of $C_{1\times3} \cdot D_{3\times3}$ are 1 and 3. Thus, the product CD will have dimensions 1×3. ❖

interdisciplinary CONNECTION

Sports Present the following problem to students to discuss and solve.

In a swimming competition among 5 teams at summer camp, 3 teams, Team A, Team B, and Team C, were vying for first place. The outcomes of 12 events for these three teams are shown in Matrix M. Teams were given 5 points for first place, 3 points for second place, and 1 point for third place in each event. This scoring is shown in Matrix N.

$M = \begin{matrix} A \\ B \\ C \end{matrix} \begin{bmatrix} 5 & 4 & 1 \\ 4 & 7 & 0 \\ 3 & 1 & 6 \end{bmatrix}$ (1st 2nd 3rd); $N = \begin{matrix} 1st \\ 2nd \\ 3rd \end{matrix} \begin{bmatrix} 5 \\ 3 \\ 1 \end{bmatrix}$ (Points)

How can you determine which team won the competition? [By finding the product MN] Who won the competition? [Team B won the competition with 41 points.]

Product Matrix

To find the product CD, pair rows of C with columns of D.

$$\underbrace{[2\ \ 0 -1]}_{C}\underbrace{\begin{bmatrix} 1 & 3 & 2 \\ 7 & -2 & 1 \\ 0 & 4 & 3 \end{bmatrix}}_{D} = \begin{bmatrix} \overbrace{[2\ \ 0 -1]\begin{bmatrix}1\\7\\0\end{bmatrix}}^{cd_{11}} & \overbrace{[2\ \ 0 -1]\begin{bmatrix}3\\-2\\4\end{bmatrix}}^{cd_{12}} & \overbrace{[2\ \ 0 -1]\begin{bmatrix}2\\1\\3\end{bmatrix}}^{cd_{13}} \end{bmatrix}$$

$$= \begin{bmatrix} \overbrace{(2)(1)+(0)(7)+(-1)(0)}^{cd_{11}} & \overbrace{(2)(3)+(0)(-2)+(-1)(4)}^{cd_{12}} & \overbrace{(2)(2)+(0)(1)+(-1)(3)}^{cd_{13}} \end{bmatrix}$$

$$= \begin{bmatrix} 2 & 2 & 1 \end{bmatrix}$$

Thus, the product CD is $[2\ \ 2\ \ 1]$.

EXAMPLE 3

Let $X = \begin{bmatrix} -1 & 4 & 3 & 5 \\ 2 & 0 & -6 & 1 \end{bmatrix}$ and $Y = \begin{bmatrix} 2 & 0 \\ 1 & 4 \\ 3 & -2 \\ -5 & 1 \end{bmatrix}$.

A What row of X and column of Y is used to determine the element xy_{21} of the product matrix XY?

B Will there be an entry xy_{23} in the product matrix XY? Explain.

C Find the product XY.

Solution▶

A To determine the element xy_{21} of the product matrix XY, row 2 of X and column 1 of Y are used.

B No. There would not be an entry xy_{23} because matrix Y does not have a column 3.

C To find the product XY, pair rows of X with columns of Y as shown.

$$\underbrace{\begin{bmatrix} -1 & 4 & 3 & 5 \\ 2 & 0 & -6 & 1 \end{bmatrix}}_{X}\underbrace{\begin{bmatrix} 2 & 0 \\ 1 & 4 \\ 3 & -2 \\ -5 & 1 \end{bmatrix}}_{Y} = \begin{bmatrix} \overbrace{[-1\ 4\ 3\ 5]\begin{bmatrix}2\\1\\3\\-5\end{bmatrix}}^{xy_{11}} & \overbrace{[-1\ 4\ 3\ 5]\begin{bmatrix}0\\4\\-2\\1\end{bmatrix}}^{xy_{12}} \\ \underbrace{[2\ 0\ -6\ 1]\begin{bmatrix}2\\1\\3\\-5\end{bmatrix}}_{xy_{21}} & \underbrace{[2\ 0\ -6\ 1]\begin{bmatrix}0\\4\\-2\\1\end{bmatrix}}_{xy_{22}} \end{bmatrix} = \underbrace{\begin{bmatrix} -14 & 15 \\ -19 & 13 \end{bmatrix}}_{XY}$$

❖

Try This Find the product YX.

Alternate Example 4

Write the two equations represented by the matrix equation.

$$\begin{bmatrix} 3 & -1 \\ -2 & 4 \end{bmatrix}\begin{bmatrix} x \\ y \end{bmatrix} = \begin{bmatrix} 7 \\ 2 \end{bmatrix}$$

$[3x - y = 7; -2x + 4y = 2]$

Alternate Example 5

The local movie theater is showing a sneak preview of a new movie on Friday, Saturday, and Sunday at 12:30 and 8:00. The prices for the performances are listed in Matrix T. The number of tickets sold at each performance is shown in Matrix N.

$$\begin{array}{c} \\ 12:30 \\ 8:00 \end{array}\begin{array}{ccc} \text{Child} & \text{Adult} & \text{Senior} \end{array}$$
$$\begin{bmatrix} \$4.00 & \$4.00 & \$4.00 \\ \$4.00 & \$7.00 & \$5.00 \end{bmatrix} = T$$

$$\begin{array}{c} \text{Child} \\ \text{Adult} \\ \text{Senior} \end{array}\begin{array}{ccc} \text{Fri.} & \text{Sat.} & \text{Sun.} \\ 10 & 120 & 100 \\ 150 & 100 & 120 \\ 100 & 80 & 60 \end{array} = N$$

A. Give the inner labels of $T \cdot N$. [**Child, Adult, Senior**]

B. Give the outer labels of $T \cdot N$. [**row labels: 12:30 and 8:00; column labels: Fri., Sat., Sun.**]

C. Find the product TN.

$$\begin{array}{c} \\ 12:30 \\ 8:00 \end{array}\begin{array}{ccc} \text{Fri.} & \text{Sat.} & \text{Sun.} \\ \$1040 & \$1200 & \$1120 \\ \$1590 & \$1580 & \$1540 \end{array}$$

D. On which day is the revenue highest? [**Saturday**]

E. Is the revenue greater at the 12:30 show or the 8:00 show? [**8:00**]

Matrix multiplication can be used to represent a system of equations as one matrix equation.

EXAMPLE 4

Write the system represented by the matrix equation.

$$\begin{bmatrix} 3 & 1 \\ 2 & -4 \end{bmatrix}\begin{bmatrix} x \\ y \end{bmatrix} = \begin{bmatrix} -12 \\ 7 \end{bmatrix}$$

Solution

Multiply the two matrices.

$$\begin{bmatrix} 3 & 1 \\ 2 & -4 \end{bmatrix}\begin{bmatrix} x \\ y \end{bmatrix} = \begin{bmatrix} -12 \\ 7 \end{bmatrix}$$

$$\begin{bmatrix} 3x + 1y \\ 2x - 4y \end{bmatrix} = \begin{bmatrix} -12 \\ 7 \end{bmatrix} \rightarrow \begin{cases} 3x + y = -12 \\ 2x - 4y = 7 \end{cases}$$

Meaning of the Product Matrix

You know that the inner dimensions of two matrices must be the same in order to find their product. Also, the outer dimensions of the two matrices become the dimensions of the product matrix. The same is true for the *labels* of two matrices that are multiplied. The outer labels become the labels of the product matrix, and the inner labels of the matrices must be the same in order for the product matrix to have meaning.

EXAMPLE 5

Nutrition Karl and Heather mix dried fruit and nuts in two snack mixes. Matrix N lists, in grams per scoop, the amounts of protein, carbohydrates, and fat for both dried fruit and nuts. Matrix G shows the combination, given in number of scoops of dried fruit and nuts, for the two mixes.

$$\begin{array}{c} \text{Protein} \\ \text{Carbohydrates} \\ \text{Fat} \end{array}\begin{array}{cc} \begin{array}{c}\text{Dried}\\\text{fruit}\end{array} & \text{Nuts} \\ 3 & 20 \\ 65 & 21 \\ 1 & 52 \end{array} = N$$

$$\begin{array}{c} \text{Dried fruit} \\ \text{Nuts} \end{array}\begin{array}{cc} \begin{array}{c}\text{Sport}\\\text{mix}\end{array} & \begin{array}{c}\text{Camp}\\\text{mix}\end{array} \\ 4 & 3 \\ 2 & 3 \end{array} = G$$

A Give the inner labels of $N \cdot G$. **B** Give the outer labels of $N \cdot G$.

C Find the product NG. **D** Which mix has more protein?

E Which mix has less fat?

Solution

A The column labels of N and the row labels of G are **Dried fruit** and **Nuts**. These are the inner labels.

B The row labels of N are **Protein, Carbohydrates**, and **Fat**, and the column labels of G are **Sport mix** and **Camp mix**. These are the outer labels. The row labels of N become the row labels of the product NG. The column labels of G become the column labels of the product NG.

$\overline{\text{R}}$ETEACHING
the
lesson

Using Cognitive Strategies Write matrix M on the chalkboard, where

$$M = \begin{bmatrix} 1 & 2 & 3 & 4 \\ 5 & 6 & 7 & 8 \\ 9 & 10 & 11 & 12 \end{bmatrix}.$$

Ask students to create a matrix N, all of whose elements are 1, for which the product NM exists. Then create another matrix P, all of whose elements are 2, for which the product MP exists. Then have students work in groups to compare the dimensions of their matrices N and P. Ask each student to find the products NM and MP. Have students compare their products with the products found by the other members of their group.

C

$$\underset{N_{3\times2}}{\begin{bmatrix} 3 & 20 \\ 65 & 21 \\ 1 & 52 \end{bmatrix}} \underset{G_{2\times2}}{\begin{bmatrix} 4 & 3 \\ 2 & 3 \end{bmatrix}} = \begin{matrix} \text{Protein} \\ \text{Carbohydrates} \\ \text{Fat} \end{matrix} \underset{NG_{3\times2}}{\begin{bmatrix} \overset{\text{Sport}}{\underset{\text{mix}}{}} & \overset{\text{Camp}}{\underset{\text{mix}}{}} \\ 52 & 69 \\ 302 & 258 \\ 108 & 159 \end{bmatrix}}$$

D Row 1 of *NG* lists amounts of protein and 69 > 52. So the Camp mix has more protein.

E Row 3 of *NG* lists amounts of fat and 108 < 159. So the Sport mix has less fat. ❖

CRITICAL
Thinking

In Example 5, would the matrix multiplication be meaningful if the row labels and column labels of matrix *G* were exchanged? Explain.

Properties of Matrix Multiplication

Exploration *Matrix Multiplication*

Graphics Calculator

Let $A = \begin{bmatrix} 1 & -2 \\ 3 & -5 \end{bmatrix}$ and $B = \begin{bmatrix} 4 & 7 \\ 1 & 2 \end{bmatrix}$.

1 Find *AB* and *BA*. Does *AB* = *BA*?

2 Let $C = \begin{bmatrix} -5 & 2 \\ -3 & 1 \end{bmatrix}$. Find *AC* and *CA*. Does *AC* = *CA*?

3 The product matrix in Step 2 is called the *identity* matrix. The identity matrix is denoted by the letter *I*. Find the indicated products to complete the following chart.

Column 1	Column 2	Column 3	Column 4	Column 5	Column 6
AI = ?	*BI* = ?	*CI* = ?	(*AB*)*C* = ?	(*CB*)*A* = ?	*BC* = ?
IA = ?	*IB* = ?	*IC* = ?	*A*(*BC*) = ?	*C*(*BA*) = ?	*CB* = ?

4 Look at Columns 1, 2, and 3. Does multiplication with the identity matrix seem to be commutative?

5 Look at Columns 4 and 5. How do parentheses affect the product? Does matrix multiplication seem to be associative?

6 Look at Column 6. Does matrix multiplication seem to be commutative?

7 Why do you think matrix *I* is called the identity matrix? ❖

SOME PROPERTIES OF MATRIX MULTIPLICATION

Let *A*, *B*, and *C* represent any $n \times n$ matrices.

Matrix multiplication is associative: (*AB*)*C* = *A*(*BC*).
If *I* is an $n \times n$ matrix and *AI* = *A* = *IA*, then *I* is the identity matrix.

Technology

A graphics calculator is needed for Exercises 31–34, 36–38, and 51. A graphics calculator may be helpful to check Exercises 23–26.

EXERCISES & PROBLEMS

Communicate

1. Describe the relevance of inner dimensions and outer dimensions to matrix multiplication.

2. What is the next step in the matrix multiplication shown in the photo below?

3. If all of the elements of two matrices are negative, can the product of the two matrices have positive elements? Can the product have negative elements? Explain.

4. Explain how to find matrix products using your graphics calculator.

5. What happens when you use your graphics calculator to multiply two matrices with inner dimensions that are not the same?

6. Describe the importance of labels in terms of the interpretation of the matrix product.

7. What is the identity matrix, and what is its significance?

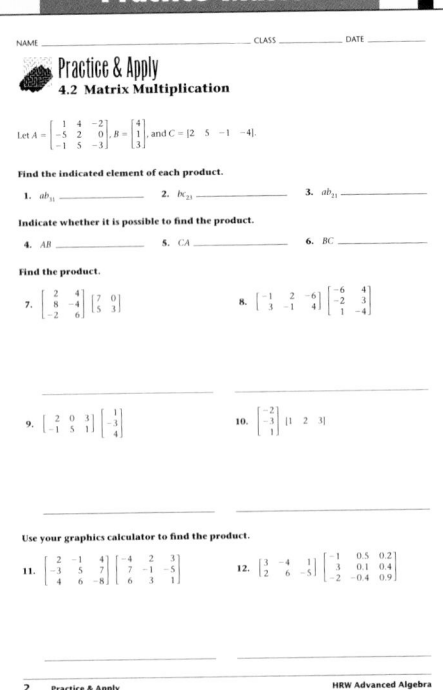

Practice & Apply

Use these matrices for Exercises 8–17.

$$A = \begin{bmatrix} 4 & -2 & 8 & 0 \\ 1 & 3 & -6 & 9 \\ -5 & 7 & 2 & 1 \end{bmatrix} \quad B = \begin{bmatrix} 7 & 6 \\ 2 & -3 \\ -1 & 8 \\ 9 & 5 \end{bmatrix} \quad C = \begin{bmatrix} 0 & 8 \\ -2 & 1 \end{bmatrix}$$

Determine which row of A and which column of B are paired to find the indicated element of the product AB.

8. ab_{12} 1st row A, 2nd col B

9. ab_{22} 2nd row A, 2nd col B

10. ab_{31} 3rd row A, 1st col B

Find the indicated element of the product AB. If the element of AB does not exist, write *does not exist* and explain.

11. ab_{21} 100

12. ab_{13} does not exist, no 3rd col B

13. ab_{11} 16

Determine whether it is possible to find the product.

14. BC yes

15. CB no

16. $(AB)C$ yes

17. $A(BC)$ yes

Sports A high school football team has played 4 games this season. Matrix S shows the number of touchdowns, extra points, and field goals scored in each game. Use the point values matrix to find the indicated scores.

Point values		Touchdowns	Extra points	Field goals
$\begin{bmatrix} 6 \\ 1 \\ 3 \end{bmatrix} = P$	Game 1	2	2	1
	Game 2	4	4	3
	Game 3	3	1	3
	Game 4	5	4	2

with $= S$

18. Find the total number of points scored in all 4 games. 122

19. Find the total score for Game 3. 28

20. Find the total number of points scored in Games 2 and 4. 77

21. Find the difference in the number of points scored in Games 2 and 3. In which of these two games were more points scored? 9; Game 2

22. Suppose matrix S included the information for all 9 games of the season. What would be the dimensions of the resulting product matrix? 9×1

Find the product.

23. $\begin{bmatrix} -1 & 3 & 5 \end{bmatrix} \begin{bmatrix} -4 \\ 2 \\ -5 \end{bmatrix}$ $[-15]$

24. $\begin{bmatrix} 1 & 5 \\ -3 & 0 \end{bmatrix} \begin{bmatrix} 3 & -2 \\ -4 & 6 \end{bmatrix}$ $\begin{bmatrix} -17 & 28 \\ -9 & 6 \end{bmatrix}$

25. $\begin{bmatrix} 2 \\ 0 \\ 6 \end{bmatrix} \begin{bmatrix} 1 & -3 & 4 \end{bmatrix}$ $\begin{bmatrix} 2 & -6 & 8 \\ 0 & 0 & 0 \\ 6 & -18 & 24 \end{bmatrix}$

26. $\begin{bmatrix} 2 & 5 & 0 \end{bmatrix} \begin{bmatrix} 8 & 1 \\ 0 & 4 \\ 2 & 5 \end{bmatrix}$ $\begin{bmatrix} 16 & 22 \end{bmatrix}$

Error Analysis

There are many opportunities for students to make errors when multiplying matrices by hand. Be sure students are very careful to multiply the correct row and column. Emphasize that row 1 is multiplied by column 1, then by column 2, then by column 3, and so on. Next, row 2 is multiplied by column 1, then by column 2, and so on. Encourage students to use their graphics calculator to check all matrix products.

Portfolio Assessment

Have students write a paragraph explaining to someone who has never performed matrix multiplication before how to multiply a 4×3 matrix and a 3×2 matrix. Instruct students to include information on the inner and outer dimensions. Ask students to list at least two real-world applications to motivate the person learning how to multiply matrices. In classes where technology is stressed, have students explain the procedure for multiplying matrices on a calculator.

Band Expenses The band director at a high school wants to raise enough money to purchase additional saxophones for the upcoming year. Matrix P gives the new and used prices for each type of saxophone. Matrix N gives the number of each type of saxophone the high school band has this year and the total number they wish to have next year.

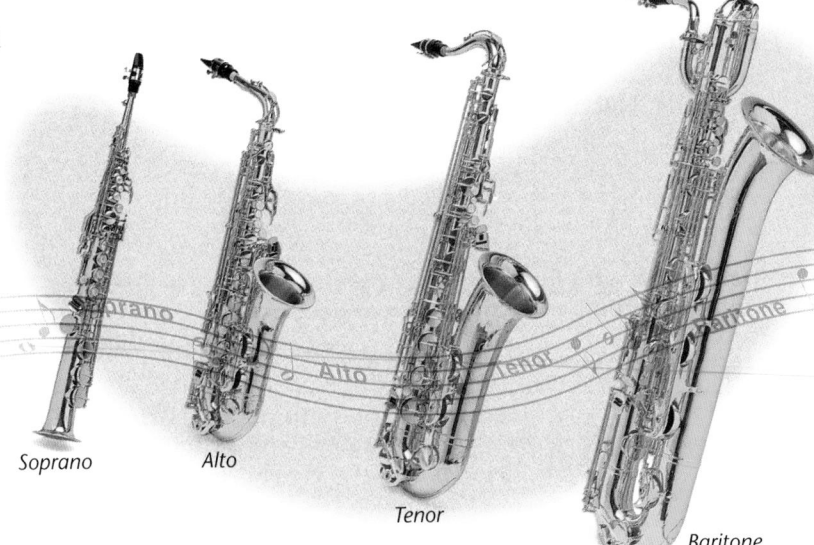

Soprano   Alto   Tenor   Baritone

	Soprano	Alto	Tenor	Baritone
New	1800	1600	3200	3800
Used	600	500	750	850

$= P$

	This year	Next year
Soprano	1	2
Alto	6	9
Tenor	2	4
Baritone	0	1

$= N$

27. Suppose the band director wants to find the value of all of the school's saxophones. Would the band director need to find the product NP or PN? Should the band director use the new values or the used values? Explain. *PN*

28. What is the present value of all the school's saxophones? $5100

29. If the school purchases used instruments to obtain the number of instruments they want to have next year, what will be the total value of their saxophones next year? $9550

30. Explain how the band director could determine the amount of money needed to purchase the saxophones for next year at the new prices.

Use your graphics calculator to find the product.

31. $\begin{bmatrix} 3 & 9 \\ 2 & -1 \end{bmatrix} \begin{bmatrix} 7 & 0 \\ 1 & 3 \end{bmatrix}$ $\begin{bmatrix} 30 & 27 \\ 13 & -3 \end{bmatrix}$

32. $\begin{bmatrix} 5 & 3 \\ 0 & 1 \end{bmatrix} \begin{bmatrix} 4 & 2 & -1 \\ 0 & 1 & 3 \end{bmatrix}$ $\begin{bmatrix} 20 & 13 & 4 \\ 0 & 1 & 3 \end{bmatrix}$

33. $\begin{bmatrix} 4 & -6 & 5 \\ 2 & 4 & -1 \end{bmatrix} \begin{bmatrix} 2 & 7 & 1 \\ -1 & 5 & 2 \\ 1 & 4 & 1 \end{bmatrix}$ $\begin{bmatrix} 19 & 18 & -3 \\ -1 & 30 & 9 \end{bmatrix}$

34. $\begin{bmatrix} 1 & 1 & -1 \\ 2 & 1 & 1 \\ 1 & -2 & 3 \end{bmatrix} \begin{bmatrix} 1 & -0.2 & 0.4 \\ -1 & 0.8 & -0.6 \\ -1 & 0.6 & -0.2 \end{bmatrix}$ $\begin{bmatrix} 1 & 0 & 0 \\ 0 & 1 & 0 \\ 0 & 0 & 1 \end{bmatrix}$

Animal Nutrition A veterinarian is mixing two brands of pet food, each with three varieties (Regular, Lite, and Growth), to create her own brand to sell from her office. The amounts of protein, fiber, and fat (in percentage per serving) are given on page 179 in matrix G for Brand 1 and in matrix H for Brand 2. The three combinations she is comparing, given in scoops per variety, are stated in matrix J.
Complete Exercises 35–38.

27. *PN*; the inner labels are the same. The used values should be used since all saxophones the school has this year are already used.

30. Find the difference between the value of all of the saxophones next year at the new price and the value of all of the saxophones this year at the new price: $34,600 − $17,800 = $16,800. It would cost $16,800 to purchase new saxophones for next year.

	Brand 1		
	Regular	**Lite**	**Growth**
Protein	0.22	0.14	0.26
Fiber	0.03	0.15	0.03
Fat	0.13	0.04	0.17

$= G$

	Brand 2		
	Regular	**Lite**	**Growth**
Protein	0.26	0.22	0.17
Fiber	0.05	0.05	0.04
Fat	0.15	0.12	0.28

$= H$

	Mix 1	**Mix 2**	**Mix 3**
Regular	1	2	1
Lite	2	1	1
Growth	1	1	2

$= J$

35. Which two matrices must be multiplied to determine the nutritional *GJ; HJ*
content of Mixes 1, 2, and 3 using Brand 1? Which two matrices
would be multiplied to find the same information using Brand 2?

36 The veterinarian wants Mix 3 to have the highest percentage of
protein and fiber per serving. Should she use Brand 1 or
Brand 2 for the mixture? Brand 1

37 The veterinarian wants Mix 1 to have the lowest percentage of
fat per serving. Determine whether Brand 1 or Brand 2
should be used in this mixture? Brand 1

38 The veterinarian would like Mix 2 to be the maintenance
pet food. Which brand should be used for this mixture?
Explain your response.

39. **Portfolio Activity**   Complete the Portfolio
Activity on page 163.

Look Back

Find the slope and the *y*-intercept of each line.   **[Lesson 1.2]**

40. $4(2x - 7) = -6$ **41.** $y = -2\left(\frac{1}{3}x + 5\right)$ $-\frac{2}{3}$; -10 **42.** $6 - \frac{2}{3}(y + 9) = 12x$ -18; 0
undef; none

43. A soap solution mixes palm oil, glycerin, and water in the ratio 1:4:8.
How much of each ingredient is there in 520 grams of the solution?
[Lesson 1.5] palm oil:40g, glycerine:160g, water:320g

**Which of the following are true for all nonzero real numbers,
a and *b*? Give counterexamples to show which are false.
[Lesson 2.2]**

44. $a \div b = b \div a$ **45.** $a + b = b + a$ true   **46.** $a - b = b - a$ **47.** $a \cdot b = b \cdot a$ true
false; $\frac{1}{2} \neq \frac{2}{1}$ false; $2 - 1 \neq 1 - 2$

Let $f(x) = 2x + 3$ and $g(x) = 5x - 2$.   **[Lesson 3.3]**

48. Find $(f \circ g)(x)$.   **49.** Find $(g \circ f)(x)$.   **50.** Is $(f \circ g)(x) = (g \circ f)(x)$? no
$10x - 1$ $10x + 13$

Look Beyond

51 Use your graphics calculator to graph each
line on the same coordinate plane. Find the
coordinates of the point of intersection. (5, 0)

$$\begin{cases} 2x + 5y = 10 \\ x + y = 5 \end{cases}$$

38. Answers may vary. One possible answer for
Mix 2 is Brand 1 because it has more fiber
and less fat, but another answer could be
Brand 2 because it has more protein.

39. **Portfolio Activity:** (1) To find the total
number of work hours, call the matrix
of volunteer groups N and the matrix of
hours worked H. Write H as a 1×4 matrix,
and find the product HN. The elements in
the 1×3 product matrix, [208 202 229],

show that neighborhood Group 1 spent 208
hours, Group 2 spent 202 hours, and Group
3 spent 229 hours working on the new
playground. (2) To find the total dollar value
of the work done, multiply the product
matrix HN by the 3×1 matrix of dollar
values V. The total dollar value, HNV, is
$7950.50.

Look Beyond

This exercise prepares students
for their work solving systems of
equations in Lesson 4.3. Many
students will recall studying this
concept in previous mathematics courses.

Systems of Two Linear Equations

PREPARE

Objectives

- Identify each type of system of two linear equations.
- Solve a system of two linear equations using elimination by adding.

- Practice Master   **4.3**
- Enrichment Master   **4.3**
- Technology Master   **4.3**
- Lesson Activity Master   **4.3**
- Quiz   **4.3**
- Spanish Resources   **4.3**

Assessing Prior Knowledge

Express each equation in slope-intercept form.

1. $3x + 4y = 12$
   $[y = -\frac{3}{4}x + 3]$

2. $x - \frac{y}{4} + 8 = 0$
   $[y = 4x + 32]$

3. $0.1y + 0.2 = x$
   $[y = 10x - 2]$

4. $\frac{5x}{4y} = 1$
   $[y = \frac{5}{4}x]$

TEACH

Discuss the fact that systems of two linear equations are used to model information in many disciplines. Ask students to work together to make a list of some of these disciplines.

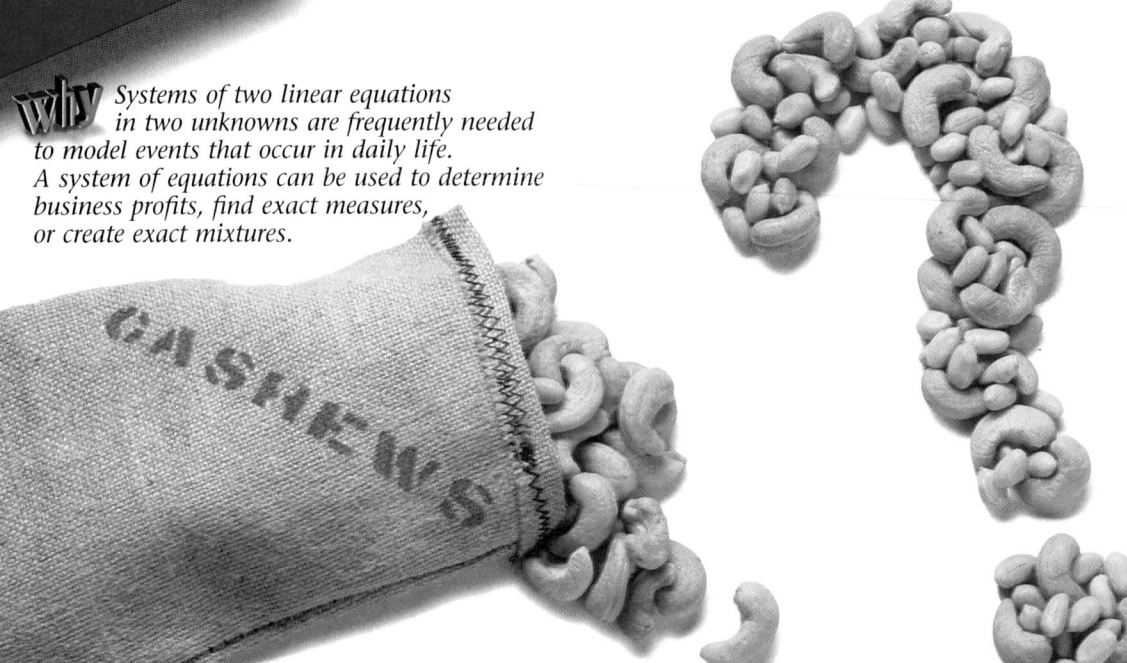

Why *Systems of two linear equations in two unknowns are frequently needed to model events that occur in daily life. A system of equations can be used to determine business profits, find exact measures, or create exact mixtures.*

Business Suppose you are a manager of a grocery store that sells bulk nuts. Cashews regularly cost $4 per pound, and peanuts regularly cost $1 per pound. To sell more cashews, you decide to create a 20-pound mixture of peanuts and cashews that will cost $3 per pound. A system of linear equations can be used to find how many pounds of cashews and how many pounds of peanuts to put in the 20-pound mixture in order to make as much money as you would by selling them separately.

Types of Systems of Linear Equations

Exploration *Intersections*

Graphics Calculator

The following exploration involves three types of systems of linear equations. For each type of system, graph each pair of linear equations on the same coordinate plane. Then answer the questions.

Type 1:
$\begin{cases} x + 2y = 5 \\ 3x + 2y = 7 \end{cases}$
$\begin{cases} 4x + 3y = 1 \\ -4x + y = -3 \end{cases}$
$\begin{cases} -x - 2y = 5 \\ -x - 3y = 4 \end{cases}$

1 Which pairs of lines intersect? How is each pair of equations alike or different?

ALTERNATIVE teaching strategy

Using Visual Models Place a coordinate grid and 2 pieces of acetate with one line on each on an overhead projector. Place the lines so that they represent each of the different types of systems of equations. As you represent each type, have students work in small groups to write the equations of the lines shown. Then have students in each group share their equations. Discuss any differences in the equations presented. It may be helpful to review the properties of parallel lines and intersecting lines prior to asking students to name the equations.

Type 2: $\begin{cases} x + y = 4 \\ 3x + 3y = 12 \end{cases}$ $\begin{cases} x - 2y = -6 \\ -4x + 8y = 24 \end{cases}$ $\begin{cases} 4x + y = -1 \\ -12x - 3y = 3 \end{cases}$

 Which pairs of lines intersect?
How is each pair of equations alike or different?

Type 3: $\begin{cases} x - 3y = -15 \\ x - 3y = 6 \end{cases}$ $\begin{cases} -2x + y = -3 \\ -4x + 2y = 14 \end{cases}$ $\begin{cases} -x + 4y = 1 \\ x - 4y = -2 \end{cases}$

3 Which pairs of lines intersect?
How is each pair of equations alike or different?

4 Describe the three types of systems of linear equations in terms of intersecting lines. ❖

Each type of system of equations in this exploration has a special name.

Type 1 is called an *independent system*.
Type 2 is called a *dependent system*.
Type 3 is called an *inconsistent system*.

One of each type of system of linear equations is shown below.

Type 1
$\begin{cases} y = \left(\dfrac{1}{2}\right)x - \dfrac{3}{2} \\ y = -x - \dfrac{1}{2} \end{cases}$

Type 2
$\begin{cases} y = \left(\dfrac{1}{2}\right)x - \dfrac{3}{2} \\ y = \left(\dfrac{1}{2}\right)x - \dfrac{3}{2} \end{cases}$

Type 3
$\begin{cases} y = \left(\dfrac{1}{2}\right)x + 1 \\ y = \left(\dfrac{1}{2}\right)x - 2 \end{cases}$

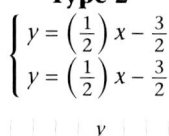 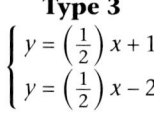

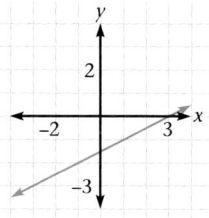

Different Slope	Same Slope	Same Slope
Same or Different *y*-intercepts	Same *y*-intercept	Different *y*-intercept
Intersecting Lines	Same Line	Parallel Lines
One Unique Solution	Infinitely Many Solutions Along This Line	No Solutions
Independent System	*Dependent* System	*Inconsistent* System

Consistent Systems

CRITICAL *Thinking*

Describe the unique solution to a system of equations in which lines have *different* slopes but the *same* *y*-intercept. What is a general characteristic of the graph of this system?

interdisciplinary
CONNECTION

Fitness Have students use a system of two linear equations to solve the following problem.

While training for the Corporate Challenge, Brielle ran 3 miles more each week than Greg.

Together, they ran 57 miles each week. How far did each run each week? [**Brielle: 30 miles, Greg: 27 miles**]

Then ask students to write problems of their own. Have them exchange and solve each other's problems.

TEACHING *tip*

Review the Systems of Equations section of your *HRW Technology Handbook*. Also review the friendly viewing window parameters for the different calculators.

Exploration Notes

Have students review the various ways their calculator can graph and find the intersections of linear functions. Remind students to solve for *y* before trying to graph a linear function on their graphics calculator.

Ongoing ASSESSMENT

4. Type 1 systems of equations intersect at one point. Type 2 systems of equations intersect at an infinite number of points. Type 3 systems of equations do not intersect.

Cooperative Learning

Have students work in pairs. Ask each student to list the number of males and females that live in his or her home. Then have each pair represent the total number of males and females in both of their homes. Each student should have a system of two linear equations. Have students exchange and solve each other's systems.

CRITICAL *Thinking*

The unique solution is the value of the common *y*-intercept. The graphs intersect at the *y*-intercept.

A **consistent system** of equations has at least one solution.

An **independent system** has a unique solution.

A **dependent system** has an infinite number of solutions along the graph of the single line.

An **inconsistent system** of equations has no solution.

EXAMPLE 1

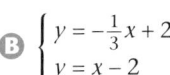

Graphics Calculator

Determine whether each system of equations is inconsistent, dependent, or independent.

A $\begin{cases} -\frac{2}{3}x - y = -5 \\ y = -\frac{2}{3}x - 3 \end{cases}$

B $\begin{cases} x + 3y = 6 \\ x - y = 2 \end{cases}$

C $\begin{cases} 3x + 5y = 7 \\ 6x = -10y + 14 \end{cases}$

D $\begin{cases} 2x + y = 5 \\ 3x - y = 2 \end{cases}$

Solution

Rewrite each equation in slope-intercept form and graph.

A $\begin{cases} y = -\frac{2}{3}x + 5 \\ y = -\frac{2}{3}x - 3 \end{cases}$

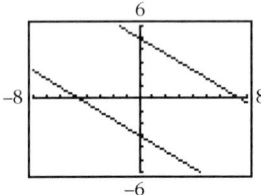

Parallel lines
Inconsistent

B $\begin{cases} y = -\frac{1}{3}x + 2 \\ y = x - 2 \end{cases}$

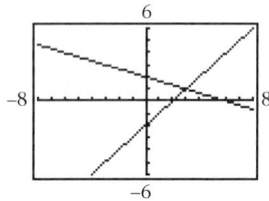

Intersecting lines
Independent

C $\begin{cases} y = -\frac{3}{5}x + \frac{7}{5} \\ y = -\frac{3}{5}x + \frac{7}{5} \end{cases}$

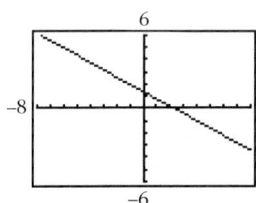

Same line
Dependent

D $\begin{cases} y = -2x + 5 \\ y = 3x - 2 \end{cases}$

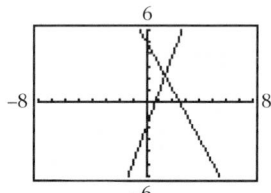

Intersecting lines
Independent

Solving Independent Systems by Elimination

You can solve an independent system of equations because it has a unique solution.

Multiplying each side of an equation by the same number results in an equivalent equation. You can solve systems of equations by using the properties of equality to eliminate a variable.

EXAMPLE 2

Nutrition Low-fat milk is 1% milk fat, and whole milk is 4% milk fat. How many ounces of each could you combine to make a 9-ounce mixture of milk with 2% milk fat?

Ⓐ Write a system of two linear equations to represent the mixture.

Ⓑ Solve the system using your graphics calculator.

Ⓒ Solve the system using elimination.

Solution

Ⓐ Let x be the number of ounces of milk with 1% milk fat.
Let y be the number of ounces of milk with 4% milk fat.

Make a table to organize the information.
Write each percentage as a decimal.

	Low-fat	Whole	Mixture
(1) **Ounces**	x	y	$x + y$
(2) **Milk fat**	$0.01x$	$0.04y$	$0.01x + 0.04y$

From the rows of the table, you can write the equations.

(1) The mixture contains 9 ounces, so $x + y = 9$.

(2) The mixture contains 2% milk fat, so $0.01x + 0.04y = (0.02)(9)$.

  Multiply both sides of this equation by 100 to get $1x + 4y = 18$.

Thus, the system of equations is $\begin{cases} x + y = 9 \\ x + 4y = 18 \end{cases}$.

Ⓑ Solve the equations in the system for y, and graph them on your graphics calculator.

$\begin{cases} y = -x + 9 \\ y = -\dfrac{1}{4}x + \dfrac{9}{2} \end{cases}$

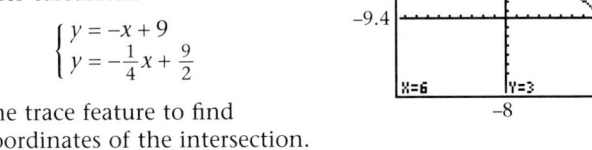
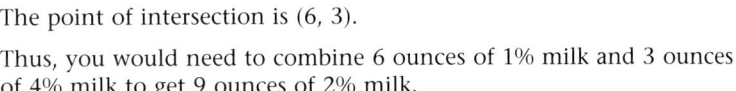

Use the trace feature to find the coordinates of the intersection. The point of intersection is (6, 3).

Thus, you would need to combine 6 ounces of 1% milk and 3 ounces of 4% milk to get 9 ounces of 2% milk.

Using Cognitive Strategies Write the equation $x + 2y = 8$ on the chalkboard. Ask students to write a linear equation that will result in

a) an independent system,
b) a dependent system, and
c) an inconsistent system

when combined with the equation $x + 2y = 8$.

Alternate Example 2

Ricky has three times more broccoli than squash. A pound of broccoli sells for $1.00 and a pound of squash sells for $1.25. He made $297.50 when he sold all the broccoli and squash at market. How many pounds of broccoli and squash did Ricky have?

A. Write a system of two linear equations to represent this information. [**Let x be pounds of broccoli and y be pounds of squash. Then $x = 3y$ and $1.00x + 1.25y = 297.50$.**]

B. Solve the system using your graphics calculator. [**210 pounds of broccoli and 70 pounds of squash**]

C. Solve the system using elimination. [**(210, 70)**]

Alternate Example 3

Describe the steps for solving each system using elimination by adding.

A $\begin{cases} 3x + y = 8 \\ \frac{1}{2}x - y = -1 \end{cases}$ [Add like terms of the two equations to eliminate y. The result is $\frac{7}{2}x = 7$. Then solve for x. Multiply both sides of the equation by $\frac{2}{7}$. The result is 2. Substitute 2 for x in either equation, and solve for y. The result is 2.]

B $\begin{cases} 4x + 3y = 24 \\ 3x + y = 13 \end{cases}$ [Multiply the second equation by -3 and add like terms of the equations to eliminate y. The result is $-5x = -15$. Solve for x. Divide both sides of the equation by -5. The result is 3. Substitute 3 for x in either equation, and solve for y. The result is 4.]

C Multiply each side of the first equation by -4. This will allow the y-terms to be eliminated when the like terms of the two equations are added.

$$\begin{cases} x + y = 9 \\ x + 4y = 18 \end{cases} \rightarrow \begin{cases} -4x - 4y = -36 \\ x + 4y = 18 \end{cases}$$

Add like terms of the two equations.

Solve for x.

$$\begin{array}{r} -4x - 4y = -36 \\ +x + 4y = +18 \\ \hline -3x = -18 \\ x = 6 \end{array}$$

To find y, substitute 6 for x in either of the original equations.

$$\begin{array}{r} x + y = 9 \\ (6) + y = 9 \\ y = 3 \end{array}$$

Check▸

The solution of a system must satisfy each equation in the system, so check the values for x and y in each original equation.

Check 1	**Check 2**
$x + y = 9$	$x + 4y = 18$
$6 + 3 \stackrel{?}{=} 9$	$6 + 4(3) \stackrel{?}{=} 18$
$9 = 9$ True	$18 = 18$ True

The solution to the system is $(6, 3)$. Thus, you would need to combine 6 ounces of 1% milk and 3 ounces of 4% milk to make a 9-ounce mixture of 2% milk. ❖

Try This Find the number of pounds of cashews and the number of pounds of peanuts needed for the mixture described on page 180.

The method used to solve the system of two linear equations in Example 2 Part **C** is called *elimination by adding*.

EXAMPLE 3

Describe the steps for solving each system using elimination by adding.

A $\begin{cases} -2x + y = 7 \\ \frac{1}{2}x - y = -1 \end{cases}$ **B** $\begin{cases} 3x + 2y = 8 \\ 2x + y = 5 \end{cases}$

Solution▸

Possible solutions are given.

A Add the left and right sides to eliminate the variable y. Then solve for x. The result is -4. Substitute -4 for x in either equation, and solve for y. The result is -1. The solution is $(-4, -1)$.

B Multiply the second equation by -2 to eliminate the variable y. Add like terms of the equations, and solve for x. The result is 2. Substitute 2 for x in either equation, and solve for y. The result is 1. The solution is $(2, 1)$. ❖

Exercises & Problems

Communicate

1. Why do you think independent and dependent systems are called consistent systems?

2. Describe the graphs of each type of system of linear equations.

3. Create three systems of equations, one to demonstrate each type of system of linear equations: inconsistent, dependent, and independent.

Explain how to solve each system using elimination by adding.

4. $\begin{cases} x + y = 2 \\ 2x - y = 7 \end{cases}$

5. $\begin{cases} 3x - 4y = 3 \\ 2x + y = -5 \end{cases}$

6. When attempting to solve by elimination, what are the results for an inconsistent system? for a dependent system?

7. How do you check your solution to a system of equations?

Practice & Apply

Postal Service A mail-order company charges for postage and handling according to the weight of the package. An order of 12 packages had a total postage and handling charge of $29.

8. Write a system of equations that can be solved to find how many light packages and how many heavy packages were in the order.

9. Graph the system you wrote in Exercise 8, and find the solution. $(7, 5)$

10. Find the solution to the system you wrote in Exercise 8 using elimination by adding. $(7, 5)$

A package that weighs less than 3 pounds is charged $2, and a package that weighs 3 pounds or more is charged $3.

$3/lb $2/lb

SHIPPING & HANDLING

8. $\begin{cases} x + y = 12 \\ 2x + 3y = 29 \end{cases}$

ASSESS

Selected Answers
Odd-numbered Exercises 9–49

Assignment Guide

Core 1–20, 23–37, 40–52

Core Plus 1–13, 15, 16, 19–30, 34–52

Technology
A graphics calculator is needed for Exercises 9, 17–22, and 25. A graphics calculator is helpful for working or checking Exercises 49 and 50.

Error Analysis

Watch for students who make errors when solving a system of equations in which there are variables on both sides of either equation, as in Exercises 33, 35, 38, and 39. Suggest that students write all of the variables on the same side of the equation and then line up like terms before using elimination to solve.

Encourage students to check their final answers by substituting the values back into *both* original equations.

Small Business A candy manufacturer wishes to mix an inexpensive candy with an expensive candy as a sales promotion. The expensive candy wholesales for $2.00 per pound and the inexpensive candy for $0.75 per pound. The manufacturer wishes to mix 1000 pounds of candy and sell the mixture for $1.35 per pound.

11. Write the system of equations that can be solved to find the amounts of each type of candy in the mixture. $x + y = 1000$ $2.00x + 0.75y = 1350$

12. How many pounds of each type of candy should be used in the mixture? 480 lb, 520 lb

13. How many pounds of each type of candy should be used if the mixture sells for $1.50 per pound? 600 lb of $2.00; 400 lb of $0.75

Estimate the solution to each system of equations.

14.
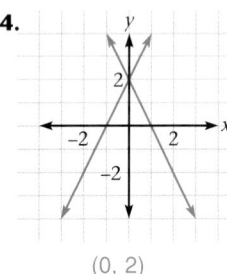
(0, 2)

15.
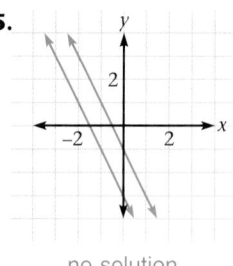
no solution

16.
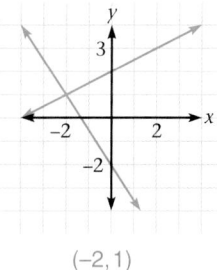
(−2, 1)

Determine whether each of the following systems is inconsistent, dependent, or independent.

17 $\begin{cases} -\dfrac{1}{2}x + y = 4 \\ x + 2y = 8 \end{cases}$ indep

18 $\begin{cases} 2x + y = 5 \\ 4x + 2y = 6 \end{cases}$ incon

19 $\begin{cases} -3x - y = 2 \\ -6x + 2y = -2 \end{cases}$ indep

20 $\begin{cases} -x + 2y = 3 \\ 2x - 4y = -6 \end{cases}$ dep

21 $\begin{cases} 3x - y = 2 \\ -3x + y = 1 \end{cases}$ incon

22 $\begin{cases} 6x - 3y = 9 \\ 3x + 7y = 47 \end{cases}$ indep

Nutrition One unit of whole-wheat flour has 13.6 grams of protein and 2.5 grams of fat. One unit of whole milk has 3.4 grams of protein and 3.7 grams of fat.

23. Write a system of equations that can be used to find the number of units of whole-wheat flour and the number of units of whole milk that must be mixed to make a dough that has 75 grams of protein and 15 grams of fat.
$13.6x + 3.4y = 75$
$2.5x + 3.7y = 15$

24. Solve the system and give, to the nearest tenth, the number of units of flour and of milk in the dough. 5.4, 0.4

25 Graph the system on your graphics calculator. Approximate, to the nearest tenth, the coordinates of the intersection. (5.4, 0.4)

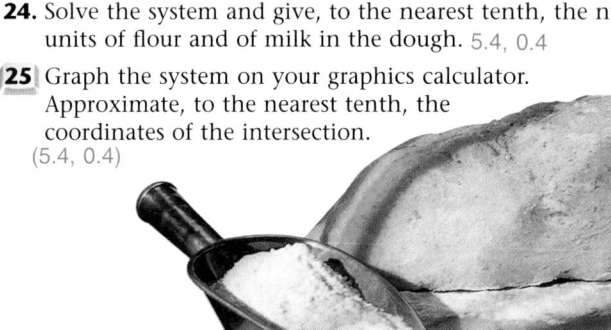

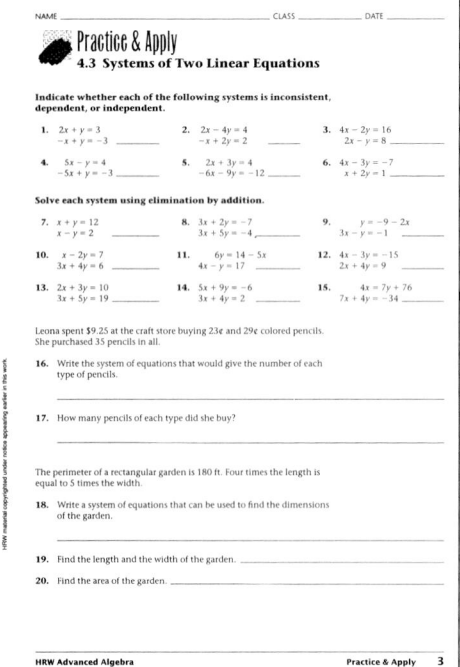

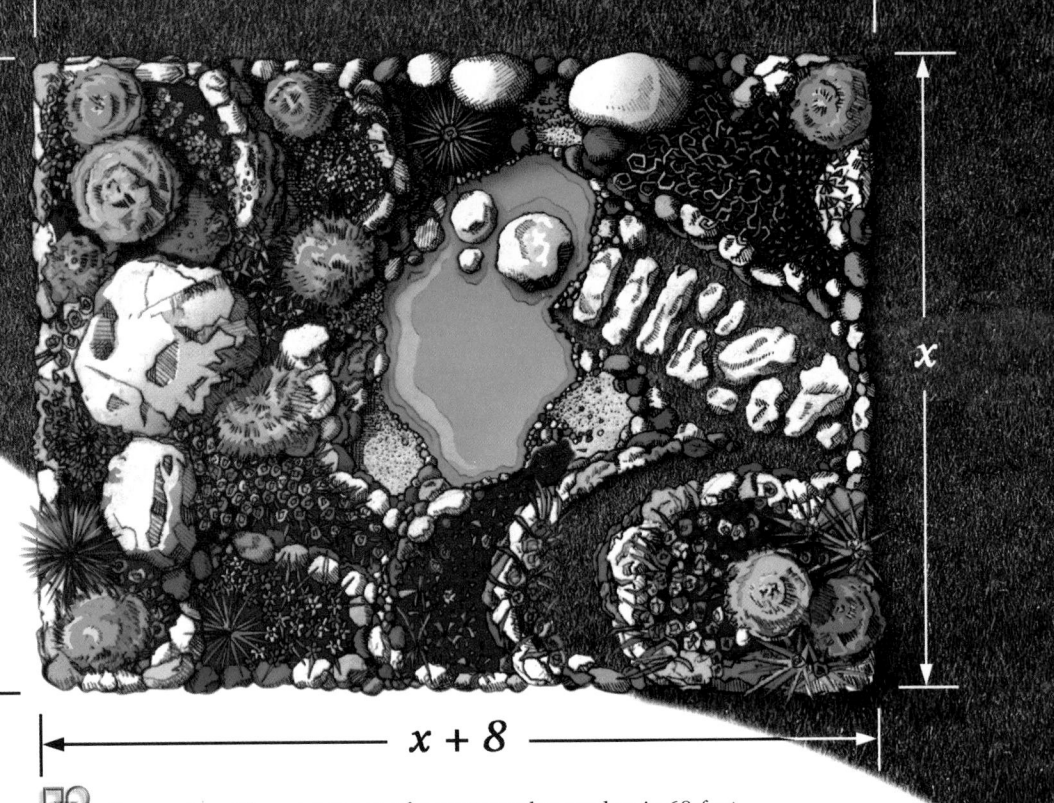

x

$\longleftarrow x + 8 \longrightarrow$

Geometry The perimeter of a rectangular garden is 60 feet.

26. Write a system of two linear equations that can be used to find the dimensions of the garden. $2x + 2y = 60$ and $y = x + 8$

27. Solve the system of equations, and find the area of the garden. (11, 19); 209 sq ft

Banking Beth has saved $4500. She would like to earn $250 per year by investing the money. She received advice about two different investment opportunities: a low-risk investment that pays 5% annually and a high-risk investment that pays 9% annually.

28. Write the system of equations that can be solved to find the amounts to invest in each type of investment. $x + y = 4500$ and $0.05x + 0.09y = 250$

29. How much should Beth invest in each type of investment to earn her annual goal? $3875 at 5%; $625 at 9%

30. How much should Beth invest in each type if she wishes to earn $325 per year? $2000 at 5%; $2500 at 9%

Solve each system of linear equations using any method.

31. $\begin{cases} x - y = 3 \\ x + y = 1 \end{cases}$ (2, −1)

32. $\begin{cases} 2x + y = 5 \\ 3x - 4y = 2 \end{cases}$ (2, 1)

33. $\begin{cases} x + y = 0 \\ y = 4 - 2x \end{cases}$ (4, −4)

34. $\begin{cases} x - 2y = 0 \\ 2x - y = 6 \end{cases}$ (4, 2)

35. $\begin{cases} 3y = 5 - x \\ x + 4y = 8 \end{cases}$ (−4, 3)

36. $\begin{cases} 5x + 4y = 2 \\ 2x + 3y = 5 \end{cases}$ (−2, 3)

37. $\begin{cases} 5x + 2y = 2 \\ 4x + 3y = 4 \end{cases}$ $\left(-\dfrac{2}{7}, \dfrac{12}{7}\right)$

38. $\begin{cases} -2x + 5y = -23 \\ 24 + 4y = 3x \end{cases}$ (4, −3)

39. $\begin{cases} 6x = 10 - 2y \\ 3x + y = 5 \end{cases}$

39. dependent system; infinite number of solutions

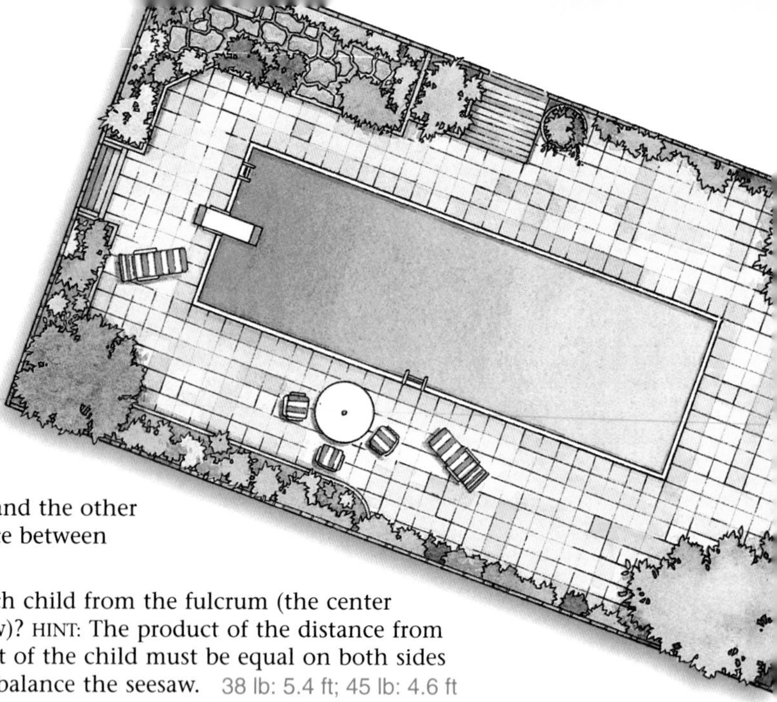

Geometry

The perimeter of a rectangular swimming pool is 100 meters. Twice the length is equal to six times the width.

40. Find the length and the width of the pool.
12.5 m; 37.5 m

41. Find the area of the pool. 468.75 m²

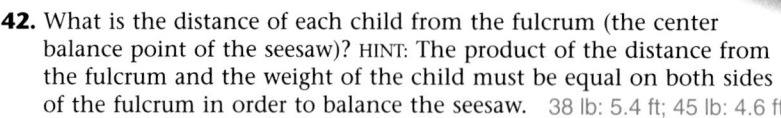

Physics Two children balance on opposite ends of a seesaw. One child weighs 45 pounds, and the other weighs 38 pounds. The distance between them is 10 feet.

42. What is the distance of each child from the fulcrum (the center balance point of the seesaw)? HINT: The product of the distance from the fulcrum and the weight of the child must be equal on both sides of the fulcrum in order to balance the seesaw. 38 lb: 5.4 ft; 45 lb: 4.6 ft

43. If the children are 15 feet apart on a seesaw, approximately how far is each child from the fulcrum? 38 lb: 8.1 ft; 45 lb: 6.9 ft

 *Look Back*

Find the linear function for each graph described below. [Lessons 1.2, 2.5]

44. Passes through the points (2, 5) and (−3, 7) $y = -\frac{2}{5}x + \frac{29}{5}$

45. Has a y-intercept of (0, 3) and contains the point (−2, 1) $y = x + 3$

46. Has a slope of 1 and contains the point (0, −4) $y = x - 4$

47. Find f^{-1} for $f(x) = 2x - 3$. **[Lesson 3.2]** $\frac{x+3}{2}$

48. Find g^{-1} for $g(x) = \frac{1}{x-2}$. **[Lesson 3.2]** $2 + \frac{1}{x}$

Find the product. [Lesson 4.2]

49. $\begin{bmatrix} 2 & -1 \\ 3 & 0 \\ -6 & 4 \end{bmatrix}\begin{bmatrix} 5 \\ -2 \end{bmatrix}$ $\begin{bmatrix} 12 \\ 15 \\ -38 \end{bmatrix}$

50. $\begin{bmatrix} 2 & 7 & 3 \\ -1 & 0 & 2 \\ 0 & -3 & 1 \end{bmatrix}\begin{bmatrix} -2 \\ 0 \\ 3 \end{bmatrix}$ $\begin{bmatrix} 5 \\ 8 \\ 3 \end{bmatrix}$

Look Beyond

These exercises prepare students for their work with coordinate geometry and transformations which will be studied in depth in Lesson 4.7.

Look Beyond

Coordinate Geometry Plot the following points on a coordinate plane. Identify the geometric figure whose vertices are plotted.

51. (−1, 2), (−3, −3), (2, −5) triangle **52.** (−4, 1), (0, 5), (2, 1), (−2, −2) quadrilateral

51. Triangle;

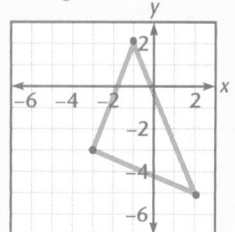

52. Quadrilateral;

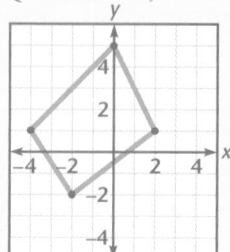

LESSON 4.4 Using Matrix Row Operations

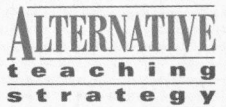

Why *Systems of linear equations are useful for a variety of reasons. Some systems of linear equations can be solved in a few steps using substitution. However, in practice, most systems are large and very complicated. You can use matrix row operations to solve systems of linear equations that would be difficult or impossible to solve by substitution.*

$$[3 \;\; -3 \;\; -6 \;\; 12] \div 3 = [1 \;\; -1 \;\; -2 \;\; 4]$$

$$\left[\begin{array}{ccc|c} 3 & -3 & -6 & 12 \\ 5 & -2 & 3 & 3 \\ -2 & 1 & -1 & 1 \end{array}\right] \xrightarrow{R_1 \div 3} \left[\begin{array}{ccc|c} 1 & -1 & -2 & 4 \\ 5 & -2 & 3 & 3 \\ -2 & 1 & -1 & 1 \end{array}\right]$$

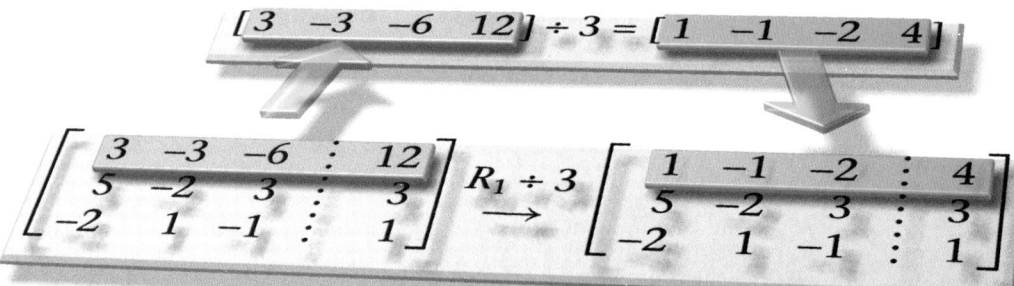

Back Substitution

How can you solve $y = 4 + x$ when you know that x is 5?

One method of solving a system of equations is called *back substitution*. **Back substitution** involves solving for one variable first and working backward to solve for the other variables.

EXAMPLE 1

Solve the given system of equations using back substitution.

$$\begin{cases} x + y + z = 6 \\ 4y + 2z = 16 \\ 3z = -6 \end{cases}$$

Solution

Use the third equation to solve for z.

$$3z = -6$$
$$z = -2$$

Substitute –2 for z in the second equation, and solve for y.

$$4y + 2z = 16$$
$$4y + 2(-2) = 16$$
$$4y = 20$$
$$y = 5$$

Finally, substitute the values of z and y into the first equation, and solve for x.

$$x + y + z = 6$$
$$x + 5 + (-2) = 6$$
$$x + 3 = 6$$
$$x = 3$$

Thus, the solution is (3, 5, –2). ❖

In Example 1, when you find that z is –2, what would happen if you substituted –2 for z into the *first* equation to solve for y?

ALTERNATIVE teaching strategy

Using Algorithms The row reduction method can be used without back substitution if it is continued until there are 1s on the main diagonal and 0s everywhere else in the constant matrix. The values of the variables can then be read from the matrix. In Example 4, continue to perform row reductions after there are zeros below the main diagonal. The steps are outlined.

1. $-R_2 \rightarrow R_2$
2. $-0.5R_3 \rightarrow R_3$
3. $R_1 + R_2 \rightarrow R_1$
4. $-3R_3 + R_1 \rightarrow R_1$
   $-5R_3 + R_2 \rightarrow R_2$

The final augmented matrix is

$$\left[\begin{array}{ccc|c} 1 & 0 & 0 & 6 \\ 0 & 1 & 0 & 12 \\ 0 & 0 & 1 & -1 \end{array}\right].$$

Read the values of x, y, and z from the fourth column. So $x = 6$, $y = 12$, and $z = -1$.

PREPARE

Objectives

- Represent a system of linear equations with an augmented matrix.
- Solve a system of linear equations using the row reduction method and back substitution.

RESOURCES

• Practice Master	4.4
• Enrichment Master	4.4
• Technology Master	4.4
• Lesson Activity Master	4.4
• Quiz	4.4
• Spanish Resources	4.4

Assessing Prior Knowledge

Find the value of y in each equation.

1. $y = 2x - z$ if $x = 4$ and $z = -2$ **[10]**
2. $z = 2y - 3x$ if $x = -6$ and $z = -1$ **[–9.5]**

TEACH

Why Point out that matrix row operations provide an organized method for solving systems of equations.

ongoing ASSESSMENT

Substitute 5 for x in $y = 4 + x$, and evaluate for y.

Alternate Example 1

Solve the given system of equations using back substitution.

$$\begin{cases} x + y + z = 5 \\ 2y - 3z = 21 \\ 3z = -9 \end{cases}$$

[$x = 2, y = 6, z = -3$]

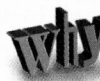

Alternate Example 2

Write the augmented matrix for this system of equations. Identify the elements in the main diagonal.

$$\begin{cases} 3x + 4y - x = 12 \\ 7x - y + 2z = -2 \\ 2x - 2y - 5z = 18 \end{cases}$$

$$\left[\begin{array}{ccc|c} 3 & 4 & -1 & 12 \\ 7 & -1 & 2 & -2 \\ 2 & -2 & -5 & 18 \end{array}\right]$$

The elements of the main diagonal are 3, −1, and −5.]

Augmented Matrix

Recall that a system of linear equations can be represented by a matrix equation. The system of linear equations from Example 1 is represented by the following matrix equation.

$$\begin{cases} x + y + z = 6 \\ 4y + 2z = 16 \\ 3z = -6 \end{cases} \rightarrow \begin{bmatrix} 1 & 1 & 1 \\ 0 & 4 & 2 \\ 0 & 0 & 3 \end{bmatrix} \begin{bmatrix} x \\ y \\ z \end{bmatrix} = \begin{bmatrix} 6 \\ 16 \\ -6 \end{bmatrix}$$

A matrix equation consists of three individual matrices.

Coefficient matrix	Variable matrix	Constant matrix
$\begin{bmatrix} 1 & 1 & 1 \\ 0 & 4 & 2 \\ 0 & 0 & 3 \end{bmatrix}$	$\begin{bmatrix} x \\ y \\ z \end{bmatrix}$	$\begin{bmatrix} 6 \\ 16 \\ -6 \end{bmatrix}$

You can also represent a system of equations with an *augmented matrix*. The **augmented matrix** is a single matrix that contains both the coefficient matrix and the constant matrix. The variable matrix is not included in this augmented matrix.

The same system of equations from Example 1 can be represented by the augmented matrix at the right. A vertical dotted line is placed between the coefficients and the constants.

$$\left[\begin{array}{ccc:c} 1 & 1 & 1 & 6 \\ 0 & 4 & 2 & 16 \\ 0 & 0 & 3 & -6 \end{array}\right]$$

The elements 1, 4, and 3 in blue are the elements in the **main diagonal** of the coefficient matrix.

Notice that a triangle of zeros is formed below the main diagonal. When the elements below the main diagonal of an augmented matrix are zeros, the system represented can be solved by back substitution.

EXAMPLE 2

Write the augmented matrix for this system of equations. Identify the elements in the main diagonal.

$$\begin{cases} -2x + y - z = 1 \\ 5x - 2y + 3z = 3 \\ x - y - 2z = -4 \end{cases}$$

Solution►

$$\begin{cases} -2x + y - z = 1 \\ 5x - 2y + 3z = 3 \\ x - y - 2z = -4 \end{cases} \rightarrow \left[\begin{array}{ccc:c} -2 & 1 & -1 & 1 \\ 5 & -2 & 3 & 3 \\ 1 & -1 & -2 & -4 \end{array}\right]$$

Thus, the elements of the main diagonal are −2, −2, and −2. ❖

ENRICHMENT Have students solve the system of four linear equations given below.

$$\begin{cases} 6x + 4y + 3z + q = 3 \\ 12x - 8y + 2q = 10 \\ 16y + 5z + 5q = 1 \\ 4x + 4y + 4z + 4q = 5 \end{cases}$$

$$\left[x = \tfrac{1}{2}, y = -\tfrac{1}{4}, z = 0, q = 1\right]$$

INCLUSION strategies **Using Visual Models** Encourage students to use highlighting markers to highlight the row or rows of a matrix that they are working with. This will help them focus on the correct row or rows.

The Chinese Counting Board

Cultural Connection: Asia Over 2000 years ago, the Chinese used a counting board to solve systems of equations. A system of equations was represented on a counting board. The counting board later evolved into the abacus.

The sticks, or rod numerals, arranged on the counting board represent this system of equations.

$$\begin{cases} 3x + 2y + z = 39 \\ 2x + 3y + z = 34 \\ x + 2y + 3z = 26 \end{cases}$$

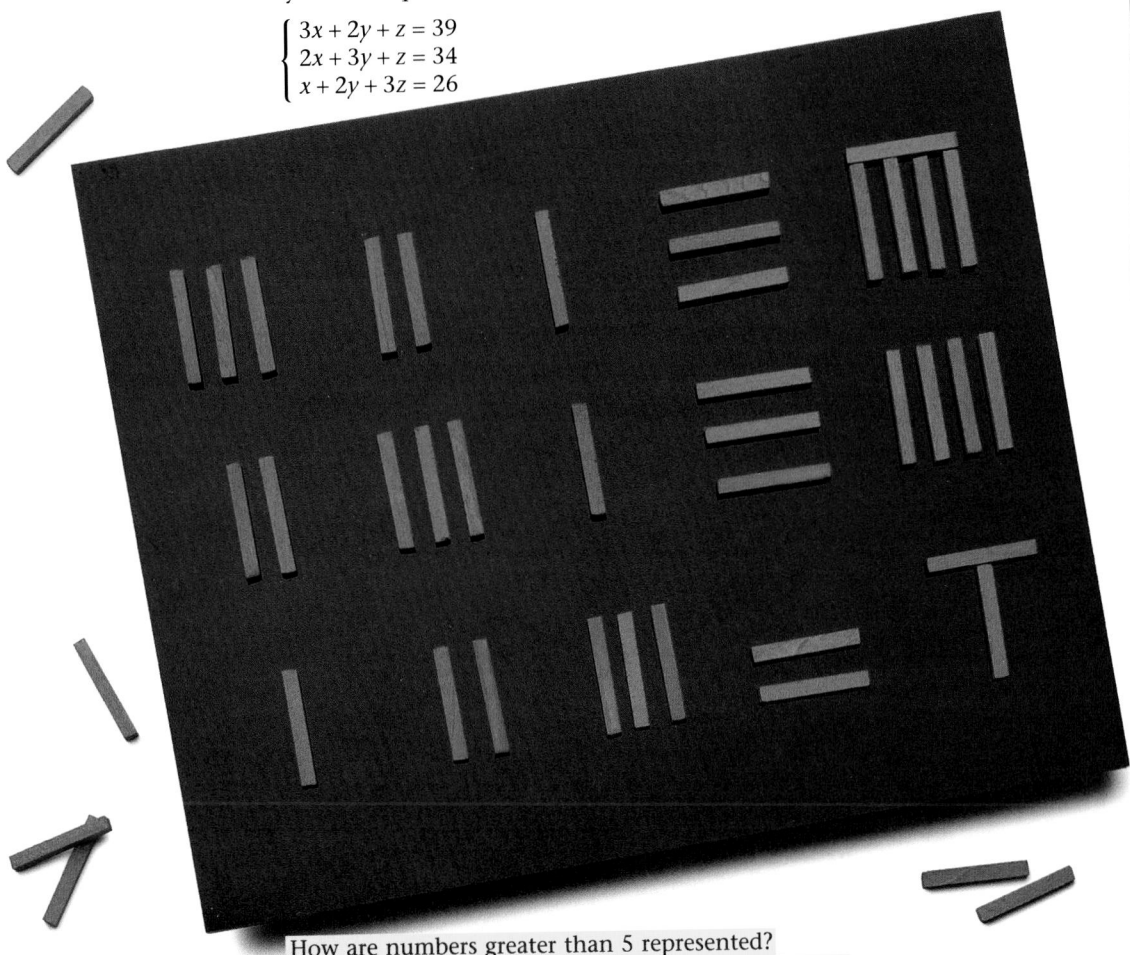

How are numbers greater than 5 represented?
How are the numerals in the tens place represented?

Try to represent other numbers as the Chinese did on the counting board.

The Chinese developed a method of solving systems of equations by moving sticks on the counting board. The Chinese method used the idea of manipulating a representation of systems of equations.
The representation of systems is the basis of using matrices to solve systems of equations today.

CRITICAL Thinking

How is the Chinese counting board like an augmented matrix?

Cooperative Learning
Have students refer back to the photograph of Chinese counting rods shown in the Chapter 2 Opener on page 61. Have students use their own materials to duplicate the Chinese counting rods and boards described. Have them use their rods and boards to represent several different systems of equations. It may be helpful for students to do research to find out more about these boards and exactly how and why they evolved into the abacus.

ongoing ASSESSMENT

Numbers in the units place that are 5 or more are represented by one horizontal rod worth 5 and vertical rods necessary to add up to the required number.

Numbers in the tens place are in a separate column from numbers in the units place. Here, each horizontal rod is worth ten.

CRITICAL Thinking

The first three columns represent the coefficient matrix; the remaining columns represent the constant matrix.

RETEACHING the lesson

Cooperative Learning Separate students into three groups, A, B, and C. Give each group the augmented matrix below.

$$\begin{bmatrix} 2 & 0 & 5 & | & 45 \\ 1 & 1 & 3 & | & 32 \\ 2 & 2 & 1 & | & 29 \end{bmatrix}$$

Have each group solve the system of equations, beginning the solution as described.

- Have each member of Group A interchange row 1 and row 2 of the matrix.
- Have each member of Group B multiply row 1 by −1, add it to row 3, and then place the result in row 3.
- Have each member of Group C begin by multiplying row 1 by $\frac{1}{2}$ and placing the result in row 1.

Have the entire class compare their results. **[5, 6, 7]**

Alternate Example 3

Perform the following row operations on this matrix with your graphics calculator.

$$\begin{bmatrix} 1 & 3 & 5 & | & 10 \\ 2 & 5 & 3 & | & 12 \\ -4 & 6 & 8 & | & 16 \end{bmatrix}$$

A. $-2R_1 + R_2 \to R_2$ [**One calculator appears like this:**

```
*Row+(-2,[A],1,2
)
[ 1   3   5   10]
[ 0  -1  -7  -8]
[ -4  6   8   16]
Ans→[B]
```

]

B. $4R_1 + R_3 \to R_3$ [**One calculator appears like this:**

```
*Row+(4,[B],1,3)
[ 1   3   5   10]
[ 0  -1  -7  -8]
[ 0  18  28  56]
```

]

Elementary Row Operations

Some systems of equations, such as the one in Example 2, cannot be solved using back substitution. However, you can change a matrix so that the elements below the main diagonal are zeros by performing *elementary row operations* on the matrix.

Elementary row operations are derived from the properties of equality for real numbers. Each time you perform a row operation on a matrix, the resulting matrix represents a system of equations that is *equivalent* to the original system.

ELEMENTARY ROW OPERATIONS

Elementary row operations are operations performed on a matrix that result in an equivalent matrix. They are:

1. Interchanging two rows
2. Multiplying all elements of a row by a nonzero constant
3. Adding a constant multiple of the elements of one row to the corresponding elements of another row

You can use *row operation notation* to keep a record of the row operations that you perform on the matrix.

Row Operation Notation	
1. Interchange rows 1 and 2.	$R_1 \leftrightarrow R_2$
2. Multiply all elements of row 3 by –2.	$-2R_3 \to R_3$
3. Multiply row 2 by 4, add row 1 and replace row 1 with the resulting row.	$4R_2 + R_1 \to R_1$

EXAMPLE 3

Graphics Calculator

Perform the following row operations on this matrix with your graphics calculator.

Ⓐ $6R_1 + R_2 \to R_2$ Ⓑ $-7R_1 + R_3 \to R_3$

$$\begin{bmatrix} 1 & -1 & 5 & \vdots & 2 \\ -6 & 3 & 1 & \vdots & 4 \\ 7 & -9 & 2 & \vdots & 1 \end{bmatrix}$$

Solution➤

Enter the matrix into your calculator. Use the command in the matrix menu that allows you to multiply a row by a constant and add the result to another row. One calculator appears like this.

Ⓐ 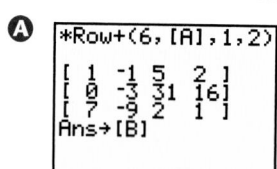   Ⓑ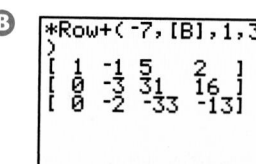

Row Reduction Method

The process of performing elementary row operations on an augmented matrix to solve a system of equations is called the **row reduction method**.

EXAMPLE 4

Solve this system of linear equations using the row reduction method and back substitution.

$$\begin{cases} -2x + y - z = 1 \\ 5x - 2y + 3z = 3 \\ x - y - 2z = -4 \end{cases}$$

Solution▸

Represent the system in an augmented matrix.

$$\begin{cases} -2x + y - z = 1 \\ 5x - 2y + 3z = 3 \\ x - y - 2z = -4 \end{cases} \longrightarrow \left[\begin{array}{ccc:c} -2 & 1 & -1 & 1 \\ 5 & -2 & 3 & 3 \\ 1 & -1 & -2 & -4 \end{array}\right]$$

It is easier to produce zeros below the main diagonal when elements of the main diagonal are 1 or –1. Interchange rows 1 and 3 to get a 1 in the first element of the main diagonal.

$$\left[\begin{array}{ccc:c} -2 & 1 & -1 & 1 \\ 5 & -2 & 3 & 3 \\ 1 & -1 & -2 & -4 \end{array}\right] \xrightarrow{R_1 \leftrightarrow R_3} \left[\begin{array}{ccc:c} 1 & -1 & -2 & -4 \\ 5 & -2 & 3 & 3 \\ -2 & 1 & -1 & 1 \end{array}\right]$$

To produce a zero in row 2 of column 1, multiply the first row by –5, and add it to the second row.

$$\left[\begin{array}{ccc:c} 1 & -1 & -2 & -4 \\ 5 & -2 & 3 & 3 \\ -2 & 1 & -1 & 1 \end{array}\right] \xrightarrow{-5R_1 + R_2 \rightarrow R_2} \left[\begin{array}{ccc:c} 1 & -1 & -2 & -4 \\ 0 & 3 & 13 & 23 \\ -2 & 1 & -1 & 1 \end{array}\right]$$

To produce a zero in row 3 of column 1, multiply the first row by 2, and add it to the third row.

$$\left[\begin{array}{ccc:c} 1 & -1 & -2 & -4 \\ 0 & 3 & 13 & 23 \\ -2 & 1 & -1 & 1 \end{array}\right] \xrightarrow{2R_1 + R_3 \rightarrow R_3} \left[\begin{array}{ccc:c} 1 & -1 & -2 & -4 \\ 0 & 3 & 13 & 23 \\ 0 & -1 & -5 & -7 \end{array}\right]$$

To get a 1 or –1 on the main diagonal, interchange rows 2 and 3.

$$\left[\begin{array}{ccc:c} 1 & -1 & -2 & -4 \\ 0 & 3 & 13 & 23 \\ 0 & -1 & -5 & -7 \end{array}\right] \xrightarrow{R_2 \leftrightarrow R_3} \left[\begin{array}{ccc:c} 1 & -1 & -2 & -4 \\ 0 & -1 & -5 & -7 \\ 0 & 3 & 13 & 23 \end{array}\right]$$

You can now produce the third zero below the main diagonal. Multiply the second row by 3, and add it to the third row.

$$\left[\begin{array}{ccc:c} 1 & -1 & -2 & -4 \\ 0 & -1 & -5 & -7 \\ 0 & 3 & 13 & 23 \end{array}\right] \xrightarrow{3R_2 + R_3 \rightarrow R_3} \left[\begin{array}{ccc:c} 1 & -1 & -2 & -4 \\ 0 & -1 & -5 & -7 \\ 0 & 0 & -2 & 2 \end{array}\right]$$

Alternate Example 4

Solve the following system of linear equations using the row reduction method and back substitution.

$$\begin{cases} 2x + 3y - z = 1 \\ x + 2y + 3z = 38 \\ -5x + y + 2z = 35 \end{cases}$$

[$x = -2, y = 5, z = 10$]

TEACHING *tip*

Have students use the method suggested in the Alternative Teaching Strategy on page 189 to solve the systems of equations in both Example 4 and Alternate Example 4. Have students compare their answers and discuss which method they prefer and why.

TEACHING tip

Emphasize that back substitution should not be used until all elements below the main diagonal of a matrix are zeros.

Cooperative Learning

Have students work in groups of three to solve

$$\begin{cases} 2x + 3y - z = -10 \\ 3x - 2y + 4z = 100 \\ -x - 5y + 2z = 45 \end{cases}$$

Ask one student in each group to use the row reduction method and back substitution with a calculator. Have the second student do the same without a calculator. Then have the third student solve the system without using matrices. Ask each group to discuss which method would be easiest if a large matrix is involved.

Aongoing ASSESSMENT

Try This

$$\begin{bmatrix} 1 & \frac{1}{3} & -\frac{1}{3} & | & \frac{19}{3} \\ 0 & 1 & -\frac{1}{2} & | & \frac{1}{4} \\ 0 & 0 & 1 & | & -\frac{3}{2} \end{bmatrix};$$
$$x = 6, y = -\frac{1}{2}, z = -\frac{3}{2}$$

Aongoing ASSESSMENT

It is easier because you simply multiply the 1 by the negative of the value in the row where you want a zero.

CRITICAL Thinking

Multiply row 2 by $-\frac{1}{3}$;
$-\frac{1}{3}R_2 \rightarrow R_2$

The augmented matrix now represents a system of equations that can be solved using back substitution.

$$\begin{bmatrix} 1 & -1 & -2 & \vdots & -4 \\ 0 & -1 & -5 & \vdots & -7 \\ 0 & 0 & -2 & \vdots & 2 \end{bmatrix} \longrightarrow \begin{cases} x - y - 2z = -4 \\ -y - 5z = -7 \\ -2z = 2 \end{cases}$$

From the third equation in this system of equations, solve for z.

$$-2z = 2$$
$$z = -1$$

Substitute –1 for z in the second equation, and solve for y.

$$-y - 5z = -7$$
$$-y - 5(-1) = -7$$
$$y = 12$$

Substitute –1 for z and 12 for y in the first equation, and solve for x.

$$x - y - 2z = -4$$
$$x - (12) - 2(-1) = -4$$
$$x = 6$$

Thus, the solution is (6, 12, –1).

Check►

Check the solution in each original equation of the system.

Check 1	**Check 2**
$-2x + y - z = 1$	$5x - 2y + 3z = 3$
$-2(6) + (12) - (-1) \overset{?}{=} 1$	$5(6) - 2(12) + 3(-1) \overset{?}{=} 3$
$1 = 1$ True	$3 = 3$ True

Check 3
$$x - y - 2z = -4$$
$$(6) - (12) - 2(-1) \overset{?}{=} -4$$
$$-4 = -4 \quad \text{True} ❖$$

Try This Solve this system of linear equations using the row reduction method and back substitution. Check your row operations with your graphics calculator.

$$\begin{cases} 3x + y - z = 19 \\ 2x - y + z = 11 \\ 4x + 2y = 23 \end{cases}$$

Why is it easier to produce zeros below the main diagonal when elements of the main diagonal are 1 or –1?

CRITICAL Thinking

Using elementary row operations, how can you produce a 1 in the second element of the main diagonal without exchanging rows?

$$\begin{bmatrix} 1 & 2 & 1 & \vdots & 3 \\ 0 & -3 & -1 & \vdots & 10 \\ 0 & -1 & 1 & \vdots & 6 \end{bmatrix}$$

The row reduction method consists of repetitive steps and can be easily computerized. This method can be used to solve a system of n equations in n unknowns, no matter how large n might be.

Exercises & Problems

Communicate

1. Explain how to use back substitution to solve $\begin{cases} 2x - 3y + z = 12 \\ 5y - 4z = 6 \\ -z = 3 \end{cases}$

Identify the coefficient matrix, the constant matrix, and the augmented matrix for each system of equations.

2. $\begin{cases} 3x - 2y + z = 6 \\ x + y = 3 \\ 2y + 4x = 0 \end{cases}$

3. $\begin{cases} x - 4y + 7z = 17 \\ 2x + y - z = -5 \\ x + 4z = 13 \end{cases}$

4. Identify the main diagonal and the system of equations represented by this augmented matrix. $\left[\begin{array}{ccc:c} -3 & -4 & 0 & 2 \\ 4 & 2 & -3 & 6 \\ -2 & 0 & 1 & -6 \end{array}\right]$

5. How is a system of linear equations solved using the row reduction method and back substitution?

Describe each row operation in words.

6. $-5R_1 + R_3 \rightarrow R_3$

7. $R_1 \leftrightarrow R_3$

8. $2R_1 \rightarrow R_1$

Practice & Apply

Cultural Connection: Asia Write the system of equations represented by each Chinese counting board.

9.

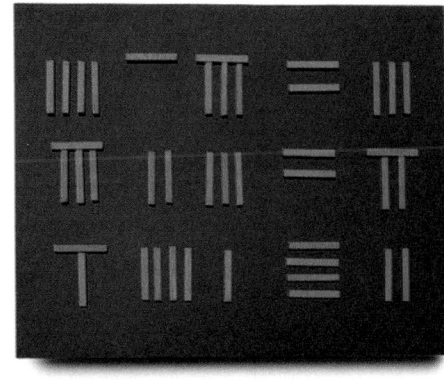

10.

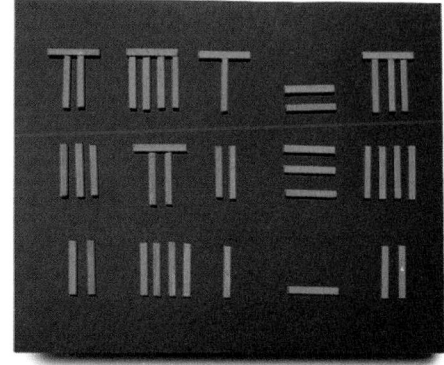

Perform the indicated row operations on matrix A.

$A = \left[\begin{array}{ccc:c} -1 & 2 & -5 & 4 \\ 3 & 7 & -2 & 3 \\ -8 & 4 & 1 & -7 \end{array}\right]$

11. $3R_1 + R_2 \rightarrow R_2$ R_2: 0 13 −17 ⋮ 15

12. $-8R_1 + R_3 \rightarrow R_3$ R_3: 0 −12 41 ⋮ −39

9. $\begin{cases} 4x + 5y + 8z = 23 \\ 8x + 2y + 3z = 27 \\ 6x + 4y + z = 42 \end{cases}$

10. $\begin{cases} 7x + 9y + 6z = 28 \\ 3x + 7y + 2z = 34 \\ 2x + 4y + z = 12 \end{cases}$

Write the augmented matrix for each system of equations.

13. $\begin{cases} x + 3y = 23 \\ 4x - 2y = -6 \end{cases}$ $\begin{bmatrix} 1 & 3 & \vdots & 23 \\ 4 & -2 & \vdots & -6 \end{bmatrix}$

14. $\begin{cases} -x + 2y - 5z = -12 \\ 2x - 3y + 7z = 19 \\ -5x - 2y + z = -10 \end{cases}$

15. $\begin{cases} 2x - 5y - z = -32 \\ -x + 4y + 2z = 34 \\ 3x + 7y - 3z = -2 \end{cases}$

Use back substitution to solve each system of equations.

16. $\begin{cases} 4x + 2y - 5z = -9 \\ 9y + z = 66 \\ 5z = 15 \end{cases}$ $(-2, 7, 3)$

17. $\begin{cases} x + 4y - 3z = -13 \\ -2y + z = 1 \\ -6z = -30 \end{cases}$ $(-6, 2, 5)$

18. $\begin{cases} 4x - 7y + 5z = -52 \\ 3y + 8z = 7 \\ -z = 1 \end{cases}$ $(-3, 5, -1)$

Human Resources An automobile service station employs attendants and mechanics. An attendant pumps gas for the entire 8-hour work day, and a mechanic is expected to spend 6 hours repairing automobiles and 2 hours pumping gas. One person can pump gas for 10 cars per hour. The service station owner wants to be able to pump gas for 320 cars per day and have a total of 24 hours of the mechanics' time per day for repairing automobiles.

19. Write a system of two linear equations in two variables to represent this situation.

20. Solve the system of linear equations using the row reduction method and back substitution. (3, 4)

21. How many attendants and how many mechanics should be hired? 3 attendants, 4 mechanics

Manufacturing A tool company manufactures pliers and scissors. In 1 hour the company uses 140 units of steel and 290 units of aluminum.

Complete Exercises 22–24.

Each pair of scissors requires 1 unit of steel and 3 units of aluminum.

22. Write a system of two linear equations to represent this situation.

23. Solve the system of linear equations using the row reduction method and back substitution. (65, 10)

24. How many pliers and how many scissors can this tool company make in one hour? 65 pliers, 10 scissors

Each pair of pliers contains 2 units of steel and 4 units of aluminum.

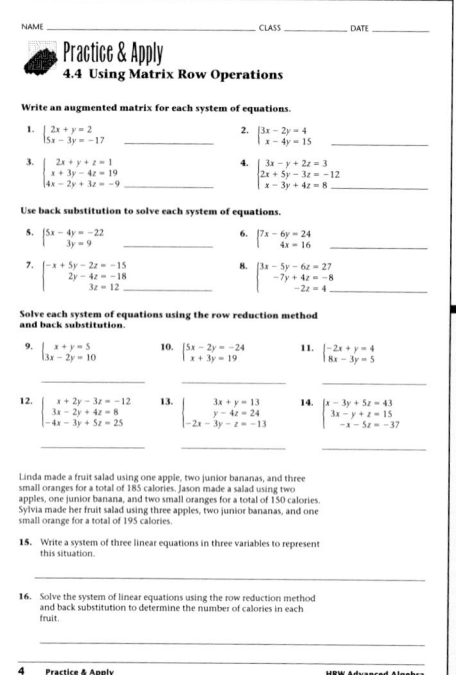

25. Use your graphics calculator to perform the indicated row operations on this matrix. $\begin{bmatrix} -1 & 4 & 9 & \vdots & 10 \\ 5 & -8 & -1 & \vdots & 7 \\ -3 & 7 & 2 & \vdots & -6 \end{bmatrix}$ $\xrightarrow{5R_1 + R_2 \to R_2}$ $\xrightarrow{-3R_1 + R_3 \to R_3}$

Solve each system of equations using the row reduction method and back substitution.

26. $\begin{cases} x + y = 7 \\ 2x - 3y = 4 \end{cases}$ (5, 2)

27. $\begin{cases} x + 3y = 23 \\ 4x - 2y = -6 \end{cases}$ (2, 7)

28. $\begin{cases} -2x + 5y = -4 \\ 3x - y = -7 \end{cases}$ (-3, -2)

29. $\begin{cases} -x + 2y - 5z = -12 \\ 2x - 3y + 7z = 19 \\ -5x - 2y + z = -10 \end{cases}$ (3, -2, 1)

30. $\begin{cases} 2x + y = 8 \\ y - 2z = -9 \\ -3x - 4y - z = 9 \end{cases}$ $\left(\frac{33}{4}, -\frac{17}{2}, \frac{1}{4}\right)$

31. $\begin{cases} x - 4y + 7z = 17 \\ 2x + y - z = -5 \\ x + 4z = 13 \end{cases}$

14. $\begin{bmatrix} -1 & 2 & -5 & \vdots & -12 \\ 2 & -3 & 7 & \vdots & 19 \\ -5 & -2 & 1 & \vdots & -10 \end{bmatrix}$

15. $\begin{bmatrix} 2 & -5 & -1 & \vdots & -32 \\ -1 & 4 & 2 & \vdots & 34 \\ 3 & 7 & -3 & \vdots & -2 \end{bmatrix}$

19. Let x and y represent the number of attendants and mechanics respectively.
$\begin{cases} 8x + 2y = 32 \\ 6y = 24 \end{cases}$

22. Let x and y represent the number of pliers and scissors respectively. $\begin{cases} 2x + y = 140 \\ 4x + 3y = 290 \end{cases}$

25. $\begin{bmatrix} -1 & 4 & 9 & \vdots & 10 \\ 0 & 12 & 44 & \vdots & 57 \\ 0 & -5 & -25 & \vdots & -36 \end{bmatrix}$

31. $\left(-\frac{17}{11}, \frac{19}{11}, \frac{40}{11}\right)$

Geography A traveler is going south along the west coast of South America through the Andes Mountains. While in Peru, the traveler spent $20 a day on housing and $30 a day on food and travel. Passing through Bolivia, the traveler spent $30 a day for housing and $20 a day on food and travel. Finally, while enjoying the long coast of Chile, the traveler spent $20 a day both for housing and for food and travel. In each country, the traveler spent $10 a day on miscellaneous items. The traveler spent a total of $220 on housing, $230 on food and travel, and $100 on miscellaneous items.

32. Write a system of three linear equations in three unknowns to find the number of days the traveler spent in each country.

33. Solve the system of linear equations using the row reduction method and back substitution. (3, 2, 5)

34. How many days did the traveler spend in each country?
    3 in Peru, 2 in Bolivia, 5 in Chile

35.  **Probability** Suppose that a certain experiment has three possible (separate) outcomes with probabilities p_1, p_2, and p_3. If $3p_1 + 18p_2 - 12p_3 = 3$ and $p_1 - 2p_2 - 2p_3 = 0$, determine the probabilities of the three outcomes. (HINT: In any experiment, the sum of the probabilities of the outcomes is 1.) $\left(\frac{2}{3}, \frac{1}{6}, \frac{1}{6}\right)$

Look Back

Tell whether each equation is linear. **[Lesson 1.1]**

36. $y + 1 = 5$ yes   **37.** $y = 4x^2 - 7$ no   **38.** $x + 2 = 0$ yes

Determine whether each relation is a function. Explain your answers. **[Lesson 2.3]**

39. {(−1, 6), (0, 6), (1, 6), (2, 6)} yes

40. {(0, 14), (1, 12), (2, 10), (3, 8), (4, 6)} yes

41. {(−2, 0), (0, 0), (2, 0), (4, 0), (6, 0)} yes

42. {(−1, 0), (0, −1), (0, 1), (1, 0)} no

Let $A = \begin{bmatrix} 3 & 2 & -2 \\ 1 & 1 & 4 \\ -1 & 2 & 3 \end{bmatrix}$ and $B = \begin{bmatrix} 2 & 3 & 1 \\ -1 & -2 & 2 \\ 4 & 1 & -4 \end{bmatrix}$. **[Lesson 4.2]**

43. Find AB.   **44.** Find BA.   **45.** Is AB equal to BA? no

Look Back

Name the multiplicative inverse of each of the following numbers, and show how it can be used to solve an equation.

46. 5 $\frac{1}{5}$ **47.** $\frac{1}{4}$ 4 **48.** $-\frac{1}{5}$ -5

---

32. Let x, y, and z represent the number of days spent in Peru, Bolivia, and Chile respectively. $\begin{cases} 20x + 30y + 20z = 220 \\ 30x + 20y + 20z = 230 \\ 10x + 10y + 10z = 100 \end{cases}$

39. function; only one y-value for each x-value

40. function; only one y-value for each x-value

41. function; only one y-value for each x-value

42. not a function; two y-values for $x = 0$

43. $\begin{bmatrix} -4 & 3 & 15 \\ 17 & 5 & -13 \\ 8 & -4 & -9 \end{bmatrix}$

44. $\begin{bmatrix} 8 & 9 & 11 \\ -7 & 0 & 0 \\ 17 & 1 & -16 \end{bmatrix}$

The answers to Exercises 46–48 can be found in Additional Answers beginning on page 842.

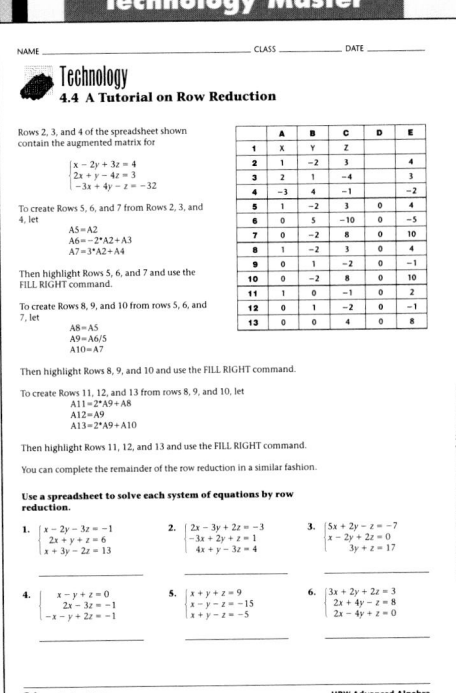

Objectives
- Find the inverse of a matrix.

- Practice Master   4.5
- Enrichment Master   4.5
- Technology Master   4.5
- Lesson Activity Master   4.5
- Quiz   4.5
- Spanish Resources   4.5

Assessing Prior Knowledge

Find the multiplicative inverse of each number.

1. -3 $\left[-\frac{1}{3}\right]$

2. $-\frac{1}{7}$ $[-7]$

3. 2.5 $[0.4]$

4. $-\frac{6}{5}$ $\left[-\frac{5}{6}\right]$

TEACH

 Review how inverse operations are used to solve equations. The exercises in Look Beyond on page 197 focus on the use of multiplicative inverses to solve equations. Ask students to suggest how inverse matrices could be used to solve matrix equations.

LESSON 4.5 The Inverse of a Matrix

why *Just as inverse operations can be used to solve equations, inverse matrices can be used to solve matrix equations that represent systems of equations.*

The statement $3 \cdot \frac{1}{3} = 1 = \frac{1}{3} \cdot 3$ is true because $\frac{1}{3}$ and 3 are multiplicative inverses and 1 is the multiplicative identity.

Recall that the matrix $I_2 = \begin{bmatrix} 1 & 0 \\ 0 & 1 \end{bmatrix}$ is the identity matrix for 2×2 matrices.

IDENTITY MATRIX

The **identity matrix**, I_n, is an $n \times n$ matrix with ones along the main diagonal and zeros elsewhere.

Let $A = \begin{bmatrix} 1 & 2 \\ 3 & 5 \end{bmatrix}$ and $A^{-1} = \begin{bmatrix} a & b \\ c & d \end{bmatrix}$.

Then $A \cdot A^{-1} = I_2 = A^{-1} \cdot A$ is

$$\begin{bmatrix} 1 & 2 \\ 3 & 5 \end{bmatrix} \begin{bmatrix} a & b \\ c & d \end{bmatrix} = \begin{bmatrix} 1 & 0 \\ 0 & 1 \end{bmatrix} = \begin{bmatrix} a & b \\ c & d \end{bmatrix} \begin{bmatrix} 1 & 2 \\ 3 & 5 \end{bmatrix}.$$

ALTERNATIVE teaching strategy

Using Cognitive Strategies After students have described how they would solve the two systems of equations presented on page 199, have them use their suggestions to find the inverse. Then have students use the same process to find the inverses of the matrices shown.

$$\begin{bmatrix} -1 & 1 \\ 8 & -6 \end{bmatrix} \begin{bmatrix} 3 & 0.5 \\ 4 & 0.5 \end{bmatrix}$$

$$\begin{bmatrix} 1 & 1 \\ 2 & 3 \end{bmatrix} \begin{bmatrix} 3 & -1 \\ -2 & 1 \end{bmatrix}$$

$$\begin{bmatrix} 0 & -4 \\ -4 & 0 \end{bmatrix} \begin{bmatrix} 0 & -0.25 \\ -0.25 & 0 \end{bmatrix}$$

Finding an Inverse Matrix

If $\begin{bmatrix} a & b \\ c & d \end{bmatrix}$ is the inverse matrix for $\begin{bmatrix} 1 & 2 \\ 3 & 5 \end{bmatrix}$, then

$$\begin{bmatrix} 1 & 2 \\ 3 & 5 \end{bmatrix} \begin{bmatrix} a & b \\ c & d \end{bmatrix} = \begin{bmatrix} 1 & 0 \\ 0 & 1 \end{bmatrix}.$$

How can you find the values for a, b, c, and d to know what the inverse matrix is? Multiply the two matrices.

$$\begin{bmatrix} 1 & 2 \\ 3 & 5 \end{bmatrix} \begin{bmatrix} a & b \\ c & d \end{bmatrix} = \begin{bmatrix} 1a + 2c & 1b + 2d \\ 3a + 5c & 3b + 5d \end{bmatrix} = \begin{bmatrix} 1 & 0 \\ 0 & 1 \end{bmatrix}$$

The result is two systems of linear equations.

$\begin{cases} 1a + 2c = 1 \\ 3a + 5c = 0 \end{cases}$ A system with variables a and c

$\begin{cases} 1b + 2d = 0 \\ 3b + 5d = 1 \end{cases}$ A system with variables b and d

Describe the steps you would take to solve for the variables a, b, c, and d.

The mathematician Abraham Adrian Albert, 1905–1972, of the University of Chicago discovered that you can solve for all four variables, a, b, c, and d, with one augmented matrix.

You can use a graphics calculator to find the inverse of a matrix. First, enter the coefficient matrix $\begin{bmatrix} 1 & 2 \\ 3 & 5 \end{bmatrix}$ as matrix A. Use the $\boxed{x^{-1}}$ key to find the inverse of matrix A. Thus, $A^{-1} = \begin{bmatrix} -5 & 2 \\ 3 & -1 \end{bmatrix}$. To check, find $A \cdot A^{-1}$ to see if the result is the identity matrix.

THE INVERSE OF A MATRIX

If A is an $n \times n$ matrix with an inverse, then A^{-1} is its **inverse matrix**, and

$$A \cdot A^{-1} = I_n = A^{-1} \cdot A.$$

ongoing ASSESSMENT

Try This

A. $A^{-1} = \begin{bmatrix} -5 & -2 \\ 8 & 3 \end{bmatrix}$

B. $\begin{bmatrix} 3 & 2 \\ -8 & -5 \end{bmatrix} \begin{bmatrix} -5 & -2 \\ 8 & 3 \end{bmatrix} = \begin{bmatrix} 1 & 0 \\ 0 & 1 \end{bmatrix}$

$= \begin{bmatrix} -5 & -2 \\ 8 & 3 \end{bmatrix} \begin{bmatrix} 3 & 2 \\ -8 & -5 \end{bmatrix}$

CRITICAL Thinking

A nonsquare matrix cannot have an inverse because the different row and column dimensions would not allow the multiplication to be defined in both directions.

ongoing ASSESSMENT

Not every square matrix has an inverse. The matrix $\begin{bmatrix} 4 & 2 \\ 6 & 3 \end{bmatrix}$ does not have an inverse matrix. The calculator gives an error message.

EXAMPLE

Graphics Calculator

Let $A = \begin{bmatrix} 3 & 2 \\ -8 & -5 \end{bmatrix}$.

Ⓐ Find A^{-1}.

Ⓑ Check that $A \cdot A^{-1} = I_2 = A^{-1} \cdot A$.

Solution▸

Ⓐ Enter the matrix $\begin{bmatrix} 3 & 2 \\ -8 & -5 \end{bmatrix}$ as matrix A.

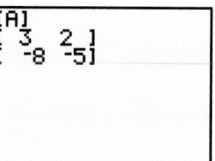

Use the $\boxed{x^{-1}}$ key to find the inverse of matrix A.

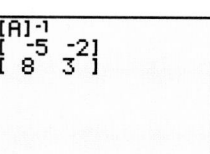

Ⓑ To check, find $A \cdot A^{-1}$ and $A^{-1} \cdot A$. Both products are the identity matrix.

Thus, $A^{-1} = \begin{bmatrix} -5 & -2 \\ 8 & 3 \end{bmatrix}$ because $A \cdot A^{-1} = I_2 = A^{-1} \cdot A$. ❖

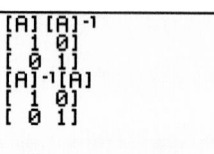

Try This Let $A = \begin{bmatrix} 3 & 2 \\ -8 & -5 \end{bmatrix}$.

Ⓐ Find A^{-1}.

Ⓑ Check that $A \cdot A^{-1} = I_2 = A^{-1} \cdot A$.

CRITICAL Thinking

Explain why a nonsquare matrix cannot have an inverse.

Do you think every square matrix has an inverse?
What happens when you try to find the inverse of the matrix $\begin{bmatrix} 4 & 2 \\ 6 & 3 \end{bmatrix}$?

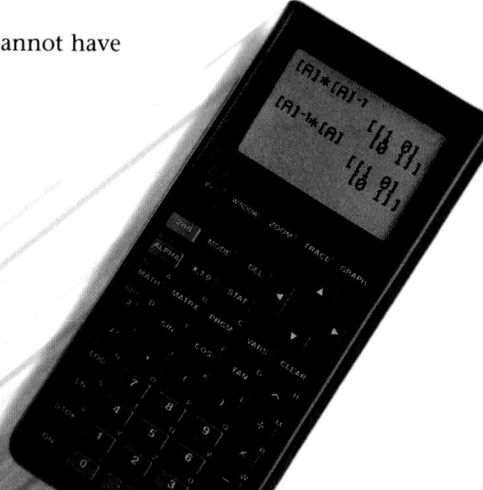

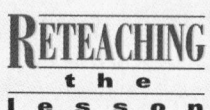

EXERCISES & PROBLEMS

Communicate

Describe the product matrix without multiplying.

1. $\begin{bmatrix} 1 & 3 \\ 2 & -1 \end{bmatrix} \begin{bmatrix} 1 & 0 \\ 0 & 1 \end{bmatrix}$

2. $\begin{bmatrix} 1 & 0 \\ 0 & 1 \end{bmatrix} \begin{bmatrix} -2 & 1 \\ 5 & 3 \end{bmatrix}$

3. Explain why the product $I_3 I_4$ cannot be found.

4. What is the notation for the 4×4 identity matrix, and what does it look like?

5. How do you find the inverse of a matrix with a graphics calculator?

Practice & Apply

Use your graphics calculator to determine which of the following matrices has an inverse.

6 $\begin{bmatrix} -2 & -2 & 1 \\ 3 & 3 & -1 \\ 2 & 3 & -1 \end{bmatrix}$ yes

7 $\begin{bmatrix} 1 & 6 & 2 \\ -2 & 3 & 5 \\ 7 & 12 & -4 \end{bmatrix}$ no

8 $\begin{bmatrix} -2 & -1 & 1 \\ 1 & -2 & 3 \\ 4 & 1 & 2 \end{bmatrix}$ yes

For Exercises 9–11, determine whether each pair of matrices are inverses.

9. $\begin{bmatrix} 4 & 0 \\ 0 & 3 \end{bmatrix}, \begin{bmatrix} \frac{1}{4} & 0 \\ 0 & \frac{1}{3} \end{bmatrix}$ yes

10. $\begin{bmatrix} 6 & 3 \\ 4 & 2 \end{bmatrix}, \begin{bmatrix} 2 & -3 \\ -4 & 6 \end{bmatrix}$ no

11. $\begin{bmatrix} 1 & 2 \\ 3 & 4 \end{bmatrix}, \begin{bmatrix} -2 & 1 \\ \frac{3}{2} & -\frac{1}{2} \end{bmatrix}$ yes

Investment Among three plans, A, B, and C, a firm invested $16,110.00 in low-risk stocks, $9016.75 in medium-risk stocks, and $5698.75 in high-risk stocks.

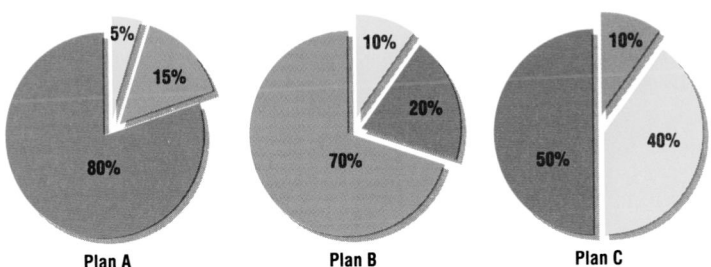

Plan A   Plan B   Plan C

- ■ Low-risk stocks
- ■ Medium-risk stocks
- □ High-risk stocks

12. Write the system of equations that you would solve to find the amount invested in each plan.

13. Write the coefficient matrix.

14 Use your graphics calculator to find the inverse of the coefficient matrix. Round numbers to the nearest hundredth.

12. Let x, y, and z represent the amount invested in Plan A, Plan B, and Plan C respectively. Then

$$\begin{cases} 0.80x + 0.20y + 0.50z = 16{,}110.00 \\ 0.15x + 0.70y + 0.10z = 9016.75 \\ 0.05x + 0.10y + 0.40z = 5698.75 \end{cases}$$

13. $\begin{bmatrix} 0.80 & 0.20 & 0.50 \\ 0.15 & 0.70 & 0.10 \\ 0.05 & 0.10 & 0.40 \end{bmatrix}$

14. $\begin{bmatrix} 0.80 & 0.20 & 0.50 \\ 0.15 & 0.70 & 0.10 \\ 0.05 & 0.10 & 0.40 \end{bmatrix}^{-1}$

$= \begin{bmatrix} 1.38 & -0.15 & -1.69 \\ -0.28 & 1.51 & -0.03 \\ -0.10 & -0.36 & 2.72 \end{bmatrix}$

ASSESS

Selected Answers
Odd-numbered Exercises 7–41

Assignment Guide
Core 1–7, 9–20, 23–41

Core Plus 1–5, 7–8, 10–14, 18–25, 29–42

Technology
A graphics calculator is needed for Exercises 6–8, 14–22, and 25–32. A graphics calculator or computer software such as *f(g) Scholar*™ may be helpful to find the inverses of matrices and to answer Exercise 42.

Error Analysis
An error message on a graphics calculator may not always mean that the matrix does not have an inverse. Encourage students to make sure they have correctly entered all elements of a matrix before assuming that it does not have an inverse.

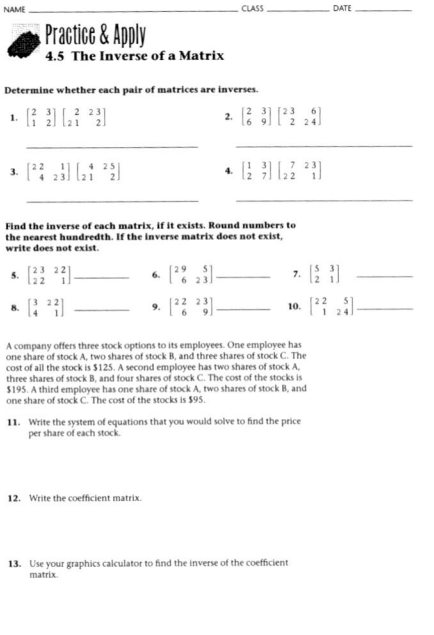

ALTERNATIVE ASSESSMENT

Authentic Assessment

The method for finding the inverse of a 2 × 2 matrix is shown on the top of page 199. Have students use this procedure as a model for finding the inverse of a 3 × 3 matrix. Have students explain the steps and give an example. Have them check the inverse by multiplying the inverse and the original matrix.

23. $\begin{cases} 2x - z = 0 \\ x - 2y + z = 0 \\ x + y + z = 180 \end{cases}$

24. $\begin{bmatrix} 2 & 0 & -1 \\ 1 & -2 & 1 \\ 1 & 1 & 1 \end{bmatrix}$

25. $\begin{bmatrix} 0.33 & 0.11 & 0.22 \\ 0 & -0.33 & 0.33 \\ -0.33 & 0.22 & 0.44 \end{bmatrix}$

Look Beyond

This exercise will help prepare students to perform the mathematical steps necessary in solving a matrix equation.

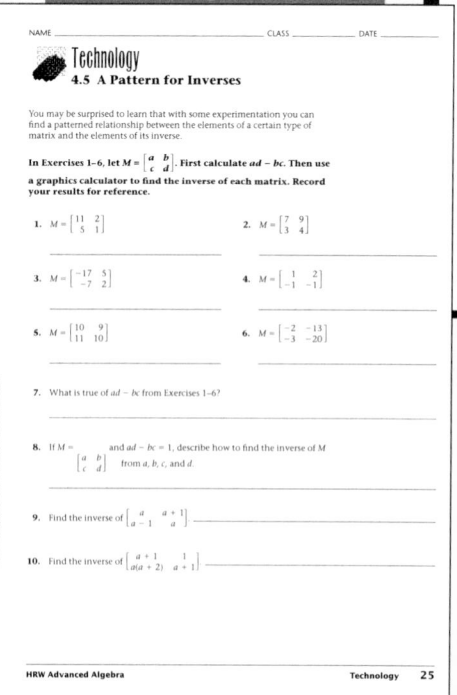

Technology Master

NAME _____ CLASS _____ DATE _____

Technology
4.5 A Pattern for Inverses

You may be surprised to learn that with some experimentation you can find a patterned relationship between the elements of a certain type of matrix and the elements of its inverse.

In Exercises 1–6, let $M = \begin{bmatrix} a & b \\ c & d \end{bmatrix}$. First calculate $ad - bc$. Then use a graphics calculator to find the inverse of each matrix. Record your results for reference.

1. $M = \begin{bmatrix} 11 & 2 \\ 5 & 1 \end{bmatrix}$ 2. $M = \begin{bmatrix} 7 & 9 \\ 3 & 4 \end{bmatrix}$

3. $M = \begin{bmatrix} -17 & 5 \\ -7 & 2 \end{bmatrix}$ 4. $M = \begin{bmatrix} 1 & 2 \\ -1 & -1 \end{bmatrix}$

5. $M = \begin{bmatrix} 10 & 9 \\ 11 & 10 \end{bmatrix}$ 6. $M = \begin{bmatrix} -2 & -13 \\ -3 & -20 \end{bmatrix}$

7. What is true of $ad - bc$ from Exercises 1–6?

8. If $M = \begin{bmatrix} a & b \\ c & d \end{bmatrix}$ and $ad - bc = 1$, describe how to find the inverse of M from a, b, c, and d.

9. Find the inverse of $\begin{bmatrix} a & a+1 \\ a-1 & a \end{bmatrix}$

10. Find the inverse of $\begin{bmatrix} a+1 & 1 \\ a(a+2) & a+1 \end{bmatrix}$

HRW Advanced Algebra   Technology   25

Find the inverse of each matrix. Round numbers to the nearest hundredth.

15. $\begin{bmatrix} 2 & 3 \\ 1 & 2 \end{bmatrix} \begin{bmatrix} 2 & -3 \\ -1 & 2 \end{bmatrix}$ 16. $\begin{bmatrix} 3 & 7 \\ 2 & 5 \end{bmatrix} \begin{bmatrix} 5 & -7 \\ -2 & 3 \end{bmatrix}$ 17. $\begin{bmatrix} 2 & -6 \\ 1 & -2 \end{bmatrix} \begin{bmatrix} -1 & 3 \\ -0.5 & 1 \end{bmatrix}$ 18. $\begin{bmatrix} 2 & 1 \\ 1 & 1 \end{bmatrix} \begin{bmatrix} 1 & -1 \\ -1 & 2 \end{bmatrix}$

19. $\begin{bmatrix} 1 & 3 \\ 2 & 7 \end{bmatrix} \begin{bmatrix} 7 & -3 \\ -2 & 1 \end{bmatrix}$ 20. $\begin{bmatrix} 5 & -7 \\ -2 & 3 \end{bmatrix} \begin{bmatrix} 3 & 7 \\ 2 & 5 \end{bmatrix}$ 21. $\begin{bmatrix} 3 & -1 \\ -4 & 2 \end{bmatrix} \begin{bmatrix} 1 & 0.5 \\ 2 & 1.5 \end{bmatrix}$ 22. $\begin{bmatrix} 4 & -3 \\ 11 & 2 \end{bmatrix} \begin{bmatrix} 0.05 & 0.07 \\ -0.27 & 0.10 \end{bmatrix}$

Geometry The largest angle of a triangle is twice the measure of the smallest angle. The measure of the remaining angle of the triangle is the average of the measures of the largest and the smallest angle.

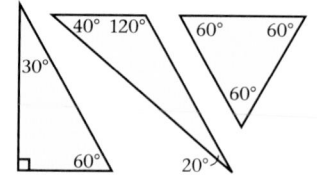

The sum of the measures of the angles of any triangle is 180°.

23. Write the system of equations that you would solve to find the measure of each angle of the triangle.

24. Write the coefficient matrix.

25. Use your graphics calculator to find the inverse of the coefficient matrix. Round numbers to the nearest hundredth.

For each matrix, find the inverse matrix, if it exists. Round numbers to the nearest hundredth. If the inverse matrix does not exist, write *does not exist*.

26. $\begin{bmatrix} 2 & -4 \\ -3 & 6 \end{bmatrix}$ does not exist   27. $\begin{bmatrix} \frac{1}{2} & 0 \\ 1 & \frac{1}{4} \end{bmatrix} \begin{bmatrix} 2 & 0 \\ -8 & 4 \end{bmatrix}$ 28. $\begin{bmatrix} 2 & -10 \\ -1 & 10 \end{bmatrix}$ 29. $\begin{bmatrix} \frac{1}{2} & \frac{1}{10} \\ \frac{3}{2} & \frac{1}{5} \end{bmatrix}$

30. $\begin{bmatrix} -2 & 2 & -1 \\ 3 & -5 & 4 \\ 5 & -6 & 4 \end{bmatrix}$ 31. $\begin{bmatrix} -2 & -1 & 1 \\ 1 & -2 & 3 \\ 4 & 1 & 2 \end{bmatrix}$ 32. $\begin{bmatrix} 1 & -1 & 2 \\ 0 & 0 & 0 \\ 2 & 1 & 2 \end{bmatrix}$ does not exist

～～ Look Back

33. Evaluate the expression $2a - 5b + b^2$ when a is 3 and b is -4. 42

34. Write the linear equation, in slope-intercept form, of a line that passes through the points $(2, -2)$ and $(0, -1)$. **[Lesson 1.2]** $y = -\frac{1}{2}x - 1$

Solve. Justify each step. [Lesson 1.5]

35. $8.91 + x = 11.09$ 2.18

36. $\frac{1}{4} = \frac{3}{4} + x$ $-\frac{1}{2}$

37. $1 + x = 7\frac{1}{2}$ $6\frac{1}{2}$

38. $-0.25x = 8$ -32

39. $-2\frac{1}{6} = 8\frac{2}{3}x$ $-\frac{1}{4}$

40. $5\frac{1}{2}x = -62$ $-11\frac{3}{11}$

41. Solve the system of linear equations $\begin{cases} 5x + 7y = 32 \\ 2x - 14y = 6 \end{cases}$. **[Lesson 4.3, 4.4]** $\left(\frac{35}{6}, \frac{17}{42} \right)$

Look Beyond ～～

42. Let $A = \begin{bmatrix} 3 & 7 \\ 2 & 5 \end{bmatrix}$. Find A^{-1}. Find the product of both sides of $A^{-1} \begin{bmatrix} 3 & 7 \\ 2 & 5 \end{bmatrix} \begin{bmatrix} x \\ y \end{bmatrix} = A^{-1} \begin{bmatrix} 4 \\ 1 \end{bmatrix}$. What is the resulting matrix equation?

28. $\begin{bmatrix} 1 & 1 \\ 0.1 & 0.2 \end{bmatrix}$

29. $\begin{bmatrix} -4 & 2 \\ 30 & -10 \end{bmatrix}$

30. $\begin{bmatrix} 4 & -2 & 3 \\ 8 & -3 & 5 \\ 7 & -2 & 4 \end{bmatrix}$

31. $\begin{bmatrix} -0.54 & 0.23 & -0.08 \\ 0.77 & -0.62 & 0.54 \\ 0.69 & -0.15 & 0.38 \end{bmatrix}$

35. Subtract Prop of Equal; $x = 2.18$

36. Subtract Prop of Equal; $x = -\frac{1}{2}$

37. Subtract Prop of Equal; $x = 6\frac{1}{2}$

38. Div Prop of Equal; $x = -32$

The answers to Exercises 39–40 and 42 can be found in Additional Answers beginning on page 842.

PREPARE

Objectives

• Use matrix algebra to solve a system of equations.

RESOURCES

• Practice Master   4.6
• Enrichment Master   4.6
• Technology Master   4.6
• Lesson Activity Master   4.6
• Quiz   4.6
• Spanish Resources   4.6

Assessing Prior Knowledge

Classify each system as independent, dependent, or inconsistent.

1. $3x + 4y = -10$
    $-6x - 8y = 20$
    [**dependent**]

2. $3x + 4y = -8$
    $6x + 8y = 16$
    [**inconsistent**]

3. $-2x + 5y = 15$
    $10x + 2y = 6$
    [**independent**]

TEACH

 Using matrix algebra and a graphics calculator or computer software such as *f(g) Scholar*™ significantly decreases the steps needed to solve a system of linear equations. Even systems with more than three equations can be quickly and easily solved using this method.

Why *Systems of equations can be solved by using matrix equations. Matrix algebra allows you to solve systems of equations in less time because there are fewer steps.*

Recall that a system of equations can be represented by a matrix equation.

$$\begin{cases} x + y + 3z = 2 \\ 2x - y - 2z = -3 \\ 3x + 2y - 2z = -13 \end{cases} \rightarrow \begin{bmatrix} 1 & 1 & 3 \\ 2 & -1 & -2 \\ 3 & 2 & -2 \end{bmatrix} \begin{bmatrix} x \\ y \\ z \end{bmatrix} = \begin{bmatrix} 2 \\ -3 \\ -13 \end{bmatrix}$$

A = the coefficient matrix $\qquad$ X = the variable matrix $\qquad$ B = the constant matrix

$$A = \begin{bmatrix} 1 & 1 & 3 \\ 2 & -1 & -2 \\ 3 & 2 & -2 \end{bmatrix} \qquad X = \begin{bmatrix} x \\ y \\ z \end{bmatrix} \qquad B = \begin{bmatrix} 2 \\ -3 \\ -13 \end{bmatrix}$$

You can now write the matrix equation using the names of the matrices.

$$\underbrace{\begin{bmatrix} 1 & 1 & 3 \\ 2 & -1 & -2 \\ 3 & 2 & -2 \end{bmatrix}}_{A} \underbrace{\begin{bmatrix} x \\ y \\ z \end{bmatrix}}_{X} = \underbrace{\begin{bmatrix} 2 \\ -3 \\ -13 \end{bmatrix}}_{B} \rightarrow AX = B$$

ALTERNATIVE teaching strategy

Hands-On Strategies
Give students the systems of equations shown below.

$$\begin{cases} 7x + 2y = 24 \\ 5x - 3y = -36 \end{cases} \qquad \begin{cases} x - 3y = -14 \\ 2x - y = 2 \end{cases}$$

Have them solve the systems first using elimination, then matrix algebra. [(0, 12), (4, 6)] Discuss students' ideas of which method is easier and why. Then have students use both methods

to solve the system of three equations in three variables shown below.

$$\begin{cases} x + 2y + z = -5 \\ 2x - 3y - 4z = 50 \\ 3x + 4y - 2z = 35 \end{cases} \quad [x = 5, y = 0, z = -10]$$

Students who use a calculator will notice that the value they get for y is $-6\text{E}{-}13$, not zero. Explain that this is due to rounding errors that occur in the operation of the calculator.

Matrix Algebra

You can solve the matrix equation $AX = B$ for X in the same way that you solve the algebraic equation $ax = b$ for x.

To solve $ax = b$, multiply both sides of the equation by the multiplicative inverse of a.	$ax = b$ $\frac{1}{a}(ax) = \frac{1}{a}(b)$
This leaves only x on the left side and the solution for x on the right side of the equation.	$x = \frac{b}{a}$
Similarly, to solve $AX = B$, multiply both sides of the equation by A^{-1}.	$AX = B$ $A^{-1}(AX) = A^{-1}(B)$
The solution for X is $A^{-1}B$.	$X = A^{-1}B$

Since X is a matrix and you are solving a matrix equation, this method is called **matrix algebra**. Since matrix multiplication is not commutative in general, you need to make sure that you multiply A^{-1} on the *left* side of the coefficient matrix, B, to obtain the product $A^{-1}B$ rather than BA^{-1}.

EXAMPLE 1

Graphics Calculator

Solve the system of linear equations using matrix algebra.

$$\begin{cases} x + y + 3z = 2 \\ 2x - y - 2z = -3 \\ 3x + 2y - 2z = -13 \end{cases}$$

Solution

Write the matrix equation, $AX = B$, that represents this system of linear equations.

$$\underbrace{\begin{bmatrix} 1 & 1 & 3 \\ 2 & -1 & -2 \\ 3 & 2 & -2 \end{bmatrix}}_{A} \underbrace{\begin{bmatrix} x \\ y \\ z \end{bmatrix}}_{X} = \underbrace{\begin{bmatrix} 2 \\ -3 \\ -13 \end{bmatrix}}_{B}$$

To find the solution, $A^{-1}B$, enter matrices A and B in your graphics calculator. Then find the product $A^{-1}B$.

```
[A]⁻¹[B]
[ -1]
[ -3]
[  2]
```

Thus, $\begin{bmatrix} x \\ y \\ z \end{bmatrix} = \begin{bmatrix} -1 \\ -3 \\ 2 \end{bmatrix}$, or x is -1, y is -3, and z is 2.

Check►
Substitute the solution in each original equation.

Equation 1
$$x + y + 3z = 2$$
$$-1 + (-3) + 3(2) \overset{?}{=} 2$$
$$2 = 2$$
True

Equation 2
$$2x - y - 2z = -3$$
$$2(-1) - (-3) - 2(2) \overset{?}{=} -3$$
$$-3 = -3$$
True

Equation 3
$$3x + 2y - 2z = -13$$
$$3(-1) + 2(-3) - 2(2) \overset{?}{=} -13$$
$$-13 = -13$$
True ❖

Try This Solve the system of equations using matrix algebra.
$$\begin{cases} -2x + 2y - z = 4 \\ 3x - 5y + 4z = -2 \\ 5x - 6y + 4z = 8 \end{cases}$$

You can solve a system of n equations in n unknowns using matrix algebra, no matter how large n is. Computers and calculators are used to speed up the process.

Graphics Calculator

EXAMPLE 2

Solve the system of equations using matrix algebra.
$$\begin{cases} 2a - c + d = 12 \\ -2a - b + 2c - d = 7 \\ 3a + 2b - c + d = -3 \\ b - c + d = 4 \end{cases}$$

Solution►
Write the matrix equation $AX = B$.

Remember to use zeros for the coefficients of missing variables.

$$\begin{bmatrix} 2 & 0 & -1 & 1 \\ -2 & -1 & 2 & -1 \\ 3 & 2 & -1 & 1 \\ 0 & 1 & -1 & 1 \end{bmatrix} \begin{bmatrix} a \\ b \\ c \\ d \end{bmatrix} = \begin{bmatrix} 12 \\ 7 \\ -3 \\ 4 \end{bmatrix}$$

To find the solution $A^{-1}B$, enter matrices A and B in your graphics calculator. Then find the product $A^{-1}B$.

```
[A]⁻¹[B]
[ .2  ]
[ -7.6]
[ 11.4]
[ 23  ]
```

Thus, $\begin{bmatrix} a \\ b \\ c \\ d \end{bmatrix} = \begin{bmatrix} 0.2 \\ -7.6 \\ 11.4 \\ 23 \end{bmatrix}$, or a is 0.2, b is –7.6, c is 11.4, and d is 23.

Aongoing
ASSESSMENT

Try This

$x = 44, y = 78, z = 64$

Alternate Example 2

Solve the system of equations using matrix algebra.
$$\begin{cases} 3a + 2b - d = 13 \\ 2a + 3c + 2d = -16 \\ a + b + c + d = -5 \\ 2b + 2c - 3d = 18 \end{cases}$$
$[a = 1, b = 2, c = -2, d = -6]$

INCLUSION **strategies**

Visual Learners Have students graph each of the following systems of equations on a coordinate grid or by using a graphics calculator.

$$\begin{cases} x + 2y = 8 \\ 2x - 3y = 2 \end{cases} \begin{cases} x + 2y = 8 \\ -2x - 4y = -16 \end{cases} \begin{cases} x + 2y = 8 \\ x + 2y = 10 \end{cases}$$

Have students describe each system as independent, dependent, or inconsistent. Then ask them to try to solve the matrix equation for each system. Ask them to record all of their findings about different systems of equations in a chart like the one below.

Independent	Dependent	Inconsistent

Students should notice that dependent and inconsistent systems both produce the same error message on a calculator.

Substitute the solution in each original equation.

Equation 1
$$2a - c + d = 12$$
$$2(0.2) - (11.4) + (23) \stackrel{?}{=} 12$$
$$12 = 12$$
True

Equation 2
$$-2a - b + 2c - d = 7$$
$$-2(0.2) - (-7.6) + 2(11.4) - (23) \stackrel{?}{=} 7$$
$$7 = 7$$
True

Equation 3
$$3a + 2b - c + d = -3$$
$$3(0.2) + 2(-7.6) - (11.4) + 23 \stackrel{?}{=} -3$$
$$-3 = -3$$
True

Equation 4
$$b - c + d = 4$$
$$-7.6 - (11.4) + 23 \stackrel{?}{=} 4$$
$$4 = 4$$
True ❖

Why is it important to check the solution in *each* original equation?

Types of Systems of Linear Equations

A dependent system of two linear equations has infinitely many solutions, and an inconsistent system has no solution. Both dependent and inconsistent systems have coefficient matrices that do not have inverses. An independent system of linear equations has a unique solution, so it will have a coefficient matrix with an inverse.

EXAMPLE 3

Graphics Calculator

Solve the system of equations using matrix algebra.
$$\begin{cases} x + 2y - z = 5 \\ 3x + 4y + 5z = 9 \\ 2x + 3y + 2z = 1 \end{cases}$$

Solution➤

Write the matrix equation $AX = B$.

$$\begin{bmatrix} 1 & 2 & -1 \\ 3 & 4 & 5 \\ 2 & 3 & 2 \end{bmatrix} \begin{bmatrix} x \\ y \\ z \end{bmatrix} = \begin{bmatrix} 5 \\ 9 \\ 1 \end{bmatrix}$$

When you try to find the inverse of this coefficient matrix, you will find that this matrix does not have an inverse. Thus, there is no *unique* solution to this system of equations. ❖

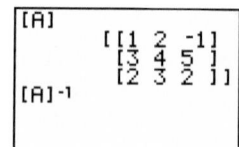

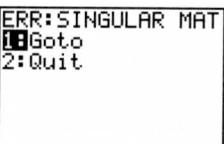

CRITICAL Thinking

Explain why the coefficient matrix for a dependent or inconsistent system does not have an inverse.

Alternate Example 3

Solve the system of equations using matrix algebra.

$$\begin{cases} x + z = 29 \\ x - 4y + 3z = -22 \\ 2x - 6y + 5z = 10 \end{cases}$$

[**no unique solution**]

TEACHING *tip*

When a calculator gives an error message indicating that a coefficient matrix does not have an inverse, the system has no unique solution. More complicated calculations must be performed to determine whether the system is dependent or inconsistent.

CRITICAL *Thinking*

An inconsistent system of equations does not have a solution. When row reduction is performed on it, a false statement results. A dependent system of equations has an infinite number of solutions. The coefficients of one equation are scalar multiples of the other(s). When row reduction is performed on the matrix, the result is a true statement $(0 = 0)$. Neither of these results can be represented by matrix equations.

RETEACHING the lesson

Cooperative Learning Have students work in pairs to write a system of two linear equations whose solution is $x = -3$ and $y = 4$. Then have them work together to write a matrix equation for their system. Ask each pair to use a graphics calculator or computer software such as *f(g) Scholar*™ to solve the matrix equation. If necessary, have students complete another example using a system of two equations. Ask each pair to share their equations and solutions with the class. Then have each pair complete the same procedure for a system of three linear equations whose solution is $x = 2$, $y = -4$, and $z = 1$.

Exercises & Problems

Communicate

1. How do you write a matrix equation to represent a system of equations?

2. What system of equations does this matrix equation represent?
$$\begin{bmatrix} 2 & -1 & 3 \\ -3 & 0 & -1 \\ 1 & -3 & 1 \end{bmatrix} \begin{bmatrix} x \\ y \\ z \end{bmatrix} = \begin{bmatrix} 4 \\ 1 \\ 5 \end{bmatrix}$$

3. Explain how to solve a system of linear equations using matrix algebra.

4. When solving a matrix equation, why is it necessary to multiply the inverse matrix, A^{-1}, on the *left* side of the constant matrix B?

Describe each step in solving the system of equations using matrix algebra.

5. $\begin{cases} 2x + 3y - 2z = 4 \\ 3x - 3y + 2z = 16 \\ 6x - 2y + 8z = 10 \end{cases}$

6. $\begin{cases} x + 2y = -6 \\ y + 2z = 11 \\ 2x + z = 16 \end{cases}$

7. What types of systems of linear equations have coefficient matrices that do not have inverses?

Practice & Apply

For Exercises 8 and 9, write the matrix equation that represents each system. Use your graphics calculator to solve it, and check your answer.

8. $\begin{cases} x - 2y + 3z = 11 \\ 4x + y - z = 4 \quad (2, -3, 1) \\ 2x - y + 3z = 10 \end{cases}$

9. $\begin{cases} x + y - z = 7 \\ 4x - y + 5z = 4 \quad \text{no solution} \\ 6x + y + 3z = 20 \end{cases}$

Entertainment One hundred twenty people attended a school play. The total amount of money collected for tickets was $1515. There were twice as many students as general patrons.

10. Write a system of equations that you would solve to find the number of students, general patrons, and senior citizens.

11. Use your graphics calculator to solve the system. Give the number of students, general patrons, and senior citizens. 70, 35, 15

12. Check your solution in the original problem.

8. $\begin{bmatrix} 1 & -2 & 3 \\ 4 & 1 & -1 \\ 2 & -1 & 3 \end{bmatrix} \begin{bmatrix} x \\ y \\ z \end{bmatrix} = \begin{bmatrix} 11 \\ 4 \\ 10 \end{bmatrix}; \begin{bmatrix} x \\ y \\ z \end{bmatrix} = \begin{bmatrix} 2 \\ -3 \\ 1 \end{bmatrix}$

9. $\begin{bmatrix} 1 & 1 & -1 \\ 4 & -1 & 5 \\ 6 & 1 & 3 \end{bmatrix} \begin{bmatrix} x \\ y \\ z \end{bmatrix} = \begin{bmatrix} 7 \\ 4 \\ 20 \end{bmatrix};$ no solution

10. $\begin{cases} x + y + z = 120 \\ 12x + 15y + 10z = 1515 \\ x = 2y \end{cases}$

12.
$$x + y + z = 120$$
$$(70) + (35) + (15) = 120 \quad \text{True}$$

$$12x + 15y + 10z = 1515$$
$$12(70) + 15(35) + 10(15) = 1515 \quad \text{True}$$

$$x = 2y$$
$$(70) = 2(35)$$

Assess

Selected Answers

Odd-numbered Exercises 9–31

Assignment Guide

Core 1–14, 16–20, 22–36

Core Plus 1–7, 9–12, 14–18, 20–38

Technology

A graphics calculator is needed for Exercises 8, 9, 11, 13–15, 17, 19–21, and 23. A graphics calculator can be helpful for checking Exercises 29–31.

Error Analysis

Not all calculators report an error message when a matrix does not have an inverse. Refer to your *HRW Technology Handbook* or the calculator manual to determine how a particular calculator responds to a noninvertible matrix.

Be sure students do not make careless errors entering the elements of a matrix into their calculator. Encourage students to double check all entries. The most common errors occur when working with negative signs. Also watch to be sure students enter a 0 for any missing terms.

For Exercises 13–15, write the matrix equation that represents each system. Use your graphics calculator to solve it, and check your answer.

13
$$\begin{cases} 3x + 6y - 6z = 9 \\ 2x - 5y + 4z = 6 \\ -x + 16y + 14z = -3 \end{cases}$$
(3, 0, 0)

14
$$\begin{cases} x + y - z = 7 \\ 4x - y + 5z = 4 \\ 2x + 2y - 3z = 0 \end{cases}$$
(−9, 30, 14)

15
$$\begin{cases} 3x + 6y - 6z = 9 \\ 2x - 5y + 4z = 6 \\ 5x + 28y - 26z = -8 \end{cases}$$
no solution

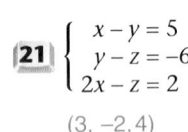 **Geometry** The largest angle of a triangle is three times the measure of the smallest angle. The measure of the remaining angle of the triangle is the average of the measures of the largest and the smallest angle.

16. Write the system of equations that you would solve to find the measure of each angle of the triangle.

17 Use your graphics calculator to solve the system. Give the measure of each angle of the triangle. 30°, 60°, 90°

18. Check your solution in the original problem.

For Exercises 19–21, write the matrix equation that represents each system. Use your graphics calculator to solve it, and check your answer.

19
$$\begin{cases} x + y + z + w = 10 \\ 2x - y + z - 3w = -9 \\ 3x + y - z - w = -2 \\ 2x - 3y + z - w = -5 \end{cases}$$
(1, 2, 3, 4)

20
$$\begin{cases} x + 2y - 6z + w = 12 \\ -2x - 3y + 9z + w = -19 \\ x + 2y - 5z + 2w = 15 \\ 2x + 4y - 12z + 3w = 24 \end{cases}$$
(2, 14, 3, 0)

21
$$\begin{cases} x - y = 5 \\ y - z = -6 \\ 2x - z = 2 \end{cases}$$
(3, −2, 4)

Investment Among three plans, A, B, and C, a firm invested $16,110.00 in low-risk stocks, $9016.75 in medium-risk stocks, and $5698.75 in high-risk stocks. How much did the firm invest in each plan?

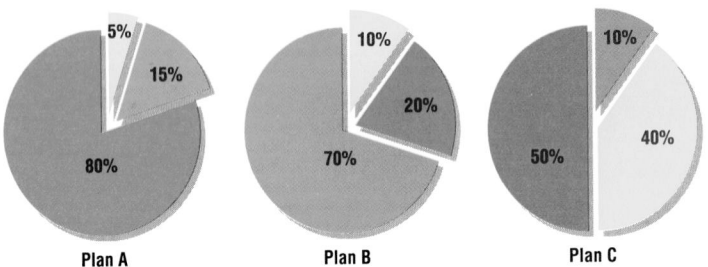

5%   15%   80%
Plan A

10%   20%   70%
Plan B

10%   40%   50%
Plan C

■ Low-risk stocks
■ Medium-risk stocks
□ High-risk stocks

22. Write the system of equations that you would solve to find the amount the firm invested in each plan. Write the matrix equation that represents this system.

23 Use your graphics calculator to solve the system. Give the amount that the firm invested in each plan.

24. Show your check of the solution in each original equation of the system.

13. $$\begin{bmatrix} 3 & 6 & -6 \\ 2 & -5 & 4 \\ -1 & 16 & 14 \end{bmatrix} \begin{bmatrix} x \\ y \\ z \end{bmatrix} = \begin{bmatrix} 9 \\ 6 \\ -3 \end{bmatrix};$$ $$\begin{bmatrix} x \\ y \\ z \end{bmatrix} = \begin{bmatrix} 3 \\ 0 \\ 0 \end{bmatrix}$$

14. $$\begin{bmatrix} 1 & 1 & -1 \\ 4 & -1 & 5 \\ 2 & 2 & -3 \end{bmatrix} \begin{bmatrix} x \\ y \\ z \end{bmatrix} = \begin{bmatrix} 7 \\ 4 \\ 0 \end{bmatrix};$$ $$\begin{bmatrix} x \\ y \\ z \end{bmatrix} = \begin{bmatrix} -9 \\ 30 \\ 14 \end{bmatrix}$$

15. $$\begin{bmatrix} 3 & 6 & -6 \\ 2 & -5 & 4 \\ 5 & 28 & -26 \end{bmatrix} \begin{bmatrix} x \\ y \\ z \end{bmatrix} = \begin{bmatrix} 9 \\ 6 \\ -8 \end{bmatrix};$$ no solution

16. $$\begin{cases} x + y + z = 180 \\ 3x - z = 0 \\ x - 2y + z = 0 \end{cases}$$

The answers to Exercises 18–24 and 29 can be found in Additional Answers beginning on page 842.

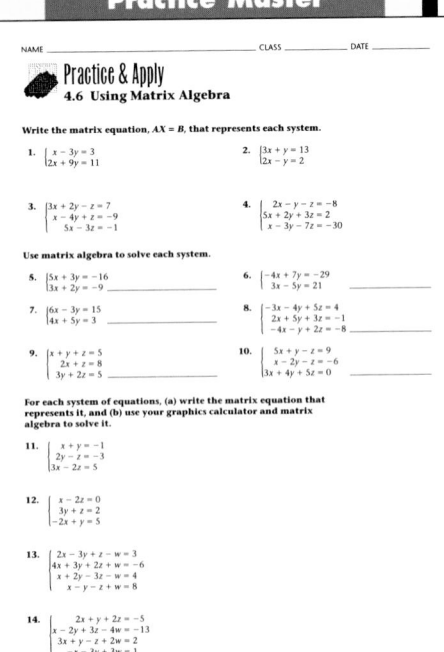

25. Determine the domain and range of the function $f(x) = |x - 5|$.
[Lesson 2.3] R; $y \geq 0$

For Exercises 26–28, let $f(x) = 2x + 3$ and let $g(x) = x^2 - 3x + 1$.
[Lessons 2.4, 3.3]

26. Find $f(a)$. $2a + 3$ **27.** Find $g \circ f$. $4x^2 + 6x + 1$ **28.** Find $f \circ g$. $2x^2 - 6x + 5$

29. Multiply $\begin{bmatrix} 5 & -4 \\ -3 & 3 \end{bmatrix} \begin{bmatrix} -5 & -2 \\ 8 & 3 \end{bmatrix}$. Show your steps.   **[Lesson 4.2]** $\begin{bmatrix} -57 & -22 \\ 39 & 15 \end{bmatrix}$

Solve the following systems of equations using the row reduction method. [Lesson 4.4]

30. $\begin{cases} x - 3y = 4 \\ -4x + 2y = 6 \end{cases}$ $\left(-\frac{13}{5}, -\frac{11}{5}\right)$ **31.** $\begin{cases} 2x - 8y = 5 \\ -3x + 12y = 8 \end{cases}$ no solution

Look Beyond

32. **Coordinate Geometry** On graph paper, graph the coordinates $A(3, 6)$, $B(5, -4)$, $C(-5, -4)$, and $D(-3, 6)$, and connect each vertex in order from A to D. What geometrical shape did you make? isosceles trapezoid

33. **Coordinate Geometry** On the same coordinate plane that you used for Exercise 32, graph the coordinates $A'(-6, 3)$, $B'(4, 5)$, $C'(4, -5)$, and $D'(-6, -3)$, and connect each vertex in order from A' to D'. Explain how this shape and the shape you made in Exercise 32 are alike and how they are different. isos trap; congruent but rotated counterclockwise by 90°

Magic Addition Squares The square at the right is a magic square because its numbers add up to the same sum whether you add them across, up and down, or diagonally.

38	93	16
27	49	71
82	5	60

Are these illustrated squares magic addition squares?

34. Find the sum of the magic square above. 147

Complete the following magic squares.

35.

75	106	59
64	80	96
101	54	85

36.

88	9	11	70
24	57	55	42
47	34	32	65
19	78	80	1

37.

45	27	29	41
31	39	37	35
36	34	32	40
30	42	44	26

38.

21	7	8	18
10	16	15	13
14	12	11	17
9	19	20	6

32.

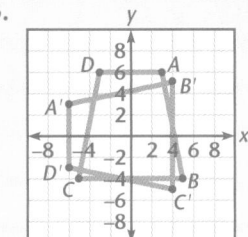

isosceles trapezoid

33.

The isosceles trapezoid is the same as in Exercise 32, but it has been rotated counterclockwise by 90°.

Exploring Transformations Using Matrices

PREPARE

Objectives

• Use matrices to represent and transform objects.

RESOURCES

• Practice Master	**4.7**
• Enrichment Master	**4.7**
• Technology Master	**4.7**
• Lesson Activity Master	**4.7**
• Quiz	**4.7**
• Spanish Resources	**4.7**

Assessing Prior Knowledge

Plot and connect each set of points. Identify the resulting polygon.

1. $P(3, 0)$, $R(-3, 0)$, $Q(0, 4)$
   [**triangle**]

2. $A(-3, 3)$, $B(-3, -3)$, $C(3, -3)$, $D(3, 3)$
   [**square**]

3. $H(6, 2)$, $I(-4, 2)$, $J(-4, -3)$, $K(6, -3)$
   [**rectangle**]

4. $X(4, 5)$, $Y(0, -5)$, $Z(-6, -5)$, $W(-2, 5)$
   [**parallelogram**]

TEACH

Discuss situations and professions in which individuals would need to transform objects. Ask students to bring in examples of designs that involve transformations of objects and designs. You may want to provide an example of such a design to get students started.

Some computer-graphics systems use matrices to manipulate objects and to transform one object into another. You can use transformation matrices to move, change, and rotate designs made out of a variety of shapes.

why *You can make a variety of designs with computer drawing programs. With matrix multiplication and transformation matrices, you can imitate the method that computers use to create graphic designs.*

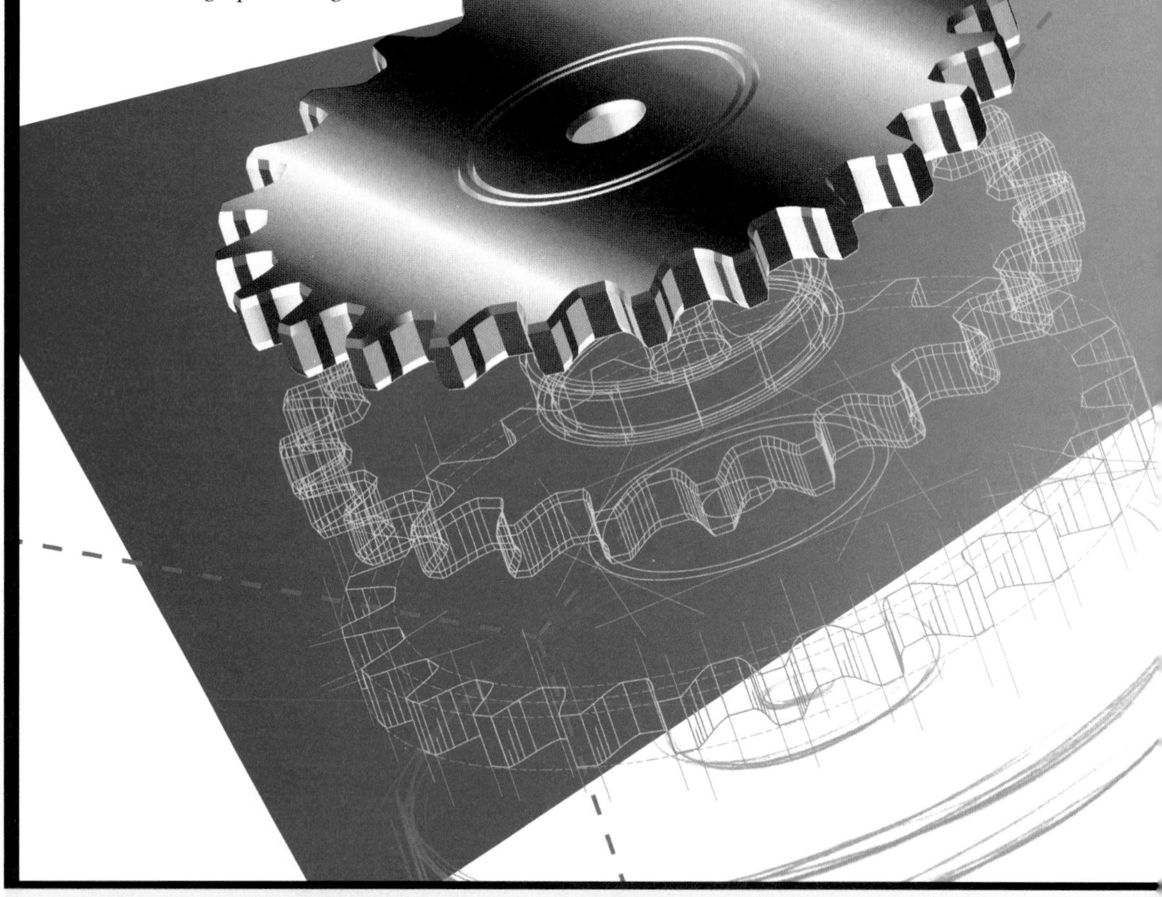

ALTERNATIVE teaching strategy

Hands-On Strategies Provide small groups of students with a clear piece of acetate. Ask them to place the piece of acetate over a coordinate grid and draw polygon *PQRS* as shown on page 211. Have them rotate the polygon 90° counterclockwise and record the coordinates of the new vertices. Then ask students to find the product of the matrices *A* and *B* given on page 211. Have them compare *BA* with the coordinates they found when they rotated polygon *PQRS*. Their answers should be the same. Have students repeat this process as many times as necessary until they grasp the concept.

Representing Objects With Matrices

COORDINATE GEOMETRY
Connection

Geometric figures, called polygons, are made up of straight lines and can be specified by the vertex points (corners) that make up the figures.
For example, the points $P(3, 5)$, $Q(6, -5)$, $R(-6, -5)$, and $S(-3, 5)$ are the corners of a trapezoid centered at the origin.

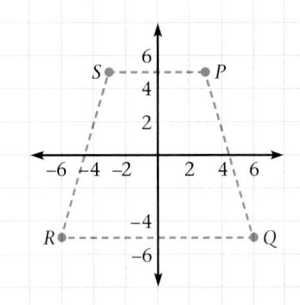

The coordinates of the vertices that describe such figures or objects are placed in a matrix, where the first row lists the *x*-coordinates and the second row lists the *y*-coordinates.

Matrix A lists the vertices of trapezoid *PQRS*.

$$A = \begin{bmatrix} 3 & 6 & -6 & -3 \\ 5 & -5 & -5 & 5 \end{bmatrix} \begin{matrix} \leftarrow x\text{-coordinates} \\ \leftarrow y\text{-coordinates} \end{matrix}$$

Transformations

 *Using Matrix Products*

Graphics Calculator

Let $B = \begin{bmatrix} 0 & -1 \\ 1 & 0 \end{bmatrix}$. Enter matrix A, which lists the vertices of trapezoid *PQRS*, and matrix B in your calculator.

1 Find the product *BA*.

2 Plot the vertices of *BA* on graph paper. Connect the vertices to form a polygon.

3 Compare the figure from *BA* with trapezoid *PQRS* shown. How are they alike? Are they congruent (exactly the same shape and same size)? Are their centers in the same location?

4 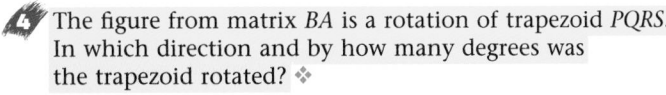 The figure from matrix *BA* is a rotation of trapezoid *PQRS*. In which direction and by how many degrees was the trapezoid rotated? ❖

Printing After Exploration 3 on page 212, ask students to write down several examples of objects that may need to be enlarged or reduced. Some examples include photographs, pictures, pages from books, and so on. Discuss ways in which these enlargements and reductions are accomplished. For example, ask students to think about a copy machine that enlarges and reduces copies. Then have students write a transformation matrix to represent each

enlargement or reduction described.

- enlargement of 125% $\begin{bmatrix} 1.25 & 0 \\ 0 & 1.25 \end{bmatrix}$

- reduction by 75% $\begin{bmatrix} 0.75 & 0 \\ 0 & 0.75 \end{bmatrix}$

- enlargement of 150% $\begin{bmatrix} 1.5 & 0 \\ 0 & 1.5 \end{bmatrix}$

- reduction by 50% $\begin{bmatrix} 0.5 & 0 \\ 0 & 0.5 \end{bmatrix}$

Cooperative Learning

Have students work in pairs to graph a polygon of their choice on a coordinate place. Then have one student rotate the figure 90° counterclockwise and note the coordinates of the vertices. Next have the other student rotate the image 90° clockwise. Have students repeat this procedure rotating 180°, 270°, and 360°. They should then explore rotations of other degrees. The pairs should describe the results of their investigations. Have students repeat this procedure three more times and describe the results.

Math Connection
Coordinate Geometry

Ask students how they could prove that *PQRS* is a trapezoid. [**prove that *PS* and *QR* are parallel by proving that their slopes are the same (they are both zero). Also have students show that the slopes of lines *RS* and *PQ* are not the same. Otherwise, the figure would be a parallelogram.**]

Exploration 1 Notes

Be sure that students notice that the trapezoid *BA* is also a trapezoid centered at the origin. Point out that the center of rotation is the origin.

Aongoing
SSESSMENT

4. The figure from matrix *BA* is rotated 90° counterclockwise.

Exploration 2 · Using Repeated Products

Graphics Calculator

1 Plot the vertices in matrix A on graph paper. Connect the vertices to form a polygon. What shape is this figure?

$$A = \begin{bmatrix} -1 & -6 & 6 & 4 \\ 3 & -1 & -5 & 3 \end{bmatrix}$$

2 Find the product BA, where $B = \begin{bmatrix} 0 & -1 \\ 1 & 0 \end{bmatrix}$.

Plot the vertices of BA on graph paper. Connect the vertices to form a polygon. Is this figure congruent to the figure represented by A? How is this figure different from the figure represented by A?

3 Find the product $B(BA)$. Plot the vertices in $B(BA)$, and draw the figure. Is this figure congruent to the figures represented by A and BA? How is this figure different from the figures represented by A and BA?

4 How do you think you could rotate the object represented by A by 270° counterclockwise? Experiment to find out.

5 Find the product BBB. Graph and describe the figure represented by the product $(BBB)A$.

6 How can you rotate an object by 90°, 180°, or 270° counterclockwise? ❖

A ongoing ASSESSMENT

6. For counterclockwise 90°, multiply by $\begin{bmatrix} 0 & -1 \\ 1 & 0 \end{bmatrix}$. For 180°, multiply by $\begin{bmatrix} -1 & 0 \\ 0 & -1 \end{bmatrix}$. For counterclockwise 270°, multiply by $\begin{bmatrix} 0 & 1 \\ -1 & 0 \end{bmatrix}$.

Exploration 3 Notes

As students work through this exploration, ask them to note the similarities and differences in matrices B, C, and D.

Exploration 3 · Enlarging and Reducing Objects

Graphics Calculator

Let $A = \begin{bmatrix} 2 & -2 & -8 & 8 \\ 4 & 4 & -3 & -3 \end{bmatrix}$, $B = \begin{bmatrix} 2 & 0 \\ 0 & 2 \end{bmatrix}$, $C = \begin{bmatrix} 3 & 0 \\ 0 & 3 \end{bmatrix}$, and $D = \begin{bmatrix} \frac{1}{2} & 0 \\ 0 & \frac{1}{2} \end{bmatrix}$.

Plot the vertices in matrix A on graph paper. Connect the vertices to form a polygon. What shape is this figure?

1 Find the product BA. Plot the vertices in BA on graph paper. Connect the vertices to form a polygon.

2 Find the product CA. Plot the vertices in CA on graph paper. Connect the vertices to form a polygon.

3 Describe how the multiplication of matrix B changed the size of the object represented by matrix A. Describe how the multiplication of matrix C changed the size of the object represented by A.

4 Find the product $D(BA)$. Plot the vertices in $D(BA)$ on graph paper. Connect the vertices to form a polygon.

5 Describe how the multiplication of matrix D changed the object represented by matrix BA.

6 If $X(CA) = A$, what is matrix X?

7 Explain how to enlarge and reduce an object represented by a matrix. ❖

A ongoing ASSESSMENT

7. To enlarge or reduce an object represented by a matrix, multiply the matrix by a 2 × 2 matrix of the form $\begin{bmatrix} a & 0 \\ 0 & a \end{bmatrix}$, where a can be any real number greater than 1 for an enlargement, and a can be any positive real number less than 1 for a reduction.

ENRICHMENT Have each student draw a polygon on a coordinate plane. Have them perform combinations of transformations on the figure and then draw the final transformed figure in a different color on the same grid. Have students exchange graphs and try to determine what transformations were used to draw the final transformed figure.

INCLUSION *strategies* **Using Manipulatives** Provide students with a coordinate grid with a clear piece of acetate anchored to the origin. Have students draw a figure to be rotated and then turn the piece of acetate the required amount to show the rotated figure. A geoboard can also be used to help students work with enlarging and reducing figures.

Exploration 4 *Reversing Transformations*

Graphics Calculator

1 Plot the vertices in matrix *A* from Exploration 3 on graph paper. Connect the vertices to form a polygon. What shape is this figure?

2 Find the product *BA*, where $B = \begin{bmatrix} 0 & -1 \\ 1 & 0 \end{bmatrix}$. Plot the vertices in *BA*. Connect the vertices to form a polygon. Describe this transformation.

3 Find $B^{-1}A$. Plot the vertices in $B^{-1}A$. Connect the vertices to form a polygon. Describe this transformation.

4 Find $BB^{-1}A$. Plot the vertices in $BB^{-1}A$. Connect the vertices to form a polygon. How does the object represented by $BB^{-1}A$ compare with the object represented by *A*?

5 Compare the transformations on *A* in Steps 2 and 3. How are they alike? How are they different?

6 Find the product *CA*, where $C = \begin{bmatrix} 2 & 0 \\ 0 & 2 \end{bmatrix}$. Plot the vertices in *CA*. Connect the vertices to form a polygon. Describe this transformation.

7 Find $C^{-1}A$. Plot the vertices in $C^{-1}A$. Connect the vertices to form a polygon. Describe this transformation.

8 Find $CC^{-1}A$. Plot the vertices in $CC^{-1}A$. Connect the vertices to form a polygon. How does the object represented by $CC^{-1}A$ compare with the object represented by *A*? ❖

In general, any transformation can be undone by multiplying by the inverse of the transformation matrix.

CRITICAL *Thinking*

Why is shrinking an object to $\frac{1}{3}$ its size the same as the inverse transformation of enlarging an object to 3 times its size?

RETEACHING the lesson

Hands-On Strategies
Have students study transformations using a figure such as a triangle or a rectangle that is centered at the origin. Give students the transformation matrices and have them draw the original figure in one color and the transformed figure in another color. Be sure that students complete a sufficient number of examples to feel comfortable with the concepts.

Exploration 4 Notes

Before beginning this exploration, review the concept of the inverse matrix. Discuss how the inverse matrix is used to solve a matrix equation. Tell students that in a similar manner they will use the inverse matrices to "undo" or reverse transformations.

Aongoing **SSESSMENT**

5. The objects represented by *BA* and $B^{-1}A$ are congruent. Multiplication by matrix *B* or matrix B^{-1} rotates the object by 90°. Matrix *B* rotates the object *counterclockwise*, while matrix B^{-1} rotates the object *clockwise*.

8. The trapezoid represented by $CC^{-1}A$ is the same as the trapezoid represented by *A*.

CRITICAL *Thinking*

The inverse of the matrix that enlarges by 3, $\begin{bmatrix} 3 & 0 \\ 0 & 3 \end{bmatrix}$, is $\begin{bmatrix} \frac{1}{3} & 0 \\ 0 & \frac{1}{3} \end{bmatrix}$. This is also the matrix that reduces by $\frac{1}{3}$.

ASSESS

Selected Answers
Odd-numbered Exercises 7–45

Assignment Guide
Core 1–8, 10–16, 19–25, 27–47

Core Plus 1–6, 8, 9, 14–22, 25–47

Technology
A graphics calculator is needed for Exercises 10–18, 23–26, 36, 45, and 47.

7.

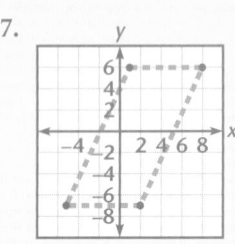

parallelogram

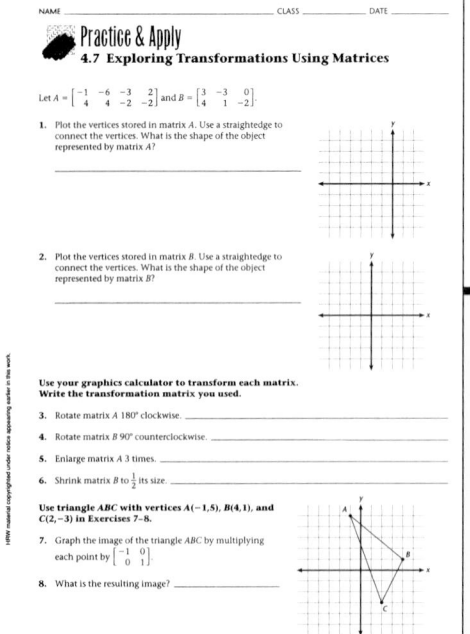
EXERCISES & PROBLEMS

Communicate

1. What matrix can be used to rotate an object 90° counterclockwise?
2. What matrix can be used to rotate an object 180° counterclockwise?
3. What matrix can be used to rotate an object 270° counterclockwise?
4. Explain how to "undo" the transformation of an object.
5. Find the matrix that can be used to rotate an object 180° clockwise.
6. What matrix can be used to enlarge an object to 5 times its size?

Practice & Apply

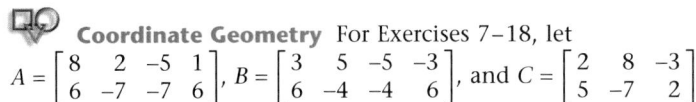

Coordinate Geometry For Exercises 7–18, let
$$A = \begin{bmatrix} 8 & 2 & -5 & 1 \\ 6 & -7 & -7 & 6 \end{bmatrix}, B = \begin{bmatrix} 3 & 5 & -5 & -3 \\ 6 & -4 & -4 & 6 \end{bmatrix}, \text{ and } C = \begin{bmatrix} 2 & 8 & -3 \\ 5 & -7 & 2 \end{bmatrix}.$$

7. Plot the vertices in A. Connect the vertices to form a polygon. What is the shape of the object represented by A? parallelogram
8. Plot the vertices in matrix B. Connect the vertices to form a polygon. What is the shape of the object represented by B? isos trapezoid
9. Plot the vertices in matrix C. Connect the vertices to form a polygon. What is the shape of the object represented by C? triangle

Perform each of the following transformations on the object represented by matrix A. Write the matrix equation you used.

10. Rotate 270° clockwise.
11. Enlarge to 5 times its size.
12. Rotate 90° counterclockwise, and shrink to $\frac{2}{3}$ of its size.
13. Enlarge to 3 times its size, and rotate 180° clockwise.

Perform each of the following transformations on the object represented by matrix B. Write the matrix equation you used.

14. Rotate 180° counterclockwise, and enlarge to 4 times its size.
15. Enlarge to 7 times its size, and shrink to $\frac{1}{2}$ of that size.
16. Rotate 90° clockwise, and rotate 180° counterclockwise.

Perform each of the following transformations on the object represented by matrix C. Write the matrix equation you used.

17. Rotate 180° counterclockwise, and enlarge to 3 times its size.
18. Reduce to $\frac{3}{4}$ of its size, and rotate 270° clockwise.

8.

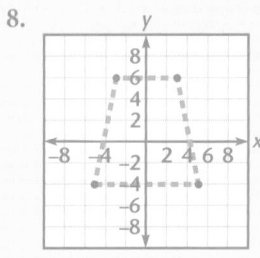

isoceles trapezoid

9.

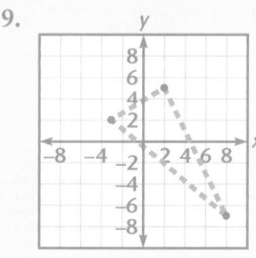

triangle

The answers to Exercises 10–18 can be found in Additional Answers beginning on page 842.

Animation A cartoonist draws the individual frames of an animation sequence on paper that is 11 inches wide and 8 inches long. Each frame is then put together and projected as an animated movie on a screen that is $5\frac{1}{2}$ feet wide and 4 feet long.

19. What matrix could be used to enlarge the cartoonist's drawings from the size of the paper to the size of the screen on which the movie is projected? $\begin{bmatrix} 6 & 0 \\ 0 & 6 \end{bmatrix}$

20. What matrix could be used to reduce the drawings from the size of the screen to the size of the paper? $\begin{bmatrix} \frac{1}{6} & 0 \\ 0 & \frac{1}{6} \end{bmatrix}$

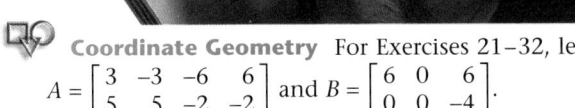

Coordinate Geometry For Exercises 21–32, let
$$A = \begin{bmatrix} 3 & -3 & -6 & 6 \\ 5 & 5 & -2 & -2 \end{bmatrix} \text{ and } B = \begin{bmatrix} 6 & 0 & 6 \\ 0 & 0 & -4 \end{bmatrix}.$$

21. Plot the vertices in A. Connect the vertices to form a polygon. What is the shape of the object represented by A? isos trap symmetric about y-axis

22. Plot the vertices in matrix B. Connect the vertices to form a polygon. What is the shape of the object represented by B? rt tri with long leg on x-axis

21.
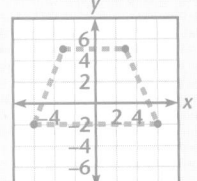

Isosceles trapezoid symmetric about the y-axis.

22.
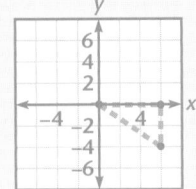

Right triangle with longer leg on the x-axis.

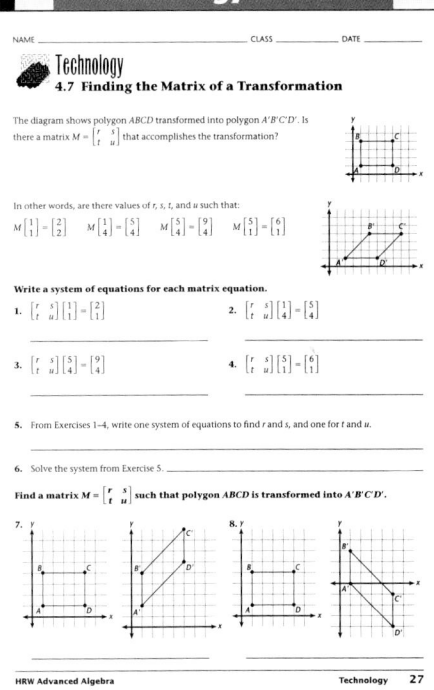

**Performance
Assessment**

Have students draw a parallelogram on the coordinate plane. Ask them to provide the coordinates of its vertices in a 2 × 4 matrix named *A*. Then have them give several different transformation matrices that will rotate the parallelogram in a clockwise and counterclockwise direction, and several that will enlarge and reduce the parallelogram. Have students find the product of each transformation matrix and matrix *A* to find the coordinates of each transformed figure.

23.

$$\begin{bmatrix} -1.42 & -5.68 & -2.84 & 5.68 \\ 5.68 & 1.42 & -5.68 & 2.84 \end{bmatrix}$$

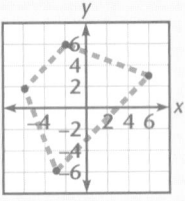

📈 **Transformations** Use your graphics calculator to multiply the indicated matrices. Graph the object represented by the product.

23. matrix *A*; $\begin{bmatrix} 0.71 & -0.71 \\ 0.71 & 0.71 \end{bmatrix}$

24. matrix *A*; $\begin{bmatrix} 0.71 & 0.71 \\ -0.71 & 0.71 \end{bmatrix}$

25. matrix *B*; $\begin{bmatrix} 0.87 & -0.5 \\ 0.5 & 0.87 \end{bmatrix}$

26. matrix *B*; $\begin{bmatrix} 0.87 & 0.5 \\ -0.5 & 0.87 \end{bmatrix}$

27. Describe the transformation in Exercise 23 on the object represented by *A*. rot CC by 45°

28. Describe the transformation in Exercise 24 on the object represented by *A*. rot C by 45°

29. Describe the transformation in Exercise 25 on the object represented by *B*. rot CC by 30°

30. Describe the transformation in Exercise 26 on the object represented by *B*. rot C by 30°

31. How are the transformations in Exercises 23 and 25 alike, and how are they different?

32. How are the transformations in Exercises 24 and 26 alike, and how are they different?

Architecture The standard scale for an architectural drawing of the floor plan of a building is 1 inch = 1 foot. In other words, each inch of the drawing represents 1 foot of the actual floor of the building.

33. Explain how to find the area of the building from the area of the floor plan.

34. Explain how to find the area of the drawing of the floor plan from the area of the building.

Landscaping A courtyard has two paths that intersect between four curved rows of hedges. The owner of the courtyard wants to have these paths paved as sidewalks. The diagram shown is an illustration of the courtyard drawn to scale, where 1 grid unit is equal to 1 foot.

35. Look at the illustration and create a matrix that represents the horizontal sidewalk.

36. Write the matrix multiplication you can use to rotate this sidewalk 90° clockwise. What is the product?

37. Let the length of one grid unit be $\frac{1}{a}$ of an inch long, where *a* is a positive, real number. Write the matrix multiplication you can use to enlarge the illustration to actual size.

38. Find the number of grid units in the diagram of the sidewalk. Find the actual sidewalk area. 96, 96 ft²

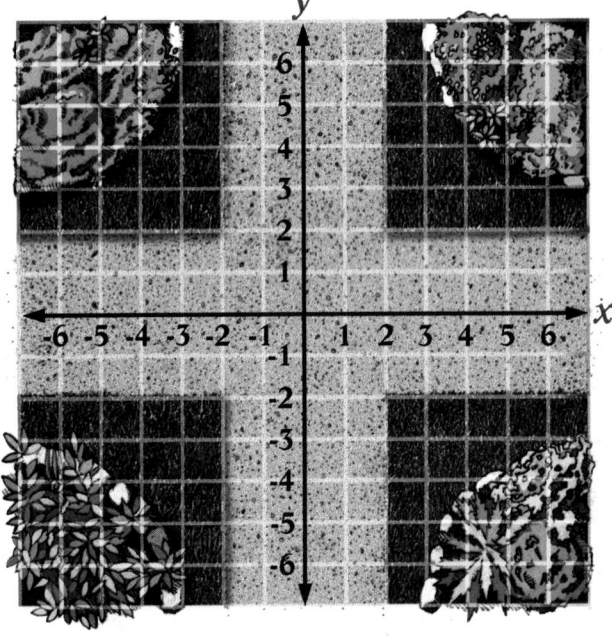

---

24. $\begin{bmatrix} 5.68 & 1.42 & -5.68 & 2.84 \\ 1.42 & 5.68 & 2.84 & -5.68 \end{bmatrix}$

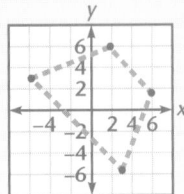

25. $\begin{bmatrix} 5.22 & 0 & 7.22 \\ 3 & 0 & -0.48 \end{bmatrix}$

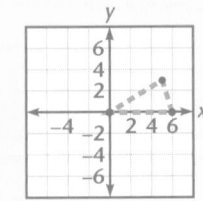

The answers to Exercises 26 and 31–37 can be found in Additional Answers beginning on page 842.

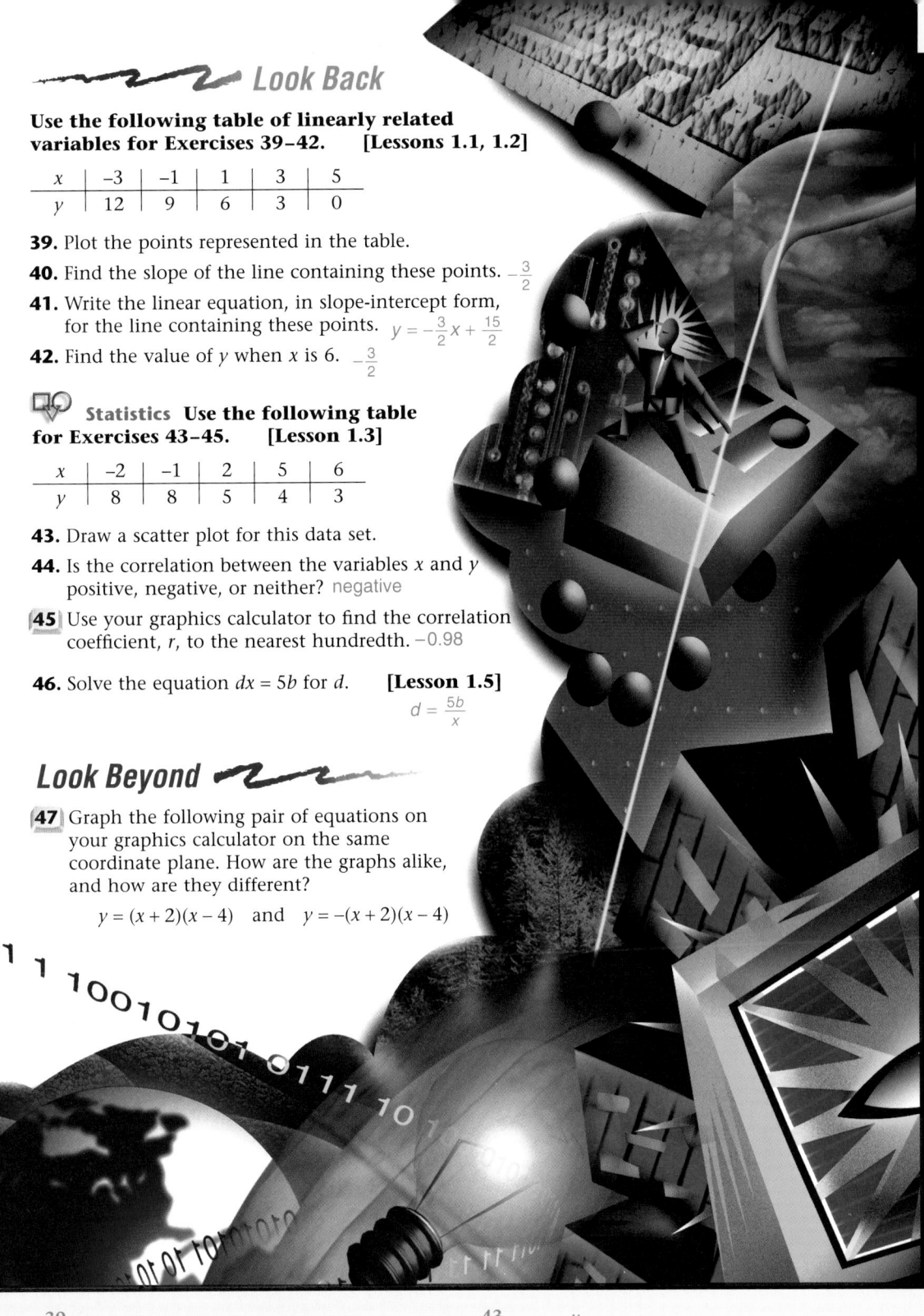

Look Back

Use the following table of linearly related variables for Exercises 39–42. [Lessons 1.1, 1.2]

x	–3	–1	1	3	5
y	12	9	6	3	0

39. Plot the points represented in the table.

40. Find the slope of the line containing these points. $-\frac{3}{2}$

41. Write the linear equation, in slope-intercept form, for the line containing these points. $y = -\frac{3}{2}x + \frac{15}{2}$

42. Find the value of y when x is 6. $-\frac{3}{2}$

Statistics Use the following table for Exercises 43–45. [Lesson 1.3]

x	–2	–1	2	5	6
y	8	8	5	4	3

43. Draw a scatter plot for this data set.

44. Is the correlation between the variables x and y positive, negative, or neither? negative

45 Use your graphics calculator to find the correlation coefficient, r, to the nearest hundredth. −0.98

46. Solve the equation $dx = 5b$ for d.   [Lesson 1.5]
$$d = \frac{5b}{x}$$

Look Beyond

47 Graph the following pair of equations on your graphics calculator on the same coordinate plane. How are the graphs alike, and how are they different?

$$y = (x + 2)(x - 4) \quad \text{and} \quad y = -(x + 2)(x - 4)$$

47. Both graphs are parabolas. One graph opens upward, and the other opens downward. They both intercept the x-axis at the points $(-2, 0)$ and $(4, 0)$.

Look Beyond

The graphs in this exercise are reflections of one another over the x-axis. This exercise begins to get students ready for quadratic functions in Chapter 5.

39.

43.

MOTIVATE

Discuss the difference between two-dimensional images and three-dimensional images. Ask students to give examples of when they have seen each type of image used.

Ask students if they have ever used 3-D glasses. Have those students who answer *yes* describe what happened when they looked through such glasses. If possible, provide students with several pairs of 3-D glasses for all students to experiment with.

Have students work in pairs. Ask each student to look at the picture and then discuss with their partner what they saw. Then have each pair share their results with the class.

EYEWITNESS MATH
3-D or not 3-D?

How can a three-dimensional image rise from a flat display of senseless swirls? When you gaze at the page the right way, your brain interprets some of the patterns in the picture as depth clues.

You can find out how these optical illusions are made, and you can make your own. To do that, you can think of stereograms as rectangular arrays of dots, so your knowledge of matrices will come in handy.

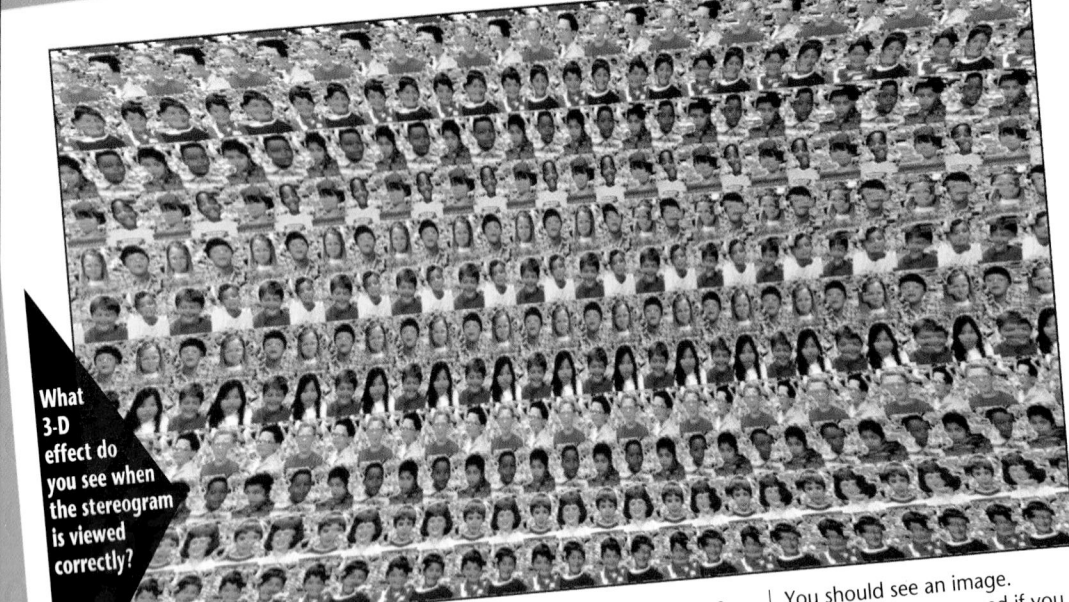

What 3-D effect do you see when the stereogram is viewed correctly?

3-D glasses won't help

In the past, you could see three-dimensional illusions only when wearing speacial glasses with red and green lenses or with polarized lenses to separate "left eye" and "right eye" images. Computers, however, now allow creators to make single 3-D images on paper.

Holding the page about one foot in front of you, blur your vision and relax your eyes, as if you were looking at something several feet away. It may help to blink one eye while your vision is blurred to split the image.

Now bring your visual attention to the middle as if you're looking through the page. Maintain your gaze for a few seconds and give your brain time to make sense of the visual field. Then gently let your gaze drift down the rectangle.

You should see an image. Don't be discouraged if you can't see depth on the first try. Some people need only a few seconds before they see it, but most people require as long as 20 minutes. Only a few cannot see it even after a half hour or so.

If your eyes converge the images on the page itself, you won't see any depth. You must let your eyes diverge "past" the page in order to see the image in the illusion. It's like looking out a window at the buildings beyond it but still being aware of what's on the window itself.

Cooperative Learning

The stereogram on page 218 contains hundreds of thousands of dots. You can make one that works with less than 2000 dots. To learn the process quickly, you will first create a nonworking model of a stereogram with only 54 dots.

A. The background of the stereogram is made by generating a few columns of random dots, and then repeating those columns over and over.

① On graph paper, outline a 6×9 matrix. Label it A.

② Flip a coin to generate 18 random zeros and ones. Enter those 18 values in the first three columns of A. (Write the ones lightly because you will erase them later.) An example is shown on the right.

$$A = \begin{array}{|ccc}
0 & 1 & 0 \\
1 & 0 & 0 \\
1 & 1 & 0 \\
0 & 1 & 1 \\
0 & 0 & 1 \\
0 & 1 & 1
\end{array}$$

③ a_{ij} stands for the element in row i and column j of matrix A. Suppose you used the rule $a_{ij} = a_{i,j-3}$ to assign values for the other 36 elements of A. How would you find the value of a_{24}?

④ Use the rule in Step 3 to fill in the rest of A.

B. Suppose you want to design your stereogram so that a T shape, such as the one shown, floats above the background. You need to change the elements within the T according to a new rule.

0	1	0	0	1	0	0	1	0
1	0	0	1	0	0	1	0	0
1	1	0	1	1	0	1	1	0
0	1	1	0	1	1	0	1	1
0	0	1	0	0	1	0	0	1
0	1	1	0	1	1	0	1	1

① Which 6 elements of A will be affected?

② Give new values to each of those 6 elements according to this rule: $a_{ij} = a_{i,j-2}$.

③ Describe how the new matrix A compares to the original matrix A created in part **A**.

C. To turn your matrix into a stereogram, just make a pictorial version of it.

Shade each square that has a zero in it. Erase all of the ones. You will be left with an array of shaded and unshaded squares.

You can also write a computer program to do the work for you.

Cooperative Learning

Have students work in pairs to complete the activities. For Activity A, have one student flip the coin while the other student records the results. Then have students work together to complete the rest of Activity A.

Have students work with their partner to complete Activities B and C. Then have each pair of students work with another pair of students and compare their results. Ask each group of four students to make any necessary corrections to their work and prepare a final stereogram to present to the class.

Discuss

Have each group of four students share their final stereogram with the class. Discuss any differences among the stereograms.

After students have had a chance to compare their work, discuss the advantages of using a computer program to do the work. Ask students how they think programmers might tell the computer how to draw a stereogram. If possible, invite a computer programmer who has created such a program to come and talk to the class.

A. **1.** Answers may vary

2. Answers may vary

3. $a_{24} = a_{21}$

4. Answers may vary

B. **1.** The elements in the T

2. Answers may vary. In the T shown, the new values are

*	*	*
0	0	1
*	0	*
*	1	*
*	1	*
*	*	*

3. Answers may vary.

C. Answers may vary.

Objectives

- Graph the solution to a system of linear inequalities.

Assessing Prior Knowledge

Graph each inequality on the coordinate plane.

1. $y < 2x - 1$

[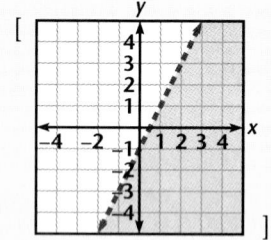]

2. $y \geq 3x + 2$

[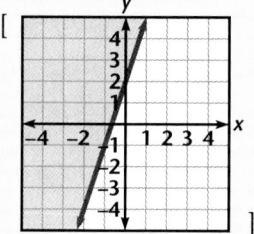]

TEACH

Discuss situations in which individuals would need to find the greatest (maximum) or least (minimum) possible values. Use the situations given on the student page to help get the discussion started.

LESSON 4.8
Exploring
Systems of Linear Inequalities

 *In the manufacturing industry, there are many situations in which a variety of items are manufactured with the least possible cost and the greatest possible profit. Dietitians want to plan meals with the greatest possible nutrient content and the least possible salt and fat content. You can use systems of linear inequalities to describe such maximization and minimization problems.*

ALTERNATIVE teaching strategy

Technology Have students use a graphics calculator to explore the graph of two inequalities. Refer to your *HRW Technology Handbook* to determine how to graph linear inequalities and systems of linear inequalities. When you progress to three inequalities, students can graph two of the inequalities using a graphics calculator, transfer the graph to their paper, and then graph the third inequality to get the final graph. Caution students to use a dashed line when graphing strict inequalities. Most calculators show only solid lines.

Exploration 1 Graphing Two Linear Inequalities

You will need two different colored pencils and graph paper for this exploration.

1 Graph the inequality $y \geq 2x - 1$, and shade the appropriate half-plane with one of the two colors.

2 On the same coordinate plane, graph the inequality $y < -\frac{2}{3}x + 2$, and shade the appropriate half-plane with the second color.

3 You should now have a graph of two linear inequalities on the same coordinate plane. Are the boundary lines solid or dashed? Why?

Describe the region that you shaded with both colors. Is it possible to view *all* of the points that could possibly be in this region of the graph? Explain why or why not.

4 Select three points from the region that you shaded with both colors. Write the ordered pairs for these points in the first column of the following table.

Substitute the x- and y-values of each point into each inequality. If it makes the inequality true, write *true* under the inequality in the table. If it makes the inequality false, write *false* under the inequality.

Point	$y \geq 2x - 1$	$y < -\frac{2}{3}x + 2$

5 Select three points from any of the regions that you did not shade with both colors. Substitute the x- and y-values into each inequality to complete the true-false table as you did in Step 4.

Point	$y \geq 2x - 1$	$y < -\frac{2}{3}x + 2$

6 Compare your table from Step 4 with those of your classmates. Did anyone choose a point that made either of the inequalities false? Why or why not?

7 Compare your table from Step 5 with those of your classmates. Did anyone choose a point that made both of the inequalities true? Why or why not?

8 The two linear inequalities you graphed in this exploration make up a system of two linear inequalities. How do you decide what areas to shade? How do you know what area contains the solution points? ❖

interdisciplinary CONNECTION

Budgets Have students solve the following problem:

Cory's two favorite cold-weather activities are playing an hour of indoor tennis and going to the movies. The only time he is free to do one of these activities is on Saturday, and he never does more than one of these activities per week during the 39 weeks of the year in which he cannot play tennis outdoors. One hour of indoor tennis costs $30.00, and one movie ticket costs $6.00. He has a maximum of $900 in his budget to spend on these activities during these 39 weeks.

1. What is the maximum number of times he can play indoor tennis? [**30**]

2. What is the maximum number of movies he can see? [**150**]

After students have solved this problem, have them write and solve other questions about budgeting money.

Again, be sure students use colors that contrast well. As suggested in the Alternative Teaching Strategy on page 220, you could have students use a graphics calculator to complete Steps 1 and 2 and then transfer these graphs to coordinate paper. Alternatively, students could complete Steps 1 and 2 on paper and use a graphics calculator to verify their results.

Aongoing
ASSESSMENT

9. **Shade each area that makes each inequality true. The intersection of all the shaded areas represents the solution to the system of inequalities.**

Exploration 2 *Graphing Three or More Linear Inequalities*

You will need three different colored pencils and graph paper for this exploration.

1 Graph the inequality $y \le -\frac{1}{2}x + 1$, and shade the appropriate half-plane with one of the three colors.

2 On the same coordinate plane, graph the inequality $y < \frac{3}{2}x + 3$, and shade the appropriate half-plane with a different color.

3 On the same coordinate plane, graph the inequality $y > -2$, and shade the appropriate half-plane with the third color.

4 You should now have a graph of three linear inequalities on the same coordinate plane. Describe the shape of the region that you shaded with all three colors.

Is it possible to view this entire region of the graph? Explain why or why not.

5 Complete the true-false table as you did in Exploration 1 using three points from the region that you shaded with all three colors.

Point	$y \le -\frac{1}{2}x + 1$	$y < \frac{3}{2}x + 3$	$y > -2$

6 Complete the true-false table using three points from any of the regions that you did *not* shade with all three colors.

Point	$y \le -\frac{1}{2}x + 1$	$y < \frac{3}{2}x + 3$	$y > -2$

7 Compare your table from Step 5 with those of your classmates. Did anyone choose a point that made any of the inequalities false? Why or why not?

8 Compare your table from Step 6 with those of your classmates. Did anyone choose a point that made *all* of the inequalities true? Why or why not?

9 The three linear inequalities you graphed in this exploration make up a system of three linear inequalities. Describe how to determine the solutions to any system of linear inequalities. ❖

ENRICHMENT Have students write and solve a system of two inequalities whose solution is a triangular-shaped bounded area. As a hint, suggest that they use an absolute-value sign in one inequality. [**Possible answer:** $y \ge |x + 1|$ **and** $y \le 4$]

INCLUSION **strategies** **Using Visual Models** Provide students with several pieces of acetate containing a line with one half-plane shaded. Some pieces should show solid lines and some dashed. Each piece should be shaded with a different color. Have students place these pieces of acetate on top of a coordinate plane to show the graphs of linear inequalities and the graphs of systems of linear inequalities. Then have students write the system of equations and the solution.

The solutions to the system of linear inequalities in Exploration 1 are located in a region that is not limited by a boundary. It is impossible to view the entire region. This region is said to be an **unbounded region**.

The solutions to the system of linear inequalities in Exploration 2 are located in a region that has boundary lines and does not extend beyond its boundary lines. This region is said to be a **bounded region**.

 *CRITICAL Thinking*

Do all systems of *two* linear inequalities have solutions that are located in an *unbounded* region on the coordinate plane? Explain why or why not.

Do all systems of *three* linear inequalities have solutions that are located in a *bounded* region on the coordinate plane? Explain why or why not.

CRITICAL Thinking

Yes, a polygon cannot be formed by 2 lines. No. Answers may vary. One possible answer is that a system of three parallel lines would not be bounded.

Exploration 3 — Applying Systems of Linear Inequalities

Nutrition

A school dietitian wants to prepare a meat-and-rice dish that contains not more than 20 grams of fat and not more than 30 milligrams of salt. Each serving of meat contains 10 grams of fat and 9 milligrams of salt, and each serving of rice contains 2 grams of fat and 6 milligrams of salt.

MAXIMUM
MINIMUM
Connection

1 Let x represent the number of servings of meat and y represent the number of servings of rice. Complete the following table with expressions in terms of x and y for each cell.

	Number of servings	Amount of fat (g)	Amount of salt (mg)
Meat	x	$10x$	
Rice	y		
Meat and rice	$x+y$		

2 Write a system of two linear inequalities.

  a. Write one linear inequality to represent the allowable amount of fat in the meat-and-rice dish.

  b. Write another linear inequality to represent the allowable amount of salt in the meat-and-rice dish.

3 Limitations on the possible values for x and y, such as "x and y cannot be negative," are called *constraints*.

Write two inequalities to represent the constraints that x and y cannot be negative.

4 Graph the system of linear inequalities you wrote in Steps 2 and 3. Use your graph to answer the remaining questions in this exploration.

Exploration 3 Notes

Encourage students to use colored pencils again if necessary. They can also use different types of shading to show the different parts. For example, students can use vertical lines to shade one region, horizontal lines to shade another region, and diagonal lines to shade the third region. After students have completed this exploration, have them share their results with the entire class.

 RETEACHING *the lesson*

Hands-On Strategies
Provide students with several pieces of clear acetate and several different colored markers with which to write on the acetate. Have them draw the graph of each inequality on a separate piece of acetate and then place all of the pieces together to show the final graph. Allow students to use the pieces of acetate as long as necessary. Then have them progress to drawing their graphs on paper, using the acetate method to check their graphs when necessary.

5 Assume that you can only use full servings and cannot use a negative number of servings of meat or rice. If 1 serving of meat is used, how many servings of rice are possible?

6 If 3 servings of rice are used, how many servings of meat are possible?

7 What is the maximum number of servings of meat possible? How many servings of rice does the maximum number of servings of meat allow? Is this reasonable for a meat-and-rice dish? Why or why not?

8 What is the maximum number of servings of rice possible? How many servings of meat does the maximum number of servings of rice allow? Is this reasonable for a meat-and-rice dish? Why or why not?

9 What is the minimum number of servings of meat and rice possible? Is this reasonable for any dish? Why or why not? ❖

EXERCISES & PROBLEMS

Communicate

1. Is it possible for a system of linear inequalities to have no solution? Explain.

2. Describe the graph of a system of linear inequalities with solutions contained in a bounded region.

3. Describe the graph of a system of linear inequalities with solutions contained in an unbounded region.

Time Management Angela works 40 hours or less per week programming computers and tutoring. She earns $20 per hour programming and $10 per hour tutoring. Angela needs to earn at least $500 per week.

4. Explain how to write a system of linear inequalities that represents the possible combinations of hours tutoring and hours programming that will meet Angela's needs.

5. What are the constraints in this situation?

6. Graph the system of linear inequalities. Are the solutions contained in a bounded or unbounded region?

7. Find a point that is a solution to the system of linear inequalities. What are the coordinates of this point? What do the coordinates of this point represent?

8. Which point in the solution region represents the best way for Angela to spend her time? Explain why you think this is the best solution.

9.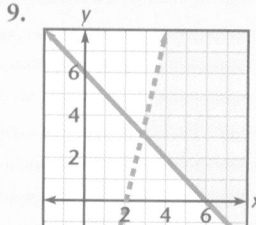

10.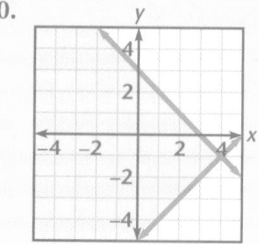

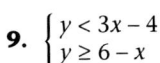
Practice & Apply

Graph the solution to each of the following systems of linear inequalities.

9. $\begin{cases} y < 3x - 4 \\ y \geq 6 - x \end{cases}$ **10.** $\begin{cases} x + y \leq 3 \\ x - y \leq 5 \end{cases}$ **11.** $\begin{cases} x \geq 0 \\ y \geq 1 \\ x + y \leq 5 \end{cases}$ **12.** $\begin{cases} y + 2x \geq 0 \\ y \geq 2x - 4 \\ y \leq 3 \end{cases}$

Entertainment A rock band sells reserved tickets and general-admission tickets to a concert. The auditorium normally holds, at most, 5000 people. There can be, at most, 3000 reserve seats and, at most, 4000 general-admission tickets sold.

13. Let x represent the number of reserved tickets, and let y represent the number of general-admission tickets. Write a system of three linear inequalities to represent the possible combinations of reserved and general-admission tickets that can be sold.

14. Graph the region of possible combinations of reserved and general-admission tickets that can be sold.

15. In order to increase the number of people that come to the concert, the auditorium adds extra folding chairs so that the capacity of the auditorium increases to 5500. Explain how this addition of chairs changes the region of possible seating combinations.

Graph the solution to each of the following systems of linear inequalities.

16. $\begin{cases} 2x - y \leq 16 \\ x + y \leq 10 \\ x \geq 0 \\ y \geq 0 \end{cases}$ **17.** $\begin{cases} 3x - y \leq 15 \\ x + 2y \leq 10 \\ x \geq 0 \\ y \geq 0 \end{cases}$ **18.** $\begin{cases} 3x - 2y \geq 4 \\ x + y > 4 \\ x - y \leq 7 \\ x \geq 0 \\ y \geq 0 \end{cases}$ **19.** $\begin{cases} x + 3y < 6 \\ 4x + 3y \geq 18 \\ x - y \leq -3 \end{cases}$

11.

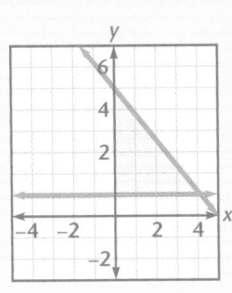

12.

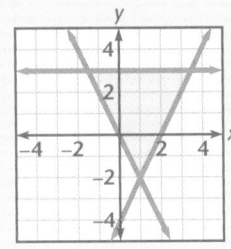

Error Analysis

Be sure students use a dashed line when necessary and that they shade on the correct half-plane. Allow students who struggle with this concept to complete as many examples as necessary of problems like those given in the Assessing Prior Knowledge section on page 220.

13. $\begin{cases} x + y \leq 5000 \\ 0 \leq x \leq 3000 \\ 0 \leq y \leq 4000 \end{cases}$

The answers to Exercises 14–19 can be found in Additional Answers beginning on page 842.

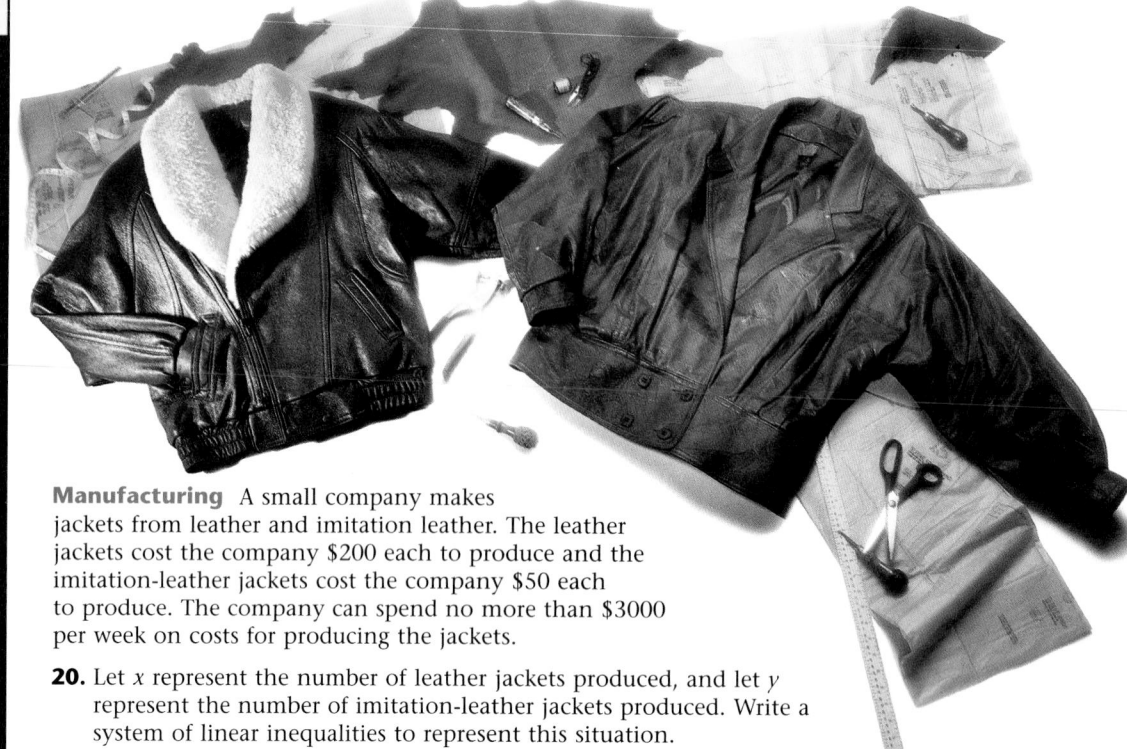

alternative ASSESSMENT

Portfolio Assessment

Have students compare Exercise sets 20–26, 27–29, and 30–32. How does considering only cost limitations affect the solution region? How does considering only profit goals affect the solution region? What kind of solution region would you expect if you considered both cost limitations and profit goals? Have students make-up a company that sells some type of product. Have them determine at least three reasonable constraints (cost and profit) for their company. Have students determine solution regions and reasonable goals for their company, and present these to the class.

20. $\begin{cases} 200x + 50y \leq 3000 \\ x \geq 0 \\ y \geq 0 \end{cases}$

Manufacturing A small company makes jackets from leather and imitation leather. The leather jackets cost the company $200 each to produce and the imitation-leather jackets cost the company $50 each to produce. The company can spend no more than $3000 per week on costs for producing the jackets.

20. Let x represent the number of leather jackets produced, and let y represent the number of imitation-leather jackets produced. Write a system of linear inequalities to represent this situation.

21. Graph the system of linear inequalities.

22. List the possible combinations of leather and imitation-leather jackets that the company might make to satisfy these conditions.

23. Are there any constraints? If so, list the constraints. If not, why not? yes; $x \geq 0$, $y \geq 0$

24. What is the greatest number of leather jackets that can be made? 15

25. What is the greatest number of imitation-leather jackets that can be made? 60

26. If the cost of making leather jackets were $150 per jacket, would the solution region increase or decrease in size? Explain your answer.
Increase; the number of poss leather jackets would increase from 15 to 20

Business The Lawn and Garden Shop sells hedge clippers with electric motors and with gasoline motors. The store wants to sell at least 45 hedge clippers per month. The profit on each electric model is $50, and the profit on each gasoline model is $40. The shop wants to earn at least $2000 per month on the sale of hedge clippers.

27. Let x represent the number of hedge clippers with electric motors, and let y represent the number of hedge clippers with gasoline motors. Write a system of linear inequalities to represent the possible combinations of each model sold.

28. Graph the region of possible combinations of each model sold.

29. Suppose the shop wants to sell at least 50 hedge clippers per month. How would this affect the maximum number of electric models and the maximum number of gasoline models that can be sold?

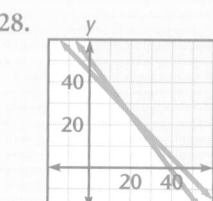

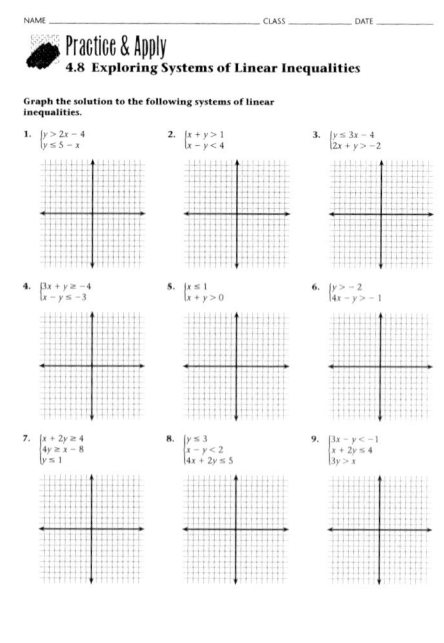
21.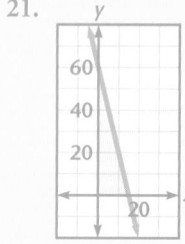

The answer to Exercise 22 can be found in Additional Answers beginning on page 842.

27. $\begin{cases} x + y \geq 45 \\ 50x + 40y \geq 2000 \\ x \geq 0 \\ y \geq 0 \end{cases}$

28.

Manufacturing The Happy Home small-appliance manufacturing company makes standard and deluxe models of a toaster oven. The company can make up to 200 ovens per week. The standard model costs the company $20 to produce, and the deluxe model costs the company $30 to produce. The Happy Home company has budgeted no more than $3600 per week to produce the ovens.

30. Let x represent the number of standard models and y represent the number of deluxe models. Write a system of linear inequalities to represent the possible combinations of standard and deluxe models that the company can make in 1 week.

31. Graph the region of possible combinations of standard and deluxe models that the company can make in 1 week.

32. Due to an increase in the rental costs of the building that the Happy Home Company is renting, the company can budget no more than $3000 per week to make the toaster ovens. Explain how this changes the region of possible combinations of standard and deluxe models that the company can make in 1 week.

Look Back

33. Graph the equation $5x - 3y = 7$. **[Lesson 1.1]**

34. Find the x- and y-intercepts of the equation $5x - 3y = 30$. **[Lesson 1.1]** $(6, 0); (0, -10)$

35. Write the equation of a line that is parallel to $2x - 4y = 26$. **[Lesson 1.1]**
Answers may vary; must have slope $\frac{1}{2}$

Solve for x in terms of y, and then solve for y in terms of x. [Lessons 1.5, 1.6]

36. $5x - 3y = 15$ **37.** $7x + 2y - 13 = 0$ **38.** $x - y > 9$

Solve for x and y. Identify each system as *independent*, *dependent*, or *inconsistent*. **[Lessons 4.3, 4.4]**

39. $\begin{cases} 4x - 2y = 3 \\ 8y - 6x = 24 \end{cases}$ **40.** $\begin{cases} 5x + y = -1 \\ 10x - y = 3 \end{cases}$ **41.** $\begin{cases} 2x + 5y = 10 \\ 2x - 5y = 0 \end{cases}$ **42.** $\begin{cases} 3x + y = 6 \\ 6x + 2y = 12 \end{cases}$

$\left(\frac{18}{5}, \frac{57}{10}\right)$; indep $\left(\frac{2}{15}, -\frac{5}{3}\right)$; indep $\left(\frac{5}{2}, 1\right)$; indep   dependent

Look Beyond

Evaluate for the given pairs of values.

43. $P = 20x + 30y$ for $(3, 5)$, $(4, 7)$, and $(5, 8)$. As x- and y-values increase, does the value of P increase or decrease? 210, 290, 340; increase

44. $C = 400p + 500q$ for $(1, 2)$, $(2, 5)$, $(3, 8)$. As p- and q-values increase, does the value of C increase or decrease? 1400, 3300, 5200; increase

45. Define the word *feasible*. Use a dictionary or your own words. What is one synonym for *feasible*?

29. The increase in the number of electric models that could be sold will be greater than the increase in the number of gasoline models. However, the solution region is still unbounded, and so there is no maximum number of either electric or gasoline models that could be sold.

30. $\begin{cases} x + y \le 200 \\ 20x + 30y \le 3600 \\ x \ge 0 \\ y \ge 0 \end{cases}$

31.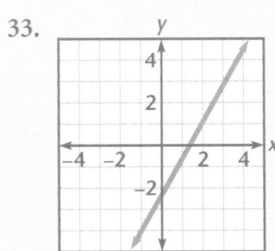

32. It decreases the region of possible combinations. The maximum number of standard models is decreased from 180 to 150, and the maximum number of deluxe models is decreased from 120 to 100.

33.

The answers to Exercises 36–38 and 45 can be found in Additional Answers beginning on page 842.

Look Beyond

These exercises prepare students for their work on linear programming.

Technology Master

NAME _____ CLASS _____ DATE _____

Technology
4.8 Bar Graphs and Linear Inequalities

If the number of bars in a bar graph is great enough, you can simulate the intersection of the solution regions of a set of linear inequalities with the use of a bar graph. The diagram shows the solution of the system.

$\begin{cases} x \ge 0 \\ y \ge 0 \\ x + y \le 7.5 \\ 3x + y \le 14 \\ x + 3y \le 18 \end{cases}$

To create the graph,

• Write each inequality involving x and y in terms of y.
• To obtain very narrow bars, increment x by 0.1. These entries will be in column A.

Enter
• Cell B2 as 7.5−A2. Cell C2 as −3*A2+14. Cell D2 as −A2/3+6.
• Using the FILL DOWN command, you can complete the data set.

	A	B	C	D
1	X	Y1	Y2	Y3
2	0.0	7.5	14.0	6.00
3	0.1	7.4	13.7	5.97
4	0.2	7.3	13.4	5.93
...				
60	5.8	1.7	−3.4	4.07
61	5.9	1.6	−3.7	4.03
62	6.0	1.5	−4.0	4.00

Use a spreadsheet to graph the solution region for each set of inequalities.

1. $\begin{cases} x \ge 0 \\ y \ge 0 \\ y \le x \\ y \le -x + 6 \end{cases}$ **2.** $\begin{cases} 0 \le x \le 6 \\ y \ge 0 \\ y \le 5x \\ 3x + 5y \le 28 \end{cases}$ **3.** $\begin{cases} x \ge 0 \\ y \ge 0 \\ y \le -0.5x + 6 \\ y \le -2x + 12 \end{cases}$

4. $\begin{cases} 0 \le x \le 6 \\ y \ge 0 \\ y \le -0.1x + 12 \\ y \le 3.6x + 1 \\ y \le -3.6x + 29 \end{cases}$ **5.** $\begin{cases} x \ge 0 \\ y \ge 0 \\ y \le 6 \\ y \le -1.5x + 9 \end{cases}$ **6.** $\begin{cases} 0 \le x \le 6 \\ 0 \le y \le 6 \\ y \le -0.6x + 13.4 \\ y \le 1.6x + 2.4 \end{cases}$

7. Explain how to make a bar graph solution region even finer than what is shown on this page.

_____

LESSON 4.9 Introduction to Linear Programming

why *People make choices based on what is feasible. For example, you cannot drive 500 miles if you have only 2 hours to get to the destination. You cannot purchase a luxury sports car if you have only $5. You can use a system of linear inequalities to describe the options that are possible for a situation in which there are one or more restrictions.*

Feasible Region

Points in the solution region of a graph of a system of linear inequalities indicate the possible, or feasible, options.

EXAMPLE 1

Business The Zoom City Skateboard Company manufactures two models of skateboards—the Rapido model and the Speed Merchant model. The dealers who sell these models want to receive from 30 to 80 Rapido skateboards and from 10 to 30 Speed Merchant skateboards per day. Quality controls and limits on the availability of high-quality ball bearings limit production to no more than 80 skateboards per day. The company makes a profit of $15 for each Rapido model sold and $8 for each Speed Merchant model sold.

Ⓐ Write a system of three linear inequalities to represent the constraints on the number of skateboards that the Zoom City Skateboard Company can produce.

Ⓑ On one coordinate plane, graph the constraints for the number of skateboards that the Zoom City Skateboard Company can produce.

Ⓒ Write a profit function for the Zoom City Skateboard Company.

Solution▸

A Let r be the number of Rapido models produced, and let s be the number of Speed Merchant models produced. Make a table showing the constraints on the number of skateboards that can be produced.

	r	s	r and s
Number produced	$30 \le r \le 80$	$10 \le s \le 30$	$r + s \le 80$

Thus, the constraints on the number of skateboards that the Zoom City Skateboard Company can produce is represented by the following system of linear inequalities.

$$\begin{cases} 30 \le r \le 80 \\ 10 \le s \le 30 \\ r + s \le 80 \end{cases}$$

B Each inequality is graphed by itself first.

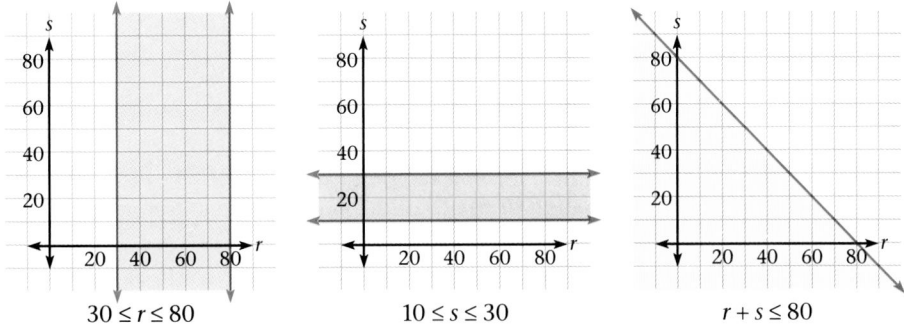

$30 \le r \le 80$ $10 \le s \le 30$ $r + s \le 80$

Since a negative number of skateboards cannot be produced, there are also the constraints $r \ge 0$ and $s \ge 0$. Graph these constraints on the same coordinate plane.

Thus, the graph of all the constraints for the number of skateboards that the Zoom City Skateboard Company can produce is the intersection of all the shaded areas.

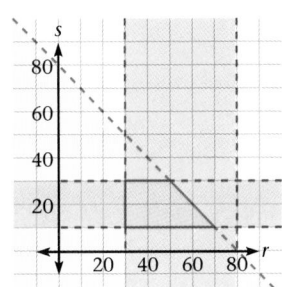

C Combine the profit information in a table.

	r	s	r and s
Profit from sales	$15r$	$8s$	$P = 15r + 8s$

Thus, the profit function for the Zoom City Skateboard Company is

$$P = 15r + 8s. \; ❖$$

In Example 1, the only values of r and s that are *feasible* for use in the profit function, $P = 15r + 8s$, are the coordinates of the points in the solution region. This region is called the **feasible region** because it contains the values of r and s that satisfy all of the constraints.

ENRICHMENT Have students work in small groups to develop revenue and cost functions for the Pro Tour Tennis Company in Alternate Example 1. They should assume that the revenue for each Pro model sold is $25 and the revenue for each Extra Light model sold is $20. (Assume that profit = revenue − cost.) [$R = 25p + 20e$; $C = 5p + 4e$]

INCLUSION **strategies** **English Language Development** Encourage students to write about the concepts presented in this lesson using their own words. Ask them to focus on the terms *feasible*, *profit*, *linear programming*, and *objective function*. These words may be difficult for students to understand. Allow students to learn from each other by working on the examples and exploration in small groups.

Alternate Example 1

The Pro Tour Tennis Company manufactures two models of tennis racquets: the Pro model and the Extra Light model. The dealers who sell these models want to receive from 50 to 70 Pro racquets and from 20 to 60 Extra Light racquets per day. The company can produce up to 120 racquets per day. The company makes a profit of $20 for each Pro model sold and $16 for each Extra Light model sold.

A. Write a system of three linear inequalities to represent the constraints on the number of tennis racquets that the Pro Tour Tennis Company can produce.

[**Let p be the number of Pro models and e be the number of Extra Light models produced.**

$$\begin{cases} 50 \le p \le 70 \\ 20 \le e \le 60 \\ p + e \le 120 \end{cases}]$$

B. On one coordinate plane, graph the constraints for the number of tennis racquets that the Pro Tour Tennis Company can produce.

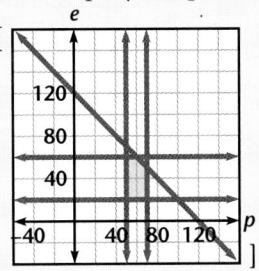

C. Write a profit function for the Pro Tour Tennis Company. [$P = 20p + 16e$]

Use Transparency ▶ 17

Objective Function

MAXIMUM MINIMUM *Connection*

A certain profit value, such as $P = 38$, can result from particular values of r and s. For example, $38 = 15r + 8s$ is true when (r, s) is $(2, 1)$ or $(-6, 16)$ or $(10, -14)$. All of these points are on the line $38 = 15r + 8s$, but none of them are in the feasible region.

What values of r and s will maximize profit? **Linear programming** involves finding a maximum or a minimum value that satisfies all of the given conditions.

The *object* is to maximize the profit function, $P = 15r + 8s$. Thus, P is called the **objective function**. Notice that the objective function, P, is a linear function in r and s.

The possible profit values are the values of P for each point in the feasible region. The largest value is the maximum profit possible.

•Exploration• *Graphing Objective Functions*

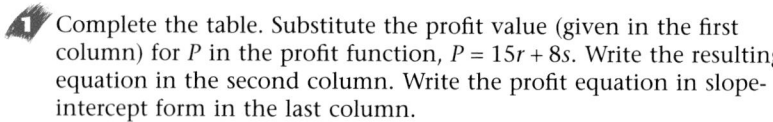

1 Complete the table. Substitute the profit value (given in the first column) for P in the profit function, $P = 15r + 8s$. Write the resulting equation in the second column. Write the profit equation in slope-intercept form in the last column.

Profit	Profit equation	Slope-intercept form
400	$400 = 15r + 8s$	$s = -\frac{15}{8}r + 50$
600		
800		
1000		
1200		

2 Look at the slopes and intercepts in these equations. Discuss how they are alike and how are they different.

3 Graph all five profit equations on the same coordinate plane. How are the graphs alike? How are the graphs different?

4 Find any point (r, s) in the feasible region from Example 1 that results in a profit $(P > 0)$ using the profit function, $P = 15r + 8s$. (Several answers are possible.)

5 Use the profit value you found in Step 4 to write a profit equation in slope-intercept form. Graph your profit equation on the same coordinate plane that you used in Step 3.

6 Suppose a profit can be obtained from the coordinates of a point in the feasible region. Do you think that the graph of the profit equation would intersect the feasible region? Why or why not? ❖

CRITICAL *Thinking*

What does it mean when the graph of an objective function does not intersect the feasible region?

EXERCISES & PROBLEMS

Communicate

1. How would the feasible region for Example 1 change if the dealer demand were changed to between 20 and 60 Rapido skateboards?

2. Suppose the profit from selling Rapido skateboards was $30 and the profit from selling Speed Merchant skateboards was $50. How would the profit function change? How would the profit lines change?

Practice & Apply

Graph the feasible region for each set of constraints. Then determine any four points in the feasible region.

3.
$$\begin{cases} x + 2y \le 12 \\ 2x - y \ge -5 \\ x \ge 0 \\ y \ge 0 \end{cases}$$

4.
$$\begin{cases} 3x + 2y \le 12 \\ x - 2y \le -2 \\ x \ge 0 \\ y \ge 0 \end{cases}$$

5.
$$\begin{cases} -3x + y \le 12 \\ 3x - 4y \le -4 \\ x \ge 0 \\ y \ge 0 \end{cases}$$

6.
$$\begin{cases} 3x + 2y \le 12 \\ x - 4y \le -8 \\ x \ge 0 \\ y \ge 0 \end{cases}$$

For each of the feasible regions in Exercises 3–6, find a value of the objective function determined by a point that you chose in the feasible region.

7. $P = 5x + 3y$

8. $C = 3x + 2y$

Manufacturing The Olympic Ski manufacturing company manufactures two types of skis and has a fabricating department and a finishing department. Downhill skis require 6 hours to fabricate and 1 hour to finish. Cross-country skis require 4 hours to fabricate and 1 hour to finish. The fabricating department has 108 hours of labor available per day. The finishing department has 24 hours of labor available per day. The company makes $40 profit on each pair of downhill skis and $30 profit on each pair of cross-country skis.

9. Create a table to organize the information given.

10. Write a system of linear inequalities to describe the constraints of this situation.

11. Write an objective function for the total profit.

12. Determine the value of the objective function for four points with integer coordinates in the feasible region.

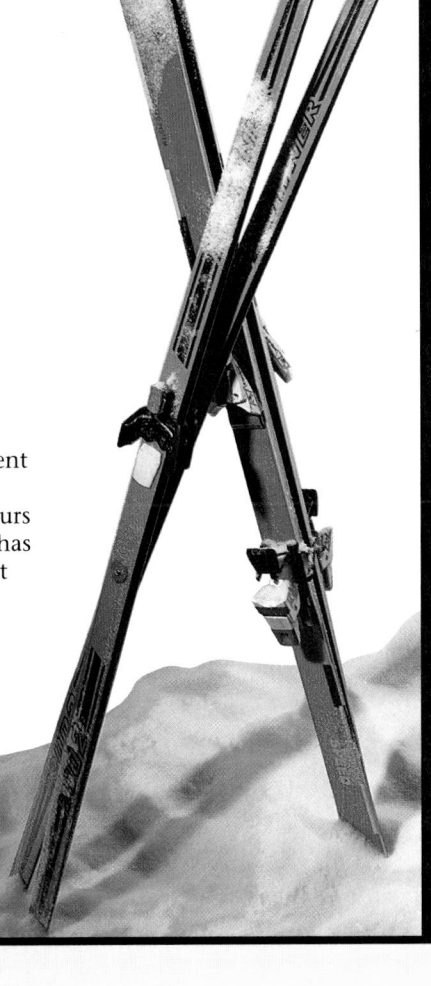

ASSESS

Selected Answers
Odd-numbered Exercises 3–33

Assignment Guide
Core 1–5, 7–35

Core Plus 1–2, 4–35

Technology
A graphics calculator is needed for Exercises 25–30 and 34. A graphics calculator can be helpful in answering Exercises 3–6.

5.
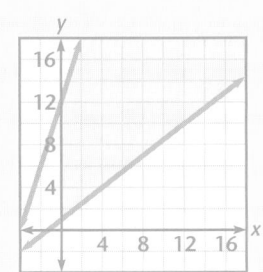

Four possible points are $(2, 4)$, $(2, 5)$, $(2, 6)$, and $(2, 7)$.

The answers to Exercises 6–12 can be found in Additional Answers beginning on page 842.

3.
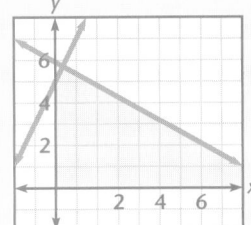

Four possible points are $(1, 1), (1, 2), (1, 3)$, and $(1, 4)$.

4.
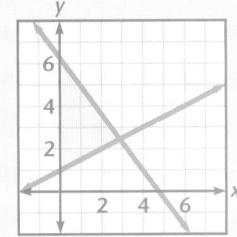

Four possible points are $(1, 2), (1, 3), (1, 4)$, and $(2, 3)$.

Watch for students who do not use the correct inequality symbol for a given situation. Encourage students to double check all of their inequalities before continuing to solve the problem.

The answers to Exercises 13, 17, and 21 can be found in Additional Answers beginning on page 842.

14. $\begin{cases} n + d \le 10 \\ 150n + 100d \le 1200 \\ n \ge 0 \\ d \ge 0 \end{cases}$

16. Answers may vary. Examples, where ordered pairs represent (n, d), are

(2, 8): $C = 1,080,000$;
(4, 6): $C = 1,260,000$;
(5, 4): $C = 1,260,000$;
(8, 0): $C = 1,440,000$;

Tourism A tourist agency can sell up to 1200 travel packages for a bowl game. The package includes air fare, weekend accommodations, and the choice of two types of flights: nonstop flight or direct flight with two stops. The nonstop flight can carry up to 150 passengers, and the direct flight can carry up to 100 passengers. The agency can locate no more than 10 planes. Each package with a nonstop flight costs $1200, and each package with a direct flight costs $900.

13. Create a table to organize the constraints given.

14. Write a system of linear inequalities to describe the constraints of this situation.

15. Write an objective function for the cost. $C = 1200(150n) + 900(100d)$

16. Determine the value of the objective function for four points with integer coordinates in the feasible region.

Farming A farmer has 90 acres available for planting millet and alfalfa. Seed costs $4 per acre for millet and $6 per acre for alfalfa. Labor costs will amount to $20 per acre for millet and $10 per acre for alfalfa. The expected income from millet is $110 per acre, and from alfalfa, $150 per acre. The farmer intends to spend no more than $480 for seed and $1400 for labor.

17. Create a table to organize the information given.

18. Write a system of linear inequalities to describe the constraints of this situation.

19. Write an objective function for the income. $I = 110m + 150a$

20. Determine the value of the objective function for four points with integer coordinates in the feasible region.

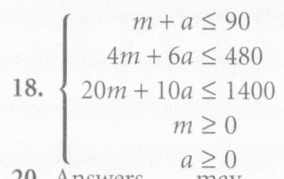

Community Planning Trenton, MI, a small Midwestern town, is trying to establish a public transportation system using 10 or fewer vans and buses. It can spend no more than $100,000 for both types of vehicles and no more than $500 per month for maintenance. Trenton can purchase a van for $10,000 and maintain it for $100 per month. The buses cost $20,000 each and can be maintained for $75 per month. Each van can carry a maximum of 15 passengers and each bus a maximum of 25 passengers.

21. Create a table to organize the information given.

22. Write a system of linear inequalities to describe the constraints of this situation.

23. Write an objective function for the total number of passengers.

24. Determine the value of the objective function for four points with integer coordinates in the feasible region.

18. $\begin{cases} m + a \le 90 \\ 4m + 6a \le 480 \\ 20m + 10a \le 1400 \\ m \ge 0 \\ a \ge 0 \end{cases}$

20. Answers may vary. Examples, where ordered pairs represent (m, a), are

(70, 0): $I = 7700$;
(0, 80): $I = 12,000$;
(30, 60): $I = 12,300$;
(50, 40): $I = 11,500$

22. $\begin{cases} v + b \le 10 \\ 10{,}000v + 20{,}000b \le 100{,}000 \\ 100v + 75b \le 500 \\ v \ge 0 \\ b \ge 0 \end{cases}$

23. $T = 15v + 25b$

24. Answers may vary. Examples, where ordered pairs represent (v, b), are (0, 0): $T = 0$; (5, 0): $T = 75$; (0, 5): $T = 125$; (2, 4): $T = 130$

Let $f(x) = 2x^2 - 5x + 1$ and $g(x) = 6x - 1$. Use your graphics calculator to graph each function. Find the domain and range of each function.   [Lesson 2.3, 2.6]

25 f R; $y \geq -2.125$ **26** g R; R   **27** $f + g$ R; $y \geq -0.125$

28 $f \cdot g$ R; R   **29** $g - f$ R; $y \leq 13.125$ **30** $f - g$ R; $y \geq -13.125$

For Exercises 31–34, let $A = \begin{bmatrix} 2 & 0 & 0 & \vdots & 8 \\ 0 & 1 & 2 & \vdots & 0 \\ 0 & 0 & 3 & \vdots & 9 \end{bmatrix}$.   [Lessons 4.4, 4.6]

31. Write the system of linear equations represented by the augmented matrix A.

32. Use back substitution to solve the system of equations represented by matrix A. $(4, -6, 3)$

33. Write the matrix equation $AX = B$ to represent this system of equations.

34 Use matrix algebra to solve the system of equations. $(4, -6, 3)$

Look Beyond

Communications A **network** is a set of connected points. Networks can be used to represent telephone systems, roadways, electrical systems, and ecological relationships such as food chains. A **directed network** shows the direction of travel along each connection.

A matrix can be used to represent a network. The zero in the first row, third column of the matrix means that there is *no* direct path from A to C. Likewise, there is *one* direct path from C to C, and there are *two* direct paths from C to D.

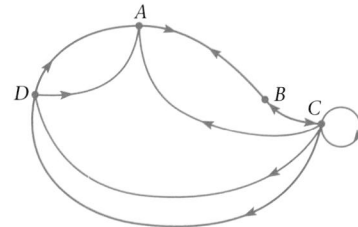

$$\begin{array}{c} \\ A \\ B \\ C \\ D \end{array} \begin{array}{cccc} A & B & C & D \\ \begin{bmatrix} 0 & 1 & 0 & 0 \\ 1 & 0 & 1 & 0 \\ 1 & 1 & 1 & 2 \\ 2 & 0 & 0 & 0 \end{bmatrix} \end{array}$$

35. Construct the matrix for the following telephone network.

Telephone System

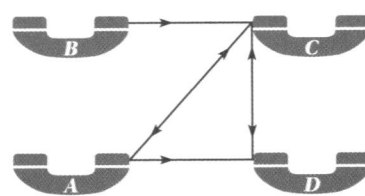

$$\begin{array}{c} \\ A \\ B \\ C \\ D \end{array} \begin{array}{cccc} A & B & C & D \\ \begin{bmatrix} 0 & 0 & 1 & 1 \\ 0 & 0 & 1 & 0 \\ 1 & 0 & 0 & 1 \\ 0 & 0 & 1 & 0 \end{bmatrix} \end{array}$$

31. $\begin{cases} 2x = 8 \\ y + 2z = 0 \\ 3z = 9 \end{cases}$

33. $\begin{bmatrix} 2 & 0 & 0 \\ 0 & 1 & 2 \\ 0 & 0 & 3 \end{bmatrix} \begin{bmatrix} x \\ y \\ z \end{bmatrix} = \begin{bmatrix} 8 \\ 0 \\ 9 \end{bmatrix}$

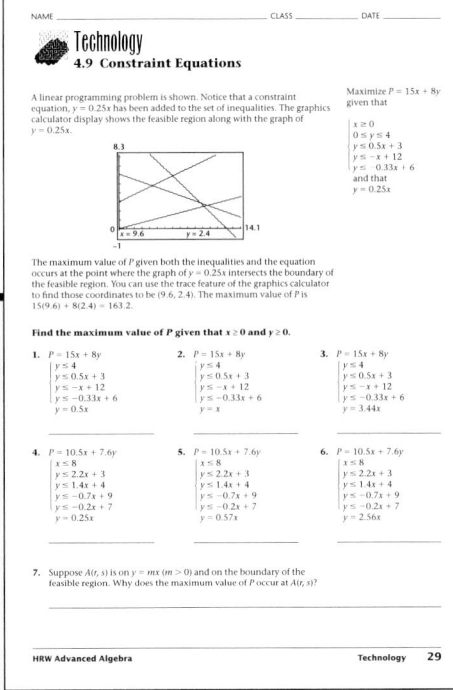

Objectives

- Find the maximum and minimum values of the objective function determined by the feasible region.

RESOURCES

- Practice Master   **4.10**
- Enrichment Master   **4.10**
- Technology Master   **4.10**
- Lesson Activity Master   **4.10**
- Quiz   **4.10**
- Spanish Resources   **4.10**

This lesson may be considered optional in the core course. See page 162A.

Assessing Prior Knowledge

Find the solution to each system of equations.

1. $\begin{cases} x - 2y = -8 \\ 3x + y = 18 \end{cases}$ [(4, 6)]

2. $\begin{cases} 4x + 2y = 2 \\ 5x - y = -22 \end{cases}$ [(−3, 7)]

3. $\begin{cases} 5x - 4y = 15 \\ 3x - 2y = 7 \end{cases}$ [(−1, −5)]

TEACH

 Have students work together to create a list of real-world items or situations that people want to maximize or minimize.

why *In Lesson 4.9, you learned that the minimum and maximum values of an objective function are obtained from the coordinates of points within the feasible region. You can use linear programming to find the exact minimum or maximum value.*

It is helpful to use a table to organize a large amount of information. A method of minimizing or maximizing the situation's objective function for points in the feasible region is based on geometry.

Exploration *Estimating Minimums and Maximums*

1 Graph the feasible region for these constraints.

$\begin{cases} x + y \le 8 \\ x - y \ge -4 \\ x \ge 0 \\ y \ge 0 \end{cases}$

MAXIMUM MINIMUM *Connection*

2 Let the objective function be $P = 2x + 3y$. Complete the table.

P	Objective equation	Slope-intercept form
6		
12		
18		
21		
24		
30		

ALTERNATIVE teaching strategy

Using Cognitive Strategies Have students find the coordinates of the point of intersection of a pair of objective equations using algebraic methods. In Example 2, one vertex of the feasible region is at (0, 0), the point at which the lines x = 0 and y = 0 intersect. The second vertex occurs

where $x + y = 8$ and $x = 0$ intersect. Substitute 0 for x so that y is 8 and the point is (0, 8). A third vertex occurs where the lines $2x + y = 10$ and $x + y = 8$ intersect. By elimination, x is 2 and y is 6, which gives (2, 6). The fourth vertex occurs at the intersection of $2x + y = 10$ and $y = 0$. Substitute 0 for y so that x is 5 and the point is (5, 0).

3 Graph the objective equations on the same coordinate plane that you used to graph the feasible region in Step 1. Which objective equation lines intersect the feasible region? Which objective equation lines do not intersect the feasible region?

4 What values of *P* are feasible? What values of *P* are *not* feasible? Write an inequality, $a \le P \le b$, where *a* and *b* are real numbers, for the values of *P* that *are* feasible.

5 Estimate the maximum value of *P* that is feasible. What part of the feasible region does the objective equation line for this maximum value of *P* intersect?

Estimate the minimum value of *P* that is feasible. What part of the feasible region does the objective equation line for this minimum value of *P* intersect?

6 Where do you think a maximum or minimum value will occur in a feasible region? ❖

Recall the Zoom City Skateboard Company from Example 1 on page 228. Using linear programming, can you find the maximum profit that the company can make?

EXAMPLE 1

Business   The Zoom City Skateboard Company manufactures two models of skateboards—the Rapido model and the Speed Merchant model. The dealers who sell these models want to receive from 30 to 80 Rapido skateboards and from 10 to 30 Speed Merchant skateboards per day. Quality controls and limits on the availability of high-quality ball bearings limit production to no more than 80 skateboards per day. The company makes a profit of $15 for each Rapido model sold and $8 for each Speed Merchant model sold. Find the maximum profit that the Zoom City Company can make under these conditions.

Solution▶

Graph the constraints on the number of skateboards that the Zoom City Skateboard Company can produce, where *r* is the number of Rapido skateboards manufactured and *s* is the number of Speed Merchant skateboards manufactured.

Constraints: $\begin{cases} 30 \le r \le 80 \\ 10 \le s \le 30 \\ r + s \le 80 \end{cases}$

Feasible Region:

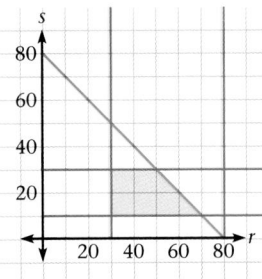

It is very beneficial to allow students to work together when studying a concept such as linear programming. Students can learn much more from each other than they can when working alone. Try to provide opportunities for students to work in groups with a wide variety of strengths. Form the groups in such a way that all students can and will be able to participate and learn most efficiently.

CRITICAL *Thinking*

Yes, because the coordinates of the vertices of the feasible region are the extreme values of x and y that are possible, or feasible, for the objective function.

The objective (profit) function is $P = 15r + 8s$. The graph shown has the profit equations that you graphed in the exploration of Lesson 4.9.

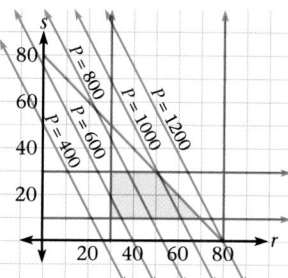

What happens to the profit values as the profit equation lines move up and to the right? Where do you think the maximum feasible profit value will occur?

When you examine the graph of the profit equation lines on the same coordinate plane as the graph of the feasible region, you can see that the maximum profit will occur between $1000 and $1200.

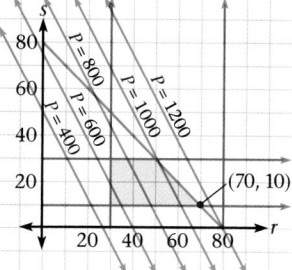

The maximum will occur at the vertex where the lines $s = 10$ and $s = 80 - r$ intersect.

Solve $\begin{cases} s = 10 \\ s = 80 - r \end{cases}$ to find the coordinates of the vertex, (70, 10).

$$(r, s) = (70, 10)$$

Thus, the maximum value of the objective profit function occurs at (70, 10). What is the maximum profit value?

To find the maximum profit, substitute 70 for r and 10 for s in the objective profit function.

$$P = 15r + 8s$$
$$= 15(70) + 8(10)$$
$$= 1130$$

The maximum profit is $1130. Thus, for the specified dealer demands and quality control limits on skateboard production, the company should manufacture 70 Rapido model and 10 Speed Merchant model skateboards per week to achieve a maximum profit of $1130 per week. ❖

CRITICAL *Thinking*

Will the maximum or minimum value of an objective function always be given by the coordinates of a vertex of the feasible region? Why or why not?

ENRICHMENT Determine the maximum value of the profit function $P = 3x + y$ under the given constraints.

$$\begin{cases} x + y \geq 10 \\ 2x + 3y \leq 120 \\ 0 \leq x \leq 30 \end{cases}$$

[**The maximum profit occurs at the vertex (30, 20) of the feasible region, yielding a profit of $P = 110$.**]

INCLUSION **strategies**

English Language Development As in Lesson 4.9, students may find many of the concepts difficult to understand. Encouraging students to write about the concepts in their own words and to share their thoughts with each other will be very helpful in overcoming this problem. Suggest that students review the explanations they wrote for the terms in Lesson 4.9 to help them get started.

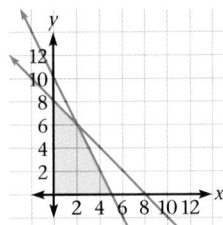

EXAMPLE 2

Find the maximum value of the following objective function under the given constraints.

Constraints: $\begin{cases} 2x + y \le 10 \\ x + y \le 8 \\ x \ge 0 \\ y \ge 0 \end{cases}$ Objective function: $C = 2x + 3y$

Solution➤

Graph the feasible region.

Find the coordinates of the vertices of the feasible region.

(0, 0), (0, 8), (5, 0), and (2, 6)

Substitute these coordinates in the objective function.

Vertex	Value of $C = 2x + 3y$
(0, 0)	$C = 2(0) + 3(0) = 0$
(0, 8)	$C = 2(0) + 3(8) = 24$
(5, 0)	$C = 2(5) + 3(0) = 10$
(2, 6)	$C = 2(2) + 3(6) = 22$

Thus, the maximum value of 24 for C is achieved when x is 0 and y is 8.

From the graph you can see that the point (0, 8) is where the maximum objective function value occurs. ❖

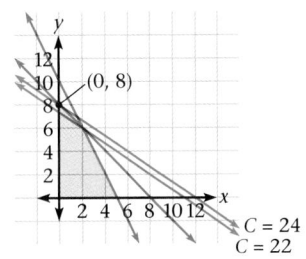

The minimum value of an objective function also occurs at a vertex. In Example 2, the minimum value for C is 0 and occurs at the point (0, 0).

Try This   Find the maximum value for $C = 4x + 3y$ with the constraints given in Example 2.

Alternate Example 3

Find the approximate maximum and minimum values of the objective function under the given constraints.

Objective function: $R = -x + 4y$

Constraints: $\begin{cases} x + y \geq 4 \\ 2x + 3y \leq 24 \\ 3x + y < 15 \\ x \geq 0 \\ y \geq 0 \end{cases}$

[maximum value: 32, minimum value: −4]

EXAMPLE 3

Graphics Calculator

Find the approximate maximum and minimum values of the objective function under the given constraints.

Constraints: $\begin{cases} u + v \geq 2 \\ 3u + 2v \leq 18 \\ 2u + v \leq 10 \\ u \geq 0 \\ v \geq 0 \end{cases}$ Objective function: $R = -u + 3v$

Solution

Use your graphics calculator to graph the feasible region. Use the trace feature to estimate the vertices.

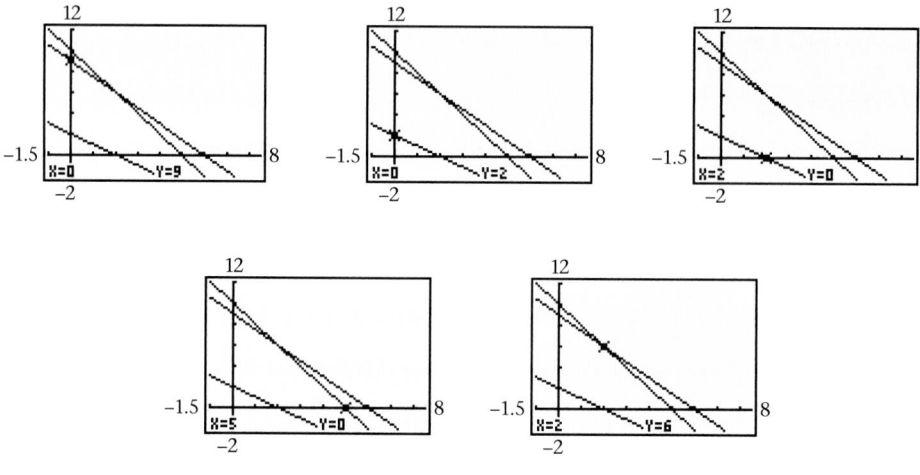

The vertices are (0, 9), (0, 2), (2, 0), (5, 0), and (2, 6). Create a table containing the coordinates of the vertices and the resulting R-values.

Vertex	Value of $R = -u + 3v$
(0, 9)	$R = -(0) + 3(9) = 27$
(0, 2)	$R = -(0) + 3(2) = 6$
(2, 0)	$R = -(2) + 3(0) = -2$
(5, 0)	$R = -(5) + 3(0) = -5$
(2, 6)	$R = -(2) + 3(6) = 16$

The maximum value of 27 for R occurs at the vertex (0, 9).
The minimum value of −5 for R occurs at the vertex (5, 0). ❖

A common method of performing linear programming with a large number of constraints and variables is the **simplex method**. There are computer programs that have been written to facilitate the use of the simplex method.

EXERCISES & PROBLEMS

Communicate

1. Where will a maximum or a minimum objective function line intersect the feasible region?

2. Explain how to find a minimum or a maximum value of an objective function subject to given constraints.

3. Explain how to use your graphics calculator to estimate the maximum or minimum of an objective function subject to given constraints.

Practice & Apply

Nutrition A school lunchroom dietitian wants to prepare a meal of meat and vegetables that has the lowest possible amount of fat and that meets the Food and Drug Administration recommended daily allowances (RDA) for iron and protein. The RDA minimum is 20 milligrams for iron and 45 grams for protein. Each 3-ounce serving of meat contains 45 grams of protein, 10 milligrams of iron, and 4 grams of fat. Each 1-cup serving of vegetables contains 9 grams of protein, 6 milligrams of iron, and 2 grams of fat. Let x represent the number of 3-ounce portions of meat, and let y represent the number of 1-cup portions of vegetables.

4. Write an inequality for each constraint in this situation. Graph the feasible region.

5. Is the feasible region bounded or unbounded? What are the vertices of the feasible region?

6. Determine the objective function. $F = 4x + 2y$

7. Find the minimum value of the objective function that satisfies the constraints. 7

ASSESS

Selected Answers
Odd-numbered Exercises 5–35

Assignment Guide
Core 1–10, 12–18, 20–39

Core Plus 1–7, 10–15, 17–39

Technology
A graphics calculator is needed for Exercises 26 and 32–39. A graphics calculator or computer software such as *f(g) Scholar*™ may be helpful to answer Exercises 8–11, 14, 16–19, and 22.

4. $\begin{cases} 45x + 9y \geq 45 \\ 10x + 6y \geq 20 \\ \quad\quad x > 0 \\ \quad\quad y > 0 \end{cases}$

5. unbounded; (0, 5), (0.5, 2.5), (2, 0)

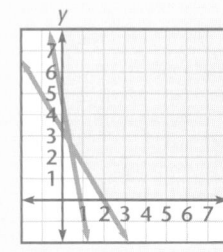

Error Analysis

Watch for students who confuse maximum and minimum values. Be sure they understand how to use the feasible region of the graph to find these values and then use the graph to double check their answers. They should also check to make sure the answers they give make sense.

Maximize each of the following objective functions under the given constraints.

8. $P = 5x + 3y$ 30
Constraints
$$\begin{cases} x + y \le 6 \\ x - y \le 4 \\ x \ge 0 \\ y \ge 0 \end{cases}$$

9. $P = 3x + y$ 4.5
Constraints
$$\begin{cases} 2x + y \le 3 \\ 3x + 4y \le 8 \\ x \ge 0 \\ y \ge 0 \end{cases}$$

10. $P = 4x + 7y$ 56
Constraints
$$\begin{cases} x + y \le 8 \\ x - y \le 2 \\ x \ge 0 \\ y \ge 0 \end{cases}$$

11. $P = 2x + 7y$ 36
Constraints
$$\begin{cases} x \ge 1 \\ 0 \le y \le 4 \\ 4x - 2y \le 8 \end{cases}$$

Small Business A carpenter makes bookcases in two sizes, large and small. It takes 6 hours to make a large bookcase and 2 hours to make a small one. The profit on a large bookcase is $50, and the profit on a small one is $20. The carpenter can spend only 24 hours per week making bookcases and must make at least two of each size per week. The carpenter wants to know how many of each size must be made per week in order to maximize profit.

12. Write and graph the feasible region.

13. Determine the objective function. $P = 50l + 20s$

14. What are the vertices of the feasible region? $(2, 2), (2, 6), \left(\frac{10}{3}, 2\right)$

15. Find the maximum value of the objective function that satisfies the constraints. 220

Minimize each of the following objective functions under the given constraints.

16. $C = 2x + y$
Constraints
$$\begin{cases} x + y \ge 6 \\ x - y \ge -4 \\ x \ge 0 \\ y \ge 0 \quad 7 \end{cases}$$

17. $C = x + y$
Constraints
$$\begin{cases} 2x + y \le 3 \\ 3x + 4y \le 12 \\ x \ge 0 \\ y \ge 0 \quad 0 \end{cases}$$

18. $C = 3x + 5y$
Constraints
$$\begin{cases} x - 2y \ge 0 \\ x + 2y \ge 8 \\ 1 \le x \le 6 \\ y \ge 0 \quad 22 \end{cases}$$

19. $C = 3x + 2y$
Constraints
$$\begin{cases} x + y \le 5 \\ y - x \ge 5 \\ 4x + y \ge -10 \\ \qquad -5 \end{cases}$$

Communications At a radio station, 6 minutes of each hour are devoted to news, and the remaining 54 minutes are devoted to other programming, such as music and commercials. Station policy requires at least 30 minutes of music per hour and at least 3 minutes of music for each minute of commercials. The radio station profits from selling commercial time, so the manager wants to find the maximum number of minutes of commercial time available each hour.

20. Write and graph the feasible region.

21. Determine the objective function. value of c

22. Find the vertices of the feasible region. (30, 0), (30, 10), (40.5, 13.5), (54, 0)

23. Find the maximum for the objective function that satisfies the constraints. 13.5

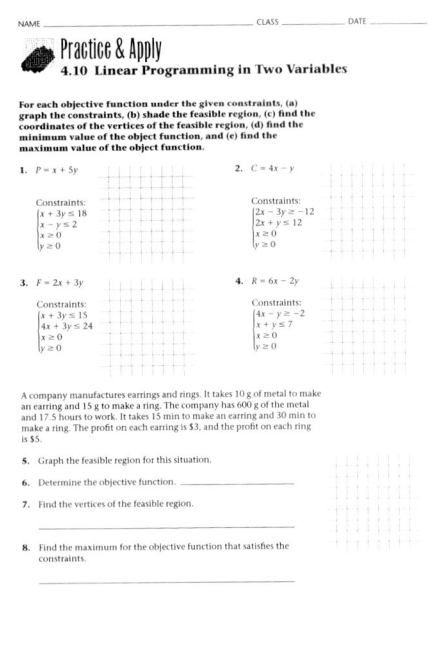
12.
$$\begin{cases} 6l + 2s \le 24 \\ l \ge 2 \\ s \ge 2 \end{cases}$$

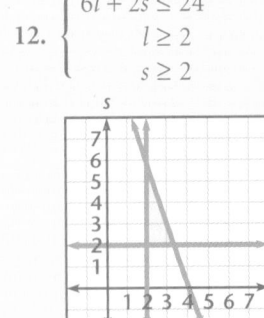

20.
$$\begin{cases} m + c \le 54 \\ m \ge 30 \\ m \ge 3c \\ c > 0 \end{cases}$$

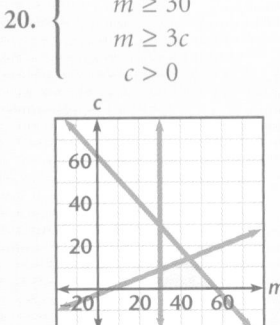

Manufacturing A tire manufacturer has 1000 units of raw rubber to use in producing radial tires for passenger cars and tractor tires. Each radial tire requires 5 units of rubber; each tire for tractors requires 20 units. Labor costs are $8 for a radial tire and $12 for a tractor tire. The manufacturer does not want to pay more than $1500 in labor costs and wants to make a profit of $10 per radial tire and $25 per tractor tire. The manager needs to know how many of each kind of tire to make in order to maximize profits.

24. Write and graph the feasible region.

25. Determine the object function. $P = 10r + 25t$

26 Use your graphics calculator to estimate the vertices of the feasible region.
(0, 0), (0, 50), (187.5, 0), (180, 5)

27. Find the maximum for the objective function that satisfies the constraints.
1925

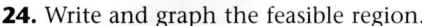 **Look Back**

Graph the following systems of linear inequalities. [Lesson 1.6]

28. $\begin{cases} -x + 2y \geq 4 \\ y \geq 3 \end{cases}$

29. $\begin{cases} 4x - y \leq -3 \\ y < -1 \end{cases}$

30. $\begin{cases} y \leq x + 4 \\ x < 4 \end{cases}$

31. $\begin{cases} y > -\frac{2}{3}x + 4 \\ y \leq \frac{4}{3}x - 2 \end{cases}$

Find the product. Use your graphics calculator to check your answer. [Lesson 4.2]

32 $[4 \quad 8 \quad 3] \begin{bmatrix} 2 \\ 0 \\ 5 \end{bmatrix}$ [23]   **33** $\begin{bmatrix} 0 & 1 \\ 3 & -8 \end{bmatrix} \begin{bmatrix} -2 & -5 \\ -5 & 34 \end{bmatrix}$ **34** $\begin{bmatrix} 7 & 3 \\ 0 & -1 \end{bmatrix} \begin{bmatrix} 5 & -2 \\ 4 & 6 \end{bmatrix}$ **35** $\begin{bmatrix} 4 & -1 \\ 3 & 2 \end{bmatrix} \begin{bmatrix} 1 & 0 \\ 5 & 3 \end{bmatrix}$

$\begin{bmatrix} 47 & 4 \\ -4 & -6 \end{bmatrix}$ $\begin{bmatrix} -1 & -3 \\ 13 & 6 \end{bmatrix}$

36 Solve $\begin{cases} x + y + z = 60 \\ 4y + 2z = 160 \\ 2x - 3z = -40 \end{cases}$ using matrix algebra. [Lesson 4.6]
(10, 30, 20)

Look Beyond

Graph each function, find the x-intercepts, and explain how the x-intercepts are related to the constants in the function.

37 $f(x) = (x - 3)(x + 2)$ **38** $f(x) = (x + 4)(x - 9)$ **39** $f(x) = x(x + 12)$

24. $\begin{cases} 5r + 20t \leq 1000 \\ 8r + 12t \leq 1500 \\ r \geq 0 \\ t > 0 \end{cases}$

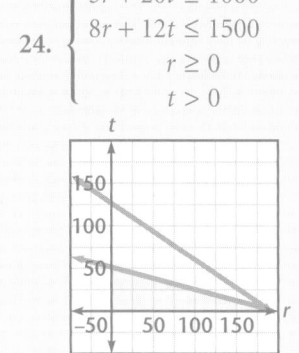

The answers to Exercises 29–31 and 37–39 can be found in Additional Answers beginning on page 842.

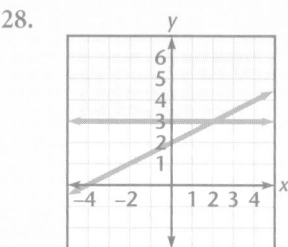
Technology Master

NAME _____ CLASS _____ DATE _____

Technology
4.10 Minimums, Maximums, and Fixed Profits

In some situations, a problem will contain a set of constraints and a profit level that is fixed, such as the situation at the right. What are the least and greatest production levels x that will yield the fixed profit level?

$\begin{cases} x \geq 0 \\ x \geq 0 \\ y \leq 2x + 2.5 \\ y \leq -0.2x + 7 \\ y \leq -1.5x + 13 \end{cases}$

The graphics calculator display shows the feasible region and the profit equation for the given problem. The least and greatest production levels give the indicated profit. These points are represented by the intersection points of the lower horizontal line.

Profit level:
$12 = x + 3y$

To find the x-coordinates of these points, use the trace feature on your graphics calculator. You will find them to be x = 0.6 and x = 7.7. If 0.6 ≤ x ≤ 7.7, the manufacturer will realize a fixed profit.

If you also want to find the values of y that correspond to the minimum and maximum values of x, you can read them from the display as you use trace key.

Find the minimum and maximum values of x that give the desired profit level P.

1. $\begin{cases} x \geq 0 \\ y \geq 0 \\ y \leq 4 \\ y \leq -1.5x + 8 \\ y \leq -0.75x + 5 \end{cases}$
Profit level:
P: y = 3

2. $\begin{cases} x \geq 0 \\ y \geq 0 \\ y \leq 4 \\ y \leq -1.5x + 8 \\ y \leq -0.75x + 5 \end{cases}$
Profit level:
P: 12 = 4x + y

3. $\begin{cases} x \geq 0 \\ y \geq 0 \\ y \leq 4 \\ y \leq -1.5x + 8 \\ y \leq -0.75x + 5 \end{cases}$
Profit level:
P: 2 = -x + y

4. $\begin{cases} x \geq 0 \\ y \geq 0 \\ y \leq x + 2 \\ y \leq -0.5x + 8 \\ y \leq -4x + 36 \end{cases}$
Profit level:
P: 36 = 2x + 9y

5. $\begin{cases} x \geq 0 \\ y \geq 0 \\ y \leq x + 2 \\ y \leq -0.5x + 8 \\ y \leq -4x + 36 \end{cases}$
Profit level:
P: 12 = -3x + 4y

6. $\begin{cases} x \geq 0 \\ y \geq 0 \\ y \leq x + 2 \\ y \leq -0.5x + 8 \\ y \leq -4x + 36 \end{cases}$
Profit level:
P: 72 = 2x + 9y

7. Suppose that no fixed profit level is given for Exercises 4–6. Instead you are given only that P = 2x + 9y. What would be the minimum and maximum profits?

30   Technology   HRW Advanced Algebra

FOCUS

Many beautiful, creative designs can be made by performing transformations on figures. As students work to create their designs, they will learn more about the usefulness of some of the concepts they studied in this chapter.

MOTIVATE

Have students look for examples of designs that they think can be recreated using transformations. Make a display of these designs and, after students have completed their work creating their own designs, have them use transformations to try to recreate some of the designs in the display.

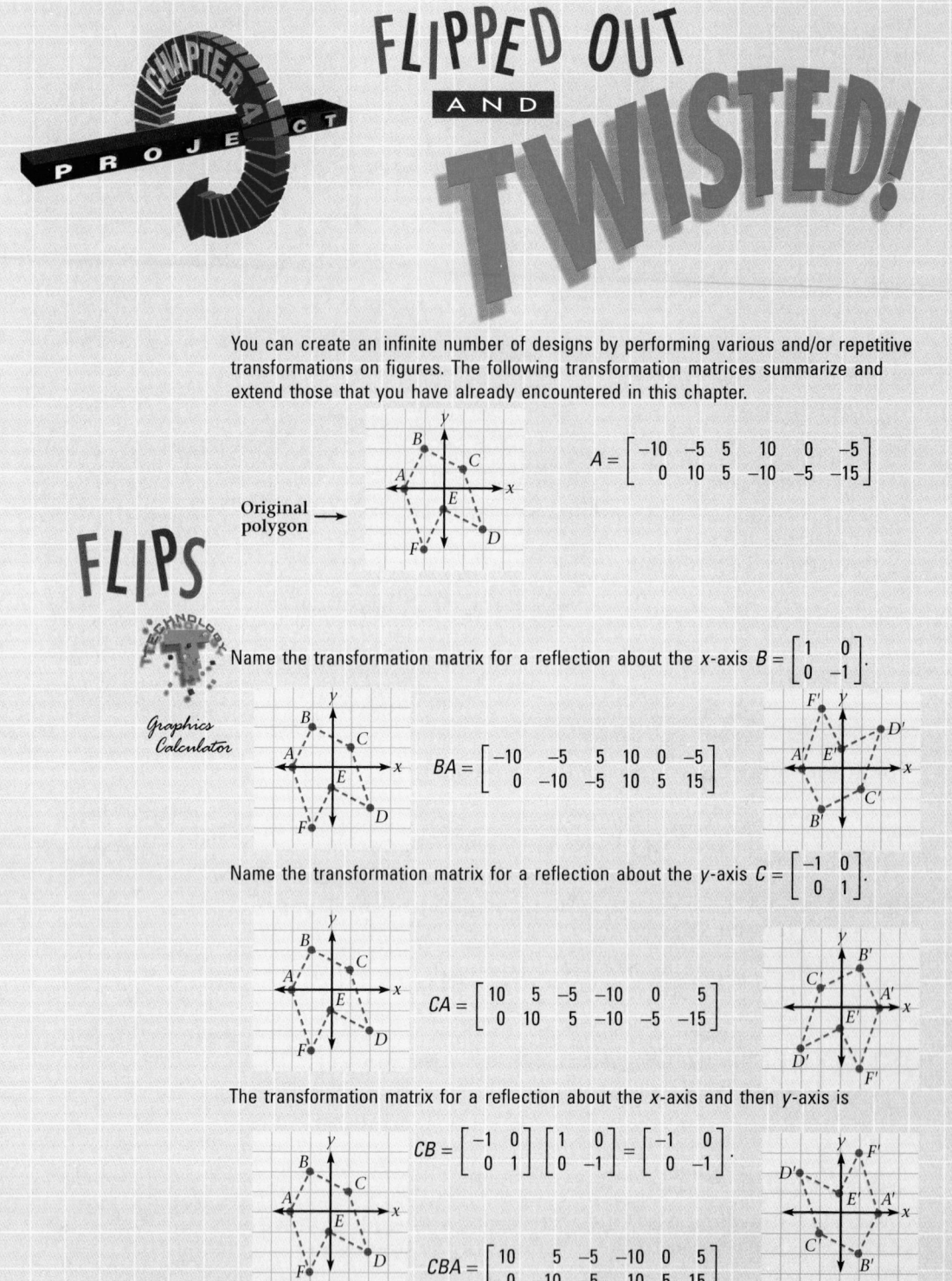

You can create an infinite number of designs by performing various and/or repetitive transformations on figures. The following transformation matrices summarize and extend those that you have already encountered in this chapter.

$$A = \begin{bmatrix} -10 & -5 & 5 & 10 & 0 & -5 \\ 0 & 10 & 5 & -10 & -5 & -15 \end{bmatrix}$$

Original polygon →

FLIPS

Name the transformation matrix for a reflection about the x-axis $B = \begin{bmatrix} 1 & 0 \\ 0 & -1 \end{bmatrix}$.

$$BA = \begin{bmatrix} -10 & -5 & 5 & 10 & 0 & -5 \\ 0 & -10 & -5 & 10 & 5 & 15 \end{bmatrix}$$

Name the transformation matrix for a reflection about the y-axis $C = \begin{bmatrix} -1 & 0 \\ 0 & 1 \end{bmatrix}$.

$$CA = \begin{bmatrix} 10 & 5 & -5 & -10 & 0 & 5 \\ 0 & 10 & 5 & -10 & -5 & -15 \end{bmatrix}$$

The transformation matrix for a reflection about the x-axis and then y-axis is

$$CB = \begin{bmatrix} -1 & 0 \\ 0 & 1 \end{bmatrix} \begin{bmatrix} 1 & 0 \\ 0 & -1 \end{bmatrix} = \begin{bmatrix} -1 & 0 \\ 0 & -1 \end{bmatrix}.$$

$$CBA = \begin{bmatrix} 10 & 5 & -5 & -10 & 0 & 5 \\ 0 & -10 & -5 & 10 & 5 & 15 \end{bmatrix}$$

TWISTS

Name the transformation matrix for a rotation by 30° clockwise $D = \begin{bmatrix} 0.86 & 0.5 \\ -0.5 & 0.86 \end{bmatrix}$.

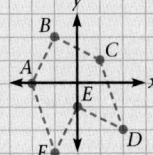

$$DA = \begin{bmatrix} -8.6 & 0.7 & 6.8 & 3.6 & -2.5 & -11.8 \\ 5.0 & 11.1 & 1.8 & -13.6 & -4.3 & -10.4 \end{bmatrix}$$

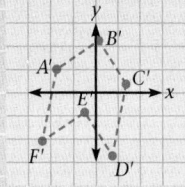

Name the transformation matrix for a rotation by 45° counterclockwise $E = \begin{bmatrix} 0.7 & -0.7 \\ 0.7 & 0.7 \end{bmatrix}$.

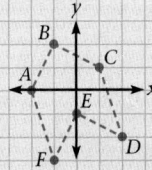

$$EA = \begin{bmatrix} -7 & -10.5 & 0 & 14 & 3.5 & 7 \\ -7 & 3.5 & 7 & 0 & -3.5 & -14 \end{bmatrix}$$

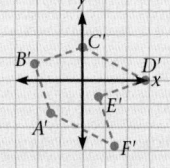

Beginning with the figure rotated 45° counterclockwise represented by *EA*, the figure is then rotated by 60° more, for a total counterclockwise rotation of the original by 105°. Name the transformation matrix for a rotation by 60° counterclockwise $F = \begin{bmatrix} 0.5 & -0.86 \\ 0.86 & 0.5 \end{bmatrix}$.

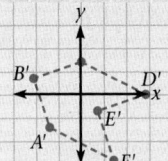

$$F(EA) = \begin{bmatrix} 2.52 & -8.26 & -6.02 & 7 & 4.76 & 15.54 \\ -9.52 & -7.28 & 3.5 & 12.04 & 1.26 & -0.98 \end{bmatrix}$$

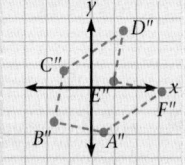

The following designs were created by using one or more of the transformation matrices that you learned in this chapter.

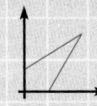

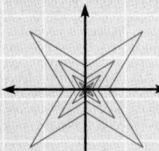

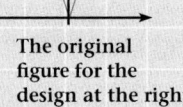

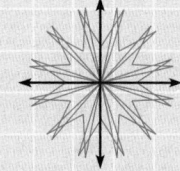

The original figure for the design at the right

The original figure for the design at the right

For this chapter project, select any shape that can be determined by its vertices and have fun making your own design. You may use any transformation matrices that you have learned or combinations, inverses, or other transformation matrices that you can discover. Keep in mind that these transformations may change the size of the figure but cannot change its shape.

Have students work in small groups to create at least one simple design. Then students may want to work in pairs to create a different design. After students feel comfortable with the procedure used to create designs, have them create at least one on their own. Make a display of the designs students choose.

DISCUSS

Have students look at the display of examples of designs they found and compare them with the designs they made themselves. Discuss any differences students notice. Then ask students what happened when they tried to recreate the examples with transformations. In closing, ask students if they think all designs can be created using transformations.

Chapter 4 Review

Vocabulary

Key Skills & Exercises

Lesson 4.1

➤ **Key Skills**

Use matrices to store and represent data.

$$C = \begin{bmatrix} \text{red} & \text{yellow} & \text{purple} \\ \text{blue} & \text{green} & \text{white} \end{bmatrix}$$

The dimensions of matrix C are 2 rows × 3 columns, or 2×3. C_{23} is the element located in row 2, column 3, or *white*.

Add and subtract matrices.

To find the sum of A and B, find the sum of each element of A and the corresponding element of B.

To subtract B from A, subtract each element of matrix B from each corresponding element of matrix A.

➤ **Exercises**

Matrix T shows the number of assessments that Mark had in 1 month. 4×3

Subject	Tests	Quizzes
Algebra	4	4
History	2	3
Chemistry	4	2
German	2	5

$$T = \begin{bmatrix} 4 & 4 \\ 2 & 3 \\ 4 & 2 \\ 2 & 5 \end{bmatrix}$$

1. What are the dimensions of matrix T?
2. Describe the data in location t_{33}.
   Chemistry quizzes
3. In matrix notation, write an expression for the total number of assessments that Mark had in algebra. $t_{12} + t_{13}$
4. Use a matrix equation to represent the following set of linear equations.
   $-3x + 2y + z = 9$, $4x - 5y + z = 2$, and $-x - 3y - z = 5$

4. $\begin{bmatrix} -3x + 2y + z \\ 4x - 5y + z \\ -x - 3y - z \end{bmatrix} = \begin{bmatrix} 9 \\ 2 \\ 5 \end{bmatrix}$

Let $A = \begin{bmatrix} -1 & 0 & 3 \\ 2 & -3 & 4 \end{bmatrix}$ and $B = \begin{bmatrix} 1 & 2 & -2 \\ -1 & 1 & -5 \end{bmatrix}$.

5. Find $A + B$. $\begin{bmatrix} 0 & 2 & 1 \\ 1 & -2 & -1 \end{bmatrix}$

6. Find $B - A$. $\begin{bmatrix} 2 & 2 & -5 \\ -3 & 4 & -9 \end{bmatrix}$

Lesson 4.2

➤ Key Skills

Perform matrix multiplication.

To find the matrix product, pair rows of the first matrix with columns of the second matrix. The inner dimensions of the two matrices must be the same.

$$\begin{bmatrix} 4 & 0 \\ -1 & 3 \end{bmatrix} \begin{bmatrix} 5 \\ -2 \end{bmatrix} = \begin{bmatrix} (4)(5) + (0)(-2) \\ (-1)(5) + (3)(-2) \end{bmatrix} = \begin{bmatrix} 20 \\ -11 \end{bmatrix}$$

➤ Exercises

Let $A = \begin{bmatrix} -1 \\ 2 \\ 0 \end{bmatrix}$, $B = \begin{bmatrix} -2 & 1 & 3 \\ 0 & \frac{1}{2} & 5 \end{bmatrix}$, $C = [-1 \quad 2 \quad 3]$, and $I = \begin{bmatrix} 1 & 0 \\ 0 & 1 \end{bmatrix}$.

Find the indicated product.

7. CA [5]

8. BA $\begin{bmatrix} 4 \\ 1 \end{bmatrix}$

9. IB $\begin{bmatrix} -2 & 1 & 3 \\ 0 & \frac{1}{2} & 5 \end{bmatrix}$

Lesson 4.3

➤ Key Skills

Identify each type of system of two linear equations.

An *independent* system of linear equations has one unique solution.

A *dependent* system of linear equations has infinitely many solutions.

An *inconsistent* system of linear equations has no solution.

Solve a system of two linear equations using elimination by adding.

$$\begin{cases} -2x + y = -2 \\ x - 3y = -4 \end{cases} \rightarrow \begin{array}{l} -6x + 3y = -6 \\ \underline{x - 3y = -4} \\ -5x = -10 \\ x = 2 \end{array}$$

Substitute 2 for x. $\begin{array}{l} -2(2) + y = -2 \\ y = 2 \end{array}$

Thus, $(2, 2)$ is the solution.

➤ Exercises

Solve each system of linear equations. Indicate whether each system is inconsistent, dependent, or independent.

10. $\begin{cases} 4x - 5y = 20 \\ -\frac{1}{2}x = -\frac{5}{8}y + \frac{5}{4} \end{cases}$ no solution; incon

11. $\begin{cases} -3x + 2y = 7 \\ 2x + 5y = 8 \end{cases}$ $(-1, 2)$; indep

12. $\begin{cases} 6x + 4y = 12 \\ 2y = 6 - 3x \end{cases}$ infinite; dep

13. $\begin{cases} 2x + y = 3 \\ 3x - 2y = 8 \end{cases}$ $(2, -1)$; indep

Lesson 4.4

➤ Key Skills

Represent a system of linear equations with an augmented matrix.

$$\begin{cases} y + 2x = 2 \\ -y - 2z = -9 \\ -3x = z + 4y \end{cases} \rightarrow \begin{cases} 2x + y + 0z = 2 \\ 0x - y - 2z = -9 \\ -3x - 4y - z = 0 \end{cases} \rightarrow \begin{bmatrix} 2 & 1 & 0 & \vdots & 2 \\ 0 & -1 & -2 & \vdots & -9 \\ -3 & -4 & -1 & \vdots & 0 \end{bmatrix}$$

Solve a system of linear equations using the row reduction method and back substitution.

$$\begin{cases} 2x + 3y - z = -7 \\ x + y - z = -4 \\ 3x - 2y - 3z = -7 \end{cases} \rightarrow \begin{bmatrix} 2 & 3 & -1 & \vdots & -7 \\ 1 & 1 & -1 & \vdots & -4 \\ 3 & -2 & -3 & \vdots & -7 \end{bmatrix} \xrightarrow{R_1 \leftrightarrow R_2} \begin{bmatrix} 1 & 1 & -1 & \vdots & -4 \\ 2 & 3 & -1 & \vdots & -7 \\ 3 & -2 & -3 & \vdots & -7 \end{bmatrix} \begin{matrix} \xrightarrow{-2R_1+R_2 \rightarrow R_2} \\ \xrightarrow{-3R_1+R_3 \rightarrow R_3} \end{matrix}$$

$$\begin{bmatrix} 1 & 1 & -1 & \vdots & -4 \\ 0 & 1 & 1 & \vdots & 1 \\ 0 & -5 & 0 & \vdots & 5 \end{bmatrix} \xrightarrow{5R_2+R_3 \rightarrow R_3} \begin{bmatrix} 1 & 1 & -1 & \vdots & -4 \\ 0 & 1 & 1 & \vdots & 1 \\ 0 & 0 & 5 & \vdots & 10 \end{bmatrix} \rightarrow \begin{cases} x + y - z = -4 \\ y + z = 1 \\ 5z = 10 \end{cases}$$

$$\begin{array}{ccc} 5z = 10 & y + z = 1 & x + y - z = -4 \\ z = 2 & y + (2) = 1 & x + (-1) - (2) = -4 \\ & y = -1 & x = -1 \end{array}$$

The solution is $(-1, -1, 2)$.

➤ Exercises

Write the augmented matrix for each system of equations. Then solve the system using row reduction and back substitution.

14. $\begin{cases} 3x - 2y + z = -5 \\ -2x + 3y - 3z = 12 \\ 3x - 2y - 2z = 4 \end{cases}$ **15.** $\begin{cases} 3x - y + 2z = 9 \\ x - 2y - 3z = -1 \\ 2x - 3y + z = 10 \end{cases}$ **16.** $\begin{cases} -3x + 2y + 2z = 9 \\ 2x - 5y - 3z = -2 \\ -6x + 3y - 4z = 4 \end{cases}$

$(0, 1, -3)$ $(1, -2, 2)$ $(-3, -2, 2)$

Lesson 4.5

➤ Key Skills

Find the inverse of a matrix.

To find the inverse of $A = \begin{bmatrix} 3 & -1 \\ -4 & 2 \end{bmatrix}$ with your graphics calculator, enter matrix A, and use the $\boxed{x^{-1}}$ key to find the inverse.

```
[A]
[ 3  -1]
[ -4  2 ]
[A]⁻¹
[ 1  .5 ]
[ 2  1.5]
```

➤ Exercises

Find the inverse of each matrix, if it exists. If it does not exist, write *does not exist*. Write elements to the nearest hundredth.

17. $\begin{bmatrix} 2 & 4 \\ 4 & 8 \end{bmatrix}$ does not exist **18.** $\begin{bmatrix} -1 & -2 \\ 2 & 3 \end{bmatrix}$ $\begin{bmatrix} 3 & 2 \\ -2 & -1 \end{bmatrix}$

19. $\begin{bmatrix} 2 & 3 & -1 \\ 4 & 6 & -2 \\ 1 & 0 & 2 \end{bmatrix}$ does not exist **20.** $\begin{bmatrix} 4 & -1 \\ -5 & 2 \end{bmatrix}$ $\begin{bmatrix} 0.67 & 0.33 \\ 1.67 & 1.33 \end{bmatrix}$

Lesson 4.6

➤ *Key Skills*

Use matrix algebra to solve a system of equations.

To solve a system of equations using matrix algebra, write a matrix equation, and solve it by multiplying both sides of the equation by the inverse of the coefficient matrix.

$$\begin{cases} x + 2y - z = -2 \\ -2x + y - 3z = -6 \\ x + y + 2z = 2 \end{cases} \rightarrow \begin{bmatrix} 1 & 2 & -1 \\ -2 & 1 & -3 \\ 1 & 1 & 2 \end{bmatrix} \begin{bmatrix} x \\ y \\ z \end{bmatrix} = \begin{bmatrix} -2 \\ -6 \\ 2 \end{bmatrix}$$

$$X = A^{-1}B \rightarrow \begin{bmatrix} x \\ y \\ z \end{bmatrix} = \begin{bmatrix} 1 \\ -1 \\ 1 \end{bmatrix} \text{ or } x = 1, y = -1, \text{ and } z = 1$$

➤ *Exercises*

For each system of equations, write the matrix equation that represents it, and solve it using matrix algebra.

21. $\begin{cases} x + y - z = -3 \\ 3x - y + 2z = -3 \\ -2x + 2y - z = 4 \end{cases}$ **22.** $\begin{cases} x + y + z = 2 \\ 2x - y + z = 6 \\ -2x + 3y - z = -8 \end{cases}$ **23.** $\begin{cases} 2x - 3y + 5z = 5 \\ x - y - 2z = 2 \\ -3x + 3y + 6z = 5 \end{cases}$

$(-2, 1, 2)$ $(2, -1, 1)$ no solution

Lesson 4.7

➤ *Key Skills*

Use matrices to represent and transform objects.

The matrix $\begin{bmatrix} 8 & -4 & 3 \\ 6 & -3 & 4 \end{bmatrix}$ represents an object with vertices (8, 6), (−4, −3), and (3, 4).

To rotate the object by 90° counterclockwise, multiply by $\begin{bmatrix} 0 & -1 \\ 1 & 0 \end{bmatrix}$.

To enlarge or reduce an object by a, multiply by $\begin{bmatrix} a & 0 \\ 0 & a \end{bmatrix}$.

To reverse a transformation, multiply by the inverse of the transformation matrix.

➤ *Exercises*

Give the vertices after the indicated transformation on

$A = \begin{bmatrix} 5 & 1 & 2 & -2 \\ 2 & -3 & 4 & -1 \end{bmatrix}$.

24. Rotate 90° counterclockwise, and enlarge to 3 times its size.

25. Rotate 180° clockwise, and reduce to $\frac{1}{2}$ of its size.

Lesson 4.8

➤ *Key Skills*

Graph the solution to a system of linear inequalities.

To graph the solutions to the system of linear inequalities $\begin{cases} y > 2x - 4 \\ x + y \leq 1 \end{cases}$,

graph each inequality and shade the intersection of the solutions to each inequality.

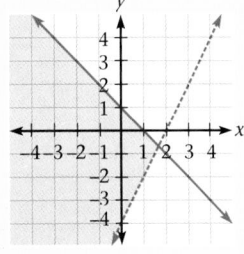

24. $\begin{bmatrix} -6 & 9 & -12 & 3 \\ 15 & 3 & 6 & -6 \end{bmatrix}$

25. $\begin{bmatrix} -2.5 & -0.5 & -1 & 1 \\ -1 & 1.5 & -2 & 0.5 \end{bmatrix}$

26.

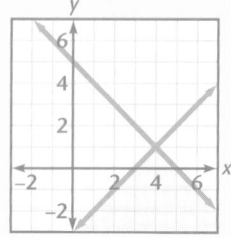

27.

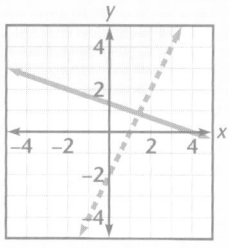

28.

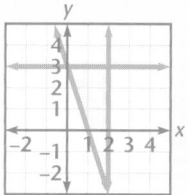

29.

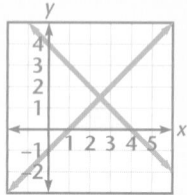

➤ **Exercises**

Graph the solution to each of the following systems of linear inequalities.

26. $\begin{cases} x + y \leq 5 \\ x - y \geq 3 \end{cases}$ **27.** $\begin{cases} 2x - y < 2 \\ x + 2y \geq 4 \end{cases}$ **28.** $\begin{cases} 3x + y \geq 3 \\ 0 \leq x \leq 2 \\ 0 \leq y \leq 3 \end{cases}$ **29.** $\begin{cases} x + y \leq 4 \\ x - y \geq 1 \\ y \geq 0 \end{cases}$

Lessons 4.9 and 4.10

➤ **Key Skills**

Graph a feasible region determined by constraints.

To graph the feasible region determined by constraints, graph the inequalities on the same coordinate plane.

$\begin{cases} x + y \leq 4 \\ x - y \leq 2 \\ x \geq 0 \\ y \geq 0 \end{cases}$

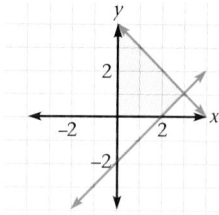

Find the maximum and minimum values of the objective function determined by the feasible region.

The vertices of the feasible region and the values of an objective function, $P = x^2 - y$, are as follows.

(0, 0): $P = 0 - 0 = 0$
(0, 4): $P = 0 - 4 = -4$
(2, 0): $P = 4 - 0 = 4$
(3, 1): $P = 9 - 1 = 8$

The maximum value of P is 8 and occurs at the point (3, 1).

➤ **Exercises**

For each of the feasible regions, find the vertices of the feasible region, and find the minimum or maximum values of the objective function that satisfies the constraints.

30. Constraints: $\begin{cases} x + 2y \geq 4 \\ 3x - 2y \geq 6 \\ y \geq 0 \\ x \leq 6 \end{cases}$

Minimize $P = 2x + 3y$.

(2.5, 0.75), (4, 0), (6, 0), (6, 6); 7.25

31. Constraints: $\begin{cases} x - y \leq 4 \\ 2x + 3y \leq 6 \\ x \geq 0 \\ y \geq 0 \end{cases}$

Maximize $P = x + 2y$.

(0, 2), (0, 0), (3, 0); 4

Applications

Fund-Raising Two hundred and ten people attended a school carnival. The total amount of money collected for tickets was $710. Prices were $5 for regular admission, $3 for students and $1 for children. The number of regular admissions was 10 more than twice children.

32. Write a system of equations that you would solve to find the number of tickets for regular admission, students, and children.

33. Write the matrix equation that represents this system of equations.

34. Give the number of tickets for each group. 70 regular, 110 student, 30 child

35 **Business** The Rollerblade Shop sells in-line skates with and without custom boots. The store wants to sell at most 90 pairs of skates per month. The in-line skates without custom boots cost $80 to produce, and the in-line skates with custom boots cost $150. The store can spend no more than $11,400 per month on in-line skates. The profit on each pair of in-line skates without custom boots is $60, and the profit on each pair of in-line skates with custom boots is $65. Find the number of each type of in-line skates to maximize the profit. 30 without 60 with

32. $\begin{cases} a + s + c = 210 \\ 5a + 3s + c = 710 \\ a = 10 + 2c \end{cases}$

33. $\begin{bmatrix} 1 & 1 & 1 \\ 5 & 3 & 1 \\ 1 & 0 & -2 \end{bmatrix} \begin{bmatrix} a \\ s \\ c \end{bmatrix} = \begin{bmatrix} 210 \\ 710 \\ 10 \end{bmatrix}$

Chapter 4 Assessment

Sports The high school basketball team has played 5 games this season. Matrix G shows the number of baskets for each game.

	Free throw	Field goals	3-point FGs
Game 1	10	26	4
Game 2	12	31	2
Game 3	9	28	3
Game 4	15	22	0
Game 5	4	35	5

$= G$

Point values

Free throws	1
Field goals	2
3-point FGs	3

$= P$

1. What are the dimensions of matrix G? 5×3
2. Describe the data in location g_{41}. 15 free throws in Game 4
3. Use matrix multiplication to find the total number of points for Game 3. 74

Solve each system of linear equations. Indicate whether each system is inconsistent, dependent, or independent.

4. $\begin{cases} 3x - 6y = 9 \\ \frac{1}{2}x = y + \frac{3}{2} \end{cases}$
infinite sol; dep

5. $\begin{cases} 2x + 3y = 1 \\ -3x + 4y = -10 \end{cases}$
$(2, -1)$; indep

6. $\begin{cases} 0.5x - 0.75y = 0.125 \\ 4x - 6y = 12 \end{cases}$
no solution; incon

Use the following system of linear equations for Items 7–9.

7. Write a matrix equation to represent the system.
8. Write the augmented matrix for the system.
9. Use matrix algebra to solve the system.

$\begin{cases} -2x + y + z = -1 \\ 2x - 2y + z = 4 \\ 2x - 3y - 2z = 1 \end{cases}$

Find the inverse of each matrix, if it exists. If it does not exist, write *does not exist*.

10. $\begin{bmatrix} 4 & 2 \\ -3 & -1 \end{bmatrix}$ $\begin{bmatrix} -0.5 & -1 \\ 1.5 & 2 \end{bmatrix}$

11. $\begin{bmatrix} 2 & 3 & 5 \\ 1 & 2 & -1 \\ 2 & 4 & -2 \end{bmatrix}$ does not exist

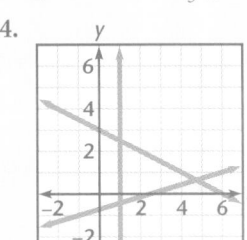 **Transformations** Let $A = \begin{bmatrix} 1 & 5 & 4 \\ 0 & 3 & -2 \end{bmatrix}$. **Give the vertices of the object after the indicated transformations on A.**

12. Rotate 90° counterclockwise, and enlarge 2 times its size.
13. Rotate 180° counterclockwise, and reduce to $\frac{1}{3}$ of its size.

For Items 14–16, use the following system of linear inequalities.

14. Graph the system of inqualities.
15. Find the vertices of the feasible region.
16. Maximize the objective function, $P = 2x + y$, that satisfies the given constraints. 12

$\begin{cases} x + 2y \le 6 \\ x - 3y \ge 3 \\ x \ge 1 \\ y \ge 0 \end{cases}$

7. $\begin{bmatrix} -2 & 1 & 1 \\ 2 & -2 & 1 \\ 2 & -3 & -2 \end{bmatrix} \begin{bmatrix} x \\ y \\ z \end{bmatrix} = \begin{bmatrix} -1 \\ 4 \\ 1 \end{bmatrix}$

8. $\begin{bmatrix} -2 & 1 & 1 & \vdots & -1 \\ 2 & -2 & 1 & \vdots & 4 \\ 2 & -3 & -2 & \vdots & 1 \end{bmatrix}$

9. $(0.8, -0.6, 1.2)$

12. $\begin{bmatrix} 0 & -6 & 4 \\ 2 & 10 & 8 \end{bmatrix}$

13. $\begin{bmatrix} -\frac{1}{3} & -\frac{5}{3} & -\frac{4}{3} \\ 0 & -1 & \frac{2}{3} \end{bmatrix}$

14.

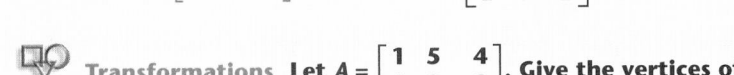

15. $(4.8, 0.6), (3, 0), (6, 0)$

COLLEGE ENTRANCE-EXAM PRACTICE

Multiple-Choice and Quantitative-Comparison Samples

The first half of the Cumulative Assessment contains two types of items found on standardized tests — multiple-choice questions and quantitative-comparison questions. Quantitative-comparison items emphasize the concepts of equality, inequality, and estimation.

Free-Response Grid Samples

The second half of the Cumulative Assessment is a free-response section. A portion of this part of the Cumulative Assessment consists of student-produced response items commonly found on college entrance exams. These questions require the use of machine-scored answer grids. You may wish to have students practice answering these items in preparation for standardized tests.

Sample answer-grid masters are available in the *Chapter Teaching Resources Booklets*.

Chapters 1–4 Cumulative Assessment

College Entrance Exam Practice

Quantitative Comparison For items 1–4 write

A if the quantity in Column A is greater than the quantity in Column B;

B if the quantity in Column B is greater than the quantity in Column A;

C if the two quantities are equal; or

D if the relationship cannot be determined from the given information.

	Column A	Column B	Answers
1. D	x $\quad |2x+1|+2=7$ $\quad y$		Ⓐ Ⓑ Ⓒ Ⓓ [Lesson 3.4]
2. C	$\left[5.001\right]-\left[3.125\right]$	$\left[\sqrt{3}\right]-\left\lfloor\sqrt{3}\right\rfloor$	Ⓐ Ⓑ Ⓒ Ⓓ [Lesson 3.5]
3. B	x, given that $\begin{bmatrix}-1 & 2 & -1\end{bmatrix}\begin{bmatrix}0\\-2\\3\end{bmatrix}=\begin{bmatrix}x\end{bmatrix}$	$\begin{bmatrix}-5 & 0\end{bmatrix}\begin{bmatrix}0\\2\end{bmatrix}=\begin{bmatrix}x\end{bmatrix}$	Ⓐ Ⓑ Ⓒ Ⓓ [Lessons 4.2]
4. B	x $\quad\begin{cases}2x+3y=8\\-3x+y=-1\end{cases}$ $\quad y$		Ⓐ Ⓑ Ⓒ Ⓓ [Lessons 4.4, 4.6]

5. What is the inverse of {(0, 1), (2, − 1), (3, 2), (4, −1)}?
d Is the inverse a function?   **[Lesson 3.2]**
 a. (0, 1), (2, −1), (3, 2), (4, −1), Yes   **b.** (0, 1), (2, −1), (3, 2), (4, −1), No
 c. (1, 0), (−1, 2), (2, 3), (−1, 4), Yes   **d.** (1, 0), (−1, 2), (2, 3), (−1, 4), No

6. What is the inverse of $f(x) = 3x - 5$?   **[Lesson 3.2]**
c **a.** $f^{-1}(x) = \frac{x}{3}+5$ **b.** $f^{-1}(x) = 3x+5$ **c.** $f^{-1}(x) = \frac{x+5}{3}$ **d.** $f^{-1}(x) = \frac{x+3}{5}$

7. What single function represents this system of parametric equations?
b **[Lesson 3.6]**
$$\begin{cases} x(t) = 2t + 1 \\ y(t) = t - 2 \end{cases}$$
 a. $y = x - 5$ **b.** $y = \frac{1}{2}x - \frac{5}{2}$ **c.** $y = x - t - 3$ **d.** $y = \frac{1}{2}x + 3$

8. If $A = \begin{bmatrix} 3 & 0 & -2 \\ -1 & 2 & 1 \\ 1 & 5 & -1 \end{bmatrix}$ and $B = \begin{bmatrix} 0 & -3 & -1 \\ -1 & 4 & 3 \\ 2 & 2 & 2 \end{bmatrix}$, what is $A - B$?   **[Lesson 4.1]**

a.

a. $\begin{bmatrix} 3 & 3 & -1 \\ 0 & -2 & -2 \\ -1 & 3 & -3 \end{bmatrix}$ **b.** $\begin{bmatrix} 3 & -3 & -3 \\ -2 & 6 & 4 \\ 3 & 7 & 1 \end{bmatrix}$ **c.** $\begin{bmatrix} 3 & -3 & -3 \\ 0 & -2 & 4 \\ -1 & 7 & 1 \end{bmatrix}$ **d.** $\begin{bmatrix} 3 & 3 & -1 \\ -2 & -2 & -2 \\ 3 & 3 & -3 \end{bmatrix}$

9 What is the inverse of $A = \begin{bmatrix} 3 & -1 \\ -4 & 2 \end{bmatrix}$?   **[Lesson 4.5]**

b.

a. $\begin{bmatrix} 3 & 1 \\ 4 & 2 \end{bmatrix}$ **b.** $\begin{bmatrix} 1 & 0.5 \\ 2 & 1.5 \end{bmatrix}$ **c.** $\begin{bmatrix} 2 & 1 \\ 4 & 3 \end{bmatrix}$ **d.** $\begin{bmatrix} 1.5 & 0.5 \\ 2 & 1 \end{bmatrix}$

10 Solve this system of equations. $\begin{cases} x + y - 2z = -1 \\ 2x + 2y + z = 3 \\ -3x - 2y - 3z = -4 \end{cases}$ **[Lesson 4.6]**

d.

a. $(0, 1, 1)$ **b.** $(-2, 2, 3)$ **c.** $(1, -2, 1)$ **d.** $(-1, 2, 1)$

11. Explain the relationship between the parametric representation of a function and the function itself.   **[Lesson 3.6]**

12. Graph the solution to the system of linear inequalities.   **[Lesson 4.8]** $\begin{cases} x + 2y \le 4 \\ -2x + 3y \ge 6 \end{cases}$

13. Let $f(x) = 3x - 5$ and $g(x) = x^2 + 1$. Find $(g \circ f)(a + 1)$.   **[Lesson 3.3]**

14. Given the pre-image points $(2, 1)$, $(-4, 2)$, $(3, 1)$, and $(-1, -2)$, determine the coordinates of the image points that are symmetric with respect to the x-axis.   **[Lesson 3.1]** $(2, -1), (-4, -2), (3, -1), (-1, 2)$

15. Let $A = \begin{bmatrix} 6 & 2 & 3 & 4 \\ 3 & -2 & 5 & 5 \end{bmatrix}$. Give the vertices of the transformed object represented by matrix A after it has been rotated $90°$ counterclockwise and reduced to $\frac{1}{2}$ its size.   **[Lesson 4.7]** $\begin{bmatrix} -1.5 & 1 & -2.5 & -2.5 \\ 3 & 1 & 1.5 & 2 \end{bmatrix}$

Free-Response Grid Exercises 16–20 may be answered using a free-response grid commonly used by standardized test services.

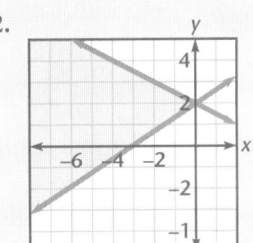

16. Let $f(x) = x + 1$ and $g(x) = x^2 - 1$.
4 Find $(f \circ g)(2)$.   **[Lesson 3.3]**

17. Find the maximum value of the objective
12 function $P = 2x + 3y$ that satisfies the given constraints.   **[Lesson 4.10]**

$$\begin{cases} x + y \le 4 \\ 2x + y \ge 2 \\ x \ge 0 \\ y \ge 0 \end{cases}$$

18. What is the slope of the linear function
9 $f(x) = 9x + 6$?   **[Lesson 2.5]**

For Exercises 19 and 20, use the following information.

Joshua wants to mix two types of candy. Type A costs \$2.50 per pound, type B costs \$4.50 per pound. Ten pounds of the combined candy mixture would cost \$37.00.   **[Lesson 4.3]**

19. How many pounds of Type A candy is in the combined mixture? 4

20. How many pounds of Type B candy is in the combined mixture? 6

11. The parametric representation of a function examines the changes in two values as a function of a third quantity.

12.

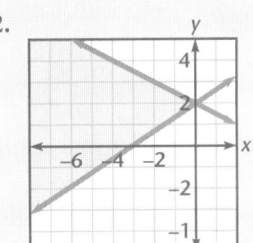

13. $(g \circ f)(a + 1) =$ $9(a + 1)^2 - 30(a + 1) + 26$ or $9a^2 - 12a + 5$

Quadratic Functions

Meeting Individual Needs

5.1 Exploring Products of Two Linear Functions

Core	Core Plus
Inclusion Strategies, p. 256	Practice Master 5.1
Reteaching the Lesson, p. 257	Enrichment, p. 255
Practice Master 5.1	Technology Master 5.1
Enrichment Master 5.1	
Technology Master 5.1	
Lesson Activity Master 5.1	
[2 days]	**[2 days]**

5.2 Solving Quadratic Equations

Core	Core Plus
Inclusion Strategies, p. 264	Practice Master 5.2
Reteaching the Lesson, p. 265	Enrichment, p. 264
Practice Master 5.2	Technology Master 5.2
Enrichment Master 5.2	Interdisciplinary
Technology Master 5.2	Connection p. 263
Lesson Activity Master 5.2	
[2 days]	**[2 days]**

5.3 Graphs of Quadratic Functions

Core	Core Plus
Inclusion Strategies, p. 271	Practice Master 5.3
Reteaching the Lesson, p. 272	Enrichment, p. 271
Practice Master 5.3	Technology Master 5.3
Enrichment Master 5.3	Interdisciplinary
Technology Master 5.3	Connection p. 270
Lesson Activity Master 5.3	
[2 days]	**[2 days]**

5.4 Completing the Square

Core	Core Plus
Inclusion Strategies, p. 278	Practice Master 5.4
Reteaching the Lesson, p. 279	Enrichment, p. 277
Practice Master 5.4	Technology Master 5.4
Enrichment Master 5.4	
Technology Master 5.4	
Lesson Activity Master 5.4	
[2 days]	**[1 day]**

5.5 Quadratic Formula

Core	Core Plus
Inclusion Strategies, p. 284	Practice Master 5.5
Reteaching the Lesson, p. 285	Enrichment, p. 284
Practice Master 5.5	Technology Master 5.5
Enrichment Master 5.5	Interdisciplinary
Technology Master 5.5	Connection, p. 283
Lesson Activity Master 5.5	Mid-Chapter Assessment
Mid-Chapter Assessment Master	Master
[2 days]	**[1 day]**

5.6 The Discriminant and Imaginary Numbers

Core	Core Plus
Inclusion Strategies, p. 290	Practice Master 5.6
Reteaching the Lesson, p. 291	Enrichment, p. 289
Practice Master 5.6	Technology Master 5.6
Enrichment Master 5.6	
Technology Master 5.6	
Lesson Activity Master 5.6	
[2 days]	**[2 days]**

5.7 Complex Numbers

Core

Inclusion Strategies, p. 296
Reteaching the Lesson, p. 297
Practice Master 5.7
Enrichment Master 5.7
Technology Master 5.7
Lesson Activity Master 5.7

[2 days]

Core Plus

Practice Master 5.7
Enrichment, p. 296
Technology Master 5.7
Interdisciplinary
    Connection, p. 295

[1 day]

5.8 Curve Fitting With Quadratic Functions

Core

Inclusion Strategies, p. 304
Reteaching the Lesson, p. 305
Practice Master 5.8
Enrichment Master 5.8
Technology Master 5.8
Lesson Activity Master 5.8

[1 day]

Core Plus

Practice Master 5.8
Enrichment, p. 303
Technology Master 5.8

[1 day]

5.9 Exploring Quadratic Inequalities

Core

Inclusion Strategies, p. 310
Reteaching the Lesson, p. 311
Practice Master 5.9
Enrichment Master 5.9
Technology Master 5.9
Lesson Activity Master 5.9

[2 days]

Core Plus

Practice Master 5.9
Enrichment, p. 309
Technology Master 5.9

[1 day]

Chapter Summary

Core

Chapter 5 Project,
    pp. 316–317
Lab Activity
Long-Term Project
Chapter Review, pp. 318–322
Chapter Assessment, p. 323
Chapter Assessment, A/B
Alternative Assessment

[4 days]

Core Plus

Chapter 5 Project,
    pp. 316–317
Lab Activity
Long-Term Project
Chapter Review,
    pp. 318–322
Chapter Assessment,
    p. 323
Chapter Assessment, A/B
Alternative Assessment

[3 days]

Hands-on Strategies

Provide students with a coordinate plane and several different parabolas in different colors drawn on separate pieces of clear acetate. Demonstrate on the overhead how to place one of the parabolas to show the function $y = x^2$. Then move the parabola to demonstrate a transformation of the function with x-intercepts of 1 and 3. Ask students if they can think of a different way to show a function with these same intercepts. Repeat this activity with different x-intercepts. Then use the other parabolas in similar demonstrations.

Cooperative Learning

GROUP ACTIVITIES	
Exploring quadratic functions	Lesson 5.0, The Portfolio Activity
Exploring properties of parabolas	Lesson 5.1, Explorations 1–4
Exploring the discriminant	Lesson 5.6, Reteaching the Lesson
Finding break-even points	Lesson 5.7, Cooperative Learning
Fitting quadratic curves	Lesson 5.8, Cooperative Learning
Finding path of a basketball	Lesson 5.P, Cooperative Learning

You may wish to have students work with partners for some of the above activities. Additional suggestions for cooperative group activities are noted in the teacher's notes in each lesson.

Multicultural

The cultural connections in this chapter include references to Europe and Africa.

CULTURAL CONNECTIONS	
Europe: "Pythogorean" Right-Triangle Theorem	Lesson 5.2
Europe: Definintion of imaginary numbers	Lesson 5.6
Africa: The al-Khowarizmi Method	Lesson 5.4

Portfolio Assessment

Below are portfolio activities for the chapter. They are listed under seven activity domains that are appropriate for portfolio development.

1. **Investigation/Exploration** One project listed in *Enrichment* on page 284 (Lesson 5.5) of the teacher's notes asks students to investigate a method or a relationship.

2. **Applications** Recommended are any of the following: small business (Lesson 5.9), sports (Lesson 5.9).

3. **Non-Routine Problems** The Portfolio Activity on page 253 asks students to use a graph of a quadratic function to help them estimate answers. Later in the chapter, students are asked to find precise answers and then compare these answers with the estimates.

4. **Project** The Chapter Project on pages 316 and 317 asks students to compare the motion of a basketball on Earth with its motion on 8 other planets. The teacher's notes suggest that students make observations beyond those requested in the project.

5. **Interdisciplinary Topics** Students may choose from physics (Lessons 5.3, 5.8, and 5.9), astronomy (Lessons 5.8 and 5.9), geometry (Lessons 5.1 and 5.5), maximum/minimum (Lessons 5.1, 5.9), coordinate geometry (Lesson 5.2), and statistics (Lesson 5.7).

6. **Writing** The Communicate questions at the beginning of the exercise sets are recommended for journal writing.

7. **Tools** Chapter 5 uses tiles and graphic calculators or computer software, such as *f(g) Scholar*™. The technology masters are recommended to measure individual students' proficiency.

Technology

Graphics calculators and computer software such as *f(g)* *Scholar*™ provide students with an opportunity to study the graphs of quadratic functions without spending the unnecessary time graphing them manually. Students can quickly graph a function and use the trace feature to locate points on the graph. With this capability, students are much more likely to experiment and therefore learn more about the usefulness and properties of the graphs of quadratic functions. For more information on the use of this technology, refer to your *HRW Technology Handbook*.

Graphics Calculator

Finding Points on a Parabola Graphics calculators are used for graphing quadratic functions and locating the coordinates of points on the graph. For instructions for using the calculate or solve feature on specific graphics calculators, please refer to your *HRW Technology Handbook*. Below are the instructions for finding the minimum of a parabola on the TI-82 or the TI-83.

Let the function be $f(x) = 2x^2 + x - 1$. Use the **CALC**, or [2nd] [GRAPH], CALCULATE 3:minimum feature to find the minimum value, or vertex, of this parabola.

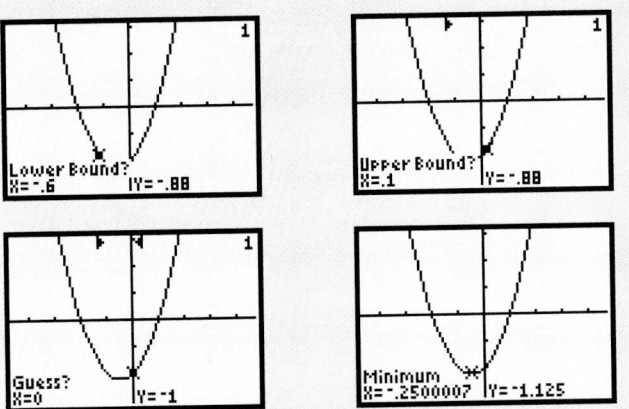

Therefore, the vertex of this quadratic function is $(-0.25, -1.125)$.

Integrated Software

f(g) *Scholar*™ is an integrated computer-based mathematics productivity tool that combines calculator, spreadsheet, and graphics capabilities to provide a dynamic and interactive environment for explorations in mathematics. Use *f(g)* *Scholar*™ for any lesson needing a spreadsheet, calculator, graphics calculator, or any combination of the three.

5 Quadratic Functions

ABOUT THE CHAPTER

Background Information

In this chapter, major emphasis is placed on the nature and number of the zeros of a quadratic function, as well as on the maximum or minimum value. The graph of a function can provide information about all of these points. The videotape of the basketball game provides an ideal visual example of the usefulness of quadratic functions.

CHAPTER RESOURCES

- Practice Masters
- Enrichment Masters
- Technology Masters
- Lesson Activity Masters
- Lab Activity Masters
- Long-Term Project Masters
- Assessments Masters
  Chapter Assessments, A/B
  Mid-Chapter Assessment
  Alternative Assessment, A/B
- Teaching Transparencies
- Spanish Resources

CHAPTER OBJECTIVES

- Write quadratic functions as a product of two linear functions.
- Approximate the minimum or maximum value of a quadratic function and the x-intercepts of its graph.
- Solve quadratic equations.
- Use the distance formula.
- Analyze graphs of quadratic functions.
- Find the vertex, axis of symmetry, and direction of opening for the graphs of quadratic functions in the form $f(x) = a(x - h)^2 + k$.

CHAPTER 5

Quadratic Functions

Quadratic functions can be used to find the area of a region, to find a distance between two locations that cannot be measured directly, or to find the height of a falling object as it descends.

ABOUT THE PHOTOS

The photographs illustrate the importance of the vertex of a parabolic curve. The athlete who has best timed when the ball will reach its highest point has the advantage in the beginning of a basketball game. The video shows, in a frame-by-frame manner, how the vertical path of the basketball is parabolic-shaped, like the graph of a quadratic function.

You are sitting in a basketball arena. The game is about to begin. The two centers are poised, and the referee is about to toss the ball vertically into the air. A video camera follows the motion of the ball as it rises vertically to a maximum height and then begins to fall.

One center, who has timed the jump perfectly, makes contact and taps the ball to a teammate.

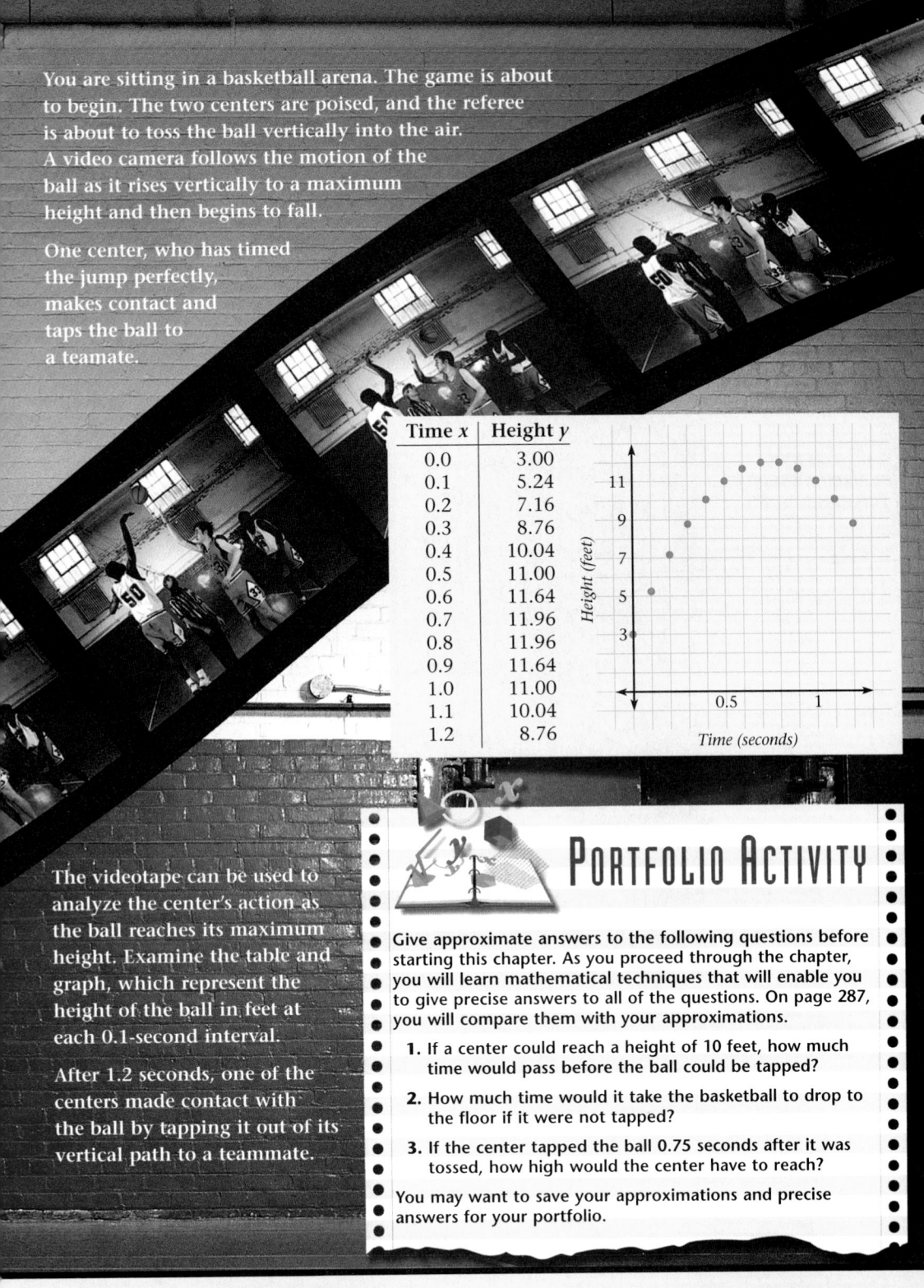

Time x	Height y
0.0	3.00
0.1	5.24
0.2	7.16
0.3	8.76
0.4	10.04
0.5	11.00
0.6	11.64
0.7	11.96
0.8	11.96
0.9	11.64
1.0	11.00
1.1	10.04
1.2	8.76

The videotape can be used to analyze the center's action as the ball reaches its maximum height. Examine the table and graph, which represent the height of the ball in feet at each 0.1-second interval.

After 1.2 seconds, one of the centers made contact with the ball by tapping it out of its vertical path to a teammate.

PORTFOLIO ACTIVITY

Give approximate answers to the following questions before starting this chapter. As you proceed through the chapter, you will learn mathematical techniques that will enable you to give precise answers to all of the questions. On page 287, you will compare them with your approximations.

1. If a center could reach a height of 10 feet, how much time would pass before the ball could be tapped?

2. How much time would it take the basketball to drop to the floor if it were not tapped?

3. If the center tapped the ball 0.75 seconds after it was tossed, how high would the center have to reach?

You may want to save your approximations and precise answers for your portfolio.

ABOUT THE CHAPTER PROJECT

The Chapter 5 Project on pages 316 and 317 extends the idea presented in the Portfolio Activity by including accelerations caused by the gravity on different planets. Students may be intrigued by comparing the path of the ball on different planets with the path of the ball on Earth.

- Use tiles to complete the square for quadratic expressions of the form $x^2 + bx + c$.
- Solve quadratic equations by completing the square for expressions of the form $ax^2 + bx + c$, where $a \neq 0$.
- Use the quadratic formula to solve quadratic equations that model real-world situations.
- Use the axis of symmetry formula to find maximum or minimum values of quadratic equations that model real-world situations.
- Determine the number of real-number solutions using the discriminant.
- Solve quadratic equations with imaginary solutions.
- Identify, operate with, and graph complex numbers.
- Find the complex roots of quadratic equations that model real-world situations.
- Write a quadratic model that fits three data points from real-world data.
- Write, solve, and graph quadratic inequalities that model real-world situations.

PORTFOLIO ACTIVITY

Have students study the data given in the table and the corresponding graph. Ask them to look for patterns. Discuss students' observations.

Have students work in small groups to answer the questions in the portfolio activity. Make certain students understand that the ball must be falling before it can be tapped. Ask each group to justify their answers using a graph or a table.

Have students determine how they could use a videotape to make a table like the one given in the text. If possible, have students actually make a videotape, construct a table, and draw a graph.

Objectives

- Write quadratic functions as a product of two linear functions that model real-world situations.
- Use the graphics calculator to approximate the minimum or maximum value of a quadratic function and the *x*-intercepts of its graph.

RESOURCES

- Practice Master   **5.1**
- Enrichment Master   **5.1**
- Technology Master   **5.1**
- Lesson Activity Master   **5.1**
- Quiz   **5.1**
- Spanish Resources   **5.1**

Assessing Prior Knowledge

Find each product.

1. $(3x - 1)(x - 4)$
   $[3x^2 - 13x + 4]$

2. $-(x + 5)(4x - 3)$
   $[-4x^2 - 17x + 15]$

TEACH

why Point out to students that the measurement "20 yards" represents the sum of the lengths of three sides of the dog pen. The existing fence forms the fourth side of the pen.

LESSON 5.1 Exploring Products of Two Linear Functions

why *The area of a rectangular pen can be modeled by a quadratic function, and a graph of the function can be used to find the maximum area of the pen.*

Centerville has just passed a new leash law. When outside, dogs must be either fenced in or on a leash. Carly has a new dog and a limited budget. She plans to buy 20 yards of fence material.

Existing fence

$w(x) = x$
$l(x) = 20 - 2x$

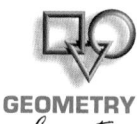
GEOMETRY *Connection*

The table shows how different widths and lengths affect the area when a 3-sided pen is built against an existing fence.

How can you find a function to model the data in the table?

Width (yd)	Length (yd)	Area (yd²)
1	18	18
2	16	32
3	14	42
4	12	48
⋮	⋮	⋮

ALTERNATIVE teaching strategy

Using Visual Models

Write a function such as $y = x(x + 3)$ on the chalkboard. Give small groups of students a piece of clear acetate. Have them draw a coordinate plane on a piece of graph paper. Have students place the piece of acetate on top of the coordinate plane and draw the graph of the function. Ask them to name

1. the *x*-intercepts, [(0, 0); (−3, 0)]
2. the maximum or minimum point, [**minimum at (−1.5, −2.25)**]

3. the values for which the function is increasing, and [*x* > −1.5]
4. the values for which the function is decreasing. [*x* < −1.5]

Then have students reflect the graph over the *x*-axis and write a function to describe the new graph. $[f(x) = -x(x + 3)]$ Have students give the information in 1–4 for this new graph. Repeat this activity for $y = (x + 2)(x + 4)$ and $y = (x - 5)(x + 1)$.

Exploration 1 Area Function

Graphics Calculator

1 Let *w* be the width function. Graph $w(x) = x$. What domain for the linear function *w* is possible in the real world?

2 Let *l* be the length function. Graph $l(x) = 20 - 2x$. What domain for the linear function *l* is possible in the real world?

3 Let *A* be the area function. Then $A = w \cdot l$, or $A(x) = w(x) \cdot l(x)$. Graph *A*. Is *A* a linear function? What domain for *A* is possible in the real world?

4 Use the trace feature to find the highest point on the curve, the *maximum point*. What is the maximum possible area? What width and length will produce the maximum area?

5 Graph all three functions on the same coordinate plane. Use the trace feature to estimate the *x*-intercepts of all the functions.

6 What is the relationship between the *x*-intercepts of the two linear functions and the *x*-intercepts of the area function? ❖

The area function in Exploration 1 is

$$A(x) = x(20 - 2x).$$

The graph of the area function is a parabola. The area function is an example of a *quadratic function*.

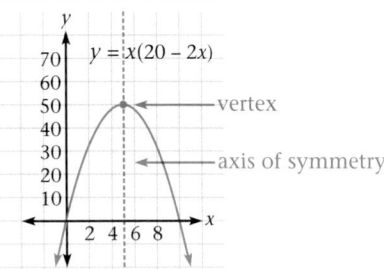

$y = x(20 - 2x)$

vertex

axis of symmetry

QUADRATIC FUNCTION

A **quadratic function** is a function that can be written in the form $f(x) = ax^2 + bx + c$, where *a*, *b*, and *c* are real numbers and $a \neq 0$. The graph of a quadratic function is a **parabola**. The turning point of a parabola is called the **vertex**. The **axis of symmetry of a parabola** goes through the vertex and divides the parabola in half.

APPLICATION

Mitch has 50 yards of fence material to enclose a rectangular area. The table shows possible widths, lengths, and resulting areas.

Width (yd)	Length (yd)	Area (yd²)
1	24	24
2	23	46
3	22	66
⋮	⋮	⋮
x	$25 - x$	$x(25 - x)$

$$\text{Area } (x) = x(25 - x)$$
$$= 25x - x^2$$
$$= -x^2 + 25x + 0$$

The area function is a quadratic function because it can be written in the form $f(x) = ax^2 + bx + c$, where *a* is –1, *b* is 25, and *c* is 0. ❖

Have students discuss how to find profit if the cost and the revenue are known. Also, have students discuss how revenue might be a quadratic function. A window that shows *x*-values from −5 to 15 will show all of the *x*-intercepts. The *y*-values can be set at −50 to 50 so that all functions can be distinguished.

Cooperative Learning

Have students work in small groups to complete this exploration. Have each group summarize their findings and share the information with the class.

TEACHING *tip*

This exploration may be more challenging than Exploration 1. Have students review material from Lessons 4.8 through 4.10 on cost, revenue, and profit before beginning this exploration. Then have students give many examples of possible ticket prices and the resulting attendance for each different ticket price. Continue this process until students notice the relationship between ticket price and attendance. Have students extend this relationship to the revenue at each different ticket price.

•Exploration 2 *Revenue Function*

Graphics Calculator

Use your graphics calculator for Steps 4–6 of this exploration.

The student council plans to run a talent show to raise money. Last year, tickets sold for $5 each and 300 people attended. This year, the student council wants to make an even bigger profit than last year. They estimate that for each dollar increase in ticket price, attendance will drop by 20 people, and for each dollar decrease in ticket price, attendance will increase by 20 people.

Let *x* be the increase or decrease in the ticket price, in dollars.

1 Complete the table below.

2 Write the ticket price, *t*(*x*), as a function of *x*. What type of function is *t*?

3 Write the attendance, *a*(*x*), as a function of *x*. What type of function is *a*?

x	Ticket price ($)	Attendance ($)	Revenue ($)
−2			
−1			
0	5	300	1500
1	6	280	1680
2			
3			
4			
x	5 + *x*		

INCLUSION strategies

Using Visual Models Have students use different colored pencils to draw the different parts of a parabola. For example, have them use a red pencil to show the point(s) where the parabola crosses the *x*-axis, a blue pencil to represent the maximum or minimum point, a green pencil to represent the part of the parabola that is increasing, and a purple pencil to represent the part of the parabola that is decreasing. Have students use these same colors for each of the different graphs that they draw, and have students use the same colors to record the ordered pairs that correspond to each set of points.

4 Write the revenue, $r(x)$, as a function of x. Graph your revenue function, r. What type of function is the revenue function, r? What is the y-intercept of the revenue function? What does the y-intercept represent?

5 Estimate the coordinates of the maximum point on the graph. How much should the student council increase the ticket price to generate the maximum revenue? What should the ticket price be in order to obtain the maximum revenue? What will the maximum revenue be?

MAXIMUM
MINIMUM
Connection

6 Graph the ticket-price function and the attendance function on the same coordinate plane. Find the x-intercept for each function. Graph the revenue function and find its x-intercepts. What is the relationship between the x-intercepts of the price and attendance functions and the x-intercepts of the revenue function?

7 What is the smallest possible realistic x-value in this exploration? What is the largest possible realistic x-value in this exploration? Explain. ❖

Exploration 3 *Increasing and Decreasing*

Graph the revenue function from Exploration 2.

1 Use the trace feature to find the values for which the revenue function is increasing.

*Graphics
Calculator*

2 Use the trace feature to find the values for which the revenue function is decreasing.

3 Where does the function stop increasing and begin to decrease? Where do you think any parabola changes from increasing to decreasing or from decreasing to increasing? ❖

Exploration 4 *Vertex and x-Intercepts*

1 Graph each function listed in the following table. For each function, find approximate answers, and complete the table.

	x-intercepts	Midpoint of x-coordinates of x-intercepts	Vertex	x-coordinate of vertex
$y = (x - 4)(x + 2)$				
$y = (x + 1)(x + 6)$				
$y = (x - 3)(x - 5)$				

2 Compare the midpoint of the x-coordinates of the x-intercepts with the x-coordinate of the vertex. How are they alike or different? What does this tell you about the vertex location in relation to the x-intercepts?

3 Use the patterns you see in the table to predict the x-intercepts and the x-coordinate of the vertex of $y = (x + 1)(x - 9)$. Use your graphics calculator to check your prediction. Were you correct? Explain. ❖

Ongoing ASSESSMENT

6. The x-intercepts of the price function and the attendance function are x-intercepts of the revenue function, -5 and 15.

Exploration 3 Notes

Use student answers to Step 3 to review the meaning of the term *vertex*.

Ongoing ASSESSMENT

3. Any parabola changes from increasing to decreasing or from decreasing to increasing at its vertex.

Exploration 4 Notes

Have students compare the results of this exploration with the results of Exploration 3.

Ongoing ASSESSMENT

2. The vertex is located on a vertical line that passes midway between the x-intercepts.

Use Transparency ▶ 18

TEACHING *tip*

Discuss how the phrase "zeros of a function" relates to the fact that the product of the factors of a quadratic equation is zero when the function crosses the *x*-axis. Understanding this relationship may help students to remember how to solve quadratic equations using factoring or graphing.

Aongoing SSESSMENT

The maximum revenue occurs at the vertex of the graph, when *x* is 5.

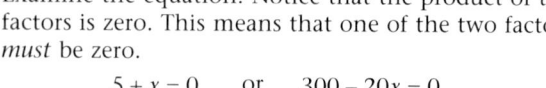

EXTENSION

Refer to the revenue function from Exploration 2.

$$r(x) = (5 + x)(300 - 20x)$$

When the revenue is zero, then $0 = (5 + x)(300 - 20x)$. How can you solve this equation?

Examine the equation. Notice that the product of the factors is zero. This means that one of the two factors *must* be zero.

$$5 + x = 0 \quad \text{or} \quad 300 - 20x = 0$$
$$-20x = -300$$
$$x = -5 \quad \text{or} \quad x = 15$$

The **zeros of a function**, f, are the values of x that make $f(x)$ equal to zero. The **roots of an equation** are the x-values corresponding to $y = 0$. Both the zeros of a function and the roots of an equation can be used to find the x-intercepts of the graph.

The zeros of the revenue function, $r(x) = (5 + x)(300 - 20x)$, are **−5** and **15**.

Thus, the x-intercepts occur when x is **−5** or when x is **15**.

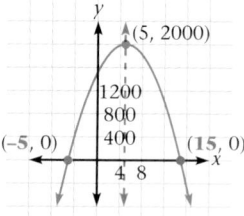

Where on the graph does the maximum revenue occur?

The x-coordinate of the vertex, x_v, is the midpoint of the two zeros, **−5** and **15**.

$$x_v = \frac{-5 + 15}{2} = 5$$

Substitute 5 for x in the revenue function to find y_v, the y-coordinate of the vertex.

$$y_v = (5 + x)(300 - 20x)$$
$$= [5 + (5)][300 - 20(5)]$$
$$= 2000$$

The vertex is (5, 2000). Thus, the maximum revenue from ticket sales, $2000, occurs when there is a $5 increase in ticket price. ❖

To find the x-intercepts for a quadratic function that is written as a product of linear functions, use the *zero product property*.

ZERO PRODUCT PROPERTY

If $ab = 0$, then either $a = 0$ or $b = 0$ must be true.

EXERCISES & PROBLEMS

Communicate

1. Explain how to find the maximum area of a rectangular pen enclosed by 60 yards of fencing.

2. How can you determine whether a function is quadratic without graphing it?

Small Business Julia sells tamales for $5.00 per dozen. At this price, she can sell about 30 dozen tamales per day. Julia wants to maximize the revenue from tamale sales. She estimates that each 50¢ increase in price will result in about 2 dozen less tamales sold, and each 50¢ decrease in price will result in about 2 dozen more tamales sold. **Complete Exercises 3 and 4.**

3. Explain how to develop a quadratic function to represent the revenue from selling tamales. What will the linear factors be?

4. How can you find the price Julia should charge for the tamales to maximize revenue?

5. What does it mean when a function is increasing or decreasing?

6. Explain how to find the x-intercepts and the vertex of a quadratic function.

ASSESS

Selected Answers
Odd-numbered problems 7–55

Assignment Guide
Core 1–11, 13–33, 36–57

Core Plus 1–6, 9–26, 30–57

Technology
A graphics calculator is needed for Exercises 18–23, 27–35, 39–40, 43–45, and 57. Computer software such as *f(g) Scholar*™ could also be used for these exercises.

Practice & Apply

Show that each function is a quadratic function by writing it in the form $f(x) = ax^2 + bx + c$ and identifying a, b, and c.

7. $f(x) = -(x - 2)(x + 6)$

8. $f(x) = (2x + 5)(3x + 1)$

9. $f(x) = 3(x - 2)(3x - 4)$

10. $f(x) = (4 - x)(7 + x)$

11. $f(x) = (x - 3)(x + 8)$

12. $f(x) = 2(x + 4)(x + 6)$

Maximum/Minimum For Exercises 13–16, find the maximum area of a rectangular pen that you can build (all four sides) with the indicated amount of fencing material.

13. 80 feet 400 sq ft

14. 100 feet 625 sq ft

15. 60 feet 225 sq ft

16. n feet $\frac{n^2}{16}$ sq ft

17. Look at your answers to Exercises 13–16. What do they all have in common? How does this compare with constructing a rectangular pen against a building, where fencing material is needed for only three sides?

7. $f(x) = -x^2 - 4x + 12; a = -1, b = -4, c = 12$

8. $f(x) = 6x^2 + 17x + 5; a = 6, b = 17, c = 5$

9. $f(x) = 9x^2 - 30x + 24;$ $a = 9,$ $b = -30,$ $c = 24$

10. $f(x) = -x^2 - 3x + 28; a = -1, b = -3, c = 28$

11. $f(x) = x^2 + 5x - 24; a = 1, b = 5, c = -24$

12. $f(x) = 2x^2 + 20x + 48; a = 2, b = 20, c = 48$

17. In Exercises 13–16, all the answers are perfect squares. So, with a four-sided pen, the maximum area occurs when the pen is in the shape of a square. For a rectangle with only 3 sides fenced, the maximum area occurs when the length is twice the width.

Error Analysis

Be sure students understand the meaning of the terms *maximum* and *minimum*. Discuss the use of these terms in everyday life. Then relate the everyday uses to the meaning of the terms when used to describe parabolas.

18. The graphs all have the same x-intercepts, $(-2, 0)$ and $(4, 0)$. All the functions are of the form $f(x) = a(x + 2)(x - 4)$, where a is not equal to 0. This is why all the functions have the same zeros. Since all the functions have the same x-intercepts, they all have the same x-coordinate for the vertex, $\frac{-2 + 4}{2} = 1$.

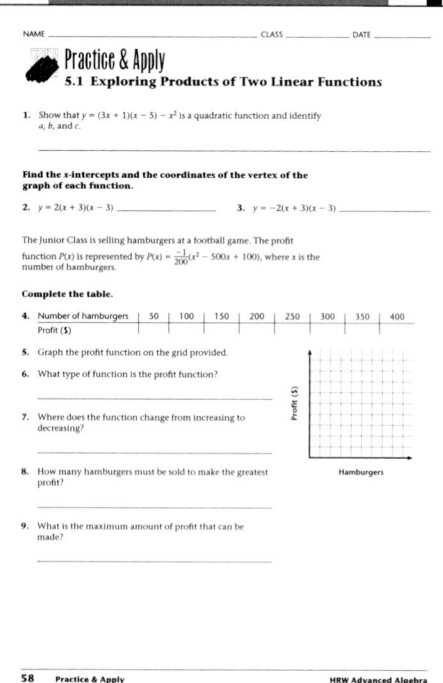

Transformations Use your graphics calculator to graph each function and answer Exercises 18–20.

$f_1(x) = (x + 2)(x - 4)$ 　 $f_2(x) = 2(x + 2)(x - 4)$ 　 $f_3(x) = \frac{1}{2}(x + 2)(x - 4)$

$f_4(x) = -(x + 2)(x - 4)$ 　 $f_5(x) = -2(x + 2)(x - 4)$ 　 $f_6(x) = -\frac{1}{2}(x + 2)(x - 4)$

18 What do all of the graphs have in common? Why?　x-intercepts $(-2, 0)$ and $(4, 0)$

19 Which graphs have a maximum point?　f_4; f_5; f_6

20 Which graphs have a minimum point?　f_1; f_2; f_3

For Exercises 21–23, let $f(x) = (2x + 5)(3x + 1)$.

21 Approximate the x-intercepts.　$(-2.5, 0)$; $(-0.3, 0)$

22 Approximate the coordinates of the vertex. Is the vertex a minimum or a maximum point?　$(-1.4, -7)$; minimum

23 Approximate the range of values of x for which f is increasing and for which f is decreasing.　increasing for $x > -1.4$; decreasing for $x < -1.4$

For Exercises 24–26, let $f(x) = (4 - x)(7 + x)$.

24. Find the zeros of f. What are the x-intercepts of the graph of f?　-7, 4; $(-7, 0)$, $(4, 0)$

25. Find the vertex of the graph. Is it a minimum or a maximum point?　$(-1.5, 30.25)$; max

26. What are the values of x for which f is increasing and for which f is decreasing?　increasing for $x < -1.5$; decreasing for $x > -1.5$

For the graph of each function, find the x-intercepts and coordinates of the vertex. Check using your graphics calculator.

27 $f(x) = (5x - 3)(2x - 1)$ 　　　**28** $f(x) = (x + 4)(x - 2)$

29 $f(x) = -(x - 2)(x + 6)$ 　　　**30** $f(x) = 3(x + 8)(-x + 9)$

31 $f(x) = -(4x + 1)(x + 4)$ 　　　**32** $f(x) = (x + 32)(x - 4)$

33 $f(x) = 2(x + 4)(x + 6)$ 　　　**34** $f(x) = (3x + 6)(5x - 1)$

35 $f(x) = (3x + 4)(2x - 5)$

Small Business A jazz club charges a $5.00 cover charge when the regular local musicians play. The manager estimates that each 50¢ increase in cover charge will decrease attendance by 10 people, and each 50¢ decrease in cover charge will increase attendance by 10 people. Currently, the average attendance is about 200.
Complete Exercises 36 and 37.

36. What should the cover charge be to maximize revenue?　$7.50

37. What will the maximum revenue be?　$1125

38. Explain why 3 must be a zero of $f(x) = (x - 3)(x + 1995)$.

39 Create a quadratic function that has x-intercepts of $(5, 0)$ and $(1, 0)$. Check your answer using a graphics calculator.

40 Create a quadratic function that has -2 and 3 as zeros. Check your answer using a graphics calculator.

---

27. $(0.6, 0)$ and $(0.5, 0)$; $(0.55, -0.025)$

28. $(-4, 0)$ and $(2, 0)$; $(-1, -9)$

29. $(2, 0)$ and $(-6, 0)$; $(-2, 16)$

30. $(-8, 0)$ and $(9, 0)$; $(0.5, 216.75)$

31. $(-0.25, 0)$ and $(-4, 0)$; $(-2.125, 14.0625)$

32. $(-32, 0)$ and $(4, 0)$; $(-14, -324)$

33. $(-6, 0)$ and $(-4, 0)$; $(-5, -2)$

34. $(-2, 0)$ and $(0.2, 0)$; $(-0.9, -18.15)$

35. $\approx (-1.33, 0)$ and $(2.5, 0)$; $\approx (0.58, -22.04)$

38. When x is 3, the function value, $f(3)$, is 0. 3 will always be a zero of the f when $(x - 3)$ is a factor of f.

39. Answers may vary but must be similar to $f(x) = a(x - 5)(x - 1)$ for any nonzero a.

40. Answers may vary but must be similar to $f(x) = a(x + 2)(x - 3)$ for any nonzero a.

Small Business Roberto and Brian decide to start a small business making carved animals. They know that the selling price will affect the number of animals that they can sell. They decide to use a mathematical model to relate the selling price, x, to the number, n, of items sold, $n(x) = -10x + 200$.

41. Complete the following table.

Selling price	Number of items sold	Income
1	190	190
2	180	360
3	170	510
x	$-10x + 200$	$-10x^2 + 200x$

42. Write income as a function of x, the selling price. $I(x) = -10x^2 + 200x$

43 Graph the income function. Estimate the selling price that will produce the maximum income. $10

44 Estimate the number of items that must be sold to produce the maximum income. 100 items

45 Estimate the maximum income. $1000

Look Back

Find the equation of the line passing through the given points. [Lesson 1.2] $y = -\frac{15}{19}x + \frac{92}{19}$

46. (7, –4) and (12, –5) $y = -\frac{1}{5}x - \frac{13}{5}$ **47.** (–4, 8) and (15, –7)

For Exercises 48–52, let $f(x) = -4x + 11$. [Lessons 1.2, 2.5]

48. Graph the function.

49. What type of function is this? linear

50. What is the x-intercept? $\left(\frac{11}{4}, 0\right)$ **51.** What is the y-intercept? (0, 11)

52. Is the function increasing or decreasing? decreasing

Let $f(x) = 3x + 2$ and $g(x) = 5x$. Find the indicated functions. [Lesson 3.3]

53. $(f \circ g)(x)$
$15x + 2$

54. $(g \circ f)(x)$
$15x + 10$

55. $(f \circ f)(x)$
$9x + 8$

56. $(g \circ g)(x)$
$25x$

Look Beyond

57 Use your graphics calculator to graph $y = x^2$. What range did you use to view the graph? Use the trace feature to approximate any x-intercepts of the graph. x-intercept (0, 0)

48. $f(x) = -4x + 11$

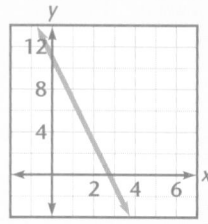

57. Answers may vary. Any range that includes the point (0, 0) and the positive y-axis will work. There is only one x-intercept, (0, 0).

Objectives

* Solve quadratic equations by taking square roots, or by squaring.
* Use the distance formula to find the distance between two points.

RESOURCES

* Practice Master   **5.2**
* Enrichment Master   **5.2**
* Technology Master   **5.2**
* Lesson Activity Master   **5.2**
* Quiz   **5.2**
* Spanish Resources   **5.2**

Assessing Prior Knowledge

Simplify.

1. $\sqrt{81}$ **[9]**

2. $-\sqrt{144}$ **[−12]**

3. $\sqrt{99}$ **[$3\sqrt{11}$]**

4. $-\sqrt{68}$ **[$-2\sqrt{17}$]**

TEACH

why Discuss several situations in which it is necessary to find a square root. For example, if you want to fence your yard and you know the square footage of the property, then you must solve an equation involving square roots to find the length of a side.

ongoing ASSESSMENT

The square root of 9 is 3.

LESSON 5.2 Solving Quadratic Equations

why *There are many methods for obtaining useful information from quadratic equations. However, in order to use any of these methods, you first need to recall the properties of square roots.*

Taking the square root of a number is the inverse of squaring a number.

What is the square root of 9, written $\sqrt{9}$?

Square Roots

There are two numbers that can be squared to get 9.

$$3^2 = 9 \quad \text{and} \quad (-3)^2 = 9$$

Does $\sqrt{9} = 3$ or $\sqrt{9} = -3$? To avoid confusion, 3 is called the **principal square root** and is indicated by the square root sign. When there is no minus sign in front of the square root sign, the principal root is implied.

$x^2 = 9$ has two solutions, 3 and –3. However, $\sqrt{9} = 3$ and $-\sqrt{9} = -3$.

ALTERNATIVE teaching strategy

Technology Introduce students to square roots by using a calculator. Students will find that all of the square roots given by the calculator are positive. They may also discover that the square root of a negative number is undefined.

Discuss different methods for solving a quadratic equation. Then, discuss the "Pythagorean"

Right-Triangle Theorem. Have students solve the equation $c^2 = a^2 + b^2$ for each of the variables. [$c = \pm\sqrt{a^2 + b^2}$; $a = \pm\sqrt{c^2 - b^2}$; $b = \pm\sqrt{c^2 - a^2}$]. Students can use the appropriate equation and their calculator or computer software, such as *f(g) Scholar*[TM], to solve problems using the "Pythagorean" Right-Triangle Theorem.

SQUARE ROOTS

Let a and b represent real numbers, with $b \geq 0$.
If $a^2 = b$, then $a = \sqrt{b}$ or $a = -\sqrt{b}$, also written $a = \pm\sqrt{b}$.

EXAMPLE 1

Solve $3(x^2 - 28) = 108$.

Solution➤

Since there is no x-term, isolate x^2 on one side. Then use the definition of square root.

$$3(x^2 - 28) = 108$$
$$x^2 - 28 = 36$$

Include both positive and negative
square roots.
$$x^2 = 64$$
$$x = \pm\sqrt{64} = \pm 8$$

Thus, the solutions to $3(x^2 - 28) = 108$ are 8 and –8. ❖

Try This Solve $3(x^2 + 15) = 120$.

EXAMPLE 2

Solve $4x^2 + 13 = 253$ for x.

Solution➤

Since there is no x-term, isolate x^2 on one side. Then use the definition of square root.

$$4x^2 + 13 = 253$$
$$4x^2 = 240$$
$$x^2 = 60$$
$$x = \pm\sqrt{60}$$

Thus, the solutions are $\sqrt{60}$ and $-\sqrt{60}$. ❖

When you have a number that is not a perfect square under the square root sign, you can sometimes simplify the number by factoring.

$$\sqrt{60} = \sqrt{4 \cdot 15}$$
$$= \sqrt{4} \cdot \sqrt{15}$$
$$= 2\sqrt{15}$$

The number $2\sqrt{15}$ is called the **simple radical form** of $\sqrt{60}$.
Since it takes fewer keystrokes to enter $\sqrt{60}$ than $2\sqrt{15}$ into a calculator for an approximate value, the simple radical form will not be used here.

Cooperative Learning

Have students work in small groups and do research to learn more about the history of the "Pythagorean" Right-Triangle Theorem. Have students find out the most recent information about the Babylonian tablet. Ask students to summarize their findings in a report.

Alternate Example 1

Solve $2(x^2 - 10) = 180$ [**±10**]

Aongoing **SSESSMENT**

Try This

±5

Alternate Example 2

Solve $3x^2 - 251 = 1$. [**$\pm\sqrt{84}$**]

TEACHING *tip*◆

Encourage students to check their answers in the original equation.

interdisciplinary
CONNECTION

Physics Have students use the "Pythagorean" Right-Triangle Theorem to solve the following problem.

An airplane is flying due west at 400 miles per hour. The wind is blowing due south at 30 miles per hour. At what speed is the plane actually flying?

$$\left[\sqrt{400^2 + 30^2} \approx 401.12 \text{ mph} \right]$$

Path of plane without wind, 400 mph

Wind,
30 mph

Resulting path of plane with wind

Alternate Example 3

Solve $(x + 7)^2 = 25$.

[**−12 and −2**]

Alternate Example 4

Solve $\sqrt{2x + 1} = -7$.

[**no solution**]

CRITICAL
Thinking

By definition, the principal square root of any positive real number is positive. Therefore, $\sqrt{x - 3}$ can never equal −5, so the equation has no solution.

Aongoing
SSESSMENT

The squares of a and $-a$, for any real number a, are equal to one another and are both a positive number. Therefore, squaring both sides of an equation can result in two different numbers appearing to be equal. By checking your answers in the original problem, you will discover if you have an unreasonable (or extraneous) result.

EXAMPLE 3

Solve $(x - 2)^2 = 64$.

Solution

Use the definition of square root and simplify.

$$(x - 2)^2 = 64$$
$$x - 2 = \pm 8$$
$$x = \pm 8 + 2$$
$$x = 8 + 2 \text{ or } x = -8 + 2$$

Thus, the solutions are 10 and −6. ❖

When you square both sides of an equation, you may obtain **extraneous roots**, or false roots. For example, −5 = 5 is false, but $(-5)^2 = 5^2$ is true. Therefore, when you *square* both sides of an equation, you *must* check your solution in the original equation.

EXAMPLE 4

Solve $\sqrt{x - 3} = -5$.

Solution

Square both sides and solve for x.

$$\sqrt{x - 3} = -5$$
$$x - 3 = (-5)^2$$
$$x - 3 = 25$$
$$x = 28$$

Check

Substitute 28 for x in the original equation and evaluate.

$$\sqrt{x - 3} = -5$$
$$\sqrt{28 - 3} \overset{?}{=} -5$$
$$\sqrt{25} \overset{?}{=} -5$$
$$5 \overset{?}{=} -5 \qquad \text{False}$$

The principal square root of 25 is 5, *not* −5. Therefore, $\sqrt{(x - 3)} = -5$ has no solution. ❖

CRITICAL
Thinking

How can you tell that the equation in Example 4 has no solution without solving the equation?

Why is it necessary to check your solution in the original equation when you square both sides of an equation?

ENRICHMENT Have students do research to find out about the history of the square root sign. Ask them to do the research individually and then work in pairs or small groups to write a report about their findings. Ask each group to share their report with the class.

INCLUSION **Using Algorithms** Students should include as many
strategies steps as necessary when solving equations. For example,

$$(x - 2)^2 = 64$$
$$\sqrt{(x - 2)^2} = \sqrt{64}$$
$$x - 2 = \pm 8$$
$$x - 2 + 2 = \pm 8 + 2$$
$$x = \pm 8 + 2$$
$$x = 8 + 2 = 10 \text{ or } x = -8 + 2 = -6$$

"Pythagorean" Right-Triangle Theorem

Cultural Connection: Asia Sometime between 1900 B.C.E. and 1600 B.C.E. in ancient Babylon (now Iraq), a table of numbers was inscribed in a clay tablet. When archaeologists discovered the tablet, now kept at Columbia University, part of the tablet was missing, so the information on the tablet remained a mystery. However, most historians now believe that the numbers in the table form sets of triples that form a special right-triangle relationship.

This relationship is commonly referred to as the *"Pythagorean" Right-Triangle Theorem.*

The Greek Mathematician Pythagorus, 585 B.C.E.–500 B.C.E.

Alternate Example 5

Find the length of the hypotenuse.

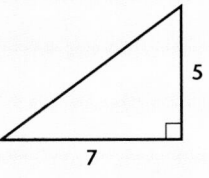

$[\sqrt{74}]$

A**ongoing ASSESSMENT**

Given leg a and hypotenuse c, rearranging the equation of the "Pythagorean" Right-Triangle Theorem gives $b^2 = c^2 - a^2$. So you can find leg b by solving $b = \sqrt{c^2 - a^2}$.

"PYTHAGOREAN" RIGHT-TRIANGLE THEOREM

The "Pythagorean" Right-Triangle Theorem states that in a right triangle, the length of the hypotenuse, c, is related to the lengths of the legs, a and b, by the equation $c^2 = a^2 + b^2$.

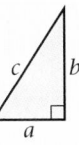

By using the definition of square root, you find $c = \pm\sqrt{a^2 + b^2}$. Since length is always positive, $c = \sqrt{a^2 + b^2}$.

EXAMPLE 5

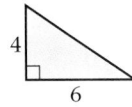

Find the length of the hypotenuse.

Solution▸

Substitute 4 and 6 in the "Pythagorean" Right-Triangle Theorem formula.

$$c^2 = a^2 + b^2$$
$$c^2 = 4^2 + 6^2$$
$$c = \sqrt{4^2 + 6^2}$$
$$c = \sqrt{52}$$

Thus, the length of c is $\sqrt{52}$, or approximately 7.2. ❖

How do you find the length of one leg of a right triangle when you know the length of the hypotenuse and the length of the other leg?

RETEACHING **the lesson**

Using Cognitive Strategies Have students work in pairs or small groups to use square roots to solve quadratic equations of increasing difficulty. Examples of the types of equations you can use are given.

1. $x^2 = 36$ $[6, -6]$

2. $2x^2 = 36$ $[\sqrt{18}, -\sqrt{18}]$

3. $(x + 1)^2 = 36$ $[5, -7]$

4. $3(x + 1)^2 = 36$ $[-1 + \sqrt{12}, -1 - \sqrt{12}]$

5. $3(x + 1)^2 + 9 = 36$ $[2, -4]$

Then have students use a similar procedure to solve radical equations.

Students can find the lengths of the legs of a right triangle by counting squares on a grid. In order to help students better understand the derivation of the distance formula, redraw the graph of the triangle and replace the coordinates given with (x_1, y_1) and (x_2, y_2). Then show students how to replace a, b, and c in the "Pythagorean" Right-Triangle Theorem with the appropriate parts of the distance theorem (a is the distance $x_2 - x_1$, b is the distance $y_2 - y_1$, and c is the distance along the hypotenuse).

Alternate Example 6

Find the distance between points $(-3, 5)$ and $(-5, 9)$. $[\sqrt{20}$, or $\approx 4.5]$

CRITICAL
Thinking

No. The coordinates may be substituted in either order. When the difference between the x- or y-values is squared, the result will be the same no matter what order you use because
$$(x_1 - x_2)^2 = (-(x_2 - x_1))^2$$
$$= (-1)^2 (x_2 - x_1)^2$$
$$= (x_2 - x_1)^2$$

Distance Formula

COORDINATE GEOMETRY
Connection

You can use the "Pythagorean" Right-Triangle Theorem to find the distance between two points on a coordinate plane.

Suppose you have two points, $A(5, 4)$ and $B(-1, -4)$, on a coordinate plane.

You can see that the horizontal distance, a, and the vertical distance, b, are the legs of a right triangle.

$a = 5 - (-1) = 6$ and $b = 4 - (-4) = 8$

The distance between the two points is the length of the hypotenuse of the right triangle.

Substitute 6 for a and 8 for b in the "Pythagorean" Right-Triangle Theorem formula.

$$c = \sqrt{a^2 + b^2}$$
$$= \sqrt{6^2 + 8^2}$$
$$= \sqrt{100}$$
$$= 10$$

The distance between points A and B is 10. Using this method to find the distance between two points, (x_1, y_1) and (x_2, y_2), you can develop the *distance formula*.

DISTANCE FORMULA

The distance, d, between the points (x_1, y_1) and (x_2, y_2) is
$$d = \sqrt{(x_2 - x_1)^2 + (y_2 - y_1)^2}.$$

EXAMPLE 6

Find the distance between points $(-3, 6)$ and $(1, -2)$.

Solution

Let $(-3, 6)$ be (x_1, y_1) and let $(1, -2)$ be (x_2, y_2). Then substitute the x- and y-coordinates into the distance formula.

$$d = \sqrt{(x_2 - x_1)^2 + (y_2 - y_1)^2}$$
$$d = \sqrt{[1 - (-3)]^2 + (-2 - 6)^2}$$
$$d = \sqrt{16 + 64}$$
$$d = \sqrt{80}$$

Thus, the distance between points $(-3, 6)$ and $(1, -2)$ is $\sqrt{80}$, or approximately 8.9. ❖

CRITICAL
Thinking

In the distance formula, does the order in which you substitute the coordinates of the points, (x_1, y_1) and (x_2, y_2), make a difference? Explain.

EXERCISES & PROBLEMS

This display below illustrates the "Pythagorean" Right-Triangle Theorem. Colored water is contained in squares attached to the sides of a right triangle. As it rotates, the water flows from the two squares opposite the hypotenuse, completely filling the square on the hypotenuse.

Communicate

1. Explain why you must include both positive and negative square roots when using the definition of square root to solve an equation.

2. When is it possible to obtain extraneous roots when solving an equation?

3. How can you tell which roots are extraneous?

4. Under what circumstances would you retain only the principal root?

5. How can you find the length of the hypotenuse of a right triangle with legs of lengths 3 and 4?

6. Explain how to find the distance between two points on the coordinate plane.

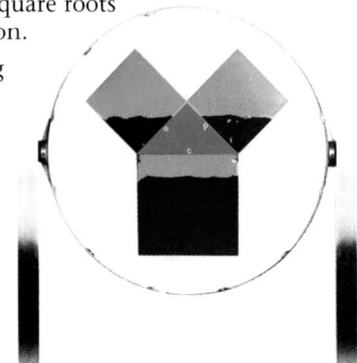

Practice & Apply

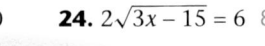

Solve and check.

7. $x^2 = 121$ ± 11

8. $x^2 = 32$ $\pm\sqrt{32}$

9. $3x^2 = 49$ $\pm\dfrac{7}{\sqrt{3}}$

10. $\frac{1}{2}x^2 = 6$ $\pm\sqrt{12}$

11. $4x^2 - 5 = 6$ $\pm\dfrac{\sqrt{11}}{2}$

12. $\frac{1}{3}(x^2 - 15) = 37$ $\pm\sqrt{126}$

13. $\frac{1}{2}(x^2 + 6) - 5 = 10$ $\pm\sqrt{24}$

14. $2(x^2 - 4) + 3 = 15$ $\pm\sqrt{10}$

15. $4(x^2 + 7) - 9 = 39$ $\pm\sqrt{5}$

16. $\sqrt{x} = -2$ no solution

17. $\sqrt{x + 3} = 7$ 46

18. $\sqrt{4x + 4} = 12$ 35

19. $-\sqrt{x} = 3$ no solution

20. $-7 + \sqrt{2x - 3} = -4$ 6

21. $15 = 3\sqrt{x}$ 25

22. $\sqrt{3x} + 4 = 5$ $\frac{1}{3}$

23. $3\sqrt{2x + 1} + 5 = 0$ no solution

24. $2\sqrt{3x - 15} = 6$ 8

Ecology The erosive force of moving water is partly determined by its velocity. The velocity, *v* (in feet per second), can be measured with a hollow L-shaped tube.

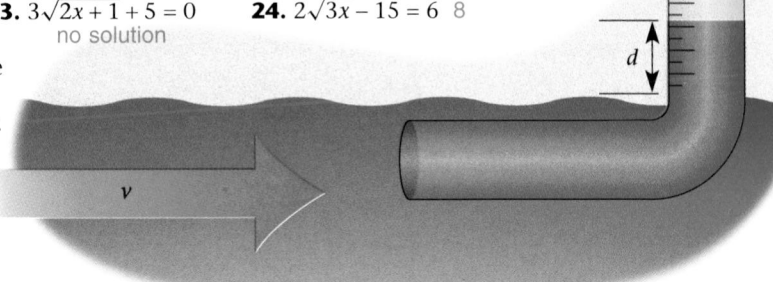

25. Approximately how fast is a stream flowing if *d* = 1.5 feet?   9.80 ft/sec

26. Approximately how fast is a stream flowing if *d* = 0.5 feet?   5.66 ft/sec

Physicists have determined that $v^2 = 64d$.

Find the missing side length of right triangle *ABC*.

27. *a* is 8 and *b* is 4. $\sqrt{80}$

28. *a* is 9 and *b* is 2. $\sqrt{85}$

29. *c* is 5 and *b* is 3.   4

30. *c* is $\sqrt{29}$ and *b* is 5.   2

31. *a* is 9 and *c* is $\sqrt{90}$.   3

32. *a* is 7 and *c* is $\sqrt{74}$.   5

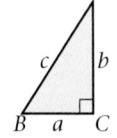

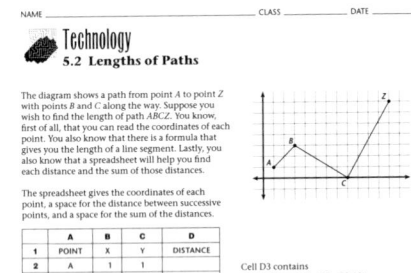

Find the distance between the points with the given coordinates.

33. $(1, 2)$ and $(4, 3)$ $\sqrt{10}$ 34. $(2, -6)$ and $(7, -5)$ $\sqrt{26}$ 35. $(4, -3)$ and $(-4, 3)$ 10

36. $(4, -3)$ and $(-1, -3)$ 5   37. $(7, 1)$ and $(6, 2)$ $\sqrt{2}$ 38. $(-2, -5)$ and $(-2, 4)$ 9

Coordinate Geometry Find the perimeter of each triangle with vertices at the given coordinates.

39. $(0, 8)$, $(-6, 0)$, and $(15, 0)$ 48   40. $(-3, -1)$, $(1, 2)$, and $(1, -1)$ 12

41. **Coordinate Geometry** Use the distance formula to show that $R(-4, 1)$, $S(5, 4)$, and $T(2, -2)$ are the coordinates of the vertices of an isosceles triangle.

42. **Coordinate Geometry** Use the distance formula and the "Pythagorean" Right-Triangle Theorem to show that $A(-4, 1)$, $B(5, 4)$, and $C(2, -2)$ are the coordinates of the vertices of a right triangle.

43. **Coordinate Geometry** Find the length of the radius of a circle that has a center $O(3, 4)$ and passes through the point $P(2, -5)$. $\sqrt{82}$

Look Back

44. Make a table of values that are linearly related.   **[Lesson 1.1]**

For Exercises 45–47, find the slope of the line passing through the indicated points, and write the equation of the line in slope-intercept form.   **[Lesson 1.2]**

45. $(-4, -2)$ and $(6, -7)$ 46. $(3, 5)$ and $(-6, 1)$ 47. $(-5, 4)$ and $(1, -7)$

Simplify.   **[Lesson 2.2]**

48. $4^5 \cdot 4^{-3}$ 16   49. $(5^4)^{-\frac{1}{2}}$ $\frac{1}{25}$ 50. $\left(\frac{x^3 y^{-2}}{x^2 y^5}\right)^{-3}$ $\frac{y^{21}}{x^3}$

Give the domain and range of each function.   **[Lesson 2.3]**

51. $f(x) = 3x^2 - 7$ R; $y \geq -7$ 52. $f(x) = \frac{6}{x}$ $x \neq 0$, $y \neq 0$ 53. $f(x) = \frac{2x^2}{(x-3)^2}$ $x \neq 3$, $y \geq 0$

54. What is an axis of symmetry?   **[Lesson 3.1]**

Look Beyond

55. Graph $f(x) = 2(x - 3)^2 + 5$. Use the trace feature to find the vertex and x-intercepts of the parabola. vertex $(3, 5)$; no x-intercepts

56. Graph $f(x) = -4(x + 5)^2 - 2$. Use the trace feature to find the vertex and x-intercepts of the parabola. vertex $(-5, -2)$; no x-intercepts

57. How are the two parabolas in Exercises 55 and 56 alike, and how are they different? Compare the similarities and differences of the two graphs with the similarities and differences of their equations. Can you see any patterns?

The answers to Exercises 41 and 42 can be found in Additional Answers beginning on page 842.

44. Answers may vary.

45. $-\frac{1}{2}$; $y = -\frac{1}{2}x - 4$

46. $\frac{4}{9}$; $y = \frac{4}{9}x + \frac{11}{3}$

47. $-\frac{11}{6}$; $y = -\frac{11}{6}x - \frac{31}{6}$

54. An axis of symmetry is a line that divides a figure into two parts that are reflections of each other.

57. Both parabolas have no x-intercepts. When the coefficient of x^2 is positive, the parabola opens upward, and when the coefficient of x^2 is negative, the parabola opens downward. Also, the x- and y-coordinates of the vertex are the same as the number subtracted from x and the number added to the squared term.

LESSON 5.3 Graphs of Quadratic Functions

why *Quadratic functions are used to model objects affected by the force of gravity, such as the free fall of an object from a certain height to the ground. Physicists usually write the quadratic function in the form $f(x) = ax^2 + bx + c$, where $a \neq 0$, because each of the terms has a specific physical interpretation. However, there is another form of a quadratic function that you can use to determine properties of the graph without graphing.*

How do the terms in a quadratic function determine the size and placement of the parabola on the coordinate plane?

Exploration Exploring Graphs

Use your graphics calculator to graph the functions and complete the given tables.

1

	$f(x) = \frac{1}{2}x^2$	$f(x) = x^2$	$f(x) = 2x^2$	$f(x) = 5x^2$
Vertex				
Axis of symmetry				

As the coefficient of x^2 increases from small numbers to large numbers, describe what happens to the shape of the parabola.

2

	$f(x) = -\frac{1}{2}x^2$	$f(x) = -x^2$	$f(x) = -2x^2$	$f(x) = -5x^2$
Vertex				
Axis of symmetry				

Describe the effect that a negative coefficient of x^2 has on the graph of a parabola.

Disregarding air resistance, all free-falling objects accelerate at the same rate, depending on the force of gravity.

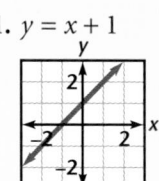
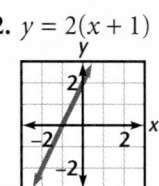

Exploration Notes

Students will have an opportunity to understand how information about a quadratic function can be obtained by writing it in the form $f(x) = a(x-h)^2 + k$.

Cooperative Learning

Have students work in pairs or small groups to complete the exploration. Students should be able to see the relationship between the form of the function and the shape of the curve. Students can help one another understand these connections.

TEACHING tip

Remind students that on some graphics calculators, the table feature can be used to obtain more precise coordinates for points on a graph. Have students discuss how they can determine the vertex and axis of symmetry by studying a table of values for a quadratic function.

Ongoing ASSESSMENT

7. If $f(x) = a(x-h)^2 + k$, the vertex is at (h, k) and the axis of symmetry is $x = h$. If $a > 0$, the parabola opens upward; if $a < 0$, the parabola opens downward. As $|a|$ increases, the parabola becomes narrower.

3

	$f(x) = x^2 + 4$	$f(x) = x^2 + 1$	$f(x) = x^2 - 2$	$f(x) = x^2 - 5$
Vertex				
Axis of symmetry				

Predict the vertex and axis of symmetry for $f(x) = x^2 + 3$. Check your answer by graphing.

4

	$f(x) = (x+2)^2$	$f(x) = (x+3)^2$	$f(x) = (x-1)^2$	$f(x) = (x-3)^2$
Vertex				
Axis of symmetry				

Predict the vertex and axis of symmetry for $f(x) = (x-6)^2$. Check your answer by graphing.

5

	$f(x) = (x-1)^2 + 3$	$f(x) = (x-1)^2 - 3$	$f(x) = (x+1)^2 + 3$	$f(x) = (x+1)^2 - 3$
Vertex				
Axis of symmetry				

Predict the vertex and the axis of symmetry for $f(x) = (x+4)^2 - 2$. Check your answer by graphing.

6

	$f(x) = 3(x-1)^2 - 2$	$f(x) = -2(x+3)^2 - 1$	$f(x) = \frac{1}{2}(x+2)^2 + 3$	$f(x) = -\frac{1}{3}(x-4)^2$
Vertex				
Axis of symmetry				

Predict the vertex and the axis of symmetry for $f(x) = -6(x-4)^2 - 5$.

7 Summarize what you discovered in each step of the exploration. ❖

Vertex of the Graph of $f(x) = a(x-h)^2 + k$

Graphics Calculator

With the trace feature on your graphics calculator, you can see that the coordinates of the vertex of $y = 2(x-3)^2 - 8$ are $(3, -8)$.

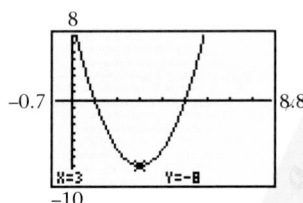

Examine the quadratic function $f(x) = 2(x-3)^2 - 8$. Since $(x-3)^2$ is the square of a number, it will always be *non-negative*. The minimum value of f will occur when $(x-3)^2$ is zero, or when x is 3.

---

interdisciplinary CONNECTION **Physics** Have students work in small groups and do research to find out about other uses of the quadratic equation in physics. Ask each group to design a poster that describes their findings and to present and describe their poster to the class. Then display the posters around your classroom.

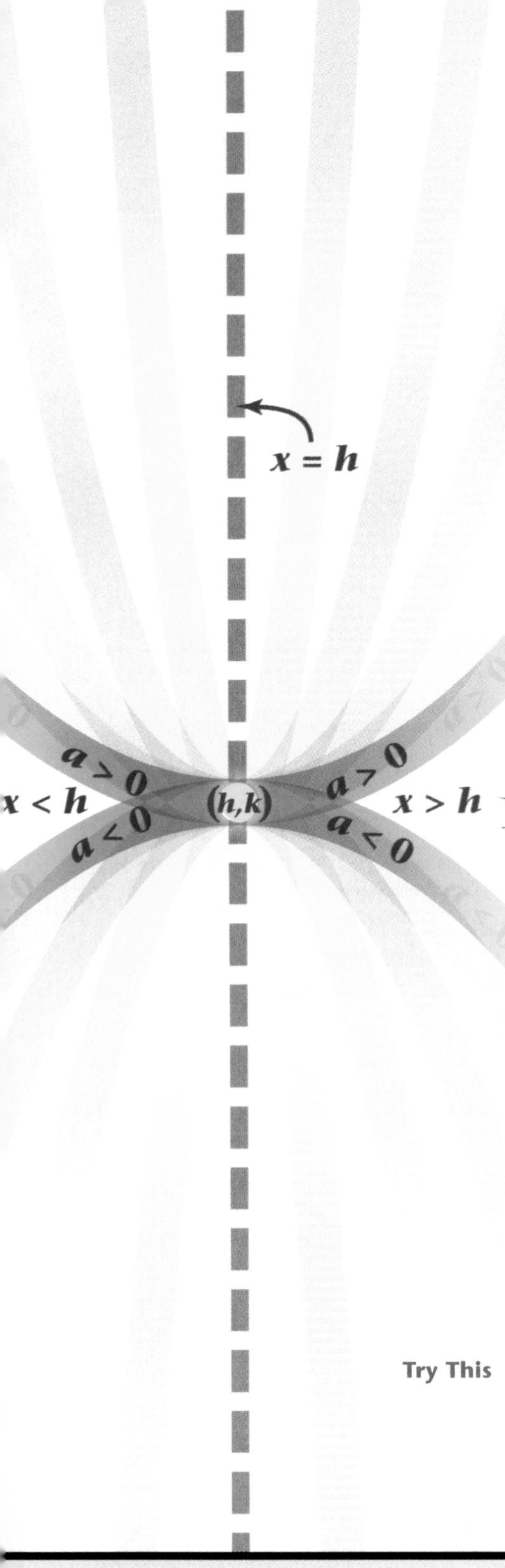

The x-coordinate of the vertex is 3. To find the y-coordinate, find $f(3)$.

$$f(x) = 2(x-3)^2 - 8$$
$$f(3) = 2(3-3)^2 - 8$$
$$= 2(0)^2 - 8$$
$$= -8$$

Thus, the coordinates of the vertex are $(3, -8)$.

In the equation $f(x) = 2(x-3)^2 - 8$, written in the form $f(x) = a(x-h)^2 + k$, what are the values of h and k? What are the values of the x- and y-coordinates of the vertex? How are the values of h and k related to the vertex?

Quadratic Functions of the Form
$$y = a(x-h)^2 + k$$

	$a > 0$	$a < 0$
Vertex	(h, k)	(h, k)
Axis of symmetry	$x = h$	$x = h$
Direction of opening	upward	downward

$x = h$

$a > 0$, $a < 0$ $x < h$ (h,k) $a > 0$, $a < 0$ $x > h$

EXAMPLE 1

Let $f(x) = 2(x+1)^2 - 10$.

A Find the coordinates of the vertex of the graph of f.

B What is the equation for the axis of symmetry?

C For what values of x is f increasing, and for what values of x is f decreasing?

Solution

A In the form $f(x) = a(x-h)^2 + k$, h is -1 and k is -10. The coordinates of the vertex are $(-1, -10)$.

B The equation for the axis of symmetry is $x = h$. The axis of symmetry is $x = -1$.

C Since a is 2, and $2 > 0$, the graph opens upward, and the vertex is a minimum point. Thus, the graph is decreasing for values of $x < -1$ and increasing for values of $x > -1$. ❖

Try This Let $f(x) = -4(x-3)^2 + 7$.

A Find the coordinates of the vertex of the graph of f.

B What is the equation for the axis of symmetry?

C For what values of x is f increasing, and for what values of x is f decreasing?

Alternate Example 2

While a rescue helicopter is 1000 feet above water, a life raft is dropped out. Its height, h, after t seconds is given approximately by the function $h(t) = -16t^2 + 1000$.

A. Use your calculator to find the time it will take the life raft to reach the water. [$\approx$ **7.9 seconds**]

B. Solve the equation to find the time it will take the life raft to reach the water. [$\sqrt{\frac{1000}{16}}$, **or** $\approx$ **7.9 seconds**]

T~EACHING~ *tip*

Ask students to investigate the meaning of the term *acceleration*. Encourage students to think about the way they use the term in everyday life and to discuss how that definition relates to the concepts presented in this lesson.

CRITICAL
Thinking

The initial vertical velocity is 0 because the life raft is released with an initial velocity of zero, in other words, without being pushed.

Applications of the Quadratic Function

EXAMPLE 2

Physics   While a rescue helicopter is 500 feet above water, a life raft is dropped out. Its approximate height, h, after t seconds is given by the function $h(t) = -16t^2 + 500$.

Ⓐ Use your calculator to find the time it will take the life raft to reach the water.

Ⓑ Solve the equation to find the time it will take the life raft to reach the water.

Solution▸

When the life raft reaches the water, its height, h, is 0.

Ⓐ Graph $y = -16x^2 + 500$. Use the trace feature to find points where y is 0.

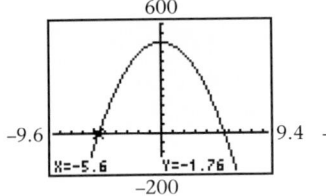 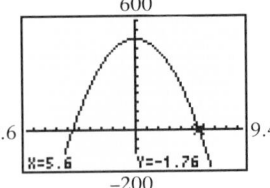

According to the graph, y is approximately 0 when x is 5.6 or −5.6. So $h(t)$ is approximately 0 when t is 5.6 or −5.6. Since time can only be positive in this situation, the life raft will reach the water after about 5.6 seconds.

Ⓑ Find the zeros of h. Consider only positive values of t.

$$0 = -16t^2 + 500$$
$$16t^2 = 500$$
$$t = \sqrt{\frac{500}{16}}, \text{ or approximately } 5.6$$

Thus, the life raft will reach the water approximately 5.6 seconds after it is dropped. ❖

Throughout this chapter, you will see how to use quadratic functions to model the height of a falling object. The rate that measures the pull of gravity on an object toward Earth, *acceleration due to gravity*, is 32 feet per second squared, or, in metric units, 9.8 meters per second squared.

Disregarding air resistance, the model for the height of a falling object affected by gravity is $h(t) = -\frac{1}{2}gt^2 + v_0 t + h_0$, where h_0 is the initial height, v_0 is the initial vertical velocity, and g is the acceleration due to gravity.

CRITICAL
Thinking

In Example 2, why is there no v_0-term in the equation $h(t) = -16t^2 + 500$, which models the falling life raft?

R~ETEACHING~
the
lesson

Guided Research Have students work through a series of explorations like those given in the exploration in the text. Before completing each table, ask students to compare the functions given and record the similarities and the differences. Then ask students to complete the tables and discuss how the differences they noted affected the graphs. After students have completed this work, give them a vertex and an axis of symmetry for the graphs of several different functions. Ask students to write an equation and to draw a graph for each function described.

EXERCISES & PROBLEMS

ASSESS

Selected Answers
Odd-numbered problems 7–59.

Assignment Guide
Core 1–9, 11–21, 23–26, 29–36, 39–43, 45–61

Core Plus 1–6, 8–16, 19–22, 25–30, 34–61

Technology
A graphics calculator is needed for Exercises 31–38, 45, 48, and 49. A graphics calculator or computer software, such as *f(g) Scholar*™, could be helpful for Exercises 12, 14, 16, 23–28, and 47.

Communicate

1. Explain how to write a quadratic function with a graph that has a vertex at (–3, 2), an axis of symmetry at $x = -3$, and that opens upward.

2. Explain how to find the coordinates of the vertex and the axis of symmetry of the graph of $f(x) = -5(x + 3)^2 - 2$.

3. Without graphing a quadratic function, how can you tell whether its graph opens upward or downward?

4. Explain how to find the *x*-intercepts of the graph of $f(x) = -3(x + 1)^2 - 7$.

5. If $h(t) = -16t^2 + 5t + 12$ models the height of a falling object, what point(s) on the graph represent(s) when the object will reach the ground? What point(s) on the graph represent(s) the maximum height of the object?

6. Suppose an object is dropped from a height of 10 feet. Explain how to write a quadratic function that models the height of the object in terms of time, *t*, in seconds.

Practice & Apply

For each function in Exercises 7–10, give the coordinates of the vertex and the equation of the axis of symmetry.

7. $f(x) = -\frac{1}{3}x^2$ $V(0,0), x = 0$

8. $f(x) = (x - 5)^2 + 13$
$V(5,13), x = 5$

9. $f(x) = 3(x + 2)^2 - 15$
$V(-2, -15), x = -2$

10. $f(x) = -\frac{1}{3}(x - 4)^2 - 6$
$V(4, -6), x = 4$

Rescue A crate of blankets and clothing is dropped without a parachute from a hovering helicopter at a height of 200 feet.

11. Use the quadratic model for the height of a falling object affected by gravity, $h(t) = -\frac{1}{2}gt^2 + v_0t + h_0$, to write an equation for the height of the crate as a function of time after it is dropped. $h(t) = -16t^2 + 200$

12. Approximately how long will it take for the crate to reach the ground?
3.5 sec

13. Write an equation for the height of the crate as a function of time after it is dropped if the helicopter is hovering 100 meters above the ground. $h(t) = -4.9t^2 + 100$

14. Approximately how long will it now take for the crate to reach the ground?   4.5 sec

23. increasing for $x > 1$; decreasing for $x < 1$

24. increasing for $x < -7$; decreasing for $x > -7$

25. increasing for $x < 9$; decreasing for $x > 9$

Recreation A child in a swimming class steps off a 12-foot-high platform and jumps into the pool.

15. Use the quadratic model for the height of a falling object affected by gravity, $h(t) = -\frac{1}{2}gt^2 + v_0t + h_0$, to write an equation for the height of the child t seconds after stepping off the platform. $h(t) = -16t^2 + 12$

16. Approximately how long will it take for the child to reach the surface of the water in the pool? Approx 0.9 sec

Give the coordinates of the vertex and the equation of the axis of symmetry.

17. $f(x) = 4(x + 3)^2 + 1$ (−3, 1), $x = -3$ 18. $f(x) = 3(x + 7)^2 - 8$ (−7, −8), $x = -7$

19. $f(x) = -(x - 1)^2$ (1, 0), $x = 1$ 20. $f(x) = \left(x + \frac{3}{4}\right)^2 + 12$ $\left(-\frac{3}{4}, 12\right)$, $x = -\frac{3}{4}$

21. $f(x) = \frac{2}{3}x^2$ (0, 0), $x = 0$ 22. $f(x) = x^2 - \frac{3}{5}$ $\left(0, -\frac{3}{5}\right)$, $x = 0$

For Exercises 23–28, give the values of x for which the function is increasing and for which the function is decreasing.

23. $f(x) = 4(x - 1)^2 + 5$ 24. $f(x) = -2(x + 7)^2 - 4$

25. $f(x) = -(x - 9)^2 + 12$ 26. $f(x) = 3 + (x - 4)^2 - 6$

27. $f(x) = \frac{1}{4} + \left(x - \frac{1}{2}\right)^2$ 28. $f(x) = \left(x + \frac{2}{3}\right)^2$

29. Create a quadratic function whose vertex (2, 3) is a minimum point.

30. Create a quadratic function whose vertex (2, 3) is a maximum point.

For each function in Exercises 31–38, find the zeros of the function. Check using your graphics calculator.

31. $f(x) = (x - 9)^2 - 1$ (8, 0), (10, 0)   32. $f(x) = (x - 9)^2$ (9, 0)

33. $f(x) = (x + 2)^2 - 4$ (−4, 0), (0, 0)   34. $f(x) = (x - 8)^2 - 4$ (6, 0), (10, 0)

35. $f(x) = 2(x - 8)^2 - 2$ (7, 0), (9, 0)   36. $f(x) = \frac{1}{2}(x - 8)^2 - 2$ (6, 0), (10, 0)

37. $f(x) = -7(x - 3)^2 + 7$ (2, 0), (4, 0)   38. $f(x) = (x - 2)^2 + 1$ none

39. Create a quadratic function of the form $f(x) = a(x - h)^2 + k$ that has exactly one x-intercept.

40. Create a quadratic function of the form $f(x) = a(x - h)^2 + k$ that has no x-intercepts.

Determine which functions are linear, quadratic, or neither.

41. $f(x) = x - 3^2$ linear   42. $f(x) = x^3 - 8$ neither   43. $f(x) = \frac{x^2 - 8}{2}$ quadratic   44. $f(x) = \frac{2}{x^2 - 8}$ neither

26. increasing for $x > 4$; decreasing for $x < 4$

27. increasing for $x > \frac{1}{2}$; f is decreasing for $x < \frac{1}{2}$

28. increasing for $x > -\frac{2}{3}$; decreasing for $x < -\frac{2}{3}$

29. Answers may vary but must be equivalent to $f(x) = a(x - 2)^2 + 3$, where $a > 0$

30. Answers may vary but must be equivalent to $f(x) = a(x - 2)^2 + 3$, where $a < 0$

39. Answers may vary but should be equivalent to $f(x) = a(x - h)^2$, where h and a are any real numbers and $a \neq 0$.

40. Answers may vary but should be equivalent to $f(x) = a(x - h)^2 + k$, where either $a > 0$ and $k > 0$ or $a < 0$ and $k < 0$.

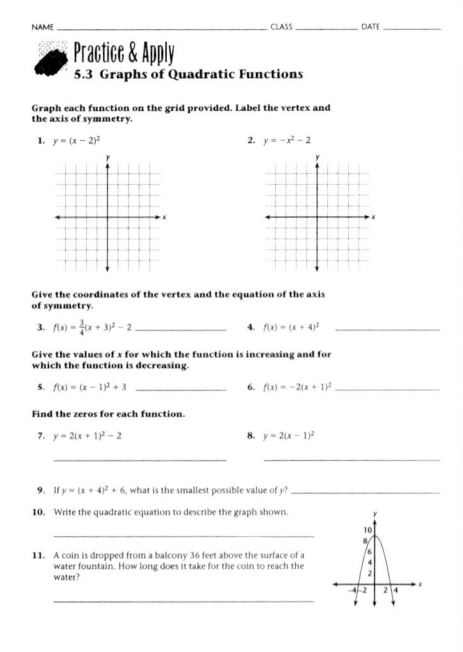

45 Write a quadratic function for each of the interlocking parabolas trapping the point $(-1, 3)$. Use your graphics calculator to check your answer.

$f(x) = (x + 1)^2 + 2$

$g(x) = -(x + 1)^2 + 4$

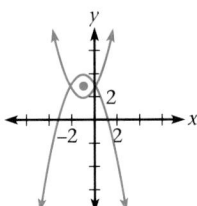

Sports A trained cliff diver stands on a cliff but does not know the height of the cliff. To find the height, he drops a pebble into the ocean and discovers that it takes approximately 3 seconds for the pebble to hit the water.

46. Create a quadratic model for the height of the pebble in terms of time, t. $h(t) = -16t^2 + h_0$

47. Approximately how high is the cliff? 144 ft

Transformations For Exercises 48 and 49, let $f(x) = (x - 5)^2 + 3$.

48 Find g, the reflected image of f over the x-axis.

49 Find h, the reflected image of f over the y-axis.

 **Look Back**

Find the product.

50. $(3x + 4)(-x - 5)$
$-3x^2 - 19x - 20$

51. $(-2x + 9)(-4x + 7)$
$8x^2 - 50x + 63$

52. $\left(\frac{1}{3}x + \frac{1}{4}\right)(-5x - 2)$
$-\frac{5}{3}x^2 - \frac{23}{12}x - \frac{1}{2}$

Solve and graph the inequality. [Lesson 1.6]

53. $2x - 4 > 12 + 5x$ $x < -\frac{16}{3}$

54. $2x - \frac{3}{4}y \geq 7$ $y \leq \frac{8}{3}x - \frac{28}{3}$

55. $3(3x + 7) - 12 \leq 8 - \left(\frac{1}{2}x + 9\right)$ $x \leq -\frac{20}{19}$

56. $-2\left(\frac{2}{3}x + 5\right) - 13 < -6y - 1$ $y < \frac{2}{9}x + \frac{11}{3}$

Solve each equation. [Lesson 5.2]

57. $-2x^2 = -16$ $\pm\sqrt{8}$

58. $-3x^2 + 15 = -6$ $\pm\sqrt{7}$

59. $32 = 2x^2 - 4$ $\pm\sqrt{18}$

Look Beyond

60. Multiply to complete the table.

Exponential form	$(x + 3)^2$	$(x - 4)^2$	$(x + 6)^2$
Product	$x^2 + 6x + 9$	$x^2 - 8x + 16$	$x^2 + 12x + 36$

61. Compare the exponential form with the product. Can you see a pattern? How can you write the product without actually multiplying?

48. $g(x) = -(x - 5)^2 - 3$

49. $h(x) = (x + 5)^2 + 3$

53. $x < -\frac{16}{3}$

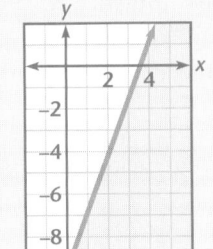

54. $y \leq \frac{8}{3}x - \frac{28}{3}$

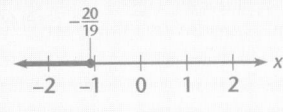

55. $x \leq -\frac{20}{19}$

56. $y < \frac{2}{9}x + \frac{11}{3}$

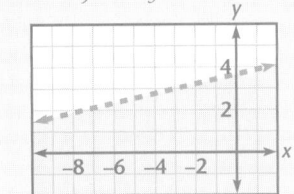

Objectives

- Use tiles to complete the square for quadratic expressions of the form $x^2 + bx + c$, where b and c are real numbers.
- Solve quadratic equations by completing the square for expressions of the form $ax^2 + bx + c$, where $a \neq 0$, and b and c are real numbers.

RESOURCES

- Practice Master   5.4
- Enrichment Master   5.4
- Technology Master   5.4
- Lesson Activity Master 5.4
- Quiz   5.4
- Spanish Resources   5.4

Assessing Prior Knowledge

Solve and check.

1. $x^2 = 100$ [± 10]

2. $2x^2 - 7 = 281$ [± 12]

3. $(x - 1)^2 - 2 = 7$ [$-2, 4$]

TEACH

 Ask students to consider the meaning of the phrase "completing the square" and then share their thoughts with the class. Many students may think of a geometric model. This model can be used to demonstrate the algebraic procedures introduced in this lesson.

Use Transparency ▶ 20

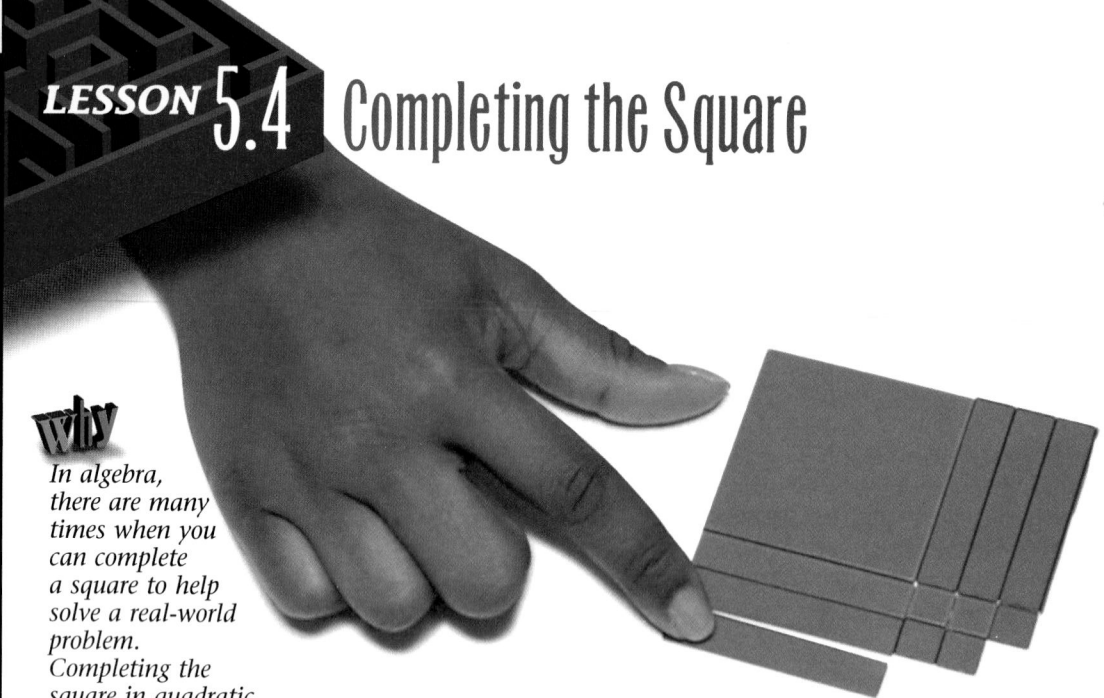

LESSON 5.4 Completing the Square

In algebra, there are many times when you can complete a square to help solve a real-world problem. Completing the square in quadratic expressions can be modeled with geometrical shapes, called tiles.

Algebra tiles can be used to model quadratic expressions.

Model x^2 with a square tile that is x units long on each side.	x^2-tile
Model x with a rectangular tile that is x units long and 1 unit wide.	x-tile
Model a constant with a square tile that is 1 unit long on each side.	unit tile

Using Tiles to Complete the Square

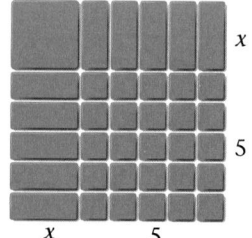

This model has 1 x^2-tile, 10 x-tiles, and 25 unit tiles.

The area of this square is the sum of the tiles, $x^2 + 10x + 25$.

Since the length of a side of this square is $x + 5$, the area can also be written $(x + 5)^2$.

This model forms a complete square. The quadratic expression is a complete square, called a **perfect-square trinomial**.

A perfect-square trinomial, such as $x^2 + 10x + 25$, can also be written $(x + 5)^2$, called the **square of a binomial**.

Technology Introduce the concept of completing the square by asking students to draw representations of the tiles shown in the text. Students can either draw these by hand or use computer software that enables them to draw figures. After students feel comfortable with the geometric model, introduce the procedure for solving quadratic equations by completing the square. Have students work through several examples in small groups, using a calculator or computer software, such as *f(g) Scholar*™, as necessary. Students can also use a calculator or computer software to find or to check the solutions. After students have completed their work, have them share the ways in which they used technology with the class.

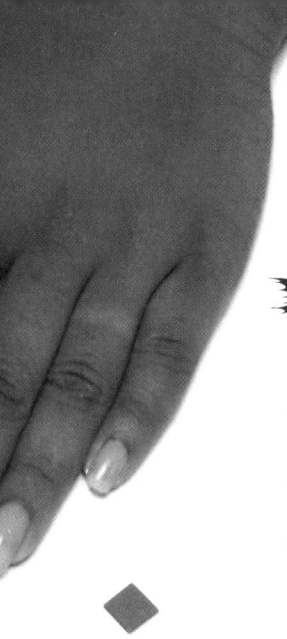

A Write the quadratic expression modeled by these tiles.

B Find the minimum number of additional unit tiles needed to form a square.

C Write the perfect-square trinomial modeled by the completed square.

Solution▶

A There are 0 unit tiles, 6 x-tiles, and 1 x^2-tile. The expression modeled is the sum of the tiles.

$$x^2 + 6x + 0, \text{ or } x^2 + 6x$$

B Take $\frac{1}{2}$ of the x-tiles and move them to an adjacent side of x^2. Now you have 3 x-tiles on two sides of x^2.

There are 3^2, or 9, unit tiles needed to fill in the missing corner, or to complete the square.

C By adding the tiles, you find that the new square models the perfect-square trinomial $x^2 + 6x + 9$. ❖

Try This

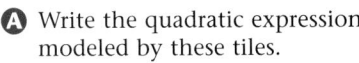

A Write the quadratic expression modeled by these tiles.

B Find the minimum number of additional unit tiles needed to form a square.

C Write the perfect-square trinomial modeled by the completed square.

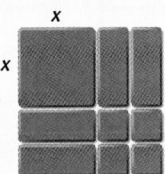
In a perfect-square trinomial, such as $x^2 + 6x + 9$, the square of half the coefficient of x, or $\left(\frac{6}{2}\right)^2$, is the constant term, 9. For example,

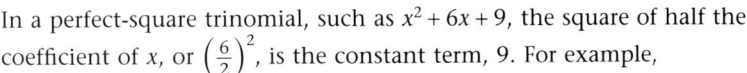

$$x^2 + 8x + 16 = (x+4)^2 \qquad x^2 + 2x + 1 = (x+1)^2$$

$$\left(\frac{8}{2}\right)^2 = 16 \qquad \left(\frac{2}{2}\right)^2 = 1$$

If $x^2 + bx + c$ is a perfect-square trinomial, then $\left(\frac{b}{2}\right)^2 = c$.

Thus, to complete the square in the expression $x^2 + bx$, add $c = \left(\frac{b}{2}\right)^2$.

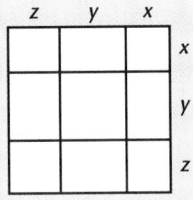

TEACHING *tip*

Emphasize the importance of dividing through by a before trying to complete the square. You may want to show students what will happen if they forget to complete this important step.

Solving by Completing the Square

You can use the technique of completing the square to solve quadratic equations.

EXAMPLE 2

Solve $x^2 + 6x - 17 = 0$ by completing the square.

Solution➤

Isolate the variable terms on one side.

$$x^2 + 6x - 17 = 0$$
$$x^2 + 6x = 17$$

Complete the square by adding $c = \left(\frac{6}{2}\right)^2 = 9$ to each side.

$$x^2 + 6x + 9 = 17 + 9$$
$$(x + 3)^2 = 26$$
$$x + 3 = \pm\sqrt{26}$$
$$x = -3 \pm \sqrt{26}$$
$$x = -3 + \sqrt{26} \text{ or } x = -3 - \sqrt{26}$$
$$x \approx 2.1 \qquad \text{or } x \approx -8.1$$

Thus, the solutions for $x^2 + 6x - 17 = 0$ are approximately 2.1 and −8.1. ❖

When you want to complete the square in a quadratic equation of the form $ax^2 + bx + c = 0$ where $a \neq 1$, first divide through by a.

EXAMPLE 3

Solve $2x^2 - 6x - 7 = 0$ by completing the square.

Solution➤

Divide through by the coefficient, 2.

$$2x^2 - 6x - 7 = 0$$
$$x^2 - 3x - \frac{7}{2} = 0$$

Isolate the variable terms on one side.

$$x^2 - 3x = \frac{7}{2}$$

Complete the square by adding $c = \left(\frac{-3}{2}\right)^2$ to each side.

$$x^2 - 3x + \left(\frac{-3}{2}\right)^2 = \frac{7}{2} + \left(\frac{-3}{2}\right)^2$$
$$\left(x - \frac{3}{2}\right)^2 = \frac{23}{4}$$
$$x - \frac{3}{2} = \pm\sqrt{\frac{23}{4}}$$
$$x = \pm\sqrt{\frac{23}{4}} + \frac{3}{2}$$
$$x = \frac{\sqrt{23}}{2} + \frac{3}{2} \text{ or } x = -\frac{\sqrt{23}}{2} + \frac{3}{2}$$
$$x = \frac{\sqrt{23} + 3}{2} \text{ or } x = \frac{-\sqrt{23} + 3}{2}$$
$$x \approx 3.9 \text{ or } x \approx -0.9$$

Thus, the solutions for $2x^2 - 6x - 7 = 0$ are approximately 3.9 and −0.9. ❖

English Language Development This lesson provides students who have a limited command of the English language with an excellent opportunity to understand the concept of quadratic expressions through the use of tiles. Make sure, however, that students still understand the terms used in this lesson. Have students write the definitions of any necessary terms or phrases in their own words, using drawings when possible. Encourage students to share their work with each other.

The al-Khowarizmi Method

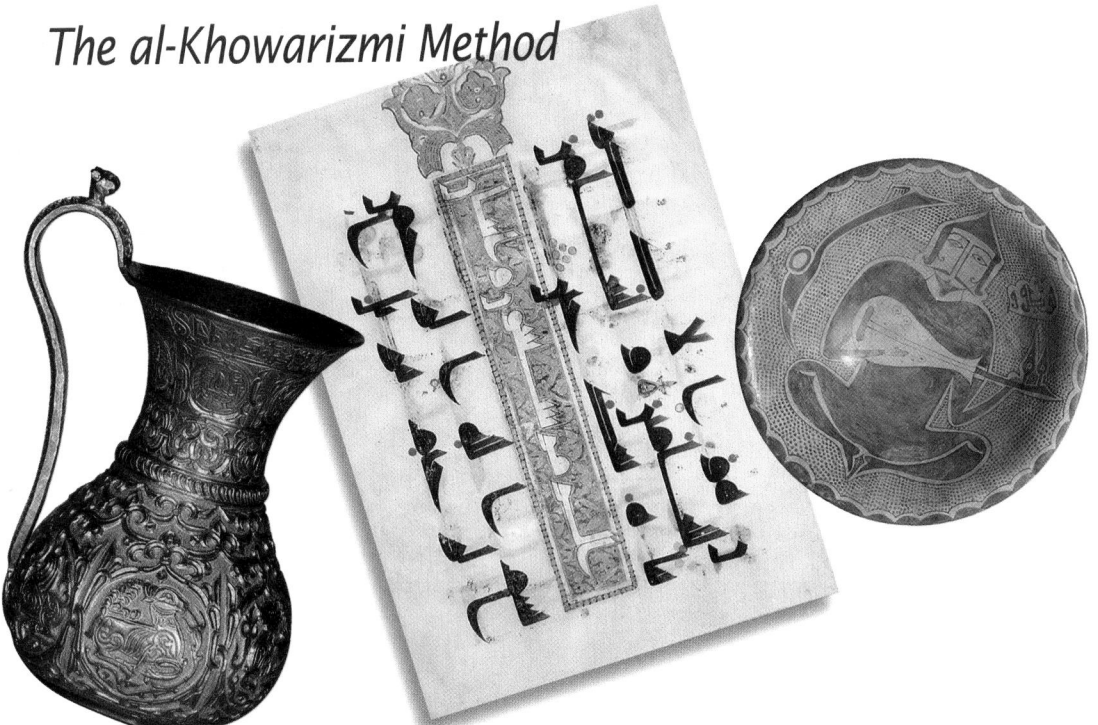

Islamic artifacts from the 9th and 10th century, (from left) a gold pitcher, a page from the Qu'ran, and a large serving bowl.

Cultural Connection: Africa One of the first algebra books, written in Arabic by al-Khowarizmi, was imported into Europe from Africa over 1000 years ago. Because *al-jabr* was part of the title of the book, the subject became known as *algebra*. This book used a method similar to tiles to complete the square and solve for *x*.

To complete the square in the equation $x^2 + 12x = 45$, the al-Khowarizmi method also begins with one square that is *x* units long on each side, and 12 rectangles that are 1 unit wide by *x* units long.

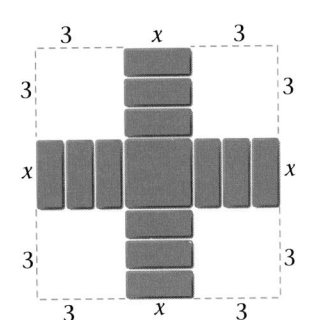

Divide the 12 rectangles equally, and arrange them on the 4 sides of the square.

To complete the square, add 9 units to each corner: 9 units × 4 corners = 36. The new area is the original constant, 45, plus 36: $45 + 36 = 81$.

Since the new area is 81, the side of the new square has length $\sqrt{81}$, or 9. From the model, you can see that the length of the side is $3 + x + 3 = 9$; therefore, *x* is 3.

You have found one solution for *x*. How can the other solution be found?

CRITICAL *Thinking*

Using Manipulatives Have students work in small groups. Ask each group to use tiles to make squares that are complete. Ask them to draw the completed squares on paper. Then ask them to draw the squares again on a separate sheet of paper, this time leaving out the unit tiles. Have groups exchange the papers of the incomplete squares. Ask students to work in their groups and use tiles to complete the squares, recording their work. Then have groups check their answers against the original squares.

TEACHING *tip*

Many students may be interested by the historical discussion of algebra presented here. Encourage students to do further research into the subject presented. Also, ask students to look for other historical facts that were developed by early Arabic mathematicians.

CRITICAL *Thinking*

Use completing the square to find the other solution for *x*.

11. $x^2 + 10x$

12. $x^2 + 12x + 36$

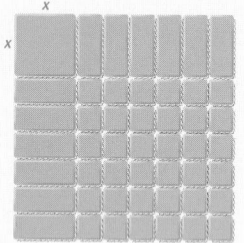

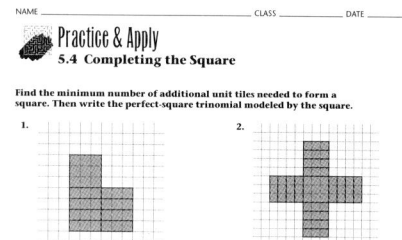
EXERCISES & PROBLEMS

Communicate

1. How are tiles used to complete the square in the quadratic expression $x^2 + 12x$?

2. Explain how to complete the square to solve the equation $x^2 + 4x - 13 = 0$.

3. Explain how to complete the square to solve the equation $2x^2 + 4x = 15$.

4. How can you solve $x^2 + 20x = 108$ using the al-Khowarizmi method?

5. Compare completing the square using tiles with completing the square using the al-Khowarizmi method. How are they alike? How are they different?

Practice & Apply

Write the quadratic expression modeled by each set of tiles.

6.

$x^2 + 8x + 16$

7.

$x^2 + 3x + 3$

8.

$x^2 + 7x + 6$

9.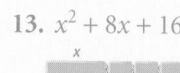

$x^2 + 7x + 6$

10.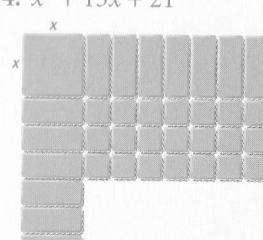

$x^2 + 7x + 7$

Draw a tile model to represent each of the following quadratic expressions.

11. $x^2 + 10x$ **12.** $x^2 + 12x + 36$ **13.** $x^2 + 8x + 16$
14. $x^2 + 13x + 21$ **15.** $x^2 + 2x + 3$ **16.** $x^2 + 6x + 14$

For Exercises 17–19, find the minimum number of additional unit tiles needed to form a perfect square. Then write the perfect-square trinomial modeled by the square.

17.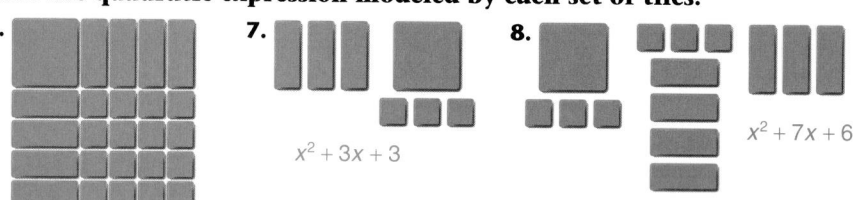

16; $x^2 + 8x + 16$

18.

4; $x^2 + 4x + 4$

19.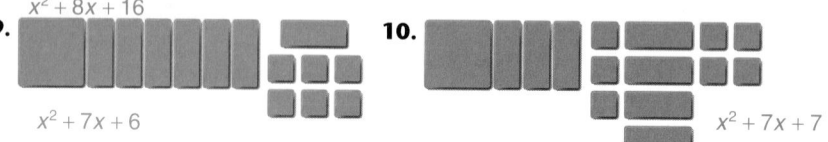

$\frac{1}{4}$; $x^2 + x + \frac{1}{4}$

13. $x^2 + 8x + 16$ **14.** $x^2 + 13x + 21$ **15.** $x^2 + 2x + 3$

16. $x^2 + 6x + 14$

Complete the square for each quadratic expression.

20. $x^2 + 6x$ $x^2 + 6x + 9$ **21.** $x^2 - 8x$ $x^2 - 8x + 16$ **22.** $x^2 - 7x$ $x^2 - 7x + \frac{49}{4}$

Solve by completing the square. Give your answers to the nearest tenth.

23. $x^2 - 8x = 3$ $-0.4, 8.4$ **24.** $x^2 + 2x = 13$ $-4.7, 2.7$ **25.** $x^2 - 5x - 1 = 4 - 3x$ $-1.4, 3.4$

26. $0 = x^2 - 6x + 3$ $0.6, 5.4$ **27.** $0 = x^2 + 7x - 26$ **28.** $0 = x^2 - 3x - 6$ $-1.4, 4.4$

29. $0 = 3x^2 - 2x - 12$ **30.** $-2x^2 + 14x + 60 = 0$ **31.** $0 = 3x^2 - 11x + 6$ $0.7, 3$
$\quad$ $-1.7, 2.4$ $\qquad\qquad$ $-3, 10$

Electronics A certain variable electric current is given by $I = t^2 - 6t + 7$, where $0 \le t \le 7$ is time in seconds and I is the electric current in amperes.

32. In approximately how many seconds will the current reach 2.6 amperes? 0.9 sec

33. In how many seconds will the current reach -2 amperes? 3 sec

34. For what values of t are the corresponding values of I negative? What is the meaning of these values? $1.6 < t < 4.4$; neg charge

Solve by completing the square. Give your answers to the nearest tenth.

$\qquad\qquad\qquad\qquad\qquad\qquad\qquad$ $-10.4, 0.4$

35. $-10 = x^2 - 8x + 2$ $2, 6$ **36.** $x^2 + 16x = 2$ $-16.1, 0.1$ **37.** $4 - x^2 = 10x$

38. $x^2 = 23 - 15x$ $-16.4, 1.4$ **39.** $8x - 2 = x^2 + 150x$ **40.** $-32x = 16 - x^2$

41. $2x^2 = 22x - 11$ $0.5, 10.5$ **42.** $4x^2 - 8 = -13x$ $-3.8, 0.5$ **43.** $2x^2 - 3x + 12 = -42$

44. Explain how to solve a quadratic equation of the form $ax^2 + bx + c = 0$ for x by completing the square.

Look Back

Let p be any positive number, let n be any negative number, and let z be zero. Write _positive_, _negative_, or _zero_ for the value of each expression. If the value cannot be determined, write _cannot be determined_.   **[Lessons 2.1, 3.4]**

45. pn neg   **46.** $p + n$ cannot   **47.** $z - n$ pos

48. $-np$ pos   **49.** $|npz|$ zero   **50.** $|-p + n|$ pos

Evaluate.   **[Lessons 2.2, 5.2]**

51. $5\sqrt{2} + 11\sqrt{2}$ $16\sqrt{2}$ **52.** $\frac{2}{3}\sqrt{5} - \frac{4}{3}\sqrt{5}$ $-\frac{2}{3}\sqrt{5}$ **53.** $2\sqrt{11}(4\sqrt{5})$ $8\sqrt{55}$ **54.** $(-\sqrt{20})(5\sqrt{2})$
$\qquad\qquad\qquad\qquad\qquad\qquad\qquad\qquad\qquad\qquad\qquad\qquad\qquad\qquad\qquad\qquad\qquad\qquad$ $-5\sqrt{40}$

55. Create a quadratic function that has -2 and 6 as zeros.   **[Lesson 5.1]**
$\quad$ Answers may vary. Sample: $f(x) = (x + 2)(x - 6)$

Look Beyond

For each function, use your graphics calculator to find the number of x-intercepts and the coordinates of the vertex.

56. $y = 2x^2 + 4x + 1$ **57.** $y = 2x^2 + 4x + 2$ **58.** $y = 2x^2 + 4x + 3$ $0; (-1, 1)$
$\quad$ $2; (-1, -1)$ $\qquad\qquad$ $1; (-1, 0)$
59. How are the graphs and equations alike? How are they different?

---

27. -9.7 and 2.7

39. -142.0 and 0

40. -0.5 and 32.5

43. no real solutions

44. Subtract c from both sides. Divide both sides by the numerical coefficient, a. Add the square of half of $\frac{b}{a}$ to both sides. Rewrite the perfect square trinomial as a binomial squared. Take the square root of both sides and solve for x.

59. The graphs are all congruent parabolas with the same axis of symmetry, $x = -1$. The graphs all have a minimum when $x = -1$. The equations are all of the form $y = 2x^2 + 4x + c$. When $c = 1$ there are two x-intercepts, when $c = 2$ there is one x-intercept, and when $c = 3$ there are no x-intercepts.

Error Analysis

Be sure students understand that the last term of a perfect-square trinomial is always positive. Use FOIL (**F**irst, **O**utside, **I**nside, **L**ast) to expand $(a - b)^2$. This will emphasize the point.

alternative ASSESSMENT

Performance Assessment

Have students write several quadratic equations that can be solved by completing the square. Include equations in which $a = 1$ and equations in which $a \ne 1$. Have students exchange and solve each others' equations.

Look Beyond

These exercises preview the next lesson by using a graph to determine the number of solutions to an equation, or the number of roots of a quadratic function.

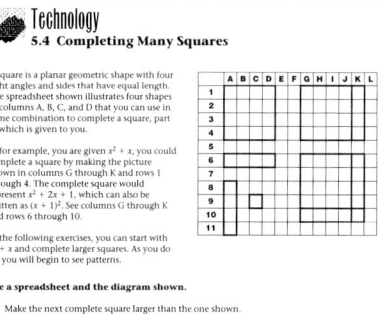

- Use the quadratic formula to solve quadratic equations that model real-world situations.
- Use the axis of symmetry formula to find maximum or minimum values of quadratic equations that model real-world situations.

RESOURCES

- Practice Master 5.5
- Enrichment Master 5.5
- Technology Master 5.5
- Lesson Activity Master 5.5
- Quiz 5.5
- Spanish Resources 5.5

Assessing Prior Knowledge

Solve for x.

1. $x^2 = 49$ [±7]

2. $x^2 - 4 = 0$ [±2]

3. $9x^2 - 25 = 0$ [$\pm\frac{5}{3}$]

4. $x^2 + 8x + 16 = 0$ [−4]

TEACH

Discuss the advantages of having one formula that can be used to solve all equations of a given type. Mention the practical advantages, such as the speed and efficiency of obtaining solutions. Mention the mathematical gains, like obtaining insight into the relationship between the coefficients of a polynomial and its zeros.

LESSON 5.5 The Quadratic Formula

why *One of the advantages of working with variables is being able to generalize. You have learned how to solve any quadratic equation by completing the square. To derive a general formula for solving quadratic equations, you can complete the square for the general quadratic equation $ax^2 + bx + c = 0$, where $a \neq 0$.*

Quadratic Formula

The *quadratic formula* is developed by following the steps for completing the square.

$$ax^2 + bx + c = 0$$

Divide each side by a.
$$x^2 + \frac{b}{a}x + \frac{c}{a} = 0$$

$$x^2 + \frac{b}{a}x = -\frac{c}{a}$$

Complete the square by adding $\left(\frac{b}{2a}\right)^2$ to each side.
$$x^2 + \frac{b}{a}x + \left(\frac{b}{2a}\right)^2 = -\frac{c}{a} + \left(\frac{b}{2a}\right)^2$$

$$\left(x + \frac{b}{2a}\right)^2 = \frac{b^2}{4a^2} - \frac{c}{a}$$

$$\left(x + \frac{b}{2a}\right)^2 = \frac{b^2 - 4ac}{4a^2}$$

$$x + \frac{b}{2a} = \pm\sqrt{\frac{b^2 - 4ac}{4a^2}}$$

Subtract $\frac{b}{2a}$ from each side.
$$x = -\frac{b}{2a} \pm \frac{\sqrt{b^2 - 4ac}}{2a}$$

$$x = \frac{-b \pm \sqrt{b^2 - 4ac}}{2a}$$

QUADRATIC FORMULA

If $ax^2 + bx + c = 0$, where $a \neq 0$, then $x = \frac{-b \pm \sqrt{b^2 - 4ac}}{2a}$.

You can use the quadratic formula to solve *any* quadratic equation.

When the number under the square root symbol is *not* a perfect square, use a decimal number approximation.

ALTERNATIVE teaching strategy

Hands-On Strategies Give students the equation $2x^2 + 5x - 8 = 0$, and ask them to solve for x by completing the square. $\left[\frac{-5 \pm \sqrt{89}}{4}\right]$ Encourage them to write out every step. Repeat this with other equations. Then ask students to work in small groups to solve the equation $ax^2 + bx + c = 0$ for x. $\left[\frac{-b \pm \sqrt{b^2 - 4ac}}{2a}\right]$

As they work, remind them to think about the equations they have just solved individually and to use the same methods to solve this general form of a quadratic equation. Provide encouragement and guidance as necessary. After students have completed their work, have them share their results with the class. Present the quadratic formula, and ask students to compare it with their results. Then have students work in groups to use the quadratic formula to solve several quadratic equations.

EXAMPLE 1

Solve $4x^2 = 8 - 3x$ for x to the nearest tenth.

Solution

Write the equation in the form $ax^2 + bx + c = 0$.

$4x^2 + 3x - 8 = 0$, where a is 4, b is 3, and c is –8.

Use the quadratic formula. Substitute the values for a, b, and c into the quadratic formula.

$$x = \frac{-b \pm \sqrt{b^2 - 4ac}}{2a}$$

$$x = \frac{-3 \pm \sqrt{3^2 - 4(4)(-8)}}{2(4)}$$

$$x \approx \frac{-3 + 11.7}{8} \text{ or } x \approx \frac{-3 - 11.7}{8}$$

$$x \approx 1.1 \qquad \text{ or } x \approx -1.8$$

Thus, the solutions for $4x^2 = 8 - 3x$ are approximately 1.1 and –1.8. ❖

Try This Solve $2x^2 - 5x = -3$ for x. Write your answer to the nearest tenth.

How can you use the quadratic formula to find the x-intercepts of a parabola?

EXAMPLE 2

Graphics Calculator

Find the x-intercepts of the parabola given by $f(x) = 2x^2 + 10x - 28$. Check your answers with your graphics calculator.

Solution

Solve $2x^2 + 10x - 28 = 0$. $2x^2 + 10x - 28 = 0$
Divide both sides by 2. $x^2 + 5x - 14 = 0$

Find the roots of $x^2 + 5x - 14 = 0$.

Substitute 1 for a, 5 for b, and –14 for c in the quadratic formula.

$$x = \frac{-b \pm \sqrt{b^2 - 4ac}}{2a}$$

$$x = \frac{-(5) \pm \sqrt{5^2 - 4(1)(-14)}}{2(1)}$$

$$x = -7 \text{ or } x = 2$$

Check

Graph $f(x) = 2x^2 + 10x - 28$, and use the trace feature.

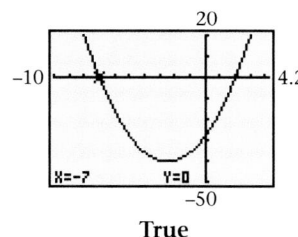

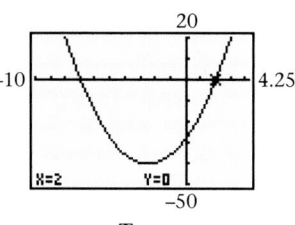

True   True

Thus, the x-intercepts are (–7, 0) and (2, 0). ❖

interdisciplinary
CONNECTION

Physics Have students solve the following problem about minimal stopping distance. The equation $d = 0.042s^2 + 1.1s$, where s is the speed in miles per hour and d is the distance in feet, describes an approximation of the minimal distance it takes for a car to stop, including the reaction time for the driver. Suppose a car stopped after 150 feet. What was the approximate speed of the car? [**about 48 mph**]

Cooperative Learning

Students can work in small groups to check the solution to Example 1. Have each group think of different ways to check the solutions to quadratic equations. Have the students in each group test their methods and share them with the class.

Alternate Example 1

Solve $16x^2 + 9 = 24x$. [**0.75**]

ongoing
ASSESSMENT

Try This

1.0 and 1.5

TEACHING *tip*

Emphasize that before the quadratic formula can be used to solve a quadratic equation, the equation must be in the form $ax^2 + bx + c = 0$.

ongoing
ASSESSMENT

You can use the quadratic formula to determine the roots of the equation. The roots of the equation are the x-coordinates of the x-intercepts.

Alternate Example 2

Find the x-intercepts of the parabola given by $f(x) = 4x^2 + 20x + 25$. Check your answer(s) with your graphics calculator. [**–2.5**]

Alternate Example 3

Use the axis of symmetry formula to find the vertex of the parabola given by $f(x) = 3x^2 + 5x - 4$. $\left[\left(\frac{-5}{6}, \frac{-73}{12}\right)\right]$ Find the vertex with your graphics calculator. $[\approx (-0.8, -6.1)]$

Vertex of a Parabola

ROOTS

MIDPOINT FORMULA

AXIS OF SYMMETRY

Recall that the axis of symmetry of a parabola goes through the vertex and divides the parabola in half. Therefore, the axis of symmetry intersects the x-axis exactly halfway between the x-intercepts of the parabola.

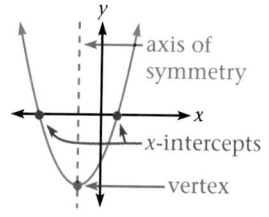

A formula for the axis of symmetry can be found using the two solutions of the quadratic formula, $x = \frac{-b \pm \sqrt{b^2 - 4ac}}{2a}$, and the midpoint formula.

$$x = \frac{\left(\frac{-b + \sqrt{b^2 - 4ac}}{2a}\right) + \left(\frac{-b - \sqrt{b^2 - 4ac}}{2a}\right)}{2} = \frac{-b}{2a}$$

Therefore, the equation of the axis of symmetry of a parabola given by a quadratic function of the form $f(x) = ax^2 + bx + c$ is $x = \frac{-b}{2a}$.

CRITICAL
Thinking

Explain why the midpoint formula and the quadratic formula are used to find the axis of symmetry formula.

EXAMPLE 3

Graphics
Calculator

Use the axis of symmetry formula to find the vertex of the parabola given by $f(x) = 4x^2 - 7x + 2$. Check your answers with your graphics calculator.

Solution

Substitute 4 for a and -7 for b in the axis of symmetry formula to find the x-coordinate of the vertex.

$$x = \frac{-b}{2a} = \frac{-(-7)}{2(4)} = \frac{7}{8} = 0.875$$

Substitute 0.875 for x in $f(x) = 4x^2 - 7x + 2$ to find the y-coordinate of the vertex.

$$f(x) = 4x^2 - 7x + 2$$
$$f(0.875) = 4(0.875)^2 - 7(0.875) + 2$$
$$= -1.0625$$

The vertex is $(0.875, -1.0625)$.

Check

Graph $f(x) = 4x^2 - 7x + 2$ with your graphics calculator.

Thus, the coordinates of the vertex are $(0.875, -1.0625)$.

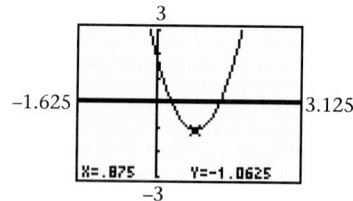

EXERCISES & PROBLEMS

Communicate

1. Explain how to solve $-2x^2 + 3x = -8$ using the quadratic formula.

2. What are the different methods you can use to find the x-intercepts of a parabola?

3. What are the different methods you can use to find the vertex of a parabola?

4. How can the axis of symmetry formula be used to find minimum and maximum values?

5. Which method do you prefer for solving quadratic equations: completing the square or using the quadratic formula? Explain why you choose that method.

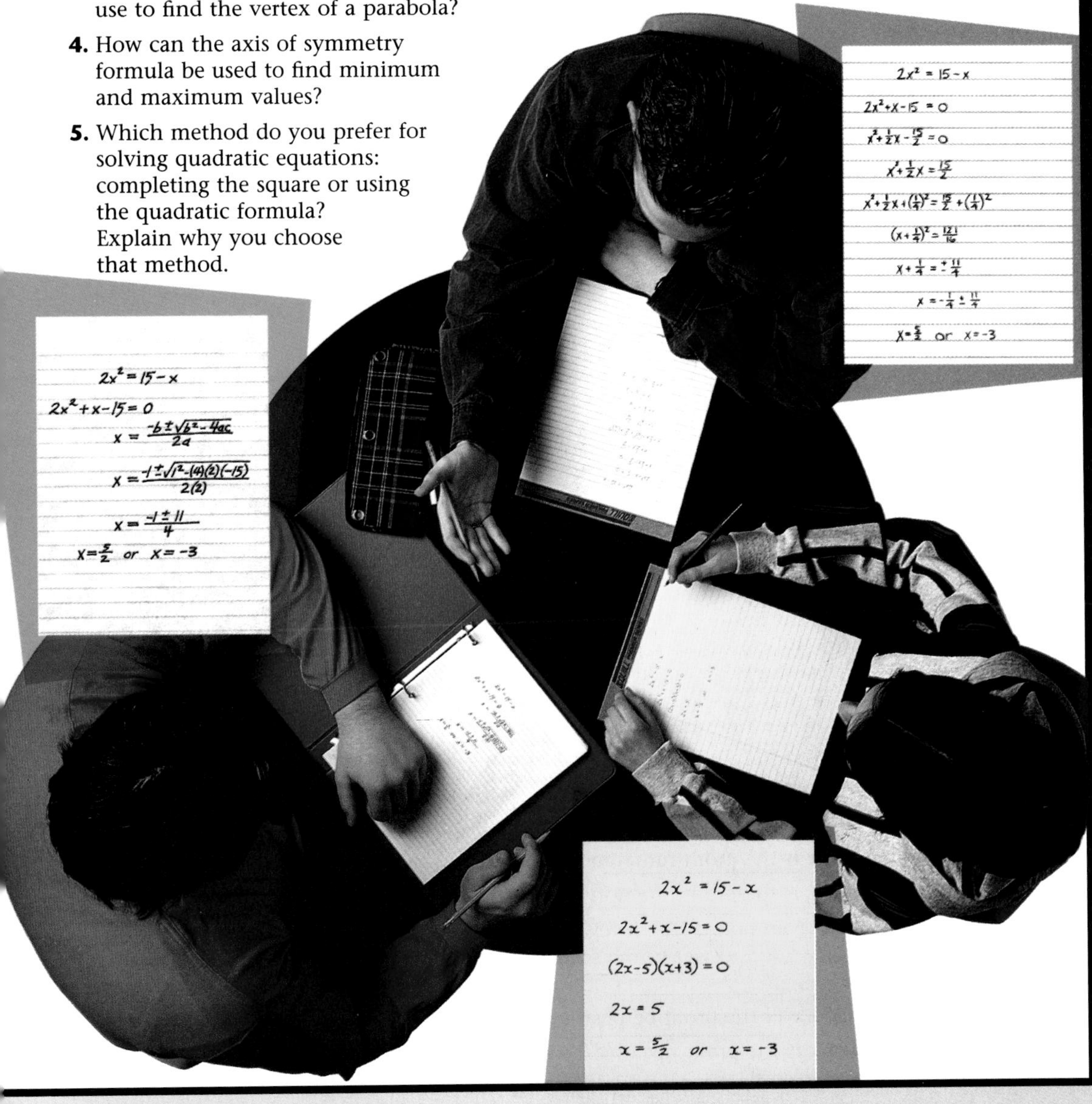

ASSESS

Selected Answers

Odd-numbered problems 7–51

Assignment Guide

Core 1–15, 18–29, 32–55

Core Plus 1–5, 11–19, 25–58

Technology

A graphics calculator is needed on Exercises 20–31 and 44–46.

$2x^2 = 15 - x$

$2x^2 + x - 15 = 0$

$x = \frac{-b \pm \sqrt{b^2 - 4ac}}{2a}$

$x = \frac{-1 \pm \sqrt{1^2 - (4)(2)(-15)}}{2(2)}$

$x = \frac{-1 \pm 11}{4}$

$x = \frac{5}{2}$ or $x = -3$

$2x^2 = 15 - x$

$2x^2 + x - 15 = 0$

$x^2 + \frac{1}{2}x - \frac{15}{2} = 0$

$x^2 + \frac{1}{2}x = \frac{15}{2}$

$x^2 + \frac{1}{2}x + \left(\frac{1}{4}\right)^2 = \frac{15}{2} + \left(\frac{1}{4}\right)^2$

$\left(x + \frac{1}{4}\right)^2 = \frac{121}{16}$

$x + \frac{1}{4} = \pm\frac{11}{4}$

$x = -\frac{1}{4} \pm \frac{11}{4}$

$x = \frac{5}{2}$ or $x = -3$

$2x^2 = 15 - x$

$2x^2 + x - 15 = 0$

$(2x - 5)(x + 3) = 0$

$2x = 5$

$x = \frac{5}{2}$ or $x = -3$

RETEACHING the lesson

Using Discussion Have students work in groups of three. Give each group a list of quadratic equations to solve. Before asking students to solve these equations, ask them to study and discuss the three different methods for solving the same equation shown on the student papers on page 285. Then ask each group to choose one of the three methods and solve each equation. If students do not choose to use the quadratic formula initially, have them use it to check their solutions. Ask each group to discuss which method they prefer and share their ideas with the class.

Watch for students who do not write quadratic equations in standard form before trying to use the quadratic formula.

20. x-intercepts: $(-1.38, 0)$ and $(0.52, 0)$;
vertex: $(-0.43, -6.29)$

21. x-intercepts: $(-2, 0)$ and $(-7, 0)$;
vertex: $(-4.5, -6.25)$

22. x-intercepts: $(-0.5, 0)$ and $(-3, 0)$;
vertex: $(-1.75, -3.125)$

23. x-intercepts: $(0.56, 0)$ and $(-3.56, 0)$;
vertex: $(-1.5, 21.25)$

24. x-intercepts: $(1.65, 0)$ and $(-3.65, 0)$;
vertex: $(-1, -21)$

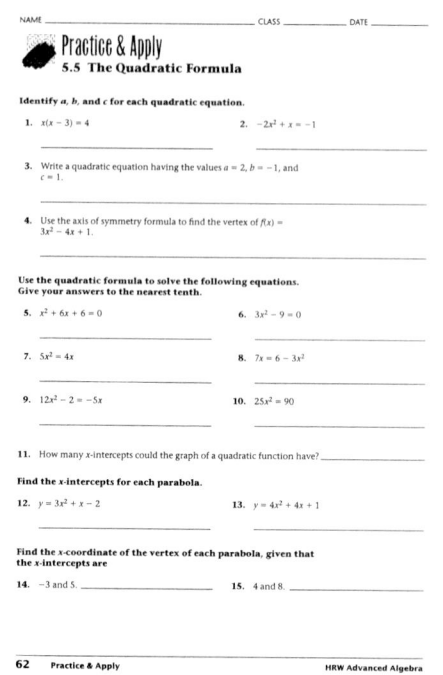

Practice & Apply

Use the quadratic formula to solve the following equations. Give your answers to the nearest tenth.

6. $x^2 + 7x + 9 = 0$ $-1.7, -5.3$
7. $x^2 + 6x = 0$ $-6, 0$
8. $(x + 1)(x - 2) = 5$ $3.2, -2.2$
9. $x^2 + 5x - 2 = 0$ $0.4, -5.4$
10. $t^2 - 9t + 5 = 0$ $8.4, 0.6$
11. $x^2 - 3x - 1 = 0$ $3.3, -0.3$
12. $x^2 + 9x - 2 = -16$ $-2, -7$
13. $x^2 - 5x - 6 = 18$ $8, -3$
14. $5x^2 + 16x - 6 = 3$ $0.5, -3.7$
15. $4x^2 = -8x - 3$ $-0.5, -1.5$
16. $3x^2 - 3 = -5x - 1$ $\frac{1}{3}, -2$
17. $x^2 + 3x = 2 - 2x$
$0.4, -5.4$

Geometry To form a rectangular rain gutter from a flat sheet of aluminum, an equal amount of aluminum is bent up on each side. The area of the cross section is 18 square inches.

18. Write a quadratic equation to model the area of the cross section. $18 = x(12 - 2x)$

19. Find the depth, x, of the gutter. 3 in.

Find the x-intercepts and the vertex for each parabola. Use your graphics calculator to check your answers.

20. $y = 7x^2 + 6x - 5$
21. $y = x^2 + 9x + 14$
22. $y = 2x^2 + 7x + 3$
23. $y = -5x^2 - 15x + 10$
24. $y = 3x^2 + 6x - 18$
25. $y = -2x^2 + 8x + 14$
26. $y = 5x^2 - 10x + 4$
27. $y = -x^2 - 6x + 2$
28. $y = 3x^2 + 21x - 4$
29. $y = -2x^2 + 3x - 1$
30. $y = 3x^2 - 18x + 22$
31. $y = -2x^2 + 8x + 13$

Business The president of a manufacturing company hires a financial consultant. The consultant's job is to create a mathematical model that describes profit in terms of items sold. The consultant analyzes the data and arrives at the function $p(x) = -2x^2 + 16x - 22$, where x represents the number of items sold (in thousands) and $p(x)$ represents the profit (in thousands of dollars).

32. What is the maximum profit that the company can make? $10,000

33. How many items must be sold to earn the maximum profit? 4000 items

34. For what values of x is the profit function increasing? $0 < x < 4$

35. For what values of x is the profit function decreasing? $x > 4$

36. What are the break-even points (the levels of production for which the profit is 0)? Give your answer to the nearest whole number of items. 1764 and 6236

37. For what values of x will the profit be positive? $1.764 < x < 6.236$

38. For what values of x is the profit negative? What does this mean?

25. x-intercepts: $(-1.32, 0)$ and $(5.32, 0)$; vertex: $(2, 22)$

26. x-intercepts: $(1.45, 0)$ and $(0.55, 0)$; vertex: $(1, -1)$

27. x-intercepts: $(0.32, 0)$ and $(-6.32, 0)$; vertex: $(-3, 11)$

28. x-intercepts: $(0.186, 0)$ and $(-7.186, 0)$; vertex: $(-3.5, -40.75)$

29. x-intercepts: $(0.5, 0)$ and $(1, 0)$; vertex: $(0.75, 0.125)$

30. x-intercepts: $(4.29, 0)$ and $(1.71, 0)$; vertex: $(3, -5)$

31. x-intercepts: $(5.24, 0)$ and $(-1.24, 0)$; vertex: $(2, 21)$

38. $p < 0$ for $0 < x < 1.764$ and $x > 6.236$. The cost is greater than the revenue for these values of x. The company is losing money.

Maximum/Minimum A professional pyrotechnician shoots fireworks vertically into the air from the ground with an initial vertical velocity of 192 feet per second. The height of the fireworks is given by $h(t) = -16t^2 + 192t$.

39. What is the maximum height reached by the fireworks? 576 ft

40. How long does it take for the fireworks to reach the maximum height? 6 sec

41. How long does it take for the fireworks to return to the ground level? 12 sec

42. Certain fireworks shoot sparks for 2.5 seconds, leaving a comet-like trail. After how many seconds might the pyrotechnician want the fireworks to begin firing in order to have sparks at the maximum height? Explain.

43.  **Portfolio Activity** Complete the Portfolio Activity on page 253. How do the approximations that you made at the beginning of the chapter compare with the answers that you just found?

Look Back

Use your graphics calculator to find the inverse of each matrix. Round numbers to the nearest hundredth. Indicate if the matrix does not have an inverse. [Lesson 4.5]

44. $\begin{bmatrix} -3 & 2 \\ 12 & 9 \end{bmatrix}$

45. $\begin{bmatrix} -1 & 4 \\ 8 & -7 \end{bmatrix}$

46. $\begin{bmatrix} 6 & -4 & 18 \\ 21 & -3 & 19 \\ 4 & 5 & -2 \end{bmatrix}$

Determine which of the following represent real numbers. [Lesson 5.2]

47. $\sqrt{13}$ yes   **48.** $\sqrt{-9}$ no   **49.** $\sqrt{0}$ yes   **50.** $\sqrt{-15}$ no

51. If you are certain that $\sqrt{x}$ represents a real number, what does this tell you about the number represented by x? $x \geq 0$

52. If you are certain that $\sqrt{x}$ does not represent a real number, what does this tell you about the number represented by x? $x < 0$

Look Beyond

Solve each equation and determine how many real-number solutions it has: 0, 1, or 2.

53. $x^2 = 19$ 2   **54.** $x^2 = 0$ 1   **55.** $x^2 = -7$ 0

56. $(x - 5)^2 = 4$ 2   **57.** $(x - 5)^2 = 0$ 1   **58.** $(x - 5)^2 = -4$ 0

The answers to Exercises 42 and 43 can be found in Additional Answers beginning on page 842.

46. $A^{-1} \approx \begin{bmatrix} -0.08 & 0.07 & -0.02 \\ 0.11 & -0.08 & 0.24 \\ 0.11 & -0.04 & 0.06 \end{bmatrix}$

44. $A^{-1} \approx \begin{bmatrix} -0.18 & 0.04 \\ 0.24 & 0.06 \end{bmatrix}$

45. $A^{-1} = \begin{bmatrix} 0.28 & 0.16 \\ 0.32 & 0.04 \end{bmatrix}$

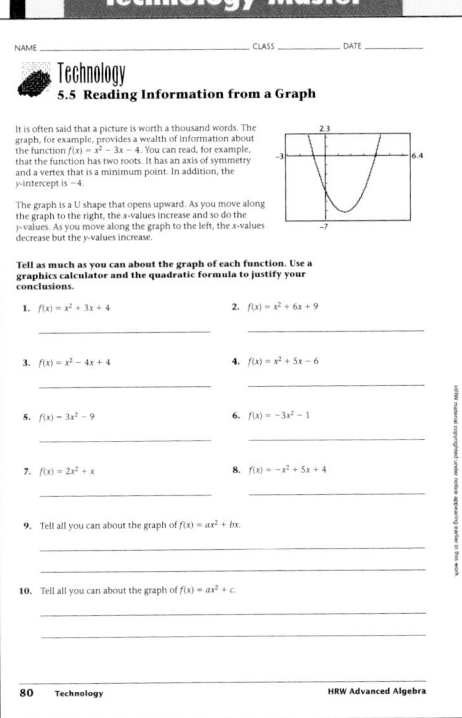

PREPARE

Objectives

- Determine the number of real-number solutions using the discriminant.
- Solve quadratic equations with imaginary-number solutions.

RESOURCES

- Practice Master   5.6
- Enrichment Master   5.6
- Technology Master   5.6
- Lesson Activity Master   5.6
- Quiz   5.6
- Spanish Resources   5.6

Assessing Prior Knowledge

Find the value of each expression.

1. $b^2 - 4ac$, if $a = -3$, $b = 4$, and $c = 2$ [**40**]

2. $b^2 - 4ac$, if $a = 2, b = 5$, and $c = 3$ [**1**]

3. $b^2 - 4ac$, if $a = -2$, $b = -6$, and $c = 1$ [**44**]

4. $b^2 - 4ac$, if $a = -3$, $b = -2$, and $c = -2$ [**−20**]

TEACH

why Ask students what they think the graph of a quadratic equation with no real solutions looks like. If they struggle with the answer, remind them that the x-intercepts give the roots of a quadratic equation. Then discuss the possibility for 1 or 2 real solutions.

Use Transparency ▶ 21

LESSON 5.6 The Discriminant and Imaginary Numbers

why The quadratic equations you have solved so far have had real-number solutions. Some quadratic equations have solutions that are not real numbers. Imaginary numbers are used to represent these numbers. You can determine whether a quadratic equation has real-number solutions by examining the discriminant.

Consider the following graphs.

$y_1 = x^2 - 6x + 5$

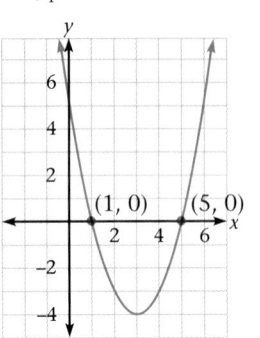

$y_2 = x^2 - 6x + 9$

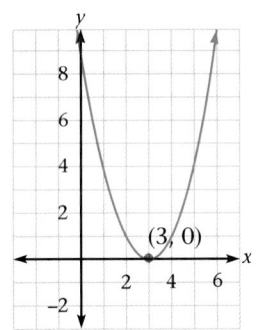

$y_3 = x^2 - 6x + 14$

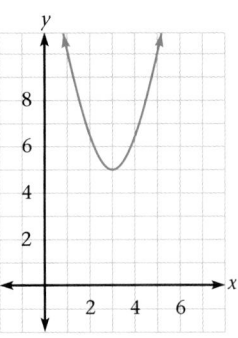

The equation $0 = x^2 - 6x + 5$ has two real solutions, 1 and 5.

The equation $0 = x^2 - 6x + 9$ has one real solution, 3.

The equation $0 = x^2 - 6x + 14$ has *no* real solutions.

$$x = \frac{-b \pm \sqrt{}}{2a}$$

Discriminant ⟶ $b^2 - 4ac$

ALTERNATIVE teaching strategy

Technology To introduce students to the three possibilities for finding the number of real solutions to a quadratic equation, have them graph the functions below using their graphics calculator or computer software such as *f(g) Scholar™*.

1. $y = x^2 - 4x + 3$ [**two real solutions**]

2. $y = x^2 - 4x + 4$ [**one real solution**]

3. $y = x^2 - 4x + 9$ [**no real solutions**]

Ask students to share their results with the class. Then have students evaluate $b^2 - 4ac$ for each of the functions, and have them record their results. [**1. 4; 2. 0; 3. −20**] Have students compare their results with the number of solutions they found when they graphed the functions.

The Discriminant

The discriminant allows you to find out how many real-number solutions a quadratic equation has without solving the equation or graphing.

If $b^2 - 4ac > 0$, then $\pm\sqrt{b^2 - 4ac}$ yields two different real numbers.

If $b^2 - 4ac = 0$, then $\pm\sqrt{b^2 - 4ac}$ is 0.

If $b^2 - 4ac < 0$, then $\pm\sqrt{b^2 - 4ac}$ is not defined for real numbers.

SOLUTIONS TO A QUADRATIC EQUATION

Let $ax^2 + bx + c = 0$ be any quadratic equation, where a, b, and c are real numbers and $a \neq 0$.
1. If $b^2 - 4ac > 0$, the equation has *two* real-number solutions.
2. If $b^2 - 4ac = 0$, the equation has *one* real-number solution.
3. If $b^2 - 4ac < 0$, the equation has *no* real-number solutions.

EXAMPLE 1

Determine how many real-number solutions exist for each equation.

Ⓐ $2x^2 + 4x + 1 = 0$ Ⓑ $2x^2 + 4x + 2 = 0$ Ⓒ $2x^2 + 4x + 3 = 0$

Solution▹

Equation	$b^2 - 4ac$	Real-number solutions
Ⓐ $2x^2 + 4x + 1 = 0$	$4^2 - 4(2)(1) = 8$ $8 > 0$	2
Ⓑ $2x^2 + 4x + 2 = 0$	$4^2 - 4(2)(2) = 0$ $0 = 0$	1
Ⓒ $2x^2 + 4x + 3 = 0$	$4^2 - 4(2)(3) = -8$ $-8 < 0$	0

❖

Examine the graphs of the equations from Example 1.

$y = 2x^2 + 4x + 1$ $y = 2x^2 + 4x + 2$ $y = 2x^2 + 4x + 3$

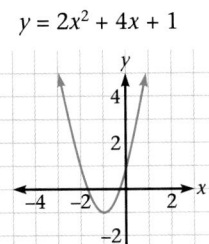 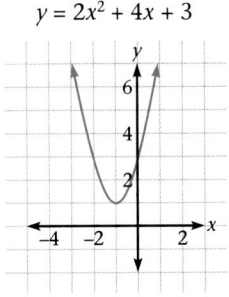

$b^2 - 4ac = 8$ $b^2 - 4ac = 0$ $b^2 - 4ac = -8$
2 *different* real roots   2 *equal* real roots   No real roots

Cooperative Learning

Have students work in pairs or small groups to do research to find out about the history of the discriminant. Ask students to summarize their findings in a report that they can share with the class.

TEACHING *tip*

Emphasize that the discriminant is $b^2 - 4ac$, not $\sqrt{b^2 - 4ac}$.

Alternate Example 1

Determine how many real-number solutions exist for each equation.
A. $3x^2 - 6x + 1 = 0$ [2]
B. $3x^2 - 6x + 3 = 0$ [1]
C. $3x^2 - 6x + 5 = 0$ [0]

TEACHING *tip*

Technology Encourage students to check the number of real solutions. Have them use a graphics calculator to draw the graphs of equations.

ENRICHMENT

Have students find information about the history of imaginary numbers.

For example, in the sixteenth century, mathematicians reluctantly began to accept the existence of imaginary numbers. Cardano noticed problems with the solution of cubic equations such as $x^3 - 15x - 4 = 0$. He presented a solution of the general cubic equation in his *Ars magma*, published in 1545 and found that he could express the solution of $x^3 - 15x - 4 = 0$ as $\sqrt[3]{2 + 11\sqrt{-1}} + \sqrt[3]{\sqrt{2 - 11\sqrt{-1}}}$. Cardano also knew, by direct substitution, that 4 is a *real* solution of the equation. Could the expression be equal to 4? Bombelli proved this to be true by showing that the expression simplifies to $\left(2 + \sqrt{-1}\right) + \left(2 - \sqrt{-1}\right) = 4$. Such examples showed the existence of imaginary numbers.

Review with students all the properties of multiplication that apply to square roots. For example, the square root of a number is equal to the product of the square roots of its factors. This property is used to show that $\sqrt{-4} = \sqrt{4}\sqrt{-1} = 2i$ and that $\sqrt{-5} = \sqrt{-1}\sqrt{5} = i\sqrt{5}$. Also, review how to find the power of a product. This concept is used to show that $(2i)^2 = 2^2 i^2 = -4$ and that $\left(i\sqrt{5}\right)^2 = i^2 \left(\sqrt{5}\right)^2 = -5$.

Alternate Example 2

Simplify.

A. $\sqrt{-81}$ $[9i]$

B. $\sqrt{-37}$ $\left[i\sqrt{37}\right]$

Ongoing ASSESSMENT

Yes; changing the order when multiplying does not change the product.

Imaginary Numbers

When you square any real number, the result is always a non-negative number. So how can you take the square root of a negative number? That is, what is the solution for $x^2 = -1$?

Cultural Connection: Europe Square roots of negative numbers puzzled mathematicians for a long time. Finally, in the sixteenth century, the Italian mathematician Girolamo Cardano, 1501–1576, developed a way to work successfully with square roots of negative numbers. Cardano defined $\sqrt{-1} \cdot \sqrt{-1} = -1$. The French mathematician Rene Descartes, 1596–1650, defined the **imaginary unit, i**, such that $i = \sqrt{-1}$ and $i^2 = -1$.

With the imaginary unit, the square root of any negative number can be defined. For example,

$$\sqrt{-4} = \sqrt{4} \cdot \sqrt{-1} = 2i \qquad \text{and} \qquad (2i)^2 = (2^2)(i^2) = (4)(-1) = -4$$
$$\sqrt{-5} = \sqrt{-1} \cdot \sqrt{5} = i\sqrt{5} \qquad \text{and} \qquad (i\sqrt{5})^2 = i^2(\sqrt{5})^2 = (-1)(5) = -5$$

IMAGINARY NUMBER

For any positive real number a, $\sqrt{-a}$ is an **imaginary number**, where $\sqrt{-a} = i\sqrt{a}$ and $(i\sqrt{a})^2 = -a$.

INCLUSION strategies

Inviting Participation

Discuss the meaning of the term *imaginary*. Begin by asking students what comes to mind when they hear the term used. Then apply their ideas to the mathematical meaning of the term. It may be helpful for students to do more research into why the term *imaginary* was originally chosen to describe $\sqrt{-1}$.

EXAMPLE 2

Simplify. **A** $\sqrt{-36}$ **B** $\sqrt{-24}$

Solution

A $\sqrt{-36} = \sqrt{36} \cdot \sqrt{-1}$
     $= 6i$

B $\sqrt{-24} = \sqrt{-1} \cdot \sqrt{24}$
           $= i\sqrt{24}$, or $2i\sqrt{6}$ ❖

Is multiplication of imaginary numbers commutative?

EXAMPLE 3

Simplify. **A** $2i \cdot 5i$ **B** $\sqrt{-8} \cdot \sqrt{-2}$

Solution

A $2i \cdot 5i = (2 \cdot 5)(i \cdot i)$
        $= 10i^2$
        $= (10)(-1)$
        $= -10$

B $\sqrt{-8} \cdot \sqrt{-2} = i\sqrt{8} \cdot i\sqrt{2}$
               $= i^2\sqrt{16}$
               $= (-1)(4)$
               $= -4$ ❖

Try This Simplify. **A** $\sqrt{-12} \cdot 3i$ **B** $2i(5i + 3i)$

Is multiplication of imaginary numbers associative?

CRITICAL *Thinking*

Recall that for all real numbers $a \geq 0$ and $b \geq 0$, $\sqrt{a} \cdot \sqrt{b} = \sqrt{ab}$. In Example 3 Part **B**, is $\sqrt{-8} \cdot \sqrt{-2}$ equal to $\sqrt{(-8)(-2)}$? Why or why not?

Exploration *Powers of i*

1 Study the following powers of *i*. What patterns do you observe?

$i = \sqrt{-1}$
$i^2 = i \cdot i = \sqrt{-1} \cdot \sqrt{-1} = -1$
$i^3 = i^2 \cdot i = -1 \cdot \sqrt{-1} = -i$
$i^4 = i^2 \cdot i^2 = -1 \cdot (-1) = 1$

$i^5 = i^4 \cdot i = 1 \cdot i = i$
$i^6 = i^4 \cdot i^2 = 1 \cdot i^2 = -1$
$i^7 = i^4 \cdot i^3 = 1 \cdot (-i) = -i$
$i^8 = i^4 \cdot i^4 = 1$

2 Complete the following table.

$i^9 = ?$	$i^{13} = ?$	$i^{17} = ?$
$i^{10} = ?$	$i^{14} = ?$	$i^{18} = ?$
$i^{11} = ?$	$i^{15} = ?$	$i^{19} = ?$
$i^{12} = ?$	$i^{16} = ?$	$i^{20} = ?$

3 What happens when the powers of *i* become greater than 4? Describe what happens as the powers of *i* increase. How can you tell when i^n is a positive real number or a negative real number? ❖

RETEACHING *the lesson*

Using Discussion Give small groups of students several different quadratic equations. Ask them to graph each equation and find the value of $b^2 - 4ac$. Then ask students to state whether the value of $b^2 - 4ac$ is greater than 0, equal to 0, or less than 0. Have each group compare their answers with the number of real solutions shown by the corresponding graph. Ask each group to write a summary of what they have discovered, and have each group share their summary with the class. Then introduce imaginary numbers, and allow students to experiment with imaginary numbers in their groups. Provide several exercises for each group to complete, and ask them to share their results with the class.

Alternate Example 3
Simplify.
A. $-4i \cdot 3i$ [12]
B. $\sqrt{-32} \cdot \sqrt{-2}$ [-8]

Ongoing ASSESSMENT

Try This
A. $-6\sqrt{3}$
B. -16

Ongoing ASSESSMENT

Yes; grouping different factors when multiplying does not change the product.

CRITICAL *Thinking*

No. $\sqrt{-8} \cdot \sqrt{-2} = i\sqrt{8} \cdot i\sqrt{2}$
               $= i^2\sqrt{16} = -4$
but $\sqrt{(-8)(-2)} = \sqrt{16} = 4$

Therefore, the expressions are not the same.

Exploration Notes

Encourage students to use the pattern they discover in Step 1 to complete the table in Step 2.

Ongoing ASSESSMENT

3. When the powers of *i* become greater than 4, the cycle of i, i^2, i^3, and i^4 is repeated. $i^{4a} = 1$ and $i^{4a-2} = -1$, where *a* is an integer. i^n is a positive real number if *n* is divisible by 4, and it is a negative real number if *n* is even but not divisible by 4.

Alternate Example 4

Use imaginary numbers to solve the equations.
A. $x^2 + 7 = 0$ [$\pm i\sqrt{7}$]
B. $x^2 + 3 = -8$ [$\pm i\sqrt{11}$]

ASSESS

Selected Answers

Odd-numbered problems 7–41.

Assignment Guide

Core 1–24, 27–32, 35–43

Core Plus 1–8, 11–14, 19–26, 30–43

Technology

A graphics calculator is needed for Exercise 43.

Error Analysis

Be sure students understand the pattern for powers of i that they discovered in the exploration. Encourage students to refer back to their exploration work as often as necessary.

Practice Master

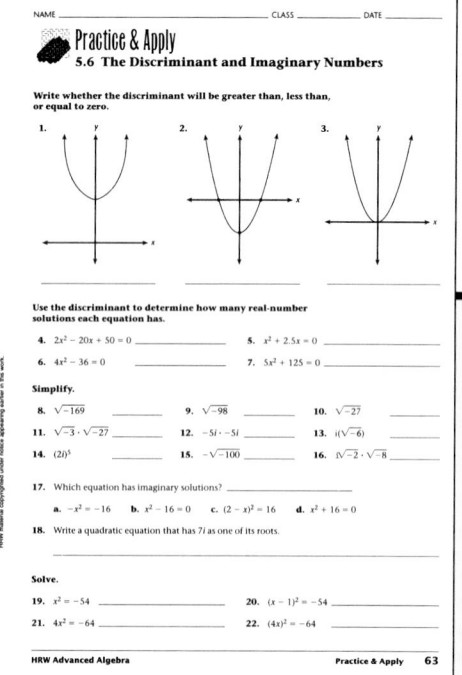

Solving Equations With Imaginary Roots

When solving equations with radicals, both positive and negative roots must be considered.

EXAMPLE 4

Use imaginary numbers to solve the equations.

Ⓐ $x^2 + 5 = 0$ Ⓑ $x^2 - 4 = -6$

Solution▶

Ⓐ $x^2 + 5 = 0$
$$x^2 = -5$$
$$x = \pm\sqrt{-5}$$
$$x = i\sqrt{5} \text{ or } x = -i\sqrt{5}$$

Ⓑ $x^2 - 4 = -6$
$$x^2 = -2$$
$$x = \pm\sqrt{-2}$$
$$x = i\sqrt{2} \text{ or } x = -i\sqrt{2}$$

Check▶

Substitute the results in the original equation to check.

Ⓐ $\left(i\sqrt{5}\right)^2 + 5 \stackrel{?}{=} 0$ $\left(-i\sqrt{5}\right)^2 + 5 \stackrel{?}{=} 0$ Ⓑ $\left(i\sqrt{2}\right)^2 - 4 \stackrel{?}{=} -6$ $\left(-i\sqrt{2}\right)^2 - 4 \stackrel{?}{=} -6$

$(-1 \cdot 5) + 5 \stackrel{?}{=} 0$ $(-1 \cdot 5) + 5 \stackrel{?}{=} 0$ $(-1 \cdot 2) - 4 \stackrel{?}{=} -6$ $(-1 \cdot 2) - 4 \stackrel{?}{=} -6$

$-5 + 5 = 0$ $-5 + 5 = 0$ $-2 - 4 = 6$ $-2 - 4 = 6$

True   True   True   True

Thus, the solutions are $i\sqrt{5}$ and $-i\sqrt{5}$.   Thus, the solutions are $i\sqrt{2}$ and $-i\sqrt{2}$. ❖

EXERCISES & PROBLEMS

Communicate

1. Explain how to determine how many real-number solutions a quadratic equation has without solving it.

2. Compare the graphs of quadratic equations having negative discriminants with graphs of quadratic equations having discriminants equal to zero. How are they alike? How are they different?

3. Explain how the square root of a negative number is defined using the imaginary unit.

4. Is it true that $\sqrt{-2} \cdot \sqrt{-3} = \sqrt{6}$? Why or why not?

5. What do all powers of i with exponents divisible by 4 have in common?

6. Explain how to solve the equation $x^2 - 4 = -9$.

Practice & Apply

Business A company determines that its profit, p, from manufacturing and selling x items is given by $p(x) = -0.05x^2 + 2x - 15$.

7. How many real-number roots does this profit function have? 2

8. What does your answer to Exercise 7 imply about the company's profit?
   $P(x) > 0$ between the two roots

Without solving, determine how many real-number solutions each equation has.

9. $x^2 - 4x + 7 = 0$ 0   **10.** $2x^2 - 5x + 1 = 0$ 2   **11.** $-x^2 + 8x - 16 = 0$ 1

12. $-3x^2 + 6x - 5 = 0$ 0   **13.** $x^2 - 7 = -6x$ 2   **14.** $-3x^2 + 6x - 5 = -4x^2 - 15$ 0

Simplify.

15. $\sqrt{-17} \cdot \sqrt{-17}$ -17 **16.** $\left(-\sqrt{-10}\right)\left(-\sqrt{-10}\right)$ -10 **17.** $(3i)(3i)$ -9 **18.** $\left(-i\sqrt{5}\right)\left(-i\sqrt{5}\right)$ -5

19. $\left(-i\sqrt{-7}\right)$ $\sqrt{7}$ **20.** $i^2 \cdot i^3$ i **21.** $4i \cdot (-6i)$ 24 **22.** $-6i \cdot (-i)$ -6

23. $-i\sqrt{3} \cdot i\sqrt{7}$ $\sqrt{21}$ **24.** $i\sqrt{7} \cdot (-2i)$ $2\sqrt{7}$ **25.** $-i\sqrt{-4} \cdot i^4$ 2 **26.** $i^3 \cdot (-2i^2)$ $-2i$

Solve.

27. $v^2 = -7$ $\pm i\sqrt{7}$ **28.** $x^2 = -20$ $\pm i\sqrt{20}$ **29.** $m^2 = -16$ $\pm 4i$ **30.** $y^2 = -18$ $\pm i\sqrt{18}$

31. $z^2 = -45$ $\pm i\sqrt{45}$ **32.** $p^2 = -19$ $\pm i\sqrt{19}$ **33.** $q^2 = -18$ $\pm i\sqrt{18}$ **34.** $r^2 = -49$ $\pm 7i$

Look Back

Solve each equation for x in terms of the other variables.
[Lesson 1.5]

35. $\frac{a}{x} = \frac{5}{e}$ $x = \frac{ae}{5}$ **36.** $\frac{sx}{r} + a = b$ $x = \frac{r}{s}(b-a)$ **37.** $\frac{x}{b} + c = d - f$ $x = b(d - f - c)$

Simplify.   [Lesson 2.2]

38. $7xy - 2yz + 3xz - 4yz + 4x - 7y$

39. $-5b^2 - 6a^2 - 2b^2 + 4a^2 + 7a + 6b + 3a$ $-2a^2 - 7b^2 + 10a + 6b$

40. $(-3a)(4b^2)(2a) + (2a^2)(4b^2) - (5ab^2)(2ab)$ $-2a^2b^2(8 + 5b)$

41. Solve $\begin{cases} 5x + y = 13 \\ -6x - 5y = -27 \end{cases}$ using matrix algebra.   **[Lesson 4.6]** $(2, 3)$

42. How can you tell the direction in which the graph of a quadratic function opens?   **[Lesson 5.3]** upward if $a > 0$, downward if $a < 0$

Look Beyond

43 Graph the function $f(x) = 3x^2 - 7x + 5$ with your graphics calculator. Estimate the x-intercept(s). What do you think the solutions to $0 = 3x^2 - 7x + 5$ would be?

38. $7xy - 6yz + 3xz + 4x - 7y$

43. There are no x-intercepts, so there are no real solutions for x; solutions are complex numbers.

PREPARE

Objectives

- Identify, operate with, and graph complex numbers.
- Find the complex roots of quadratic equations that model real-world situations.

RESOURCES

- Practice Master   5.7
- Enrichment Master   5.7
- Technology Master   5.7
- Lesoon Activity Master 5.7
- Quiz   5.7
- Spanish Resources   5.7

Assessing Prior Knowledge

Simplify.

1. $(3 + 4a) + (2 + 5a)$ $[5 + 9a]$

2. $(4 + 2x) - (7 + 3x)$ $[-3 - x]$

3. $-3y(4 + 3y)$ $[-12y - 9y^2]$

4. $(5 - 2m)(-3 + 4m)$ $[-15 + 26m - 8m^2]$

TEACH

Complex numbers are very useful in many different fields. Many students are very interested in fractals, such as the Julia sets shown on this page. Ask students to study the fractals shown and discuss where they have seen fractals used.

LESSON 5.7 Complex Numbers

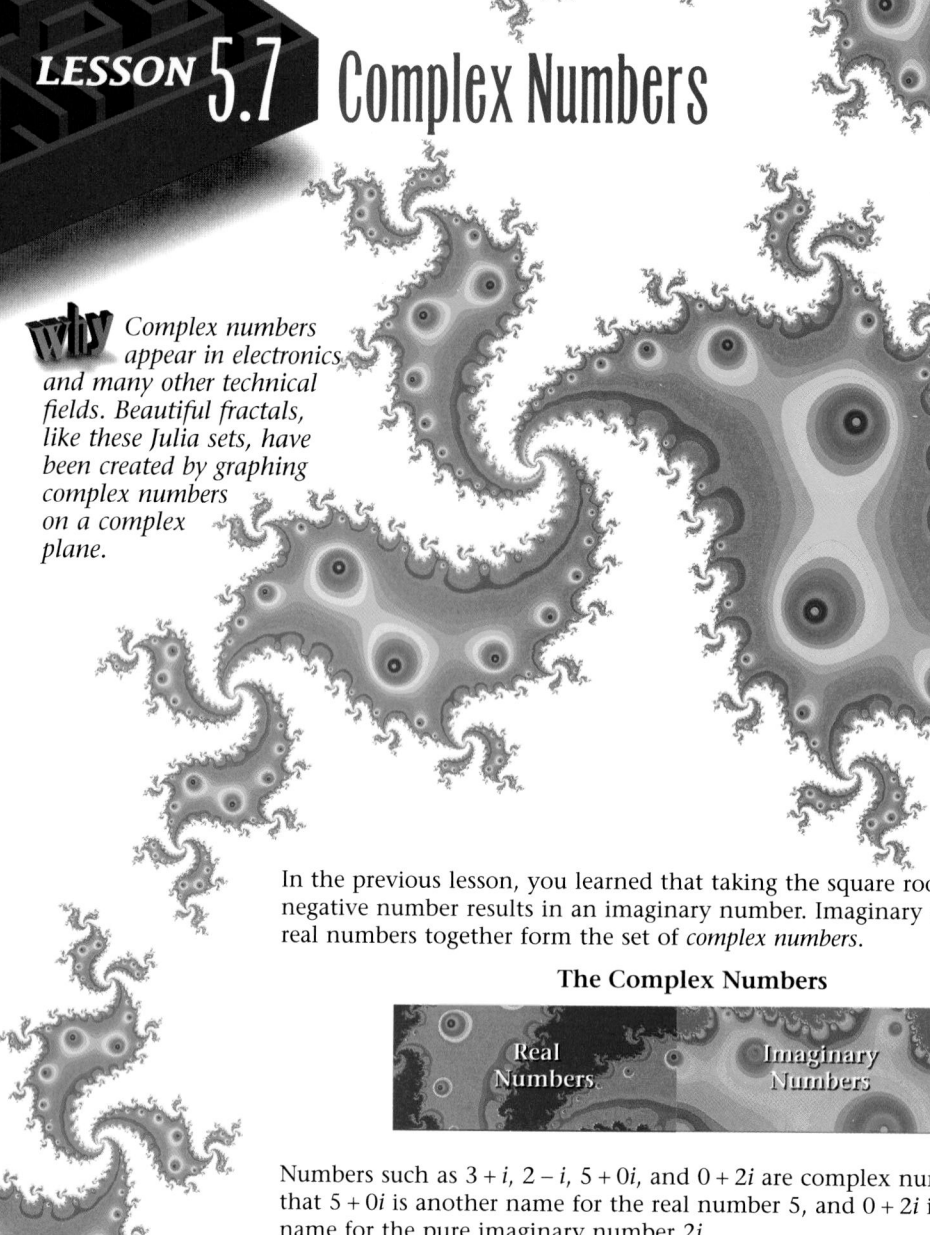

Why Complex numbers appear in electronics and many other technical fields. Beautiful fractals, like these Julia sets, have been created by graphing complex numbers on a complex plane.

In the previous lesson, you learned that taking the square root of a negative number results in an imaginary number. Imaginary numbers and real numbers together form the set of *complex numbers*.

The Complex Numbers

Numbers such as $3 + i$, $2 - i$, $5 + 0i$, and $0 + 2i$ are complex numbers. Notice that $5 + 0i$ is another name for the real number 5, and $0 + 2i$ is another name for the pure imaginary number $2i$.

COMPLEX NUMBER

A **complex number** is a number that can be written in the form $a + bi$, where a and b are real numbers and i is the imaginary unit.

Two complex numbers are equal if their real parts are equal and their imaginary parts are equal. That means for real numbers a, b, c, and d, if $a = c$ and $b = d$, then the complex numbers $a + bi$ and $c + di$ are equal.

Using Visual Models Have students write the real part of a complex number using one color, such as purple, and the imaginary part using a different color, such as red. It may help to show the two parts of a complex number, writing "real numbers" in purple and "imaginary numbers" in red. As students work through examples that show operations with complex numbers, have them continue to use purple for real numbers and red for imaginary numbers.

Alternate Example 1

Find the values of x and y such that $2x + 16i = 6 + 4yi$.
$[x = 3, y = 4]$

CRITICAL
Thinking

If $a, b, c,$ and d are *real* numbers, then it follows that if $a + bi = c + di$, then $a = c$ and $b = d$.

Alternate Example 2

Simplify.
A. $(4 + 5i) + (1 - 6i)$ $[5 - i]$
B. $(3 - 8i) - (2 + 3i)$ $[1 - 11i]$

ongoing
ASSESSMENT

Yes, grouping different factors when adding complex numbers does not affect the sum

Alternate Example 3

Simplify.
A. $-3i(6 + 5i)$ $[15 - 18i]$
B. $(-2 - 3i)(5 + 4i)$ $[2 - 23i]$

ongoing
ASSESSMENT

Yes, a factor can be distributed to both terms of a complex number to find the correct product.

EXAMPLE 1

Find the values of x and y such that $3x + 10i = 12 + 5yi$.

Solution

Real parts are equal.

$3x = 12$
$x = 4$

Imaginary parts are equal.

$10i = 5yi$
$10 = 5y$
$2 = y$

Thus, the value for x is 4 and the value for y is 2. ❖

CRITICAL
Thinking

Is it always true that if $a + bi$ and $c + di$ are equal, then $a = c$ and $b = d$? Explain.

Operating With Complex Numbers

Operating with complex numbers is very similar to operating with binomials. The imaginary unit i is treated as a variable. To add and subtract complex numbers, you combine like terms.

EXAMPLE 2

Simplify.   A $(3 + 4i) + (7 - 5i)$ B $(2 - 9i) - (6 + 3i)$

A $(3 + 4i) + (7 - 5i)$
$= (3 + 7) + (4i - 5i)$
$= 10 - i$

B $(2 - 9i) - (6 + 3i)$
$= (2 - 6) + (-9i - 3i)$
$= -4 - 12i$ ❖

Is addition of complex numbers associative?

Multiplying complex numbers is like multiplying binomials.

EXAMPLE 3

Simplify.   A $-2i(5 + 3i)$ B $(5 - 2i)(-4 + 3i)$

Solution

A $-2i(5 + 3i)$
$= -10i + (-6i^2)$
$= -10i + (-6)(-1)$
$= -10i + 6$
$= 6 - 10i$

B $(5 - 2i)(-4 + 3i)$
$= -20 + 15i + 8i - 6i^2$
$= -20 + 23i - (6)(-1)$
$= -20 + 23i + 6$
$= -14 + 23i$ ❖

Does the distributive property apply to complex numbers?

interdisciplinary
CONNECTION

Electronics Three basic components are considered in a simplified electric current: the flow of the current, I, given in amps, the resistance to that flow, Z, called impedance and given in ohms, and the electromotive force, V, called voltage and given in volts. The formula $V = ZI$ gives the relationship between these three components. The current, impedance, and voltage are often expressed as complex numbers. Electrical engineers often represent imaginary numbers using j instead of i. That is, $j = \sqrt{-1}$ and $j^2 = -1$. They also write the j before any coefficients instead of after the coefficient. For example, engineers would write the complex number $4 + 2i$ as $4 + j2$. Have students use this information to solve the following problem.

Find the voltage if the current is $(10 + j6)$ amps and the impedance is $(5 - j2)$ ohms.
$[(62 + j10) \text{ volts}]$

Solving Quadratic Equations With Complex Roots

With complex numbers, you can find the solutions to a quadratic equation that has a discriminant less than zero. The solutions to this type of quadratic equation are complex numbers.

EXAMPLE 4

Solve $3x^2 - 7x + 5 = 0$. Write the solutions in simplest $a + bi$ form.

Solution

Identify a, b, and c, and substitute these values into the quadratic formula.

$$x = \dfrac{-b \pm \sqrt{b^2 - 4ac}}{2a}$$

a is 3, b is -7, and c is 5.

$$= \dfrac{-(-7) \pm \sqrt{(-7)^2 - 4(3)(5)}}{2(3)}$$

$$= \dfrac{7 \pm \sqrt{-11}}{6}$$

$$x = \dfrac{7 + \sqrt{-11}}{6} \text{ or } x = \dfrac{7 - \sqrt{-11}}{6}$$

$$x = \dfrac{7}{6} + \dfrac{\sqrt{11}}{6}i \text{ or } x = \dfrac{7}{6} - \dfrac{\sqrt{11}}{6}i \qquad \diamond$$

Try This   Solve $2x^2 - 4x + 5 = 0$. Write the solutions in simplest $a + bi$ form.

EXAMPLE 5

Graphics Calculator

A manufacturing company is considering manufacturing and selling a new product. How can the financial analyst for this company find out whether it would be profitable for the company to proceed with plans to launch the new product? The variable x represents the number (in hundreds) of items manufactured and sold.

One model describes the cost (in thousands of dollars) of manufacturing the product (money going out from the company).

Cost model:
$c(x) = 4x + 65$

The other model describes the revenue (in thousands of dollars) from selling the product (money coming into the company).

Revenue model:
$r(x) = -x^2 + 20x$

Ⓐ Find the break-even points, where cost equals revenue. Use your graphics calculator to check your answer.

Ⓑ Should the company proceed with plans to launch its new product? Explain why or why not.

ENRICHMENT Have students create a fractal of their own. If computer software is available that students can use to do this, encourage them to print out several designs to share with the class. Such fractals make an excellent bulletin board display.

INCLUSION **strategies** **English Language Development** Discuss the meaning of the term *complex*. Begin by asking students what they think of when they hear the term used. Then apply their ideas to the mathematical meaning of the term. It may be helpful for students to do more research into why the term *complex* was originally chosen.

Cooperative Learning

Technology Have students work in pairs or small groups to find a cost model and a revenue model that will show one or more break-even points. Groups can use graphics calculators or computer software such as *f(g) Scholar*™ to change the coefficients and constants until they get two equations that will work.

TEACHING *tip*

Be sure students understand that actual break-even points will be real numbers. Emphasize the concept that complex solutions indicate that the graphs do not intersect.

Solution▸

Ⓐ The break-even point is the point at which cost and revenue are equal — the point(s) of intersection of the cost and revenue graphs.

To find the exact break-even point(s), solve $4x + 65 = -x^2 + 20x$.

$$4x + 65 = -x^2 + 20x$$
$$x^2 - 16x + 65 = 0$$
$$x = \frac{16 \pm \sqrt{16^2 - 4(1)(65)}}{2(1)}$$
$$x = 8 + i \text{ or } x = 8 - i$$

Since the solutions of the equation are not real numbers, the two graphs do not intersect. There are no break-even points.

Use your graphics calculator to graph the cost function and revenue function on the same coordinate plane. Use the trace feature to find the point(s) of intersection. It appears that a break-even point occurs at an approximate *x*-value of 8.

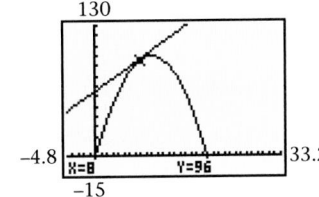

However, if you zoom in on the graphs around this point, you can see that they do *not* intersect. That is why the solutions to $4x + 65 = -x^2 + 20x$ are complex numbers.

Thus, there is no break-even point for the company.

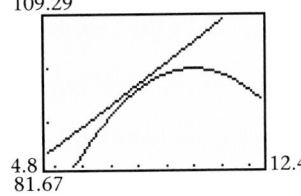

Ⓑ Since the cost function is always above the revenue function, cost will always exceed revenue. The company should not launch its new product unless it can find a way to decrease cost or increase revenue so that revenue will exceed cost. ❖

RETEACHING
the
lesson

Independent Learning Have students name the real part and the imaginary part of each complex number.

1. $2 + 4i$ [**real: 2, imaginary: 4i**]

2. $0 - 3i$ [**real: 0; imaginary: −3i**]

3. $0 + 6i$ [**real : 0; imaginary: 6i**]

4. $7 + 0i$ [**real: 7; imaginary: 0i**]

5. $-1 - 0i$ [**real: −1; imaginary: 0i**]

Ask students if any of the numbers can be simplified. [**2. −3i; 3. 6i; 4. 7; 5. −1**] Then ask if any of the numbers are actually real and why. [**Yes, 4 and 5 are real since the imaginary part of these numbers is 0i = 0.**] Progress to operations on complex numbers. First, have students simplify a similar expression that contains a variable other than *i*. Then have students simplify the same expression with *i* instead of the variable.

Complex Conjugates

Examine the solutions for Example 4. How are the real parts of the solutions alike or different? How are the imaginary parts alike or different?

The solutions in Example 4 are called *complex conjugates.*

COMPLEX CONJUGATES

The complex numbers $a + bi$ and $a - bi$ are **complex conjugates.** The complex conjugate of $a + bi$ is denoted $\overline{a + bi}$.

Examine the solutions in Examples 4 and 5. What is the relationship between the two solutions found by using the quadratic formula?

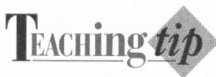

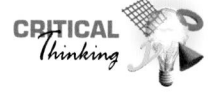

EXAMPLE 6

Find $\overline{-3 + 7i}$.

Solution

The complex conjugate of $-3 + 7i$ will have the *same real* part, -3.

The complex conjugate of $-3 + 7i$ will have the *opposite imaginary* part, $-7i$.

Thus, $\overline{-3 + 7i}$ is $-3 - 7i$. ❖

CRITICAL *Thinking* Is the product of complex conjugates always a real number? Explain.

Graphing Complex Numbers

Complex numbers are graphed on the *complex plane.* On the **complex plane,** the horizontal axis is called the **real axis** and the vertical axis is called the **imaginary axis.**

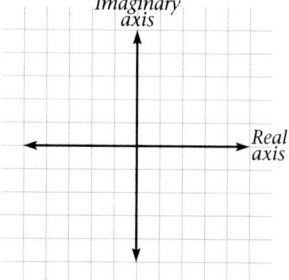

To graph the complex number $a + bi$, plot the point (a, b) on the complex plane. For example, the point $(3, 4)$ represents the complex number $3 + 4i$.

ongoing ASSESSMENT

The real parts of the solutions are the same. The imaginary parts have opposite signs.

TEACHING *tip*

Be sure students realize that the only difference between a complex number and its conjugate is the sign of the imaginary part.

ongoing ASSESSMENT

The two solutions are complex conjugates.

Alternate Example 6

Find $\overline{-4 - 3i}$ [$\mathbf{-4 + 3i}$]

CRITICAL *Thinking*

Yes, the product of complex conjugates is always a real number since $(a + bi)(a - bi) = a^2 - b^2i^2 = a^2 + b^2$.

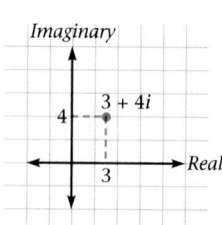

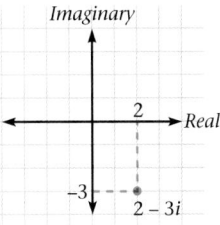

EXAMPLE 7

Plot $2 - 3i$ on the complex plane.

Solution➤

The real part is 2; locate 2 on the real axis.

The imaginary part is -3; locate -3 on the imaginary axis.

Plot the point on the complex plane. ❖

Try This   Plot $-3 + 4i$ on the complex plane.

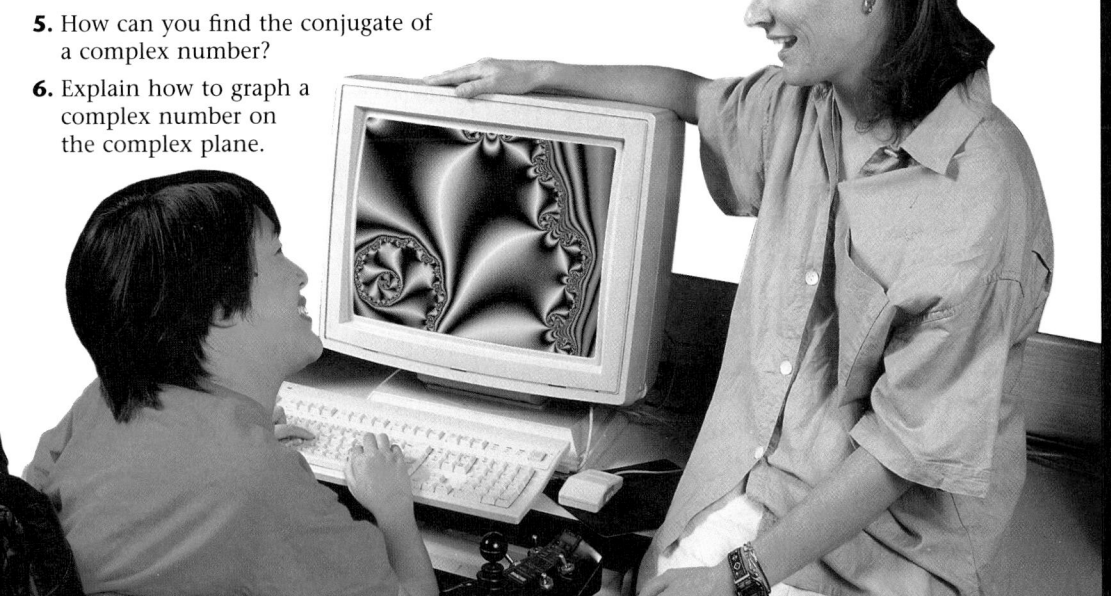

EXERCISES & PROBLEMS

Communicate

1. Why do the complex numbers include both imaginary and real numbers?

2. How can you solve $2x + 5i = 9 + 20yi$ for x and y?

3. Explain how to add, subtract, and multiply complex numbers.

4. If the graph of $f(x) = ax^2 + bx + c$ has no x-intercepts, what type of solutions will the equation $ax^2 + bx + c = 0$ have? How are these solutions related to each other?

5. How can you find the conjugate of a complex number?

6. Explain how to graph a complex number on the complex plane.

Alternate Example 7

Plot $-2 - 5i$ on the complex plane.

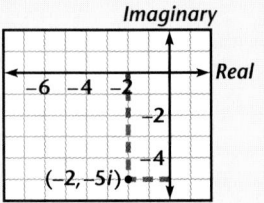

ongoing

ASSESSMENT

Try This

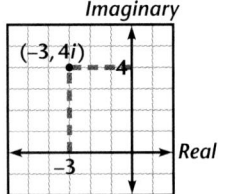

ASSESS

Selected Answers

Odd-numbered problems 7–59

Assignment Guide

Core 1–13, 16–21, 26–29, 34–39, 42–62

Core Plus 1–7, 11–17, 18–25, 29–33, 37–62

Technology

A graphics calculator is needed for Exercise 60.

Remind students to simplify i^2, that $i = \sqrt{-1}$, and that therefore $i^2 = -1$.

7. $x^2 - 8x + 17$
$= (4 + i)^2 - 8(4 + i) + 17$
$= 16 + 8i - 1 - 32 - 8i + 17$
$= 33 - 33 + (8 - 8)i$
$= 0$

8. $-\dfrac{5}{2} + \dfrac{\sqrt{7}}{2}i, -\dfrac{5}{2} - \dfrac{\sqrt{7}}{2}i$

9. $\dfrac{5}{6} + \dfrac{\sqrt{23}}{6}i, \dfrac{5}{6} - \dfrac{\sqrt{23}}{6}i$

10. $\dfrac{7}{2} + \dfrac{\sqrt{3}}{2}i, \dfrac{7}{2} - \dfrac{\sqrt{3}}{2}i$

11. $\dfrac{1}{2} + \dfrac{\sqrt{15}}{10}i, \dfrac{1}{2} - \dfrac{\sqrt{15}}{10}i$

12. $\dfrac{5}{2} + \dfrac{\sqrt{5}}{2}i, \dfrac{5}{2} - \dfrac{\sqrt{5}}{2}i$

13. $\dfrac{3}{2} + \dfrac{\sqrt{19}}{2}i, \dfrac{3}{2} - \dfrac{\sqrt{19}}{2}i$

14. $4 + i\sqrt{3}, 4 - i\sqrt{3}$

15. $\dfrac{3}{2} + \dfrac{1}{2}i, \dfrac{3}{2} - \dfrac{1}{2}i$

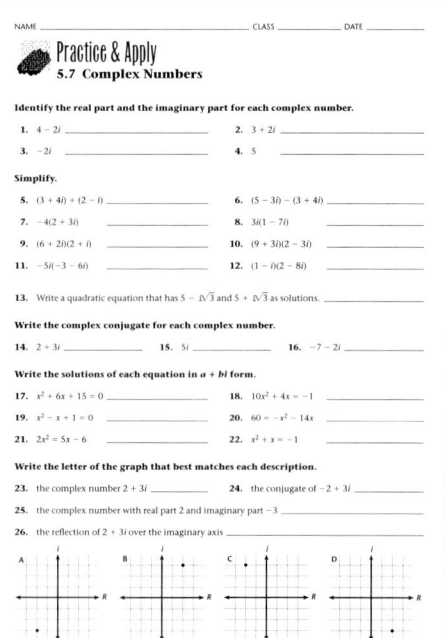

Practice & Apply

7. Use substitution to show that $4 + i$ is a solution of $x^2 - 8x + 17 = 0$.

Write the solutions of each equation in $a + bi$ form.

8. $x^2 + 5x + 8 = 0$ 9. $3x^2 - 5x + 4 = 0$ 10. $x^2 - 7x = -13$ 11. $5x^2 - 5x + 2 = 0$
12. $-2x^2 + 10x - 15 = 0$ 13. $x^2 - 3x = -7$ 14. $-x^2 + 8x - 19 = 0$ 15. $2x^2 - 6x = -5$

Refer to Example 5 in the lesson. Modify the cost model as indicated and repeat parts Ⓐ and Ⓑ of the example.

16. **Business** The fixed cost of \$65 is reduced to \$64: $c(x) = 4x + 64$.

17. **Business** The fixed cost of \$65 is reduced to \$55: $c(x) = 4x + 55$.

Identify the real part and the imaginary part for each complex number.

18. $-5 + 6i$ 19. $0 + 6i$ 20. 8 21. $4i$

Write the complex conjugate for each complex number.

22. $2 + 3i$ $2 - 3i$ 23. $-5 + i$ $-5 - i$ 24. $-4 - i$ $-4 + i$ 25. $8 - 3i$ $8 + 3i$

Plot each number on a separate complex plane.

26. $3 - 5i$ 27. $-2 + 7i$ 28. $5 + i$ 29. $-7 - 2i$
30. -4 31. $6i$ 32. $-2i$ 33. 3

Plot each complex number and its conjugate, together, on a complex plane.

34. $5 + 3i$ 35. $-2 + 4i$ 36. $-3 - 5i$ 37. $2 - 3i$
38. $5 - 2i$ 39. $7 + 6i$ 40. $-8 + 5i$ 41. $-3 - 2i$

42. **Transformations** Describe the relationship between the graphs of a complex number and its conjugate in terms of a translation, rotation, or reflection.   reflection about real axis

43. Describe the location of real numbers, such as 2, –5, and $\sqrt{7}$, when graphed on the complex plane.   points on the real axis

44. Describe the location of pure imaginary numbers, such as $2i$, $-5i$, and $i\sqrt{7}$, when graphed on the complex plane.   points on the imaginary axis

45. Identify the complex numbers graphed on the complex plane at the right. $-2 + 4i, 3 - i, -3 - 3i$

46. Identify the coordinates of the conjugate for each of the complex numbers graphed at the right. Describe the relationship between the locations of a complex number and its conjugate.

47. Find the product of each pair of conjugates from Exercise 46, and graph the conjugate pairs and their products. How is the location of the product related to the location of its complex factors?

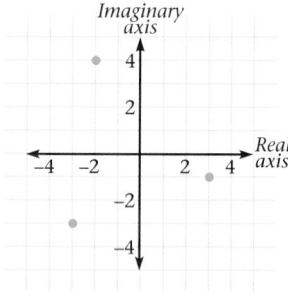

16. **A.** The break-even point occurs when $x = 8$ or when 800 items are sold.
B. Since there is only one solution, they only break even. They do not make a profit, so they would probably not launch the product in this case.

B. Two break-even points mean that in between these x-values, revenue exceeds cost. The company could find the x-value that maximizes their revenue (between 500 and 1100 items). They would probably launch the product in this case.

17. **A.** There are two break-even points: when $x = 5$ and when $x = 11$, or when 500 items are sold and when 1100 items are sold.

The answers for Exercises 18–21, 26–41, 46, and 47 can be found in Additional Answers beginning on page 842.

 **Look Back**

For Exercises 48–51, y varies directly as x. **[Lesson 1.4]**

48. If y is –10 when x is 5, find y when x is 3. –6

49. If y is 6 when x is $-\frac{1}{3}$, find y when x is –1. 18

50. If y is 2.7 when x is 3, find y when x is –5. –4.5

51. If y is $4\frac{1}{2}$ when x is 2, find y when x is 7. 15.75

52. Business Management Gene ordered 8 prints for his new restaurant. Each unframed print costs $50, and each framed print costs $98. The total cost of the prints was $640. How many framed prints did Gene buy? **[Lesson 1.5]** 5

Graph each function and its inverse on the same coordinate plane. **[Lesson 3.2]**

53. $f(x) = 2x - 3$ **54.** $f(x) = -3x + 5$ **55.** $f(x) = -x$

Let $f(x) = 3x - 5$. **[Lesson 3.3]**

56. Find $f^{-1}(x)$. $\frac{x+5}{3}$ **57.** Find $f \circ f^{-1}$. x **58.** Find $f^{-1} \circ f$. x

59. Ecology Waste-Not recycling center pays 40¢ per pound of aluminum and 25¢ per 100 pounds of newspaper. Linda took a 460-pound load of aluminum and newspaper to the Waste-Not recycling center and collected $21.82. How many pounds of aluminum and how many pounds of newspaper did Linda have? **[Lessons 4.3, 4.4, 4.6]**

52 lb alum, 408 lb newspaper

 **Look Beyond**

 Statistics Use the following table for Exercises 60–62.

x	–5	–3	–1	1	3	5
y	–2	1	5	6	4	1

60 Use your graphics calculator to plot the points (x, y) from the table on a scatter plot. Graph the line of best fit on the scatter plot. Use the trace feature on the line of best fit to predict the y-value for an x-value of 8. approx 5.36

61. Plot the points on paper on a coordinate plane. Connect all of the points, in order, for values of x ranging from –5 to 5. Describe the shape of this graph. Look at your graph on paper and predict the y-value for an x-value of 8.

62. Did you predict the same number in Exercise 61 that you did in Exercise 60? Why or why not?

53. $f(x) = 2x - 3; f^{-1}(x) = \frac{x+3}{2}$

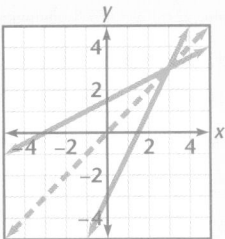

54. $f(x) = -3x + 5; f^{-1}(x) = \frac{5-x}{3}$

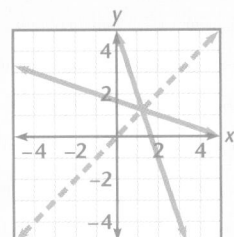

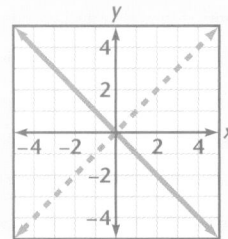
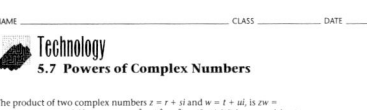

Objectives

- Write a quadratic model that fits three data points from real-world data.

RESOURCES

• Practice Master	**5.8**
• Enrichment Master	**5.8**
• Technology Master	**5.8**
• Lesson Activity Master	**5.8**
• Quiz	**5.8**
• Spanish Resources	**5.8**

Assessing Prior Knowledge

Use a graphics calculator to find the inverse of each matrix.

1. $\begin{bmatrix} 3 & 5 \\ 1 & 2 \end{bmatrix}$ $\left[\begin{bmatrix} 2 & -5 \\ -1 & 3 \end{bmatrix}\right]$

2. $\begin{bmatrix} -1 & 0 \\ 0 & 1 \end{bmatrix}$ $\left[\begin{bmatrix} -1 & 0 \\ 0 & 1 \end{bmatrix}\right]$

3. $\begin{bmatrix} 4 & -1 & 3 \\ 2 & -2 & 4 \\ 3 & -1 & 1 \end{bmatrix}$

   $\left[\begin{bmatrix} 0.2 & -0.2 & 0.2 \\ 1 & -0.5 & -1 \\ 0.4 & 0.1 & -0.6 \end{bmatrix}\right]$

TEACH

 A model can be used to show the path that an object will follow. Ask students to think of objects that might follow a path that can be modeled by a parabola.

Aongoing ASSESSMENT

$a > 0$

LESSON 5.8 Curve Fitting With Quadratic Functions

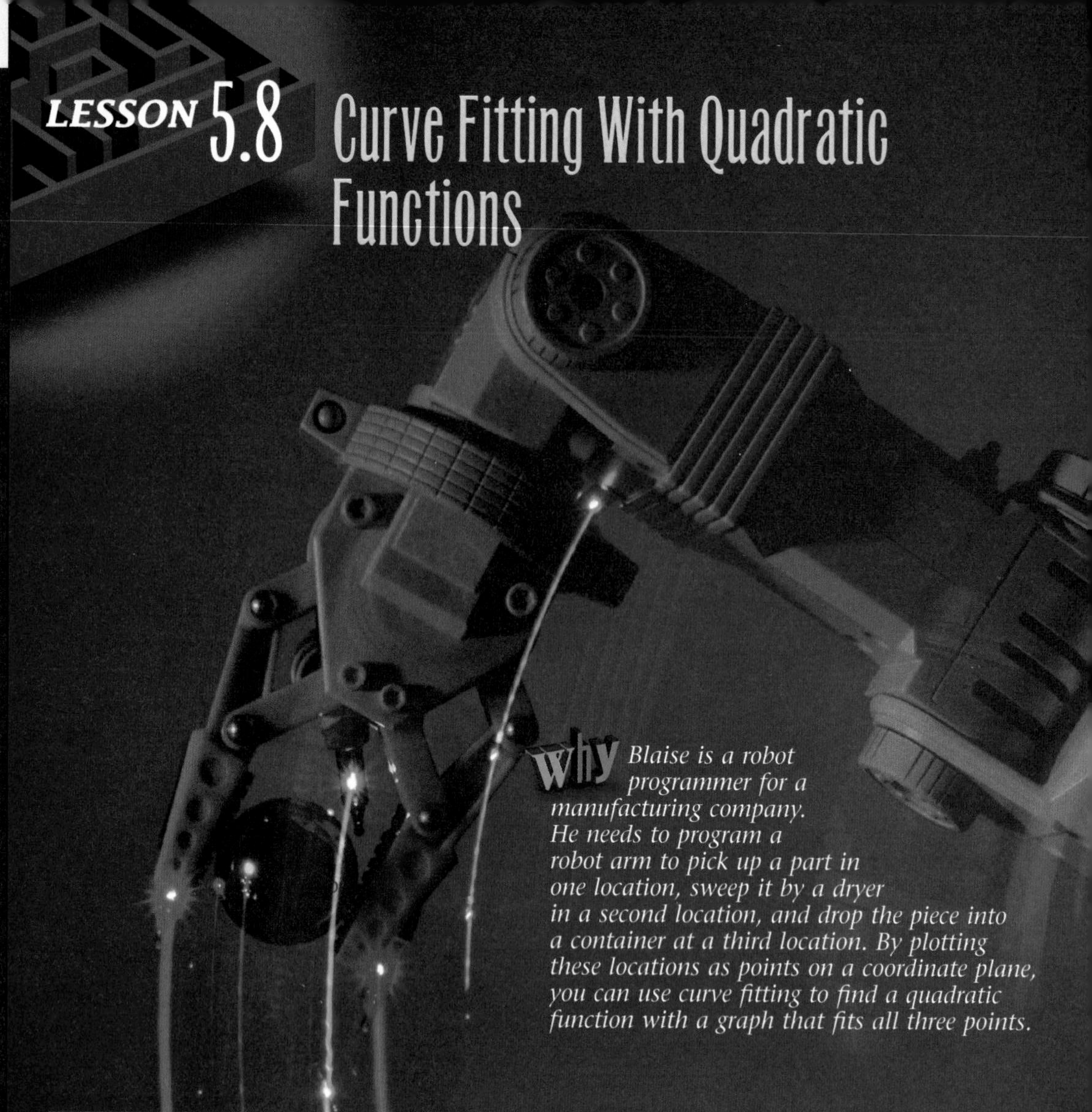

why *Blaise is a robot programmer for a manufacturing company. He needs to program a robot arm to pick up a part in one location, sweep it by a dryer in a second location, and drop the piece into a container at a third location. By plotting these locations as points on a coordinate plane, you can use curve fitting to find a quadratic function with a graph that fits all three points.*

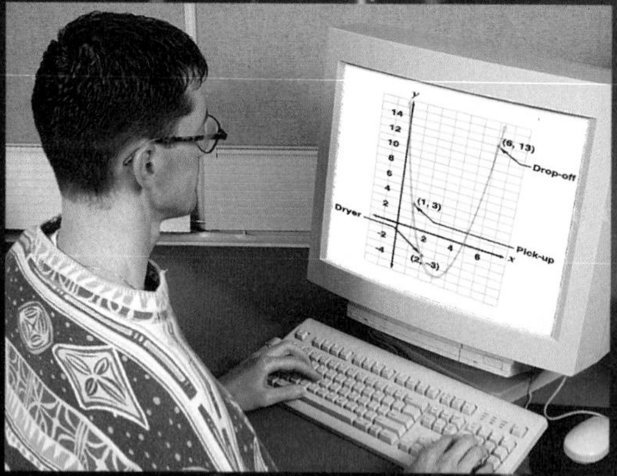

Blaise first draws a sketch of the assembly room. He then plots the points (1, 3), (2, −3), and (6, 13) to represent the locations through which the robot's arm must pass. Since the path is U-shaped, Blaise will search for a quadratic equation to fit the points.

The *y*-value decreases at the second point, then increases at the third point. Therefore, the minimum point should be somewhere between the second and third points.

Since the graph opens upward, is $a > 0$ or is $a < 0$?

ALTERNATIVE teaching strategy

Hands-On Strategies Give small groups of students a set of three points to graph, such as (1, 3), (3, −1), and (5, 3). Ask them to draw a rough graph of a parabola that will pass through all three points. Then instruct students to find the *x*-intercepts of the graph. [(2, 0), (4, 0)] Have them use the *x*-intercepts to write an equation to describe their graph, and use a graphics calculator to view their graph. Have students make adjustments to their graph, if necessary, and write a new equation. Students should continue to make adjustments until they find the correct graph and equation. Have students repeat this activity as many times as necessary until they feel comfortable with the concept. It may be helpful to have students use different colored pencils while making their adjustments each time. This will enable students to look back on their work and analyze the steps they took to get to the correct function. [$f(x) = x^2 - 6x + 8$]

EXAMPLE 1

Graphics Calculator

Find a quadratic model that fits the points (1, 3), (2, −3), and (6, 13), through which the robot's arm must pass.

Solution

To identify the coefficients a, b, and c for the model $y = ax^2 + bx + c$, substitute the values of x and y from the given points into the equation $y = ax^2 + bx + c$.

Point	x	y	$y = ax^2 + bx + c$
(1, 3)	1	3	$3 = a(1)^2 + b(1) + c$
(2, −3)	2	−3	$-3 = a(2)^2 + b(2) + c$
(6, 13)	6	13	$13 = a(6)^2 + b(6) + c$

This creates a system of equations in three unknowns.

$$\begin{cases} a + b + c = 3 \\ 4a + 2b + c = -3 \\ 36a + 6b + c = 13 \end{cases}$$

Solve the system of equations using matrix algebra.

$$\begin{bmatrix} 1 & 1 & 1 \\ 4 & 2 & 1 \\ 36 & 6 & 1 \end{bmatrix} \begin{bmatrix} a \\ b \\ c \end{bmatrix} = \begin{bmatrix} 3 \\ -3 \\ 13 \end{bmatrix} \quad \leftarrow AX = B$$

$$\begin{bmatrix} a \\ b \\ c \end{bmatrix} = \begin{bmatrix} 2 \\ -12 \\ 13 \end{bmatrix} \quad \leftarrow X = A^{-1}B \rightarrow$$

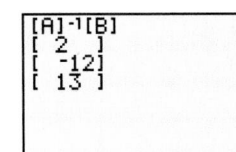

Substituting a, b, and c, the model is $y = 2x^2 - 12x + 13$.

Check

Make sure that all three points are on the graph of $y = 2x^2 - 12x + 13$.

Point (1, 3)

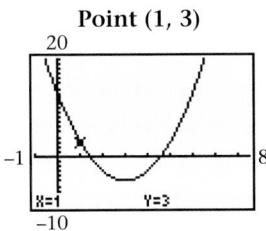

Point (2, −3)

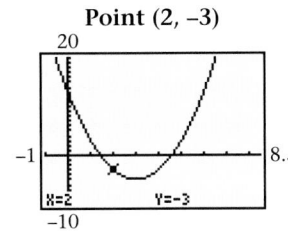

Point (6, 13)

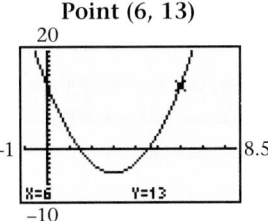

Thus, the quadratic function that fits the points (1, 3), (2, −3), and (6, 13) is $y = 2x^2 - 12x + 13$. ❖

Try This   Find a quadratic function that fits the points (2, −3), (4, 2), and (6, −2).

Alternate Example 1

Find a quadratic model that fits the points (1, 4), (2, −3), and (6, 9). [$y = 2x^2 - 13x + 15$]

TEACHING *tip*

You may want to review using matrices to solve systems of equations. Start with a system of two equations, and progress to a system of three equations.

ongoing ASSESSMENT

Try This

$y = -1.125x^2 + 9.25x - 17$

ENRICHMENT

Ask students to try to fit a quadratic curve to each of the following sets of points.

1. (2, −3), (7, 6)

2. (1, 0), (2, −6), (3, 4), (4, 7)

Students will find that an infinite number of quadratic curves will fit the points in 1, while *no* quadratic curve will fit all 4 points in 2.

Substitute 4 for x and 5 for y in the equation $y = x^2 + 2x + c$, and solve for c. [$c = -19$]

Alterate Example 2

Imagine an experiment conducted on planet X in which an object is propelled vertically from a height h_0 above the surface of the planet and with an initial velocity v_0. The height is measured at three different points in time, and the data are recorded in the following table.

t (sec)	h (ft)
1	9
2	10
3	5

A. Find a quadratic function for the data.
[$y = -3x^2 + 10x + 2$]
B. Find an approximation for the acceleration due to gravity on Planet X. [**6 feet per second squared**]
C. Find an approximation for the initial height and the initial velocity of the object.[**initial height: 2 feet; initial velocity: 10 feet per second**]

If the point (4, 5) is on the parabola with the equation $y = x^2 + 2x + c$, how can you find c?

Motion on the Moon is governed by the same laws of physics that govern motion on Earth, but each has its own acceleration due to gravity. On Earth, the acceleration due to gravity is 32 feet per second squared. Find a quadratic function based on experimental data to approximate the acceleration due to gravity on the Moon.

EXAMPLE 2

Astronomy Imagine an experiment conducted on the Moon in which an object is propelled vertically from a height, h_0, above the surface of the Moon and with an initial velocity, v_0. The height is measured at three different points in time, and the data is recorded in a table.

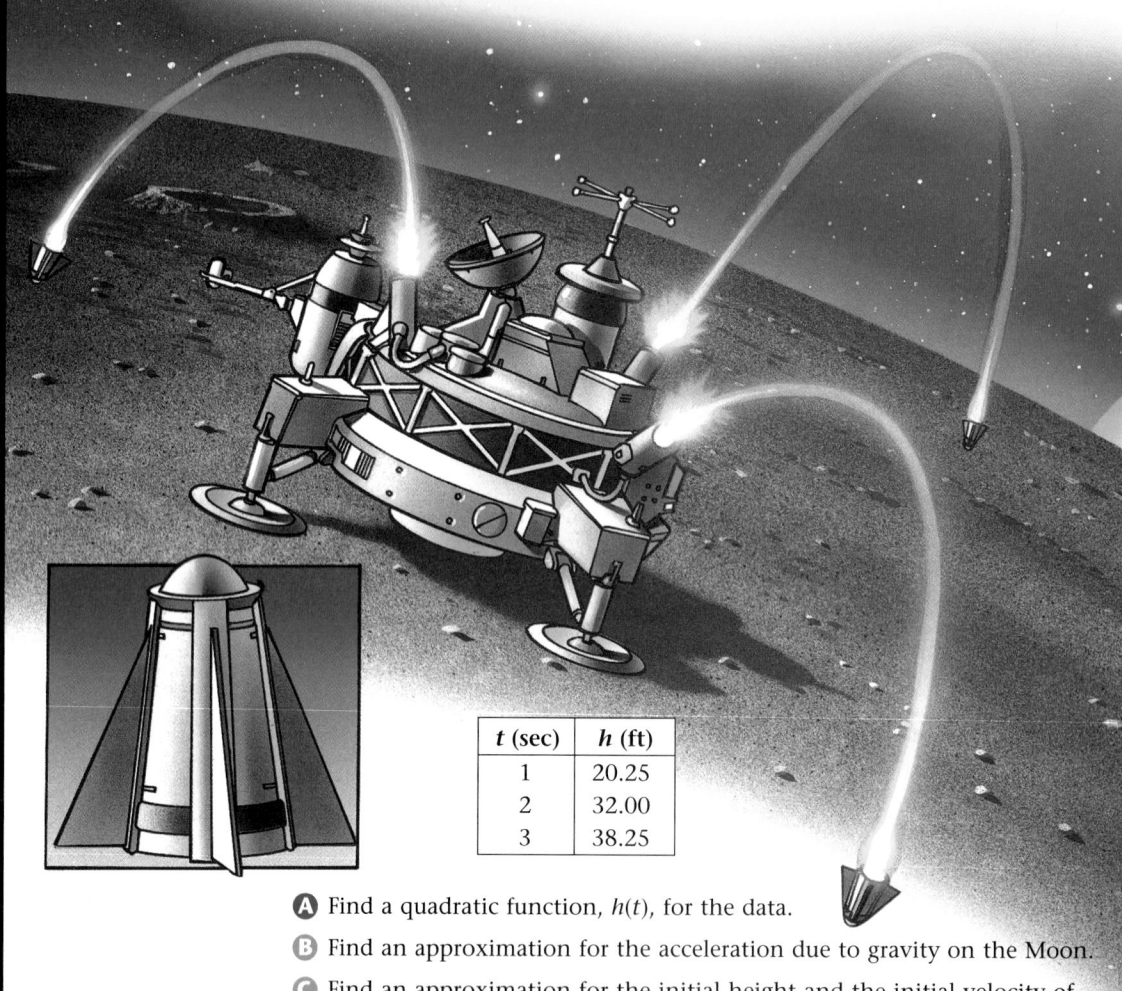

t (sec)	h (ft)
1	20.25
2	32.00
3	38.25

Ⓐ Find a quadratic function, $h(t)$, for the data.

Ⓑ Find an approximation for the acceleration due to gravity on the Moon.

Ⓒ Find an approximation for the initial height and the initial velocity of the object.

INCLUSION strategies **English Language Development** Students with limited command of English will find the graphs in this lesson to be very helpful. Be sure students are able to describe the concepts presented in this lesson using their own words. Allow students who have mastered English to work with those with limited English skills.

Solution➤

Ⓐ Substitute the coordinates of the data points into $h(t) = at^2 + bt + c$ to obtain a system of equations in three unknowns.

$$20.25 = a(1)^2 + b(1) + c$$
$$32.00 = a(2)^2 + b(2) + c \quad \rightarrow \quad \begin{cases} a + b + c = 20.25 \\ 4a + 2b + c = 32.00 \\ 9a + 3b + c = 38.25 \end{cases}$$
$$38.25 = a(3)^2 + b(3) + c$$

Set up a matrix equation, and solve for the variable matrix.

$$\begin{bmatrix} 1 & 1 & 1 \\ 4 & 2 & 1 \\ 9 & 3 & 1 \end{bmatrix} \begin{bmatrix} a \\ b \\ c \end{bmatrix} = \begin{bmatrix} 20.25 \\ 32.00 \\ 28.25 \end{bmatrix} \leftarrow AX = B$$

$$\begin{bmatrix} a \\ b \\ c \end{bmatrix} = \begin{bmatrix} -2.75 \\ 20 \\ 3 \end{bmatrix} \leftarrow X = A^{-1}B$$

Thus, according to the experimental data, the quadratic function for the height of a falling object on the Moon is $h(t) = -2.75t^2 + 20t + 3$.

Ⓑ The general quadratic model for the height of a falling object as a function of time is $h(t) = -\frac{1}{2}gt^2 + v_0t + h_0$. The coefficient of t^2 is $-\frac{1}{2}g$, where g is the acceleration due to gravity.

Since the coefficient of t^2 is –2.75, solve $-\frac{1}{2}g = -2.75$ to find g.

$$-\frac{1}{2}g = -2.75$$
$$g = (-2)(-2.75) = 5.5$$

Thus, the acceleration due to gravity on the Moon, based on experimental data, is 5.5 feet per second squared.

Ⓒ Since $h(t) = -2.75t^2 + 20t + 3$, the initial height, h_0, is 3 feet, and the initial velocity, v_0, is 20 feet per second. ❖

The answers to Example 2 are approximations because they were obtained through experimentation.

RETEACHING *the lesson*

Technology Give small groups of students several different sets of three points. Have them use the statistics feature on their graphics calculator to find a quadratic equation to fit the points. Have students use the calculator to graph the equation and use the trace or table feature to be sure the points are actually on the graph. Students should repeat this activity as necessary.

ASSESS

Selected Answers

Odd-numbered problems 5–29.

Assignment Guide

Core 1–17, 19–31

Core Plus 1–12, 15–31

Technology

A graphics calculator is needed for Exercises 5–21, 30, and 31.

EXERCISES & PROBLEMS

Communicate

Let the points (1, 6), (–3, 8), and (5, –2) be on a parabola.

1. Explain how to write a system of equations representing the points.

2. Explain how to find a quadratic function that fits the points.

3. How can you check your solution with and without a graphics calculator?

Practice & Apply

4. If the point (2, 3) is on a parabola, which of the following equations must be true?

$y = 2x^2 + 3x + c$ 　　　 $y = x^2 + 2x + 3$ 　　　 $2 = 9a + 3b + c$

$3 = 4x^2 + 2x + c$ 　　　 $\boxed{3 = 4a + 2b + c}$

t (sec)	h (ft)
1	19
2	21
3	11

Astronomy Imagine an experiment conducted on Mars in which an object is propelled vertically from a height h_0 above the surface of the planet and with an initial velocity v_0. The height is measured at three different points in time, and the data are recorded in the table shown.

5. Find a quadratic function that fits the data. $h(t) = -6t^2 + 20t + 5$

6. Find an approximation for the acceleration due to gravity on Mars. 12 ft/sec^2

7. Find an approximation for the initial height and the initial velocity of the object. $h_0 = 5$ft, $v_0 = 20$ ft/sec

8. What is the approximate maximum height reached by the object?

9. Approximately how long will it take for the object to reach its maximum height?

10. How long after the object is propelled from its initial height will it take for the object to return to the surface of Mars?

11. Use your function to predict the height of the object when t is 2.5.

12. Use your function to predict the time(s) when the object is at a height of 16 feet.

8. 21.67 feet

9. 1.67 seconds

10. 3.57 seconds

11. 17.5 feet

12. 0.69 seconds and 2.64 seconds

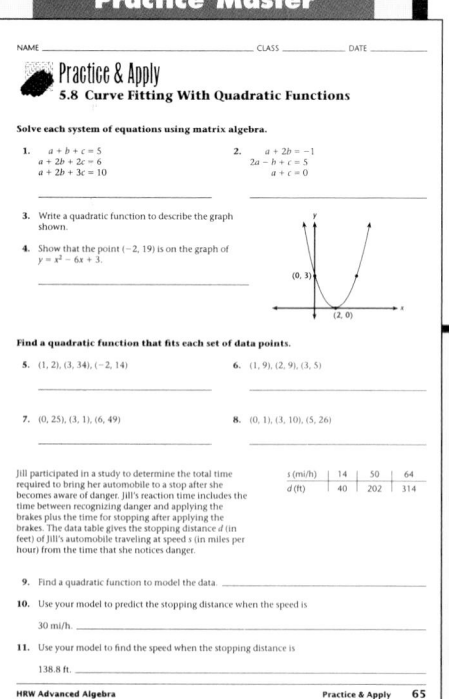

Find a quadratic function that fits each set of data points in Exercises 13–18.

$y = x^2 + 3x - 5$

13 (1, –1), (2, 5), (3, 13)

14 (1, 13), (4, 7), (5, –3) $y = -2x^2 + 8x + 7$

15 (2, 18), (6, 10), (8, –6)

16 (–2, –7), (1, 8), (2, 21) $y = 2x^2 + 7x - 1$

17 (0, 4), (1, 5), (3, 25)

18 (–3, 7), (–1, –5), (6, 16) $y = x^2 - 2x - 8$

$y = 3x^2 - 2x + 4$

Business Management A company keeps track of items produced and profit over several months. Three data points are shown in the table.

Number of items produced	Profit
5	$51
10	$56
15	$11

19 Use the data in the table to create a quadratic function that describes the profit as a function of the number of items produced. $y = -x^2 + 16x - 4$

20 Use your function to predict the maximum profit possible. $60

21 Use your function to predict the level of production that will maximize profit. 8 items

Look Back

22. Solve the equation $(x - 3)^2 - 7 = 0$. **[Lesson 5.2]** $3 \pm \sqrt{7}$

For Exercises 23–25, let $f(x) = x^2 - 2x - 15$.

23. Find the coordinates of the vertex of the graph of f. (1, –16) Describe the method that you used. **[Lesson 5.3]**

24. Find the zeros of f by completing the square. **[Lesson 5.4]** 5, –3

25. Find the zeros of f by using the quadratic formula. **[Lesson 5.5]** 5, –3

26. The equation $0 = x^2 + 20x + c$ is known to have a negative discriminant. What does this tell you about the number and type of solutions of the equation? **[Lesson 5.6]** there are two complex conjugate solutions

27. What does the negative discriminant of the function $f(x) = x^2 + 2x + c$ tell you about its graph? **[Lesson 5.6]** no x-intercepts

28. Given that the function $f(x) = x^2 + 2x + c$ has a negative discriminant, find three different values for c. **[Lesson 5.6]** Answers will vary; any $c > 1$

29. Given that the function $f(x) = x^2 + 2x + c$ has a negative discriminant, describe *all* possible values for c. **[Lesson 5.6]** $c > 1$

Look Beyond

30 Use your graphics calculator to graph the function $f(x) = 2x^2 + 4x - 5$. Use the trace feature to approximate the zeros of f. –2.9, 0.9

31 For what approximate values of x will f be greater than zero? $x < -2.9, x > 0.9$
For what approximate values of x will f be less than zero? $-2.9 < x < 0.9$

15. $y = -x^2 + 6x + 10$

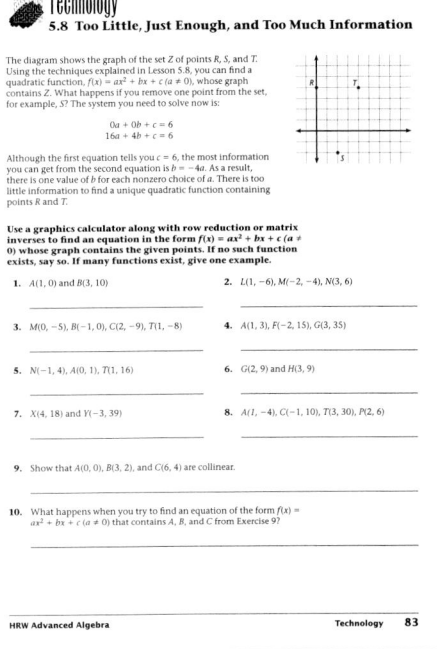

PREPARE

Objectives

- Write, solve, and graph quadratic inequalities that model real-world situations.

RESOURCES

- Practice Master 5.9
- Enrichment Master 5.9
- Technology Master 5.9
- Lesson Activity Master 5.9
- Quiz 5.9
- Spanish Resources 5.9

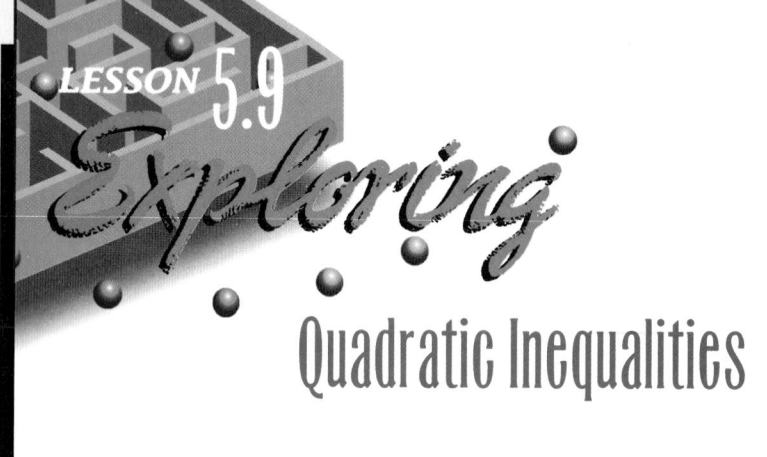

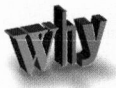

Assessing Prior Knowledge

Find the zeros of each function.

1. $f(x) = (x + 1)(2x - 5)$ $\left[\mathbf{-1, \frac{5}{2}}\right]$

2. $f(x) = x^2 - 9$ [**±3**]

3. $f(x) = x^2 - 2x + 1$ [**1**]

4. $f(x) = x^2 - 2x - 3$ [**−1, 3**]

TEACH

why In Lesson 5.7, the break-even point was described and illustrated as the intersection of a linear cost function and a quadratic revenue function. Now, from the viewpoint of the profit function, profit and loss are interpreted in terms of quadratic inequalities.

LESSON 5.9

Exploring

Quadratic Inequalities

why *Kate produces and sells novelty T-shirts. Quadratic inequalities can be used to determine the levels of production that will produce a profit.*

A **quadratic inequality** is an inequality that contains one or more quadratic expressions.

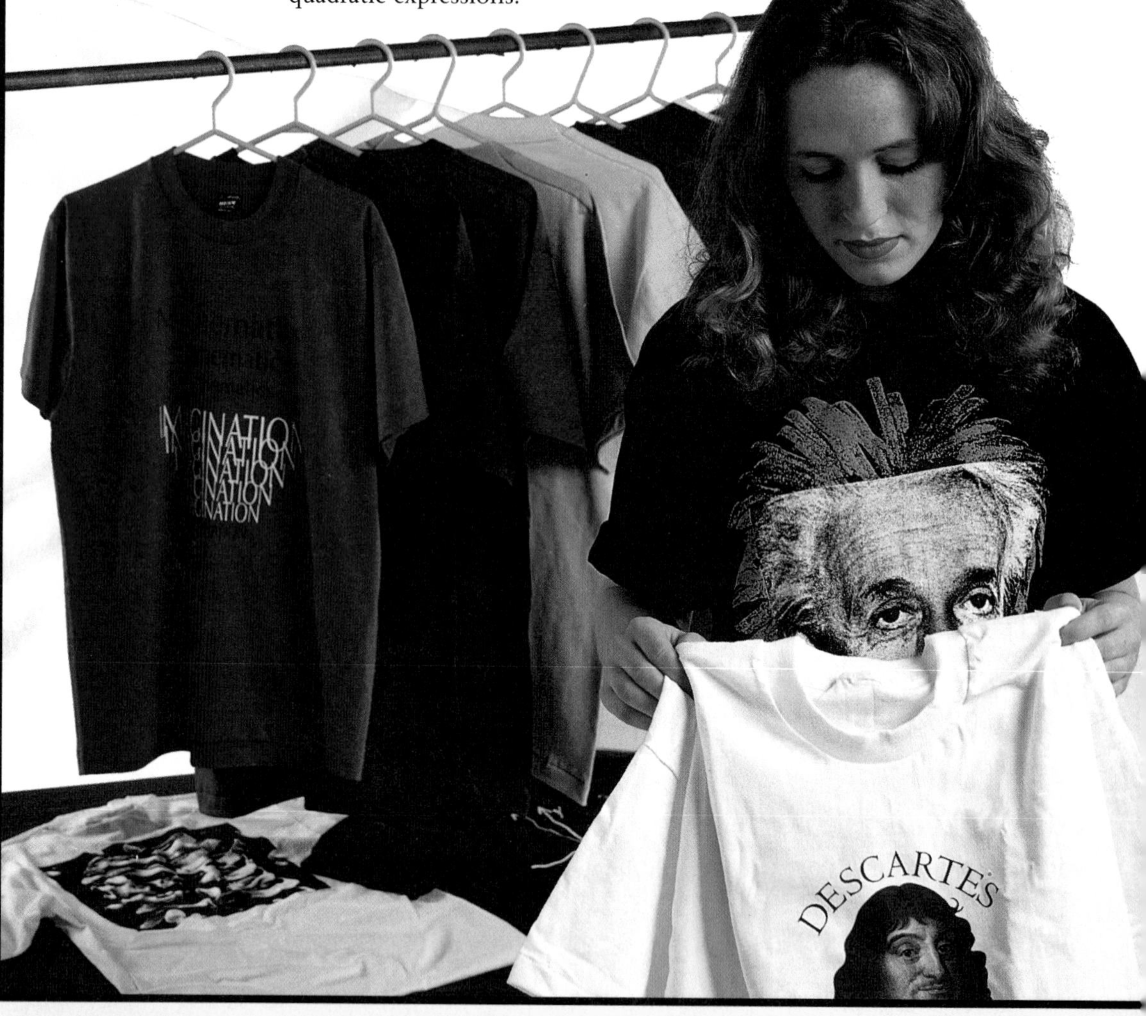

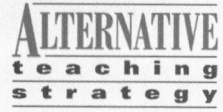

ALTERNATIVE teaching strategy

Using Visual Models Have students use factoring to find the solution region. For example, consider the inequality $x^2 - 2x - 3 < 0$. The expression $x^2 - 2x - 3$ can be written in factored form as $(x - 3)(x + 1)$. So, for $x^2 - 2x - 3 < 0$, $(x - 3)(x + 1) < 0$. Since the product must be negative, the following statement must be true.

$$x - 3 < 0 \text{ and } x + 1 > 0 \text{ or}$$
$$x - 3 > 0 \text{ and } x + 1 < 0,$$

so $x < 3$ and $x > -1$ or $x > 3$ and $x < -1$.

Since the second part of the statement *cannot* be true, the solution must be $x < 3$ and $x > -1$, or $-1 < x < 3$. Have students draw a line graph of the solution and check these x values in the original inequality. Ask students to repeat this procedure for $x^2 - 2x - 3 > 0$. [$\mathbf{x < -1}$ **or** $\mathbf{x > 3}$] Have students compare and discuss the differences between the solutions to the two inequalities.

Solving Quadratic Inequalities

Exploration 1 Finding Solution Regions

Use your graphics calculator to complete Step 4 of the following exploration.

Graphics Calculator

1 Find the zeros of $f(x) = x^2 - 2x - 3$. Plot the zeros on a number line. Notice that the two points divide the number line into 3 regions.

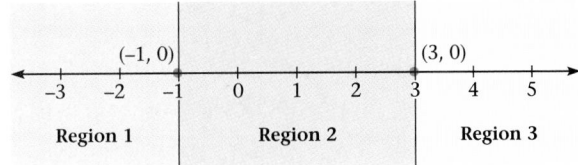

2 Choose three numbers from each region, and complete the following table.

Region 1		Region 2		Region 3	
x	$x^2 - 2x - 3$	x	$x^2 - 2x - 3$	x	$x^2 - 2x - 3$
−4	21				

3 Which regions contain values of x that result in $x^2 - 2x - 3 < 0$? Which regions contain values of x that result in $x^2 - 2x - 3 > 0$?

4 Graph the function $f(x) = x^2 - 2x - 3$. Choose a window wide enough to include all of the x-values that you chose in Step 2 for each region. Use the trace feature to answer the following questions.

For which values of x is $f(x) > 0$? For which values of x is $f(x) < 0$?

How do your answers here compare with your answers from Step 3?

5 Explain how to find the values of x for which $x^2 - 2x - 3 > 0$ with and without a graphics calculator. Explain how to find the values of x for which $x^2 - 2x - 3 < 0$ with and without a graphics calculator.

6 Explain the difference between the graph of the solutions to $x^2 - 2x - 3 < 0$ and the graph of the solutions to $x^2 - 2x - 3 \leq 0$. Explain the difference between the graph of solutions to $x^2 - 2x - 3 > 0$ and the graph of solutions to $x^2 - 2x - 3 \geq 0$. ❖

ENRICHMENT

Students can use the shade feature on their graphics calculator to graph two inequalities. Review the Systems of Linear Inequalities section of your *HRW Technology Handbook*. Have students write their own situation and graph the quadratic inequalities appropriate for their problem.

Exploration 1 Notes
Students should choose at least 3 numbers from each region. Encourage students to choose as many points as they believe are necessary to be able to complete the exploration.

ongoing ASSESSMENT

5. $x^2 - 2x - 3 > 0$: With a graphics calculator, use the trace feature to find the x-intercepts and, from the graph, determine values of x for which the graph is above the x-axis. Without a graphics calculator, find the zeros of the function, and test x-values on either side of the zeros to determine where the function is greater than zero. $x^2 - 2x - 3 < 0$: The methods are similar except, on the graphics calculator, trace to determine the values of x for which the graph is below the x-axis. Using the algebraic method, test values of x to determine where the function is less than zero.

6. The graph of the solutions to $x^2 - 2x - 3 < 0$ does not include the zeros of $x^2 - 2x - 3 = 0$, whereas the graph of the solutions to $x^2 - 2x - 3 \leq 0$ does include the zeros. Similarly, the graph of the solutions to $x^2 - 2x - 3 > 0$ does not include the zeros, while the graph of the solutions to $x^2 - 2x - 3 \geq 0$ does.

Cooperative Learning

Have students reexamine the application and Exploration 2 with real data. Divide students into groups and present each group with a list of tasks. The first task is to gather data. Coach the groups on the kind of data they need to find. Find several small stores who are willing to share information about their business. Have each group choose a store. Students visit the proprietor and ask for the information on fixed and productions costs. if usable incremental information is available, it may be included; otherwise, assign a hypothetical value. The next set of tasks is to have members of the group determine the cost and revenue functions and provide the financial analysis shown in the application and Exploration 2.

Exploration 2 Notes

Have students work in small groups to complete the exploration. Emphasize the difference between the graph students draw in Step 1 and the graph they draw in Step 3. The graph in Step 1 includes a parabola and a line. The graph in Step 3 is a graph on a number line that shows the three regions created by the equation $10x = 0.1x^2 + 5x + 40$.

APPLICATION

Small Business

Recall Kate's T-shirt business. You can use a quadratic inequality to determine the levels of production that will produce a profit.

Let x represent the number of T-shirts sold. If Kate decides to charge $10 per T-shirt, the revenue (money coming in), R, is given by

$$R(x) = 10x.$$

To find a function for the cost (money going out), C, in a week, estimate

1. the fixed costs (including rent, electricity, etc.) per week,

2. the cost to produce each T-shirt, and

3. an *incremental cost* to account for increases in cost that accompany increased levels of production.

Let the fixed costs be $40 per week, the cost to produce each T-shirt be $5, and the incremental cost be $0.1x^2$.

Then the total cost, C, is $C(x) = 0.1x^2 + 5x + 40$.

If the revenue, $R(x)$, is greater than the cost, $C(x)$, then there is a profit. The goal is to find the values of x for which $R(x) > C(x)$, or for which $10x > 0.1x^2 + 5x + 40$. ❖

·Exploration 2 *Finding the Solution Region*

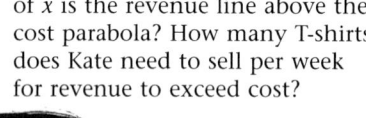
Graphics Calculator

Use your graphics calculator to complete Steps 1 and 2 of this exploration.

1 Graph the revenue function and the cost function from the application on the same coordinate plane.

2 Use the trace feature to find the points of intersection. For what values of x is the revenue line above the cost parabola? How many T-shirts does Kate need to sell per week for revenue to exceed cost?

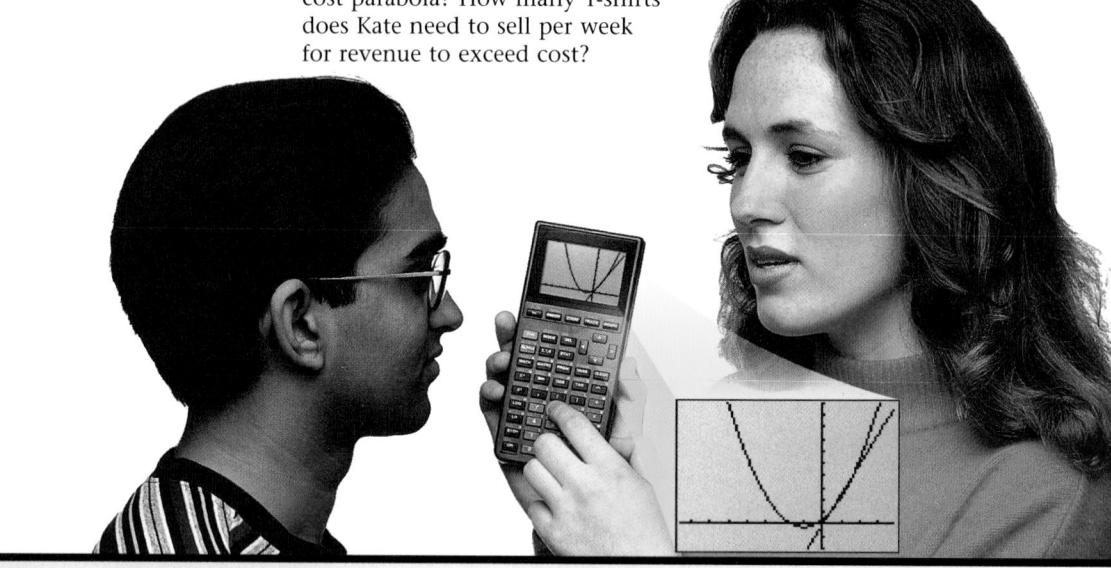

INCLUSION
strategies

English Language Development Encourage students to write definitions for the terms *cost*, *revenue*, *incremental cost*, and *profit function* in their own words. Have students share their definitions with one another. Be sure students understand how each graph presented in this lesson relates to these terms. It may be helpful for students to work in small groups to write a problem using these terms.

3 Without using a graphics calculator, solve $10x = 0.1x^2 + 5x + 40$ for x. Plot the solutions on a number line. What three regions do the points create?

4 Complete the following table.

x	Revenue $R(x) = 10x$	Cost $C(x) = 0.1x^2 + 5x + 40$	True or false: $R(x) > C(x)$
Region 1 $x < ?$	0			
Region 2 ? < x < ?	20			
	30			
Region 3 $x > ?$	50			

Which region(s) contain(s) function values for $R(x)$ and $C(x)$ such that $R(x) > C(x)$?

5 Explain how to find values of x for which $R(x) > C(x)$ with and without a graphics calculator. Explain how to find values of x for which $R(x) < C(x)$ with and without a graphics calculator. ❖

CRITICAL *Thinking*

Do all x-values in a region result in either $R(x) > C(x)$ or $R(x) < C(x)$? Explain.

EXTENSION

MAXIMUM MINIMUM *Connection*

You can create the *profit function* from the fact that profit will be the amount remaining after revenue is received and costs are paid. In other words, profit = revenue − cost.

$$P(x) = R(x) - C(x)$$
$$P(x) = 10x - (0.1x^2 + 5x + 40)$$
$$P(x) = -0.1x^2 + 5x - 40$$

Graph of P(x)

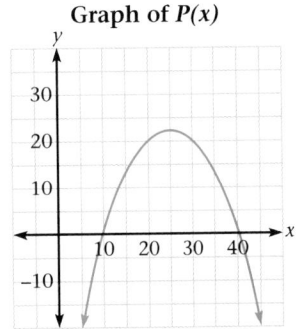

Will the maximum possible profit be at the vertex of the profit function, $P(x)$? Why or why not?

$P(x) = -0.1x^2 + 5x - 40$, where a is −0.1, b is 5, and c is −40

The x-coordinate of the vertex is $\frac{-b}{2a} = 25$.

Substitute 25 for x in the profit function to find the y-coordinate of the vertex.

$$y = -0.1x^2(25)^2 + 5(25) + 40$$
$$y = 22.5$$

Thus, the maximum profit, $22.50, results from selling 25 T-shirts per week. ❖

RETEACHING *the lesson*

Using Patterns Review with students the concept of graphing linear inequalities. Give them several examples to complete in small groups. Then change each of the equations to a quadratic inequality. Instruct students to apply the concepts used to graph linear inequalities to graph the quadratic inequalities.

---

5. $R(x) > C(x)$: With a graphics calculator, graph the two functions $Y_1 = R(x)$ and $Y_2 = C(x)$. Use the trace feature to find their points of intersection. By observation, record the region(s) on either side of or between the points of intersection for which the graph of $R(x)$ lies above the graph of $C(x)$.

Without graphics calculator, find the points of intersection by solving the equation $R(x) = C(x)$. Then test x-values in each region to determine where $R(x) > C(x)$. The values of x for which $R(x) < C(x)$ are found in a similar way, except that, with a graphics calculator, you can observe where $R(x)$ is below $C(x)$.

CRITICAL *Thinking*

Yes. Since $R(x) - C(x)$ is a quadratic function, it has at most two zeros, and, except at the zeros where $R(x) = C(x)$, is either greater than zero or less than zero. If the value of $R(x) - C(x)$ changes from positive to negative within a region, then the function crosses the x-axis, which contradicts the fact that this is a quadratic with at most two zeros.

Yes. Since $a < 0$, $P(x)$ is a parabola that opens downward, so the maximum occurs at the vertex.

ASSESS

Selected Answers

Odd-numbered problems 7–59

Assignment Guide

Core 1–29, 31–39, 42–61

Core Plus 1–27, 29–35, 39–65

Technology

A graphics calculator is needed for Exercises 15–17, 24, 26–30, 32–33, and 61–64.

EXERCISES & PROBLEMS

Communicate

1. Explain how to find the solutions to $x^2 - 2x - 8 > 0$ with a graphics calculator.

2. Explain how to find the solutions to $x^2 - 2x - 8 > 0$ without a graphics calculator.

3. Explain how to solve $x^2 - 2x - 8 < 4x + 3$.

4. Describe the graph of $x^2 - 2x - 8 < 4x + 3$.

5. Explain how to create a quadratic function $f(x)$ for which $f > 0$ when $-1 < x < 4$.

6. Refer to Exploration 2. If Kate decides to sell the T-shirts for $12 each, what would be the new revenue function, $R(x)$? What would be the new profit function, $P(x)$?

Practice & Apply

Small Business Jennifer and Miguel produce and sell x key chains per day. The graphs of the revenue, cost, and profit functions are shown. Use the graphs to find approximate answers to Exercises 7–12.

7. What will the revenue be if they sell 4 key chains? *approx $28*

8. Will they have a profit or a loss if they sell 9 key chains? *loss*

9. How many key chains must they sell to break even? *2 or 8*

10. How many key chains must they sell to earn a profit? *3, 4, 5, 6, or 7*

11. How many key chains must they sell to earn the maximum profit? *5*

12. What is the maximum profit? *about $5*

13. If the selling price is $7, what is the revenue function? *$R(x) = 7x$*

14. If Jennifer and Miguel have a fixed cost of $8, a production cost of $2 per key chain, and an incremental cost of $0.5x^2$, what is their cost function? *$C(x) = 0.5x^2 + 2x + 8$*

15. Use your answers to Exercises 13 and 14 to graph the revenue, cost, and profit functions on the same coordinate plane. Compare your graphs to the ones provided for Exercises 7–12. How are they alike? How are they different? *$P(x) = -0.5x^2 + 5x - 8$; same graphs*

16. Find the break-even points. *$x = 2, x = 8$*

17. Find the maximum profit possible. Find the number of key chains that must be sold to earn the maximum profit. *$4.50, 5 key chains*

18. According to your answers to Exercise 17, would you suggest that Jennifer and Miguel continue their key-chain business? Why or why not?

18. Answers may vary. A sample result: since the profit is very low, it is probably not worth their continuing the key-chain business.

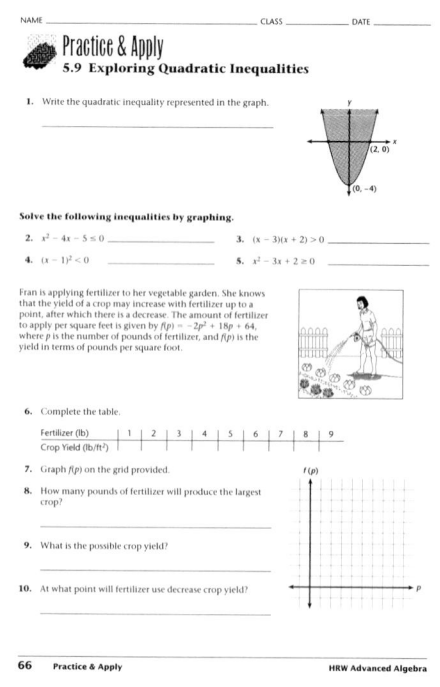

Sales Jon is a traveling sales representative for a wholesaler who sells radios. The price per radio varies based on the number of radios purchased. Beginning with a price of $124 for one radio, the price for each radio is reduced by $1 for each additional radio purchased.

19. Copy and complete the table.

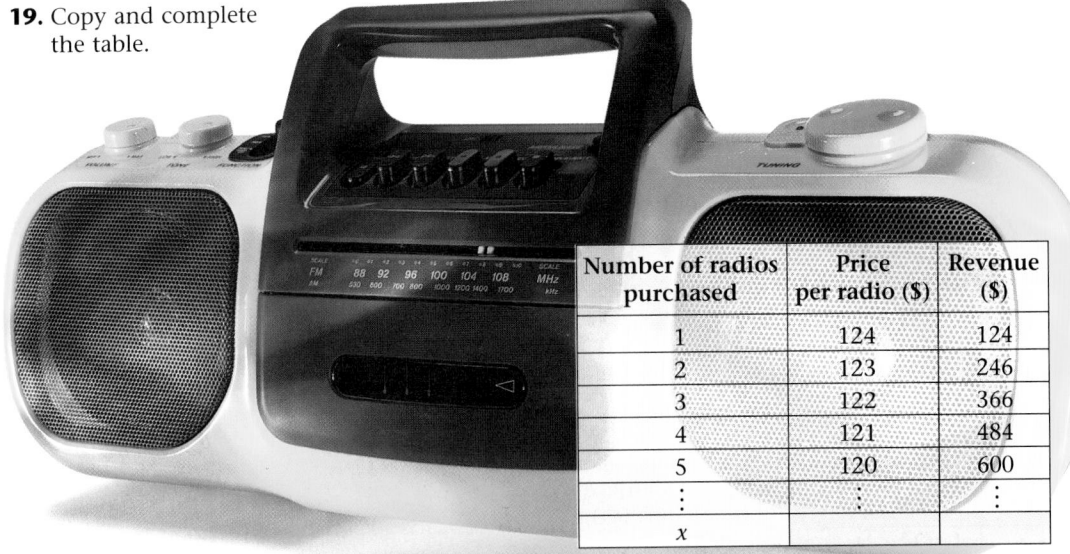

Number of radios purchased	Price per radio ($)	Revenue ($)
1	124	124
2	123	246
3	122	366
4	121	484
5	120	600
⋮	⋮	⋮
x		

20. What is the equation for the revenue function? Is the revenue function linear or quadratic? $R(x) = -x^2 + 125x$; quadratic

21. What is the maximum revenue for a given sale? $3906

22. How many radios must be sold to attain the maximum revenue? 62 or 63

23. Assume that it costs the wholesaler $68 to produce each radio, plus an average of $128 in fixed costs (travel expenses, phone calls, etc.) per sale. Based on these two factors alone, what is the cost function? Is the cost function linear or quadratic? $C(x) = 68x + 128$; linear

24 Graph the revenue and cost functions on the same coordinate plane. How many radios does Jon need to sell for the revenue to be greater than cost? between 3 and 54

25. What is the profit function? $P(x) = -x^2 + 57x - 128$

26 Graph the profit function on the same coordinate plane with the revenue and cost functions. How many radios does Jon need to sell to make a profit? between 3 and 54

27 What is the maximum profit for a given sale? How many radios must be sold to earn the maximum profit? $684, 28 or 29 radios

Solve the following inequalities. Use your graphics calculator to check your answers.

28 $-x^2 + 6x - 5 > 0$ **29** $2x^2 - 3x - 20 < 0$ **30** $-3x^2 + 7x - 1 < 0$

31. Sports At the beginning of a basketball game, the referee tosses the ball vertically into the air. Its height (in feet) after t seconds is given by $h(t) = -16t^2 + 24t + 5$. During what time interval (to the nearest tenth of a second) is the height of the ball above 9 feet? $0.2 \leq t \leq 1.3$

19. Price: $124 - (x - 1)$
    Revenue: $x[124 - (x - 1)]$

28. $1 < x < 5$

29. $-\frac{5}{2} < x < 4$

30. $x < \frac{7}{6} - \frac{\sqrt{37}}{6}$ or $x > \frac{7}{6} + \frac{\sqrt{37}}{6}$, or approximately $x < 0.15$ or $x > 2.18$

Error Analysis

Encourage students always to check several points from each region in the original inequalities to be sure they determine the correct solution region.

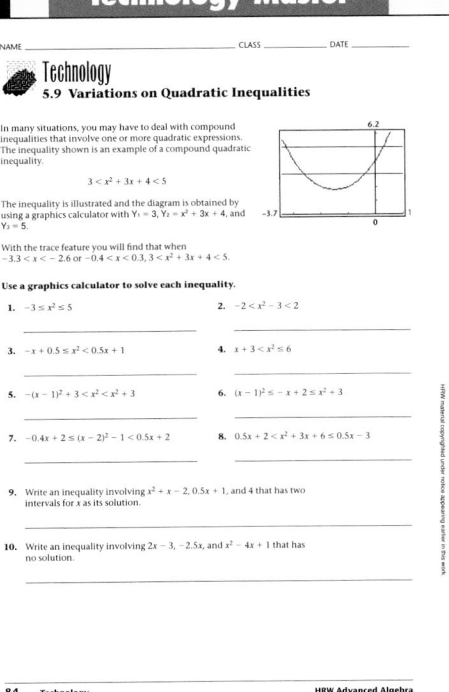

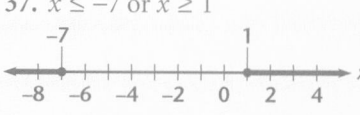

Performance Assessment

Have each student write exercises similar to those in Exercises 36–41. Be sure students use several different inequality symbols. Have students exchange, graph, and solve each other's inequalities.

36. $x < \dfrac{1 - \sqrt{13}}{2}$ or $x > \dfrac{1 + \sqrt{13}}{2}$, or $\approx x < -1.30$ or $x > 2.30$

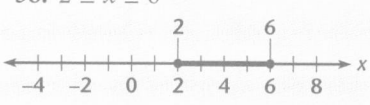

−1.3   2.3

37. $x \leq -7$ or $x \geq 1$

−7   1

38. $2 \leq x \leq 6$

2   6

39. $\dfrac{3 \pm \sqrt{33}}{2}$, or $\approx x < -1.4$ or $x > 4.4$

−1.4   4.4

Small Business Suppose the profit, p, for selling x bumper stickers is given by $p(x) = -0.1x^2 + 8x - 50$.

32 What is the minimum number of bumper stickers that must be sold to make a profit?   7

33 Is it possible for the profit to be greater than $100? Explain. yes

Free Fall An object is dropped from a height of 1000 feet. Its height (in feet) after t seconds is given by $h(t) = -16t^2 + 1000$.

34. For approximately how long will the height of the object be above 500 feet? ≈ 5.6 sec

35. For what approximate values of t will the height of the object be below 500 feet? $t > 5.6$

Solve the following inequalities. Graph your solutions on a number line.

36. $x^2 - x - 3 > 0$

37. $x^2 + 6x - 7 \geq 0$

38. $x^2 - 8x + 12 \leq 0$

39. $-x^2 + 3x + 6 < 0$

40. $x^2 + 4x - 1 > 8$

41. $x^2 + 5x - 7 < 4x$

42. Transportation The stopping distance of a car on a particular type of road surface is given by $d = 0.05s^2 + 11s$, where s is the speed of the car in miles per hour and d is the stopping distance in feet for a reaction time of $\frac{3}{4}$ of a second. For what speeds is the stopping distance less than two car lengths (assume a car length of 14.5 feet)?
$0 < s < 2.61$ mph

Physics The length, l (in feet), of a pendulum is related to the time, t (in seconds), required for one complete swing by the formula $l \approx 0.81t^2$.

43. For what values of t is $l > 2$? $t > 1.57$ sec

44. For what values of t is $l < 5$? $0 < t < 2.48$ sec

Small Business Suppose the cost, c, of producing x handmade leather sandals is given by $c(x) = 0.4x^2 + 5x + 100$. **Complete Exercises 45 and 46.**

45. For what values of x is the cost, c, less than a budgeted amount of $1000?

46. For what values of x does the cost, c, exceed the budgeted amount of $1000? $x > 41.6$, or 42 or more items

47. Create a quadratic function for which $f(x) > 0$ for the values of x between 2 and 6 only.

48. Write a quadratic inequality whose solution is $2 < x < 5$.

49. Write a quadratic inequality whose solution is $x < 3$ or $x > 7$.

---

40. $\dfrac{-4 \pm \sqrt{52}}{2}$ or $\approx x < -5.6$ or $x > 1.6$

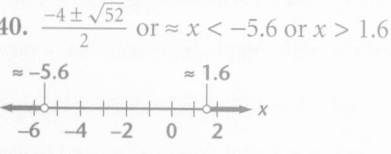

≈ -5.6 ≈ 1.6

41. $\dfrac{-1 \pm \sqrt{29}}{2}$, or $\approx -3.2 < x < 2.2$

≈ -3.2 ≈ 2.2

45. $0 < x < 42$ sandals

47. Answers may vary but should be equivalent to $f(x) = a(x^2 - 8x + 12) > 0$, $a < 0$

48. Answers may vary but should be equivalent to $f(x) = a(x^2 - 7x + 10) < 0$, $a > 0$

49. Answers may vary but should be equivalent to $f(x) = a(x^2 - 10x + 21) > 0$, $a > 0$, or $f(x) = a(x - 3)(x - 7) < 0$, $a < 0$.

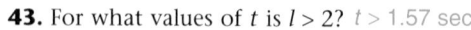
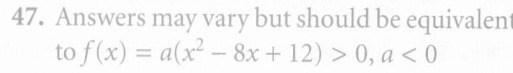

Look Back

Astronomy A group of scientists tested the effect of weightlessness on an astronaut's reaction time. After 5 hours of weightlessness, it took the astronaut 0.7 second to respond to a flashing light by pressing a button. After 13 hours of weightlessness, the astronaut's reaction time was 2 seconds.   **[Lesson 1.5]**

Weightless astronaut Robert Cabana, pilot of the Shuttle Discovery.

50. Assume that the relation is linear. Graph the relation.

51. Write an equation for the relation. Use w for the number of hours of weightlessness and t for the reaction time in seconds. $t = 0.1625w - 0.1125$

52. Use the equation you wrote for Exercise 51 to predict the astronaut's reaction time after 24 hours of weightlessness. 3.79 sec

53. Use the equation you wrote for Exercise 51 to predict the number of hours of weightlessness that causes a reaction time of 1 minute. about 370 hr

Graph each of the following equations. State whether or not it is a function.   [Lesson 2.3]

54. $y = |x|$ yes   **55.** $x = |y|$ no   **56.** $y = -|x|$ yes

Evaluate the following expressions, and state your answers in simplest terms.   [Lesson 5.7]

57. $(8 - 2i)(6 + 3i)$ **58.** $(3 - i)t$ $3t - it$ **59.** $(4 - 3i)(6i + 7)$ $46 + 3i$

60. Find a function for the parabola that contains the points $(1, 5)$, $(2, 6)$, and $(3, 11)$.   **[Lesson 5.8]** $y = 2x^2 - 5x + 8$

Look Beyond

61. Let $f(x) = x^3 - 2x^2 + 3x + d$. If the graph of f contains the point $(1, 9)$, find the value of d. Use your graphics calculator to verify your answer by graphing f and tracing to determine whether the point $(1, 9)$ lies on the graph. $d = 7$

62. Plot the function $f(x) = (x + 2)(x - 1)(x - 4)$, and find the x-intercepts. $(-2, 0)$, $(1, 0)$, $(4, 0)$

63. Plot the function $f(x) = (x + 3)(x - 5)(x + 1)$, and find the x-intercepts. $(-3, 0)$, $(-1, 0)$, $(5, 0)$

64. Plot the function $f(x) = (x - 2)(x + 1)(x - 7)$, and find the x-intercepts. $(-1, 0)$, $(2, 0)$, $(7, 0)$

65. Look for a connection between the location of the x-intercepts and the linear factors of the functions. What do you see?

55.

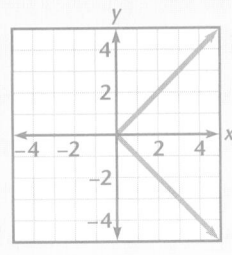

not a function

56.

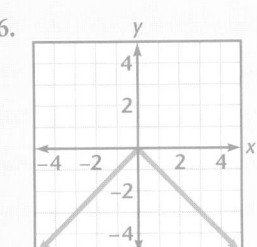

a function

57. $54 + 12i$

65. The x-intercepts of the functions are the zeros of the linear factors of each function.

Look Beyond

These exercises explore the relationship between the number of factors of a polynomial function and the number of x-intercepts in its graph.

50.

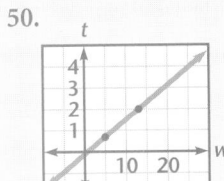

54.

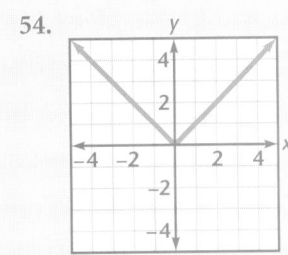

a function

FOCUS

Describe the path of a basketball on several different planets with a quadratic function.

MOTIVATE

Before students complete the table in Part 1, ask them the following questions.

- What affects the motion of a basketball on each planet? [**the acceleration due to gravity, which varies from planet to planet**]
- The quadratic model for the motion of the basketball is $h(t) = \frac{1}{2}gt^2 + 24t + 3$ What does g represent? [**the gravity at the surface of the planet in feet per second**]

After students have completed the table in Part 1, have them compare the quadratic models for each planet using their graphics calculator to graph all of the functions on the same coordinate axes. You may want to have students duplicate the graphs on a piece of graph paper, using a different color for each function.

INTERPLANETARY BASKETBALL

The Chapter Opener describes the vertical motion of a basketball at the beginning of a game. The referee is holding the ball at a height of 3 feet above the floor. He tosses the ball upward at a velocity of 24 feet per second. The ball is affected by the acceleration due to gravity at the Earth's surface, which is approximately 32 feet per second squared.

The motion of the basketball can be described by the quadratic model

$$h(t) = -16t^2 + 24t + 3.$$

In this project, you will compare the motion of the basketball on Earth with its motion on each of the other eight planets. On each planet, assume that the initial height of the ball is 3 feet and that the initial velocity of the ball is 24 feet per second. However, the acceleration due to gravity near the surface of each planet is different from that on Earth.

The following table contains the acceleration due to gravity near the surface of each planet in terms of that on Earth.

1. Complete the table.

Planet	Gravity at surface (as a fraction of Earth's)	Gravity at surface (in ft/sec²)	Quadratic Model: $h(t) = -\frac{1}{2}gt^2 + 24t + 3$
Mercury	0.37		
Venus	0.88		
Earth	1.00	32 ft/sec²	$h(t) = -16t^2 + 24t + 3$
Mars	0.38		
Jupiter	2.64		
Saturn	1.15		
Uranus	1.15		
Neptune	1.12		
Pluto	0.04		

1.

Planet	Gravity at surface (as a fraction of Earth's)	Gravity at surface (in ft/sec²)	Quadratic model: $h(h) = -\frac{1}{2}gt^2 + 24t + 3$
Mercury	0.37	11.84	$h(t) = -5.92t^2 + 24t + 3$
Venus	0.88	28.16	$h(t) = -14.08t^2 + 24t + 3$
Earth	1.00	32	$h(t) = -16t^2 + 24t + 3$
Mars	0.38	12.16	$h(t) = -6.08t^2 + 24t + 3$
Jupiter	2.64	84.48	$h(t) = -42.24t^2 + 24t + 3$
Saturn	1.15	36.8	$h(t) = -18.4^2 + 24t + 3$
Uranus	1.15	36.8	$h(t) = -18.4t^2 + 24t + 3$
Neptune	1.12	35.84	$h(t) = -17.92t^2 + 24t + 3$
Pluto	0.04	1.28	$h(t) = -0.64t^2 + 24t + 3$

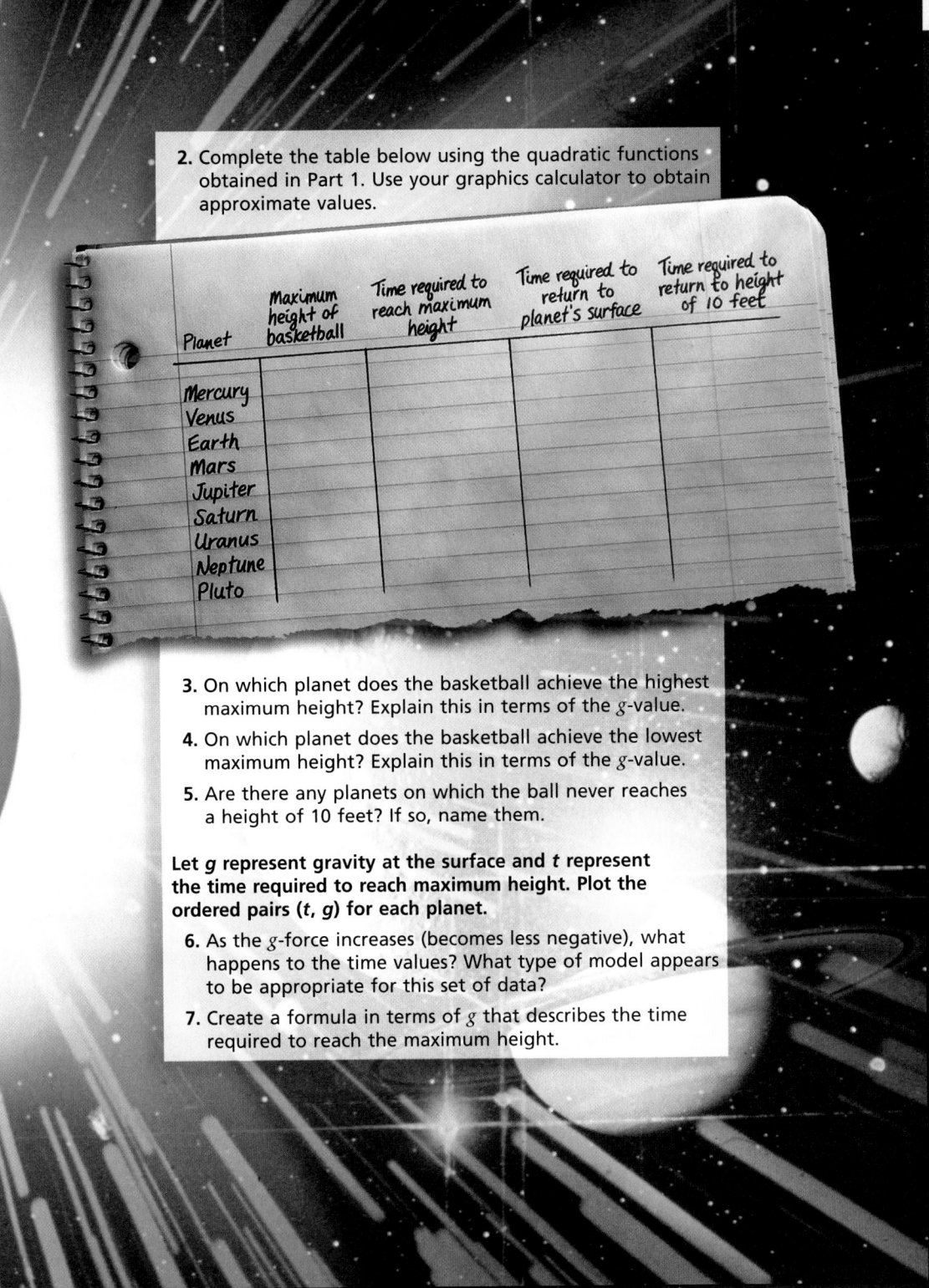

2. Complete the table below using the quadratic functions obtained in Part 1. Use your graphics calculator to obtain approximate values.

Planet	Maximum height of basketball	Time required to reach maximum height	Time required to return to planet's surface	Time required to return to height of 10 feet
Mercury				
Venus				
Earth				
Mars				
Jupiter				
Saturn				
Uranus				
Neptune				
Pluto				

3. On which planet does the basketball achieve the highest maximum height? Explain this in terms of the g-value.

4. On which planet does the basketball achieve the lowest maximum height? Explain this in terms of the g-value.

5. Are there any planets on which the ball never reaches a height of 10 feet? If so, name them.

Let _g_ represent gravity at the surface and _t_ represent the time required to reach maximum height. Plot the ordered pairs (_t_, _g_) for each planet.

6. As the g-force increases (becomes less negative), what happens to the time values? What type of model appears to be appropriate for this set of data?

7. Create a formula in terms of g that describes the time required to reach the maximum height.

2.

Planet	Maximum height of basketball (in feet)	Time required to reach maximum height (in seconds)	Time required to return to planets surface (in seconds)	Time required to return to height of 10 feet (in seconds)
Mercury	27.32	2.027	4.175	3.738
Venus	13.23	0.852	1.822	1.331
Earth	12	0.750	1.616	1.104
Mars	26.68	1.974	4.069	3.630
Jupiter	6.41	0.284	0.674	no solution
Saturn	10.83	0.652	1.419	0.864
Uranus	10.83	0.652	1.419	0.864
Neptune	11.04	0.670	1.454	0.910
Pluto	228	18.75	37.625	37.206

Chapter 5 Review

Vocabulary

Key Skills & Exercises

Lesson 5.1

➤ **Key Skills**

Write quadratic functions as a product of two linear functions that model real-world situations.

The x-intercepts or the zeros of the quadratic function are the same as the x-intercepts of the two linear functions.

The x-coordinate of the vertex is the midpoint of the zeros of the quadratic function.

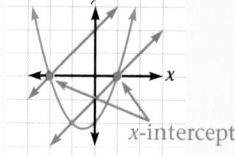

Use the graphics calculator to approximate the minimum or maximum value of a quadratic function and the x-intercepts of its graph.

Use the trace feature to approximate the coordinates of the vertex, which is the maximum or minimum point on the parabola, and to find the x-intercepts.

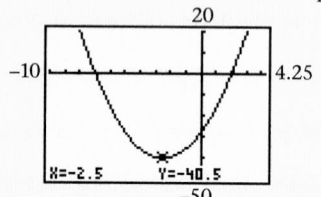

➤ **Exercises**

Let $f(x) = (1 - 2x)(2x + 3)$.

1. Show that $f(x)$ is a quadratic function by writing it in the form $f(x) = ax^2 + bx + c$, and identify a, b, and c.

2. Use your graphics calculator to approximate the x-intercepts.

3. Use your graphics calculator to approximate the coordinates of the vertex. Is the vertex a minimum or maximum point?

4. What are the values of x for which f is increasing and for which f is decreasing?

Lesson 5.2

➤ Key Skills

Solve quadratic equations by taking square roots, or by squaring.

To solve $(x-3)^2 = 36$ for x, take the square root of both sides and simplify.

To solve $\sqrt{(x-4)} = 2$ for x, square both sides and solve for x. When you square both sides of an equation, you may obtain extraneous, or false, roots.

Use the distance formula to find the distance between two points.

The distance, d, between the points (x_1, y_1) and (x_2, y_2) is $d = \sqrt{(x_2 - x_1)^2 + (y_2 - y_1)^2}$. The "Pythagorean" Right-Triangle Theorem states that in a right triangle, the lengths of the legs, a and b, are related to the length of the hypotenuse, c, by $a^2 + b^2 = c^2$.

➤ Exercises

Solve and check.

5. $\frac{1}{2}(x-1)^2 = 18$ -5 or 7

6. $4\sqrt{2x-1} - 5 = 0$ $\frac{41}{32}$

7. Find the distance between the points $(-1, 2)$ and $(3, -3)$.

$\sqrt{41}$

Find the missing side length of right triangle *ABC*.

8. a is 5 and b is 2. $\sqrt{29}$

9. c is $\sqrt{35}$ and b is 4. $\sqrt{19}$

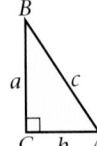

Lesson 5.3

➤ Key Skills

Analyze graphs of quadratic functions to identify the transformations that result from changing the terms of the function.

As the coefficient of x^2 increases from small numbers to large numbers, the graph of the parabola becomes narrower.

Find the vertex, axis of symmetry, and direction of opening for the graphs of quadratic functions in the form $f(x) = a(x-h)^2 + k$.

	$a > 0$	$a < 0$
Vertex	(h, k)	(h, k)
Axis of symmetry	$x = h$	$x = h$
Direction of opening	upward	downward

➤ Exercises

Let $f(x) = 3(x+2)^2 - 5$.

10. Find the coordinates of the vertex. $(-2, -5)$

11. What is the equation for the axis of symmetry? $x = -2$

12. What is the direction of opening of the graph of f? upward

13. What happens to the graph of f when the coefficient of x^2 is changed from 3 to $\frac{1}{3}$? The graph widens.

14.

20. x-intercepts: $(0.25, 0)$
and $(-1, 0)$;
vertex: $(-0.375, -1.563)$

21. x-intercepts: $(-0.608, 0)$
and $(4.108, 0)$;
vertex: $(1.75, -11.125)$

Lesson 5.4

➤ **Key Skills**

Use tiles to complete the square for quadratic expressions of the form $x^2 + bx + c$, where b and c are real numbers.

The quadratic expression $x^2 + 4x + 4$ can be modeled by 1 x^2-tile, 4 x-tiles, and 4 unit tiles. If $x^2 + bx + c$ is a perfect-square trinomial, then $\left(\frac{b}{2}\right)^2 = c$.

Solve quadratic equations by completing the square for expressions of the form $ax^2 + bx + c$, where $a \neq 0$, and b and c are real numbers.

$$3x^2 + 4x - 7 = 0$$
$$x^2 + \frac{4}{3}x = \frac{7}{3}$$
$$x^2 + \frac{4}{3}x + \left(\frac{1}{2} \cdot \frac{4}{3}\right)^2 = \frac{7}{3} + \left(\frac{1}{2} \cdot \frac{4}{3}\right)^2$$
$$x^2 + \frac{4}{3}x + \left(\frac{4}{9}\right)^2 = \frac{7}{3} + \frac{4}{9}$$
$$\left(x + \frac{2}{3}\right)^2 = \frac{25}{9}$$
$$x + \frac{2}{3} = \pm\frac{5}{3}$$
$$x = 1 \text{ or } x = -\frac{7}{3}$$

➤ **Exercises**

14. Draw a model to represent $x^2 + 3x + 5$.
15. Complete the square for $x^2 - 9x$. $x^2 - 9x + \frac{81}{4}$

Solve by completing the square. Give your answers to the nearest tenth.

16. $x^2 - 5x = 3$ $-0.5; 5.5$ **17.** $6 - 2x^2 = 8x$ $-4.6; 0.6$

Lesson 5.5

➤ **Key Skills**

Use the quadratic formula to solve quadratic equations that model real-world situations.

If $ax^2 + bx + c = 0$, where $a \neq 0$, then $x = \frac{-b \pm \sqrt{b^2 - 4ac}}{2a}$.
You can use the quadratic formula to solve *any* quadratic equation.

Use the axis of symmetry formula to find maximum or minimum values of quadratic equations that model real-world situations.

The axis of symmetry of a parabola goes through the vertex and divides the parabola equally in half. The equation of the axis of symmetry of a parabola given by a quadratic function of the form $f(x) = ax^2 + bx + c$ is $x = \frac{-b}{2a}$.

➤ **Exercises**

Use the quadratic formula to solve each of the following equations. Give your answers to the nearest tenth.

18. $2x^2 + 3 = -5x$ $-1.5; -1$ **19.** $3x^2 + 7x = 0$ $0; -2.3$

Find the x-intercepts and the vertex for each parabola. Give your answer to the nearest thousandth. Use your graphics calculator to check your answers.

20. $y = 4x^2 + 3x - 1$ **21.** $y = 2x^2 - 7x - 5$

Lesson 5.6

➤ Key Skills

Determine the number of real-number solutions using the discriminant.

The discriminant allows you to find out how many real-number solutions that a quadratic equation has without solving the equation or graphing.

1. If $b^2 - 4ac > 0$, the equation has *two different* real-number solutions.
2. If $b^2 - 4ac = 0$, the equation has *two equal* real-number solutions.
3. If $b^2 - 4ac < 0$, the equation has *no* real-number solutions.

Solve quadratic equations with imaginary-number solutions.

Imaginary numbers are used to represent solutions to quadratic equations that are not real numbers.

For any positive real number a, $\sqrt{-a}$ is an imaginary number.

$$\sqrt{-a} = i\sqrt{a} \text{ and } (i\sqrt{a})^2 = -a.$$

Solve.
$$x^2 + 7 = 0$$
$$x^2 = -7$$
$$x = \pm i\sqrt{7}$$

25. -5

31.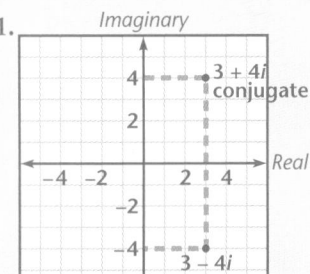

➤ Exercises

Without solving, determine how many real-number solutions each equation has.

22. $x^2 - 3x + 5 = 0$ ⁰
23. $2x^2 - 5x + 2 = 0$ ₂
24. $-3x^2 + 6x - 3 = 0$ ₁

Simplify.

25. $\sqrt{-5} \cdot \sqrt{-5}$
26. $3i \cdot (-2i)$ 6
27. $4i^2 \cdot (-5i)$ 20i
28. Solve $x^2 = -13$. $\pm i\sqrt{13}$

Lesson 5.7

➤ Key Skills

Identify, operate with, and graph complex numbers.

A *complex number* is a number that can be written in the form $a + bi$, where a and b are real numbers, and i is the imaginary unit.

A complex number can be graphed on the complex plane, which consists of a real axis and an imaginary axis.

To add and subtract complex numbers, combine like terms. Multiply complex numbers just as you would multiply binomials.

The complex numbers $a + bi$ and $a - bi$ are complex conjugates.

Find the complex roots of quadratic equations that model real-world situations.

Use the quadratic formula to solve $3x^2 - 2x + 1 = 0$.

$$x = \frac{-(-2) \pm \sqrt{(-2)^2 - 4(3)(1)}}{2(3)}$$
$$x = \frac{2 \pm \sqrt{-8}}{6}$$
$$x = \frac{2 \pm 2i\sqrt{2}}{6} = \frac{1 \pm i\sqrt{2}}{3}$$

➤ Exercises

29. Write the solutions for $-5x^2 + 3x - 1 = 0$ in $a + bi$ form. $\frac{3}{10} \pm \frac{\sqrt{11}}{10}i$
30. Identify the real part and the imaginary point of the complex number $-7 - 9i$. real: -7; imag: $-9i$
31. Plot $-4i + 3$ and its conjugate together on a complex plane.

Lesson 5.8

➤ **Key Skills**

Write a quadratic model that fits three data points from real-world data.

To find a quadratic model that fits the points (1, 0), (2, 3), and (−1, 6), substitute the values of x and y from the given points into the equation $y = ax^2 + bx + c$ and solve for the three unknowns a, b, and c.

$$\begin{cases} a + b + c = 0 \\ 4a + 2b + c = 3 \\ a - b + c = 6 \end{cases}$$

$a = 2$, $b = -3$, $c = 1$

Substituting the values a, b, and c, the quadratic model that fits the points is $y = 2x^2 - 3x + 1$.

➤ **Exercises**

Find a quadratic model that fits each set of data points.

32 (0, −1), (1, −2), (−1, −6) $y = -3x^2 + 2x - 1$

33 (0, −1), (1, −4), (−1, 6) $y = 2x^2 - 5x - 1$

Lesson 5.9

➤ **Key Skills**

Write, solve, and graph quadratic inequalities that model real-world situations.

To solve $x^2 - x - 6 > 0$, plot the zeros of $f(x) = x^2 - x - 6$ on a number line. Since $f(x) > 0$ for $x < -2$ and for $x > 3$, the graph of the solution is shown as follows.

➤ **Exercises**

Solve and graph the following inequalities.

34. $x^2 - 2x - 8 \le 0$ $-2 \le x \le 4$

35. $x^2 + x - 12 > 0$ $x < -4$ or $x > 3$

Applications

Business Management A dinner theater charges $15 per ticket. The manager estimates that each $1 increase in ticket price will decrease the attendance by 10 people, and each $1 decrease in ticket price will increase the attendance by 10 people. Currently the average attendance is about 200 people.

36. What should the ticket price be to maximize revenue? $17.50

37. What will the maximum revenue be? $3062.50

Sales The mathematical model that describes profit in terms of items sold for the Brentwood Skate Patrol is $p(x) = -2x^2 + 12x - 10$, where x represents the number of items sold (in thousands) and $p(x)$ represents the profit (in thousands of dollars).

38. What is the maximum profit that the company can make? $8000

39. How many items must be sold to earn the maximum profit?   3000

40. For what value of x is the profit function increasing? $x < 3000$

41. For what values of x is the profit decreasing? $x > 3000$

34. $-2 \le x \le 4$

35. $x < -4$ or $x > 3$

Chapter 5 Assessment

For Items 1–3, let $f(x) = (1 - x)(2x + 1)$.

1. What is the shape of the graph of f? parabola, opens downward

2. What are the x-intercepts of the graph of f? $(-0.5, 0), (1, 0)$

3. Maximum/Minimum What are the coordinates of the vertex of the graph of f? Is it a maximum or a minimum point?

4. Geometry Explain how to find the length of the hypotenuse of a right triangle with legs of lengths 2 and 3.

5. Solve $\frac{1}{3}(x - 3)^2 = 2$ for x.

6. Coordinate Geometry Find the distance between the points $(3, -2)$ and $(1, 4)$.

For Items 7–10, let $f(x) = -4(x - 1)^2 + 3$.

7. Explain how to find the x-intercepts. What are they?

8. Find the coordinates of the vertex. $(1, 3)$

9. What is the equation for the axis of symmetry? $x = 1$

10. For what values of x is f increasing, and for what values of x is f decreasing? increasing for $x < 1$; decreasing for $x > 1$

11. Solve $5 - 2x^2 = 8x$ by completing the square. Give your answer to the nearest tenth. 0.5 or -4.5

12. Solve $4x - 2x^2 = 5$ using the quadratic formula. Give your answer to the nearest tenth. $1 \pm 1.2i$

13. Find the x-intercepts and the vertex of the graph of $y = 4x^2 + 7x - 3$. Give your answer to the nearest thousandth.

14. Without solving, determine the number of real-number solutions for $x^2 - 7x - 9 = 0$. 2

15. Simplify $\sqrt{-5} \cdot (-3i^2)$. $3i\sqrt{5}$

16. Solve $x^2 + 5 = 0$. $\pm i\sqrt{5}$

17. Write the solutions in $a + bi$ form for $-5x^2 + x - 3 = 0$. $0.1 \pm 0.768i$

18. What is the complex conjugate for $3i + 1$? $1 - 3i$

19. Find a quadratic model that fits the points $(0, 3)$, $(1, -1)$, and $(-1, 5)$. $y = -x^2 - 3x + 3$

20. Solve $x^2 - x - 6 \leq 0$. $-2 \leq x \leq 3$

Physics A package is dropped from an airplane from a height of 3000 feet without a parachute.

21. Use the quadratic model for the height of a falling object affected by gravity, $h(t) = -\frac{1}{2}gt^2 + v_0t + h_0$, to write an equation for the height of the package after it is dropped as a function of time. $h(t) = -16t^2 + 3000$

22. How long will it take the package to reach the ground? 13.69 sec

23. **Business** The Printer Connection manufactures and sells boxes. The cost model that describes the cost of manufacturing x million boxes is $c(x) = 35 - 5x$. The revenue model, which describes the revenue from selling x million boxes, is $r(x) = -x^2 + 10x$. Find the break-even points, where the cost equals the revenue. 2,890,000 items, or 12,110,000 items

6. $\sqrt{40}$

7. Let $f(x) = 0$ and solve for x. $-4(x - 1)^2 + 3 = 0$, $x = 1 \pm \frac{\sqrt{3}}{2}$ so that the x-intercepts are $(1 + \frac{\sqrt{3}}{2}, 0)$ and $(1 - \frac{\sqrt{3}}{2}, 0)$

13. $(0.356, 0)$
$(-2.106, 0)$
$(-0.875, -6.063)$

3. $(0.25, 1.125)$, maximum

4. $\sqrt{2^2 + 3^2} = \sqrt{13}$

5. $3 \pm \sqrt{6}$

6 **P**olynomial Functions

CHAPTER

Meeting Individual Needs

6.1 Exploring Products of Linear Functions

Core	Core Plus
Reteaching the Lesson, p. 328 Inclusion Strategies, p. 327 Lesson Activity Master 6.1 Practice Master 6.1 Enrichment Master 6.1 Technology Master 6.1	Practice Master 6.1 Enrichment, p. 327 Technology Master 6.1
[2 days]	**[1 day]**

6.2 The Factored Form of a Polynomial

Core	Core Plus
Reteaching the Lesson, p. 335 Inclusion Strategies, p. 334 Lesson Activity Master 6.2 Practice Master 6.2 Enrichment Master 6.2 Technology Master 6.2	Practice Master 6.2 Enrichment, p. 334 Technology Master 6.2 Interdisciplinary Connection, p. 333
[2 days]	**[2 days]**

6.3 Dividing Polynomials

Core	Core Plus
Reteaching the Lesson, p. 343 Inclusion Strategies, p. 342 Lesson Activity Master 6.3 Practice Master 6.3 Enrichment Master 6.3 Technology Master 6.3 Mid-Chapter Assessment Master	Practice Master 6.3 Enrichment, p. 342 Technology Master 6.3 Mid-Chapter Assessment Master
[2 days]	**[1 day]**

6.4 Exploring Polynomial Function Behavior

Core	Core Plus
Reteaching the Lesson p. 352 Inclusion Strategies p. 351 Lesson Activity Master 6.4 Practice Master 6.4 Enrichment Master 6.4 Technology Master 6.4	Practice Master 6.4 Enrichment, p. 351 Technology Master 6.4
[Optional*]	**[1 day]**

6.5 Applications of Polynomial Functions

Core	Core Plus
Reteaching the Lesson, p. 361 Inclusion Strategies, p. 360 Lesson Activity Master 6.5 Practice Master 6.5 Enrichment Master 6.5 Technology Master 6.5	Practice Master 6.5 Enrichment, p. 360 Technology Master 6.5 Interdisciplinary Connection, p. 359
[2 days]	**[2 days]**

Chapter Summary

Core	Core Plus

Eyewitness Math, pp. 356–357
Lab Activity
Long-Term Project
Chapter Review, pp. 368–370
Chapter Assessment, p. 371
Chapter Assessment, A/B
Alternative Assessment
Cumulative Assessment,
   pp. 372–373
	Eyewitness Math,
	   pp. 356–357
	Chapter 6 Project,
	   pp. 366–367
	Lab Activity
	Long-Term Project
	Chapter Review,
	   pp. 368–370
	Chapter Assessment,
	   p. 371
	Chapter Assessment, A/B
	Alternative Assessment
	Cumulative Assessment,
	   pp. 372–373

[3 days]	**[3 days]**

Reading Strategies

Encourage students to read through the chapter quickly to see what is covered and to identify lessons that could be difficult. Then they should read each lesson in detail. Later, students can work with partners or in small groups to discuss what they read. When reading mathematics, students should be prepared to read the material several times and work the examples during the course of their reading. They should be able to ask intelligent questions about material they do not understand. They should also be able to explain the material they have read to others.

Visual Strategies

Throughout the chapter, students are asked to graph polynomial functions using a graphics calculator. Graphics calculators and computer software, like $f(g)$ *Scholar*™, provide students with a quick and easy way to graph these functions. However, it is also beneficial for students to use the graphics calculator or computer software only for computation and to sketch the graph on graph paper by hand.

They can draw the graph using a colored pencil and then use a different color to draw important points on the graph, such as the turning point(s) or the point(s) of intersection, stressing the x-intercepts. Highlighting these points in this way will help to emphasize their importance. Labeling the points helps to connect the specific points with the terms used to describe them.

Hands-on Strategies

The Portfolio Activity asks students to make a box from a 12 inch by 12 inch piece of cardboard. The volume of this box can be expressed using a polynomial function. Encourage students to use several different pieces of cardboard of this size to make several different boxes. They then write the expression used to find the volume of each box. Ask students to look at all of their expressions and use them to write the polynomial function requested in the Portfolio Activity. This activity can be repeated with pieces of cardboard that have different dimensions, such as the 8 inch by 16 inch piece suggested on page 358 or the 5 inch by 18 inch piece suggested in Try This on page 359 (both in Lesson 6.5).

Cooperative Learning

GROUP ACTIVITIES	
Exploring polynomial functions	Chapter Opener, The Portfolio Activity
Zeros of a function	Lesson 6.1, Explorations 1, 2
Multiple zeros	Lesson 6.4, Exploration 2
Scream Machine	Eyewitness Math
Taylor Series	Chapter 6 Project

You may wish to have students work with partners for some of the above activities. Additional suggestions for cooperative group activities are noted in the teacher's notes in each lesson.

Multicultural

The cultural connections in this chapter include references to Europe and Asia.

CULTURAL CONNECTIONS	
Europe: Fundamental Theorem of Algebra	Lesson 6.3
Asia: Khayyam methods	Lesson 6.5

Portfolio Assessment

Below are portfolio activities for the chapter. They are listed under seven activity domains that are appropriate for portfolio development.

1. **Investigation/Exploration** Two projects listed in the teacher's notes ask students to investigate a method and/or a relationship. They are *Enrichment* on page 334 (Lesson 6.2) and *Enrichment* on page 342 (Lesson 6.3).

2. **Applications** Any of the following are recommended: thermodynamics (Lesson 6.2), economics (Lesson 6.4), banking (Lesson 6.5), radioactive decay (Lesson 6.5), and from the teacher's notes, *Enrichment* on page 360 (Lesson 6.5) and *Interdisciplinary Connection*: Nature on page 359 (Lesson 6.5).

3. **Non-Routine Problems** The Portfolio Activity on page 325 asks students to investigate the possibilities for the volume of a box made from a piece of cardboard of a given size. Students find a polynomial function to describe the possible volumes for the box.

4. **Project** The Chapter Project on pages 366 and 367 introduces students to the Taylor series for sines. Students use graphics calculators and hand-drawn graphs to explore the relationship between polynomial functions and the sine function.

5. **Interdisciplinary Topics** Students may choose from the following: physics (Lessons 6.2 and 6.4), geography (Lesson 6.3), economics (Lessons 6.4), and from the teacher's notes *Interdisciplinary Connection*: Engineering on page 333 (Lesson 6.2).

6. **Writing** The portfolio assessment on page 365 suggests an excellent opportunity for journal writing. Also recommended are the Communicate questions at the beginning of the exercise sets.

7. **Tools** Chapter 6 uses cardboard, graphic calculators or computer software, such as *f(g) Scholar*™, graph paper, and colored pencils. The technology masters are recommended to measure individual student's proficiency.

Technology

Graphics calculators and computer software such as *f(g)* *Scholar*™ provide students with an opportunity to study the graphs of polynomial functions without spending time graphing them manually. Students can quickly graph a function and use the TRACE feature and other available features to find points on the graph. With this capability, students are much more likely to experiment and therefore learn more about the usefulness of the graphs of polynomial functions. For more information on the use of technology, refer to the *HRW Technology Handbook*.

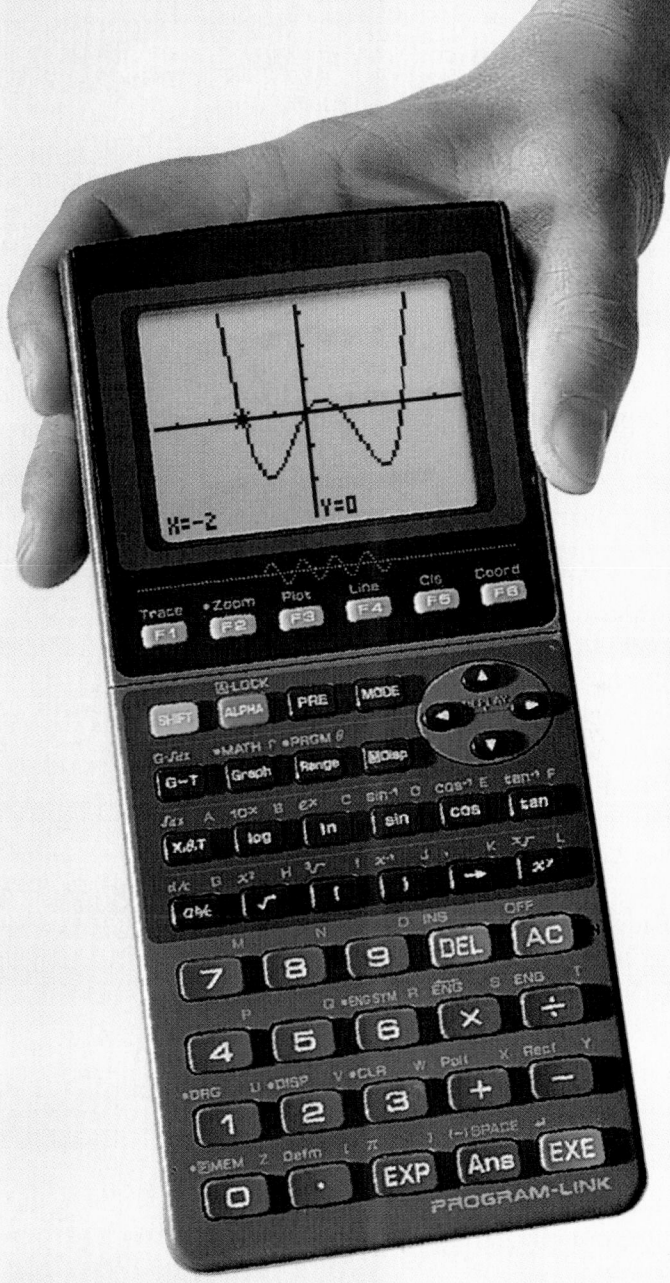

Graphics Calculator

Finding Roots of Polyomial Functions Graphics calculators are used for graphing polynomial functions and finding the coordinates of points on the graph. Instructions given below show how to use the CALC feature on the TI-82. For instructions on using other graphics calculators, refer to the *HRW Technology Handbook*.

1. Enter the appropriate polynomial function, set the window, and graph.

2. Press `2nd` `CALC` to display the CALCULATE menu. To find a zero of the function to a tolerance of 1E-5, choose 2: root from the menu. The graph is then displayed along with the prompt "Lower Bound?:" and the current position of the cursor. Press the ◄ or ► keys to move the cursor to the appropriate *x*-value for the lower bound and then press `ENTER`. A > then appears on the screen to indicate the lower bound.

3. Use the same process to set the upper bound. A < will appear on the screen to indicate the upper bound.

4. The prompt "Guess?" then appears on the screen. Move the cursor to a point that is between the bounds and close to the zero of the function. This helps the calculator find the zero more quickly. Press `ENTER` and the display shows "Root" and the *x*- and *y*-values.

5. Repeat steps 1–4 for any additional zero of the function.

A similar procedure can be followed to find any maximum or minimum values of functions. Choose 3: minimum or 4: maximum from the CALCULATE menu and set the lower bound, upper bound, and guess as described above. The *x*- and *y*-values for the appropriate point will be given.

Integrated Software

f(g) Scholar™ is an integrated computer-based mathematics productivity tool that combines calculator, spreadsheet, and graphics capabilities to provide a dynamic and interactive environment for explorations in mathematics. Use *f(g) Scholar*™ for any lesson needing a spreadsheet, calculator, graphics calculator, or any combination of the three.

6 Polynomial Functions

ABOUT THE CHAPTER

Background Information

This page gives a brief introduction to the history of polynomial functions. More detailed information about the method of Omar Khayyam for finding zeros of a cubic function is given in Lesson 6.5. You may want to have students do research to find out more about the other mathematicians mentioned here.

CHAPTER RESOURCES

- Practice Masters
- Enrichment Masters
- Technology Masters
- Lesson Activity Masters
- Lab Activity Masters
- Long-Term Project Masters
- Assessments Masters
  Chapter Assessments, A/B
  Mid-Chapter Assessment
  Alternative Assessment, A/B
- Teaching Transparencies
- Cumulative Assessment
- Spanish Resources

CHAPTER OBJECTIVES

- Determine realistic domain restrictions of a volume function.
- Define and examine the two forms of polynomial functions.
- Write the factored form of a polynomial.
- Define and use the Factor Theorem of Algebra to find the zeros of polynomial functions.
- Define and use the division algorithm for polynomials.

CHAPTER 6

LESSONS

6.1 *Exploring* Products of Linear Functions

6.2 The Factored Form of a Polynomial

6.3 Dividing Polynomials

6.4 *Exploring* Polynomial Function Behavior

Eyewitness Math
Scream Machine

6.5 Applications of Polynomial Functions

Chapter Project
The Taylor Series

Polynomial Functions

Operations with polynomials were developed by Islamic mathematicians of the Middle Ages to help solve problems in inheritance law, hydraulic engineering, and commerce. The first algebra book to reach Europe was written in central Asia by al-Khowarizmi. Abu Kamil of Egypt was the first to use polynomials of degrees higher than 2. Al-Karaji and al-Samaw'al, both of Baghdad, developed the rules for exponents, division of polynomials, and a form of synthetic division. They thought so highly of their subject that al-Karaji named his book *The Marvelous*, and al-Samaw'al named his book on polynomials *The Shining*. Al-Samaw'al was only 19 years old when he wrote this important book.

▲ Sculptural relief from Baghdad, Iraq

At-Tusi's Arabic rendition ▶ of Euclid's proof of the "Pythagorean" Right-Triangle Theorem, 1258 C.E.

ABOUT THE PHOTOS

Ancient Arabic mathematical artifacts and artwork are displayed on these two pages. The page of mathematics will help students realize that much of the mathematics we know now was developed in Asia Minor. The artwork will show students how important symmetry was to these people.

PORTFOLIO ACTIVITY

Construct an open box from a piece of 12-inch-by-12-inch cardboard. Cut out squares of equal size from the corners of the piece of cardboard, and fold up the sides. The size of the cut-out squares determines the volume of the box.

Express the volume of the box as a polynomial function in terms of the length of the sides of the cut-out squares. The polynomial describes the effect that the size of the cut-out squares will have on the volume and what the maximum volume will be.

You will be asked to complete this activity on page 365. You may include it in your portfolio.

x

12"

12"

- Define and use the Fundamental Theorem of Algebra to find the zeros of a polynomial function.
- Determine and classify the behavior of single zeros and multiple zeros of a polynomial function.
- Determine and classify the behavior and basic shape of polynomial functions of varying degrees.
- Determine and analyze polynomial functions that model real-world situations.
- Use the algebraic tool of variable substitution to simplify a polynomial function.

PORTFOLIO ACTIVITY

This activity provides new insight into the volume of a rectangular prism. You may want to have students work with numbers before progressing to the use of x for the height of the box.

Have students work in small groups and discuss how they can solve this problem. Ask each group to record the ideas discussed and compile a list of possible solutions. Have each group share its final list with the class. Have students record this list so that they can refer to it when asked to complete the activity in Lesson 6.5.

As an extension, ask students to complete the activity using pieces of cardboard that are of different sizes. Encourage students to choose square as well as rectangular pieces of cardboard. Ask them to summarize the results of using the different sizes of cardboard.

ABOUT THE CHAPTER PROJECT

The Chapter 6 project on pages 366–367 introduces students to a special type of polynomial function called the Taylor series for sines. Students use graphics calculators, graph paper, and colored pencils to explore this special series. Students should be encouraged to investigate this subject more thoroughly on their own.

Exploring Products of Linear Functions

PREPARE

Objectives

- Determine realistic domain restrictions of a volume function.
- Define and examine the two forms of polynomial functions.

RESOURCES

• Practice Master	**6.1**
• Enrichment Master	**6.1**
• Technology Master	**6.1**
• Lesson Activity Master	**6.1**
• Quiz	**6.1**
• Spanish Resources	**6.1**

Assessing Prior Knowledge

The width and length of the base of a carton are 5 feet and 8 feet, respectively. The height of the carton is 3 feet.

1. What is the perimeter of the base of the carton? [**26 ft**]

2. What is the area of the base of the carton? [**40 ft²**]

3. What is the volume of the carton? [**120 ft³**]

TEACH

Volume is a concept that is used in many different situations. Ask students to name several situations or professions in which volume might need to be found. Have students describe why volume is important in the situation or profession they name.

You have seen that area is modeled by the product of two linear functions, or a quadratic function. A different type of function is needed to model volume.

Judy wants to build a compost bin shaped like a rectangular box. Because of constraints on the amount of material available to build the bin, the perimeter of its base cannot exceed 18 feet.

Judy plans to rotate the compost from the bottom up with a shovel that is 2 feet wide. Therefore, the width must be at least 2 feet. If the compost bin is too tall, rotating the compost will be difficult. Therefore, Judy wants the height to be only 1 foot greater than the width.

Using Visual Models Have students work in small groups and draw several labeled pictures of possible compost bins that Judy can build. Suggest that they use the table on page 327 to get started. However, emphasize that the dimensions given in the table are not the only ones they should use. Have students study their pictures and draw new pictures, if necessary, as they work through Exploration 1. Then have students use what they have learned in previous mathematics courses to determine the roots of each of the six polynomial functions given in Exploration 2. Have them use colored pencils to show these roots on a coordinate plane. Then ask students to graph each function on their graphics calculator or on computer software such as *f(g) Scholar*™ to check their roots against the roots shown on the graph.

Volume Function

The volume of the box is the product of the width, length, and height, written $V = w \cdot l \cdot h$.

If the width is 2 feet and the perimeter is 18 feet, then the length is 7 feet. Why?

Examine the following table, which shows some dimensions of the compost bin and the resulting volumes.

Width (ft)	Length (ft)	Height (ft)	Volume (ft³)
2	7	3	42
3	6	4	72
4	5	5	100
5	4	6	120
⋮	⋮	⋮	⋮
x	$9 - x$	$x + 1$	$x(9-x)(x+1)$

Exploration 1 Zeros and the Realistic Domain

Graphics Calculator

1 Graph the width function, $w(x) = x$, using a viewing window that shows the x- and y-intercepts. What is the zero of the width function? What restriction on x is necessary for a realistic width?

2 Since the perimeter is 18 feet, the length must be less than 9 feet, or $l(x) = 9 - x$, where x is the width. Graph the length function using a viewing window that shows the x- and y-intercepts. What is the zero of the length function? What restriction on x is necessary for a realistic length?

3 Graph the height function, $h(x) = x + 1$, using a viewing window that shows the x- and y-intercepts. What is the zero of the height function? What restriction on x is necessary for a realistic height?

4 Graph the volume function, $V(x) = w(x) \cdot l(x) \cdot h(x)$, or $V(x) = x(9-x)(x+1)$. Is V a linear function, quadratic function, or neither? Explain. What are the zeros of the volume function?

5 When a function models a real-world situation, the domain is restricted to x-values that are meaningful, and is called the **realistic domain** of that function. What is the realistic domain of V?

MAXIMUM
MINIMUM
Connection

6 Use the trace feature to approximate, to the nearest tenth, the maximum value of the volume function. What length, width, and height correspond to the maximum volume? Are these reasonable dimensions for a compost bin? Explain. ❖

CRITICAL
Thinking

Find negative values of x that make V positive. Why are these values not included in the realistic domain of V?

Exploration 2 Notes

Be sure students understand that they obtain the same zeros from the expanded form as they do from the factored form of a polynomial function.

ASSESSMENT

2. The zeros are the same as the *r*-value for each polynomial.

3. The highest power of *x is* the same as the number of linear factors in each polynomial.

CRITICAL
Thinking

Changing the constant will change the width of the graph of the function. There are no changes in the *zeros* of the function because when $f(x) = 0$, for any constant C, $C \cdot f(x)$ still equals zero.

TEACHING *tip*

Point out that C in the factored form is the same as a_n in the expanded form.

Linear Factors

The volume function in Exploration 1 is a *polynomial function*.

A **polynomial function** is any function that can be expressed as a product of linear factors multiplied by a constant. A **linear factor** of a function is any factor of the form $x - r$, where r can be a real or a complex number.

•Exploration 2 Linear Factors and Zeros

Graphics Calculator

Graph each polynomial function using your graphics calculator. Make sure to use a viewing window that includes the *x*- and *y*-intercepts of each function. Complete Steps 1–3 for each function.

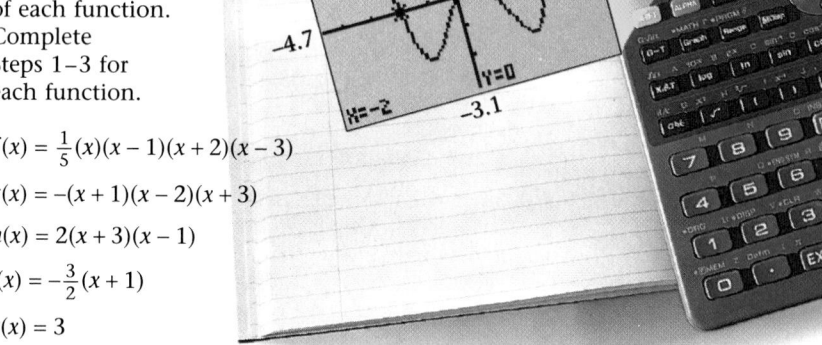

$f(x) = \frac{1}{5}(x)(x - 1)(x + 2)(x - 3)$

$g(x) = -(x + 1)(x - 2)(x + 3)$

$h(x) = 2(x + 3)(x - 1)$

$j(x) = -\frac{3}{2}(x + 1)$

$k(x) = 3$

$l(x) = (x - 3)(x + 1)$

1 How many linear factors are there? What is the constant?

2 Find the zeros of the function from its graph. How are these zeros related to the value of *r* in each linear factor?

3 Multiply the factors to expand each polynomial. How is the highest power of *x* related to the number of linear factors? ❖

CRITICAL
Thinking

How do you think a change in the constant affects the shape of the graph? Does changing the constant change the zeros of the function? Explain.

FACTORED FORM OF A POLYNOMIAL FUNCTION

If f is any polynomial function, then the factored form of f is $f(x) = C(x - r_1)(x - r_2) \ldots (x - r_{n-1})(x - r_n)$, where the subscripted *r*-values, $r_1, r_2, \ldots, r_n$, are the roots, or zeros, of the function and C is a constant.

RETEACHING
the
lesson

Using Cognitive Strategies Have each student write three polynomial functions in factored form: one with two factors, one with three factors, and one with four factors. Then have students exchange and graph each other's functions. Have students use two different methods to find the zeros of each function. Students should make sure they get the same zeros us-

ing both methods. Have each student multiply the factors in each polynomial to write the expanded form of each function. As a class, discuss the results for several functions of each type. Ask students to record any observations they make during the discussion. Conclude the discussion by asking students to share their observations. Work together as a class, and use these observations to make a final list to summarize the concepts presented in this lesson.

In the factored form of a polynomial, the constant, C, vertically stretches or shrinks the graph without affecting the roots of the polynomial.

EXPANDED FORM OF A POLYNOMIAL FUNCTION
If f is any polynomial function, then f can be written in the expanded form $f(x) = a_n x^n + a_{n-1} x^{n-1} + \ldots + a_2 x^2 + a_1 x + a_0$, where the subscripted a-terms are the *coefficients*.

In the expanded form, the coefficient a_n is the **leading coefficient** of the polynomial, and the coefficient a_0 is the constant term.

The **degree of the polynomial** is the highest power of x with a nonzero coefficient in expanded form. It is equal to the total number of linear factors in factored form.

The terms of a polynomial in expanded form are usually written in decreasing order of the degree of the variable. For example, $2x^3 - x^4 + 5$ is usually written $-x^4 + 2x^3 + 5$.

Polynomials of the first four degrees have special names. They are listed below.

Linear (1st degree):	$a_1 x + a_0$
Quadratic (2nd degree):	$a_2 x^2 + a_1 x + a_0$
Cubic (3rd degree):	$a_3 x^3 + a_2 x^2 + a_1 x + a_0$
Quartic (4th degree):	$a_4 x^4 + a_3 x^3 + a_2 x^2 + a_1 x + a_0$

EXERCISES & PROBLEMS

Communicate

1. In your own words, define *polynomial function*.

2. How do the linear factors of a polynomial function relate to the roots, or zeros, of the function?

3. Describe the factored form of a polynomial function, including the meaning of C and r_n. How do you determine the degree of a polynomial function in factored form?

4. Describe the expanded form of a polynomial function, including the meaning of a_n and a_0. How do you determine the degree of a polynomial function in expanded form?

5. Use the definition of a polynomial function to explain why a quadratic function is a polynomial function.

6. Use the definition of a polynomial function to explain why both constant and nonconstant linear functions are polynomial functions.

ASSESS

Selected Answers
Odd-numbered Exercises 7–35

Assignment Guide
Core 1–9, 11–13, 15–42

Core Plus 1–6, 9, 10, 13–42

Technology
A graphics calculator is needed for Exercises 15–20 and 22–28.

Error Analysis

Be sure students use the appropriate friendly viewing window on their calculator to find all zeros. Students should determine the number of zeros of a function *before* graphing so that they can be sure they have found every zero.

17. A squared linear factor causes the graph to touch but not cross the x-axis.

20.

w	0	1	2	3	4	5
l	2	1	0	−1	−2	−3
h	−1	0	1	2	3	4
V	0	0	0	−6	−24	−60

21. The width, w, must be between 1 and 2, so the length, l, must be between 0 and 1, and the height, h, must be between 0 and 1. All the possible width values do not give realistic volume values since some width values cause the length or the height to be negative.

Practice & Apply

Determine the degree of each polynomial function.

7. $f(x) = (x + 6)(3x - 1)(x - 3)(2x - 7)$ 4

8. $f(x) = 7x^5 + 3x^3 - 2x + 4$ 5

9. $f(x) = -4x^2 + 3x^3 - 5x^6 + 4$ 6

10. $f(x) = 2(x + 1)^3 - 5x + 4x^2$ 3

Determine the zeros of each polynomial function.

11. $f(x) = 8(x - 1)(x + 4)(x - 7)$ 1, −4, 7

12. $f(x) = (x + 6)(x - 1)(x - 3)(x - 7)$ −6, 1, 3, 7

13. $f(x) = (x + 2)(x - 1)(x - 3)(x + 7)$ −2, 1, 3, −7

14. $f(x) = (x - 4)(x - 1)$ 4, 1

For Exercises 15–18, graph $f(x) = (x - 3)^2(x + 1)$.

15 What are the zeros of this function? 3, −1

16 At which zero(s) does the graph *cross* the x-axis? *touch* the x-axis? crosses: −1; touches: 3

17 How does a squared linear factor affect the graph of a function?

18 For what domain is the function greater than zero? Find the approximate maximum value of the function between the zeros. $f(x) > 0$ when $-1 < x < 3$ and when $x > 3$ maximum: 9.5

Geometry The perimeter of the base of a crate cannot exceed 4 meters. The height of the crate is 1 meter less than the width.
Complete Exercises 19–22.

19 Write a function for the volume of the crate in terms of the width. Graph this volume function. $V(w) = w(2 - w)(w - 1)$

20 Use the trace feature to make a table of values for the volume of the crate. Use width values of 0, 1, 2, 3, 4, and 5 meters. Include in your table the corresponding values of the length and height.

21. Which values of the width give realistic volumes? What are the corresponding values of the length and height for the values of the width that result in realistic volumes? Explain why all of the possible width values do not give realistic volumes.

22 **Maximum/Minimum** Find the dimensions that result in the maximum volume. Give the maximum volume to the nearest tenth of a meter.
$w = 1.6$ m, $h = 0.6$ m, $l = 0.4$ m; $V_{max} = 0.4$ m^3

23 Use your graphics calculator to compare the graphs of $f(x) = (x - 3)(x - 5)$, $g(x) = (x - 3)^2(x - 5)$, and $h(x) = (x - 3)^3(x - 5)$. How do the zeros of these functions compare? Do the graphs have the same shape? Explain.

23. All functions have the same zeros, 3 and 5, but their graphs have different shapes. The graph of f is a parabola, the graph of g is a cubic, and the graph of h is a quartic.

24. The graph gains a zero as another factor is included.

25. There are no turns in the graph of f, one turn in the graph of g, two turns in the graph of h, and three turns in the graph of k. The number of turns is one less than the degree of the polynomial.

Practice Master

NAME ______________ CLASS ________ DATE ________

Practice & Apply
6.1 Exploring Products of Linear Functions

Identify whether the function is a polynomial function, a linear function, or neither.

1. $f(x) = x^2 + x - 1$ ________
2. $f(x) = \frac{(x - 3)(x^2 - 1)}{x}$ ________
3. $f(x) = 3x - 2$ ________
4. $f(x) = (2x - 1)(x + 2)$ ________

Use the function $f(x) = -(x + 2)(x - 3)(x - 1)$ to answer the following exercises.

5. What is the degree of the function? ________
6. What are the zeros of the function? ________
7. Write the function in expanded form. ________
8. What is the leading coefficient of the function? ________
9. What is the constant term of the function? ________

Determine the degree of each polynomial function.

10. $f(x) = x(x - 5)^2(x + 2)$ ________
11. $f(x) = 4x^2 - 5x^3 + 1$ ________
12. $f(x) = x(1 - x^4)$ ________
13. $f(x) = x^2(x - 4)^4$ ________

Determine the degree of each function and create an equation of a polynomial function having the same zeros as the functions graphed.

14.
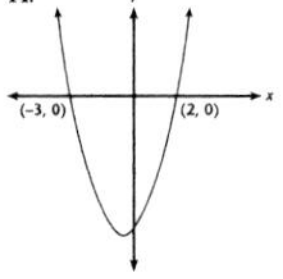

____

15.
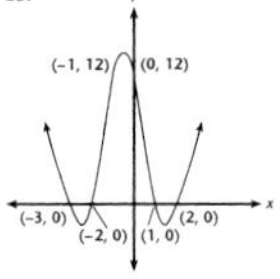

____

16.
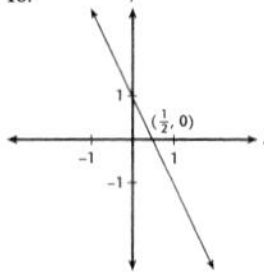

____

110 Practice & Apply HRW Advanced Algebra

For Exercises 24 and 25, use your graphics calculator to examine the graphs of polynomial functions *f*, *g*, *h*, and *k*. Use the same viewing window for each graph.

$f(x) = x - 1$ $\quad$ $g(x) = (x - 1)(x + 2)$

$h(x) = (x - 1)(x + 2)(x - 3)$ $\quad$ $k(x) = (x - 1)(x + 2)(x - 3)(x + 4)$

24 How does the graph of each function change as an additional linear factor is included?

25 How many turns are there in the graph of each function? How does the number of turns compare with the degree of each polynomial?

For Exercises 26–28, graph $f(x) = (x - 1)^2(x + 2)(x - 3)$ and $g(x) = (x - 1)^3(x + 2)(x - 3)$.

26 What are the zeros of each function? *f*: 1, −2, 3; *g*: 1, −2, 3

27 Graph each function. How are the graphs alike? How are they different?

28 How does a cubed linear factor affect the graph of a function?

Look Back

Let $A = \begin{bmatrix} 1 & -1 \\ 0 & 2 \end{bmatrix}$ and $B = \begin{bmatrix} -1 & 4 \\ 5 & 3 \end{bmatrix}$. Find each value. [Lessons 4.1, 4.2]

29. $B - A$ **30.** $2A$ **31.** BA **32.** $-AB$ **33.** $B - 3A$ **34.** B^2

35. Give an example of a perfect-square trinomial. **[Lesson 5.4]**

36. Rewrite your perfect-square trinomial from Exercise 35 as a binomial squared. **[Lesson 5.4]**

Look Beyond

For Exercises 37–40, use the following pairs of linear factors.
$(x + 3)(x - 3)$ $\quad$ $(x + 5)(x - 5)$ $\quad$ $(x + 2)(x - 2)$ $\quad$ $(x + 8)(x - 8)$

37. How is each pair of linear factors alike or different?

38. Is each expanded product a linear or quadratic expression? quadratic

39. Is there an *x*-term in each expanded product? no

40. How does the constant term in each expanded product compare with the *r*-values in each corresponding pair of linear factors? the constant term is $-r^2$

41. Examine the pattern in the expanded products you found in Exercises 37–40. Use this pattern to write the product of $(x + a)(x - a)$ in terms of *x* and *a*. $x^2 - a^2$

42. Use your answer to Exercise 41 to find the linear factors of each expression.

a. $x^2 - 16$ **b.** $x^2 - 36$ **c.** $x^2 - 49$ **d.** $x^2 - 256$ $(x + 16)(x - 16)$

27. The zeros of *f* and *g* are the same; 1, −2 and 3. *f* just touches the *x*-axis at 1 while *g* crosses the *x*-axis at 1. The graph of *f* has three turning points, two local minimums and a local maximum, while the graph of *g* has only 2 turning points, one minimum and 1 maximum.

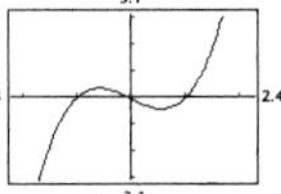

28. A cubed linear factor flattens the graph of the function at this crossing point zero.

29. $\begin{bmatrix} -2 & 5 \\ 5 & 1 \end{bmatrix}$ $\quad$ 30. $\begin{bmatrix} 2 & -2 \\ 0 & 4 \end{bmatrix}$

31. $\begin{bmatrix} -1 & 9 \\ 5 & 1 \end{bmatrix}$ $\quad$ 32. $\begin{bmatrix} 6 & -1 \\ -10 & -6 \end{bmatrix}$

33. $\begin{bmatrix} -4 & 7 \\ 5 & -3 \end{bmatrix}$ $\quad$ 34. $\begin{bmatrix} 21 & 8 \\ 10 & 29 \end{bmatrix}$

alternative ASSESSMENT

Portfolio Assessment

Ask students to describe how to convert a polynomial function expressed in factored form to a polynomial function expressed in expanded form. Have them provide three examples of this procedure and identify the leading coefficient, the constant term, and the degree for each polynomial. Also ask students to find the roots of each function.

35. Answers may vary; should be of the form $x^2 \pm 2ax + a^2$

36. Answers may vary; should be of the form $(x \pm a)^2$

The answers to Exercises 37 and 42 can be found in Additional Answers beginning on page 842.

Look Beyond

These exercises preview factoring a polynomial, which is presented in the next lesson.

Technology Master

NAME ____________ CLASS ________ DATE ________

Technology
6.1 Symmetric Zeros

Suppose that $f_a(x) = (x + a)(x)(x - a)$ and that a is a nonnegative real number. If, for example, $a = 1$, $f_a(x) = (x + 1)(x)(x - 1)$. The graph has three zeros, two of which are an equal distance from the origin. Notice that the graph is symmetric about the origin and that if you expand the polynomial $(x + 1)(x)(x - 1)$ you will get $x^3 - x$, a polynomial with only odd powers of x.

With a graphics calculator, you can classify all the different graphs that arise from different values of a.

Use a graphics calculator to graph $f_a(x) = (x + a)(x)(x - a)$ for each value of a.

1. $a = 0$ **2.** $a = 1.5$ **3.** $a = 2$

4. Classify and describe the graphs of $f_a(x) = (x + a)(x)(x - a)$.

Let $g_{a,b}(x) = (x + b)(x + a)(x)(x - a)(x - b)$. Use a graphics calculator to graph g for each value of a and b.

5. $a = 0; b = 0$ **6.** $a = 1.5; b = 0$ **7.** $a = 2; b = 1$

8. Describe the graph of $g_{a,b}(x) = (x + b)(x + a)(x)(x - a)(x - b)$ if a and b are not 0 and $a = b$.

9. Describe the graph of $g_{a,b}(x) = (x + b)(x + a)(x)(x - a)(x - b)$ if a and b are not 0 and $a \neq b$.

10. Describe the graph of $h_{a,b}(x) = (x + b)(x + a)(x - a)(x - b)$.

- Write the factored form of a polynomial.
- Define and use the Factor Theorem of Algebra to find the zeros of polynomial functions.

RESOURCES

- Practice Master   6.2
- Enrichment Master   6.2
- Technology Master   6.2
- Lesson Activity Master   6.2
- Quiz   6.2
- Spanish Resources   6.2

Assessing Prior Knowledge

Solve each quadratic equation using the quadratic formula.

1. $x^2 - 4x - 21 = 0$ [**-3, 7**]

2. $4x^2 - 8x - 32 = 0$ [**-2, 4**]

3. $3x^2 + 6x = 0$ [**-2, 0**]

4. $6x^2 - 150 = 0$ [**-5, 5**]

TEACH

Point out that the polynomial function $h(t) = -16t^2 + 32t + 240$ represents just one of many real-world situations that can be modeled using a polynomial. Create a list of other polynomial functions that model real-world situations.

LESSON 6.2 The Factored Form of a Polynomial

Polynomial functions are used in many practical applications. These range from the abstract theory of atomic and molecular structure to more practical problems that arise in investment planning. Important information may be obtained from the roots of a polynomial. These can be determined by converting the expanded form of the polynomial to its factored form.

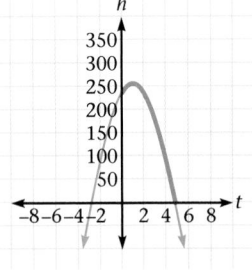

Physics   Jean throws a pebble from a 240-foot-high bridge with an initial upward velocity of 32 feet per second. The quadratic function $h(t) = -16t^2 + 32t + 240$ models the height of the pebble as the time changes. When the function h is written in expanded form, it has a leading coefficient of -16, and its degree is 2. The function can also be expressed in factored form as $h(t) = -16(t - r_1)(t - r_2)$.

Using Visual Models
Write several monomials, such as x, $2x$, $3x$, and $4x$, on separate index cards. Write several binomials, such as $(x - 3)$, $(x + 2)$, $(2x + 3)$, and $(3x - 4)$, on separate index cards. Ask students to choose a monomial card and a binomial card, and use the cards to complete the steps below.

Record the expressions on your cards. These are the factors. Then find the product.

1. Write "factored form" by the factored expression.

2. Write "expanded form" by the product.

3. Use the product to write a polynomial function.

4. Find the zeros of the function.

Repeat the steps with different combinations of cards, such as two binomial cards, or one monomial card and two binomial cards.

The graphs show that the zeros of $h(t) = -16t^2 + 32t + 240$ are -3 and 5

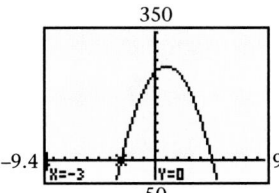 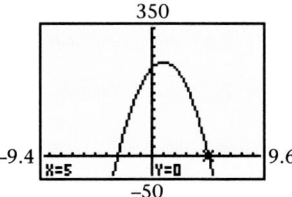

Since the zeros of h are the same as the r-values of h in factored form, rewrite the equation.

$$h(t) = C(t - r_1)(t - r_2)$$
$$h(t) = -16(t - 5)(t - (-3))$$
$$= -16(t - 5)(t + 3)$$

To check whether the factored form of h is equivalent to the expanded form, multiply the factors of h.

$$h(t) = -16(t - 5)(t + 3)$$
$$= -16(t^2 - 5t + 3t - 15)$$
$$= -16t^2 + 32t + 240$$

Use the r-values of h to find the time, t, when the pebble will hit the water. The time for the pebble to fall must be positive, so 5 is the only realistic solution. Thus, it will take 5 seconds for the pebble to hit the water.

What is the realistic domain of h?

Special Products

Exploration 1 *Difference of Two Squares*

Graphics Calculator

1 Find each product.

$(x + 1)(x - 1)$ $(x + 4)(x - 4)$ $(5x + 2)(5x - 2)$

What pattern do you notice between the linear factors and the expanded products? Write a general rule to describe this pattern. Use your rule to multiply $(a + b)(a - b)$, where a and b are any numbers.

2 Graph $f(x) = x^2 - 1$, $g(x) = x^2 - 16$, and $h(x) = 25x^2 - 4$ one at a time. How are the zeros of these functions related to the r-values in the linear factors in Step 1?

3 Use the pattern you developed in Step 1 to write $f(x) = x^2 - a^2$ in factored form. What are the zeros of the function $f(x) = x^2 - a^2$? ❖

The special product in Exploration 1 is called the **difference of two squares**.

In Lesson 5.4 you learned how to complete the square to form a *perfect-square trinomial*. A polynomial of the form $x^2 + 2kx + k^2$, with the constant $k \neq 0$, can always be factored as the square of the binomial, $(x + k)^2$.

Ongoing ASSESSMENT

$0 < h \leq 5$

Exploration 1 Notes

Once students work with the product $(a + b)(a - b) = a^2 - b^2$, some begin to confuse it with the expansion of $(a - b)^2$ and write this product as $a^2 - b^2$. It often helps to review the product of $(a - b)^2$ and point out the differences between the binomial square and the difference of two squares.

Ongoing ASSESSMENT

1. The product is the square of the first term minus the square of the second term.
$$(a + b)(a - b) = a^2 - b^2$$

3. $f(x) = x^2 - a^2$
$$= (x + a) \cdot (x - a).$$
The zeros of f are $\pm a$.

---

interdisciplinary CONNECTION

Engineering One end of a horizontal beam is built into a wall and the other end of the beam rests on a support with the weight of the beam distributed uniformly along its length. The function that describes this situation is $f(x) = -x^4 + 24x^3 - 135x^2$, where x is the distance, in meters, from the wall to any point on the beam, and the value of the function is the distance, in hundredths of a millimeter, from the x-axis to the beam caused by the sag.

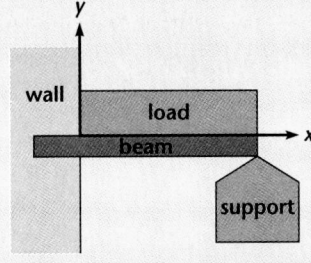

Find the zeros of this function, and explain the meaning of each. [**The zeros are 0, 9, and 15. There is no sag at the wall, 9 meters from the wall, or 15 meters from the wall.**]

EXAMPLE 1

Write each polynomial in factored form.

A $f(x) = 3x^3 - 12x$ **B** $g(x) = 2x^4 - 12x^3 + 18x^2$

Solution➤

A First, factor out the greatest common factor, or GCF, $3x$.

$$f(x) = 3x^3 - 12x$$
$$= 3x(x^2 - 4)$$

Then, factor $x^2 - 4$, a difference of two squares.

$$x^2 - 4 = (x - 2)(x + 2) \quad \rightarrow \quad f(x) = 3x(x - 2)(x + 2)$$

Since all of the factors are linear, $f(x) = 3x(x - 2)(x + 2)$ is the factored form of f.

B First, factor out the GCF, $2x^2$.

$$g(x) = 2x^4 - 12x^3 + 18x^2$$
$$= 2x^2(x^2 - 6x + 9)$$

Then, factor $x^2 - 6x + 9$, a perfect-square trinomial.

$$x^2 - 6x + 9 = (x - 3)^2 \quad \rightarrow \quad g(x) = 2x^2(x - 3)^2$$

Thus, $g(x) = 2x^2(x - 3)^2$ is the factored form of g. ❖

Other Factorable Trinomials

A trinomial written in the form $x^2 + bx + c$ can be factored quickly if you can find two numbers whose product is c and whose sum is b. This process of *factoring by inspection* is illustrated below.

If $c < 0$, then the constant parts of the linear factors have *different* signs.

$x^2 - 3x - 10$	$x^2 + 5x - 6$
$-10 < 0$, so the factors have different signs	$-6 < 0$, so the factors have different signs
↓ ↓	↓ ↓
$(x - 5)(x + 2)$	$(x + 6)(x - 1)$

If $c > 0$, then both constant parts of the linear factors have the *same* sign as *b*.

$x^2 + 11x + 10$	$x^2 - 7x + 10$
$11 > 0$, so the factors have positive signs	$-7 < 0$, so the factors have negative signs
↓ ↓	↓ ↓
$(x + 10)(x + 1)$	$(x - 5)(x - 2)$

TEACHING *tip*

Review perfect-square trinomials before asking students to complete Example 1.

Alternate Example 1

Write each polynomial in factored form.

A. $f(x) = 5x^3 - 45x$
    $[f(x) = 5x(x - 3)(x + 3)]$
B. $g(x) = 3x^4 + 24x^3 + 48x^2$
    $[g(x) = 3x^2(x + 4)^2]$

TEACHING *tip*

The method for factoring trinomials that is presented here is sometimes called the trial-and-error method, because there are cases in which several different combinations will need to be considered before the correct combination is found.

ENRICHMENT   Have students work in small groups to answer the question below.

What are the factored forms of $x^3 - a^3$ and $x^3 + a^3$? HINT: One of the factors of $x^3 - a^3$ is $x - a$ and one of the factors of $x^3 + a^3$ is $x + a$. The other factor in both cases is a trinomial. $[x^3 - a^3 = (x - a)(x^2 + ax + a^2); x^3 + a^3 = (x + a)(x^2 - ax + a^2)]$

INCLUSION **strategies**   **English Language Development** The use of graphs in this lesson provides students with many opportunities to communicate and learn visually. However, be sure students can describe the concepts presented in their own words. Have students keep a list of each term or phrase presented. They should define each term or phrase using their own words. Then have students work in pairs to discuss each other's lists.

EXAMPLE 2

Factor each trinomial.

A $x^2 + 7x + 12$　　**B** $x^2 - 13x + 12$　　**C** $x^2 + 4x - 12$　　**D** $x^2 - x - 12$

Solution➤

Given $x^2 + bx + c$, find two numbers whose product is c and whose sum is b

A

Product is c.	Sum is b.
$3 \times 4 = 12$	$3 + 4 = 7$
↓	↓
$x^2 + 7x + 12$	$x^2 + 7x + 12$

Thus, $x^2 + 7x + 12 = (x + 3)(x + 4)$.

B

Product is c.	Sum is b.
$-12 \times (-1) = 12$	$-12 + (-1) = -13$
↓	↓
$x^2 - 13x + 12$	$x^2 - 13x + 12$

Thus, $x^2 - 13x + 12 = (x - 12)(x - 1)$.

C

Product is c.	Sum is b.
$-2 \times 6 = -12$	$-2 + 6 = 4$
↓	↓
$x^2 + 4x - 12$	$x^2 + 4x - 12$

Thus, $x^2 + 4x - 12 = (x - 2)(x + 6)$.

D

Product is c.	Sum is b.
$-4 \times 3 = -12$	$-4 + 3 = -1$
↓	↓
$x^2 - x - 12$	$x^2 - x - 12$

Thus, $x^2 - x - 12 = (x - 4)(x + 3)$. ❖

Try This　Factor the trinomials.

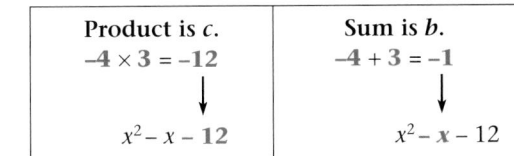

A $x^2 - 13x + 22$　　**B** $x^2 - 9x - 22$　　**C** $x^2 + 13x + 22$　　**D** $x^2 + 9x - 22$

In Example 2, each trinomial is *factored over the integers*. A polynomial is **factored over the integers** when it is expressed as a product of two or more factors having integral coefficients.

Alternate Example 2

Factor each trinomial.

A. $x^2 + 6x + 8$
　　$[(x + 2)(x + 4)]$

B. $x^2 - 12x + 20$
　　$[(x - 10)(x - 2)]$

C. $x^2 + 2x - 15$
　　$[(x - 3)(x + 5)]$

D. $x^2 - 2x - 24$
　　$[(x - 6)(x + 4)]$

ongoing
ASSESSMENT

Try This

A. $(x - 11)(x - 2)$
B. $(x - 11)(x + 2)$
C. $(x + 11)(x + 2)$
D. $(x + 11)(x - 2)$

RETEACHING the lesson

Using Patterns Provide students with a list of polynomials and one factor for each polynomial. A sample list is given below.

1. $x^2 - 9 = (x - 3)(\underline{\quad})$

2. $x^2 + 8x + 16 = (x + 4)(\underline{\quad})$

3. $x^2 - 3x - 4 = (x - 4)(\underline{\quad})$

Have students work in pairs or small groups to find the missing factor for each polynomial.

[**1.** $(x + 3)$　**2.** $(x + 4)$　**3.** $(x + 1)$]

Alternate Example 3

Use the Factor Theorem to write each polynomial in factored form.

A. $f(x) = 2x^3 - 32x$
   $[f(x) = 2x(x + 4)(x - 4)]$
B. $g(x) = 3x^4 + 12x^3 + 12x^2$
   $[g(x) = 3x^2(x + 2)^2]$

The polynomial $x^2 + 1$ cannot be factored over the rational numbers. Such a polynomial is called a prime polynomial with respect to the rationals. It is not prime with respect to the *complex numbers*, however, since $x^2 + 1 = x^2 - i^2 = (x + i)(x - i)$.

Recall from Lesson 6.1 that if the factored form of a polynomial has n linear factors, then the polynomial is of the nth degree. The Fundamental Theorem of Algebra, which you will study in the next lesson, guarantees that any nth degree polynomial always has n linear factors (real or complex).

Each linear factor produces a root. Repeated linear factors produce multiple roots, or **multiple zeros**. For example, the factored form of $f(x) = x^5 - 6x^4 + 9x^3$ is $f(x) = (x - 0)(x - 0)(x - 0)(x - 3)(x - 3)$. The distinct roots are zero and three. The *multiplicity* of zero is 3 and the multiplicity of three is 2. Counting multiplicities, this 5th degree polynomial has 5 roots: 3 zeros and 2 threes.

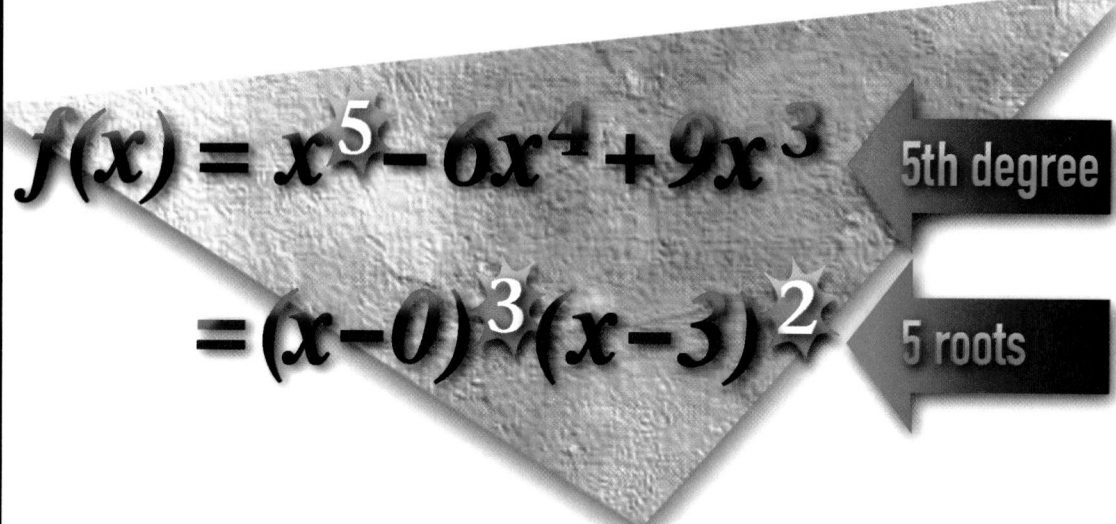

The Factor Theorem

The *Factor Theorem* summarizes the relationship between the factors of a polynomial function and its zeros.

THE FACTOR THEOREM

If f is any polynomial function, $x - a$ is a factor of f if and only if a is a zero of f. Also, a is a zero of f if $x - a$ divides f exactly.

EXAMPLE 3

Use the Factor Theorem to write each polynomial in factored form.

A $f(x) = 3x^3 - 12x$ **B** $g(x) = 2x^4 - 12x^3 + 18x^2$

Graphics Calculator

Solution

A Graph $f(x) = 3x^3 - 12x$, and find the zeros of f.

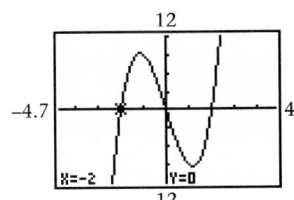

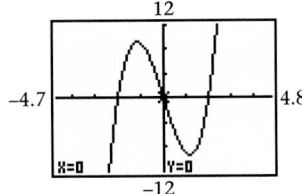

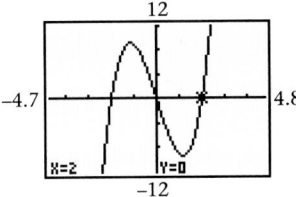

The zeros of f and the r-values in the factored form are -2, 0, and 2.

The leading coefficient, a_3, is 3. Therefore, in factored form, the constant factor, C, is 3.

$$f(x) = C(x - r_1)(x - r_2)(x - r_3)$$
$$f(x) = 3(x - 0)(x - (-2))(x - 2)$$
$$f(x) = 3x(x + 2)(x - 2)$$

B Graph $g(x) = 2x^4 - 12x^3 + 18x^2$, and find the zeros of g.

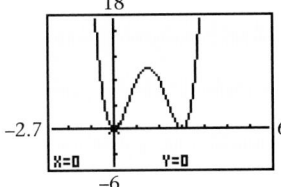

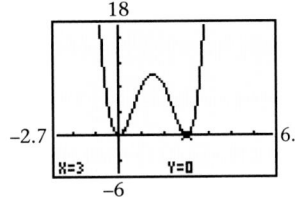

Notice that the graph of g only *touches* the x-axis at the two zeros, 0 and 3. When the graph *touches* (not crosses) the x-axis at a turn in the graph, the zero is a repeated zero of *even* multiplicity. By contrast, at a zero of *odd* multiplicity, the graph *crosses* the x-axis. You will further investigate this in Lesson 6.4.

The leading coefficient, a_4, is 2. Therefore, in factored form, the constant factor, C, is 2.

$$g(x) = C(x - r_1)(x - r_2)(x - r_3)(x - r_4)$$
$$g(x) = 2(x - 0)(x - 0)(x - 3)(x - 3)$$
$$g(x) = 2x^2(x - 3)^2$$

The quartic function, g, has 4 roots because the roots 0 and 3 have double multiplicity and are each counted twice. ❖

Try This Use the Factor Theorem to write each polynomial in factored form.

A $f(x) = 2x^3 - 18x$ **B** $g(x) = 3x^5 + 12x^4 + 12x^3$

TEACHING *tip*

Discuss the advantages of using technology to find the factors of a polynomial. Then ask students how to find the factors of the functions on this page without using technology.

Aongoing **SSESSMENT**

Try This

A. $f(x) = 2x(x - 3)(x + 3)$

B. $g(x) = 3x^3(x + 2)^2$

Use Transparency ▶ 22

Exploration 2 Notes

You may wish to conduct a short review of the quadratic formula before having students complete Exploration 2.

A ongoing
ASSESSMENT

2. They are the same.

3. The product is $\frac{f(x)}{2}$.

5. Use the quadratic formula to find the roots of $g(x)$, $\frac{1}{5}$ and $-\frac{5}{2}$. So $\left(x - \frac{1}{5}\right)\left(x + \frac{5}{2}\right)$ is the factored form of $\frac{1}{10}g(x)$. The factored form of g is $g(x) = 10\left(x - \frac{1}{5}\right)\left(x + \frac{5}{2}\right)$. You can multiply through by 10 to write g factored over the integers: $g(x) = (5x - 1)(2x + 5)$.

CRITICAL
Thinking

a_n in the expanded form of a quadratic polynomial function $(n = 2)$ is the same as C in the factored form of a quadratic polynomial function.

Alternate Example 4

Use the quadratic formula and the value of C to write $g(x) = 3x^2 - 5x - 2$ in factored form. $[g(x) = 3\left(x + \frac{1}{3}\right)(x - 2)]$

A ongoing
ASSESSMENT

The graphs are the same.

Exploration 2 Factors of Quadratic Functions

Graphics Calculator

Graph $f(x) = 2x^2 - 5x - 3$.

1 Use the graph to find the zeros of f.

2 Use the quadratic formula to find the roots of f. Name these roots r_1 and r_2. How do these roots compare with the zeros you found in Step 1?

3 Find the product $(x - r_1)(x - r_2)$. How does this product compare with f? Find a number C such that $C(x - r_1)(x - r_2) = f(x)$. How does C compare with the leading coefficient of f?

4 Divide every term in f by its leading coefficient, 2. Use the quadratic formula to find the roots of the resulting quadratic function. How do the values of these roots compare with the roots you found in Step 2?

5 How can you use the quadratic formula to write $g(x) = 10x^2 + 23x - 5$ in factored form? ❖

CRITICAL
Thinking

What is the relationship between a_n in the expanded form and C in the factored form of a quadratic polynomial function?

EXAMPLE 4

Graphics Calculator

Use the quadratic formula to write $f(x) = 2x^2 + 7x + 3$ in factored form.

Solution

Recall the quadratic formula.

$$x = \frac{-b \pm \sqrt{b^2 - 4ac}}{2a}$$

Substitute 2 for a, 7 for b, and 3 for c.

$$x = \frac{-(7) \pm \sqrt{(7)^2 - 4(2)(3)}}{2(2)}$$

$$x = \frac{-7 \pm 5}{4}$$

$$x = \frac{-7 + 5}{4} \text{ or } x = \frac{-7 - 5}{4}$$

$$x = -\frac{1}{2} \text{ or } x = -3$$

The roots of f are -3 and $-\frac{1}{2}$, and the leading coefficient is 2.

Thus, in factored form, $f(x) = 2\left(x + \frac{1}{2}\right)(x + 3)$. ❖

The function in Example 4 can be written in factored form over the integers. Multiply the constant factor, 2, and the first linear factor, $x + \frac{1}{2}$. The result is $f(x) = (2x + 1)(x + 3)$.

Graph $f(x) = 2\left(x + \frac{1}{2}\right)(x + 3)$ and $f(x) = (2x + 1)(x + 3)$ on the same coordinate plane. What do you notice?

Exercises & Problems

Communicate

1. Why is the factored form of a polynomial useful?

2. What is the relationship between the linear factors of a polynomial and the zeros of a polynomial function?

3. What is a perfect-square trinomial? How do you factor a perfect-square trinomial? How do you factor the difference of two perfect squares?

4. When a quadratic function can be factored over the rational numbers, can it be factored over the integers? Explain.

5. When the discriminant of the quadratic formula is not a perfect square, can the quadratic function be factored over the integers? Explain.

Practice & Apply

Write each polynomial expression in factored form.

6. $x^2 + 3x - 4$ $(x+4)(x-1)$

7. $x^2 + 5x + 6$ $(x+2)(x+3)$

8. $x^2 + 9x + 14$ $(x+2)(x+7)$

9. $x^2 - 2x - 15$ $(x+3)(x-5)$

10. $x^2 + 2x - 15$ $(x-3)(x+5)$

11. $2x^2 - 5x - 3$ $2(x+\frac{1}{2})(x-3)$

12. $-3x^2 + 7x - 2$ $-3(x-\frac{1}{3})(x-2)$

13. $6x^2 - x - 2$ $6(x-\frac{2}{3})(x+\frac{1}{2})$

14. $x^2 - 4x + 4$ $(x-2)^2$

15. $x^2 + 10x + 25$ $(x+5)^2$

16. $x^2 - 2x + 1$ $(x-1)^2$

17. $x^2 + 6x + 9$ $(x+3)^2$

18. $x^2 - 9$ $(x+3)(x-3)$

19. $x^2 - 16$

20. $4x^2 - 1$ $4(x+\frac{1}{2})(x-\frac{1}{2})$

21. $64x^2 - 4$

Find the zeros of each function from its graph. Then write each function in factored form.

22. $f(x) = 3x^2 + 9x - 12$

23. $f(x) = 2x^3 + 18x^2 + 28x$

24. $f(x) = 3x^3 - 6x^2 - 45x$

25. $f(x) = -4x^2 - 16x - 12$

26. $f(x) = 3x^3 - 45x^2 + 168x$

27. $f(x) = 3x^2 + 3x - 18$

28. $f(x) = -x^2 + 25$

29. $f(x) = -2x^3 - 10x^2 + 12x$

30. $f(x) = -3x^2 + 9x + 120$

31. $f(x) = x^4 - x^3 - 12x^2$

32. Graph $f(x) = 10x^2 - 31x + 15$, and find the zeros of the function. Write f in factored form. $\frac{3}{5}, \frac{5}{2}$; $f(x) = 10(x - \frac{3}{5})(x - \frac{5}{2})$

33. Thermodynamics A hot-air balloon is rising with a constant velocity of 2 feet per second. When it reaches a height of 200 feet, it begins to accelerate at a rate of 2 feet per second squared. Thereafter, its height can be found using the function $f(t) = 2t^2 + 2t + 200$. After how many seconds will it reach a height of 523 feet? $\approx$ 12.2 seconds

---

19. $(x+4)(x-4)$

21. $64(x+\frac{1}{4})(x-\frac{1}{4})$

22. zeros at -4 and 1; $f(x) = 3(x+4)(x-1)$

23. zeros at $-7, -2,$ and 0;
$f(x) = 2x(x+2)(x+7)$

24. zeros at 0, 5, and -3;
$f(x) = 3x(x+3)(x-5)$

25. zeros at -3 and -1; $f(x) = -4(x+3)(x+1)$

26. zeros at 0, 7, and 8; $f(x) = 3x(x-7)(x-8)$

27. zeros at -3 and 2; $f(x) = 3(x+3)(x-2)$

28. zeros at 5 and -5; $f(x) = -(x-5)(x+5)$

29. zeros at 0, -6, and 1;
$f(x) = -2x(x+6)(x-1)$

30. zeros at 8 and -5; $f(x) = -3(x-8)(x+5)$

31. zeros at -3, 0, and 4;
$f(x) = x^2(x+3)(x-4)$

Performance
Assessment

Have students give the factors of $x^2 - a^2$, $x^2 - 2ax + a^2$, and $x^2 + (a + b)x + ab$.
$[x^2 - a^2 = (x - a)(x + a);$
$x^2 - 2ax + a^2 = (x - a)^2;$
$x^2 + (a + b)x + ab =$
$(x + a)(x + b)]$

Then have students provide a numerical example of each type of quadratic polynomial, including its factors.

35. Answers should be of the form $f(x) = a(x - 2)(x + 3)$ or $a(x^2 + x - 6)$, where a is a non-zero constant.

Look Beyond

These exercises review the concept of long division. A similar procedure that can be used to divide polynomials is presented in the next lesson.

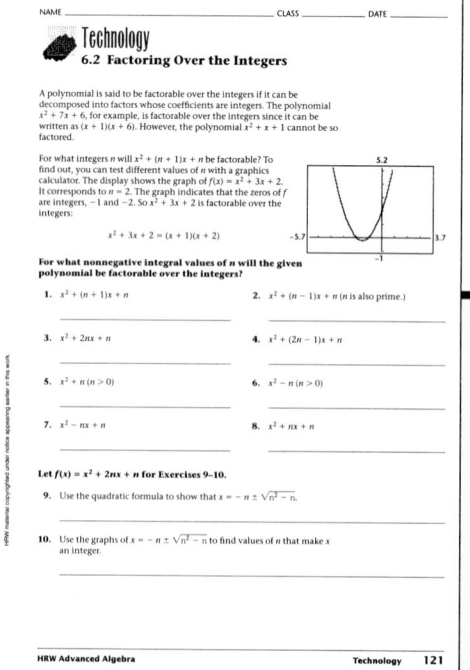

34. If $x - 2$ is a linear factor of a polynomial function, name at least one zero of that polynomial. 2

35. Write a quadratic function with zeros at 2 and -3. Then use the quadratic formula to verify that the roots are 2 and -3.

36. Write a quadratic function with zeros at 4 and -1. Then use the quadratic formula to verify that the roots are 4 and -1.

37. Write a quadratic function with only one zero at 2. Then use the quadratic formula to verify that the only root is 2.

38. Write a polynomial function with two roots at -3 and two roots at 1. Describe the graph, including the zeros of the function.

39. Write a different polynomial function for Exercise 38 that has the same zeros.

Look Back

40. Write the system of equations represented by this matrix equation. $\begin{cases} -3x + 3y = 3 \\ 5x + y = 2 \end{cases}$ **[Lesson 4.1]** $\begin{bmatrix} -3 & 3 \\ 5 & 1 \end{bmatrix}\begin{bmatrix} x \\ y \end{bmatrix} = \begin{bmatrix} 3 \\ 2 \end{bmatrix}$

41. Solve the system of equations you wrote for Exercise 40 by elimination. Check your solutions by graphing the system on your graphics calculator.   **[Lesson 4.3]** $x = \frac{1}{6}, y = \frac{7}{6}$

42. Use your graphics calculator to find the inverse of matrix A. Round entries to the nearest hundredth.   **[Lesson 4.5]** $A = \begin{bmatrix} -3 & 3 \\ 5 & 1 \end{bmatrix}$ $A^{-1} = \begin{bmatrix} -0.06 & 0.17 \\ 0.28 & 0.17 \end{bmatrix}$

43. Solve the system of equations you wrote for Exercise 40 using matrix algebra. Round answers to the nearest hundredth.   **[Lesson 4.6]** $x \approx 0.17$ $y \approx 1.17$

Physics The path of a dust particle moving away from a wall in a gentle breeze is modeled by the formula $d(x) = -x^2 + 3x$, where d is the height in feet of the particle from the floor, and x is the distance in feet from the wall. Use your graphics calculator to graph d.   **[Lesson 5.2]**

44. Does the particle ever touch the floor? If so, where? yes; at $x = 0$ and $x = 3$

45. What is the realistic domain of d? $0 \leq x \leq 3$

46. At what approximate distance (or distances) from the wall is the dust particle when it is 2 feet above the floor? 1 ft and 2 ft

Look Beyond

47. Use long division to divide 11,536 by 28. Show your steps.

48. Examine your division steps in your answer to Exercise 47. Describe the steps that are repeated in the process of long division.

36. Answers should be of the form $f(x) = a(x - 4)(x + 1)$ or $a(x^2 - 3x - 4)$, where a is a non-zero constant.

37. Answers should be of the form $f(x) = a(x - 2)^2$ or $a(x^2 - 4x + 4)$, where a is a constant.

38. Answers should be of the form $f(x) = a(x + 3)^2(x - 1)^2$ or $f(x) = a(x^4 + 4x^3 - 2x^2 - 12x + 9)$, where

a is a constant. The graph of the quartic function touches the x-axis at $x = -3$ and $x = 1$, the zeros of the function.

39. Answers should be of the form $f(x) = b(x + 3)^2(x - 1)^2$ or $f(x) = b(x^4 + 4x^3 - 2x^2 - 12x + 9)$, where b is a constant and $b \neq a$ from Exercise 38.

The answers to Exercises 47 and 48 can be found in Additional Answers beginning on page 842.

LESSON 6.3 Dividing Polynomials

 *If you know the zeros of a polynomial, you can write the polynomial in factored form. If a graph cannot be used to determine all of the zeros, division can be used to determine the remaining factors.*

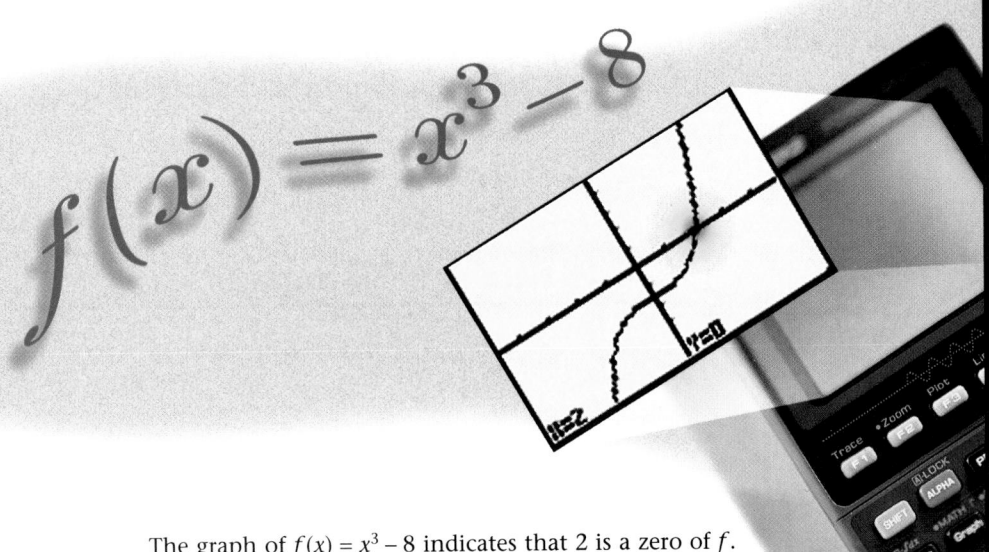

$$f(x) = x^3 - 8$$

The graph of $f(x) = x^3 - 8$ indicates that 2 is a zero of f. The Factor Theorem states that if 2 is a zero of f, then $x - 2$ will divide f exactly. How can you find the quotient of $x^3 - 8$ divided by $x - 2$?

The Division Algorithm

You have used the *division algorithm* to divide 17 by 3.

When 17 is divided by 3, the quotient is 5 and the remainder is 2. This division can be written as $17 = 5 \times 3 + 2$	$\begin{array}{r} 5 \ \ \text{R2} \\ 3\overline{)17} \\ -15 \\ \hline 2 \end{array}$
When a number, such as 3, divides a number, such as 18, *exactly*, the quotient is an integer, 6, and the remainder is 0. This division can be written as $18 = 6 \times 3 + 0$.	$\begin{array}{r} 6 \ \ \text{R0} \\ 3\overline{)18} \\ -18 \\ \hline 0 \end{array}$

ALTERNATIVE teaching strategy

Cooperative Learning Provide students with several transparencies that they can use when dividing polynomials. An example of what can be shown on each transparency is given below.

$$x\_\_ \overline{)\_\_ x^3 \_\_\_ x^2 \_\_\_ x \_\_\_}$$

Each blank should be filled in with either a plus (+) sign, a minus (−) sign, or an integer. Give small groups of students a polynomial and a factor. Have the groups show how the polynomial can be divided by the factor chosen. Have students use an appropriate transparency as they work. Ask each group to present its work to the class. Provide similar transparencies for students' work with the quadratic formula and synthetic division.

PREPARE

Objectives

- Define and use the division algorithm for polynomials.
- Define and use the Fundamental Theorem of Algebra to find the zeros of a polynomial function.

RESOURCES

- Practice Master   6.3
- Enrichment Master   6.3
- Technology Master   6.3
- Lesson Activity Master 6.3
- Quiz   6.3
- Spanish Resources   6.3

Assessing Prior Knowledge

Arrange each polynomial in descending order of x. If a term is missing, write that term with a coefficient of 0.

1. $x^2 + 3 - 4x + x^4$
   $[x^4 + 0x^3 + x^2 - 4x + 3]$

2. $-1000 + x^3$
   $[x^3 + 0x^2 + 0x - 1000]$

3. $6 - 3x + 2x^3$
   $[2x^3 + 0x^2 - 3x + 6]$

4. $x^2 + 2 + 3x^4$
   $[3x^4 + 0x^3 + x^2 + 0x + 2]$

TEACH

 Emphasize that some polynomials cannot be easily factored. Create a class list of polynomials that cannot be factored by inspection. Point out to students that there can be a remainder when one polynomial is divided by another.

Alternate Example 1

Use the division algorithm to divide $f(x) = x^3 - 216$ by $x - 6$.
$[f(x) = (x - 6)(x^2 + 6x + 36)]$

TEACHING tip

Be sure students understand that missing powers of x are inserted with a coefficient of 0 before beginning the division process.

CRITICAL Thinking

Conjecture: $(x + a)$ is one factor of $x^3 + a^3$.
$x^3 + a^3 = (x + a)(x^2 - ax + a^2)$

$$\text{If } 3\overline{)17}^{\;5\text{ R}2}\text{, then } 17 = 5 \times 3 + 2$$

The division algorithm states that when a is divided by b, the result can be written as $a = Qb + R$, where Q is the quotient and R is the remainder.

$$P = Q \times D + R$$

The division algorithm can be extended to polynomials. If a polynomial, P, is divided by a polynomial, D, of lower degree, the result is a quotient, Q, with remainder, R.

THE DIVISION ALGORITHM

If the polynomial P is divided by the polynomial D, then $P = DQ + R$, where Q is the quotient polynomial, and R is the remainder polynomial of degree less than D. If R is 0, then D divides P *exactly*.

EXAMPLE 1

Use the division algorithm to divide $x^3 - 8$ by $x - 2$.

Solution➤

$$\begin{array}{rl}
 & x^2 + 2x + 4 \\
x - 2 & \overline{)x^3 + 0x^2 + 0x - 8} \\
\text{Subtract} & \underline{(x^3 - 2x^2)} \\
 & 2x^2 + 0x \\
\text{Subtract} & \underline{(2x^2 - 4x)} \\
 & 4x - 8 \\
\text{Subtract} & \underline{(4x - 8)} \\
 & 0
\end{array}$$

The remainder is zero, so $x - 2$ divides $x^3 - 8$ exactly.

$$x^3 - 8 = (x - 2)(x^2 + 2x + 4)$$ ❖

The quotient $x^2 + 2x + 4$ cannot be factored over the rationals because the discriminant of the quadratic formula, $b^2 - 4ac$, is less than zero. Therefore, the roots are complex and $x^2 + 2x + 4$ is prime with respect to the rationals. Thus, P factored over the rationals is $(x - 2)(x^2 + 2x + 4)$.

The expression $x^3 - 8$ is an example of a difference of two cubes, x^3 and 8. In general, the **difference of two cubes** can be written as the product of two factors, $x^3 - b^3 = (x - b)(x^2 + bx + b^2)$.

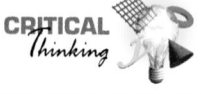

CRITICAL Thinking

Make a conjecture about how the *sum of two cubes* is factored. Use the division algorithm to write a general rule for factoring the sum of two cubes over the rationals.

ENRICHMENT Have some students do research about the history of long division, some students do art, and some find real-world examples where long division is necessary. Then have students work in pairs to make a poster to display some of their findings.

INCLUSION strategies

Using Models Provide students with ample visual examples that require little or no explanation. For example, have students divide $x^2 - 4x + 4$ by $x - 2$, $x^2 - 5x + 6$ by $x - 3$, and $2x^2 - 5x - 3$ by $2x + 1$. Then have students explain each step of each example using their own words.

The Fundamental Theorem Of Algebra

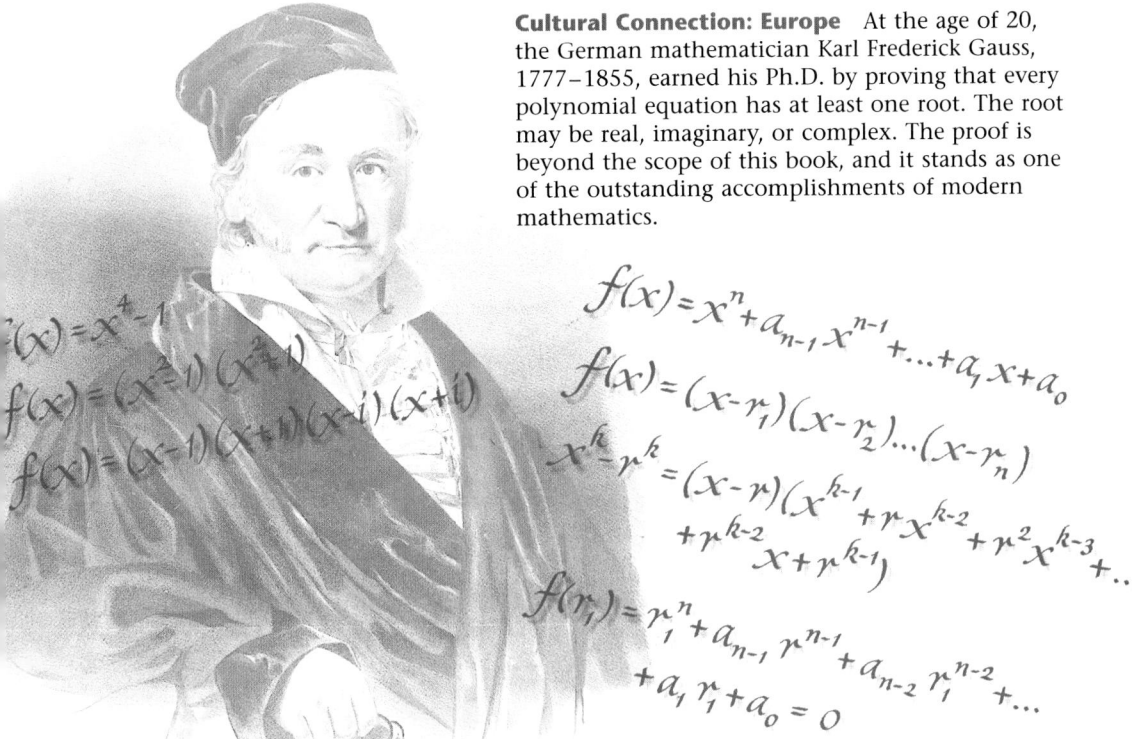

Cultural Connection: Europe At the age of 20, the German mathematician Karl Frederick Gauss, 1777–1855, earned his Ph.D. by proving that every polynomial equation has at least one root. The root may be real, imaginary, or complex. The proof is beyond the scope of this book, and it stands as one of the outstanding accomplishments of modern mathematics.

$$f(x) = x^n + a_{n-1}x^{n-1} + \ldots + a_1x + a_0$$

$$f(x) = (x - r_1)(x - r_2)\ldots(x - r_n)$$

$$x^k - r^k = (x - r)(x^{k-1} + rx^{k-2} + r^2x^{k-3} + \ldots + r^{k-2}x + r^{k-1})$$

$$f(r_1) = r_1^n + a_{n-1}r_1^{n-1} + a_{n-2}r_1^{n-2} + \ldots + a_1r_1 + a_0 = 0$$

The theorem that Gauss proved is called the *Fundamental Theorem of Algebra*.

THE FUNDAMENTAL THEOREM OF ALGEBRA
Every polynomial function of degree $n > 0$ has at least one complex zero.

Two important results follow from this theorem. The first result gives the number of zeros that the polynomial has.

THE NUMBER OF POLYNOMIAL ROOTS
Every polynomial $P_n(x)$ of degree n can be written as a factored product of exactly n linear factors, corresponding to exactly n zeros (counting multiplicities).
$$P_n(x) = C(x - r_1)(x - r_2)\ldots(x - r_{n-1})(x - r_n)$$

If a polynomial with real coefficients has one complex root, r, then its complex conjugate, $\bar{r}$, is also a root. Therefore, a polynomial of odd degree n may have at most an even number, $n - 1$, of complex factors. This implies the second result: *Every polynomial of odd degree has at least one real root.*

TEACHING tip

You may wish to have students do research to find out more about the history of the Fundamental Theorem of Algebra and the people who worked on its proof. Have students share their findings with the class.

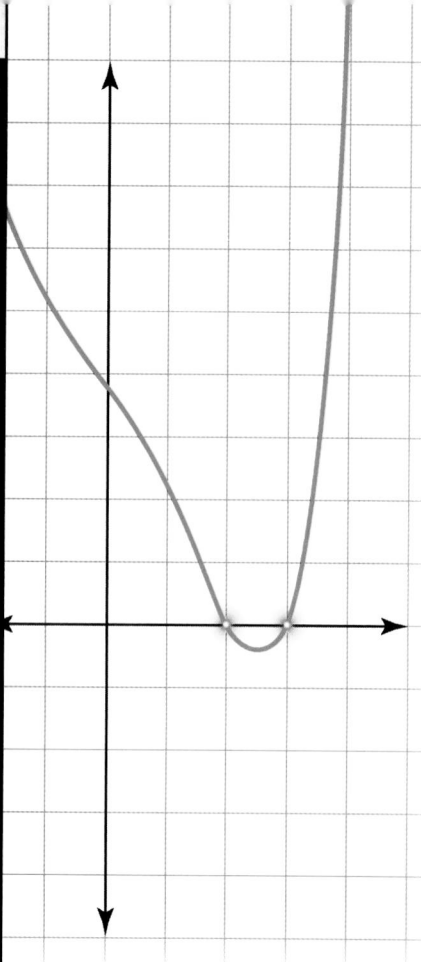

EXAMPLE 2

Ⓐ Determine the real zeros of $f(x) = x^4 - 2x^3 - 4x^2 - 7x + 30$.

Ⓑ Use the division algorithm to factor f over the complex numbers.

Solution

Ⓐ Graph $f(x) = x^4 - 2x^3 - 4x^2 - 7x + 30$.

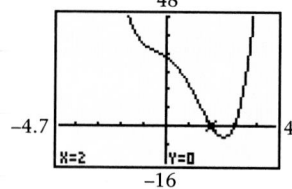 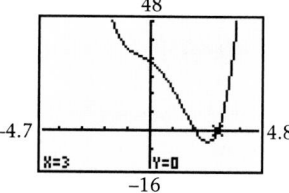

You can see from the graph that f has two real zeros at 2 and 3.

Ⓑ Since $f(x) = x^4 - 2x^3 - 4x^2 - 7x + 30$ is of degree 4 and the graph crosses the x-axis only two times, at 2 and 3, f also has two nonreal complex roots. Use the division algorithm to factor out $x - 2$ and $x - 3$. The result is $f(x) = (x - 2)(x - 3)(x^2 + 3x + 5)$, which is factored over the rationals. However, the trinomial can be factored over the complex numbers using the quadratic formula.

$$x = \frac{-b \pm \sqrt{b^2 - 4ac}}{2a}$$

$$x = \frac{-3 \pm \sqrt{(3)^2 - 4(1)(5)}}{2(1)}$$

$$x = \frac{-3 + i\sqrt{11}}{2} \text{ or } x = \frac{-3 - i\sqrt{11}}{2}$$

Notice that these complex roots are conjugates. Therefore, the factored form of f over the complex numbers is written as follows.

$$f(x) = (x - 2)(x - 3)\left(x - \frac{-3 + i\sqrt{11}}{2}\right)\left(x - \frac{-3 - i\sqrt{11}}{2}\right) \quad ❖$$

It is possible to simplify the factored form of f by clearing fractions from the complex factors.

$$f(x) = \frac{1}{4}(x - 2)(x - 3)(2)\left(x - \frac{-3 + i\sqrt{11}}{2}\right)(2)\left(x - \frac{-3 - i\sqrt{11}}{2}\right)$$

$$= \frac{1}{4}(x - 2)(x - 3)\left(2x + 3 - i\sqrt{11}\right)\left(2x + 3 + i\sqrt{11}\right)$$

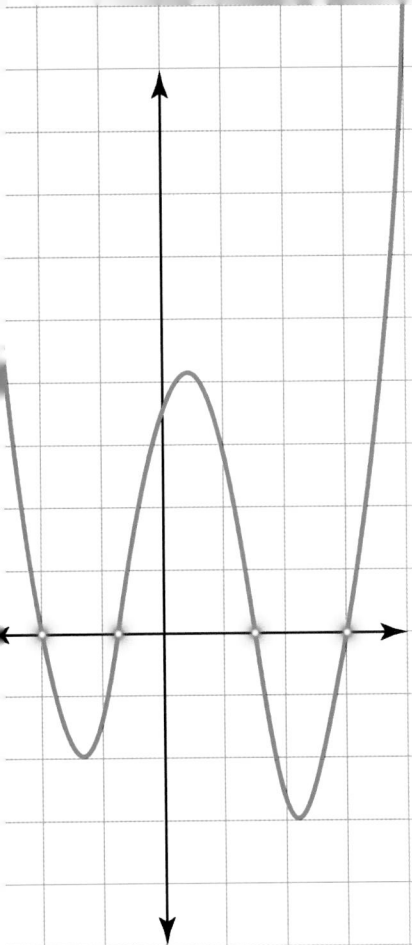

EXAMPLE 3

Factor $f(x) = 10x^4 - 17x^3 - 65x^2 + 54x + 72$ over the integers.

Solution▶

Graph $f(x) = 10x^4 - 17x^3 - 65x^2 + 54x + 72$, and find the real zeros.

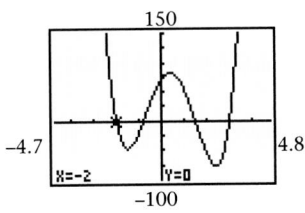

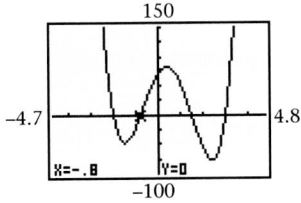

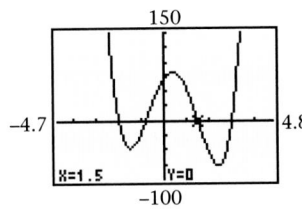

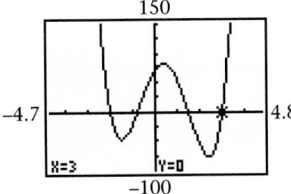

The real zeros are -2, -0.8 or $-\frac{4}{5}$, 1.5 or $\frac{3}{2}$, and 3.

The leading coefficient of f is 10.

The four real zeros are:

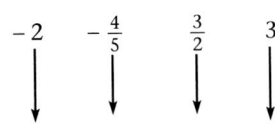

$$-2 \qquad -\frac{4}{5} \qquad \frac{3}{2} \qquad 3$$

Thus, the factored form of f is:

$$10(x + 2)\left(x + \frac{4}{5}\right)\left(x - \frac{3}{2}\right)(x - 3)$$

Observe that the denominators in the fractions, 2 and 5, are factors of $C = 10$. Factor **10** as **5 · 2**, and use the associative property.

$$f(x) = \mathbf{10}(x + 2)\left(x + \tfrac{4}{5}\right)\left(x - \tfrac{3}{2}\right)(x - 3)$$

$$f(x) = (x + 2)\left[(\mathbf{5})\left(x + \tfrac{4}{5}\right)\right]\left[(\mathbf{2})\left(x - \tfrac{3}{2}\right)\right](x - 3)$$

$$f(x) = (x + 2)(5x + 4)(2x - 3)(x - 3)$$

Thus, $f(x) = 10x^4 - 17x^3 - 65x^2 + 54x + 72$ factored over the integers is $f(x) = (x + 2)(5x + 4)(2x - 3)(x - 3)$. ❖

Try This Factor $f(x) = 4x^4 - 8x^3 - 53x^2 + 30x + 27$ over the integers.

Synthetic Division

A compact form of division, called **synthetic division**, can be used to divide polynomials without writing out all of the steps of long division.

EXAMPLE 4

Use synthetic division to divide $11x^3 - 27x + 2x^4 - 10$ by $x + 5$.

Solution

Write $11x^3 - 27x + 2x^4 - 10$ in descending degree order. Use a zero to represent the coefficient of the x^2 term.	$2x^4 + 11x^3 + 0x^2 - 27x - 10$
Write the coefficients of the polynomial with the r-value of the divisor, $x - r$, on the left. In this case, $x - r$ is $x - (-5)$, so the r-value is -5.	$-5 \mid$ 2 11 0 -27 -10
Write the first coefficient again below the line. Multiply the r-value, -5, by the number below the line, 2. Write the product, -10, under the next coefficient, 11.	$-5 \mid$ 2 11 0 -27 -10 $\downarrow$ -10 2
Add the second column, and write the sum below the line. Again, multiply the r-value, -5, by the new number below the line, 1. Write the product, -5, under the next coefficient, 0.	$-5 \mid$ 2 11 0 -27 -10 $\downarrow$ -10 -5 2 1
Continue this pattern. The numbers below the line are the coefficients of the quotient polynomial. The number at the far right below the line is the remainder.	$-5 \mid$ 2 11 0 -27 -10 $\downarrow$ -10 -5 25 10 2 1 -5 $-2 \mid$ 0 $\downarrow$ $\downarrow$ $\downarrow$ $\downarrow$ $2x^3 + 1x^2 - 5x - 2$

Thus, $(11x^3 - 27x + 2x^4 - 10) \div (x + 5) = 2x^3 + 1x^2 - 5x - 2$. ❖

You can verify that the quotient, $2x^3 + 1x^2 - 5x - 2$, is correct by graphing $y = (x + 5)(2x^3 + x^2 - 5x - 2)$ and $y = 11x^3 - 27x + 2x^4 - 10$ on the same coordinate plane. Why is it important to use a zero to represent the coefficient of x^2 in Example 4?

Try This   Use synthetic division to divide $3 + 16x^2 - 5x + 6x^3$ by $x + 3$.

Exercises & Problems

Communicate

1. How do you find the zeros of a polynomial function from its graph?

2. Explain how to use a graph to write a polynomial function as a product of its linear factors.

3. How many linear factors will a polynomial of degree n have? Justify your response.

4. List the possible numbers of complex zeros that a polynomial of degree n may have if n is even. For each number listed, how many real zeros are possible?

5. List the possible numbers of complex zeros that a polynomial of degree n may have if n is odd. For each number listed, how many real zeros are possible?

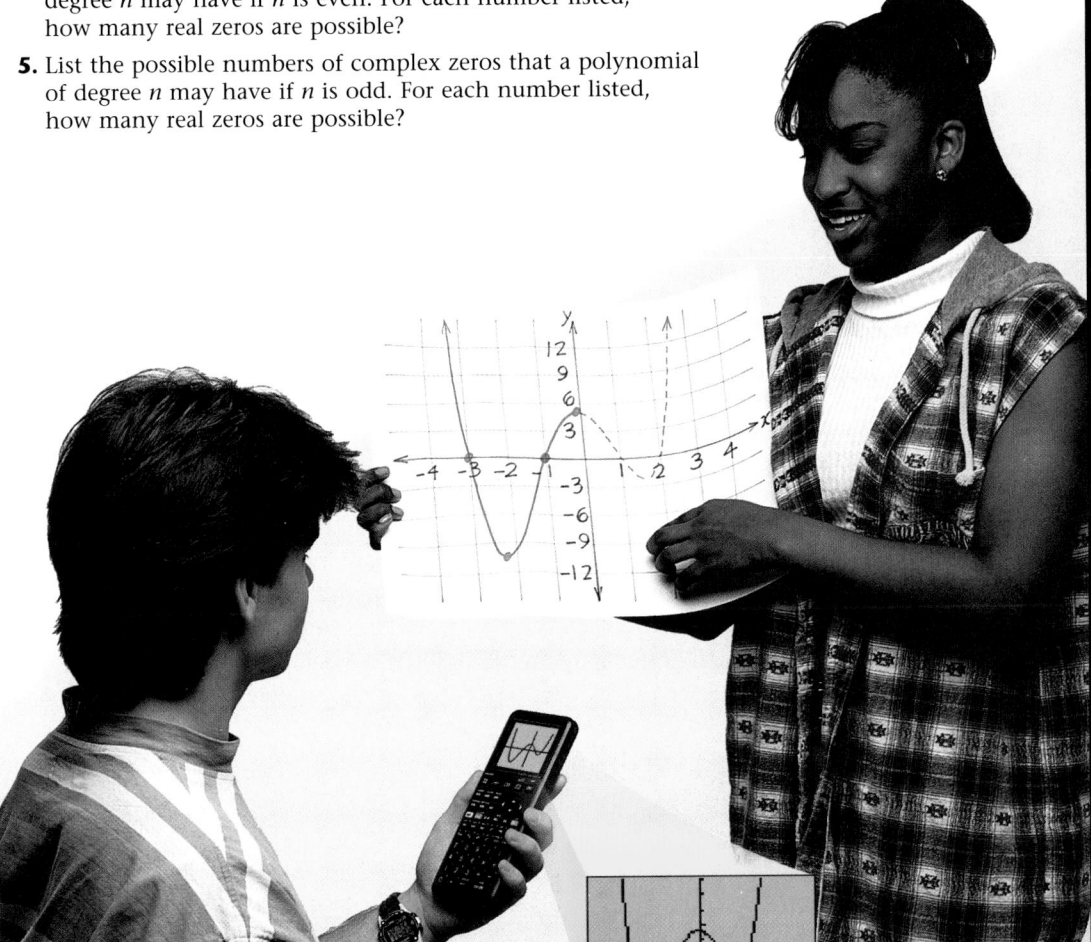

ASSESS

Selected Answers
Odd-numbered Exercises 7–37

Assignment Guide
Core 1–37

Core Plus 1–38

Technology
A graphics calculator is needed for Exercises 12, 17, 18, 21–24, and 29.

Error Analysis

Watch for students who choose the incorrect sign for the divisor when performing synthetic division. Remind students that if the divisor is in the form $x - r$ before dividing, the number used in the synthetic division process is r.

Practice & Apply

6. If $x - 2$ is a factor of a polynomial, name a zero of that polynomial. 2

7. If $x^2 - 5x - 14$ is a factor of a polynomial, find two zeros of the polynomial. 7, −2

8. What is the largest number of distinct zeros that a fourth-degree polynomial can have? the smallest it can have? Explain. 4, 0

9. Can a fourth-degree polynomial have 2 real zeros and 2 complex zeros? Explain. yes

The numbers –2 and 1 are zeros of the polynomial $f(x) = x^3 - 6x^2 - 9x + 14$. Complete Exercises 10–12.

10. Find the other zero(s) using long division. 7

11. Find the other zero(s) using synthetic division. 7

12. Find the other zero(s) using the graph of f. 7

13. If $x - 2$ is a factor of P, a polynomial of degree 7, will the quotient of $P \div (x - 2)$ be a polynomial? If so, what will be its degree? yes; 6

For Exercises 14 and 15, let f be a polynomial function that has a quadratic factor whose discriminant is negative.

14. How many zeros of f are associated with this quadratic factor? no *real* zeros

15. Are the zeros real numbers or complex numbers? complex

16. Write two different polynomial functions of degree 3 that have a single zero at the x-value 5.

Geography A mountain ridge is drawn to scale on the wall of a tourist hut in West Virginia. The ridge line has approximately the same shape as the graph of the function $f(x) = -x^4 + 3x^2 - 3x + 6$ between x-values of –2.5 and 2.0. Use your graphics calculator to graph f.

17. The zeros of the function represent sea level. What are the approximate zeros of f? −2.32, 1.79

18. Use the trace feature to find the approximate maximum value of f. 12.24

19. Find a vertical scale factor corresponding to this ridge that has a maximum height of 3200 feet above sea level. 1 unit represents ≈ 261.4 ft

20. If a tunnel goes straight through the mountain at sea level, how long is the tunnel? approx. 1075 ft

8. 4; 0; a fourth degree polynomial can have as many as 4 distinct zeros or as few as 0 distinct zeros.

9. Yes. The complex zeros will be conjugates.

16. Answers may vary. The general form should be $p(x) = a(x - 5)(x - r_1)(x - r_2)$, where a is any real number ≠ 0, and r_1 and $r_2 \neq 5$.

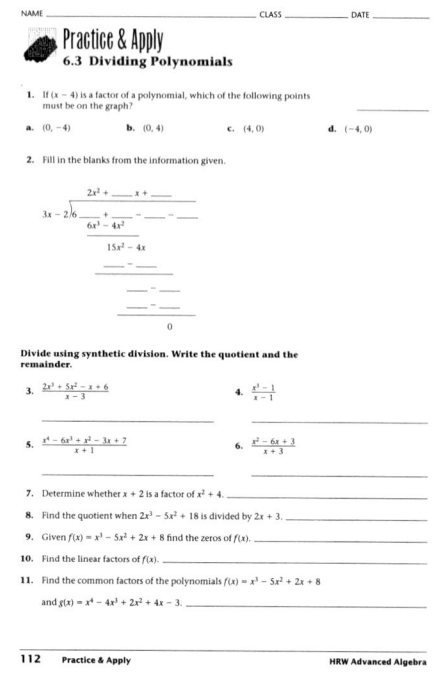

21. Use synthetic division to divide $x^4 - 19x^3 + 83x^2 + 139x - 924$ by $x + 3$.

22. Factor the polynomial $f(x) = 6x^4 - 49x^3 + 84x^2 + 181x - 462$ by graphing. $f(x) \approx 6(x + 2)(x - 3)(x - 3.5)(x - 3.7)$

23. Find the common factor(s) of the two polynomials $f(x) = x^3 + 3x^2 - 18x - 40$ and $g(x) = x^3 - 9x^2 - 4x + 96$. $(x - 4)$

24. Find the common factor(s) of the two polynomials $f(x) = x^3 + 3x^2 - 4x - 12$ and $g(x) = x^4 + x^3 - 14x^2 - 8x + 48$. $(x - 2)$, $(x + 3)$

25. Can the product of a third-degree polynomial with distinct real zeros and a fourth-degree polynomial with distinct real zeros be a polynomial that has all zeros with multiplicity 2? no Explain.

26. Write a quadratic polynomial function with real zeros at 2 and −3 whose y-intercept is −12. $f(x) = 2(x - 2)(x + 3)$

27. Write a quadratic polynomial function with real zeros at 2 and −3 whose y-intercept is 12. $f(x) = -2(x - 2)(x + 3)$

28. Find two quadratic functions that have a single real zero at 2.

For Exercises 29–31, let $f(x) = 10x^2 - 31x + 15$.

29. Find the zeros of f. $\frac{3}{5}, \frac{5}{2}$

30. Find the linear factors associated with the zeros of f. $\left(x - \frac{3}{5}\right), \left(x - \frac{5}{2}\right)$

31. Write f in factored form over the integers. $f(x) = (5x - 3)(2x - 5)$

Look Back

32. Solve the inequality $x + 3 \leq 3(x - 1)$. **[Lesson 1.6]** $x \geq 3$

33. Write the expression $\frac{3}{x - 2} - \frac{5}{x + 3}$ as a single fraction. **[Lesson 2.1]** $\frac{-2x + 19}{(x - 2)(x + 3)}$

34. Simplify $\frac{x - 3}{4x + 12} \div \frac{2x - 6}{4}$. **[Lesson 2.1]** $\frac{1}{2(x + 3)}$

35. Is the set of points $\{(2, 3), (3, 5), (4, 7)\}$ a function? Explain.
[Lesson 2.3] yes, each x-value is associated with a unique y-value.

For Exercises 36 and 37, let $f(x) = \frac{1}{x}$ and $g(x) = \sqrt{x + 1}$.

36. Find $f^{-1}(x)$ and $g^{-1}(x)$. **[Lesson 3.2]** $f^{-1}(x) = \frac{1}{x}, x \neq 0$; $g^{-1}(x) = x^2 - 1, x \geq -1$

37. Find $(f \circ g)(x)$ and $(g \circ f)(x)$. **[Lesson 3.3]**

$(f \circ g)(x) = \frac{1}{\sqrt{x + 1}}, x > -1, (g \circ f)(x) = \sqrt{\frac{x + 1}{x}}, x \neq 0, x \geq -1$

Look Beyond

38. Find values for y that satisfy the equation $x^2 + y^2 = 9$ for the x-values 0, 1, 2, and 3. Based on your results, will the graph of this equation be the graph of a function? How does this graph compare with the graphs of $y = \sqrt{-x^2 + 9}$ and $y = -\sqrt{-x^2 + 9}$, for $-3 \leq x \leq 3$?

21. $x^3 - 22x^2 + 149x - 308$

25. No. Since the product of the polynomials will have 7 linear factors, it is impossible for each linear factor to appear twice (which would result in a multiplicity of 2). The number of linear factors must be even in order to have the possibility of all factors having a multiplicity of 2.

28. The functions should be of the form $p(x) = C(x - 2)^2$, for 2 different values of C.

38. $x^2 + y^2 = 9 \rightarrow y = \pm\sqrt{9 - x^2}$

x	y
0	±3
1	±$\sqrt{8}$
2	±$\sqrt{5}$
3	0

The graph of $x^2 + y^2 = 9$ will not be the graph of a function. Together the graphs of $y = \sqrt{-x^2 + 9}$ and $y = -\sqrt{-x^2 + 9}$, for $|x| \leq 3$, are the same as the graph of $x^2 + y^2 = 9$.

PREPARE

Objectives

- Determine and classify the behavior of single zeros and multiple zeros of a polynomial function.
- Determine and classify the behavior and basic shape of polynomial functions of varying degrees.

RESOURCES

- Practice Master   **6.4**
- Enrichment Master   **6.4**
- Technology Master   **6.4**
- Lesson Activity Master **6.4**
- Quiz   **6.4**
- Spanish Resources   **6.4**

This lesson may be considered optional in the core course. See page 324A.

Assessing Prior Knowledge

Use your graphics calculator to graph each function. Determine the real zeros of each function.

1. $f(x) = 3x^3$ [0]

2. $g(x) = x^2 + 2x + 1$ [−1]

3. $h(x) = 2x^2 − 8$ [−2, 2]

4. $P(x) = x^3 − 27$ [3]

TEACH

Ask students why only a portion of the usual function domain is included in real-world situations. Have students give some examples of such situations, stating the reason for the restricted domain.

Exploring Polynomial Function Behavior

When a polynomial function describes a real-world situation, such as the volume function in Lesson 6.1, its realistic domain may include only a portion of the usual function domain. Examining the behavior of a polynomial function often provides insight into the problem being modeled.

Business A high school chess club is having a bake sale to earn money. From previous bake sales, it has been found that cookie sales are the most profitable. A function used to model the profit from selling cookies is $P(x) = x^3 + 9.25x^2 + 3x − 99$, where x is the number of cookies sold in units of 100. How many cookies must the chess club sell to make a profit? Is this a reasonable number of cookies to sell?

From the factored form, $P(x) = \left(x − \frac{11}{4}\right)(x + 6)^2$, −6 is a zero of multiplicity 2, and $\frac{11}{4}$, or 2.75, is a zero of multiplicity 1.

ALTERNATIVE teaching strategy

Using Visual Models Have students work in small groups to graph the equations given in Exploration 1. Have one student graph the first equation in one color, have a second student graph the second equation in a second color, and so on until all of the graphs in each step have been graphed on the same coordinate plane.

Then have students complete Steps 1–3. Have them duplicate the table on page 352 on a clear piece of acetate. Then have students fill in the table and complete Steps 4 and 5. Have each group place its transparency on the overhead projector and give their answers to Steps 4 and 5, using the entries in the table to justify the answers. Have students complete Exploration 2 in a similar manner.

The graph of P is positive for x-values greater than 2.75. Since x is the number of cookies sold in 100s, the students will begin to show a profit after 275 cookies are sold.

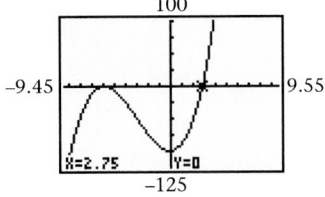

If no cookies are sold (x is 0), the club will lose $99.

Notice the points on the graph where P has a peak, (–6, 0), or a valley, (0, –99). These points are called **turning points**. At a peak, the graph changes from increasing to decreasing. At a valley, the change is from decreasing to increasing.

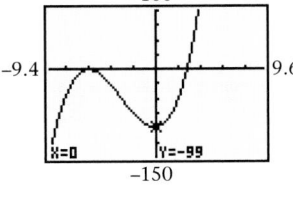

At a **crossing point**, the value of the function changes signs. Note that a crossing point must be a zero of P, but not every zero is a crossing point. Knowledge of the turning points and crossing points of the graph gives important information about the polynomial function.

CRITICAL *Thinking*
Explain why each zero must be either a turning point or a crossing point but cannot be both.

•Exploration 1 *Even- and Odd-Degree Functions*

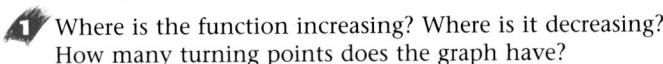

Graphics Calculator

Graph each of the following sets of polynomial functions using a viewing window that includes the x- and y-intercepts of each function.
Trace each curve from the far left of the viewing window to the right.
Complete Steps 1–3 for each set of functions, and enter your results and other requested information in the table that follows the steps.

Linear: $f(x) = x$, $g(x) = x - 2$, and $h(x) = -x$

Quadratic: $f(x) = x^2$, $g(x) = x(x - 2)$, and $h(x) = -x^2$

Cubic: $f(x) = x^3$, $g(x) = x^2(x - 2)$, $h(x) = x(x - 2)(x + 4)$, and $k(x) = -x^3$

Quartic: $f(x) = x^4$, $g(x) = x^3(x - 2)$, $h(x) = x^2(x - 2)(x + 4)$, $k(x) = x(x - 2)(x + 4)(x - 6)$, and $m(x) = -x^4$

1 Where is the function increasing? Where is it decreasing? How many turning points does the graph have?

2 How many zeros does each function have? Specify whether each zero is a turning point or a crossing point.

3 What is the effect of the sign of the leading coefficient on the shape of the graph at the far left of the viewing window? What is the effect of the leading coefficient's sign on the shape of the graph at the far right of the viewing window?

Cooperative Learning

Have students work in pairs to complete the explorations. Then have two pairs of students compare and discuss their work, making any necessary corrections.

CRITICAL *Thinking*

If the graph of a polynomial $p(x)$ crosses the x-axis at $x = r$, then $p(r) = 0$. Hence, r is a zero of p by definition. The graph of a polynomial may also have a local maximum or minimum which touches the x-axis and at which the graph changes from increasing to decreasing or from decreasing to increasing. At this value of x, the function is also equal to zero and is therefore a turning point zero. Since a graph cannot both turn and cross the x-axis at the same time, all zeros are either crossing or turning point zeros, but not both.

Exploration 1 Notes

As students graph the functions, be sure they enter the parentheses in the correct place and do not confuse subtraction signs with negative signs. Have students use their completed table from page 352 to explain their answers to Steps 4 and 5.

Use your answers to complete the table.

	Degree 1: linear	Degree 2: quadratic	Degree 3: cubic	Degree 4: quartic
Domain	All reals	All reals	All reals	All reals
Range				
Number of distinct zeros				
Number of turning points				
Number of crossing points				
Far left and far right behavior				

Use the table to complete Steps 4 and 5.

4 What do the quadratic and quartic functions have in common? What do you think will be true for all polynomial functions of *even* degree?

5 What do the linear and cubic functions have in common? What do you think will be true for all polynomial functions of *odd* degree? ❖

In Exploration 2 you will investigate the effect of multiple zeros on the graph of a quartic polynomial function.

•Exploration 2 *Multiple Zeros*

Graphics Calculator

1 Graph each function.

$f_1(x) = (x - 3)^4$

$f_2(x) = (x - 1)(x - 3)^3$

$f_3(x) = (x - 1)^2(x - 3)^2$

$f_4(x) = (x - 1)^2(x - 3)(x + 2)$

2 How many turning and crossing points does each graph have? Which zeros are turning points, and which zeros are crossing points?

3 How is the evenness or oddness of the multiplicity of a zero related to whether it is a turning point or a crossing point?

4 Construct a quartic polynomial that has only 2 crossing points. Determine the number of turning points it has. Can any turning points be zeros?

5 Construct a quartic polynomial that has 2 turning points at its zeros. Determine what other turning points (if any) it has.

EXERCISES & PROBLEMS

Communicate

Let $f(x) = (x - 2)(x + 4)(x - 5)$. Complete Exercises 1–5.

1. What are the zeros and the degree of f?

2. Is f always an increasing function? Why or why not?

3. Which direction (upward or downward) will the ends of the graph turn?

4. Describe what happens at a turning point. Does f have any zeros that are turning points?

5. Describe what happens at a crossing point. How many crossing points does the graph of f have?

6. Can a fourth-degree polynomial function have turning points but no crossing points? Explain your answer.

Practice & Apply

7. Write an example of a polynomial in factored form that is of degree 5 and has 3 crossing-point zeros and 1 turning-point zero.

8. Write an example of a polynomial in factored form that is of degree 7 and has 3 turning-point zeros and 1 crossing-point zero.

9. Write a polynomial in factored form that is of degree 3 with a left end that turns up and a right end that turns down.

10. Write a polynomial in factored form that turns up at both ends, is of degree 6, and has 2 zeros of multiplicity 3 at –4 and 7.

11. Write a polynomial in expanded form that is of degree 7 and turns down at the right end.

Let $f(x) = x^4 + 6x^3 - 21x^2 - 72x + 168$. **Use your graphics calculator to graph f. Make sure that all of the x- and y-intercepts are included in your viewing window.**

12 Estimate all of the zeros of f to the nearest tenth. $-7.0, -3.9, 2.1, 2.9$

13. Use the approximate zeros of f to factor f into its linear factors. $f(x) = (x + 7.0)(x + 3.9)(x - 2.1)(x - 2.9)$

14 How does the graph of the approximate factored form of f compare with the graph of f? the graphs are essentially the same

15 Where are the turning points on the graph of f? Where is f increasing? Where is f decreasing?

---

7. Answers may vary. Answers must be similar to $f(x) = a(x - 1)(x - 2)(x - 3)(x - 4)^2$

8. Answers may vary. Answers must be similar to $f(x) = a(x - 1)^2(x - 2)^2(x - 3)^2(x - 4)$ for some real number a.

9. Answers may vary. Answers must be similar to $f(x) = -ax(x - 1)(x - 2)$ for some positive real number a.

10. Answers may vary. Answers must be similar to $f(x) = a(x + 4)^3(x - 7)^3$ for some positive real number a.

11. Answers may vary. Answers must be similar to $f(x) = -ax^7 + \ldots$ for some positive real number a, and where no term has degree greater than 7.

Physics Suppose a charged particle is moving in one direction in response to an electrical field. Let x represent the position of the charged particle. Within the interval $-2.5 \leq x \leq 2.5$, the speed (in meters per second) of the particle as a function of its position, x, is given by $v(x) = x^4 - 8x^2 + 16$.

16 Graph v. Where do the turning points of the speed function occur? Describe how the speed changes as each turning point is passed. $(-2,0)$, $(0,16)$, $(2,0)$

17 Use the graph to factor $v(x)$ into linear factors. Is the speed ever negative? Explain what this would mean.
$v(x) = (x + 2)^2(x - 2)^2$, no

18 **Maximum/Minimum** Where does the maximum speed occur? What is the maximum speed?
x is 0; $v(0) = 16$ m/sec

For Exercises 19–22, let $g(x) = x^5 + x^4 - 40x^3 - 3x^2 + 316x - 336$. Use your graphics calculator to graph g. Make sure that all of the x- and y-intercepts are included in your viewing window.

19 Estimate all of the zeros of g. -5.9, -3.7, 1.4, 2.1, 5.1

20 Use the approximate zeros of g to factor g into its linear factors. $g(x) \approx (x + 5.9)(x + 3.7)(x - 1.4)(x - 2.1)(x - 5.1)$

21 How does the graph of the approximate linear factors of g compare with the graph of g? The graphs are essentially the same

22 Where are the turning points on the graph of g? Where is g increasing? Where is g decreasing?

For Exercises 23–26, let $f(x) = x^4 + 5x^3 - 2x^2 + 5x - 3$. Use your graphics calculator to graph f. Make sure that all of the x- and y-intercepts are included in your viewing window.

23 Estimate all of the zeros of f. (HINT: $(x - i)$ is a linear factor of f). -5.5, 0.5, i, $-i$

24 Use the approximate zeros of f to factor f into its linear factors. $f(x) \approx (x + 5.5)(x - 0.5)(x + i)(x - i)$

25 How does the graph of the approximate factored form of f compare with the graph of f? The graphs are essentially the same

26 Where are the turning points on the graph of f? Where is f increasing? Where is f decreasing?

A student claims that the graph at the right is the graph of $f(x) = (x - 2)(x - 3)(x + 4)^2(x + 7)$. Complete Exercises 27–30.

27. Give at least two reasons why the graph shown is *not* the graph of f.

28 Graph f. Compare your viewing window with the viewing window shown, and list any similarities and differences.

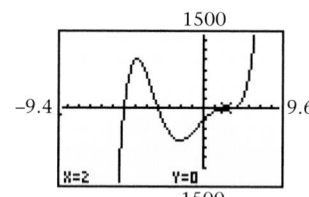

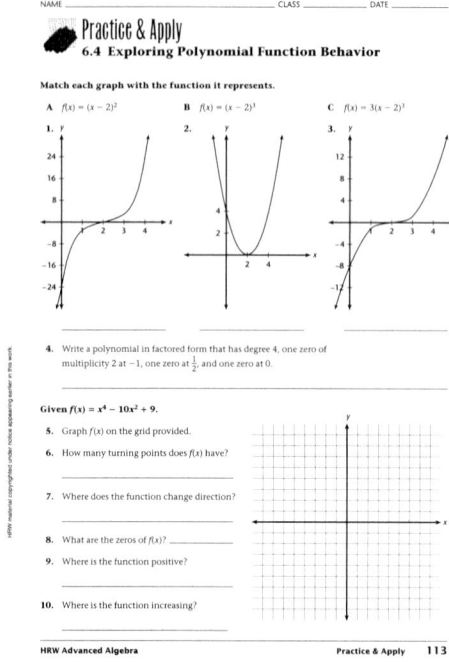

22. g is increasing $x < -5.0$, $-1.7 < x < 1.8$, and $x > 4.1$. g is decreasing $-5.0 < x < -1.7$, $1.8 < x < 4.1$. The turning points are at about $(-5.0, 509.4)$, $(-1.7, -691.2)$, $(1.8, 19.3)$, $(4.2, -407.7)$.

26. f is decreasing for $x < -4.1$ and increasing for $x > -4.1$. The turning point is $(-4.1, -119.2)$.

27. $x + 4$ is a factor of even multiplicity, so the graph should have a turning point at $x = -4$, not a crossing point. y should be positive when x is between -4 and 2.

28. Differences are listed in the answer to Exercise 27. Both graphs head down on the left and up on the right.

29 Where is f increasing? Where is f decreasing? Where is the graph shown increasing? Where is it decreasing? How do your responses for each graph compare?

30 List the turning and crossing points for each graph.

31. Which of the following polynomials is graphed at the right? Explain why.
  a. $f(x) = (x-4)^3(x+5)^2$ *d*
  b. $f(x) = x(x-4)^2(x+5)^2$
  c. $f(x) = (x-4)(x+5)^4$
  d. $f(x) = (x-4)^2(x+5)^3$
  e. $f(x) = (x-4)^3(x+5)$

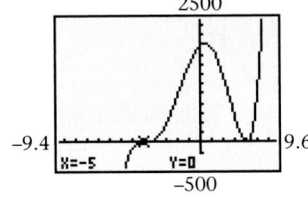

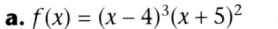 *Look Back*

For Exercises 32–34, let $f(x) = -2x^2 - 3x + 5$. Use your graphics calculator to graph f. Make sure your viewing window includes the x- and y-intercepts.

32 **Maximum/Minimum** Find the maximum value of f.
[Lesson 5.3] 6.125

33 Find the zeros of f.   **[Lesson 6.1]** −2.5, 1

34. Write f as a product of its linear factors.   **[Lesson 6.1]** $f(x) = -2(x + \frac{5}{2})(x-1)$

For Exercises 35–37, let $f(x) = 2x^2 + 8x - 7$. Use your graphics calculator to graph f. Make sure your viewing window includes the x- and y-intercepts.

35 **Maximum/Minimum** Find the minimum value of f.
[Lesson 5.3] −15

36 Approximate the zeros of f to the nearest tenth.   **[Lesson 6.1]** −4.7, 0.7

37. Use the approximate zeros of f to write f as a product of its linear factors.   **[Lesson 6.1]** $f(x) = 2(x + 4.7)(x - 0.7)$

Look Beyond

38. Economics The interest rate on a new bank account deposit is 7% compounded (paid out) annually. How much interest will be paid on an initial deposit of $200 after the first year? after 3 years? What is the total amount in the account after 3 years if no other deposits are made?

39 Use the statistics features of your graphics calculator to find the function of best fit for the table at the right. $f(x) = 3^x$

x	$f(x)$
1	3
2	9
3	27
4	81
5	243

29. f is increasing for $x < -6$, $-4 < x < -0.4$, and $x > 2.6$. f is decreasing for $-6 < x < -4$, $-0.4 < x < 2.6$. The function shown is increasing for $x < -6$ and $x > -2$. It is decreasing for $-6 < x < -2$. The responses are different for x-values between -6 and 2.6.

30. f has a turning point at $(-6, 293)$, $(-4, 0)$, $(-0.4, 698)$, and $(2.5, -101)$. It has crossing points at $(-7, 0)$, $(2, 0)$, and $(3, 0)$. The

function shown has turning points at the x-values -6 and -2. It has crossing points at $(-7, 0)$, $(-4, 0)$, and $(2, 0)$.

31. Since $x = 4$ is a turning-point zero, the function must have an even multiplicity on the factor $x - 4$. Since $x = -5$ is a crossing point zero, the function must have an odd multiplicity on the factor $x + 5$. Therefore the polynomial in d is the only one that could correspond to the graph shown.

Ask students the following questions.

- What roller coasters have you ridden on?
- Which one did you like best? Why?
- If you were designing a roller coaster, what would you like to include in the ride? Why?
- What restrictions do you think exist as to how roller coasters can be built? Why do you think these restrictions are important?

Encourage students to utilize a library or an on-line service to find out more about the construction of roller coasters and any building restrictions that actually do exist. Have students share the results of their research with the class.

EYEWITNESS MATH

SCREAM MACHINE

Coasting to Records

Highest, fastest, steepest in the worl

On June 9, 1988, Cedar Point's Magnum XL-200 roller coaster was certified as the fastest roller coaster in the world, with the longest vertical drop. As a result, it was listed in the 1990 edition of the Guinness Book of World Records. To enter the record books, the coaster needed two witnesses. One was Lt. Gov. Paul Leonard, who observed the ride's top speed 72 miles per hour and vertical drop of 194 feet, 8 inches.

NOTES:

1–2. One example sets the height of the hill 1 as 225 feet, hill 2 at 100 feet, and hill 3 at 49 feet.

Bottom of hill 1	
Theoretical speed	120 feet per second
Loss of speed (5%)	6 feet per second
Actual speed:	114 feet per second (≈ 78 miles per hour)

Top of hill 2	
Theoretical speed	34 feet per second
Loss of speed (5%)	5.7 ≈ 6 feet per second
Actual speed	28 feet per second (≈ 19 miles per hour)

Bottom of hill 2	
Theoretical speed	108 feet per second
Loss of speed (5%)	5.4 ≈ 5 feet per second
Actual speed	103 feet per second (≈ 70 miles per hour)

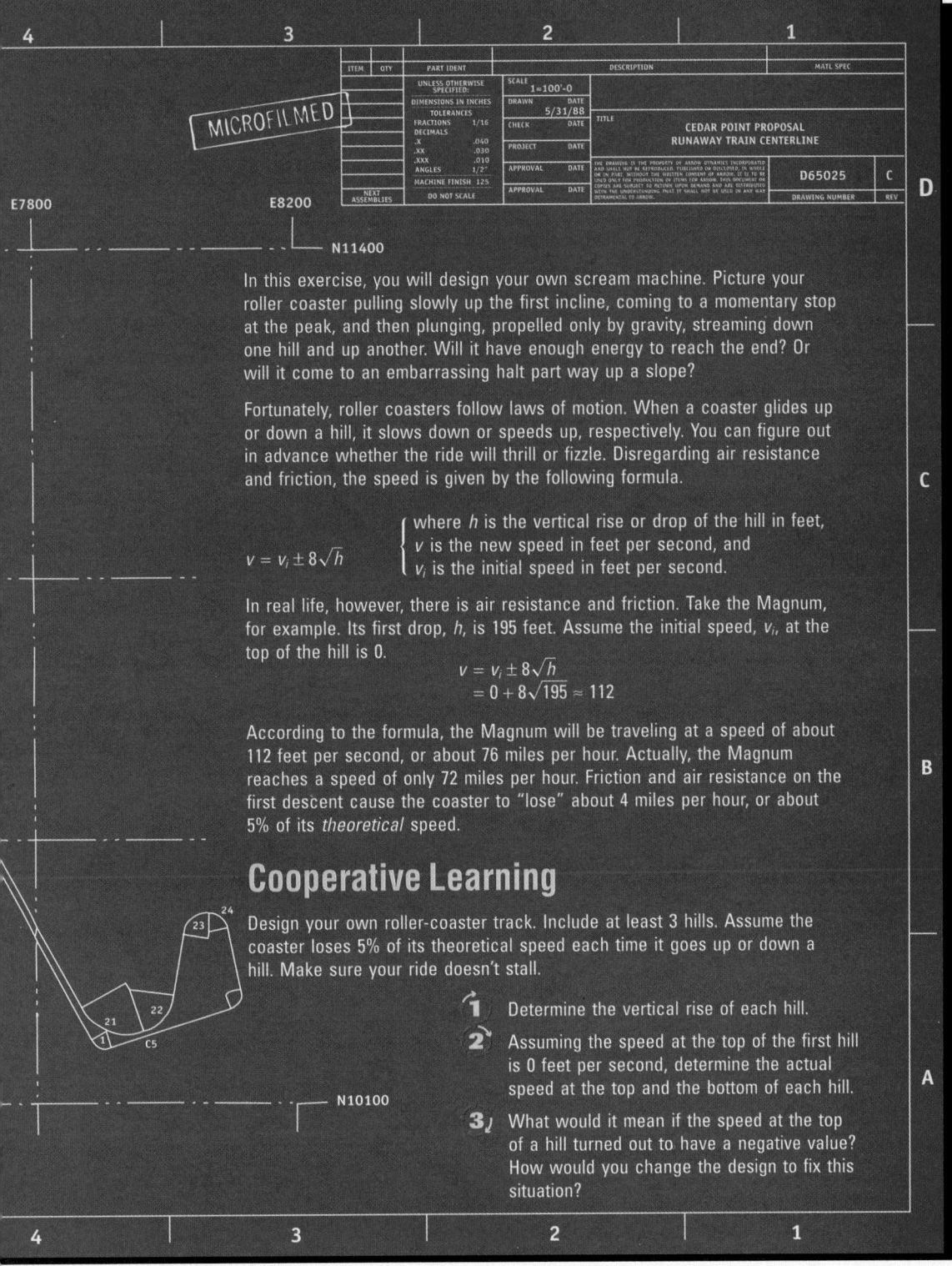

In this exercise, you will design your own scream machine. Picture your roller coaster pulling slowly up the first incline, coming to a momentary stop at the peak, and then plunging, propelled only by gravity, streaming down one hill and up another. Will it have enough energy to reach the end? Or will it come to an embarrassing halt part way up a slope?

Fortunately, roller coasters follow laws of motion. When a coaster glides up or down a hill, it slows down or speeds up, respectively. You can figure out in advance whether the ride will thrill or fizzle. Disregarding air resistance and friction, the speed is given by the following formula.

$$v = v_i \pm 8\sqrt{h}$$
where h is the vertical rise or drop of the hill in feet,
v is the new speed in feet per second, and
v_i is the initial speed in feet per second.

In real life, however, there is air resistance and friction. Take the Magnum, for example. Its first drop, h, is 195 feet. Assume the initial speed, v_i, at the top of the hill is 0.

$$v = v_i \pm 8\sqrt{h}$$
$$= 0 + 8\sqrt{195} \approx 112$$

According to the formula, the Magnum will be traveling at a speed of about 112 feet per second, or about 76 miles per hour. Actually, the Magnum reaches a speed of only 72 miles per hour. Friction and air resistance on the first descent cause the coaster to "lose" about 4 miles per hour, or about 5% of its *theoretical* speed.

Cooperative Learning

Design your own roller-coaster track. Include at least 3 hills. Assume the coaster loses 5% of its theoretical speed each time it goes up or down a hill. Make sure your ride doesn't stall.

1 Determine the vertical rise of each hill.

2 Assuming the speed at the top of the first hill is 0 feet per second, determine the actual speed at the top and the bottom of each hill.

3 What would it mean if the speed at the top of a hill turned out to have a negative value? How would you change the design to fix this situation?

Cooperative Learning

Have students work in pairs to design a roller-coaster track. Encourage each pair to discuss several possibilities for a track before actually deciding upon the final design. Suggest that students use both drawings and descriptions as they work on their designs.

After each pair has completed their design, ask them to explain and justify their design in a short report that they can share with the class.

Discuss

Have each pair of students share their designs and report with the class. Discuss the similarities and differences in the designs.

Ask students to share their answers to the questions in Step 3 with the class. Ask students how they used the answer to this question as they completed their design.

Discuss the possibility of using a computer to help with the design process. Ask students if they are familiar with any specific computer software that could be used. If a demonstration of this software is possible, allow students to show the class how they think it can be used.

Top of hill 3	
Theoretical speed	47 feet per second
Loss of speed (5%)	5.15 ≈ 5 feet per second
Actual speed:	42 feet per second (≈ 29 miles per hour)

Bottom of hill 3	
Theoretical speed	98 feet per second
Loss of speed (5%)	4.9 ≈ 5 feet per second
Actual speed:	93 feet per second (≈ 63 miles per hour)

3. A negative speed would mean the coaster would not make it up the hill. You could fix that by making that hill lower or by making the first hill higher.

Objectives

- Determine and analyze polynomial functions that model real-world situations.
- Use the algebraic tool of variable substitution to simplify a polynomial function.

R**ESOURCES**

- Practice Master   **6.5**
- Enrichment Master   **6.5**
- Technology Master   **6.5**
- Lesson Activity Master   **6.5**
- Quiz   **6.5**
- Spanish Resources   **6.5**

Assessing Prior Knowledge

A missile is fired from a nozzle that is 10 feet above the ground. Its initial velocity is 500 feet per second. Its height, $h(t)$, t seconds after being launched, is given by $h(t) = -16t^2 + 500t + 10$.

1. When will the missile reach its maximum height?
   [≈ **15.6 seconds**]

2. How high will the missile be at its maximum height?
   [≈ **3916 feet**]

T**EACH**

Many examples of applications of polynomial functions are given here. Ask students to suggest how polynomial functions might be used in these areas.

Math Connection
Geometry

Students may benefit from drawing a three-dimensional model of the box described.

Use Transparency ▶ 24

LESSON **6.5** Applications of Polynomial Functions

Many real-world problems in fields such as economics, physics, data analysis, and the arts can be solved by examining an appropriate polynomial function. For instance, constructing a cardboard box with the maximum volume possible can be analyzed using a polynomial function.

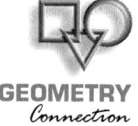
GEOMETRY
Connection

To construct an open box, cut small squares from the corners of a rectangular piece of cardboard, and fold up the sides that were created by removing the cut-out squares.

Let the sides of each square piece be x inches long. Then the volume of the box is a polynomial function in terms of x, the height of the box.

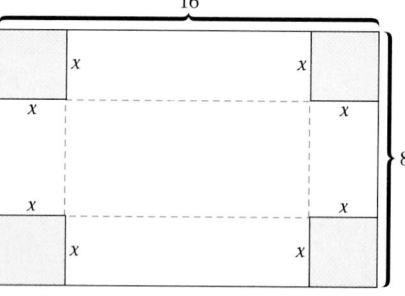

Volume = length × width × height
$V = (16 - 2x)(8 - 2x)(x)$

A**LTERNATIVE**
t e a c h i n g
s t r a t e g y

Using Visual Strategies Have students work in small groups. Ask each group to draw a graph of the polynomial function given in Example 1. Then ask a volunteer from one group to draw their graph on a coordinate plane transparency on an overhead projector. Ask the student to use different colors to indicate the crossing points and turning points of the graph. Have students describe the significance of these points in relationship to this problem. They should notice that the important values on this graph are those in which $0 \le x \le 4$. This means that the height must be between 0 inches and 4 inches. Also, the turning point in this section of the graph represents the maximum volume, which is about 98.5 cubic inches. Have students continue to work in their groups as they use a similar approach to look at other examples presented in this lesson.

MAXIMUM MINIMUM *Connection*

EXAMPLE 1

A rectangular piece of cardboard measures 8 inches by 16 inches. Find the maximum volume of the box to the nearest tenth. How large are the cut-out squares that produce the maximum volume?

Solution▸

Graph the volume function that models the cardboard box, $V = (16 - 2x)(8 - 2x)(x)$.

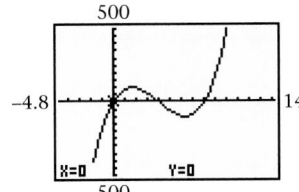
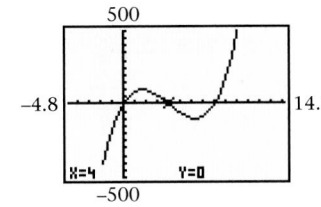
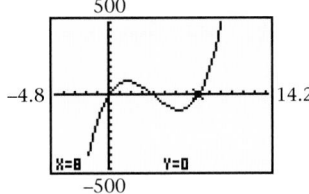

The zeros of V are 0, 4, and 8.

When $x \leq 0$, there are no squares cut from the cardboard.

When $x \geq 4$, there is no width to the box.

Thus, the value of x must be between 0 inches and 4 inches. The realistic domain of V is $0 < x < 4$.

Find the maximum value of V for $0 < x < 4$.

The maximum volume of the open-top box is approximately 98.5 cubic inches This occurs when squares about 1.7 inches by 1.7 inches are removed from each corner of the cardboard.

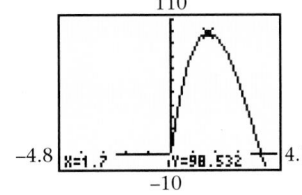

Try This A rectangular piece of cardboard measures 5 inches by 18 inches. Find the maximum volume of the box to the nearest tenth. How large are the cut-out squares that produce the maximum volume?

CRITICAL *Thinking*

In Example 1, are there any x-values outside of the realistic domain that make V positive? Why are these not included in the realistic domain of V?

EXAMPLE 2

Banking Sharon is planning to save money for her first year of college. At the end of her freshman year in high school, Sharon plans to deposit $300 in an account that has an annual interest rate of 5%.

Ⓐ By the end of her senior year in high school, how much money will she have from her original investment of $300?

Ⓑ If she invests $300 at the end of *each* year, how much money will she have by the end of her senior year in high school?

Ⓒ To accumulate $1400 for her first-year college tuition, at what interest rate should she invest $300 each year?

interdisciplinary
CONNECTION

Nature A cross section of a bee's honeycomb is a mosaic of hexagons. In this mosaic, three hexagons meet at each vertex. The mosaic starts with one hexagon. The hexagon is surrounded by a "circle" of six more hexagons, that "circle" is surrounded by another "circle" of 12 hexagons, and so on. Let n represent the number of circles with the center hexagon being circle 1.

1. The total number of hexagons in 2 circles is 7. What is the total in 3? 4? [**19, 37**]

2. Find a polynomial equation to indicate the total number of hexagons, h, as a function of the total number of "circles," n. [$h(n) = 3n^2 - 3n + 1$]

3. Find the total number of hexagons in a honeycomb with 12 "circles" [**397**]

Alternate Example 2

Alex is planning to save money for his first year of college. At the end of his freshman year in high school, Alex plans to deposit $500 in an account that has an annual interest rate of 6%.

A. By the end of his senior year in high school, how much money will he have from his original investment of $500? [**$595.51**]

B. If he invests $500 at the end of each year, how much money will he have by the end of his senior year in high school? [**$2187.31**]

C. To accumulate $2,250 for his first-year college tuition, at what interest rate should he invest $500 each year? [**about 7.9%**]

Cooperative Learning

Have students work in small groups as they study the examples. First, have students work through each example individually. Then have them discuss their work with their group. Encourage students to discuss any questions they may have about each example, saving for class discussion any questions that they cannot answer in their group for class discussion. Ask students to record the important concepts that they discuss in their group work. Have each group share their discussions with the class and ask any necessary questions.

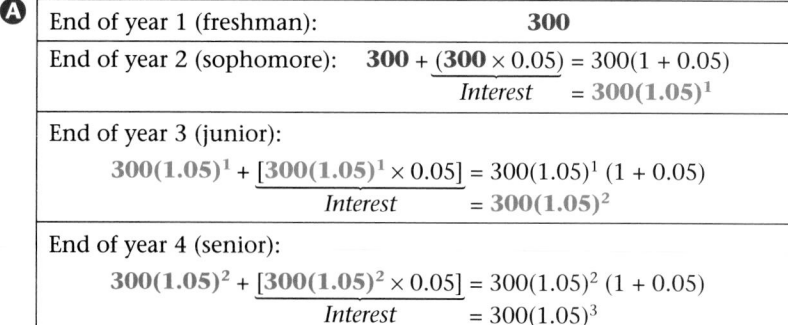

A

End of year 1 (freshman):	300
End of year 2 (sophomore):	$300 + \underbrace{(300 \times 0.05)}_{Interest} = 300(1 + 0.05)$ $= 300(1.05)^1$
End of year 3 (junior): $300(1.05)^1 + \underbrace{[300(1.05)^1 \times 0.05]}_{Interest} = 300(1.05)^1(1 + 0.05)$ $= 300(1.05)^2$	
End of year 4 (senior): $300(1.05)^2 + \underbrace{[300(1.05)^2 \times 0.05]}_{Interest} = 300(1.05)^2(1 + 0.05)$ $= 300(1.05)^3$	

The amount of Sharon's initial investment at the end of her senior year will be $300(1.05)^3$, or $347.28.

B

End of year 1 (freshman):	300
End of year 2 (sophomore):	$300(1.05)^1 + 300$
End of year 3 (junior):	$300(1.05)^2 + 300(1.05)^1 + 300$
End of year 4 (senior): $300(1.05)^3 + 300(1.05)^2 + 300(1.05)^1 + 300 = 1293.04$	

If Sharon invests $300 at the end of *each* year, she will have $1293.04 by the end of her senior year of high school.

C If Sharon wants four annual deposits of $300 to be worth $1400 by the end of her senior year, she will need to invest her money at a rate higher than 5%.

To determine what the interest rate should be, substitute r for 0.05.

$$300(1 + r)^3 + 300(1 + r)^2 + 300(1 + r) + 300 = 1400$$

Simplify this equation by letting $1 + r$ be the variable R.

$$300R^3 + 300R^2 + 300R + 300 = 1400$$

Approximate the zeros of the related polynomial function $f(R) = 300R^3 + 300R^2 + 300R - 1100$.

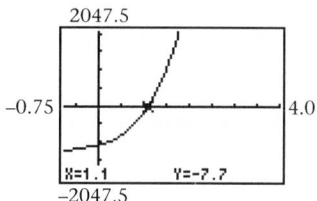

The value for R is about 1.1.

$$R = 1 + r$$
$$1.1 = 1 + r$$
$$0.1 = r$$

Since r is about 0.1, Sharon needs to invest her money at a rate of about 10% to have $1400 by the end of her senior year. ❖

ENRICHMENT Have students do research to find out about different accounts available at several different local banks. Have them use this information to write several problems that can be solved using polynomial functions. Then have students exchange and solve the problems.

INCLUSION strategies **Tactile-Kinesthetic Strategies** Some students may have difficulty visualizing a flat piece of cardboard being made into a box. Use a flat piece of cardboard to demonstrate this process. Then allow students to duplicate this procedure as necessary.

How could the value of *r* in Example 2 have been found without using the substitution $R = r + 1$? Would this be more or less complicated than the method used?

CRITICAL
Thinking

If the interest rate remained at 5%, how much more money would Sharon need to save each year to have $1400 by the end of her senior year?

Khayyam Methods

Cultural Connection: Asia Omar Khayyam, 1050–1122, is usually remembered as a poet and author of the *Rubaiyat*. In his native Persia (modern-day Iran), Khayyam is also remembered as a prominent scientist and mathematician. Like most scholars of his time and place, he became politically active, serving his ruler, Malik-Shah, in a variety of ways.

Khayyam developed a method for finding the zeros of a cubic polynomial function of the form $f(x) = x^3 - bx - a$, where *a* and *b* are real numbers. His method involved using two curves, $y = -\frac{1}{\sqrt{b}}x^2$ and $y^2 = x^2 + \frac{a}{b}x$, which he was already familiar with. Example 3 shows Khayyam's method.

RETEACHING
the
lesson

Using Manipulatives Write several different types of problems on index cards and place the cards in a small index-card file box. You may want to have students write some or all of the problems. Then have students work in small groups and choose one problem at a time to solve. Encourage each group to use whatever method its members like best. After each group has solved one problem, ask them to share their solutions and strategies with the class. Encourage the rest of the class to ask questions about the strategies and solution. Repeat this activity as many times as is necessary.

TEACHING tip

Be sure students understand that the Khayyam method involves finding the intersection of the graphs of two equations in order to find the roots of a cubic equation.

Graphics Calculator

EXAMPLE 3

Find the roots of $x^3 = 7x + 6$ using the Khayyam method.

Solution

Write the related polynomial function in the form $f(x) = x^3 - bx - a$, and identify a and b.

$$f(x) = x^3 - 7x - 6; \ b \text{ is } 7 \text{ and } a \text{ is } 6$$

The Khayyam method uses the two equations, $y = -\dfrac{1}{\sqrt{b}}x^2$ and $y^2 = x^2 + \dfrac{a}{b}x$, to find the roots of a cubic polynomial of the form $f(x) = x^3 - bx - a$. Substitute 6 for a and 7 for b in these two equations.

$$y = -\dfrac{1}{\sqrt{7}}x^2 \text{ and } y^2 = x^2 + \dfrac{6}{7}x$$

Graph these Khayyam curves. The graph of $y^2 = x^2 + \dfrac{6}{7}x$ is obtained by plotting both $y = \sqrt{x^2 + \dfrac{6}{7}x}$ and $y = -\sqrt{x^2 + \dfrac{6}{7}x}$.

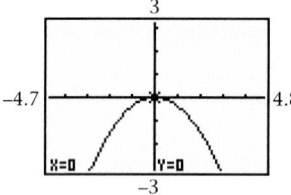

$$y = -\dfrac{1}{\sqrt{7}}x^2$$

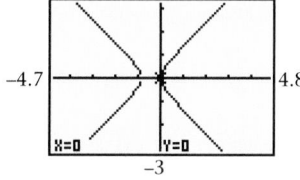

$$y^2 = x^2 + \dfrac{6}{7}x$$

Find the nonzero x-coordinates of the intersection points of these graphs.

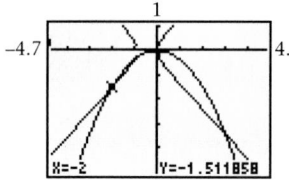

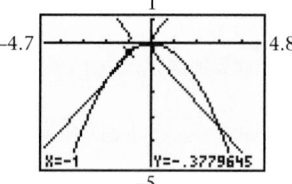

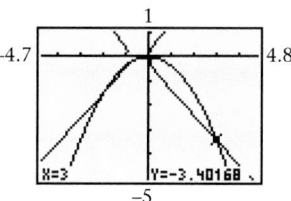

The nonzero x-coordinates of the intersection points are -2, -1, and 3. These are the roots of the original cubic equation, $x^3 = 7x + 6$, and the zeros of the related cubic function, $f(x) = x^3 - 7x - 6$.

Check

Graph $f(x) = x^3 - 7x - 6$, and verify that the zeros of this function are -2, -1, and 3.

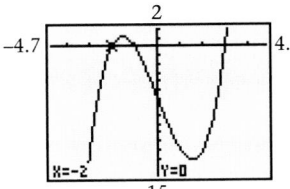 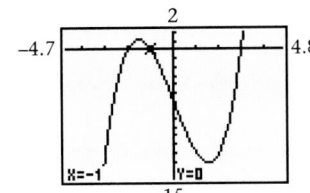 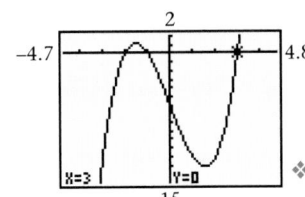

EXERCISES & PROBLEMS

Communicate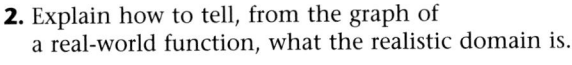

1. If a piece of cardboard is 6 inches by 4 inches, what would be the volume as a function of the length of the sides of the square cut from each corner?

2. Explain how to tell, from the graph of a real-world function, what the realistic domain is.

3. How large a square could you cut out of each corner, when making a box, if the piece of cardboard were 12 inches by 12 inches?

4. Explain how to determine possible volumes for the box described in Exercise 1.

5. Describe the behavior of the volume function of Exercise 1.

6. In Example 2, how much would Sharon have accumulated in 3 years if she had been able to save $400 in her freshman year in high school?

ASSESS

Selected Answers
Odd-numbered Exercises 7–29

Assignment Guide
Core 1–29

Core Plus 1–30

Technology
A graphics calculator is needed for Exercises 14–15, 17–18, 28–29.

Make sure that students write the correct equations to solve each problem. Before students begin to solve a problem, have them determine which example the problem resembles, and have students use the solution shown in the example as a guide.

Practice & Apply

Economics William is able to deposit $200, $300, $400, and $500 at the end of his freshman, sophomore, junior, and senior years of high school, respectively.

7. How much money will he have by the end of his senior year if the money is invested at 6% interest? $1499.28

8. If his total tuition for two semesters at the local community college is $560, how much of his savings will he have left over after he pays the tuition? $939.28

9. If William's total tuition at a state university is $2000 for the first two semesters, will his investment return be enough to pay for the tuition if he delays college for one year? Explain. no

Economics Calinda is able to deposit $400, $300, $200, and $100 at the end of her freshman, sophomore, junior, and senior years of high school, respectively.

10. At what approximate interest rate must Calinda invest her money to cover her college tuition costs of $1100? 5%

11. Suppose that the money is invested at the interest rate you found in Exercise 10, but she deposits $425, $350, $275, and $200 at the end of each high school year. How much money will be left over after she pays the $1100 tuition? $266.62

12. Using the investment amounts from Exercise 11, what interest rate would Calinda need if the tuition were increased by $400? approx 11%

Geometry A piece of cardboard measures 8 inches by 6 inches. Complete Exercises 13–15.

13 What would be the volume function in terms of the length, x, of the sides of the squares cut from each corner? Graph the volume function, making sure that the x- and y-intercepts are included. $V(x) = (8 - 2x)(6 - 2x)(x)$

14 What is the realistic domain of the volume function? $0 < x < 3$

15 Maximum/Minimum What is the maximum volume of the box to the nearest tenth? What size of square should be removed from each corner? volume 24.2 cubic inches, 1.1 in. squares

9. No; $(1499.98)(1.06) = 1589.24,$ and $2000 - $1589.29 = 410.76. William will need another $410.76 to pay for tuition at a state university.

Banking Gary is able to deposit $A, $B, $C, and $D at the end of his freshman, sophomore, junior, and senior years of high school, respectively.

16. Write the polynomial function in terms of R, where $R = 1 + r$, for these investments. $f(R) = AR^3 + BR^2 + CR + D$

17 Graph the polynomial function if the amounts deposited each year are $100, $200, $400, and $800, respectively, and he wants to earn $1600. What interest rate will most closely yield this amount? 9%

18 Graph the polynomial function if the amounts deposited each year are $800, $400, $200, and $100, respectively, and he wants to earn $1600. What interest rate will most closely yield this amount? 3%

19. From the results of Exercises 17 and 18, is it better if $A < B < C < D$ or if $A > B > C > D$? Why? $A > B > C > D$

20. **Portfolio Activity** Complete the Portfolio Activity described on page 325. $V(x) = (12 - 2x)^2 x$, where x is the length

Look Back

Let $f(x) = 5x + 4$ and $g(x) = 2 - x$. Find each of the following functions. [Lesson 3.3]

21. $(f \circ g)(x)$ **22.** $(g \circ f)(x)$ **23.** $(f \circ f)(x)$ **24.** $(g \circ g)(x)$
   $-5x + 14$ $-5x - 2$ $25x + 24$ x

Use synthetic division to find the quotient.   [Lesson 6.3]

25. $\dfrac{3x^4 - 4x^2 + 2x - 1}{x - 1}$ $3x^3 + 3x^2 - x + 1$ **26.** $\dfrac{x^4 + 4x^3 + 5x^2 - 5x - 14}{x + 2}$ $x^3 + 2x^2 + x - 7$

27. Write a polynomial of degree 5 that has three crossing-point zeros at -3, 4, and 5 and one turning-point zero at 1.   **[Lesson 6.4]**

28 Use your graphics calculator to graph $f(x) = -3x^2 + 4x - 5$, and describe its behavior.   **[Lesson 6.4]**

29 Describe the behavior of the function $f(x) = -3(x + 1)^2(x - 1)^3(x - 2)^4$.   **[Lesson 6.4]**

Look Beyond

30. **Radioactive Decay** A radioactive substance has a half-life (the period of time in which half of the material decays) of 5 years. If you begin with 20 grams of the substance, how much of the substance is left after 10 years? after 20 years? 5 g; 1.25 g

19. It is better if $A > B > C > D$, since with a lower rate of interest he still earns his $1600.

27. One possible answer is $p(x) = (x + 3)(x - 4)(x - 5)(x - 1)^2$.

28. $f(x) = -3x^2 + 4x - 5; f(0) = -5$ The graph is a parabola, opening down, vertex at $(0.67, -3.67)$.

29. The zeros of f are -1, 1, 2. The turning points are $x = -1$ and $x = 2$ and the crossing point is $x = 1$. The graph heads upward at the left and downward at the right.

FOCUS

This project explores the use of nonperiodic polynomial functions with a restricted domain to approximate the graph of a periodic function. Students need to use either a graphics calculator, or computer software such as *f(g) Scholar*™, and graph paper and colored pencils to explore these graphs.

MOTIVATE

Review the graph of the sine function in Chapter 3. Ask students to complete the following activities.

• Make a list of the similarities between the graph of the sine function and the graph of a polynomial function.

• Make a list of the differences between the graph of the sine function and the graph of a polynomial function.

• Share your lists with your class. Create a final list that describes the similarities and differences.

You have been studying many different polynomial functions. You will now graph and find patterns in a *specific* set of polynomial functions. The patterns that you discover in these functions and in the shape of their graphs will be extended. By combining an infinite number of these specific polynomial functions, you can create a function that is no longer polynomial.

Materials needed:

1. several pieces of graph paper
2. colored pencils
3. graphics calculator

The Project:

Recall that $n! = n(n-1)(n-2)\cdots(3)(2)(1)$. For example, $3! = 3 \cdot 2 \cdot 1 = 6$ and $5! = 5 \cdot 4 \cdot 3 \cdot 2 \cdot 1 = 120$.

1. Using a full piece of graph paper, draw your coordinate axes so that x and y range from -10 to 10, and graph $s_1(x) = x$ using a colored pencil.

2. Use your graphics calculator to graph $s_2(x) = x - \dfrac{x^3}{3!}$. Make a table of values, and transfer the graph to your graph paper. Use the same coordinate axes but a different colored pencil.

3. Repeat Step 2 for each of the following functions.

   a. $s_3(x) = x - \dfrac{x^3}{3!} + \dfrac{x^5}{5!}$

   b. $s_4(x) = x - \dfrac{x^3}{3!} + \dfrac{x^5}{5!} - \dfrac{x^7}{7!}$

   c. $s_5(x) = x - \dfrac{x^3}{3!} + \dfrac{x^5}{5!} - \dfrac{x^7}{7!} + \dfrac{x^9}{9!}$

   d. $s_6(x) = x - \dfrac{x^3}{3!} + \dfrac{x^5}{5!} - \dfrac{x^7}{7!} + \dfrac{x^9}{9!} - \dfrac{x^{11}}{11!}$

   e. $s_7(x) = x - \dfrac{x^3}{3!} + \dfrac{x^5}{5!} - \dfrac{x^7}{7!} + \dfrac{x^9}{9!} - \dfrac{x^{11}}{11!} + \dfrac{x^{13}}{13!}$

These functions are each *partial sums* of the **Taylor series for sines**, which has an unlimited number of terms.

$$\sin(x) = x - \frac{x^3}{3!} + \frac{x^5}{5!} - \frac{x^7}{7!} + \frac{x^9}{9!} - \frac{x^{11}}{11!} + \frac{x^{13}}{13!} - \cdots$$

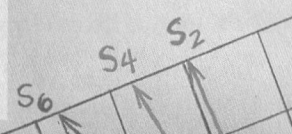

1. Each successive partial sum of the Taylor series more closely resembles the graph of $\sin x$. For all partial sums, the Taylor series resembles the sine function for $-\dfrac{\pi}{2} < x < \dfrac{\pi}{2}$.

2. As each additional term is added to the Taylor series, a closer fit to the sine function is made between the domain of $-\dfrac{3}{2}\pi \le x \le \dfrac{3}{2}\pi$.

3. The turning points at $x = \pm\dfrac{\pi}{2}$ are the same for all these graphs. The turning points beyond the domain $-\dfrac{\pi}{2} < x < \dfrac{\pi}{2}$ vary until $s_7(x)$ when the turning points at $x = \pm\dfrac{3\pi}{2}$ approach those of the sine function.

The Results:

1. Graph the sine function: $f(x) = \sin(x)$. How does each successive partial sum of the Taylor series compare to the sine function?

2. As each additional term is added, describe how the polynomial curve begins to resemble the behavior of the nonpolynomial curve, $f(x) = \sin(x)$? In what portion of the domain does the resemblance occur?

3. How do the turning points of each successive polynomial function compare to those of the sine function?

4. How do the zeros of each successive polynomial function compare to those of the sine function?

5. About how many terms must be in the partial sum before the polynomial function matches four turns of the sine function?

6. Use the trace feature to find $\sin\left(\frac{\pi}{2}\right)$. About how many terms must be in the partial sum before the polynomial function approximates $\sin\left(\frac{\pi}{2}\right)$ to two decimal places?

The graph of s_7 is shown on the same coordinate plane as the graph of $f(x) = \sin(x)$.

Cooperative Learning

Have students work in small groups to complete the project. First, ask each student to work individually to draw the graphs. Next, have members of each group compare their graphs and agree upon a final graph. Then have them work together to answer the questions. After they have completed the project, ask each group to write a paragraph summarizing what they have learned as they have worked on this project.

DISCUSS

Have students compare their final graphs with the one shown at the bottom of this page. Their graphs through Part(c) of Step 3 should be the same as the one shown in the text. Ask each group to share their answers to the steps in the *Results* section and the summary paragraph suggested in the Cooperative Learning section above. Conduct a class discuss about these answers and summaries.

As an extension to this project, have students do research to find other examples of Taylor Series.

4. All graphs have a zero at 0. The partial sums have other zeros corresponding to the sine function, at $y = \pm\pi$, for s_5 and up.

5. about 7 terms

6. about 3 terms

6 Chapter Review

Chapter 6 Review

Vocabulary

Key Skills & Exercises

Lesson 6.1

➤ Key Skills

Determine realistic domain restrictions of a volume function.

The restriction on the domain of the volume function, $V(x) = w(x) \cdot l(x) \cdot h(x)$, is determined by the realistic constraints on each dimension: width, length, and height.

Define and examine the two forms of polynomial functions.

The factored form of a polynomial is $C(x - r_1)(x - r_2) \ldots (x - r_{n-1})(x - r_n)$, where C is a constant, $x - r$ is a linear factor, and r is a zero of f.

The expanded form of a polynomial is $a_n x^n + a_{n-1} x^{n-1} + \cdots + a_2 x^2 + a_1 x + a_0$, where a_n is the leading coefficient.

➤ Exercises

For Exercises 1 and 2, let $f(x) = (x + 2)(3x + 5)(3 - x)(2x + 1)(x - 6)$.

1. Determine the degree of f. 5

2. Determine the zeros of f. $-2, -\frac{5}{3}, 3, -\frac{1}{2}, 6$

For Exercises 3–7, use your graphics calculator to graph the function $f(x) = (x - 3)^2(2x + 5)(1 + x)^2(7 - 5x)$.

3 What are the zero(s) of this function? 3, −2.5, −1, 1.4

4 At which zero(s) does the graph *cross* the *x*-axis? −2.5, 1.4

5 At which zero(s) does the graph only *touch* the *x*-axis? 3, −1

6 For what domain is the function greater than zero? $-2.5 < x < -1$; $-1 < x < 1.4$

7 Approximate the maximum values of the function between the zeros. 440.6; 384.2

Lesson 6.2

➤ Key Skills

Write the factored form of a polynomial.

Difference of two squares:
$$(x^2 - a^2) = (x - a)(x + a)$$
Perfect-square trinomial:
$$(x + k)^2 = x^2 + 2k + k^2$$
Factoring by inspection:
$$x^2 + bx + c = (x + m)(x + n),$$
where $m \cdot n = c$ and $m + n = b$

Define and use the Factor Theorem to find the zeros of polynomial functions.

For any polynomial function f, $x - a$ is a factor of f if and only if a is a zero of f. Also, a is a zero of f if $x - a$ divides f exactly.

➤ Exercises

8 Write $2x^2 + x - 6$ in factored form. $2(x + 2)(x - \frac{3}{2})$

9. Find the exact zeros of $f(x) = 3x^4 + 3x^3 - 36x^2$. Then write f in factored form. $f(x) = 3x^2(x - 3)(x + 4)$

10. Write a quadratic function with zeros at $-\frac{1}{2}$ and 3.

10. Answers may vary but should be of the form $f(x) = C\left(x + \frac{1}{2}\right)(x - 3)$, where $C \neq 0$.

Lesson 6.3

➤ Key Skills

Define and use the division algorithm for polynomials.

If the polynomial P is divided by the polynomial D, then $P = QD + R$, where Q is the quotient polynomial and R is the remainder polynomial of degree less than D. If R is 0, then D divides P *exactly*.

Synthetic division is a shorthand notation for dividing polynomials.

Define and use the Fundamental Theorem of Algebra to find the zeros of a polynomial function.

Every polynomial function $P_n(x)$ of degree $n > 0$ has at least one complex zero.

Every polynomial $P_n(x)$ of degree n can be written as a factored product of exactly n linear factors, corresponding to exactly n zeros (counting multiplicities).

➤ Exercises

The numbers –2 and 2 are zeros of $f(x) = x^5 + x^4 - 3x^3 - 3x^2 - 4x - 4$.

11. Find the other real zero(s) of f using synthetic division. -1

12. Write f in factored form over the complex numbers. $f(x) = (x + 2)(x - 2)(x + 1)(x + i)(x - i)$

13. Write a quadratic polynomial function that has zeros at –2 and 1 and whose graph has a y-intercept at –4. $f(x) = 2(x + 2)(x - 1)$

Lesson 6.4

➤ Key Skills

Determine and classify the behavior of single zeros and multiple zeros of a polynomial function.

A repeated zero is called a multiple zero. When a zero occurs n times, the zero has a multiplicity of n.
Zeros of an odd order are crossing points.
Zeros of an even order are turning points.

Determine and classify the behavior and basic shape of polynomial function of varying degrees.

At a *peak turning point* the function changes from increasing to decreasing.
At a *valley turning point* the function changes from decreasing to increasing.
At a *crossing point*, the function value changes signs.

14. Answers may vary but should be of the form $f(x) = C(x-3)(x+2)^2$ $(x+1)^2(x-2)^2$

15. Answers may vary but should be of the form $f(x) = -C(x+1)^2(x-1)^3$, where $C > 0$.

19. turning points:
$(-1.21, 6.78)$ and
$(0.38, -9.30)$
increasing: $x < -1.21$ and $x > 0.38$;
decreasing:
$-1.21 < x < 0.38$

20. $V(x) = (12 - 2x)(8 - 2x)x$

> ## Exercises

14. Write a polynomial function in factored form that has degree 7, 3 turning-point zeros, and 1 crossing-point zero.

15. Write a polynomial function in factored form of degree 5 that has one zero of multiplicity 2 at –1, one zero of multiplicity 3 at 1, and whose graph turns upward at the left end.

Let $f(x) = 8x^3 + 10x^2 - 11x - 7$.

16 Estimate all the zeros of f. $-1.7, -0.5, 1$

17. Use the approximate zeros to write f in factored form. $f(x) = 8(x + 1.7)(x + 0.5)(x - 1)$

18 Compare the graph of the approximate factored form of f with the graph of f. graphs are essentially the same

19 Where are the turning points of f? Where is f increasing? Where is f decreasing?

Lesson 6.5

> ### Key Skills

Determine and analyze polynomial functions that model real-world situations.

An example of a real-world problem that can be analyzed using a polynomial function is constructing an open-top box.

Use the algebraic tool of variable substitution to simplify a polynomial function.

To simplify the equation
$300(1 + r)^2 + 300(1 + r) + 3 = 1400$
let $1 + r$ be the variable R.
Solve the resulting equation,
$300R^2 + 300R + 3 = 1400$, for R.
Then solve the equation $1 + r = R$, to find the original r.

> ### Exercises

A piece of cardboard measures 12 inches by 8 inches.

20. Write a function for the volume of the box created from the cardboard, in terms of the length of the sides of the squares cut from each corner.

21 Graph the function from Exercise 20. What is the realistic domain of the volume function? $0 < x < 4$

22 What is the maximum volume of the box? approx. 67.6 in^3

23. The perimeter of the base of a crate cannot exceed 6 feet. The height is 1 foot less than twice the width. Write a function for the volume of the crate in terms of the width. $V(x) = x(3 - x)(2x - 1)$

Applications

Investment While in high school, Andrew saved money for his first year of college. At the end of his freshman year, he deposited $500 into an account that earns 6% simple interest, compounded annually.

24. By the end of his sophomore year, how much will Andrew have in his account? $530

25. If he deposits another $500 at the end of *each* year, how much money will Andrew have by the end of his sophomore year? his junior year? his senior year? $1030, $1591.80, $2187.31

26. To accumulate $2650 for his first semester of college tuition at a state university, at what interest rate should he invest the $500 *each* year? 19%

Chapter 6 Assessment

For Items 1–6, let $f(x) = (x + 1)^3(2x + 3)(2 - x)^2$.

1. Determine the degree of f. 6

2. What are the zeros of f? $-1, -\frac{3}{2}, 2$

3. At which zero(s) does the graph *cross* the x-axis? $-1, -\frac{3}{2}$

4. At which zero(s) does the graph only *touch* the x-axis? 2

5. For what domain is the function greater than zero? $x < -1.5, -1 < x < 2$, and $x > 2$

6. Find the maximum value of the function between the zeros. approx 40

7. Write $15x^2 + 11x - 14$ in factored form. $15\left(x - \frac{2}{3}\right)\left(x + \frac{7}{5}\right)$

8. Use your graphics calculator to find the zeros of $f(x) = x^5 - x^4 + x^3 - x^2$. Then write f in factored form over the complex numbers. $0, 1; x^2(x - 1)(x - i)(x + i)$

9. Write a quadratic function that has zeros at –2 and 9. $f(x) = C(x + 2)(x - 9)$

10. Write a quadratic function that has real zeros at –1 and 3 and whose y-intercept is 6. $f(x) = -2(x + 1)(x - 3)$

The numbers 3 and –2 are zeros of the polynomial function $f(x) = x^6 - x^5 - 5x^4 - x^3 - 8x^2 + 2x + 12$.

11. Find the other real zero using synthetic division. (HINT: –1 is a zero.) 1

12. Write f in factored form over the complex numbers. $f(x) = (x - 3)(x + 2)(x - 1)(x + 1)\left(x - i\sqrt{2}\right)\left(x + i\sqrt{2}\right)$

Let $f(x) = (3x - 2)^2(x + 1)(1 - x)^3(x - 4)^2$.

13. What are the zeros and the degree of f? $\frac{2}{3}, -1, 1, 4; 8$

14. Describe what happens at a turning point. Does f have any zeros that are turning points? What are they?

15. Describe what happens at a crossing point. How many crossing points does f have? What are they?

16. **Geometry** The perimeter of the base of a crate cannot exceed 8 feet. The height is 1 foot more than twice the width. Write a function for the volume of the crate in terms of the width. $V(w) = w(4 - w)(2w + 1)$

Geometry A piece of cardboard measures 16 inches by 20 inches.

17. Write a function for the volume of a box in terms of the length of the sides of the squares cut from each corner. $V(x) = x(16 - 2x)(20 - 2x)$

18. Graph the volume function you wrote for Item 17. What is the realistic domain for the volume function? $0 < x < 8$

19. **Maximum/Minimum** What is the maximum volume of the box? approx 420 in²

Investment While in high school, Jenny saved money for her first year of college. At the end of her freshman year, she deposited $400 in an account that earns 6% simple interest, compounded annually.

20. By the end of her sophomore year, how much will Jenny have in the account? $424

21. If she deposits another $400 at the end of *each* year after her freshman year, how much money will Jenny have by the end of her sophomore year? her junior year? her senior year? $824; $1273.44; $1749.85

14. At a turning point, $f(x)$ changes direction; at a peak the polynomial changes from increasing to decreasing; at a valley the polynomial changes from decreasing to increasing; yes; $x = \frac{2}{3}$ and $x = 4$.

15. At a crossing point in the graph of $f(x)$ changes sign; 2; $x = -1$ and $x = 1$

COLLEGE ENTRANCE-EXAM PRACTICE

Multiple-Choice and Quantitative-Comparison Samples

The first half of the Cumulative Assessment contains two types of items found on standardized tests — multiple-choice questions and quantitative-comparison questions. Quantitative-comparison items emphasize the concepts of equality, inequality, and estimation.

Free-Response Grid Samples

The second half of the Cumulative Assessment is a free-response section. A portion of this part of the Cumulative Assessment consists of student-produced response items commonly found on college entrance exams. These questions require the use of machine-scored answer grids. You may wish to have students practice answering these items in preparation for standardized tests.

Sample answer-grid masters are available in the *Chapter Teaching Resources Booklets*.

Chapters 1–6 Cumulative Assessment

College Entrance Exam Practice

Quantitative Comparison For Items 1–4, write

A if the quantity in Column A is greater than the quantity in Column B;

B if the quantity in Column B is greater than the quantity in Column A;

C if the quantities are equal; or

D if the relationship cannot be determined from the given information.

	Column A	Column B	Answers
1. B	$\frac{1}{5}(x-2)^2 = 1$		Ⓐ Ⓑ Ⓒ Ⓓ [Lesson 5.2]
	The absolute value of the sum of the solutions	The absolute value of the difference of the solutions	
2. B	The distance between $(-2, -1)$ and $(-5, 3)$	The y-coordinate of the vertex of the parabola given by $f(x) = 2x^2 + 4x + 12$	Ⓐ Ⓑ Ⓒ Ⓓ [Lesson 5.2, 5.5]
3. C	The degree of the polynomial $f(x) = (x + 1)^2(x - 1)(2 - x)$	The largest zero of the polynomial $g(x) = x^3 - 4x^2 - x + 4$	Ⓐ Ⓑ Ⓒ Ⓓ [Lessons 6.1, 6.3]
4. A	$f(x) = (x - 5)^2(2x - 3)$		Ⓐ Ⓑ Ⓒ Ⓓ [Lesson 6.4]
	The number of turning points in the graph of f	The number of crossing points in the graph of f	

5. If $A = \begin{bmatrix} 2 & 3 & 8 \\ -4 & 5 & -8 \\ 0 & 6 & -5 \end{bmatrix}$ and $B = \begin{bmatrix} 0 & 3 & -1 \\ -1 & 4 & 3 \\ 2 & -7 & 2 \end{bmatrix}$, what is $A - B$?   **[Lesson 4.1]**

b

a. $\begin{bmatrix} 2 & 0 & 7 \\ -5 & 1 & -5 \\ -2 & -1 & -3 \end{bmatrix}$ **b.** $\begin{bmatrix} 2 & 0 & 9 \\ -3 & 1 & -11 \\ -2 & 13 & -7 \end{bmatrix}$ **c.** $\begin{bmatrix} -2 & 0 & -9 \\ -5 & -1 & -5 \\ 2 & -13 & -3 \end{bmatrix}$ **d.** $\begin{bmatrix} -2 & 0 & -9 \\ 3 & -1 & 11 \\ 2 & -13 & 7 \end{bmatrix}$

6. Let $f(x) = (3 - 2x)(4x + 3)$. Find the approximate coordinates of the
a   vertex. Is the vertex a minimum or maximum point?   **[Lesson 5.1]**
    a. (0.375, 10.125), maximum   **b.** (0.375, 10.125), minimum
    c. (0.75, 10.125), maximum   **d.** (0.375, 7.875), maximum

7. What is the inverse of $\{(5, 1), (5, 2), (3, 3), (4, -1)\}$? Is the inverse a
c   function?   **[Lesson 3.2]**
    a. (5, 1), (5, 2), (3, 3), (4, -1); No   **b.** (5, 1), (5, 2), (3, 3), (4, -1); Yes
    c. (1, 5), (2, 5), (3, 3), (-1, 4); Yes   **d.** (1, 5), (2, 5), (3, 3), (-1, 4); No

8. What single function represents this system of parametric equations?
d   **[Lesson 3.6]**

$$\begin{cases} x(t) = 2t \\ y(t) = t + 1 \end{cases}$$

    a. $y = -\frac{1}{2}x + 1$ **b.** $y = 2x + 1$ **c.** $y = -2x + 1$ **d.** $y = \frac{1}{2}x + 1$

9. Find the x-intercepts of $y = 5x^2 - 8x - 4$.   **[Lesson 5.1]**
d   **a.** $\left(\frac{2}{5}, 0\right)$ and $(-2, 0)$ **b.** $\left(\frac{2}{5}, 0\right)$ and $(2, 0)$ **c.** $\left(-\frac{2}{5}, 0\right)$ and $(-2, 0)$ **d.** $\left(-\frac{2}{5}, 0\right)$ and $(2, 0)$

10. Which correlation coefficient is most likely to apply to the
d   relationship between the amount of sunscreen applied to the skin and
    the amount of damage to the skin caused by the sun?   **[Lesson 1.3]**
    a. $r = 0.65$ **b.** $r = 0.43$ **c.** $r = -0.03$ **d.** $r = -0.88$

11. What are the approximate zeros of $f(x) = -2x^2 + 4x + 1$? What is the
    vertex?   **[Lesson 5.1]** $-0.2, 2.2$, Vertex:$(1,3)$

12. Let $f(x) = (5 - 2x)(3x + 2)$. What are the values of x for which f is
    increasing and for which f is decreasing?   **[Lesson 5.3]**

13. Write the solutions in $a + bi$ form for $-3x^2 + x - 4 = 0$.   **[Lesson 5.7]**

14. In factored form over the integers, write a polynomial with real zeros
    at -1, 0, and 4 and imaginary zeros at $2i$ and $-2i$.   **[Lesson 6.3]** $f(x) = x(x + 1)(x - 4)(x^2 + 4)$

**Use the functions f and g to find the combinations $(f - g)(x)$ and
$(f \div g)(x)$.**   **[Lesson 2.6]**

15. $f(x) = -2x^2 + 4$, $g(x) = x + 3$ **16.** $f(x) = \frac{1}{x+3}$, $g(x) = \frac{2}{x^2-9}$

Free-Response Grid The following questions may be answered using a
free-response grid commonly used by standardized test services.

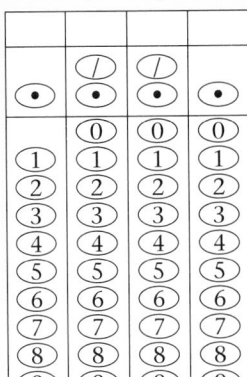

17. A package is dropped from the top of a
    building 500 feet high. Using the quadratic
    model for the height of a falling object
    affected by gravity, $h(x) = -16t^2 + v_0t + h_0$, find
    how long it will take the package to reach
    the ground.   **[Lesson 5.3]** ≈ 5.6 seconds

18. Solve $6x - 2x^2 = -3$ for x using the quadratic
    formula. Give your answers to the nearest
    tenth.   **[Lesson 5.5]** $-0.4, 3.4$

19. Let $f(x) = -4x^2 - 8x + 15$. What is the
    maximum value of $f(x)$?   **[Lesson 5.1]** 19

20. Let $f(x) = (x + 3)^3(x - 2)^2(x + 1)^2$. At which zero
    does the graph cross the x-axis?   **[Lesson 6.4]** -3

12. increasing: $x < 0.9$
    decreasing: $x > 0.9$

13. $\frac{1}{6} + \frac{1}{6}i\sqrt{47}$ and
    $\frac{1}{6} - \frac{1}{6}i\sqrt{47}$

14. $f(x) = x(x + 1)(x - 4)(x^2 + 4)$

15. $(f - g)(x) = -2x^2 - x + 1$;
    $(f \div g)(x) = \frac{-2x^2 + 4}{x + 3}$,
    $x \neq -3$

16. $(f - g)(x) = \frac{x-5}{x^2-9}$, $x \neq \pm 3$;
    $(f \div g)(x) = \frac{x-3}{2}$, $x \neq \pm 3$

CHAPTER 7

Exponential and Logarithmic Functions

Meeting Individual Needs

7.1 Exploring Population Growth

Core	Core Plus
Inclusion Strategies, p. 378	Practice Master 7.1
Reteaching the Lesson, p. 379	Enrichment, p. 378
Practice Master 7.1	Technology Master 7.1
Enrichment Master 7.1	Interdisciplinary
Technology Master 7.1	Connection, p. 377
Lesson Activity Master 7.1	
[2 days]	**[1 day]**

7.2 The Exponential Function

Core	Core Plus
Inclusion Strategies, p. 384	Practice Master 7.2
Reteaching the Lesson, p. 385	Enrichment, p. 384
Practice Master 7.2	Technology Master 7.2
Enrichment Master 7.2	
Technology Master 7.2	
Lesson Activity Master 7.2	
[2 days]	**[2 days]**

7.3 Logarithmic Functions

Core	Core Plus
Inclusion Strategies, p. 392	Practice Master 7.3
Reteaching the Lesson, p. 393	Enrichment, p. 391
Practice Master 7.3	Technology Master 7.3
Enrichment Master 7.3	
Technology Master 7.3	
Lesson Activity Master 7.3	
[2 days]	**[1 day]**

7.4 Exploring Properties of Logarithmic Functions

Core	Core Plus
Inclusion Strategies, p. 398	Practice Master 7.4
Reteaching the Lesson, p. 399	Enrichment, p. 398
Practice Master 7.4	Technology Master 7.4
Enrichment Master 7.4	Interdisciplinary
Technology Master 7.4	Connection, p. 397
Lesson Activity Master 7.4	Mid-Chapter Assessment
Mid-Chapter Assessment Master	Master
[2 days]	**[2 days]**

7.5 Common Logarithms

Core	Core Plus
Inclusion Strategies, p. 403	Practice Master 7.5
Reteaching the Lesson, p. 404	Enrichment, p. 403
Practice Master 7.5	Technology Master 7.5
Enrichment Master 7.5	
Technology Master 7.5	
Lesson Activity Master 7.5	
[2 days]	**[1 day]**

7.6 The Natural Base, e

Core	Core Plus
Inclusion Strategies, p. 409	Practice Master 7.6
Reteaching the Lesson, p. 410	Enrichment, p. 408
Practice Master 7.6	Technology Master 7.6
Enrichment Master 7.6	Interdisciplinary
Technology Master 7.6	Connection, p. 408
Lesson Activity Master 7.6	
[2 days]	**[2 days]**

7.7 Solving Exponential and Logarithmic Equations

Core

Inclusion Strategies, p. 417
Reteaching the Lesson, p. 418
Practice Master 7.7
Enrichment Master 7.7
Technology Master 7.7
Lesson Activity Master 7.7

[2 days]

Core Plus

Practice Master 7.7
Enrichment, p. 417
Technology Master 7.7
Lesson Activity Master 7.7

[2 days]

Chapter Summary

Core

Eyewitness Math, pp. 414–415
Chapter 7 Project,
    pp. 422–423
Lab Activity
Long-Term Project
Chapter Review, pp. 424–426
Chapter Assessment, p. 427
Chapter Assessment, A/B
Alternative Assessment

[4 days]

Core Plus

Eyewitness Math,
    pp. 414–415
Chapter 7 Project,
    pp. 422–423
Lab Activity
Long-Term Project
Chapter Review,
    pp. 424–426
Chapter Assessment,
    p. 427
Chapter Assessment, A/B
Alternative Assessment

[3 days]

Reading Strategies

Students should carefully read the lesson once before the class studies that material and read it again after the class, to prepare for working the exercises. If students read the lessons carefully before studying them in class, they will be able to ask more pertinent questions and contribute more to the discussions. Emphasize to students that they may need to read a paragraph or sentence several times before it is fully understood.

Visual Strategies

Encourage students to look at the graphs in the chapter and discuss how they differ from graphs in earlier chapters. Encourage students to explore as many characteristics of exponential and logarithmic functions as they can by experimenting with graphics calculators.

Hands-On Strategies

Many phenomena in the physical world exhibit characteristics that can be modeled by exponential or logarithmic functions. The bouncing ball experiment in the Chapter Project provides an example of data that can be modeled by the concepts in this chapter.

Cooperative Learning

GROUP ACTIVITIES	
Modeling population growth	Lesson 7.1, Explorations 1–2
Properties of logarithms	Lesson 7.4, Exploration 3; Exercises 43–45
Exponential functions	Eyewitness Math: Meet *e* in St. Louis
Exponential growth or decay	Chapter Project

You may wish to have students work with partners for some of the above activities. Additional suggestions for cooperative group activities are noted in the teacher's notes in each lesson.

Multicultural

The cultural connections in this chapter include references to Asia and Europe.

CULTURAL CONNECTIONS	
Asia: Babylonian Compound Interest	Portfolio Activity
Asia: Population Growth	Lesson 7.1, Exercises 34–38
Europe: Logarithmic Functions, Napier's rods	Lesson 7.5, Introduction

Portfolio Assessment

Below are portfolio activities for the chapter. They are listed under seven activity domains that are appropriate for portfolio development.

1. **Investigation/Exploration** Modeling population growth is the objective of Explorations 1–3 in Lesson 7.1. Powers of 2 is the focus of the exploration in Lesson 7.3. The properties of logarithms are studied in Explorations 1–3 in Lesson 7.4. Approaching *e* is the focus of the exploration in Lesson 7.6. Exponential functions are also investigated in the Eyewitness Math.

2. **Applications** Recommended are any of the following: demographics, Lesson 7.1, Exercises 42–44; accounting, Lesson 7.2, Exercises 36–38; molecular chemistry, Lesson 7.3, Exercises 30–33; optics, Lesson 7.4, Exercises 43–45; acoustics, Lesson 7.5, Exercises 20–21; radioactive decay, Lesson 7.6, Exercise 47; biology, Lesson 7.7, Exercise 34.

3. **Non-Routine Problems** The Meet *e* in St. Louis problem from Eyewitness Math presents the students with non-traditional but interesting examples of exponential functions.

4. **Project Exponential Bouncing Ball Experiment** See pages 422 and 423. Students are asked to perform an experiment to demonstrate that the height of successive bounces decreases exponentially as the number of bounces increases.

5. **Interdisciplinary Topics** Students may choose from the following: space science, Lesson 7.1, Exercises 21–25; transformations, Lesson 7.2, Exercises 17–22.

6. **Writing** Communicate exercises of the type where a student is asked to describe basic concepts of exponential and logarithmic functions, their characteristics and their graphs offer excellent writing selections for the portfolio.

7. **Tools** Chapter 7 uses graphics calculators to assimilate data and model that data with various logarithmic and exponential functions. To measure the individual student's proficiency, it is recommended that the student complete selected worksheets from the Technology Masters.

Technology

Technology is a powerful tool that makes the study of exponential and logarithmic functions interesting and enjoyable. Graphs can be quickly made and compared to determine the most effective way to communicate results. The calculator enables students to generate exponential growth patterns by repeating the application of the multiplier.

Graphics Calculator

The graphing calculator permits a way to see the overall growth or decay patterns. If the technology is used to model real-world exponential growth or decay functions, students can quickly and easily see the shape of the graph for these functions, work with the restrictions of domain and range, and associate the real-world situation with the exponential function and its graph. If data is used, fitting the regression line and analyzing the resulting equation helps students acquaint themselves with the function. Students benefit from being able to visualize functions. By tracing along the graph or by using a table feature, students can quickly determine approximate values of the function for various values from the domain.

The graphics calculator enables the student to gain a holistic grasp of a whole family of functions. When graphing the family of exponential functions, $f(x) = ab^x$, students can alter values of a and b and examine the effects that the changes have on the basic function. Many Explorations in the text lead the student to discover generalities about functions by experimenting with the calculator.

The graphing calculator can also be used to solve exponential and logarithmic equations. The left side of the equation is entered as Y1 = *left side* and the right side of the equation is entered as Y2 = *right side*. This plots both expressions simultaneously on the same coordinate plane. Points of intersection of the two graphs are the solutions to the equation.

$f(x) = e^x$

$f(x) = \ln x$

The values can be found using the calculator solve feature or by using trace and zoom. Refer to your *HRW Technology Handbook* for specific instructions on finding intersection points on various calculators.

Integrated Software

f(g) Scholar™ is an integrated computer-based mathematics productivity tool that combines calculator, spreadsheet, and graphics capabilities to provide a dynamic and interactive environment for explorations in mathematics. Use *f(g) Scholar*™ for any lesson needing a spreadsheet, calculator, graphics calculator, or any combination of the three.

Exponential and Logarithmic Functions

CHAPTER 7

Exponential and Logarithmic Functions

The functions you have studied so far have been polynomial functions. They are defined as sums of powers of *x*. *Exponential* and *logarithmic* functions, which you will study in this chapter, are called *transcendental* functions because they transcend, or go beyond, the standard polynomial functions. These exponential and logarithmic functions model many real-world events, such as population growth, radioactive decay, atmospheric pressure, human memory, loudness of sounds, and intensity of earthquakes.

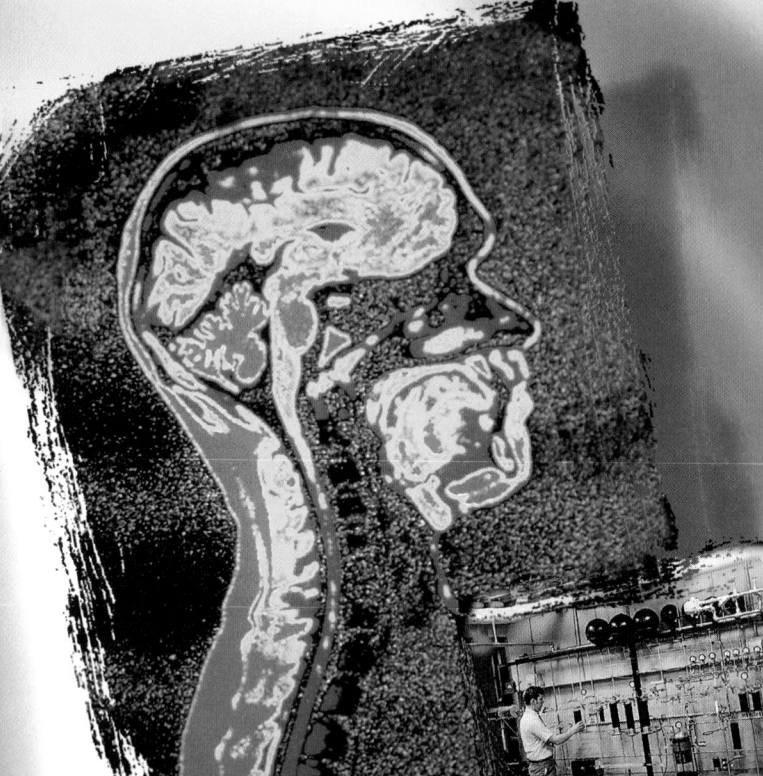

ABOUT THE CHAPTER

Background Information

Exponential functions are significantly different from polynomial functions. Exponential growth, for example, increases much faster than polynomial growth. These exponential growth functions have graphs that are steeply inclined.

RESOURCES

- Practice Masters
- Enrichment Masters
- Technology Masters
- Lesson Activity Masters
- Lab Activity Masters
- Long-Term Project Masters
- Assessments Masters
  Chapter Assessments, A/B
  Mid-Chapter Assessment
  Alternative Assessment, A/B
- Teaching Transparencies
- Spanish Resources

CHAPTER OBJECTIVES

- Model the growth or decline of a population with a graphics calculator by repeatedly applying the multiplier.
- Identify the behavior of exponential functions by inspection and by graphing.
- Determine the growth of funds under various compounding methods.
- Identify the exponential and logarithmic functions as inverse functions.
- Determine equivalent forms for exponential and logarithmic equations.
- Identify the product, quotient, and power properties of logarithms.

ABOUT THE PHOTOS

Charles F. Richter, 1900–1985, an American physicist, set up the scale for measuring the intensity of earthquakes in 1935 based on the logarithmic scale. The photos on page 375 show the results of two earthquakes with very high Richter numbers. The intensities of light and sound are also measured using a logarithmic scale. One photo on this page shows an image of the human neurological system, the system which allows human beings to perceive light and sound.

In this chapter, you will investigate the nature and behavior of exponential and logarithmic functions. You will use your graphics calculator to construct exponential functions that model real-world phenomena.

- Simplify expressions and solve equations involving logarithms.
- Identify and use the common logarithmic function.
- Write equivalent logarithmic and exponential equations.
- Evaluate expressions involving the natural number, e, and identify the relationship between the natural logarithmic and exponential functions.
- Model growth and decay processes with natural exponential functions.
- Solve logarithmic and exponential equations by graphing.
- Use the exponential-log inverse properties to solve logarithmic and exponential equations.

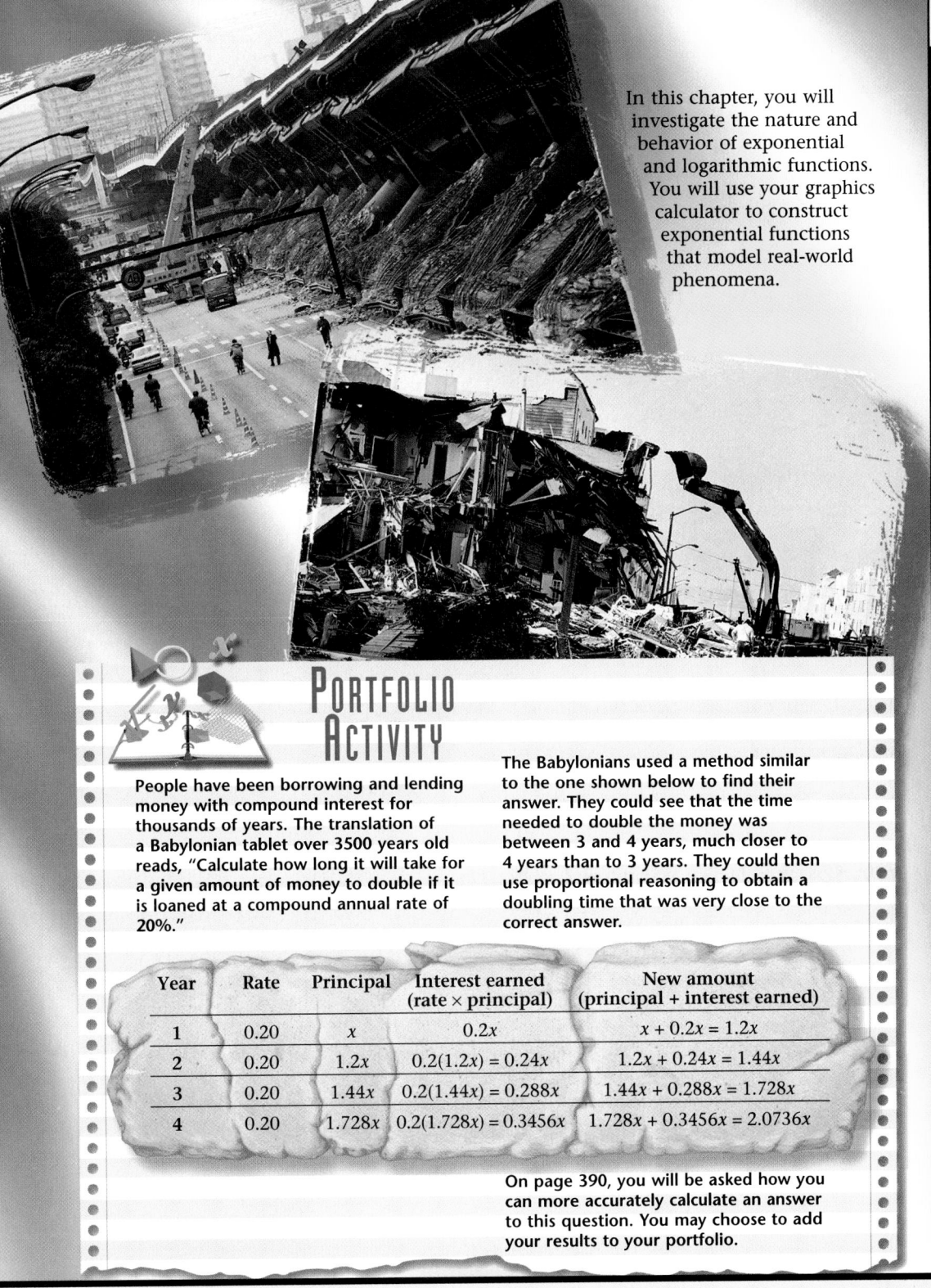

PORTFOLIO ACTIVITY

People have been borrowing and lending money with compound interest for thousands of years. The translation of a Babylonian tablet over 3500 years old reads, "Calculate how long it will take for a given amount of money to double if it is loaned at a compound annual rate of 20%."

The Babylonians used a method similar to the one shown below to find their answer. They could see that the time needed to double the money was between 3 and 4 years, much closer to 4 years than to 3 years. They could then use proportional reasoning to obtain a doubling time that was very close to the correct answer.

Year	Rate	Principal	Interest earned (rate × principal)	New amount (principal + interest earned)
1	0.20	x	$0.2x$	$x + 0.2x = 1.2x$
2	0.20	$1.2x$	$0.2(1.2x) = 0.24x$	$1.2x + 0.24x = 1.44x$
3	0.20	$1.44x$	$0.2(1.44x) = 0.288x$	$1.44x + 0.288x = 1.728x$
4	0.20	$1.728x$	$0.2(1.728x) = 0.3456x$	$1.728x + 0.3456x = 2.0736x$

On page 390, you will be asked how you can more accurately calculate an answer to this question. You may choose to add your results to your portfolio.

PORTFOLIO ACTIVITY

In this activity, students will explore the Babylonian method of proportional reasoning to solve the problem of compounding interest on a loan. Students will see that this method has its shortcomings and will then explore methods on their graphics calculators that are more accurate.

Organize students into groups of five. Each group should use a different value for the principle; for example, $x = 1000$, 2000, or 3000. Within each group, students should compute t in the formula on page 386 for a different value of n: $n = 1$, 2, 3, and 4. They will see that doubling times for many values of x are the same when n is held constant. However, different values of n give significantly different answers.

ABOUT THE CHAPTER PROJECT

In the Chapter Project on pages 422 and 423, students will perform a variation of the bouncing ball experiment from Chapter 2 by dropping a ball from three different heights. They will observe the height of the first, second and third bounce for each dropping height. Students will try to discover a mathematical pattern in their data.

PREPARE

Objectives

• Model the growth or decline of a population with a graphics calculator by repeatedly applying the multiplier.

RESOURCES

• Practice Master	7.1
• Enrichment Master	7.1
• Technology Master	7.1
• Lesson Activity Master	7.1
• Quiz	7.1
• Spanish Resources	7.1

Assessing Prior Knowledge

1. Simplify the expression, $3x(2 - y)$. [$6x - 3xy$]

2. Use the distributive property to factor out the greatest common factor of $8xy + 4x^2 - 16xy^2$. [$4x(2y + x - 4y^2)$]

TEACH

Discuss, in non-mathematical terms, how certain natural phenomena increase exponentially. Have students consider situations in the real world that involve exponential growth.

Exploring

Population Growth

Why *Many natural processes involve growth or decay of a specific quantity. Often the rate of increase or decrease is proportional to the amount present. What type of function models this kind of increase or decrease over time? The increasing world population is modeled by this kind of growth function, which has properties that are different from functions you have studied so far.*

Modeling Population Growth

Nicki is keeping track of bacteria growth in a culture in her school's laboratory. She places a prepared slide containing a culture with approximately 100 bacteria under a microscope.

After 1 hour, Nicki observes that the size of the culture has doubled, so that there are approximately 200 bacteria in the culture.

After 2 hours, there are 400 bacteria.

After 3 hours, there are 800 bacteria, and so on.

You can find a mathematical model to describe this bacteria growth.

ALTERNATIVE teaching strategy

Hands-On Strategies

Have the students take a piece of notebook paper and fold it in half. Students should record the number of regions and fold the paper in half again. Have them record the number of regions. Have students continue this process and keep a running total at each step. Have students determine the multiplier that is used at each step.

Exploration 1 Bacteria Growth

Graphics Calculator

1 To model the bacteria's growth using a calculator, first enter 100.

This is the value of the bacteria population at the start of the experiment, or at 0 hours.

```
100
          100
```

2 The number of bacteria doubles every hour. So multiply the value entered by 2, by pressing [×] [2] [ENTER].

This gives the number of bacteria after 1 hour.

```
100
          100
Ans*2
          200
```

3 Press [ENTER] to repeat the doubling operation.

The new display is the bacteria population after 2 hours.

```
100
          100
Ans*2
          200
          400
```

4 Continue pressing [ENTER] to find the bacteria population at each successive hour for 20 hours. Record each result in a table.

Hour, x	0	1	2	3	4	5	⋯
Population, y	100	200	400				⋯

5 What will the bacteria population be after 10 hours? after 20 hours?

6 When will the bacteria population exceed 1 million? 10 million? 100 million?

7 What number is multiplied by each population value in each calculator step to obtain the next hour's population? ❖

Your answer in Step 7 of Exploration 1 is the **multiplier** for the bacteria population growth.

interdisciplinary CONNECTION

Literature Find and read the tale of Hercules and the Hydra. In this tale, each time Hercules cuts off one of the Hydra's heads, two heads grow in its place. Write out a table of values for the number of heads the Hydra grows after each cut. Assume that with each cut, Hercules cuts off all of the Hydra's heads.

TEACHING tip

Review with students how to enter data repeatedly on the graphics calculator. The instructions given in the text are intended to be general and not calculator specific. Some calculators may use a different sequence of steps to perform this process.

Exploration 1 Notes

Have the students write a brief description, in outline form, of the steps they went through to find the results for Step 5. Have students do the same for Step 6. Have students describe the similarities and differences in the methods used.

Aongoing SSESSMENT

7. 2

Between the years 1990 and 2017, and between the years 1991 and 2018, the population approximately doubled. In fact, the population doubles every 27 years. This provides a very quick method for computing the population in 2017 + 27, or 2044, 2071, and so on.

Aongoing SSESSMENT

6. 1.026

Demographics   In the mid-1980s, a large Southwestern city began a growth spurt due to a significant relocation of computer-related industries into its metropolitan area.

In 1990, the population of the city was 465,622 and growing at a rate of about 2.6% per year. There is a mathematical model to demonstrate the population growth of this city.

•Exploration 2 Urban Population Growth

Graphics Calculator

1 To model the city growth enter the 1990 population, 465,622, in your calculator.

```
465622
              465622
```

2 A constant growth rate of 2.6% means that the population after 1 year, in 1991, will increase by 0.026 times the population in 1990.

Let x represent the 1990 population. The 1991 population will be

$$x + 0.026x = x(1 + 0.026) = 1.026x.$$

Multiply the 1990 population by 1.026.

This gives the projected 1991 population.

```
465622
              465622
Ans*1.026
         477728.172
```

3 Continue by pressing ENTER to find the projected population for each successive year until the year 2000. Record each result in a table.

Round your population values to the nearest whole number, since it is not realistic to count a fraction of a person.

Year, x	1990	1991	1992	1993	1994	1995
Population, y	465,622	477,728				

4 What is the projected population for the year 2000?

5 When does this model predict that the population will exceed 1 million people?

6  What number is multiplied by each year's projected population in each calculator step to obtain the next year's projected population? ❖

Your answer in Step 6 of Exploration 2 is the multiplier for the urban population growth.

ENRICHMENT   In reality, populations do not increase as regularly as many examples suggest. Natural and human-made disasters can cause sudden drops or leveling-off periods. For example, epidemics, earthquakes, overuse of natural resources, and wars can all cause irregularities in the growth of populations. Have students write a short presentation on one cause of irregularity in population growth (or decline).

INCLUSION **strategies**   **Using Visual Strategies** Students may benefit by graphing on paper the data in Explorations 1 and 2. In this way, students will see how sudden the increase is in Exploration 1 and how slow the increase is in Exploration 2.

In Exploration 2, what happens if you use 0.026 as the multiplier instead of 1.026? Why?

Assume that in 1990 the population of a city was x and decreasing at a rate of about 2.6% per year. The 1991 population is given by the following equation.

$$x - 0.026x = x(1 - 0.026) = 0.974x$$

When a population *decreases*, the growth rate is negative. Thus, the growth rate is *subtracted* from the original population, and the multiplier is less than 1.

Exploration 3 *Regression Models*

Graphics Calculator

1 Use your graphics calculator to graph the ordered pairs in the bacteria growth table from Exploration 1.

2 Using the linear regression feature, find the linear function that best fits these data points. What is the correlation coefficient, r?

3 Using the power regression feature, find the power function that best fits these data points. What is the correlation coefficient, r?

4 Using the exponential regression feature, find the exponential function that best fits these data points. What is the correlation coefficient, r?

5 Compare the three types of *best fit* models. Which type of regression model results in the correlation coefficient closest to 1? What type of function best models the bacteria population growth?

6 Repeat Steps 2–5 for the data in the urban population table from Exploration 2. What type of function best models this population growth? ❖

Using Cognitive Strategies Summarize Explorations 1 and 2 and define a *beginning value* and a *multiplier*. Then take any beginning value and any multiplier, repeatedly apply the multiplier, and the result is an exponential function. Have the students do this with three different pairs of numbers. Have the students write a short description of the general situation that occurs and the process that causes this situation.

CRITICAL
Thinking

Using 0.026 as the multiplier yields a decreasing function. In fact, any multiplier less than 1 will yield a decreasing function, and any multiplier greater than 1 will yield an increasing function.

TEACHING *tip*

Refer to the Statistics section of your *HRW Technology Handbook*. Notice that different calculators come equipped to find different types of regression models.

Exploration 3 Notes

Review with students the use of the statistical features on their graphics calculator. Have each student make a list of the different types of regression models their calculator is able to compute.

Cooperative Learning

Have the students go to the library in teams of two: one student will find population data for the United States over a period of 8 decades. Each group should use a different 80-year period. The other student will enter the data in the graphics calculator, and determine the exponential function that models this data. Have each group find the multiplier used when the data set uses population values taken every ten years. Have students compare their multipliers.

Ongoing ASSESSMENT

5. exponential

6. exponential

ASSESS

Selected Answers

Odd-numbered Exercises 7–53

Assignment Guide

Core 1–16, 21–55

Core-Plus 1–5, 13–55

Technology

A graphics calculator is needed for Exercises 22–25, 27–29, 31–33, 35–44, and 51–55.

Error Analysis

Remind students that to change percentages to decimals, they need to drop the percent symbol and move the decimal point two places to the left.

Practice Master

NAME ______________ CLASS ________ DATE ________

Practice & Apply
7.1 Exploring Population Growth

According to the Environmental Protection Agency (EPA), the air quality standard for the pollutant ozone is 0.12 parts per million. The Philadelphia metropolitan area released an average of 8.40 parts per million of ozone in 1990. The area is required to decrease emissions at a projected constant rate of 35% per year.

1. Complete the table by finding the projected ozone level for each successive year until the year 2000. Round the ozone level to the nearest hundredth.

Year x	1990	1991	1992	1993	1994	1995
Ozone Level y						
Year x	1996	1997	1998	1999	2000	
Ozone Level y						

2. According to the table, what is the projected ozone level for 1998? ________
3. When will the ozone level meet the air quality standard of 0.12 parts per million? ________
4. In each calculator step, what number is multiplied by each year's projected ozone level to obtain the next year's projected ozone level? ________

The number of cable television subscribers in the United States on January 1, 1970 was 2500 and was growing at a rate of 7% per year.

5. What is the growth rate? ________
6. What is the multiplier for this growth rate? ________
7. Find, to the nearest whole number, the projected population of cable television subscribers after 20 years. ________
8. Find the function describing the growth of the number of cable television subscribers.

____

The retail trade sales totaled $957 billion in the United States in 1980 and was growing at a rate of 7.5% per year.

9. Find the function that models the growth of retail trade sales. ________
10. To the nearest billion, what is the projected total sales for the year 2000? ________

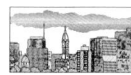

EXERCISES & PROBLEMS

Communicate

1. Describe the method used in Explorations 1 and 2 for modeling population growth.
2. Describe how the regression feature may be used to find a model for population growth.
3. Explain how the multiplier is determined in population growth.
4. Compare the graphs of the two exponential regression models from Exploration 3. How do the respective multipliers affect the graphs?
5. What is happening to a population if the multiplier is less than 1?

Practice & Apply

Identify the multiplier for each growth rate.

6. 4% 1.04
7. –3.7% 0.963
8. 0.9% 1.009
9. –31% 0.69
10. 140% 2.4
11. 3% 1.03
12. –2.01% 0.9799
13. 1.6% 1.016
14. –0.001% 0.99999
15. 2.17% 1.0217
16. 1.73% 1.0173
17. 0.105% 1.00105
18. 12.01% 1.1201
19. –10.8% 0.892
20. 27% 1.27

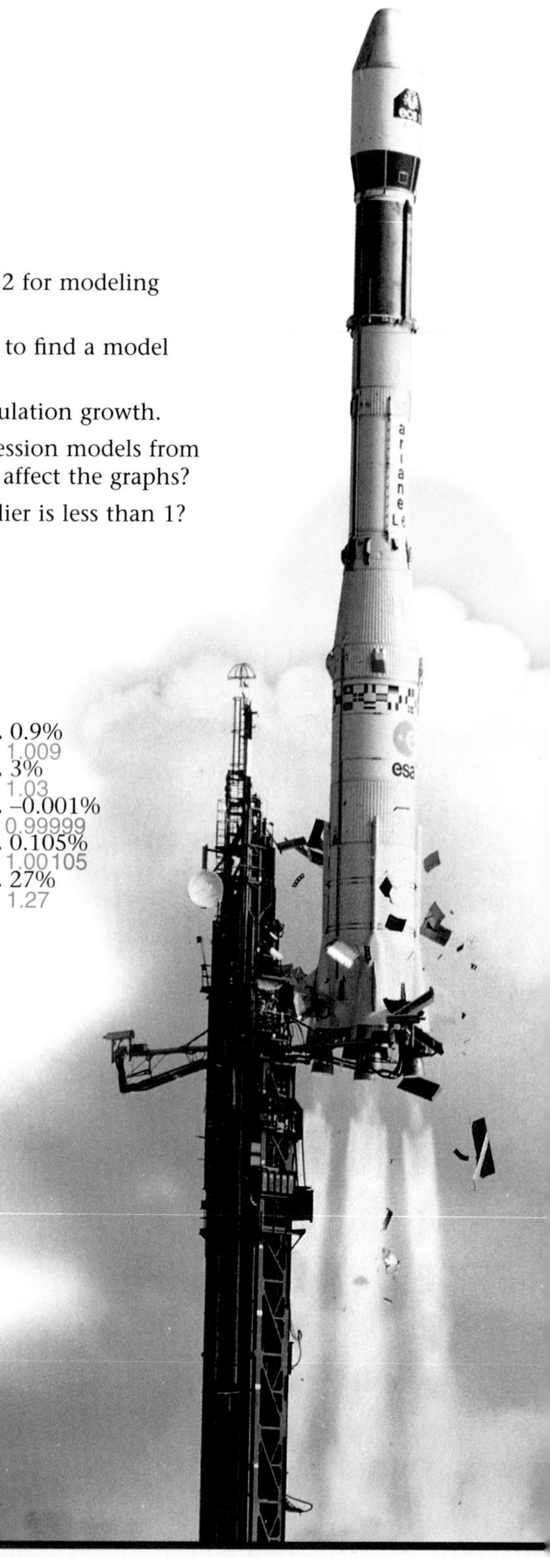

Space Science The first stage of the Saturn 5 rocket that propelled astronauts to the Moon burned about 8% of its available fuel every 15 seconds, and carried about 600,000 gallons at liftoff.

21. What is the multiplier for the amount of fuel left in the first stage after 15 seconds? 0.92

22 How much first-stage fuel remains after 2 minutes? ≈ 307,931 gal

23 How much first-stage fuel remains after $2\frac{3}{4}$ minutes? ≈ 239,782 gal

24 About how long after liftoff was half of the first-stage fuel used? 2 to 2.25 min

25 When was $\frac{3}{4}$ of the first-stage fuel used? 4 to 4.25 min

Investment $1000 is invested at 6% interest compounded once per year.

26. What is the growth rate for the $1000 invested? What is the multiplier for this growth rate? 0.06; 1.06

27 What will the value of the original investment be after 10 years? $1790.85

28 How long will it take for the original investment to double? ≈ 12 yr

29 Find the function that models the growth of the original investment. $y = 1000(1.06)^x$

Demographics The population of Sacramento, California, was 369,365 in 1990 and was growing at a rate of about 2.5% per year.

30. What was the multiplier for Sacramento's population growth? 1.025

31 What was the projected population of Sacramento for the year 2000? 472,818

32 Find the function that models Sacramento's projected population growth. $y = 369,365(1.025)^x$

33 During what year will Sacramento's projected population double? 2018

Demographics The 1990 population of the United States was 248,709,873 and was growing at a rate of 0.6%. The 1990 population of Indonesia was 191,256,000 and was growing at a rate of 1.9%.

34. What is the growth multiplier for each country? U.S.A.:1.006 Indonesia: 1.019

35 Find the projected population of the United States for the year 2000. 264,041,890

36 Find the projected population of Indonesia for the year 2000. 230,864,368

37 Find and graph the functions for the populations of both the United States and Indonesia on the same coordinate plane. Compare the two graphs.

38 During what year will the projected population of Indonesia surpass that of the United States? 2010

The United States has a land area of 3,536,278 square miles. Indonesia has a land area of 741,052 square miles.

Medan

Sibolga

SUMATERA

Bukittinggi

Padang

Palembang

INDIAN OCEAN

Jakarta

Purwakarta

Bandung JAWA

BALI

JAVA SEA

Surabaya

Waingapu

Kupang

Balikpapan

Banjarmasin

SULAWESI

Buru

SERAM

Padang

TIMOR

HALMAHERA

PACIFIC OCEAN

MALUKU

ARAFURA SEA

INDONESIA

37. The graph of Indonesia's population starts far below the graph of the United States population, but because it increases much more rapidly, it quickly overtakes the United State's graph. The function for the United State's population growth is $y = 248,709,873(1.006)^x$, and the function for Indonesia's population growth is $y = 191,256,000(1.019)^x$.

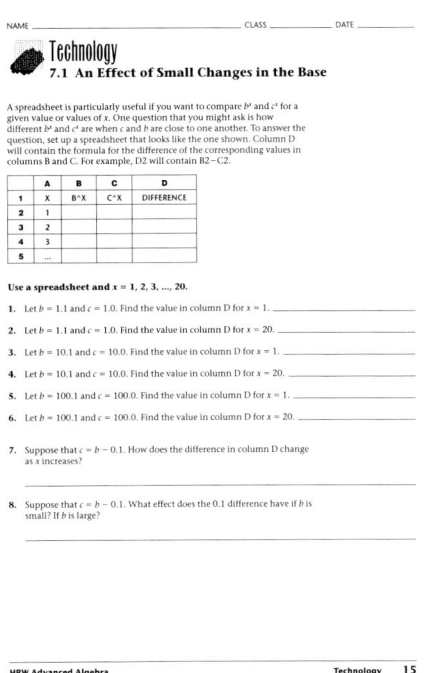

Cultural Connection: Asia In 1972, U.N. figures showed the population of the Republic of Korea (R.O.K.) as 32,530,000 with a growth rate of 2.2%.

39 At that rate, what would be the expected population of the R.O.K. in 1994? 52,505,376

40 The *1995 World Almanac* lists the 1994 R.O.K. population as 45,083,000. What is the difference between the actual population and the predicted population based on the 1972 rate of increase? 7,422,376

41 Determine the average growth rate for the R.O.K. from 1972 to 1994 using the actual populations from these two years. How does this growth rate compare with the 1972 growth rate? ≈ 1.5%; much slower

Demographics Using 1980 and 1990 census information, find the 1980 and 1990 populations of the community you live in.

42 Determine the growth rate for your community using the 1980 and 1990 populations. Answers may vary.

43 Find the function that models your community's projected population growth. Answers may vary.

44 Find the projected population for your community for the year 2000 and for the year 2010. Explain whether or not you believe this growth rate is accurate for your community. Answers may vary.

Look Back

Let $f(x) = 3x^2 - 2x + 1$. Find each value, and simplify.   [Lesson 2.4]

45. $f(2.4)$ 13.48   **46.** $f(-a)$ $3a^2 + 2a + 1$ **47.** $f(b + 3)$ $3b^2 + 16b + 22$

48. $f(x - 1)$ $3x^2 - 8x + 6$ **49.** $f(x - h)$ **50.** $\frac{f(x-h)-f(x)}{h}$ $3h - 6x + 2$

$3x^2 - 6hx + 3h^2 - 2x + 2h + 1$

Solve each system using matrix algebra. [Lesson 4.6]

51 $\begin{cases} y = 3x + 1 \\ y = x - 1 \end{cases}$ **52** $\begin{cases} y = \frac{1}{2}x + 3 \\ y = 3x + 7 \end{cases}$ **53** $\begin{cases} y = x - 2 \\ y = -2x + 1 \end{cases}$ **54** $\begin{cases} y = \frac{x+4}{3} \\ y = -x + 4 \end{cases}$

(−1, −2)   (−1.6, 2.2)   (1, −1)   (2, 2)

Look Beyond

55 **Investment**   Suppose your savings account has a balance of $177.25, and you started the account 9 years ago with $100. If you never withdrew or deposited any money during this time, what is the annual interest rate at the bank? ≈ 6.57%

Chunchŏn
Ŭijonbu   Kangnŭ
Sŏul
Inch'ŏn   Yŏngdŭngpo
Sǔwon

Chungju
HUANG   Chonan
HAI   Taejŏn

Taejon
Yŏngdong   Pohan
Kimchŏn   Kyo
Kunsan   Chŏnju   Ulsa

Chinju   Masan
Kwangju   Pu
Samchonpo
Chungmu
Mokpo

REPUBLIC
OF KOREA

Look Beyond

This exercise encourages students to look at the inverse of the exponential growth process. In discussing this exercise, review the concept of inverse functions.

LESSON 7.2 The Exponential Function

In Lesson 7.1, bacteria growth and urban population growth were each modeled by an exponential function. Properties of the exponential function provide the connection between exponential functions and the growth and decay processes.

Objectives

- Identify the behavior of exponential functions by inspection and by graphing.
- Determine the growth of funds under various compounding methods.

RESOURCES

- Practice Master   7.2
- Enrichment Master   7.2
- Technology Master   7.2
- Lesson Activity Master   7.2
- Quiz   7.2
- Spanish Resources   7.2

Compound interest is a percentage of the balance that is added periodically to the preceding balance, resulting in a new, larger balance. Each time interest is compounded, the amount of interest added is based on the latest balance. Compound interest payments are often referred to as *interest paid on interest*.

EXAMPLE 1

Investment

Suppose the parents in a family want to have $10,000 available for each of their children by the time each child reaches his or her 18th birthday. At each child's 3rd birthday, the parents deposit $3000 in a fund for college education. The fund earns 8.5% interest compounded annually. Will $10,000 be available for college expenses on the child's 18th birthday?

Spreadsheet

Solution▶

Make a table or use a spreadsheet to show the growth of the college fund. The interest rate is assumed to be fixed, so each year the multiplier is 1.085.

	C6		=3000*(1.085)^A6	
				Investment
	A	**B**	**C**	**D**
1	Years	Child's Age	Balance ($)	
2	x		y	
3	0	3	$3000.00	
4	1	4	$3255.00	
5	2	5	$3531.68	
6	3	6	$3831.87	
7	4	7	$4157.58	
8				
9	n	n + 3	3000(1.085)^n	

Assessing Prior Knowledge

1. Express $a^{0.3}$ using fractional exponents. $\left[a^{\frac{3}{10}} \right]$

2. Express $a^{\frac{3}{10}}$ using integer exponents. $\left[\sqrt[10]{a^3} \right]$

TEACH

Polynomial functions describe the data from situations like free fall and volume. Exponential functions describe the data from situations like population growth and investment.

Alternate Example 1

Suppose you want to have 1 million dollars by the time you retire in 50 years. Suppose also that interest rates will be stable at 6% during this entire time. Will a $60,000 investment now gross 1 million dollars by your retirement? [**yes, you will have approximately 1.10520925 million dollars, or $1,105,209.25, in 50 years**]

ALTERNATIVE **teaching strategy**

Using Symbols Some students have difficulty with the formalism of the notation $f(x) = ab^x$. Show them how the explorations of Lesson 7.1 can be written using this notation. The bacterial growth model can be written as $f(x) = 100 \cdot 2^x$, and the urban population growth can be written as $f(x) = 465,622 \cdot (1.026)^x$.

A ongoing
SSESSMENT

$f(x) = 3000 \cdot (1.085)^x$

CRITICAL *Thinking*

Since $0 \cdot c = 0$ for any real number c, $f(x) = 0 \cdot b^x = 0$ is a constant function, $f(x) = 0$. A constant function is not an exponential function.

Since $1^x = 1$ for all real values of x, $f(x) = a \cdot 1^x = a$ is a constant function, $f(x) = a$, for all real values of x. A constant function is not an exponential function.

If $b < 0$, then the function is undefined for many values of x. For example, when x is an even integer or a rational number $\frac{p}{q}$, where q is even, b^x is undefined.

TEACHING*tip*

Technology Have students use their graphics calculators to graph a variety of exponential functions (for example, 2^x, 3^x, 5^x, 10^x). This should help give students an intuitive feel for growth functions. Then have students graph $\left(\frac{1}{2}\right)^x$, $\left(\frac{1}{3}\right)^x$, $\left(\frac{1}{5}\right)^x$, and $\left(\frac{1}{10}\right)^x$. This should help them visualize decay functions.

Use Transparency 25

When a child is 18 years old, the number of times the interest has been compounded is $n = 15$. Thus, the accumulated balance is

$$3000(1.085)^{15} = 10{,}199.23.$$

Thus, each child will have over $10,000 on his or her 18th birthday. ❖

In Example 1, the expression for the yearly balance is $3000(1.085)^n$. Notice that a constant *base*, 1.085, is raised to a variable power, n, and that this entire value is multiplied by a constant, 3000. A function that consists of a constant base, b, raised to a variable power, x, which is then multiplied by a constant, a, is called an *exponential function*.

$$f(x) = ab^x$$

CONSTANT   EXPONENT   BASE

EXPONENTIAL FUNCTION

The function $f(x) = ab^x$, where a and b are real numbers such that $a \neq 0$, $b > 0$, and $b \neq 1$, and x is any real number, is an **exponential function**. The domain of f is the set of real numbers. The number b is called the **base** of the exponential function.

What is the exponential function that models the investment plan in Example 1?

CRITICAL *Thinking*

What happens to the values of an exponential function when $a = 0$? What happens when $b = 1$? Recall that $b^{\frac{1}{n}} = \sqrt[n]{b}$. What happens when $b < 0$?

Exponential Growth or Decay

In most applications of exponential functions, the constant a in $f(x) = ab^x$ is positive.

When $a > 0$ and $b > 1$, the function is that of **exponential growth.**

When $a > 0$ and $0 < b < 1$, the function is that of **exponential decay.**

When a is positive, the range of $f(x) = ab^x$ is the set of positive real numbers.

Examine the exponential function $f(x) = 3(2)^x$, which has a base of 2. As x increases, f increases.

Thus, f is an exponential growth function.

The domain is the set of real numbers, and the range is the set of all positive real numbers.

When x is 0, $f(0) = 3$. In general, for $f(x) = ab^x$, when $x = 0$, $f(0) = a$.

Graph showing: $f(x) = 3(2)^x$ with points $(1, 6)$, $(0, 3)$, $(-1, 1.5)$

ENRICHMENT Organize students into groups of three. Encourage them to imagine how much money they will be earning per year after they finish college. Designate 10% as savings for retirement. They should then use their calculators to determine how much they will have at retirement.

INCLUSION *strategies* **Using Cognitive Strategies** To connect this section with the previous section, take Explorations 1 and 2 from Section 7.1 and show how these explorations are modeled by functions of the form $f(x) = ab^x$. For each example in this lesson, write the result in similar functional notation. Make it clear to students how a and b are derived from the examples.

Graphics Calculator

EXAMPLE 2

Let $f(x) = 2\left(\frac{1}{2}\right)^x$.

A Explain why f is an exponential function.

B Find the y-intercept of f.

C Is f a function of exponential growth or exponential decay?

D Graph f. Describe how f behaves as values of x become larger.

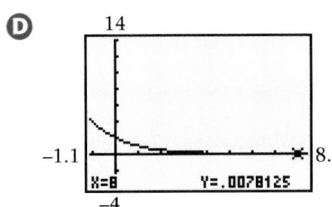

Solution▸

A By definition of exponential function, f is an exponential function because the variable, x, is an exponent; the base, $\frac{1}{2}$, is positive and is not equal to 1; and the constant, 2, is not equal to 0.

B Since $f(0) = 2$, the y-intercept of f is 2.

C Since the base of f, $\frac{1}{2}$, is between 0 and 1, f is an exponential decay function.

D

As the x-values become larger, the values of f get closer and closer to 0. No matter how large x becomes, the value of f never reaches 0. ❖

Try This Graph $g(x) = 5(3)^x$. Is g a function of exponential growth or exponential decay? Describe how g behaves as values of x become larger.

In Example 2, the graph of the exponential decay function $f(x) = 2\left(\frac{1}{2}\right)^x$ gets closer and closer to the x-axis without reaching it. In this graph, the x-axis is called an **asymptote** of f. Exponential growth and decay functions generally have the x-axis as an asymptote.

Recall that $\left(\frac{1}{2}\right)^x$ may be written as $\frac{1}{2^x}$ or 2^{-x}. In a similar way, any exponential decay function may be written as $f(x) = ab^{-x}$, where $b > 1$.

TEACHING tip

Write out the expansions of the compound interest formula for different values of n.

$$f(x) = P \cdot \left(1 + \frac{r}{1}\right)^{1t} = \ldots$$

$$f(x) = P \cdot \left(1 + \frac{r}{2}\right)^{2t} = \ldots$$

$$f(x) = P \cdot \left(1 + \frac{r}{3}\right)^{3t} = \ldots$$

See whether the students can detect the emerging pattern.

Alternate Example 3

$500 is invested at an annual percentage rate of 10%.

A. Find the balance after 10 years if the interest is compounded annually, quarterly, and daily. [**$1296.87, $1342.53, $1358.95**]

B. Which compound method gives the best return on the initial investment? [**daily**]

Cooperative Learning

Divide the class into groups of four. Designate one student the local banker, another the local business person, and let the other students be depositors. The banker charges 5% interest on deposits. The banker loans money to the local business person at 8% interest. Using play money, have students carry out a number of transactions, and discuss why the interest on loans is higher than the interest on deposits.

Compound Interest

Simple interest is the interest on the original amount of money, the **principal**, P. **Compound interest** is the interest on the principal *plus* the interest on any interest already earned. If the annual percentage rate, r, is compounded once yearly, the amount after t years is given by $A(t) = P(1 + r)^t$.

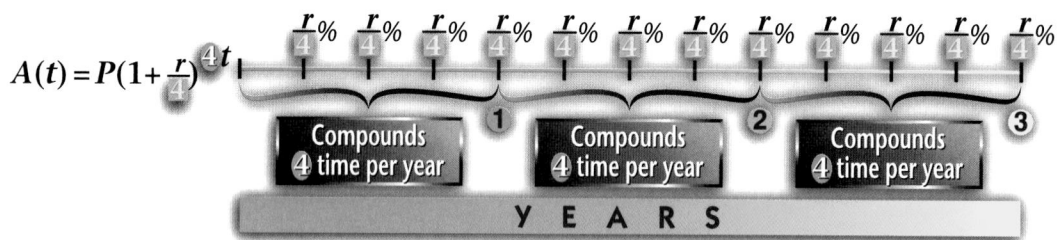

$$A(t) = P(1+r)^t$$

The base, $1 + r$, is called the growth rate, and the principal (initial) amount, P, is the y-intercept.

In most situations, interest is compounded more than once per year.

Suppose interest is compounded 4 times per year, or compounded quarterly. Then the **compounding period** is $\frac{1}{4}$ of a year. The interest rate for each compounding period is $\frac{1}{4}$ of the annual rate, r, or $\frac{r}{4}$. In t years, the number of times interest is compounded is $4t$, so the total amount after t years is $A(t) = P\left(1 + \frac{r}{4}\right)^{4t}$.

$$A(t) = P\left(1 + \frac{r}{4}\right)^{4t}$$

In general, if n is the number of times per year that the interest is compounded, the total amount after t years is

$$A(t) = P\left(1 + \frac{r}{n}\right)^{nt}.$$

In this exponential function the base, or *growth*, is $1 + \frac{r}{n}$.

EXAMPLE 3

Graphics Calculator

$100 is invested at an annual rate of 5%.

Ⓐ Find the balances after 10 years if the interest is compounded annually, quarterly, and daily.

Ⓑ Which compounding method gives the best return on the initial investment?

Solution▶

A The principal, P, is \$100, the annual rate, r, is 0.05, and the time of investment, t, is 10 years.

Since n is the number of times interest is compounded in 1 year,

$n = 1$ for an annual compounding period,
$n = 4$ for a quarterly compounding period, and
$n = 365$ for a daily compounding period.

Compounding period	$A(t) = 100\left(1 + \frac{0.05}{n}\right)^{nt}$	Final amount
Annual ($n = 1$)	$A(10) = 100\left(1 + \frac{0.05}{1}\right)^{1 \times 10}$	\$162.89
Quarterly ($n = 4$)	$A(10) = 100\left(1 + \frac{0.05}{4}\right)^{4 \times 10}$	\$164.36
Daily ($n = 365$)	$A(10) = 100\left(1 + \frac{0.05}{365}\right)^{365 \times 10}$	\$164.87

B Since \$164.87 is greater than the other amounts, daily compounding gives the best return on the principal. ❖

Try This Find the balances after 10 years if \$100 is invested at an annual percentage rate of 6.5% compounded annually, quarterly, and daily.

From Example 3, you can see that a larger number of compounding periods results in a higher return over the same period of time. However, the difference is relatively small, even after 10 years.

What is the effect of a change in the interest rate? The following table lists the resulting balances after 10 years for a principal of \$100 invested at different annual percentage rates and compounded *daily*.

Interest rate	$A(r) = 100\left(1 + \frac{r}{365}\right)^{3650}$	Final amount
5%	$A(0.05) = \$100\left(1 + \frac{0.05}{365}\right)^{3650}$	\$164.87
5.5%	$A(0.055) = \$100\left(1 + \frac{0.055}{365}\right)^{3650}$	\$173.32
6%	$A(0.06) = \$100\left(1 + \frac{0.06}{365}\right)^{3650}$	\$182.20

Notice that a 6% interest rate yields a significantly higher return over a 10-year period than a 5% rate.

CRITICAL Thinking Why do you think increasing the annual interest rate by 1% has a greater effect on the return after 10 years than increasing the number of compounding periods? Would this be true after 100 years? Explain.

ASSESS

Selected Answers

Odd-numbered Exercises 7–37, 41–43

Assignment Guide

Core 1–13, 15–30, 35–52

Core Plus 1–10, 13–22, 28–53

Technology

A calculator is needed for Exercises 11–16, 21–35, 37–38, 40–43, and 53.

Error Analysis

Caution students about the use of parentheses when keying functions into the graphics calculator.

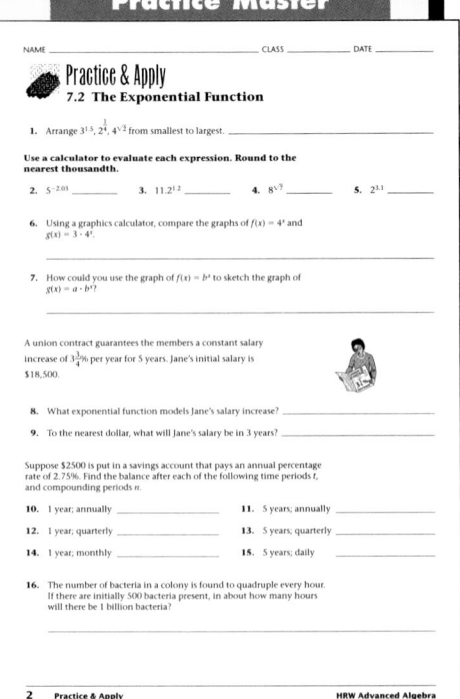

EXERCISES & PROBLEMS

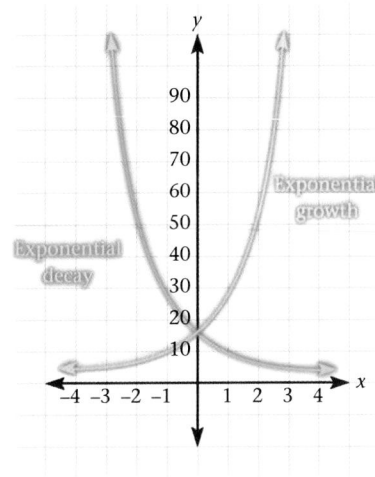

Communicate

1. Describe at least three characteristics of an exponential function.

2. Explain the difference between exponential growth and exponential decay functions.

3. Explain what an asymptote is.

4. What are the domain and range of an exponential function?

5. How does an exponential function differ from a polynomial function?

6. Explain why $f(x) = \left(\frac{1}{4}\right)^x$ is equivalent to $f(x) = 4^{-x}$.

Practice & Apply

Classify each function as polynomial or exponential. Explain your response.

7. $f(x) = x^3$ **8.** $f(x) = 4^x$ **9.** $f(x) = x + 1$ **10.** $f(x) = \pi^x$
polynomial   exponential   polynomial   exponential

Use a calculator to evaluate each expression to the nearest thousandth.

11 $2^{3.45}$ 10.928   **12** $4^{\sqrt{3}}$ 11.036   **13** $3^{-4.56}$ 0.007   **14** $1.4^{6.38}$ 8.557

Graph each pair of functions to determine whether they are equivalent. Explain your result.

15 $f(x) = 3^{-x}$ and $g(x) = \left(\frac{1}{3}\right)^x$ **16** $h(x) = 7^{-x}$ and $j(x) = \left(\frac{1}{7}\right)^{-x}$
   Equivalent   Not equivalent, $7^{-x} = \left(\frac{1}{7}\right)^x$

Transformations Graph each pair of functions. Explain how the graph of the function in the right column can be obtained from the graph of the function in the left column.   shift g_1 2 units to the right

17. $f_1(x) = 3^x$ $f_2(x) = 3^x + 2$ **18.** $g_1(x) = \left(\frac{1}{3}\right)^x$ $g_2(x) = \left(\frac{1}{3}\right)^{x-2}$
    shift f_1 up 2 units

19. $h_1(x) = 7^{-x}$ $h_2(x) = 7^{-x} - 3$ **20.** $j_1(x) = \left(\frac{1}{7}\right)^{-x}$ $j_2(x) = \left(\frac{1}{7}\right)^{-x+3}$
    shift h_1 3 units down

21 **Transformations** Compare the graphs of $f(x) = 3^x$ and $g(x) = 3^x + 5$. Describe how you can use the graph of $f(x) = b^x$ to sketch the graph of $g(x) = b^x + c$.

22 **Transformations** Compare the graphs of $f(x) = 3^x$ and $g(x) = 3^{(x+5)}$. Describe how you can use the graph of $f(x) = b^x$ to sketch the graph of $h(x) = b^{(x+c)}$.

20. Shift j_1 3 units to the right.

21. The graph of $g(x) = 3^x + 5$ is the graph of $f(x) = 3^x$ shifted 5 units upward. For example, the y-intercept of f is 1, and of g is $1 + 5$ or 6. To sketch the graph of $g(x) = b^x + c$ from the graph of $f(x) = b^x$, shift the graph of f upward c units for $c > 0$, and downward $|c|$ units for $c < 0$.

22. The graph of $g(x) = 3^{(x+5)}$ is the graph of $f(x) = 3^x$ shifted 5 units to the left. For example, f passes through $(0, 1)$, and g passes through $(0 - 5, 1)$, or $(-5, 1)$. To sketch the graph of $h(x) = b^{(x+c)}$ from the graph of $f(x) = b^x$, shift the graph of f to the left c units for $c > 0$, and to the right $|c|$ units for $c < 0$.

Investment Suppose $1000 is invested at 3.5% interest. Determine how much the investment is worth after each of the following time periods, t, and compounding periods, n.

23 1 year; annually
$1035

24 10 years; annually
$1410.60

25 20 years; annually
$1989.79

26 1 year; quarterly
$1035.46

27 10 years; quarterly
$1416.91

28 20 years; quarterly
$2007.63

29 1 year; monthly
$1035.57

30 10 years; monthly
$1418.34

31 20 years; monthly
$2011.70

32 1 year; daily
$1035.62

33 10 years; daily
$1419.04

34 20 years; daily
$2013.69

35 Biology Suppose the number of bacteria in a colony triples every 24 hours. If the colony begins with 50 bacteria, how many bacteria are present after 10 days? How many days must pass before there are 1,000,000 bacteria? 2,952,450; 9 days, 21 minutes

Accounting Suppose the used car below sold for $10,500 when new.

36. By what percentage does this car's value decrease each year? 25%

37 Determine the value of the car 3 years after it was purchased. $4429.69

38 Graph v. Determine the approximate age of the car in the photo. ≈ 4.5 years

After t years, the value of this car decreases, or depreciates, according to the function
$v(t) = 10,500\left(\frac{3}{4}\right)^t.$

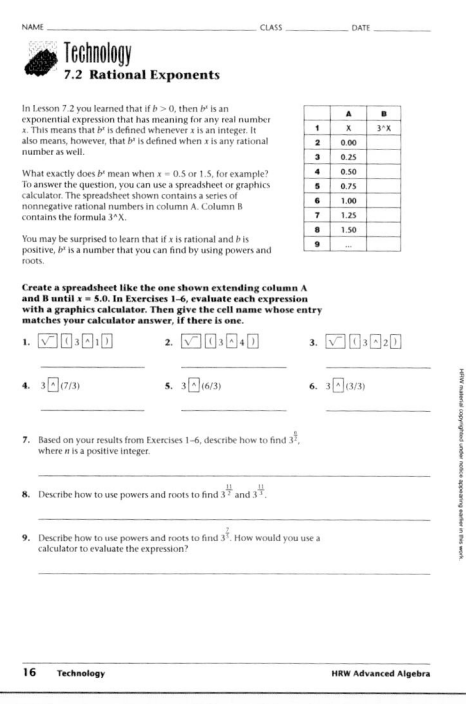

39. Since the money increased by 1.728 times in 3 years and 2.0736 times in 4 years, the Babylonians might have reasoned that the difference of $2 - 1.728 = 0.272$ is to the whole increase of $2.0736 - 1.728 = 0.3456$ as the time past 3 years to double the money is to the time difference of 1 year, or $\frac{0.272}{0.3456} = \frac{x}{1}$, which gives $x \approx 0.787$. This means that the money will double in about $3 + 0.787$, or 3.787 years. This is just an approximation, however, because this kind of proportional reasoning assumes that the function is linear. Another way is to graph the function $y = 1.2^x$, (counting the principal as 1), and then trace to find x for $y = 2$, which represents a doubling of the principal. Tracing gives an estimate of 3.802 years, which is close to the estimate by proportional reasoning, but is more accurate. Answers may vary on student comparisons with their own method and the Babylonian method.

Look Beyond

This problem looks ahead to the graphs of the inverse of exponential functions, logarithmic functions, which will be studied in the next lesson.

39.  **Portfolio Activity** Refer to the Portfolio Activity on page 375. Explain how the Babylonians used proportional reasoning to compute compound interest. Explain how you would solve the problem described on the Babylonian tablet. Compare your method with that of the Babylonians.

Economics Suppose the annual rate of inflation averages 3% over the next 10 years. With this rate of inflation, the approximate cost, C, of goods or services during any year in the following decade will be given by $C(t) = P(1.03)^t$ such that $0 \le t \le 10$, where t is time in years and P is the present cost.

40 If a tune-up for your car presently costs $40, estimate how much it will cost 3 years from now. $43.71

41 Estimate what the tune-up will cost 10 years from now. $53.76

42 If you live in an apartment for which the monthly rent is $300, about how much will the rent be 3 years from now? $327.82

43 Estimate what the rent will be 10 years from now. $403.17

Look Back

Solve for x.   [Lesson 3.4]

44. $|-2x| = -1$ **45.** $5x - 3 = |x + 12|$ **46.** $|12x - 47| = |18x + 45|$
no solution   3.75 $\frac{1}{15}, -\frac{46}{3}$

Graph each of the following systems of inequalities. [Lesson 4.8]

47. $\begin{cases} 2x - y \ge 3 \\ 5x + 2y < 3 \end{cases}$ **48.** $\begin{cases} x + 3y > 4 \\ -4x - y \le 0 \end{cases}$ **49.** $\begin{cases} x - 2y \ge 1 \\ 2x + y < -1 \\ -x - y > 2 \end{cases}$

Graph each of the following quadratic inequalities. [Lesson 5.9]

50. $y \ge 3x^2 - 4$ **51.** $y \ge -x^2 + 4x - 12$ **52.** $x^2 - y > 3$

Look Beyond

53 Graph $y = 4^x$ and its inverse, $x = 4^y$, on the same coordinate plane. What are the domain and range of $x = 4^y$? Is $x = 4^y$ a function? Explain your responses.

47.

48.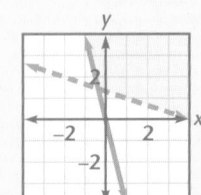

49.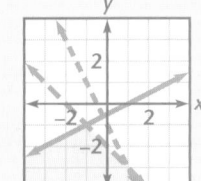

The answers to Exercises 50–53 can be found in Additional Answers beginning on page 842.

LESSON 7.3 Logarithmic Functions

PREPARE

Objectives

- Identify the exponential and logarithmic functions as inverse functions.
- Determine equivalent forms for exponential and logarithmic equations.

RESOURCES

- Practice Master   7.3
- Enrichment Master   7.3
- Technology Master   7.3
- Lesson Activity Master   7.3
- Quiz   7.3
- Spanish Resources   7.3

You have explored functions and their inverse relations in previous lessons. The inverse of an exponential function is itself a function that is often used in a wide variety of applications.

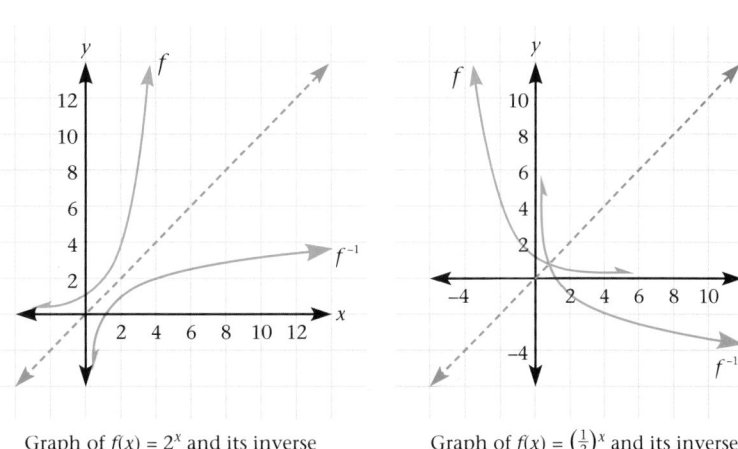

Graph of $f(x) = 2^x$ and its inverse   Graph of $f(x) = \left(\frac{1}{2}\right)^x$ and its inverse

The graph of f, shown in **blue**, has been reflected over the line $y = x$. Thus, the reflected curve, shown in **red**, is the inverse relation of f.

Use the vertical line test. Is the inverse of $f(x) = 2^x$ a function?

Exploration *Powers of 2*

 Complete the following tables.

Table 1

x	1	2	3	4	5	6
$y = 2^x$	2	4				

Table 2

y						
$x = 2^y$	2	4	8	16	32	64

 Describe the relationship between Table 1 and Table 2. ❖

Notice that the y-values in Table 2 are all exponents. What is the precise relationship between the table values? To solve $x = 2^y$ for y, a new expression is needed. This expression is called a *logarithm*. By definition, the exponent, y, is the **logarithm** of x to the base 2, or $y = \log_2 x$. This is the inverse function of $f(x) = 2^x$.

Assessing Prior Knowledge

1. Find the inverse relation for $f(x) = x^2$. $\left[f^{-1}(x) = +\sqrt{x} \text{ or } f^{-1}(x) = -\sqrt{x}\right]$

2. Find the inverse of $f(x) = 2x + 3$. $\left[f^{-1}(x) = \frac{1}{2}(x - 3)\right]$

TEACH

To find the value of an investment after n years, you can use an exponential function. How do you find the inverse of an exponential function?

Ongoing ASSESSMENT

Yes

ALTERNATIVE teaching strategy

Using Patterns Reverse the exponential explorations from Lesson 7.1. Bacteria growth, for example, can be approached as follows: Ask, After how many hours does the population double? triple? quadruple? The population growth for cities can be viewed in the same way. How many years does it take for the population to double? triple? quadruple?

ENRICHMENT

The procedure for solving problems on page 392 depends on the following property of inverse functions: if f and g are inverse functions, then $f(g(x)) = x$ and $g(f(x)) = x$. Have the students look back over what they have learned about linear and polynomial functions. Have them find a few of each type of function and their inverses.

Exploration Note

Table 2 can be completed analytically, e.g. think: 2 to what power gives 2, 2 to what power gives 4, and so on.

ongoing
ASSESSMENT

2. The rows are the same.

Alternate Example 1

A. Write $3^8 = 6561$ in logarithmic form. [$\log_3 6561 = 8$]

B. Write $\log_7 49 = 2$ in exponential form. [$7^2 = 49$]

CRITICAL
Thinking

Any number raised to the zero power is 1. Therefore, if b is any non-zero number, zero is the power of b that equals 1. In other words, $b^0 = 1$. If b is 0, b^0 is not defined.

Alternate Example 2

Solve each equation.

A. $y = \log_2 64$ [$y = 6$]

B. $y = \log_{49} 7$ $\left[y = \frac{1}{2} \right]$

Use Transparency ▶ **27**

Any equation written in exponential form can also be written in an equivalent logarithmic form.

3 is the power of **2** that gives **8**.
The base **2** to the power of **3** is **8**.

3 is the power of **2** that gives **8**.
The log base **2** of **8** is **3**.

LOGARITHMIC FUNCTION

The **logarithmic function** $y = \log_b x$ is the inverse of the exponential function $y = b^x$, where $b \neq 1$ and $b > 0$.

$$y = \log_b x \text{ if and only if } x = b^y$$

EXAMPLE 1

Ⓐ Write $5^3 = 125$ in logarithmic form.

Ⓑ Write $\log_3 81 = 4$ in exponential form.

Solution

Ⓐ 3 is the power of 5 that gives 125, or $3 = \log_5 125$.

Ⓑ 4 is the power of 3 that gives 81, or $3^4 = 81$. ❖

The following table lists some equivalent exponential and logarithmic equations.

Exponential equation	$2^5 = 32$	$10^3 = 1000$	$3^{-2} = \frac{1}{9}$	$16^{\frac{1}{2}} = 4$
Logarithmic equation	$\log_2 32 = 5$	$\log_{10} 1000 = 3$	$\log_3 \frac{1}{9} = -2$	$\log_{16} 4 = \frac{1}{2}$

CRITICAL
Thinking

Explain why $\log_b 1 = 0$ for any positive base $b \neq 1$.

EXAMPLE 2

Solve each equation.

Ⓐ $y = \log_{10} 100$ 　　　　　　　　Ⓑ $y = \log_{125} 5$

Solution

Ⓐ $y = \log_{10} 100$ is the same as $10^y = 100$. Since 10^2 is 100, $y = 2$.

Ⓑ $y = \log_{125} 5$ is the same as $125^y = 5$. Since $125 = 5^3$ and $5 = 125^{\frac{1}{3}}$, $y = \frac{1}{3}$. ❖

INCLUSION
strategies

Using Algorithms Students will benefit by writing out the procedure, step-by-step, for solving a simple logarithmic equation.

1. For $y = \log_{10} 100$, treat each side as the exponent of an exponential expression with 10 as the base.

2. On the right-hand side, $10^{\log_{10}}$ cancel each other out, since they are inverse operations. This leaves $10^y = 100$.

3. Apply knowledge of the basic rules for working with exponents to solve for y.

Compare the graphs of $y = 2^x$ and $y = \log_2 x$. The two functions are inverses, and, as you can see, the line $y = x$ is a line of symmetry for the two graphs.

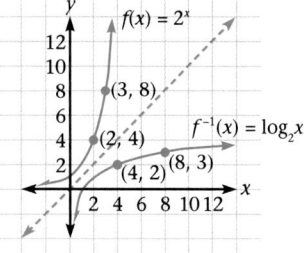

In general, for any base b, if $f(x) = b^x$ and $g(x) = \log_b x$, then f and g are inverse functions.

The domain of logarithmic functions is all $x > 0$, while the range is all real numbers.

This leads to two important properties of exponentials and logarithms.

EXPONENTIAL-LOG INVERSE PROPERTIES

$$b^{\log_b x} = x \text{ for } x > 0$$
$$\log_b b^x = x \text{ for all } x$$

If $b^x = b^y$, then $x = y$. Similarly, if $\log_b x = \log_b y$, then $x = y$.

CRITICAL *Thinking*

Is the converse, if $x = y$ (where $x > 0$ and $y > 0$), then $\log_b x = \log_b y$, true? Explain.

The exponential-log inverse properties can be used when solving equations involving logarithmic equations.

EXAMPLE 3

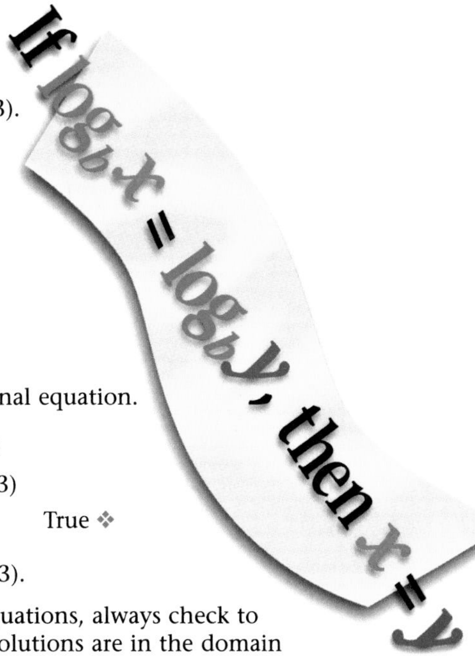

Solve $\log_2(5x - 1) = \log_2(4x + 3)$.

Solution▶

If $\log_2(5x - 1) = \log_2(4x + 3)$, then $5x - 1 = 4x + 3$.
Solve $5x - 1 = 4x + 3$ for x.

$$5x - 1 = 4x + 3$$
$$x = 4$$

Check▶

Substitute 4 for x in the original equation.

$$\log_2(5x - 1) = \log_2(4x + 3)$$
$$\log_2(5(4) - 1) \overset{?}{=} \log_2(4(4) + 3)$$
$$\log_2 19 = \log_2 19 \qquad \text{True} ❖$$

Try This   Solve $\log_{10}(3x - 1) = \log_{10}(x + 3)$.

When solving logarithmic equations, always check to make sure that the possible solutions are in the domain of the original equation.

If $\log_b x = \log_b y$, then $x = y$

CRITICAL *Thinking*

Yes, but only when $x > 0$ and $y > 0$. Then $\log_b x$ and $\log_b y$ are defined. By substitution, $\log_b x = \log_b y$. This works because $y = \log_b x$ is a function, so for any value of x in its domain, there can be only one value of y.

Alternate Example 3

Solve.
$\log_3(2x + 3) = \log_3(5x - 9)$.
[$x = 4$]

Aongoing
SSESSMENT

Try This

$x = 2$

EXAMPLE 4

Solve $\log_{10}(x - 2) = \log_{10}(2x + 1)$.

Solution▶

If $\log_{10}(x - 2) = \log_{10}(2x + 1)$, then $x - 2 = 2x + 1$.
Solve $x - 2 = 2x + 1$ for x.

$$x - 2 = 2x + 1$$
$$-x = 3, \text{ or } x = -3$$

Check▶

Substitute -3 for x in the original equation.

$$\log_{10}(x - 2) = \log_{10}(2x + 1)$$
$$\log_{10}(-3 - 2) \stackrel{?}{=} \log_{10}(2(-3) + 1)$$
$$\log_{10}(-5) \stackrel{?}{=} \log_{10}(-5) \quad \text{Appears to be true}$$

However, this equation is invalid because only logarithms of positive numbers are defined. Therefore, the equation has no solution. ❖

EXERCISES & PROBLEMS

Communicate

1. How are logarithmic functions and exponential functions related?
2. Describe the meaning of $\log_b x$ where $b \neq 1$ and $x > 0$.
3. What are the domain and range of a logarithmic function?
4. What is the value of $\log_b 1$? Explain.
5. Explain why a logarithm is actually an exponent.

Practice & Apply

Write an equivalent logarithmic equation.

6. $16^{\frac{1}{4}} = 2$ $\log_{16} 2 = \frac{1}{4}$

7. $5^4 = 625$ $\log_5 625 = 4$

8. $\left(\frac{1}{3}\right)^2 = \frac{1}{9}$ $\log_{\frac{1}{3}}\left(\frac{1}{9}\right) = 2$

9. $2^{-3} = \frac{1}{8}$ $\log_2 \frac{1}{8} = -3$

10. $27^{\frac{1}{3}} = 3$ $\log_{27} 3 = \frac{1}{3}$

11. $7^3 = 343$ $\log_7 343 = 3$

12. $\left(\frac{1}{4}\right)^{-3} = 64$ $\log_{\frac{1}{4}} 64 = -3$

13. $6^{-2} = \frac{1}{36}$ $\log_6\left(\frac{1}{36}\right) = -2$

22.

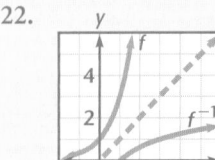

x	-1	0	1
$f(x) = 3^x$	$\frac{1}{3}$	1	3

x	$\frac{1}{3}$	1	3
$f^{-1}(x) = \log_3 x$	-1	0	1

Write an equivalent exponential equation.

14. $\log_6 36 = 2$ $6^2 = 36$

15. $\log_{10} 1000 = 3$ $10^3 = 1000$

16. $-2 = \log_2\left(\frac{1}{4}\right)$ $2^{-2} = \frac{1}{4}$

17. $3 = \log_3 27$ $3^3 = 27$

18. $\log_5 125 = 3$ $5^3 = 125$

19. $\log_{10} 0.001 = -3$ $10^{-3} = 0.001$

20. $-4 = \log_3 \frac{1}{81}$ $3^{-4} = \frac{1}{81}$

21. $4 = \log_5 625$ $5^4 = 625$

22. Graph $f(x) = 3^x$. Then sketch f^{-1} on the same coordinate plane. Write a table of values that illustrates the relationship between f and f^{-1}.

23. Graph $f(x) = 3^{-x}$. Then sketch f^{-1} on the same coordinate plane. Write a table of values that illustrates the relationship between f and f^{-1}.

Solve each equation, and check your answers.

24. $\log_3(2x + 1) = \log_3(x + 3)$ 2

25. $\log_{10}(3x - 3) = \log_{10}(x + 3)$ 3

26. $\log_5(2x + 3) = \log_5(x - 1)$ no solution

27. $\log_2(5x - 10) = \log_2(3x - 5)$ 2.5

28. $\log_{10}(x^2 - x) = \log_{10}(4x - 6)$ 2 or 3

29. $\log_4(x^2 - 2x) = \log_4(3x - 6)$ 3

Molecular Chemistry In chemistry, pH is defined as pH $= -\log_{10}[H^+]$, where $[H^+]$ is the hydrogen ion concentration in moles per liter.

30. Write $[H^+]$ in terms of the pH by writing pH $= -\log_{10}[H^+]$ in exponential form. $[H^+] = 10^{-pH}$

31. What is $[H^+]$ if the pH is 9, the pH level of bile? 1×10^{-9}

32. What is $[H^+]$ if the pH is 7, the pH level of distilled water? 1×10^{-7}

33. What is $[H^+]$ if the pH is 3, the pH level of grapefruit juice? 1×10^{-3}

 **Look Back**

34. Write a linear equation with slope 4 and y-intercept $(0, 3)$ **[Lesson 1.2]** $y = 4x + 3$

35. Find the inverse of the matrix $\begin{bmatrix} 2 & 3 \\ 1 & -4 \end{bmatrix}$. **[Lesson 4.5]**

36. Solve the quadratic equation $x^2 - 6x + 9 = 0$. **[Lessons 5.2, 5.4]** $x = 3$

37. Name the two solutions of the equation $x^2 + 1 = 0$. **[Lesson 5.7]** $\pm i$

38. What are two important consequences of the Fundamental Theorem of Algebra? **[Lesson 6.3]**

39. If an interest rate is 7.3%, what is the growth multiplier? **[Lesson 7.1]**   1.073

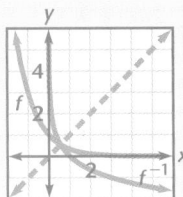 **Look Beyond**

40. Solve the exponential equation $3^x = 8$. $x \approx 1.89$

23.

x	-1	0	1
$f(x) = 3^{-x}$	3	1	$\frac{1}{3}$

x	3	1	$\frac{1}{3}$
$f^{-1}(x) = -\log_3 x$	-1	0	1

35. $\begin{bmatrix} 0.36 & 0.27 \\ 0.09 & 0.18 \end{bmatrix}$

- Identify the product, quotient, and power properties of logarithms.
- Simplify expressions and solve equations involving logarithms.

RESOURCES

Assessing Prior Knowledge

Simplify each expression.

1. $2^3 \times 2^4$ $[2^7]$

2. $\frac{5^8}{5^5}$ $[5^3]$

3. $\left((3)^2\right)^3$ $[3^6]$

TEACH

Many times, simplifying a mathematical statement can make its solution easier to find. The properties of exponential and logarithmic functions can be used to simplify solving equations.

LESSON 7.4

Exploring
Properties of Logarithmic Functions

why *In equations containing more than one logarithm, such as $\log_{10} I = -0.245(\log_{10} d) + 3$, which relates light intensity to depth below a water surface, it is often possible to combine all the logarithms into a single logarithmic expression. The converted form may yield additional information or be easier to solve for any variable that the equation contains.*

$$\log_{10} I = -0.245\,\log_{10} d + 3$$

Recall that the logarithmic function with base b is the inverse of the exponential function with base b. In other words,

$$y = b^x \text{ and } x = \log_b y \text{ are equivalent.}$$

Since a logarithm is an exponent, it is reasonable that each property of exponents you discovered in Lesson 2.2 should have a corresponding logarithmic property. For example, recall that $b^0 = 1$ for all numbers $b \neq 0$. Therefore, $\log_b 1 = 0$.

Also, recall that $\log_{10} 10^2 = 2$ and $\log_2 2^5 = 5$. In general, $\log_b b^n = n$. This is because logarithmic and exponential functions are inverses of each other.

ALTERNATIVE
teaching
strategy

Using Visual Strategies Graph $f(x) = \log_{10} x$. Draw three vertical line segments: L_2 connects $(2, 0)$ and $(2, \log_{10} 2)$; L_4 connects $(4, 0)$ and $(4, \log_{10} 4)$: L_8 connects $(8, 0)$ and $(8, \log_{10} 8)$. Show students that the length of L_8 is the sum of the lengths of L_2 and L_4, which means $\log_{10} 8 = \log_{10} 2 + \log_{10} 4$. The length of L_4 is the length of L_8 minus L_2, which means $\log_{10} 4 = \log_{10} 8 - \log_{10} 2 = \log_{10}\left(\frac{8}{2}\right)$. The length of L_8 is three times the length of L_2, which means $3 \log_{10} 2 = \log_{10} 2^3$.

•Exploration 1 *Product Property of Logarithms*

1 Complete the table.

y	2	4	8	16	32	64	128
$\log_2 y$	1	2	3				

2 Find each of the following pairs of values and compare them.
   a. $\log_2(2 \cdot 4)$ and $\log_2 2 + \log_2 4$
   b. $\log_2(2 \cdot 8)$ and $\log_2 2 + \log_2 8$
   c. $\log_2(8 \cdot 4)$ and $\log_2 8 + \log_2 4$

3 Make a conjecture about the relationship between $\log_2(ab)$ and $\log_2 a + \log_2 b$.

4 Explain how your conjecture can be used with values in the table you completed in Step 1 to evaluate $\log_2 512$. ❖

In Exploration 1, you discovered the product property for base 2 logarithms. This property of logarithms is true for any positive base $b \neq 1$.

PRODUCT PROPERTY OF LOGARITHMS

For any two positive numbers m and n, the **product property** states that
$$\log_b(mn) = \log_b m + \log_b n.$$

EXTENSION

Suppose you know that $\log_2 10 \approx 3.322$. This fact and the product property of logarithms can be used to find $\log_2 40$.

$$\begin{aligned}
\log_2 40 &= \log_2(4 \cdot 10) \\
&= \log_2 4 + \log_2 10 \\
&\approx 2.000 + 3.322 \\
&\approx 5.322 \quad ❖
\end{aligned}$$

$$\log_{10}\left(Id^{0.245}\right) = 3$$

interdisciplinary CONNECTION
Astronomy The difference between the apparent brightness of a star, m, and its true brightness, M, is given by the equation $m - M = 5\log_{10}\left(\frac{r}{10}\right)$ where r is its distance in kiloparsecs from the Earth.

A. Given $m = 20$ and $r = 100$ find M. [15]
B. Given $m = 5$ and $M = 10$ find r. [1]

A ongoing **ASSESSMENT**

4. Answers may vary. One possible answer is that $512 = 8(64)$, so
$$\begin{aligned}
\log_2 512 &= \log_2 8(64) \\
&= \log_2 8 + \log_2 64 \\
&= 3 + 6 = 9.
\end{aligned}$$

TEACHING *tip*

Show students that they can prove the product property of logarithms from the product property of exponents.

$$m \times n = b^{\log_b m} \times b^{\log_b n}$$
$$m \times n = b^{(\log_b m + \log_b n)}$$

Apply $\log_b$ to both sides.

$$\log_b(m \times n) = \log_b b^{(\log_b m + \log_b n)}$$
$$\log_b(m \times n) = \log_b m + \log_b n$$

TEACHING *tip*

Technology Graph both $\log(x + 2)$ and $\log x + \log 2$ on the same coordinate plane. Students should see that these functions are not the same. Graph $\log\left(\frac{x}{2}\right)$ and $\frac{\log x}{\log 2}$. This graph shows that the logarithm of a quotient is not the quotient of the logarithms. Finally, graphing $\log x^2$ and $(\log x)^2$ shows that the logarithm of a power is not the power of a logarithm.

Exploration 2 Notes

This exploration leads students to discover the quotient property for logarithms by generalizing from three concrete examples.

2. **An informal way to state this is: The logarithm of a quotient is the difference of the logarithms of the numerator and the denominator.**

TEACHING *tip*

Show students that they can prove the quotient property of logarithms from the quotient property of exponents.

$$\frac{m}{n} = \frac{b^{\log_b m}}{b^{\log_b n}} = b^{\log_b m - \log_b n}$$

Apply $\log_b$ to both sides.

$$\log_b \left(\frac{m}{n}\right) = \log_b m - \log_b n$$

•Exploration 2 *Quotient Property of Logarithms*

1 Find each of the following values and compare them.

a. $\log_2 \left(\frac{16}{2}\right)$ and $\log_2 16 - \log_2 2$ **b.** $\log_2 \left(\frac{4}{2}\right)$ and $\log_2 4 - \log_2 2$

c. $\log_2 \left(\frac{16}{4}\right)$ and $\log_2 16 - \log_2 4$

2 How can you express $\log_2 \left(\frac{a}{b}\right)$ in terms of $\log_2 a$ and $\log_2 b$? In your own words, define what you think the quotient property of logarithms is.

3 Use your answer from Step 2 to evaluate $\log_3 \left(\frac{243}{9}\right)$, given that $\log_3 243 = 5$. ❖

In Exploration 2, you discovered the quotient property for base 2 logarithms. This property of logarithms is true for any positive base $b \neq 1$.

QUOTIENT PROPERTY OF LOGARITHMS
For any two positive numbers m and n, the **quotient property** states that
$$\log_b \frac{m}{n} = \log_b m - \log_b n.$$

EXTENSION

Suppose you know that $\log_3 18 \approx 2.631$. This fact and the quotient property of logarithms can be used to find $\log_3 6$.

$$\begin{aligned}
\log_3 6 &= \log_3 \left(\frac{18}{3}\right) \\
&= \log_3 18 - \log_3 3 \\
&\approx 2.631 - 1 \\
&\approx 1.631 \quad ❖
\end{aligned}$$

ENRICHMENT Have students visit the library to find examples of logarithmic equations in many different areas. For example, they might look in books about astronomy, optics, archeology, geology, or chemistry. Students may also explore periodicals and journals, such as *Scientific American*, *Science*, *Physics Teacher*, and so on. They should write a short report summarizing an example they found.

INCLUSION **strategies** **Using Patterns** Some students prefer to view a function as a table of data, others prefer a graph, while others prefer an equation. Take the optics example on page 399 and draw a diagram of the rays of sunlight penetrating the ocean surface and refracting to deep levels. Then show a graph of the equation given. Examine each way of looking at the function. Explain the advantages and disadvantages of each representation.

Exploration 3 Power Property Of Logarithms

1 Notice that, using the product property of logarithms,
$$\log_b 6^2 = \log_b(6 \cdot 6) = \log_b 6 + \log_b 6 = 2 \log_b 6.$$
Use the product property of logarithms to show that each of the following statements is true.

    a. $\log_b 3^2 = 2 \log_b 3$ **b.** $\log_b 5^3 = 3 \log_b 5$ **c.** $\log_b 7^4 = 4 \log_b 7$

2 Let p be any real number. Use your results from Step 1 to complete the statement $\log_b m^p =$ _____.

3 Define, in your own words, this power property of logarithms.

4 Use this power property of logarithms to explain why $\log_{10} 100^4 = 8$. ❖

POWER PROPERTY OF LOGARITHMS

For any positive number m and any real number p, the **power property** states that
$$\log_b m^p = p \log_b m.$$

APPLICATION

Optics   The luminous intensity, I, of sunlight that reaches levels below the surface of an ocean varies exponentially with the depth, d, below the surface of the water.

A formula for the intensity near the surface of the water is $\log_{10} I = -0.245(\log_{10} d) + 3$, where d is measured in meters and I in candelas.

To find the luminous intensity, I, at various depths, d, first use the properties of logarithms to write the intensity as a function of depth.

$$\log_{10} I = -0.245 \log_{10} d + 3$$
$$\log_{10} I + 0.245 \log_{10} d = 3$$
$$\log_{10} I + \log_{10} d^{0.245} = 3$$
$$\log_{10} (Id^{0.245}) = 3$$
$$Id^{0.245} = 10^3$$
$$I = \frac{1000}{d^{0.245}} = 1000d^{-0.245}$$

To find the luminous intensity at various depths, make a table.

Depth (meters)	Intensity $I(d) = 1000\, d^{-0.245}$ (candelas)
2	$1000 \cdot 2^{-0.245} = 843.8$
4	$1000 \cdot 4^{-0.245} = 712.0$
6	$1000 \cdot 6^{-0.245} = 644.7$
8	$1000 \cdot 8^{-0.245} = 600.8$
10	$1000 \cdot 10^{-0.245} = 568.9$

❖

CRITICAL *Thinking*   Why is the formula $I(d) = 1000d^{-0.245}$ invalid at the lake surface?

RETEACHING *the lesson*   **Using Cognitive Strategies** Summarize the properties for logarithms along with their equivalent properties for exponents. To make it more concrete, substitute numbers into the equations.

Exploration 3 Notes

This exploration leads students to discover the power property for logarithms by generalizing from three concrete examples.

A**ongoing** SSESSMENT

3. The logarithm of a number raised to a power is the power times the logarithm of the number.

CRITICAL *Thinking*

At the lake surface, $d = 0$, so $0^{-0.245} = \frac{1}{0^{0.245}} = \frac{1}{0}$, which is not defined.

Assignment Guide

Core   1–13,   16–21,   24–32, 34–38, 42–53

Core Plus 1–6, 11–15, 20–27, 31–33, 37–54

Technology

Graphics calculators are not needed in this lesson. However, calculators can be used throughout the lesson to test the properties and to check results.

Error Analysis

Students may have problems recognizing which rule applies to which step in Exercises 42 and 46–48. Encourage students to write the steps vertically with the rules stated to the right of each step.

Practice Master

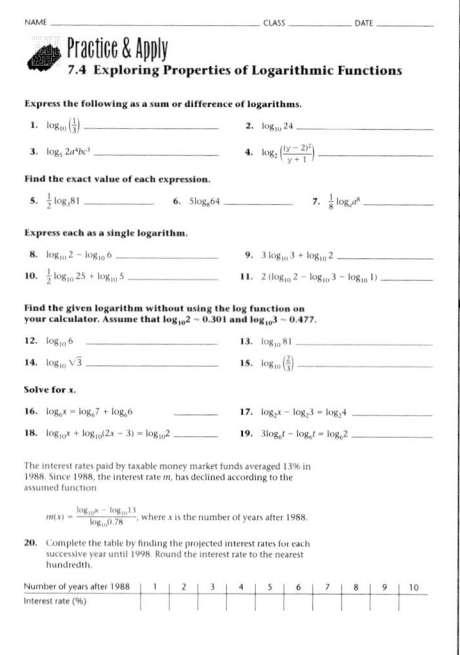

EXERCISES & PROBLEMS

Communicate

1. Explain the product property of logarithms.
2. Explain the quotient property of logarithms.
3. Explain the power property of logarithms.
4. Explain how to write $\log_5 uvw$ as a sum of logarithms.
5. Describe how to simplify $\log_3 x(x+3)^2$.
6. Describe how to express $\log_4(x+3)^2 - \log_4 3(x+3)$ with a single logarithm.

Practice & Apply

Express the following expressions as a sum of logarithms.

7. $\log_3(5 \cdot 8)$ $\log_3 5 + \log_3 8$

8. $\log_2 xy$ $\log_2 x + \log_2 y$

9. $\log_2 x^2 y$

10. $\log_{10} x(x+2)$ $\log_{10} x + \log_{10}(x+2)$

11. $\log_5(3 \cdot 5 \cdot 7)$ $\log_5 3 + \log_5 5 + \log_5 7$

12. $\log_{10}(2 \cdot 3 \cdot 4 \cdot 5)$ $\log_{10} 2 + \log_{10} 3 + \log_{10} 4 + \log_{10} 5$

13. $\log_2 vxyz$ $\log_2 v + \log_2 x + \log_2 y + \log_2 z$

14. $\log_7 x^3 y^2 z$

15. $\log_{10} x(x+2)^4 (x-2)^3$

Express each of the following expressions as a difference of logarithms.

16. $\log_{10}\left(\frac{16}{9}\right)$ $\log_{10} 16 - \log_{10} 9$

17. $\log_2\left(\frac{80}{15}\right)$ $\log_2 80 - \log_2 15$

18. $\log_3\left(\frac{4}{15}\right)$ $\log_3 4 - \log_3 15$

19. $\log_6\left(\frac{xy}{z}\right)$ $\log_6 xy - \log_6 z$

20. $\log_{10}\left(\frac{x}{yz}\right)$ $\log_{10} x - \log_{10}(yz)$

21. $\log_5\left(\frac{2x}{3y}\right)$ $\log_5 2x - \log_5 3y$

22. $\log_{10}\left(\frac{x^3}{y^2}\right)$ $\log_{10} x^3 - \log_{10} y^2$

23. $\log_2\left(\frac{uv^4}{w^5}\right)$ $\log_2 uv^4 - \log_2 w^5$

Find the exact value of each of the following expressions.

24. $6\log_5 25$ 12

25. $2\log_3 27$ 6

26. $3\log_4\left(\frac{1}{16}\right)$ −6

27. $4\log_x x^5$ 20

Write each of the following expressions with a single logarithm.

28. $\log_2 5 + \log_2 7$ $\log_2 35$

29. $\log_{10} 14 - \log_{10} 7$ $\log_{10} 2$

30. $\log_5 3 + \log_5 6 + \log_5 9$ $\log_5 162$

31. $\log_2 12 - \log_2 6 + \log_2 3$ $\log_2 6$

32. $\log_7 2 - \log_7 3 - \log_7 4$ $\log_7\left(\frac{1}{6}\right)$

33. $\log_4 a + \log_4 b - \log_4 c$ $\log_4\left(\frac{ab}{c}\right)$

9. $\log_2 x^2 + \log_2 y = 2\log_2 x + \log_2 y = \log_2 x + \log_2 x + \log_2 y$

14. $\log_7 x^3 + \log_7 y^2 + \log_7 z$
$= 3\log_7 x + 2\log_7 y + \log_7 z$
$= \log_7 x + \log_7 x + \log_7 x + \log_7 y + \log_7 y + \log_7 z$

15. $\log_{10} x + 4\log_{10}(x+2) + 3\log_{10}(x-2)$
$= \log_{10} x + \log_{10}(x+2) + \log_{10}(x+2) + \log_{10}(x+2) + \log_{10}(x+2) + \log_{10}(x-2) + \log_{10}(x-2) + \log_{10}(x-2)$

For Exercises 34–41, assume that $\log_{10} 2 \approx 0.301$, $\log_{10} 3 \approx 0.477$, and $\log_{10} 5 \approx 0.699$. Evaluate each of the following expressions without using your calculator.

34. $\log_{10} 4$ 0.602 **35.** $\log_{10} 15$ 1.176 **36.** $\log_{10} 18$ 1.255 **37.** $\log_{10}\left(\frac{2}{5}\right)$ –0.398

38. $\log_{10} 2^{50}$ 15.05 **39.** $\log_{10} 125$ 2.097 **40.** $\log_{10} 7.5$ 0.875 **41.** $\log_{10}\left(\frac{1}{2}\right)$ –0.301

42. Show that $\log_b \frac{x}{y} = -\log_b \frac{y}{x}$.

Optics Refer to the Application on page 399.

43. Find luminous intensities of sunlight, I, at depths of 1, 3, 5, 7, and 9 meters.

44. Graph the ordered pairs (d, I) from the application on page 399 and Exercise 43.

45. At what depth can one see only 50% of the illumination seen at a depth of 1 meter?
approx. 16.9 m

Solve for x.

46. $\log_3 x = 3\log_3 2 + \log_3 5$ 40
47. $\log_{10}(x^2 - 4) - \log_{10}(x - 2) = \log_{10}(2x - 3)$ 5
48. $\log_2 x + \log_2(x - 1) = \log_2(x + 4)$ $1 + \sqrt{5}$

DEEP DIVER

A diver tests and inspects a submersible before making a deep dive.

Look Back

Find the image points that are symmetric to each set with respect to the x-axis. [Lesson 3.5]

49. $\{(2, 3), (3, -8), (-7, -6)\}$
$\{(2, -3), (3, 8), (-7, 6)\}$

50. $\{(-8, 3), (4, -2), (-2, -3)\}$
$\{(-8, -3), (4, 2), (-2, 3)\}$

Minimize each of the following objective functions under the given constraints. [Lesson 4.10]

51. $\begin{cases} x + 5y \geq 8 \\ y - 3x \leq 14 \\ x \geq 0 \\ y \geq 0 \end{cases}$ $C = 2x - 5y$ $C = -70$

52. $\begin{cases} x - y \geq 12 \\ 7y - x \leq 2 \\ x \geq 0 \\ y \geq 0 \end{cases}$ $C = x + 4y$ $C = 12$

53. Investment $100 is invested at an annual percentage rate of 5%. Find the balances after 10 years if the interest is compounded annually, quarterly, and daily. **[Lesson 7.2]**
$162.89; $164.36; $164.87

Look Beyond

54. Solve $2e^{3x} = 5$ for x. $x \approx 0.31$

42. $\log_b \frac{x}{y} = \log_b x - \log_b y$
$= -(\log_b y - \log_b x)$
$= -\log_b \frac{y}{x}$

43. (answers in candelas) 1: 1000, 3: 764.0, 5: 674.1, 7: 620.8, 9: 583.73

44.

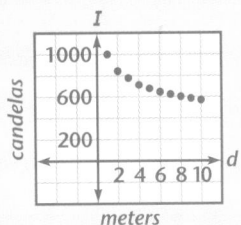

candelas
1000
600
200
2 4 6 8 10
meters

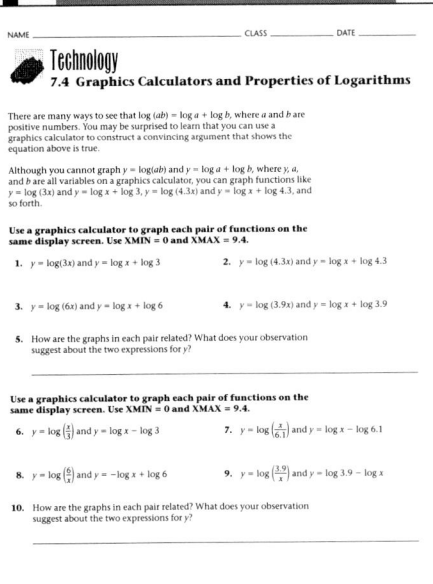

Objectives

- Identify and use the common logarithmic function.
- Write equivalent logarithmic and exponential equations.

RESOURCES

- Practice Master   7.5
- Enrichment Master   7.5
- Technology Master   7.5
- Lesson Activity Master   7.5
- Quiz   7.5
- Spanish Resources   7.5

Assessing Prior Knowledge

1. $10^0 = ?$
    $10^1 = ?$
    $10^2 = ?$

[1, 10, 100]

2. $10^{-1} = ?$
    $10^{-2} = ?$

$\left[\dfrac{1}{10}, \dfrac{1}{100}\right]$

TEACH

In order to appreciate both the mathematics and the technological advances, it is beneficial for students to understand how these computations were done before technology was available.

LESSON 7.5 Common Logarithms

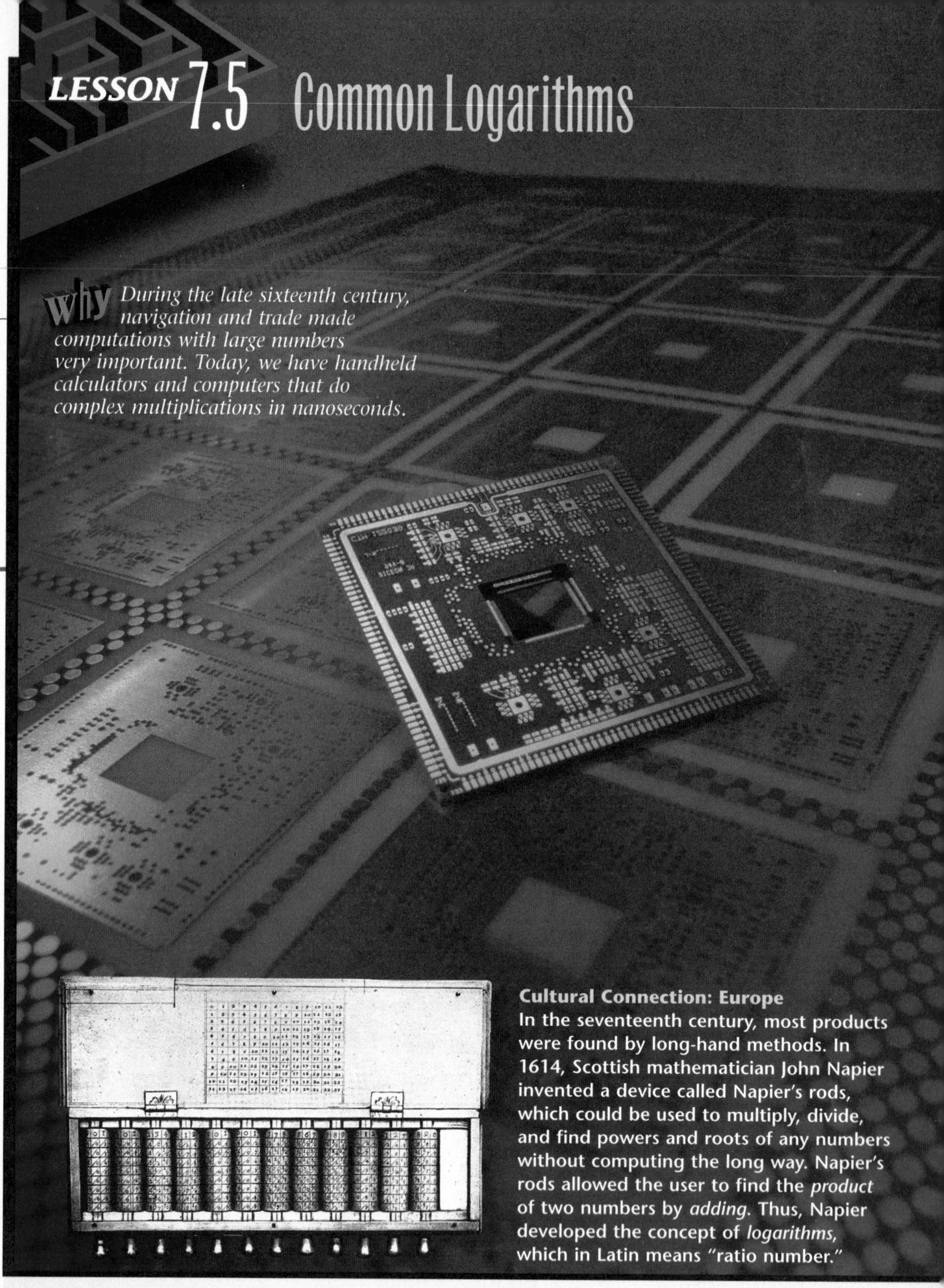

why *During the late sixteenth century, navigation and trade made computations with large numbers very important. Today, we have handheld calculators and computers that do complex multiplications in nanoseconds.*

Cultural Connection: Europe
In the seventeenth century, most products were found by long-hand methods. In 1614, Scottish mathematician John Napier invented a device called Napier's rods, which could be used to multiply, divide, and find powers and roots of any numbers without computing the long way. Napier's rods allowed the user to find the *product* of two numbers by *adding*. Thus, Napier developed the concept of *logarithms*, which in Latin means "ratio number."

Using Algorithms
Make a list of the definitions and rules given in the earlier sections of this chapter. These should include the product, quotient, and power rules for logarithms and the definition of logarithm in terms of the exponential function. Have students write the same rules for the common logarithm.

The logarithmic function with base 10, $f(x) = \log_{10} x$, is called the **common logarithmic function**. When the logarithm is written without a base, it is assumed that the base is 10. On most calculators this function is designated with the [LOG] key.

The equation $y = \log x$ is equivalent to $x = 10^y$. Thus, $\log 1000 = 3$ because $10^3 = 1000$.

A table of values for $y = 10^x$ and a table of values for $y = \log x$ are shown below.

x	$y = 10^x$
–2	0.01
–1	0.1
0	1
1	10
2	100

x	$y = \log x$
0.01	–2
0.1	–1
1	0
10	1
100	2

Notice that the x- and y-values in the two tables are exchanged, since $y = 10^x$ and $y = \log x$ are inverse functions. The graph of $y = \log x$ is a reflection of $y = 10^x$ across the line $y = x$.

What are the domain and range of $y = \log x$?

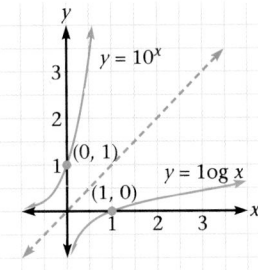

EXAMPLE 1

Write an equivalent logarithmic or exponential equation.

A $\log 10,000 = 4$

B $10^2 = 100$

Solution

A $\log 10,000 = 4$

The power of 10 that equals 10,000 is 4.

Thus, $10^4 = 10,000$.

B $10^2 = 100$

2 is the power of 10 that equals 100.

Thus, $2 = \log 100$. ❖

Working with logarithms in Napier's day was long and hard work. With today's technology, the level of accuracy you can easily produce is limited only by the real-world meaning of the result.

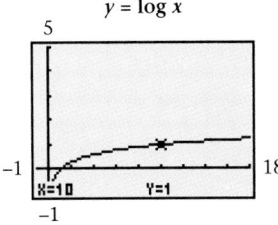

ENRICHMENT Give students the three functions $y = 2^x$, $y = e^x$, and $y = \log x$. Have students write these functions across the top of a page of note paper and write a function that is a translation, a reflection, and the inverse for each function.

INCLUSION **strategies** **Hands-On Strategies**
Graph paper can be very useful here. 10^x and $\log_{10} x$ are easy to sketch on graph paper and allow the student to slow down enough to see the real nature of these functions. Part of the challenge of this exercise is to choose the scale wisely. This requires the student to gain a general understanding of these functions before plotting points.

Alternate Example 2

Graph $y = \log x$ and evaluate each expression using the graph.
A. $\log 850$ [**2.93**]
B. The number whose log is 0.963 [**9.18**]

ASSESS

Selected Answers
Odd-numbered Exercises 7–49

Assignment Guide
Core 1–12, 14–18, 20–50

Core Plus 1–5, 10–13, 17–50

Technology
Graphics calculators are needed for Exercises 6–19, 21–32, 34–36, 40–41, and 50.

Error Analysis
Students often have difficulty determining the domain and range of various functions. Students must be careful to select a friendly viewing window that permits them to see the complete graph.

EXAMPLE 2

Graphics Calculator

Graph $y = \log x$, and evaluate each expression using the graph.

Ⓐ $\log 629$ Ⓑ the number whose log is 1.782

Solution▸

Ⓐ Find the *y*-value when *x* is 629.

Thus, $\log 629 \approx 2.80\%$.

Ⓑ Find the *x*-value when *y* is 1.782.

Since $\log x \approx 1.782$ when $x = 60.5$, the number whose log is 1.782 is approximately 60.5.

EXERCISES & PROBLEMS

Communicate

1. What did Napier's rods allow the user to do?
2. What base does the common logarithmic function have?
3. Name the inverse function of $f(x) = \log x$.
4. Compare the domain and range of $y = \log x$ with the domain and range of $y = 10^x$.
5. Explain how to use your calculator to find common logs.

Practice & Apply

Use your calculator to evaluate each logarithm. Round to the nearest hundredth.

6. $\log 35$ 1.54
7. $\log 2$ 0.30
8. $\log 0.135$ −0.87
9. $\log \pi$ 0.50
10. $\log 432$ 2.64
11. $\log \left(\frac{2}{27} \right)$ −1.13
12. $\log 0.0032$ −2.49
13. $\log \sqrt{\pi}$ 0.25

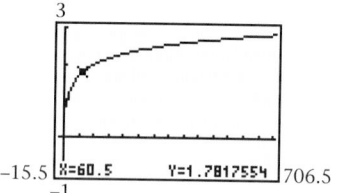

RETEACHING the lesson

Hands-On Strategies Have students use graph paper for this exercise. The left-most vertical line will be the *y*-axis and the bottom-most horizontal line will be the *x*-axis. Starting from the left, the marks along the *x*-axis will represent 1, 10, 100, 1000, and so on. Starting from the bottom, the marks along the *y*-axis will represent 0, 1, 2, 3, 4, and so on. Use these axes to graph the function $f(x) = \log_{10} x$. Have students describe the graph and explain why the graph is shaped in this way.

Use your calculator to find the number whose common log is the given value. Round to the nearest hundredth.

14 1.3
19.95

15 –1
0.10

16 4.5
31,622.78

17 0.017
1.04

18 0.3
2.00

19 –0.76
0.17

Acoustics The decibel, D (named after Alexander Graham Bell), is the unit of measure for the intensity, i, of sound, and is defined by $D(i) = 10 \log \left(\frac{i}{i_0} \right)$, where i_0 is the least intensity that can be heard by the human ear (approximately 10^{-16} watts per meter squared).

20. Find the loudness, in decibels, of conversational speech that is 1 million times as intense as the minimum, i_0.
60 W/m^2

21 How much more intense than the minimum intensity is the sound of an automobile with a decibel level of 55?
$\approx$ 316,228 times

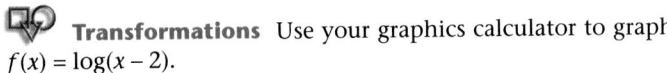 **Transformations** Use your graphics calculator to graph $f(x) = \log(x - 2)$.

Alexander Graham Bell, 1847-1922, inventor of the telephone, the audiometer (an early hearing-aid) and wax recorders for phonographs.

22 Find the domain and range of f. $x > 2$; R

23 How does the graph of f compare with that of $y = \log x$?
The graph of f is the graph of $y=\log x$ shifted 2 units right

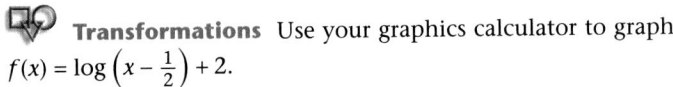 **Transformations** Use your graphics calculator to graph $f(x) = \log \left(x - \frac{1}{2} \right) + 2$.

24 What is the y-intercept? there is no y-intercept

25 Find the domain and range of f. $x > 0.5$; R

26 How does the graph of f compare with that of $y = \log x$?
The graph of f is the graph of $y=\log x$ shifted 0.5 units right and 2 units up

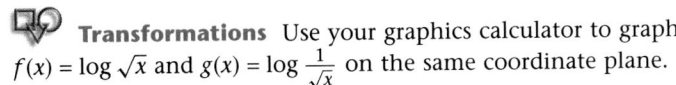 **Transformations** Use your graphics calculator to graph $f(x) = \log \sqrt{x}$ and $g(x) = \log \frac{1}{\sqrt{x}}$ on the same coordinate plane.

27 What is the axis of symmetry between the two graphs? the x-axis

28 How do f and g appear to be related? They are reflections across the x-axis

29 How are the graphs of f and g related to that of $y = \log x$?

Molecular Chemistry In chemistry, pH is defined as pH = $- \log[H^+]$, where $[H^+]$ is the hydrogen ion concentration in moles per liter. For tomato juice, $[H^+]$ is about 6.3×10^{-5}. Complete Exercises 30–36.

30 Find the pH of tomato juice. $\approx$ 4.2

31 An alkaline solution has 7 < pH ≤ 14. Determine the corresponding allowable range of $[H^+]$ for alkaline substances. 10^{-14} mol / L ≤ $[H^+]$ < 10^{-7} mol / L

29. Because $\log \sqrt{x} = \log x^{0.5} = 0.5 \log x$, the graph of f is a vertical compression of the graph of $y = \log x$ by a factor of 0.5. Because $\log \frac{1}{\sqrt{x}} = \log x^{-0.5} = -0.5 \log x$, the graph of g is a vertical compression of the graph of $y = \log x$ by a factor of 0.5, combined with a reflection across the x-axis.

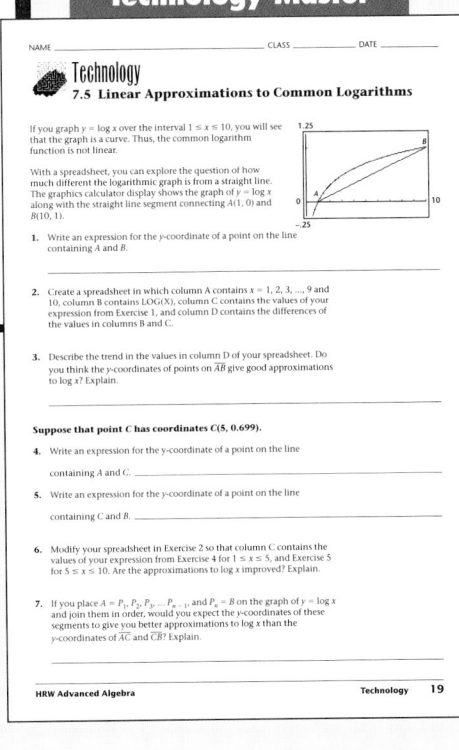

36. Acids and bases neutralize each other. Because milk of magnesia is alkaline, it works to neutralize the effects of acids. So, its effects are those of an anti-acid, or an antacid.

37. Let $y = \log_b x$. Then $b^y = x$. Taking the base 10 logarithm of both sides, $\log b^y = \log x$, or $y \log b = \log x$. So, $y = \frac{\log x}{\log b}$, and by substitution, $\log_b x = \frac{\log x}{\log b}$.

Look Beyond

This exercise helps the student visualize the natural exponential number and the graph of its function, which are introduced in the next section.

32 An acidic solution has $0 \le \text{pH} < 7$, and a neutral solution has pH = 7. Determine the corresponding allowable range of [H⁺] for acidic substances. 10^{-7} mol / L $< $ [H⁺] ≤ 1 mol / L

33. Are tomatoes acidic, neutral, or alkaline? acidic

34 Eggs have a [H⁺] of about 1.6×10^{-8}. Find their pH. Determine whether eggs are alkaline, acidic, or neutral. ≈ 7.80; alkaline

35 Milk of magnesia has a [H⁺] of 3.2×10^{-11}. Find its pH. Determine whether milk of magnesia is alkaline, acidic, or neutral. ≈ 10.49; alkaline

36 Using the results of Exercise 35, explain why milk of magnesia is called an antacid (anti-acid).

37. Prove that if a, b, and x are positive real numbers such that $b \ne 1$, then $\log_b x = \frac{\log x}{\log b}$. This is referred to as the **change of base formula**.

38. Use your calculator and the *change of base formula* shown in Exercise 37 to find $\log_2 15$. ≈ 3.91

Radioactive Decay Radioactive carbon-14 is used to date objects that are up to 50,000 years old. It has a half-life of about 5700 years, which means that a skeleton having half of its original amount of carbon-14 is about 5700 years old. The amount of carbon-14 present after t years is $A(t) = A_0 10^{-kt}$, where A_0 is the initial amount and $k = 5.3 \times 10^{-5}$.

39. Solve the decay equation for t in terms of $\frac{A}{A_0}$ and k. $t = \frac{-\log\left(\frac{A}{A_0}\right)}{k}$

40 How old is an object that has 65% of its original carbon-14? ≈ 3500 yr

41 How old is an object that has 15% of its original carbon-14? ≈ 15,500 yr

Look Back

Simplify. [Lessons 5.6, 5.7]

42. $\left(-i^3 \sqrt{-4}\right)$ −2

43. $(4 + 2i)(-5 - 7i)$
−6 − 38i

44. $(-3i)(4 - 6i)$ −18 − 12i

Identify the multiplier for each growth rate. [Lesson 7.1]

45. 4% 1.04   **46.** −3.7% 0.963   **47.** 0.9% 1.009   **48.** −31% 0.69   **49.** 140% 2.4

Look Beyond

50 Graph $f(x) = e^x$, and describe the graph.

50. The graph looks similar to the graph of $f(x) = 2^x$, except that it increases a little more rapidly. The graph always increases as x increases, has the x-axis as a horizontal asymptote, and has y-intercept (0, 1). The domain is the set of all real numbers, and the range is the set of all positive real numbers.

$e = 1 + \dfrac{1}{1} + \dfrac{1}{1} + \dfrac{1}{2 \cdot 1} + \dfrac{1}{3 \cdot 2 \cdot 1} + \dfrac{1}{4 \cdot 3 \cdot 2 \cdot 1} + \dfrac{1}{5 \cdot 4 \cdot 3 \cdot 2 \cdot 1}$

$e = 2.718281828459045235360287471352$

Why *Recall from Lesson 7.2 that the more frequently interest is compounded, the higher the investment return will be. The highest possible investment return for a principal amount, P, at an annual interest rate, r, is one that is continuously compounded. The formula for this type of investment includes a special number called the natural number. The natural number, e, is used in many business and scientific models.*

The Natural Exponential Function

Like π, the number e is a special number.

Exploration Approaching e

Graphics Calculator

Let $f(n) = 1 + \dfrac{1}{n}$ and $g(n) = \left(1 + \dfrac{1}{n}\right)^n$.

Use your calculator to evaluate each of the following.

1 $f(1)$ and $g(1)$

2 $f(10)$ and $g(10)$

3 $f(100)$ and $g(100)$

4 $f(1000)$ and $g(1000)$

5 $f(10{,}000)$ and $g(10{,}000)$

6 What number does g approach as values of n become larger? ❖

You can see in the exploration that even though the values of f approach 1 as values of n become larger, the values of g do not approach 1. As values of n become larger, the values of g approach a number represented by e, which is approximately 2.718.

ALTERNATIVE teaching strategy

Using Patterns The number pattern that approximates the value of e is illustrated in the lesson opener. This interesting pattern can be generalized to explore values of the natural exponential function. Have students approximate e by generating and adding several terms of the pattern.

$e = 1 + \dfrac{1}{1} + \dfrac{1}{2 \cdot 1} + \dfrac{1}{3 \cdot 2 \cdot 1} + \dfrac{1}{4 \cdot 3 \cdot 2 \cdot 1}$
$\qquad + \dfrac{1}{5 \cdot 4 \cdot 3 \cdot 2 \cdot 1} + \dots$

Students may then explore the related function by selecting a value for x, substituting, and adding the terms for an approximation.

$e^x = 1 + x + \dfrac{x^2}{2!} + \dfrac{x^3}{3!} + \dfrac{x^4}{4!} + \dots$

Technology Refer to your *HRW Technology Handbook* for instructions on computing factorials on various calculators.

Cooperative Learning

Have the students form pairs. One student will use the calculator to compute:

$1 + \frac{1}{1!}$

$1 + \frac{1}{1!} + \frac{1}{2!}$

$1 + \frac{1}{1!} + \frac{1}{2!} + \frac{1}{3!}$

and so on up to $\frac{1}{10!}$

The other student will compile a table of the results.

Alternate Example 1

Suppose you invest $2000 at an interest rate of 6%.

A. How much is the investment worth after 8 years if it is compounded quarterly? [**$3220.65**]

B. How much is the investment worth after 8 years if the interest is compounded continuously? [**$3232.15**]

C. Which type yields the larger return? [**B**]

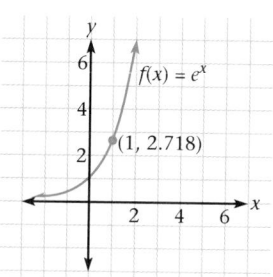

The exponential function, $f(x) = e^x$, with **natural number e**, as the base, is graphed here.

This function, $f(x) = e^x$, is called the **natural exponential function**.

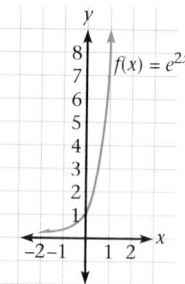

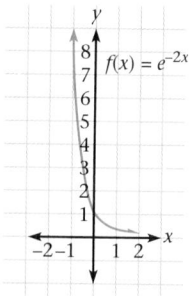

The graph of $f(x) = e^{ax}$, where $a > 0$, is an exponential growth function.

The graph of $f(x) = e^{ax}$, where $a < 0$, is an exponential decay function.

Investment In today's banking industry, interest is usually compounded continuously. The formulas for compounding interest periodically and compounding interest continuously are compared below.

Periodic compounding	**Continuous compounding**
$A = P\left(1 + \frac{r}{n}\right)^{nt}$	$A = Pe^{rt}$

In both formulas, P is the principal amount invested, r is the annual interest rate, and t is the number of years of investment. In the formula for periodic compounding, n is the number of times interest is compounded per year.

EXAMPLE 1

Suppose you invest $1000 at an interest rate of 7.6%.

A How much is the investment worth after 8 years if the interest is compounded quarterly?

B How much is the investment worth after 8 years if the interest is compounded continuously?

C Compare the results of Parts **A** and **B**. Which type of compound interest yields the larger return?

interdisciplinary
CONNECTION

History The history of the number e and other irrational numbers, like π and $\sqrt{2}$, is long and fascinating. Suggest to the students that they go to the library and research the history of some irrational number. When did it first appear and in what field?

ENRICHMENT

Two apparently different definitions of e are given on page 407. The first is in terms of factorials.

$$1 + \frac{1}{1!} + \frac{1}{2!} + \frac{1}{3!} + \ldots$$

The other is in terms of n, $g(n) = \left(1 + \frac{1}{n}\right)^n$. Students may compare these two definitions.

Solution➤

In this example, P is 1000, r is 0.076, and t is 8.

Ⓐ When interest is compounded quarterly, n is 4.

$$A = P\left(1 + \frac{r}{n}\right)^{nt} = 1000\left(1 + \frac{0.076}{4}\right)^{4\cdot8} \approx 1826.31$$

Ⓑ When interest is compounded continuously, $A = Pe^{rt}$.

$$A = Pe^{rt} = 1000e^{0.076\cdot8} \approx 1836.75$$

Ⓒ Investing a principal amount at a given interest rate for a certain number of years will always result in a greater return if the interest is compounded continuously. ❖

The Natural Logarithmic Function

The inverse of the natural exponential function, $y = e^x$, is the **natural logarithmic function**, $y = \log_e x$. The natural logarithmic function is usually written $y = \ln x$, where ln is the abbreviation for *natural log*.

When you need to find the time, t, that it will take for an investment that is growing continuously to reach a specific amount, use a logarithmic function.

EXAMPLE 2

Investment Suppose you invest $1500 at an annual interest rate of 8.5%.

Ⓐ How long does it take to double your investment if the interest is compounded quarterly?

Ⓑ How long does it take to double your investment if the interest is compounded continuously?

Solution➤

P is 1500, r is 0.085, and A is double the principal, or $3000.

Ⓐ Substitute 4 for n, and solve for t.

$$A = P\left(1 + \frac{r}{n}\right)^{nt}$$
$$3000 = 1500\left(1 + \frac{0.085}{4}\right)^{4t}$$
$$2 = 1.02125^{4t}$$
$$\ln 2 = \ln(1.02125)^{4t}$$
$$\ln 2 = 4t\ln(1.02125)$$
$$\frac{\ln 2}{4\ln(1.02125)} = t$$
$$8.24 \approx t$$

Thus, it will take almost 8 years and 3 months to double $1500 if 8.5% interest is compounded quarterly.

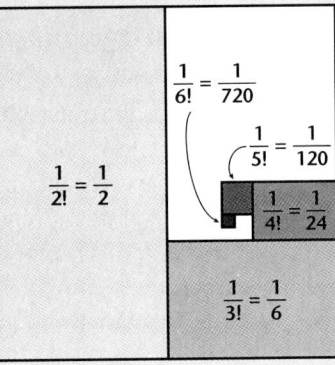

TEACHING tip

Technology Refer to your *HRW Technology Handbook* for instructions on graphing e^x and $\ln x$.

Yes. The general formula for doubling time is $2P = Pe^{rt}$. Dividing both sides by P gives $2 = e^{rt}$, which simplifies (by taking natural logarithms of both sides) to $t = \frac{\ln 2}{r}$, which depends only on the interest rate.

Alternate Example 3

How much of a 100 gram sample of radioactive strontium-90 will remain after 100 years? [**6.27 grams**]

Aongoing
SSESSMENT

Try This

$N = 40e^{-0.0277 \cdot 35} = 15.17$ grams

B Use $A = Pe^{rt}$, and solve for t.

$$A = Pe^{rt}$$
$$3000 = 1500e^{0.085t}$$
$$2 = e^{0.085t}$$
$$0.085t = \ln 2$$
$$t = \frac{\ln 2}{0.085}$$
$$t \approx 8.15$$

Thus, it will take almost 8 years and 2 months to double $1500 at 8.5%, if the interest is compounded continuously. ❖

CRITICAL
Thinking

It is sometimes said that the doubling time depends only on the interest rate, not on the principal. Is this true? Explain.

Exponential Growth and Decay

Pictured above is the flame test for strontium-90.

Radioactive decay is one example of exponential decay. Radioactive substances contain unstable atoms that change to more stable ones. This results in an exponential decrease in the amount of original substance present. The equation that models exponential decay is

$$N = N_0 e^{-kt}, \text{ where } \begin{cases} N_0 \text{ is the amount at time } t = 0, \\ N \text{ is the amount at time } t, \text{ and} \\ k \text{ is a positive constant that depends} \\ \text{on the substance.} \end{cases}$$

EXAMPLE 3

Science   The amount of radioactive strontium-90 remaining after t years decreases according to the formula $N = N_0 e^{-0.0277t}$. How much of a 40-gram sample will remain after 25 years?

Solution

To find N, the amount remaining after 25 years, substitute 40 for the initial amount, N_0, and 25 for t in $N = N_0 e^{-0.0277t}$.

$$N = N_0 e^{-0.0277t}$$
$$N = 40e^{-0.0277(25)}$$
$$N \approx 20.01$$

After 25 years, about 20 grams of the original 40 grams of strontium-90 will remain. This means that strontium-90 has a *half-life* of about 25 years. ❖

Try This   How much of the 40-gram sample of strontium-90 remains after 35 years?

RETEACHING
the
lesson

Using Tables Use the calculator to compute successive values of e^x and $\ln x$ to illustrate the behavior of these functions. Using the e^x key and the $\ln x$ key, have the students construct a table of values for the two functions. The progressively steeper increase of e^x and the progressively more gradual increase in the values of $\ln x$ show the inverse characteristics of these two functions.

Exercises & Problems

Communicate

1. What is the *natural* exponential base? Give its value correct to the nearest thousandth.

2. What kind of exponential growth and decay models have *e* as a base?

3. How are the natural logarithmic function and the natural exponential function related?

4. Give the formula that describes continuous exponential growth.

5. Give the formula that describes continuous exponential decay.

6. Explain what is meant by the half-life of a radioactive substance.

Practice & Apply

Use your calculator to evaluate each expression to the nearest hundredth.

7. $e^{-1.3}$ 0.27

8. e^3 20.09

9. $\ln 4.78$ 1.56

10. $\ln 10$ 2.30

11. e^π 23.14

12. $\ln \pi$ 1.14

13. $e^{\sqrt{2}}$ 4.11

14. $\ln \sqrt{5}$ 0.80

Write an equivalent logarithmic or exponential equation.

15. $e^{0.5} = 1.65$ $\ln 1.65 = 0.5$

16. $\ln 2 = 0.69$ $e^{0.69} = 2$

17. $e^{\frac{1}{3}} = 1.40$ $\ln 1.40 = \frac{1}{3}$

18. $\ln \pi = 1.14$ $e^{1.14} = \pi$

When solving $5^x = 30$, Geoff took the log of both sides, Jean took the ln of both sides, and Tanu'e took the $\log_5$ of both sides.

19. Explain why $\log 5^x \neq x$ and $\ln 5^x \neq x$.

20. What could be a disadvantage of the method Tanu'e used?

19. $\log_b 5^x = x$ only when $b = 5$, because logarithmic and exponential functions are inverses only when they have the same base. The base of 5^x is 5, but the bases of the common log and natural log are 10 and e, respectively.

20. Taking the $\log_5$ of both sides gives $x = \log_5 30$, which she would have to change to an expression involving common or natural logarithms in order to find the value using a calculator or standard tables.

Assess

Selected Answers
Odd-numbered Exercises 7–57

Assignment Guide
Core 1–12, 15–24, 27–57

Core Plus 1–6, 11–20, 23–58

Technology
Graphics calculators are needed for Exercises 7–14, 21–26, 28–33, 36–47, 49–51.

Error Analysis
Remind students to use friendly viewing windows when approximating values on their graphics calculators.

Have the students go to the library and look up elevations for 10 major cities around the world. Students should use the formula given for Exercises 48–51 to compute the standard atmospheric pressure at each city.

Solve each of the following equations for x without graphing.

21 $35^x = 30$ 0.96   **22** $1.3^x = 8$ 7.93   **23** $3^{-x} = 17$ –2.58

24 $36^{2x} = 20$ 0.42   **25** $0.42^{-x} = 7$ 2.24   **26** $2^{-\frac{1}{3}x} = 10$ –9.97

27. Solve the general equation $b^x = c$ for x in terms of b and c. $x = \frac{\log c}{\log b}$ or $x = \frac{\ln c}{\ln b}$

28 **Investment** How long will it take an investment of $800 to triple in value at 5% interest compounded continuously? $\approx$ 22 yr

29 Graph $f(x) = e^{-2x}$, e^{-x}, e^x, and e^{2x} on the same coordinate plane. What controls the rate of growth or decay in these functions?

Economics The demand, x, for a certain product is related to its price, P, by the function $P = 30 + 0.3e^{0.012x}$.

30 Find the price if the demand is 700 units.   1364

31 Find the price if the demand is 800 units.   4459

32 What can you conclude about the connection between price and demand for this product?

33 **Transformations** Graph $f(x) = e^x$ and $g(x) = e^{-x}$ on the same coordinate plane. What is the axis of symmetry of this pair of graphs? By what kind of transformation are f and g related?

Transformations For Exercises 34 and 35, graph $f(x) = \ln x^2$ and $g(x) = \ln \frac{1}{x^2}$ on the same coordinate plane.

34 Are there any lines of symmetry? Explain.

35 What transformation (if any) relates f to g?
They are reflections about the x-axis

Use your graphics calculator to graph $f(x) = \ln x + 2$.

36 Find the x-intercept of f. approx.(0.14,0)

37 Compare the graph of f with that of $g(x) = \ln x$.
graph of f is graph of g shifted 2 units up.

Use your graphics calculator to graph $f(x) = \ln x - 2$.

38 Find the x-intercept of f. approx. (7.39,0)

39 Compare the graph of f with that of $g(x) = \ln x$.
graph of f is graph of g shifted 2 units down.

40 **Investment** Use your graphics calculator to compare the growth of an investment of $2000 in two different accounts. One account earns 3% interest and the other earns 5% interest, and both are compounded continuously over 20 years.

Use your graphics calculator to graph $f(x) = 5e^{-2x}$.

41 Find and graph the inverse function. $f^{-1}(x) = -0.5\ln\left(\frac{x}{5}\right)$

42 What are the domain and range of f and f^{-1}? f:R; $y > 0$; f^{-1}:$x > 0$; R

43 What is the x-intercept of f and of f^{-1}? f^{-1}:(5, 0), f: no x-intercept

29. When the coefficient of x is positive, the graph is of exponential growth. When the coefficient of x is negative, the graph is of exponential decay. The greater the absolute value of the coefficient, the greater the rate of growth or decay.

32. The greater the demand, the greater the price. Price grows exponentially relative to demand.

33. y-axis; they are reflections with respect to the y-axis.

34. f and g are each symmetric to the y-axis; f and g are reflections of each other with respect to the x-axis

40. 3%: $3644.24; 5%: $5436.56

Investment Use your graphics calculator for Exercises 44–46.
Assume that all interest rates are compounded continuously.

44 How long will it take an investment of $5000 to double if the annual percentage rate is 6%? ≈ 11.55 years

45 How long will it take an investment to double at 10% interest? ≈ 6.93 years

46 If it takes a certain amount of money 3.7 years to double, at what interest rate was the money invested? ≈ 18.7%

47 Radioactive Decay
A radioactive element decays according to the decay function $A = A_0 e^{-(0.04t)}$, where A_0 is the amount of the substance initially present, and t is time in years from the present. How long will it take a sample of this element to decay to one-half its original amount? ≈ 17.33 years

Meteorology Under normal conditions, the atmospheric pressure at height h, in feet is given by $P = 1013e^{-0.0000385h}$, where P is the pressure in millibars.

48. At sea level, h is 0. What is the standard atmospheric pressure at sea level? 1013 millibars

49 Find the pressure at 5000 feet.

50 Find the pressure at 1500 feet.

51 Find the height above (or below) sea level for your school. What is the normal atmospheric pressure at your school? Answers may vary

A technician prepares a weather balloon.

Look Back

Describe the behavior of each of the given functions.
[Lesson 3.4]

52. $f(x) = \frac{1}{2}|x|$ **53.** $g(x) = -|x|$ **54.** $h(x) = |x - 3|$

Factor each of the polynomial expressions.   [Lesson 6.2]

55. $x^2 - 3x - 10$ **56.** $3x^2 - 6x + 3$ **57.** $x^2 - 49$
$(x - 5)(x + 2)$ $3(x - 1)^2$ $(x + 7)(x - 7)$

Look Beyond

58 Solve $\ln x + \ln(x + 2) = 5$ by graphing. $x \approx 11.22$

52. The graph of $f(x) = \frac{1}{2}|x|$ is the graph of $f(x) = |x|$ vertically compressed by a factor of 0.5. So, it still has the vertex $(0, 0)$, but it opens wider for the same domain.

53. The graph of $g(x) = -|x|$ is the graph of $f(x) = |x|$ reflected over the x-axis. So, it still has the vertex $(0, 0)$ and the same shape, but it opens downward instead of upward.

54. The graph of $g(x) = |x - 3|$ is the graph of $f(x) = |x|$ shifted 3 units to the right. So, it has the same shape and direction, but the vertex is now $(3, 0)$.

Look Beyond

This problem employs many of the properties of logarithmic expressions. It also prepares students for solving more complex logarithmic equations using the graphics calculator.

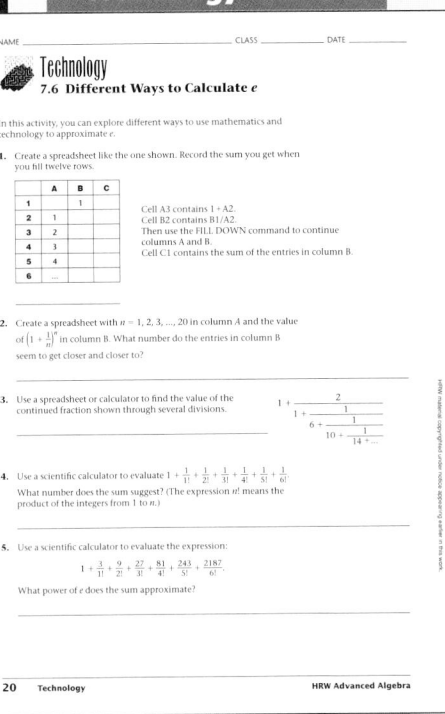

Focus

An inverted *catenary* curve is a mathematics abstraction that is based on the sum of e^x functions. This stable and artistically sweeping curve is used for the structure of the St. Louis Gateway Arch. Students will have the opportunity to explore the function, graph the curve, examine its unusual properties, and experiment with some transformations. Since the curve is similar to a parabola, students will also compare a parabola with a catenary curve.

Motivate

Before studying the catenary, have students experiment on a graphics calculator or computer with the function e^x and the sum function $e^x + e^{-x}$. Have them also do research to learn about the Gateway Arch in St. Louis and its construction.

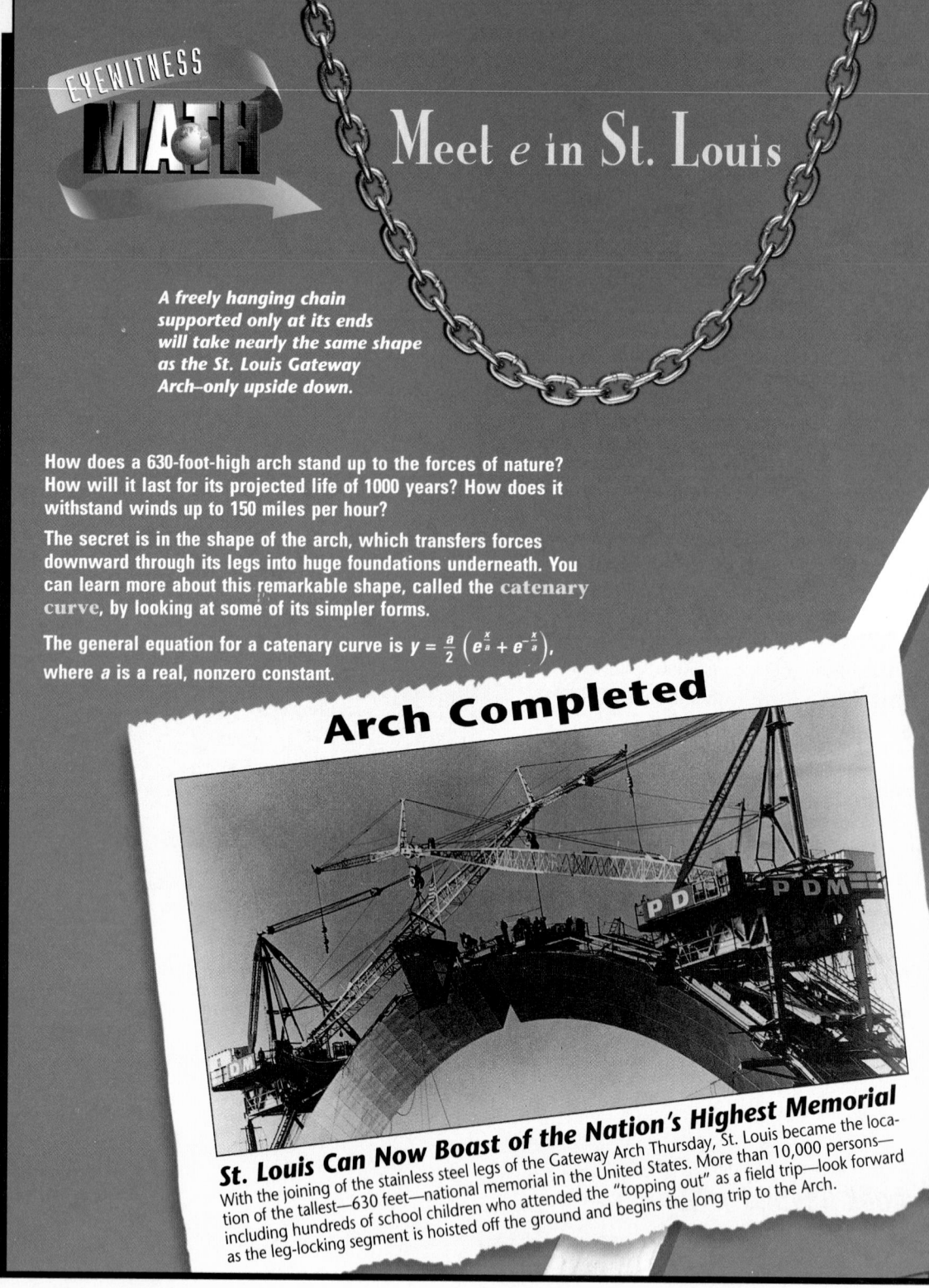

Meet *e* in St. Louis

A freely hanging chain supported only at its ends will take nearly the same shape as the St. Louis Gateway Arch—only upside down.

How does a 630-foot-high arch stand up to the forces of nature? How will it last for its projected life of 1000 years? How does it withstand winds up to 150 miles per hour?

The secret is in the shape of the arch, which transfers forces downward through its legs into huge foundations underneath. You can learn more about this remarkable shape, called the **catenary curve,** by looking at some of its simpler forms.

The general equation for a catenary curve is $y = \frac{a}{2}\left(e^{\frac{x}{a}} + e^{-\frac{x}{a}}\right)$, where *a* is a real, nonzero constant.

Arch Completed

St. Louis Can Now Boast of the Nation's Highest Memorial

With the joining of the stainless steel legs of the Gateway Arch Thursday, St. Louis became the location of the tallest—630 feet—national memorial in the United States. More than 10,000 persons—including hundreds of school children who attended the "topping out" as a field trip—look forward as the leg-locking segment is hoisted off the ground and begins the long trip to the Arch.

A1. $y = e^{\frac{x}{2}} + e^{\frac{-x}{2}}$

A2.

x	0	1	2	3	4	5
y	2.00	2.26	3.09	4.70	7.52	12.26

A3. $y = 2.26$ for $x = -1$. This is the same value as for $x = 1$.

A4. If you substitute $-x$ for x in the equation, you get $y = e^{\frac{-x}{2}} + e^{\frac{x}{2}}$, which is equivalent to the original equation.

B1. The curve looks like a parabola-type curve that opens up and has a vertex at $(0, 2)$.

B2. Multiply the function by -1; $y = -\left(e^{\frac{-x}{2}} + e^{\frac{x}{2}}\right)$.

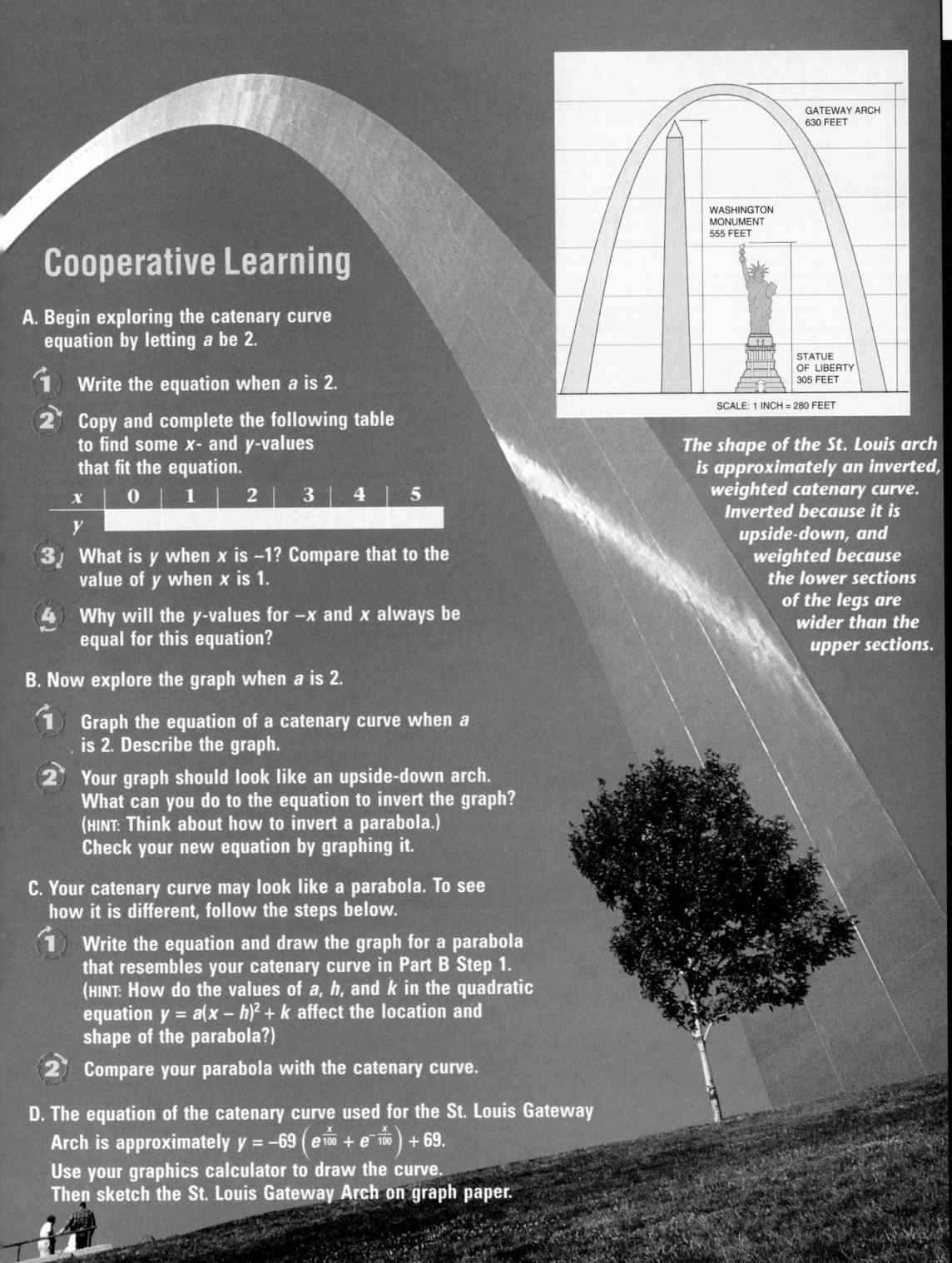

Cooperative Learning

A. Begin exploring the catenary curve equation by letting *a* be 2.

1 Write the equation when *a* is 2.

2 Copy and complete the following table to find some *x*- and *y*-values that fit the equation.

x	0	1	2	3	4	5
y						

3 What is *y* when *x* is –1? Compare that to the value of *y* when *x* is 1.

4 Why will the *y*-values for –*x* and *x* always be equal for this equation?

B. Now explore the graph when *a* is 2.

1 Graph the equation of a catenary curve when *a* is 2. Describe the graph.

2 Your graph should look like an upside-down arch. What can you do to the equation to invert the graph? (HINT: Think about how to invert a parabola.) Check your new equation by graphing it.

C. Your catenary curve may look like a parabola. To see how it is different, follow the steps below.

1 Write the equation and draw the graph for a parabola that resembles your catenary curve in Part B Step 1. (HINT: How do the values of *a*, *h*, and *k* in the quadratic equation $y = a(x - h)^2 + k$ affect the location and shape of the parabola?)

2 Compare your parabola with the catenary curve.

D. The equation of the catenary curve used for the St. Louis Gateway Arch is approximately $y = -69 \left(e^{\frac{x}{100}} + e^{-\frac{x}{100}} \right) + 69$.
Use your graphics calculator to draw the curve. Then sketch the St. Louis Gateway Arch on graph paper.

GATEWAY ARCH
630 FEET

WASHINGTON MONUMENT
555 FEET

STATUE OF LIBERTY
305 FEET

SCALE: 1 INCH = 280 FEET

The shape of the St. Louis arch is approximately an inverted, weighted catenary curve. Inverted because it is upside-down, and weighted because the lower sections of the legs are wider than the upper sections.

Cooperative Learning

Have groups of students use graphics calculators to explore the graphs of functions based on the function $f(x) = b \left(e^{\frac{x}{a}} + e^{\frac{-x}{a}} \right) + c$. As each group discovers more about the behavior of the function and its transformations, have the group record that information and prepare a presentation of their findings to the class.

DISCUSS

Discuss the results from the Cooperative Learning task above. Have students make conjectures and generalizations from their observations. Then discuss how the function formula for the basic catenary curve must be changed to translate the graph of the function horizontally.

Additional discussion can focus on the function e^x. The history, application, and calculation of this function can stimulate interest and motivate research projects. One such aspect of *e* is the series approximation:

$$e \approx 1 + \frac{1}{1!} + \frac{1}{2!} + \frac{1}{3!} + \ldots + \frac{1}{n!} \ldots$$

$$e \approx 2.718$$

C1. Answers may vary. A possible parabola, since the vertex is (0, 2) and the curve is very wide, is $y = 0.5x^2 + 2$.

C2. Answers may vary. Using the answer given in C1, the catenary is wider than the parabola near the vertex, but as *x* gets larger, the catenary becomes narrower than the parabola.

D. Answers may vary.

Objectives

- Solve logarithmic and exponential equations by graphing.
- Use the exponential-log inverse properties to solve logarithmic and exponential equations.

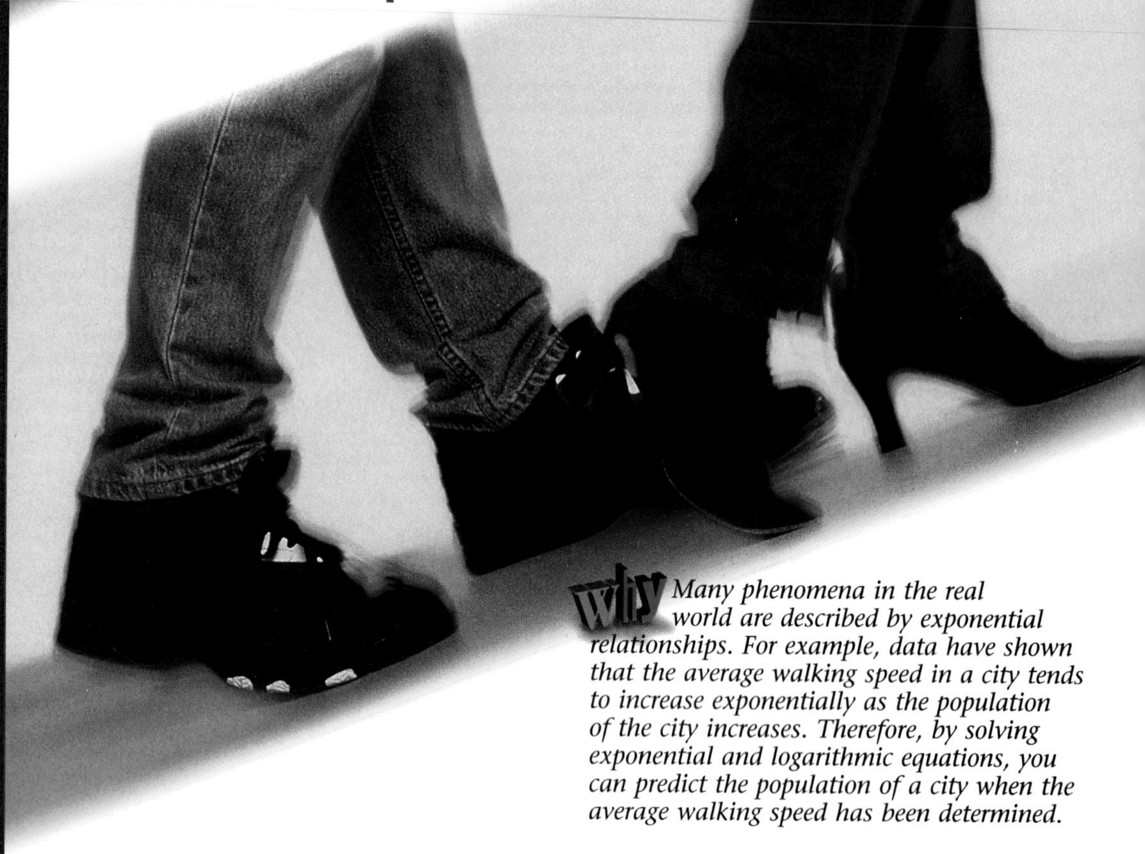

RESOURCES

- Practice Master   7.7
- Enrichment Master   7.7
- Technology Master   7.7
- Lesson Activity Master   7.7
- Quiz   7.7
- Spanish Resources   7.7

Assessing Prior Knowledge

1. $\log_2 2^4 =?$ [4]

2. $\log_5 25 =?$ [2]

3. $2^{\log_2 7.4} =?$ [7.4]

4. $e^{\ln 2} =?$ [2]

TEACH

 Discuss the fact that formulas are commonly used in the physical and social sciences. Formulas allow us to discover new information by substituting given information into the formula. Encourage students to recall other examples of formulas that they studied in science courses.

LESSON 7.7 Solving Exponential and Logarithmic Equations

Many phenomena in the real world are described by exponential relationships. For example, data have shown that the average walking speed in a city tends to increase exponentially as the population of the city increases. Therefore, by solving exponential and logarithmic equations, you can predict the population of a city when the average walking speed has been determined.

The graphics calculator makes it possible to solve many problems involving exponents and logs. Keep the following properties in mind as you solve each equation you encounter.

1. Basic Definition $y = \log_b x$ if and only if $b^y = x$

2. Product Property $\log_b mn = \log_b m + \log_b n$

3. Quotient Property $\log_b \left(\dfrac{m}{n} \right) = \log_b m - \log_b n$

4. Power Property $\log_b m^p = p \log_b m$

5. Inverse Properties $b^{\log_b x} = x$ for $x > 0$, $\log_b b^x = x$ for all x

6. Equality Property   If $\log_b m = \log_b n$, then $m = n$.

 **Using Models** Students understand the rules more clearly if they understand why the rules work the way they do. Present an example of each rule using numbers. Encourage students to create their own examples. For example, consider $\log_2(32) = \log_2(8 \cdot 4) = \log_2 8 + \log_2 4$.

Alternate Example 1

Solve $\log x + \log(x - 9) = 1$ for x by graphics calculator and by using the exponential-log properties. [$x = 10$]

EXAMPLE 1

Graphics Calculator

Solve $\log x + \log(x - 3) = 1$ for x:

A by graphing.

B by using the exponential-log properties.

Solution

A Graph $y = \log x + \log(x - 3)$.

Use the trace feature to find the value of x when y is 1.

From the graph, you can see that the solution is 5.

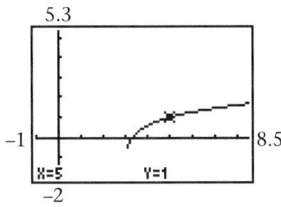

B Apply the product property of logarithms.
Apply the definition of logarithms.
Apply the equality property.
Solve the resulting quadratic equation.

$$\log x + \log(x - 3) = 1$$
$$\log x(x - 3) = 1$$
$$\log x(x - 3) = \log 10$$
$$x(x - 3) = 10$$
$$x^2 - 3x - 10 = 0$$
$$(x - 5)(x + 2) = 0$$
$$x = 5 \text{ or } x = -2$$

Since the domain of a logarithmic function cannot be negative, the only solution is 5. Notice that this is the same solution found by graphing.

Check

$$\log x + \log(x - 3) = 1$$
$$\log(5) + \log(5 - 3) \overset{?}{=} 1$$
$$\log(5 \cdot 2) \overset{?}{=} 1$$
$$\log 10 = 1 \quad \text{True} \; \diamond$$

Try This Solve the logarithmic equation $\ln(x) + \ln(x - 4) = 3$ by graphing and by using the exponential-log properties.

TEACHING tip

Have students do Example 1 in pairs. One student can do the problem using the graphics calculator and the other by applying the rules for exponents and logarithms. Then students can compare results for accuracy and increased understanding.

A ongoing ASSESSMENT

Try This

6.9

TEACHING tip

This is a good section to acquaint, or reacquaint, students with the idea of proofs or formal derivations. The solutions to these exercises can be written out step-by-step in a vertical column with reasons written out to the right. Each reason should refer to one of the rules given at the beginning of the lesson.

ENRICHMENT Take the function given in Example 3 and convert the result to miles per hour. Using a graphics calculator, graph this function for populations up to 20 million. Do the walking speeds for small towns seem reasonable? What about a population of 1 or a population of 2? What about a population near 20 million? Have students discuss their findings.

INCLUSION strategies **English Language Development** It may help students if the equation solving process is described in terms of a goal, such as isolating x on one side of the equality sign. Each step is a matter of undoing whatever the given operation did. Exponential functions undo whatever a logarithmic function did, and logarithmic functions undo whatever an exponential function did.

TEACHING tip

For Example 2, the students can graph $4e^{3x-5}$ and trace to determine when it is equal to 72. Alternatively, one student can graph $4e^{3x-5} - 72$ and determine when it is equal to zero.

Alternate Example 2

Solve $5e^{2x} = 100$ by using the exponential-log properties.

$$\left[x = \frac{\ln 20}{2} \right]$$

Alternate Example 3

What is the predicted population in a city where the average walking speed is 6.53 feet per second? **[51,300]**

EXAMPLE 2

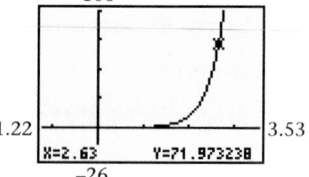

Graphics Calculator

Solve $4e^{3x-5} = 72$ for x:

Ⓐ by graphing.

Ⓑ by using the exponential-log properties.

Solution➤

Ⓐ Graph $y = 4e^{3x-5}$.

Use the trace feature to find the value of x when y is 72.

Using the zoom feature, you can find that x is approximately 2.6 when y is approximately 72.

[Graph showing exponential curve with window: top 100, left −1.22, right 3.53, bottom −26, with X=2.63 Y=71.973238]

Ⓑ Divide both sides of the equation by 4.

$$4e^{3x-5} = 72$$
$$e^{3x-5} = 18$$

Take the ln of both sides.
Use the logarithmic-exponential inverse property.
Solve the resulting linear equation.

$$\ln(e^{3x-5}) = \ln 18$$
$$3x - 5 = \ln 18$$
$$3x = \ln 18 + 5$$
$$x \approx 2.63$$

Notice that the graphing and nongraphing methods yield the same approximate solution for x.

Check➤

$$4e^{3x-5} = 72$$
$$4e^{3(2.63)-5} \stackrel{?}{=} 72$$
$$4(17.99) \approx 72 \quad \text{True} \; ❖$$

EXAMPLE 3

Stress Suppose the average walking speed as a function of city population, p, is $s(p) = 1.38 \log(p) + 0.03$. What is the predicted population in a city where the average walking speed is 7.45 feet per second?

Solution➤

To solve for p, you need a way to isolate p. This may be done by first isolating $\log p$. Then p may be obtained using the logarithmic-exponential inverse properties.

First solve for $\log p$.

$$s = 1.38 \log p + 0.03$$
$$\frac{s - 0.03}{1.38} = \log p$$

Apply the definition of logs to isolate p.

$$10^{\left(\frac{s-0.03}{1.38} \right)} = p$$

Substitute 7.45 for s.

$$10^{\frac{7.42}{1.38}} = p$$
$$238,128.6 \approx p$$

Thus, in a city where the average walking speed is 7.45 feet per second, the predicted population is approximately 240,000. ❖

RETEACHING the lesson

Inviting Participation
Write out the solution to an equation on the board. Ask the students to supply the reasons for each step and what rule is used. Then engage them in a discussion about the goal of solving such equations and the most important rules of exponents and logarithms that are used at crucial steps.

Try this   Find the walking speed for each city.

City Name	Population
New York City, NY	7,323,000
Phoenix, AZ	983,000
Cleveland, OH	506,000
Sacramento, CA	369,000
St. Petersburg, FL	240,000
Garland, TX	180,000
Laie, HA	6,000

CRITICAL
Thinking

Explain how to determine the average walking speed in any city for which you know the population.

EXERCISES & PROBLEMS

Communicate

1. Explain how an exponential equation may be solved by graphing.
2. Outline the method for solving an exponential equation without graphing.
3. Explain how a logarithmic equation may be solved by graphing.
4. Outline the method for solving a logarithmic equation without graphing.
5. Describe how the inverse relation between exponential and logarithmic functions can be used to solve real-world problems.

ongoing
ASSESSMENT

Try This

City	Walking Speed
NY	9.5
Pho	8.3
Cle	7.9
Sac	7.7
St. P	7.5
Gar	7.3
Lai	5.2

CRITICAL
Thinking

If you know the population, substitute the population for p in $s = 1.38 \log p + 0.03$, and find s.

ASSESS

Selected Answers

Odd-numbered Exercises 7–45

Assignment Guide

Core 1–9, 11–22, 27–45

Core Plus 1–5, 8–17, 21–46

Technology

A graphics calculator is needed for Exercises 11–26, 28–31, 34–36, 38–39, and 46.

Error Analysis

Only equations with integer solutions can be done without a calculator. Students can recognize these equations if they recall the most common powers. Have them review basic powers, such as 2^2, 2^3, 2^4, as well as simple powers for bases 3, 4, and 5.

Practice & Apply

Solve each of the following equations.

6. $3^x = 3^4$ **7.** $3^{2x} = 81$ **8.** $\log_3 x = 2$ **9.** $\ln(2x - 3) = \ln 21$ **10.** $4^{x+1} = \frac{1}{128}$
   4   2   9   12   −4.5

Use your calculator to solve the following exponential equations to the nearest hundredth.

11. $2^x = 35$ 5.13   **12.** $8^x = 5$ 0.77   **13.** $e^{2x} = 20$ 1.50

Use your calculator to solve the following logarithmic equations to the nearest tenth.

14. $\log x = 5.2$ **15.** $\ln 2x = 3$ **16.** $\ln x^2 = 8$
   158,489.3   10.0   −54.6 or 54.6

17. Accounting How many years will it take an investment to triple itself when interest is compounded continuously at 6.5%?   ≈ 16.9 years

Use your calculator to solve each equation to the nearest hundredth.

18. $5 \ln(2x - 3) = 1.8$ **19.** $\ln \sqrt{x} = 7.5$ **20.** $5\left(1 + e^{\frac{x}{3}}\right) = 82$
     2.22   3,269,017.37   8.20

21. $3^{x+2} = 5^{x-1}$ **22.** $\ln\left(3\sqrt{x}\right) = \sqrt{\ln x}$ **23.** $\log x^3 = (\log x)^3$
     7.45   no solution $x = 1$ or $x \approx 0.02$ or $x \approx 53.96$

24. $\log(\log x) = 4$ **25.** $e^x - e^{-x} = 5$ **26.** $7^{x-1} - 2^{x+3} = 0$
    $10^{10,000}$ 1.65   3.21

Psychology Educational psychologists sometimes use mathematical models of memory. Suppose a group of students takes a chemistry test, and after a time, t (in months), they take an equivalent form of the same test. The mathematical model $a(t) = 82 - 12 \log(t + 1)$, where a is the average score at time t, represents a function describing the students' retention of the material. **Complete Exercises 27–29.**

27. What was the average score when the students first took the test?   82

28. What was the average score after 6 months?   ≈ 72

29. After how many months will the average score be 60?   ≈ 67 months

30. Solve the exponential equation $10^{2x} + 75 = 150$ by graphing. $x \approx 0.94$

31. Solve the exponential equation $10^{-2x} = 75$ by graphing. $x \approx -0.94$

32. Solve the equation of Exercise 30 without graphing. Compare the results. $x \approx 0.94$, same

33. Solve the equation of Exercise 31 without graphing. Compare the results. $x \approx -0.94$, same

34. Biology A colony of bacteria is growing according to the growth function $w(t) = \dfrac{2.70}{1 + 0.45e^{-(0.05t)}}$, where w is the weight in grams of the colony, and t is the time from the present in hours. How much does the colony weigh at present? Is it possible to double its original weight? Explain.   ≈ 1.86g; no. The population of the bacteria colony never exceeds 2.7 g.

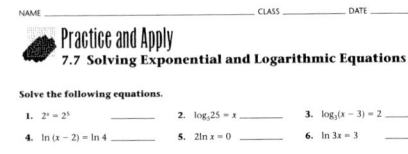
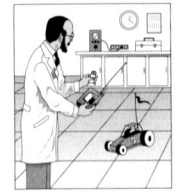
40.

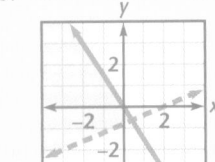

41.

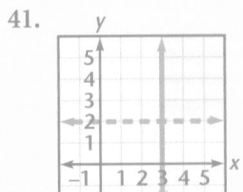

Archaeology The process of measuring C-14 in an organism to determine its age is known as radiocarbon dating. Most of the carbon in Earth's atmosphere is the isotope C-12, but a small amount is the radioactive isotope C-14. When an organism dies, the amount of C-12 it contains remains the same, but C-14 with a half-life of 5730 years decays. Archaeologists use the half-life decay function $N(t) = N_0 \left(\frac{1}{2}\right)^{\frac{t}{h}}$, where h is the half-life of a substance. If a sample initially contains 50 grams of C-14, how many grams would remain after the given number of years?

35 20,000 years   **36** 45,000 years
   ≈ 4.45 g   ≈ 0.22 g

Disease Control During a past epidemic, the number of infected people increased exponentially with time, or $f(t) = ab^t$, where f is the number of infected people at time t, in days. Suppose that at the onset (0 days) there are 30 people infected, and 6 days after the onset there are 300 people infected.

37. Specify the function, f, by determining a and b. $f(t) = 30(1.4678)^t$

38 Predict the number of people who will be infected at the end of 2 weeks. ≈ 6463 people

39 Approximately when will the number of infected people reach 10,000? ≈ 15 days, 3 to 4hr

Look Back

Graph each of the following systems of linear inequalities. [Lesson 4.8]

40. $\begin{cases} 2x - 5y < 4 \\ -3x \geq 2y \end{cases}$ **41.** $\begin{cases} -x \leq -3 \\ y - 5 > -3 \end{cases}$ **42.** $\begin{cases} 7 < 2x - y \\ -y \geq 0 \end{cases}$ **43.** $\begin{cases} -y + 3 \leq 12 \\ -x < y + 8 \end{cases}$

Solve each of the following equations for x.   [Lesson 5.2]

44. $x^2 - 3 = 46$ ±7   **45.** $x + 4 = \sqrt{12}$ ≈ −0.54

Look Beyond

46 Graph the functions $f(x) = \sin x$ and $g(x) = \cos x$. What does the shape of these functions look like? Use the **SIN** and **COS** keys on your calculator.

42.

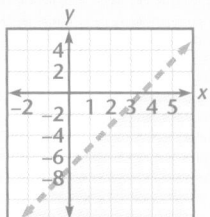

43.

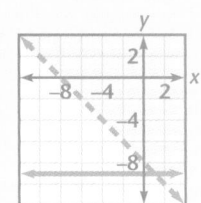

46. The graphs of the functions have the shape of repeating waves.

FOCUS

In this project, students will drop a ball and record the heights of successive bounces. They will seek to discover an exponential function that models the data they gather.

MOTIVATE

Some intuition is needed to find the mathematical function underlying the observed data. The central intuition is, if the first bounce is a certain fraction of the original height, wouldn't the second bounce be the same fraction of the first bounce? The second bounce is no different from dropping the ball from the first bounce height.

CHAPTER PROJECT 9

Exponential

bouncing ball
EXPERIMENT

Graphics Calculator

Students should work in groups of two to complete this project. Each team will record data on the height of successive bounces of a ball dropped from different heights. The object is to demonstrate that the height of successive bounces decreases exponentially as the number of bounces increases.

Materials

- meter sticks (or other means of measuring in centimeters)
- bouncing ball (tennis ball, racquetball, handball, etc.)
- smooth, uncarpeted floor

The Experiment

Each team will drop the ball from 3 different heights. On each drop, measure the height to which the ball rebounds after each of 3 bounces. Estimate each bounce height as well as you can. Repeat the experiment at each height 10 times. Then find the average value for each bounce height. Record your average data in a table.

Initial height	Bounce 1 height	Bounce 2 height	Bounce 3 height
200 cm			
150 cm			
100 cm			

1. Enter and graph the data for each initial height, with the bounce number as the x-coordinate and the bounce height as the y-coordinate. Let the "zero" bounce correspond to the initial height.

2. Use the exponential regression feature on your graphics calculator to find a best-fit model for the data. What is the base of the exponential function that best fits the data points? What is the correlation coefficient, r?

In the Chapter 2 Project, it was demonstrated that the height of the first bounce is a definite fraction, k, of the initial height, h_0. That is,

$$\frac{\text{bounce 1 height}}{\text{initial height}} = k.$$

If this holds true for successive bounces, then the heights should be $h_0 k$, $h_0 k^2$, ..., $h_0 k^n$. The bounce heights should correspond to the exponential function $f(x) = h_0 k^x$ for $x = 0, 1, 2, ..., n$.

3. For each initial height, measure the value of k using the initial height and the average height of the first bounce.

4. Compare your measured k for each initial height with the base obtained by exponential regression.

5. Do your data and calculations support the conclusion that the height of a bouncing ball decreases exponentially? Explain your reasoning.

Cooperative Learning

Students will work in pairs. One will drop the ball while the other observes and records the height of each bounce. As in any scientific experiment, the data will not conform perfectly to any known mathematical function. Students can plot these points on a graphing calculator and test for an equation of best fit.

Discuss

The accuracy of the outcome depends on many factors. Did the ball have irregularities on its surface? Was it a perfect sphere? Was the floor smooth and flat? Statistical methods try to compensate for these irregularities.

Chapter 7 Review

Vocabulary

Key Skills & Exercises

Lesson 7.1

➤ Key Skills

Model the growth or decline of a population with a graphics calculator by repeatedly applying the multiplier.

A rate of 4.5% represents a multiplier of 1.045. To model the growth of $500 invested at 4.5% interest compounded annually, first enter the principal, $500. Then multiply 500 by 1.045. Press ENTER repeatedly to watch the money grow year by year.

➤ Exercises

Give the multiplier for each growth rate.

1. 7.1% 1.071   **2.** –11% 0.89   **3.** 475% 5.75   **4.** –0.03% 0.9997

Lesson 7.2

➤ Key Skills

Identify the behavior of exponential functions by inspection and by graphing.

An exponential function is $f(x) = ab^x$, with $a > 0$, $b > 0$, and $b \neq 1$.
When $b > 1$, the graph of f shows exponential growth, and when $0 < b < 1$, the graph shows exponential decay.

Determine the growth of funds under various compounding methods.

To find the balance after 15 years if $1000 is invested at 8.5% compounded monthly, substitute 1000 for P, 15 for t, 0.085 for r, and 12 for n, and evaluate $A(t) = P\left(1 + \frac{r}{n}\right)^{nt}$.

➤ Exercises

For each exponential function, give the y-intercept and indicate whether the function represents exponential growth or decay.

5. $f(x) = 0.81^x$ **6.** $f(x) = 3(2.5)^x$ **7.** $f(x) = 1.5\left(\frac{5}{6}\right)^x$ **8.** $f(x) = 2(3)^{-x}$

5. (0, 1); decay

6. (0, 3); growth

7. (0, 1.5); decay

8. (0, 2); decay

Lesson 7.3

➤ Key Skills

Identify the exponential and logarithmic functions as inverse functions.

If $f(x) = b^x$, then $f^{-1}(x) = \log_b x$.

If $f(x) = \log_b x$, then $f^{-1}(x) = b^x$.

The domain of logarithmic functions is all $x > 0$, and the range is all real numbers.

Determine equivalent forms for exponential and logarithmic equations.

Since a logarithm is an *exponent*, the equivalent logarithmic expression for the exponential expression $7^3 = 343$ is $\log_7 343 = 3$.

➤ Exercises

Give an equivalent logarithmic or exponential equation.

9. $6^4 = 1296$ $\log_6 1296 = 4$

10. $\log 100{,}000 = 5$ $10^5 = 100{,}000$

11. $5^{-4} = 0.0016$ $\log_5 0.0016 = -4$

12. $-3 = \log_4\left(\frac{1}{64}\right)$ $4^{-3} = \frac{1}{64}$

13. Solve $A = 10^{-20t}$ for t in terms of A. $t = -\frac{\log A}{20}$

Lesson 7.4

➤ Key Skills

Identify the product, quotient, and power properties of logarithms.

For $x > 0$, $y > 0$, $b > 0$, and $b \ne 1$:

$$\log_b xy = \log_b x + \log_b y$$
$$\log_b \frac{x}{y} = \log_b x - \log_b y$$
$$\log_b x^r = r \log_b x$$

Simplify expressions and solve equations involving logarithms.

Express $\log m - \log n + p \log q$ with a single logarithm.

$$\log m - \log n + p \log q$$
$$= \log \frac{m}{n} + \log q^p$$
$$= \log \frac{mq^p}{n}$$

➤ Exercises

Evaluate each expression without using your calculator. Assume that $\log 4 \approx 0.602$, $\log 5 \approx 0.699$, and $\log 6 \approx 0.778$.

14. $\log 20$ 1.301

15. $\log 120$ 2.079

16. $\log \frac{4}{5}$ -0.097

17. $\log 125$ 2.097

18. Solve $3 \log 2x + \log 0.5 = \log 12 - \log 3$ for x. 1

Lesson 7.5

➤ Key Skills

Identify and use the common logarithmic function.

$y = \log_{10} x = \log x$ if and only if $x = 10^y$

Write equivalent logarithmic and exponential equations.

$\log 100{,}000 = 5 \leftrightarrow 10^5 = 100{,}000$

$10^3 = 1000 \leftrightarrow 3 = \log 1000$

➤ Exercises

Use your calculator to evaluate each logarithm. Round to the nearest hundredth.

19. $\log 53$ 1.72

20. $\log 3$ 0.48

21. $\log 0.513$ -0.29

22. $\log \pi$ 0.50

Use your calculator to find the number whose common log is the given value. Round to the nearest hundredth.

23. 3.1 1258.93

24. -0.2 0.63

25. 4.5 $31{,}622.78$

26. 0.071 1.18

27. $-\pi$ $7.22 \times 10^{-4} \approx 0$

Lesson 7.6

➤ Key Skills

Evaluate expressions involving the natural number, *e*, and identify the relationship between the natural logarithmic and exponential functions.

Find $e^{5.2}$ to 2 decimal places, and write a logarithmic equation that is equivalent to your result.

$e^{5.2} \approx 181.27$. The equivalent logarithmic equation is $\ln 181.27 = 5.2$.

Model growth and decay processes with natural exponential functions.

To find how long it will take $5000 to double at 6.8% interest compounded continuously, substitute 10,000 for *A*, 5000 for *P*, and 0.068 for *r* in the equation $A(t) = Pe^{rt}$, and solve for *t*.

➤ Exercises

Evaluate each expression to the nearest hundredth. Then use your result to write an equivalent exponential or logarithmic equation.

28 e^4 54.60
   $\ln 54.60 = 4$

29 $\ln 148.41$
   5.00; $e^{5.00} = 148.41$

30 e^{-2} 0.14
   $\ln 0.14 = -2$

31 $\ln 1.28$ 0.25
   $e^{0.25} = 1.28$

Lesson 7.7

➤ Key Skills

Solve logarithmic and exponential equations by graphing.

To solve $2 \log 4(x + 7) - \log(2x + 14) = 2$, graph $y = 2 \log 4(x + 7) - \log(2x + 14)$, and trace to find the value of *x* when *y* is 2.

Use the exponential-log inverse properties to solve logarithmic and exponential equations.

$b^{\log_b x} = x$ for $x > 0$, and $\log_b b^x = x$
$$\log 8(x + 7) = 2$$
$$8(x + 7) = 10^2$$
$$x = 5.5$$

➤ Exercises

Solve each equation for *x*. If *x* is not an integer, estimate *x* to the nearest hundredth.

32 $2 \ln 5 - \ln x = 1$
   $x = \frac{25}{e} \approx 9.20$

33 $4 + 3e^{5+2x} = 13$
   $x = \frac{\ln 3 - 5}{2} \approx -1.95$

34 $\log x^3 = 2.4$
   $x = 10^{0.8} \approx 6.31$

35 $\ln(x^2 - 3x + 4) = \ln 2$
   $x = 1$ or $x = 2$

Applications

Investment An investor has $11,000 invested at 7.5% compounded quarterly and $11,500 at 7.1% compounded continuously.

36 How much will the $11,000 be worth in 10 years? $23,125.84

37 How much will the $11,500 be worth in 10 years? $23,390.90

38 Graph and trace to estimate how long it will take for the investments to have equal value. Estimate this value to the nearest month. 13 years, 5 months $29,900 value

Demographics The population of Cleveland, Ohio, was 505,616 in 1990 and was declining at a rate of 1.3% per year.

39. What was the multiplier for Cleveland's population decline? 0.987

40 Find the projected population of Cleveland for the year 2000. 443,601

41 Find the function that models Cleveland's projected population growth. $505,616(0.987)^x$

42 During what year will Cleveland's projected population be one-half of the 1990 population? 2042

Chapter 7 Assessment

Economics The gross domestic product (GDP) of a country is an estimate of the goods and services that it produces in a year. In 1991, some economists estimated that the Republic of Singapore had a GDP of $38 billion and was growing at a rate of 6.5%, and that the Philippines had a GDP of $47 billion and was growing at a rate of 0.1%.

1 If an economist predicts that Singapore's growth rate would remain the same for the subsequent 5 years, what would be the prediction, to the nearest billion dollars, for the 1996 Singapore GDP? $52 billion

2 At a 6.5% growth rate, how long, to the nearest year, would it take for the Singapore GDP to double? 11 yr

3 Graph and trace to estimate during which year the GDP of Singapore would surpass that of the Philippines at the predicted growth rates. 1994

Write an equivalent logarithmic or exponential equation.

4. $\log 0.01 = -2$ $10^{-2} = 0.01$

5. $8^5 = 32{,}768$ $\log_8 32{,}768 = 5$

6. $\ln 54.6 = 4$ $e^4 = 54.6$

7. $e^{-3} = 0.05$ $\ln 0.05 = -3$

8. $10^{1.5} = 31.62$ $\log 31.62 = 1.5$

9. $\log_2 512 = 9$ $2^9 = 512$

10. $\ln c = d$ $e^d = c$

11. $e^{x+y} = z$ $\ln z = x + y$

Write each expression with a single logarithm.

12. $\log 16 - 3 \log 2$ $\log 2$

13. $\ln 10 + \ln 6 - \ln 4$ $\ln 15$

14. $5 \ln b - 2 \ln 3b$ $\ln \frac{b^3}{9}$

Solve each equation. When necessary, round to the nearest hundredth.

15. $-8.5 = -5 \log(2t + 3)$ $t = 23.56$

16. $\ln(x + 3) = \ln(x^2 + x - 6)$ $x = 3$

17. $9 \cdot 3^{-x} = 81$ $x = -2$

18 $0.2e^{-x-4} = 0.11$ $x = -3.40$

19. Using the estimates $\ln 3 = 1.099$ and $\ln 5 = 1.609$, write an expression for $\ln 45$. $\ln 45 = 2\ln 3 + \ln 5 \approx 3.807$

Radioactive Decay Radioactive decay follows the general exponential decay function $A(t) = A_0 e^{-kt}$, where $A(t)$ is the amount of a substance remaining at time t, A_0 is the amount initially present, and k is a constant that depends on the element or isotope involved. The gas radon has a half-life of about 3.8 days.

20 Find the constant, k, for radon, for t in days. Estimate to the nearest hundredth.   0.18

21 A closed lab is contaminated by radon. If the radon level is initially 5 times above the safe level, about how many days must the lab be closed before it is safe?   9 days

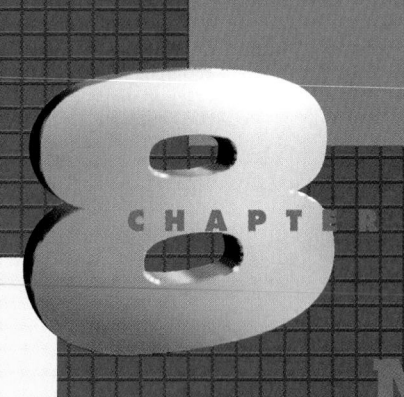

CHAPTER 8 Trigonometric Functions

Meeting Individual Needs

8.1 Exploring Special Right Triangles

Core

Inclusion Strategies, p. 432
Reteaching the Lesson, p. 433
Practice Master 8.1
Enrichment Master 8.1
Technology Master 8.1
Lesson Activity Master 8.1

[2 days]

Core Plus

Practice Master 8.1
Enrichment, p. 432
Technology Master 8.1
Lesson Activity Master 8.1
Interdisciplinary
    Connection, p. 431

[2 days]

8.2 The Unit Circle

Core

Inclusion Strategies, p. 440
Reteaching the Lesson, p. 441
Practice Master 8.2
Enrichment Master 8.2
Technology Master 8.2
Lesson Activity Master 8.2

[2 days]

Core Plus

Practice Master 8.2
Enrichment, p. 440
Technology Master 8.2
Lesson Activity Master 8.2
Interdisciplinary
    Connection, p. 439

[2 days]

8.3 Trigonometric Applications

Core

Inclusion Strategies, p. 447
Reteaching the Lesson, p. 448
Practice Master 8.3
Enrichment Master 8.3
Technology Master 8.3
Lesson Activity Master 8.3

[2 days]

Core Plus

Practice Master 8.3
Enrichment, p. 447
Technology Master 8.3
Lesson Activity Master 8.3

[2 days]

8.4 Exploring the Sine and Cosine Graphs

Core

Inclusion Strategies, p. 455
Reteaching the Lesson, p. 456
Practice Master 8.4
Enrichment Master 8.4
Technology Master 8.4
Lesson Activity Master 8.4
Mid-Chapter Assessment Master

[2 days]

Core Plus

Practice Master 8.4
Enrichment, p. 455
Technology Master 8.4
Interdisciplinary
    Connection, p. 454
Mid-Chapter Assessment
    Master

[1 day]

8.5 Radian Measure

Core

Inclusion Strategies, p. 461
Reteaching the Lesson, p. 462
Practice Master 8.5
Enrichment Master 8.5
Technology Master 8.5
Lesson Activity Master 8.5

[3 days]

Core Plus

Practice Master 8.5
Enrichment, p. 460
Technology Master 8.5
Lesson Activity Master 8.5

[2 days]

8.6 Arc Length and Sector Area

Core

Inclusion Strategies, p. 470
Reteaching the Lesson, p. 471
Practice Master 8.6
Enrichment Master 8.6
Technology Master 8.6
Lesson Activity Master 8.6

[Optional*]

Core Plus

Practice Master 8.6
Enrichment, p. 470
Technology Master 8.6
Lesson Activity Master 8.6
Interdisciplinary
    Connection, p. 469

[1 day]

8.7 Applications of Periodic Functions

Core

Inclusion Strategies, p. 477
Reteaching the Lesson, p. 478
Practice Master 8.7
Enrichment Master 8.7
Technology Master 8.7
Lesson Activity Master 8.7

[Optional*]

Core Plus

Practice Master 8.7
Enrichment, p. 476
Technology Master 8.7

[1 day]

Chapter Summary

Core

Chapter 8 Project,
    pp. 482–483
Lab Activity
Long-Term Project
Chapter Review, pp. 484–488
Chapter Assessment, p. 489
Chapter Assessment, A/B
Alternative Assessment
Cumulative Assessment,
    pp. 490–491

[3 days]

Core Plus

Chapter 8 Project,
    pp. 482–483
Lab Activity
Long-Term Project
Chapter Review,
    pp. 484–488
Chapter Assessment,
    p. 489
Chapter Assessment, A/B
Alternative Assessment
Cumulative Assessment,
    pp. 490–491

[3 days]

Reading Strategies

Before reading each lesson, students should skim all the lessons in the chapter to get an idea of how the material in the lessons relates to the overall subject of the chapter. As the students skim the lessons, they should pay close attention to the applications represented in the explorations, examples, and exercises to see how the information relates to the real-world problems. Encourage students to discuss their perceptions of the practical value of learning this chapter.

Visual Strategies

Direct the students to look at the pictures and associated graphs in the chapter opener. Encourage them to speculate about how mathematics might be useful in studying such phenomena. Also have them look at the graphs in the chapter and discuss how they differ from graphs in earlier chapters. What asptect of these graphs makes them different from graphs in earlier chapters? Can they imagine how such functions could model the phenomena in the photographs?

Hands-on Strategies

This chapter presents the students with a number of opportunities to become actively involved with mathematics in the physical world. Several applications problems, such as construction, cycling, and business, could be performed as experiments.

Cooperative Learning

You may wish to have students work with partners for some of the above activities. Additional suggestions for cooperative group activities are noted in the teacher's notes in each lesson.

Multicultural

The cultural connections in this chapter include references to Asia and Africa.

Portfolio Assessment

Below are portfolio activities for the chapter. They are listed under seven activity domains that are appropriate for portfolio development.

1. **Investigation/Exploration** The study of special triangles is the focus of Explorations 1 and 2 in Lesson 8.1. Transformations of the sine and cosine curves is the objective of Explorations 1–5 in Lesson 8.4. Radian measures of common rotations are studied in the exploration of Lesson 8.5. Trigonometric functions and the daylight around the world is the focus of the Chapter Project.

2. **Applications** Recommended are any of the following: construction, Lesson 8.1, Exercise 18; transportation, Lesson 8.2, Exercises 44–45; cycling, Lesson 8.3, Exercises 36–41; music, Lesson 8.4, Exercises 24–26; astronomy, Lesson 8.5, Exercise 58; business, Lesson 8.6, Exercises 9–11; physiology, Lesson 8.7, Exercises 25–28.

3. **Non-Routine Problems** Lesson 8.6, Exercises 45–46 provide students with the opportunity to copy congruent figures on dot paper.

4. **Project Daylight Around the World** See pages 482–483. Students are asked to represent the similarities and differences in daylight in several areas of the world.

5. **Interdisciplinary Topics** Students may choose from the following: geometry, Lesson 8.1, Exercises 16 and 17; physics, Lesson 8.6, Exercise 21.

6. **Writing** Communicate exercises where a student is asked to describe basic concepts of trigonometric functions, the unit circle, and transformations of circular functions, their characteristics, and their graphs, offer excellent writing selections for the portfolio.

7. **Tools** In Chapter 8, graphics calculators are used to assimilate data and model that data with various trigonometric functions. To measure the individual student's proficiency, it is recommended that he or she complete selected worksheets from the Technology Masters.

Technology

Technology can simplify computational and graphing processes. This allows the student to focus attention on the target concepts and avoids the time-consuming task of creating several graphs.

Graphics Calculators

One of the most significant advantages of a graphics calculator or computer software is its ability to instantaneously display graphs and transformations of these graphs.

A family of functions, such as those based on the fundamental function $\sin x$, can be generated by varying the parameters in the transformation function $f(x) = a + b \sin c(x - d)$. Students experiment on a graphics calculator to identify the effects caused by changing each of the values a, b, c, and d.

Graphing Transformations

To view trigonometric transformations, make certain that the viewing window includes at least one complete cycle of the function. For instance, the functions $g(x) = (0.5) \sin x$ and $h(x) = 2 \sin x$ could use the TI-82 friendly viewing window shown. The calculator is in degree mode. In degree mode a friendly viewing window is −470 to 470.

The maximum value of g is $\frac{1}{2}$ and the maximum value of h is 2. Both occur exactly at 90°, or at $\frac{\pi}{2}$ radians. Notice that the viewing window includes both the maximum and the minimum as well as several cycles of each function.

If these functions were graphed in radian mode, since $2 \cdot \pi$ is about 6.3, one friendly viewing window for the TI-82 is −9.4 to 9.4. The same maximum value for g is found by using the solve capabilities of this calculator. However, the x-value is not exact and is difficult to relate to a familiar representation of an angle. Even when this x-value is divided by π, it is difficult to see the connection between the radian number and a familiar representation of the angle.

To avoid this inconvenience, especially when students are trying to find *exact* radian values, set the calculator in degree mode and use a *friendly viewing window*. Once the degree measure has been found, it can be converted to radians by multiplying by $\frac{\pi}{180}$ and simplifying the resulting fraction. The result will be an exact radian value. Some calculators will even display the result in fraction form.

Integrated Software

f(g) Scholar™ is an integrated computer-based mathematics productivity tool that combines calculator, spreadsheet, and graphics capabilities to provide a dynamic and interactive environment for explorations in mathematics. Use *f(g) Scholar*™ for any lesson needing a spreadsheet, calculator, graphics calculator, or any combination of the three.

8 Trigonometric Functions

Trigonometric Functions

ABOUT THE CHAPTER

Background Information

Trigonometry has existed for over 3000 years. In its earliest form, it dealt with the relationships between the sides and angles of triangles. In modern mathematics, the relationships between sides and angle of triangles and angles defined by the unit circle are expressed as functions.

RESOURCES

- Practice Masters
- Enrichment Masters
- Technology Masters
- Lesson Activity Masters
- Lab Activity Masters
- Long-Term Project Masters
- Assessments Masters
  Chapter Assessments, A/B
  Mid-Chapter Assessment
  Alternative Assessment, A/B
- Teaching Transparencies
- Cumulative Assessment
- Spanish Resources

CHAPTER OBJECTIVES

- Identify the trigonometric ratios in special right triangles.
- Apply the special right triangle relationships to find missing lengths of sides of special triangles.
- Identify the coordinates of a point (x, y) on the unit circle given an angle in standard position using the relationship $(x, y) = (\cos\theta, \sin\theta)$.
- Identify angles and their coterminal angles in standard position.

LESSONS

8.1 *Exploring* Special Right Triangles

8.2 The Unit Circle

8.3 Trigonometric Applications

8.4 *Exploring* Sine and Cosine Graphs

8.5 Radian Measure

8.6 Arc Length and Sector Area

8.7 Applications of Periodic Functions

Chapter Project
Daylight Around the World

The word *trigonometry* comes from the Greek words *trigonon*, meaning "triangle," and *metria*, meaning "measurement." Originally, trigonometry dealt with the measurement of the sides and angles of triangles. The ancient Egyptians used ratios in triangles to solve problems associated with land measurement and to construct pyramids. One of the earliest known uses of trigonometry is an Egyptian table that shows the relationship between the time of day and the length of the shadow cast by a vertical stick. The Egyptians knew that this shadow was longer in the morning, decreased to a minimum at noon, and increased again until sundown. Trigonometry has been used in astronomy for over 3500 years.

ABOUT THE PHOTOS

Trigonometry is based on right triangles whose hypotenuses coincide with the radius of the unit circle. When a sequence of the altitudes of the inscribed triangles are graphed on a coordinate plane, a smooth wave-like curve is created.

Trigonometry was used by early Egyptians to measure time based on the position of the sun. In this way, measurements were made that assisted in constructing the pyramids.

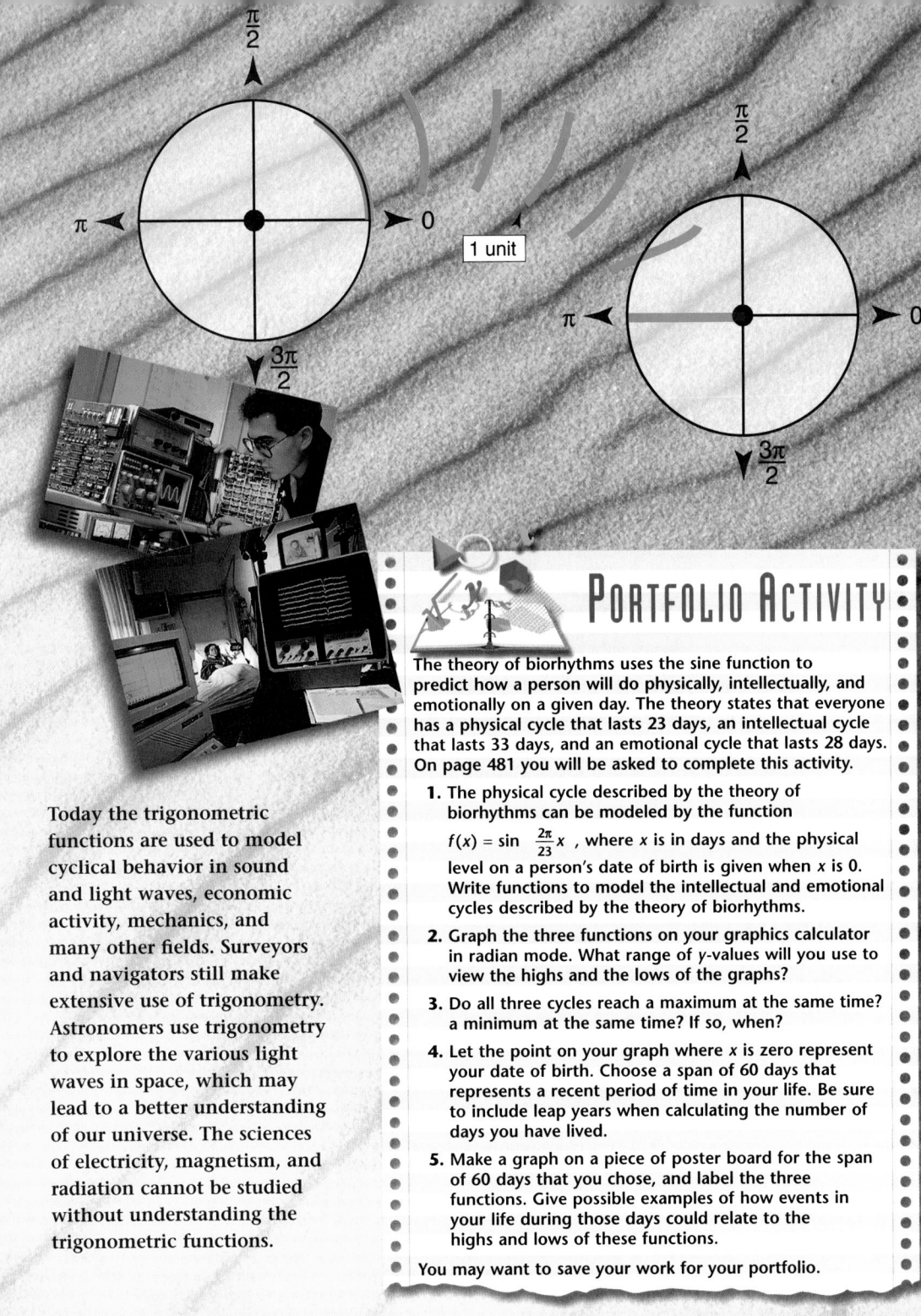

1 unit

Today the trigonometric functions are used to model cyclical behavior in sound and light waves, economic activity, mechanics, and many other fields. Surveyors and navigators still make extensive use of trigonometry. Astronomers use trigonometry to explore the various light waves in space, which may lead to a better understanding of our universe. The sciences of electricity, magnetism, and radiation cannot be studied without understanding the trigonometric functions.

PORTFOLIO ACTIVITY

The theory of biorhythms uses the sine function to predict how a person will do physically, intellectually, and emotionally on a given day. The theory states that everyone has a physical cycle that lasts 23 days, an intellectual cycle that lasts 33 days, and an emotional cycle that lasts 28 days. On page 481 you will be asked to complete this activity.

1. The physical cycle described by the theory of biorhythms can be modeled by the function

    $f(x) = \sin \dfrac{2\pi}{23}x$, where x is in days and the physical level on a person's date of birth is given when x is 0. Write functions to model the intellectual and emotional cycles described by the theory of biorhythms.

2. Graph the three functions on your graphics calculator in radian mode. What range of y-values will you use to view the highs and the lows of the graphs?

3. Do all three cycles reach a maximum at the same time? a minimum at the same time? If so, when?

4. Let the point on your graph where x is zero represent your date of birth. Choose a span of 60 days that represents a recent period of time in your life. Be sure to include leap years when calculating the number of days you have lived.

5. Make a graph on a piece of poster board for the span of 60 days that you chose, and label the three functions. Give possible examples of how events in your life during those days could relate to the highs and lows of these functions.

You may want to save your work for your portfolio.

ABOUT THE CHAPTER PROJECT

For the Chapter Project on pages 482 and 483, students will compile information on times of sunrise and sunset for a number of cities all over the world. Students will graph these results over time for each city in each location. This activity will give students an authentic example of several cyclic curves.

- Determine sides or angles of right triangles using trigonometric functions or their inverses.
- Determine, from the principal values of the inverse trigonometric functions, any angle in standard position.
- Identify how a, b, c, and d in $f(\theta) = a + b \sin c(\theta - d)$ and $f(\theta) = a + b \cos c(\theta - d)$ transform the graphs of $f(\theta) = \sin \theta$ and $f(\theta) = \cos \theta$.
- Convert degree to radian measure, and radian to degree measure.
- Graph and identify properties of trigonometric functions in radian measure.
- Find the arc length and sector area determined by the central angle of a circle.
- Model applications with circular functions of the form $f(x) = a + b \cos c(x - d)$ or $f(x) = a + b \sin c(x - d)$

PORTFOLIO ACTIVITY

In this Exercise, students will explore the theory of biorhythms and apply it to their own lives. This theory illustrates one example of how mathematical patterns occur in the natural world.

Organize students into groups of three. One student will prepare physical cycle charts for all three students; the second, intellectual charts; the third, emotional charts.

Discuss with students the possible practical applications of the theory of biorhythms. Could students better organize their time to allow for these patterns? Could organizations be organized to capitalize on its employees' cycles?

PREPARE

Objectives

- Identify the trigonometric ratios in special right triangles.
- Apply the special right triangle relationships to find missing lengths of sides of special triangles.

- Practice Master   8.1
- Enrichment Master   8.1
- Technology Master   8.1
- Lesson Activity Master   8.1
- Quiz   8.1
- Spanish Resources   8.1

Assessing Prior Knowledge

Let a and b be the legs of a right triangle and c its hypotenuse. Find

1. c if $a = 1, b = 1$ $[\sqrt{2}]$
2. b if $a = 1, c = 2$ $[\sqrt{3}]$

TEACH

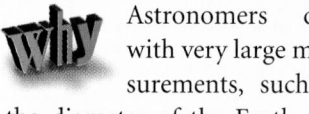

Astronomers deal with very large measurements, such as the diameter of the Earth and the distances from our planet to the stars. These distances cannot be measured directly, so indirect methods must be utilized. Special right triangle relationships can be used to find these measurements.

Exploration 1 Notes

Have students complete Exploration 1 in groups of three. Each student should complete Steps 1–4 individually. Then students can discuss Steps 5–8 as a group.

Exploring Special Right Triangles

Why *Sometimes it is necessary to find a measurement indirectly. Surveyors use special right-triangle relationships to find heights or distances without directly measuring them.*

A right triangle containing angles of 30° and 60° or containing two 45° angles is known as a special right triangle. For Explorations 1 and 2, you will need a protractor and a centimeter ruler to explore special right triangles.

Exploration 1 The 45-45-90 Triangle

1 Draw a horizontal line across the width of your paper. Construct a 45° angle with the horizontal line such that the rays of the angle on your paper are as long as possible.

2 Draw three parallel lines, each perpendicular to the original horizontal line, to form three similar triangles with the 45° angle as shown.

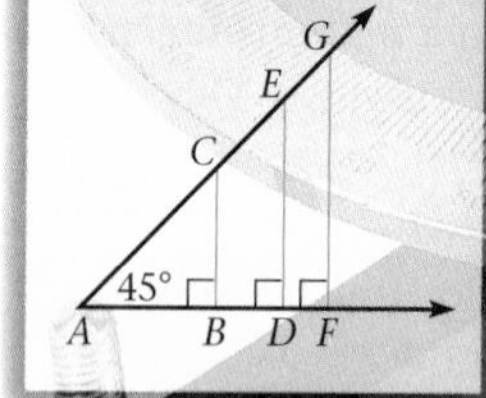

ALTERNATIVE teaching strategy

Hands-On Strategies
Have students stand at one corner of the classroom. Beginning from this position, have students measure two adjacent sides of the classroom with a yardstick. Have students measure the shorter wall first. Then have students measure the length of the shorter wall along the longer wall and mark the point. Ask them to find the length of the diagonal that connects the shorter wall with the mark on the longer wall. Have students measure the diagonal connecting these two points with the yardstick to check their results.

3 The *measure of* ∠A is abbreviated m∠A.
Find the measure of each angle in the three triangles,
and complete the following table.

Triangle ABC	Triangle ADE	Triangle AFG
m∠A = 45°	m∠A = 45°	m∠A = 45°
m∠B = ?	m∠D = ?	m∠F = ?
m∠C = ?	m∠E = ?	m∠G = ?

4 Use a ruler to find the length, in centimeters,
of the sides of each right triangle you drew,
and complete the following table.

	Triangle ABC	Triangle ADE	Triangle AFG
Length of vertical leg			
Length of horizontal leg			
Length of hypotenuse			
Length of vertical leg / Length of hypotenuse			
Length of horizontal leg / Length of hypotenuse			

5 How does the $\frac{\text{length of vertical leg}}{\text{length of hypotenuse}}$ ratio for each triangle compare?

How does the $\frac{\text{length of horizontal leg}}{\text{length of hypotenuse}}$ ratio for each triangle compare?

6 What do you think the measure of angle X in triangle XYZ is? Explain.

7 Use the "Pythagorean" Right-Triangle Theorem to find the length of the hypotenuse in triangle XYZ. Then find the value of the ratio $\frac{XY}{XZ}$. How does this ratio compare with the $\frac{\text{length of horizontal leg}}{\text{length of hypotenuse}}$ ratio in each triangle that you drew in Step 1?

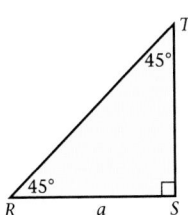

8 Let triangle RST be a 45-45-90 triangle, and let RS be a. Draw triangle RST, and label sides $\overline{RT}$ and $\overline{ST}$ with the lengths of their sides in terms of a. ❖

TEACHING tip

Students should find the ratios for all three triangles to be very similar. They should realize that when comparing similar 45-45-90 triangles, the corresponding ratios between the lengths of the legs and the length of the hypotenuse will always be the same. Suggest that students draw larger diagrams to see the relationships.

interdisciplinary CONNECTION

Astronomy The distances to the Moon, planets, and nearby stars were all determined using simple trigonometry. The distance to the Moon is one leg of a right triangle with the radius of the Earth as the other leg. The angle *a* is determined by noting the position of the Moon with respect to background stars. For measuring the distance to planets and nearby stars, the shorter leg of the triangle is the diameter of the Earth's orbit around the Sun.

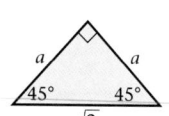

EXTENSION

The length of the hypotenuse of a 45-45-90 triangle is 8. To find the lengths of the legs of the right triangle, use the 45-45-90 triangle relationships you found in Exploration 1.

GEOMETRY *Connection*

The legs of a 45-45-90 triangle are equal in length. Let a represent the length of each leg. Then the length of the hypotenuse is $a\sqrt{2}$.

Since you are given the length of the hypotenuse, 8, solve for a in the equation $a\sqrt{2} = 8$.

$$a\sqrt{2} = 8$$
$$2a^2 = 64$$
$$a^2 = 32$$
$$a = \pm\sqrt{32}$$

Thus, the length of each leg of the 45-45-90 triangle is $\sqrt{32}$, or approximately 5.7. ❖

• Exploration 2 *The 30-60-90 Triangle*

1 Draw a horizontal line across the width of your paper. Construct a 30° angle with the horizontal line such that the rays of the angle on your paper are as long as possible.

2 Draw three parallel lines, each perpendicular to the original horizontal line, to form three similar triangles with the 30° angle as shown.

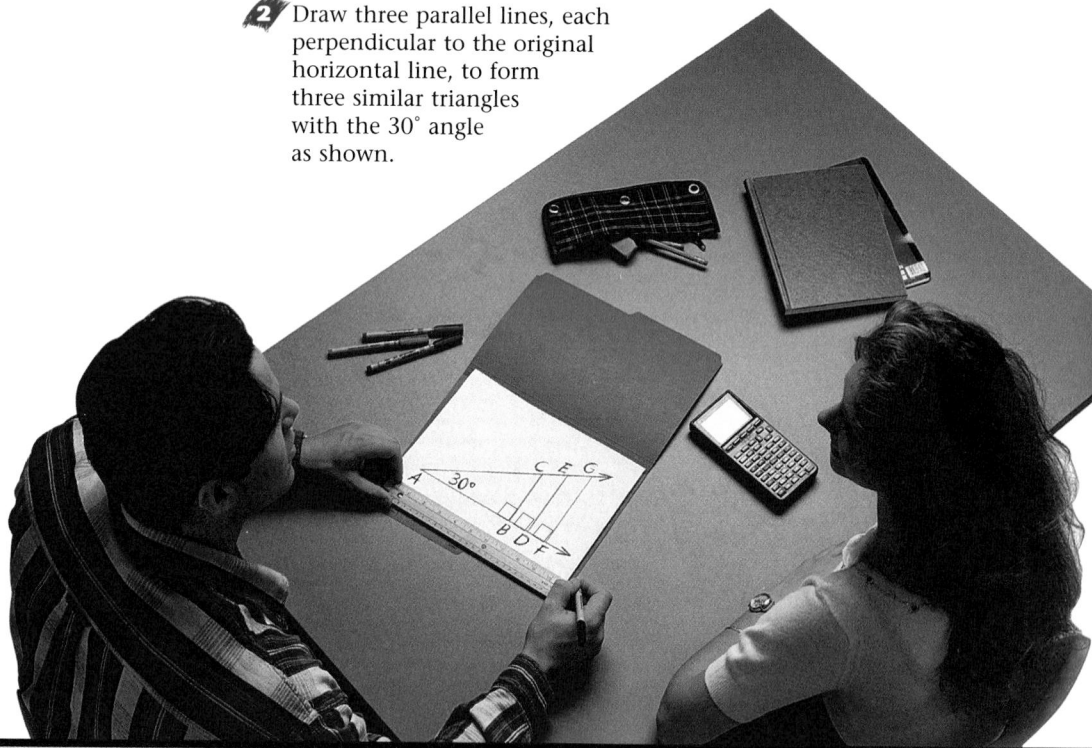

ENRICHMENT The history of mathematics in Babylonia and in ancient Greece shows the origin of the split between pure and applied mathematics. The "Pythagorean" Right-Triangle Theorem is considered part of pure, or theoretical, mathematics. The applications of the standard triangles in construction, physics, and other areas are considered to be applied mathematics. Have students obtain more information about this subject.

INCLUSION **strategies** **Using Models** Students may have problems with the ideas of limits and asymptotes. Have them draw a series of right triangles with measures of 30°, 20°, 10°, and 5° keeping the length of the shorter leg constant. Have students measure the length of the longer leg in each case. Encourage students to speculate about the length of this leg as the angle continues to get smaller.

3 Find the measure of each angle in the three triangles, and complete the following table.

Triangle ABC	Triangle ADE	Triangle AFG
m∠A = 30°	m∠A = 30°	m∠A = 30°
m∠B = ?	m∠D = ?	m∠F = ?
m∠C = ?	m∠E = ?	m∠G = ?

4 Use a ruler to find the length, in centimeters, of the sides of each right triangle you drew, and complete the following table.

	Triangle ABC	Triangle ADE	Triangle AFG
Length of vertical leg			
Length of horizontal leg			
Length of hypotenuse			
Length of vertical leg / Length of hypotenuse			
Length of horizontal leg / Length of hypotenuse			

5 How does the $\frac{\text{length of vertical leg}}{\text{length of hypotenuse}}$ ratio for each

triangle compare?

How does the $\frac{\text{length of horizontal leg}}{\text{length of hypotenuse}}$ ratio for each

triangle compare?

6 What do you think the measure of angle X in triangle XYZ is? Why?

7 Use the "Pythagorean" Right-Triangle Theorem to find the missing length in triangle XYZ. Then find the value of $\frac{XY}{XZ}$. How does this ratio compare with

the $\frac{\text{length of horizontal leg}}{\text{length of hypotenuse}}$ ratio in

each triangle that you drew in Step 1?

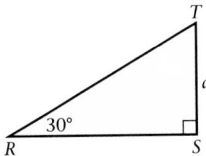

8 Let RST be a 30-60-90 triangle, and let ST be a. Draw triangle RST, and label sides $\overline{RS}$ and $\overline{RT}$ with the lengths of their sides in terms of a. ❖

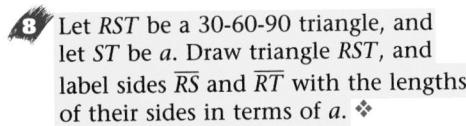

Using Visual Models Have students draw a 30-60-90 triangle using 1 as the length of the leg adjacent to the 60° angle, and 2 as the length of the hypotenuse. Have students measure the length of the other leg. Then have students use the "Pythagorean" Right-Triangle Theorem or the 30-60-90 triangle relationships to find this length. Students should compare the two values and comment on their findings.

A ongoing SSESSMENT

6. 30°; The ratio of the vertical leg to the hypotenuse is the same as that of the triangles in Steps 3 and 4, so the angles opposite them must be the same. In all of them, the angle opposite the vertical (shorter) leg has a measure of 30°.

8. $RS = a\sqrt{3}$; $RT = 2a$.

Geometry All 30-60-90 triangles are similar. Similar triangles have identical ratios of the lengths of their sides.

CRITICAL
Thinking

$a < a\sqrt{3} < 2a$; the length of the hypotenuse is always represented by $2a$, since

$$(a)^2 + (a\sqrt{3})^2 = a^2 + a^2 \cdot 3$$
$$= 4a^2,$$

which is the square of $2a$.

EXTENSION

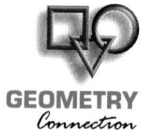

GEOMETRY
Connection

The length of the hypotenuse in a 30-60-90 triangle is 12. To find the lengths of the legs of the right triangle, use the 30-60-90 triangle relationships that you found in Exploration 2.

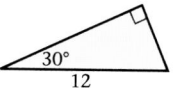

Let a represent the length of the leg opposite the 30° angle. Then the hypotenuse is $2a$, and the length of the leg opposite the 60° angle is $a\sqrt{3}$.

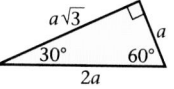

Since you are given the length of the hypotenuse, 12, solve for a in the equation $2a = 12$.

$$2a = 12$$
$$a = 6$$

Thus, the length of the leg opposite the 30° angle is 6, and the length of the leg opposite the 60° angle, the hypotenuse, is $6\sqrt{3}$, or approximately 10.4. ❖

CRITICAL
Thinking

List, in order from least to greatest, the lengths (a, $2a$, and $a\sqrt{3}$) of the legs of a 30-60-90 triangle. Which value always represents the length of the hypotenuse? Explain.

The ratios you found in the right triangles in Explorations 1 and 2 are called *trigonometric ratios*. The three basic trigonometric ratios are defined for the acute angles in *any* right triangle. They are

$\sin A$ read "the sine of angle A"
$\cos A$ read "the cosine of angle A"
$\tan A$ read "the tangent of angle A."

TRIGONOMETRIC RATIOS

$$\sin A = \frac{\text{length of side opposite } \angle A}{\text{length of hypotenuse}} = \frac{a}{c}$$

$$\cos A = \frac{\text{length of side adjacent } \angle A}{\text{length of hypotenuse}} = \frac{b}{c}$$

$$\tan A = \frac{\text{length of side opposite } \angle A}{\text{length of side adjacent } \angle A} = \frac{a}{b}$$

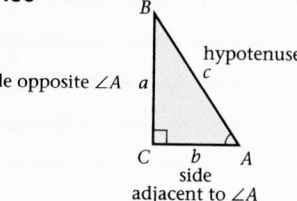

The ratios are sometimes abbreviated.

$$\sin A = \frac{\text{opp}}{\text{hyp}} \qquad \cos A = \frac{\text{adj}}{\text{hyp}} \qquad \tan A = \frac{\text{opp}}{\text{adj}}$$

EXTENSION

To find cos 45°, you can use the
45-45-90 triangle relationships.

$$\cos 45° = \frac{\text{adj}}{\text{hyp}} = \frac{a}{a\sqrt{2}}, \text{ or } \frac{1}{\sqrt{2}}$$

To find sin 60° and tan 30°, use the
30-60-90 triangle relationships.

$$\sin 60° = \frac{\text{opp}}{\text{hyp}} = \frac{a\sqrt{3}}{2a}, \text{ or } \frac{\sqrt{3}}{2}$$
$$\tan 30° = \frac{\text{opp}}{\text{adj}} = \frac{a}{a\sqrt{3}}, \text{ or } \frac{1}{\sqrt{3}} \text{ ❖}$$

Show that $\frac{1}{\sqrt{2}}$ is equivalent to $\frac{\sqrt{2}}{2}$ and $\frac{1}{\sqrt{3}}$ is equivalent to $\frac{\sqrt{3}}{3}$.

When you use the special right-triangle relationships to find the sines,
cosines, and tangents of special angles, the results are exact. When you
use your calculator, the results are approximate.

$$\cos 45° \approx 0.71, \text{ to the nearest hundredth}$$
$$\sin 60° \approx 0.87, \text{ to the nearest hundredth}$$
$$\tan 30° \approx 0.58, \text{ to the nearest hundredth}$$

What calculator keystrokes are used to find these values?

The sine, cosine, and tangent ratios can be used to find measures of the
unknown sides and angles of any right triangle.

EXERCISES & PROBLEMS

Communicate

1. Suppose that one leg of a 45-45-90 triangle is 5.
Explain how to find the other two sides.

2. Suppose that the shorter leg of a 30-60-90 triangle is 4.
Explain how to find the other two sides.

3. How can you use the triangle relationships of a 45-45-90 triangle to
find sin 45°?

4. How can you use the triangle relationships of a 30-60-90 triangle to
find cos 30°?

5. How can the words *ratio* and *proportion* be used to describe all
30-60-90 triangles?

Aongoing **SSESSMENT**

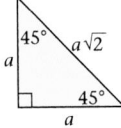

$$\frac{1}{\sqrt{2}} = \frac{1}{\sqrt{2}} \cdot \frac{\sqrt{2}}{\sqrt{2}} = \frac{\sqrt{2}}{(\sqrt{2})^2} = \frac{\sqrt{2}}{2}$$
$$\frac{1}{\sqrt{3}} = \frac{1}{\sqrt{3}} \cdot \frac{\sqrt{3}}{\sqrt{3}} = \frac{\sqrt{3}}{(\sqrt{3})^2} = \frac{\sqrt{3}}{3}$$

Aongoing **SSESSMENT**

The exact keystrokes will
depend on the calculator. The
calculator should be in degree
mode and keystrokes should
be similar to

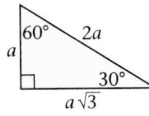

ASSESS

Selected Answers
Odd-numbered Exercises 7–31

Assignment Guide
Core 1–12, 14–31
Core Plus 1–5, 9–35

Technology
Graphics calculators are needed
for Exercises 32–34.

Error Analysis
Students should be careful to use
parentheses to enclose the angles of trigonometric functions.
If parentheses are omitted, the
calculator may misinterpret the
order of operations and the angle intended for the trigonometric function.

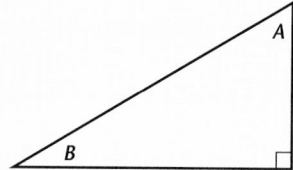
Performance Assessment

1. Have students use the triangle below. Students should measure the lengths of the sides and find the sine, cosine, and tangent of angle A. What is the measure of angle A?

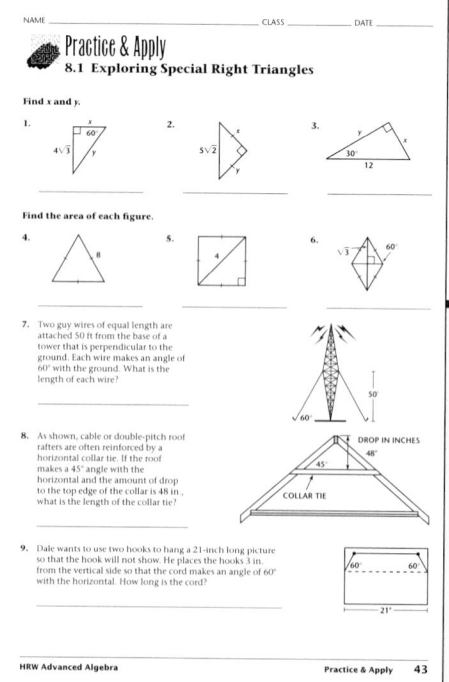

[sin A = 0.87, cos A = 0.5, tan A = 1.73, m∠A = 60°]

2. Ask students to explain how the "Pythagorean" Right-Triangle Theorem helps to evaluate the trigonometric ratios of the special triangles studied in this lesson.

Practice & Apply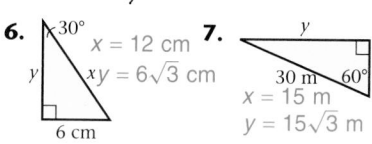

Find x and y.

6.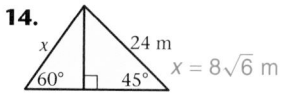
30° x = 12 cm
y   xy = 6√3 cm
6 cm

7. y
30 m 60° x
x = 15 m
y = 15√3 m

8. 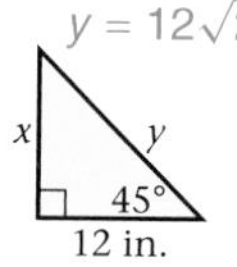 x = 12 in.
y = 12√2 in.
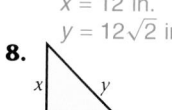
x   y
45°
12 in.

9. 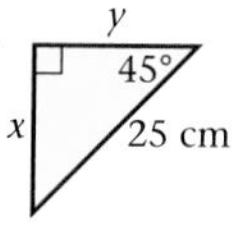 y
45°
x   25 cm
x = 25/√2 cm
y = 25/√2 cm

10. 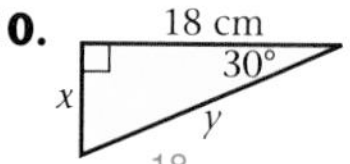 18 cm
30°
x
y
x = 18/√3 cm
y = 36/√3 cm

11.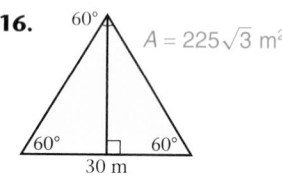
y   x
45°
2√2 in.
x = 2√2 in.
y = 4 in.

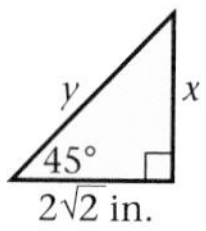

12. 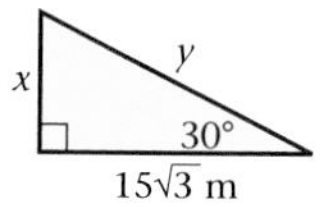 x   y
30°
15√3 m
x = 15 m
y = 30 m

13. 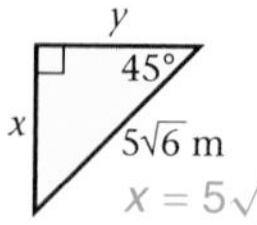 y
45°
x   5√6 m
x = 5√3 m
y = 5√3 m

Find x.

14. x   24 m
60° 45°
x = 8√6 m

15. 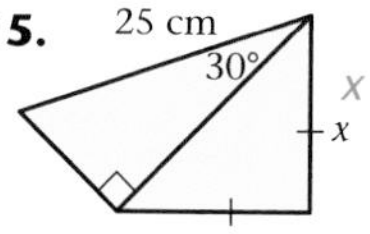 25 cm
30°
x = 6.25√6 cm
x

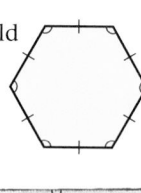 **Geometry** Find the area of each figure.

16. 60°
A = 225√3 m²
60° 60°
30 m

17. 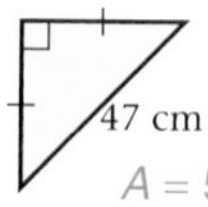
47 cm
A = 552.25 cm²

18. **Construction** The city park manager would like to pave an area in the shape of a regular hexagon with sides 20 feet long. If it will cost $15 per square-foot unit, find the cost to pave the hexagonal area.
approx. $15,600

A regular hexagon is a 6-sided polygon with all sides equal in length and all angles having the same measure.

Home Improvement
Mary and Chris want to build a right-triangular deck off the back of their house. They would like the hypotenuse of the deck to be 20 feet long, and they would like both sides of the deck to be equal in length.

19. Find the length of the sides of the deck. approx 14 ft, 2 in.

20. Find the area of the deck. 100 ft²

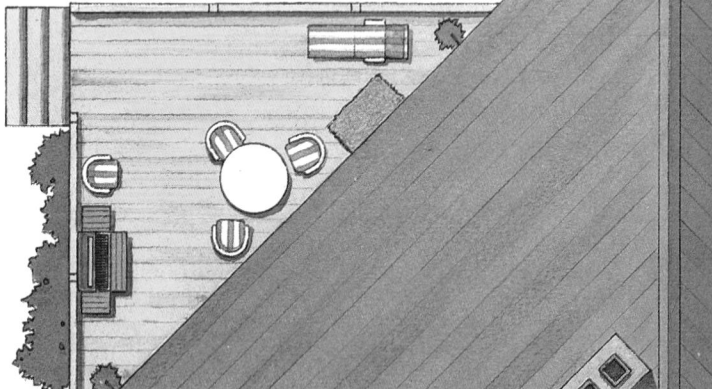

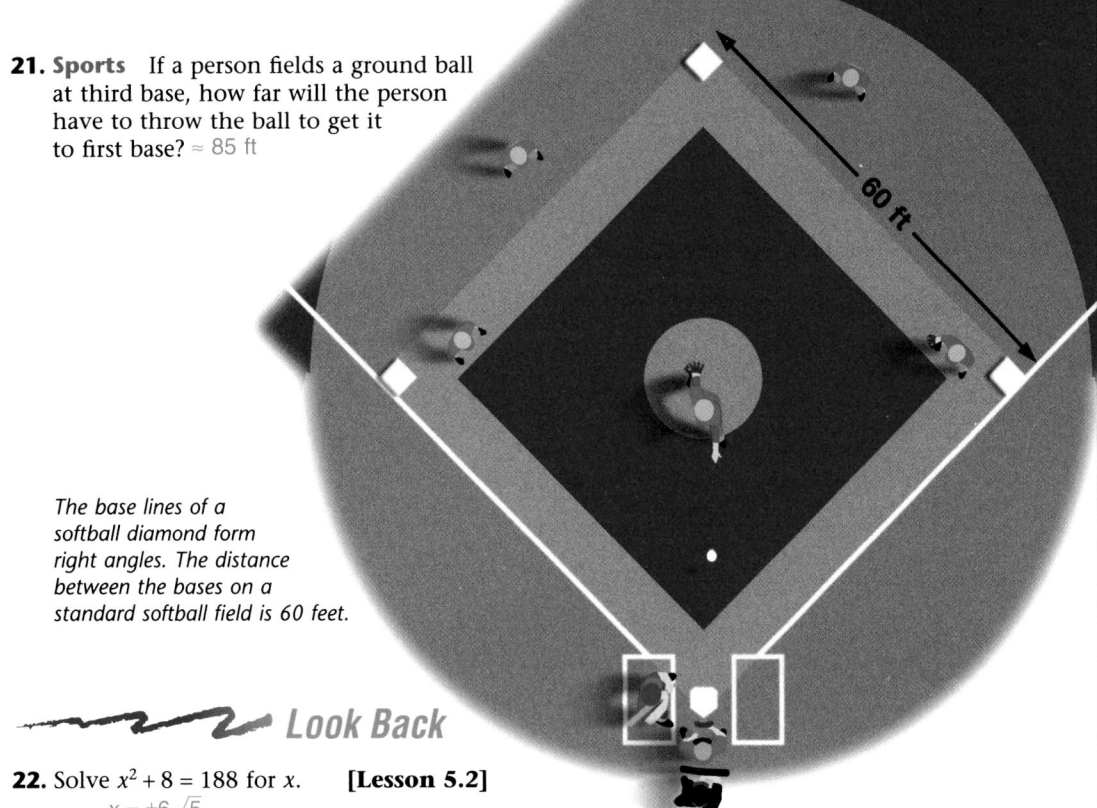

21. Sports If a person fields a ground ball at third base, how far will the person have to throw the ball to get it to first base? ≈ 85 ft

60 ft

The base lines of a softball diamond form right angles. The distance between the bases on a standard softball field is 60 feet.

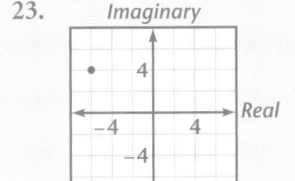 **Look Back**

22. Solve $x^2 + 8 = 188$ for x.   **[Lesson 5.2]**
$x = \pm 6\sqrt{5}$

Plot each number on a separate complex plane.   **[Lesson 5.7]**

23. $-6 + 4i$ **24.** $8 + 9i$ **25.** $-3 - 4i$

Determine the degree of each polynomial function.   **[Lesson 6.1]**

26. $f(x) = -5(x-4)(x+7)(x+1)^2$ 4   **27.** $f(x) = 7(x^2+3)(x-3)^2$ 4

28. $f(x) = 2x^5 - (x^2-9)(x^3+4)$ 5   **29.** $f(x) = 8$ 0

Write the polynomial in factored form.   **[Lesson 6.1]**

30. $f(x) = 2x^3 - 18x$ **31.** $f(x) = 3x^3 - 7x^2 + 2x$
$f(x) = 2x(x+3)(x-3)$ $f(x) = 3x(x-2)(x-\frac{1}{3})$

Look Beyond

32 Graph $y = \sqrt{1-x^2}$ and $y = -\sqrt{1-x^2}$ on the same coordinate plane. Describe the shape of the two graphs. A circle centered at the origin with the radius 1.

33 Find the x- and y-intercepts of both graphs. x-intercepts:(−1, 0), (1, 0)
y-intercepts:(0, −1), (0, 1)

34 Graph $y = \sqrt{1-x^2}$ and $y = x$ on the same coordinate plane. Find the coordinates of the point of intersection of these two graphs. (≈ 0.71, ≈ 0.71)

35. How do the coordinates of the point of intersection from Exercise 34 compare with the values of cos 45° and sin 45°? Appear to be the same

Look Beyond

These exercises prepare students for graphing and evaluating trigonometric values for angles placed on the coordinate plane.

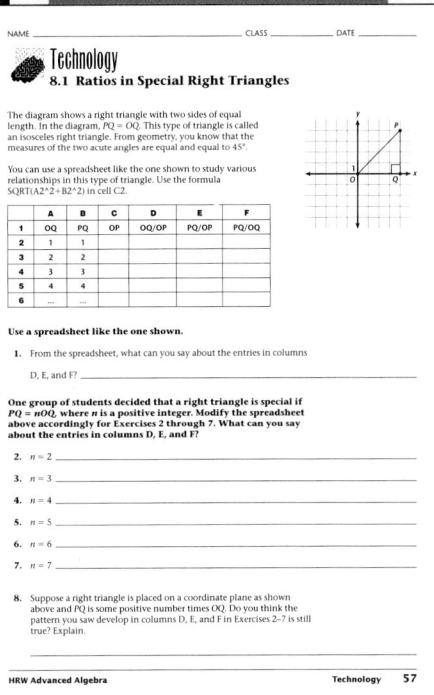

23. *Imaginary*

24. *Imaginary*

25. *Imaginary*

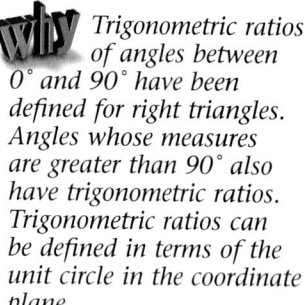
LESSON 8.2 The Unit Circle

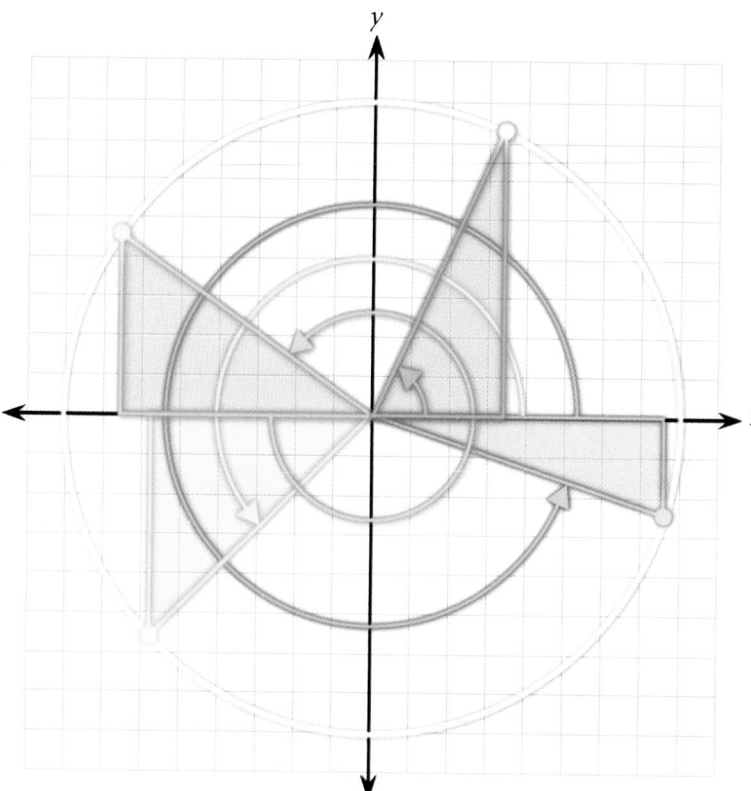

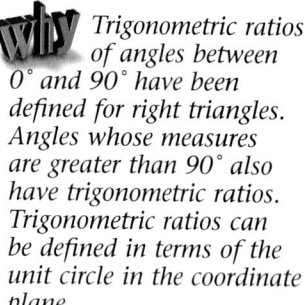 *Trigonometric ratios of angles between $0°$ and $90°$ have been defined for right triangles. Angles whose measures are greater than $90°$ also have trigonometric ratios. Trigonometric ratios can be defined in terms of the unit circle in the coordinate plane.*

An angle in the coordinate plane is in **standard position** if its vertex is at the origin and its **initial side** is on the positive x-axis.

The measure of the angle is the amount of rotation from the initial side to the **terminal side**.

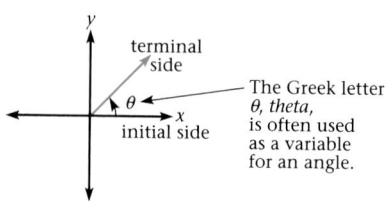

The Greek letter θ, theta, is often used as a variable for an angle.

COORDINATE GEOMETRY *Connection*

The **unit circle** is a circle with a radius of 1 centered at the origin of the coordinate plane.

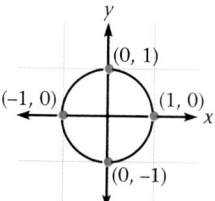

Suppose that an angle is in standard position and its terminal side has been rotated through the entire circle. Where is its terminal side now?

 **teaching strategy**

Technology Have students use a square viewing window to graph the unit circle on a graphics calculator as $y_1 = \sqrt{1 - x^2}$ and $y_2 = -\sqrt{1 - x^2}$. Then have them graph each of the following functions, one at a time. Students should use their intersect or trace and zoom calculator features to determine the coordinates of the points of intersection between each function and the unit circle. The x-coordinate of the intersect

point is the cosine of the angle, and the y-coordinate is the sine of the angle. Have students use these coordinates to find the angle represented by using the inverse sine and cosine buttons on their calculators ([COS⁻¹] and [SIN⁻¹] on most calculators).

$y_3 = \left(\frac{1}{\sqrt{3}}\right) x$ [**30°**] $y_3 = -\left(\sqrt{3}\right) x$ [**150°**]

$y_3 = x$ [**45°**] $y_3 = -x$ [**135°**]

$y_3 = \left(\sqrt{3}\right) x$ [**60°**] $y_3 = -\left(\frac{1}{\sqrt{3}}\right) x$ [**120°**]

Degree Measures in the Unit Circle

One quarter rotation is 90°.	One half rotation is 180°.	One full rotation is 360°.
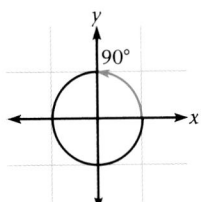	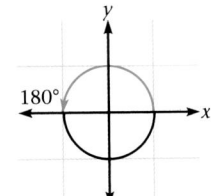	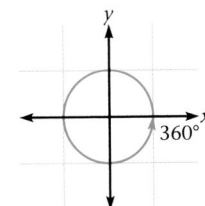

How many degrees is one-eighth of a full rotation? What fraction of a full rotation is one degree (1°)?

EXAMPLE 1

Draw each angle in standard position.

A 15°   **B** 75°   **C** 210°   **D** 345°

Solution

A One quarter rotation is 90°, and 15° is $\frac{1}{6}$ of 90°.

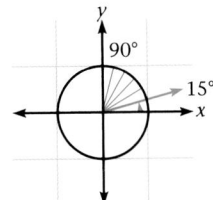

B One quarter rotation is 90°, and 75° is $\frac{5}{6}$ of 90°, or 15° less than 90°.

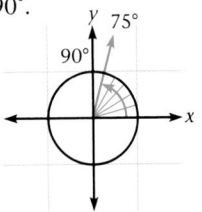

C One half rotation is 180°, and 210° is 30° more than 180°.

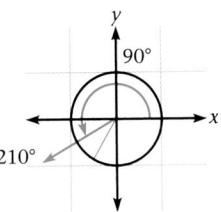

D One full rotation is 360°, and 345° is 15° less than 360°.

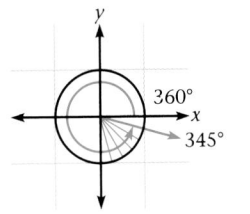

Try This   Draw each angle in standard position.

A 105°   **B** 135°   **C** 240°   **D** 315°

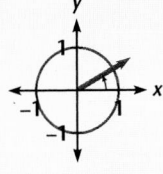 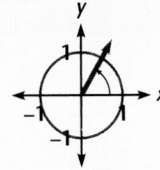
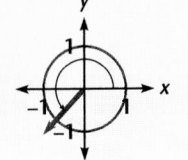 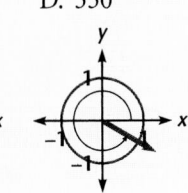
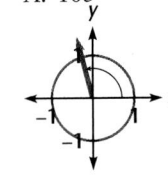 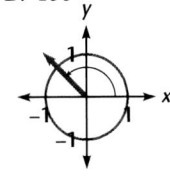
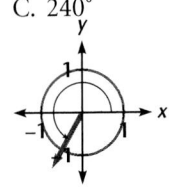 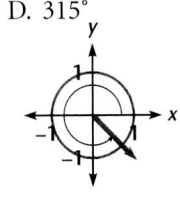
interdisciplinary CONNECTION

History of Science One of the most important events in the history of science was the development of analytic geometry. The individual associated with this historic event was René Descartes. Many objects and events in the physical world have geometric properties. Analytic geometry combines a coordinate plane and algebraic equations to describe geometric figures. This idea brought together Euclidean geometry of ancient Greece and algebra from Asia Minor.

Coterminal Angles

Angles in standard position can have positive or negative measure.

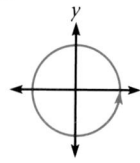

If the rotation from the initial side is measured in the *counterclockwise* direction, then the angle has a **positive measure.**

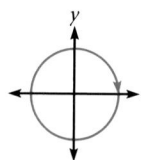

If the rotation from the initial side is measured in the *clockwise* direction, then the angle has a **negative measure.**

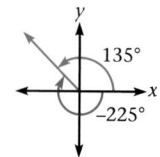

Angles that share the same terminal side are **coterminal angles**. For example, angles of 135° and −225° are coterminal.

Since 135 + 360 = 495, angles of 135° and 495° are also coterminal.

EXAMPLE 2

Find a coterminal angle for each angle.

A 235°   **B** −285°   **C** 390°

Solution

A Sketch an angle of 235°.

235 − 360 = −125

Thus, angles of 235° and −125° are coterminal.

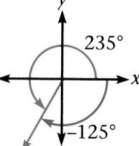 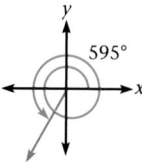

Since 235 + 360 = 595, angles of 235° and 595° are also coterminal.

B Sketch an angle of −285°.

−285 + 360 = 75

Thus, angles of −285° and 75° are coterminal.

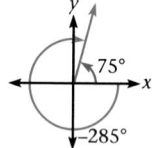 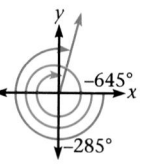

Since −285 − 360 = −645, angles of −285° and −645° are also coterminal.

C Sketch an angle of 390°.

390 – 360 = 30

Thus, angles of 30° and 390° are coterminal.

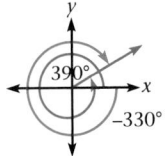

Since 390 – 360 – 360 = –330, angles of 390° and –330° are also coterminal. ❖

 **CRITICAL Thinking**

How many angles are coterminal with any given angle?

Consider a 45° angle in standard position intersecting the unit circle. The terminal side intersects the unit circle at a point $P(x, y)$, and forms a 45-45-90 triangle. The legs of the triangle are equal in length. By the "Pythagorean" Right-Triangle Theorem, the length of the hypotenuse is $\sqrt{a^2 + a^2} = \sqrt{2a^2} = a\sqrt{2}$.

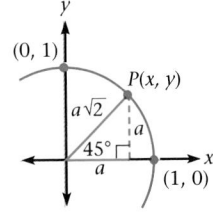

Since the hypotenuse of the triangle is also the radius of the unit circle, you know that $a\sqrt{2}$ is 1. To find the coordinates of P, solve for a.

$$a\sqrt{2} = 1$$
$$a = \frac{1}{\sqrt{2}}, \text{ or } \frac{\sqrt{2}}{2}$$

The coordinates of P are $\left(\frac{\sqrt{2}}{2}, \frac{\sqrt{2}}{2}\right)$. In Lesson 8.1, you learned that the cosine of 45° is $\frac{\text{length of side adjacent 45° angle}}{\text{length of hypotenuse}}$.

In this case, you can see that $\cos 45°$ is $\frac{\sqrt{2}}{2}$, the x-coordinate of P.

Why is the sine of 45° the y-coordinate of P?

The unit circle can be used to formulate the definition of sine and cosine for *any* angle.

COORDINATES OF POINTS ON THE UNIT CIRCLE

For all angles θ in standard position, the coordinates of the point where the terminal side of θ intersects the unit circle are $(\cos \theta, \sin \theta)$.

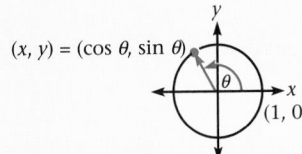

 **CRITICAL Thinking**

Explain how the unit circle can be used to show that $\tan \theta = \frac{\sin \theta}{\cos \theta}$ for any angle θ.

CRITICAL Thinking

Adding any integral multiple of 360° to a given angle gives an angle that is coterminal with the given angle. Therefore, there are *infinitely many* angles coterminal with a given angle.

ongoing ASSESSMENT

The value of the y-coordinate of P is a. The sine of 45° is $\frac{\text{length of side opposite 45° angle}}{\text{length of hypotenuse}}$ or $\frac{a}{a\sqrt{2}} = \frac{1}{\sqrt{2}}$. But in a unit circle, the hypotenuse is 1, so $1 = a\sqrt{2}$, or $a = \frac{1}{\sqrt{2}}$. Therefore a is the sine of 45°, and since a is the y-coordinate of P, the sine of 45° is the y-coordinate of P.

CRITICAL Thinking

The tangent of θ is given by $\frac{\text{side opposite } \theta}{\text{side adjacent } \theta}$. On the unit circle, the side opposite θ has length equal to the y-coordinate, which is $\sin \theta$. Also, the side adjacent θ has length equal to the x-coordinate, which is $\cos \theta$. So, $\tan \theta = \frac{\sin \theta}{\cos \theta}$.

Use Transparency ▶ **29**

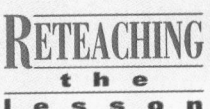 **RETEACHING the lesson**

Using Visual Models
Have students draw a unit circle where the terminal side of an angle of 60° intersects the circle at point $P(x, y)$. Have students find the coordinates of point P using the relationship $(x, y) = (\cos \theta, \sin \theta)$. Have students write a paragraph explaining why the $\cos 60°$ is the x-coordinate of P, and the $\sin 60°$ is its y-coordinate.

Aongoing
SSESSMENT

For each special angle θ and the ordered pair (x, y) shown on the unit circle, x represents $\cos \theta$ and y represents $\sin \theta$.

CRITICAL
Thinking

An angle with measure $0° < \theta < 90°$ intersects the unit circle at a positive x-coordinate and a positive y-coordinate.

An angle with measure $90° < \theta < 180°$ intersects the unit circle at a negative x-coordinate and a positive y-coordinate.

An angle with measure $180° < \theta < 270°$ intersects the unit circle at a negative x-coordinate and a negative y-coordinate.

An angle with measure $270° < \theta < 360°$ intersects the unit circle at a positive x-coordinate and a negative y-coordinate.

Alternate Example 3

Find x and y to the nearest hundredth where $x = \cos 25°$ and $y = \sin 25°$. [$x = 0.91$, $y = 0.42$]

Aongoing
SSESSMENT

Try This

$(x, y) \approx (0.26, 0.97)$

Special Angles in the Unit Circle

In Lesson 8.1, you worked with special angles from $0°$ to $90°$. The special angles from $0°$ to $360°$ are shown in standard position intersecting the unit circle below. The coordinates of the intersecting points are indicated.

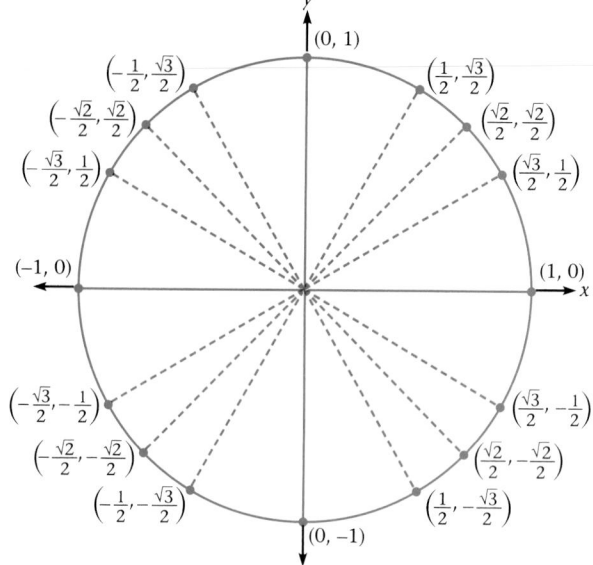

Use the unit circle shown to make a table that includes each special angle and its sine and cosine.

CRITICAL
Thinking

Determine the signs (positive or negative) of the coordinates where the terminal side of each of the following angles in standard position intersects the unit circle.

- An angle with measure $0° < \theta < 90°$
- An angle with measure $90° < \theta < 180°$
- An angle with measure $180° < \theta < 270°$
- An angle with measure $270° < \theta < 360°$

EXAMPLE 3

Graphics Calculator

Find x and y to the nearest hundredth.

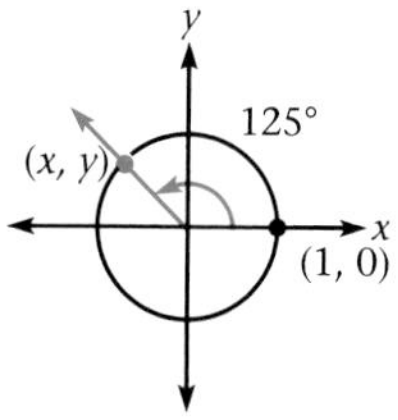

Solution▶

Since x and y are the coordinates of a point on the unit circle, x is $\cos 125°$ and y is $\sin 125°$.

$$x = \cos 125° \approx -0.57$$
$$y = \sin 125° \approx 0.82 ❖$$

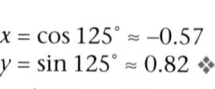

Try This Find the coordinates of the point where the terminal side of an angle of $75°$ intersects the unit circle.

EXERCISES & PROBLEMS

Communicate

1. If θ is an angle in standard position, explain why the coordinates (x, y) of the point where the terminal side of θ intersects the unit circle are $(\cos \theta, \sin \theta)$.

2. How do you find the coordinates of the point where the terminal side of a 60° angle intersects the unit circle?

3. Why is tan 90° undefined?

4. Explain how to find sin 360°, cos 360°, and tan 360° without using a calculator.

Practice & Apply

5. Express two-thirds of a full rotation in degrees. 240°

6. Express three-fourths of a full rotation in degrees. 270°

7. Express one-eighth of a full rotation in degrees. 45°

Draw each angle in standard position.

8. 260°	**9.** 132°	**10.** −125°	**11.** 318°
12. −245°	**13.** 54°	**14.** 94°	**15.** −315°
16. −12°	**17.** 890°	**18.** −490°	**19.** −756°
20. 45°	**21.** 60°	**22.** 90°	**23.** −30°
24. 135°	**25.** 210°	**26.** 300°	**27.** 152°
28. −340°	**29.** −114°	**30.** 745°	**31.** −400°

32. Let the coordinates of the point where θ intersects the unit circle be (−0.8, −0.6). Find tan θ. 0.75

Navigation In navigation, angles are often measured as clockwise rotations from due north. The measure of the angle, θ, is called the bearing.

33. Draw diagrams showing the paths a ship would take when sailing at bearings of 60°, 135°, and 200°.

34. Express the paths taken by the ship in Exercise 33 as angles in standard position.

The answers to Exercises 8–31 can be found in Additional Answers beginning on page 842.

34. Bearing of 60°: −330° (or 30°); bearing of 135°: −45° (or 315°); bearing of 200°: −110° (or 250°)

33.

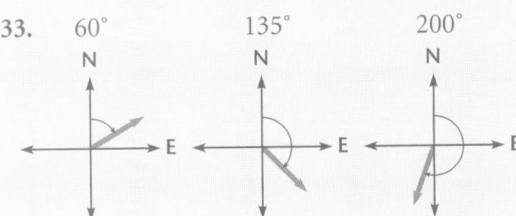

ASSESS

Selected Answers
Odd-numbered Exercises 5–49

Assignment Guide
Core 1–26, 32–40, 44–52

Core Plus 1–6, 16–34, 40–53

Technology
A graphics calculator is needed for Exercises 35–43, 46, and 51–53.

Error Analysis
When using compass bearing, 0 is located at the top for North and measurement progresses clockwise. This is different from standard trigonometric graphs. Students should be aware of the differences and have some time to reorient before working compass bearing exercises.

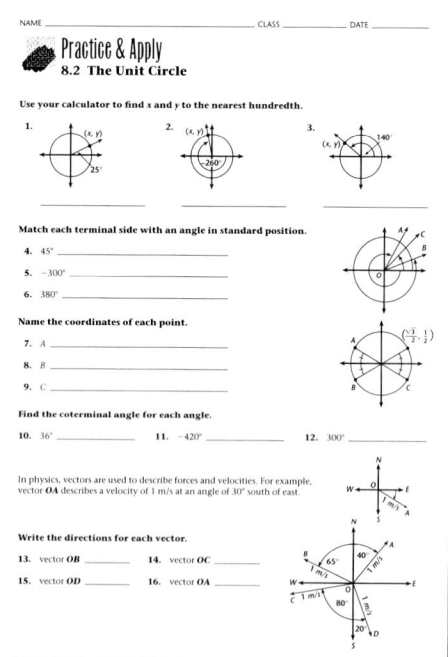
Authentic Assessment

Have the students do an actual surveying project with a protractor and a tape measure. Students can measure the height of a building or tree by standing at some distance from the base of the object so that the angle from the student to the top of the object is 30°, 45°, or 60°. Using the measurement of the base and the angle, students can determine the height of the object.

Practice Master

Use your calculator to find the coordinates (x, y) on the unit circle to the nearest hundredth.

35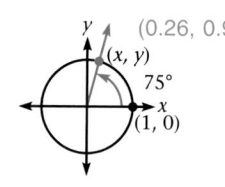
210° (−0.87, −0.5)

36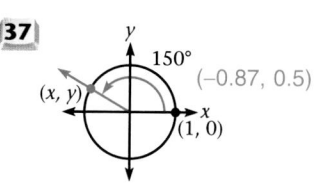
(0.26, 0.97) 75°

37 (−0.87, 0.5)
150°

38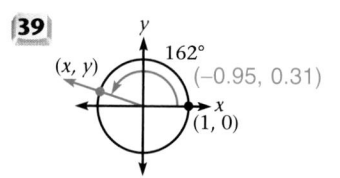
325° (0.82, −0.57)

39 162° (−0.95, 0.31)

40 250° (−0.34, −0.94)

41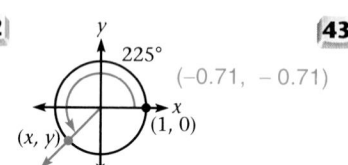
282° (0.21, −0.98)

42 225° (−0.71, −0.71)

43 (x, y) 65° (0.42, 0.91)

Transportation The freeway loop around a city is roughly circular. Christie and Sharmila drive through an angle of about 140° on the loop daily on their way to work. One morning, they hear that there is an accident between where they usually get on and get off the loop. They decide to go the opposite way.

44. Through what angle must they drive if they go the 220° other way? What portion of the entire loop is this? ≈ 61%

45. If they usually drive 6.5 miles on the loop, about how far will they have to drive if they go the other way? (HINT: What portion of the loop is 6.5 miles?)
≈10.2 miles

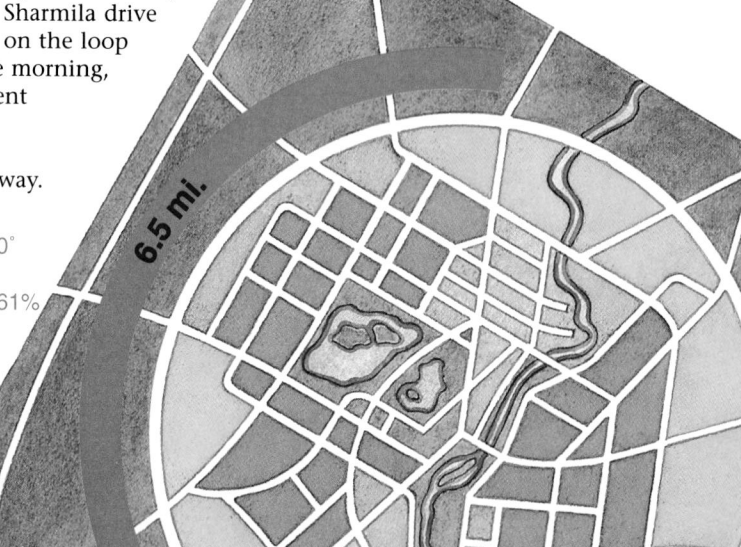

6.5 mi.

Look Back

Health A doctor is studying the relationship between smoking and cholesterol levels. The doctor randomly chooses several patients who smoke. The results are shown in the following table.

Cigarettes per day	15	21	40	28	25	32	45
Cholesterol level	190	210	250	245	220	228	235

46. 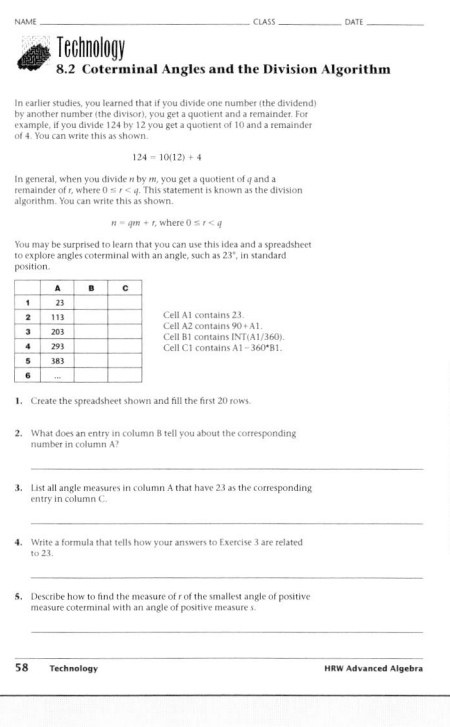 **Statistics** Find the equation of the line of best fit and the linear correlation coefficient for the data. **[Lesson 1.3]** $y \approx 1.59x + 178.61$, $r \approx 0.80$

47. Use your equation to estimate the cholesterol level of a person who smokes 36 cigarettes per day. **[Lesson 1.3]** ≈ 236

48. Can the doctor conclude that a person's cholesterol level can be lowered by cutting down on the number of cigarettes smoked per day? Explain.

Use the distance formula for Exercises 49 and 50. [Lesson 5.2]

49. **Coordinate Geometry** Show that the triangle with vertices $A(-2, 5)$, $B(4, 13)$, and $C(10, 5)$ is an isosceles triangle.

50. **Coordinate Geometry** Find y so that the point $(1, y)$ is equidistant from the points $(6, 17)$ and $(-4, -7)$. $y = 5$

Look Beyond

51. Use your graphics calculator to find $\cos 40°$. ≈ 0.766

52. Use your graphics calculator to find the inverse cosine $(\cos^{-1})$ of your answer to Exercise 51. $40°$

53. Use your graphics calculator to find the inverse sine $(\sin^{-1})$ of $\frac{\sqrt{3}}{2}$. $60°$

48. There is a fairly strong positive correlation between cigarettes smoked per day and cholesterol level. The doctor could conclude that cutting down on smoking may reduce cholesterol. Still, the doctor will need to do more studies on whether the cholesterol levels will actually drop when smoking is reduced, since this was not the exact purpose of the experiment.

49. $AB = \sqrt{(-2-4)^2 + (5-13)^2} = 10$, $AC = \sqrt{(-2-10)^2 + (5-5)^2} = 12$, and $BC = \sqrt{(4-10)^2 + (13-5)^2} = 10$. Since $AB = BC$, triangle ABC is an isosceles triangle.

PREPARE

Objectives

- Determine sides or angles of right triangles using trigonometric functions or their inverses.
- Determine, from the principal values of the inverse trigonometric functions, any angle in standard position.

RESOURCES

- Practice Master   8.3
- Enrichment Master   8.3
- Technology Master   8.3
- Lesson Activity Master   8.3
- Quiz   8.3
- Spanish Resources   8.3

Assessing Prior Knowledge

Find the inverse of each function.

A. $f(x) = 3x - 8$

$\left[f^{-1}(x) = \dfrac{x+8}{3} \right]$

B. $f(x) = 2e^x$

$\left[f^{-1}(x) = \ln\left(\dfrac{x}{2}\right) \right]$

TEACH

 If you know one side of a right triangle and one other angle, all the other measurements can be found. This is one tool used to determine indirectly a hard to measure distance.

Alternate Example 1

$m\angle A = 48°$ and the length of c is 30 centimeters.

A. Find a. [22.29 cm]
B. Find b. [20.07 cm]

LESSON 8.3 Trigonometric Applications

Brass astrolabe, 1295-96, used by Islamic astronomers

 *You can use the trigonometric ratios as functions to solve problems involving right triangles.*

Cultural Connection: Asia
Nearly all of trigonometry as it is known today was fully developed in the eastern civilizations before the Renaissance in Europe. Trigonometry went hand in hand with astronomy. Many of the famous Islamic astronomers came from Central Asia, Africa, and Spain, as well as from the Arab countries. Within the Islamic empire there were Jewish, Christian, and Muslim astronomers. Indian and Chinese astronomers also made contributions to trigonometry.

EXAMPLE 1

Graphics Calculator

Use triangle ABC.

Ⓐ Find a.   Ⓑ Find b.

Solution

Ⓐ Use the sine ratio to find a because it involves the hypotenuse and a, the side opposite the angle of $35°$.

$$\sin 35° = \frac{a}{20}$$

Multiply both sides by 20. $20 \sin 35° = a$
Use your calculator. $11.47 \approx a$

Thus, a is approximately 11.47 centimeters. This seems reasonable since the length of the hypotenuse is 20 centimeters.

Ⓑ Use the cosine ratio to find b because it involves the hypotenuse and b, the side adjacent to the $35°$ angle.

$$\cos 35° = \frac{b}{20}$$

Multiply both sides by 20. $20 \cos 35° = b$
Use your calculator. $16.38 \approx b$

Thus, b is approximately 16.38 centimeters. ❖

CRITICAL Thinking

Once you find the value of a in Example 1, describe another method you can use to find b.

ALTERNATIVE **teaching strategy**

Using Discussion Review the concept of inverse functions. Remind students that an inverse is a function only when each value in its domain has exactly one value in the range. Discuss how an inverse of the sine or cosine would allow someone to find an angle when given a side. Emphasize that since $\sin(30°) = \sin(390°)$, the domain of the sine function must be restricted if it is to have an inverse that is a function.

Inverse Trigonometric Functions

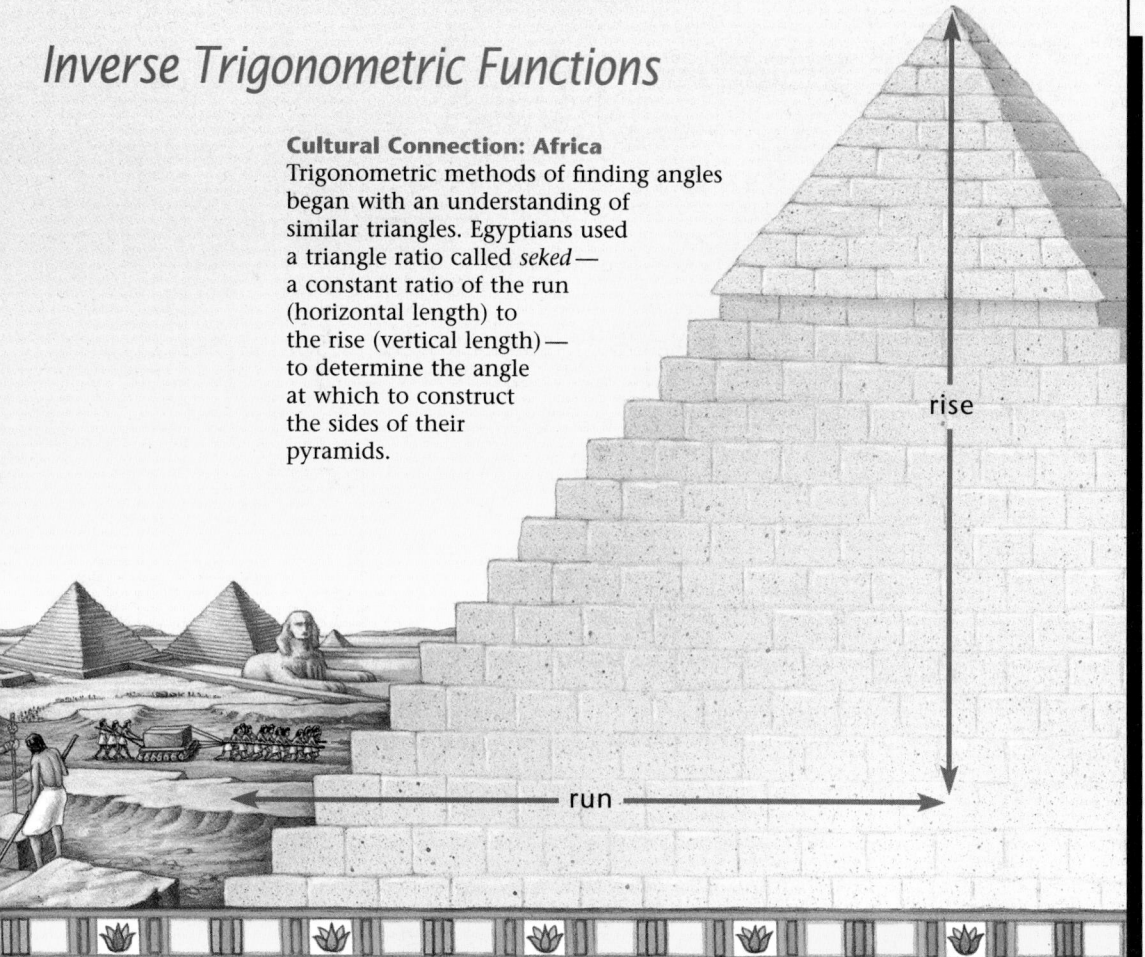

Cultural Connection: Africa
Trigonometric methods of finding angles began with an understanding of similar triangles. Egyptians used a triangle ratio called *seked*— a constant ratio of the run (horizontal length) to the rise (vertical length)— to determine the angle at which to construct the sides of their pyramids.

rise

run

CRITICAL
Thinking

Once you know a and c, use the "Pythagorean" Right-Triangle Theorem to find b:
$c^2 = a^2 + b^2 \rightarrow 20^2 \approx 11.47^2 + b^2$,
or $b \approx \sqrt{20^2 - 11.47^2}$.
So, $b \approx 16.38$.

Today, we use the tangent ratio, the reciprocal of the *seked*, to measure the slope of an incline. In right triangle *ABC*, the rise is 4 feet and the run is 3 feet. Thus, the tangent of angle *A* is $\frac{4}{3}$. When you know the tangent ratio of an angle, you can find the measure of that angle by using the *inverse* relation of the tangent.

In right triangle *ABC*, tan *A* is $\frac{4}{3}$, sin *A* is $\frac{4}{5}$, and cos *A* is $\frac{3}{5}$.

Trigonometric ratios also have *inverse* relations.

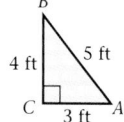

The *angle whose tangent is* $\frac{4}{3}$ is written **tan⁻¹** $\left(\frac{4}{3}\right)$. Thus, $\tan^{-1}\left(\frac{4}{3}\right) = A$.

The *angle whose sine is* $\frac{4}{5}$ is written **sin⁻¹** $\left(\frac{4}{5}\right)$. Thus, $\sin^{-1}\left(\frac{4}{5}\right) = A$.

The *angle whose cosine is* $\frac{3}{5}$ is written **cos⁻¹** $\left(\frac{3}{5}\right)$. Thus, $\cos^{-1}\left(\frac{3}{5}\right) = A$.

Recall the special angles in the unit circle. What is $\cos^{-1}\left(\frac{\sqrt{3}}{2}\right)$?

Aongoing
SSESSMENT

30°

ENRICHMENT Have students use the *x*-axis to represent radians and the *y*-axis to represent sin(*x*). Ask students to plot a number of points computed by calculator. Students should note that for values of $x > 2\pi$, the graph begins to repeat. Explain that this phenomenon creates a problem when finding the inverse of sin(*x*).

INCLUSION **Using Visual Models** The **strategies** inverse of the sine can be determined by drawing a triangle. For example, to find the $\sin^{-1}(0.7)$, construct a right triangle with hypotenuse = 1 and one leg = 0.7. Complete the drawing of the triangle, and then use a protractor to measure the angle.

Alternate Example 2

Let the length of $BC = 5$ and $CA = 7$. Find $m\angle A$ in the right triangle ABC (where C is the right angle). [$\approx 35.5°$]

A^{ongoing}**SSESSMENT**

An infinite number; any angle coterminal with 45° or 315° has $\frac{\sqrt{2}}{2}$ for its cosine.

TEACHING tip

Draw each trigonometric function on an overhead transparency overlaid on a coordinate grid. Then, take the function and reflect it with respect to the line $y = x$. Ask students whether the graph shown is a function. Remind students to use the vertical-line test. This will help students understand the domain restrictions on the inverse trigonometric functions.

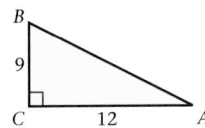

EXAMPLE 2

Find $m\angle A$ in $\triangle ABC$.

Solution➤

The length of the side opposite $\angle A$ is 9, and the length of the side adjacent to $\angle A$ is 12. Use the inverse tangent.

If $\tan A = \frac{9}{12}$, then $A = \tan^{-1}\left(\frac{9}{12}\right)$.

Use your calculator in degree mode.

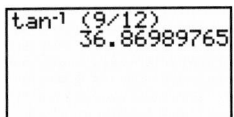

Thus, the measure of angle A is approximately 37°. ❖

Notice that $y = \cos x$ is a function because each x-value (angle) has exactly one y-value (cosine ratio). However, the inverse, $y = \cos^{-1} x$, is not a function because each x-value (cosine ratio) has more than one y-value (angle). For example,

$$\cos^{-1}\left(\frac{\sqrt{2}}{2}\right) = 45° \text{ and } \cos^{-1}\left(\frac{\sqrt{2}}{2}\right) = 315°.$$

Considering angles of any measure, how many values are there for $\cos^{-1}\left(\frac{\sqrt{2}}{2}\right)$?

Principal Values

A graphics calculator graphs and evaluates only functions. An inverse trigonometric relation can become a function if the range of the inverse function is restricted. The numbers in the restricted range are called the **principal values** of the inverse relation.

The principal values for the inverse cosine, inverse sine, and inverse tangent functions are given below.

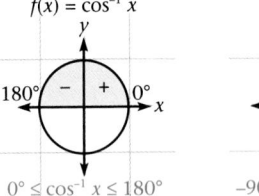

$0° \leq \cos^{-1} x \leq 180°$

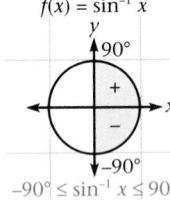

$-90° \leq \sin^{-1} x \leq 90°$

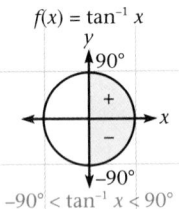

$-90° < \tan^{-1} x < 90°$

RETEACHING *the lesson*

Hands-On Strategies

Have the students construct several right triangles and measure their sides. Then have them compute the ratios of the lengths of pairs of sides. With a protractor, have students measure the angles. Explain that these ratios are the sine, cosine, or tangent depending on which sides of the triangle are used for the ratios.

CRITICAL
Thinking

Why are the principal values for $f(x) = \sin^{-1} x$ and $f(x) = \tan^{-1} x$ different from those for $f(x) = \cos^{-1} x$?

EXAMPLE 3

Graphics Calculator

Find $\tan^{-1}(-0.6)$ to the nearest degree for $90° < \theta < 270°$.

Solution

First, find the principal value for $\tan^{-1}(-0.6)$.

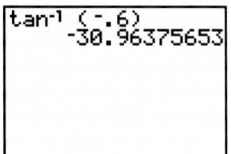

```
tan-1 (-.6)
    -30.96375653
```

The principal value is approximately $-31°$.

Tangent function values are negative in Quadrants II and IV, so the angle between $90°$ and $270°$ whose tangent is -0.6 is in Quadrant II.

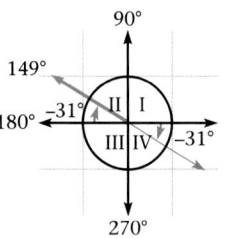

$$180 - 31 = 149$$

Thus, for $90° < \theta < 270°$, $\tan^{-1}(-0.6) \approx 149°$. ❖

Try This Find $\cos^{-1}(0.9)$ to the nearest degree for $180° < \theta < 360°$.

Applications of Trigonometric Functions

EXAMPLE 4

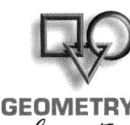

GEOMETRY
Connection

Find the area of the triangular piece of land between the lake and the road.

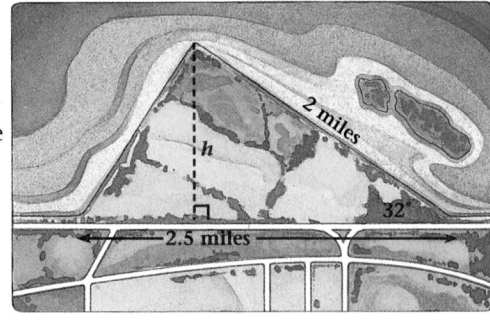

Solution

The formula for the area of a triangle is $A = \frac{1}{2}bh$. You are given that the length of the base, b, is 2.5 miles. Use the sine ratio to find h.

$$\sin 32° = \frac{h}{2}$$
$$2 \sin 32° = h$$

Substitute $2 \sin 32°$ for h and 2.5 for b in the area formula.

$$A = \frac{1}{2}bh$$
$$A = \frac{1}{2}(2.5)(2 \sin 32°)$$
$$A \approx 1.325$$

Thus, the area of the triangular piece of land is approximately 1.325 square miles. ❖

CRITICAL
Thinking

The principal values for $\sin^{-1} x$ and $\tan^{-1} x$ ensure that there is only one angle for each value in their domains, that is, that they are functions. If $\cos^{-1} x$ had principal values between $90°$ and $-90°$, there would be *two* different angles for each value of x, and $\cos^{-1} x$ would not be a function. For example, $\cos^{-1} \frac{\sqrt{2}}{2}$ could be either $45°$ or $-45°$. With the principal values $0° \le \cos^{-1} x \le 180°$, however, $\cos^{-1} \frac{\sqrt{2}}{2}$ will have only one value, $45°$.

Alternate Example 3

Find $\tan^{-1}(-0.8)$ to the nearest degree for $90° < \theta < 270°$. [$\approx 141°$]

Aongoing
SSESSMENT

Try This

$334°$

Alternate Example 4

Find the area of a triangular piece of land with a road at the base 10 miles long, a road on the right of the base 8 miles long, and an angle between the roads of $40°$. [≈ 25.7 square miles]

Math Connection

Geometry Students see that they can use the formula $A = \frac{1}{2}ab \sin C$ to find the area of a triangle when they know two sides of the triangle and the angle between them.

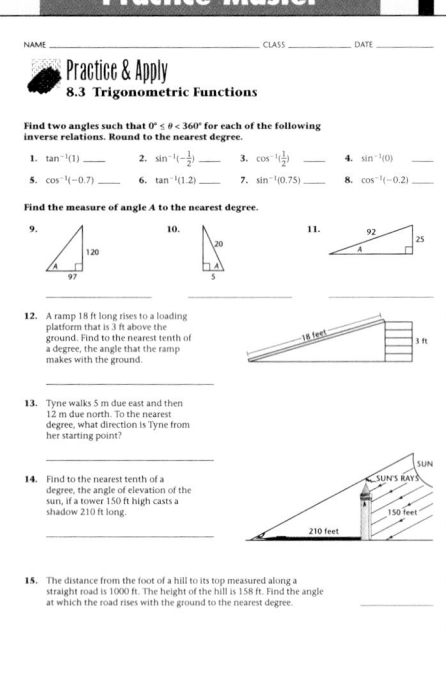

EXERCISES & PROBLEMS

Communicate

1. Explain how to find the missing lengths of the sides of a right triangle when you know the measure of one of the acute angles and the length of one side.

2. Explain how to find the measures of the acute angles of a right triangle when you know the lengths of the sides.

3. How do you find the measure of the angle whose terminal side intersects the unit circle at the point (–0.88, 0.47)?

4. Why do the inverse trigonometric relations have principal values?

5. If cos 58° is approximately 0.1192, how can you find another angle between 0° and 360° whose cosine is approximately 0.1192?

6. What are the domains of the functions $f(x) = \cos^{-1} x$, $f(x) = \sin^{-1} x$, and $f(x) = \tan^{-1} x$?

Practice & Apply

Find two angles between 0° and 360° for each of the following inverse *relations*.

7. $\tan^{-1}(-1)$ 135°, 315° 8. $\sin^{-1}\left(-\dfrac{\sqrt{3}}{2}\right)$ 240°, 300° 9. $\cos^{-1}\left(-\dfrac{1}{2}\right)$ 120°, 240°

10. $\tan^{-1}(\sqrt{3})$ 60°, 240° 11. $\cos^{-1}\left(-\dfrac{\sqrt{3}}{2}\right)$ 150°, 210° 12. $\sin^{-1}\left(\dfrac{\sqrt{2}}{2}\right)$ 45°, 135°

13. $\tan^{-1}\left(-\dfrac{\sqrt{3}}{3}\right)$ 150°, 330° 14. $\sin^{-1}\left(\dfrac{1}{2}\right)$ 30°, 150° 15. $\cos^{-1}\left(\dfrac{\sqrt{2}}{2}\right)$ 45°, 315°

16. $\tan^{-1} 0$ 0°, 180° 17. $\sin^{-1}\left(-\dfrac{1}{2}\right)$ 210°, 330° 18. $\cos^{-1}\left(\dfrac{\sqrt{3}}{2}\right)$ 30°, 330°

19. Find an angle θ other than 90° such that tan is undefined. answers may vary but should be equivalent to any angle of the form 90° + 180°·n, n an integer.

Safety Robin needs to climb a $6\dfrac{1}{2}$-foot wall using an 8-foot ladder. For safety purposes, the boss warned Robin not to lean the ladder at an angle to the ground that is greater than 60°, and always to let the top of the ladder hang over the top of the wall.

20. What angle does the ladder make with the ground if Robin allows 1 foot of the ladder to hang over the wall? ≈ 68°

21. What length of the ladder (to the nearest inch) should hang over the wall to create an angle of 60° with the ground? 6 inches

Laser Surgery A doctor plans to conduct laser surgery on a tumor lying 5.4 centimeters below the patient's skin. There is an organ directly above the tumor, and the doctor moves the source of the beam over 8.5 centimeters.

22. At what angle to the skin must the doctor aim the laser to reach the tumor? ≈ 32.4°

23. How far will the laser beam travel through the patient's body to reach the tumor? ≈ 10.1 cm

Indicate whether each statement can possibly be true. Explain your answers.

24. $\cos^{-1}\left(\frac{1}{2}\right) = \sin^{-1}\left(-\frac{\sqrt{3}}{2}\right)$

25. $\cos^{-1} 1 = \sin^{-1}(-1)$

26. $\cos^{-1}\left(-\frac{\sqrt{2}}{2}\right) = \sin^{-1}\left(\frac{\sqrt{3}}{2}\right)$

27. $\tan^{-1}\left(\frac{\sqrt{3}}{3}\right) = \cos^{-1}\left(\frac{1}{2}\right)$

28. $\tan^{-1} 1 = \sin^{-1}\left(-\frac{\sqrt{2}}{2}\right)$

29. $\tan^{-1} 0 = \cos^{-1} 0$

30. $\tan^{-1}\left(-\sqrt{3}\right) = \cos^{-1}\left(\frac{1}{2}\right)$

31. $\sin^{-1}\left(\frac{\sqrt{3}}{2}\right) = \tan^{-1}\left(\sqrt{3}\right)$

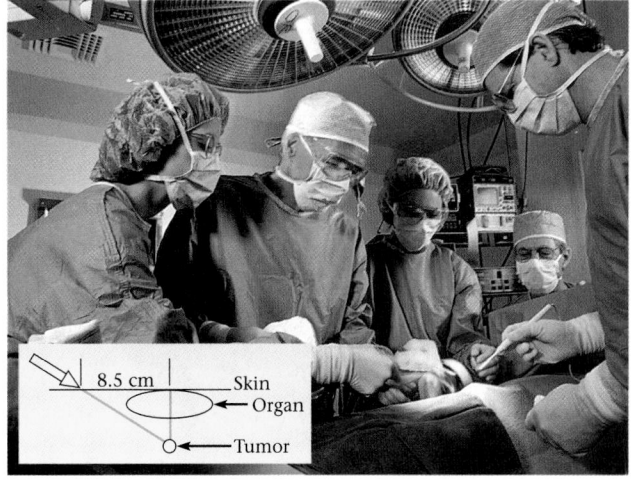

Aviation Commercial airline pilots fly at an altitude of about $6\frac{1}{2}$ miles. To make a gentle descent when landing, pilots begin descending toward the airport when they are still far away.

32. If the pilot begins descending at a constant angle 186 miles from the airport, what angle will the plane's path make with the runway? ≈ 2°

33. If the plane's path is to make an angle of 5° with the ground, how far from the airport must the pilot begin descending? ≈ 74 miles

You can substitute $\cos\theta$ for x and $\sin\theta$ for y in the equation for the unit circle, $x^2 + y^2 = 1$, to obtain the trigonometric identity $\cos^2\theta + \sin^2\theta = 1$. Use this identity for Exercises 34 and 35.

34. Find all values for $\cos\theta$ such that $\sin\theta$ is 0.96. −0.28 (Quad II) or 0.28 (Quad I)

35. Find all values for $\sin\theta$ such that $\cos\theta$ is $\frac{\sqrt{3}}{2}$. $\frac{1}{2}$ (Quad I) or $-\frac{1}{2}$ (Quad IV)

24. True for any angle coterminal with 300°. For example, $\cos 300° = \frac{1}{2}$ and $\sin 300° = -\frac{\sqrt{3}}{2}$.

25. Never true; the cosine is 1 only when the sine is 0, and the sine is −1 only when cosine is 0.

26. Never true; the cosine would only take on the given value for angles coterminal with 135°, or 225° and the sine would only take on the given value for angles coterminal

with 60° or 120°.

27. Never true; the cosine would only take on the given value for angles coterminal with 60° or 300°, and the tangent would only take on the given value for angles coterminal with 30° or 330°.

28. True for any angle coterminal with 225°. For example, $\tan 225° = 1$ and $\sin 225° = -\frac{\sqrt{2}}{2}$.

29. Never true; the tangent is 0 only when the cosine is ±1, and the cosine is 0 only when the tangent is undefined.

30. True for any angle coterminal with 300°. For example, $\tan 300° = -\sqrt{3}$ and $\cos 300° = \frac{1}{2}$.

31. True for any angle coterminal with 60°. For example, $\sin 60° = \frac{\sqrt{3}}{2}$ and $\tan 60° = \sqrt{3}$.

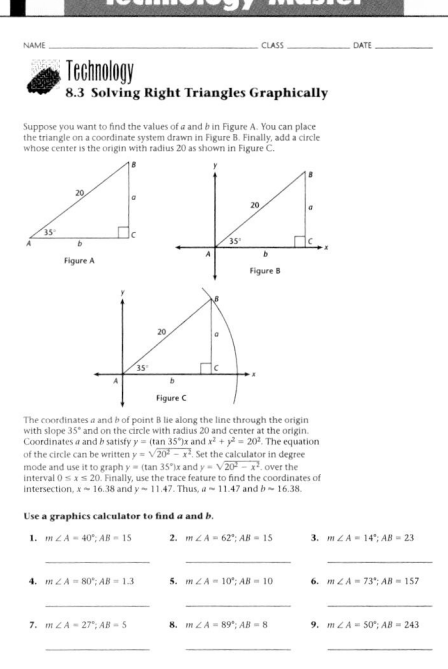

Cycling The wheels of a bicycle have a diameter of 26 inches. The valve on the back tire starts at a point 13 inches above the ground and turns counterclockwise. **Complete Exercises 36–39.**

36. How high off the ground will the valve be after the tire has rotated 40°? Through how many degrees will the tire have to rotate before the valve next reaches this height? ≈ 21.4 in., 140°

37. How high off the ground will the valve be after the tire has rotated 220°? ≈ 4.60 in. Through how many degrees will the tire have to rotate before the valve next reaches this height? 320°

38. If the valve is 20 inches off the ground, find two possible angles the tire has rotated.
any angles co-terminal with 32.6° or 147.4°

39. If the valve is 6 inches off the ground, find two possible angles the tire has rotated.
any angles co-terminal with 212.6° or 327.4°

40. Let the coordinates of the point where θ intersects the unit circle be (0.96, 0.28). Find θ to the nearest degree. 16°

41. Let the coordinates of the point where θ intersects the unit circle be (0.39, 0.92). Find θ to the nearest degree. 67°

← Valve

13 in.

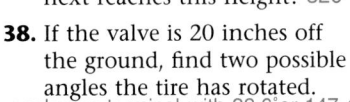 **Look Back**

Let $f(x) = x^5 + 3.5x^4 - 41x^3 - 52.5x^2 + 225x$. Use your graphics calculator to graph f.   [Lesson 6.4]

42 Find all the zeros of f. –7.5, – 3,0, 2,5

43. Use the zeros of f to write f in factored form. $f(x) = (x + 7.5)(x + 3)(x)(x - 2)(x - 5)$

44 Where is f increasing? Where is f decreasing?

45. Identify the domain and range of $f(x) = 6^x$.   [Lesson 7.2]
R; y > 0

Use your calculator to solve each equation. Round your answer to the nearest hundredth. [Lesson 7.6]

46 $4^x = 35$ $x ≈ 2.56$ **47** $\ln x^2 = 4$ $x ≈ ± 7.39$ **48** $10\ln(2x + 3) = 1.8$ $x ≈ - 0.90$

Look Beyond

This prepares students for the concept of transformations of trigonometric functions, which they will study in the next lesson.

Look Beyond

49 Use your graphics calculator (in degree mode) to graph $f(x) = \sin x$ and $f(x) = \cos x$ on the same coordinate plane. How are the two graphs alike? How are they different?

44. f is increasing for $x < -6.1$, $-1.8 < x < 1.1$, and $x > 4.0$; f is decreasing for $-6.1 < x < -1.8$, and for $1.1 < x < 4.0$. (note: all values are rounded to the nearest tenth)

49. Both graphs have the same shape, the same domain, and the same range. The graph of $f(x) = \sin x$ is the graph of $f(x) = \cos x$ shifted 90° to the right.

LESSON 8.4

Exploring

Sine and Cosine Graphs

why *In the real world, many things come in cycles, or periods, like spring, summer, fall, and winter. The functions needed to model such periodic events must possess the same characteristics.*

Regardless of what time you set a clock, every 12 hours the hands of the clock will point to the same numerals. A 12-hour interval is one *cycle* of the clock, and 12 hours is the *period* of the clock. Functions that behave in a similar manner are periodic functions. The sine and cosine are examples of periodic functions.

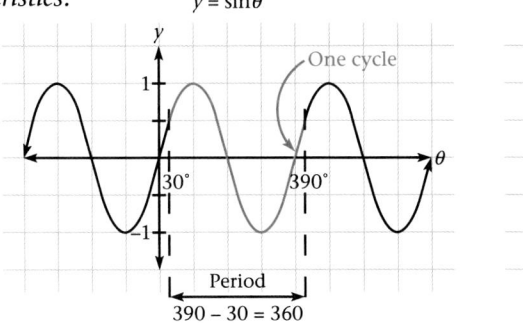

$y = \sin\theta$

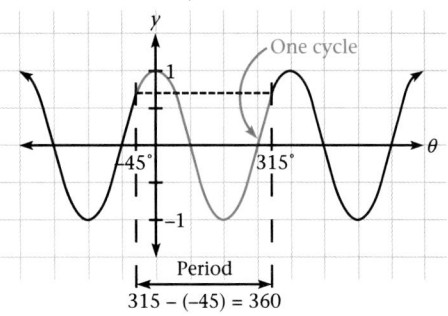

$y = \cos\theta$

The **period** of the sine and cosine functions is 360°.

Exploration 1 *Vertical Shift*

Graphics Calculator

Graph each function, and complete the table.

Function	Domain	Range	Period
$y = \sin\theta$	All real numbers	$-1 \le y \le 1$	360°
❶ $y = 2 + \sin\theta$			
❷ $y = -3 + \sin\theta$			

❸ Predict the domain, range, and period of $y = \frac{1}{2} + \sin\theta$. Check your results by graphing.

❹ What happens to the graph of $y = a + \sin\theta$ when a is changed?

❺ Repeat Steps 1–4 with the cosine function. ❖

ALTERNATIVE teaching strategy

Using Algorithms
Discuss the general functions of the forms $f(\theta) = a + b\sin c(\theta - d)$ and $f(\theta) = a + b\cos c(\theta - d)$. Explain what effect each of the constants a, b, c, and d have on the shape of the graph. Have students substitute 3 or 4 different values in for each variable and explain how each value affects the graph of the function.

PREPARE

Objectives
- Identify how a, b, c, and d in $f(\theta) = a + b\sin c(\theta - d)$ and $f(\theta) = a + b\cos c(\theta - d)$ transform the graphs of $f(\theta = \sin\theta$ and $f(\theta) = \cos\theta$.

RESOURCES

- Practice Master   8.4
- Enrichment Master   8.4
- Technology Master   8.4
- Lesson Activity Master   8.4
- Quiz   8.4
- Spanish Resources   8.4

Assessing Prior Knowledge
Find the vertex of the parabola described by each quadratic function.
A. $f(x) = (x + 1)^2 - 2$
    $[(-1, -2)]$
B. $f(x) = -(x - 3)^2 + 4$ $[(3, 4)]$

TEACH

why Most trigonometric functions, no matter how complex, can be expressed as transformations of sine and cosine functions.

Exploration 1 Notes
Students should notice that the domain remains unchanged as the range changes.

ongoing ASSESSMENT

4. For $a > 0$, the graph of $y = \sin\theta$ is translated upward a units. For $a < 0$, the graph of $y = \sin\theta$ is translated downward $|a|$ units.

Use Transparency ▶ 31

Students should see that the size of the range changes with the value of *b*.

4. For *b* > 1, the graph of *y* = sin *θ* is stretched vertically *b* units. For 0 < *b* < 1, the effect is a vertical compression of the graph.

Exploration 3 Notes

Changing the sign of the function reflects the graph with respect to the *x*-axis. The domain, range, and period of the function remain the same.

2. The graphs are reflections of one another with respect to the *x*-axis.

Exploration 4 Notes

These transformations stretch or compress the curve along the *x*-axis. The domain and range remain unchanged, but the period changes.

4. For *c* < 1, the period of the graph stretches horizontally from 360° to $\frac{360°}{c}$. For *c* > 1, the period compresses horizontally from 360° to $\frac{360°}{c}$.

6. The graphs of *y* = sin(−*θ*) and *y* = −sin *θ* are the same. The graph of *y* = −cos *θ* is a reflection of the graph of *y* = cos(−*θ*) with respect to the *x*-axis.

Exploration 2 Vertical Stretch

Graphics Calculator

Graph each function, and complete the table.

Function	Domain	Range	Period
$y = \sin \theta$	All real numbers	$-1 \le y \le 1$	360°
$y = \frac{1}{2}\sin \theta$			
$y = 3 \sin \theta$			

3 Predict the domain, range, and period of $y = 5 \sin \theta$. Check your results by graphing.

4 What happens to the graph of $y = b \sin \theta$, $b > 0$, when *b* gets larger?

5 Repeat Steps 1–4 with the cosine function. ❖

Exploration 3 Reflections

1 Graph $y = \sin \theta$ and $y = -\sin \theta$. How are the graphs alike or different?

2 Predict what the relationship between $y = \cos \theta$ and $y = -\cos \theta$ will be. Check by graphing. ❖

If $y = b \sin \theta$ or $y = b \cos \theta$, $|b|$ is the **amplitude**.

Exploration 4 Horizontal Stretch

Graph each function, and complete the table.

Function	Domain	Range	Period
$y = \sin \theta$	All real numbers	$-1 \le y \le 1$	360°
$y = \sin \frac{1}{2}\theta$			
$y = \sin 3\theta$			

3 Predict the domain, range, and period of $y = \sin \frac{1}{4}\theta$. Check your results by graphing.

4 The period of the function is $\frac{360°}{c}$. What happens to the graph of $y = \sin c\theta$ as *c* is changed?

5 Repeat Steps 1–4 with the cosine function.

6 What is the relationship between $y = -\sin \theta$ and $y = \sin(-\theta)$? between $y = -\cos \theta$ and $y = \cos(-\theta)$? ❖

interdisciplinary **Astronomy** A sine curve can be used to model the orbits of Jupiter's moons if they are viewed edge-on. Io takes 1.77 days to orbit Jupiter. Assume the distance of Io from the center of Jupiter is 421,770 km. Use 421,770 km as the amplitude and 1.77 days as the period of Io's sine curve. Assume also the distances for Europa, Ganymede, and Cal- listo from the center of Jupiter are 671,000 km at 3.55 days, 1,070,000 km at 7.15 days, and 1,885,000 km at 16.69 days, respectively. Assume that all the moons begin their orbits simultaneously. Have students use poster board and draw the graphs of the four moons on one coordinate plane. Have them label the *x*-axis in *days*, and the *y*-axis in *kilometers*. Let $0 \le x \le 20$ for each sine curve.

TRANSFORMATIONS
Connection

1 Determine how a change in d affects the graph of $y = \sin(\theta - d)$.

2 Determine how a change in d affects the graph of $y = \cos(\theta - d)$.
d is called the **phase shift** of the function. ❖

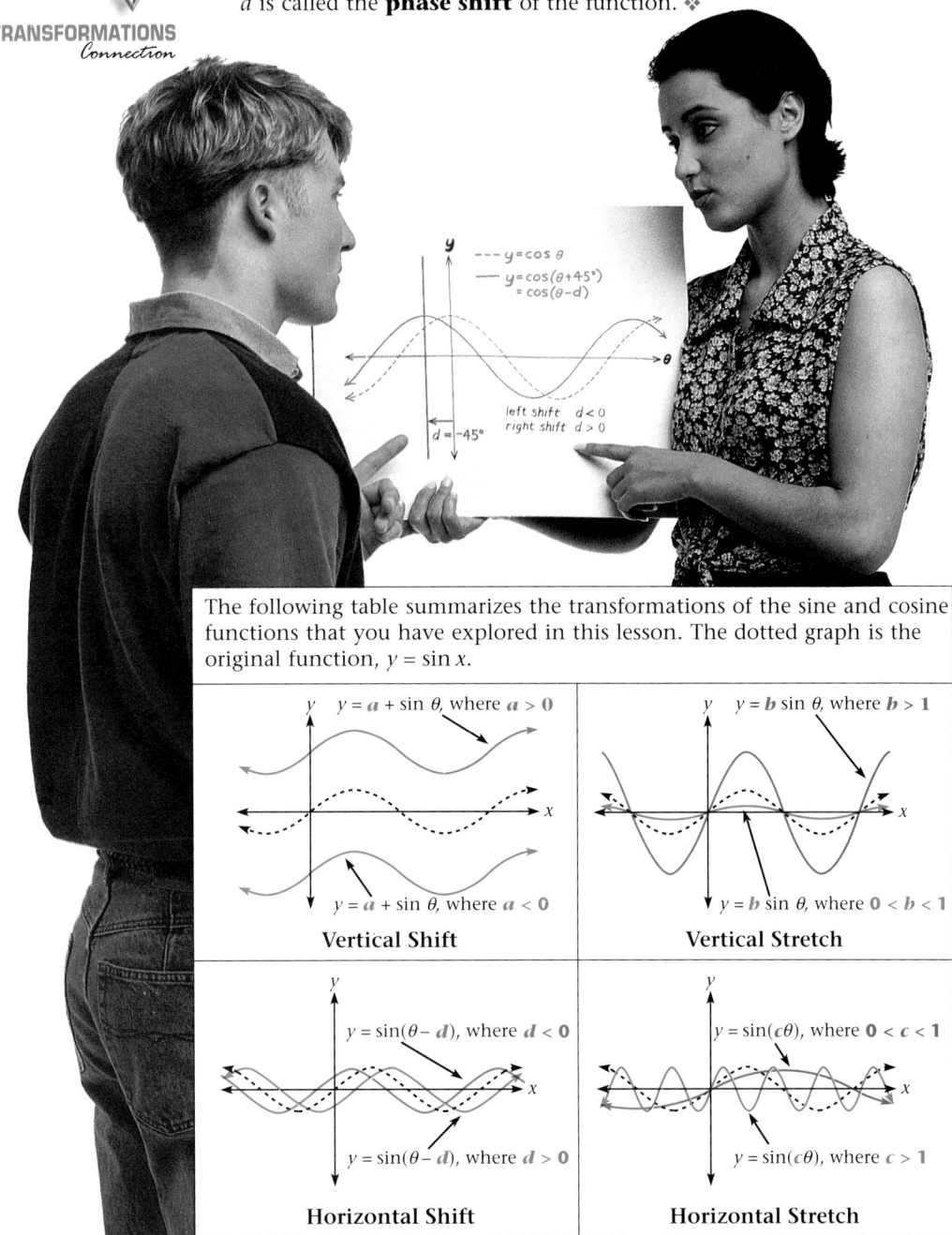

The following table summarizes the transformations of the sine and cosine functions that you have explored in this lesson. The dotted graph is the original function, $y = \sin x$.

$y = a + \sin\theta$, where $a > 0$ / $y = a + \sin\theta$, where $a < 0$ **Vertical Shift**	$y = b\sin\theta$, where $b > 1$ / $y = b\sin\theta$, where $0 < b < 1$ **Vertical Stretch**
$y = \sin(\theta - d)$, where $d < 0$ / $y = \sin(\theta - d)$, where $d > 0$ **Horizontal Shift**	$y = \sin(c\theta)$, where $0 < c < 1$ / $y = \sin(c\theta)$, where $c > 1$ **Horizontal Stretch**

Exploration 5 Notes
Explain to the students that d moves the curve to the right or left without changing the shape, domain, or range of the function.

A **ongoing**
SSESSMENT

1. For $d > 0$, the graph of $y = \sin(\theta - d)$ is the graph of $y = \sin\theta$ translated d units to the right. For $d < 0$, the graph of $y = sin(\theta - d)$ is the graph of $y = \sin\theta$ translated $|d|$ units to the left.

2. For $d > 0$, the graph of $y = \cos(\theta - d)$ is the graph of $y = \cos\theta$ translated d units to the right. For $d < 0$, the graph of $y = \cos(\theta - d)$ is the graph of $y = \cos\theta$ translated $|d|$ units to the left.

Math Connection

Transformations Most of these trigonometric transformations have equivalents in polynomial transformations. For $f(x) = a(x - h)^2 + k$, h gives a horizontal shift, k a vertical shift, and a stretches. If a is negative, the function is reflected.

Cooperative Learning

Working in pairs, students can develop flashcards for a variety of transformations of the sine curve. Students can ask each other questions related to the horizontal and vertical shifts, the amplitude, period, domain, and the phase shift of each function. Exact answers are not necessary. Approximate answers will help develop student intuition.

ENRICHMENT Ask students to identify the transformations that do not change the x-intercepts and the transformations that do change the x-intercepts. Have students find the transformations that do not change the range and the transformations that do change the range. Have students name the transformations that change the shape of the curve and those that do not change the shape of the curve.

INCLUSION **Hands-On Strategies**
strategies Have students use a graphics calculator to compute a table of values for several transformations of the sine function. Have them plot these points on graph paper, and identify the domain, range, and period for each function.

ASSESS

Selected Answers

Odd-numbered Exercises 7–35

Assignment Guide

Core 1–14, 17–38

Core Plus 1–7, 12–38

Technology

Graphics calculators are needed for Exercises 27–29, and 31–35.

Error Analysis

Students may select a viewing window which fails to show enough of the graph. Remind students that 360° is the period for untransformed sine and cosine functions. The *x*-range should be at least 2 × 360°.

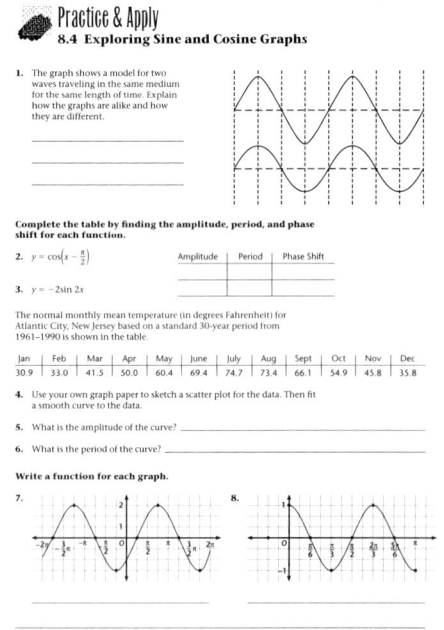
EXERCISES & PROBLEMS

Communicate

1. Given that $\sin 30° = \frac{1}{2}$, how can you use the period of the function to find another angle whose sine is $\frac{1}{2}$?

Compare the graph of each function with the graph of $f(\theta) = \cos\theta$. Explain how the graphs are alike and how they are different.

2. $g(\theta) = -3\cos\theta$

3. $g(\theta) = \cos\frac{1}{2}\theta$

4. $g(\theta) = 5 + \cos\theta$

5. $g(\theta) = -\cos\frac{1}{2}\theta$

Practice & Apply

Write a function of the form $y = \cos(c\theta)$ for each graph.

6.

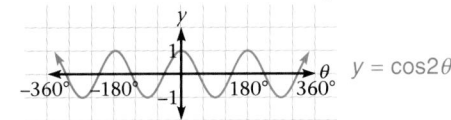

$y = \cos 2\theta$

7.

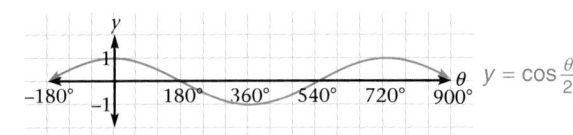

$y = \cos\frac{\theta}{2}$

Identify the vertical shift, amplitude, period, and phase shift for each function. Then graph the function.

8. $f(\theta) = 4\sin\theta$

9. $g(\theta) = -4\sin\theta$

10. $h(\theta) = 2 + \cos\theta$

11. $k(\theta) = \cos(\theta + 4)$

12. $f(\theta) = \cos(-2\theta)$

13. $g(\theta) = 3.5 + \sin(\theta + 120)$

14. $h(\theta) = 3\cos(\theta - 30)$

15. $m(\theta) = 5.1\sin\theta$

16. $n(\theta) = 4 - \cos\frac{1}{2}(\theta - 270)$

Radio Transmission The signals sent by radio stations travel in waves.

17. What changes in the waves broadcast by FM stations? What stays the same?

18. Find out what FM stands for. Explain why this name is appropriate.

FM radio wave

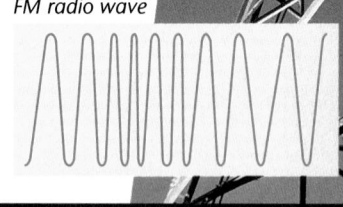

RETEACHING the lesson

Technology Illustrate the effects of the variables *a*, *b*, *c*, and *d* on the shape of the general sine and cosine graphs by graphing these functions on the graphing calculator for varying values of *a*, *b*, *c*, and *d*.

The answers to Exercises 8–16, and 19 can be found in Additional Answers beginning on page 842.

17. The period of the wave changes, but the amplitude stays the same.

18. Answers may vary but should be similar to the following. FM stands for *frequency modulation* because radio waves travel at the same speed; changing the period of the waves changes the frequency that the crests of the wave pass at a given point. Halving the period doubles the frequency, and so on.

Biology When a full-grown giraffe is standing up, its heart must pump blood upward about 10 feet to reach the brain. The graphs shown here are of the blood pressure near a giraffe's head when the giraffe is lying down and when the giraffe is standing up.

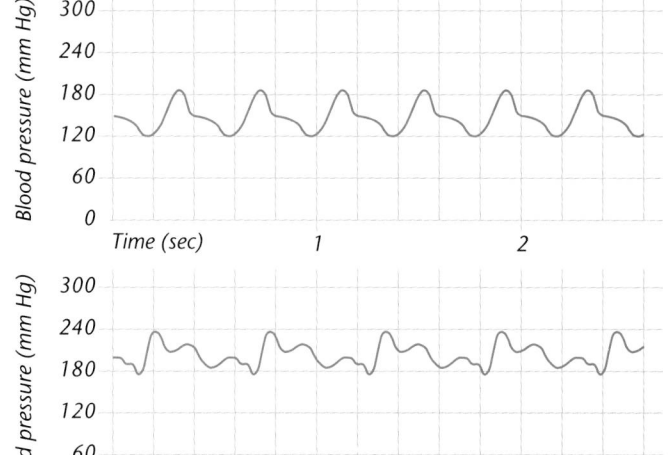

19. What is different about the two graphs? Why? Standing: period is 0.4 seconds; range, between 120 and 200 mmHg.

20. Find the period and range of each graph. Lying: period is 0.6 seconds; range, between 170 and 240 mmHg.

21. Find the number of times the giraffe's heart beats per minute when it is lying down and when it is standing up.

100 beats/minute   150 beats/minute

Write a function of the form $y = a \cos \theta$ for each graph.

22.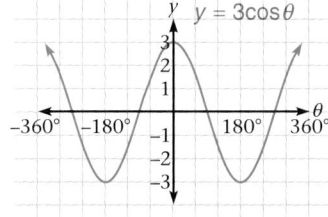

23.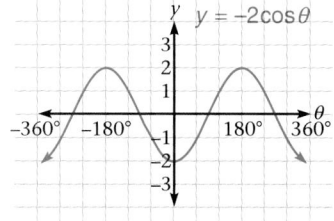

Music Since sound is generated by vibrations and travels in waves, it can be modeled with circular functions. When an object vibrates twice as fast, it produces a sound with a pitch one octave higher than before. Amplitude is associated with loudness. Suppose a sound can be modeled with the function $g(\theta) = 15 \sin(200\theta)$.

24. Write a function for a sound with the same pitch as the sound modeled by g but 3 times as loud. $g(\theta) = 45\sin(200\theta)$

25. Write a function for a sound with the same volume as the sound modeled by g but one octave higher. $g(\theta) = 15\sin(400\theta)$

26. Write a function for a sound with the same volume as the sound modeled by g but two octaves higher. $g(\theta) = 15\sin(800\theta)$

Technology Master

NAME _____ CLASS _____ DATE _____

Technology
8.4 The Graph of the Tangent Function

In Lesson 8.4, you explored the graphs of $y = \sin x$ and $y = \cos x$ and variations of those functions by using a graphics calculator. In this activity, you carry out a similar exploration for the function $y = \tan x$.

Use a graphics calculator in degree mode.

1. Graph $y = \tan x$ for $-360° \le x \le 360°$.

2. Briefly describe the calculator display in your own words.

3. Write a formula for those values of x for which $\tan x = 0$.

4. Write a formula for those values of x for which $\tan x$ is undefined.

5. Find the domain and range of $y = \tan x$.

6. Is $y = \tan x$ periodic? What is its period?

7. Does the tangent function have an amplitude? Explain.

8. Find $\tan 1170°$, if it exists.

60   Technology   HRW Advanced Algebra

28. The graphs of $f(\theta) = \cos \frac{1}{2}\theta$ and $g(\theta) = -\cos \frac{1}{2}\theta$ are reflections of one another with respect to the x-axis. The graph of $h(\theta) = -\sin \frac{1}{2}\theta$ is the reflection over the x-axis of the graph of $f(\theta) = \sin \frac{1}{2}\theta$.

29. The graphs are the same. $g(\theta) = \cos(90° - \theta) = \cos(-(\theta - 90°)) = \cos(\theta - 90°)$ since cosine is an even function. Thus, the graph of g is a shift of the graph of $\cos \theta$ by 90° to the right, which is the same as the graph of $\sin \theta$. For example, $\cos(90° - 90°) = 1 = \sin 90°$.

30. The cosine function $y = \cos \theta$ only takes on values between −1 and 1. Therefore, there can be no angle θ such that the cosine of that angle is 2.

Look Beyond

These problems refresh students' memories about the radius, circumference, and area of a circle in preparation for the concept of radian measures of angles.

27 Let $f(\theta) = \cos \theta$. Find $f(60°)$. Use the period of the cosine function to find three other values of θ such that $f(\theta) = f(60°)$. Use your graphics calculator to check your answer. $f(60°) = 0.5$ Answers will vary but be equivalent to $f(60° + 360°n) = 0.5$ for three different values of n.

28 **Transformations** Graph $f(\theta) = \cos \frac{1}{2}\theta$ and $g(\theta) = -\cos \frac{1}{2}\theta$ on the same coordinate plane. Compare the graphs. Predict what the graph of $h(\theta) = -\sin \frac{1}{2}\theta$ will look like. Check using your calculator.

29 Graph $f(\theta) = \sin \theta$ and $g(\theta) = \cos(90° - \theta)$ on the same coordinate plane. Compare the graphs. What happens? Why?

30. Explain why there is no value for θ such that $\cos \theta = 2$.

Look Back

Industrial Engineering An industrial engineer is interested in the efficiency of the workers in a factory. The engineer takes a random sample of the workers and compares how long they have worked in the factory and their output per hour. **Complete Exercises 31–35.**

Years worked	2.5	1.9	0.25	8.5	6.0	0.75	4.5	1.5
Output per hour	93	85	24	103	102	54	98	73

31 Find a logarithmic model for the data and the correlation coefficient for the model. **[Lesson 7.3]** $y \approx 62.29 + 23.34 \ln x$; $r \approx 0.97$

32 According to the model, what happens to the output per hour if the worker stays at the factory? What might cause this trend? How might management take this into account? **[Lesson 7.3]**

33 Use your model to predict the units per hour produced by a worker who has worked at the factory for 3.5 years. **[Lesson 7.3]** ≈ 91.5 units per hour

34 Use your model to predict how long a worker must work at the factory to have an output of 70 units per hour. **[Lesson 7.3]** ≈ 1.4 years

35 Find a linear model for the data. Would a linear model or a logarithmic model be more appropriate for this situation? Explain. **[Lesson 1.3]**

36. Find sin 90°, cos 90°, and tan 90° without using a calculator. Check with your calculator. **[Lesson 8.2]**
$\sin 90° = 1$; $\cos 90° = 0$; $\tan 90°$ is undefined

Look Beyond

37. **Geometry** What is the circumference of a circle of radius 1? 2π units

38. **Geometry** What is the area of a circle of radius 1? What is the area of a semicircle of radius 1? π square units; $\frac{\pi}{2}$ square units

32. The output increases rapidly at first, and then the output tends to level off. The leveling off could be caused by workers who learn new skills rapidly at first. However, once they become proficient, it becomes increasingly harder to improve their output. Management might decide to increase wages in alignment with this curve.

35. $y \approx 7.53x + 54.60$; $r \approx 0.78$. The logarithmic model is better, since it has a correlation coefficient of about 0.97, while the linear model has a correlation coefficient of about 0.78.

LESSON 8.5 Radian Measure

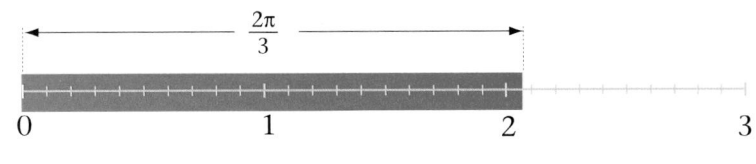

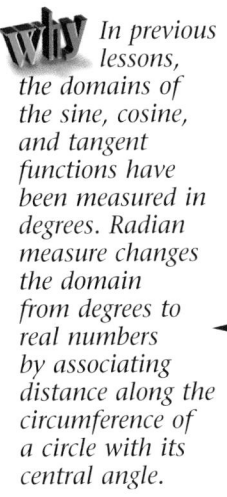 *In previous lessons, the domains of the sine, cosine, and tangent functions have been measured in degrees. Radian measure changes the domain from degrees to real numbers by associating distance along the circumference of a circle with its central angle.*

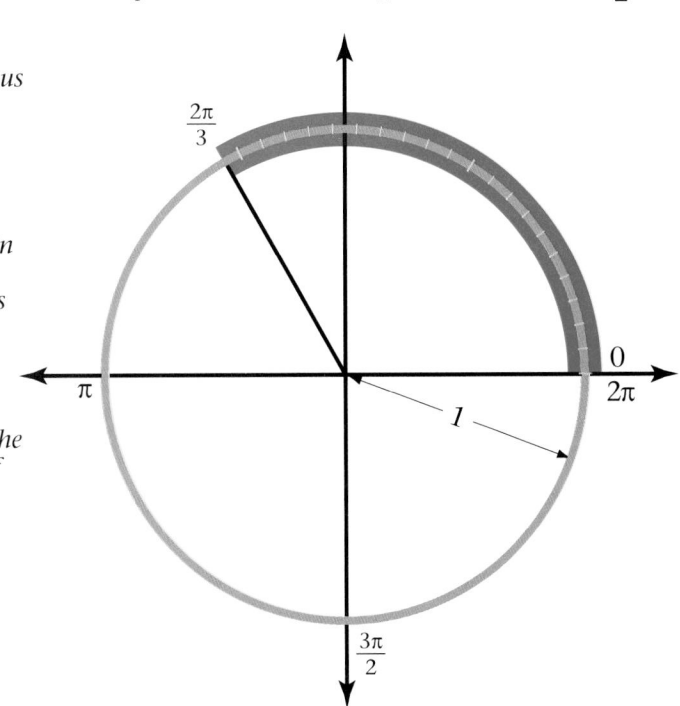

Radian Measure

GEOMETRY
Connection

The circumference of a circle with radius r is $2\pi r$. Since the radius of the unit circle is 1, the circumference of the unit circle is 2π. *Radians* are used to measure the length of an arc associated with a given rotation.

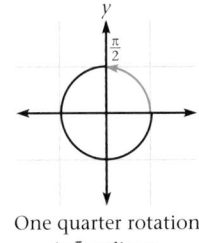

One quarter rotation is $\frac{\pi}{2}$ radians.

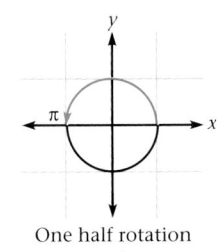

One half rotation is π radians.

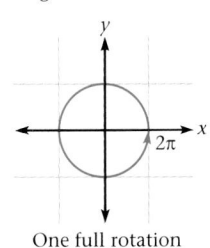

One full rotation is 2π radians.

Students should be able to move back and forth between the two systems of measurement for special angles quickly and easily.

4. $\frac{1}{12}$ of a rotation is $\frac{2\pi}{12} = \frac{\pi}{6}$ radians, $\frac{5}{12}$ of a rotation is $\frac{5 \cdot 2\pi}{12} = \frac{5\pi}{6}$ radians, and $\frac{n}{12}$ of a rotation is $\frac{n \cdot 2\pi}{12} = \frac{n\pi}{6}$ radians.

Cooperative Learning

Have students make flashcards for the special angles between 0 and 2π radians. Have them write the degree measure on one side of the card and the equivalent radian measure on the other side. Students can then quiz each other in pairs until each knows the equivalent measures well.

Math Connection

Geometry Review geometric concepts including *arc*, *sector*, *central angle*, *arc length* and *sector area* formulas for a given central angle.

CRITICAL *Thinking*

0 to π, or 0 to 3.14 radians

•Exploration• *Radian Measures of Common Rotations*

Complete the following circles by writing the radian measures for each of the indicated rotations. Reduce fractions to lowest terms.

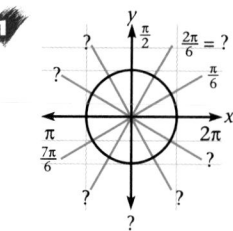 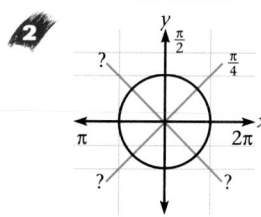

 **3** What is the degree measure for $\frac{1}{6}$ of a full rotation?
What is the radian measure for $\frac{1}{6}$ of a full rotation?

4 How can you find the radian measure for $\frac{1}{12}$ of a full rotation? for $\frac{5}{12}$ of a full rotation? for $\frac{n}{12}$ of a full rotation?

5 Use your calculator to find decimal approximations for each of the radian measures you found in Steps 1 and 2. Write the approximations to complete this circle. ❖

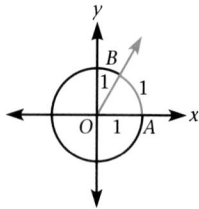

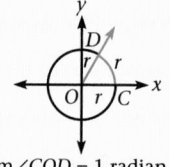 **GEOMETRY** *Connection*

A central angle is an angle with its vertex at the center of a circle. Examine the central angle *AOB* in the unit circle. Notice that the arc intercepted by the central angle *AOB* in the unit circle is equal in length to the radius, 1.

Angle *AOB* has a measure of 1 *radian*.

RADIAN MEASURE

The measure of a central angle where the radii and the intercepted arc are the same length is defined to be 1 **radian.**

m∠COD = 1 radian

 **CRITICAL** *Thinking*

The calculator gives inverse cosine values of 0° to 180° when in degree mode. What radian measures will the calculator give for inverse cosine?

ENRICHMENT Have the students use their graphics calculator on radian mode to compare the graph of $\sin x$ with the graph of $\cos(x + \pi)$ and the graph of $\sin(x + \pi)$ with the graph of $\cos x$. Encourage students to speculate about the possibility of expressing all cosine functions in terms of sine functions. Have students write a rule that converts sine functions to cosine functions. Then have students test their rule using several different functions and their graphics calculator.

Converting Measures of Rotation

EXAMPLE 1

Convert each degree measure to radian measure.

A 45°   **B** 190°

Solution▶

Degree measure can be converted to radian measure using proportional reasoning. The degree measure is part of the entire circle of 360°.

A 45° is $\frac{45}{360}$, or $\frac{1}{8}$, of the circle.

Since the entire circle is 2π radians,

45° is $\frac{1}{8}(2\pi) = \frac{\pi}{4} \approx 0.785$ radians.

B 190° is $\frac{190}{360}(2\pi) = \frac{19\pi}{18} \approx 3.316$ radians.

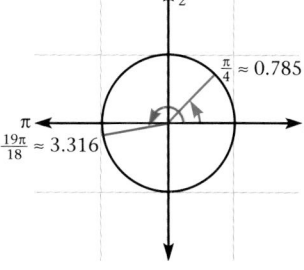

Try This Convert each degree measure to radian measure.

A 60°   **B** 217°

EXAMPLE 2

Convert each radian measure to degree measure.

A $\frac{5\pi}{6}$ radians   **B** 2.5 radians   **C** 4.2 radians

Solution▶

To convert radians to degrees, recall that π radians is 180°, so 1 radian is $\frac{180°}{\pi}$.

A Since π radians is 180°, $\frac{5}{6}\pi = \frac{5}{6} \cdot 180 = 150$.

So, $\frac{5\pi}{6}$ radians is 150°.

B Since 1 radian is $\frac{180°}{\pi}$, 2.5 radians is $2.5 \cdot \frac{180}{\pi} = \frac{450}{\pi} \approx 143°$.

C 4.2 radians = $4.2 \cdot \frac{180}{\pi} = \frac{756}{\pi} \approx 240°$. ❖

Try This Convert each radian measure to degree measure.

A $\frac{2\pi}{3}$ radians   **B** 1.14 radians   **C** 5.26 radians

---

Alternate Example 1

Convert each degree measure to radian measure.

A. 30°
   $\left[\frac{\pi}{6} \approx 0.524 \text{ radians}\right]$

B. 280°
   $\left[\frac{14\pi}{9} \approx 4.887 \text{ radians}\right]$

A ongoing ASSESSMENT

Try This

A. $\frac{\pi}{3} \approx 1.047$ radians

B. $\frac{217\pi}{180} \approx 3.787$ radians

Alternate Example 2

Convert each radian measure to degree measure.

A. 1.4 radians [$\approx$ **80°**]

B. $\frac{7\pi}{8}$ radians [$\approx$ **158°**]

C. 5.0 radians [$\approx$ **286°**]

A ongoing ASSESSMENT

Try This

A. 120°

B. $\approx 65°$

C. $\approx 301°$

Graphing in Radian Mode

The following graphs of $y = \cos x$ and $y = \sin x$ with domain in radian measure are shown.

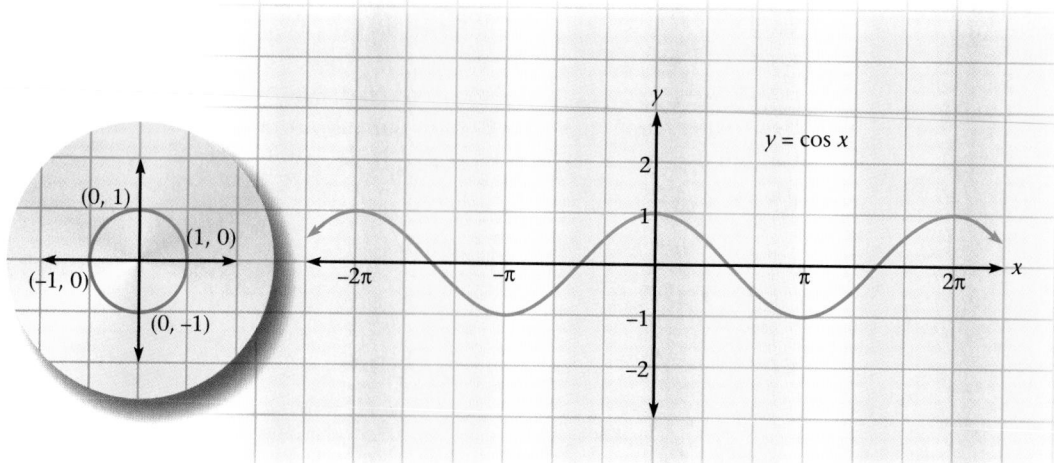

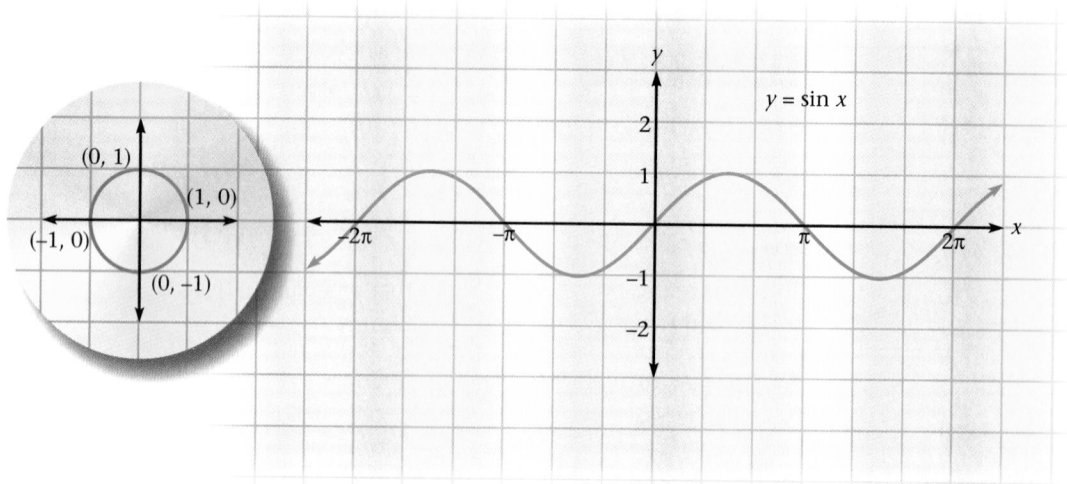

What is the period in radians of the functions $y = \cos x$ and $y = \sin x$?

The general formula for the sine function that combines the transformations you explored in Lesson 8.4 is given by

$$y = a + b \sin c(x - d).$$

In Example 3, you will determine the effects that a, b, c, and d have on the graph in radian mode.

Hands-On Strategies

Draw several angles in standard position inscribed within a unit circle. For each angle, use a protractor to find its degree measure. To find the radian measure, measure the part of the circumference intersected by the central angle.

EXAMPLE 3

TRANSFORMATIONS
Connection

Let $f(x) = 7 + 5 \sin \frac{1}{2}\left(x - \frac{\pi}{3}\right)$. Explain the effect of each of the numbers 7, 5, $\frac{1}{2}$, and $\frac{\pi}{3}$. Then graph the function.

Solution

Graph

$$y = 7 + 5 \sin \frac{1}{2}\left(x - \frac{\pi}{3}\right)$$

The equation $f(x) = 7 + 5 \sin \frac{1}{2}\left(x - \frac{\pi}{3}\right)$ is in the form $y = a + b \sin c(x - d)$. Thus a is **7** b is **5**, c is $\frac{1}{2}$, and d is $\frac{\pi}{3}$.

Since a is **7**, the graph of $y = \sin x$ is shifted up by **7** units. In radians, $y = \sin x$ has a period of 2π. Since c is $\frac{1}{2}$, f has a period of $\frac{2\pi}{c} = \frac{2\pi}{\frac{1}{2}}$, or 4π. This represents a horizontal stretch by a factor of **2**.

Since d is positive, f is shifted right $\frac{\pi}{3}$ units. Since the graph of f is shifted to the right $\frac{\pi}{3}$ units and stretched horizontally by a factor of 2, the maximum of f occurs at $\frac{\pi}{3}$ units to the right of $2\left(\frac{\pi}{2}\right)$, or at the x-value $\frac{4\pi}{3}$. The amplitude, b, is **5**, so the maximum y-value is **7 + 5**, or 12. The minimum of f occurs at the x-value $\frac{-2\pi}{3}$. The minimum y-value is **7 – 5**, or 2.

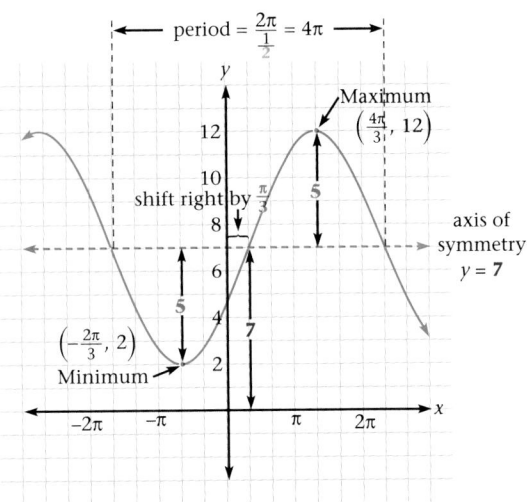

Let $y = 2 + 4 \sin\left(x - \frac{\pi}{3}\right)$. Explain the effect of each of the numbers 2, 4, and $\frac{\pi}{3}$. Then graph the function.

[The graph is shifted up by 2 units; the amplitude is 4 so the maximum y-value is 6 and the minimum y-value is –2; the period is 2π; and the graph is shifted to the right by $\frac{\pi}{3}$ radians.

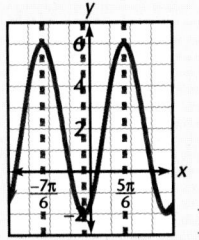

]

A. Write the function, in the form $f(x) = a + b \cos c(x - d)$ for the following graph.

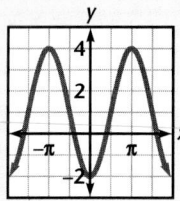

$[f(x) = 1 + 3\cos(x + \pi)]$

B. Identify the domain and range for the function. [**domain is all real numbers, the range is** $-2 \le y \le 4$]

Graphics Calculator

EXAMPLE 4

Ⓐ Write a function, in the form $f(x) = a + b \cos c(x - d)$, for the following graph. Check using your graphics calculator.

Ⓑ Identify the domain and range for the function.

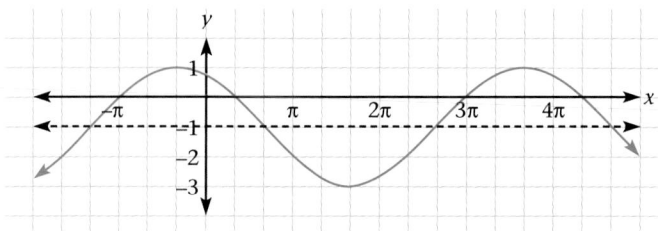

Solution

Ⓐ The graph is shifted down 1 unit, so the vertical shift, a, is -1. Also, the graph is symmetrical with respect to the line $y = -1$.

The vertical stretch or amplitude, b, is half of the difference between the maximum y-value, 1, and the minimum y-value, -3. Therefore, $b = \frac{1 - (-3)}{2}$, or 2.

A complete period occurs between $-\frac{\pi}{3}$ and $\frac{11\pi}{3}$, so the period is $\frac{11\pi}{3} - (-\frac{\pi}{3})$, or 4π. Solve $4\pi = \frac{2\pi}{c}$ for c.

$$4\pi = \frac{2\pi}{c}$$
$$4\pi \cdot c = 2\pi$$
$$c = \frac{2\pi}{4\pi}$$
$$c = \frac{1}{2}$$

The graph of $f(x) = \cos x$ has a maximum y-value when x is 0. The graph in this example has a maximum y-value when x is $-\frac{\pi}{3}$, a shift of $\frac{\pi}{3}$ units to the left. So, the horizontal phase shift, d, is $-\frac{\pi}{3}$.

Therefore, the function for the graph, in the form $f(x) = a + b \cos c(x - d)$, is $f(x) = -1 + 2 \cos \frac{1}{2}\left(x + \frac{\pi}{3}\right)$.

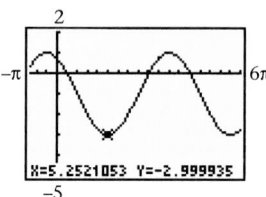

Ⓑ The domain is all real numbers. The range is $-3 \le y \le 1$. ❖

CRITICAL Thinking

How can you write a function in the form $f(x) = a + b \sin c(x - d)$ for the graph in Example 4?

CRITICAL *Thinking*

The vertical shift, amplitude, and period remain the same as the cosine function, so a is -1, b is 2, and c is $\frac{1}{2}$. The sine function is the cosine function shifted horizontally by π radians, so d becomes $-\frac{4\pi}{3}$. This gives the equation $f(x) = -1 + 2 \sin \frac{1}{2}\left(x + \frac{4\pi}{3}\right)$.

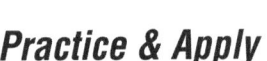

EXERCISES & PROBLEMS

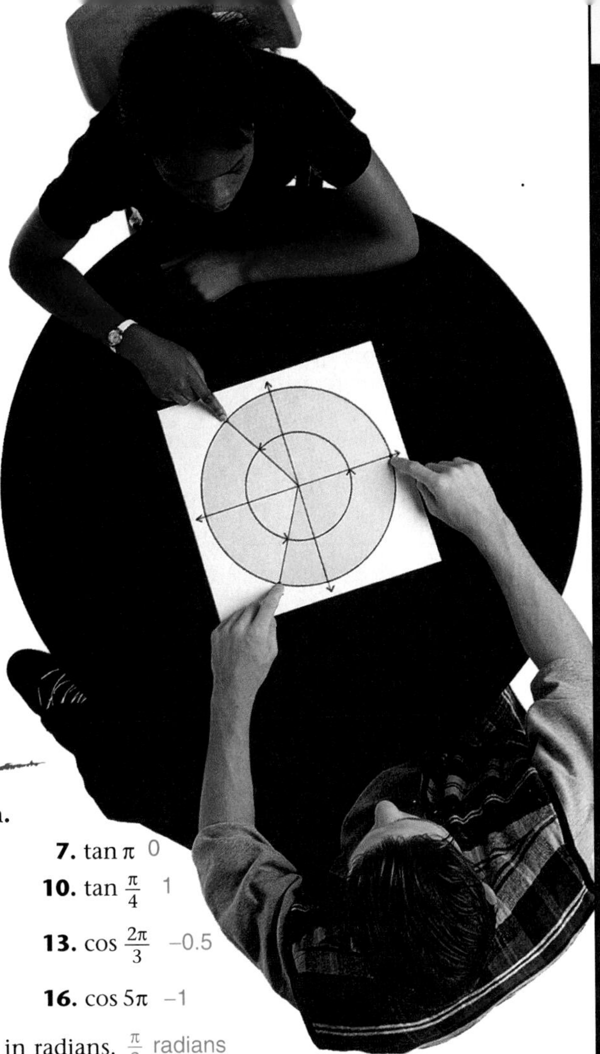

Communicate

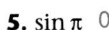

1. Explain the radian measure of an angle. Compare radian and degree measures. How are they alike? How are they different?

2. How can you find the measure of one-third of a rotation in radians?

3. How can you convert degrees to radians? How can you convert radians to degrees?

4. What is the phase shift, in radians, from $f(x) = \cos x$ to $g(x) = \sin x$?

Practice & Apply

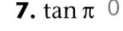

Give the exact value for each expression.

5. $\sin \pi$ 0 **6.** $\cos \pi$ -1 **7.** $\tan \pi$ 0

8. $\cos \frac{\pi}{3}$ 0.5 **9.** $\sin \frac{\pi}{6}$ 0.5 **10.** $\tan \frac{\pi}{4}$ 1

11. $\tan \frac{7\pi}{6}$ $\frac{1}{\sqrt{3}}$ **12.** $\cos \frac{5\pi}{3}$ 0.5 **13.** $\cos \frac{2\pi}{3}$ -0.5

14. $\cos \frac{7\pi}{4}$ $\frac{1}{\sqrt{2}}$ **15.** $\tan \frac{9\pi}{4}$ 1 **16.** $\cos 5\pi$ -1

17. Express one-fourth of a complete rotation in radians. $\frac{\pi}{2}$ radians

18. Express three-fourths of a complete rotation in radians. $\frac{3\pi}{2}$ radians

19. Express one-eighth of a complete rotation in radians. $\frac{\pi}{4}$ radians

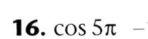

Draw each angle and convert radian to degree measure and degree to radian measure.

20. $\frac{\pi}{5}$ radians **21.** $-90°$ **22.** $\frac{2\pi}{3}$ radians **23.** $270°$

24. $-\frac{3\pi}{4}$ radians **25.** $450°$ **26.** 6.28 radians **27.** $158°$

28. 1 radian **29.** $-85°$ **30.** -10 radians **31.** $-1440°$

Find each value to the nearest hundredth using your calculator in radian mode.

32 $\cos\left(\frac{\pi}{9}\right) \approx 0.94$ **33** $\tan 1.3 \approx 3.60$ **34** $\sin 0.7 \approx 0.64$ **35** $\cos 2 \approx -0.42$

36 $\tan \pi$ 0 **37** $\tan\left(\frac{3\pi}{20}\right) \approx 0.51$ **38** $\cos 0.75 \approx 0.73$ **39** $\sin\left(\frac{5\pi}{7}\right) \approx 0.78$

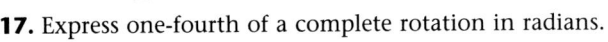

20. $\frac{\pi}{5}$ radians $= 36°$

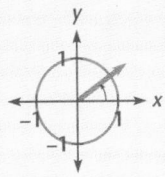

21. $-90° = -\frac{\pi}{2}$ radians

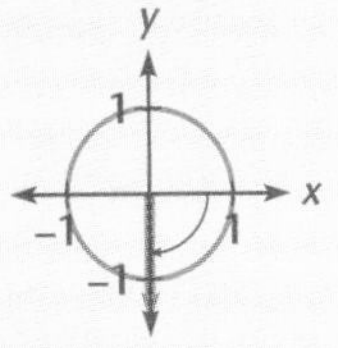

22. $\frac{2\pi}{3}$ radians $= 120°$

23. $270° = \frac{3\pi}{2}$ radians

ASSESS

Selected Answers

Odd-numbered Exercises 5–67

Assignment Guide

Core 1–13, 17–27, 32–37, 40–46, 49–69

Core Plus 1–4, 10–19, 26–31, 35–39, 43–69

Technology

Graphics calculators are needed for Exercises 32–53. Graphics calculators would be helpful for Exercises 59 and 61.

TEACHING tip

Make certain students have their calculators in radian mode before performing the exercises in this lesson. Also, explain that some scientific calculators have an additional angle measure called *gradians.* Tell students that 100 gradians = 90° = $\frac{\pi}{2}$ radians. Gradians are generally referred to as *grads.*

Error Analysis

Degree mode and radian mode will give very different results. Students must be careful to have the calculator in the correct mode.

The answers to Exercises 24–31 can be found in Additional Answers beginning on page 842.

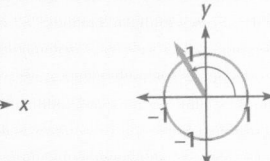

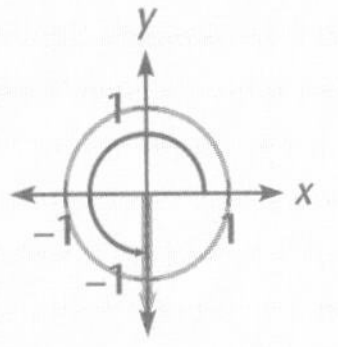

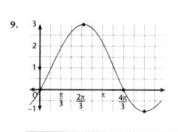

alternative ASSESSMENT

Portfolio Assessment

Have students write a short essay explaining why two different units of measure exist and when it is best to use each. Students should then choose a trigonometric function, and graph it using their graphics calculator. Ask students to graph the function using both the degree and the radian mode of the calculator. Have students point out the differences between tracing and calculating on the two graphs.

40. vertical shift: none; amplitude: 1; period: 6π; phase shift: none

41. vertical shift: none; amplitude: 1; period: 2; phase shift: none

---

Identify the vertical shift, amplitude, period (in radians), and phase shift for each function. Then sketch the graph. Use your graphics calculator to check each graph.

40 $f(x) = \cos \frac{1}{3}x$

41 $f(x) = \sin \pi x$

42 $f(x) = 1.6 \cos 3x$

43 $f(x) = 3 + 4 \sin x$

44 $f(x) = -3 + \sin(x - \pi)$

45 $f(x) = 5 \sin \frac{1}{5}x$

46 $f(x) = 4.5 + \cos \left(x - \frac{\pi}{2}\right)$

47 $f(x) = 3 \cos \frac{\pi}{2}x$

48 $f(x) = 2.5 + 7.5 \sin \frac{3}{2}x$

Write a function in the form $f(x) = a + b \cos c(x - d)$ for each graph. Use your graphics calculator to check each function.

49
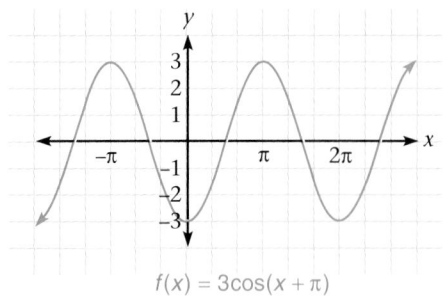
$f(x) = 3\cos(x + \pi)$

50
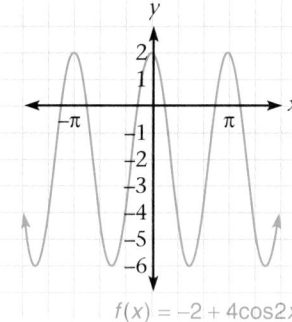
$f(x) = -2 + 4\cos 2x$

51
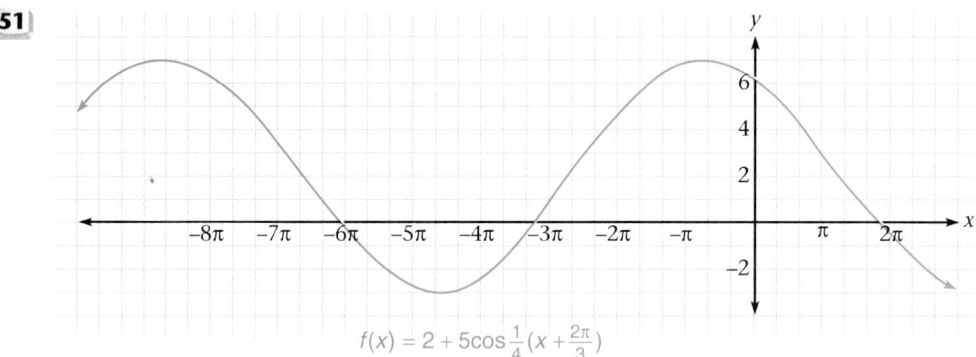
$f(x) = 2 + 5\cos \frac{1}{4}(x + \frac{2\pi}{3})$

52 Find three values for x other than $\frac{\pi}{2}$ that are not in the domain of the function $f(x) = \tan x$. Write your answers in radians. Answers will vary but must be similar to $\frac{\pi}{2} + n\pi$ for 3 different values of n.

53 Let $g(x) = \tan x$. Find $g\left(\frac{\pi}{4}\right)$. Use the period of the tangent function to find three other values of x such that $g(x) = g\left(\frac{\pi}{4}\right)$. Check using your calculator. $g(\frac{\pi}{4}) = 1$; Answers will vary but must be similar to $\frac{\pi}{4} + n\pi$ for 3 different values of n.

54. Find x in $\triangle ABC$.

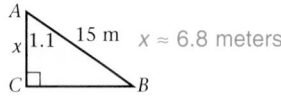
$x \approx 6.8$ meters

55. Find y in radians in $\triangle XYZ$.

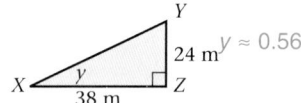
$y \approx 0.56$

 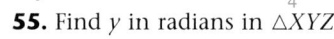
---

42. vertical shift: none; amplitude: 1.6; period: $\frac{2\pi}{3}$; phase shift: none

43. vertical shift: upward 3 units; amplitude: 4; period: 2π; phase shift: none

44. vertical shift: 3 units downward; amplitude: 1; period: 2π; phase shift: π units to the right

45. vertical shift: none; amplitude: 5; period: 10π; phase shift: none

46. vertical shift: 4.5 units upward; amplitude: 1; period: 2π; phase shift: $\frac{\pi}{2}$ units to the right

47. vertical shift: none; amplitude: 3; period: 4; phase shift: none

48. vertical shift: 2.5 units upward; amplitude: 7.5; period: $\frac{4\pi}{3}$; phase shift: none

56. Analog Systems Through how many radians does the minute hand of an analog clock turn in 45 minutes? in 1 hour and 15 minutes? $\frac{-3\pi}{2}$; $\frac{-5\pi}{2}$

57. Analog Systems Through how many radians does the minute hand of an analog clock turn in 50 minutes? in 1 hour and 20 minutes? $\frac{-5\pi}{3}$; $\frac{-8\pi}{3}$

58. Astronomy Through what angle does a person on the equator rotate in 1 hour? Write your answer in degrees and in radians. $15°$, $\frac{\pi}{12}$ radians

~~~ Look Back

Medical Research A doctor is studying how fast a certain disease spreads. The number of new cases diagnosed in a given week in the weeks after the initial cases are reported is taken from several different highly populated areas.

Weeks after first case	3	7	2	9	3	8	1	7	10
New cases diagnosed	42	67	37	84	40	75	34	70	94

59. Statistics Find a linear model and the correlation coefficient for the data. **[Lesson 1.3]** $y \approx 6.71x + 23.03$; $r \approx 0.993$

60. Use the linear model to predict the number of new cases diagnosed in week 12. **[Lesson 1.3]** ≈ 104

61. Find an exponential model and the correlation coefficient for the data. **[Lesson 7.2]** $y \approx 29.47\,(1.12)^x$; $r \approx 0.998$

62. Use the exponential model to predict the number of new cases diagnosed in week 12. **[Lesson 7.2]** ≈ 115

63. Which model do you think is more appropriate? Explain why. exponential, r is closer to 1

Let $f(x) = \frac{x^2 - 5}{10}$. **[Lesson 3.2]**

64. Find the inverse of f. $f^{-1}(x) = \pm\sqrt{10x + 5}$

65. Is the inverse of f a function? Explain. No, for most values of x, there are 2 values of f^{-1}.

66. Find three angles that are coterminal with $125°$. **[Lesson 8.2]** Answers may vary but must be equivalent to $125° + 360°n$ for 3 different values of n

67. Given any angle θ, find three coterminal angles. **[Lesson 8.2]** Answers may vary but must be equivalent to $\theta + 360°n$ for 3 different values of n

Look Beyond ~~~

68. Geometry Find the circumference of Circle M. 12π cm

69. Geometry Find the area of Circle M. 36π cm^2

$r = 6$ cm

Look Beyond

These problems review the relationship between circumference and area of a circle to prepare students for the concepts of arc length and sector area.

Objectives

- Find the arc length and sector area determined by the central angle of a circle.

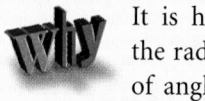

RESOURCES

- Practice Master   **8.6**
- Enrichment Master   **8.6**
- Technology Master   **8.6**
- Lesson Activity Master **8.6**
- Quiz   **8.6**
- Spanish Resources   **8.6**

This lesson may be considered optional in the core course. See page 428A.

Assessing Prior Knowledge

Find the circumference of a circle with the following radii.

1. 1   [2π]

2. 1.5   [3π]

3. 2.3   [4.6π]

TEACH

It is helpful to use the radian measures of angles in solving many real-life problems. This system of angle measurement offers a direct connection between distance and angle of rotation.

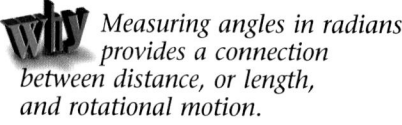

LESSON 8.6 Arc Length and Sector Area

why *Measuring angles in radians provides a connection between distance, or length, and rotational motion.*

A CD rotates at different speeds depending upon where the laser is reading on the disc.
In CD players, the laser must read the digital information at a constant rate regardless of where the information is located in the disc.
How does this compare with how information is "read" on a standard turntable?

Arc Length

Recall the definition of a radian. When the length, d, of an arc determined by a central angle, θ, is equal to the radius of the circle, r, the measure of θ is 1 radian.

For a complete circle, the central angle is 2π and the circumference, or arc length, is $2\pi r$. Use proportional reasoning to find the arc length, d, when the central angle, θ, is less than 2π.

$$\frac{\theta}{2\pi} = \frac{d}{2\pi r}, \text{ or } d = r\theta$$

ARC LENGTH

The length, d, of $\overset{\frown}{AB}$ intercepted by central angle θ, measured in radians, in a circle with radius, r, is

$$d = r\theta.$$

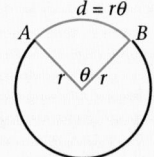

In the unit circle, the arc length, d, is equal to θ since the radius, r, is 1.

Cooperative Learning Before you present any information from this lesson, have students work in groups to try to solve the following problem. It may help students to draw a representation of the situation. Have each group keep a record of the different methods they used in approaching this problem. Then present the material in the lesson, and have students work in

groups to solve the problem. Have each group write a report describing what they learned in trying to solve this problem.

The president of a local bakery wants to build his swimming pool in the shape of a slice of pie. It should represent one-sixth of a pie. The arc length is 30 feet. If the pool will be 10 feet deep, how much water will be required to fill it? [≈ **4297 cubic feet**]

EXAMPLE 1

Find the length of $\overwidehat{AB}$ when θ is in radians.

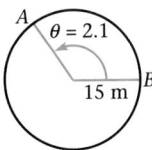

Solution▸

The central angle is given in radians. Substitute 2.1 for θ and 15 for r in the formula for the arc length.

$$d = r\theta$$
$$d = (15)(2.1) = 31.5$$

The length of the radius is given in meters. Thus, the length of the arc is 31.5 meters. ❖

Try This Find the radius of the circle if the length of the arc intercepted by $\frac{1}{4}$ a rotation is 3π.

Once you know how to find the length of an arc, you can estimate the velocity of an object moving in a circular path. Velocity is usually measured in units of distance per time, such as miles per hour or meters per second. By converting an angle measure to a distance or length measure, you can find the velocity of an object traveling in circular motion.

EXAMPLE 2

Sports Chris is located at the center of a circular raceway. The track has a radius of 500 meters. There is a 69° angle between two billboards on the outer edge of the track. Using a stopwatch, Chris notices that it takes 4.6 seconds for a car to travel from one billboard to the next. Find the velocity of the car moving around the track.

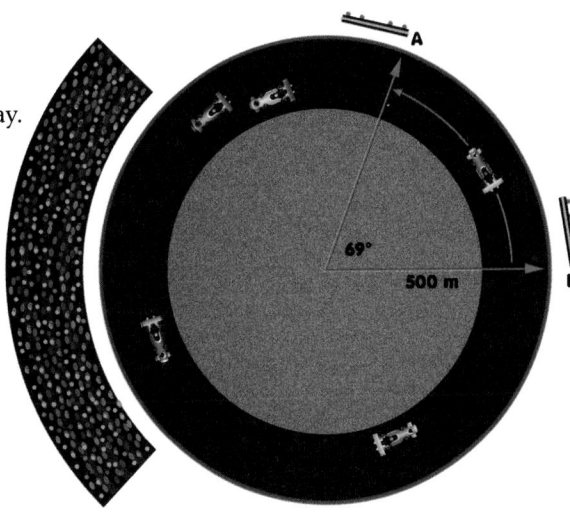

Solution▸

The car traveled 69° in 4.6 seconds. The distance traveled is the length of $\overwidehat{AB}$. To find the length of $\overwidehat{AB}$, convert 69° to radians.

$$\frac{69}{360} \cdot 2\pi \approx 1.2 \text{ radians}$$

Substitute 1.2 for θ and 500 for r.

$$d = r\theta$$
$$d \approx (500)(1.2) \approx 600$$

Thus, the velocity of the car is approximately $\frac{600 \text{ meters}}{4.6 \text{ seconds}}$, or 130.4 meters per second. ❖

Alternate Example 1

Find the length of an arc $\overwidehat{AB}$ where the angle between A and B is 0.8 radians and the radius is 21 meters. [**16.8 meters**]

Aongoing ASSESSMENT

Try This

$r = 6$

TEACHING tip

Explain to students that rounding is performed in Example 2 when converting from degrees to radians and when finding the arc length, d. This is done only to demonstrate the method of solution clearly. In general, rounding is performed only *after* finding the desired value. When rounding in this way, the speed of the car in Example 2 is approximately 130.9 meters per second.

Alternate Example 2

Find the velocity of a car moving around a circular auto raceway. The radius of the circular track is 500 meters. The angle between two billboards on the outer edge of the track is 62°. It takes a car 3.2 seconds to travel from one billboard to the next. [**169.1 meters per second**]

interdisciplinary **Astronomy** Planets travel around the sun in elliptical orbits. As they travel closer to the sun, their velocity is greater and they cover a longer distance than when they are far away from the sun. However, regardless of where the planet is, the sector area it sweeps out in one unit of time remains constant. This is Kepler's third law of planetary motion. Have students find information about the velocity and distance covered of one of the planets. Students need at least 10 pairs of data at points spread out at different positions along the planet's orbit. Have students prove Kepler's third law using this data for one planet. Then have students do research to determine *why* the orbits of the planets have this characteristic.

Alternate Example 3

Find the velocity, in meters per second, of a point 3 centimeters from the center of a record that is turning at 45 revolutions per minute. [≈ **0.14 meters per second**]

CRITICAL Thinking

The definition of radian measure is based directly on distance around a circle, while degree measure is not. So, radian measure is more convenient for finding distance and velocity in rotational motion.

Math Connection

Geometry When the central angle is proportionately less than 2π, the sector area is proportionately less than πr^2.

Cooperative Learning

In groups, have students study Examples 2 and 3. Have them generate and answer at least three additional questions based on the original information. Students should then ask three "What if…" questions that alter the given information one piece at a time. They should then explore what occurs when these pieces of information are changed. These 2 techniques teach students how to extend problems by exploring, in greater depth, both the mathematics and the topic of the application.

EXAMPLE 3

Physics Find the velocity, in meters per second, of a point 11 centimeters from the center of a record that is turning at $33\frac{1}{3}$ revolutions per minute.

Solution▸

Each revolution is 2π radians, so $33\frac{1}{3}$ revolutions is

$\left(33\frac{1}{3}\right)(2\pi)$ radians, or approximately 209.44 radians.

Substitute 209.44 for θ and 11 for r.

$d = r\theta$
$d \approx (11)(209.44) \approx 2303.8.$

$$\frac{2303.8 \text{ cm}}{\text{min}} \times \frac{1 \text{ m}}{100 \text{ cm}} \times \frac{1 \text{ min}}{60 \text{ sec}} \approx 0.38 \text{ m/sec}$$

Thus, the point is moving approximately 2303.8 centimeters per minute, or approximately 0.38 meters per second. ❖

CRITICAL Thinking

Why is radian measure more convenient than degree measure to find the distance traveled in rotational motion?

Sector Area

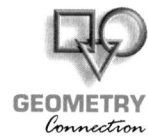

GEOMETRY Connection

For a complete circle, the area is πr^2 and the central angle is 2π. What happens when the central angle is less than 2π?

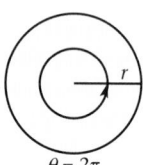

$\theta = 2\pi$

When the measure of a central angle is not a complete rotation, the area formed is a *sector*.

A **sector of a circle** is an area of a circle enclosed by a central angle, θ, and the arc intercepted by the central angle.

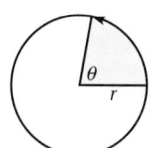

ENRICHMENT Have students explain the formula $A = \frac{1}{2}\theta r^2$, based on the formula for the area of a circle, $A = \pi r^2$, and the definition of arc length. Encourage students to describe the steps they followed in this process.

INCLUSION strategies **Independent Learning** Ask students to develop a solution to the following problem: For two cities that are in different latitudes and in equal longitudes, find the surface distance between them.

Since the area of a complete circle is πr^2, the area of the sector is

$$\frac{\theta}{2\pi} = \frac{A}{\pi r^2}, \text{ or } A = \frac{1}{2}\theta r^2.$$

The area of any sector can be found when the central angle and radius are given.

AREA OF A SECTOR

The area of sector *AOB* with central angle θ, measured in radians, in a circle with radius *r* is

$$A = \frac{1}{2}\theta r^2.$$

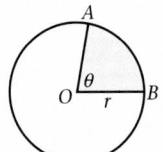

EXAMPLE 4

Find the area of the unshaded sector *AOB*.

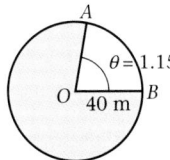

Solution

The central angle is 1.15 radians.

$$A = \frac{1}{2}\theta r^2$$

Substitute 1.15 for θ and 40 for *r*. $= \frac{1}{2}(1.15)(40)^2 = 920$

Thus, the area of the sector is 920 square meters. ❖

Try This   Find the area of the shaded sector in Example 4.

EXERCISES & PROBLEMS

Communicate

1. How is the area of a sector related to the area of a circle?
2. Describe how to find the length of an arc on a circle with a radius of 13.
3. Explain how to find the velocity of an object rotating in a circular path.
4. Explain how to find the area of a sector of a circle.

Using Patterns A carnival ride company wants to build a Ferris-wheel ride. With a radius of 40 feet and 12 seats evenly spaced, have students find the distance, along the curve of the Ferris-wheel, between the seats. Have students repeat the problem with 13 seats, with 14 seats, and so on. Have students determine the maximum number of seats the ferris-wheel can have

if no one's feet should be touching any one's head. Students should justify their responses. [12: 20.9 feet; 13: 19.3 feet; 14: 18.0 feet; 15: 16.8 feet, 16: 15.7 feet; 17: 14.8 feet; 18: 14.0 feet; 19: 13.2 feet; 20: 12.6 feet; etc.; answers may vary but should be greater than 30 seats and fewer than 45. With 30 seats, there are 8.4 feet between seats. With 45 seats, there are 5.6 feet between seats.]

Alternate Example 4

Consider a circle of radius 3 miles. For a central angle of 2.1 radians, find the area of the sector bounded by this angle and the circle. [**9.45 square miles**]

ongoing
ASSESSMENT

Try This

$1600\pi - 920$ square meters $\approx$ 4106.55 square meters

ASSESS

Selected Answers
Odd-numbered Exercises 5–43

Assignment Guide
Core 1–15, 18–26, 28–46

Core Plus 1–12, 15–46

Technology
A graphics calculator is needed for Exercise 35.

Error Analysis
Students often confuse the formula for circumference with the formula for area of a circle. Remind students that in one formula the variable is squared. This should be associated with area. Have students estimate their answer before solving the problem. This will give students an immediate check of their answer.

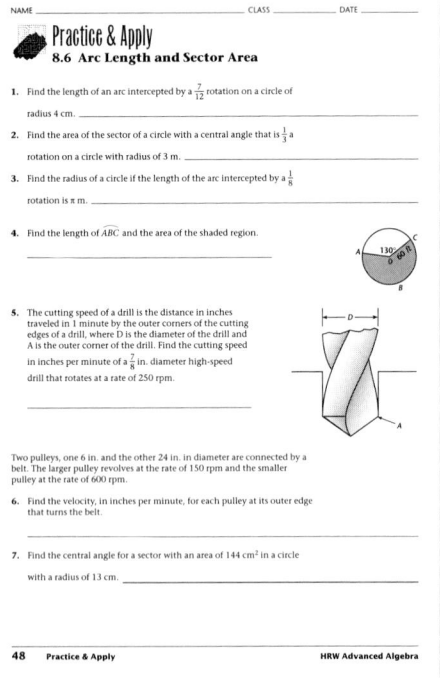

Alternative ASSESSMENT

Performance Assessment

Have students write one or two paragraphs describing what a measure of 1 radian is, and how they would determine the area of a sector of a circle given the radius and the central angle.

Practice & Apply

5. Find the length of $\overarc{AB}$ and the area of sector *AOB*.
length: 56 m
area: 1960 m²
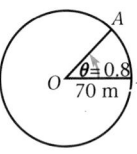

6. Find the length of $\overarc{CD}$ and the area of sector *COD*.
length: ≈ 63 in.
area: ≈ 754 in.²
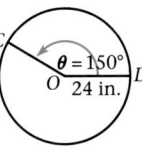

7. Find the length of $\overarc{ABC}$ and the area of the shaded sector.
length: ≈ 1833 m
area: ≈ 366,655 m²
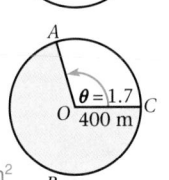

8. Find the length of $\overarc{AB}$ and the area of sector *AOB*.
length: 175.5 m
area: 5704 m²
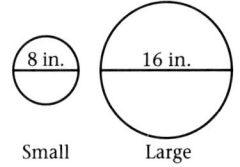

Business All pizzas at Gino's Pizzeria are cut into 8 equal sector-shaped pieces.
Complete Exercises 9–11.

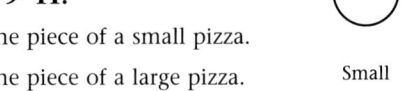
Small   Large

9. Find the area of one piece of a small pizza.
$2\pi \approx 6.28$ in.²
10. Find the area of one piece of a large pizza.
$8\pi \approx 25.13$ in.²
11. If $1.00 is charged for one piece of a small pizza, how much should be charged for one piece of a large pizza? Explain.

12. Find the length of $\overarc{ABC}$ and the area of the shaded sector.
length: ≈ 69 meters;
area: 415 meters²
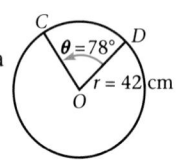

13. Find the length of $\overarc{CD}$ and the area of sector *COD*.
length: ≈ 57 cm;
area: 1201 cm²

14. Find the central angle that intercepts an arc length of 24 centimeters on a circle with a radius of 15 centimeters. 1.6 radians

15. Find the central angle for a sector with an area of 18 centimeters squared in a circle with a radius of 10 centimeters. 0.36 radians

16. Find the central angle that intercepts an arc length of 70 meters on a circle with a radius of 32 meters. 2.1875 radians

17. Find the central angle for a sector with an area of 125 centimeters squared in a circle with a radius of 30 centimeters. ≈ 0.28 radians

Laser Technology A CD player rotates the CD at different speeds depending upon where the laser is reading on the disc.

18. When the laser is at the outer edge of the CD, the disc is rotating at 200 revolutions per minute, its slowest speed. Find the velocity, in centimeters per second, at which a point on the outer edge of a CD is moving when the CD player is at its slowest speed. ≈ 63 cm/sec

19. Find the velocity, in centimeters per second, at which a point on the outer edge of a CD is moving when the CD is rotating at 240 revolutions per minute. ≈ 75 cm/sec

11. $4.00; one piece of a large pizza has 4 times the area, and thus 4 times the ingredients, of one piece of a small pizza.

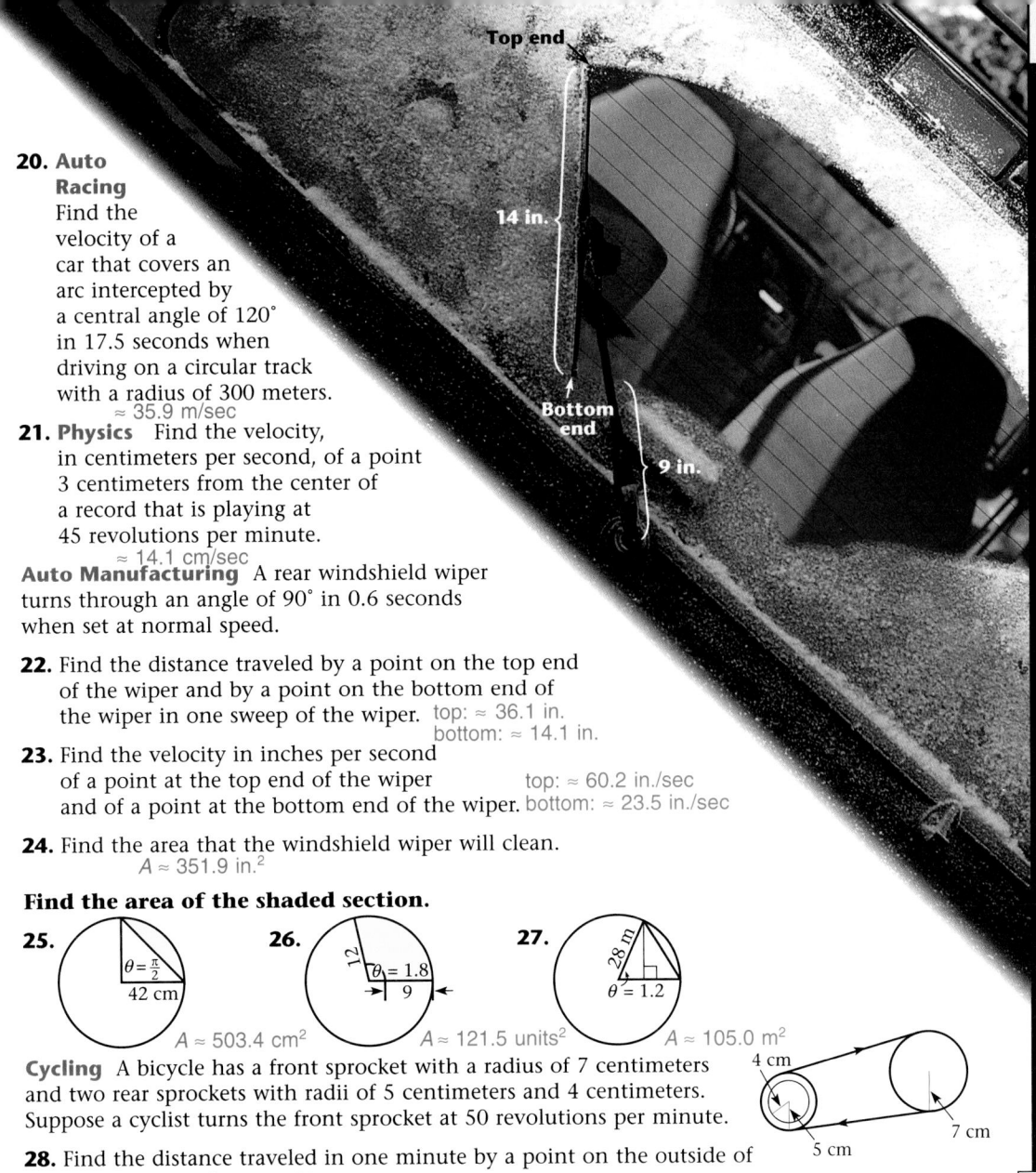

20. Auto Racing Find the velocity of a car that covers an arc intercepted by a central angle of 120° in 17.5 seconds when driving on a circular track with a radius of 300 meters.
≈ 35.9 m/sec

21. Physics Find the velocity, in centimeters per second, of a point 3 centimeters from the center of a record that is playing at 45 revolutions per minute.
≈ 14.1 cm/sec

Auto Manufacturing A rear windshield wiper turns through an angle of 90° in 0.6 seconds when set at normal speed.

22. Find the distance traveled by a point on the top end of the wiper and by a point on the bottom end of the wiper in one sweep of the wiper. top: ≈ 36.1 in.
bottom: ≈ 14.1 in.

23. Find the velocity in inches per second of a point at the top end of the wiper top: ≈ 60.2 in./sec
and of a point at the bottom end of the wiper. bottom: ≈ 23.5 in./sec

24. Find the area that the windshield wiper will clean.
A ≈ 351.9 in.²

Find the area of the shaded section.

25. $\theta = \frac{\pi}{2}$ 42 cm
A ≈ 503.4 cm²

26. 12 $\theta = 1.8$ 9
A ≈ 121.5 units²

27. 28 m $\theta = 1.2$
A ≈ 105.0 m²

Cycling A bicycle has a front sprocket with a radius of 7 centimeters and two rear sprockets with radii of 5 centimeters and 4 centimeters. Suppose a cyclist turns the front sprocket at 50 revolutions per minute.

4 cm   5 cm   7 cm

28. Find the distance traveled in one minute by a point on the outside of the front sprocket. ≈ 2199 cm

29. Find the velocity of a point on the outside of the front sprocket in centimeters per second. ≈ 36.7 cm/sec

30. Find the velocity, in centimeters per second, of a point on the outside of the rear sprocket, which has a radius of 5 centimeters. How many revolutions per minute will the sprocket turn? velocity≈ 36.7 cm/sec
rpm=70

31. Find the velocity, in centimeters per second, of a point on the outside of the rear sprocket, which has a radius of 4 centimeters. How many revolutions per minute will the sprocket turn? velocity ≈ 36.7 cm/sec, 87.5 rpm

39. 337°

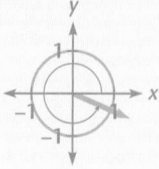

Answers may vary but must be similar to $337 + 360n$.

40. −118°

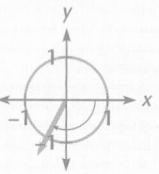

Answers may vary but must be similar to $-118 + 360n$.

Look Beyond

These exercises introduce students to the idea of rotating a geometric figure without changing its size or shape.

32. Astronomy The Earth completes one rotation about its axis in 24 hours. If the radius of the Earth is 6370 kilometers, find the velocity, in kilometers per hour, of a person standing on the equator.
$\approx$ 1668 km/hour

Entertainment The outer 14 feet of the Space Needle Restaurant in Seattle rotates once every 58 minutes.

33. Find the area of the rotating part of the restaurant. approx 7938.8 ft²

34. Find the velocity in feet per minute of a person sitting by the window at this restaurant.
$\approx$ 10.5 ft/min

 **Look Back**

35 Solve the system of equations using matrix algebra. **[Lesson 4.6]** (3, 5, −1)

$$\begin{cases} 2x + 3y + 5z = 16 \\ -4x + 2y + 3z = -5 \\ 3x - y - z = 5 \end{cases}$$

36. Find the zeros of the function $f(x) = x^3 + 5x^2 - 8x - 12$. **[Lesson 6.2]** −6, −1, 2

Solve each equation for x. [Lesson 7.7]

37. $5e^{2x-1} = 60$ $x = \frac{\ln 12 + 1}{2}$

38. $2\log_3 x + \log_3 9 = 4$
$x = 3$

Draw each angle in standard position, and find the measure of a coterminal angle. [Lesson 8.2]

39. 337°   **40.** −118°   **41.** $\frac{3\pi}{4}$ **42.** $-\frac{4\pi}{3}$

Convert each radian measure to degree measure. [Lesson 8.5]

43. $\frac{5\pi}{12}$ radians 75°

44. 5.85 radians $\approx$ 335°

Look Beyond

Finish drawing the figure on the right so that it is congruent to the figure on the left.

45.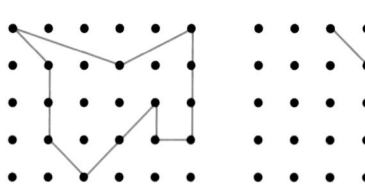

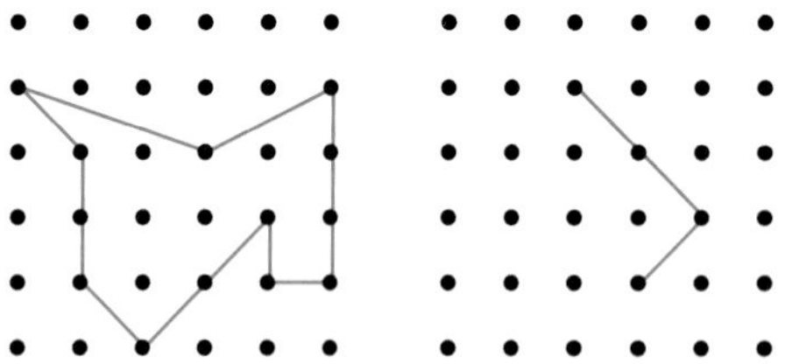

46.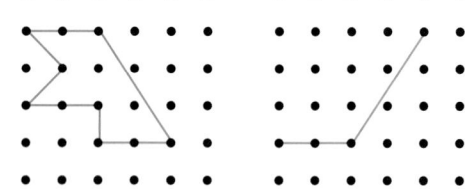

41. $\frac{3\pi}{4}$ radians

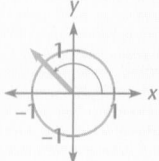

Answers may vary but must be similar to $\frac{3\pi}{4} + 2\pi n$.

42. $-\frac{4\pi}{3}$ radians

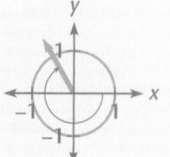

Answers may vary but must be similar to $-\frac{4\pi}{3} + 2\pi n$.

The answers to Exercises 45 and 46 can be found in Additional Answers beginning on page 842.

Applications of Periodic Functions

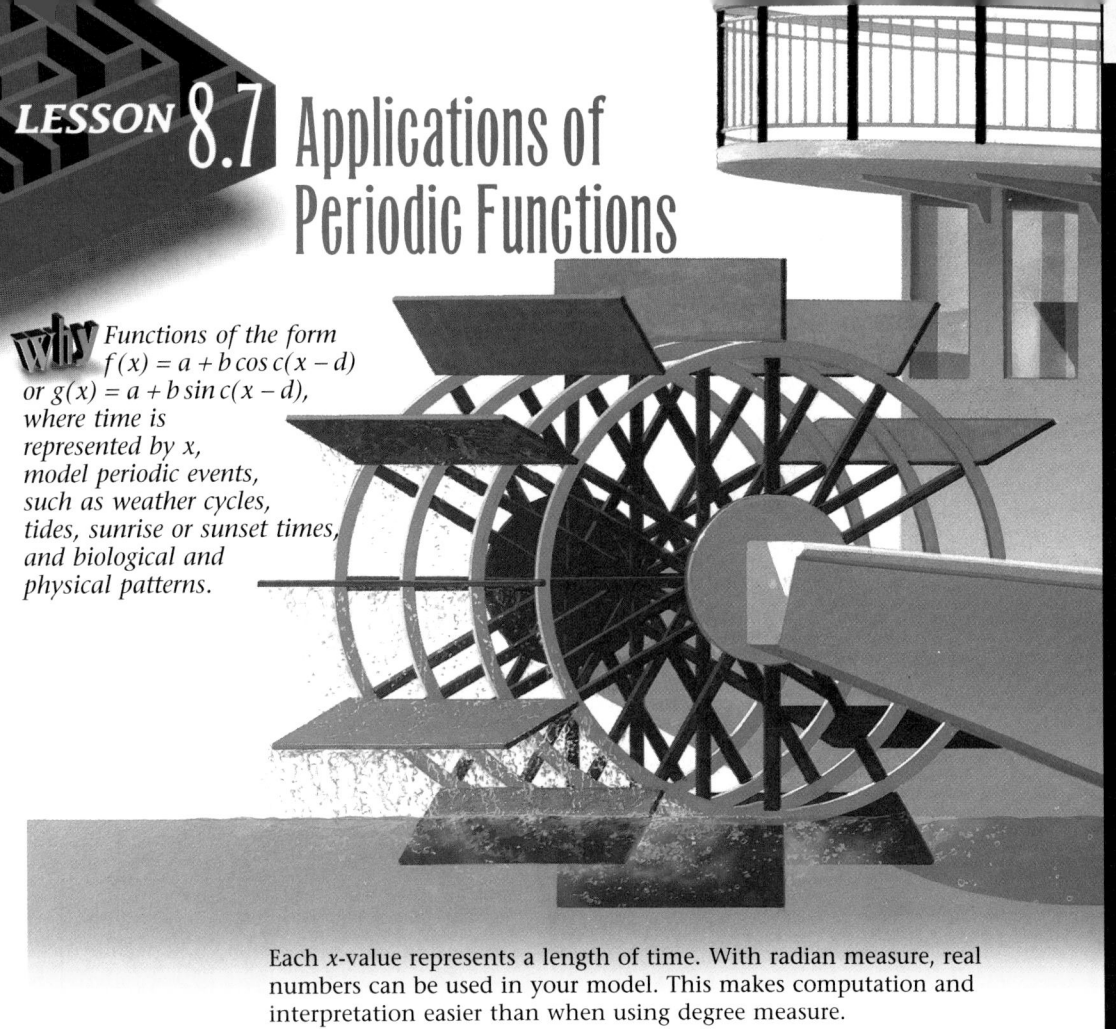

 *Functions of the form*
$f(x) = a + b \cos c(x - d)$
or $g(x) = a + b \sin c(x - d)$,
where time is
represented by x,
model periodic events,
such as weather cycles,
tides, sunrise or sunset times,
and biological and
physical patterns.

Each *x*-value represents a length of time. With radian measure, real numbers can be used in your model. This makes computation and interpretation easier than when using degree measure.

EXAMPLE 1

 *Graphics Calculator*

A model for the motion of a point on the outside edge of the paddle wheel of a steamboat is $f(x) = 8 + 11 \cos\left(\frac{\pi x}{10}\right)$, where *x* is the time in seconds after the wheel starts, and $f(x)$ is the height in feet above the surface of the water. Use a graphics calculator in radian mode to answer the following questions. In each case, look back and see how your answer relates to the numbers in the model.

Ⓐ How far underwater does the point go?

Ⓑ What is the diameter of the wheel?

Ⓒ How high above the water is the axle of the wheel?

Ⓓ How long does it take for the wheel to make one complete turn?

Ⓔ How high is the wheel after 5 seconds?

Ⓕ How can you adjust the function to model the motion of the point when the boat is going faster?

ALTERNATIVE **teaching strategy**

Technology On a graphics calculator in radian mode, have students graph a basic sine function. Then give them a function of the form $f(x) = a + b \sin cx$ such as $f(x) = 6 + 12 \sin\left(\frac{\pi x}{2}\right)$. Have students describe how the basic sine curve changes to represent the given function. Ask students to determine the maximum and minimum points on the graph, the period of the function, and the first positive value of *x* for which $f(x)$ is –6. Ask students why it helps to graph the basic sine function first. [**maximum: 18; minimum: –6; period is 4; x = 3; to see the shape of the graph**]

PREPARE

Objectives

• Model applications with circular functions of the form
$f(x) = a + b \cos c(x - d)$ or
$f(x) = a + b \sin c(x - d)$.

RESOURCES

• Practice Master   8.7
• Enrichment Master   8.7
• Technology Master   8.7
• Lesson Activity Master   8.7
• Quiz   8.7
• Spanish Resources   8.7

This lesson may be considered optional in the core course. See page 428A.

Assessing Prior Knowledge

In the following function $f(x) = 5 + 2 \cos 3(x + \pi)$, identify the vertical shift, horizontal phase shift, amplitude, and period.
[**up 5 units, –π (π units to the left), vertical stretch of 2 (3 to 7), $\frac{2\pi}{3}$**]

TEACH

 Encourage students to think up cyclical patterns that exist in the real world. Many are connected with the revolution of the Earth about the Sun and the rotation of the Earth on its own axis.

Let $f(x) = 4 + 10 \cos\left(\frac{\pi x}{8}\right)$ where x is time in seconds and $f(x)$ is the height in feet above the surface of the water. Use the situation in the text in Example 1, and answer the questions listed at the bottom of page 475.

[A. 6 feet

B. 20 feet

C. 4 feet

D. 16 seconds

E. ≈ 0.2 feet

F. decrease the period by decreasing the denominator of the fraction $\frac{\pi x}{8}$]

Cooperative Learning

Examples 1 and 2 can be worked best by putting students in pairs. Each student can work three of the six questions. They should discuss their answers to make sure they are consistent and both understand the entire process.

ongoing ASSESSMENT

The period describes how long the paddle wheel takes to make one turn: the shorter the period, the faster the wheel is turning.

Solution►

Graph the function $f(x) = 8 + 11 \cos\left(\frac{\pi x}{10}\right)$.

Ⓐ The y-coordinate of each of the lowest points on the graph is –3. Since $f(x)$ is the height in feet above the surface of the water, the point on the wheel goes 3 feet underwater.

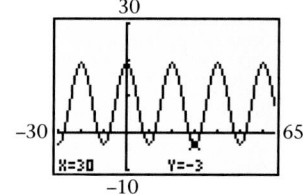

Ⓑ The y-coordinate of the highest points on the graph is 19. Thus, the point reaches a maximum height of 19 feet above the water. The diameter of the wheel must be the distance between the highest point, 19, and lowest point, –3. Thus, the diameter of the wheel is 22 feet.

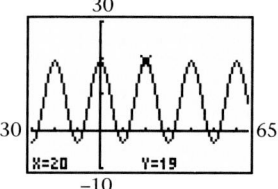

Ⓒ The axle is at the middle of the wheel. Therefore, the axle is the midpoint of the diameter.

$$\frac{19 + (-3)}{2} = 8$$

Thus, the axle is 8 feet above water.

Ⓓ The period of the graph is 20. Since x is time in seconds, the wheel completes one full turn every 20 seconds.

Ⓔ $f(5)$ gives the height of the wheel after 5 seconds. Find $f(5)$. Trace the graph until the x-coordinate is 5, or substitute 5 for x in $f(x) = 8 + 11 \cos\left(\frac{\pi x}{10}\right)$. Since $f(5)$ is 8, the point is 8 feet above the water after 5 seconds.

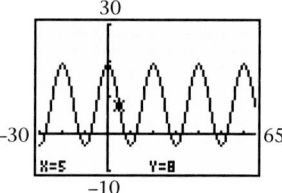

Ⓕ A shorter period means the wheel completes a cycle in fewer seconds and is going faster. How do you decrease the period? Experiment.

Graph $f(x) = 8 + 11 \cos\left(\frac{\pi x}{8}\right)$.

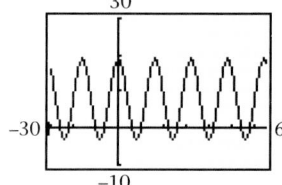

Then graph $f(x) = 8 + 11 \cos\left(\frac{\pi x}{12}\right)$.

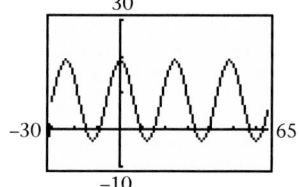

Decreasing the denominator of $\frac{\pi x}{10}$ decreases the period and models the motion of the paddle wheel when the boat is going faster. ❖

How does the period of the graph describe the paddle wheel on the steamboat?

ENRICHMENT

The sun's angular height above the horizon at noon plotted against each day of the year gives a sine curve. So does the sun's angular distance from the meridian throughout 1 day. Have students perform research to determine the data, and graph these data for both aspects of the sun. Have students determine the amplitude and the period of each graph and the function that describes the graph.

EXAMPLE 2

Graphics Calculator

Jesse wants to vacation at a resort that is not very crowded. A travel agent gives Jesse a graph showing the population throughout the year for a certain resort. The population graph is modeled by $g(x) = 7.7 - 3.2 \cos\left(\frac{\pi x}{6}\right)$, where x is the month of the year, $x = 1$ represents January, and $g(x)$ is the population, in thousands, of the town that month. Use a graphics calculator in radian mode for this example. Look back and see how each answer relates to the model.

A What type of resort town might this be? Explain.

B What seems to be the permanent population of the town?

C Predict the population of the town in July.

D During which months of the year will the population be about 7700?

E What is the period of the function? What does this represent?

F How might you adjust the function to represent the effects of a large company beginning a business in the town?

Solution▶

Graph the function on your graphics calculator.

A Since the population is at its highest when the *x*-coordinate is 6, in June, the resort is probably a summer resort — perhaps near a lake or beach.

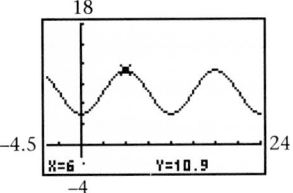

B The permanent population would be the minimum value on the graph — the people who live there. The minimum value on this graph is 4.5. Thus, the permanent population of the town is about 4500 people.

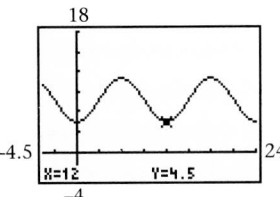

 **Using Visual Models**

strategies Have students graph $y = a + b \sin c(x - d)$ on the graphics calculator for $a = 0$, $b = 1$, $c = 1$, and $d = 0$. Then graph the function for each of the following values, one variable at a time:

$a = 1, -1$

$b = \frac{1}{2}, 2$

$c = \frac{1}{2}, 2$

$d = \pi, -\pi$

Students should make a drawing of each of these functions on 3" × 5" index cards. Then, have students organize these cards in whatever manner helps them to see what each variable does.

Alternate Example 2

Let $g(x) = 4 - 4 \cos\left(\frac{\pi x}{6}\right)$. Use the situation in the text in Example 2, and answer the questions on page 477.

[A. **Summer Resort**

B. **The permanent population is 0**

C. **≈ 7464 people in July**

D. **during May and June**

E. **12, the 12 months of the year**

F. **Increase the value of** *a* **(which is 4 in the function** *g***)**]

C To predict the population of the town in July, find $g(7)$. Trace the graph until the x-coordinate is 7, or substitute 7 for x in $g(x) = 7.7 - 3.2 \cos\left(\frac{\pi x}{6}\right)$.

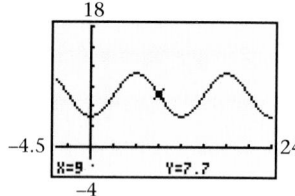

$$g(7) \approx 10.5$$

Thus, there are about 10,500 people in the town during July.

D To find the months that the population will be about 7700, find the values of x for which $f(x)$ is 7.7. The x-coordinates are 3 and 9. Thus, the population in the town is about 7700 in March and in September.

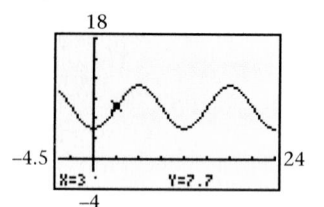

E The period of the function is 12 because the population follows yearly cycles of 12 months.

F Consider how a new business may affect the population. A new business would probably increase the permanent population without affecting the increase of population in the summer. How can you move the entire graph up without changing the shape of the graph?

Experiment. Change the first number in the function to a larger number, such as 10.7.

$$f(x) = 10.7 - 3.2 \cos\left(\frac{\pi x}{6}\right)$$

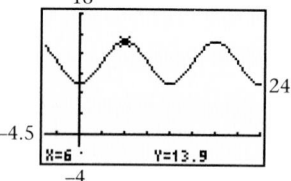

Then change the first number in the function to a smaller number, such as 3.7.

$$f(x) = 3.7 - 3.2 \cos\left(\frac{\pi x}{6}\right)$$

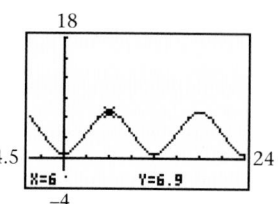

Notice that changing the first number to a larger number will move the function up to account for a larger permanent population without changing the amount of fluctuation. ❖

The graphics calculator has made working with trigonometric models much easier. However, you must be sure what the variables represent in order to interpret these models effectively.

EXERCISES & PROBLEMS

Communicate

1. Which measure of rotation (radians or degrees) is best to use in functions that model periodic changes over time? Why?

2. Describe the changes to the graph of $f(\theta) = a + \sin\theta$ as a gets larger or smaller.

3. Describe the changes to the graph of $f(\theta) = b\sin\theta$ as b gets larger or smaller.

4. Describe the changes to the graph of $f(\theta) = \sin(c\theta)$ as c gets larger or smaller.

Practice & Apply

Use a calculator in radian mode to graph the function $f(x) = 4 + 7\sin\left(\frac{\pi x}{3}\right)$.

5 Find the maximum and minimum values of f.   max: 11, min: −3

6 Find $f(6)$. 4

7 Find $f(20)$. ≈ 10.06

8 Find two values of x such that $f(x)$ is 11.

9 Find the period of the function.
6

Ferris Wheel The distance above the ground for the last person to get on a Ferris wheel before it starts can be modeled with the function $f(x) = 27 - 25\cos\left(\frac{\pi x}{20}\right)$, where x is the number of seconds after the Ferris wheel starts, and f is the height of the person from the ground in feet after x seconds.

10 What is the highest point reached by the person riding this Ferris wheel? How long does it take to reach that height? max: 52 ft, time=20 sec

11 What is the height of a person 12 seconds after the Ferris wheel starts? When will the person next reach that height?

12 Find two times when the person is 27 feet above the ground.

13 Find the diameter of the Ferris wheel. 50 ft

Selected Answers
Odd-numbered Exercises 5–37 and 41–45

Assignment Guide
Core 1–45

Core Plus 1–47

ASSESS

Technology
Graphics calculators are needed for Exercises 5–38 and 42–43.

Error Analysis
Make certain students have their calculators in radian mode for these exercises. Students often confuse the concepts of the difference in the minimum and minimum values and the period.

8. Answers may vary but must be equivalent to $1.5 + 6n$ for two different integer values of n.

11. 34.73 feet; after 16 additional seconds, 28 seconds total

12. Answers may vary but must be equivalent to $10 + 20n$ seconds for 2 different non-negative, integer values of n.

Authentic Assessment

Have students describe the periodicity of the sundial. Have them explain what will happen to the graph of the function describing the shadows on a sundial if a, b, or c changes.

14. Winter resort, because the population of the town is highest in January.

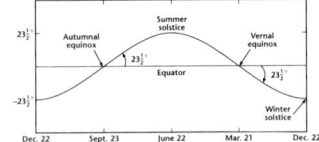

NAME _____ CLASS _____ DATE _____

Practice & Apply
8.7 Applications of Periodic Functions

The yearly change in seasons can be modeled by a sine function as shown. In the diagram, the sky appears flat. The ecliptic, the apparent path of the sun, is shown as a curve that crosses the equator at the two equinoxes.

1. Find the domain, range, and period of the function. _____
2. At what angle does the ecliptic intersect the equator? _____
3. When is the sun highest in the sky? _____
4. Describe the path of the sun from the vernal equinox to the winter solstice.

The range of a projectile is modeled by $y = \frac{(2200)^2}{32} \sin 2x$ where 2200 ft/s is the initial velocity, x is the angle of elevation, and y is the range of a projectile.

5. Use a graphics calculator to sketch the graph on the grid provided.
6. Identify the amplitude. _____
7. Give the period. _____

The sales of a seasonal product are modeled by $y = 74 + 40 \sin \frac{\pi}{6}x$, where y is measured in thousandths of units and x is the time in months, with $x = 1$ corresponding to January.

8. Use a graphics calculator to sketch the graph of the given function.
9. Find the amplitude, period, and vertical shift of the function. _____

HRW Advanced Algebra   Practice & Apply   49

Employment The number of people employed in a resort town can be modeled with the function $g(x) = 5.2 + 1.5 \sin\left(\frac{\pi x}{6} + 1\right)$, where x is the month of the year (beginning with 1 for January), and $g(x)$ is the number of people (in thousands) employed in the town that month.

14 What type of resort might this be? Explain.

15 How many people are permanently employed in the town? ≈ 3700

16 Predict the number of people employed in February. ≈ 6533

17 Find two months when there are about 4500 people employed in the town. May and September

18 Adjust the function to model the employment if a major business in the resort town closes.
Answers may vary, but all functions should have the number 5.2 decreased.

Use a graphics calculator in radian mode to graph the function $f(x) = 58\sin(4\pi x)$.

19 Find the maximum and minimum values of f. 58, −58

20 Find $f(1)$. $-1.2 \times 10^{-11} \approx 0$ **21** Find $f(1.1)$. ≈ 55.2 **22** Find $f(1.5)$. $1.2 \times 10^{-11} \approx 0$

23 Find two values such that $f(x) = -3$. Answers may vary, but must be similar to $-0.004 + 0.5n$ for 2 different values of n.

24 Find the period of the function. 0.5 radians

Physiology The amount of air in a resting person's lungs can be modeled by the function $a(x) = 0.8 + 0.7 \sin\left(\frac{2\pi x}{3}\right)$, where x is the time in seconds after the measurement is started, and $a(x)$ is the amount of air in the person's lungs in liters.

25 Find two times when the person's lungs have 0.7 liters of air in them.
Answers may vary, but must be similar to $-0.068 + 3n$ or $1.57 + 3n$ for 2 different values of n.
26 How much air will the person's lungs have in them after 10 seconds? ≈ 1.41 liters

27 How many breaths per minute is the person taking? 20

28 Adjust the function to model the amount of air in the person's lungs just after the person has been running.

Temperature The temperature in an air-conditioned office on a hot day can be modeled by the function $t(x) = 67 + 1.5 \cos\left(\frac{\pi x}{12}\right)$, where x is the time in minutes after the air conditioner turns on, and $t(x)$ is the temperature in degrees Fahrenheit after x minutes.

29 How long does the air conditioner run after turning on? Explain. 12 min

30 Find the maximum and minimum temperatures in the office building. max:68.5°, 65.5°: min

31 Find the temperature 10 minutes after the air conditioner turns on. $\approx 65.7°$

32 Find two times when the temperature is 67°.

33 Adjust the function to model the temperature in the building when the thermostat is set to a higher temperature.

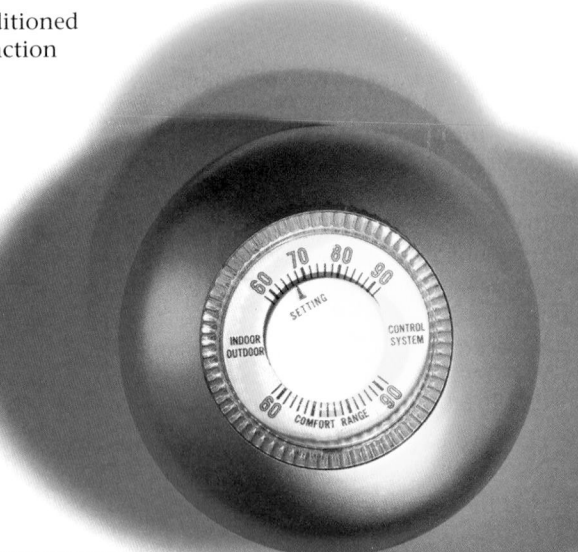

28. Answers may vary. A possible answer is: a person breathes harder and deeper after running, so the amplitude will have to be increased from 0.7 to a higher amount. Also, the vertical shift will have to be increased from the current value of 0.8 to compensate for the change in amplitude. The vertical shift should always be greater than the amplitude, or you will get negative values for the volume.

32. Answers may vary but must be similar to $6 + 12n$ for 2 different non-negative integer values of n.

33. Answers may vary, but all functions must have an increase in the number 67.

Physics A spring with a weight attached is suspended from a board. The weight is pulled and then released. The height of the weight above the floor can be modeled by the function $h(x) = 15 - 9\cos(2\pi x)$, where x is the number of seconds after the weight is released, and $h(x)$ is the height of the weight in centimeters at that time.

34 What was the height of the weight before it was pulled down? How far was it pulled down?
15 cm, 9 cm

35 Find the height of the weight 0.5 seconds after it was released. 24 cm

36 Find two times when the height of the weight is 15 centimeters. Answers may vary, but must be similar to $0.25 + 0.5n$ seconds for 2 different values of n.

37 Find the period of the function. What does the period represent?

38 Adjust the function to model the height of the weight when the weight is pulled farther down before being released.

39. 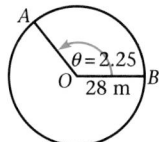 **Portfolio Activity** Complete the Portfolio Activity on page 429.

 **Look Back**

Geometry The surface area of a right cylinder is $A = 2\pi r^2 + 2\pi rh$. **[Lesson 1.5]** $h = \frac{A - 2\pi r^2}{2\pi r}$

40. Solve the surface area formula for h in terms of r and A.

41. Find the height of a right cylinder with a surface area of 125π cm^2 and a radius of 5 cm. 7.5 cm

Demographics The population of Monterey is 6740 and is increasing at the rate of 3.7% a year. **[Lesson 7.2]**

42 Find the projected population of Monterey in 10 years. 9693

43 How many years will it take for the projected population of Monterey to double? $\approx$ 19 years, 1 month

44. Find the length of $\overset{\frown}{AB}$.
[Lesson 8.6] 63 meters

45. Find the area of sector AOB.
[Lesson 8.6] 882 m^2

$\theta = 2.25$, O — 28 m — B

Look Beyond

46. Write the reciprocal of the cosine function. $f(x) = \frac{1}{\cos x}$

47. Graph the cosine function and its reciprocal function on the same axes. Compare their graphs. How are they alike? How are they different?

37. The period is 1 second; it represents the time that it takes the spring to go from maximum height to maximum height or from minimum height to minimum height.

38. Answers may vary, but all should have the number "9" increased.

The answers to Exercises 39 and 47 can be found in Additional Answers beginning on page 842.

Look Beyond

These exercises introduce students to the secant function, which will be covered in Chapter 14.

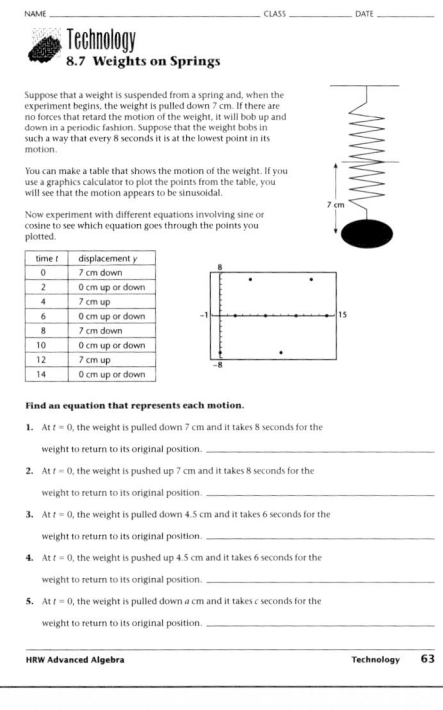

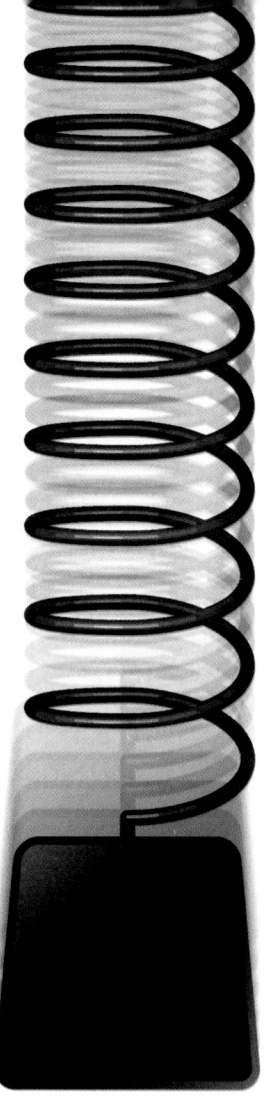

Daylight Around the World

FOCUS

Students will be asked to compile data on times of sunrise and sunset over a period of 1 year for their own city, and then do the same for several other cities. For each city, students will graph these results and discover that they will have a sine curve.

MOTIVATE

In this project, students see one of the earliest known and most natural examples of a sine curve. Questions about amplitude and period will have a very concrete and tangible meaning.

1. Choose about 20 to 30 dates, spread out evenly throughout the calendar year. Include the dates for the spring equinox, fall equinox, summer solstice, and winter solstice. Begin with the spring equinox.

2. Find the times for sunrise and sunset in your local area on these 20 to 30 days that you chose. To find the times, you may consult an almanac, a library, or a local TV or radio station.

3. Let the spring equinox date begin at 0 on your horizontal axis. Plot your data on a graph, and connect the points with a smooth curve. An example of the axes for the graph is given.

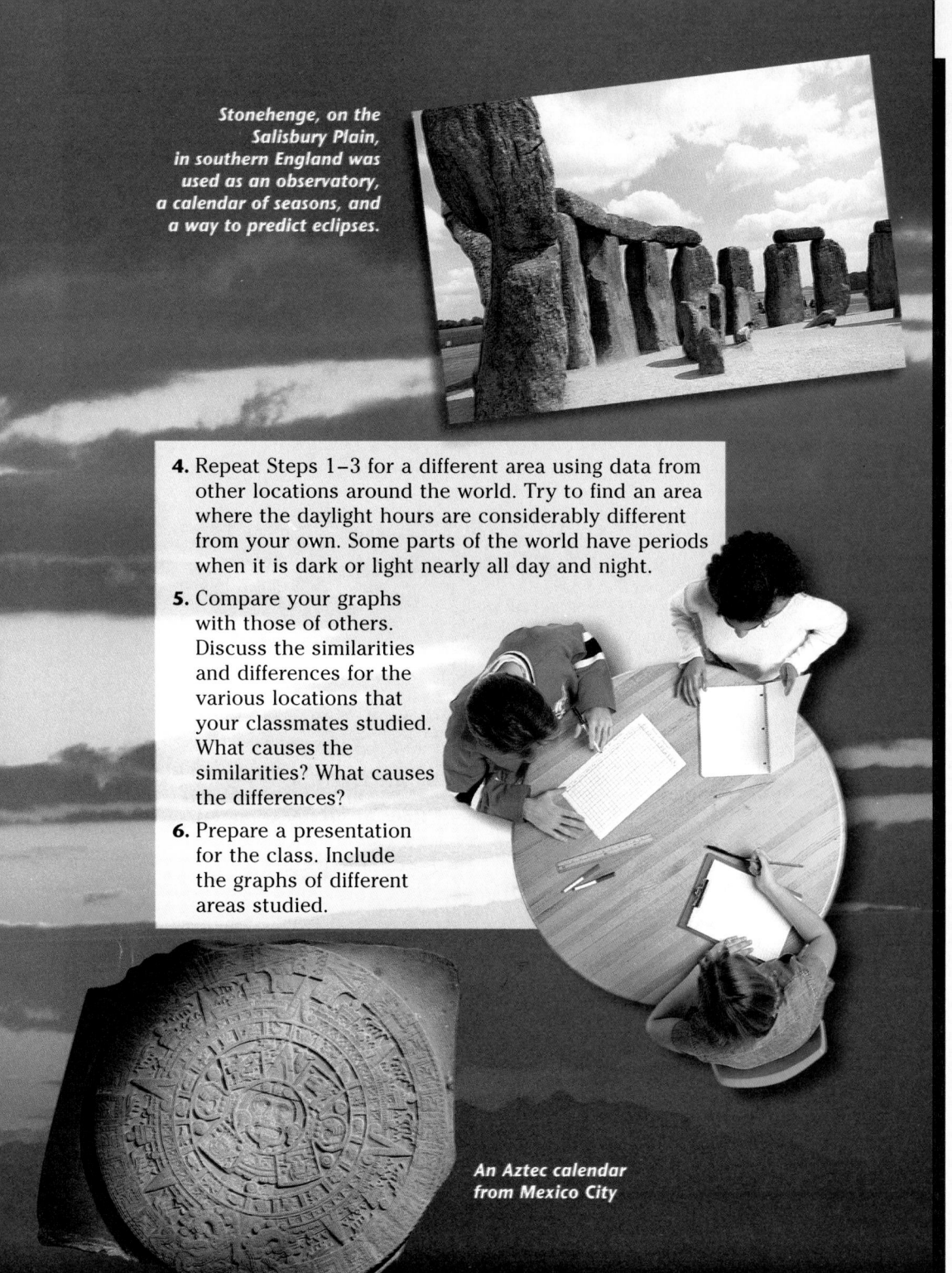

Stonehenge, on the Salisbury Plain, in southern England was used as an observatory, a calendar of seasons, and a way to predict eclipses.

4. Repeat Steps 1–3 for a different area using data from other locations around the world. Try to find an area where the daylight hours are considerably different from your own. Some parts of the world have periods when it is dark or light nearly all day and night.

5. Compare your graphs with those of others. Discuss the similarities and differences for the various locations that your classmates studied. What causes the similarities? What causes the differences?

6. Prepare a presentation for the class. Include the graphs of different areas studied.

An Aztec calendar from Mexico City

For Steps 4 and 5, organize students into groups of four. Students who previously lived in another city should be encouraged to choose that city for their data collection. After students have constructed their own graphs, they should discuss Step 5 as a group.

Discuss

Students can extend and apply this project by constructing similar graphs for other planets. For example, Neptune's axis is tilted 29°, its day is 18 hours long and its year is the same length as ≈ 60,188 of our days.

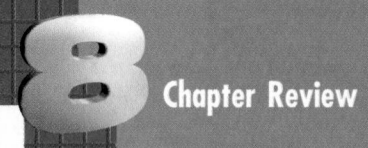

Chapter 8 Review

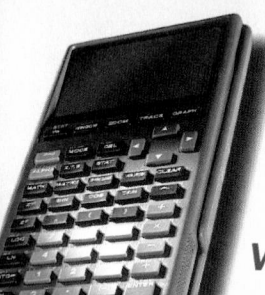

Vocabulary

30-60-90 triangle	432	negative measure	440
45-45-90 triangle	430	period	453
amplitude	454	phase shift	455
arc length	468	positive measure	440
area of a sector	471	principal values	448
central angle	460	radian measure	460
cosine of an angle	434	sector of a circle	470
coterminal angles	440	sine of an angle	434
initial side of an angle	438	standard position of an angle	438
inverse cosine relation	447	tangent of an angle	434
inverse sine relation	447	terminal side of an angle	438
inverse tangent relation	447	unit circle	438

Key Skills & Exercises

Lesson 8.1

➤ Key Skills

Identify the trigonometric ratios in special right triangles.

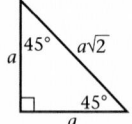

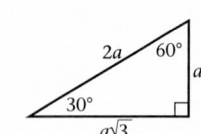

$$\sin 45^\circ = \frac{\text{opp}}{\text{hyp}} = \frac{a}{a\sqrt{2}} = \frac{1}{\sqrt{2}}, \text{ or } \frac{\sqrt{2}}{2}$$

$$\cos 60^\circ = \frac{\text{adj}}{\text{hyp}} = \frac{a}{2a} = \frac{1}{2}$$

Apply the special right triangle relationships to find missing lengths of sides of special triangles.

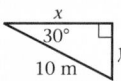

To find x and y, use the 30-60-90 triangle relationships. The lengths of the legs in terms of a are a and $a\sqrt{3}$. The hypotenuse, $2a$, is 10. Therefore, a is 5 and $a\sqrt{3}$ is $5\sqrt{3}$. Thus, $x = 5\sqrt{3}$, and $y = 5$.

➤ Exercises

Use the special right triangle relationships to find x and y.

1.

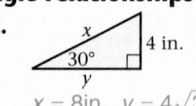

$x = y$
$= \frac{8}{\sqrt{2}}$ cm

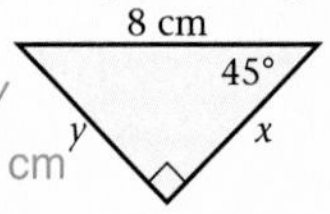

2.

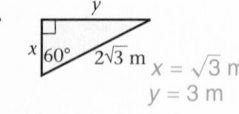

$x = 8$in., $y = 4\sqrt{3}$ in.

3.

$x = \sqrt{3}$ m
$y = 3$ m

Lesson 8.2

➤ Key Skills

Identify the coordinates of a point (x, y) on the unit circle given an angle in standard position using the relationship $(x, y) = (\cos\theta, \sin\theta)$.

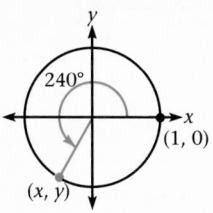

Find x and y.
Since (x, y) is on the unit circle,
$x = \cos 240°$ and $y = \sin 240°$.
So, $x = -\frac{1}{2}$ and $y = -\frac{\sqrt{3}}{2}$.

Identify angles and their coterminal angles in standard position.

Draw an angle of 135° in standard position. Find a negative angle that is coterminal with it.
The angle 135° is halfway between one-quarter rotation, 90°, and one-half rotation, 180°. Since $135 - 360 = -225$, the angle $-225°$ is coterminal with the angle 135°.

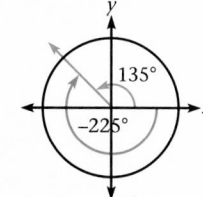

➤ Exercises

Find x and y to the nearest hundredth.

4.
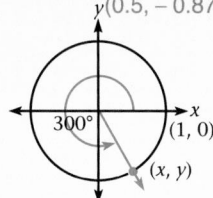

5.
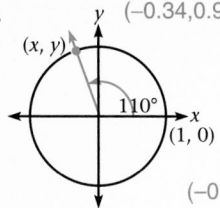

6.

7. Give two positive and two negative angles that are coterminal with 57°.
Answers may vary, but must be similar to $57° + 360°n$ for 2 different values of n.

Lesson 8.3

➤ Key Skills

Determine sides or angles of right triangles using trigonometric functions or their inverses.

Find $m\angle A$. Then use this value to find a.

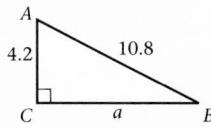

$m\angle A = \cos^{-1}\left(\frac{4.2}{10.8}\right) \approx 67.1°$
$\tan 67.1° = \frac{a}{4.2}$, so $a \approx 9.9$

Determine, from the principal values of the inverse trigonometric functions, any angle in standard position.

Find $\cos^{-1}(-0.4)$ for $180° < \theta < 360°$.
The principal value is about 114° and lies in Quadrant II. The desired value lies in Quadrant III.
Since $360 - 114 = 246$, $\theta \approx 246°$.

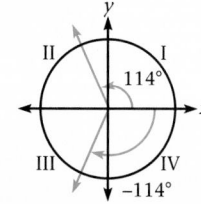

13. vertical shift: 2; amplitude: 1; period: 360°; phase shift: none

14. vertical shift: none; amplitude: 1; period: 360°; phase shift: 90° to the right

15. vertical shift: none; amplitude: 3; period: 360°; phase shift: none

16. vertical shift: none; amplitude: 1; period: 180°; phase shift: none

17. 390°

➤ **Exercises**

Find two angles between 0° and 360°, inclusive, for each inverse relation.
                                             225°, 315°   135°, 225°

8. $\cos^{-1} 1$ 0°, 360°   **9.** $\tan^{-1} 1$ 45°, 225° **10.** $\sin^{-1}\left(-\frac{\sqrt{2}}{2}\right)$ **11.** $\cos^{-1}\left(-\frac{\sqrt{2}}{2}\right)$

12. In $\triangle ABC$, $m\angle C = 90°$, $b = 6.5$, and $a = 9.7$. Find $m\angle B$ to the nearest degree. Then use this value and the sine or cosine function to find c to the nearest tenth. $m\angle B = 34°$; $c = 11.7$

Lesson 8.4

➤ **Key Skills**

Identify how a, b, c, and d in $f(\theta) = a + b\sin c(\theta - d)$ and $f(\theta) = a + b\cos c(\theta - d)$ transform the graphs of $f(\theta) = \sin\theta$ and $f(\theta) = \cos\theta$.

In general, for the graph of $f(\theta) = a + b\sin c(\theta - d)$,

- a shifts the graph vertically by a;
- b stretches the graph vertically by a factor of b, resulting in an amplitude of b;
- c changes the period to $\frac{360}{c}$ for degrees;
- d shifts the graph horizontally by d units, resulting in a phase shift of d.

Also, the graph of $-f(\theta)$ is a reflection of the graph of $f(\theta)$ across the x-axis.

➤ **Exercises**

Identify the vertical shift, amplitude, period, and phase shift for each function.

13. $f(\theta) = 2 + \sin\theta$ **14.** $g(\theta) = \cos(\theta - 90°)$
15. $h(\theta) = 3\cos\theta$ **16.** $k(\theta) = \sin 2\theta$

Lesson 8.5

➤ **Key Skills**

Convert degree to radian measure, and radian to degree measure.

Convert 2.4 radians to degree measure. Because π radians $= 180°$,
$2.4 \cdot \pi$ radians $= 2.4 \cdot 180°$,
or 2.4 radians $= 2.4 \cdot \frac{180°}{\pi} \approx 137.5°$.
Similarly, to convert from degrees to radian measure, multiply by $\frac{\pi}{180°}$.

Graph and identify properties of trigonometric functions in radian measure.

What are the period and phase shift of the graph of $\sin 4(\theta + \pi)$?
The period is $\frac{2\pi}{4} = \frac{\pi}{2}$ radians, and the phase shift is $-\pi$ radians (or π radians to the left).

➤ **Exercises**

Convert radian to degree measures (to the nearest tenth), and degree to radian measure.

17. $\frac{13\pi}{6}$ radians   **18.** 147° 2.6 radians **19.** 4 radians 229.2° **20.** 240° 4.2 radians

21. Write the equation of a cosine function with a vertical shift of 1 unit, an amplitude of 2, a period of $\frac{5\pi}{2}$ radians, and a phase shift of π radians to the left. Then sketch the graph.

21. $f(x) = 1 + 2\cos(0.8(x + \pi))$

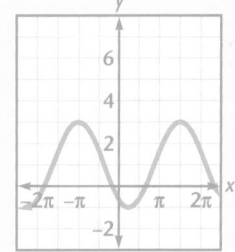

Lesson 8.6

> ## Key Skills

Find the arc length and sector area determined by the central angle of a circle.

Find the arc length and sector area determined by a central angle of $\frac{2\pi}{3}$ radians in a circle with a radius of 9 centimeters.

$$\text{arc length} = r\theta = 9\left(\frac{2\pi}{3}\right) \approx 18.8 \text{ cm}$$

$$\text{sector area} = \frac{1}{2}\theta r^2 = \left(\frac{1}{2}\right)\left(\frac{2\pi}{3}\right)(81) \approx 84.8 \text{ cm}^2$$

> ## Exercises

Find the arc length and sector area (to the nearest tenth) determined by the given central angle and radius of each circle.

22. 4.4 radians; 10 inches
Length: 44 in. ; Area 220 in.^{2}

23. π radians; 5.5 meters
Length: 17.3 meters
Area: 47.5 m ^{2}

Lesson 8.7

> ## Key Skills

Model applications with circular functions of the form $f(x) = a + b \cos c(x - d)$ or $f(x) = a + b \sin c(x - d)$.

The height (in centimeters) of a piston in a cylinder after t seconds is given by $h(x) = 8 + 6 \sin 4\pi t$. Graph and trace to find the maximum height and the period. The maximum height is 14 centimeters. By tracing to the next peak, you can see that the period is 0.5 seconds.

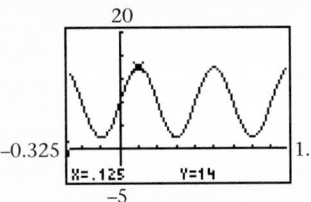

> ## Exercises

Use a calculator in radian mode to graph $f(x) = 30 + 9.2 \cos \frac{\pi x}{6}$.

24 Find the maximum and minimum values of f.   max: 39.2; min:20.8

25 Find $f(22)$.   34.6

26 Find two values of x such that $f(x)$ is 25.4.
Answers may vary but must be similar to $4 + 12n$ or $8 + 12n$ for 2 different integers n.

Applications

Agriculture An irrigation system consists of a 500-foot pipe with wheels. The pipe is fixed on one end and swings through an arc of 150°. The system irrigates from 100 feet out from the fixed end to the far end of the pipe.

27. How far does the far end of the pipe move, to the nearest foot? 1309 ft

28. To the nearest tenth, how many acres will the system water, given that an acre is 43,560 square feet? (Remember that the system doesn't deliver water for the first 100 feet.) 7.2 acres

29. To water the land equally, a sprayer 500 feet from the center must deliver water how many times as fast as a sprayer 100 feet from the center? 5

30. $y = 74 + 8 \sin \frac{\pi}{12} (x - 10)$,
or
$y = 74 + 8 \sin \frac{\pi}{12} (x + 14)$

31.

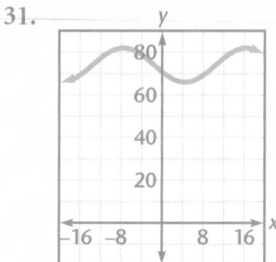

At noon, the temperature is 78°

32. about 10:30 A.M. and 9:30 P.M.

34. $d = \frac{18}{\cos 30°} = 12\sqrt{3}$ ft
≈ 20 ft, 10 in.

36. Approximately 252,336 stades, or about 25,233.6 miles

37. Answers may vary; inaccurate assumptions about the cities being on the same meridian and rounding error are two possible reasons for the differences.

38–41. Depending on rounding, days can vary by 1 day in either direction.

Meteorology Suppose the temperature during a week changes in a steady manner as follows: each day, there is a high temperature of 82° at 4 P.M. and a low temperature of 66° at 4 A.M.

30. Find a transformation of the sine function in form $y = a + b \sin c(x - d)$ that fits these conditions, where x is the number of hours from an initial time of 0. For example, when x is 0, it is midnight on Saturday (the beginning of the week), and when x is 28, it is 4 A.M. on Monday.

31. Sketch the function. Find the temperature at noon on any day.

32. Find the times in one 24-hour period when the temperature is 75°.

Construction A carpenter wants to build a roof with a slope of 30° above the horizontal. The horizontal distance from the outer edge of the roof to the center of the span is 18 feet.

33. How high must the carpenter build the vertical support for the center of the roof? Give an exact value and a value to the nearest inch. $6\sqrt{3}$ft $\approx$ 10ft, 5 in.

34. Give an expression involving a cosine function for the distance, d, along the deck of the roof. Then give an exact value for this expression and a value to the nearest inch.

35. If the carpenter is thinking about increasing the slope to 50°, how many times as much plywood (to the nearest hundredth) will be needed to deck the roof? 1.35

History The circumference of the Earth was computed by the Greek mathematician Eratosthenes (276–195 B.C.E.). He knew that the distance from Aswân, Egypt, to Alexandria, Egypt, was 5000 stades, or approximately 500 miles. He believed that the cities were on the same meridian and observed the Sun at noon on the summer solstice at both cities. At Aswân, the Sun cast no shadow. At Alexandria, the Sun cast a shadow at an angle of about 7 degrees, 8 minutes.

36. Find Eratosthenes' measurement of the circumference of the Earth in stades (based on the Greek stadium) and in miles.

37. How does this value compare with the value used today? Explain why the differences in the two values might occur.

Meteorology The average daily temperature for a city can be approximated by a circular function. Assume that the average temperature of a city (in degrees Fahrenheit) is given by $t(d) = 46 + 27 \sin \frac{2\pi}{365} (t - 111)$, where t is the day of the year. For your answers, assume that it is not a leap year.

38 What is the coldest average temperature, and on which date does it occur? 19°, Jan. 20

39 What is the warmest average temperature, and on which date does it occur? 73°, July 21

40 Give the two days when the average temperature is closest to 60°. May 23, September 19

41 On about what date in the fall does the average temperature first drop below freezing? November 21

Chapter 8 Assessment

Use the special right triangle relationships to find x and y.

1.

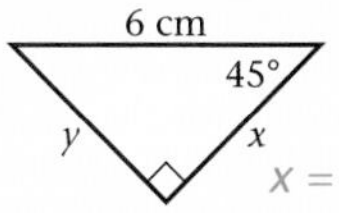

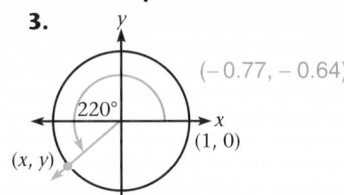

6 cm   45°

$x = 3\sqrt{2}$ cm
$y = 3\sqrt{2}$ cm

2.

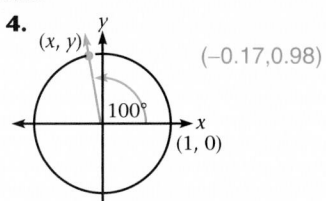

30°   8 in.

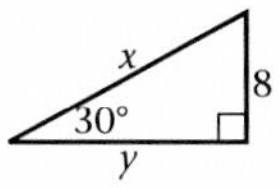

$x = 16$ in.
$y = 8\sqrt{3}$ in.

Find x and y to the nearest hundredth.

3.

220°   (1, 0)   (x, y)

(−0.77, −0.64)

4.

(x, y)   100°   (1, 0)

(−0.17, 0.98)

5. Give two positive and two negative angles that are coterminal with 112°. Answers may vary but must be similar to: positive: 112° + 360°n, negative: 112° − 360°n for positive integers n.

Find two angles between 0° and 360°, inclusive, for each inverse relation.

6. $\sin^{-1} 0$ 0°, 360°   **7.** $\tan^{-1} \dfrac{\sqrt{3}}{3}$ 30°, 210°   **8.** $\cos^{-1}\left(-\dfrac{\sqrt{2}}{2}\right)$ 135°, 225°   **9.** $\sin^{-1}\left(-\dfrac{\sqrt{3}}{2}\right)$ 240°, 300°

Write the function in the form $f(\theta) = a + b \cos c(\theta - d)$ for the following graph.

$f(x) = -2 + 3\cos 0.5(x + 120)$

10.

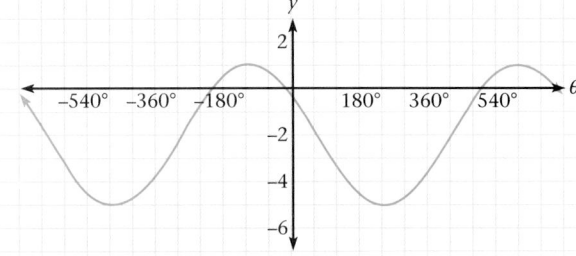

11. Write the equation of a cosine function with a vertical shift of 2 units, an amplitude of 1, a period of π radians, and a phase shift of $\frac{\pi}{4}$ radians to the right. Then sketch the graph.

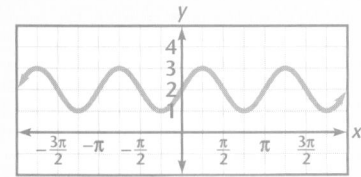

Find the arc length and sector area (to the nearest tenth) determined by the given central angle and radius of each circle.

12. 3.7 radians; 14 centimeters   **13.** $\dfrac{3\pi}{2}$ radians; 48 inches

length: 51.8 cm   area: 362.6 cm²   length: 226.2 in.   area: 5428.7 in²

Use a calculator in radian mode to graph $f(x) = 12 - 6.2 \cos \frac{\pi x}{4}$.

14 Find the maximum and minimum values of f. 18.2, 5.8

15 Find $f(8)$. 5.8

16 Find two values of x such that $f(x)$ is 18.2. Answers may vary but must be similar to 4 + 8n for 2 different integers n.

Chapters 1–8 Cumulative Assessment

COLLEGE ENTRANCE-EXAM PRACTICE

Multiple-Choice and Quantitative-Comparison Samples

The first half of the Cumulative Assessment contains two types of items found on standardized tests — multiple-choice questions and quantitative-comparison questions. Quantitative-comparison items emphasize the concepts of equality, inequality, and estimation.

Free-Response Grid Samples

The second half of the Cumulative Assessment is a free-response section. A portion of this part of the Cumulative Assessment consists of student-produced response items commonly found on college entrance exams. These questions require the use of machine-scored answer grids. You may wish to have students practice answering these items in preparation for standardized tests.

Sample answer-grid masters are available in the *Chapter Teaching Resources Booklets.*

College Entrance Exam Practice

Quantitative Comparison For Items 1–6, write

A if the quantity in Column A is greater than the quantity in Column B;
B if the quantity in Column B is greater than the quantity in Column A;
C if the two quantities are equal; or
D if the relationship cannot be determined from the information given.

	Column A	Column B	Answers
1. B	$\frac{5\pi}{3}$ radians	$305°$	(A) (B) (C) (D) [Lesson 8.5]
2. A	The number of real-number solutions $2x^2 - 3x - 3 = 0$	$4x^2 - 4x + 1 = 0$	(A) (B) (C) (D) [Lesson 5.6]
3. D	$\ln x$	$\log x$	(A) (B) (C) (D) [Lesson 7.4]
4. A	$f(g(x))$ $f(x) = x^2$ and $g(x) = x^2 + 2$ $g(f(x))$		(A) (B) (C) (D) [Lesson 3.3]
5. B	The period of $f(\theta) = 2 + 2\cos 5\theta$	$f(\theta) = -3 \sin \frac{\theta}{2}$	(A) (B) (C) (D) [Lesson 8.5]
6. D	The total multiplicity of the real zeros of		(A) (B) (C) (D) [Lesson 6.4]

7. If A is a 2×4 matrix, B is a 4×2 matrix, and C is a 4×4 matrix, which
^{C} one of the following products is not defined?   **[Lesson 4.2]**
  a. AB **b.** AC **c.** CA **d.** CB
8. If $a + bi = 4 - 3i$, what is the conjugate of $a + bi$?   **[Lesson 5.7]**
^{C} **a.** $-4 - 3i$ **b.** $-4 + 3i$ **c.** $4 + 3i$ **d.** $4 - 3i$
9. Which of the following is equivalent to $\lceil -3.01 \rceil - \lceil -1.99 \rceil$? **[Lesson 3.5]**
^{d} **a.** -1 **b.** 1 **c.** -1.02 **d.** -2

10. Referring to triangle ABC at the right,
d which of the following is *not* true?   **[Lesson 8.3]**

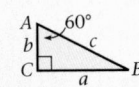

a. $c = \dfrac{a}{\cos 30°}$ **b.** $a = b\sqrt{3}$ **c.** $b = \dfrac{c}{2}$ **d.** $c = a \sin 60°$

11. For $f(x) = 2x^2 - 1$, which expression represents $f(t - 1)$? **[Lesson 2.4]**
a   **a.** $2t^2 - 4t + 1$ **b.** $2(t - 1)^2$ **c.** $2t^2 - 4t + 2$ **d.** $2t^2 - 2$

12. Which one of the following expressions is not equivalent to $\log 72$?
b   **[Lesson 7.5]**

    a. $3 \log 2 + 2 \log 3$ **b.** $\log 18 + 3 \log 2$

    c. $2 \log 12 - \log 2$ **d.** $2 \log 6 + \log 2$

Find the product.   **[Lesson 4.2]**

13. $\begin{bmatrix} 1 & 0 \\ -2 & 1 \end{bmatrix} \begin{bmatrix} 0 & 2 \\ 3 & -1 \end{bmatrix}$ **14.** $\begin{bmatrix} -1 & 0 & 1 \\ 2 & 1 & -2 \end{bmatrix} \begin{bmatrix} 1 \\ 2 \\ 3 \end{bmatrix}$ **15.** $\begin{bmatrix} 2 \\ 1 \\ 3 \\ 0 \end{bmatrix} [3 \ \ 4 \ \ 0 \ \ -2]$

Use the functions g and h to find the combinations $(g - h)(x)$ and $(g + h)(x)$.   **[Lesson 2.6]**

16. $g(x) = 2x + 3$, $h(x) = 6 + x - x^2$ **17.** $g(x) = \dfrac{x}{x^2 - 9}$, $h(x) = \dfrac{2}{x + 3}$

Use your graphics calculator to find the zeros of each function. Then write each function in factored form over the integers.   **[Lesson 6.2]**

18 $f(x) = 3x^2 - 15x + 18$ **19** $f(x) = x^3 + 2x^2 - x - 2$ **20** $f(x) = 4x^3 + 8x^2 - 29x + 12$

Simplify.   **[Lesson 5.6]**

21. $i^3 \cdot i^5$ =1   **22.** $i\sqrt{5} \left(-i\sqrt{-5}\right)$ 5i   **23.** $\left(-i^3\sqrt{3}\right)\left(i^2\sqrt{12}\right)$ −6i

Landscaping A landscaper is selecting trees to plant behind a house. The owner has a mountain view from the deck and does not want the trees to block the view. The trees will be planted 100 feet behind the house, at which point the ground level is 15 feet below eye-level from the deck.   **[Lesson 8.3]**

24. If the angle of elevation from eye-level on the deck to the bottom of the view that is to be kept clear is 10°, what is the maximum mature height, to the nearest foot, of the trees that can be chosen? 33 ft

25. If the view from the highest level of the house is 12 feet higher than that from the deck, and trees would have to be over 43 feet tall to begin blocking this view, what is the angle of elevation to the view, to the nearest degree, from the upper level of the house? 9°

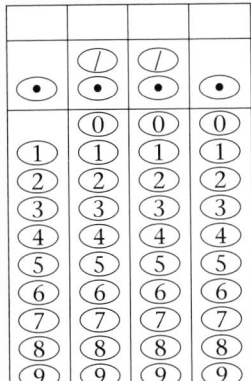

Free-Response Grid The following questions may be answered using a free-response grid commonly used by standardized test services.

Investment If $2500 is invested at 6.9%, determine the worth of the investment, to the nearest dollar, in 12 years with each of the following compounding periods.
[Lesson 7.2]
                                  $5708
26. yearly $5568 **27.** monthly   **28.** daily $5721

29 Solve the equation $5e^{(3-x)} = 2$ for x. Answer to the nearest hundredth.   **[Lesson 7.7]**
            3.92

13. $\begin{bmatrix} 0 & 2 \\ 3 & -5 \end{bmatrix}$

14. $\begin{bmatrix} 2 \\ -2 \end{bmatrix}$

15. $\begin{bmatrix} 6 & 8 & 0 & -4 \\ 3 & 4 & 0 & -2 \\ 9 & 12 & 0 & -6 \\ 0 & 0 & 0 & 0 \end{bmatrix}$

16. $(g - h)(x) = x^2 + x - 3$;
$(g + h)(x) = -x^2 + 3x + 9$

17. $(g - h)(x) = \dfrac{-x + 6}{x^2 - 9}; x \neq 3, -3$
$(g + h)(x) = \dfrac{3x - 6}{x^2 - 9}; x \neq 3, -3$

18. $x = 2, x = 3; f(x) = 3(x - 2)(x - 3)$

19. $x = -2, x = -1, x = 1$;
$f(x) = (x + 2)(x + 1)(x - 1)$

20. $x = -4, x = 0.5, x = 1.5$;
$f(x) = (x + 4)(2x - 1)(2x - 3)$

CHAPTER 9 Rational Functions

Meeting Individual Needs

9.1 Inverse Variation

Core

Inclusion Strategies, p. 495
Reteaching the Lesson, p. 496
Practice Master 9.1
Enrichment Master 9.1
Technology Master 9.1
Lesson Activity Master 9.1

[2 days]

Core Plus

Practice Master 9.1
Enrichment, p. 495
Technology Master 9.1

[1 day]

9.2 Simple Reciprocal Functions

Core

Inclusion Strategies, p. 504
Reteaching the Lesson, p. 505
Practice Master 9.2
Enrichment Master 9.2
Technology Master 9.2
Lesson Activity Master 9.2

[2 days]

Core Plus

Practice Master 9.2
Enrichment, p. 503
Technology Master 9.2

[1 day]

9.3 Exploring Reciprocals of Polynomial Functions

Core

Inclusion Strategies, p. 511
Reteaching the Lesson, p. 512
Practice Master 9.3
Enrichment Master 9.3
Technology Master 9.3
Lesson Activity Master 9.3
Mid-Chapter Assessment Master

[2 days]

Core Plus

Practice Master 9.3
Enrichment, p. 510
Technology Master 9.3
Mid-Chapter Assessment
    Master

[2 days]

9.4 Quotients of Polynomial Functions

Core

Inclusion Strategies, p. 518
Reteaching the Lesson, p. 519
Practice Master 9.4
Enrichment Master 9.4
Technology Master 9.4
Lesson Activity Master 9.4

[Optional*]

Core Plus

Practice Master 9.4
Enrichment, p. 518
Technology Master 9.4

[1 day]

9.5 Solving Rational Equations

Core

Inclusion Strategies, p. 525
Reteaching the Lesson, p. 526
Practice Master 9.5
Enrichment Master 9.5
Technology Master 9.5
Lesson Activity Master 9.5

[2 days]

Core Plus

Practice Master 9.5
Enrichment, p. 525
Technology Master 9.5

[1 day]

Chapter Summary

Core

Eyewitness Math pp. 522–523
Chapter 9 Project,
    pp. 528–529
Lab Activity
Long-Term Project
Chapter Review, pp. 530–532
Chapter Assessment, p. 533
Chapter Assessment, A/B
Alternative Assessment

[3 days]

Core Plus

Eyewitness Math
    pp. 522–523
Chapter 9 Project,
    pp. 528–529
Lab Activity
Long-Term Project
Chapter Review,
    pp. 530–532
Chapter Assessment,
    p. 533
Chapter Assessment, A/B
Alternative Assessment

[3 days]

Reading Strategies

Before reading each lesson, students should examine the word problems in the exercises of each lesson. This will provide an idea of the sort of applications that justify the study of this lesson. As students read each lesson, they can try to determine the type of exercise and problem that relates to the lesson. They can then try to characterize the general method used in solving the class of problems. Students can add applications they find personally interesting to their portfolio. This will develop a portfolio that reflects their individual interests.

Visual Strategies

Direct the students to look at the pictures and associated graphs in the chapter opener. Encourage them to speculate about how mathematics might be useful in studying such phenomena. Also have them look at the graphs in the chapter and discuss how they differ from graphs in earlier chapters. Can they imagine how such functions could model the phenomena in the photographs?

Hands-on Strategies

This chapter presents the students with a number of opportunities to become actively involved with mathematics in the physical world. Geometry connections enable students to draw, or physically construct, models of concepts.

Cooperative Learning

GROUP ACTIVITIES	
Rational functions	Lesson 9.2, Example 4
Even and odd functions	Lesson 9.3, Exploration 1
Secret messages and rational functions	Eyewitness Math: How Secret is Secret?
Modeling the stock market	Chapter Project

You may wish to have students work with partners for some of the above activities. Additional suggestions for cooperative group activities are noted in the teacher's notes in each lesson.

Multicultural

The cultural connections in this chapter include references to Europe and Asia.

CULTURAL CONNECTIONS	
Europe: Halley's comet	Lesson 9.1, Introduction
Asia: Mahavira's rational function problem	Lesson 9.5, Exercise 25

Portfolio Assessment

Below are portfolio activities for the chapter listed. They are under seven activity domains that are appropriate for portfolio development.

1. **Investigation/Exploration** Even and odd functions is the focus of Exploration 1, and graphing reciprocal quadratic functions is the focus of Exploration 2 in Lesson 9.3; modeling the stock market is the objective of the Chapter Project.

2. **Applications Recommended** To be included are any of the following: construction, Lesson 9.1, Exercise 18–19; business, Lesson 9.2, Exercises 13–14; electronics, Lesson 9.4, Exercises 24–25; travel, Lesson 9.5, Exercises 29–30.

3. **Non-Routine Problems** The How Secret is Secret? problem from Eyewitness Math, and the "Modeling the Stock Market" activity from the Chapter Project present the students with non-traditional but interesting representations and applications of rational functions.

4. **Project Modeling the Stock Market** See pages 528–529. Students are asked to predict the future performance of different stocks.

5. **Interdisciplinary Topics** Students may choose from the following: geometry, Lesson 9.1, Exercise 14; physics, Lesson 9.1, Exercises 16–17; chemistry, Lesson 9.2, Exercises 25–27.

6. **Writing** Communicate exercises of the type in which a student is asked to describe basic concepts of rational equations and reciprocal functions, their characteristics, and their graphs offer excellent writing selections for the portfolio.

7. **Tools** Chapter 9 uses graphics calculators to assimilate data and model that data with various rational functions. To measure the individual student's proficiency, it is recommended that he or she complete selected worksheets from the Technology Masters.

Technology

Graphics Calculators

Graphing Rational Functions The use of the graphics calculator for discovery is especially important with rational functions. Students can quickly see the shape and form of the graphs and use these visualizations to compare the graphs of many functions. This allows the student to study and explore the behavior and fundamental concepts of rational functions. Students can see the vertical asymptotes of the functions, and they can actually see how the factors of the polynomial in the denominator affect these asymptotes.

For example, the graph of the function $y = \frac{x+2}{(x-1)(x+3)}$ is shown. Using a TI-82 friendly viewing window, the vertical asymptotes, -3 and 1, are obvious.

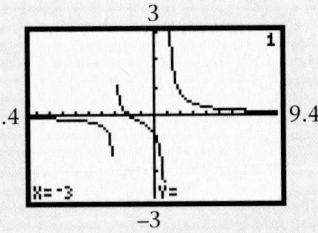

Identifying Limits Students can also comprehend fundamental ideas about limits when they trace far to the left or right of a rational function when finding the horizontal asymptotes. For example, using the function f above, the student can see that as x gets larger, y approaches 0 from above and as x gets smaller, y approaches 0 from below. Thus, the horizontal asymptote is $y = 0$.

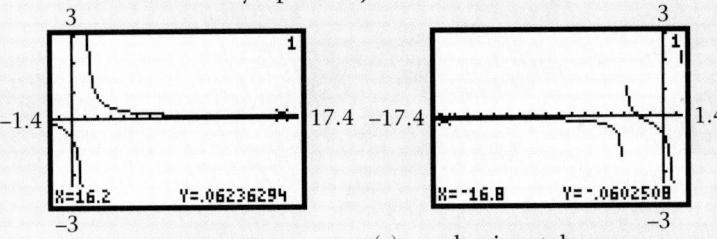

Now, the notation $f(x) = \frac{p(x)}{q(x)} \rightarrow$ horizontal asymptote as $|x| \rightarrow \infty$ has a meaning that can be visually displayed.

Calculator Drawbacks Calculators do present some drawbacks, but they can help students see the need for algebraic manipulations and factoring. When both the numerator and the denominator have a factor in common, a hole exists in the rational function, and the calculator does not show this point of discontinuity clearly. However, when the student traces along the graph of the function, the hole's x-value has an undefined y-value. For example, with the function $f(x) = \frac{x+2}{(x-1)(x+2)}$, the graph appears to be a rational function with one vertical asymptote and no other points of discontinuity. However, when the graph of the function is traced to the x-value -2, where the hole exists, no y-value is given.

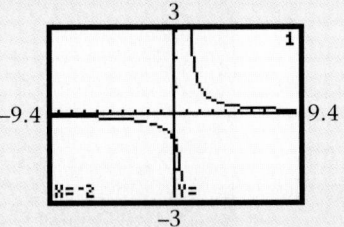

Also, vertical asymptotes do not always appear in the same way. The view is dependent on the x-values chosen for the viewing window. Using the function $f(x) = \frac{x+2}{(x-1)(x+3)}$, notice the change in the appearance of the vertical asymptotes as the x-values of the viewing window are slightly changed.

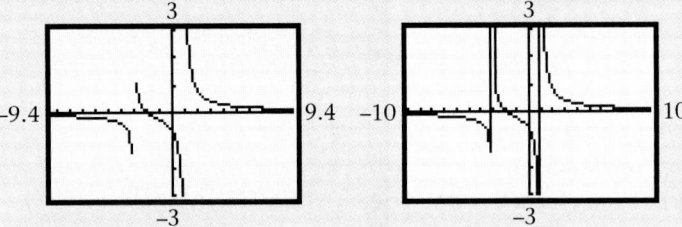

Despite these drawbacks, the graphics calculator is an effective tool for exploring and analyzing families of rational functions.

Integrated Software

f(g) Scholar™ is an integrated computer-based mathematics productivity tool that combines calculator, spreadsheet, and graphics capabilities to provide a dynamic and interactive environment for explorations in mathematics. Use *f(g) Scholar*™ for any lesson needing a spreadsheet, calculator, graphics calculator, or any combination of the three.

9 Rational Functions

CHAPTER 9

Rational Functions

ABOUT THE CHAPTER

Background Information

Rational functions are functions described as the quotient of polynomials. They are defined as long as the denominator does not equal zero. Graphs of *rational* functions have at least one break in continuity at the *x*-value that causes the denominator of the function to equal zero. Inverse variations are useful examples of rational functions.

CHAPTER RESOURCES

- Practice Masters
- Enrichment Masters
- Technology Masters
- Lesson Activity Masters
- Lab Activity Masters
- Long-Term Project Masters
- Assessments Masters
  - Chapter Assessments, A/B
  - Mid-Chapter Assessment
  - Alternative Assessment, A/B
- Teaching Transparencies
- Spanish Resources

CHAPTER OBJECTIVES

- Given an inverse variation relationship, find the constant of variation, and write the equation of variation.
- Identify the graph, and write the equation of the vertical and horizontal asymptotes of a rational function that is the quotient of two linear functions.

LESSONS

9.1 Inverse Variation

9.2 Simple Reciprocal Functions

9.3 *Exploring* Reciprocals of Polynomial Functions

9.4 Quotients of Polynomial Functions

Eyewitness Math
How Secret is Secret?

9.5 Solving Rational Equations

Chapter Project
Modeling the Stock Market

In Chapter 6, you studied polynomial functions. In this chapter, you will study functions that are reciprocals and quotients of polynomial functions. These functions, called *rational functions*, have characteristics that are quite different from polynomial functions. Rational functions can be used to model real-world phenomena such as light intensity, radio transmission, Boyle's law of gases, and electrical currents.

ABOUT THE PHOTOS

In preparing to take a photograph, a photographer does not set the camera's aperture and shutter speed to match the intensity of the light *source*. Instead, the settings are based on the amount of light that *reaches* the subject of the photograph. This amount of light is inversely proportional to the square of the distance from the light source to the subject. The expression of this relationship is one common application of a rational function.

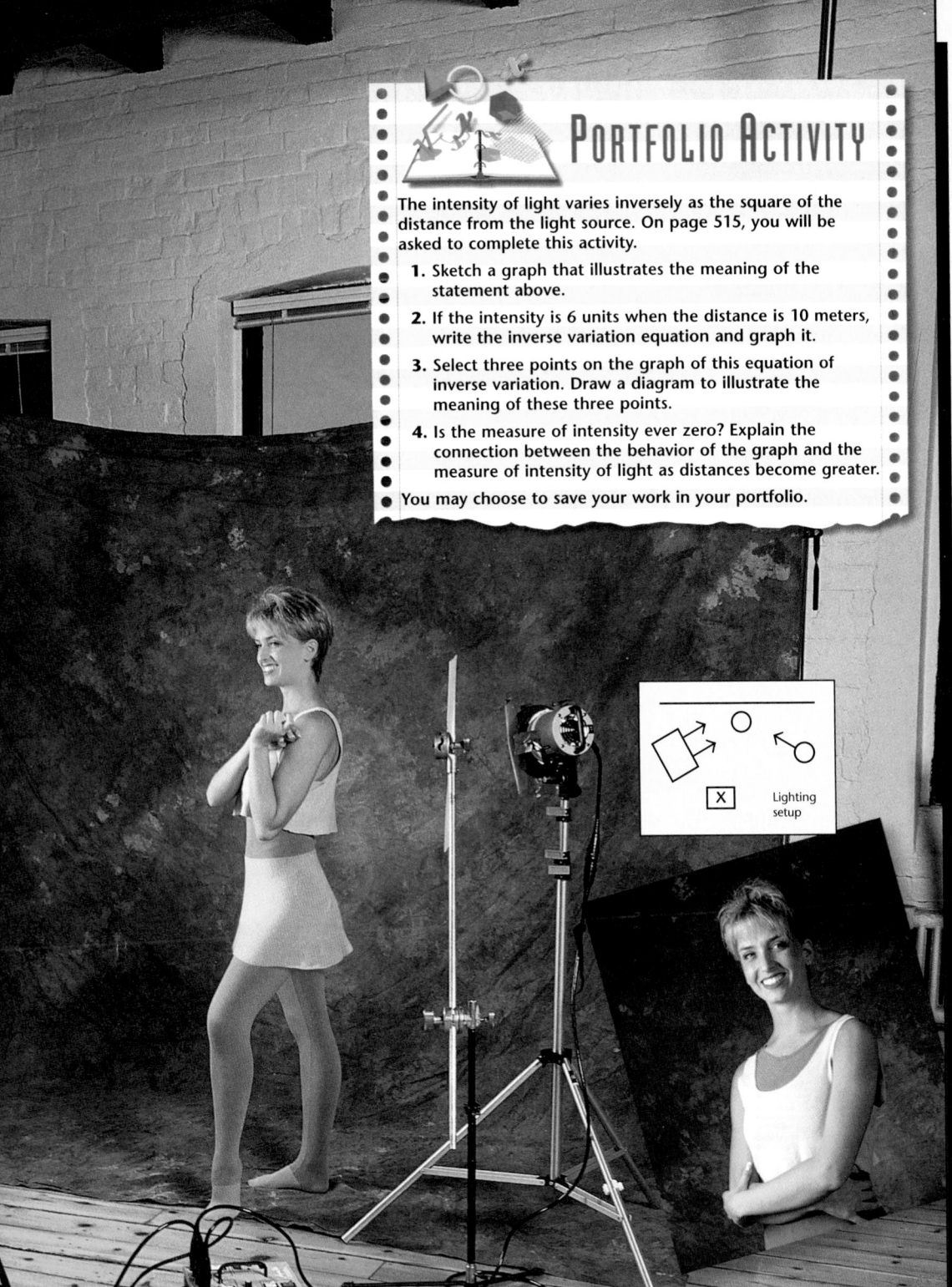

Lighting setup

About the Chapter Project

In the Chapter Project on pages 528 and 529, students will be asked to explore mathematical models for predicting the future behavior of two companies in the stock market. Students will try to develop a winning investment strategy for maximizing their return over a period of 8 years.

- Determine whether a function is even, odd, or neither.
- Graph a function that is the reciprocal of a polynomial, and find its vertical asymptotes.
- Find the vertical and horizontal asymptotes and the domain of a rational function in which the degree of the numerator is not greater than the degree of the denominator.
- Combine rational functions to get a single rational function.
- Solve equations that contain rational expressions.

Portfolio Activity

In Parts 1 and 2 of this activity, students are asked to make a graph to represent a physical phenomenon that can be expressed as a rational function. This graph should help students visualize the relationship between the variables in an inverse variation function. By using specific values for the variables, students can determine the constant of variation and write the equation.

Have students explore the relationship between the ratio of distances and the ratio of intensities by using three actual points on the graph of the function found in Part 2. Remind students that two expressions that are equal to the same constant are equal to each other.

Have students compare and contrast their graphs and sequence of steps and discuss any differences.

Objectives

• Given an inverse variation relationship, find the constant of variation, and write the equation of variation.

RESOURCES

• Practice Master   9.1
• Enrichment Master   9.1
• Technology Master   9.1
• Lesson Activity Master   9.1
• Quiz   9.1
• Spanish Resources   9.1

Assessing Prior Knowledge

Find the constant of variation in each equation of direct variation.

A. $\frac{y}{3x} = 6$ [18]

B. $\frac{x}{4y} = 5$ $\left[\frac{1}{20}\right]$

TEACH

As in direct variation, inverse variation involves the relationship between two variables. In an inverse variation, however, the absolute value of one variable decreases as the absolute value of the other variable increases.

LESSON 9.1 Inverse Variation

 *In contrast to direct variation, which you studied in Chapter 1, inverse variation describes situations in which one variable decreases as the other variable increases.*

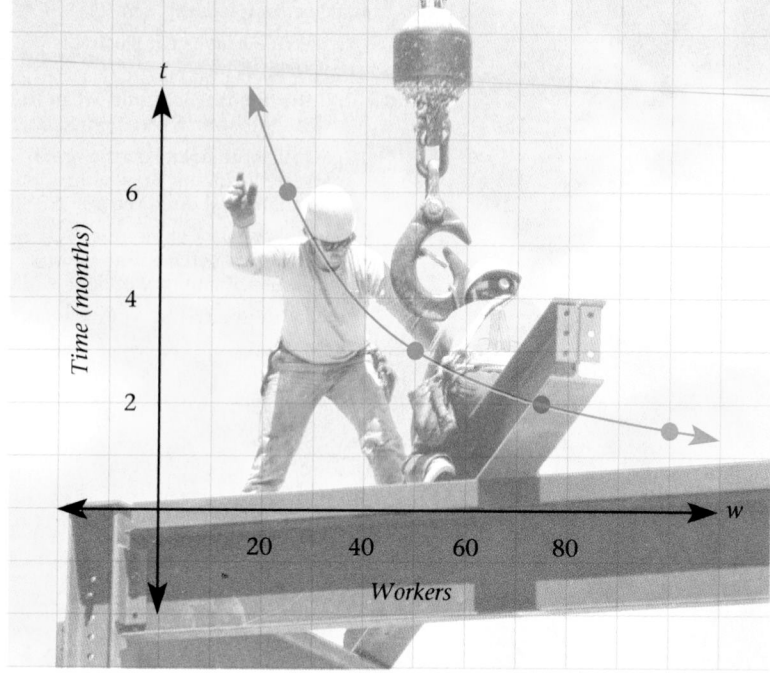

Business Management

The foreman of a construction company must hire workers to complete a certain job. He estimates that 50 workers can complete the job in approximately 3 months. The following table shows the relationship between the number of workers, w, and the time, t, in months needed to complete the job.

w	25	50	75	100
t	6	3	2	$\frac{3}{2}$

Examine the table. Notice that the product wt is always equal to 150. This relationship can be written in three different ways:

$$wt = 150, \qquad t = \frac{150}{w}, \qquad \text{or} \qquad w = \frac{150}{t}.$$

In this relationship, the variable t *varies inversely* as the variable w. As t increases, w decreases, but the product wt is always equal to the constant 150.

INVERSE VARIATION

In an **inverse variation**, y varies inversely as x. That is, $xy = k$, or $y = \frac{k}{x}$, where k is the **constant of variation**, $k \neq 0$, and $x \neq 0$.

ALTERNATIVE
teaching strategy

Hands-On Strategies
Applications in science of inverse variation relationships can be visualized using various models. To demonstrate the relationship between the variables in Boyle's law, presented in Exercises 20–21, provide students with inflated balloons. Have students press the balloons carefully between their hands. Students can feel increased pressure as the volume decreases. To demonstrate the relationship be-

tween the intensity of light and the distance from the light source, presented in the Chapter Opener and in Exercises 26–28, students can use a light source and a piece of poster board. The intensity of light will noticeably decrease as distance from the source to the poster board increases. Have students develop other demonstrations to illustrate the relationships between the variables for any of the functions in this lesson.

Suppose you know that x varies inversely as y, and y is 9 when x is 4. How can you find the value of y when x is 18?

Since y is 9 and x is 4, $xy = 36$. Thus, the constant of variation, k, is 36. Substitute 18 for x in the equation $xy = 36$ and solve for y.

$$18y = 36$$
$$y = 2$$

Thus, when x is 18, the value of y is 2.

Graphics Calculator

Use your graphics calculator to examine the graph of the inverse variation equation described above, $y = \frac{36}{x}$.

Notice that when x is 4, y is 9.

What is the value of y when x is 3?
What is the constant of variation?

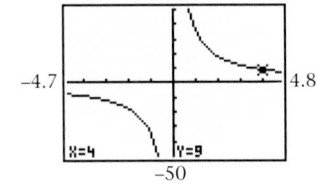

Cultural Connection: Europe The astronomer Edmund Halley, 1656–1742, asked Sir Isaac Newton, 1642–1727, to describe the shape of a planet's orbit around the sun. Newton had shown that *the sun attracts a planet with a force of gravity inversely proportional to the square of its distance from the sun*. Newton proved that the orbits of planets would be ellipses. It followed that the paths of comets would also be elliptical. This enabled Halley to predict the path and location of the comet that would later be named for him. Once considered "signs of heavenly wrath," comets were revealed by the work of Newton and Halley to be astronomical bodies that obey laws of universal gravitation.

Halley's comet most recently passed by the Earth and Sun in late 1985 and early 1986 moving at a speed of more than 80,000 miles per hour. Temperatures within the coma of Halley's Comet have been measured to be about 330 K ± 20 K.

Newton

Halley

ENRICHMENT One of the most common mathematical relationships in physics is the inverse square relationship, which is common to light, sound, and gravity. Encourage interested students to prepare a captioned photo essay on the wide range of natural phenomena that follow this pattern.

INCLUSION strategies **Using Tables** Have students construct a table of positive integer values for the inverse variation $y = \frac{1}{x}$, for $x = 1, 2, \ldots, 10$. Then have students plot these points on graph paper and describe any patterns they observe orally.

ongoing ASSESSMENT

y is 12 when x is 3; k is 36

TEACHING tip

Remind students that x cannot equal zero because division by zero is undefined. The constant, k, cannot equal zero for a different reason. If k is zero, the function becomes the constant function $y = 0$, not an inverse variation function.

Cooperative Learning

Have students work in groups of six. They will need packs of 3×5 inch index cards. Have three students count the cards while one of the other students times them. Record the time in seconds. Now, have 2 students count the cards while another student times them and records the time. Repeat this counting and timing procedure three more times with 1, 4, and 5 students counting the cards. Have students make a graph of their data letting the horizontal axis represent the number of students counting and the vertical axis represent the time it takes to count. Students should see that it takes less time to count the cards as more students participate in counting. Have students use the statistics feature on their graphics calculators to determine this inverse variation function. Then have each group compare their graph and equation with the other groups. Discuss why outliers might occur, and other factors that could affect the inverse variation equations and graphs.

Alternate Example 1

The distance from Earth to the nearest star, Alpha Centauri, is 4.3 light-years, which means that a spacecraft traveling at the speed of light would take 4.3 years to get to Alpha Centari. Complete the following table for a trip from Earth to Alpha Centauri. Let the rate, r, represent the fraction of the speed of light. Let t be the time in years.

r	1	$[\frac{1}{2}]$	$\frac{1}{4}$	$[\frac{1}{8}]$	$\frac{1}{10}$
t	4.3	8.6	[17.2]	34.4	[43]

Use the table to write the inverse variation equation for this trip.

$$\left[t = \frac{4.3}{r} \right]$$

Use the statistical regression feature on your graphics calculator to check your inverse variation equation.

[

```
PwrReg
y=a*x^b
a=4.3
b=-1
r=-1
```
]

TEACHING *tip*

Have students graph the following inverse variation functions on one coordinate grid.

$$y = \frac{1}{x}, y = \frac{2}{x}, y = \frac{5}{x}$$

Have students compare the graphs. Then, have students repeat the procedure with $y = -\frac{1}{x}, y = -\frac{2}{x}, y = -\frac{5}{x}$.

Students should notice that the graphs of the first group of functions are in the first and third quadrants. The others are in the second and fourth quadrants.

EXAMPLE 1

Travel The equation $d = rt$ gives the distance traveled in time t at a constant rate r. The time required to travel a given distance in a car varies inversely as the car's average rate of speed.

Graphics Calculator

A Write the inverse variation equation, and construct a table of values for a 200-mile trip.

B Use the statistical regression mode of your graphics calculator to check your answer.

Solution

A The distance, 200 miles, is the constant of variation, k.

Thus, the inverse variation equation is $t = \frac{200}{r}$, where t is time in hours, and r is the average rate of speed in miles per hour and cannot equal zero.

Make a table of values for different average speeds.

r (mph)	20	40	60	80	100
t (hours)	10	5	$3\frac{1}{3}$	$2\frac{1}{2}$	2

B Graph the points represented in the table on your graphics calculator.

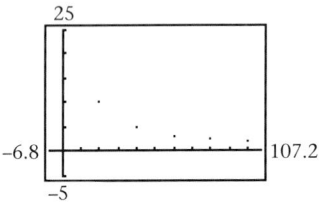

STATISTICS *Connection*

Calculate the power regression for this data.

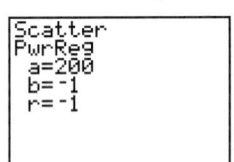 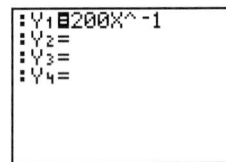

Notice that a is the constant of variation, 200, and b is the power of x, −1. The inverse variation equation $y = ax^b$, is $y = 200x^{-1}$, or $y = \frac{200}{x}$.

The correlation coefficient, r, is −1. Therefore, the points represented in the table fit the graph of the inverse variation function $y = \frac{200}{x}$ exactly.

Since the power regression equation fits exactly, you can use the trace feature to check the values in the table.

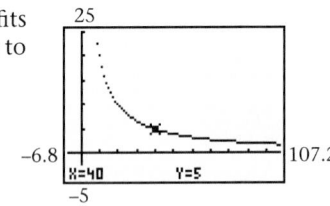

RETEACHING *the lesson*

Hands-On Strategies Have students work in groups to design and conduct a fulcrum experiment similar to Exercises 16 and 17. Materials could include two items of differing weights, a meter stick as a lever, and a fulcrum made from a piece of cardboard that is bent in the middle. Have students use a scale to weigh one of the objects. After balancing the objects on their lever and determining the length of the objects from the fulcrum, have students calculate the weight of the second object. Use the scale to confirm this weight. This experiment can be conducted several times using different items each time. Have students write the inverse variation equation for each trial in the experiment. How do these equations compare?

How long would it take each vehicle to travel this 20-mile route when each travels at a different rate?

TEACHING tip

Have students assign reasonable average speeds to the vehicles over the 20-mile route shown. For example, the following could be used as sample average speeds: bike, 8 mph; sports car 40 mph; school bus, 27 mph; and the dump truck, 22 mph. Have students find the time each vehicle will take. Have them choose other reasonable speeds and compare the new times with their other values.

You can see from the table in Example 1 that as the car's average speed increases, the time it takes to travel a given distance decreases. In other words, the *larger r* becomes, the *smaller t* becomes. How is this relationship different from direct variation?

In any inverse variation table, you can find the constant of variation *and* any missing term. Examine the following table.

x	2	5	10	40
y	50	20	10	?

The product xy is 100, the constant of variation. Let y_1 represent the missing y-value. Then $40y_1 = 100$, or $y_1 = 2.5$.

CRITICAL Thinking

If x_1 and y_1 are any pair of values in an inverse variation table, what is their relationship to the product xy?

EXAMPLE 2

Electronics

Ohm's law states that with a constant voltage, the current, I, flowing through a wire varies inversely as the resistance, R, of the wire. If the current is 6 amperes when the resistance is 20 ohms, find the constant of variation (voltage), and write the equation of inverse variation.

Solution

If I varies inversely as R, then $IR = k$. Substitute 6 for I and 20 for R.

$$IR = k$$
$$(6)(20) = k$$
$$120 = k$$

Thus, the variation equation is $I = \frac{120}{R}$. ❖

Try This   What is the current when the resistance is 8 ohms? What is the voltage?

A ongoing ASSESSMENT

In direct variation, as one variable gets larger, the other variable gets larger. In the inverse variation equation $t = \frac{200}{r}$, as r gets larger, t gets smaller.

CRITICAL Thinking

The same, $x_1y_1 = xy$

Alternate Example 2

In an electrical circuit, if voltage is constant, the current, I, varies inversely with the resistance, R, of the wire. Find the constant of variation, and complete the following table of values.

R	5	10	12	20	30	60
I	12	[6]	[5]	[3]	[2]	[1]

$[k = 60]$

A ongoing ASSESSMENT

Try This

$I = 15$ amperes; $V = k = 120$ volts

Alternate Example 3

Suppose y varies inversely as the square of x, and y is 10 when x is 2.

Find the constant of variation, and write the equation of inverse variation. $[k = 40, y = \frac{40}{x^2}]$

CRITICAL
Thinking

The value of y varies directly with x and inversely with z. So, as x increases (assuming that z remains constant), y increases. As z increases (assuming that x remains constant), y decreases.

ASSESS

Sometimes y varies inversely as a *power* of x, or $x^n y = k$ and $y = \frac{k}{x^n}$. Given this type of inverse variation, you can still find the constant of variation from just one pair of terms.

EXAMPLE 3

Suppose y varies inversely as the cube of x, and y is 2 when x is 4. Find the constant of variation, and write the equation of inverse variation.

Solution▶

If y varies inversely as the cube of x, then $x^3 y = k$. Substitute 4 for x and 2 for y.

$$x^3 y = k$$
$$(4)^3(2) = k$$
$$128 = k$$

Thus, the equation of inverse variation is $x^3 y = 128$, or $y = \frac{128}{x^3}$. ❖

CRITICAL
Thinking

Let x, y, and z be variables and k be a constant. Explain how the equation $y = k\left(\frac{x}{z}\right)$ combines direct and inverse variation.

EXERCISES & PROBLEMS

Communicate 〰

1. Write an inverse variation equation with a constant of variation of 10.

2. If k is a constant, which equation does *not* express an inverse variation? Explain.

$$y = \frac{k}{x^2} \qquad y = \frac{k}{x} \qquad y = \frac{x}{k} \qquad xy = k$$

3. In the following table, y varies inversely as x. Explain how to find the constant of variation using any pair of values.

x	3	6	9	12
y	4	2	$1\frac{1}{3}$	1

4. How can you use the statistical regression feature on your graphics calculator to verify that a table of values represents an inverse variation function?

5. Is the equation $y = \frac{k}{\sqrt{x}}$ an inverse variation equation? Explain.

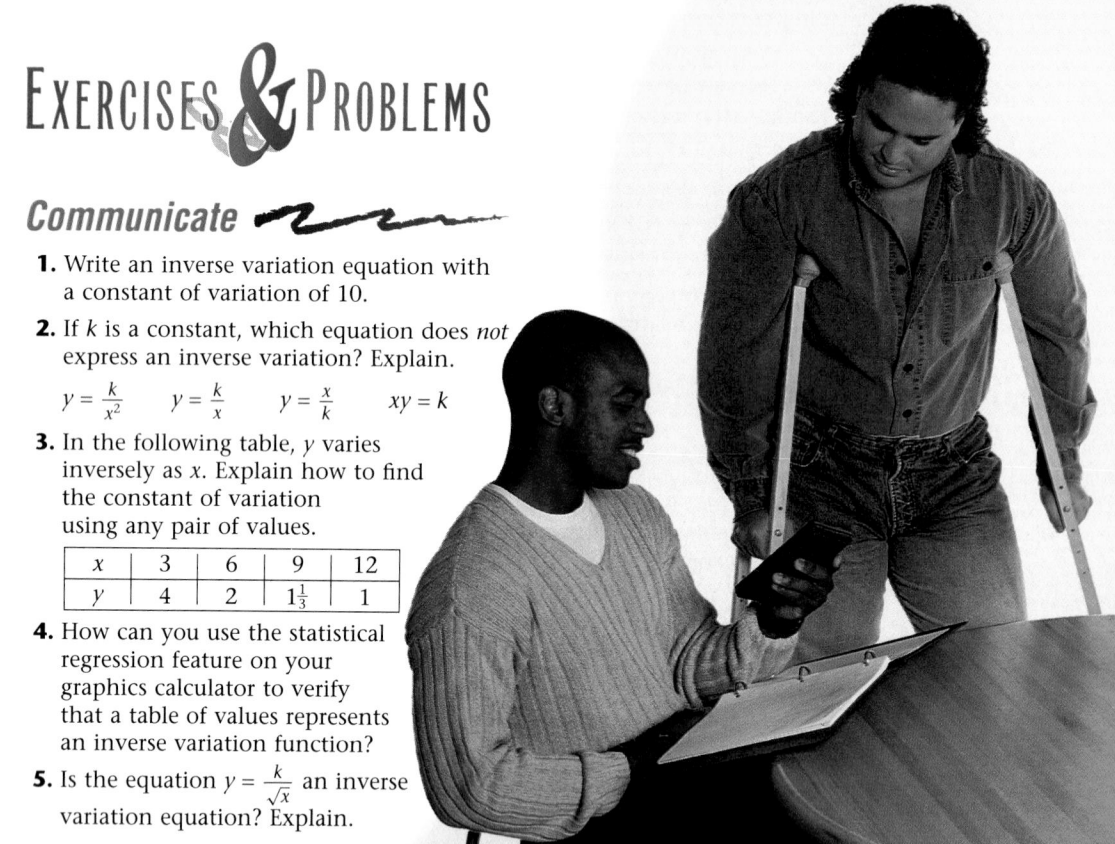

Practice & Apply

In the following table, y varies inversely as the square of x.
Complete Exercises 6–9.

x	5	10	?
y	4	?	0.25

6. Find the constant of variation. $k = 100$ **7.** Write the equation of inverse variation. $y = \frac{100}{x^2}$

8. Complete the table. $x = \pm 20,\ y = 1$

9 Check your answers to Exercises 6–8 using the statistical regression feature on your calculator. What equation and value of r do you find?

10. In the following table, y varies inversely as x. Complete the table.

x	8	10	? 15	a	$8a$
y	5	? 4	$2\frac{2}{3}$	$?\frac{40}{a}$	$?\frac{5}{a}$

Find the constant of variation, and write the equation of inverse variation.

11. y varies inversely as x, and y is 20 when x is 3. $k = 60,\ y = \frac{60}{x}$

12. y varies inversely as the square of x, and y is 12 when x is 3. $k = 108,\ y = \frac{108}{x^2}$

13. y varies inversely as the cube of x, and y is 2 when x is 3. $k = 54,\ y = \frac{54}{x^3}$

14. **Geometry** If the area of a rectangle is held constant, the length varies inversely as the width. Write the inverse variation equation, and construct a table of values for a rectangular area of 24 square meters. $l = \frac{24}{w}$

15. Cycling A bicycle wheel with a 26-inch diameter takes 10 revolutions to go a certain distance. How many revolutions will a wheel with a 20-inch diameter complete while covering the same distance? 13 revolutions

Physics If a seesaw is balanced, each person's distance from the fulcrum varies inversely as his or her weight.
A 120-pound person sits 5 feet away from a seesaw fulcrum.

16. Where must a 100-pound person sit to balance the seesaw?
6 ft from the fulcrum

17. What is the constant of variation?
$k = 600$

Construction The time required to do a certain carpentry job varies inversely as the number of people working. It takes 6 hours for 9 carpenters to complete a job.

18. How long would it take 12 carpenters to complete the job? 4.5 hr

19. Write the equation of inverse variation, and sketch its graph.

9. $y = 100x^{-2} = \frac{100}{x^2},\ r = -1$

19.

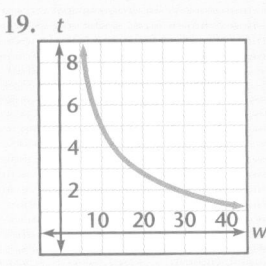

$t = \frac{54}{w}$ where w is the number of workers

22. Since $y_1 = \dfrac{k}{x_1}$ and $y_2 = \dfrac{k}{x_2}$ for the same value of k, then $x_1 y_1 = x_2 y_2$, or $\dfrac{y_1}{y_2} = \dfrac{x_2}{x_1}$

Chemistry According to Boyle's law, the volume, V, of a fixed amount of gas (at constant temperature) is inversely proportional to the pressure, P, of the gas. **Complete Exercises 20 and 21.**

20. If the gas volume is 250 cubic feet when the pressure is 60 pounds per square inch, find the constant of variation, and write the equation of inverse variation. $k = 15{,}000;\ V = \dfrac{15{,}000}{P}$

21. What pressure would be necessary to maintain the gas volume at 200 cubic feet? $75\ \text{lb/in}^2$

22. Suppose y varies inversely with x. Show that $\dfrac{y_1}{y_2} = \dfrac{x_2}{x_1}$, where $(x_1,\ y_1)$ and $(x_2,\ y_2)$ are ordered pairs of the relation.

23. Suppose y varies inversely as the square of x. If $(x_1,\ y_1)$ and $(x_2,\ y_2)$ are ordered pairs of the relation, what expression of x_1 and x_2 is equal to $\dfrac{y_1}{y_2}$? $\dfrac{x_2^2}{x_1^2}$

Physics According to Newton's law of gravitation, the weight of an object varies inversely as the square of the distance from the object to the center of the Earth (the radius of Earth is approximately 4000 miles).

24. If an astronaut weighs 125 pounds on Earth, what will the astronaut weigh at a height of 50 miles above the Earth's surface before going into orbit? $121.93\ \text{lb}$

25. If an astronaut weighs 160 pounds when 70 miles above the Earth's surface, how much does the astronaut weigh on Earth? $165.65\ \text{lb}$

Optics The intensity of light, I, measured in lux, from a light bulb varies inversely as the square of the distance, d, from the light bulb. Suppose I is 100 lux when d is 4.5 meters. **Complete Exercises 26–28.**

$506.25\ \text{lux}$
26. Find the intensity at a distance of 2 meters.

27. Find the intensity at a distance of $5.06\ \text{lux}$ 20 meters.

28. Sketch the graph of the inverse variation equation. $I = \dfrac{2025}{d^2}$

29. Suppose y varies inversely as x. What is the effect on y when x is tripled? y is divided by 3

30. Suppose y varies inversely as x. What is the effect on y when x is multiplied by c? y is divided by c

31. Show that if x varies inversely as y, and y varies inversely as z, then x varies directly as z.

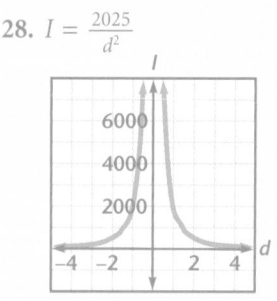

28. $I = \dfrac{2025}{d^2}$

31. If $x = \dfrac{k_1}{y}$ and $y = \dfrac{k_2}{z}$, then $y = \dfrac{k_1}{x}$ or $\dfrac{k_1}{x} = \dfrac{k_2}{z}$. Multiply both sides by xy, $x = \dfrac{k_1}{k_2} z$. Since $\dfrac{k_1}{k_2}$ is just some constant, let $k = \dfrac{k_1}{k_2}$ so that $x = kz$. Therefore, x varies directly as z.

Medicine The intensity of radiation varies inversely as the square of the patient's distance from the source. Suppose the intensity is 3 milliroentgens per hour at 10 meters and 12 milliroentgens per hour at 5 meters.

32. Find the constant of variation. $k = 300$

33. Write the inverse variation equation. $I = \dfrac{300}{d^2}$

Economics When the price of gold goes up, the prices of stocks tend to go down. Some economists believe that stock prices vary inversely with the price of gold.

34. Plot the measure of the price of stocks (Dow Jones Industrial Average) versus the price of gold for any one-month period. Is there an inverse relationship?
Answers may vary

35. Write an inverse variation equation that best fits your data. Do you think these two variables are exactly inversely related? Explain.
Answers may vary

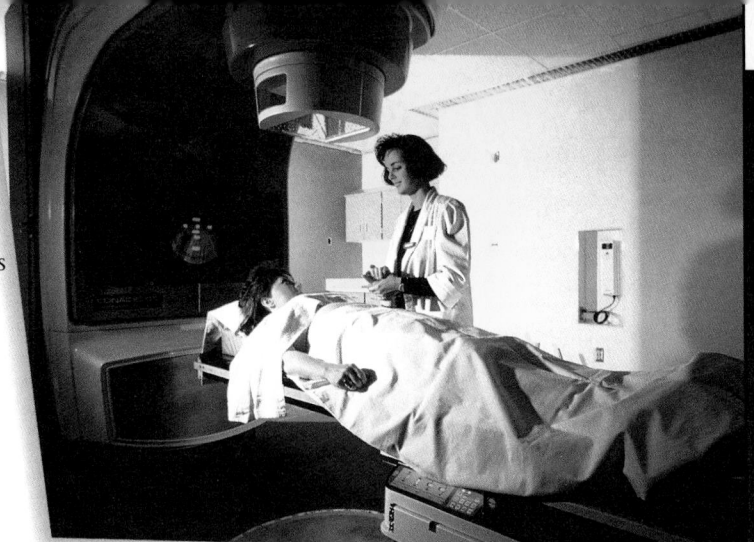

Tumors are sometimes treated with gamma rays from a radioactive source such as cobalt 60.

Look Back

Use your graphics calculator to graph each parametric system. Then combine each system into one linear function. Graph the function to check your result. **[Lesson 3.6]**

36 $\begin{cases} x(t) = 3t \\ y(t) = t - 2 \end{cases}$ $y = \dfrac{x}{3} - 2$

37 $\begin{cases} x(t) = t + 4 \\ y(t) = 2t + 1 \end{cases}$ $y = 2x - 7$

38 $\begin{cases} x(t) = 3t + 1 \\ y(t) = -t \end{cases}$ $y = -\dfrac{x}{3} + \dfrac{1}{3}$

39 Solve the system of equations using matrix algebra. **[Lesson 4.6]** $\begin{cases} 2y + x = 8 \\ y - 2x = 3 \end{cases}$ $\left(\dfrac{2}{5}, 3\dfrac{4}{5}\right)$, or (0.4, 3.8)

40. Find the discriminant in the quadratic function $f(x) = 2x^2 - 3x - 1$. 17 **[Lesson 5.6]**

41. Name three points on the unit circle. **[Lesson 8.2]**

Answers may vary. Examples are $\left(\dfrac{\sqrt{2}}{2}, \dfrac{\sqrt{2}}{2}\right)$, $\left(-\dfrac{1}{2}, -\dfrac{\sqrt{3}}{2}\right)$, (0, 1)

In general, (cos x, sin x) where $0° \le x \le 360°$

Look Beyond

42. Construct a table for the function $f(x) = \dfrac{1}{x + 2}$. Find the rate of change, $\dfrac{f(x_2) - f(x_1)}{x_2 - x_1}$, between $x_1 = 2$ and $x_2 = 3$. Do the same for other pairs of points. What pattern do you notice?

42.

x	1	2	3	4	5	6
$y = \dfrac{1}{x + 2}$	$\dfrac{1}{3}$	$\dfrac{1}{4}$	$\dfrac{1}{5}$	$\dfrac{1}{6}$	$\dfrac{1}{7}$	$\dfrac{1}{8}$

$$\dfrac{f(x_2) - f(x_1)}{x_2 - x_1} = \dfrac{\frac{1}{5} - \frac{1}{4}}{3 - 2} = -\dfrac{1}{20}$$

$$\dfrac{\frac{1}{4} - \frac{1}{3}}{2 - 1} = -\dfrac{1}{12}, \quad \dfrac{\frac{1}{6} - \frac{1}{5}}{4 - 3} = -\dfrac{1}{30},$$

$$\dfrac{\frac{1}{7} - \frac{1}{6}}{5 - 4} = -\dfrac{1}{42}$$

The rate of change of the function f approaches zero as x increases.

Look Beyond

This problem introduces students to the concept of increasing and decreasing functions and the connection between a non-linear function and the slope, or rate of change, of that function.

Technology Master

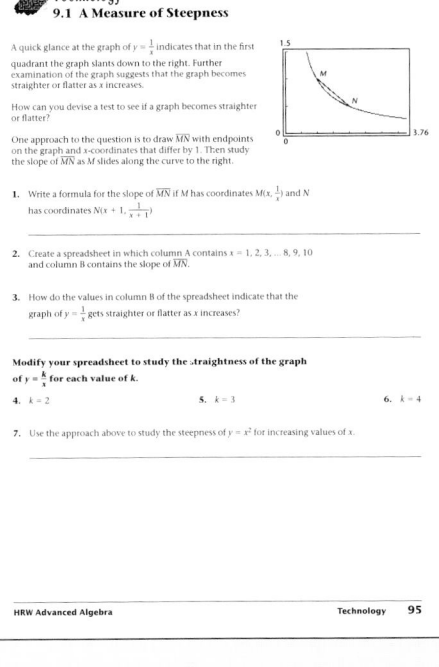

NAME _____ CLASS _____ DATE _____

Technology
9.1 A Measure of Steepness

A quick glance at the graph of $y = \frac{1}{x}$ indicates that in the first quadrant the graph slants down to the right. Further examination of the graph suggests that the graph becomes straighter or flatter as x increases.

How can you devise a test to see if a graph becomes straighter or flatter?

One approach to the question is to draw $\overline{MN}$ with endpoints on the graph and x-coordinates that differ by 1. Then study the slope of $\overline{MN}$ as M slides along the curve to the right.

1. Write a formula for the slope of $\overline{MN}$ if M has coordinates $M(x, \frac{1}{x})$ and N has coordinates $N(x + 1, \frac{1}{x + 1})$

2. Create a spreadsheet in which column A contains $x = 1, 2, 3, \dots 8, 9, 10$ and column B contains the slope of $\overline{MN}$.

3. How do the values in column B of the spreadsheet indicate that the graph of $y = \frac{1}{x}$ gets straighter or flatter as x increases?

Modify your spreadsheet to study the straightness of the graph of $y = \frac{k}{x}$ for each value of k.

4. $k = 2$ 5. $k = 3$ 6. $k = 4$

7. Use the approach above to study the steepness of $y = x^2$ for increasing values of x.

HRW Advanced Algebra   Technology   95

Objectives

• Identify the graph, and write the equation of the vertical and horizontal asymptotes of a rational function that is the quotient of two linear functions.

RESOURCES

• Practice Master   9.2
• Enrichment Master   9.2
• Technology Master   9.2
• Lesson Activity Master 9.2
• Quiz   9.2
• Spanish Resources   9.2

Assessing Prior Knowledge

Evaluate each fraction for the given values of x.

A. $\frac{1}{x+3}$ when x is $-7, -5, -3,$ $-1,$ and 1

$[-\frac{1}{4}, -\frac{1}{2}, \text{undefined}, \frac{1}{2}, \frac{1}{4}]$

B. $\frac{3x+1}{2x-3}$ when x is $-1, 1, \frac{3}{2},$ $2,$ and 4

$[\frac{2}{5}, -4, \text{undefined}, 7, \frac{13}{5}]$

TEACH

 Any function that is defined as the quotient of two linear functions can be expressed as a simple reciprocal function plus some constant. These functions are rational functions and are defined for all x-values that do not make the denominator of the fraction zero.

LESSON 9.2 Simple Reciprocal Functions

 Reciprocals of linear functions model real-world applications of subjects as diverse as chemical engineering and marketing.

A rational *number* is the quotient of two integers. Similarly, a rational *function* is the quotient of two polynomials.

RATIONAL FUNCTION

The function $f(x) = \frac{p(x)}{q(x)}$, where both the numerator, $p(x)$, and denominator, $q(x)$, are polynomials, is called a **rational function.** The domain of a rational function is all real numbers, x, such that the denominator $q(x) \neq 0$.

Reciprocal Functions

Recall that the graph of the linear function $g(x) = x$ is a line through the origin that divides the first and the third quadrants in half. It is known as the identity function. What is the shape of the graph of the function defined by the reciprocal of g, $f(x) = \frac{1}{g(x)}$, or $\frac{1}{x}$?

Examine the graph of the *basic reciprocal function*, $f(x) = \frac{1}{x}$.

Notice that when x is zero, $f(x)$ is undefined. Therefore, the domain of f is all real numbers except zero. Recall from Chapter 7 that an asymptote is a line that a graph approaches but never touches. Observe from the graph that the y-axis, or the line $x = 0$, is a *vertical asymptote* for the graph of $f(x) = \frac{1}{x}$.

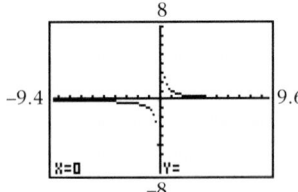

Also, $f(x) = \frac{1}{x}$ can never equal zero. Therefore, the range of f is all real numbers except zero. The x-axis, or the line $y = 0$, is a *horizontal asymptote* for the graph of f.

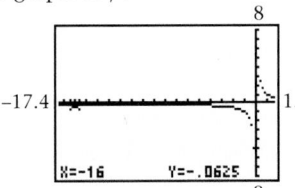   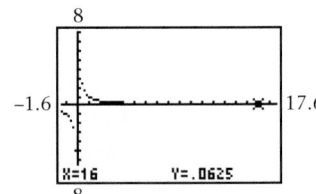

Using the trace feature, you can see that the value of y never reaches zero.

ALTERNATIVE teaching strategy

Using Visual Models On separate pieces of graph paper, have students graph a linear function, such as $f(x) = x - 2$, and the reciprocal of this linear function. Have students use a pencil (representing the vertical-line test) to show that

the graph of the linear function is continuous and can be traced with no break in the graph. However, the graph of the reciprocal function has a break in its graph where the function is undefined. Emphasize the importance of locating this break, or point of discontinuity, on the graph of a reciprocal function.

The graph of $f(x) = \frac{1}{x}$ is called a *hyperbola*. You will learn more about hyperbolas in the next chapter. How does the function $f(x) = \frac{1}{x}$ compare with the definition of inverse variation in Lesson 9.1?

The graph of $f(x) = \frac{1}{x}$ is symmetric with respect to the line $y = x$. Knowing the position of the asymptotes and the line of symmetry makes it easy to sketch the graph of f.

Transformations of Reciprocal Functions

Variations of the basic reciprocal function, $f(x) = \frac{1}{x}$, can be investigated using transformations.

TRANSFORMATIONS
Connection

EXAMPLE 1

Compare the graph of $g(x) = \frac{1}{x-2}$ with the graph of the basic reciprocal function $f(x) = \frac{1}{x}$.

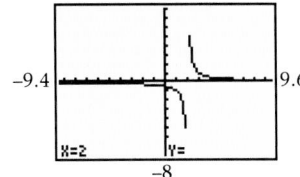

Solution▸

Graphics Calculator

The graph of $g(x) = \frac{1}{x-2}$ is a translation of the graph of $f(x) = \frac{1}{x}$ shifted 2 units to the right.

Since the graph is shifted 2 units to the right, the vertical asymptote is also shifted 2 units to the right. Thus, the vertical asymptote of $g(x) = \frac{1}{x-2}$ is $x = 2$. The horizontal asymptote is still $y = 0$. ❖

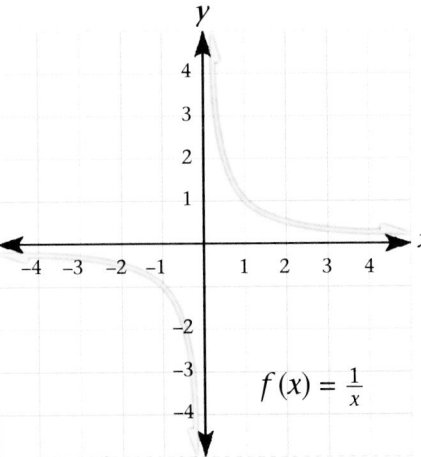

$f(x) = \frac{1}{x}$

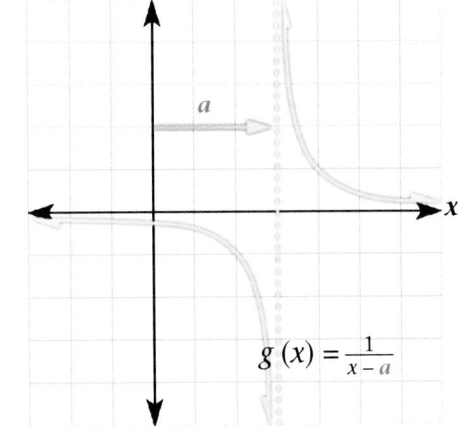

$g(x) = \frac{1}{x-a}$

CRITICAL
Thinking

Use the two graphs above to describe the effect of the variable a on the function $g(x) = \frac{1}{x-a}$.

TEACHING *tip*

Have students explain why division by zero is undefined. One possible explanation is as follows: If $\frac{x}{0} = y$ for x and $y \neq 0$, then $x = 0 \cdot y = 0$, which contradicts the fact that $x \neq 0$.

Alternate Example 1

Use your graphics calculator to compare the graph of $h(x) = \frac{1}{x+3}$ with the graph of the basic reciprocal function, $f(x) = \frac{1}{x}$. [h **is a horizontal translation of f shifted 3 units to the left. The vertical asymptote of f is $x = 0$, so the vertical asymptote of h is $x = -3$. The horizontal asymptotes stay the same.**]

CRITICAL
Thinking

In general, if $a > 0$, a shifts the graph of f horizontally to the right by a units; if $a < 0$, a shifts the graph of f horizontally to the left by $|a|$ units. The vertical asymptote of f is $x = 0$, so the vertical asymptote of g is $x = a$.

Use Transparency ▶ **33**

ENRICHMENT Have students who have mastered graphing rational functions use the following table to test several functions for symmetry with respect to the x-axis, the y-axis, and the origin. First, have the students write three rational functions and graph each function. Name the vertical asymptote for each function. Then, test each function for symmetry.

Replace	If the shape of the graph is	the function is symmetric with respect to the
x by $-x$	unchanged	y-axis
y by $-y$	unchanged	x-axis

Have students write a conclusion summarizing their findings.

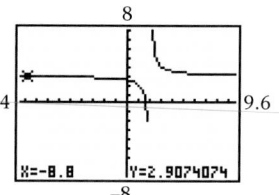

TRANSFORMATIONS
Connection

EXAMPLE 2

Compare the graph of $h(x) = \frac{1}{x-2} + 3$ with the graph of the basic reciprocal function, $f(x) = \frac{1}{x}$.

Solution►

The graph of $h(x) = \frac{1}{x-2} + 3$ is a translation of $f(x) = \frac{1}{x}$ shifted 2 units to the right and 3 units upward.

Since the graph is shifted 2 units to the right, the vertical asymptote is $x = 2$. Since the graph is shifted 3 units up, the horizontal asymptote is $y = 3$. ❖

Use the graphs below to describe how the horizontal and vertical asymptotes of $h(x) = \frac{1}{x-a} + b$ are related to the constants *a* and *b*.

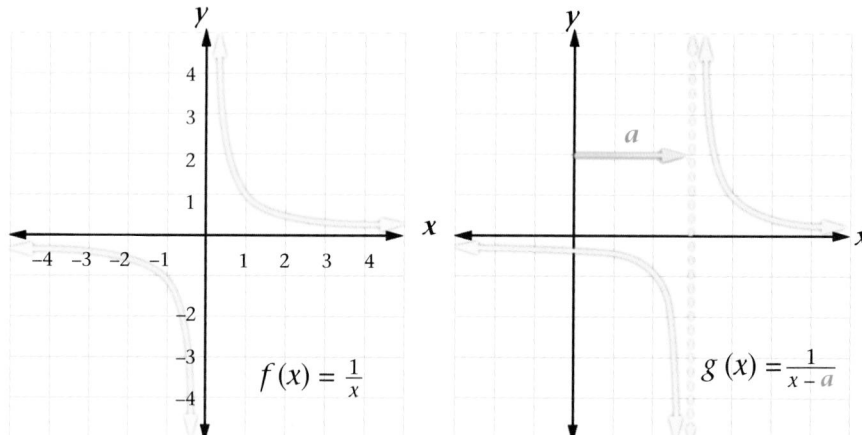

$$f(x) = \frac{1}{x}$$

$$g(x) = \frac{1}{x-a}$$

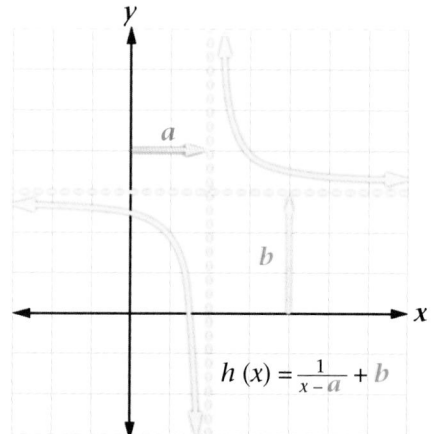

$$h(x) = \frac{1}{x-a} + b$$

Using Visual Models
Have each student place a transparency sheet over a piece of graph paper that has the axes labeled. Have them graph the rational function $f(x) = \frac{1}{x-a} + b$ on the transparency for $a = 2$ and $b = 3$. Then have students work in pairs to move the transparency to represent changes in *a* and *b*. Students may check their changes and the resulting graphs using their graphics calculators.

A function that is the quotient of two linear functions has a graph that is similar in shape to the graph of a reciprocal function.

EXAMPLE 3

Graph $j(x) = \frac{3x-5}{x-2}$, and use the graph to find the horizontal and vertical asymptotes.

Graphics Calculator

Solution▸

When x is 2, j is not defined. So, the vertical asymptote is $x = 2$.

Use the trace feature to find the horizontal asymptote. The value of j approaches 3 but never reaches it. Therefore, the horizontal asymptote is $y = 3$. ❖

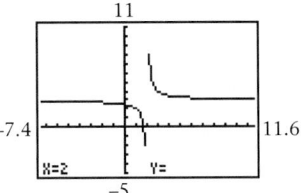

Notice that the graphs in Examples 2 and 3 are the same. That is, $j(x) = \frac{3x-5}{x-2}$ and $h(x) = \frac{1}{x-2} + 3$ are equivalent functions. When you divide $3x - 5$ by $x - 2$, the result is a quotient of 3 and a remainder of $\frac{1}{x-2}$.

The function, j, can be written in the form $j(x) = \frac{1}{x-2} + 3$. In this form, you can determine, without graphing, that the graph is a translation of the basic reciprocal function, $y = \frac{1}{x}$, by 2 units to the right and 3 units upward.

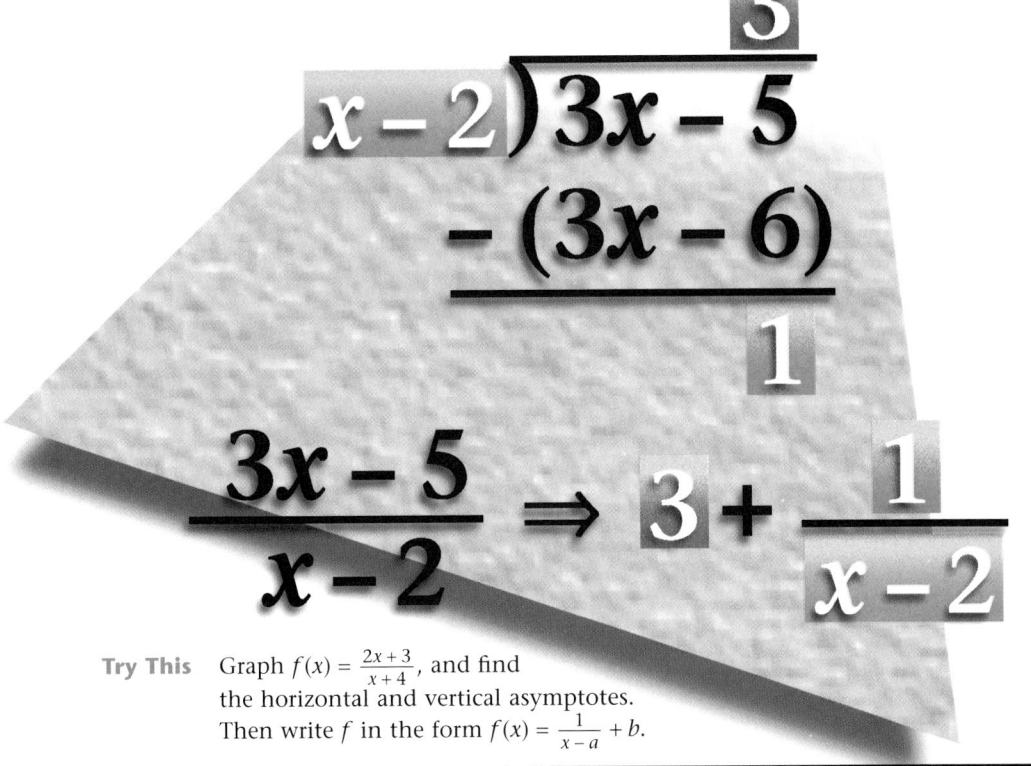

Try This Graph $f(x) = \frac{2x+3}{x+4}$, and find the horizontal and vertical asymptotes. Then write f in the form $f(x) = \frac{1}{x-a} + b$.

Alternate Example 3

Graph $k(x) = \frac{5x+1}{x+1}$, and use the graph to find the horizontal and vertical asymptotes.

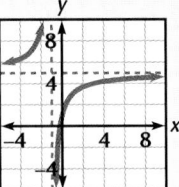

[**vertical,** $x = -1$; **horizontal,** $y = 5$]

A ongoing SSESSMENT

Try This

horizontal asymptote is $y = 2$; vertical asymptote is $x = -4$; $f(x) = -5\left(\frac{1}{x+4}\right) + 2$

TEACHING *tip*

Graphics calculators can be deceiving in the way they display the vertical asymptotes of rational functions. Refer to your *HRW Technology Handbook* for instructions on the best way to view vertical asymptotes on each calculator. Caution students to use a friendly viewing window and trace, or zoom on any suspected asymptotes. Explain to students the different ways calculators display the coordinates of a point for an x-value that is not in the domain (vertical asymptote).

RETEACHING the lesson

Using Visual Models Have students use graph paper to sketch several different rectangles, all with an area of 24 square units. Have them use their sketches to predict what the limits (asymptotes) are of the length and width. Students should be able to predict that as either length or width increases, the other variable decreases until it approaches, but does not reach, zero. Have students graph the function $f(x) = \frac{24}{x}$ and compare the asymptote of this graph with their predictions.

Applications of Reciprocal Functions

Reciprocal functions can be used to model mixture problems.

EXAMPLE 4

Chemistry Pure water is mixed with 10 milliliters of a 50% acid solution.

Ⓐ Write a function for the concentration, C, of acid in the mixture in terms of the amount, x, of pure water added.

Ⓑ Graph the concentration function, and explain what the behavior of the graph represents.

Solution▶

Ⓐ The acid concentration, C, is the original amount of acid in 10 milliliters of solution, 50%, divided by the total amount of the solution, 10 milliliters plus the added water, 10 milliliters + x.

$$C(x) = \frac{50\% \text{ of } 10 \text{ mL}}{10 \text{ mL} + \text{water added}}$$
$$= \frac{0.5(10)}{10 + x} = \frac{5}{10 + x}$$

Thus, $C(x) = \frac{5}{10 + x}$, where x is the amount of pure water added and $C(x)$ is the concentration of acid.

Ⓑ The graph of C for the realistic domain of $x > 0$ is shown.

The concentration of acid decreases as the amount of water added increases, but there is always some acid in the mixture. ❖

Water added (mL)

EXERCISES & PROBLEMS

Communicate

Describe the graph of each function. Then graph the function to check your answer.

1. $f(x) = \frac{1}{x+2}$ **2.** $f(x) = \frac{1}{x} - 4$ **3.** $f(x) = \frac{1}{x-5} + 2$ **4.** $f(x) = \frac{1}{x+1} - 6$

For Exercises 5 and 6, let $f(x) = \frac{2x-5}{x-3}$.

5. Use polynomial division to write f in the form $f(x) = \frac{1}{x-a} + b$.

6. Use your results from Exercise 5 to describe the graph of f, including the vertical and horizontal asymptotes. Then graph $f(x) = \frac{2x-5}{x-3}$ to check your answer.

Practice & Apply

Use your graphics calculator to graph each function. Identify the domain and the range for each function, and find its asymptotes.

7. $f(x) = \frac{1}{x-4}$ **8.** $f(x) = -\frac{1}{x+5}$ **9.** $f(x) = \frac{x+2}{x+1}$

10. $f(x) = \frac{1}{x-3} + 4$ **11.** $f(x) = \frac{3x-1}{x}$ **12.** $f(x) = \frac{x+2}{x-2}$

Business If a merchant wants to sell an item for $100, he can mark the regular price as $125 and then sell it at a 20% discount. This puts the price back at $100. In other words, he increases the intended price by 25% and then decreases it by 20% to recover the intended price. The rational function $d(m) = \frac{100m}{m+100}$, where m is the percentage markup and $d(m)$ is the sale percentage discount, always returns the price to the one the merchant originally planned.

13. Show that $d(25) = 20$.

14. Graph d. Explain what value d approaches as m gets very large ($m > 5000$). What does this value represent in the real world?

Find each of the following. Explain your results.

15. $d(10)$ 9.09% **16.** $d(15)$ 13.04% **17.** $d(35)$ 25.93% **18.** $d(50)$ 33.33%

19. How do you think your answers to Exercises 15–18 might affect the marketing strategy of the merchant?

20. Find m if $d(m) = 30$. Explain the meaning of this result. $m \approx 42.86$

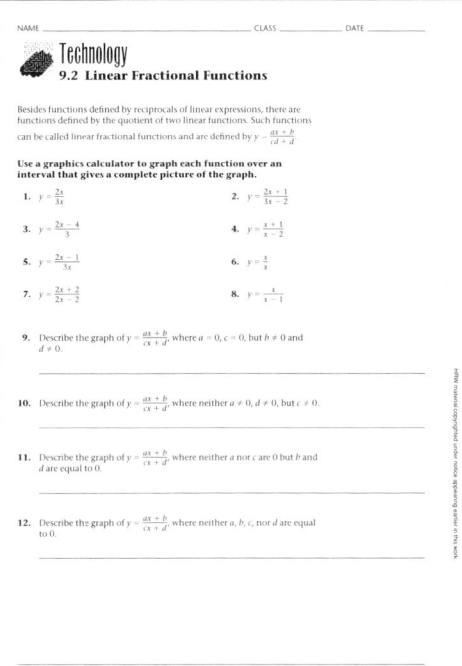

Merchants wants to sell an item for the highest possible price. However, they know that buyers will respond better if offered a discount. It is common for merchants first to mark up the price of an item, and then to offer a discount on the item.

	Domain	Range	Vert. Asym.	Horiz. Asym.
7.	$x \neq 4$	$y \neq 0$	$x = 4$	$y = 0$
8.	$x \neq -5$	$y \neq 0$	$x = -5$	$y = 0$
9.	$x \neq -1$	$y \neq 1$	$x = -1$	$y = 1$
10.	$x \neq 3$	$y \neq 4$	$x = 3$	$y = 4$
11.	$x \neq 0$	$y \neq 3$	$x = 0$	$y = 3$
12.	$x \neq 2$	$y \neq 1$	$x = 2$	$y = 1$

13. $d(25) = \frac{100(25)}{25+100} = \frac{2500}{125} = 20$

14. When m is greater than 5000, d approaches 100. No matter how much an item is marked-up, the largest possible discount is 100%.

19. The merchant might make the item seem valuable and the discount seem generous by a 50% markup and a $33\frac{1}{3}$% discount.

20. $m \approx 42.86$; If the merchant wants a sale percent discount of 30%, he should mark-up the regular price by about 43%.

ASSESS

Selected Answers
Odd-numbered Exercises 7–35

Assessment Guide
Core 1–11, 13–35

Core Plus 1–6, 10–37

Technology
Graphics calculators are necessary for Exercises 1–4, 6–12, 14–18, 20–23, 26–27, and 36–37.

Error Analysis
Vertical asymptotes are not easily identified on the graphics calculator. When $x = a$ is suspected of being an asymptote, have students compute $f(a)$ as a check.

Authentic Assessment

Have students use a road map to determine the distance from their home to a destination of their choice. Using the function $\left[t = \frac{d}{r}\right]$ and their individual values for d, have students compute travel times for several different speed limits.

21. **Transformations** Use your graphics calculator to graph $f(x) = \frac{a}{x}$ for $a = 1, 2, 3$, and 4 on the same coordinate plane. How does the graph of f change as a gets larger?
The graphs move farther from the origin as a increases.

22. **Transformations** Use your graphics calculator to graph $g(x) = \frac{1}{x-a}$ for $a = 1, 2, 3$, and 4 on the same coordinate plane. Describe how the graphs are related.
horizontal shift by a units to the right

23. **Transformations** Use your graphics calculator to graph $h(x) = -\frac{1}{x+a}$ for $a = 1, 2, 3$, and 4 on the same coordinate plane. Describe how the graphs are related. graphs are reflections of $\frac{1}{x}$ with respect to the x-axis with horizontal shift by a units to the left.

24. Find the inverse of $f(x) = \frac{1}{x-1}$. Graph f and its inverse on the same coordinate plane. How do they compare? $f^{-1}(x) = \frac{1}{x} + 1$

Chemistry Water is mixed with 25 milliliters of a 40% acid solution.

25. Write the acid concentration as a function of the amount of water added. $C(x) = \frac{10}{25+x}$

26. Graph the function on your graphics calculator, and describe its behavior.

27. What is the concentration when 10 milliliters of water have been added? 28.57%

Look Back

Graph the feasible region for each set of constraints. Then determine any four points in the feasible region. [Lesson 4.9]

28. $\begin{cases} 2x - y \le 1 \\ x + y < 7 \\ x \ge 0 \\ y \ge 0 \end{cases}$

29. $\begin{cases} x + 3y \le 12 \\ x - 2y \ge -6 \\ x \ge 0 \\ y \ge 0 \end{cases}$

30. Find the zeros of $f(x) = x^2 - 3x - 18$. **[Lesson 5.2]**
−3, 6

Write each of the following expressions as a sum or difference of logarithms. **[Lesson 7.6]**

31. $\log(6 \cdot 3)$ 32. $\ln(2 \cdot 4)$ 33. $\log \frac{17}{8}$ 34. $\log\left(\frac{xz}{y}\right)$
$\log 6 + \log 3$ $\ln 2 + \ln 4$ $\log 17 - \log 8$ $\log x + \log z - \log y$

35. y varies inversely as the square of x, and y is 3 when x is 4. Find the constant of variation, and write the equation of inverse variation. **[Lesson 9.1]** $k = 48$; $y = \frac{48}{x^2}$

Look Beyond

Look Beyond

These problems introduce the students to the ideas of even and odd functions. Students also begin to see how increasing the degree of the denominator affects the graph of the rational function and changes the number of asymptotes.

36. Use your graphics calculator to graph $f(x) = \frac{1}{x^2}$.
Does the graph have a line of symmetry? If so, where? yes; the y-axis

37. Use your graphics calculator to graph $f(x) = \frac{1}{x^2 - 4}$.
Find its asymptotes. $y = 0$, $x = 2$, $x = -2$

24. $f^{-1}(x) = \frac{1}{x} + 1$

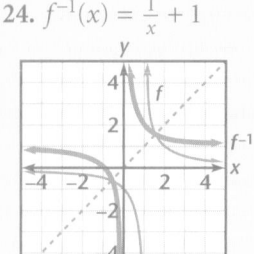

f and f^{-1} are symmetric with respect to the line $y = x$. f has a horizontal asymptote at $y = 0$ and a vertical asymptote at $x = 1$, while f^{-1} has a vertical asymptote at $x = 0$ and a horizontal asymptote at $y = 1$.

The answers for Exercises 26, 28 and 29 can be found in Additional Answers beginning on page 842.

Exploring Reciprocals of Polynomial Functions

why *Reciprocals of polynomial functions of degree 2 or higher are used to model the magnitude of light intensity and the gravitational force on a satellite. The shape of their graphs is strongly influenced by the zeros of the polynomial function.*

•Exploration 1 *Even and Odd Functions*

Graphics Calculator

Graph each function, and use the trace feature to complete the table.

	$f(-3)$	$f(-2)$	$f(-1)$	$f(0)$	$f(1)$	$f(2)$	$f(3)$
$f(x) = x^2$							
$f(x) = x^3$							
$f(x) = x^4$							
$f(x) = x^5$							
$f(x) = x^4 + x^2$							
$f(x) = 3x^5 + x$							
$f(x) = 5x^4 - 3x^2$							
$f(x) = x^4 - 2$							

1 For which of the functions in the table is the statement $f(-x) = f(x)$ true? Examine these functions. What do these functions have in common?

2 For which of the functions in the table is the statement $f(-x) = -f(x)$ true? Examine these functions. What do these functions have in common?

3 How can you determine whether a function has the property $f(-x) = f(x)$ without graphing?

4 How can you determine whether a function has the property $f(-x) = -f(x)$ without graphing? ❖

EVEN AND ODD FUNCTIONS
A function with the property $f(-x) = f(x)$ is called an **even function**.
A function with the property $f(-x) = -f(x)$ is called an **odd function**.

Using Visual Models Three types of symmetry can be used to characterize graphs: symmetry with respect to the *x*-axis, the *y*-axis, and the origin. Graph $f(x) = x^2$ and show that it is symmetric about the *y*-axis. Show that $f(x) = x^3$ is symmetric about the origin. Graph $x = y^2$. Show that it is symmetric about the *x*-axis even though it is not a function. Show that the graph of $x^2 + y^2 = 1$ exhibits all three types of symmetry, but it is also not a function.

PREPARE

Objectives
- Determine whether a function is even, odd, or neither.
- Graph a function that is the reciprocal of a polynomial, and find its vertical asymptotes.

RESOURCES

• Practice Master	9.3
• Enrichment Master	9.3
• Technology Master	9.3
• Lesson Activity Master	9.3
• Quiz	9.3
• Spanish Resources	9.3

Assessing Prior Knowledge
Factor each expression.
A. $x^2 - x - 2$
   $[(x + 1)(x - 2)]$
B. $2x^2 + 5x - 3$
   $[(x + 3)(2x - 1)]$
C. $2x^2 - 9x - 5$
   $[(2x + 1)(x - 5)]$

TEACH

why Another rational function is the reciprocal of polynomial functions with a degree of 2 or more.

Exploration 1 Notes
This exploration develops the concept of odd and even functions by exploring a number of functions.

ongoing ASSESSMENT
3. If all the powers of x are even, then $f(-x) = f(x)$.

4. If all the powers of x are odd, then $f(-x) = -f(x)$.

CRITICAL
Thinking

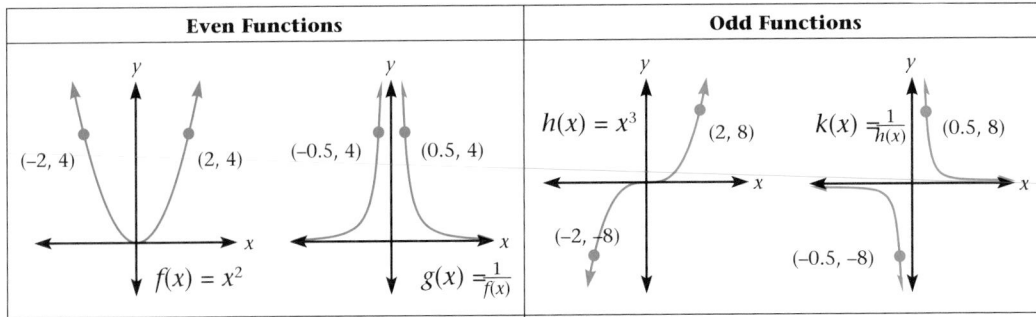

Is there an axis of symmetry for the graph of an even function? Is there an axis of symmetry for the graph of an odd function? Explain.

Even Functions		Odd Functions	
(−2, 4) (2, 4) $f(x) = x^2$	(−0.5, 4) (0.5, 4) $g(x) = \frac{1}{f(x)}$	$h(x) = x^3$ (2, 8) (−2, −8)	$k(x) = \frac{1}{h(x)}$ (0.5, 8) (−0.5, −8)
For any point (*a*, *b*) on the graph of an **even function,** the point (−*a*, *b*) is also on the graph.		For any point (*a*, *b*) on the graph of an **odd function,** the point (−*a*, −*b*) is also on the graph.	

Consider the reciprocal functions $f(x) = x^2$ and $g(x) = \frac{1}{x^2}$.

Notice that $g(x) = \frac{1}{x^2}$ can also be written $g(x) = \frac{1}{f(x)}$.

Examine the function and the graph of $g(x) = \frac{1}{x^2}$.

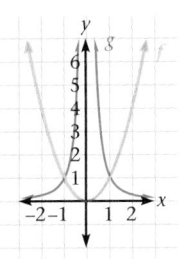

The domain of *g* is all real numbers except zero. As the absolute value of *x* gets larger, the value of *g* gets closer to zero, but never reaches it. Therefore, the *x*-axis is a horizontal asymptote of the graph of *g*.

Examine the table of values for *g*(*x*).

x	−100	−10	−3	−2	−1	0	1	2	3	10	100
g(*x*)	$\frac{1}{10,000}$	$\frac{1}{100}$	$\frac{1}{9}$	$\frac{1}{4}$	1	—	1	$\frac{1}{4}$	$\frac{1}{9}$	$\frac{1}{100}$	$\frac{1}{10,000}$

As *x* approaches zero through negative values, *g* increases. As *x* moves away from zero through positive values, *g* decreases. Thus, the function is increasing on the negative *x*-axis and decreasing on the positive *x*-axis. What is the vertical asymptote for *g*?

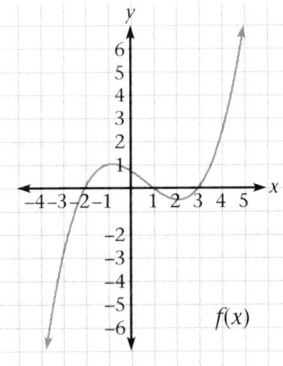

f(*x*)

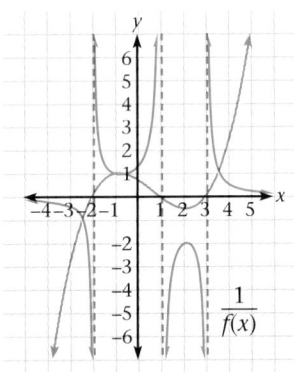

$\frac{1}{f(x)}$

Exploration 2 Graphs of Reciprocal Quadratic Functions

Graph each quadratic function, and use the trace feature to complete the table.

		Zeros of f	x-coordinate of the vertex
1	$f(x) = (x + 4)(x - 3)$		
2	$f(x) = (x - 2)(x + 1)$		
3	$f(x) = (x + 3)(x + 3)$		
4	$f(x) = (x - 4)(x - 4)$		

Now graph the reciprocal of each quadratic function, and complete the table.

		Vertical asymptotes of f	Maximum point of f between the asymptotes (if any)
5	$f(x) = \frac{1}{(x + 4)(x - 3)}$		
6	$f(x) = \frac{1}{(x - 2)(x + 1)}$		
7	$f(x) = \frac{1}{(x + 3)(x + 3)}$		
8	$f(x) = \frac{1}{(x - 4)(x - 4)}$		

Compare the graphs of each quadratic function of the form $(x + a)(x + b)$, where $a \neq b$, with its reciprocal function, and answer Questions 9 and 10.

9. How are the zeros of the quadratic function related to the vertical asymptotes of its reciprocal function?

10. How is the x-coordinate of the vertex of the quadratic function related to the x-coordinate of the maximum point between the asymptotes of its reciprocal function?

Compare the graphs of each quadratic function of the form $(x + a)(x + b)$, where $a = b$, with its reciprocal function, and answer Questions 11 and 12.

11. How are the zeros of the quadratic function related to the vertical asymptotes of its reciprocal function?

12. When a quadratic function has a zero of multiplicity 2 at its vertex, does its reciprocal function have a maximum value? Explain. ❖

INCLUSION strategies

Using Visual Models Have the students use graph paper to graph two simple functions, one odd function and one even function. Have students choose some integer a. Then have students use a straightedge to connect the points $(a, f(a))$ and $(-a, f(-a))$ in both graphs. Have students measure the length of this line in each graph. This should give students an additional visual model for even and odd symmetry.

Cooperative Learning

Have students work in pairs to develop a table, like the table shown below, that can be used to summarize the results of the Extension on page 512.

$f(x) = x^3 - x^2 - 6x$	$g(x) = \dfrac{1}{x^3 - x^2 - 6x}$
-2, 0, 3 are zeros of f	**-2, 0, 3** are vertical asymptotes of g and are not in the domain of g
(-1.1, 4.059) is a turning point and a local maximum point of f.	**(-1.1, 0.246)** is a local minimum point of g.
(1.8, -8.208) is a turning point and a local minimum point of f.	**(1.8, -0.122)** is a local maximum point of g.

Now, have each group graph $f(x) = x^3 + 3x^2 - 4x$ and $g(x) = \dfrac{1}{x^3 + 3x^2 - 4x}$. Have students construct a table similar to the table they constructed for the Extension. Then have each group discuss the similarities and differences between f and g in the Extension and f and g given above. How did the table help these similarities and differences to become more obvious?

Have students make-up a pair of functions and predict the results (zeros, asymptotes, turning points, local minimum and maximum points) before graphing the functions. Then, have students graph the functions and test their predictions.

Graphics Calculator

Compare the reciprocal functions $f(x) = x^3 - x^2 - 6x$ and $g(x) = \dfrac{1}{x^3 - x^2 - 6x}$.

The factored form of f is $f(x) = x(x + 2)(x - 3)$, so the zeros of f are -2, 0, and 3.

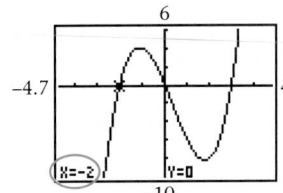 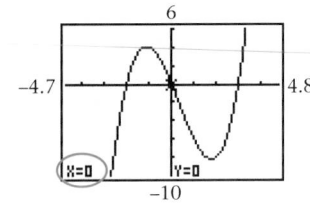 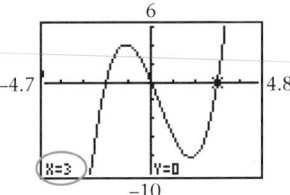

The vertical asymptotes of g are $x = 0$, $x = -2$, and $x = 3$, the same as the zeros of f.

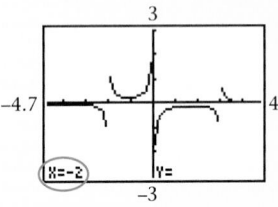 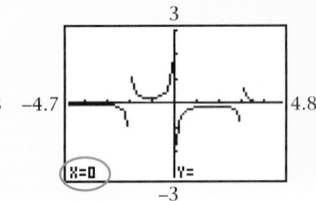 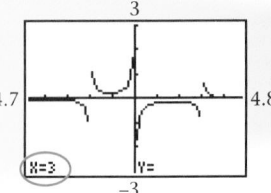

Therefore, the domain of g is all real numbers except -2, 0, and 3.

MAXIMUM MINIMUM *Connection*

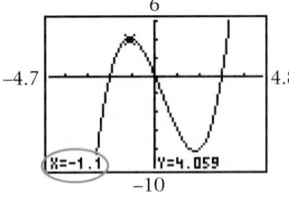 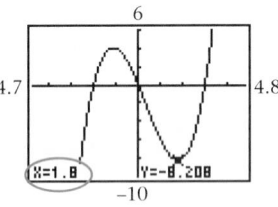

The turning points of f are approximately (-1.1, 4.059) and (1.8, -8.208).

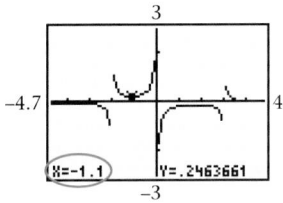 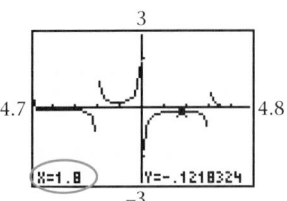

The minimum and maximum points of g between the vertical asymptotes also occur at the approximate x-values of -1.1 and 1.8.

RETEACHING *the* lesson

Cooperative Learning

Have students work in pairs. As one student graphs a polynomial function, have the other student graph its reciprocal. It will be helpful for students to com-

pare the two graphs visually. Have both students make lists of characteristics of both the quadratic function and its graph and the reciprocal function and its graph. Then, have students discuss and fine-tune their lists. These lists can be presented for the entire class to discuss.

The function g is decreasing where f is increasing. Likewise, g is increasing where f is decreasing. ❖

Compare the reciprocal functions $f(x) = x(x-1)(x+2)$ and $g(x) = \dfrac{1}{x(x-1)(x+2)}$.

CRITICAL
Thinking

Explain why the turning points of a polynomial function are maximum or minimum points of its reciprocal function.

Applications of Reciprocal Polynomial Functions

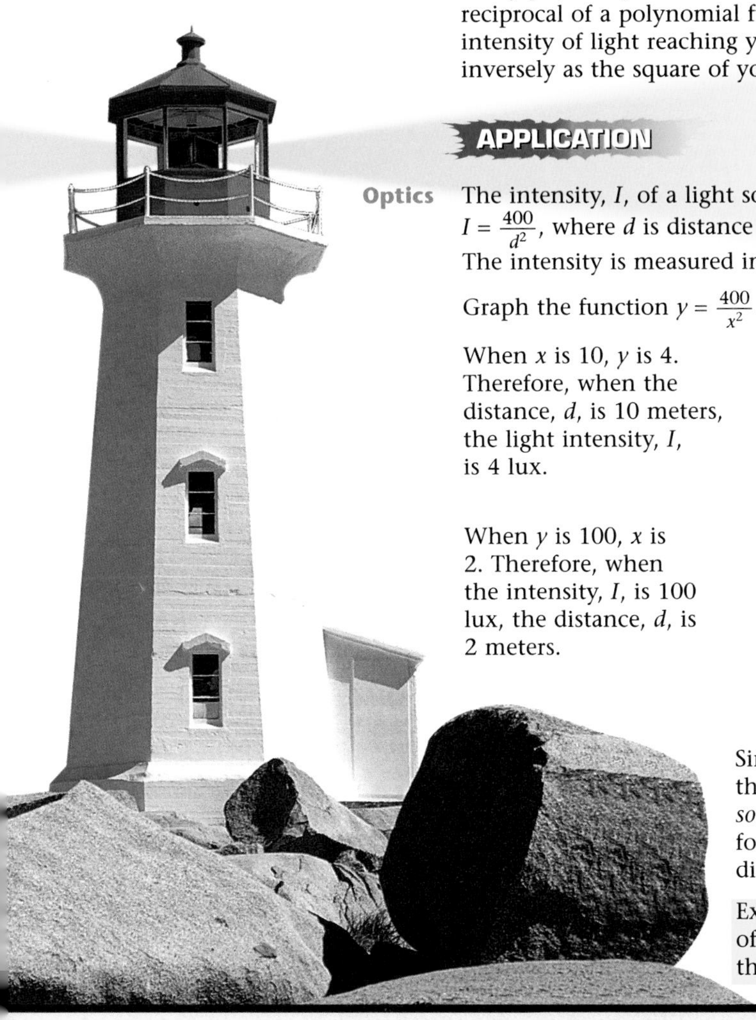

Many physical phenomena can be modeled by the reciprocal of a polynomial function. For example, the intensity of light reaching you from a light source varies inversely as the square of your distance from the source.

APPLICATION

Optics The intensity, I, of a light source is given by the function $I = \dfrac{400}{d^2}$, where d is distance in meters from the source. The intensity is measured in lux (lumens per square meter).

Graph the function $y = \dfrac{400}{x^2}$.

When x is 10, y is 4. Therefore, when the distance, d, is 10 meters, the light intensity, I, is 4 lux.

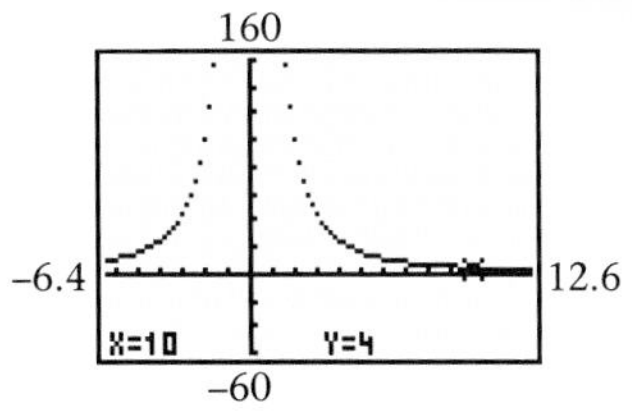

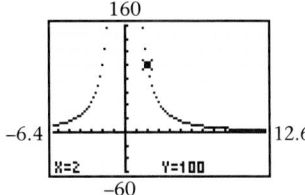

When y is 100, x is 2. Therefore, when the intensity, I, is 100 lux, the distance, d, is 2 meters.

Since the graph of y never reaches the x-axis, there will always be *some* positive measure of intensity for the illumination, even at great distances. ❖

Explain the asymptotic behavior of the intensity of light using the stars as an example.

EXERCISES & PROBLEMS

Communicate

1. Explain how to find the vertical asymptotes of the function $f(x) = \dfrac{1}{x^2 - 6x - 7}$ without graphing.

2. Write a rational function with a quadratic function in the denominator that has only one vertical asymptote.

3. What is the domain of the function $f(x) = \dfrac{1}{x^2 - 1}$?

4. Explain how the graph of $f(x) = -\dfrac{1}{x^2}$ may be obtained from that of $f(x) = \dfrac{1}{x^2}$.

5. What are the properties of even and odd functions? Give an example of each.

6. Write a rational function with vertical asymptotes at $x = -4$, $x = 1$, and $x = 3$.

Practice & Apply

Determine whether each of the following functions is even, odd or neither.

7. $f(x) = \dfrac{1}{x^2 + 2x - 3}$ neither

8. $f(x) = \dfrac{1}{x^2 + 2x}$ neither

9. $f(x) = \dfrac{1}{x^2 - 3}$ even

10. $f(x) = \dfrac{1}{x^3 + 2x}$ odd

11. Write a rational function that has a vertical asymptote at $x = -2$ and a horizontal asymptote at $y = -3$.
Answers may vary but should be similar to $f(x) = \dfrac{a}{x + 2} - 3$

12. Write a rational function with a cubic denominator that has vertical asymptotes at $x = -4$, $x = 1$, and $x = 3$.
Answers may vary but should be similar to $f(x) = \dfrac{a}{(x + 4)(x - 1)(x - 3)}$

Graph and compare each function with its reciprocal function.

13. $f(x) = -(x^2 - 3)$

14. $f(x) = x^2 + 3$

15. $f(x) = x^2 + 6x + 9$

16. $f(x) = -(x^2 - 16)$

17. $f(x) = x^3$

18. $f(x) = -(x^3 + 4x - 5)$

19. $f(x) = -(x^3 + 2)$

20. $f(x) = x^4$

21. $f(x) = x^5$

Sketch the graph of the reciprocal of each polynomial function.

22. $f(x) = x^2 - 3x + 4$

23. $f(x) = x^3 + 5x^2 + 6x$

For Exercises 24–26, let $f(x) = x^2$, $g(x) = \dfrac{1}{x^2}$, $h(x) = x^3$, and $i(x) = \dfrac{1}{x^3}$.

24. Graph each function. Determine whether each function is even, odd, or neither. Find the inverse of each function.

25. Which functions have inverses that are functions? How does this compare with whether the function is even, odd, or neither?

	Vertical Asymptote	Horizontal Asymptote	Odd or Even?
13.	$x = \pm\sqrt{3}$	$y = 0$	even
14.	none	$y = 0$	even
15.	$x = -3$	$y = 0$	neither
16.	$x = \pm 4$	$y = 0$	even
17.	$x = 0$	$y = 0$	odd
18.	$x = 1$	$y = 0$	neither
19.	$x = -\sqrt[3]{2}$	$y = 0$	neither
20.	$x = 0$	$y = 0$	even
21.	$x = 0$	$y = 0$	odd

The answers to Exercises 22–25, and 30 can be found in Additional Answers beginning on page 842.

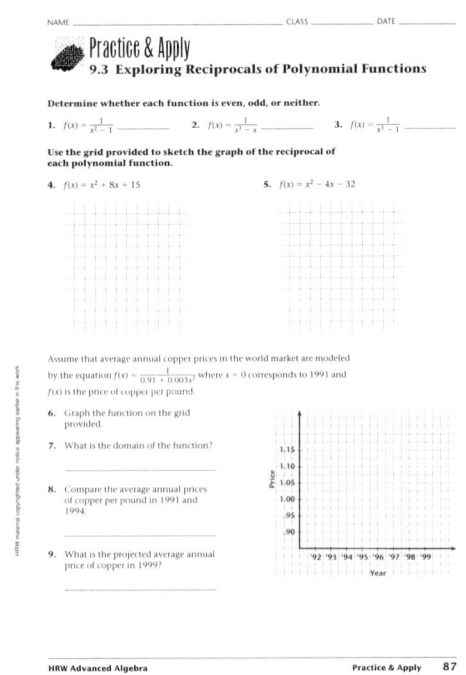

26 Determine the domain restrictions that make each an invertible function. Graph each inverse function. How do f^{-1} and g^{-1} compare? How do h^{-1} and i^{-1} compare?

27 **Transformations** Use your graphics calculator to graph $f(x) = \frac{a}{x^2}$ for $a = 2, 3, 4,$ and 5 on the same coordinate plane. graph moves farther away from the origin as a increases Describe the effect of an increase of a on the graph of the function.

28. **Maximum/Minimum** What is the maximum point on the graph $f(x) = \frac{1}{x^2 + 2}$? How is it related to the intervals where the graph is increasing or decreasing?

29 **Transformations** Use your graphics calculator to graph $f(x) = \frac{1}{x^2 + a}$ for $a = 2, 3, 4,$ and 5 on the same coordinate plane. graph moves closer to the x-axis as a increases Describe the effect of an increase of a on the graph of the function.

30. **Portfolio Activity** Complete the Portfolio Activity on page 493.

Physics The function $f(d) = \frac{1.29 \times 10^9}{d^2}$ describes the gravitational force $f(d)$ of the Earth on a spacecraft a distance, d, in miles, from the center of the Earth. Choose an appropriate viewing window and graph the function to include the domain $0 \le x \le 15,000$.

31 Compare the force at 7000 miles with the force at 12,000 miles from the center of the Earth.
$f(7000) = 26.3$
$f(12,000) = 8.96$

32 The diameter of the Earth is approximately 7927 miles. What is the gravitational force, to the nearest tenth, on the surface of the Earth? 20.5

33. Explain what the asymptote represents in terms of the gravitational force. As the distance from the Earth becomes large, the gravitational force approaches zero.

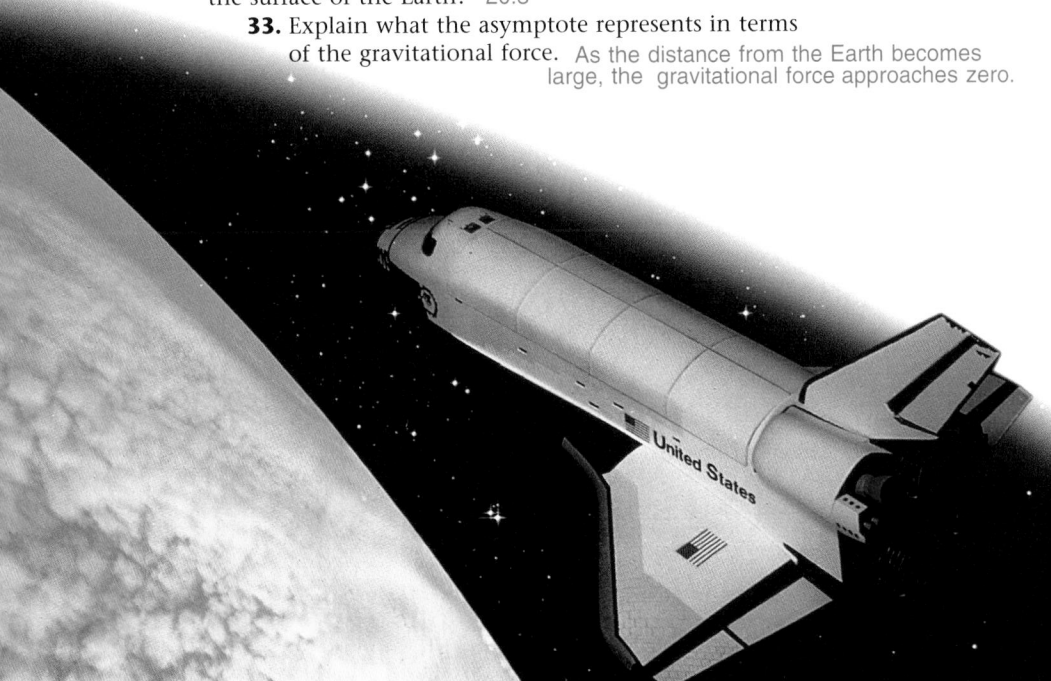

26. If $x \ge 0$, then f has an inverse function; if $x > 0$, then g has an inverse function. There is no restriction on the domains of h and i except that, for i, $x \ne 0$. As f^{-1} gets large, g^{-1} approaches zero. h^{-1} and i^{-1} have a similar relationship.

28. $(0, 0.5)$ The graph is increasing when $x < 0$ and decreasing when $x > 0$.

The answers to Exercise 30 can be found in Additional Answers beginning on page 842.

Business The total revenue function for a fine-service-ware company between the years 1980 and 1990 can be approximated by $R(t) = \dfrac{1}{2t^2 - 3.5t + 4}$, where t represents the number of years after 1980, and $R(t)$ represents the total revenue in billions of dollars. Graph the revenue function.

34 What was the revenue, to the nearest dollar, in the year 1988? $9,615,385

35 When is the revenue increasing? When is it decreasing? increasing   decreasing
$0 \le t < 0.875$; $t > 0.875$

36 What is the maximum revenue, to the nearest dollar? $405,063,291

Look Back

37. Divide $\dfrac{x^3 - x^2 - 3x + 2}{x - 2}$. **[Lesson 6.3]** $x^2 + x - 1$

38. What are the zeros of the function $f(x) = x^3 - x^2 - 3x + 2$? **[Lesson 6.3]**
$2; \approx 0.6; \approx -1.6$

Find two angles between 0° and 360° for each of the following inverse relations.   [Lesson 8.3]

39. $\tan^{-1}(1)$ **40.** $\sin^{-1}\left(-\dfrac{1}{2}\right)$ **41.** $\cos^{-1}\left(\dfrac{\sqrt{3}}{2}\right)$ **42.** $\tan^{-1}\left(\sqrt{3}\right)$
   45°, 225°   210°, 330°   30°, 330°   60°, 240°

43. Graph the function $f(x) = \dfrac{1}{x+3} - 2$.   **[Lesson 9.2]**

Look Beyond

44 Graph $f(x) = \dfrac{1}{x-2}$ and $g(x) = \dfrac{(x+1)}{(x-2)(x+1)}$. How do the graphs of f and g compare?

45 Find the horizontal asymptote for the function $f(x) = \dfrac{4x^3 - x^2 + 5}{2x^3 - 9}$. $y = 2$

46 Graph the function $f(x) = \dfrac{3x^2 - 5x - 8}{x - 2}$. Does it have any asymptotes? yes; $f(2)$ is undefined
Explain.   and $(x-2)$ is not a factor of $3x^2 - 5x - 8$, so $x = 2$ is a vertical asymptote.

43.

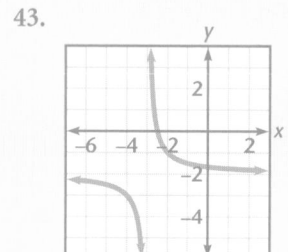

44. The graphs are the same except at $x = -1$; $f(-1) = \frac{1}{3}$, but $g(-1)$ is undefined, so the graph of g has an undefined spot at $x = -1$.

LESSON 9.4 Quotients of Polynomial Functions

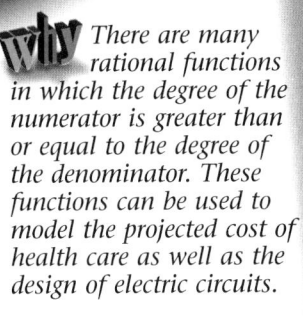

Why *There are many rational functions in which the degree of the numerator is greater than or equal to the degree of the denominator. These functions can be used to model the projected cost of health care as well as the design of electric circuits.*

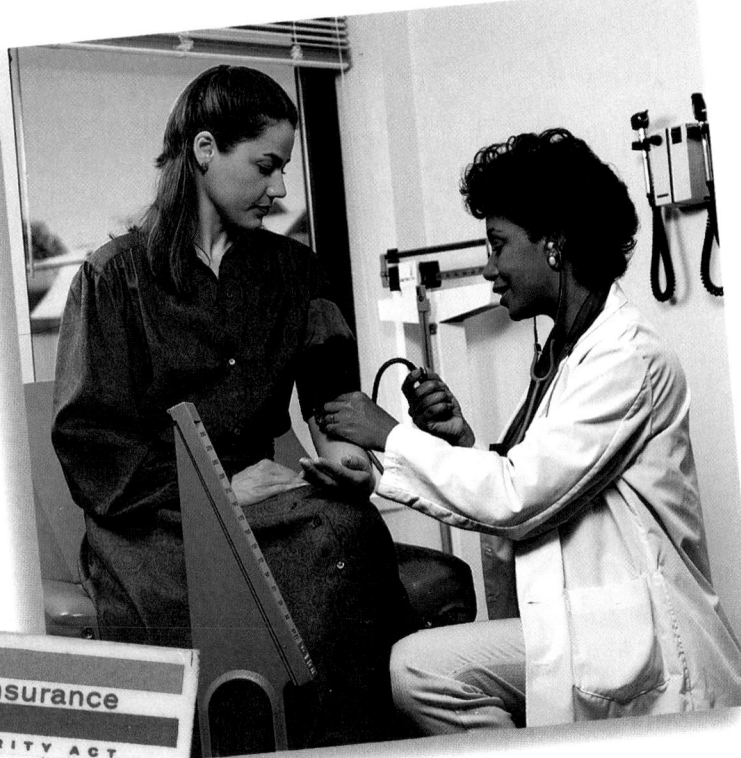

To analyze a rational function, factor the denominator over the integers, if possible. This will enable you to find the vertical asymptotes.

EXAMPLE 1

Find the vertical and horizontal asymptotes of the function $f(x) = \dfrac{3x^2 + x - 2}{x^2 + 2x - 3}$.

Solution

To find the vertical asymptotes, factor the denominator.

$$f(x) = \frac{3x^2 + x - 2}{x^2 + 2x - 3} = \frac{3x^2 + x - 2}{(x+3)(x-1)}$$

ALTERNATIVE teaching strategy

Cooperative Learning Have pairs of students work through several examples and exercises. Have one student find the asymptotes by using a graphics calculator and the other student find the asymptotes by analytical methods. Then have students compare results. Be sure students switch methods several times.

PREPARE

Objectives

- Find the vertical and horizontal asymptotes and the domain of a rational function in which the degree of the numerator is not greater than the degree of the denominator.

RESOURCES

- Practice Master   9.4
- Enrichment Master   9.4
- Technology Master   9.4
- Lesson Activity Master 9.4
- Quiz   9.4
- Spanish Resources   9.4

This lesson may be considered optional in the core course. See page 492A.

Assessing Prior Knowledge

Divide each rational expression.

A. $\dfrac{2x^2 + 2x + 1}{x^2 - x - 2}$ $\left[2 + \dfrac{4x + 5}{x^2 - x - 2} \right]$

B. $\dfrac{4x^2 + 4x + 5}{2x^2 + 3}$ $\left[2 + \dfrac{4x - 1}{2x^2 + 3} \right]$

TEACH

Why When analyzing rational functions, it is often helpful to factor the polynomials in the numerator and the denominator.

Alternate Example 1

Find the vertical and horizontal asymptotes of the function $f(x) = \dfrac{4x^2 - 3x + 1}{x^2 - x - 2}$ [vertical asymptotes at $x = 2, -1$; horizontal asymptote at $y = 4$]

Alternate Example 2

Simplify each expression

A. $\dfrac{x^2 + 3x + 2}{x^2 + x - 2}$

$\left[\dfrac{x+1}{x-1} \text{ for } x \neq -2 \text{ and } x \neq 1 \right]$

B. $\dfrac{x^2 + 6x - 7}{x^2 + 2x - 3}$

$\left[\dfrac{x+7}{x+3} \text{ for } x \neq -3 \text{ and } x \neq 1 \right]$

Since f is undefined when x is -3 or when x is 1, you can expect to find vertical asymptotes at $x = -3$ and at $x = 1$. The graph of f shows that $x = -3$ and $x = 1$ are vertical asymptotes.

To find the horizontal asymptotes, you can use the trace feature on your graphics calculator.

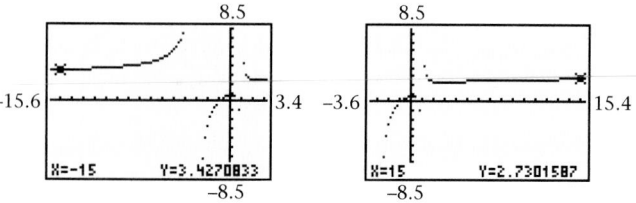

Notice that as the absolute value of x gets larger, the value of f approaches 3. Thus, f has a horizontal asymptote at $y = 3$.

You can also find the horizontal asymptote by dividing the polynomials.

$$x^2 + 2x - 3 \overline{\smash{\big)}\ \begin{aligned} &\quad\quad 3 \\ &3x^2 + \ x - 2 \\ &\underline{-(3x^2 + 6x - 9)} \\ &\quad\quad\ -5x + 7 \end{aligned}}$$

So, $\dfrac{3x^2 + x - 2}{x^2 + 2x - 3}$ is equal to $3 + \dfrac{-5x + 7}{x^2 + 2x - 3}$. As the absolute value of x gets larger, the value of the remainder, $\dfrac{-5x + 7}{x^2 + 2x - 3}$, gets closer to zero, and the horizontal asymptote is $y = 3$. ❖

Try This Find the vertical and horizontal asymptotes of the function $f(x) = \dfrac{2x^2 - 5x + 4}{x^2 - 2x - 3}$.

Sometimes it is possible to factor the numerator and denominator to simplify the expression.

EXAMPLE 2

Simplify each expression.

Ⓐ $\dfrac{x^2 - 9}{x^2 + x - 6}$

Ⓑ $\dfrac{x^2 + x - 2}{x^2 - 3x + 2}$

Solution

Ⓐ $\dfrac{x^2 - 9}{x^2 + x - 6} = \dfrac{(x+3)(x-3)}{(x+3)(x-2)} = \dfrac{x-3}{x-2}$, for $x \neq -3$ and $x \neq 2$

Ⓑ $\dfrac{x^2 + x - 2}{x^2 - 3x + 2} = \dfrac{(x-1)(x+2)}{(x-1)(x-2)} = \dfrac{x+2}{x-2}$, for $x \neq 1$ and $x \neq 2$ ❖

When you simplify a function, notice that the domain of that function is still restricted by the original expression.

ENRICHMENT Students may wonder why rational functions in which the degree of the numerator is equal to or less than the degree of the denominator are considered separately from other rational functions. Using their graphics calculators, have students investigate the behavior of several kinds of rational functions and compare their observations with the behavior of the rational functions considered in this lesson.

INCLUSION strategies **Using Patterns** Have students compare the graphs of each of the following functions: $f(x) = \dfrac{1}{x}$, $f(x) = -x$, $f(x) = \dfrac{1}{x^2}$, $f(x) = -\dfrac{1}{x^2}$, $f(x) = \dfrac{1}{x^3}$, $f(x) = -\dfrac{1}{x^3}$, $f(x) = \dfrac{1}{x^4}$, $f(x) = -\dfrac{1}{x^4}$. Have students discuss how rational functions might be divided into different classes. Encourage students to draw generalizations about these different classes. Have them note any similarities in asymptotes and symmetries.

EXAMPLE 3

Find the vertical and horizontal asymptotes of the function
$f(x) = \frac{x+3}{x^2+4x+3}$.

Solution►

To find the vertical asymptotes, factor the denominator.

$$f(x) = \frac{x+3}{x^2+4x+3} = \frac{x+3}{(x+3)(x+1)} = \left(\frac{x+3}{x+3}\right)\left(\frac{1}{x+1}\right)$$

The zeros of the denominator are –3 and –1. Therefore, you might expect the function to have vertical asymptotes at $x = -3$ and $x = -1$. However, the factor $\frac{x+3}{x+3}$ is equal to 1 for all x-values *except* –3.

Thus, the graphs of $g(x) = \frac{1}{x+1}$ and $f(x) = \left(\frac{x+3}{x+3}\right)\left(\frac{1}{x+1}\right)$ are exactly the same *except* at $x = -3$, where a *hole in the graph* of f occurs.

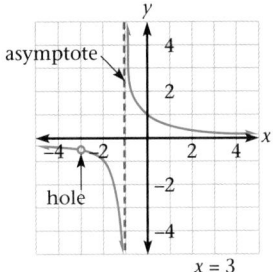

A *hole in the graph* indicates that the function is not defined for that x-value and there is no vertical asymptote at that x-value.

Thus, the graph of $f(x) = \frac{x+3}{x^2+4x+3}$ has only one vertical asymptote at $x = -1$.

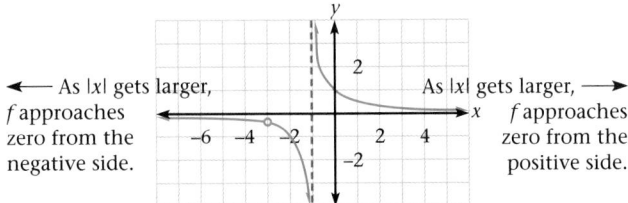

◄— As |x| gets larger, f approaches zero from the negative side.

As |x| gets larger, —► f approaches zero from the positive side.

The x-axis is the horizontal asymptote of $f(x) = \frac{x+3}{x^2+4x+3}$. ❖

HOLE IN THE GRAPH

For any real number a, when $x - a$ is a factor of *both* the numerator and the denominator of a function, the graph of the function has a **hole in the graph** at $x = a$.

When $x - a$ is a factor of *only* the denominator of a function, the graph of the function has a *vertical asymptote* at $x = a$.

CRITICAL
Thinking

For a rational function in which the degree of the numerator is less than that of the denominator, is the x-axis always a horizontal asymptote? Explain.

RETEACHING
the
lesson

Using Cognitive Strategies Have students create their own exercises by working from the solution to the problem. For example, $\frac{x+1}{x-3} = \frac{(x+1)(x+2)}{(x-3)(x+2)} = \frac{x^2+3x+2}{x^2-x-6}$ leads to the problem: Find the asymptotes and domain of the function $f(x) = \frac{x^2+3x+2}{x^2-x-6}$. Have students exchange problems and solve them.

Alternate Example 3

Find the vertical and horizontal asymptotes and any holes for the function $f(x) = \frac{x-4}{x^2-16}$ [**vertical: $x = -4$; horizontal: $y = 0$; hole: $x = 4$**]

Cooperative Learning

Have students work in groups of three to discuss the Communicate Exercises. The goal of each group is to be able to determine when a function has a vertical asymptote and when it has a hole, or a *removable discontinuity*.

CRITICAL
Thinking

Yes, if the degree of the numerator is less than that of the denominator, then as x becomes very large the value of the denominator (with the higher power of x) will become much larger than the numerator, so the value of the rational expression will approach 0.

Use Transparency ► 35

ASSESS

Selected Answers

Odd-numbered Exercises 9–39

Assessment Guide

Core 1–12, 14–20, 23–40

Core Plus 1–7, 10–13, 19–41

Technology

Graphics calculators are needed for Exercises 14–22, 25–26, 28–30, 36–39, 41.

Error Analysis

In simplifying rational functions, remind students not to disregard common factors between the numerator and the denominators.

EXERCISES & PROBLEMS

Communicate

Determine the vertical asymptotes of each rational function.

1. $f(x) = \dfrac{x+2}{(x-6)(x-1)}$

2. $f(x) = \dfrac{(x-3)(x-5)}{x^2-25}$

3. When a rational function has a horizontal asymptote other than the x-axis, what is true about the degree of the numerator and the degree of the denominator?

4. Write a rational function that has a hole in its graph at $x = -2$.

5. At what point does the function $f(x) = \dfrac{x+4}{x^2-16}$ have a hole in its graph?

6. How is a hole in a graph different from a vertical asymptote?

7. How can you determine the horizontal asymptote(s) of the function $f(x) = \dfrac{x^2-1}{x^2-9}$?

Practice & Apply

Simplify each rational function.

8. $f(x) = \dfrac{x^2-4}{x^2+7x+10}$

9. $f(x) = \dfrac{x^2+6x+9}{x^2+x-6}$

10. $f(x) = \dfrac{x^2+3x+2}{x^2-2x-8}$

11. $f(x) = \dfrac{x^2+2x+1}{(x+1)^3}$

12. $f(x) = \dfrac{x^2-16}{2x^2+11x+12}$

13. $f(x) = \dfrac{6x^2-x-1}{4x^2-1}$

Find the holes and the vertical and horizontal asymptotes for each function. Give the domain for each function.

14. $f(x) = \dfrac{x^2-5x}{4x^2+7x-2}$

15. $f(x) = \dfrac{2x}{x^2+2x+1}$

16. $f(x) = \dfrac{x}{x^2-1}$

17. $f(x) = -\dfrac{3x}{x^2+3x+2}$

18. $f(x) = \dfrac{x-3}{2x^2-5x-3}$

19. $f(x) = \dfrac{x^2-25}{x^3+2x^2-15x}$

20. $f(x) = \dfrac{x^2-x}{x^3-2x^2}$

21. $f(x) = \dfrac{x(x-1)(x+4)}{(x+1)(x+2)(x+3)}$

22. $f(x) = \dfrac{x^2-5x+4}{x^2+2x-3}$

23. Does the function $f(x) = \dfrac{x-1}{(x-1)^2}$ have a hole in it? why or why not?

Electronics The total resistance in a circuit with two adjustable resistors in parallel is given by $\dfrac{1}{R} = \dfrac{1}{R_1} + \dfrac{1}{R_2}$, where R_1 and R_2 are the individual resistances in ohms.

24. Write a function for $R(x)$, if $R_1(x) = \dfrac{10}{x}$ and $R_2(x) = \dfrac{x}{10}$. $R(X) = \dfrac{10x}{x^2+100}$

25. Graph $R(x)$ for $1 \le x \le 20$. What is the maximum value of the total resistance over this interval? 0.5 ohms

8. $f(x) = \dfrac{x-2}{x+5}, x \neq -5, -2$

9. $f(x) = \dfrac{x+3}{x-2}, x \neq 2, -3$

10. $f(x) = \dfrac{x+1}{x-4}, x \neq 4, -2$

11. $f(x) = \dfrac{1}{x+1}, x \neq -1$

12. $f(x) = \dfrac{x-4}{2x+3}, x \neq -4, -\dfrac{3}{2}$

13. $f(x) = \dfrac{3x+1}{2x+1}, x \neq \pm\dfrac{1}{2}$

The answers to Exercises 14–23 can be found in Additional Answers beginning on page 842.

Health Care The following table contains the total projected cost, C, (in billions of dollars) of Medicare for the U.S. population from 1995 through 1999. Let $t = 0$ represent the year 1995.

Time, t	0	1	2	3	4
Cost, C	107	120.1	134.2	149.3	165.4

The projected Medicare cost for the U.S. population aged 65 and older is $P(t) = 0.14t^2 + 0.5t + 32$. **Complete Exercises 26–28.**

26  **Statistics** Use the statistical regression feature of your calculator to find the quadratic equation of best fit for the cost, C, of Medicare. What is the equation? $C(t) = 0.5t^2 + 12.6t + 107$

27. Write a rational function that represents the fraction of the total Medicare cost for persons aged 65 or older as a portion of the total cost of Medicare during the years 1995 to 1999. $F(t) = \dfrac{0.14t^2 + 0.5t + 32}{0.5t^2 + 12.6t + 107}$

28 Graph the function you wrote for Exercise 27 for $0 \le t \le 4$. What does the graph tell you about trends in Medicare during this period? The percentage of Medicare spent on people over 65 is decreasing

29 What is the horizontal asymptote in the function $f(x) = \dfrac{ax^2}{px^2}$, where $a \ne 0$ and $p \ne 0$ (HINT: Substitute different values for a and p until you notice a pattern.)? $y = \dfrac{a}{p}$

30 Does the graph of $f(x) = \dfrac{2x+3}{x^2+9}$ cross its horizontal asymptote? If so, at what point(s)? Yes; $(-1.5, 0)$

Look Back

31. Use long division to divide $(2x^3 + x^2 - 6x + 4) \div (x^2 - 2x + 1)$.
[Lesson 6.3] $2x + 5 + \dfrac{2x - 1}{x^2 - 2x + 1}$

Give the exact value for each of the following expressions. [Lesson 8.5]

32. $\sin\left(\dfrac{\pi}{4}\right)$ $\dfrac{1}{\sqrt{2}}$ **33.** $\tan\left(\dfrac{\pi}{3}\right)$ $\sqrt{3}$ **34.** $\cos(5\pi)$ -1 **35.** $\sin\left(-\dfrac{9\pi}{4}\right)$ $-\dfrac{1}{\sqrt{2}}$

Use a calculator in radian mode to graph the function $f(x) = 3 + 5\cos\left(\dfrac{\pi x}{3}\right)$. Complete Exercises 36–39. [Lesson 8.7]

36 Find the maximum and minimum values of the function. 8 ; –2

37 Find $f(4)$. 0.5

38 Find two values of x such that $f(x) = 7$. Answers may vary.

39 Find the period of f. 6 radians

40. Graph the function $f(x) = \dfrac{1}{x+4}$. **[Lesson 9.2]**

Look Beyond

41 Use your graphics calculator to find all the solutions to the rational equation $1.4 = \dfrac{(x+3)(x-1)}{(x^2-1)}$. $x = 4$

38. Answers may vary but must be similar to $x \approx 0.6 + 6n$, or $x \approx 5.4 + 6n$ for 2 different integer values of n.

40.

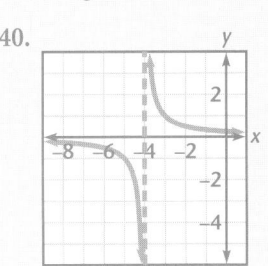

EYEWITNESS MATH

HOW SECRET IS SECRET?

Biggest Division a Giant Leap in Math

BY GINA KOLATA

Connection Machine® is one of the most powerful high-performance computers in the world.

New York, June 20, 1990—In a mathematical feat that seemed impossible a year ago, a group of several hundred researchers using about 1000 computers has broken a 155-digit number down into three smaller numbers that cannot be further divided.

The latest finding could be the first serious threat to systems used by banks and other organizations to encode secret data before transmission, cryptography experts said yesterday.

These systems are based on huge numbers that cannot be easily factored, or divided, into two numbers that cannot be divided further.

In making these codes, engineers have to strike a delicate balance when they select the numbers used to scramble messages. If they choose a number that is easy to factor, the code can be broken. If they make the number much larger, and much harder to factor, it takes much longer for the calculations used to scramble a message.

For most applications outside the realm of national security, cryptographers have settled on numbers that are about 150 digits long.

Dr. Mark Manasse of the Digital Equipment Corporation's Systems Research Center in Palo Alto, Calif., calculates that if a computer could perform a billion divisions a second, it would take 10 to the 60th years, or 1 with 60 zeros after it, to factor the number simply by trying out every smaller number that might divide into it easily. But with a newly discovered factoring method and with a world-wide collaborative effort, the number was cracked in a few months.

Factoring a 155-Digit Number:

13,407,807,929,
942,597,099,574,
024,998,205,846,
127,479,365,820,
592,393,377,723,
561,443,721,764,
030,073,546,976,
801,874,298,166,
903,427,690,031,
858,186,486,050,
853,753,882,811,
946,569,946,433,
649,006,084,097

equals
2,424,833

times
7,455,602,825,
647,884,208,337,
395,736,200,454,
918,783,366,342,
657

times
741,640,062,627,
530,801,524,787,
141,901,937,474,
059,940,781,097,
519,023,905,821,
306,144,415,759,
504,705,008,092,
818,711,693,940,
737

You can use small numbers to get an idea of how code systems like the one described in the article work. Before you begin, you need a key (a number p that is not prime). You will first use p to encode a secret message and then use p and its factors to decode the message.

$X \rightarrow$ (secret message) → Sender uses p to encode the secret message. → Y → (coded message) → Receiver uses p and its factors to decode the message. → X → (secret message)

If you were sending actual high-security messages, you would obtain a value for *p* by finding two gigantic prime numbers and multiplying them. Since you don't have a supercomputer to multiply and divide numbers with hundreds of digits, you need to use numbers small enough to handle on a calculator.

Cooperative Learning

For this activity, use 55 for *p*. You will use a special algorithm to encode and decode the message. To keep things simple, let your secret message be a two-digit number between 11 and 50.

A. To encode the secret message, follow these steps.

(1) Calculate X^3.

(2) Divide X^3 by 55. Multiply the whole-number part of the quotient by 55. Subtract that product from X^3. The difference is the remainder. Use the remainder as the coded message, *Y*.

(3) If your calculator can display only 8 digits, then *Y* cannot be more than 13. If it can display 10 digits, *Y* cannot be more than 26. If *Y* is too large, start over at Step 1 with a different value for *X*.

Let *X* be 42.

$$X^3 = 74{,}088$$
$$74{,}088 \div 55 = 1347.054545\ldots$$
$$1347 \times 55 = 74{,}085$$
$$74{,}088 - 74{,}085 = 3$$
$$Y = 3$$

B. To decode the secret message, use the factors of 55. Follow these steps.

(1) Find Y^3. Divide Y^3 by 5. Multiply the whole-number part of the quotient by 5, and subtract this product from Y^3. Call the remainder *a*.

(2) Find Y^7. Divide Y^7 by 11. Multiply the whole-number part of the quotient by 11, and subtract this product from Y^7. Call the remainder *b*.

(3) Evaluate $11a + 45b$, and divide by 55. Multiply the whole-number part of the quotient by 55, and subtract this product from the value of $11a + 45b$. How does the remainder compare to *X*, the secret "message"?

Y^3 is 27.

$$27 \div 5 = 5.4$$
$$5 \times 5 = 25$$
$$27 - 25 = 2 \rightarrow a = 2$$

Y^7 is 2187.

$$2187 \div 11 = 198.8181\ldots$$
$$198 \times 11 = 2178$$
$$2187 - 2178 = 9 \rightarrow b = 9$$

$$11a + 45b = 427$$
$$427 \div 55 = 7.7636\ldots$$
$$7 \times 55 = 385$$
$$427 - 385 = 42$$

The Enigma Machine was used in World War II for breaking coded messages.

C. Explain how you used the factors of *p* to encode or decode the message. Why would it be important for the factors of *p* to be secret?

Try using the code system to send a one-word message to a friend. Use *p* = 55 as the key.

How do you think coding systems may be affected if efficient methods for factoring very large composite numbers are found?

C. Sample answer: The factors are essential for decoding the message. Someone who could factor *p* would be able to decode a secret message not intended for him or her. This could have serious consequences if the message contained sensitive information. It could jeopardize a person's or a country's security or cause large amounts of money to be lost.

Cooperative Learning

Students can use numbers to stand for letters, for example, 11 = A and 12 = B, and so on. They can then make a small message and send it as a numerical code. Remind students that they will need to generate numbers whose values for *Y* do not exceed the size limits for their calculators or spreadsheets. Once a code is made, another activity for the class would be to break the code.

Discuss

You may wish to introduce terms and notation used in modular arithmetic. For example, $6 \equiv 68921 \pmod{55}$ means that 68921 and 6 are equivalent in a counting system in which you go back to 1 each time you reach 55 (instead of going on to 56). Also, if you divide 68921 by 55 the remainder is 6. This is also the remainder for 6 divided by 55. Modular notation provides a convenient way to account for classes of numbers with the same reminder.

Students can also find out more about codes from literature such as *The Gold Bug* by Edgar Allen Poe or *The Adventure of the Dancing Men* by Arthur Conan Doyle.

Objectives

RESOURCES

- Practice Master   9.5
- Enrichment Master   9.5
- Technology Master   9.5
- Lesson Activity Master   9.5
- Quiz   9.5
- Spanish Resources   9.5

Assessing Prior Knowledge

Add these fractions and simplify.

A. $\frac{2}{3} + \frac{5}{8}$ $\left[\frac{31}{24}\right]$

B. $\frac{2}{x} + \frac{2x}{(x+1)}$ $\left[\frac{2x^2 + 2x + 2}{x^2 + x}\right]$

TEACH

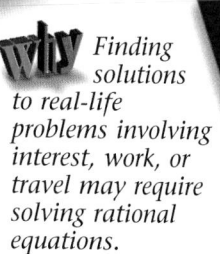 Rational functions can model many situations, like the relationship between distance and time, and the relationship between the force of gravity on Earth and the distance between the Earth and the sun.

Alternate Example 1

Donald runs 5 miles and bikes 5 miles. He bikes 15 miles per hour faster than he runs.

A. Write a rational function, in terms of running speed, for the total time it takes Donald to complete his workout. $\left[T(r) = \frac{10r + 75}{r^2 + 15r}\right]$

B. At what speeds must Donald run and bike to complete the workout in 0.7 hours? [run: 10 mph; bike: 25 mph]

LESSON 9.5   Solving Rational Equations

 *Finding solutions to real-life problems involving interest, work, or travel may require solving rational equations.*

Sometimes you can simplify rational expressions by finding a common denominator.

▶ EXAMPLE 1

Sports   In training for a triathlon, Rachel swims $\frac{1}{2}$ mile and runs 6 miles. She runs about 5 miles per hour faster than she swims.

Ⓐ Write a rational function, in terms of swimming velocity, for the total time it takes Rachel to complete her workout.

Ⓑ At what velocities must Rachel swim and run to complete the workout in 1 hour?

Solution▶

Ⓐ Let t be the total time it takes Rachel to complete her workout. Then $t = t_s + t_r$, where t_s is her swimming time and t_r is her running time. Use the relationship distance equals rate times time, or $d = rt$, to find Rachel's swimming time and running time.

Swimming time, t_s:
Rachel swims $\frac{1}{2}$ mile, so substitute $\frac{1}{2}$ for d and solve for t_s.

$$d = rt_s$$
$$\frac{1}{2} = rt_s$$
$$t_s = \frac{1}{2r}$$

Running time, t_r:
Rachel runs 6 miles, so substitute 6 for d. She runs about 5 miles per hour faster than she swims, so use $r + 5$ for rate. Solve for t_r.

$$d = rt$$
$$6 = (r + 5)t_r$$
$$t_r = \frac{6}{r + 5}$$

Total workout time $t = t_s + t_r$:

$$t = \frac{1}{2r} + \frac{6}{r + 5}$$
$$t = \frac{1}{2r}\frac{(r + 5)}{(r + 5)} + \frac{6}{(r + 5)}\frac{(2r)}{(2r)}$$
$$t = \frac{13r + 5}{2r^2 + 10r}$$

ALTERNATIVE teaching strategy

Using Cognitive Strategies Begin by adding simple rational functions and work up to more difficult ones. A method taught in this way can help students develop better problem-solving strategies.

$$\frac{1}{x} + \frac{1}{x} = \left[\frac{2}{x}\right]$$
$$\frac{2}{x} + \frac{1}{x^2} = \left[\frac{2x + 1}{x^2}\right]$$
$$\frac{2}{x + 1} + \frac{1}{x^2} = \left[\frac{2x^2 + x + 1}{x^3 + x^2}\right]$$
$$\frac{2}{x + 1} + \frac{3}{x - 2} = \left[\frac{5x - 1}{x^2 - x - 2}\right]$$
$$\frac{2}{x^2 + x} + \frac{3}{2x + 2} = \left[\frac{3x + 4}{2x^2 + 2x}\right]$$

B Graph $y = \frac{13x + 5}{2x(x + 5)}$. When y is 1, x is 2.5.

Thus, if Rachel swims 2.5 miles per hour and runs 2.5 + 5, 7.5 miles per hour, she can complete the workout in 1 hour. ❖

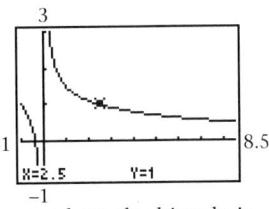

Try This Rachel decides to add biking to her workout. Her speed on the bicycle is about 12 miles per hour faster than her swimming speed. How far will she be able to bike if she completes her new workout in 2 hours and her swimming and running times do not change?

Sometimes you can solve a rational equation by converting it into a polynomial equation and finding the roots.

Consider the equation from Example 1, Part **B** $\frac{13r + 5}{2r^2 + 10r} = 1$, where $r \geq 0$.

Multiply both sides of the equation by the denominator, $2r^2 + 10r$.

$$\frac{13r + 5}{2r^2 + 10r} = 1$$
$$13r + 5 = 2r^2 + 10r$$

Combine like terms.

$$2r^2 - 3r - 5 = 0$$

Use the quadratic formula.

$$r = \frac{3 \pm 7}{4}$$
$$r = 2.5 \text{ or } r = -1$$

Since the rate must be positive, her rate must be 2.5 miles per hours as found in Example 1.

EXAMPLE 2

Solve $\frac{x}{x - 3} + \frac{2x}{x + 3} = \frac{18}{x^2 - 9}$ for x.

Solution▸

Find the lowest common denominator: LCD = $(x - 3)(x + 3) = x^2 - 9$.
Solve for x.

$$\left(\frac{x}{x - 3} + \frac{2x}{x + 3}\right)(x - 3)(x + 3) = \left(\frac{18}{x^2 - 9}\right)(x^2 - 9)$$
$$x(x + 3) + 2x(x - 3) = 18$$
$$x^2 + 3x + 2x^2 - 6x = 18$$
$$3x^2 - 3x - 18 = 0$$
$$3(x - 3)(x + 2) = 0$$
$$x = 3 \text{ or } x = -2$$

Check▸

If x is 3, the expression $\frac{x}{x - 3}$ in the original equation is undefined; therefore, x cannot be 3.

Substitute –2 for x in the original equation.

$$\frac{(-2)}{(-2) - 3} + \frac{2(-2)}{(-2) + 3} \stackrel{?}{=} \frac{18}{(-2)^2 - 9}$$
$$-\frac{18}{5} = -\frac{18}{5} \quad \text{True}$$

Thus, –2 is the only solution. ❖

Recall from Chapter 5 that roots such as "3" in Example 2 are called extraneous roots.

CRITICAL *Thinking*

Why do extraneous roots sometimes appear in the process of solving rational equations?

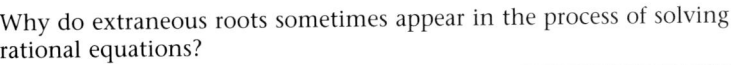

Aongoing
ASSESSMENT

Try This

14.5 miles

Alternate Example 2

Solve $\frac{2x}{x - 2} - \frac{x}{x + 4} = \frac{-25}{x^2 + 2x - 8}$.

[$x = -5$]

Cooperative Learning

Have students work in small groups to write their own real world problems similar in style to Example 1. Encourage students to make the situation fit their own abilities and needs. Make sure each student in every group understands how to express the problem as a rational equation and solve the problem. You may wish to have groups explain their situation and their problem-solving steps to the class.

CRITICAL *Thinking*

Multiplying both sides of an equation by a variable expression may introduce extraneous roots. For example,

$$x - 3 = 0 \quad \text{(root 3)}$$
$$x(x - 3) = 0x$$
$$x^2 - 3x = 0$$
$$x(x - 3) = 0 \quad \text{(roots: 0 and 3)}$$

ENRICHMENT Encourage students to create their own rational equations to solve. Motivated students might enjoy the challenge of creating rational equations with extraneous roots. Encourage students to use real-world setting for their equations. Have students trade and solve one another's problems.

INCLUSION *strategies* **Using Discussion** Encourage students to explain the goals involved in solving rational equations. Students should see that one goal is to reduce the problem to an equation they already know how to solve.

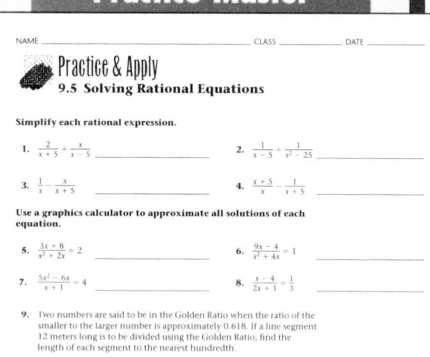
EXERCISES & PROBLEMS

Communicate

1. Explain what rational equations are.

2. How are separate rational expressions combined?

3. How can some rational equations be solved by algebraic means?

4. Explain what an extraneous root of a rational equation is.

5. Name two reasons why it is important to check possible roots of rational equations.

Practice & Apply

Simplify the following rational expressions.

6. $\frac{1}{x-2} + \frac{x}{x+2}$

7. $\frac{x}{x+3} - \frac{2x}{2x+3}$

8. $\frac{1}{x^2} + \frac{x^2}{x^2+9}$

9. $\frac{1}{x-3} + \frac{1}{x^2-9}$

10. $\frac{2}{x} - \frac{x-1}{x^3+3}$

11. $\frac{1}{x} + \frac{1}{x^2} + \frac{1}{x^3}$

Use your graphics calculator to approximate all the solutions of the following equations.

12. $\frac{12x+5}{x^2-5} = 2$ 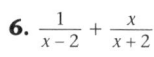 –1.06, 7.06

13. $\frac{12x+5}{x^2+5} = 2$ 0.45, 5.55

14. $\frac{3x+7}{x^2+x+1} = 4$ –1, 0.75

15. $\frac{3x+7}{x^2+x-1} = 4$ –1.79, 1.54

16. $\frac{x^2-3}{x^2+3} = -\frac{1}{2}$ –1, 1

17. $\frac{x^2+3}{x^2-3} = \frac{3}{2}$ –3.87, 3.87

18. $\frac{x+9}{x^3+27} = 1$ –2.75

19. $\frac{x^2+x+3}{x+2} = 2$ –0.62, 1.62

20. $\frac{x^2+2x-5}{3x+1} = 1$ –2, 3

21. Write a rational function that represents the sum of a positive number and its reciprocal. $F(x) = x + \frac{1}{x} = \frac{x^2+1}{x}$ for $x > 0$

22. Graph the function you wrote for Exercise 21 for $x > 0$. What is the vertical asymptote? $x = 0$

23.  **Maximum/Minimum** Use the trace feature on your calculator to find the minimum of the function you wrote for Exercise 21 where $x > 0$. What x-value corresponds to this minimum? min: $y = 2$ at $x = 1$

RETEACHING the lesson

Using Technology One method to avoid lengthy calculations is to use technology. In many cases, the rational equation can be solved by representing each side of the equation as separate functions. These functions can be graphed simultaneously. The solution to the equation is the intersection of the graphs.

6. $\frac{x^2-x+2}{x^2-4}$

7. $\frac{-3x}{2x^2+9x+9}$

8. $\frac{x^4+x^2+9}{x^4+9x^2}$

9. $\frac{x+4}{x^2-9}$

10. $\frac{2x^3-x^2+x+6}{x^4+3x}$

11. $\frac{x^2+x+1}{x^3}$

24 **Time Management** Simone can do a job in x hours. Michael can do the same job in $x+1$ hours, and Faye can do the same job in $x-1$ hours. If they can complete the job working together in 10 hours, how long would it take each person to complete the job individually? (HINT: Simone does $\frac{10}{x}$ of the job per hour when they work together.) Simone: 30 hours; Michael:31 hours; Faye:29 hours

25. **Cultural Connection: Asia** A ninth-century Indian mathematician, Mahavira, posed the following problem.

There are 4 pipes leading into a well. Individually, the pipes can fill the well (in order) in $\frac{1}{2}$, $\frac{1}{3}$, $\frac{1}{4}$, and $\frac{1}{5}$ of a day. How long would it take for the pipes to fill the well if they were all working simultaneously, and what fraction of the well would be filled by each pipe?
$\frac{1}{14}$ of a day ; $\frac{1}{7}$, $\frac{3}{14}$, $\frac{2}{7}$, $\frac{5}{14}$

Solve the following rational equations by algebraic means.

26. $\frac{6}{x-2} = \frac{5}{x-3}$ $x = 8$

27. $\frac{3}{x^2-4} = \frac{-2}{5x+10}$ $x = -5.5$

28. $\frac{2x-1}{2} - \frac{x+2}{2x+5} = \frac{6x-5}{6}$ $x = -1$

Travel Suppose that you travel by car 20 miles at r miles per hour and then 30 miles at $r+15$ miles per hour. The total time for the trip is $t_{total} = \frac{20}{r} + \frac{30}{r+15}$. The average speed for the trip is the total distance divided by the total time.

29. Express the average speed, $s(r)$, as a rational function. $s(r) = \frac{r(r+15)}{r+6}$

30 Graph the average speed for $50 \le r \le 70$. What is the average speed when r is 60?
68.18 mph

Look Back

31 Graph $f(x) = 2x^2 - 3x - 5$. Find the vertex to the nearest hundredth. **[Lesson 5.3]** (0.75, −6.13)

32. Solve $\log(x+2) = 1$. **[Lesson 7.6]** $x = 8$

33. If $\sin x = 0.8$ and $\cos x = 0.6$, find the value of $\tan x$. **[Lesson 8.2]**
$\tan x = 1.33$

Identify the vertical shift, amplitude, period, and phase shift for each of the following functions. **[Lesson 8.4]**

34. $f(\theta) = 3\cos\theta$

35. $g(\theta) = -4\sin 2\theta$

36. $h(\theta) = 1 + 2\cos\theta$

37. $k(\theta) = 2 + \sin(\theta + 120^\circ)$

Look Beyond

38 Graph the following equations and compare their graphs. How are they alike, and how are they different?

$$9x^2 + 25y^2 = 36 \qquad x^2 + y^2 = 36 \qquad \frac{x^2}{25} - \frac{y^2}{81} = 1$$

	Vertical Shift	Amplitude	Period	Phase Shift
34.	0	3	360°	0
35.	0	4	180°	90°
36.	1	2	360°	0
37.	2	1	360°	−120°

38. In all three equations the sum or difference of an x^2 term and a y^2 term is equal to a constant. The resulting graphs are different: $9x^2 + 25y^2 = 36$ is an ellipse centered at (0, 0), $x^2 + y^2 = 36$ is a circle centered at (0, 0), and $\frac{x^2}{25} = \frac{y^2}{81} = 1$ is a hyperbola centered at (0, 0).

alternative ASSESSMENT

Performance Assessment

Have small groups of students develop a hypothetical time-management problem such as Exercise 24. A sample situation might be the task of cleaning the chalkboards. Have students perform the task as a group and take a group time. Then check the results by individually performing the task. After data is collected, have each group write the rational equation that models their group, letting one person's time be x and letting the others in the group have times in terms of x. Have each student solve for x and check their solutions against the actual times.

Look Beyond

The equations listed are equations for conic sections. These exercises tie together the concepts of rational functions and conic equations.

Technology Master

NAME ______________ CLASS ________ DATE ________

Technology
9.5 Checking Your Work with a Graphics Calculator

To find the sum of $\frac{3}{x+2}$ and $\frac{-2}{x+3}$, you would work as follows.

$$\frac{3}{x+2} + \frac{-2}{x+3} = \frac{3}{x+2}\cdot\frac{x+3}{x+3} + \frac{-2}{x+3}\cdot\frac{x+2}{x+2}$$
$$= \frac{3(x+3) + (-2)(x+2)}{(x+2)(x+3)}$$
$$= \frac{3x+9-2x-4}{(x+2)(x+3)}$$
$$= \frac{x+5}{(x+2)(x+3)}$$
$$= \frac{x+5}{x^2+5x+6}$$

If you graph $y = \frac{3}{x+2} + \frac{-2}{x+3}$ and $y = \frac{x+5}{x^2+5x+6}$ on the same coordinate plane, you should see that the graphs coincide.

Use pencil and paper to find each sum or difference. Then use a graphics calculator to check your answers.

1. $\frac{8}{3x-2} - \frac{2}{3x-1}$ ______

2. $\frac{x+3}{x+1} + \frac{2}{x+1}$ ______

3. $\frac{4}{3x+2} + \frac{-2}{2x-1}$ ______

4. $\frac{-2x+1}{x+3} - \frac{2x}{2x-1}$ ______

Use pencil and paper to solve each equation. Then use a graphics calculator to check your answers.

5. $\frac{8x}{3x-2} = -4$ ______

6. $\frac{3x+1}{4x-2} = 7$ ______

7. $\frac{x-1}{6} = \frac{4}{x+1}$ ______

8. $\frac{x+1}{-6} = \frac{4}{x+1}$ ______

HRW Advanced Algebra   Technology   99

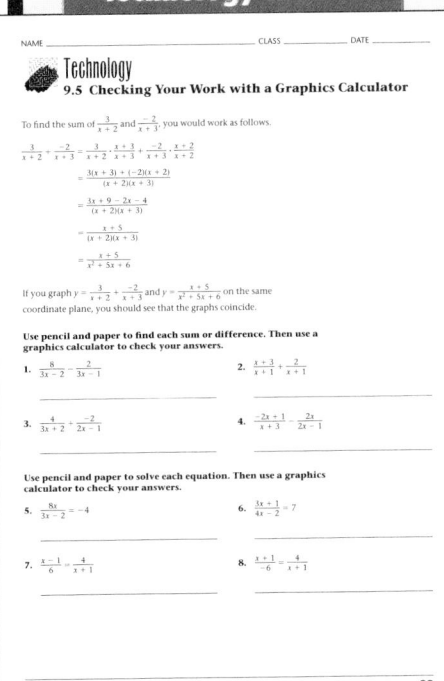

Modeling

the

Stock Market

Students should work in groups of three or four to complete the project. Each group will use mathematical models of the predicted performances of two different stocks to develop an investment strategy that maximizes the investment value growth. The models are constructed by combining exponential, trigonometric, and rational functions. Use radians.

Company A is determined by a financial analyst to have potentially large profits. However, there is also a high risk that the price of the stock may drop sharply.

The analyst determines that Company B is not as likely to make very high profits, but its profits are not as likely to drop sharply since its sales are to a broad base of regular customers.

Company A is a "high-risk, high-return" stock, while Company B is a "low-risk, moderate-return" stock.

The analyst uses the following mathematical models for the projected stock prices, P_A and P_B, for the two companies.

Company A: $P_A(t) = 5e^{-\left(\frac{t}{6}\right)} \cos 2t + \dfrac{3t^2 + 17t + 1}{t+5}$

Company B: $P_B(t) = 5e^{-\left(\frac{t}{6}\right)} \cos t + \dfrac{3t^2 + 1}{t+5}$

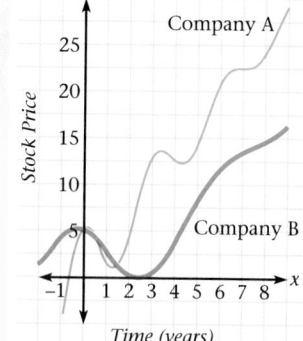

EET JOURNAL.

SHARON, PENNSYLVANIA

any, Inc. All Rights Reserved.

AUGUST 15, 1995

Company A Reports High Profits—Investors Warned of the Risks

CENTS

You will need tokens, such as counting chips or pennies, that can be used to represent shares of stock per group.

What Is the Best Investment Strategy?

1. Each team begins at $t = 0$ with 25 shares of stock in each company. Find the identical initial value of each team's investment. For simplicity, assume that the analyst's model closely predicts the actual performance of the two stocks.

2. Find the value of each stock every 6 months for 8 years. Record your data in a table.

3. Each team may sell stock in one company and buy stock in the other every 6 months during the eight years.

$$\text{Number of shares bought} = \frac{(\text{Price of shares sold})(\text{Number of shares sold})}{\text{Price of shares bought}}$$

Round the number of shares bought to the nearest whole number. Keep track of the number of shares held by each team in each company.

4. Each team may use any method to decide on its investment strategy. At the end of 8 years, the team with the largest investment value wins.

Chapter Review

Chapter 9 Review

Vocabulary

constant of variation	494	inverse variation	494
even function	510	odd function	510
hole in a graph	519	rational function	502

Key Skills & Exercises

Lesson 9.1

➤ **Key Skills**

Given an inverse variation relationship, find the constant of variation, and write the equation of variation.

If y varies inversely as x, then $xy = k$ or $y = \frac{k}{x}$, where k is the constant of variation.

➤ **Exercises**

Determine the constant of variation, and write the equation of variation.

1. y varies inversely as x, and y is 75 when x is 10. $k = 750;\ y = \frac{750}{x}$

2. In the following table, y varies inversely as x.

x	2	3	4	6
y	36	24	18	12

$k = 72;\ y = \frac{72}{x}$

Lesson 9.2

➤ **Key Skills**

Identify the graph, and write the equation of the vertical and horizontal asymptotes of a rational function that is the quotient of two linear functions.

The graph of $f(x) = \frac{2x + 7}{x + 3}$ is a hyperbola with a vertical asymptote $x = -3$, because when x is -3, the denominator of the rational function is 0. The horizontal asymptote is $y = 2$, because $(2x + 7) \div (x + 3) = 2 + \frac{1}{x + 3}$, which approaches a value of 2 as the absolute value of x gets larger.

➤ **Exercises**

Draw the graph and identify the vertical and horizontal asymptotes of the function.

3 $f(x) = \frac{1}{x + 3}$ **4** $f(x) = \frac{1}{x - 1} + 3$ **5** $f(x) = \frac{x + 2}{x - 1}$

3.
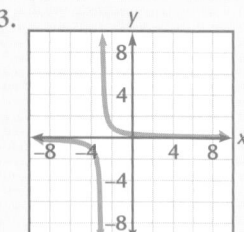

Asymptotes: $x = -3$
$y = 0$

4.
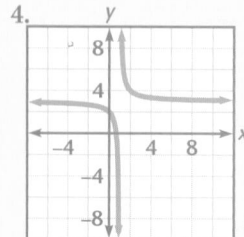

Asymptotes: $x = 1$
$y = 3$

5.
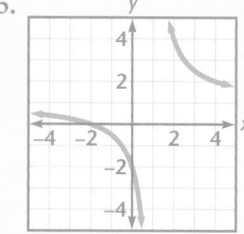

Asymptotes: $x = 1$
$y = 1$

Lesson 9.3

➤ Key Skills

Determine whether a function is even, odd, or neither.

even function: $f(-x) = f(x)$
odd function: $f(-x) = -f(x)$

Graph a function that is the reciprocal of a polynomial, and find its vertical asymptotes.

The function $f(x) = \frac{1}{(x-a)(x-b)}$ has vertical asymptotes $x = a$ and $x = b$.

➤ Exercises

Determine whether the function is even, odd, or neither.

6. $f(x) = \frac{1}{x^2 - 4}$ even

7. $f(x) = \frac{1}{x^3 - 2}$ neither

Graph each function, and find its vertical asymptotes.

8. $f(x) = -\frac{1}{x^2 - 9}$ $x = 3$; $x = -3$

9. $f(x) = \frac{1}{x^2 - x - 6}$ $x = 3$, $x = -2$

Lesson 9.4

➤ Key Skills

Find the vertical and horizontal asymptotes and the domain of a rational function in which the degree of the numerator is not greater than the degree of the denominator.

If possible, factor the denominator to find restrictions on the domain. Then try to simplify the function by factoring the numerator and cancelling common factors. If $x + a$ is a factor of the denominator in the *simplified* function, there is a vertical asymptote $x = -a$.

If the degree of the denominator is greater than the degree of the numerator, the horizontal asymptote is $y = 0$ (the x-axis). If the degrees of the numerator and denominator are equal, then the horizontal asymptote is $y = \frac{a}{b}$, where a and b are the coefficients of the highest powers of x in the numerator and denominator, respectively.

➤ Exercises

Find the holes, vertical and horizontal asymptotes, and give the domain of the function.

10. $f(x) = \frac{3x}{x^2 - x - 2}$

11. $f(x) = \frac{x^2 - 3x}{x^2 + x - 12}$

12. $f(x) = \frac{4x^2 - 9}{2x^2 - 5x + 3}$

Lesson 9.5

➤ Key Skills

Combine rational functions to get a single rational function.

Determine the common denominator of all the rational expressions, and express each with this common denominator. Then combine numerators to form one rational expression.

Solve equations that contain rational expressions.

Multiply both sides of the equation by the least common denominator of all the rational expressions. Set the resulting polynomial equal to zero, and find its zeros. Each of these zeros is a root if it is in the domain of the original function and if it is not extraneous.

	Holes	Vert. Asym.	Horiz. Asym.	Domain
10.	none	$x = -1, x = 2$	$y = 0$	$x \neq -1, 2$
11.	$x = 3$	$x = -4$	$y = 1$	$x \neq -4, 3$
12.	$x = 1.5$	$x = 1$	$y = 2$	$x \neq 1, 1.5$

➤ *Exercises*

Simplify.

13. $\dfrac{1}{x-4} + \dfrac{2x-5}{x^2-16} \quad \dfrac{3x-1}{x^2-16}$

14. $\dfrac{5}{x^2} + \dfrac{2}{x} - \dfrac{x+1}{x^2+2} \quad \dfrac{x^3+4x^2+4x+10}{x^4+2x^2}$

Find all the solutions for the equation.

15. $\dfrac{7x-3}{x^2-3} = 5 \quad -1, 2.4$

16. $\dfrac{x}{x-3} - \dfrac{7}{x+5} = \dfrac{24}{x^2+2x-15} \quad -1$

Application

Electronics The total resistance in a circuit with two resistors connected in parallel with resistances R_1 and R_2 is given by $\dfrac{1}{R} = \dfrac{1}{R_1} + \dfrac{1}{R_2}$.

17. Show how to change this equation to the form $R = \dfrac{R_1 R_2}{R_1 + R_2}$.

18. If the total resistance in a circuit is 1.5 ohms and one resistor has a resistance of 2.5 ohms, what is the resistance of the second resistor? 3.75 ohms

Physics According to Newton's Law of Gravitation, the weight of an object varies inversely as the square of the distance from the object to the center of the Earth (the radius of Earth is approximately 4000 miles).

19. If an astronaut weighs 175 pounds on Earth, what will the astronaut weigh at a height of 60 miles above the Earth's surface before going into orbit? 169.87 lb

20. If an astronaut weighs 145 pounds 80 miles above the Earth's surface, how much does the astronaut weigh on Earth? 150.86 lb

21. Cycling A bicycle wheel with a 26-inch diameter takes ten revolutions to go a certain distance, d. How many revolutions will a 20-inch diameter wheel make in covering twice that distance, $2d$? 26 revolutions

Sports Liam plans to spend a certain amount of time every day getting in shape for the school track team. He will run 3 miles and ride his bike 6 miles. He rides 6 miles per hour faster than he runs.

22. How fast must Liam run and bike in order to finish his workout in 1 hour? run: 6 mph ; bike: 12 mph

23. If Liam decides to increase his running distance by 2 miles, how long will it take him to complete his workout? $1\frac{1}{3}$ hr, or 1 hr, 20 min

24. In addition to increasing his running distance, Liam decides to increase his biking distance by 4 miles. How long will it take for Liam to complete his workout? $1\frac{2}{3}$ hr, or 1 hr, 40 min

25. After working out for several months, Liam intends to increase his running speed. His biking speed will remain the same. How fast must Liam run in order to complete the workout in Exercise 24 in 1 hour, 30 minutes? Is this reasonable? Explain. 7.5 mph

17. Multiply every term in the equation $\frac{1}{R} = \frac{1}{R_1} + \frac{1}{R_2}$ by RR_1R_2. So $R_1R_2 = RR_2 + RR_1$, or $R_1R_2 = R(R_1 + R_2)$, or finally $R = \dfrac{R_1R_2}{R_1 + R_2}$

Chapter 9 Assessment

Determine the constant of variation, and write the equation of variation.

1. y varies inversely as the square root of x, and y is 10 when x is 16. $k = 40$, $y = \frac{40}{\sqrt{x}}$

2. y varies inversely as x, and y is 0.5 when x is 16. $k = 8$, $y = \frac{8}{x}$

3. If the time required to paint a building varies inversely as the number of painters, and it takes 4 painters 5 days to do the job, how many painters would be needed to complete the job in 2 days?   10 painters

Draw the graph, and identify the vertical and horizontal asymptotes of the function.

4. $f(x) = \frac{1}{x-5}$ **5.** $f(x) = \frac{1}{x+1} - 3$ **6.** $f(x) = \frac{2x+1}{x}$

7. Explain how the graph of $g(x) = \frac{1}{x+3} + 1$ compares with the graph of $f(x) = \frac{1}{x}$.   g is translated 3 units to the left and 1 unit up

Graph each function, and find its vertical asymptotes.

8. $f(x) = \frac{1}{x^2-1}$ $x = 1, x = -1$ **9.** $f(x) = -\frac{1}{x^2+2x-3}$ $x = -3, x = 1$

Find the holes, the vertical and horizontal asymptotes, and the domain of the function.

10. $f(x) = \frac{5x}{x^2-9x+20}$ **11.** $f(x) = \frac{x^2+3x}{x^2-2x-15}$

Simplify.

12. $\frac{3}{x^2+2x} - \frac{5}{x^2-4}$ $\frac{-2x-6}{x(x^2-4)}$ **13.** $\frac{3}{x} - \frac{1}{x^2} + \frac{x}{x+1}$ $\frac{x^3+3x^2+2x-1}{x^2(x+1)}$

Find all the solutions for the equation.

14. $\frac{3x-1}{x^2+2} = -1$ $-2.62, -0.38$ **15.** $\frac{x}{x-2} + \frac{2}{x+3} = \frac{10}{x^2+x-6}$ -7

Music The pitch of a note is determined by the frequency of its sound wave. This frequency, f, is equal to the velocity, v, of the sound through the air divided by the wavelength, w. In other words, $f = \frac{v}{w}$. If the frequency doubles, the pitch rises by one octave.

16. If the velocity of a sound wave is 1056 feet per second, graph the relationship between frequency and wavelength for $w > 0$.

17. What happens to the pitch of a note as the wavelength increases?   pitch decreases

18. If the wavelength is divided by 2, what happens to the pitch of the note?   pitch rises by one octave

19. If the oboe tunes the orchestra by playing an A-note with a frequency of 440 cycles per second, what is the frequency of the A-note that is one octave higher?   880 cycles per second

20. **Sports** Nick's morning workout is a combination of jogging and riding his bicycle. He takes a total of 1 hour to jog 2 miles and bike 8 miles. His speed on the bicycle is 8 miles per hour faster than his jogging speed. Find Nick's approximate jogging speed and bicycling speed.   jogging speed: ≈ 5.1 mph   biking speed: ≈ 13.1 mph

10. Holes: none
Asymptotes:
$x = 4$
$x = 5$
$y = 0$
Domain: $x \neq 4, x \neq 5$

11. Holes: $x = -3$
Asymptotes:
$x = 5$
$y = 0$
Domain: $x \neq -3, x \neq 5$

16.
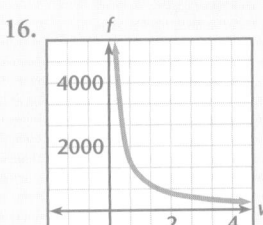

4.
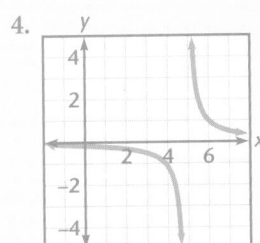

Asymptotes: $x = 5$
$y = 0$

5.
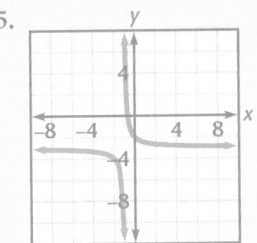

Asymptotes: $x = -1$
$y = -3$

6.
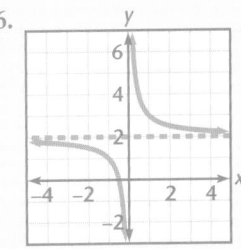

Asymptotes: $x = 0$
$y = 2$

10 CHAPTER

Analytic Geometry

Meeting Individual Needs

10.1 Parabolas

Core

Inclusion Strategies, p. 538
Reteaching the Lesson, p. 539
Practice Master 10.1
Enrichment Master 10.1
Technology Master 10.1
Lesson Activity Master 10.1

[2 days]

Core Plus

Practice Master 10.1
Enrichment, p. 538
Technology Master 10.1
Lesson Activity Master
  10.1
Interdisciplinary
  Connection, p. 537

[2 days]

10.2 Circles

Core

Inclusion Strategies, p. 546
Reteaching the Lesson, p. 547
Practice Master 10.2
Enrichment Master 10.2
Technology Master 10.2
Lesson Activity Master 10.2

[2 days]

Core Plus

Practice Master 10.2
Enrichment, p. 546
Technology Master 10.2
Lesson Activity Master
  10.2
Interdisciplinary
  Connection, p. 545

[2 days]

10.3 Ellipses

Core

Inclusion Strategies, p. 553
Reteaching the Lesson, p. 554
Practice Master 10.3
Enrichment Master 10.3
Technology Master 10.3
Lesson Activity Master 10.3

[Optional*]

Core Plus

Practice Master 10.3
Enrichment, p. 553
Technology Master 10.3
Lesson Activity Master
  10.3
Interdisciplinary
  Connection, p. 552

[2 days]

10.4 Hyperbolas

Core

Inclusion Strategies, p. 563
Reteaching the Lesson, p. 564
Practice Master 10.4
Enrichment Master 10.4
Technology Master 10.4
Lesson Activity Master 10.4
Mid-Chapter Assessment Master

[Optional*]

Core Plus

Practice Master 10.4
Enrichment, p. 562
Technology Master 10.4
Lesson Activity Master
  10.4
Interdisciplinary
  Connection, p. 561
Mid-Chapter Assessment
  Master

[2 days]

10.5 Solving Non-Linear Systems of Equations

Core

Inclusion Strategies, p. 570
Reteaching the Lesson, p. 571
Practice Master 10.5
Enrichment Master 10.5
Technology Master 10.5
Lesson Activity Master 10.5

[2 days]

Core Plus

Practice Master 10.5
Enrichment, p. 570
Technology Master 10.5
Lesson Activity Master
  10.5

[1 day]

10.6 Exploring Parametric Representation of Conic Sections

Core

Inclusion Strategies, p. 580
Reteaching the Lesson, p. 581
Practice Master 10.6
Enrichment Master 10.6
Technology Master 10.6
Lesson Activity Master 10.6

[Optional*]

Core Plus

Practice Master 10.6
Enrichment, p. 579
Technology Master 10.6
Lesson Activity Master
  10.6

[2 days]

Chapter Summary

Core	Core Plus
Chapter 10 Project, pp. 586–587	Chapter 10 Project, pp. 586–587
Lab Activity	Lab Activity
Long-Term Project	Long-Term Project
Chapter Review, pp. 588–590	Chapter Review, pp. 588–590
Chapter Assessment, p. 591	Chapter Assessment, p. 591
Chapter Assessment, A/B	Chapter Assessment, A/B
Alternative Assessment	Alternative Assessment
Cumulative Assessment, pp. 592–593	Cumulative Assessment, pp. 592–593
[3 days]	**[3 days]**

Reading Strategies

As students read the lesson they will find material that they do not understand, they will encounter problems that relate to material in a different lesson of a different book, or they will come across methods and topics they would like to pursue further. In each case, it is important that students have writing paper available. They can also jot down related thoughts, work exercises, or record questions. An attempt to rely on memory will often mean that the thought is lost. Recommend that students have a small notebook handy in which they can write thoughts about anything that occurs to them in the course of reading the material in the lessons.

Visual Strategies

Direct the students to look at the pictures and associated graphs in the chapter opener. Encourage them to speculate about where mathematics might be useful in studying such phenomena. Also have them look at the graphs in the chapter and discuss how they differ from graphs in earlier chapters. What aspect of these graphs makes them different from graphs in earlier chapters? Can they imagine how such functions could model the phenomena in the photographs?

Cooperative Learning

GROUP ACTIVITIES	
Graphing equations and functions	Lesson 10.1, Exploration
The standard equation of a circle	Lesson 10.2, Example 4
The standard equation of an ellipse	Lesson 10.3, Example 3
The equation of a hyperbola	Lesson 10.4, Example 3
Non-linear system of equations	Lesson 10.5, Example 4

You may wish to have students work with partners for some of the above activities. Additional suggestions for cooperative group activities are noted in the teacher's notes in each lesson.

Multicultural

The cultural connections in this chapter include references to Europe.

CULTURAL CONNECTIONS	
Europe: Instruments for drawing conic sections	Lesson 10.5

Portfolio Assessment

Below are portfolio activities for the chapter. They are listed under seven activity domains that are appropriate for portfolio development.

1. **Investigation/Exploration** Graphing parabolas is the focus of the exploration in Lesson 10.1; graphing other conic sections — circles, ellipses, and hyperbolas is the focus of Lessons 2 through 4.

2. **Applications Recommended** To be included are any of the following: transportation, Lesson 10.2, Exercise 22; astronomy, Lesson 10.3, Exercises 13–15 and 18–19; marine biology, Lesson 10.4, Exercises 17–19; navigation, Lesson 10.4, Exercises 30–31; sports, Lesson 10.6, Exercises 19–20.

3. **Non-Routine Problems** radio transmission, Lesson 10.2, Example 4; astronomy, Lesson 10.3, Example 3; marine research, Lesson 10.4, Example 3; and sports, Lesson 10.5, Exercises 19–20.

4. **Project Navigational Systems** Students apply their knowledge of hyperbolas to locate ships in distress at sea and to divert the closest ship in the area to assit it.

5. **Interdisciplinary Topics** geometry, Lesson 10.1, Exercise 28; geology, Lesson 10.2, Exercises 25–27; geometry, Lesson 10.5, Exercises 13–14 and 21–24.

6. **Writing** Communicate exercises in which the students generalize basic concepts of each conic section studied, offer excellent writing selections for the portfolio.

7. **Tools** Chapter 10 uses graphics calculators to assimilate data and model that data to help solve systems of parametric equations. To measure the individual student's proficiency, it is recommended that he or she complete selected worksheets from the Technology Masters.

Technology

Graphics Calculator

The graphics calculator is designed primarily to graph functions. Most conic sections are relations, not functions. To graph a conic section on a graphics calculator, the equation must be solved for y and graphed in two parts. For instance, the parabola $x = y^2 - 2y + 3$ can be solved for y by completing the square so that $y = \pm\sqrt{x - 2} + 2$. By entering the positive and negative cases individually, the two functions $Y1 = \sqrt{x - 2} + 2$ and $Y2 = -\sqrt{x - 2} + 2$ are displayed on the same coordinate plane. The parabola shown below is in a square viewing window on the TI-82.

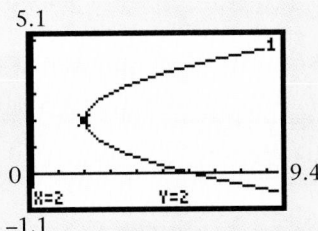

Students may need some coaching when they are tracing or solving for particular values. They will need to be aware of the location of the point on the combined graphs. They must identify which function, Y1 or Y2, they need to trace to determine a point. The point shown on the graph above, the vertex $(2, 2)$, is a point that exists on both functions.

Avoiding Distortion Another important requirement for using the graphics calculator is to view conic sections with a *square viewing window*. A circle can easily appear as an ellipse and an ellipse can appear as a circle when the actual distance between consecutive x-values is different from the distance between consecutive y-values. It is also important to know the *pixel dimension* on each calculator to find this square viewing window. Refer to the *HRW Technology Handbook* or the individual calculator manuals to determine these measurements.

On the TI-82, the dimensions of the physical viewing window are 94 pixels by 62 pixels. The circle $x^2 + y^2 = 4$ is shown on the calculators below. The calculator on the left has a non-square viewing window, −4.7 to 4.7 by −4 to 4. The calculator on the right has a square viewing window −4.7 to 4.7 by −3.1 to 3.1. Notice the difference a few pixels can make on the image. The following modifications can be made to the square viewing window to view the circle if its center is translated 4 units right and 3 units down, $(x - 4)^2 + (y + 3)^2 = 4$. The square viewing window shown above will be modified by adding 4 to both −4.7 and 4.7, and by subtracting 3 from both 3.1 and −3.1.

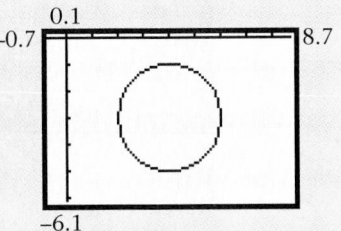

Notice that the x- and y-values for the viewing window are transformed in much the same way that the center of the circle is translated. This presents yet another opportunity to have students practice translations. Have students practice translations to their square viewing window as they study the transformation of each conic section.

Integrated Software

$f(g)$ **Scholar™** is an integrated computer-based mathematics productivity tool that combines calculator, spreadsheet, and graphics capabilities to provide a dynamic and interactive environment for explorations in mathematics. Use $f(g)$ **Scholar™** for any lesson needing a spreadsheet, calculator, graphics calculator, or any combination of the three.

CHAPTER 10

Analytic Geometry

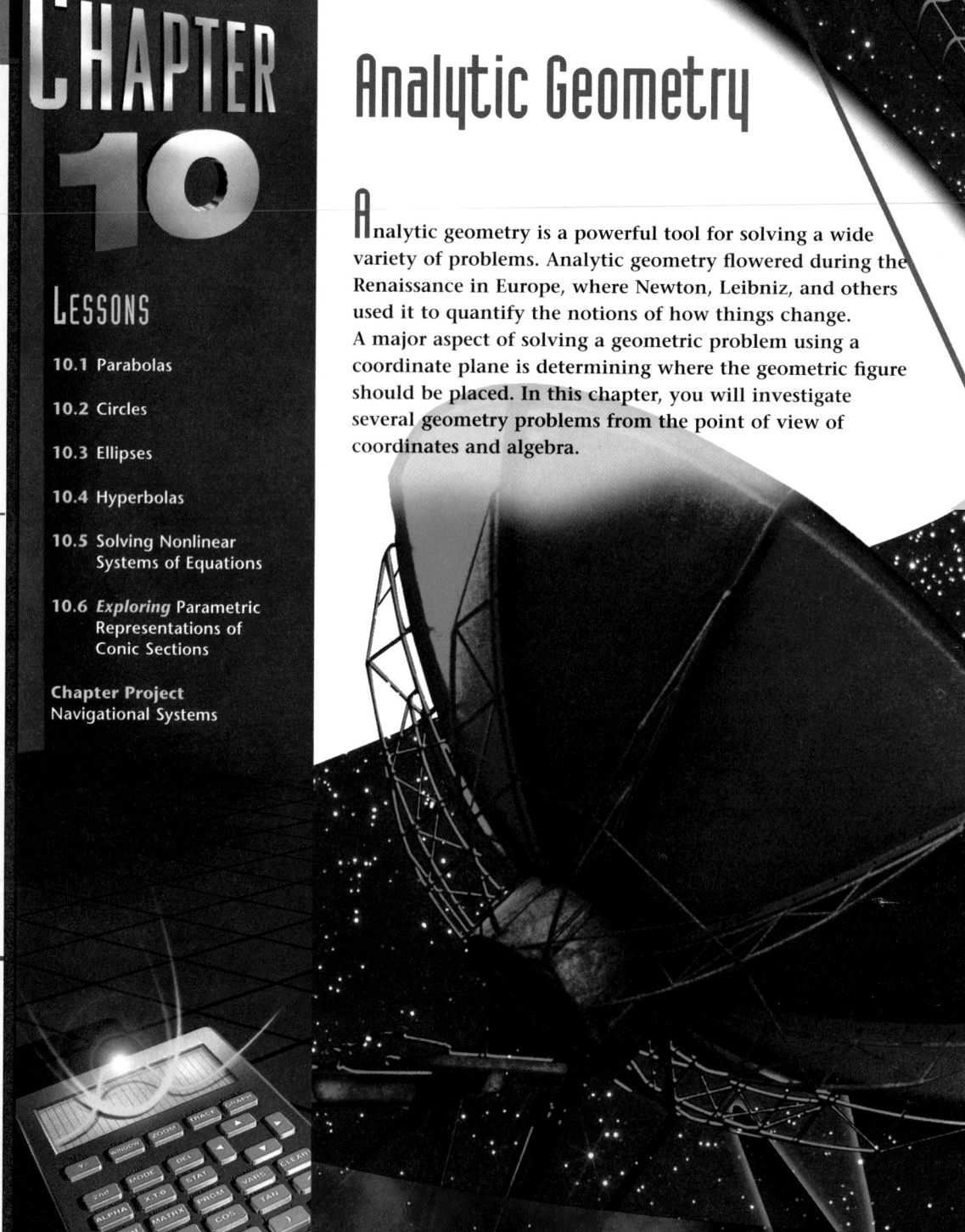

Analytic geometry is a powerful tool for solving a wide variety of problems. Analytic geometry flowered during the Renaissance in Europe, where Newton, Leibniz, and others used it to quantify the notions of how things change. A major aspect of solving a geometric problem using a coordinate plane is determining where the geometric figure should be placed. In this chapter, you will investigate several geometry problems from the point of view of coordinates and algebra.

ABOUT THE CHAPTER

Background Information

The conic sections were first discovered in ancient Greece. During the Renaissance in Europe, René Descartes formulated a system of coordinates. By placing the conic sections on the coordinate plane, Descartes and others developed analytic geometry.

CHAPTER RESOURCES

- Practice Masters
- Enrichment Masters
- Technology Masters
- Lesson Activity Masters
- Lab Activity Masters
- Long-Term Project Masters
- Assessments Masters
  Chapter Assessments, A/B
  Mid-Chapter Assessment
  Alternative Assessment, A/B
- Teaching Transparencies
- Cumulative Assessment
- Spanish Resources

CHAPTER OBJECTIVES

- Write an equation of a parabola when given any two of the following: focus, directrix, vertex.
- Identify the vertex, focus, and directrix of a parabola from its equation; then sketch a graph.
- Write the equation of a circle when given the coordinates of the center and the length of the radius.
- Determine the center and radius of a circle when given an equation of the circle.

ABOUT THE PHOTOS

The circle, ellipse, parabola and hyperbola are all created by taking cross sections of two cones joined at their vertices. By intersecting a plane with these cones at varying angles, all the different conic sections will result. Analytic geometry is used to describe such real-world phenomena as the paths of the planets, the shape of a satellite antenna, and the arches in bridge supports.

This chapter can be introduced by group or class discussion of the images on pages 534 and 535. See how many images students can identify. As each image is located, relate the image to the appropriate analytical geometry connection. Students may wish to write a short paper describing the connections between analytic geometry and real-world applications. They can use the images on these pages as a starting point.

PORTFOLIO ACTIVITY

A major breakthrough in the study of our universe occurred when Johannes Kepler (1571–1630) discovered that the planets travel along elliptical paths around the Sun. The following activities will require some research on the geometrical nature of the solar system. On page 559, you will be asked to complete this activity.

1. In Lesson 10.3, an equation is developed for the path of Mars around the Sun. Write equations for the paths of the other planets in our solar system around the Sun.

2. Comets can also have elliptical orbits around the Sun. Two famous comets are Halley's comet and the comet Kohoutek (kah-who-teck). Find the equation for the orbit of Halley's comet. The comet Kohoutek has not yet reached its farthest distance from the Sun.

You may choose to save your work for your portfolio.

- Determine the coordinates of the center, vertices, co-vertices, and foci when given an equation of an ellipse with either a horizontal or vertical major axis.
- Determine the coordinates of the center and foci, and find the lengths of the axes when given an equation of a hyperbola.
- Determine the solutions of a system consisting of one first-degree equation and one second-degree equation.
- Determine the solutions of a system consisting of two second-degree equations.
- Write the rectangular and parametric forms of the equation of a circle.
- Write the rectangular and parametric forms of the equation of an ellipse.

PORTFOLIO ACTIVITY

In this activity, students study the solar system. They can determine the orbits of all the planets and two major comets in our solar system.

Organize students into groups of four. One student will find orbiting data for the orbits of Mercury, Venus, and Earth. A second student will find data for Ceres, Jupiter, and Saturn. The third student will find data for Uranus, Neptune, and Pluto. The fourth student will find data for Halley's comet. Students can work together to research the comet Kohoutek. The group can present the equations of the paths of the planets in the solar system to the class. Students can determine when Halley's comet and the comet Kahoutek will next be visible from Earth. Students should also comment on any unusual properties of the orbit of Pluto.

ABOUT THE CHAPTER PROJECT

The Chapter Project on pages 586 and 587 gives students an opportunity to find the coordinates of a ship in distress. Students will use the equations of hyperbolas to locate this ship using data acquired by 4 monitoring stations. The procedures students will follow in this project are very similar to the actual methods used by the U.S. Coast Guard when they receive a distress signal from a ship at sea.

Objectives

- Write an equation of a parabola when given any two of the following: focus, directrix, vertex.
- Identify the vertex, focus, and directrix of a parabola from its equation; then sketch a graph.

RESOURCES

- Practice Master   10.1
- Enrichment Master   10.1
- Technology Master   10.1
- Lesson Activity Master   10.1
- Quiz   10.1
- Spanish Resources   10.1

Assessing Prior Knowledge

Solve for y.

1. $\frac{3}{2}a = 5ay$ $\left[y = \frac{3}{10}\right]$

2. $25a = 5ay$ $[y = 5]$

3. Write $y^2 + 10y + 25$ as the square of a binomial. $[(y + 5)^2]$

TEACH

 The equation of a parabola can be used to obtain the precise shape of a telescope mirror, a reflector for a headlight, or an antenna for a satellite receiver. If the shape of a headlight were not an exact parabola, for example, the light emitted by the headlight would be diffused instead of focused.

LESSON 10.1 Parabolas

Suppose you want to design a TV satellite antenna or a headlight for a car. What basic properties do you think it should have? What kind of shape would have those properties? TV satellite antennas and headlights on cars have a parabolic shape.

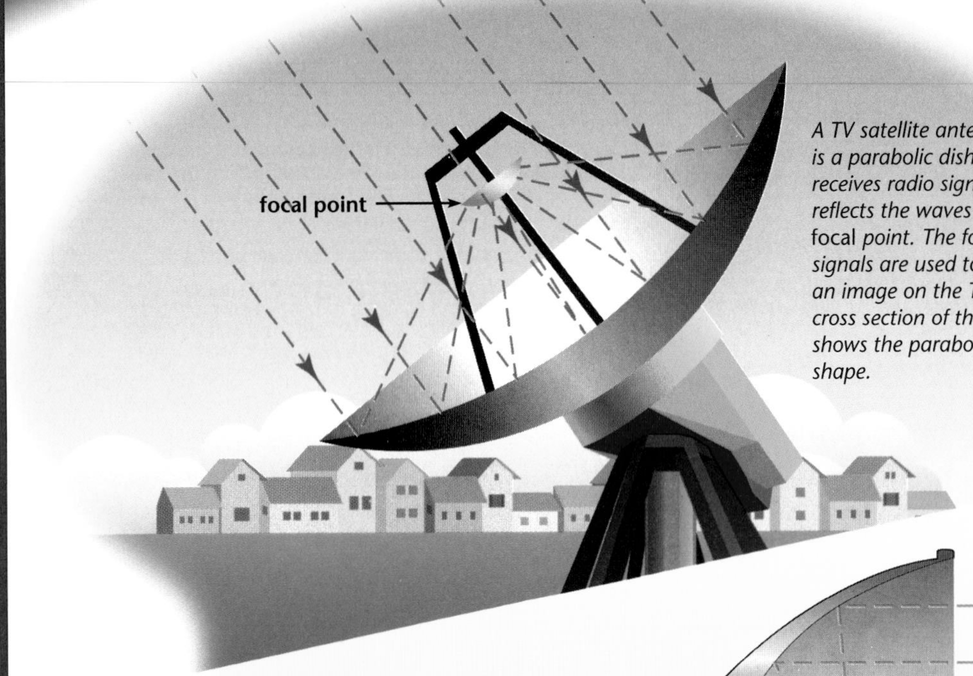

focal point →

A TV satellite antenna is a parabolic dish that receives radio signals and reflects the waves to a focal point. The focused signals are used to create an image on the TV. This cross section of the dish shows the parabolic shape.

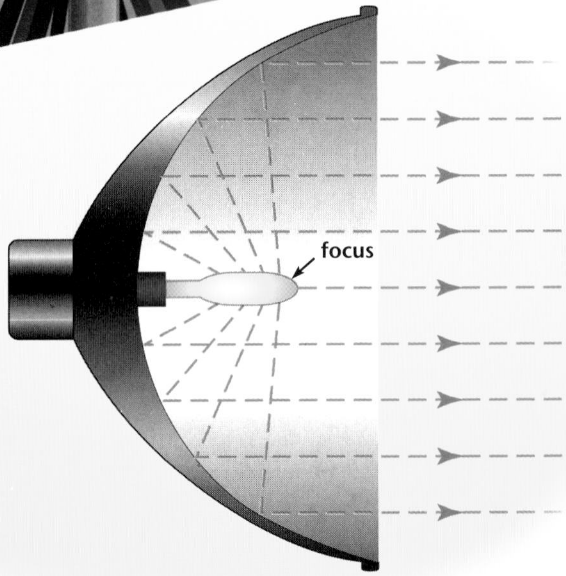

The cross section of a headlight is also a parabola. The source of light is the focus. The focus is placed so that the light reflects off the parabola in parallel rays and the headlight emits a straight beam of light.

focus

In Chapter 5, you learned that the graph of $y = ax^2$, where a is any real number except zero, is a parabola with its vertex at $(0, 0)$. What happens when x and y are interchanged?

Technology Have students use a graphics calculator as a tool for finding the equation of a parabola. Explain that in order to obtain a complete graph on the graphics calculator screen, students may need to try several sets of values for the x- and y-ranges. Students should use the trace function to find the approximate coordinates of the vertex of the parabola. With the help of the graphics calculator, students can analyze the behavior of the parabola, and find the vertex, axis of symmetry, and direction of the opening of the parabola.

•Exploration• *Graphing Equations and Functions*

*Graphics
Calculator*

1 Solve the equation $x = y^2$ for y. Graph the equation for $y > 0$. Is this graph a parabola? Explain.

2 Graph the equations $y = -\sqrt{x}$ and $y = \sqrt{x}$ on the same coordinate plane. What is the shape of these two graphs together?

3 Explain why $x = ay^2$ is not a function. How can you graph $x = ay^2$ on your graphics calculator? ❖

TRANSFORMATIONS
Connection

In Chapter 5, you studied quadratic functions of the form $f(x) = a(x - h)^2 + k$, where a is any real number except zero, and (h, k) is the vertex. This form is a translation of the graph of $y = ax^2$.

The quadratic equation $y = a(x - h)^2 + k$ is the standard form of a parabola with a *vertical* axis of symmetry. The standard form of a parabola with a *horizontal* axis of symmetry is $x = a(y - k)^2 + h$.

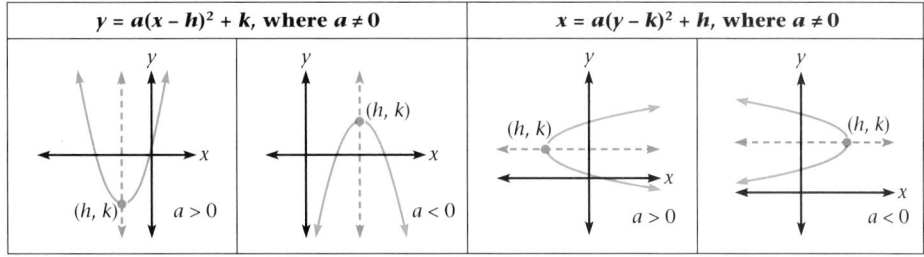

$y = a(x - h)^2 + k$, where $a \neq 0$		$x = a(y - k)^2 + h$, where $a \neq 0$	
$a > 0$	$a < 0$	$a > 0$	$a < 0$

PARABOLA

A **parabola** is a set of coplanar points, each of which is the same distance from a fixed line, called the **directrix**, as it is from a fixed point not on the line, called the **focus**.

The definition of a parabola includes all curves of parabolic shape, not just those with vertical or horizontal axes of symmetry. However, in this text, you will only examine the equations of parabolas whose graphs have horizontal or vertical equations of symmetry.

The vertex of a parabola lies on the axis of symmetry midway between the focus, F, and the directrix. By definition, $PF = PD$, where P is any point on the parabola and D is the point on the directrix such that PD is perpendicular to the directrix.

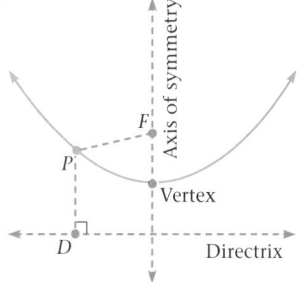

CRITICAL
Thinking

Explain why the angle between the directrix and the axis of symmetry is 90°.

interdisciplinary
CONNECTION

Optics In designing a telescope mirror, parallel light rays from a distant object are gathered. The rays are brought to a single point, or focus, where an eyepiece magnifies the image. In reflected light, the angle of incidence is equal to the angle of reflection.

This concept offers another definition of *parabola*. If possible, borrow a telescope from the science department and have students observe the mechanisms involved. Then have students design their own telescope using the concepts of focus and reflection.

You can find the equation of any parabola using the definition of a parabola and the distance formula.

EXAMPLE 1

A Find the equation of a parabola with focus $F(3, 7)$ and directrix $y = 4$.

B Find the coordinates of the vertex.

Solution▶

A Let $P(x, y)$ be any point on the parabola, and let $D(x, 4)$ be the point on the directrix with the same x-coordinate as $P(x, y)$.

Then, by definition of the parabola, $PF = PD$.

Use the distance formula to write an equation for the lengths of PF and PD. Then solve the equation for y.

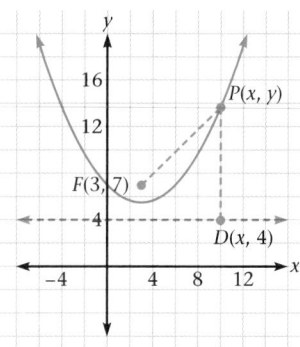

$$PF = PD$$
$$\sqrt{(x - 3)^2 + (y - 7)^2} = \sqrt{(x - x)^2 + (y - 4)^2}$$
$$(x - 3)^2 + (y - 7)^2 = (y - 4)^2$$
$$(x - 3)^2 = (y - 4)^2 - (y - 7)^2$$
$$(x - 3)^2 = (y^2 - 8y + 16) - (y^2 - 14y + 49)$$
$$(x - 3)^2 = 6y - 33$$
$$(x - 3)^2 + 33 = 6y$$
$$\frac{1}{6}(x - 3)^2 + \frac{11}{2} = y$$

Thus, the equation of the parabola with focus $F(3, 7)$ and directrix $y = 4$ is $y = \frac{1}{6}(x - 3)^2 + \frac{11}{2}$.

B The vertex of a parabola written in the form $y = a(x - h)^2 + k$ is (h, k). Thus, the vertex of $y = \frac{1}{6}(x - 3)^2 + \frac{11}{2}$ is $\left(3, \frac{11}{2} \right)$. ❖

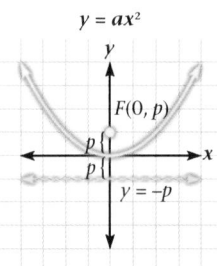

CRITICAL *Thinking*

Describe a quick way to find the vertex since you know it is halfway between the point $(3, 7)$ and the line $y = 4$.

Examine the graph of $y = ax^2$, where a is any real number except zero. How can you determine the coordinates of the focus?

$y = ax^2$

The focus must be somewhere on the axis of symmetry. Since the axis of symmetry is the line $x = 0$, the focus must have coordinates $(0, p)$ for some number $p \neq 0$.

By definition, p is the distance between the vertex and the focus, *and* the distance between the vertex and the directrix. Therefore, the equation of the directrix is $y = -p$.

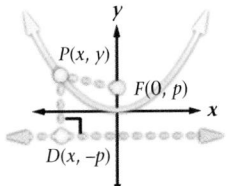

Now let $P(x, y)$ be any point on the parabola. By the definition of a parabola, the distance between $P(x, y)$ and the focus, $F(0, p)$, is the same as the distance between $P(x, y)$ and the point $D(x, -p)$ on the directrix.

$$PF = PD$$
$$\sqrt{(x-0)^2 + (y-p)^2} = \sqrt{(x-x)^2 + (y+p)^2}$$
$$x^2 + (y-p)^2 = (y+p)^2$$
$$x^2 = (y+p)^2 - (y-p)^2$$
$$x^2 = y^2 + 2yp + p^2 - y^2 + 2yp - p^2$$
$$x^2 = 4py$$
$$y = \frac{x^2}{4p}$$

Since $y = ax^2$ in the original function, substitute ax^2 for y.

$$ax^2 = \frac{x^2}{4p}$$
$$a = \frac{1}{4p} \rightarrow p = \frac{1}{4a}$$

Thus, a parabola, $y = ax^2$, has its vertex at the origin and its focus and directrix located p, or $\frac{1}{4a}$, units from the vertex.

EXAMPLE 2

Find the coordinates of the focus and the equation of the directrix for each parabola.

A $y = 0.5x^2$

B $y = \frac{1}{6}(x-5)^2 + 2$

Solution

A To find p, substitute 0.5 for a in the equation $p = \frac{1}{4a}$.

$$p = \frac{1}{4(0.5)} = \frac{1}{2}$$

Since the vertex is at $(0, 0)$, the focus is $\left(0, \frac{1}{2}\right)$ and the directrix is the line $y = -\frac{1}{2}$.

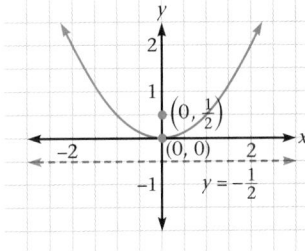

B To find p, substitute $\frac{1}{6}$ for a in the equation $p = \frac{1}{4a}$.

$$p = \frac{1}{4\left(\frac{1}{6}\right)} = \frac{1}{\frac{2}{3}} = \frac{3}{2}$$

Since the vertex is at $(5, 2)$, the focus is $\left(5, \frac{7}{2}\right)$ and the directrix is the line $y = \frac{1}{2}$. ❖

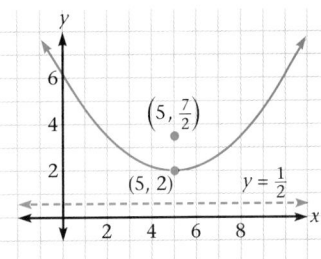

Technology On a graphics calculator, have students graph two equations of a parabola, one horizontal and one vertical. Remind students to solve the equation of a horizontal parabola for y *before* graphing, since graphics calculators can only graph functions. Have students graph both functions to obtain a complete graph of the horizontal parabola. Have students analyze each graph. Ask them if they can approximate the values of the coordinates of the vertex and focus of each parabola, and where the directrix is. Have students write the equation for each parabola in standard form. Emphasize that algebraic techniques often help in finding exact answers. [$x = -\frac{1}{4}(y-4)^2 + 8$, focus: (7, 4); vertex: (8, 4); directrix: $x = 9$]

Lesson 10.1 **539**

Alternate Example 2

Find the coordinates of the focus and the equation of the directrix for each parabola.

A. $y = 2x^2$ [focus: $\left(0, \frac{1}{8}\right)$; directrix: $y = -\frac{1}{8}$]

B. $y = 3(x-2)^2 - 4$ [focus: $\left(2, -\frac{47}{12}\right)$; directrix: $y = -\frac{49}{12}$]

Alternate Example 3

A. Write the equation of the parabola $2x + y^2 - 2y = 1$ in standard form.
$$\left[x = -\frac{1}{2}(y-1)^2 + 1\right]$$

B. Identify the vertex, axis of symmetry, focus, directrix, and direction of opening for the graph. [vertex: $(1, 1)$; axis of sym: $y = 1$; focus: $(\frac{1}{2}, 1)$; dir: $x = \frac{3}{2}$; since $-\frac{1}{2} < 0$, parabola opens to the left]

The following table summarizes the properties of a parabola.

	Standard Equation of a Parabola	
	$y = a(x - h)^2 + k$	$x = a(y - k)^2 + h$
Vertex	(h, k)	(h, k)
Axis of Symmetry	$x = h$	$y = k$
Focus	$\left(h, k + \frac{1}{4a}\right)$	$\left(h + \frac{1}{4a}, k\right)$
Directrix	$y = k - \frac{1}{4a}$	$x = h - \frac{1}{4a}$
$a > 0$	Opens upward	Opens to the right
$a < 0$	Opens downward	Opens to the left

EXAMPLE 3

Ⓐ Write the equation of the parabola $4x + y^2 - 6y = 9$ in standard form.

Ⓑ Identify the vertex, axis of symmetry, focus, directrix, and direction of opening for the graph of $4x + y^2 - 6y = 9$. Then graph $4x + y^2 - 6y = 9$.

Solution►

Ⓐ Since the standard form is $x = a(y - k)^2 + h$, complete the square on y in the equation $4x + y^2 - 6y = 9$ and solve for x.

$$4x + y^2 - 6y = 9$$
$$y^2 - 6y = 9 - 4x$$
$$y^2 - 6y + 9 = 9 - 4x + 9$$
$$(y - 3)^2 = 18 - 4x$$
$$(y - 3)^2 - 18 = -4x$$
$$-\frac{1}{4}(y - 3)^2 + \frac{9}{2} = x, \text{ or } x = -\frac{1}{4}(y - 3)^2 + \frac{9}{2}$$

Thus, the standard equation of the parabola is $x = -\frac{1}{4}(y - 3)^2 + \frac{9}{2}$.

Ⓑ In the equation $x = -\frac{1}{4}(y - 3)^2 + \frac{9}{2}$, a is $-\frac{1}{4}$, h is $\frac{9}{2}$, and k is 3.

The **vertex**, (h, k), is $\left(\frac{9}{2}, 3\right)$; the **axis of symmetry**, $y = k$, is $y = 3$; the **focus**, $\left(h + \frac{1}{4a}, k\right)$, is $\left(\frac{7}{2}, 3\right)$; and the **directrix**, $x = h - \frac{1}{4a}$, is $x = \frac{11}{2}$.

Since $a < 0$, the parabola opens to the left.

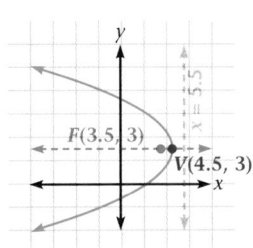

Try This **Ⓐ** Write the equation of the parabola $y^2 + 4y - 8x = -36$ in standard form.

Ⓑ Identify the vertex, axis of symmetry, focus, directrix, and direction of opening for the graph of $y^2 + 4y - 8x = -36$. Then graph $x^2 + 4x - 8y = -36$.

How can you determine the direction of opening of a parabola by looking at its equation in standard form?

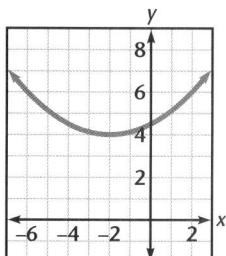
A. $x = \frac{1}{8}(y + 2)^2 + 4$

B. vertex: $(4, -2)$; axis of sym: $y = -2$; focus: $(6, -2)$; dir: $x = 2$; opens to the right

Graph of $x^2 + 4x - 8y = -36$ is shown below.

EXAMPLE 4

Satellite Write an equation for the cross section of a parabolic signal receiver that has its focus 1 foot from its vertex.

Solution▸

Place the vertex at the origin, $(0, 0)$. Then $h = 0$ and $k = 0$. The equation of the parabola has the form $y = ax^2$, where $a > 0$.

Since the focus, $\left(h, k + \frac{1}{4a}\right)$, is $(0, 1)$,

$k + \frac{1}{4a} = 1$. Substitute 0 for k, and solve for a.

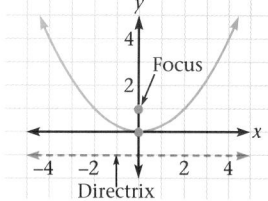

$$k + \frac{1}{4a} = 1$$
$$0 + \frac{1}{4a} = 1$$
$$\frac{1}{4} = a$$

Thus, an equation for the cross section of the satellite antenna that has its focus 1 foot from its vertex is $y = \frac{1}{4}x^2$. ❖

CRITICAL *Thinking*

How can you use the definition of a parabola to determine whether the focus of the parabola represented by $y = \frac{1}{4}x^2$ is really 1 foot from its vertex?

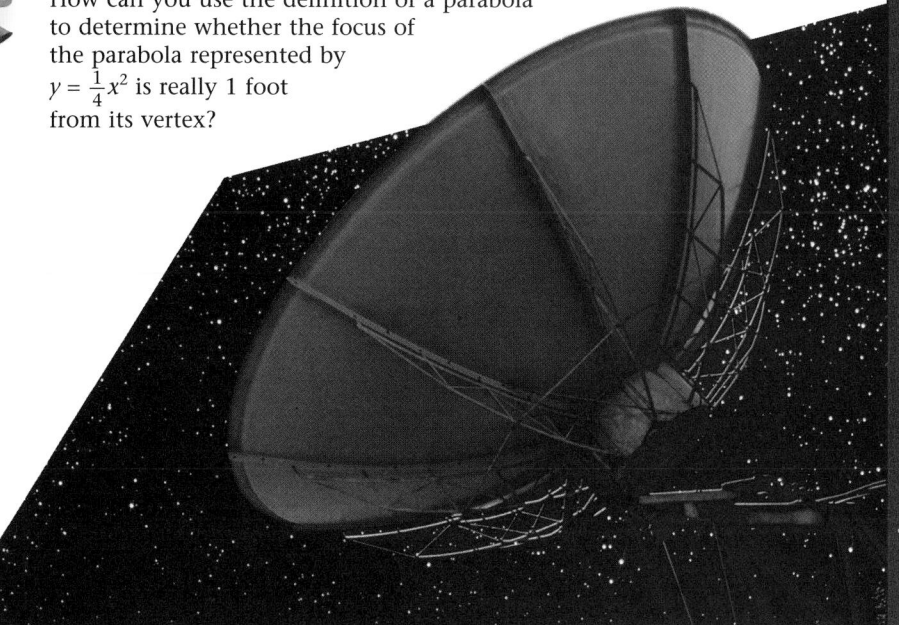

Note: The function that is graphed is different from the equation in Part A since students do not know how to solve this equation for y and graph the resulting two functions.

If the equation is written $y = a(x - h)^2 + k$, then if $a > 0$ the parabola opens upward, if $a < 0$ it opens downward. If the equation is written $x = a(y - k)^2 + h$, then if $a > 0$ the parabola opens to the right, if $a < 0$ it opens to the left.

Alternate Example 4

Write an equation for the cross section of a parabolic signal receiver that has its focus 2 feet from its vertex. $\left[y = \frac{1}{8}x^2\right]$

CRITICAL *Thinking*

$a(x - h)^2 + k = \frac{1}{4}x^2$; so $h = 0$, $k = 0$ and $a = \frac{1}{4}$. Since $\frac{1}{4a} = 1$, the focus is at $(1, 0)$.

Use Transparency ▶ **37**

Assignment Guide

Core 1–9, 11–16, 19–30

Core Plus 1–4, 7–10, 14–31

Technology

Graphics calculators are needed for Exercises 21, 24, 27, and 31. To graph a relation, the relation has to be broken into 2 parts, both of which are functions. Point out that both functions must be graphed to get a complete graph of the parabola.

Error Analysis

When solving the Exercises 19–24, students may choose too large a domain and come up with meaningless results. Remind students that the domain must have physical meaning. Students may have to try several domains before finding a reasonable solution.

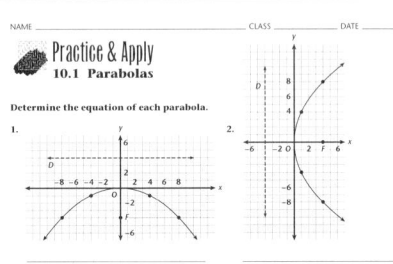

Practice Master

EXERCISES & PROBLEMS

Communicate

1. What is the relationship between (h, k) of $y = a(x - h)^2 + k$ and the coordinates of the focus and directrix?

2. What is the relationship between (h, k) of $x = a(y - k)^2 + h$ and the coordinates of the focus and directrix?

3. Explain how to find the vertex, focus, and directrix for the parabola represented by $5x - y^2 - 12y = 21$.

4. How can you determine the direction of opening of a parabola when given the focus and directrix?

Practice & Apply

Determine an equation of the parabola with the following properties.

5. Focus at $(-3, 0)$ and directrix $x = 3$ $\quad x = -\frac{1}{12}y^2$

6. Focus at $(0, -6)$ and directrix $y = 6$ $\quad y = -\frac{1}{24}x^2$

7. Focus at $(4, 0)$ and vertex at $(1, 0)$ $\quad x = \frac{1}{12}y^2 + 1$

8. Vertex at $(3, -2)$ and focus at $(3, 1)$ $\quad y = \frac{1}{12}(x - 3)^2 - 2$

9. Focus at $(5, -2)$ and directrix $x = -1$ $\quad x = \frac{1}{12}(y + 2)^2 + 2$

10. Focus at $(-3, 4)$ and directrix $y = 7$ $\quad y = -\frac{1}{6}(x + 3)^2 + \frac{11}{2}$

Determine the vertex, focus, and directrix for each parabola. Sketch each parabola.

11. $y^2 = 8x$ 12. $y^2 = -3x$

13. $x^2 = -5y$ 14. $x^2 = 12y$

15. $x^2 + 4x - 6y = -10$ 16. $x^2 - 8x - y + 20 = 0$

17. $-14x + 2y^2 - 8y = 20$ 18. $4x + y^2 + 3y = -5$

Pyrotechnics Fireworks are fired from ground level. Their path is a parabola opening downward. The fireworks reach a height of 1200 meters and travel 15,000 meters horizontally.

19. Find the coordinates of the vertex. (7500, 1200)

20. Use the coordinates of the vertex to write an equation for the parabola.

21. Graph the equation, and approximate the horizontal distance of the fireworks from the firing site when the missile first reaches a height of 500 meters. ≈ 1772 m

The answers to Exercises 11–18 and 20 can be found in Additional Answers beginning on page 842.

Civil Engineering Ideally, the main cables of a suspension bridge are parabolic to support a uniform horizontal load.

22. Place the vertex of the parabola at the origin. Find the coordinates of a point at the top of a tower. (187.5, 60); (−187.5, 60)

23. Use the coordinates you found in Exercise 22 to find the value for a in $y = ax^2$. Then write an equation for the parabola. 0.0017; $y = 0.0017x^2$

24 Graph the equation and approximate the height of the point on the cable that is 50 feet (horizontally) from one of the towers. ≈ 72 feet above water level

For Exercises 25–27, let
$4x + y^2 + 3y = 5$.

25. Write the equation in standard form, and identify the vertex and the direction of opening for the parabola.

26. Solve the equation for y.

27 Graph both equations from Exercise 26 on your graphics calculator. What is the shape of the graphs together? What viewing window did you use to view both graphs?

～～ *Look Back*

28. **Geometry** A rectangular piece of ground is to be enclosed on three sides by 160 feet of fencing. The fourth side is the side of a barn. Find the dimensions of the enclosure that maximize the area. **[Lesson 5.1]**
40 ft by 80 ft

29. **Metalworking** The diameter of a circular ventilation duct is 140 millimeters. It is joined to a square-duct system. To ensure smooth air flow, the areas of the circle and square sections must be equal. What should be the length of a side of the square section, x? **[Lesson 5.2]** 124.07 mm

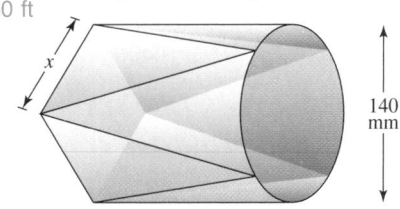

30. Determine the vertical and horizontal asymptotes and any holes for the function $f(x) = \dfrac{x^2 - 2x - 3}{x^2 - 9}$. **[Lesson 9.4]** vertical: $x = -3$, horizontal: $y = 1$
hole: $x = 3$

Look Beyond ～～

31 Let $y = 2(x - 3)^2 + 4$ and $x = \frac{1}{4}(y - 3)^2 + 1$. Use your graphics calculator to graph both parabolas on the same coordinate plane. Find the points of intersection. (2.22, 5.21), (4.13, 6.54)

25. $x = -\dfrac{1}{4}\left(y + \dfrac{3}{2}\right)^2 + \dfrac{29}{16}$; vertex: $\left(\dfrac{29}{16}, -\dfrac{3}{2}\right)$;
opens toward the left

26. $y = -\dfrac{3}{2} \pm \sqrt{\dfrac{29}{4} - 4x}$

27. A parabola; the viewing window answers may vary but should include x-values between −2 and 2 and y-values between −5 and 2.

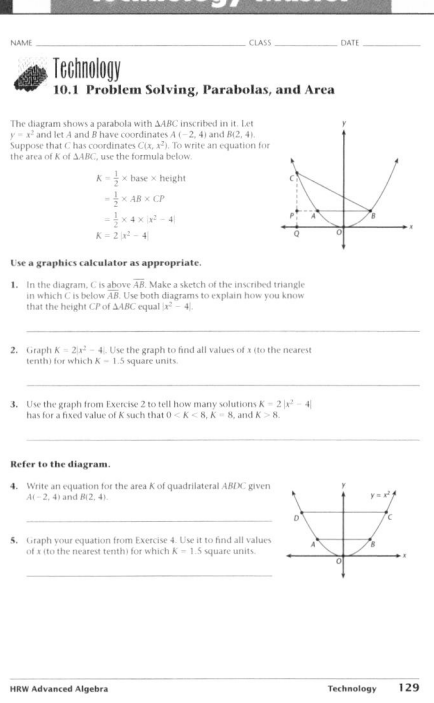

Objectives

- Write the equation of a circle when given the coordinates of the center and the length of the radius.
- Determine the center and radius of a circle when given an equation of the circle.

- Practice Master   10.2
- Enrichment Master   10.2
- Technology Master   10.2
- Lesson Activity Master   10.2
- Quiz   10.2
- Spanish Resources   10.2

Assessing Prior Knowledge

1. Factor $\quad y^2 + 12y + 36$.
   $$\left[(y + 6)^2\right]$$

2. Determine the distance between the following pairs of points.

a. (1, 2) and (3, 5) **[3.6 units]**

b. (1, 2) and (x, y)
   $$\left[\sqrt{(y-2)^2 + (x-1)^2}\right]$$

TEACH

 Circles can be found all around us. For example, a radar device sweeps out a circular area when detecting an object, and some planets of our solar system follow near-circular orbits around the sun. Mathematics helps us study circles that we cannot directly touch and measure.

why *Radio signals travel from a radio tower in all directions. The approximate horizontal or vertical region that can receive the signals can be described with a circle.*

*A radar (**ra**dio **d**etecting **a**nd **r**anging) device sweeps out a large circular area to detect a reflecting object, such as an aircraft, and to determine its direction, distance, height, or speed.*

Recall from geometry the definition of a circle.

CIRCLE

A **circle** is a set of coplanar points, each of which is the same distance from a fixed point in the plane called the **center**. The segment from the center of the circle to any point on the circle is called the **radius** of the circle.

ALTERNATIVE teaching strategy

Technology Have students graph a circle such as $(x-1)^2 + (y-4)^2 = 25$ on a graphics calculator. Remind students that in order to graph the equation of the circle on a graphics calculator, they must solve the equation for y, and graph both functions to see the complete graph on the calculator screen. Have them identify the center of the circle (1, 4). Using the trace function, read off the (x, y) coordinates of several points on the circumference such as (−2, 8). Then use the distance formula to find the distance of these points from the center. Students should realize that the distance from any point on the circumference to the center is equal to the radius of the circle.

Equation of a Circle

The equation of a circle comes directly from the distance formula. When you know the coordinates of the center and the length of its radius, you can use the distance formula to find an equation for the circle.

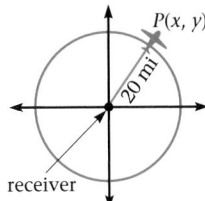

Suppose an airplane is tracked by radar, and the airplane is 20 miles from the radar receiver. The distance formula can be used to write an equation that describes the possible locations of the airplane.

Set up a coordinate system with the receiver at the origin. Let $P(x, y)$ be any point 20 miles from the receiver. Then use the distance formula.

$$\sqrt{(x-0)^2 + (y-0)^2} = 20$$
$$(x-0)^2 + (y-0)^2 = 20^2$$
$$x^2 + y^2 = 400$$

You can see from the equation that if r is the radius of a circle, then the equation of the circle with center at the origin is $x^2 + y^2 = r^2$.

What happens to the equation when the center is translated from the origin to the point (0, 1)?

EXAMPLE 1

TRANSFORMATIONS *Connection*

Use the distance formula to find the equation of the circle with its center at $C(3, 2)$ and a radius of 4.

Solution

Let $P(x, y)$ be any point on the circle. The distance from $C(3, 2)$ to $P(x, y)$ is 4, the length of the radius.

$$CP = 4$$
$$\sqrt{(x-3)^2 + (y-2)^2} = 4$$
$$(x-3)^2 + (y-2)^2 = 16$$

Thus, the equation of the circle with its center at $C(3, 2)$ and a radius of 4 is $(x-3)^2 + (y-2)^2 = 16$. ❖

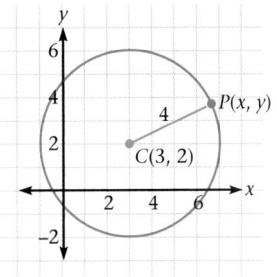

Examine the equation of the circle in Example 1. How are the constants, 3, 2, and 16, related to the center and radius of the circle?

Suppose you have a circle with its center at $C(h, k)$ and a radius of r.

Then the distance from $C(h, k)$ to any point $P(x, y)$ on the circle is $\sqrt{(x-h)^2 + (y-k)^2} = r$, or $(x-h)^2 + (y-k)^2 = r^2$.

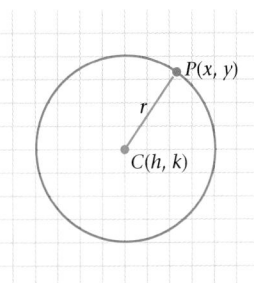

History of Science In ancient Greece, the astronomer Ptolemy developed a model of the universe with the Earth at the center and each of the planets moving around the Earth. It was believed that since these celestial objects belonged to the "heavenly realm" they must travel in perfect circular paths, since the circle was the only perfect geometric object. This model did not fit with observed motion in the night sky. Smaller circles, or epicycles, were located on the circumference of larger circles. Have interested students research the development of the model of the universe, beginning with Ptolemy's model, and concluding with the currently accepted model.

A ongoing ASSESSMENT

The equation becomes $x^2 + (y-1)^2 = r^2$.

Alternate Example 1

Use the distance formula to find the equation of the circle with its center at (1, 1) and radius of 2.
$$\left[(x-1)^2 + (y-1)^2 = 4\right]$$

A ongoing ASSESSMENT

The center is translated to the points (3, 2). The radius is 4, which squared is 16.

Cooperative Learning

Have groups of three research the history of conic sections. One student can read about Ptolemy, the second student can read about Euclid, and the third can read about Apollonius. Then students can organize their results and give a class presentation.

Use Transparency ▶ **38**

Alternate Example 2

Write the equation, in standard form, of the circle with center at $(2, -3)$ and radius 5. $[(x - 2)^2 + (y + 3)^2 = 25]$

Alternate Example 3

Find the center and the radius of the circle represented by the equation $x^2 + y^2 + 2x + 4y = 44$. [center at $(-1, -2)$; radius: 7]

STANDARD EQUATION OF A CIRCLE
The standard equation of a circle in the coordinate plane is
$$(x - h)^2 + (y - k)^2 = r^2,$$
where (h, k) is the center and $r > 0$ is the radius.

EXAMPLE 2

Write an equation, in standard form, for the circle shown.

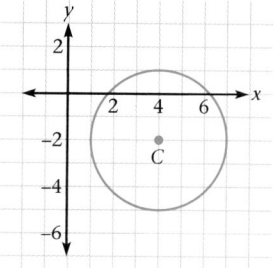

Solution➤

The center is $C(4, -2)$, and the radius, r, is 3.

Substitute 4 for h, −2 for k, and 3 for r in the equation $(x - h)^2 + (y - k)^2 = r^2$. Then simplify.

$$(x - h)^2 + (y - k)^2 = r^2$$
$$(x - 4)^2 + [y - (-2)]^2 = 3^2$$
$$(x - 4)^2 + (y + 2)^2 = 9$$

Thus, the standard equation of the circle is $(x - 4)^2 + (y + 2)^2 = 9$. ❖

Notice that both variables, x and y, are squared in the equation of a circle. Sometimes you need to complete the squares on *both* variables in order to write the equation of a circle in the form $(x - h)^2 + (y - k)^2 = r^2$.

EXAMPLE 3

Find the center and the radius of the circle represented by the equation $x^2 + y^2 + 4x - 10y = 7$. Then graph the circle.

Solution➤

Complete the squares on x and y, and write the equation in the form $(x - h)^2 + (y - k)^2 = r^2$.

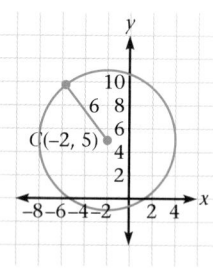

$$x^2 + y^2 + 4x - 10y = 7$$
$$(x^2 + 4x) + (y^2 - 10y) = 7$$
$$(x^2 + 4x + 4) + (y^2 - 10y + 25) = 7 + 4 + 25$$
$$(x + 2)^2 + (y - 5)^2 = 36$$

Thus, the center is $(-2, 5)$, and the radius is $\sqrt{36}$, or 6.

Sketch the graph. ❖

 **ENRICHMENT** Derive the standard equation for a circle from the definition of *distance* and the "Pythagorean" Right-Triangle Theorem. Discuss the idea of *deductive reasoning* and where students might use it in everyday life.

INCLUSION strategies **Using Models** Have students select a standard equation of a circle. Ask them to draw the circle on graph paper. By counting the squares of the grid that are within the circle, students can estimate the area within the circle. They should use the formula for the area of a circle, compute the area, and compare it with their estimate.

Applications of Circles

EXAMPLE 4

Radio Transmission A radio tower located 25 miles east and 30 miles south of Lorne's rural home emits a radio signal in all directions. The tower's signal is guaranteed to reach all receivers within 50 miles. Let the position of Lorne's home be the origin of the coordinate system.

Ⓐ Write an equation that describes the locations farthest from the radio tower within the guaranteed receiving distance.

Ⓑ Make a graph that illustrates the locations farthest from the radio tower within the guaranteed receiving distance and the location of Lorne's home. Is Lorne's home within the receiving range?

Solution➤

Ⓐ The tower is 25 miles east and 30 miles south of the origin. On a map, east is to the right, and south is downward. Therefore, the center of the circular receiving range is (25, −30). The radius, r, is 50 miles.

Thus, the equation for the guaranteed receiving distance is $(x - 25)^2 + (y + 30)^2 = 2500$.

Ⓑ The graph of the circle can be seen in the illustration below.

Lorne's home is located at the origin. The region of the circle contains the origin. Thus, Lorne's home is within the receiving range.

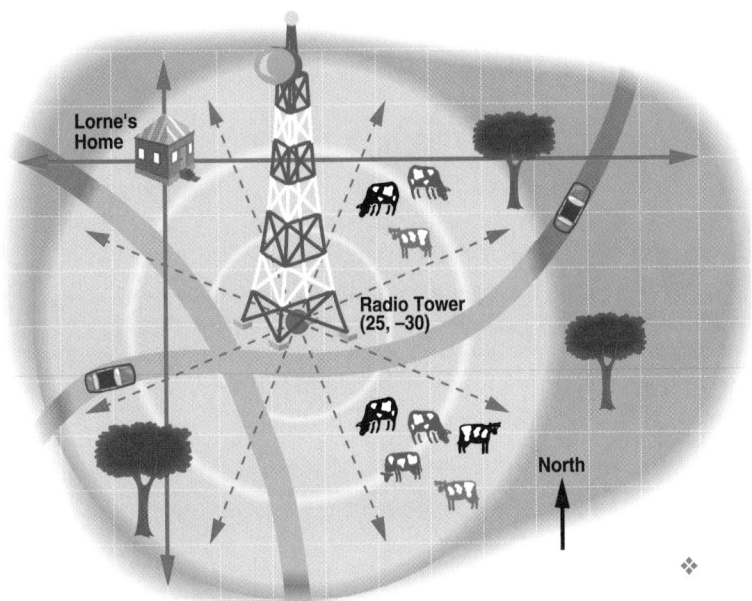

 **CRITICAL Thinking** What is the greatest distance that Lorne can go from his house and still be within the receiving range of the radio tower?

Use Transparency ▶ 39

CRITICAL Thinking

about 89 miles

 **RETEACHING the lesson**

Using Visual Models On graph paper, have students draw a circle with center at the origin and a radius of 4. Ask students to give an estimate of the coordinates of several points on the circle. Now, give students a point, such as (5, −3), and ask them to write the standard form of the equation of a circle that passes through that point and whose center is at the origin. Students should use the distance formula to find the radius. They should then use the equation of a circle with center in the origin. $[r^2 = 34, x^2 + y^2 = 34]$

Solve the equation for *y*. (*y* = 5 − *x*) Solve the equation of a circle for *y*.

Alternate Example 5

Graph $(x-3)^2 + (y+1)^2 = 16$ using your graphics calculator. Remind students to use inspection to determine the center and radius of the circle. $y = -1 \pm \sqrt{16 - (x-3)^2}$; radius: 4; center: (3, −1)]

A onogoing
SSESSMENT

Try This

$y = 7 \pm \sqrt{25 - (x+4)^2}$; radius: 5; center: (−4, 7)

What is the first step you take when graphing the equation $x + y = 5$ on your graphics calculator? What do you think is the first step in graphing the equation of a circle on the graphics calculator?

EXAMPLE 5

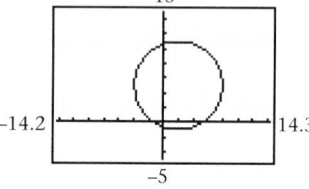

Graphics Calculator

Graph $(x-2)^2 + (y-5)^2 = 36$ using your graphics calculator.

Solution

By inspection, you can see that this is a circle centered at (2, 5) with a radius of 6. To graph the equation using a graphics calculator, solve $(x-2)^2 + (y-5)^2 = 36$ for *y*.

$$(x-2)^2 + (y-5)^2 = 36$$
$$(y-5)^2 = -(x-2)^2 + 36$$
$$y - 5 = \pm\sqrt{-(x-2)^2 + 36}$$
$$y = 5 \pm \sqrt{-(x-2)^2 + 36}$$

When you solve for *y*, you obtain two functions.

$$y_1 = 5 + \sqrt{-(x-2)^2 + 36} \text{ and } y_2 = 5 - \sqrt{-(x-2)^2 + 36}$$

Graph both functions on the same coordinate plane.

To display a *true* circle, each axis should be set with the number of units proportional to its length in pixels. On some calculators, this is called the *square* viewing window. ❖

Try This Graph $(x+4)^2 + (y-7)^2 = 25$ using your graphics calculator.

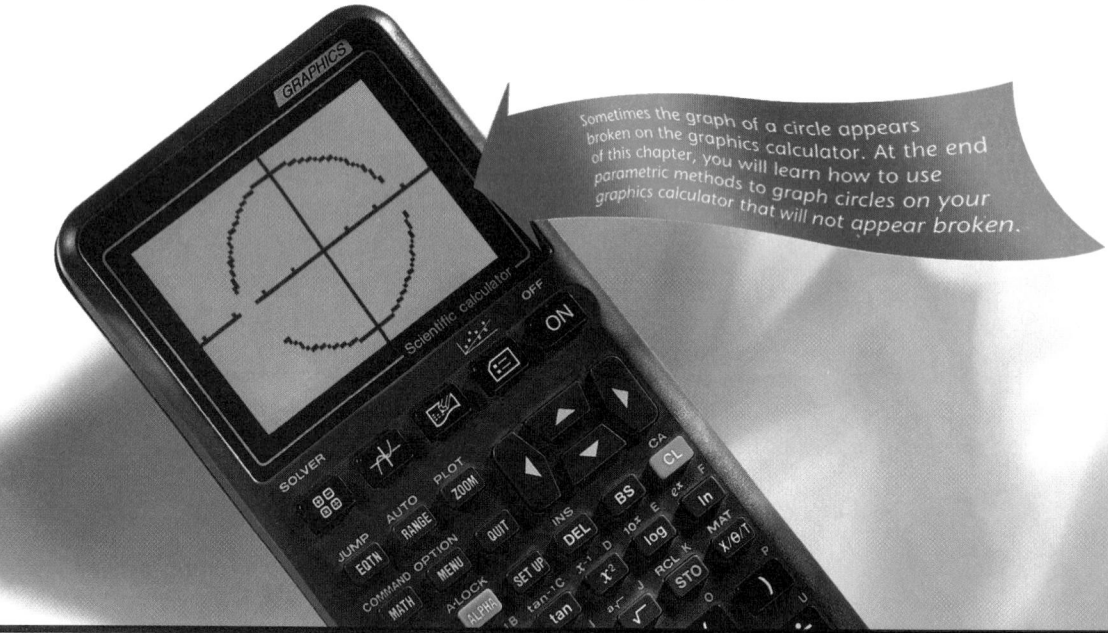

Sometimes the graph of a circle appears broken on the graphics calculator. At the end of this chapter, you will learn how to use parametric methods to graph circles on your graphics calculator that will not appear broken.

EXERCISES & PROBLEMS

Communicate

1. Explain how to find the equation of a circle when you know the coordinates of the center and the length of the radius.

2. How can you find the equation of a circle that is graphed on a coordinate plane?

3. Explain how to find the coordinates of the center and the length of the radius of a circle represented by the equation $x^2 + y^2 - 6x + 8y - 24 = 0$.

4. How can you graph the circle represented by $(x - 1)^2 + (y + 4)^2 = 16$ using your graphics calculator?

Practice & Apply

Write the standard equation of the circle with center C and radius r.

5. $C(3, 2), r = 4$

6. $C(-5, 2), r = 7$

7. $C(6, -1), r = 4$ $(x - 6)^2 + (y + 1)^2 = 16$

8. $C(-3, -4), r = \frac{7}{2}$

9. $C\left(\frac{2}{3}, -\frac{3}{5}\right), r = \sqrt{8}$

10. $C(-2, 8), r = 9$ $(x + 2)^2 + (y - 8)^2 = 81$

11. $C(1, 1), r = \sqrt{3}$
$(x - 1)^2 + (y - 1)^2 = 3$

12. $C(2, -5), r = 4.3$
$(x - 2)^2 + (y + 5)^2 = 18.49$

13. $C(-7, 1), r = \frac{3}{8}$ $(x + 7)^2 + (y - 1)^2 = \frac{9}{64}$

Determine the center and radius of the circle represented by each equation. Check by graphing on your graphics calculator.

14. $x^2 - 8x + y^2 + 4y + 16 = 0$

15. $x^2 - 6x + y^2 - 10y - 2 = 0$ $C:(3, 5);\ r = 6$

16. $x^2 + y^2 + 4x + 6y + 4 = 0$

17. $6x^2 + 6y^2 - 12x + 36y + 24 = 0$ $C:(1, -3);\ r = \sqrt{6}$

18. $x^2 + y^2 + 6x - 14y - 42 = 0$

19. $x^2 + y^2 + 8x - 4y - 5 = 0$ $C:(-4, 2);\ r = 5$

20. $x^2 + y^2 - 12x - 2y - 8 = 0$

21. $4x^2 + 4y^2 + 16x - 8y = 24$ $C:(-2, 1);\ r = \sqrt{11}$

22. **Transportation** A two-lane highway goes through a semicircular tunnel that is 14 feet high at the top. If each lane is 12 feet wide, how high is the tunnel at the edge of each lane?
7.2 feet

23. 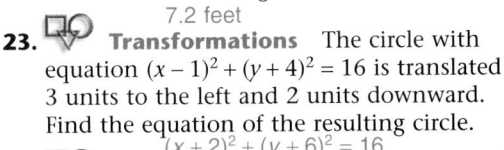 **Transformations** The circle with equation $(x - 1)^2 + (y + 4)^2 = 16$ is translated 3 units to the left and 2 units downward. Find the equation of the resulting circle.
$(x + 2)^2 + (y + 6)^2 = 16$

24. 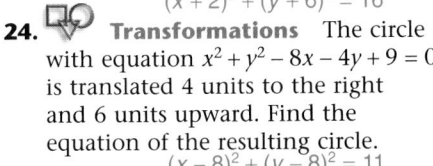 **Transformations** The circle with equation $x^2 + y^2 - 8x - 4y + 9 = 0$ is translated 4 units to the right and 6 units upward. Find the equation of the resulting circle.
$(x - 8)^2 + (y - 8)^2 = 11$

14 ft

5. $(x - 3)^2 + (y - 2)^2 = 16$

6. $(x + 5)^2 + (y - 2)^2 = 49$

8. $(x + 3)^2 + (y + 4)^2 = \frac{49}{4}$

9. $\left(x - \frac{2}{3}\right)^2 + \left(y + \frac{3}{5}\right)^2 = 8$

14. $C(4, -2), r = 2$

16. $C(-2, -3), r = 3$

18. $C(-3, 7), r = 10$

20. $C(6, 1), r = \sqrt{45}$

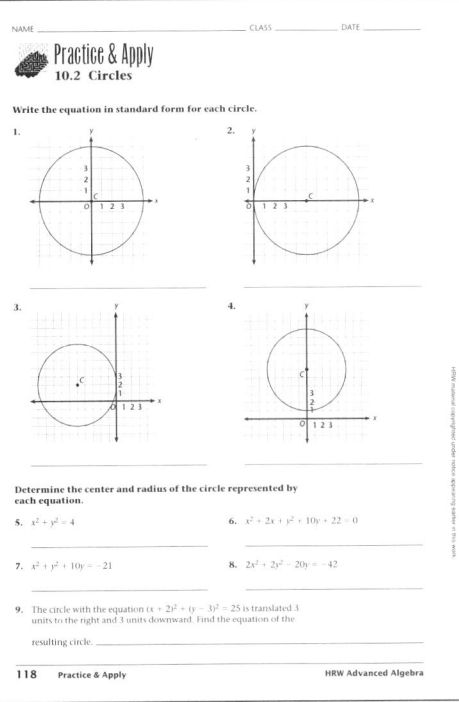

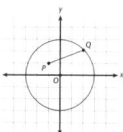

Alternative ASSESSMENT

Authentic Assessment

Communications satellites have been used for many years in long distance communications. Have students find out the approximate distance above the Earth that a communications satellite is placed. [**about 22,300 miles**] Explain that these satellites follow circular geosynchronous orbits around the Earth. Ask students to find out the radius of the Earth [**about 3960 miles**] and use this data to find an equation of the circular orbit of a communications satellite around the Earth. $\left[x^2 + y^2 = 26{,}260^2\right]$

Look Beyond

This exercise hints at the equation for the ellipse. Students can use graphics calculators to explore the shape of this graph using integer values for c.

Geology When an earthquake occurs, energy waves radiate in concentric circles from the point above where the quake occurred, called the *epicenter*. Stations with seismographs record the level of that energy and how long the energy took to reach the station.

25. Suppose one station determines that the epicenter of an earthquake is about 100 miles from the station. Find an equation for the possible locations of the epicenter using (0, 0) as the location of the station. $x^2 + y^2 = 10{,}000$

26. A second station, 120 miles east and 160 miles south of the first station, shows the epicenter to be about 135 miles away. Find a second equation for the possible locations of the epicenter. $(x - 120)^2 + (y + 160)^2 = 18{,}225$

27 Graph your equations from Exercises 25 and 26, and find the possible locations of the epicenter to the nearest whole number. (96, −27), (−1, −100)

Radio A radio tower that is 65 miles west and 45 miles north of Claudine emits a signal that is guaranteed to reach all receivers within 75 miles.

28. Write an equation for all the points that are at the farthest guaranteed receiving distance from the tower. $(x + 65)^2 + (y - 45)^2 = 5625$

29. Is Claudine within the guaranteed receiving range? Explain.
No, she is over 79 miles from the tower.

 **Look Back**

Simplify the following rational expressions.   [Lesson 2.1]

30. $\dfrac{1}{x-2} + \dfrac{4}{x^2-4}$ $\dfrac{x+6}{x^2-4}$, $x \neq \pm 2$ **31.** $\dfrac{1}{x} + \dfrac{2x}{x^3} - \dfrac{3x^3}{x^5}$ $\dfrac{x-1}{x^2}$, $x \neq 0$ **32.** $\dfrac{1}{x+3} - 4\left(\dfrac{1}{2x^2-18}\right)$
$\dfrac{x-5}{x^2-9}$, $x \neq \pm 3$

Draw an angle in standard position with approximately the given radian measure of rotation.   [Lesson 8.5]

33. 4.3 radians   **34.** $\dfrac{2\pi}{5}$ radians   **35.** $-\dfrac{3}{2}$ radians   **36.** $\dfrac{13\pi}{12}$ radians

Look Beyond

37 In this lesson you have seen that the graph of $x^2 + y^2 = c^2$, where c is a positive constant, is a circle centered at the origin. Use your graphics calculator to describe the shape of the graph of $x^2 - y^2 = c^2$, where c is a positive constant.

33.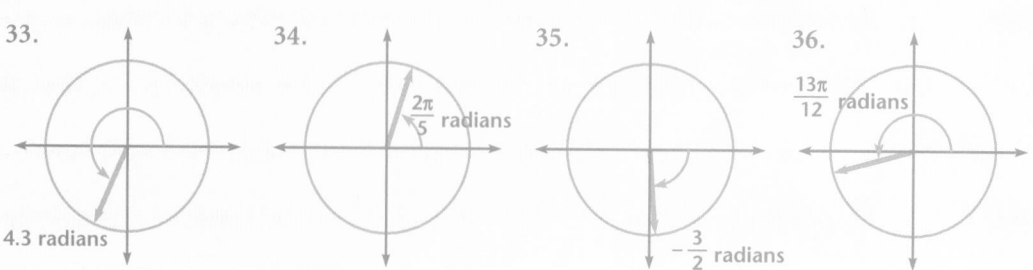

4.3 radians

34.

$\dfrac{2\pi}{5}$ radians

35.

$-\dfrac{3}{2}$ radians

36.

$\dfrac{13\pi}{12}$ radians

37. The graph has two symmetrical parts, it is a hyperbola.

LESSON 10.3 Ellipses

Why *All of the planets orbit around the Sun in an elliptical path. Satellites are placed in oval, or elliptical, paths around the Earth to perform specific tasks. You can use an equation of an ellipse to model these elliptical orbits and paths.*

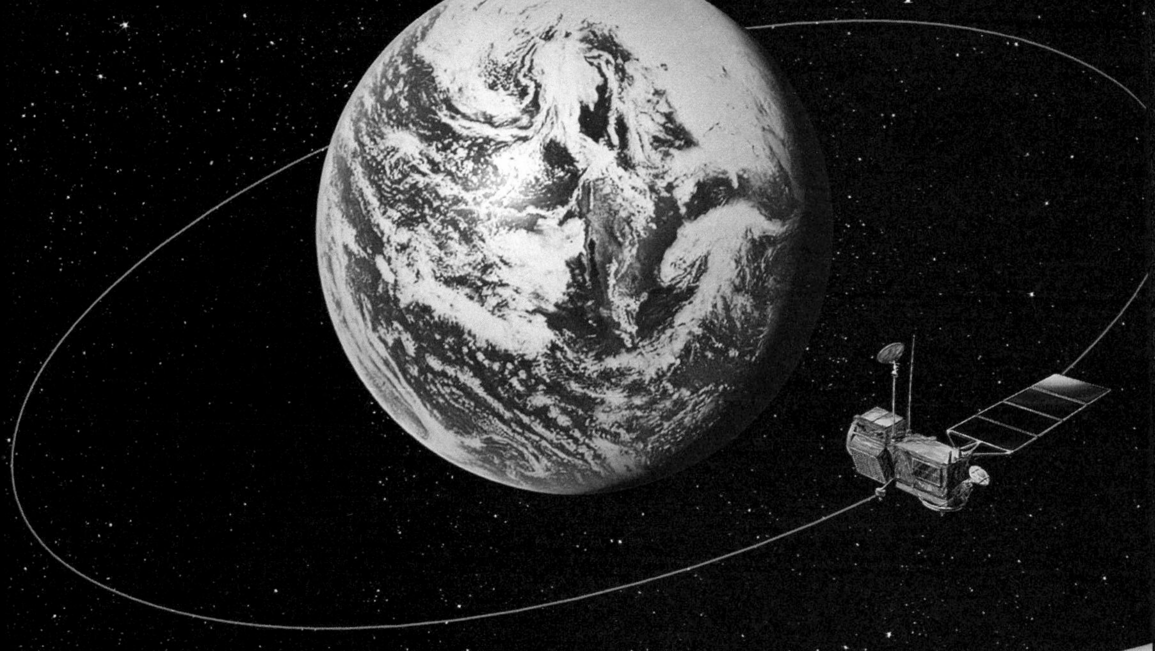

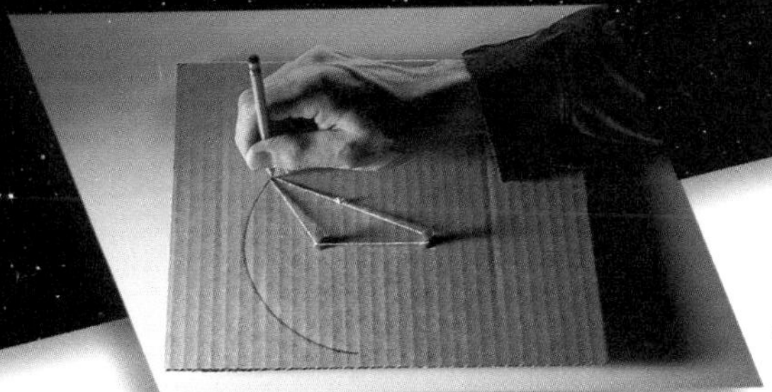

To draw an ellipse, you only need a piece of cardboard, two pushpins, and a piece of string.

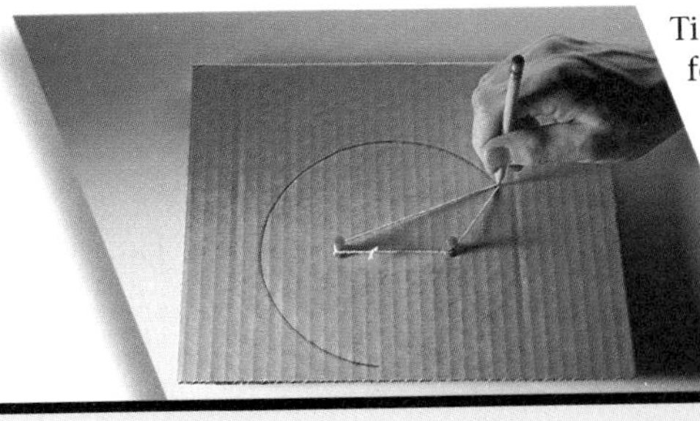

Tie the two ends of the string together to form a loop. Press the two pushpins into the cardboard and place the loop of string around them, as shown. Using a pencil with the tip on the paper, pull the string loop tight and draw a curve all the way around the pushpins until your pencil is back to the original point.

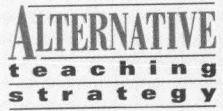

ALTERNATIVE teaching strategy

Technology Have students graph the ellipse $x^2 + 2x + 5y^2 - 20y = -1$ on their graphics calculator. Students should first solve the equation for y. Then, they should graph both functions to get a complete graph of the ellipse. On graph paper, have students sketch the graph shown on the calculator screen. Ask students to indicate the x- and y-axes. Have them describe the ellipse — the approximate coordinates of its center, vertices, and direction of the major and minor axes. Ask students to write the equation in standard form. $\left[\frac{(x+1)^2}{20} + \frac{(y-2)^2}{4} = 1\right]$

PREPARE

Objectives

- Determine the coordinates of the center, vertices, co-vertices, and foci when given an equation of an ellipse with either a horizontal or vertical major axis.

RESOURCES

• Practice Master	10.3
• Enrichment Master	10.3
• Technology Master	10.3
• Lesson Activity Master	10.3
• Quiz	10.3
• Spanish Resources	10.3

This lesson may be considered optional in the core course. See page 534A.

Assessing Prior Knowledge

Complete the square to solve each equation for x.

A. $x^2 = 5 - 4x$ **[1, −5]**

B. $3 = x^2 - 2x$ **[−1, 3]**

C. $x^2 = 7 - 6x$ **[1, −7]**

D. $x^2 = 4x - 6$ **[$2 \pm i\sqrt{2}$]**

TEACH

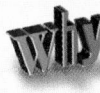

Why Mathematicians have known about ellipses since the third century B.C.E. However, the concept of the ellipse was not used until the sixteenth century when the Polish astronomer Copernicus discovered that the orbits of the planets were elliptical. The equation of an ellipse can be used to interpret the elliptical orbits of the planets.

The construction of an ellipse may help you to understand the definition of an ellipse. The part of the string that covers the distance from the pencil point to each of the two pushpins is a constant length. Therefore, the sum of the distances from any point on the ellipse (the pencil point) to each of the two fixed points (pushpins) is constant.

ELLIPSE

An **ellipse** is a set of coplanar points such that the sum of the distances from two fixed points in the plane, called **foci**, to any point on the ellipse is a constant.

 **CRITICAL** *Thinking*

If the pushpins are 10 inches apart and the loop of string has a total length of 26 inches, find the greatest length across the resulting ellipse.

Equation of an Ellipse

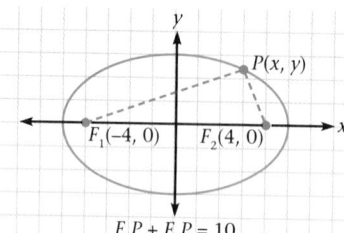

$F_1P + F_2P = 10$

Examine the ellipse shown. The center of the ellipse is located at the origin. The foci are $F_1(-4, 0)$ and $F_2(4, 0)$. Let $P(x, y)$ be any point on the ellipse. When you know the coordinates of the foci and the sum of the distances F_1P and F_2P, you can find the equation of the ellipse using the distance formula.

Given that $F_1P + F_2P = 10$, use the distance formula to express $F_1P + F_2P = 10$ in terms of the coordinates of each point. Then simplify.

$$F_1P + F_2P = 10$$
$$\sqrt{(x - (-4))^2 + (y - 0)^2} + \sqrt{(x - 4)^2 + (y - 0)^2} = 10$$
$$\sqrt{(x + 4)^2 + y^2} = 10 - \sqrt{(x - 4)^2 + y^2}$$

Square both sides, and simplify.

$$\left(\sqrt{(x + 4)^2 + y^2}\right)^2 = \left(10 - \sqrt{(x - 4)^2 + y^2}\right)^2$$
$$(x + 4)^2 + y^2 = 100 - 20\sqrt{(x - 4)^2 + y^2} + \left[(x - 4)^2 + y^2\right]$$
$$16x - 100 = -20\sqrt{(x - 4)^2 + y^2}$$
$$4x - 25 = -5\sqrt{(x - 4)^2 + y^2}$$

Square both sides again.

$$(4x - 25)^2 = \left(-5\sqrt{(x - 4)^2 + y^2}\right)^2$$
$$16x^2 - 200x + 625 = 25[(x - 4)^2 + y^2]$$
$$16x^2 - 200x + 625 = 25x^2 - 200x + 400 + 25y^2$$
$$225 = 9x^2 + 25y^2$$
$$1 = \frac{x^2}{25} + \frac{y^2}{9}$$

Thus, the equation of the ellipse centered at the origin with foci at $(-4, 0)$ and $(4, 0)$ and with a constant sum of 10 is

$$\frac{x^2}{25} + \frac{y^2}{9} = 1.$$

 **interdisciplinary** **CONNECTION**

Astronomy The planets of our solar system follow almost circular or elliptical orbits around the sun. Planets like Mercury and Mars follow elliptical orbits. Have students use an encyclopedia or other sources to find out which planets follow elliptical orbits. Ask students to find out the distance between the sun and each of those planets. Explain that the sun is located in one of the two foci of these planets orbits.

An ellipse has two axes of symmetry. The longer axis, connecting the vertices, is the **major axis**. The shorter axis, connecting the co-vertices, is the **minor axis**. The intersection of the major and minor axes is the center of the ellipse.

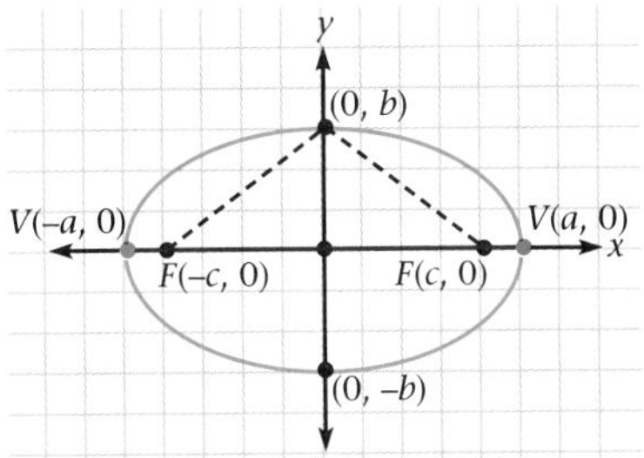

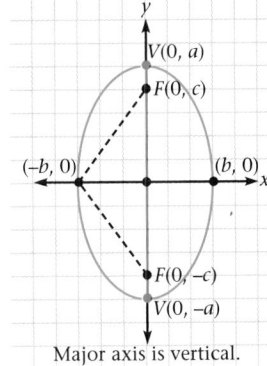

Major axis is horizontal.

Major axis is vertical.

The distance between vertices, or the length of the major axis, is $2a$. The distance between co-vertices, or the length of the minor axis, is $2b$. The distance from the center of the ellipse to either focus is c.

Notice a right triangle is formed inside the ellipse by connecting a focus and a vertex with the center.

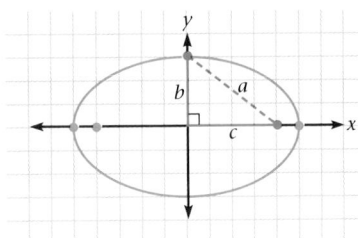

CRITICAL
Thinking

Write an equation for the relationship between a, b, and c using the "Pythagorean" Right-Triangle Theorem. Of a, b, and c, which is always the *longest*? Why?

Equation of a Translated Ellipse

A translated ellipse has its center at (h, k).

Standard Equation of an Ellipse, where $a > b > 0$	
$\dfrac{(x-h)^2}{a^2} + \dfrac{(y-k)^2}{b^2} = 1$	Major axis is horizontal. Center is (h, k).
$\dfrac{(x-h)^2}{b^2} + \dfrac{(y-k)^2}{a^2} = 1$	Major axis is vertical. Center is (h, k).

Describe a way to determine, from the standard equation of an ellipse, whether the major axis is horizontal or vertical.

CRITICAL *Thinking*

$b^2 + c^2 = a^2$; a is the longest because it is the hypotenuse of a right triangle.

A**ongoing** **SSESSMENT**

If the constant that divides the $(x - h)^2$ term is greater than the constant that divides the $(y - k)^2$ term, then the major axis is horizontal. If the constant that divides the $(x - h)^2$ term is less than the constant that divides the $(y - k)^2$ term, then the major axis is vertical.

A. Write an equation for the ellipse shown.

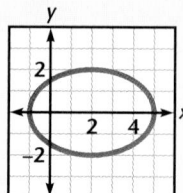

$$\left[\frac{(x-2)^2}{9} + \frac{y^2}{4} = 1\right]$$

B. Find the coordinates of the foci. [$(2 - \sqrt{5}, 0)$ and $(2 + \sqrt{5}, 0) \approx (-0.24, 0)$ and $(4.24, 0)$]

A ongoing ASSESSMENT

They are on the major axis $x = 3$ and they are inside the ellipse because $2 + \sqrt{7} < 6$ and $2 - \sqrt{7} > -2$, the y-values of the vertices.

TEACHING tip

Point out that the relationship $b^2 + c^2 = a^2$ for ellipses is derived from the "Pythagorean" Right-Triangle Theorem.

Alternate Example 2

The equation $25x^2 - 150x + 16y^2 + 32y = 159$ represents an ellipse.

A. Write the equation of the ellipse in standard form.
$$\left[\frac{(x-3)^2}{16} + \frac{(y+1)^2}{25} = 1\right]$$

B. Find the coordinates of the center, vertices, co-vertices, and the foci.
[Center: $(3, -1)$; vertices: $(3, 4)$ and $(3, -6)$; co-vertices: $(7, -1)$ and $(-1, -1)$; foci: $(3, 2)$ and $(3, -4)$]

EXAMPLE 1

Ⓐ Write an equation for the ellipse shown.

Ⓑ Find the coordinates of the foci.

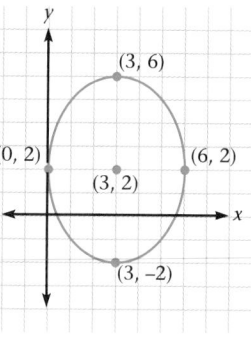

Solution➤

Ⓐ The center, (h, k), is $(3, 2)$.

The length of the major axis, $2a$, is 8. Therefore, a is 4.

The length of the minor axis, $2b$, is 6. Therefore, b is 3.

Since the ellipse has a vertical major axis, the equation will be of the form $\frac{(x-h)^2}{b^2} + \frac{(y-k)^2}{a^2} = 1$.

Substitute values of h, k, a, and b in the equation.

Thus, the equation of the ellipse shown is $\frac{(x-3)^2}{9} + \frac{(y-2)^2}{16} = 1$.

Ⓑ To find the coordinates of the foci, use the relationship $b^2 + c^2 = a^2$ for ellipses. Substitute 4 for a and 3 for b.

$$b^2 + c^2 = a^2$$
$$9 + c^2 = 16$$
$$c^2 = 7$$
$$c = \pm\sqrt{7}$$

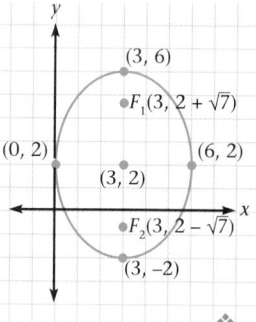

Thus, the foci are located $\sqrt{7}$ units above and $\sqrt{7}$ units below the center, $(3, 2)$.

$(3, 2 + \sqrt{7})$ and $(3, 2 - \sqrt{7})$

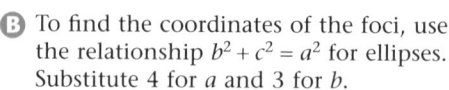

Explain why the coordinates $(3, 2 + \sqrt{7})$ and $(3, 2 - \sqrt{7})$ for the foci in Example 1 are reasonable.

When graphing an ellipse whose equation is not in standard form, complete the square to write the equation in standard form.

EXAMPLE 2

The equation $x^2 + 2x + 4y^2 - 24y = -1$ represents an ellipse.

Ⓐ Write the equation of the ellipse in standard form.

Ⓑ Find the coordinates of the center, the vertices, the co-vertices, and the foci. Then graph the ellipse.

RETEACHING the lesson

Using Patterns Examine five different ellipses. In each, determine the ratio of the length of the minor axis to the length of the major axis. Compare these ratios with the shapes of the ellipses. Display the results in a table with the column headings: major axis, minor axis, $\frac{\text{minor axis}}{\text{major axis}}$, graph.

Solution

Ⓐ To write the equation in standard form, complete the squares on x and y.

$$x^2 + 2x + 4y^2 - 24y = -1$$
$$(x^2 + 2x + 1) + 4(y^2 - 6y + 9) = -1 + 1 + 36$$
$$(x + 1)^2 + 4(y - 3)^2 = 36$$
$$\frac{(x + 1)^2}{36} + \frac{(y - 3)^2}{9} = 1$$

Ⓑ The center, (h, k), is $(-1, 3)$.

Look at the denominators, 36 and 9, in the equation of the ellipse. The larger denominator is a^2. Since the x-term has the larger denominator, the major axis is horizontal. Therefore, a is 6, and the vertices are located 6 units to the left and 6 units to the right of the center, $(-1, 3)$.

$$\text{Vertex 1: } (-1 + 6, 3) = (5, 3)$$
$$\text{Vertex 2: } (-1 - 6, 3) = (-7, 3)$$

Since b^2 is 9, b is 3. The co-vertices are located 3 units below and 3 units above the center, $(-1, 3)$.

$$\text{Co-vertex 1: } (-1, 3 + 3) = (-1, 6)$$
$$\text{Co-vertex 2: } (-1, 3 - 3) = (-1, 0)$$

To find the coordinates of the foci, use the relationship $b^2 + c^2 = a^2$ for ellipses. Substitute 6 for a and 3 for b in the equation.

$$b^2 + c^2 = a^2$$
$$9 + c^2 = 36$$
$$c^2 = 27$$
$$c = \pm\sqrt{27}$$

Thus, the foci are located $\sqrt{27}$ units to the left and $\sqrt{27}$ units to the right of the center, $(-1, 3)$.

Focus 1: $(-1 - \sqrt{27}, 3)$ Focus 2: $(-1 + \sqrt{27}, 3)$.

Plot the vertices, $(5, 3)$ and $(-7, 3)$, and the co-vertices, $(-1, 6)$ and $(-1, 0)$.

Then sketch the graph of the ellipse through these points. ❖

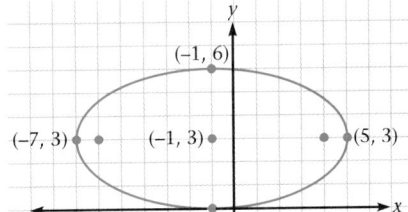

ongoing ASSESSMENT

Try This

A. $\frac{(x + 2)^2}{9} + \frac{(y + 1)^2}{4} = 1$

B.   center: $(-2, -1)$;
     vertices: $(1, -1)$ and
     $(-5, -1)$;
     co-vertices: $(-2, 1)$ and
     $(-2, -3)$;
     foci: $(-2 + \sqrt{5}, -1)$ and
     $(-2 - \sqrt{5}, -1)$ or
     $\approx (0.236, -1)$ and
     $(-4.236, -1)$

Try This   The equation $4x^2 + 16x + 9y^2 + 18y = 11$ represents an ellipse.

Ⓐ Write the equation of this ellipse in standard form.

Ⓑ Find the coordinates of the center, the vertices, the co-vertices, and the foci. Then graph the ellipse.

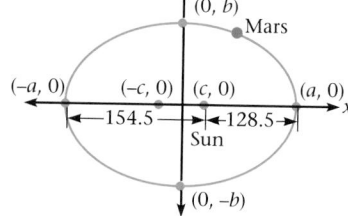

Alternate Example 3

The closest Mercury gets to the sun is 29 million miles, and its farthest distance from the sun is 43 million miles.

A. Assuming that the center of the ellipse is the origin, write the equation of the elliptical path.
$$\left[\frac{x^2}{1296} + \frac{y^2}{1247} = 1\right]$$

B. Find the distance from the sun to the other focus. [**14 million miles**]

CRITICAL *Thinking*

In the equation, $a = 141.5$ and $b = 140.9$. Therefore, the major and minor axes of the elliptical path of Mars are almost the same. A circle is an ellipse where the major and minor axes are equal.

Use Transparency ▶ **40**

Applications of Ellipses

An equation of an ellipse can be used to describe the path of a planet as it orbits the Sun.

EXAMPLE 3

Astronomy Each planet orbits the Sun in an elliptical path, with the Sun at a focus of the ellipse. The closest Mars gets to the Sun is 128.5 million miles, and its farthest distance from the Sun is 154.5 million miles.

Ⓐ Assuming that the center of the ellipse is the origin, write an equation for the elliptical path.

Ⓑ Find the distance from the Sun to the other focus.

Solution▸

Ⓐ Make a rough sketch. Let the Sun be at $(c, 0)$. Then Mars is closest to or farthest from to the Sun when its path crosses the x-axis.

In millions of miles, $a - c = 128.5$ and $c - (-a) = 154.5$.

Adding like terms of the equations, $2a = 283$, so a is 141.5. Thus, c is 13.

Substitute 141.5 for a and 13 for c in the equation $b^2 + c^2 = a^2$ for ellipses to find that b is 140.9.

Thus, an equation for the path of Mars around the Sun in millions of miles is $\frac{x^2}{141.5^2} + \frac{y^2}{140.9^2} = 1$, or $\frac{x^2}{20,022.25} + \frac{y^2}{19,853.25} = 1$.

Ⓑ The distance from the Sun, $(c, 0)$, to the other focus, $(-c, 0)$, is $2c$, or 26 million miles. ❖

CRITICAL *Thinking*

Examine the equation for the elliptical path of Mars around the Sun. Explain why people believed planets orbited the Sun in *circular* paths until Kepler discovered that the paths were elliptical.

EXERCISES & PROBLEMS

Communicate

1. How can you find an equation of an ellipse when you know the coordinates of the foci and the constant sum?

2. What is the equation of an ellipse when the center is located at (h, k)?

3. What is the difference between the equation of an ellipse whose major axis is horizontal and the equation of an ellipse whose major axis is vertical?

4. Explain how to find an equation of the ellipse with foci 6 units apart and a constant sum of 10.

Practice & Apply

Write an equation for each of the following ellipses.

5. Vertices at (4, 0) and (–4, 0); co-vertices at (0, 2) and (0, –2) $\frac{x^2}{16} + \frac{y^2}{4} = 1$

6. Vertices at (7, 0) and (–7, 0); co-vertices at (0, 5) and (0, –5) $\frac{x^2}{49} + \frac{y^2}{25} = 1$

7. Vertices at (6, 0) and (–6, 0); co-vertices at (0, 3) and (0, –3) $\frac{x^2}{36} + \frac{y^2}{9} = 1$

8. Vertices at $\left(0, \frac{11}{2}\right)$ and $\left(0, -\frac{11}{2}\right)$; co-vertices at $\left(\frac{5}{3}, 0\right)$ and $\left(-\frac{5}{3}, 0\right)$ $\frac{9x^2}{25} + \frac{4y^2}{121} = 1$

Determine the vertices and co-vertices for each ellipse.

9. $x^2 - 6x + 2y^2 + 8y + 1 = 0$

10. $5x^2 + 20x + 4y^2 + 24y + 36 = 0$

11. $2x^2 + 6y^2 - 6x - 12y = \frac{3}{2}$

12. $16x^2 + 4y^2 - 96x + 8y + 84 = 0$
   $C:(1, -1), (5, -1); V:(3, -5), (3, 3)$

Astronomy The Moon orbits Earth in an elliptical path with Earth at one focus. The major axis of the orbit is 774,000 kilometers, and the minor axis is 773,000 kilometers.

13. Using (0, 0) as the center of the ellipse, write an equation for the orbit of the Moon. $\frac{x^2}{387,000^2} + \frac{y^2}{386,500^2} = 1$

14. How far from Earth is the Moon when the Moon comes closest to Earth? 367,330 km

15. How far from Earth is the Moon when the Moon is farthest from Earth? 406,670 km

9. Vertices: $(-1, -2)$, $(7, -2)$; Co-vertices: $(3, -2 - \sqrt{8}), (3, -2 + \sqrt{8})$

10. Vertices: $(-2, -3 + \sqrt{5})$, $(-2, -3 - \sqrt{5})$; Co-vertices: $(-4, -3), (0, -3)$

11. Vertices: $\left(\frac{3}{2} - \sqrt{6}, 1\right), \left(\frac{3}{2} + \sqrt{6}, 1\right)$; Co-vertices: $\left(\frac{3}{2}, 1 - \sqrt{2}\right), \left(\frac{3}{2}, 1 + \sqrt{2}\right)$

Assignment Guide

Core 1–7, 9–45

Core Plus 1–4, 7–8, 11–47

Technology

Graphics calculators are needed for Exercises 24–26, and 44–47.

Error Analysis

When students use a graphics calculator, it may take them several trials to find an appropriate range and domain. For ellipses a good technique is to find the center by inspection and make that the center of the viewing window.

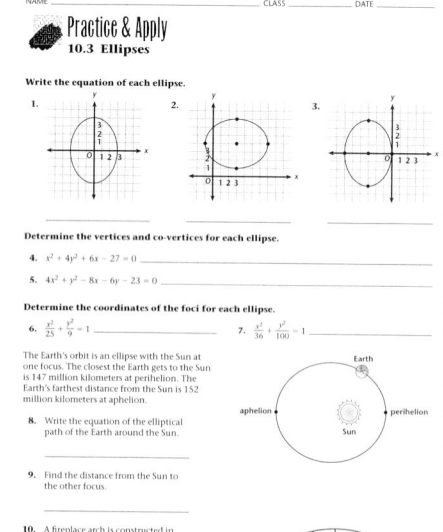

ASSESSMENT

Alternative

Performance Assessment

Have students write one or two paragraphs describing the properties of an ellipse. Ask them to indicate the standard form of the equation of the ellipse, and how to determine the coordinates of its center, vertices, co-vertices, and foci from the equation.

Architecture A hall in a museum is an elliptical room. The reflective property of an ellipse makes it possible for someone standing at one focus to hear what people standing at the other focus are saying, even if they whisper. The hall is 96 feet long and 48 feet wide.

16. Write an equation for the ellipse.

17. How far from the end of the hall are the foci? 6.4 ft

Statuary Hall in the United States Capitol is not elliptical in shape. Nevertheless, it is possible for two people standing at specific positions to hear each other whisper. What characteristics of the room might make this possible?

Astronomy A satellite orbits Earth in an elliptical path, with Earth at one focus. The closest the satellite gets to Earth is 7000 miles, and its farthest distance from Earth is 22,000 miles.

18. Using (0, 0) as the center of the ellipse, write an equation for the orbit of the satellite. $\dfrac{x^2}{14{,}500^2} + \dfrac{y^2}{12{,}400^2} = 1$

19. Find the distance from Earth to the other focus. 15,000 mi

Determine the vertices and co-vertices for each ellipse.

20. $x^2 + 4y^2 - 2x + 40y + 100 = 0$ **21.** $4x^2 + 25y^2 + 16x + 50y - 59 = 0$

22. $4x^2 + 9y^2 - 24x + 18y + 9 = 0$ **23.** $4x^2 + 9y^2 - 8x - 54y + 49 = 0$

$V{:}(0,\ -1),\ (6,\ -1)\ ;\ C{:}(3,\ 1),\ (3,\ -3)$ $V{:}(-2,\ 3),\ (4,\ 3)\ ;\ C{:}(1,\ 5),\ (1,\ 1)$

Solve each equation for *y*. Then use your graphics calculator to graph both solutions on the same coordinate plane, and verify the coordinates of the vertices and co-vertices.

24 $\dfrac{x^2}{4} + \dfrac{y^2}{1} = 1$ **25** $16x^2 + 9y^2 = 144$ **26** $\dfrac{(x+3)^2}{25} + \dfrac{(y-2)^2}{16} = 1$

27. Find the equation of the ellipse with foci at (2, −3) and (2, 5) and whose constant sum is 10. $\dfrac{(x-2)^2}{9} + \dfrac{(y-1)^2}{25} = 1$

28. Find the equation of the ellipse with foci at (0, 2) and (0, 8) and whose constant sum is 10. $\dfrac{x^2}{16} + \dfrac{(y-5)^2}{25} = 1$

29. Find the equation of the ellipse with foci at (−3, 1) and (6, 1) and whose constant sum is 14. $\dfrac{(x-1.5)^2}{49} + \dfrac{(y-1)^2}{28.75} = 1$

30. Write an equation for the set of all points, the sum of whose distances from points (2, 3) and (−4, 3) is always 12. $\dfrac{(x+1)^2}{36} + \dfrac{(y-3)^2}{27} = 1$

16. $\dfrac{x^2}{2304} + \dfrac{y^2}{576} = 1$

20. Vertices: (0, −5), (2, −5); Co-vertices: (1, −4.5), (1, −5.5)

21. Vertices: (−7, −1), (3, −1); Co-vertices: (−2, 1), (−2, −3)

24. $y = \pm\sqrt{\dfrac{4 - x^2}{4}}$; vertices (2, 0), (−2, 0); Co-vertices: (0, 1), (0, −1)

25. $y = \pm\sqrt{\dfrac{144 - 16x^2}{9}}$; vertices (0, 4), (0, −4); Co-vertices: (3, 0), (−3, 0)

26. $y = 2 \pm\sqrt{\dfrac{400 - 16(x+3)^2}{25}}$; vertices (2, 2), (−8, 2); Co-vertices: (−3, 6), (−3, −2)

31. Anatomy An artificial forearm is to be constructed so that each cross-section is an ellipse with the foci where the centers of the bones in a real forearm would be. If (0, 0) is the center of the artificial arm, find the equation of the ellipse when the foci are 10 centimeters apart and the width of the arm is 12 centimeters. $\frac{x^2}{36} + \frac{y^2}{11} = 1$

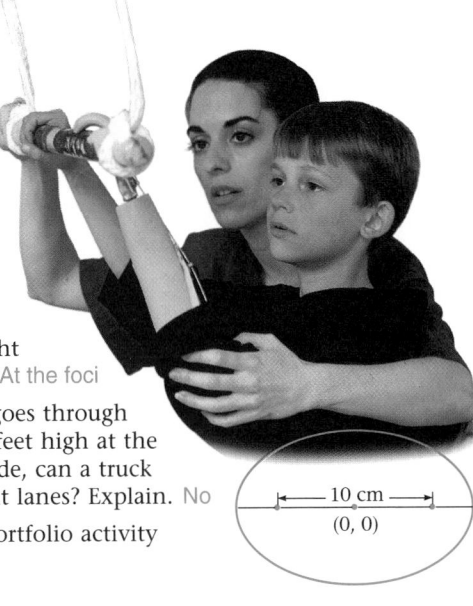

32. Acoustics Consider constructing a concert auditorium using the idea of an ellipse. How might stages or sound and light equipment be placed? At the foci

33. Transportation A one-way four-lane highway goes through a tunnel in the shape of a semi-ellipse that is 15 feet high at the center and 60 feet wide. If each lane is 12 feet wide, can a truck that is 12 feet high drive in the far left or far right lanes? Explain. No

34. 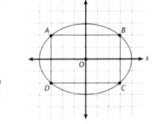 **Portfolio Activity** Complete the portfolio activity described on page 535.

|← 10 cm →|
(0, 0)

Look Back

Let a, b, and c represent real numbers. Identify the property for real numbers shown. [Lesson 2.1]

Assoc. prop. of addition
35. $(a+b)+c = a+(b+c)$ **36.** $a+0 = a = 0+a$ Additive identity
Comm. prop. of multiplication
37. $ab = ba$ **38.** ab is a real number. closure for multiplication

Show that functions are associative with respect to composition. [Lesson 3.3]

39. $f(x) = -2x$, $g(x) = 1 + \frac{2}{x}$, and $h(x) = 4 - x^2$

40. $f(x) = x^2 + 3$, $g(x) = 3x - 4$, and $h(x) = 7 - x$

Simplify. [Lesson 3.4]

41. $\lceil 5.1 \rceil + \lceil \sqrt{5} \rceil$ 9   **42.** $\lfloor -3.08 \rfloor - \lceil 6.27 \rceil$ -10   **43.** $\lceil 12.4 \rceil - \lceil 13.9 \rceil$ 0

Use your graphics calculator to graph each parametric system. Then combine each system into one linear function. Graph the function to check your result. [Lesson 3.6]

44. $\begin{cases} x(t) = 3 - t \\ y(t) = 4t \end{cases}$ $y = -4x + 12$ **45.** $\begin{cases} x(t) = t - 5 \\ y(t) = t + 1 \end{cases}$ $y = x + 6$

Look Beyond

46. Solve $x^2 - y^2 = 1$ for y, and graph both resulting equations on one coordinate plane. Describe the graph. $y = \pm\sqrt{x^2 - 1}$

47. Solve $y^2 - x^2 = 1$ for y, and graph both resulting equations on one coordinate plane. Describe the graph. $y = \pm\sqrt{x^2 + 1}$

48. Compare the two graphs in Exercises 46 and 47. How are the two graphs alike and how are they different?

The answers to Exercises 33–34, 39–40, and 48 can be found in Additional Answers beginning on page 842.

Look Beyond

Exercises 46–48 has students work with a hyperbola using their graphics calculators.

Technology Master

NAME _____ CLASS _____ DATE _____

Technology
10.3 Maximizing Area

The diagram shows the graph of $9x^2 + 16y^2 = 144$. Inscribed in the ellipse is rectangle $ABCD$. Below is a problem you can investigate with a graphics calculator.

Of all rectangles that you can inscribe in the graph of $9x^2 + 16y^2 = 144$, which one has the greatest or maximum area?

To solve the problem, notice that the area of rectangle $ABCD$ is given by $(2x)(2y) = 4(xy)$, where x and y are the coordinates of a point on the ellipse in the first quadrant.

Suppose that $9x^2 + 16y^2 = 144$.

1. Write an equation, in terms of x, for the area of rectangle $ABCD$ by using the equation Area = $4(xy)$.

2. Graph the equation you wrote in Exercise 1 over the interval $0 \le x \le 9.4$.

3. From the graph, find x (to the nearest tenth) that maximizes the area of rectangle $ABCD$ and the maximum area of $ABCD$.

Suppose that $4x^2 + 25y^2 = 100$.

4. Write an equation for the area of rectangle $ABCD$. Graph the equation you wrote over the interval $0 \le x \le 9.4$.

5. From the graph, find x (to the nearest tenth) that maximizes the area of rectangle $ABCD$ and the maximum area of $ABCD$.

6. Suppose that $ax^2 + by^2 = ab$. Make a conjecture about the maximum area of $ABCD$.

HRW Advanced Algebra   Technology **131**

Objectives

• Determine the coordinates of the center and foci, and find the lengths of the axes when given an equation of a hyperbola.

RESOURCES

• Practice Master	**10.4**
• Enrichment Master	**10.4**
• Technology Master	**10.4**
• Lesson Activity Master	**10.4**
• Quiz	**10.4**
• Spanish Resources	**10.4**

This lesson may be considered optional in the core course. See page 534A.

Assessing Prior Knowledge

Write each equation of an ellipse in standard form.

1. $9x^2 + 16y^2 = 144$

$$\left[\frac{x^2}{16} + \frac{y^2}{9} = 1\right]$$

2. $49x^2 + 81y^2 = 3969$

$$\left[\frac{x^2}{81} + \frac{y^2}{49} = 1\right]$$

TEACH

Hyperbolas have different applications, such as locating whales, sunken ships, and other objects on the ocean floor. Hyperbolas and ellipses are similar conic sections, but hyperbolas have a *constant* difference between the foci and any point on the graph.

LESSON 10.4 Hyperbolas

Many search methods use fixed sound detectors and the definition of the hyperbola. Hyperbolas are used to locate the positions of whales, sunken ships, fallen airplanes, and other objects in the ocean. They can also be used to locate electronically tagged animals.

The definition of a hyperbola is similar to that of an ellipse. Instead of a constant *sum* of distances from the foci to a point on the graph, the hyperbola has a constant *difference* between these distances.

HYPERBOLA

A **hyperbola** is a set of coplanar points such that the difference of the distances from two fixed points in the plane, called **foci**, to any point on the hyperbola is a constant.

Technology Have students graph the hyperbola represented by the equation $3x^2 - y^2 = 36$ using similar methods to those used to graph other conic sections. First ask students to solve the equation for y. Then have them graph both functions on the calculator. Remind them that they may have to try several domains before they are able to see a complete graph. Have students determine the center of the hyperbola and whether its transverse axis is horizontal or vertical. Have students complete the squares for x and y to determine the equation in its standard form. Ask them to compare the standard form of the equation and the graph. Finally, have them determine the coordinates of the foci.

Equation of a Hyperbola

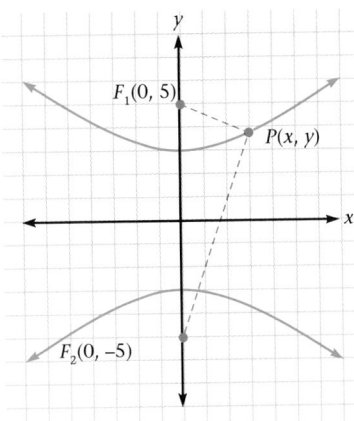

Examine the hyperbola shown. The center of the hyperbola is located at the origin. The foci are $F_1(0, 5)$ and $F_2(0, -5)$. Let $P(x, y)$ be any point on the hyperbola.

When you know the coordinates of the foci and the constant difference between the distances F_1P and F_2P, you can determine the equation of the hyperbola using the distance formula.

Suppose $|F_1P - F_2P| = 6$. Since it is not possible to determine which distance, F_1P or F_2P, is greater, use the absolute value of their difference, $|F_1P - F_2P|$.

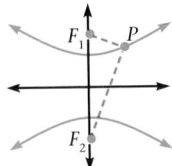 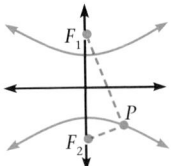

$$|F_1P - F_2P| = 6$$
$$\pm(F_1P - F_2P) = 6$$
$$F_1P - F_2P = \pm 6$$
$$\sqrt{(x-0)^2 + (y-5)^2} - \sqrt{(x-0)^2 + (y+5)^2} = \pm 6$$
$$\sqrt{x^2 + (y-5)^2} = \pm 6 + \sqrt{x^2 + (y+5)^2}$$

Square both sides.
$$x^2 + (y-5)^2 = 36 \pm 12\sqrt{x^2 + (y+5)^2} + [x^2 + (y+5)^2]$$
$$x^2 + y^2 - 10y + 25 = 36 \pm 12\sqrt{x^2 + (y+5)^2} + x^2 + y^2 + 10y + 25$$
$$-20y - 36 = \pm 12\sqrt{x^2 + (y+5)^2}$$
$$-\frac{5}{3}y - 3 = \pm\sqrt{x^2 + (y+5)^2}$$

Square both sides again.
$$\frac{25}{9}y^2 + 10y + 9 = x^2 + (y+5)^2$$
$$\frac{25}{9}y^2 + 10y + 9 = x^2 + y^2 + 10y + 25$$
$$\frac{16}{9}y^2 - x^2 = 16$$
$$\frac{y^2}{9} - \frac{x^2}{16} = 1$$

Thus, the equation of the hyperbola with foci $F_1(0, 5)$ and $F_2(0, -5)$ and with a constant difference of 6 is
$$\frac{y^2}{9} - \frac{x^2}{16} = 1.$$

interdisciplinary CONNECTION

Optics A hyperbolic mirror is designed to detect light rays and direct them to a focal point. Then these rays are directed to the other focal point of the hyperbola. Borrow several hyperbolic mirrors from the science department. Have students work in groups where each group designs and carries out its own experiment. Have each group write a description of their experiment, its setup, the desired goals, and their results. Have students comment on how they would extend their experiment if they were able to do so.

By the definition of a hyperbola and the graph, the difference between the distance from F_1 to $(0, a)$ and the distance from F_2 to $(0, a)$ is constant. So, let the constant be k, for some number k,

$$|\sqrt{(0-0)^2 + (c-a)^2} - \sqrt{(0-0)^2 + [c-(-a)]^2}| = k$$
$$|\sqrt{(c-a)^2} - \sqrt{(c+a)^2}| = k$$
$$|(c-a) - (c+a)| = k$$
$$|c-a-c-a| = k$$
$$|-2a| = k$$
$$2a = k$$

Thus, the constant difference is $2a$.

A hyperbola has two axes, the **transverse axis** and the **conjugate axis**, which intersect at the center. The **transverse axis** connects the vertices. The **conjugate axis** is perpendicular to the transverse axis. Also, the transverse axis is *not* always longer than the conjugate axis.

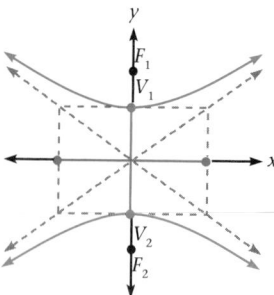

Transverse axis is vertical.

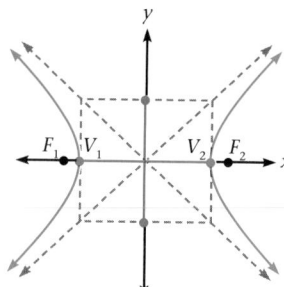

Transverse axis is horizontal.

The dashed lines through the center of a hyperbola are called **asymptotes of a hyperbola**. The hyperbola gets closer and closer to these asymptotes, but never touches them. The transverse axis and the conjugate axis define a rectangle with the asymptotes as diagonals.

The length of the transverse axis is $2a$.

The length of the conjugate axis is $2b$.

The distance between foci is $2c$.

The variables a, b, and c are related by $a^2 + b^2 = c^2$.

The constant difference of the distances from the foci to any point on the hyperbola is $2a$.

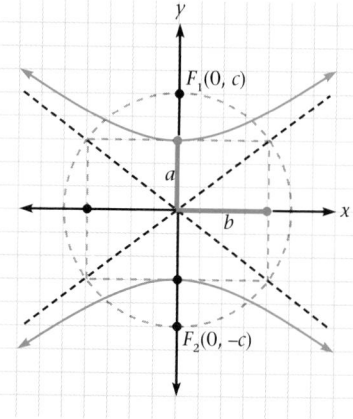

CRITICAL *Thinking*

Use the graph above and the definition of a hyperbola to prove that the constant difference is $2a$.

Equation of a Translated Hyperbola

The center of a translated hyperbola is (h, k). The standard form of a hyperbola centered at (h, k) is given in the following table.

Standard Equation of a Hyperbola, where $a > 0$ and $b > 0$		
y^2-term is negative	$\dfrac{(x-h)^2}{a^2} - \dfrac{(y-k)^2}{b^2} = 1$	Transverse axis is horizontal through the center, (h, k).
x^2-term is negative	$\dfrac{(y-k)^2}{a^2} - \dfrac{(x-h)^2}{b^2} = 1$	Transverse axis is vertical through the center, (h, k).

ENRICHMENT Generalize the proof on page 561 so that it is a general proof of the equation of a hyperbola with any axis of symmetry (not just horizontal or vertical). Let $F_1 = (f_1, f_2)$ and $F_2 = (e_1, e_2)$ and rewrite the proof.

Describe a way to determine—from the equation of a hyperbola in standard form—whether the transverse axis is horizontal or vertical.

EXAMPLE 1

A Identify the coordinates of the center and the endpoints of the axes for the hyperbola represented by $\frac{(x+4)^2}{36} - \frac{(y-1)^2}{4} = 1$. Then sketch the graph.

B Find the coordinates of the foci.

Solution

A The center of the hyperbola, (h, k), is $(-4, 1)$. The y-term is negative, so the transverse axis is horizontal.

Since a^2 is 36, a is 6. Therefore, the coordinates of the vertices are 6 units to the left and 6 units to the right of the center, $(-4, 1)$.

$$(-4 - 6, 1) = (-10, 1)$$
$$(-4 + 6, 1) = (2, 1)$$

Since b^2 is 4, b is 2. Therefore, the endpoints of the conjugate axis are 2 units above and 2 units below the center, $(-4, 1)$.

$$(-4, 1 + 2) = (-4, 3)$$
$$(-4, 1 - 2) = (-4, -1)$$

To sketch the graph, plot the endpoints of both axes, and draw the rectangle formed by these points. Draw the diagonals of the rectangle. Extend these diagonals to form the asymptotes of the hyperbola. Then sketch the hyperbola.

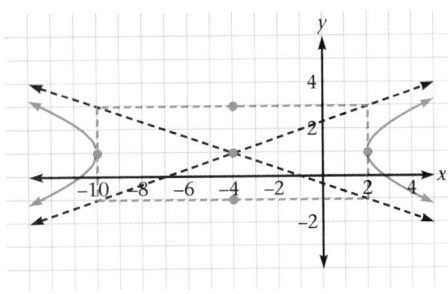

B To find the coordinates of the foci, use the relationship $c^2 = a^2 + b^2$.

Substitute 6 for a and 2 for b.

$$c^2 = a^2 + b^2$$
$$c^2 = 6^2 + 2^2$$
$$c = \pm\sqrt{40}$$

Thus, the coordinates of the foci are $\sqrt{40}$ units to the left and $\sqrt{40}$ to the right of the center, $(-4, 1)$.

Focus 1: $(-4 - \sqrt{40}, 1) \approx (-10.3, 1)$ and
Focus 2: $(-4 + \sqrt{40}, 1) \approx (2.3, 1)$ ❖

ONGOING ASSESSMENT

If the x^2 term is negative, then the transverse axis is vertical; if the y^2 term is negative, the transverse axis is horizontal.

Alternate Example 1

A. Identify the coordinates of the center and the endpoints of the axes for the hyperbola represented by $\frac{(x+2)^2}{25} - \frac{(y-1)^2}{9} = 1$. **[center: $(-2, 1)$, endpoints: transverse axis: $(-7, 1)$, $(3, 1)$; conjugate axis: $(-2, 4)$, $(-2, -2)$]**

B. Find the coordinates of the foci. **[F_1: $(-2 + \sqrt{34}, 1)$, F_2: $(-2 - \sqrt{34}, 1)$, or F_1: $\approx (3.83, 1)$, F_2: $\approx (-7.83, 1)$]**

Use Transparency 41

INCLUSION strategies

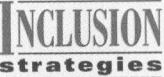

Using Visual Models Hyperbolic equations can be graphed by solving the equation for y. For instance, the equation $\frac{(x+4)^2}{36} - \frac{(y-1)^2}{4} = 1$ can be solved for y, and both resulting equations can be graphed.

$$y_1 = 1 + \sqrt{\frac{(x+4)^2 - 36}{9}}$$
$$y_1 = 1 - \sqrt{\frac{(x+4)^2 - 36}{9}}$$

Have each student graph these functions simultaneously on their graphics calculator to visualize the hyperbola.

The equation $y^2 + 4y - 4x^2 + 32x = 76$ represents a hyperbola.

A. Write the equation of the hyperbola in standard form.
$$\left[\frac{(y+2)^2}{16} - \frac{(x-4)^2}{4} = 1 \right]$$

B. Identify the coordinates of the center, endpoints of the axes, and the foci. [center: (4, −2), endpoints of transverse axis: (4, −6) and (4, 2); endpoints of conjugate axis: (6, −2), (2, −2), F_1: $(4, -2 + \sqrt{20})$, F_2: $(4, -2 - \sqrt{20})$]

Alternate Example 3

To locate a whale in the ocean, two microphones are placed under water 6000 feet apart. One microphone picks up a whale's noise 0.8 seconds after the other microphone picks up the same noise. Since the speed of sound in water is about 5000 feet per second, one microphone is (0.8)(5000), or 4000, feet farther from the whale than the other microphone.

A. Use the definition of a hyperbola and the distance formula to find an equation for the possible locations of the whale. $\left[\frac{x^2}{4} - \frac{y^2}{5} = 1 \right]$

B. What is the closest distance that the whale could be to either microphone? [1000 feet]

Sometimes you need to complete the squares on both x and y in order to write the equation of a hyperbola in the standard form.

EXAMPLE 2

The equation $y^2 + 6y - 2x^2 + 4x = -3$ represents a hyperbola.

Ⓐ Write the equation of the hyperbola in standard form.

Ⓑ Identify the coordinates of the center, the endpoints of the axes, and the foci.

Solution▶

Ⓐ Complete the squares and write the equation in standard form.

$$y^2 + 6y - 2x^2 + 4x = -3$$
$$(y^2 + 6y + 9) - 2(x^2 - 2x + 1) = -3 + 9 - 2$$
$$(y + 3)^2 - 2(x - 1)^2 = 4$$
$$\frac{(y+3)^2}{4} - \frac{(x-1)^2}{2} = 1$$

Ⓑ The center of the hyperbola, (h, k), is $(1, -3)$.

The x-term is negative, so the transverse axis is vertical.

a is 2, so the coordinates of the vertices are $(1, -1)$ and $(1, -5)$.

b is $\sqrt{2}$, so the endpoints of the conjugate axis are $(1 - \sqrt{2}, -3)$ and $(1 + \sqrt{2}, -3)$.

Using the relationship $c^2 = a^2 + b^2$, the coordinates of the foci are $(1, -3 + \sqrt{6})$ and $(1, -3 - \sqrt{6})$. ❖

Applications of Hyperbolas

Hyperbolas can be used to locate objects underwater.

EXAMPLE 3

Marine Research To locate a whale in the ocean, two microphones are placed under water 8000 feet apart. One microphone picks up a whale's noise 0.4 seconds after the other microphone picks up the same noise. Since the speed of sound in water is about 5000 feet per second, one microphone is (0.4)(5000), or 2000, feet farther from the whale than the other microphone.

Ⓐ Find an equation for the possible locations of the whale.

Ⓑ What is the closest distance that the whale could be to either microphone?

Technology Students can sometimes understand the concept of the hyperbola by using graphics calculators. On the graphics calculator, have students graph the functions

$$y_1 = k + \sqrt{\frac{(ab)^2 - b^2(x-h)^2}{-a^2}}$$

$$y_1 = k - \sqrt{\frac{(ab)^2 - b^2(x-h)^2}{-a^2}}$$

for a variety of values of a, b, h, and k. For example, let $a = 1$, $b = 2$, $h = 3$, and $k = -1$. Then let $a = 2$, $b = 3$, $h = -1$, and $k = 2$. Discuss the effect of changing these values for each graph.

Solution

A The whale could be any point that is 2000 feet farther from one microphone than the other. In other words, the whale is located on a hyperbola containing points in which the constant difference from the foci is 2000.

Make a graph in units of 1000s. Place the center of the hyperbola at the origin. The microphones (foci) are 8000 feet apart, so place the foci 4000 feet, or 4 units, from the center. Let the vertices be on the x-axis. Then $F_1(-4, 0)$ and $F_2(4, 0)$ are the foci.

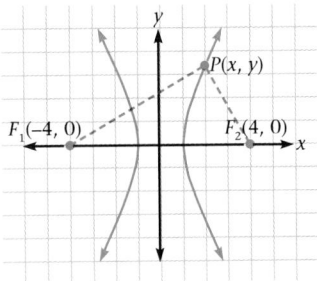

Let $P(x, y)$ be any point on the hyperbola so that the constant difference, $PF_1 - PF_2$, is 2.

Since the constant difference, $2a$, is 2, $a = 1$. Since the foci are $(-4, 0)$ and $(4, 0)$, $c = 4$.

To find b, substitute 1 for a and 4 for c in $a^2 + b^2 = c^2$.

$$a^2 + b^2 = c^2$$
$$1^2 + b^2 = 4^2$$
$$b^2 = 15$$
$$b = \sqrt{15}$$

Thus, the possible locations of the whale are described by the hyperbola $x^2 - \dfrac{y^2}{15} = 1$, where the x- and y-values are given in thousands of feet.

B The closest the whale could be to either microphone is on a vertex of the hyperbola. The distance from a vertex to a focus is 3 units. Thus, the closest the whale could be to the microphone that picks up the sound is 3000 feet. ❖

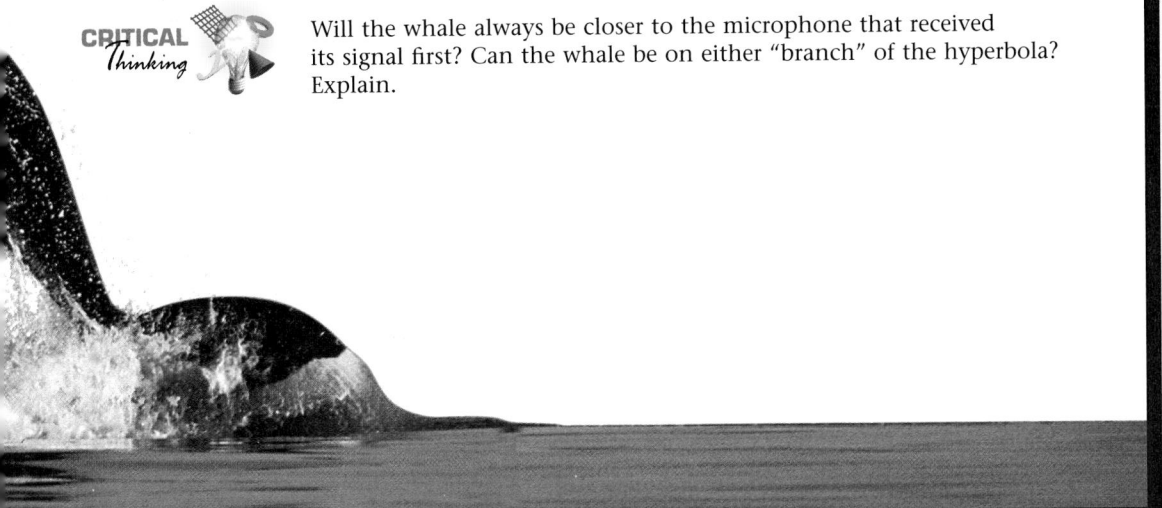

CRITICAL *Thinking*

Will the whale always be closer to the microphone that received its signal first? Can the whale be on either "branch" of the hyperbola? Explain.

CRITICAL *Thinking*

yes; yes; Since distance = speed × time, and the speed of sound is constant, the whale will always be closer to the microphone that received its signal first. The whale can be on either branch of the hyperbola, because in the simplification of the equation it is arbitrary which arm the point $P(x, y)$ is placed on.

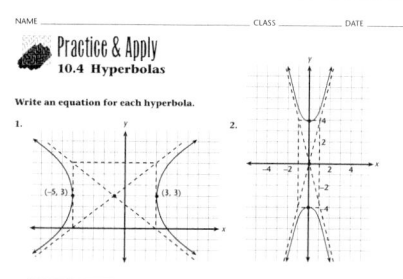
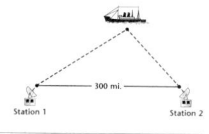
EXERCISES & PROBLEMS

Communicate

1. How can you determine, from its equation in standard form, whether the transverse axis of a hyperbola is horizontal or vertical?

2. How can you find the equation of a hyperbola when you know the coordinates of the vertices and foci?

3. Explain how to find the equation of a hyperbola when you know the coordinates of the foci and the constant difference.

4. Describe the relationship between the distances from the center of a hyperbola to the foci, from the center of a hyperbola to the vertex, and from the center of a hyperbola to one end of the conjugate axis.

5. Describe the position of the asymptotes in relation to the hyperbola.

6. If one microphone picks up the sound of a whale in 3 seconds and a second microphone located 25,000 feet from the first picks up the same sound in 4 seconds, describe all possible locations of the whale.

Practice & Apply

Write an equation for each of the following hyperbolas.

7. Transverse axis from $(7, 0)$ to $(-7, 0)$; conjugate axis from $(0, -5)$ to $(0, 5)$ $\frac{x^2}{49} - \frac{y^2}{25} = 1$

8. Transverse axis from $(0, -5)$ to $(0, 5)$; conjugate axis from $(3, 0)$ to $(-3, 0)$ $\frac{y^2}{25} - \frac{x^2}{9} = 1$

9. Transverse axis from $(-2, 0)$ to $(2, 0)$; conjugate axis from $\left(0, -\sqrt{7}\right)$ to $\left(0, \sqrt{7}\right)$ $\frac{x^2}{4} - \frac{y^2}{7} = 1$

10. Transverse axis from $(0, -3)$ to $(0, 3)$; conjugate axis from $(-\sqrt{13}, 0)$ to $(\sqrt{13}, 0)$ $\frac{y^2}{9} - \frac{x^2}{13} = 1$

11. Foci at $(-1, 0)$ and $(4, 0)$ and a constant difference of 3 $\frac{(x-1.5)^2}{2.25} - \frac{y^2}{4} = 1$

12. Foci at $(3, 2)$ and $(3, -4)$ and a constant difference of 5 $\frac{(y+1)^2}{6.25} - \frac{(x-3)^2}{2.75} = 1$

Write an equation for each of the following hyperbolas.

13. Vertices at $(0, -1)$ and $(6, -1)$; foci at $(-1, -1)$ and $(7, -1)$ $\frac{(x-3)^2}{9} - \frac{(y+1)^2}{7} = 1$

14. Vertices at $(-5, -4)$ and $(1, -4)$; foci at $(-2 - \sqrt{13}, -4)$ and $(-2 + \sqrt{13}, -4)$ $\frac{(x+2)^2}{9} - \frac{(y+4)^2}{4} = 1$

15. Vertices at $(1, -2)$ and $(1, -6)$; foci at $\left(1, -4 - \sqrt{7}\right)$ and $\left(1, -4 + \sqrt{7}\right)$ $\frac{(y+4)^2}{4} - \frac{(x-1)^2}{3} = 1$

16. Vertices at $\left(-6, \frac{1}{2}\right)$ and $\left(-6, \frac{7}{2}\right)$; foci at $(-6, 0)$ and $(-6, 4)$ $\frac{4(y-2)^2}{9} - \frac{4(x+6)^2}{7} = 1$

Marine Biology

To locate a whale, two microphones are placed 5000 feet apart in the ocean. One microphone picks up a whale's sound 0.6 second after the other microphone picks up the same sound.

17. Find the equation of the hyperbola that describes the possible locations of the whale.

18. What is the closest that the whale could be to either microphone?

19. How could the location of the whale be determined more exactly?

For Exercises 20–29, write the standard equation. Identify the coordinates of the center, the lengths of the axes, and the exact coordinates of the foci. Then graph the hyperbola.

20. $x^2 - y^2 - 12x + 8y - 5 = 0$

21. $x^2 - 2y^2 + 6x + 8y - 3 = 0$

22. $25x^2 - 4y^2 - 150x - 16y + 109 = 0$

23. $x^2 - y^2 - 2x - 4y - 4 = 0$

24. $x^2 - y^2 - 4x - 2y - 6 = 0$

25. $-x^2 + y^2 - 2x - 12y + 31 = 0$

26. $3x^2 - y^2 - 12x - 16y - 3 = 0$

27. $2y^2 - 4x^2 + 12x + 8y - 3 = 0$

28. $10y^2 - 4x^2 - 30y + 20x - 1 = 0$

29. $3x^2 - 3y^2 - 3x - 3y - 1 = 0$

Navigation The long-range navigational system (LORAN) uses hyperbolas to find the location of ships. Radio signals that travel 980 feet per microsecond (μsec) are transmitted from LORAN stations. Suppose LORAN Station 2 is 250 miles directly east of LORAN Station 1. A ship is sailing parallel to the line through the LORAN stations. A signal from Station 2 reaches the ship 500 microseconds before the signal from Station 1.

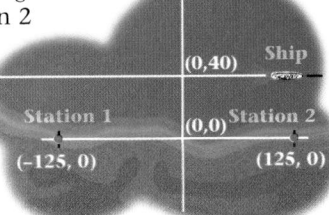

30. Write the standard equation of the hyperbola that describes the possible locations of the ship. (5280 feet = 1 mile)

31. If the ship is sailing 40 miles north of a line connecting the LORAN stations, use your equation from Exercise 30 to find the coordinates of the location of the ship. (49.1, 40)

17. $\dfrac{x^2}{2.25} - \dfrac{y^2}{4} = 1$, where x and y are in thousands of feet.

18. 1000 feet

19. Use another set of microphones so that a second hyperbola could be determined; the intersection of these two hyperbolas would give more information about the location of the whale.

20. $\dfrac{(x-6)^2}{25} - \dfrac{(y-4)^2}{25} = 1$; center: (6, 4); transverse axis: 10; conjugate axis: 10; F_1: $(6 + \sqrt{50}, 4)$, F_2: $(6 - \sqrt{50}, 4)$

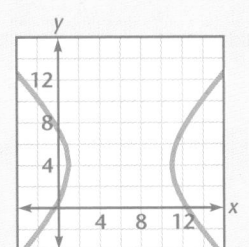

33. Answers may vary but should include x-values of −8 to 0 and y-values of −4 to 8.

34. endpoints of the transverse axis: $(-2, 2)$ and $(-4, 2)$; endpoints of conjugate axis: $(-3, 4)$ and $(-3, 0)$

35. $y = 2x + 8$ and $y = -2x - 4$; number of units depends on the viewing window; after about 4 units from the center, the asymptotes appear to touch the hyperbola.

Look Beyond

This exercise introduces students to solving a system of 2 degree-two equations in 2 variables. The calculator is used so that students can see the points of intersection. This helps students understand why there can be four unique solutions to systems of this type.

For Exercises 32 and 33, let $\frac{(x+3)^2}{1} - \frac{(y-2)^2}{4} = 1$.

32. Solve the equation for y. $y = 2 \pm \sqrt{4(x+3)^2 - 4}$

33. Graph both equations from Exercise 32 on your graphics calculator. What viewing window did you use to view the hyperbola?

34. Find the endpoints of the transverse and conjugate axes.

35. Use the information you found in Exercise 34 to find the equations of the asymptotes of this hyperbola. Graph these lines with the hyperbola on your graphics calculator. Approximately how many units from the center of the hyperbola are the asymptotes where they *appear* to touch the hyperbola?

Look Back

Give the exact value for each of the following expressions. [Lesson 8.5]

36. $\cos \frac{\pi}{3}$ $\frac{1}{2}$

37. $\tan \frac{5\pi}{4}$ 1

38. $\sin \left(-\frac{7\pi}{2}\right)$ 1

39. $\cos (-2\pi)$ 1

40. $\sin \frac{3\pi}{4}$ $\frac{1}{\sqrt{2}}$ or $\frac{\sqrt{2}}{2}$

41. $\tan \left(-\frac{2\pi}{3}\right)$ $\sqrt{3}$

Sports The distance around the track at the Indianapolis Speedway is 2.5 miles. A person with a stopwatch can quickly find a driver's average speed per lap using the formula $d = rt$, where d is the distance, r is the speed, and t is the time. **[Lesson 9.1]**

42. Write an equation for the average speed, y, in miles per hour that a driver achieves on a lap completed in x seconds. $y = \frac{9000}{x}$

43. Graph the equation you wrote for Exercise 42. What is the average speed of a driver who completes a lap in 50 seconds? 180 miles per hour

44. How fast must a driver complete a lap to achieve an average speed of 200 miles per hour? 45 seconds

Determine whether each function is even, odd, or neither. [Lesson 9.3]

45. $f(x) = \frac{1}{2x^2 - 9}$ even

46. $f(x) = \frac{1}{x^3 + 2x}$ odd

47. $f(x) = \frac{1}{x^4 + 5x^3}$ neither

Look Beyond

For Exercises 48–50, let $2x^2 + 5y^2 = 22$ and $3x^2 - y^2 = -1$.

48. What two figures are represented by these equations? ellipse and hyperbola

49. Solve both equations for y. $y = \pm \sqrt{\frac{2}{5}(11 - x^2)}$; $y = \pm \sqrt{3x^2 + 1}$

50. Graph all four equations from Exercise 49 on your graphics calculator. Use a viewing window that allows you to see all the points of intersection between the four equations. Find all the points of intersection. $(1, 2), (1, -2), (-1, 2), (-1, -2)$

LESSON 10.5 Solving Nonlinear Systems of Equations

 *Some problems can be modeled by a system of equations that includes nonlinear equations. These systems are usually solved by graphing or by elimination.*

hyperbola

parabola

The graphs of conic sections can be represented by passing a plane through a hollow double cone.

ellipse

Circles, parabolas, ellipses, and hyperbolas are examples of *conic sections*. The general equation of a **conic section** can be written in the form

$$Ax^2 + Bxy + Cy^2 + Dx + Ey + F = 0,$$

where A, B, C, D, E, and F are not all zero.

circle

Standard Equations of Conic Sections		
Conic section	**Center at (h, k)**	
Parabola	$y = a(x - h)^2 + k$ $x = a(y - k)^2 + h$	where $a \neq 0$
Circle	$(x - h)^2 + (y - k)^2 = r^2$	where $r > 0$
Ellipse	$\frac{(x-h)^2}{a^2} + \frac{(y-k)^2}{b^2} = 1$ $\frac{(x-h)^2}{b^2} + \frac{(y-k)^2}{a^2} = 1$	where $a > b > 0$
Hyperbola	$\frac{(x-h)^2}{a^2} - \frac{(y-k)^2}{b^2} = 1$ $\frac{(y-k)^2}{a^2} - \frac{(x-h)^2}{b^2} = 1$	where $a > 0$ and $b > 0$

ALTERNATIVE teaching strategy

Using Symbols Ask students to familiarize themselves with the equations of the different conic sections. This will help them determine the kind of graph that corresponds to each equation. Students should be able to sketch the system of equations and determine the number of solutions and the approximate range of values for the solutions.

PREPARE

Objectives

- Determine the solutions of a system consisting of one first-degree equation and one second-degree equation.
- Determine the solutions of a system consisting of two second-degree equations.

RESOURCES

- Practice Master   10.5
- Enrichment Master   10.5
- Technology Master   10.5
- Lesson Activity Master   10.5
- Quiz   10.5
- Spanish Resources   10.5

Assessing Prior Knowledge

Solve each equation for a.

1. $9ay = 45y$ $[a = 5]$

2. $\frac{3}{2}x = 6ax$ $[a = \frac{1}{4}]$

3. $7ya = -63y$ $[a = -9]$

TEACH

 Solving two second degree equations or one first degree equation and one second degree equation is no harder than solving systems of linear equations. Graphing methods illustrate the intersection points and help students understand why there are four possible solutions.

CRITICAL
Thinking

No; in a second degree equation for each value of y there can be at most 2 values for x.

Alternate Example 1

$$\begin{cases} x^2 - 7x - y = -13 \\ x + y = 5 \end{cases}$$

A. Solve the nonlinear system using analytical methods [(2, 3), (4, 1)]

B. Solve the nonlinear system by graphing.

[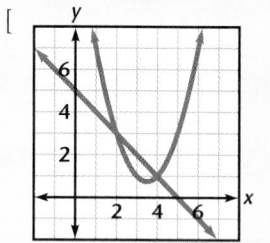]

$$\text{T}\text{EACHING } \textit{tip}$$

Point out to students that in Example 1 the y-values can also be found by substituting the x-values into the quadratic equation.

There are many possible intersections that can occur in systems of one first-degree equation and one second-degree equation.

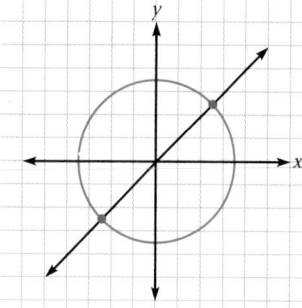 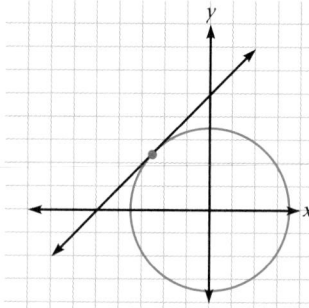 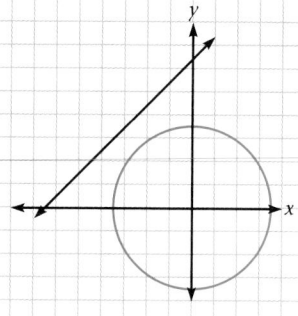

Identify the graph of a system with two solutions, the graph of a system with only one solution, and the graph of a system with no real solution.

CRITICAL
Thinking

Is there any system of one first-degree equation and one second-degree equation that has more than two solutions? Explain.

EXAMPLE 1

Graphics Calculator

A Solve the nonlinear system.

B Solve by graphing the system.

$$\begin{cases} x^2 - 6x - y = -9 \\ x + y = 5 \end{cases}$$

Solution

A Solve the linear equation for y.

$$x + y = 5$$
$$y = 5 - x$$

Then substitute the expression for y in the first equation.

$$x^2 - 6x - y = -9$$
$$x^2 - 6x - (5 - x) = -9$$
$$x^2 - 5x - 5 = -9$$
$$x^2 - 5x + 4 = 0$$
$$(x - 1)(x - 4) = 0$$
$$x = 1 \text{ or } x = 4$$

To find the corresponding y-values, substitute the x-values 1 and 4 in the linear equation.

$$x + y = 5 \qquad\qquad x + y = 5$$
Substitute 1 for x. $\quad 1 + y = 5 \qquad$ Substitute 4 for x. $\quad 4 + y = 5$
$$y = 4 \qquad\qquad\qquad\qquad y = 1$$

Thus, the solutions are (1, 4) and (4, 1).

ENRICHMENT Emphasize the importance of familiarity with the standard forms of conic sections. These formulas help determine the conic section represented by an equation. Have students write the standard form for any conic section. Then ask students to write the standard form of the equation of a circle, given its center and a point on the circle. Have students explain the process they used for this activity.

INCLUSION
strategies

Technology Have students use a graphics calculator to find the approximate solutions to the following system. Ask students to choose range parameter values for x and y that would produce complete graphs of the equations. Then students can find the exact solutions algebraically and compare them with the solutions obtained by graphing.

$$y = (x + 4)^2 - 11 \qquad y = x - 5$$
$$[(-5, -10), (-2, -7)]$$

Ⓑ Solve each equation for y, and graph them on your graphics calculator.

$$\begin{cases} y = x^2 - 6x + 9 \\ y = -x + 5 \end{cases}$$

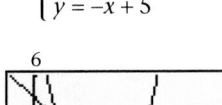

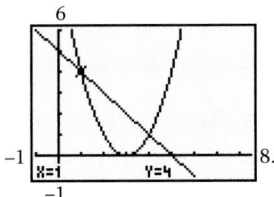 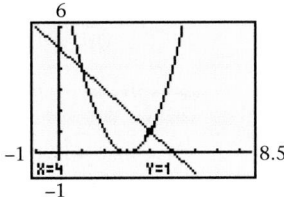

This system consists of a parabola and a line that intersect at $(1, 4)$ and $(4, 1)$. ❖

Systems of Two Second-Degree Equations

There are many possible intersections that occur in systems of two second-degree equations.

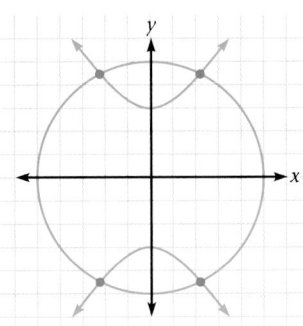 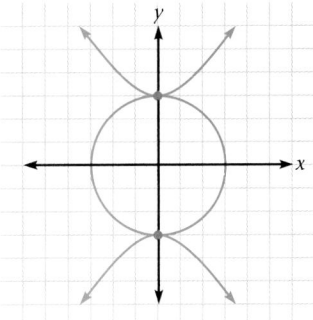

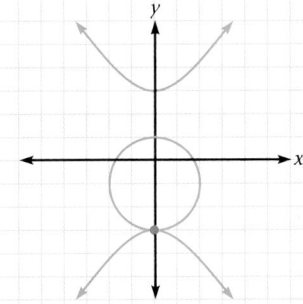 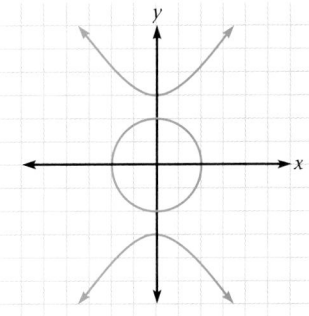

Identify the graph of a system with four solutions, the graph of a system with two solutions, the graph of a system with only one solution, and the graph of a system with no real solution.

Is there any system of two second-degree equations that has exactly three solutions? Explain.

An ellipse and a hyperbola may have 0, 1, 2, 3, or 4 intersections.

T<small>EACHING</small> *tip*

Have students write each equation in Alternate Example 2 as two functions. Then the four functions can be graphed on the same coordinate plane using their graphics calculators. Review with students how to find points of intersection on their graphics calculators.

Alternate Example 2

Solve the nonlinear system using algebra and by graphing.

$$\begin{cases} x^2 - 3y^2 = -8 \\ x^2 + 2y^2 = 12 \end{cases}$$

$[(2, 2),(2, -2),(-2, 2),(-2, -2)$

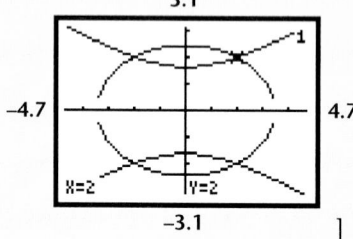

]

The system $\begin{cases} x^2 - 2y^2 = 17 \\ 2x^2 + y^2 = 54 \end{cases}$ consists of a hyperbola and an ellipse.

What are the possible intersections that can occur between an ellipse and a hyperbola?

EXAMPLE 2

Ⓐ Solve the nonlinear system using algebra.

Ⓑ Solve the nonlinear system by graphing.

$$\begin{cases} x^2 - 2y^2 = 17 \\ 2x^2 + y^2 = 54 \end{cases}$$

Graphics Calculator

Solution➤

Ⓐ Use the elimination-by-addition method to eliminate the variable x.

Multiply the first equation by –2.

$$\begin{cases} -2(x^2 - 2y^2) = -2(17) \\ 2x^2 + y^2 = 54 \end{cases} \rightarrow \begin{cases} -2x^2 + 4y^2 = -34 \\ 2x^2 + y^2 = 54 \end{cases}$$

Add both sides of the equations in the system.

$$\begin{array}{r} -2x^2 + 4y^2 = -34 \\ 2x^2 + y^2 = 54 \\ \hline 5y^2 = 20 \end{array}$$

Solve the resulting equation for y.

$$5y^2 = 20$$
$$y^2 = 4$$
$$y = \pm\sqrt{4}$$
$$y = 2 \text{ or } y = -2$$

Substitute 2 and –2 for y in either original equation to obtain the corresponding values of x.

For $y = 2$:

$$x^2 - 2y^2 = 17$$
$$x^2 - 2(2)^2 = 17$$
$$x^2 - 8 = 17$$
$$x^2 = 25$$
$$x = \pm\sqrt{25}$$
$$x = 5 \text{ or } x = -5$$

The two points of intersection corresponding to $y = 2$ are $(5, 2)$ and $(-5, 2)$.

For $y = -2$:

$$x^2 - 2y^2 = 17$$
$$x^2 - 2(-2)^2 = 17$$
$$x^2 - 8 = 17$$
$$x^2 = 25$$
$$x = \pm\sqrt{25}$$
$$x = 5 \text{ or } x = -5$$

The two points of intersection corresponding to $y = -2$ are $(5, -2)$ and $(-5, -2)$.

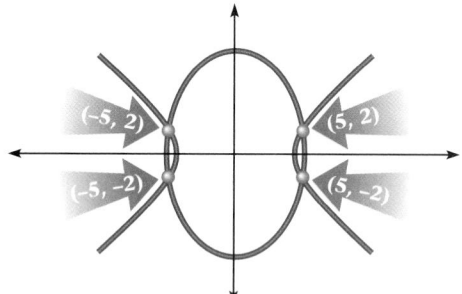

B Solve each equation for y. Write each equation as two functions, and graph all four functions on the same coordinate plane. NOTE: Some calculators will show gaps, or incomplete graphs, of some curves.

$$x^2 - 2y^2 = 17 \begin{cases} y = \sqrt{-\frac{1}{2}(-x^2 + 17)} \\ y = -\sqrt{-\frac{1}{2}(-x^2 + 17)} \end{cases}$$

$$2x^2 + y^2 = 54 \begin{cases} y = \sqrt{-2x^2 + 54} \\ y = -\sqrt{-2x^2 + 54} \end{cases}$$

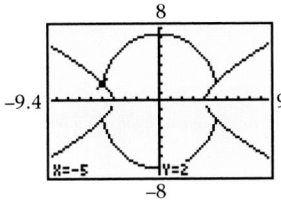

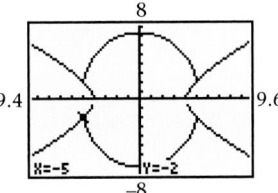

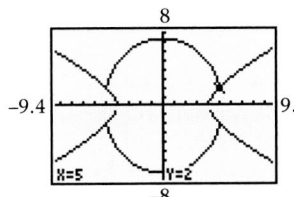

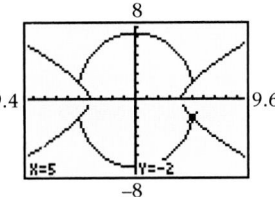

The graphs intersect at $(-5, 2)$, $(-5, -2)$, $(5, 2)$, and $(5, -2)$. ❖

Some systems of nonlinear equations do not have solutions.

EXAMPLE 3

Graphics Calculator

A Solve the nonlinear system.

B Solve by graphing the system.

$$\begin{cases} x^2 + y^2 = 4 \\ x^2 - y^2 = 9 \end{cases}$$

Solution▶

A Use the elimination-by-addition method to eliminate one of the variables.

$$\begin{cases} x^2 + y^2 = 4 \\ x^2 - y^2 = 9 \end{cases}$$
$$2x^2 = 13$$
$$x = \pm\sqrt{\frac{13}{2}}$$

Substitute $\sqrt{\frac{13}{2}}$ and $-\sqrt{\frac{13}{2}}$ for x in either original equation to find the corresponding y-values.

$$x^2 + y^2 = 4$$
$$\left(\sqrt{\frac{13}{2}}\right)^2 + y^2 = 4$$
$$y = \pm\sqrt{4 - \frac{13}{2}}$$
$$y = \pm i\sqrt{2.5}$$

$$x^2 + y^2 = 4$$
$$\left(-\sqrt{\frac{13}{2}}\right)^2 + y^2 = 4$$
$$y = \pm\sqrt{4 - \frac{13}{2}}$$
$$y = \pm i\sqrt{2.5}$$

All of the corresponding y-values are complex numbers. Thus, the system has no real solutions.

Alternate Example 3

$$\begin{cases} x^2 + y^2 = 9 \\ x^2 - y^2 = 16 \end{cases}$$

A. Solve the nonlinear system using elimination by adding. $\left[\left(\frac{5}{\sqrt{2}}, i\sqrt{\frac{7}{2}}\right),\right.$
$\left(\frac{5}{\sqrt{2}}, -i\sqrt{\frac{7}{2}}\right),$
$\left(-\frac{5}{\sqrt{2}}, i\sqrt{\frac{7}{2}}\right),$
$\left.\left(-\frac{5}{\sqrt{2}}, -i\sqrt{\frac{7}{2}}\right)\right]$

B. Solve the nonlinear system using your graphics calculator. [**The solutions from Part A are not real. Graphically, this indicates that the graphs will not intersect.**]

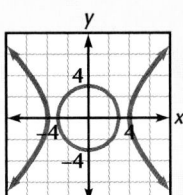

Teaching tip

Explain to students that Dürer believed that good art had mathematical rules of proportion. To draw the human figure according to such rules, he needed mathematical instruments to construct these ideal proportions.

B Solve each equation for y, and graph all resulting functions on the same coordinate plane.

$$x^2 + y^2 = 4 \begin{cases} y = \sqrt{4 - x^2} \\ y = -\sqrt{4 - x^2} \end{cases} \qquad x^2 - y^2 = 9 \begin{cases} y = \sqrt{x^2 - 9} \\ y = -\sqrt{x^2 - 9} \end{cases}$$

The circle and hyperbola do not intersect. Thus, the system has no real solution. ❖

Cultural Connection: Europe During the Renaissance, the German artist Albrecht Dürer (1471–1528) created many woodcuts, paintings, and copper engravings. Although he became well known for his artistic accomplishments, few people recognize the connections he made between art and mathematics. Dürer introduced new methods for constructing conic sections.

He called the ellipse *Eierlinie*, which means "egg-shaped line", and the hyperbola *Gabellinie*, meaning "fork-shaped line". Dürer also invented tools for drawing conic sections.

Applications of Nonlinear Systems of Equations

EXAMPLE 4

A rectangular beam is to be cut from a wooden log that is 12 inches in diameter. What are the dimensions of the cross section of the beam if its height is twice as long as its width?

Solution▸

Write the system of equations. Place the circular cross section of the log at the origin. Use the equation of a circle to represent a log that is 12 inches in diameter.

$$x^2 + y^2 = 36$$

The height of the beam is twice as long as its width

$$h = 2w$$

Since the center of the circle is $(0, 0)$, the width of the beam can be represented by $2x$ and the height by $2y$.

$$h = 2w$$
$$(2y) = 2(2x), \text{ or } y = 2x$$

The following system of equations models the problem.

$$\begin{cases} x^2 + y^2 = 36 \\ y = 2x \end{cases}$$

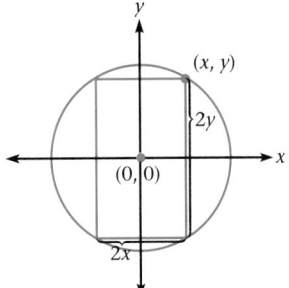

Substitute $2x$ for y in the first equation and solve for x.

$$x^2 + y^2 = 36$$
$$x^2 + (2x)^2 = 36$$
$$5x^2 = 36$$
$$x^2 = \frac{36}{5}$$
$$x = \pm\sqrt{\frac{36}{5}} = \pm\frac{6}{\sqrt{5}}$$

Find y. $y = 2x$
$$y = 2\left(\pm\frac{6}{\sqrt{5}}\right) = \pm\frac{12}{\sqrt{5}}$$

Since the width is represented by $2x$ and the length is represented by $2y$, the dimensions of the cross section of the beam are $\frac{12}{\sqrt{5}}$ inches by $\frac{24}{\sqrt{5}}$, or approximately 5.37 inches by 10.73 inches. ❖

Alternate Example 4

A rectangular beam is to be cut from a wooden log that is 14 inches in diameter. What are the dimensions of the cross section of the beam if its height is twice its width? $\left[\begin{cases} x^2 + y^2 = 7^2 \\ y = 2x \end{cases}\right.$

$w = 2x = 2\left(\dfrac{7}{\sqrt{5}}\right) = \dfrac{14}{\sqrt{5}}$, or ≈

6.26 in. $l = 2w = 2\left(\dfrac{14}{\sqrt{5}}\right) = \dfrac{28}{\sqrt{5}}$, or ≈ 12.52 in. The dimensions are $\dfrac{14}{\sqrt{5}}$ in. by $\dfrac{28}{\sqrt{5}}$ in.]

Assess

Selected Answers

Odd-numbered Exercises 7–31

Assignment Guide

Core 1–10, 12–18, 21–32

Core Plus 1–5, 8–14, 17–33

Technology

Graphics calculators are needed for Exercises 6–11, 15–20, 25–27, and 31.

Error Analysis

Students should realize that the values of the range parameters affect the solutions that they will obtain when using a graphics calculator. Students must be careful to use friendly viewing windows and any special solving features their calculators have. Remind students that some viewing windows will result in graphs of conic sections appearing to have "gaps" at the vertices. This should not affect their solutions.

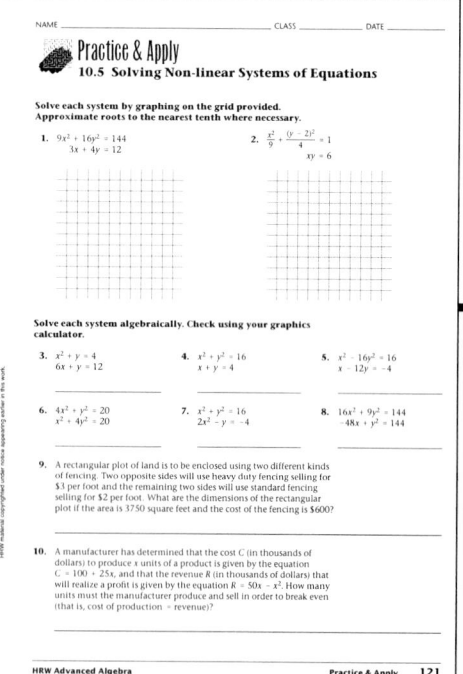

EXERCISES & PROBLEMS

Communicate

1. Describe all the different ways that the graphs of one first-degree equation and one second-degree equation can intersect.

2. Describe all the different ways that the graphs of two second-degree equations can intersect.

3. Is it possible for the graphs of two second-degree equations to intersect at three points? Explain.

4. Suppose a system consists of a hyperbola and one of its asymptotes. What is the maximum number of solutions to this system? Explain.

5. Describe the graph of a system of nonlinear equations with only non-real solutions.

Practice & Apply

Solve each system algebraically. Check using your graphics calculator.

6. $\begin{cases} x^2 + y^2 = 4 \\ x - 2y = 4 \end{cases}$ $(1.6, -1.2)$ $(0, -2)$

7. $\begin{cases} x - y = 2 \\ x^2 + y^2 = 20 \end{cases}$ $(-2, -4)$ $(4, 2)$

8. $\begin{cases} -x^2 + y = -5 \\ 3x + 2y = 10 \end{cases}$ $(-4, 11)$ $(2.5, 1.25)$

9. $\begin{cases} -x^2 + y = -3 \\ x^2 + y^2 = 9 \end{cases}$

10. $\begin{cases} x^2 - 2y^2 = 7 \\ x^2 + y^2 = 25 \end{cases}$

11. $\begin{cases} x + y = 4 \\ y = \frac{x^2}{2} \end{cases}$ $(2, 2)$ $(-4, 8)$

12. A theorem states that a curve of degree n will intersect a curve of degree m in at most nm points. State the maximum number of points of intersection of the curves in each system of Exercises 6–11.

13. **Geometry** A rectangle has a perimeter of 28 inches, and the length of a diagonal is 10 inches. Write a system of nonlinear equations to represent this situation, and find the dimensions. $2l + 2w = 28$; $l^2 + w^2 = 100$; $l = 8$ in.; $w = 6$ in.

14. **Geometry** A piece of cardboard in the shape of a rectangle has an area of 130 square inches. A square with 4-inch-long sides is cut from each corner of the piece of cardboard, and an open-top box is made by folding up the ends and sides. The volume of the box is 40 cubic inches. What are the dimensions of the original piece of cardboard? 10 in. by 13 in.

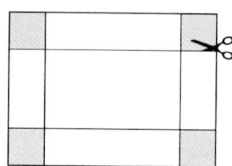

Solve each system algebraically. Check using your graphics calculator.

15. $\begin{cases} x^2 + y^2 = 14 \\ x^2 - y^2 = 4 \end{cases}$

16. $\begin{cases} x^2 - 4y^2 = 19 \\ 2x^2 - y^2 = 52 \end{cases}$

17. $\begin{cases} x^2 + y^2 = 9 \\ 16x^2 - 4y^2 = 64 \end{cases}$ $(\sqrt{5}, 2), (\sqrt{5}, -2)$ $(-\sqrt{5}, 2), (-\sqrt{5}, -2)$

18. $\begin{cases} y - x^2 = 0 \\ x^2 = 20 - y^2 \end{cases}$

19. $\begin{cases} x^2 + 4y^2 = 20 \\ y^2 = 4 \end{cases}$

20. $\begin{cases} x^2 + y^2 = 16 \\ 9x^2 - 4y^2 = 36 \end{cases}$

9. $(0, -3), (\sqrt{5}, 2), (-\sqrt{5}, 2)$

10. $(\sqrt{19}, \sqrt{6}), (\sqrt{19}, -\sqrt{6}), (-\sqrt{19}, \sqrt{6}), (-\sqrt{19}, -\sqrt{6})$

12. 6:2, 7:2, 8:2, 9:4, 10:4, 11:2

15. $(3, \sqrt{5}), (3, -\sqrt{5}), (-3, \sqrt{5}), (-3, -\sqrt{5})$

16. $(3\sqrt{3}, \sqrt{2}), (3\sqrt{3}, -\sqrt{2}), (-3\sqrt{3}, \sqrt{2}), (-3\sqrt{3}, -\sqrt{2})$

18. $(2, 4), (-2, 4)$

19. $(2, 2), (2, -2), (-2, 2), (-2, -2)$

20. $\left(\sqrt{\frac{100}{13}}, \sqrt{\frac{108}{13}}\right), \left(\sqrt{\frac{100}{13}}, -\sqrt{\frac{108}{13}}\right), \left(-\sqrt{\frac{100}{13}}, \sqrt{\frac{108}{13}}\right), \left(-\sqrt{\frac{100}{13}}, -\sqrt{\frac{108}{13}}\right)$

For Exercises 21–24, write a system of nonlinear equations to represent the situation. Then find the dimensions of the figure.

21. 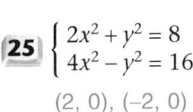 **Geometry** The area of a rectangle is 300 square yards, and the length of a diagonal is 25 yards. $lw = 300,\ l^2 + w^2 = 625$
$l = 20$ yd, $w = 15$ yd

22. **Geometry** A right triangle has a hypotenuse of 50 centimeters and an area of 600 square centimeters. Find the lengths of the legs of the triangle. $a^2 + b^2 = 2500,\ \frac{1}{2}ab = 600,$
$a = 30$ cm, $b = 40$ cm

23. **Geometry** The area of a rectangle is 120 square centimeters, and the perimeter is 44 centimeters. $lw = 120,\ 2l + 2w = 44,$ $l = 12$ cm, $w = 10$ cm

24. An open-ended cylindrical pipe has an outside surface area of 2700π square centimeters and a volume of 4050π cubic centimeters. Find the radius and the length of the pipe.
$r = 3$ cm; $l = 450$ cm

Solve each system algebraically. Then solve using your graphics calculator.

25. $\begin{cases} 2x^2 + y^2 = 8 \\ 4x^2 - y^2 = 16 \end{cases}$
(2, 0), (–2, 0)

26. $\begin{cases} \frac{x^2}{4} + y^2 = 1 \\ -\frac{x^2}{4} + y^2 = 1 \end{cases}$
(0, –1), (0, 1)

27. $\begin{cases} x^2 - y^2 - 1 = 0 \\ x^2 + y^2 - 4x = -3 \end{cases}$
(1, 0)

Look Back

28. Define a function. **[Lesson 2.3]**

29. Is $(f \circ g)(x)$ the same as $(g \circ f)(x)$? Explain. **[Lesson 3.3]**

30. Solve the system of linear equations using the row-reduction method. **[Lesson 4.4]** $\begin{cases} 3x + 2y = 9 \\ 8x - 3y = 15 \end{cases}$

31. Find the inverse of A. If it is not possible, explain why. **[Lesson 4.5]** $A = \begin{bmatrix} -1 & 6 & 2 & 0 \\ 5 & 2 & -3 & 1 \\ 6 & 6 & -2 & 7 \end{bmatrix}$ Not possible since A is not a square matrix

32. Which equation, $x^2 + y^2 = 9$ or $5x^2 + 2y^2 = 9$, represents a circle? Explain. **[Lesson 10.2]**
$x^2 + y^2 = 9$; The coefficient of x^2 and y^2 are the same for a circle

Look Beyond

33. Set your graphics calculator in parametric and radian mode, and use the following viewing window.

Tmin = 0	Tmax = 2π	Tstep = $\frac{\pi}{30}$
Xmin = –4.8	Xmax = 4.7	Xscl = 1
Ymin = –3.2	Ymax = 3.1	Yscl = 1

Graph the parametric system of equations $\begin{cases} x(t) = 3\cos t \\ y(t) = \sin t \end{cases}$ · ellipse; $\frac{x^2}{9} + \frac{y^2}{1} = 1$

What is the geometric figure? What is the rectangular equation for this figure?

28. A function is a relation between two variables in which for each element in the domain there is at most one corresponding element in the range.

29. Not always; $(f \circ g)(x) = (g \circ f)(x)$ when $f(x)$ and $g(x)$ are inverse functions.

30. $\left(\frac{57}{25}, \frac{27}{25}\right)$, or (2.28, 1.08)

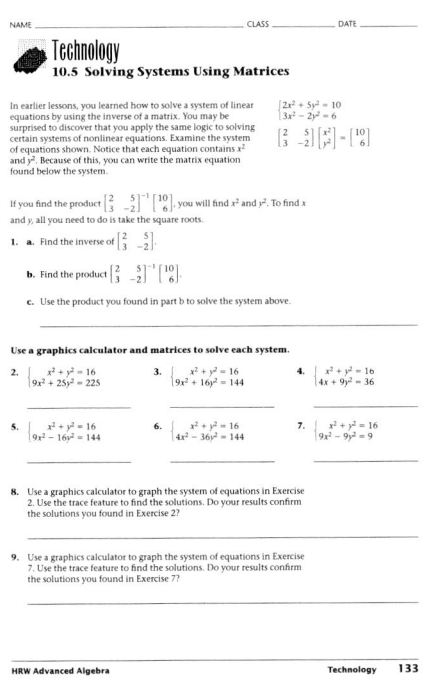

TEACH

 Parametric representations of functions allow you to describe and graph the actual path of a moving object. Parametric representations of conic sections allow you to see the vertical and horizontal motion of any object described by a conic section. You can view the path of a football or a planet, a falling pebble or a moving ship.

Exploring

Parametric Representations of Conic Sections

Why *So far, you have been graphing conic sections using rectangular coordinates. Conic sections can also be graphed using trigonometric functions in parametric mode. Parametric representation allows you to study intersections as a function of time.*

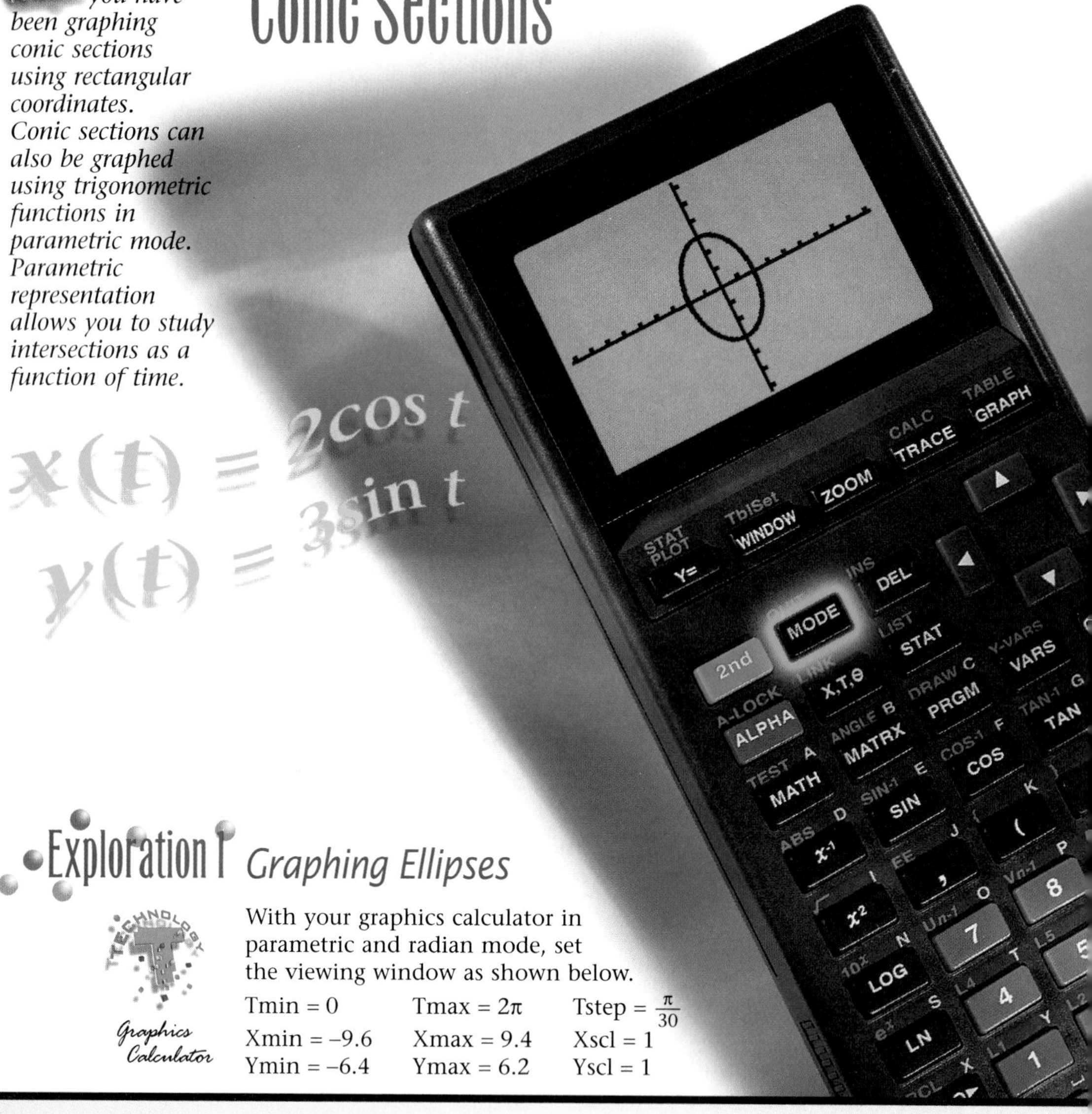

$$x(t) = 2\cos t$$
$$y(t) = 3\sin t$$

• Exploration 1 *Graphing Ellipses*

Graphics Calculator

With your graphics calculator in parametric and radian mode, set the viewing window as shown below.

Tmin = 0	Tmax = 2π	Tstep = $\frac{\pi}{30}$
Xmin = −9.6	Xmax = 9.4	Xscl = 1
Ymin = −6.4	Ymax = 6.2	Yscl = 1

ALTERNATIVE teaching strategy

Using Algorithms Let students know that trigonometric identities can be used to determine the shape of the graph. Students could use the trigonometric identity $\sin^2 t + \cos^2 t = 1$ to find out whether the equations given in Exploration 1 represent an ellipse. Students could also use this identity and write an equation in terms of only sine or cosine.

1 Graph this system of parametric equations. What is the shape of the graph?

$$\begin{cases} x(t) = 2\cos t \\ y(t) = 3\sin t \end{cases}$$

2 Find the lengths of the major and minor axes.

3 What are the coordinates of the center?

4 Use your results from Steps 1–3 to write the equation of the graph in the standard rectangular form.

Keep your calculator in parametric mode with the same viewing window. Complete the table by repeating Steps 1–4 for the following systems of parametric equations.

	Length of major axis	Length of minor axis	Coordinates of center	Equation in rectangular form
5 $\begin{cases} x(t) = 3\cos t \\ y(t) = 3\sin t \end{cases}$				
6 $\begin{cases} x(t) = 3\cos t \\ y(t) = 3.5\sin t \end{cases}$				
7 $\begin{cases} x(t) = 3.5\cos t \\ y(t) = 3\sin t \end{cases}$				
8 $\begin{cases} x(t) = 4\cos t \\ y(t) = 4\sin t \end{cases}$				

9 Compare the graphs of the parametric equations in Steps 5–8. How are they all alike?

10 Predict whether the major axis of the figure defined by

$$\begin{cases} x(t) = 4\cos t \\ y(t) = 3\sin t \end{cases}$$

will be horizontal or vertical. Predict the type of figure, the coordinates of the origin, and the lengths of the major and minor axes. Graph the equations in parametric mode. Were your predictions correct? Explain.

11 Write the system of parametric equations that defines an ellipse centered at the origin with a horizontal major axis of length 5 and a minor axis of length 3.

12 Write the system of parametric equations that defines a circle centered at the origin with a radius of 6. ❖

CRITICAL Thinking

Are all circles ellipses? Are all ellipses circles? Explain.

In Exploration 1 you were given systems of parametric equations, and you were asked to write the equations in rectangular form using the graph of the conic section. Can you think of a method using the trigonometric identity $\cos^2 t + \sin^2 t = 1$ for changing a system of parametric equations to a single equation in rectangular form?

ENRICHMENT To see complete graphs on a graphics calculator screen, it may be necessary to try different parametric values for x and y. Give students the equation of a circle, such as $x^2 + y^2 = 81$, and have them find a set of parameters for x and y that will make the graph of the given equation appear as a circle on the screen of the graphics calculator. Remind students to use the pixel dimensions of their calculator viewing window.

Exploration 1 Notes
This exploration illustrates that the ellipse can be represented by trigonometric functions. Students can identify the type of conic section represented simply by viewing its graph.

Cooperative Learning
Students can work in groups of four to complete Exploration 1 Each student can work one of the Steps 5–8. Then, as a group, students can discuss the questions in Steps 9–12.

Ongoing ASSESSMENT

11. $x(t) = 2.5\cos t,$
    $y(t) = 1.5\sin t$

12. $x(t) = 6\cos t,$
    $y(t) = 6\sin t$

CRITICAL Thinking

All circles are ellipses that have major and minor axes of the same length. All ellipses are *not* circles.

Exploration 2 Notes

This exploration helps students identify the orientation of the parabolas represented by a system of parametric equations. Suggest that students make a table of values for x, t, and y to help them imagine the type of graph they should see on their calculators.

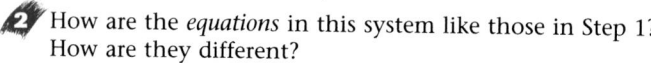
EXTENSION

Consider this system of parametric equations.

$$\begin{cases} x(t) = 2 \cos t \\ y(t) = 3 \sin t \end{cases}$$

Isolate $\sin t$ and $\cos t$.

$$\begin{cases} x(t) = 2 \cos t \\ y(t) = 3 \sin t \end{cases} \rightarrow \begin{cases} \frac{x(t)}{2} = \cos t \\ \frac{y(t)}{3} = \sin t \end{cases}$$

Square both sides of each equation.

$$\begin{cases} \frac{x^2(t)}{4} = \cos^2 t \\ \frac{y^2(t)}{9} = \sin^2 t \end{cases}$$

Add each side of the equations to obtain one equation.

$$\frac{x^2(t)}{4} + \frac{y^2(t)}{9} = \cos^2 t + \sin^2 t$$

Recall from the unit circle in Chapter 8 that $\cos^2 t + \sin^2 t = 1$. You can use this trigonometric relationship to eliminate the parameter, t.

Simplify.

$$\frac{x^2}{4} + \frac{y^2}{9} = 1 \; \diamond$$

Is this the same equation you wrote for Step 4 in Exploration 1?

•Exploration 2 *Transformations and Translations*

Set your viewing window as shown below.

Tmin = −3	Tmax = 3	Tstep = 0.1
Xmin = −76	Xmax = 76	Xscl = 10
Ymin = −50	Ymax = 50	Yscl = 10

Graphics Calculator

1 Graph this system of parametric equations. What is the shape of this graph?

$$\begin{cases} x(t) = 20t \\ y(t) = 20t - 16t^2 \end{cases}$$

2 How are the *equations* in this system like those in Step 1? How are they different?

$$\begin{cases} x(t) = 20t \\ y(t) = 20t + 16t^2 \end{cases}$$

Graph the system on the same coordinate plane with the system from Step 1. How are these two *graphs* alike? How are they different?

3 How are the *equations* in this system like those in Step 1? How are they different?

$$\begin{cases} x(t) = 20t - 16t^2 \\ y(t) = 20t \end{cases}$$

Graph the system on the same coordinate plane with the system from Step 1. How are these two *graphs* alike? How are they different?

INCLUSION **strategies**

Hands-On Strategies

Have students make a table like the one below to help them determine whether the points are on a circle. Have students use their calculators to approximate values to the nearest tenth. Then ask students to plot the points on graph paper to observe the shape of the system.

$x(t) = 6 \cos t$
$y(t) = 6 \sin t$

θ	x	y
0	[6]	[0]
$\frac{\pi}{4}$	[$3\sqrt{2}$]	[$3\sqrt{2}$]
$\frac{\pi}{2}$	[0]	[6]
$\frac{3\pi}{4}$		
π		
$\frac{3\pi}{2}$		
⋮		
$\frac{7\pi}{4}$	[$3\sqrt{2}$]	[$-3\sqrt{2}$]

4 Examine the graph shown. Compare this graph with the graph of the system given in Step 3.

Write a system of parametric equations that you think would represent this graph. Graph the system you wrote.

Is the graph of the system you wrote like the graph shown?

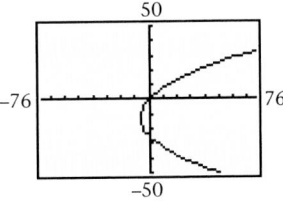

For Steps 5–8, match each pair of systems of parametric equations with its graph, **A**, **B**, **C**, or **D**. Using the same settings on your graphics calculator in parametric mode, graph the systems to check your answers.

5
$$\begin{cases} x(t) = 10t - 16t^2 \\ y(t) = 10t \end{cases}$$
$$\begin{cases} x(t) = 20t \\ y(t) = 20t - 16t^2 \end{cases}$$

6
$$\begin{cases} x(t) = 20t - 16t^2 \\ y(t) = 20t \end{cases}$$
$$\begin{cases} x(t) = 30t + 16t^2 \\ y(t) = 30t \end{cases}$$

7
$$\begin{cases} x(t) = 30t \\ y(t) = 30t - 16t^2 \end{cases}$$
$$\begin{cases} x(t) = 20t \\ y(t) = 20t - 16t^2 \end{cases}$$

8
$$\begin{cases} x(t) = 20t \\ y(t) = 20t - 16t^2 \end{cases}$$
$$\begin{cases} x(t) = 10t \\ y(t) = 10t + 16t^2 \end{cases}$$

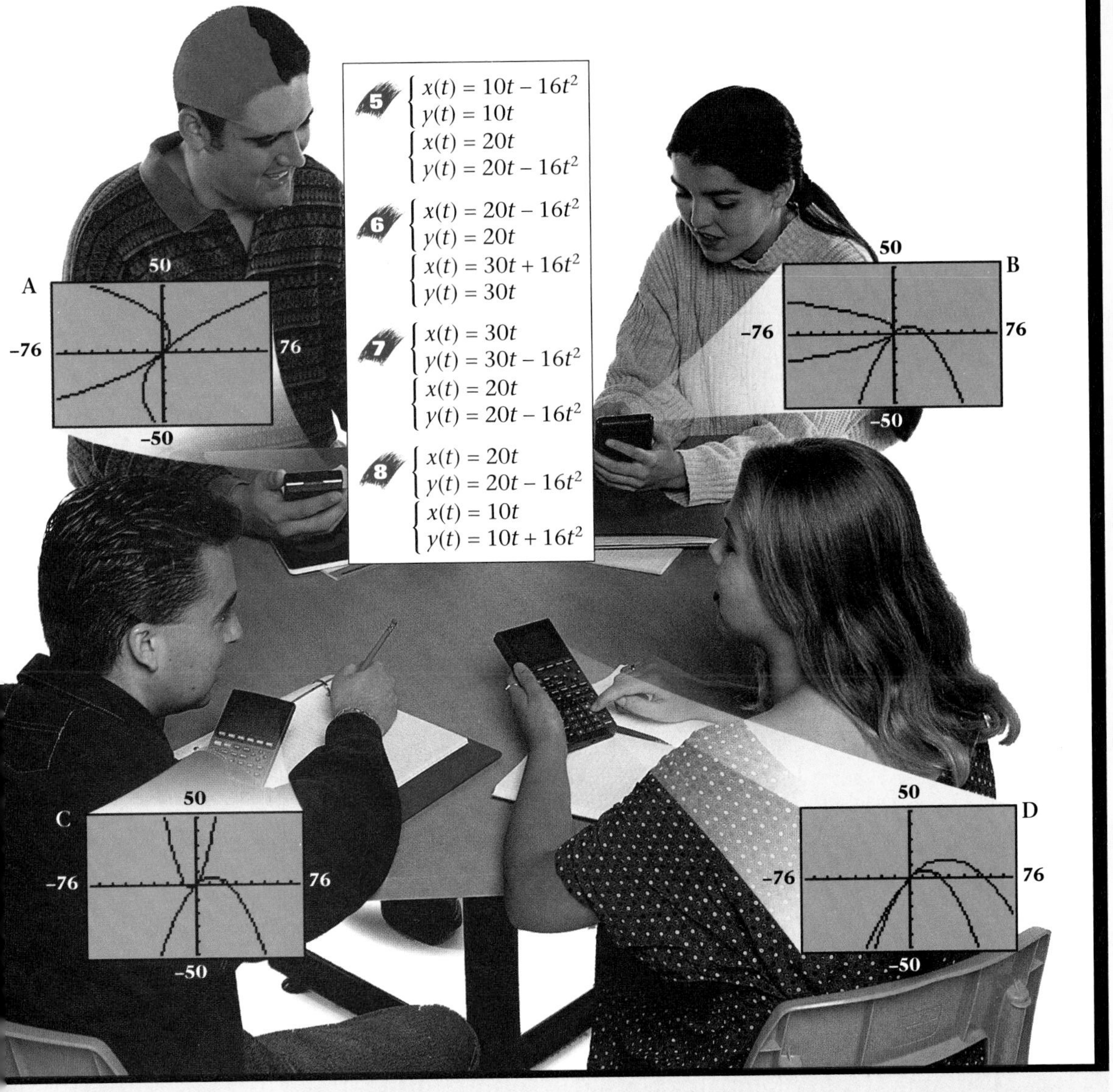

A

B

C

D

RETEACHING the lesson

Technology Have students use a graphics calculator in parametric mode to graph the following system of equations.

$x(t) = 4 \cos t$
$y(t) = 2 \sin t$

Ask students to describe the shape of the graph. Have them give you the coordinates of the center and the standard rectangular form of the equation. $\left[\dfrac{x^2}{16} + \dfrac{y^2}{4} = 1 \right]$

Exploration 3 Notes

This exploration helps students determine the starting point and the direction of motion of the graph of a system of parametric equations. Students should be able to make the connection between the different parts of the equations and how each part affects the resulting graph.

A ongoing ASSESSMENT

5. The function, sine or cosine, and whether the coefficient is negative or positive or negative affects the starting point of the graph. The direction of the graph is controlled by whether $x(t)$ is in terms of $\cos t$ or $\sin t$.

Exploration 4 Notes

This exploration helps students determine how the coordinates of the center of the graph are related to parametric equations. By viewing several graphs, students can see the transformations and translations of the functions in the system.

A ongoing ASSESSMENT

1. The radius of the circle is determined by the constant coefficient a, in the system $x(t) = a \cos t$ and $y(t) = a \sin t$.

5. The x-coordinate is the constant term in $x(t)$; the y-coordinate is the constant term in $y(t)$.

Exploration 3 Plotting Graphs

Graph each system of equations one at a time, and watch as the graph is drawn. Complete the table by recording the starting point of each graph and the direction (clockwise or counterclockwise) in which the graph is drawn. Let t range from 0 to 2π, t step $= \frac{\pi}{6}$, x from −18.8 to 19.2, x scl = 1, and y from −12.4 to 12.6, y scl = 1.

		Starting point	Direction
1	$\begin{cases} x(t) = 10 \cos t \\ y(t) = 10 \sin t \end{cases}$		
2	$\begin{cases} x(t) = -10 \cos t \\ y(t) = -10 \sin t \end{cases}$		
3	$\begin{cases} x(t) = 10 \sin t \\ y(t) = 10 \cos t \end{cases}$		
4	$\begin{cases} x(t) = -10 \sin t \\ y(t) = -10 \cos t \end{cases}$		

5. What part of the parametric equations affects the starting point of the plotted graph? What part of the parametric equations affects the direction in which the graph is plotted? ❖

Exploration 4 Transforming Ellipses

Graphics Calculator

Set your viewing window as shown below.

Tmin = 0	Tmax = 2π	Tstep = $\frac{\pi}{30}$
Xmin = −38	Xmax = 38	Xscl = 5
Ymin = −26	Ymax = 25	Yscl = 5

1. Graph the systems $\begin{cases} x(t) = 5 \cos t \\ y(t) = 5 \sin t \end{cases}$ and $\begin{cases} x(t) = 10 \cos t \\ y(t) = 10 \sin t \end{cases}$ on the same coordinate plane. How are the graphs different? How is the radius of each circle related to the equations?

The center of an ellipse is the intersection of the major and minor axes. Use the trace feature to complete the following table.

		Coordinates of center
2	$\begin{cases} x(t) = 10 \cos t + 4 \\ y(t) = 5 \sin t - 9 \end{cases}$	
3	$\begin{cases} x(t) = 15 \cos t - 10 \\ y(t) = 5 \sin t + 8 \end{cases}$	
4	$\begin{cases} x(t) = 10 \cos t - 8 \\ y(t) = 20 \sin t - 3 \end{cases}$	

5. Explain how the coordinates of the center of the graph are related to the parametric equations. ❖

EXERCISES & PROBLEMS

Communicate

1. Describe a general form of parametric equations that defines an ellipse.

2. Describe a general form of parametric equations that defines a circle.

3. How can you change a system of parametric equations for an ellipse, circle, or parabola into a rectangular equation?

4. Explain how to transform and translate the graph of a parabola described by a system of parametric equations.

5. What part of a system of parametric equations for a circle affects the starting point of the graphing and the direction in which it is graphed?

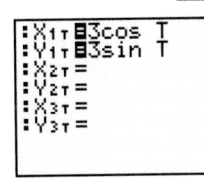 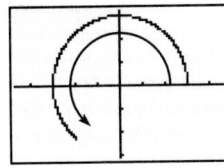

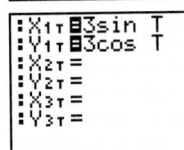 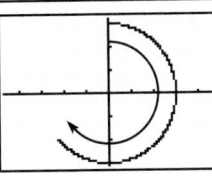

ASSESS

Selected Answers
Odd-numbered Exercises 7–37

Assignment Guide
Core 1–10, 12–17, 19–31, 34–40

Core Plus 1–5, 8–12, 15–27, 31–40

Technology
Graphics calculators are needed for Exercises 6–18, 20–24, and 27–36.

Error Analysis
For Exercise 35, point out that the differences among the approximate answers found by students is a result of each student using a different set of parameters for x- and y-values.

Practice & Apply

Write a parametric representation for each of the following equations. Name the geometric object that is represented. Check using your graphics calculator.

6 $x^2 + y^2 = 9$

7 $x^2 + y^2 = 25$

8 $x^2 + y^2 = 8$

9 $x^2 + y^2 = 21$

10 $\frac{x^2}{4} + \frac{y^2}{9} = 1$

11 $\frac{x^2}{9} + \frac{y^2}{9} = 1$

12 Use your graphics calculator to graph the translated ellipse. Write the equation in rectangular form.

$$\begin{cases} x(t) = \cos t + 2 \\ y(t) = 2 \sin t - 1 \end{cases} \quad \frac{(x-2)^2}{1} + \frac{(y+1)^2}{4} = 1$$

Write a parametric representation for each of the following equations. Name the geometric object that is represented. Check using your graphics calculator.

13 $\frac{x^2}{9} + \frac{y^2}{4} = 1$

14 $4x^2 + y^2 = 16$

15 $\frac{y^2}{9} = 1 - \frac{x^2}{25}$

16 $\frac{(x-2)^2}{16} + \frac{(y+3)^2}{9} = 1$

17 $\frac{(x+4)^2}{25} + \frac{(y-2)^2}{4} = 1$

18 $16(x-3)^2 + 9(y+5)^2 = 144$

6. $x(t) = 3 \cos t, y(t) = 3 \sin t$; circle

7. $x(t) = 5 \cos t, y(t) = 5 \sin t$; circle

8. $x(t) = \sqrt{8} \cos t, y(t) = \sqrt{8} \sin t$; circle

9. $x(t) = \sqrt{21} \cos t, y(t) = \sqrt{21} \sin t$; circle

10. $x(t) = 2 \cos t, y(t) = 3 \sin t$; ellipse

11. $x(t) = 3 \cos t, y(t) = 3 \sin t$; circle

13. $x(t) = 3 \cos t, y(t) = 2 \sin t$; ellipse

14. $x(t) = 2 \cos t, y(t) = 4 \sin t$; ellipse

15. $x(t) = 5 \cos t, y(t) = 3 \sin t$; ellipse

16. $x(t) = 4 \cos t + 2\, y(t) = 3 \sin t - 3$, ellipse

17. $x(t) = 5 \cos t - 4, y(t) = 2 \sin t + 2$, ellipse

18. $x(t) = 3 \cos t + 3, y(t) = 4 \sin t - 5$, ellipse

Portfolio Assessment

Have students write a paragraph explaining the difference between the rectangular and the parametric forms of the equation for a circle and an ellipse. Have them identify what kind of information they use to determine the type of geometric figure that is represented by a system of parametric equations.

Sports A softball is hit from home plate. If home plate is located at (0, 0) on the coordinate plane, then the position of the ball after t seconds is given by the following system of parametric equations.

$$\begin{cases} x(t) = 60t \\ y(t) = 60t - 16t^2 \end{cases}$$

19. Write an equation in rectangular form for the path of the ball. $y = x - \frac{x^2}{225}$

20 There is a wall that is 20 feet high and 200 feet from home plate. Will the ball fall short of the wall, hit the wall, or go over the wall? Explain. go over the wall; when $x = 200$, $y \approx 22.2$

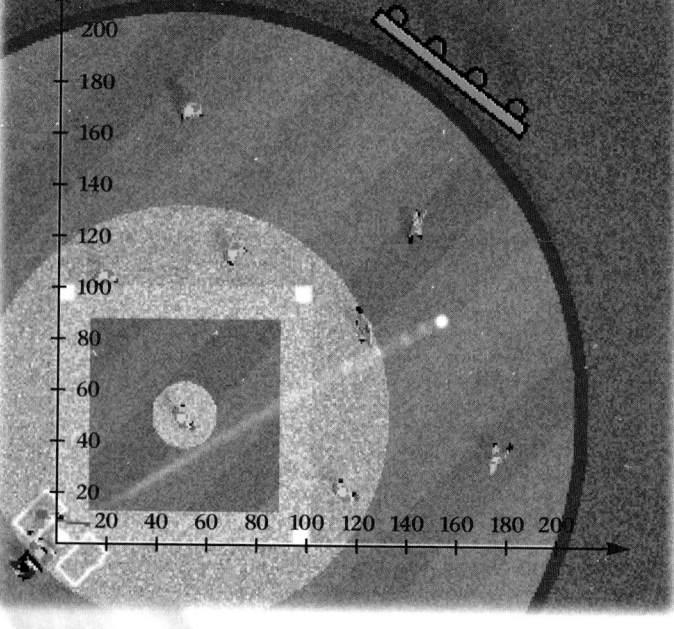

Write a parametric representation for each of the following equations. Name the geometric object that is represented. Check using your graphics calculator.

21 $(x - 2)^2 + (y - 3)^2 = 1$

22 $12 - (x + 2)^2 = (y + 3)^2$

23 $\dfrac{(y + 2)^2}{9} + \dfrac{(x - 3)^2}{16} = 1$

24 $\dfrac{(x + 2)^2}{16} + \dfrac{(y - 3)^2}{25} = 1$

Free Fall A person throws a rock horizontally into a lake from a vertical cliff that is 50 meters high. If the rock is thrown at 20 meters per second, then the position of the rock after t seconds is described by the following system of parametric equations.

$$\begin{cases} x(t) = 20t \\ y(t) = 50 - 4.9t^2 \end{cases}$$

25. Write an equation in rectangular form for the path of the rock. $y = 50 - \frac{4.9x^2}{400}$

26. How far out into the lake will the rock hit the water? 63.9 m

27 Use your graphics calculator in parametric mode to graph this system. Use the trace feature to check your result from Exercise 26. How long will it take for the rock to hit the water?

21. $x(t) = \cos t + 2, y(t) = \sin t + 3$; circle

22. $x(t) = \sqrt{12} \cos t - 2, y(t) = \sqrt{12} \sin t + 3$; circle

23. $x(t) = 4 \cos t + 3, y(t) = 3 \sin t - 2$; ellipse

24. $x(t) = 4 \cos t - 2, y(t) = 5 \sin t + 3$; ellipse

27. approximately 3.2 seconds

NAME _____ CLASS _____ DATE _____

Practice & Apply

10.6 Exploring Parametric Representations of Conic Sections

Write a parametric representation for each equation.

1. $x^2 + y^2 = 25$ 2. $\frac{x^2}{25} + \frac{y^2}{16} = 1$ 3. $y = 4x^2 - x$

Write a rectangular equation for each system of parametric equations. Name the geometric object it represents.

4. $x(t) = 4 \cos t$ 5. $x(t) = 5t - 50t^2$ 6. $x(t) = 4\cos t$
   $y(t) = 5 \sin t$ $y(t) = 5t$ $y(t) = 4\sin t$

Write a rectangular equation for each system of parametric equations. Name the geometric object represented. If the object is a circle, find the coordinates of the center and the length of the radius. If the object is an ellipse, find the coordinates of the center, and the vertices.

7. $x(t) = 3 + 4\cos t$ 8. $x(t) = -3 + \cos t$
   $y(t) = 5 + 4\sin t$ $y(t) = 5 + 4\sin t$

Write a parametric representation for each equation.

9. $x^2 + (y - 3)^2 = 25$ 10. $\frac{x^2}{25} + (y + 3)^2 = 1$

11. $\frac{y^2}{25} = 1 - \frac{x^2}{9}$ 12. $9x^2 + y^2 = 36$

For Exercises 28–33, write a rectangular equation for each system of parametric equations. Identify the geometric object represented.

28 $\begin{cases} x(t) = 6\cos t \\ y(t) = 3\sin t \end{cases}$ **29** $\begin{cases} x(t) = 2\cos t \\ y(t) = 2\sin t \end{cases}$ **30** $\begin{cases} x(t) = 2\cos t + 4 \\ y(t) = 3\sin t - 3 \end{cases}$

31 $\begin{cases} x(t) = 2\cos t + 1 \\ y(t) = 4\sin t - 2 \end{cases}$ **32** $\begin{cases} x(t) = t^2 + 2 \\ y(t) = t \end{cases}$ **33** $\begin{cases} x(t) = t \\ y(t) = 5t^2 \end{cases}$

34 Graph and describe the two following systems of parametric equations. Determine whether the two graphs intersect.

$\begin{cases} x(t) = 3\cos t \\ y(t) = 3\sin t \end{cases}$ $\begin{cases} x(t) = 3\cos t + 2 \\ y(t) = 3\sin t - 1 \end{cases}$ Two circles that intersect in two points

35 Graph and describe the two following systems of parametric equations. Determine whether the two graphs intersect.

$\begin{cases} x(t) = 2.6\cos t \\ y(t) = 2.28\sin t \end{cases}$ $\begin{cases} x(t) = \dfrac{2}{\cos t} \\ y(t) = 2\tan t \end{cases}$ an ellipse and a hyperbola that intersect in four points

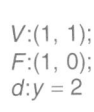

Look Back

36 Find the focus, directrix, and vertex of the parabola $x^2 - 2x + 4y - 3 = 0$. Check using your graphics calculator. **[Lesson 10.1]**
V:(1, 1); F:(1, 0); d:$y = 2$

37 Find the center, foci, vertices, and co-vertices of the ellipse $9x^2 + 4y^2 + 36x - 24y + 36 = 0$. Check using your graphics calculator. **[Lesson 10.3]**

38. Find the center, foci, and vertices of the hyperbola $x^2 - 9y^2 + 2x - 54y - 81 = 0$. Determine the asymptotes of this hyperbola. **[Lesson 10.4]**

Look Beyond

39. Suppose you have a gum-ball machine with blue and yellow gum balls in it, and there are twice as many blue gum balls as yellow ones. What are the chances that you will get a blue gum ball? $\dfrac{2}{3}$

40. What is the next number in the sequence 1, 4, 2, 5, 3, 6, 4, 7, ...? 5

Look Beyond

Exercises 39–40 introduce students to probability and sequences.

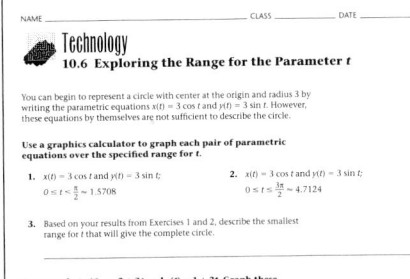

28. $\dfrac{x^2}{36} + \dfrac{y^2}{9} = 1$; ellipse

29. $x^2 + y^2 = 4$; circle

30. $\dfrac{(x-4)^2}{4} + \dfrac{(y+3)^2}{9} = 1$; ellipse

31. $\dfrac{(x-1)^2}{4} + \dfrac{(y+2)^2}{16} = 1$; ellipse

32. $x = y^2 + 2$; parabola

33. $y = 5x^2$; parabola

37. $C(-2, 3)$, F: $(-2, 3 + \sqrt{5})$, $(-2, 3 - \sqrt{5})$, V: $(-4, 3)$, $(0, 3)$, $(-2, 0)$, $(-2, 6)$

38. $C(-1, -3)$; F: $(0.05, -3)$, $(-2.05, -3)$; V: $(0, -3)$, $(-2, -3)$, $(-1, -\frac{10}{3})$, $(-1, -\frac{8}{3})$; asymptotes: $y = \frac{1}{3}x - \frac{8}{3}$ and $y = -\frac{1}{3}x - \frac{10}{3}$

FOCUS

The focus of this project is on the use of analytic geometry, specifically the hyperbola, as a practical tool in navigation and search and rescue missions.

MOTIVATE

This project can be motivated by an introduction to navigation. Guest speakers who have experience with navigation, either on ships or aircraft, can be invited to tell storeis of sea or air navigation. Nautical maps can be posted. This project also offers the opportunity to coordinate with the English department for recommending stories of shipwrecks and search and rescue missions. Have students review the relationship between the mathematics of analytical geometry and the mathematics used in navigation.

NAVIGATIONAL SYSTEMS

Some navigational systems are based on the equations of hyperbolas. For example, suppose a ship in distress at sea emits a radio signal, and two fixed navigational radio stations receive the signal at different times. The stations can then measure the difference between the times the signal arrives at each station. With this time differential and the fact that radio waves travel at about 186,000 miles per second (the speed of light), the monitors at the radio stations can determine that the ship is on a particular type of curve. If there is a second set of radio stations, the monitors at those stations can locate the ship on another set of curves. The exact position of the ship will be one of the points where these curves intersect.

There are two pairs of radio stations. Stations A and B are located at the points (0, 100) and (0, −100). Stations C and D are located at (300, 0) and (500, 0). Station B receives the distress signal 0.74 milliseconds before Station A receives it, and Station C receives the signal 0.74 milliseconds before Station D receives it.

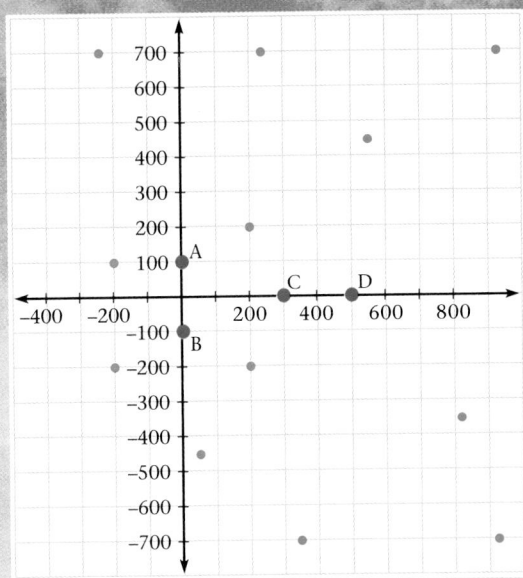

The red points in the coordinate plane represent ships in the ocean near Stations A, B, C, and D.

Which ship is in distress? What are the coordinates of this ship? If the monitors at the radio stations want to contact the ship closest to the one in distress, which ship should they contact? What are the coordinates of this ship, and what is its distance from the ship in distress?

Cooperative Learning

Have students work together to answer the questions on page 587. Then ask students to extend the problem by developing additional problems based on information from the graph. Another group activity would be to use the graph provided to create a rescue story. The story must include the mathematics that was used to locate and intercept a ship in distress.

Discuss

In groups, discuss and organize a step-by-step procedure for using a hyperbola to locate a point on a graph. Then exchange the stories from the various groups and apply the procedure to determine whether the mathematics in their stories is accurate. Groups that exchanged stories can then exchange ideas about the accuracy of the mathematics in the stories. Finally, the groups rewrite their stories incorporating the suggestions.

The ship at $(200, -200)$ is in distress. They should contact the ship at $(50, -450)$. Its distance is about 292 miles from the ship in distress.

Chapter 10 Review

Vocabulary

asymptotes of a hyperbola	562	hyperbola	560
center of a circle	544	major axis	553
conic section	569	minor axis	553
circle	544	parabola	537
conjugate axis	562	radius	544
directrix	537	standard equation of a circle	546
ellipse	552	standard equation of a hyperbola	562
foci of an ellipse	552	standard equation of a parabola	540
foci of a hyperbola	560	standard equation of an ellipse	553
focus of a parabola	537	transverse axis	562

Key Skills & Exercises

Lesson 10.1

> #### Key Skills

Write an equation of a parabola when given any two of the following: focus, directrix, vertex.

A parabola is a set of points in a plane that is the same distance from a fixed line called the *directrix*, as it is from a fixed point not on the line, called the *focus*. The *vertex* of a parabola lies midway between the focus and the directrix.

Identify the vertex, focus, and directrix of a parabola from its standard equation; then, sketch a graph.

	$y = a(x - h)^2 + k$	$x = a(y - k)^2 + h$
Vertex	(h, k)	(h, k)
Focus	$\left(h, k + \frac{1}{4a}\right)$	$\left(h + \frac{1}{4a}, k\right)$
Directrix	$y = k - \frac{1}{4a}$	$x = h - \frac{1}{4a}$

> #### Exercises

Determine an equation for the parabola with the indicated properties.

1. Focus at $(2, -3)$ and directrix $x = -1$ $x = \frac{1}{6}(y + 3)^2 + \frac{1}{2}$
2. Focus at $(0, 3)$ and vertex at $(0, 1)$ $y = \frac{1}{8}x^2 + 1$
3. Vertex at $(0, -2)$ and directrix $y = -4$ $y = \frac{1}{8}x^2 - 2$
4. Identify the vertex, focus, and directrix of the graph of $y^2 - 6y - x + 13 = 0$. Then sketch the graph.
   $V{:}(4, 3)$; $F{:}(4.25, 3)$; $d{:}x = 3.75$

4. V: $(4, 3)$; F: $(4.25, 3)$; directrix: $x = 3.75$

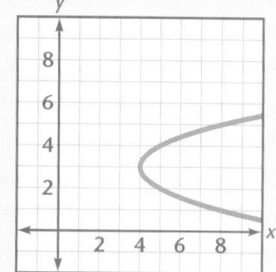

Lesson 10.2

➤ Key Skills

Write the equation of a circle when given the coordinates of the center and the length of the radius.

The standard equation of a circle with center at (h, k) and a radius of r is $(x - h)^2 + (y - k)^2 = r^2$.

Determine the center and radius of a circle when given an equation of the circle.

Complete the squares on x and y, and write the equation in standard form. The center is at (h, k) and the radius is r.

➤ Exercises

Write the standard equation of the circle with center C and radius r.

5. $C(1, 4)$, $r = 5$ $\quad (x - 1)^2 + (y - 4)^2 = 25$ **6.** $C\left(-\frac{1}{2}, -2\right)$, $r = \sqrt{11}$ $\quad \left(x + \frac{1}{2}\right)^2 + (y + 2)^2 = 11$

Determine the center and the radius of each circle.

7. $x^2 + y^2 - 10x + 8y + 5 = 0$
$\quad C:(5, -4); \ r = 6$

8. $x^2 + y^2 + 4x - 5 = 0$
$\quad C:(-2, 0); \ r = 3$

Lesson 10.3

➤ Key Skills

Determine the coordinates of the center, vertices, co-vertices, and foci when given an equation of an ellipse with either a horizontal or vertical major axis.

Complete the squares on x and y, and write the equation in the form $\frac{(x - h)^2}{a^2} + \frac{(y - k)^2}{b^2} = 1$ or $\frac{(x - h)^2}{b^2} + \frac{(y - k)^2}{a^2} = 1$ with $a > b > 0$. The center is at (h, k); the vertices are a units from the center; the co-vertices are b units from the center; and the foci are on the major axis at a distance c from the center, where $c^2 = a^2 - b^2$.

➤ Exercises

Give the coordinates of the center, vertices, co-vertices, and foci of the ellipse whose equation is given.

9. $\frac{(x - 4)^2}{9} + \frac{(y + 1)^2}{25} = 1$
$\quad C:(4, -1); \ V:(4, 4), (4, -6)$
$\quad Co - V:(1, -1), (7, -1),$
$\quad F:(4, 3), (4, -5)$

10. $x^2 + 4y^2 + 10x + 24y + 45 = 0$
$\quad C: (-5, -3); \ V: (-9, -3), (-1, -3);$
$\quad Co - V: (-5, -1), (-5, -5);$
$\quad F: (-5 - \sqrt{12}, -3), (-5 + \sqrt{12}, -3)$

Lesson 10.4

➤ Key Skills

Determine the coordinates of the center and foci, and find the lengths of the axes when given an equation of a hyperbola.

Complete the squares on x and y and write the equation in the form $\frac{(x - h)^2}{a^2} - \frac{(y - k)^2}{b^2} = 1$ or $\frac{(y - k)^2}{a^2} - \frac{(x - h)^2}{b^2} = 1$. The center is at (h, k); the transverse axis has a length of $2a$; the length of the conjugate axis is $2b$; andthe foci are on the transverse axis at a distance c from the center, where $c^2 = a^2 + b^2$.

➤ Exercises

Give the coordinates of the center, the foci, and the lengths of the axes of the hyperbola given by each of the following equations.

11. $\frac{(x + 5)^2}{4} - \frac{(y - 4)^2}{9} = 1$

12. $36y^2 - 4x^2 + 216y - 40x + 80 = 0$

11. $C: (-5, 4); F: (-5 + \sqrt{13}, 4), (-5 - \sqrt{13}, 4);$
$\quad$ length of axes: 4 and 6

12. $C: (-5, -3);$
$\quad F: (-5, -3 - \sqrt{40}),$
$\quad (-5, -3 + \sqrt{40});$ length of axes: 4 and 12

Lesson 10.5

➤ *Key Skills*

Determine the intersections of a system consisting of one first-degree equation and one second-degree equation.

Solve the first-degree equation for one variable. Then substitute this solution in the second-degree equation and solve it to find the intersections of the two graphs.

Determine the intersections of a system consisting of two second-degree equations.

Use the elimination-by-addition method to eliminate one variable. Solve the resulting equation for the second variable. Substitute to find the values for the first variable at the points of intersection.

➤ *Exercises*

Find all the points of intersection by solving the system.

13. $\begin{cases} x^2 + 2y^2 = 33 \\ 3x + 2y = -11 \end{cases}$ $(-1, \ -4), (-5, \ 2)$ 14. $\begin{cases} x^2 + 2y^2 = 33 \\ x^2 + y^2 + 2x = 19 \end{cases}$ $(1, \ 4), (1, \ -4), (-5, \ 2), (-5, \ -2)$

Lesson 10.6

➤ *Key Skills*

Write the rectangular and parametric forms of the equation of a circle.

Rectangular form: $(x - h)^2 + (y - k)^2 = r^2$

Parametric form: $\begin{cases} x(t) = r \cos t + h \\ y(t) = r \sin t + k \end{cases}$

Write the rectangular and parametric forms of the equation of an ellipse.

Rectangular form: $\frac{(x-h)^2}{a^2} + \frac{(y-k)^2}{b^2} = 1$

Parametric form: $\begin{cases} x(t) = a \cos t + h \\ y(t) = b \sin t + k \end{cases}$

➤ *Exercises*

Write the rectangular equation for each parametric system.

15. $\begin{cases} x(t) = 4 \cos t \\ y(t) = 4 \sin t + 3 \end{cases}$ $x^2 + (y - 3)^2 = 16$ 16. $\begin{cases} x(t) = 5 \cos t - 2 \\ y(t) = 3 \sin t + 1 \end{cases}$ $\frac{(x+2)^2}{25} + \frac{(y-1)^2}{9} = 1$

Write a parametric representation for each equation, and name the geometric object that is represented.

17. $\frac{(x-1)^2}{25} + \frac{y^2}{16} = 1$ $x(t) = 5 \cos t + 1$ $y(t) = 4 \sin t$; ellipse 18. $x^2 + y^2 = 16$ $x(t) = 4 \cos t$ $y(t) = 4 \sin t$; circle

Applications

Astronomy The Earth orbits the Sun in an elliptical path, with the Sun at one focus of the ellipse. The closest the Earth gets to the Sun (perigee) is 91.4 million miles. The apogee (farthest from the Sun) for the Earth's orbit is 94.6 million miles.

19. Assuming that the center of the ellipse is at the origin, write an equation for the elliptical path of the Earth around the Sun. $\frac{x^2}{8649} + \frac{y^2}{8646.44} = 1$

20. Find the distance from the Sun to the other focus. 3.2 million miles

Chapter 10 Assessment

Determine the equation of the parabola with the given properties.

1. Vertex at $(0, 0)$ and focus at $(0, -2)$ $y = -\frac{1}{8}x^2$

2. Focus at $(-2, 1)$ and directrix $x = -3$ $x = \frac{1}{2}(y - 1)^2 - \frac{5}{2}$

3. Directrix at $y = 5$ and vertex at $(3, 7)$ $y = \frac{1}{8}(x - 3)^2 + 7$

4. Identify the vertex, the focus, and the directrix of the graph of $x^2 - 2x + y + 10 = 0$. Then sketch the parabola. $V{:}(1, -9)$; $F{:}(1, -9.25)$; $d{:}y = -8.75$

Write the standard equation of the circle with center C and radius r.

5. $C(-2, 3)$, $r = 7$
$(x + 2)^2 + (y - 3)^2 = 49$

6. $C(25, -5)$, $r = \sqrt{5}$
$(x - 25)^2 + (y + 5)^2 = 5$

Determine the center and radius of the circle represented by each equation.

7. $x^2 + y^2 - 2y - 8 = 0$ $C{:}(0, 1)$; $r = 3$ **8.** $x^2 + y^2 + 8x - 6y = 0$ $C{:}(-4, 3)$; $r = 5$

Give the coordinates of the center, the vertices, the co-vertices, and the foci of the ellipse given by each equation.

9. $9x^2 + 4y^2 = 36$

10. $25x^2 + 9y^2 - 200x + 18y + 184 = 0$

Astronomy Jupiter orbits the sun in an elliptical path, with the sun at one focus of the ellipse. The distance between Jupiter and the sun varies between a maximum of 507.0 million miles and a minimum of 460.6 million miles.

11. Write the equation of Jupiter's path around the sun if the center of the ellipse is at $(0, 0)$.

12. Find the distance between the sun and the other focus. 46.4 million miles

Give the coordinates of the center, the foci, and the lengths of the axes of the hyperbola given by each equation.

13. $x^2 - y^2 = 16$

14. $9x^2 - 4y^2 + 90x + 32y + 197 = 0$

Marine Biology To locate a whale, two microphones are placed 6000 feet apart in the ocean. One microphone picks up a whale's noise 0.5 seconds after the other microphone picks up the same noise. The speed of sound in water is about 5000 feet per second.

15. Find the equation of the hyperbola that describes the possible locations of the whale.

16. What is the closest the whale could be to either microphone. 1750 ft

Find all the points of intersection by solving the system.

17. $\begin{cases} y = x^2 + 8x + 9 \\ x - y = -3 \end{cases}$ $(-1, 2)$, $(-6, -3)$

18. $\begin{cases} 7x^2 - 5y^2 + 20y = 3 \\ 21x^2 + 5y^2 = 209 \end{cases}$ $(2, 5)$, $(-2, 5)$, $(3, -2)$, $(-3, -2)$

19. Change the parametric system of equations $\begin{cases} x(t) = 2 \cos t + 5 \\ y(t) = 2 \sin t - 3 \end{cases}$ to an $(x - 5)^2 + (y + 3)^2 = 4$; circle
equation in rectangular form, and indicate the geometric object it represents.

20. Write a system of parametric equations for $x^2 + y^2 = 1$. $x(t) = \cos t$; $y(t) = \sin t$

4. V: $(1, -9)$; F: $(1, -9.25)$; directrix: $y = -8.75$

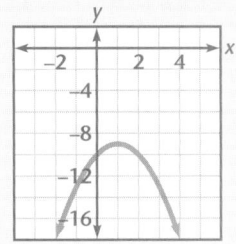

9. C: $(0, 0)$; V: $(0, 3)$, $(0, -3)$; Co-V: $(-2, 0)$, $(2, 0)$; F: $(0, \sqrt{5})$, $(0, -\sqrt{5})$

10. C: $(4, -1)$; V: $(4, -6)$, $(4, 4)$; Co-V: $(1, -1)$, $(7, -1)$; F: $(4, -5)$, $(4, 3)$

11. $\dfrac{x^2}{(483.8)^2} - \dfrac{y^2}{(483.2)^2} = 1$

13. C: $(0, 0)$; F: $(-\sqrt{32}, 0)$, $(\sqrt{32}, 0)$; lengths of axes are both 8

14. C: $(-5, 4)$; F: $(-5, 4 + \sqrt{13})$, $(-5, 4 - \sqrt{13})$; lengths of axes: 6 and 4

15. $\dfrac{x^2}{1.5625} - \dfrac{y^2}{7.4375} = 1$, where x and y values are in thousands of feet

Chapters 1–10 Cumulative Assessment

College Entrance-Exam Practice

Multiple-Choice and Quantitative-Comparison Samples

The first half of the Cumulative Assessment contains two types of items found on standardized tests — multiple-choice questions and quantitative-comparison questions. Quantitative-comparison items emphasize the concepts of equality, inequality, and estimation.

Free-Response Grid Samples

The second half of the Cumulative Assessment is a free-response section. A portion of this part of the Cumulative Assessment consists of student-produced response items commonly found on college entrance exams. These questions require the use of machine-scored answer grids. You may wish to have students practice answering these items in preparation for standardized tests.

Sample answer-grid masters are available in the *Chapter Teaching Resources Booklets*.

College Entrance Exam Practice

Quantitative Comparison For Items 1–5, write

A if the quantity in Column A is greater than the quantity in Column B;

B if the quantity in Column B is greater than the quantity in Column A;

C if the quantities are equal; or

D if the relationship cannot be determined from the given information.

	Column A	Column B	Answers
1. A	6^{-2}	-2^6	(A) (B) (C) (D) [Lesson 2.2]
2. A	$f(x) = x^2 - 5, g(x) = 3x$ $(f \circ g)(3)$	$f(x) = x^2 - 5, g(x) = 3x$ $(g \circ f)(3)$	(A) (B) (C) (D) [Lesson 3.3]
3. B	The x-coordinate of the maximum point on the graph of $y = -(x + 4)^2 + 3$	The x-coordinate of the minimum point on the graph of $y = 2(x - 5)^2 - 1$	(A) (B) (C) (D) [Lesson 5.3]
4. C	The value of $f(-2)$ if $f(x) = x^4 - x^3 - 7x^2 + x + 6$	The value of $f(3)$ if $f(x) = x^4 - x^3 - 7x^2 + x + 6$	(A) (B) (C) (D) [Lesson 2.3]
5. A	The value of x	The value of y	(A) (B) (C) (D) [Lesson 8.1]

6. The line with a slope of $-\frac{1}{2}$ and containing the point (4, 1) is *not*
C   represented by which of the following equations? **[Lesson 1.2]**

    a. $y = -\frac{1}{2}x + 3$ **b.** $x + 2y = 6$ **c.** $2y - 6 = x$ **d.** $2x - 12 = -4y$

7. If $f(x) = \frac{2x + 5}{4}$, then what is $f^{-1}(x)$?   **[Lesson 3.2]**
C

    a. $\frac{2x - 5}{-4}$ **b.** $2x - 5$ **c.** $\frac{4x - 5}{2}$ **d.** $4x + 5$

8. What is the degree of $f(x) = (x - 1)^2(3x + 5)(1 - x)^4$? **[Lesson 6.1]**
C   **a.** 3   **b.** 6   **c.** 7   **d.** 2

9. How many years (to the nearest whole year) will it take $3000 to
b   double at 10% interest compounded continuously? **[Lesson 7.4]**

    a. 6   **b.** 7   **c.** 9   **d.** 10

10. How many vertical asymptotes does the function

d

$f(x) = \frac{(x+2)(x-3)}{x(x-1)(x+5)}$ have? **[Lesson 9.4]**

   a. none   **b.** 1   **c.** 2   **d.** 3

11. Use graphing and composition to show that $f(x) = \frac{3x-15}{2}$ and

$g(x) = \frac{2}{3}x + 5$ are inverse functions.   **[Lessons 3.2, 3.3]**

12. Combine the parametric system $\begin{cases} x(t) = 2t - 3 \\ y(t) = -5t \end{cases}$ into

one linear function. **[Lesson 3.6]** $5x + 2y = -15$

13. Write a matrix equation to represent this system of linear equations, and use matrix algebra to solve it.
  $\begin{cases} -2x + y - z = -7 \\ x - 4z = -11 \\ 3x - y + 5z = 20 \end{cases}$

[Lesson 4.6]

14. Find the value of the discriminant for $2x^2 + 3x + 8 = 0$, and use it to determine how many real-number solutions exist.   **[Lesson 5.6]**   −55; no real solutions

15. Write a function that has zeros of 0, −2, and 3.   **[Lesson 6.2]**

16. Identify the amplitude and period of $f(x) = 2 + \cos\left(2x - \frac{\pi}{4}\right)$.   amp: 1; *period:* π

  [Lesson 8.5]

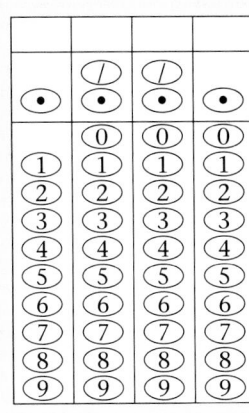

Free-Response Grid The following questions may be answered using a free-response grid commonly used by standardized test services.

17. Solve $\log_x 16 = 2$ for x.   **[Lesson 7.7]**   4

18. Find the x-coordinate of the center of the ellipse whose equation is $9x^2 + 4y^2 - 16y - 20 = 0$.   **[Lesson 10.3]**   0

19. To the nearest hundredth, what is the x-coordinate of P in the following graph? **[Lesson 8.2]**   −0.42

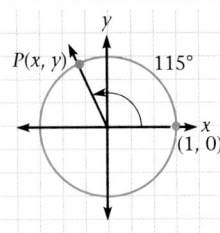

20. What is the sum of the solutions of $0 = 2x^2 - 5x - 3$?   **[Lesson 5.5]**   2.5

11. $f \circ g(x) = x$ and $(g \circ f)(x) = x$

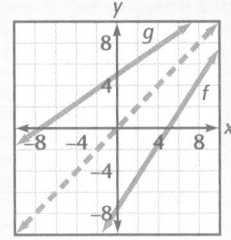

13. $\begin{bmatrix} -2 & 1 & -1 \\ 1 & 0 & -4 \\ 3 & -1 & 5 \end{bmatrix} \begin{bmatrix} x \\ y \\ z \end{bmatrix} = \begin{bmatrix} -7 \\ -11 \\ 20 \end{bmatrix}$

$x = 1, y = -2, z = 3$

15. Answers may vary but should be equivalent to $f(x) = Cx^p(x+2)^q(x-3)^r$, where C is a nonzero constant and p, q, and r are positive integers.

CHAPTER 11

Discrete Mathematics: Counting Principles and Probability

Meeting Individual Needs

11.1 Exploring Probability

Core

Inclusion Strategies, p. 598
Reteaching the Lesson, p. 599
Practice Master 11.1
Enrichment Master 11.1
Technology Master 11.1
Lesson Activity Master 11.1

[2 days]

Core Plus

Practice Master 11.1
Enrichment, p. 598
Technology Master 11.1
Lesson Activity Master
  11.1
Interdisciplinary
  Connection, p. 597

[2 days]

11.2 Counting and Permutations

Core

Inclusion Strategies, p. 607
Reteaching the Lesson, p. 608
Practice Master 11.2
Enrichment Master 11.2
Technology Master 11.2
Lesson Activity Master 11.2

[2 days]

Core Plus

Practice Master 11.2
Enrichment, p. 606
Technology Master 11.2
Lesson Activity Master
  11.2

[1 day]

11.3 Special Permutations

Core

Inclusion Strategies, p. 613
Reteaching the Lesson, p. 614
Practice Master 11.3
Enrichment Master 11.3
Technology Master 11.3
Lesson Activity Master 11.3

[2 days]

Core Plus

Practice Master 11.3
Enrichment, p. 613
Technology Master 11.3
Lesson Activity Master
  11.3

[1 day]

11.4 Counting and Combinations

Core

Inclusion Strategies, p. 622
Reteaching the Lesson, p. 623
Practice Master 11.4
Enrichment Master 11.4
Technology Master 11.4
Lesson Activity Master 11.4
Interdisciplinary Connection,
  p. 621
Mid-Chapter Assessment Master

[2 days]

Core Plus

Practice Master 11.4
Enrichment, p. 622
Technology Master 11.4
Lesson Activity Master
  11.4
Mid-Chapter Assessment
  Master

[2 days]

11.5 Independent and Mutually Exclusive Events

Core

Inclusion Strategies, p. 629
Reteaching the Lesson, p. 630
Practice Master 11.5
Enrichment Master 11.5
Technology Master 11.5
Lesson Activity Master 11.5
Interdisciplinary Connection,
  p. 628

[2 days]

Core Plus

Practice Master 11.5
Enrichment, p. 629
Technology Master 11.5
Lesson Activity Master
  11.5

[2 days]

11.6 Exploring Conditional Probability

Core

Inclusion Strategies, p. 637
Reteaching the Lesson, p. 637
Practice Master 11.6
Enrichment Master 11.6
Technology Master 11.6
Lesson Activity Master 11.6

[Optional*]

Core Plus

Practice Master 11.6
Enrichment, p. 636
Technology Master 11.6
Lesson Activity Master
  11.6

[1 day]

11.7 Simulation Methods

Core

Inclusion Strategies, p. 642
Reteaching the Lesson, p. 643
Practice Master 11.7
Enrichment Master 11.7
Technology Master 11.7
Lesson Activity Master 11.7
Interdisciplinary Connection,
    p. 641

[Optional*]

Core Plus

Practice Master 11.7
Enrichment, p. 642
Technology Master 11.7
Lesson Activity Master
    11.7

[2 days]

Chapter Summary

Core

Eyewitness Math pp. 618–619
Lab Activity
Long-Term Project
Chapter Review, pp. 648–650
Chapter Assessment, p. 651
Chapter Assessment, A/B
Alternative Assessment

[4 days]

Core Plus

Eyewitness Math pp. 618–
    619
Chapter 11 Project p. 647
Lab Activity
Long-Term Project
Chapter Review,
    pp. 648–650
Chapter Assessment,
    p. 651
Chapter Assessment, A/B
Alternative Assessment

[3 days]

Reading Strategies

The Chapter Review provides a vocabulary list that the students should examine before reading the chapter. Suggest that they keep these words in mind as they read. Students should also compile a list of definitions for any other words they encounter that they do not understand. Some words that have an imprecise meaning in daily English will have a very specific meaning when related to mathematics. Students should write down words that are used in a way that is different from what is familiar. They should then make an effort to use these words in discussion groups to be sure that they are using the terms correctly.

Visual Strategies

Direct the students to examine the pictures and associated graphs in the chapter opener. Encourage them to speculate about where mathematics might be useful in studying such phenomena. Also have them look at the graphs in the chapter and discuss how they differ from graphs in earlier chapters.

Hands-on Strategies

This chapter presents the students with a number of opportunities to become actively involved with mathematics in the physical world. Coin tossing and dice-rolling experiments give students a chance to see probability in action.

Cooperative Learning

GROUP ACTIVITIES	
Experimental probability	Lesson 11.1, Exploration 1
Games and probability	Eyewitness Math: Let's Make a Deal
Independent events	Lesson 11.5, Example 1
Dependence of events	Lesson 11.6, Exploration 1
Permutations	Chapter Project

You may wish to have students work with partners for some of the above activities. Additional suggestions for cooperative group activities are noted in the teacher's notes in each lesson.

Multicultural

The cultural connections in this chapter include references to The Americas.

CULTURAL CONNECTIONS	
The Americas: The Navaho Code Talkers	Lesson 11.3, Exercise 19

Portfolio Assessment

Below are portfolio activities for the chapter. They are listed under seven activity domains that are appropriate for portfolio development.

1. **Investigation/Exploration** Experimental probability is the focus of Exploration 1 in Lesson 11.1; theoretical probability is the focus of Exploration 2 in Lesson 11.1; computing probability with an area model is the focus of Exploration 3 in Lesson 11.1; dependence of events is the objective of the exploration in Lesson 11.6; permutations is the objective of the Chapter Project.

2. **Applications Recommended** To be included are any of the following: opinion polls, Lesson 11.1, Exercises 14–18; testing, Lesson 11.2, Exercises 6–7; manufacturing, Lesson 11.3, Exercise 20; stock exchange, Lesson 11.4, Exercise 20; sports, Lesson 11.5, Exercise 38; statistics, Lesson 11.6, Exercises 12–13; real estate, Lesson 11.7, Exercises 16–18.

3. **Non-Routine Problems** The Let's Make a Deal problem from Eyewitness Math gives the student an opportunity to use probability to solve a controversial mathematics problem.

4. **Project A-Mazing Search** Students are asked to find the probability of accessing a student's record by using two different sequences.

5. **Interdisciplinary Topics** Students may choose from the following: biology, Lesson 11.1, Exercise 38; earth science, Lesson 11.2, Exercise 23; geometry, Lesson 11.4, Exercise 24.

6. **Writing** Communicate exercises of the type in which a student is asked to describe basic concepts of experimental and theoretical probability, permutations and combinations, and real world applications of probability offer excellent writing opportunities for the portfolio.

7. **Tools** Chapter 11 uses graphics calculators to assimilate data and model that data to help perform counting procedures and simulations. To measure the individual student's proficiency, it is recommended that he or she complete selected worksheets from the Technology Masters.

Technology

Technology such as scientific calculators, calculators with statistics functions, graphics calculators, and computers can play a significant role in working with the material of this chapter. Any calculation device, however, has its limits. For example, the $n!$ key on several popular calculators has an upper limit of $n = 69$. Another calculator requires n to be less than 449. If the number of digits exceeds the limits of the calculator, the output is displayed in exponential notation. If the domain of the function extends beyond the mechanical limit of the calculator, it reports an overflow error. Because of the way that $_nC_r$ and $_nP_r$ are defined in terms of factorials, they too have limited domains and ranges. Students must appreciate these limits and have some idea of what to do if the problem contains numbers beyond the limits of their computational tools. This is where the differences between mathematics and computation become apparent.

```
13!
        6227020800
68!
    2.480035542E96
```

If the students understand the basic principles and mathematical formulas, they can transcend the computational limit. Make sure that they know that there is an approximation for $n!$ that will allow students to carry on past the limits of their calculator. Stirling's Approximation is expressed by the following formula.

$$n! \cong n^n e^{-n} \sqrt{2\pi n}$$

The computation still requires a calculator, but it allows a reasonable approximation of the factorial, combination, and permutation function when n gets large. Ask the students to consider that their calculator might be already using this approximation for large values of n, and to compare the values it gives from its function key with the values they get from using the formulas and the Stirling approximation.

The random number generator feature of calculators and computers has its own limitation. No digital or binary logic device or circuit can produce a sequence of numbers that are truly random in the purest mathematical sense. In fact, if you replace all the batteries and reset the calculator, when the calculator is restarted, some calculators will produce the same sequence of random numbers. It will continue to do this each time it is completely restarted in this fashion. Ask the student to consider what effect this fact may have on any experiment that uses random numbers generated by hardware.

Integrated Software

f(g) Scholar™ is an integrated computer-based mathematics productivity tool that combines calculator, spreadsheet, and graphics capabilities to provide a dynamic and interactive environment for explorations in mathematics. Use *f(g) Scholar*™ for any lesson needing a spreadsheet, calculator, graphics calculator, or any combination of the three.

11 Discrete Mathematics: Counting Principles and Probability

ABOUT THE CHAPTER

Background Information

An interest in counting and probability may have originated with early games of chance and the human mind's love of puzzles. Today this branch of applied mathematics is widely used in our society, for everything from sports to subatomic physics.

CHAPTER RESOURCES

- Practice Masters
- Enrichment Masters
- Technology Masters
- Lesson Activity Masters
- Lab Activity Masters
- Long-Term Project Masters
- Assessments Masters
  Chapter Assessments, A/B
  Mid-Chapter Assessment
  Alternative Assessment, A/B
- Teaching Transparencies
- Spanish Resources

CHAPTER OBJECTIVES

- Determine the theoretical probability of an event when given an appropriate sample space.
- Determine the experimental probability of an event when given the results of an experiment.
- Use the Fundamental Principle of Counting to determine how many ways a decision can be made.
- Determine the number of permutations of n distinct objects taken r at a time.
- Find the number of distinct permutations of n objects of which r_1 objects are alike and r_2 objects are alike.

CHAPTER 11

Discrete Mathematics: Counting Principles and Probability

LESSONS

11.1 *Exploring* Probability

11.2 Counting and Permutations

11.3 Special Permutations

Eyewitness Math
Let's Make a Deal

11.4 Counting and Combinations

11.5 Independent and Mutually Exclusive Events

11.6 *Exploring* Conditional Probability

11.7 Simulation Methods

Chapter Project
A-Mazing Search

Probability deals with the laws of chance. It is a numerical value that measures the likeliness that an event will occur. Statistics and probability are closely related. Probability enables you to use statistics to make inferences about populations or decisions about hypotheses. You can use probability to make inferences about a sample, while you can use statistics to make inferences about an entire population.

69% of Canadians say they cope with stress by taking a coffee or tea break.

1.9% of the U.S. population live on farms

ABOUT THE PHOTOS

One aspect of probability is the concept of randomness. The computer printout demonstrates how a computer can continuously print numbers that appear to have no identifiable pattern. These random numbers are often used for simulations. The sentences included on the pages show claims that are often based on statistics. People need to educate themselves about statistics because deceptive or false conclusions can be made to sound legitimate by using statistical language.

56% of Americans don't participate in sports because they don't feel they have the time.

46% of all American teens own or use a home computer.

- Find the number of circular permutations of *n* objects.
- Determine the number of combinations of *n* objects taken *r* at a time.
- Use the method of counting combinations with the Fundamental Principle of Counting to determine how many ways a decision can be made.
- Find the probability that two independent events, A and B, will both occur.
- Find the probability that either event A or event B will occur.
- Determine the probability of an event occurring that depends on the probability of another event occurring.
- Use a random number simulation to approximate the probability of an event.

Portfolio Activity

In this activity, the students can practice solving problems in probability by using counting procedures.

In groups of three, have one student select 5 disks (or pieces of paper, suitably marked) at random from another student, who is holding them up like playing cards. The third student records the selection. Exchange roles and repeat. This should continue long enough for students to see how much time it takes to answer these questions experimentally. Have them approximate answers with the data they have.

To extend the activity, students should work as a group to answer the questions, using the material in this chapter. They should then compare the theoretical-probability answers with their experimental approximations.

Portfolio Activity

Suppose you have a collection of compact discs from the categories rock-and-roll, folk, classical, country, jazz, and blues. If you have a multiple CD changer set on random play, can you determine the probability that a particular type of music or a particular song will be played next?

You will be asked to solve this problem on page 639, and you may wish to include your work in your portfolio.

DISK SELECTION

PROGRAM

SPIRAL 1 DISK ALL DISKS PROGRAM

RANDOM

DISK 1 DISK 2 DISK 3 DISK 4 DISK 5

About the Chapter Project

In the Chapter Project on page 647, students will explore the development of security codes for access to confidential records. They will decide how several different codes can be assigned, depending upon the system chosen. Any such system must have enough codes available so that each student can be assigned one. It is always wise to have extras available.

Exploring Probability

PREPARE

Objectives

- Determine the theoretical probability of an event when given an appropriate sample space.
- Determine the experimental probability of an event when given the results of an experiment.

- Practice Master   **11.1**
- Enrichment Master   **11.1**
- Technology Master   **11.1**
- Lesson Activity Master   **11.1**
- Quiz   **11.1**
- Spanish Resources   **11.1**

Assessing Prior Knowledge

1. What is the ratio of $\frac{1}{2}$ to $\frac{1}{4}$?
   $\left[\frac{2}{1}\right]$

2. How many sides are there in an octagon? [8]

TEACH

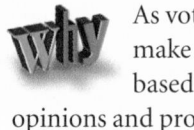

As voters, we have to make many choices based not only on opinions and promises, but also on probabilities. Challenge the students to find these hidden probabilities as they decide how they will vote.

Cooperative Learning

Have the students work in pairs to design a survey question for the class. Depending upon class size, the result could be a 10–15 question survey. Compile the questions, and have each student respond to the survey. Then have each group calculate the percentage who responded one way or another to their group's question.

USA Facts
How statistics shape our lives
Soundbites vs. stories
Where high school and college grads get the most information

H.S. grad — 53% Television, 30% Newspaper, 17% All others

College grad — 27% Television, 31% All others, 42% Newspapers

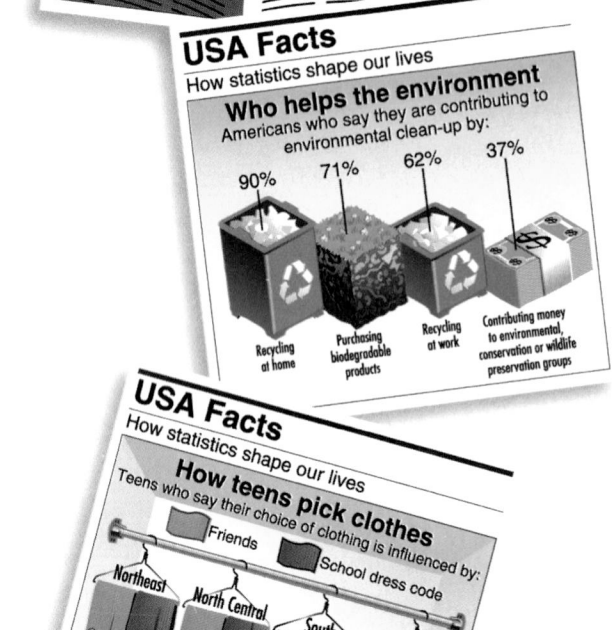

USA Facts
How statistics shape our lives
Who helps the environment
Americans who say they are contributing to environmental clean-up by:

90% Recycling at home, 71% Purchasing biodegradable products, 62% Recycling at work, 37% Contributing money to environmental, conservation or wildlife preservation groups

USA Facts
How statistics shape our lives
How teens pick clothes
Teens who say their choice of clothing is influenced by:

Friends — School dress code
Northeast 49% 41%, North Central 34%, South 46% 47%, West 29%
65% 54%

wly *You can find percentages about almost any topic imaginable, such as fashion, entertainment, environmental and health issues, and current trends. Percentages are only one of the many ways to express probabilities.*

According to a recent Harris Survey, 42% of the college graduates surveyed receive most of their information from newspapers. This means that if you question one of the college graduates surveyed, there is a 42% chance, or *probability*, that he or she gets most of his or her information from newspapers.

Probabilities that are based on the results of research, surveys, or experiments are called *experimental probabilities*. Probabilities that are determined in advance are called *theoretical probabilities*.

Using the results of a survey, many people make assumptions about an entire group, or *population*, surveyed. In the Harris Survey, the population polled is a particular group of college graduates. The results of the survey apply only to the college graduates that participated in the survey.

ALTERNATIVE teaching strategy

Guided Research
Have students bring in newspaper clippings that use or abuse probabilities. Discuss these examples in class. For each example, identify *population*, *sample space*, *outcomes*, *events*, and *probabilities*.

•Exploration 1 Experimental Probability

1 Toss a coin 50 times, and record the number of times the result is "heads."

2 Write the ratio of the number of heads recorded to the number of times the coin was tossed. Use this ratio to estimate the probability that heads will appear on the next toss. Since this probability is based on the results of an experiment, it is an *experimental probability*.

3 Collect the results of all the students in your class. Based on the data collected, how can the experimental probability that the next toss will be heads be determined? What is it?

4 When a coin is tossed 50 times, how many heads would you expect? How does your experimental probability compare with your expectation? Would the experimental results be closer to your expectation if you tossed the coin more times? Explain. ❖

EXPERIMENTAL PROBABILITY

The **experimental probability** for the occurrence of a particular event, E, is

$$P(E) = \frac{\text{number of times the event occurred}}{\text{number of times the experiment was conducted}}.$$

Each time an experiment— such as tossing a coin— is performed, the result is called an **outcome**.

Experiment: Tossing two coins
Sample space:
$S = \{HH, HT, TH, TT\}$

HH   *HT*

TH   *TT*

A set, S, of outcomes of an experiment, where *any* outcome of the experiment corresponds to exactly one element of S, is called a **sample space** of the experiment.

A set that includes elements chosen from a sample space is called an **event**.

One possible event: *HT*

interdisciplinary CONNECTION

Physics In atomic physics, the outcomes from collisions in high-energy accelerators are used to study the structure of the atom. In designing an experiment to test a certain hypothesis, the number of possible outcomes, events, and their probability are used to determine how many collisions, and how long a period of time, an experiment will take.

A sample space is used to find the *theoretical probability* of an event.

THEORETICAL PROBABILITY

When *all* the elements of a sample space are *equally likely*, the **theoretical probability** that an event, E, will occur is given by

$$P(E) = \frac{\text{number of elements in the event}}{\text{number of elements in the sample space}}.$$

Consider two possible sample spaces when a cube with faces numbered 1 through 6 is tossed.

$$S_1 = \{1, 2, 3, 4, 5, 6\} \qquad S_2 = \{\text{even, odd}\}$$

Find $P(even)$.

Using $S_1 = \{1, 2, 3, 4, 5, 6\}$, the number of elements in the event **even** is 3. The number of elements in the sample space is 6.

So $P(even) = \frac{3}{6} = \frac{1}{2}$.

Using $S_2 = \{\text{even, odd}\}$, the number of elements in the event **even** is 1. The number of elements in the sample space is 2.

So $P(even) = \frac{1}{2}$.

Find $P(6)$.

Using $S_1 = \{1, 2, 3, 4, 5, 6\}$, the number of elements in the event **6** is 1. The number of elements in the sample space is 6.

So $P(6) = \frac{1}{6}$.

The event **6** is not an element of the sample space $S_2 = \{\text{even, odd}\}$. Therefore, the number of elements in the event 6 is zero when using this sample space.

So $P(6) = \frac{0}{2} = 0$.

Are the events in sample space $S_3 = \{\text{less than 6, 6}\}$ equally likely? Explain why you cannot find the theoretical probability of $P(even)$ or $P(6)$ using this sample space.

ENRICHMENT Discuss the meaning of the following ideas:

1. an event with a probability of 0,

2. an event with a probability of 1, and

3. all probabilities have values between 0 and 1.

In each case, distinguish between theoretical and experimental probability. Encourage students to come up with examples of each.

INCLUSION strategies **Using Visual Models** Students may understand these concepts better if they convert probabilities into pie charts. Since the probabilities of the events must add up to 1, or the entire pie, this requires students to define events carefully.

 **Exploration 2** *Theoretical Probability*

Suppose two coins, a penny and a dime, are tossed. One possible sample space for this experiment is $S = \{HH, HT, TH, TT\}$.

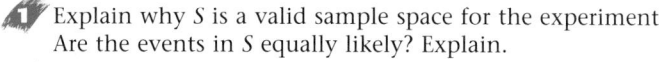 Explain why S is a valid sample space for the experiment. Are the events in S equally likely? Explain.

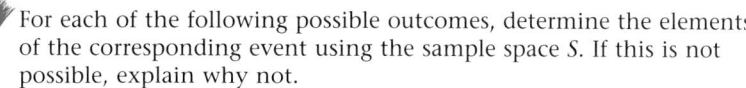 For each of the following possible outcomes, determine the elements of the corresponding event using the sample space S. If this is not possible, explain why not.

a. Tossing 2 heads   **b.** Tossing at most 1 head

c. Tossing at least 1 head   **d.** Tossing a head or a tail

e. Tossing a head and a tail **f.** Not tossing a head

g. Tossing 2 tails   **h.** Tossing 3 heads

3 Are any of the events in Step 2 the same? Explain.

4 Find the theoretical probability for each event in Step 2.

5 Write a sample space for tossing one coin twice. Repeat Steps 2–4. 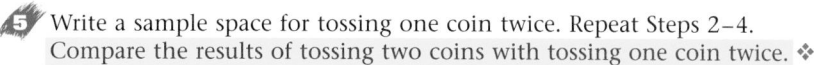 Compare the results of tossing two coins with tossing one coin twice. ❖

 **CRITICAL** *Thinking*

Is the set $S = \{0 \text{ heads}, 1 \text{ head}, 2 \text{ heads}\}$ a sample space for the experiment in Exploration 2? If so, are the elements of S equally likely? Explain.

The probability, P, of an event is a number between 0 and 1 inclusive.

$$0 \le P \le 1$$

The closer P is to 0, the less likely an event is to occur.

What can you say about an event that has a probability close to 1?

In a particular experiment with a valid sample space, the total probabilities of each event in the sample space must add up to 1.

APPLICATION

People in the United States Living Below the Federal Poverty Level

Bureau of the Census, U.S. Department of Commerce, 1990

Age group	Number
Under 16	12,342,000
16–21	3,351,000
22–44	10,170,000
45–54	2,002,000
55–59	963,000
60–64	1,098,000
65 and over	3,658,000
Total people	**33,584,000**

What is the probability that a randomly selected person living in poverty in the United States is under 16 years old?

Out of 33,584,000 total people in the United States living below the federal poverty level, 12,342,000 are under the age of 16.

$$P = \frac{12,342,000}{33,584,000} \approx 0.3675$$

Thus, there is about a 37% chance that a person living in poverty in the United States is under 16 years old.

Is this an experimental or a theoretical probability?

 **RETEACHING** *the lesson*

Using Manipulatives Theoretical probability is a predictor of events happening, based on knowing mathematically the total number of possible outcomes. If a die is tossed, the theory predicts that one of every six tosses will result in a 3. However, if students toss a die 24 times and record the results, there may or may not be four 3's in those outcomes. The experimental probability is the result of several repetitions of the specific experiment.

This introduces a more concrete application of theoretical probability. It is somewhat artificial, because it assumes that the dart player is throwing randomly. Applications often require special conditions to keep the mathematical model from becoming too complicated.

TEACHING *tip*

Point out some of the factors why random tossing cannot be involved in Exploration 3. Those factors might include the skill of the player, the eyesight of the player, and the coordination of the player. Ask students to suggest any other factors that might skew the recorded data from what the mathematics predicts.

Aongoing ASSESSMENT

3. **The probability of hitting a bull's-eye is theoretical.**

•Exploration 3 *Computing Probability With an Area Model*

GEOMETRY *Connection*

A circular dartboard has a diameter of 18 inches. The center circle, or bull's-eye, has a diameter of $\frac{1}{2}$ inch. Assume that a dart is equally likely to land anywhere on the dartboard.

1 Find the area of the dartboard. This value represents the possible outcomes.

2 Find the area of the bull's-eye. This value represents a "success."

3 Use the results of Steps 1 and 2 to find the probability of a successful outcome. Is this a theoretical or an experimental probability?

Now assume a player who is not as precise at throwing darts has an equally likely chance of hitting anywhere on the wall, 20 feet long and 8 feet high, that contains the dartboard.

4 What is the probability of a successful outcome for this player? ❖

APPLICATION

Area models can also be used to solve problems when the number of elements in the sample space is infinite. For example, suppose that Eduardo will check his electronic mail sometime between 1:00 P.M. and 2:00 P.M. Assuming all times are equally likely, what is the probability that Eduardo will check his mail between 1:30 P.M. and 1:40 P.M.?

You can represent the time between 1:00 P.M. and 2:00 P.M. as a rectangular region and the the time between 1:30 P.M. and 1:40 P.M. as the shaded part of this region.

Since the event represents $\frac{1}{6}$ of the total area, the probability that Eduardo will check his mail between 1:30 P.M. and 1:40 P.M. is $\frac{1}{6}$, or about 17%.

	1:00
	1:10
	1:20
	1:30
	1:40
	1:50
	2:00

EXERCISES & PROBLEMS

Communicate

1. Describe what is meant by the term *experiment*. Give two examples of experiments that are not presented in the lesson.

2. Describe what is meant by the phrase *probability of an event*.

3. Suppose E is an event for which $P(E) = \frac{3}{4}$. What does this statement mean? What is the probability that E will *not* occur? Explain.

4. Explain the difference between experimental probability and theoretical probability.

5. A basketball player is said to be an 85% free-throw shooter. Is this experimental or theoretical probability? Explain.

6. Johanna claims that she has a 50% chance of winning the lottery—either she wins or she does not win. Is her reasoning correct? Explain.

7. Why must the outcomes in a sample space be equally likely when determining theoretical probabilities?

Practice & Apply

For each experiment described in Exercises 8–11, give a sample space with equally likely events.

8. The answers to three true-false questions Answers may vary

9. The number of male children and female children in a family with three children Answers may vary

10. Box 1 contains cards individually numbered 1, 2, and 3. Box 2 contains a green marble (*G*) and a red marble (*R*). Two items are drawn, one item from Box 1 and the other item from Box 2. {1G,1R, 2G, 2R, 3G, 3R }

11. Two 6-sided number cubes are tossed. The sum of the top numbers is recorded. Answers may vary.
Sample: {even sum, odd sum}

12. A box contains 25 computer parts, some of which are defective. Two parts are randomly chosen from the box and tested. If the part is defective, *D* is recorded, and if it is not defective, *N* is recorded. Is it possible to have a sample space with equally likely events for this experiment? Explain.

13. Leisure A circular dartboard has a diameter of 19 inches. The center circle, or bull's-eye, has a diameter of $\frac{1}{2}$ inch. Assume that a dart is equally likely to land anywhere on the dartboard. Find the probability that the dart will miss the bull's-eye but land on the dart board. 99.93%

The elements in sets given for Exercises 8–12 may be listed in any order.

12. No, the elements in sample space $S = \{DD,DN,ND,NN\}$ are not necessarily equally likely.

8. Sample answer: $S = \{$FFF, FFT, FTF, TFF, TTF, TFT, FTT, TTT$\}$

9. Sample answer: $S = \{$MMM, MMF, MFM, FMM, FFM, FMF, MFF, FFF$\}$

ASSESS

Selected Answers

Odd-numbered Exercises 9–43

Assessment Guide

Core 1–44

Core-Plus 1–45

Technology

Graphics calculators are not needed for this lesson. The probability of various events in one sample, as in Exercises 23–26, can be stored in memory. Each new probability is then found by entering the number of successes, and then dividing by the total.

Error Analysis

A common error when determining the probability of an event is to neglect part of the total number of elements in the sample space. For example, in tossing two coins, one must include both HT and TH as well as TT and HH.

Portfolio Assessment

Have students write a short essay defining the following terms: *outcome, experiment, event, experimental probability, theoretical probability,* and *sample space.*

Opinion Polls Suppose a gun-control survey that included 200 men and 200 women had the results shown. Find the probability for each of the following events.

	For controls	Against controls	Total
Men	80	120	200
Women	106	94	200
Total	186	214	400

14. A person chosen randomly is for gun controls. 46.5%

15. A person chosen randomly is a man. 50%

16. A person chosen randomly is for gun controls and is a woman. 26.5%

17. A person chosen randomly is for gun controls or is a man. 76.5%

18. A woman chosen randomly is for gun controls. 53%

Education The state universities in a particular state have summarized their students' verbal SAT scores in a table. Find the probability that a randomly selected university student in this state has the given SAT score.

SAT score	Number
200–299	4150
300–399	17,562
400–499	25,116
500–599	8020
600–699	5235
700–799	814
Total students	60,897

19. 300–399 28.8%

20. At most 399 35.7%

21. At least 400 64.3%

22. 200–299 or 600–699 15.4%

Physiology The table displays the fingerprint ridge counts for 960 randomly selected males. Find the probability that a randomly selected male has the following fingerprint ridge count.

Ridge count	Number
0–19	15
20–39	19
40–59	38
60–79	51
80–99	88
100–119	116
120–139	107
140–159	128
160–179	152
180–199	104
200–219	77
220–239	45
240–259	12
260–279	7
280–299	1
Total males	960

23. At most 179 74.4%

24. At least 180 25.6%

25. At least 120 65.9%

26. From 120 to 179 or from 220 to 259 46.3%

Human fingerprints have many characteristics including ridge count.

NAME _____ CLASS _____ DATE _____

Practice & Apply
11.1 Exploring Probability

For each experiment, give a sample space with equally likely events and a sample space containing events that are not equally likely.

1. The number of heads and tails when a dime and a nickel are tossed. (Represent heads with H and tails with T.)

2. The number of kinds of sandwiches that can be made with the ingredients shown in the chart.

Bread	Meat	Toppings
Pumpernickel (P)	Ham (H)	Cheese (C)
White (W)	Tuna (T)	Lettuce (L)
		Sprouts (S)

3. The larger square shown has a side of 20 inches. The smaller square has a side of 5 inches. A point is chosen at random. Assume that the point is equally likely to be anywhere within the larger square. Find the probability that the point is in the shaded region.

For a group of people in their 30s who were once daily smokers or still are, the table shows the age when they first started smoking.

Age When First Smoked	
Age	Percent
0–11	16.0
12–13	21.0
14–15	26.0
16–17	19.5
18–19	9.0
20–24	7.0
25–29	1.0
30–39	0.5

Find the probability that a randomly selected person who participated in the study started smoking at the following ages.

4. After age 12 _____

5. Over age 20 _____

6. Under age 20 _____

7. Between ages 12 and 17 _____

HRW Advanced Algebra   Practice & Apply   **1**

Sports A baseball pitcher is practicing a pitch by throwing at the yellow target shown. The black square has 4-inch sides. The white circle has a 7-inch radius. Assuming there is an equally likely chance of hitting any part of the yellow target (but not missing it), find the probability that a pitch will hit the following part of the target.

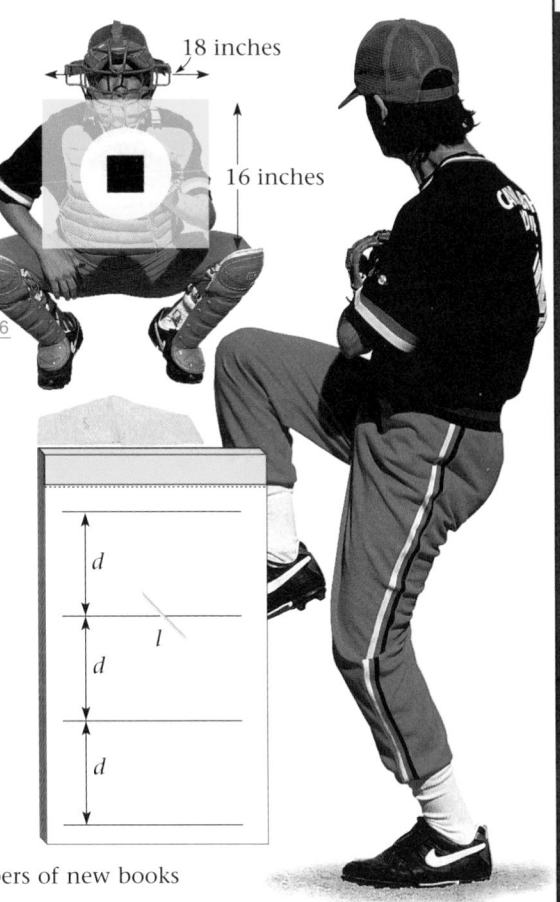

18 inches

16 inches

27. The black square $\frac{1}{18}$

28. Anywhere in the circle $\frac{49\pi}{288}$

29. In the circle but not in the black square $\frac{49\pi - 16}{288}$

30. Not in the black square $\frac{17}{18}$

31. Nowhere in the circle $1 - \frac{49\pi}{288}$

Buffon's Problem Get a toothpick and measure its length. Call its length l. Draw parallel lines at distance d apart, where $l < d$.

32. By randomly dropping the toothpick on the paper 20 times, find the probability that the toothpick will intersect one of the lines. *Answers may vary*

33. This problem was first considered and solved by the French naturalist Comte de Buffon. His solution is the value $\frac{2l}{\pi d}$. How does your approximation compare with this value? *Answers may vary*

34. If you made more trials, do you think your value would approach Buffon's value? Explain. *Answers may vary*

d

l

d

d

Publishing The following table shows the numbers of new books published in a year.

Subject	Number	Subjects	Number
Agriculture	468	Mechanics	1684
Art	1119	Medicine	2710
Biography	1890	Music	275
Business	1298	Philosophy/Psychology	1611
Education	973	Poetry and Drama	821
Fiction	4199	Religion	2059
History	2107	Science/Math	2427
Home Economics	721	Sociology/Economics	5508
Juvenile	4555	Sports/Recreation	893
Language	521	Technology	2089
Law	967	Travel	425
Literature	1903	**Total books**	41,223

Find the probability that a randomly selected book published in this year is on the following subject.

35. Literature 4.6%

36. Biography 4.6%

37. Art or travel 3.7%

34. Yes, the greater the number of trials, the greater is the probability that the value approaches Buffon's value, which is the theoretical probability.

38. Biology The Punnett square at the lower right shows the probabilities of color genes for the offspring of two purple pea plants, each parent with the color genes W and w. If an offspring plant contains at least one W, it will be purple. Only if the plant contains both ww genes will it be white. Assuming all options are equally likely, what is the probability that these two pea plants will produce a white pea plant?

$\frac{1}{4}$, or 25%

A purple pea plant

	W	w
W	WW	Ww
w	wW	ww

Look Back

39. Consider the function $I(t) = \sqrt{3.8 + t^{1.8}}$. $I(t)$ represents the number of independent: t employees of a company that will become ill, and t represents the dependent: I number of days after the beginning of a flu epidemic. Identify the independent and dependent variables.   **[Lesson 2.3]**

Let $f(x) = 3x + 10$.

40. Find the inverse of f.   **[Lesson 3.2]** $f^{-1}(x) = \frac{x-10}{3}$

41. Show that your answer to Exercise 40 is the inverse by using composition of functions.   **[Lesson 3.3]**

42. Let $f(x) = x + 7$ and $g(x) = 3x^2 - 1$. Write an expression for $(f \circ g)(x)$.   **[Lesson 3.3]** $3x^2 + 6$

43. Use matrix algebra to solve this system of equations. $\begin{cases} 5x - 4y = 12 \\ 2x + 2y = -15 \end{cases}$ $(-2, -5.5)$ **[Lesson 4.6]**

44. Suppose the real number a is a zero of even multiplicity of a polynomial function f. Describe the behavior of the graph of f as values of x approach a.   **[Lesson 6.4]** The graph remains on the same side of the x-axis

Look Beyond

45. A game consists of tossing a 4-sided number cube shaped like a tetrahedron. Each side is a different color—purple, pink, green, or yellow. Before the toss, you must guess which color will be on bottom after the toss. A correct guess is worth 24 points. An incorrect guess is worth 0 points. It costs 1 point to play the game. If you play the game 20 times, how many points do you think you have the best chance of ending with? 100 points

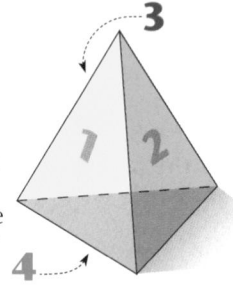

Use Transparency ▶ 42

Look Beyond

This exercise looks at mathematical ways of counting that are simpler than listing all the possible outcomes and counting the successes by hand.

41. $(f \circ f^{-1})(x) = 3\left(\frac{x-10}{3}\right) + 10 = x$ and

$(f^{-1} \circ f)(x) = \frac{(3x+10)-10}{3} = x$

LESSON 11.2 Counting and Permutations

why *In how many different ways can a film reviewer select 5 of 12 popular films to review? Answering this question involves new counting techniques.*

PREPARE

Objectives

- Use the Fundamental Principle of Counting to determine how many ways a decision can be made.
- Determine the number of permutations of n distinct objects taken r at a time.

RESOURCES

- Practice Master   **11.2**
- Enrichment Master   **11.2**
- Technology Master   **11.2**
- Lesson Activity Master   **11.2**
- Quiz   **11.2**
- Spanish Resources   **11.2**

Assessing Prior Knowledge

1. Find $3^2 \times 2^3$. [**72**]

2. Find 5! [**120**]

TEACH

why Theoretical probabilities often require the use of counting techniques.

EXAMPLE 1

A pet store has 2 brands of fish tanks and 3 brands of filter systems for fish tanks. How many different filtered fish tanks are possible?

Solution

There are two decisions to make.

Decision 1: Should I select tank 1 or tank 2? There are 2 ways to make this decision.

Decision 2: Should I select filter system A, filter system B, or filter system C? There are 3 ways to make this decision.

Alternate Example 1

A clothing store has three styles of pants and five styles of shirts in your color and size. How many different outfits are possible? [**15**]

ALTERNATIVE teaching strategy

Hands-On Strategies

Pick some game that is rich in counting and permutations, such as popular card games. Illustrate each concept with specific examples from these card games, and then generalize to the mathematical formulas.

Alternate Example 2

How many different ways can 5 people be seated in 3 chairs with 2 people left standing? [**60**]

Cooperative Learning

Have the students form groups of 3. Provide each group with a supply of 3 different items, such as paper clips, pennies, and small pieces of paper. Each group will need 27 of each item. Provide a large work surface for each group and have them arrange the items in all possible ways, like clip-clip-clip, clip-clip-penny, and so on. Then have them count all the ways to arrange these. There will be 27.

To count the number of ways the two decisions can be made, make a **tree diagram**, a visual display of all the possible choices.

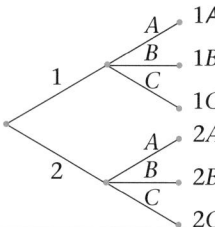

Counting the points on the right-end branches, you can see that there are 6 ways of making the two decisions. Notice that you can also find the number of possible ways of making all the decisions by multiplying 3 and 2. Thus, there are 6 different filtered fish tanks possible. ❖

This solution suggests the Fundamental Principle of Counting.

Fundamental Principle of Counting

When a sequence of decisions is to be made, the number of possible ways to make *all* the decisions can be found by multiplying together the number of ways to make each decision.

THE FUNDAMENTAL PRINCIPLE OF COUNTING

If there are m ways of making Decision 1 and n ways of making Decision 2, then there are $m \times n$ ways to make both decisions.

Sometimes you need to use the Fundamental Principle of Counting several times.

EXAMPLE 2

How many different ways can 4 people be seated in a row?

Solution►

Model the row of 4 seats. Think about how many ways you can fill the seats starting at the left.

Any one of the 4 people	Any one of the 3 remaining	Any one of the 2 remaining	The 1 person remaining

There are 4 ways to choose the first person. Once this decision is made, there are 3 ways to choose the second person. By the Fundamental Principle of Counting, there are $4 \times 3 = 12$ ways of choosing the first 2 people.

Continuing this process, there are $4 \cdot 3 \cdot 2 \cdot 1 = 24$ distinct ways of seating 4 people in a row. ❖

Try This In how many different orders can you play 5 CDs?

Notice the pattern in the multiplication $4 \cdot 3 \cdot 2 \cdot 1$ from Example 2.

$$4 \cdot 3 \cdot 2 \cdot 1 = 4 \cdot (4-1) \cdot (4-2) \cdot (4-3)$$

This pattern of multiplication beginning with 4 can be denoted 4!, read *4 factorial*.

In general, *n factorial* is $n! = n \cdot (n-1) \cdot (n-2) \cdot \ldots \cdot 3 \cdot 2 \cdot 1$ for any positive integer n.

Evaluate 5!   Evaluate 6!

Permutations

In Example 2 you found that there are 24 distinct ways to arrange 4 people in a row of 4 seats. In this arrangement, the specific *order* is important. When the order of an arrangement is important, the arrangement is called a *permutation*.

PERMUTATION

A **permutation** of a set of distinct objects is an arrangement of the objects into a specific order.

In Example 2 there are 4 people to arrange in 4 seats. This can be considered *a permutation of 4 objects taken 4 at a time*, symbolized $P(4, 4)$ or $_4P_4$.

$$_4P_4 = 4! = 4 \cdot 3 \cdot 2 \cdot 1 = 24$$

In general, for any positive integer n, a permutation of n objects taken n at a time is

$$_nP_n = n! = n \cdot (n-1) \cdot (n-2) \cdot \ldots \cdot 3 \cdot 2 \cdot 1.$$

Aongoing**SSESSMENT**

Try This

120

Aongoing**SSESSMENT**

120; 720

The number of permutations of n distinct objects taken r at a time is denoted by $P(n, r)$ or $_nP_r$. Here, r has to be less than or equal to n. Why?

Suppose 25 people enter a raffle in which there will be a first, a second, and a third prize. The number of ways this can happen is the number of permutations of 25 objects (people) chosen 3 at a time (3 prizes), or $_{25}P_3$.

By the Fundamental Principle of Counting, there are $25 \cdot 24 \cdot 23$, or 13,800, ways the prizes can be awarded. The product $25 \cdot 24 \cdot 23$ is the same as $\frac{25!}{22!}$.

$$
\begin{aligned}
25 \cdot 24 \cdot 23 &= 25 \cdot 24 \cdot 23 \cdot 1 \\
&= 25 \cdot 24 \cdot 23 \cdot \frac{22!}{22!} \\
&= \frac{25 \cdot 24 \cdot 23 \cdot 22!}{22!} \\
&= \frac{25!}{22!}
\end{aligned}
$$

Since 22! is $(25 - 3)!$, $_{25}P_3 = \frac{25!}{(25 - 3)!}$.

The formula for the number of permutations of n distinct objects taken r at a time, denoted $P(n, r)$ or $_nP_r$, can be generalized by replacing 25 with n and 3 with r.

$$_nP_r = \frac{n!}{(n - r)!}$$

EXAMPLE 3

Graphics Calculator

List all the permutations of the letters $ABCD$, taking them 2 at a time. Then verify your count using the permutation formula and a calculator.

Solution►

The permutations are organized in a table.

AB	BA	CA	DA
AC	BC	CB	DB
AD	BD	CD	DC

Since order matters, the arrangements AB and BA are counted as two distinct elements.

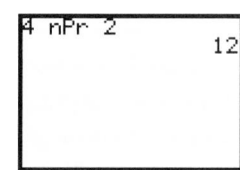

Using the permutation formula,
$_4P_2 = \frac{4!}{(4-2)!} = \frac{4!}{2!} = \frac{4 \cdot 3 \cdot 2 \cdot 1}{2 \cdot 1} = 4 \cdot 3 = 12$

Thus, there are 12 permutations of the letters $ABCD$, taking them 2 at a time. ❖

Try This Use the permutation formula to find the number of ways a film reviewer can select 5 films to review from a list of 12 films.

CRITICAL Thinking

Why do you think 0! is defined to be equal to 1?

EXERCISES & PROBLEMS

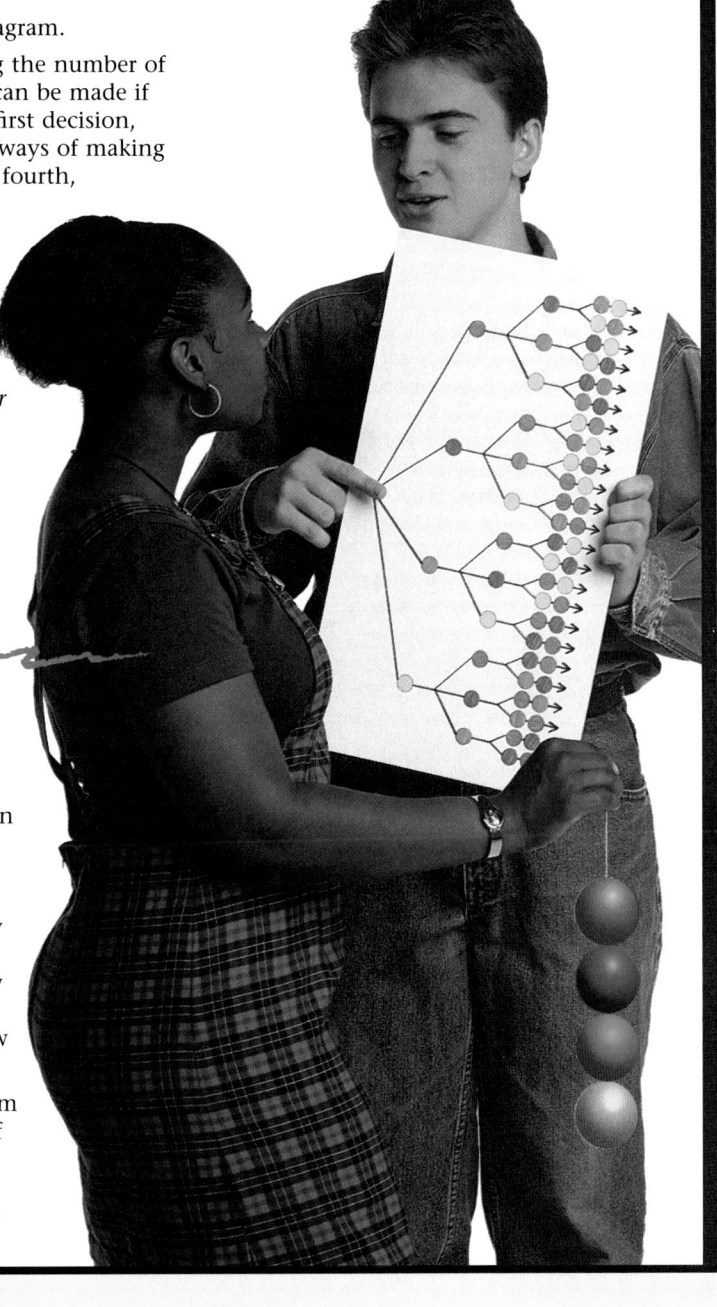

Communicate

1. Explain the purpose of a tree diagram.

2. Specify a procedure for counting the number of ways a sequence of 5 decisions can be made if there are a ways of making the first decision, b ways of making the second, c ways of making the third, d ways of making the fourth, and e ways of making the fifth.

3. Describe the meaning of the word *permutation*.

4. Give four permutations of the letters *ABCDE*, taking 3 letters at a time.

5. Explain why the integers n and r in the permutation formula, $_nP_r = \dfrac{n!}{(n-r)!}$, cannot be interchanged to $_nP_r = \dfrac{r!}{(r-n)!}$.

Practice & Apply

6. **Testing** An educational testing service plans to include 3 new questions on one of its standardized exams. One literature question will be chosen from a group of 20 possibilities, one music question will be chosen from a group of 30 possibilities, and one astronomy question will be chosen from a group of 15 possibilities. In how many different ways can the testing service include the 3 new questions? 9000

7. **Testing** A multiple-choice exam consists of 15 questions, each of which has 4 possible answers. How many different ways can all 15 questions on the exam be answered? $4^{15} = 1{,}073{,}741{,}824$

ASSESS

Selected Answers
Odd-Numbered Exercises 7–31

Assignment Guide
Core 1–32

Core Plus 1–32

Technology
Make sure students learn how their calculator computes permutations. Pay particular attention to the order of numerical entries and function keystrokes. Have students practice comparing their answers with direct calculations using the factorial key.

Error Analysis
Students will have their biggest problems when translating a word problem into the calculation needed. They may wish to use the simple $n!$ for the answers, but overlook the fact that in permutations they must account for the significance of the number taken at a time. This means dividing by $(n-r)!$.

TEACHING *tip*

The number of possible license plates increases to 962,391,456 when repetition of letters and numbers is allowed.

8. **Sports**   In a track meet, 7 runners compete for first, second, and third places. How many different outcomes are possible if there are no ties? $\frac{7!}{4!} = 210$

9. Compute the value of $_8P_3$, and explain what that value represents.
$\frac{8!}{5!} = 336$; there are 336 ways to arrange 3 objects from a set of 8.

Find the number of permutations of the letters PQRSTUV under the following circumstances.

10. Taking 5 letters at a time 2520   **11.** Taking 2 letters at a time 42

12. Taking 1 letter at a time 7   **13.** Using all the letters at one time 5040

14. Find the number of permutations of the word *NEPAL*, taking 3 letters at a time. 60

15. How many different 3-digit numbers are possible if no number can start with 0? 900

16. How many different 4-digit numbers are possible if no number can start with 0? 9000

17. In how many different ways can 5 seniors, 3 juniors, 4 sophomores, and 3 freshmen be seated in a row if a senior must be seated at each end? 124,540,416,000

18. A car holds 3 people in the front seat and 3 people in the back seat. In how many ways can 6 people be seated in the car? In how many ways can 6 people be seated if 2 people *must* sit together? $6! = 720$; $(8 \cdot 4!) = 192$

Logic Logicians make truth tables in order to study the validity of arguments. Arguments are composed of individual statements, each of which can be either true or false. A table for an argument consisting of statements p and q is shown.

Statement p	Statement q	Argument consisting of p and q
T	T	T T
T	F	T F
F	T	F T
F	F	F F

How do you think the number of possible license plates is affected when letters and numbers are repeated?

19. Make a table, and find the number of different arguments consisting of the three statements p, q, and r. 8

20. Make a table, and find the number of different arguments consisting of the four statements p, q, r, and s. 16

21. What pattern can you see in the number of different arguments consisting of n statements, where n is any positive integer? 2^n

22. **Automobile Registration**
How many different license plates can be made if each plate consists of 2 letters followed by 2 digits (1 through 9) followed by 3 letters? Assume that letters and numbers *cannot* be repeated.
568,339,200

23. Earth Science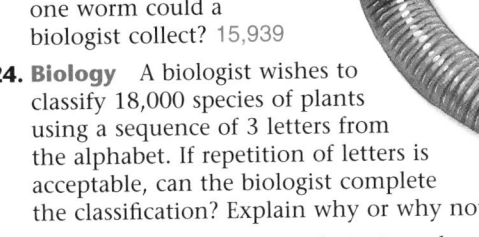
In a particular region of the desert, there are 11 species of flies, 23 species of moths, 7 species of beetles, and 9 species of worms. How many different samples of one fly, one moth, one beetle, and one worm could a biologist collect? 15,939

24. Biology A biologist wishes to classify 18,000 species of plants using a sequence of 3 letters from the alphabet. If repetition of letters is acceptable, can the biologist complete the classification? Explain why or why not. No

25. Zip Codes How many 5-digit zip codes are possible? Assume 00000 is not a valid zip code. 99,999

26. Social Security A social security number has 9 digits, and numbers may be repeated. How many social security numbers are possible? Assume a social security number consisting of 9 zeros is not valid. 999,999,999

 *Look Back*

27. Identify the system of linear equations as independent, dependent, or inconsistent. **[Lesson 4.3]**
$$\begin{cases} 2x - 7y = 15 \\ 5x - 4y = 10 \end{cases}$$ Independent

28. The revenue earned from selling x number of items is given by the revenue function $R(x) = 6x$. The cost of producing x number of items is $C(x) = -0.1x^2 + 5x + 40$. Find the number(s) of items that will produce a profit (when revenue exceeds cost). **[Lesson 5.9]** x is 16 or more

29. Write $\log_5 625 = 4$ as an exponential equation. **[Lesson 7.3]** $5^4 = 625$

30. Find two angles between 0° and 360° for the inverse relation $\tan^{-1}\left(\frac{\sqrt{3}}{3}\right)$. **[Lesson 8.3]** 30°, 210°

31. Solve the nonlinear system. **[Lesson 10.5]** $(-2, 5), (5, 26)$ $\begin{cases} y = x^2 + 1 \\ 3x - y = -11 \end{cases}$

Look Beyond

32. List all the ways that 2 of the letters ABC can be chosen if the order in which the letters are chosen is *not* important. *AB, AC, BC*

24. No; the total number of 3-letter combinations is 26^3, or 17,576. This is not enough to cover 18,000 species.

Look Beyond

Permutations require that order matters. What if some of the objects in the selection pool are identical (not distinct)? This exercise looks at this situation.

- Find the number of distinct permutations of n objects of which r_1 objects are alike and r_2 objects are alike.
- Find the number of circular permutations of n objects.

RESOURCES

• Practice Master	**11.3**
• Enrichment Master	**11.3**
• Technology Master	**11.3**
• Lesson Activity Master	**11.3**
• Quiz	**11.3**
• Spanish Resources	**11.3**

Assessing Prior Knowledge

1. Evaluate $P(4, 3)$. [**24**]

2. Evaluate 6!. [**720**]

TEACH

Sometimes the sets of elements that we try to arrange have identical elements, as well as different ones. The number of distinct permutations must take the identical elements into account.

Cooperative Learning

Have the students form groups of three. Have each group consider numbers 1, 2, and 3 and write all the possible 3-digit numbers, using each digit only once. The result will be 123, 132, 213, 231, 312, and 321, or 6 possibilities. Then give them the digits 1, 2, and 2. The only possibilities here are 122, 212, and 221. Be sure to relate this to $\frac{n!}{r!}$ seen on page 613.

LESSON 11.3 Special Permutations

why *Permutations are certain arrangements of distinct objects. Sometimes an arrangement contains objects that are not actually distinct.*

The letters *OHIO* can be considered unique if the *O*s can be distinguished from each other. Using subscripts, the letters O_1HIO_2 can be arranged in 4!, or 24, different ways.

Column 1	Column 2	Column 3	Column 4
O_1 H I O_2	H O_1 I O_2	O_2 H I O_1	H O_2 I O_1
O_1 H O_2 I	H O_1 O_2 I	O_2 H O_1 I	H O_2 O_1 I
O_1 I H O_2	H I O_1 O_2	O_2 I H O_1	H I O_2 O_1
O_1 I O_2 H	I O_1 H O_2	O_2 I O_1 H	I O_2 H O_1
O_1 O_2 H I	I O_1 O_2 H	O_2 O_1 H I	I O_2 O_1 H
O_1 O_2 I H	I H O_1 O_2	O_2 O_1 I H	I H O_2 O_1

When the *O*s are considered indistinguishable, O_1HIO_2 and O_2HIO_1 both represent the arrangement *OHIO*. In fact, if the *O*s are not distinct, every arrangement in Columns 3 and 4 of the table above is a duplicate of an arrangement in Columns 1 and 2.

There are only 12 ways to arrange the letters *OHIO* if the *O*s are not distinct.

$$\frac{24}{2} = \frac{4!}{2!} = 12$$

ALTERNATIVE teaching strategy

Using Manipulatives

Until now, all of the counting has involved the use of distinct items. Now, some of the items will be repeated. For instance, the first blue chip in a list is not different from a second blue chip, and so on. Show this difference with red, blue, and green chips.

There are *two* indistinguishable objects in *OHIO*, and the permutation formula is divided by 2!. This leads to a formula for distinct permutations of indistinguishable objects.

DISTINCT PERMUTATIONS OF INDISTINGUISHABLE OBJECTS

The number of distinct permutations of n objects, of which r objects are alike, is

$$\frac{n!}{r!}, \text{ where } r \leq n.$$

The number of distinct permutations of n objects of which r_1 are alike, r_2 are alike, r_3 are alike, ... , and r_k are alike is

$$\frac{n!}{r_1!\,r_2!\,r_3!\ldots r_k!}.$$

EXAMPLE 1

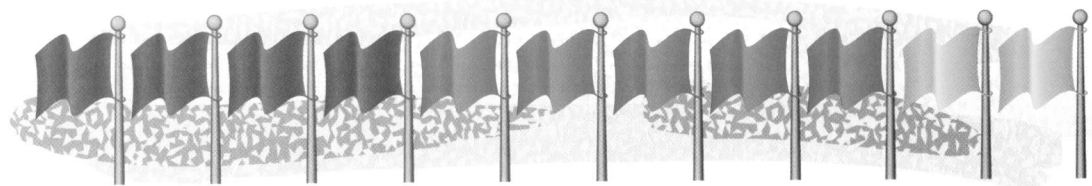

How many distinct ways can these flags be displayed in a row?

Solution▶

There are a total of 11 flags. If there is no repetition of objects, the number of permutations is 11!. However, the **4 red** flags, the **5 green** flags, and the **2 yellow** flags are indistinguishable. Thus, the number of *distinct* permutations is

$$\frac{11!}{4!\,5!\,2!} = \frac{11 \cdot 10 \cdot 9 \cdot 8 \cdot 7 \cdot 6 \cdot 5 \cdot 4 \cdot 3 \cdot 2 \cdot 1}{(4 \cdot 3 \cdot 2 \cdot 1) \cdot (5 \cdot 4 \cdot 3 \cdot 2 \cdot 1) \cdot (2 \cdot 1)} = 6930$$

Thus, there are 6930 distinct ways to display the flags. ❖

Try This   How many distinct permutations can be made from the letters *PEPPER*?

In all the examples of permutations, the arrangements have been made in rows. What happens when an arrangement is circular?

To find the number of ways that the letters *ABCD* can be arranged in a circle, draw them on a circle, and compare them to arrangements in a row. One possibility is shown.

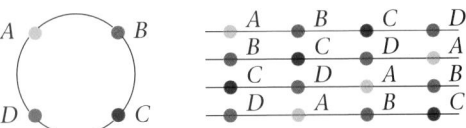

Think about cutting the circle open and straightening it into a line. Notice that the permutations *ABCD*, *BCDA*, *CDAB*, and *DABC* are *not* distinct when they appear on the circle but *are* distinct when they appear in a row.

For this 1 circular arrangement of 4 objects, there are 4 row arrangements.

Alternate Example 1

In how many distinct ways can the flags in Example 1 be displayed if you use only the 9 red and green flags? [126]

Aongoing Assessment

Try This

$\frac{6!}{3!\,2!}$ or 60

Use Transparency ▶ 43

ENRICHMENT Begin with a square and draw its two diagonals. See whether the students can discover how many ways the numerals 1, 2, 3, 4, and 5 can be set, one at each vertex and one at the intersection of the diagonals.

INCLUSION strategies **Technology** Have students consider how they might write a short computer program to calculate both distinct permutations of indistinguishable objects and circular permutations.

There are 4 linear permutations for each circular permutation.

Alternate Example 2

How many ways can all the hearts in a deck of cards be arranged in a circle? [479,001,600]

To find the number of circular arrangements of the letters *ABCD*, divide the number of row arrangements by 4. Why?

$$\frac{4!}{4} = \frac{4 \cdot 3 \cdot 2 \cdot 1}{4} = 3 \cdot 2 \cdot 1 = 3!, \text{ or } 6$$

CIRCULAR PERMUTATIONS

The number of circular permutations of *n* objects is $(n-1)!$.

EXAMPLE 2

In how many ways can 7 types of appetizers be arranged on a circular tray?

Solution

Use the formula for circular permutations, where *n* is 7.

$$(7-1)! = 6!$$
$$= 720$$

Thus, there are 720 distinct ways that 7 types of appetizers can be arranged on a circular tray. ❖

*CRITICAL
Thinking*
$\frac{(n-1)!}{2}$

CRITICAL
Thinking

Suppose several items are placed on a key ring and the ring can be turned so that each view, front and back, represents a distinct arrangement. What is the number of permutations of *n* objects on this ring?

RETEACHING
the
lesson

Hands-On Strategies
Take sets of 5 coins (3 pennies and 2 nickels) and arrange them in all possible row patterns (a roll of each coin should be enough.) Have the students keep careful count of the different rows. Relate this number to $\frac{n!}{r_1! \cdot r_2!}$.

EXERCISES & PROBLEMS

Communicate

1. Explain what is meant by a permutation of a set of objects.

2. Is it possible to use the formula $\frac{n!}{r_1! \cdot r_2! \cdot \,\cdots\, \cdot r_k!}$ to count the number of permutations of n distinguishable objects taken all at once? Explain.

3. Do you think the formula for counting the number of circular permutations of n distinct objects is valid for counting the number of permutations of n distinct objects placed around a square? Explain.

4. Describe the relationship between the variables r and n in the permutation formula for indistinguishable objects, $\frac{n!}{r_1! \cdot r_2! \cdot \,\cdots\, \cdot r_k!}$.

Practice & Apply

5. If all the letters are used, how many different permutations are possible using the letters of the word *TRIANGLE*? $8! = 40{,}320$

6. If all the letters are used, how many different permutations are possible using the letters of the word *MISSISSIPPI*? $\frac{11!}{4!\,4!\,2!} = 34{,}650$

7. If all the letters are used, how many different permutations are possible using the letters of the word *NOON*? $\frac{4!}{2!\,2!} = 6$

8. In how many ways can the numbers on a clock face be arranged, if the positions on the clock (top, bottom, and so on) do not matter?
$(12-1)! = 11! = 39{,}916{,}800$

9. In how many distinct ways can the letters of the word *ARRANGEMENT* be arranged? 2,494,800

10. **Postal Codes** In Canada, a postal code is an arrangement of 6 alternating letters (except the letter O) and numbers, beginning with a letter. How many different zip codes are possible in Canada? 15,625,000

Selected Answers
Odd-Numbered Exercises 5–29

Assignment Guide
Core 1–30

Core Plus 1–31

Technology
Graphics calculators are needed for Exercise 26. For students with some experience in programming a calculator, have the students program a function that finds circular permutations when n and r are input.

Error Analysis
The most common errors will occur when students do not carefully count the number of repeated items in a set. Instead, students might attempt to use one of the simpler formulas from earlier sections.

Portfolio Assessment

Have the students create a chart that describes various counting methods presented in this lesson. These include distinct permutations of indistinguishable objects, and circular permutations.

11. In how many ways can 3 men and 3 women be seated in a circle? $(6 - 1)! = 5! = 120$

12. In how many ways can 3 men and 3 women be seated in a row? $6! = 720$

13. In how many ways can 3 men and 3 women be seated in a circle if the men and women must be seated alternately? 24

14. In how many ways can 3 men and 3 women be seated in a row if the men and women must be seated alternately? 72

15. Find a formula for the total number of different arrangements of n keys that are possible on a key ring. $\frac{(n-1)!}{2}$

16. The English alphabet consists of 26 letters. How many different 26-letter words are possible if letters can be repeated? if letters cannot be repeated? $26^{26} \approx 6.2 \times 10^{36}$ $26! \approx 4.0 \times 10^{26}$

17. Linguistics   A palindrome is a word that is spelled the same backward as it is forward. The longest English palindrome is the word *REDIVIDER*. How many 3-letter palindromes are possible using the English alphabet? (The palindromes do not have to make sense.) 676

18. By computing, find the relationship between the following pairs.

$P(\overset{6}{3}, 3)$ and $P(\overset{6}{3}, 2)$ $P(\overset{24}{4}, 4)$ and $P(\overset{24}{4}, 3)$
$P(\overset{120}{5}, 5)$ and $P(\overset{120}{5}, 4)$ $P(\overset{720}{6}, 6)$ and $P(\overset{720}{6}, 5)$

Generalize this relationship using n and r.

19. Cultural Connection: The Americas
During World War II, 29 members of the Navaho Nation, known as the Navaho Code Talkers, developed a code used by the United States armed forces. The code used the Navaho language, which has 36 distinct sounds, 4 of which are vowel sounds. Each word has to have at least 1 vowel sound. How many 2-sound words are possible in the Navaho language? How many 3-sound words are possible in the Navaho language?

The Code Talkers began eight weeks of training in 1942 at Camp Elliot, north of San Diego. Even Marine experts couldn't break their code.

20. Manufacturing   In its assembly process, a manufacturer uses 4 stages. The first stage has 3 assembly lines, the second stage has 2 assembly lines, the third stage has 6 assembly lines, and the last stage has only 1 assembly line. How many distinct paths are there through all of the assembly stages? 36

21. FBI   Each year the FBI receives about 16,000 job applications. If the Bureau hires about 600 of the applicants, what is the probability that an applicant will be accepted for employment by the FBI? 3.75%

22. Coding   The symbols ◆, ✳, ❤, ✳, ✚, and ✕ are to be used to develop a code. How many distinct arrangements can be formed if the symbols must be used 2 at a time with no repetition? 2 at a time with repetition allowed? 3 at a time with and without repetition? 4 at a time with and without repetition? 30; 36; 216; 120; 1296; 360

18. $P(n, n) = P(n, n - 1)$.

19. 272; 13,888

NAME _____ CLASS _____ DATE _____

Practice & Apply
11.3 Special Permutations

1. If all the letters are used, how many different permutations are there of the letters of the word *ELEVEN*?

2. How many different signals, each consisting of 5 flags hung in a row, can be formed using 2 red flags, 1 blue flag, and 2 green flags?

3. The points *A*, *B*, *C*, and *D* lie on circle O. Using 3 points as vertices of a triangle, how many different inscribed triangles can be constructed?

4. How many ways can 12 wagons form a train in the shape of a circle to protect a village?

5. If there are 40,320 different seating arrangements around a round table, how many people are there?

6. In how many distinct ways can the digits of the number 11,432 be arranged?

7. In how many distinct ways can 10 books be arranged on a shelf, if 5 books are the same?

8. A rock display contains 36 rocks of which 8 are granite, 4 are limestone, 4 are slate, and 3 are marble. How many ways can the rocks be displayed in a row if the same types of rocks are not distinct?

9. A coin collection contains 25 coins. If 10 coin sets contain 2 identical coins, how many distinct ways can the 25 coins be arranged in a row?

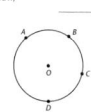

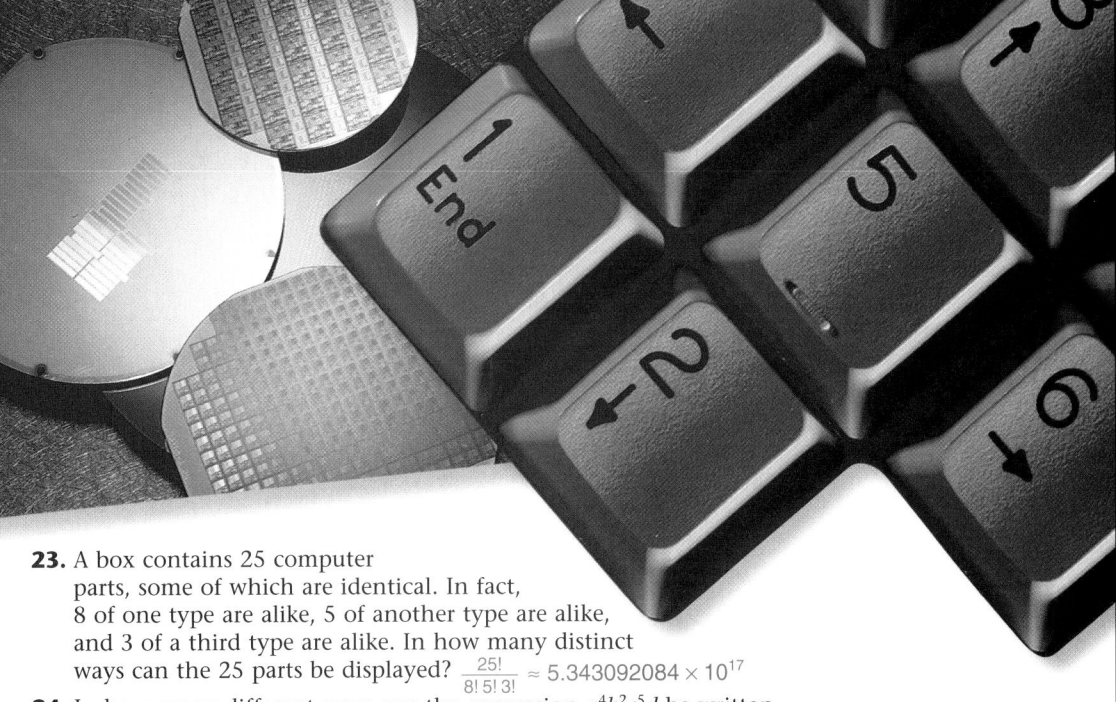

23. A box contains 25 computer parts, some of which are identical. In fact, 8 of one type are alike, 5 of another type are alike, and 3 of a third type are alike. In how many distinct ways can the 25 parts be displayed? $\frac{25!}{8!\,5!\,3!} \approx 5.343092084 \times 10^{17}$

24. In how many different ways can the expression $a^4b^2c^5d$ be written without exponents? 83,160

Look Back

25. Write an example of a polynomial function in factored form that is of degree 5 and has 3 crossing-point zeros and 1 turning-point zero. **[Lesson 6.4]** Example: $P(x) = (x-2)^2(x+1)(x-9)(x+7)$

26 A rectangular piece of cardboard measures 12 inches by 16 inches. Find the maximum volume, to the nearest tenth, of a box created by cutting out square corners of the same size and folding up the sides. **[Lesson 6.5]** 194.1 in.³

Express each of the following as a sum of logarithms. [Lesson 7.4]

27. $\log_6(x^2 - 3x)$ $\log_6 x + \log_6(x-3)$ **28.** $\log_2(x^2 - x - 12)$ $\log_2(x-4) + \log_2(x+3)$

Identify the vertical shift, amplitude, period, and phase shift for each function.   [Lesson 8.4]

29. $f(x) = 6\cos(-2\theta)$ 0; 6; 180°; 0°   **30.** $f(x) = -4 + \sin(\theta + 30°)$
$\quad$ −4; 1; 360°; −30°

Look Beyond

31. A jar contains 15 cookies, 5 of which are shortbread, 4 of which are chocolate chip, and 6 of which are oatmeal. If you randomly choose 2 of the cookies, what do you think is the probability that both will be oatmeal? Explain. $\frac{1}{7}$, or $\approx 14\%$

31. $\frac{1}{7}$, or $\approx 14\%$. First, there is a 6 out of 15 chance of selecting oatmeal. Then, assuming oatmeal was chosen the first time, there is a 5 out of 14 chance of selecting oatmeal. Multiply these two probabilities to find the chances for picking oatmeal both times: $\frac{6}{15} \cdot \frac{5}{14} = \frac{1}{7}$.

Let's Make a Deal

Let's Make a Deal
was a popular 1960s game
show hosted by Monte Hall.

FOCUS

An actual controversy over a columnist's answer to a strategy question provides the basis for using probability theory to settle a dispute. Students use a simulation and examine a table of outcomes to determine who is right and why conclusions based on logical analysis are sometimes less appropriate in reality than intuitive judgments. This is especially true in studying probability.

MOTIVATE

Before students read the excerpt from *Ask Marilyn*, make sure they understand the game show situation. Perhaps play a few rounds to demonstrate. Ask students if the question in the column accurately describes the game show situation. They may notice that the question does not state that the host must always show a losing door. You can discuss this now or leave it for the discussion of the activities.

You have reached the final round of the TV game show *Let's Make a Deal*. Behind one of the three numbered doors is a new car. Behind each of the other two doors is a goat. You have chosen door number 1. The host, Monte Hall, knows what is behind each door. He opens door number 3 to show you a goat. Then he pops the question, "Do you want to change your mind?"

What would you do? Would you stick with door number 1 or switch to door number 2?

Many contestants faced such a dilemma on the show, which ran for over 25 years. In 1990, a question based on this situation was submitted to a columnist, Marilyn vos Savant, who is reported to have the highest IQ in the world.

Behind one of these doors is a new car.

Should you stick with number 1 or switch to number 2?

Ask Marilyn

BY MARILYN VOS SAVANT

Suppose you're on a game show, and you're given the choice of three doors: Behind one door is a car; behind the others, goats. You pick a door, say No. 1, and the host, who knows what's behind the doors, opens another door, say No. 3, which has a goat. He then says to you, "Do you want to pick door No. 2?" Is it to your advantage to switch your choice?

—Craig F. Whitaker
Columbus, Md.

Yes; you should switch. The first door has a one-third chance of winning, but the second door has a two-thirds chance. Here's a good way to visualize what happened. Suppose there are a million doors, and you pick door No. 1. Then the host, who knows what's behind the doors and will always avoid the one with the prize, opens them all except door #777,777. You'd switch to that door pretty fast, wouldn't you?

Nearly a year later, an article in the *New York Times* reported that Marilyn vos Savant had received about 10,000 letters in response to her answer. Most of the letters disagreed with her. Many came from mathematicians and scientists with arguments like this one.

> ....You blew it! Let me explain: If one door is shown to be a loser, that information changes the probability of either remaining choice — neither of which has any reason to be more likely — to ½. As a professional mathematician, I'm very concerned with the general public's lack of mathematical skills. Please help by confessing your error and, in the future, be more careful....

Cooperative Learning

1. Before you analyze the problem in detail, explain which strategy you think is the best and why.

2. Use cards to model the situation. Try 10 games in which you stick with your choice. Then try 10 games in which you switch. Compare the results.

3. Make a table with the column headings shown, and complete it using all possible options.

Door with car	Door chosen	Door shown	Result if you switch	Result if you stick
1	1	2 or 3	Lose	Win
1	2	3	Win	
1	3			
2				

According to this table, should you switch? Explain.

4. Suppose you wrote the responses for Marilyn vos Savant's column. Write how you would answer the question about the game show.

Divide into groups and discuss the following situation.

• What if the host does not always have to offer a switch, but can do so whenever he feels like it. Would that affect the strategy? Why or why not?

• Why do you think that a problem like this could cause so many mathematicians to disagree?

Discuss

The situation is simple but ambiguous. Since the problem can have different interpretations, it is easy for some solutions to have subtle flaws. The original problem is not well defined, because it doesn't require the host to reveal a door in *every* case. In fact, in the actual game show, the host would use psychology to try to mislead contestants. For theoretical probability to be effective, it must be defined precisely. When the subtleties of psychological manipulation enter in, probability is no longer exact. Students should discuss the difference in mathematics between probabilities that are exact and those that are less exact, and how the probabilities can be determined.

3.

Door with car	Door you choose	Door you are shown	Result if you switch	Result if you stick
1	1	2 or 3	lose	win
1	2	3	win	lose
1	3	2	win	lose
2	1	3	win	lose
2	2	1 or 3	lose	win
2	3	1	win	lose
3	1	2	win	lose
3	2	1	win	lose
3	3	1 or 2	lose	win

In the 9 possible cases, if you always switch your choice, you win 6 times and lose 3 times. If you always stick with your first choice, you win only 3 times and lose 6 times. So, if you always switch, you should win about $\frac{2}{3}$ of the time. If you always stick, you should win about $\frac{1}{3}$ of the time.

Objectives

- Determine the number of combinations of *n* objects taken *r* at a time.
- Use the method for counting combinations with the Fundamental Principle of Counting to determine how many ways a decision can be made.

RESOURCES

- Practice Master   **11.4**
- Enrichment Master   **11.4**
- Technology Master   **11.4**
- Lesson Activity Master   **11.4**
- Quiz   **11.4**
- Spanish Resources   **11.4**

Assessing Prior Knowledge

Evaluate $\frac{6!}{4! \cdot 2!}$. [15]

TEACH

why Sometimes the order of arranging things is not important. For example, dog, cat, and rabbit are the same as rabbit, dog, and cat. They are different from pig, dog, and cat. Counting the ways that groups can be formed where order does not matter is called counting combinations.

LESSON 11.4 Counting and Combinations

Different selections from a set of objects, without regard to order, are called **combinations**.

The difference between a permutation and a combination is that

1. in a permutation, order matters, and

2. in a combination, order does not matter.

why **You have learned** how to determine the number of ways to arrange some objects from a larger set of objects by counting the number of permutations. However, sometimes you may not be concerned with the various order of arrangements, but only with the number of ways to choose the objects without regard to order.

Consider the numbers 1, 2, 3, and 4. The numbers 12 and 21 consist of the numbers 1 and 2, but the number 12 is not the same as the number 21.

Permutations		Order matters
12	21	12 is not the same as 21.
13	31	13 is not the same as 31.
14	41	14 is not the same as 41.
23	32	23 is not the same as 32.
24	42	24 is not the same as 42.
34	43	34 is not the same as 43.

Consider pennies, nickles, dimes, and quarters. A penny and a nickel together make 6¢ regardless of the order in which they are counted.

Combinations	Order does not matter
	is the same as
	is the same as
	is the same as
	is the same as
	is the same as
	is the same as

Thus, there are 12 permutations of 4 objects taken 2 at a time, but there are only 6 combinations of 4 objects taken 2 at a time.

ALTERNATIVE teaching strategy

Using Models Out of 5 different brands of laundry soap, decide how many different combinations of 3 brands a shopper could put in a cart. Looking in the cart, there is no difference in contents if the shopper has Brand 1, Brand 2, and Brand 4, or Brand 2, Brand 4, and Brand 1.

The cash register receipt will show identical purchases. However, a selection of Brand 1, Brand 2, and Brand 5 will result in a different receipt. This real-world example shows that the order with which the items are put into the cart makes *no* difference in the grocery bill if they are the same 3 items.

Counting Combinations

The notations $C(n, r)$, $_nC_r$, and $\binom{n}{r}$ all indicate combinations of n objects taken r at a time, or choices of r objects from a set of n objects. All of these notations can be read "n choose r."

Recall that $_nP_r = \frac{n!}{(n-r)!}$. In other words, there are $\frac{n!}{(n-r)!}$ permutations of n objects taken r at a time. Since a group of r objects can be arranged in $r!$ ways, you know that for each combination of objects taken r at a time, there are $r!$ permutations. Therefore, to determine the number of combinations of n objects taken r at a time, divide the number of permutations of n objects taken r at a time by $r!$.

NUMBER OF COMBINATIONS

The number of combinations of n distinct objects taken r at a time is
$$C(n, r) = {_nC_r} = \binom{n}{r} = \frac{n!}{r!(n-r)!}.$$

Graphics Calculator

EXAMPLE 1

Find and interpret each value. Then check your answer using your graphics calculator.

Ⓐ $_6C_2$ **Ⓑ** $_9C_1$

Solution

Ⓐ
$$_6C_2 = \frac{6!}{2!(6-2)!}$$
$$= \frac{6!}{2!\,4!}$$
$$= \frac{6 \cdot 5}{2!}$$
$$= 15$$

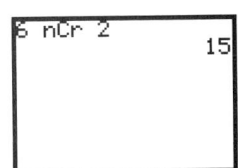

There are 15 ways to choose 2 objects from 6 objects.

Ⓑ
$$_9C_1 = \frac{9!}{1!(9-1)!}$$
$$= \frac{9!}{1!\,8!}$$
$$= \frac{9 \cdot 8!}{8!}$$
$$= 9$$

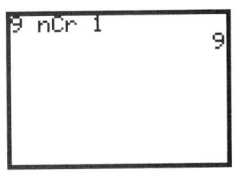

There are 9 ways to choose 1 object from 9 objects. ❖

Try This Find and interpret each value. Then check your answers using your graphics calculator.

Ⓐ $_5C_4$ **Ⓑ** $_{10}C_8$

---

interdisciplinary

CONNECTION

Marketing Retailers often combine several items in one package for quick sale. If a certain retailer wants to sell off a stock of eight different colors of ladies hosiery in packages of four each, discuss how many different sale-package combinations are possible. Assume that equal numbers of each size are available. [70]

Graphics Calculator

EXAMPLE 2

A How many ways are there to choose a committee of 3 people from 5 people?

B How many ways are there to choose a chairperson, then a secretary, and finally a treasurer from a group of 5 people?

Solution

A The 3 people chosen for the committee will be committee members regardless of the order in which they are chosen. Find $_5C_3$.

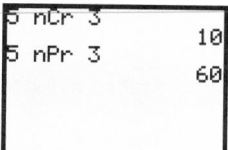

$$_5C_3 = \frac{5!}{3!(5-3)!} = \frac{5!}{3!\,2!} = 10$$

Thus, there are 10 ways to choose a committee of 3 people from 5 people.

B The order in which a person is chosen determines the position that the person takes. Find $_5P_3$.

$$_5P_3 = \frac{5!}{(5-3)!} = \frac{5!}{2!} = 60$$

Thus, there are 60 ways to choose a chairperson, then a secretary, and finally a treasurer from a group of 5 people. ❖

Try This A raffle will give away 3 identical prizes. In how many ways can the 3 prizes be awarded to 3 of the 25 people participating in the raffle?

CRITICAL Thinking

In the raffle described above, what is the probability that a particular person will win a prize?

Combinations and the Fundamental Principle of Counting

The method for counting combinations, along with the Fundamental Principle of Counting, can be used to solve many problems.

EXAMPLE 3

A college student needs to buy 2 health books, 3 political science books, and 1 reference book. How many different ways can the books be chosen from 7 health books, 5 political science books, and 3 reference books?

Solution

There are three decisions to be made.

Decision 1: Which 2 of the 7 health books should be chosen?

$$_7C_2 = \frac{7!}{2!(7-2)!} = 21$$

There are 21 ways to make this decision.

Decision 2: Which 3 of the 5 political science books should be chosen?

$$_5C_3 = \frac{5!}{3!(5-3)!} = 10$$

There are 10 ways to make this decision.

Decision 3: Which 1 of the 3 reference books should be chosen?

$$_3C_1 = \frac{3!}{1!(3-1)!} = 3$$

There are 3 ways to make this decision.

By the Fundamental Principle of Counting, the number of ways to make all three decisions is $21 \times 10 \times 3$, or 630. ❖

Try This For her hiking club, Connie wants to buy 4 backpacks, 6 sleeping bags, 3 pairs of boots, and 3 compact tents. In how many ways can she make her selections from 7 backpacks, 6 sleeping bags, 8 pairs of boots, and 5 tents?

Graphics Calculator

EXAMPLE 4

What is the probability that if 6 people are randomly chosen from the survey shown below, 2 believe and 4 do not believe that life exists in other parts of the galaxy?

Solution▶

In this experiment, there are 6 people to be randomly chosen from an equally likely sample space of 25 people. This can be done in $_{25}C_6$ ways.

To find the number of combinations in which 2 of these 6 people believe and 4 of these 6 people do not believe, consider both parts of the selection.

Part 1: There are $_{17}C_2$ ways that 2 of the 17 *believers* can be chosen.

Part 2: There are $_8C_4$ ways that 4 of the 8 *nonbelievers* can be chosen.

Therefore, to find the probability of selecting 2 believers and 4 nonbelievers out of 6 randomly chosen people from the survey, find

$$\frac{(_{17}C_2)\,(_8C_4)}{_{25}C_6}.$$

```
(17 nCr 2)(8 nCr
4)/(25 nCr 6)
        .0537549407
```

Thus, the probability that 2 out of 6 randomly chosen people from this survey believe that life exists in other parts of the galaxy and 4 do not is approximately 5%. ❖

Do you believe in Aliens?

25 people were randomly surveyed
17 say YES---8 say NO

Try This

$_7C_4 \times _6C_6 \times _8C_3 \times _5C_3 =$
19,600

Alternate Example 4

A recent poll of 100 people on the street in New York City regarding their opinion of the likelihood of snow in Southern Mississippi next winter found that 34 people think it is possible and 66 thought it was not possible. If this poll is representative of the population at large (discuss whether or not it is), what is the probability that out of 10 of the people surveyed, 5 will say "good possibility" and 5 will say "no way"? $\left[\frac{_{34}C_5 \times _{66}C_5}{_{100}C_{10}} \approx 14\%\right]$

R**ETEACHING**
the
lesson

Using Cognitive Strategies One of the simplest ways to teach the Fundamental Counting Principle is to think of the number of slots that need to be filled and then to think of how many ways each slot can be filled. A telephone number has 7 slots. The first slot cannot be a zero or a one, so 8 choices remain. Ten choices are possible for the second slot, etc. Then we simply multiply the numbers to get the total. So the number of telephone numbers would be $8 \cdot 10 \cdot 10 \cdot 10 \cdot 10 \cdot 10 \cdot 10$ or 8,000,000. If we wanted no common digits in a run of 4 numbers we would have $10 \cdot 9 \cdot 8 \cdot 7$ or 5040. The formulas in this lesson produce the same results.

EXAMPLE 5

What is the probability that exactly 2 heads will appear when 5 coins are tossed?

Solution▸

A sample space contains all the ways that 5 coins can land. The total number of ways is 2^5, or 32.

The number of ways of getting exactly 2 heads with 5 coins is $_5C_2$.

Thus, the probability that exactly 2 heads will appear when 5 coins are tossed is $P(\text{exactly 2 heads}) = \frac{_5C_2}{2^5} = \frac{10}{32} = \frac{5}{16}$, or 31.25%. ❖

EXERCISES & PROBLEMS

Communicate

1. Describe the difference between a permutation and a combination.

2. Compare the formulas for counting permutations and counting combinations. How are they alike, and how are they different?

3. Give four examples of combinations of the letters *ABCDE*, using 3 letters at a time.

4. Explain why n and r in the combination formula $_nC_r = \frac{n!}{r!(n-r)!}$ *cannot* be interchanged to $_nC_r = \frac{r!}{n!(r-n)!}$.

5. Show that for any set of n elements, there is only one way to choose them all at one time.

6. Use the formula for counting combinations to show that $_8C_5 = {_8C_3}$, $_9C_2 = {_9C_7}$, and $_{12}C_8 = {_{12}C_4}$.

Practice & Apply

Evaluate each permutation or combination.

7. $_8P_5$ 6720

8. $_{12}C_4$ 495

9. $_9C_5$ 126

10. $_{10}P_6$ 151,200

11. $_7P_3$ 210

12. $_{11}C_8$ 165

For Exercises 13–15, find the number of ways each of the following objects may be chosen, and write the combination or permutation you used to find the answer.

13. Seven out of 12 people are chosen for a committee. $_{12}C_7 = 792$

14. First-place, second-place, third-place, and fourth-place awards are given among 9 people. $_9P_4 = 3024$

15. A 13-card hand is received from a deck of 52 cards. $_{52}C_{13} = 635,013,559,600$

16. A jar contains 18 marbles, 6 of which are blue and 12 of which are green. In how many ways can 5 marbles be chosen so that 2 are blue and 3 are green? 3300

17. In how many ways can 4 cars be parked in 9 parking spaces? 3024

18. How many games are played in a league with 8 teams if each team plays each other team once? 28

19. Quality Control A shipment of 45 computers contains 5 that are defective. What is the probability that out of 10 computers chosen randomly, 3 of them are defective and 7 are nondefective? $\frac{_5C_3 \cdot _{40}C_7}{_{45}C_{10}} \approx 5.84\%$

20. Stock Exchange Of 125 stocks listed on the American Stock Exchange, 80 advanced, 30 declined, and 15 remained the same. In how many different ways could this have happened? $_{125}C_{80} \cdot _{45}C_{30} \approx 7.58 \times 10^{45}$

21. A newsstand has newspapers from Houston, Miami, and Los Angeles. In how many ways can 5 papers be chosen from these newspapers so that 3 are from Houston, 2 are from Miami, and none are from Los Angeles? 9555

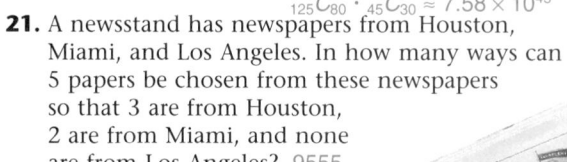

Basketball Pat Riley, a longtime NBA coach, once said that he would try every combination of the 12 players on his team in order to find a 5-person winning team.

22. Assume that the position of a player does not matter. If Coach Riley were really to do this, how many teams would he try? 792

23. How many combinations would Coach Riley have if the 6 guards had to play guard, the 4 forwards had to play forward, and the 2 centers had to play center? (Assume that each team has 2 guards, 2 forwards, and 1 center.) 180

24. **Geometry** A set of 5 parallel lines crosses another set of 8 parallel lines. How many parallelograms are formed? 280

25. Solve the equation $\binom{n+1}{3} = 2 \cdot \binom{n}{2}$ for n. 5

26. Solve the equation $\binom{n+2}{4} = 6 \cdot \binom{n}{2}$ for n. 7

27. How many of the subsets of the set $\{1, 3, 5, 7\}$ do not contain the number 5? 8

Look Back

28. Identify the domain and range of   D:$\{-2, -1, 0, 1, 2\}$; R:$\{4, 8, 12, 16, 20\}$
$f = \{(-2, 4), (-1, 8), (0, 12), (1, 16), (2, 20)\}$.   **[Lesson 2.4]**

29. Explain why you cannot find the product AB for matrices A and B.
[Lesson 4.2]   The number of columns of A does not equal the number of rows of B

$$A = \begin{bmatrix} 4 & 2 & 7 \\ 0 & -2 & 6 \\ 1 & 1 & -3 \end{bmatrix} \text{ and } B = \begin{bmatrix} 3 & -5 & -14 \\ 2 & 0 & -8 \end{bmatrix}$$

30. Explain how the graph of the function $g(x) = (x + 4)^2 + 3$ is related to the graph of the function $f(x) = x^2$.   **[Lesson 5.3]**
g is translated 4 units left and 3 units up from f.

Simplify each rational function.   **[Lesson 9.4]**

31. $f(x) = \dfrac{x-3}{2x^2 - 5x - 3}$ $\dfrac{1}{2x+1}$; $x \neq 3, -\dfrac{1}{2}$ 32. $f(x) = \dfrac{x^2 - 5x + 4}{x^2 + 2x - 3}$ $\dfrac{x-4}{x+3}$; $x \neq 1, -3$

Look Beyond

33. Specify the most reasonable next term in the sequence of numbers $-18, -10, -2, 6, 14, \ldots$ Explain your answer. 22

33. 22; each term in the sequence of numbers is 8 more than the preceding term.

Independent and Mutually Exclusive Events

PREPARE

Objectives

- Find the probability that two independent events, A and B, will both occur.
- Find the probability that either event A or event B will occur.

RESOURCES

- Practice Master   **11.5**
- Enrichment Master   **11.5**
- Technology Master   **11.5**
- Lesson Activity Master   **11.5**
- Quiz   **11.5**
- Spanish Resources   **11.5**

 *It is often necessary to understand the effect that one event has on a second event in order to solve problems involving probability.*

Suppose you draw the queen of diamonds from a full deck of 52 cards. Is it possible to draw the queen of diamonds from that deck of cards on the second draw?

If you return the queen of diamonds to the deck of cards and shuffle them before you draw a second time, it *is* possible to draw the queen of diamonds on the second draw. Because the occurrence of the first event has no effect on the occurrence of the second, these are called **independent events**.

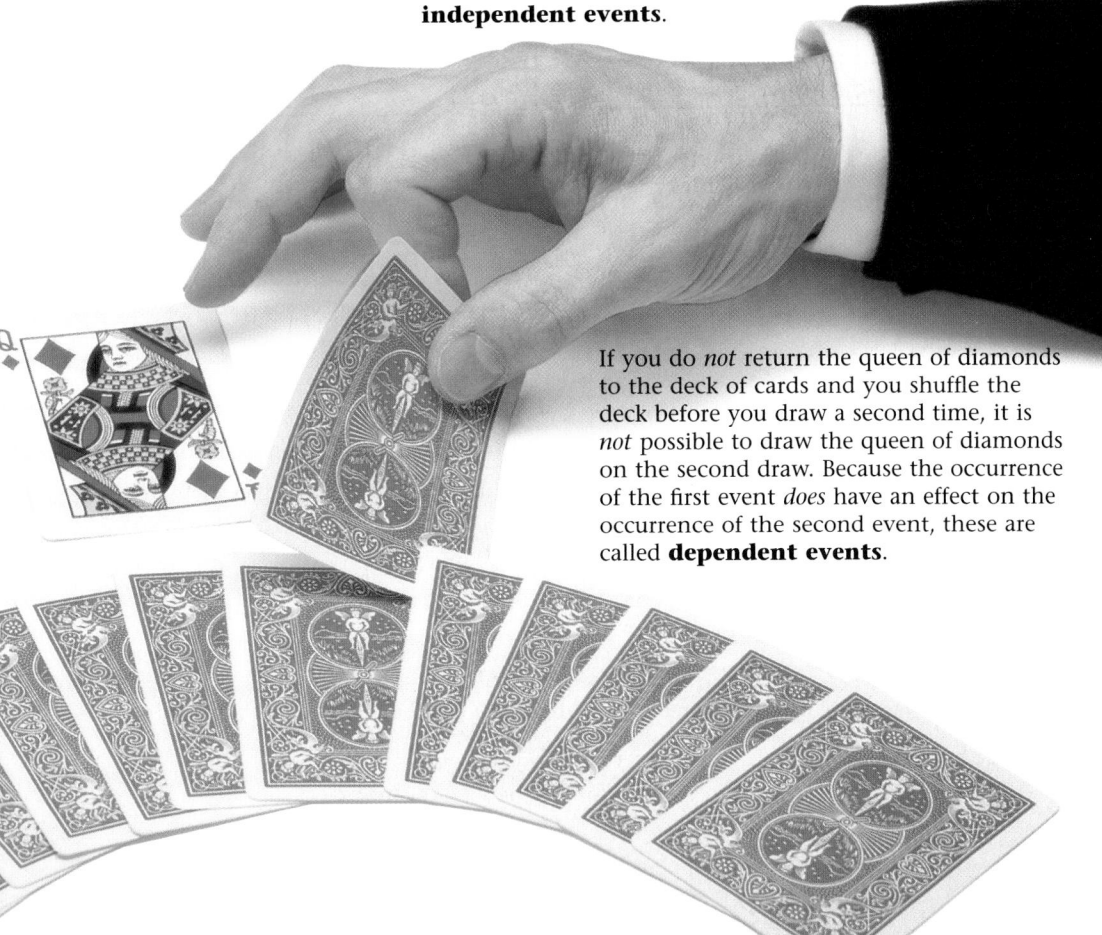

If you do *not* return the queen of diamonds to the deck of cards and you shuffle the deck before you draw a second time, it is *not* possible to draw the queen of diamonds on the second draw. Because the occurrence of the first event *does* have an effect on the occurrence of the second event, these are called **dependent events**.

Assessing Prior Knowledge

1. Find $_7C_2$. What does it represent? [**21; the number of ways to choose 2 things out of 7 things when order does not matter**]

2. Find $_7P_2$. What does it represent? [**42; the number of ways to choose 2 things out of 7 things when order does matter**]

TEACH

 In computing probabilities it is useful to know whether one event influences another. For example, in drawing a second marble from a bag of 5 red and 5 blue marbles, the probability of drawing a red one depends upon whether the first one drawn was red or blue.

ALTERNATIVE
t e a c h i n g
s t r a t e g y

Using Manipulatives
Bring in a small sack that contains 4 red and 4 blue crayons. Have a student draw one crayon and then take a second one. Record the number of times that the second one was blue. Do this 50 times. See whether there is a significant difference between when the first

one was red and when the first one was blue. Compute the experimental probabilities of each for comparison later. Point out that the second draw was affected by the first one not being put back into the bag. Thus these are mutually exclusive events, and the second is dependent upon the first outcome.

Aongoing
SSESSMENT

The number showing on the blue cube does not depend on the number showing on the red cube.

Use Transparency 44

Independent Events

Consider the experiment of tossing two 6-sided number cubes (one red cube and one blue cube) and observing the number on the top face of each cube. These events are independent. Why?

A sample space for this experiment can be represented with 36 ordered pairs, where the first coordinate represents the red cube and the second coordinate represents the blue cube.

What is the probability that the number rolled on the red cube is greater than 3 *and* the number rolled on the blue cube is greater than 4?

You can count the possibilities from the diagram.

(1, 1)	(1, 2)	(1, 3)	(1, 4)	(1, 5)	(1, 6)
(2, 1)	(2, 2)	(2, 3)	(2, 4)	(2, 5)	(2, 6)
(3, 1)	(3, 2)	(3, 3)	(3, 4)	(3, 5)	(3, 6)
(4, 1)	(4, 2)	(4, 3)	(4, 4)	(4, 5)	(4, 6)
(5, 1)	(5, 2)	(5, 3)	(5, 4)	(5, 5)	(5, 6)
(6, 1)	(6, 2)	(6, 3)	(6, 4)	(6, 5)	(6, 6)

(5, 6)

- There are 18 outcomes in which the red cube is greater than 3.
- There are 12 outcomes in which the blue cube is greater than 4.
- There are 6 outcomes in which the red cube is greater than 3 *and* the blue cube is greater than 4.

Therefore, $P(\text{red} > 3) = \frac{18}{36} = \frac{1}{2}$, $P(\text{blue} > 4) = \frac{12}{36} = \frac{1}{3}$, and

$$P(\text{red} > 3 \text{ and blue} > 4) = \frac{6}{36} = \frac{1}{6}.$$

Notice that $P(\text{red} > 3 \text{ and blue} > 4) = \frac{1}{6} = \frac{1}{2} \cdot \frac{1}{3}$
$$= P(\text{red} > 3) \cdot P(\text{blue} > 4).$$

This leads to the rule for independent events.

PROBABILITY OF (A and B)

If A and B are independent events, then $P(A \text{ and } B) = P(A) \cdot P(B)$.

interdisciplinary **Manufacturing** Quality control specialists use small samples of material or components used in manufacturing a product to determine whether or not to accept a shipment from a supplier. The size of the sample, compared with the size of the total shipment, is a very important factor in this process. The product testing and the mathematics provide the basis for decisions in this field. If a quality control supervisor is available, invite the individual to describe to students the mathematics that is used.

CONNECTION

EXAMPLE 1

Consumer Awareness A track coach buys 3 stopwatches. If 1 out of every 200 stopwatches is defective, what is the probability that all 3 of the new stopwatches are defective?

Solution➤

Assume that all the stopwatches are constructed independently of each other. Therefore, the defectiveness of one stopwatch does not affect the defectiveness of any other stopwatch. Then,

P(all 3 stopwatches are defective)
= P(1st is defective *and* 2nd is defective *and* 3rd is defective)
= P(1st is defective) · P(2nd is defective) · P(3rd is defective)
= $\frac{1}{200} \cdot \frac{1}{200} \cdot \frac{1}{200} = \frac{1}{8,000,000}$

Thus, all 3 stopwatches will be defective about 1 out of 8,000,000 times that 3 stopwatches are purchased. ❖

Try This There are 5 nominees for best actress, 2 of which are for comedy roles. There are 5 nominees for best actor, 3 of which are for comedy roles. Assuming that each nominee has an equal chance of winning, find the probability that the winner for best actress and the winner for best actor both will have comedy roles.

Consider another experiment of tossing the same 6-sided number cubes. What is the probability that either the number rolled on the red cube is greater than 3 *or* the number rolled on the blue cube is greater than 4?

Are these events independent? Explain.

(1, 1)	(1, 2)	(1, 3)	(1, 4)	(1, 5)	(1, 6)
(2, 1)	(2, 2)	(2, 3)	(2, 4)	(2, 5)	(2, 6)
(3, 1)	(3, 2)	(3, 3)	(3, 4)	(3, 5)	(3, 6)
(4, 1)	(4, 2)	(4, 3)	(4, 4)	(4, 5)	(4, 6)
(5, 1)	(5, 2)	(5, 3)	(5, 4)	(5, 5)	(5, 6)
(6, 1)	(6, 2)	(6, 3)	(6, 4)	(6, 5)	(6, 6)

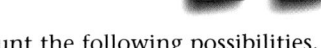

You can count the following possibilities.

$$P(\text{red} > 3 \text{ or blue} > 4) = \frac{24}{36} = \frac{2}{3}$$

There are 18 ways to roll a number greater than 3 on the red number cube. There are 12 ways to roll a number greater than 4 on the blue number cube.

Alternate Example 1

An auto parts store buys 5 remanufactured units of a particular part for its monthly sales demand. If history has shown that 30 out of 500 of this particular remanufactured part are defective, what is the probability that all 5 are defective? [$\frac{243}{312,500,000}$, or near zero]

Aongoing **SSESSMENT**

Try This

24%

Aongoing **SSESSMENT**

The events are independent. The outcome of the first roll does not change the outcome of the second.

ENRICHMENT Consider the engine control system of your car. A computer monitors many different values that represent the driving conditions at that instant, and decides what to instruct the many systems to do. Research how these systems work.

INCLUSION strategies **Guided Research** As a library assignment, have students look up a recent automobile recall campaign. Have them determine how many complaints or problems there were that prompted the recall or could be expected per 1000 vehicles produced. Is there an industry standard? What is it, and what other factors might affect the answer?

TEACHING *tip*

Point out that in these sorts of probability calculations, the number of possible outcomes is reduced by one on each subsequent trial if the trials do not include replacement of the item drawn.

Cooperate Learning

Have the students break into groups of 5. Have each group decide how many ways 2 from their group could be chosen to represent the group at a school meeting. Point out that the second person chosen is not independent of the first choice made.

Use Transparency ▶ 45

Notice that there are 6 ways to roll both red > 3 and blue > 4. Therefore, if you count red > 3 and blue > 4 separately, you count the overlapping, or *intersecting*, 6 ways twice.

Therefore, $P(\text{red} > 3 \text{ or blue} > 4)$

$$= P(\text{red} > 3) + P(\text{blue} > 4) - P(\text{red} > 3 \text{ and blue} > 4)$$
$$= \frac{1}{2} + \frac{1}{3} - \frac{1}{6}$$
$$= \frac{2}{3}$$

You can think of this as the probability that the first event will occur plus the probability that the second event will occur minus the probability that both will occur at the same time (the overlap of the first two probabilities).

This leads to a formula for finding the probability of *A or B*.

PROBABILITY OF (*A or B*)

If *A* and *B* are any events, then
$$P(A \text{ or } B) = P(A) + P(B) - P(A \text{ and } B).$$

Mutually Exclusive Events

Is it possible for the number on the blue cube to be less than 2 *and* greater than 4? Since the number on the cube cannot be less than 2 and greater than 4 at the same time, these are *mutually exclusive events*. When two events cannot occur at the same time, these events are said to be **mutually exclusive events**.

Find the probability that, when a red and a blue 6-sided number cube are tossed, the number on the blue cube is less than 2 *or* is greater than 4 (regardless of the number rolled on the red cube). You can count the possibilities.

	(1, 1)	(1, 2)	(1, 3)	(1, 4)	(1, 5)	(1, 6)
	(2, 1)	(2, 2)	(2, 3)	(2, 4)	(2, 5)	(2, 6)
	(3, 1)	(3, 2)	(3, 3)	(3, 4)	(3, 5)	(3, 6)
	(4, 1)	(4, 2)	(4, 3)	(4, 4)	(4, 5)	(4, 6)
	(5, 1)	(5, 2)	(5, 3)	(5, 4)	(5, 5)	(5, 6)
	(6, 1)	(6, 2)	(6, 3)	(6, 4)	(6, 5)	(6, 6)

RETEACHING
t h e
l e s s o n

Using Cognitive Strategies Go back to the Fundamental Counting Principle. If there are 2 choices to be made with replacement from 4 possibilities, the total possible ways would be 4 · 4 or 16. This illustrates independent events. If there is no replacement, the count is 4 · 3 or 12. No replacement illustrates mutually exclusive events.

$P(\text{blue} < 2 \text{ or blue} > 4)$

$= P(\text{blue} < 2) + P(\text{blue} > 4) - P(\text{blue} < 2 \text{ and blue} > 4)$

$= \dfrac{6}{36} + \dfrac{12}{36} - 0$

$= \dfrac{18}{36}, \text{ or } \dfrac{1}{2}$

CRITICAL *Thinking*

Explain why it is *always* true that if A and B are mutually exclusive events, then $P(A \text{ or } B) = P(A) + P(B)$.

EXAMPLE 2

There are 9 people in a high school art class, 3 of whom are going on a field trip. What is the probability that, if 4 students in the art class are randomly chosen, either exactly 1 person is going on the field trip *or* exactly 2 people are going on the field trip?

Solution▶

Let $E_1 = \{\text{exactly 1 is going}\}$, and let $E_2 = \{\text{exactly 2 are going}\}$. The events E_1 and E_2 are mutually exclusive because if one event occurs, the other cannot.

It is impossible to select exactly 1 student who is going *and* exactly 2 students who are going, $P(E_1 \text{ and } E_2) = 0$. Since you need to find $P(E_1 \text{ or } E_2)$, it is only necessary to find $P(E_1)$ and $P(E_2)$.

The number of ways to choose 4 of the 9 students in the class is $_9C_4$.

To find $P(E_1)$, use combinations. The number of ways to choose exactly 1 of the 3 students who are going is $_3C_1$. The number of ways to choose exactly 3 of the remaining 6 students who are not going is $_6C_3$.

$$P(E_1) = \frac{(_3C_1)(_6C_3)}{_9C_4} = \frac{3 \cdot 20}{126} = \frac{60}{126} = \frac{10}{21}$$

To find $P(E_2)$, use combinations again. The number of ways to choose exactly 2 of the 3 students who are going is $_3C_2$. The number of ways to choose exactly 2 of the remaining 6 students who are not going is $_6C_2$.

$$P(E_1) = \frac{(_3C_2)(_6C_2)}{_9C_4} = \frac{3 \cdot 15}{126}$$
$$= \frac{45}{126} = \frac{5}{14}$$

Thus, $P(E_1 \text{ or } E_2) = P(E_1) + P(E_2) + 0$
$$= \frac{10}{21} + \frac{5}{14}$$
$$= \frac{5}{6} \approx 0.83 \; \diamond$$

CRITICAL *Thinking*

If A and B are mutually exclusive, then they cannot occur simultaneously and so $P(A \text{ and } B)$ equals 0.

Alternate Example 2

There are 20 people who are employed in an office, 5 of whom must go to a board meeting. What is the probability that out of 7 people selected at random, exactly 2 or exactly 3 will be selected to attend? [**56%**]

TEACHING *tip*▶

In calculating the probability of 2 mutually exclusive events occurring, simply add the probabilities of each one occurring.

Assignment Guide

Core 1–47

Core-Plus 1–48

Technology

Graphics calculators are needed for Exercise 46. Calculators may also be used for exercises involving $n!$, $_nC_r$, and $_nP_r$. The use of the calculator will simplify calculations and will enable students to concentrate on the probability concepts being covered.

Error Analysis

A very common error is that students fail to notice the difference between *and* and *or*. *And* often indicates multiplication of probabilities and *or* indicates addition.

EXERCISES & PROBLEMS

Communicate

1. Explain what is meant by *mutually exclusive events*.
2. Explain what is meant by *independent events*.
3. Explain how mutually exclusive events are different from independent events.

Entertainment   A blue and a red number cube are tossed.

4. Give an example of two events that are mutually exclusive.
5. Give an example of two events that are independent.
6. Give an example of two events that are dependent.
7. Explain why $P(A$ and $B) = 0$ when A and B are mutually exclusive events.

Practice & Apply

Let $P(A) = 0.40$ and $P(B) = 0.36$, where A and B are independent.

8. Find $P(A$ and $B)$. 0.144   9. Find $P(A$ or $B)$. 0.616

Let $P(A) = 0.40$ and $P(B) = 0.36$, where A and B are mutually exclusive.

10. Find $P(A$ and $B)$. 0   11. Find $P(A$ or $B)$. 0.76

Let $P(A) = 0.35$ and $P(B) = 0.26$, where A and B are independent.

12. Find $P(A$ and $B)$. 0.091   13. Find $P(A$ or $B)$. 0.519

Let $P(A) = 0.35$ and $P(B) = 0.26$, where A and B are mutually exclusive.

14. Find $P(A$ and $B)$. 0   15. Find $P(A$ or $B)$. 0.61

After 2 number cubes are tossed, the resulting numbers on the top face of the cubes are observed. Identify each event as independent, mutually exclusive, or dependent.

16. {sum = 5} *and* {sum ≥ 2} independent   17. {sum is even} *and* {sum > 4} independent

18. {sum = 5} *and* {sum ≤ 2} mutually exclusive   19. {sum is even} *and* {sum is 5} mutually exclusive

After 2 number cubes are tossed, the resulting numbers on the top face of the cubes are observed. Find each probability.

20. {sum is 5} *and* {sum ≥ 2} $\frac{1}{9}$ 21. {sum is even} *or* {sum > 4} $\frac{17}{18}$

22. {sum is 5} *and* {sum ≤ 2} 0   23. {sum is even} *or* {sum is 5} $\frac{11}{18}$

Genetics Identical twins come from the same egg and are the same gender. Fraternal twins come from different eggs and can differ in gender. Assume that the theoretical probability that twins will be fraternal is $\frac{2}{3}$ and that the theoretical probability that a baby will be female is $\frac{1}{2}$. Find the following probabilities.

24. P (twins will be identical *and* both twins will be female) $\frac{1}{6}$

25. P (twins will be identical *or* both twins will be female)

26. P (twins will be fraternal *and* both twins will be male) $\frac{1}{6}$

27. P (twins will be fraternal *or* both twins will be male)

28. P (both twins will be of the same gender) $\frac{2}{3}$

Fraternal twins

Identical twins

According to the Youth Sports Institute of Michigan State University, a recent survey of 16-year-olds showed that 78% go out on dates and 22% play team sports.

29. Are dating and playing sports independent events? Explain. yes, they do not affect one another

30. Find the probability that of all youths polled, one randomly selected youth goes out on dates *and* plays sports. 17.16%

31. Find the probability that of all youths polled, one randomly selected youth goes out on dates *or* plays sports. 82.84%

At a particular university, 65% of all entering freshmen will graduate and 20% will participate in some form of musical activity.

32. Are graduating and participating in musical activities independent events? Explain. yes, they do not affect one another

33. Find the probability that an entering freshman will graduate *and* participate in music. 13%

34. Find the probability that an entering freshman will graduate *and* will not participate in music. 52%

35. Find the probability that an entering freshman will not graduate *and* will participate in music. 7%

36. Find the probability that an entering freshman will not graduate *and* will not participate in music. 28%

37. How do your results to Exercises 33–36 compare?

38. **Baseball** Find the probability that a baseball team will get 3 consecutive hits when the next 3 batters have batting averages of 0.225, 0.215, and 0.314, respectively. 1.5%

25. $\frac{1}{2}$

27. $\frac{5}{6}$

37. The sum of the probabilities of all the situations occurring is 1.

Security Suppose that a business security system consists of four units: a motion detector, a heat detector, a magnetic detector, and a visual spotter. After sending trained agents through the system, the following probabilities were determined.

- The probability of escaping the motion detector is 0.2.
- The probability of escaping the heat detector is 0.3.
- The probability of escaping the magnetic detector is 0.4.
- The probability of escaping the visual spotter is 0.6.

39. If the heat detector and the motion detector are independent, what is the probability that a trained agent can go through these two systems undetected?

40. If the magnetic detector and the visual spotter are independent, what is the probability that a trained agent can go through these two systems undetected?

41. If each unit acts independently, what is the probability that a trained agent can go through the entire system undetected? 1.44%

Look Back

Find the area of each figure. **[Lesson 8.1]**

42. $\frac{2025\sqrt{3}}{4} \approx$ 876.85 ft^2

60° 60°
45 ft

43. 6 in.

$54\sqrt{3} \approx$ 93.53 in.^{2}

44. 9 cm^2
6 cm

45. Find the central angle that intercepts an arc length of 48 inches on a circle with a radius of 15 inches. **[Lesson 8.6]** $\approx 183°$

46. Compare the graphs of $f(x) = \frac{1}{x}$ and $g(x) = \frac{1}{x-4}$. How are they alike and how are they different? **[Lesson 9.3]**

47. Solve $\frac{2y+5}{3y-2} = \frac{4y+3}{6y-1}$. **[Lesson 9.5]** $y = -\frac{1}{27}$

Look Beyond

48. What is the most likely next number in the sequence of numbers $\frac{2}{3}$, 1, $\frac{3}{2}$, $\frac{9}{4}$, ... ? What is the relationship between consecutive numbers in the sequence? $\frac{27}{8}$; each term is found by multiplying the previous term by $\frac{3}{2}$.

39. 6%

40. 24%

46. Both are hyperbolas; the graph of $g(x) = \frac{1}{x-4}$ is the translation of the graph of $f(x) = \frac{1}{x}$ by 4 units to the right.

Look Beyond

This exercise has students exploring sequences. These will be studied in more depth in Chapter 12.

LESSON 11.6

Exploring Conditional Probability

why *In a sequence of events, the probability of one event often depends on the occurrence of a previous event. This type of dependence of events affects the probability that both events will occur.*

• Exploration • *Dependence of Events*

A mechanic has removed the six spark plugs from a car. Four are good and two are defective. The mechanic selects one plug and without replacing it selects a second plug.

1 Labeling the good plugs as g_1, g_2, g_3, and g_4, and the bad plugs as b_1 and b_2, copy and complete the following table of possible outcomes.

g_1g_2	g_2g_1	g_3g_1	g_4g_1	b_1g_1	b_2g_1
g_1g_3	g_2g_3				
g_1g_4	g_2g_4				
g_1b_1	g_2b_1				
g_1b_2					

2 How many possible ways are there to select two plugs? What is the probability that the mechanic will select two good plugs?

Using Cognitive Strategies The goal of this lesson is to determine the probability that both of two events occur. The word *both* indicates A *and* B. The *and* is the operative word, and it means multiplication. Therefore, the probabilities of each A and B must be multiplied to reach the total probability. Emphasize the words *both* and *and*. In Exercise 6 in this lesson, the expression "… it will also be a king." means "… *and* it will be a king." Thus, this exercise is one in which the probabilities will be multiplied.

PREPARE

Objectives

• Determine the probability of an event occurring that depends on the probability of another event occurring.

RESOURCES

• Practice Master   **11.6**
• Enrichment Master   **11.6**
• Technology Master   **11.6**
• Lesson Activity Master   **11.6**
• Quiz   **11.6**
• Spanish Resources   **11.6**

This lesson may be considered optional in the core course. See page 594A.

Assessing Prior Knowledge

1. Evaluate $_5C_3$. [**10**]

2. Evaluate and simplify $\frac{_5C_3 \cdot _4C_2}{_{10}C_6}$ $\left[\frac{2}{7}\right]$

TEACH

 Sometimes, in determining the probability of an event happening, it is necessary to know the chances of a prior event occurring. In such cases, the second event's chances depend on whether or not the first event has occurred.

Cooperative Learning

In groups of three, each student should find or compose a conditional probability situation from real life and convince the other two that it is correct. For example, $E_1 = it$ *rains*, $E_2 = I$ *forgot my umbrella*, $P(I \text{ get wet}) = ?$

Exploration 1 Notes

In this exploration, students will begin to see how knowledge of the first event's outcome affects the possible choices for the second event. This illustrates the concept of *conditional probability*.

Get six identical spark plugs and secretly mark two of them, in a way only you can spot, as bad. Have each student play the role of the mechanic in both selections and record the results. Show how the results begin to converge on the predictions.

Aongoing SSESSMENT

5. There is a greater probability of selecting two good plugs if you know that the first plug is good.

---

3 Suppose the mechanic selects one good spark plug. Copy and complete the table of possible ways that the second plug can be selected.

g_1g_2	g_2g_1	g_3g_1	g_4g_1
g_1g_3			

4 How many possible ways are there to select a second plug once a good plug is selected first? What is the probability that the mechanic will select two good plugs once it is known that the first plug selected is good?

5 Compare the probability of selecting two good plugs with the probability of selecting two good plugs once the first plug selected is known to be good. How are they different and how are they alike? ❖

In the exploration, you examined the probability that, if the first spark plug is good, both plugs will be good. This is a **conditional probability** because one event depends on the occurrence of the other event.

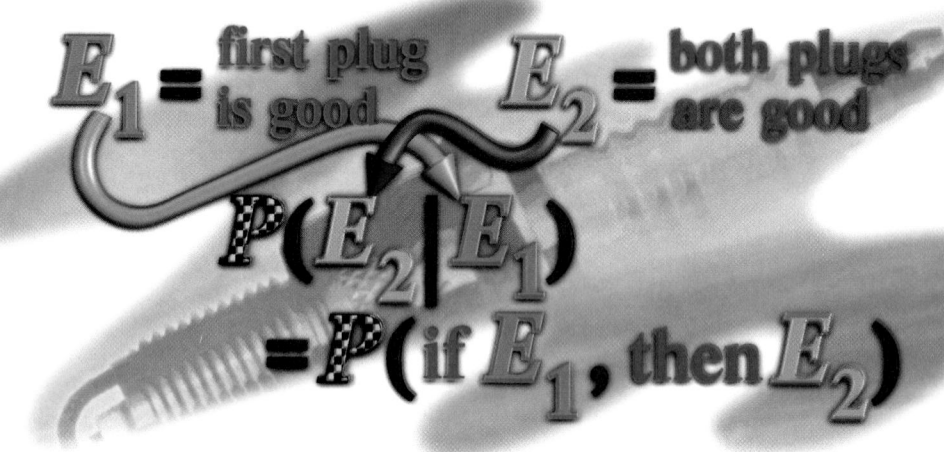

$E_1 = $ first plug is good $E_2 = $ both plugs are good

$$P(E_2|E_1) = P(\text{if } E_1, \text{ then } E_2)$$

Using conditional probability notation, you can write this conditional probability as follows:

$$P(\text{both plugs are good} \mid \text{the first plug is good})$$

The vertical line (|) stands for the words *given that*.

CONDITIONAL PROBABILITY

The conditional probability of event E_2, given event E_1, is

$$P(E_2|E_1) = \frac{P(E_1 \text{ and } E_2)}{P(E_1)}, \text{ where } P(E_1) \neq 0.$$

ENRICHMENT Life insurance companies charge premiums based on factors such as age, gender, and health-related habits. Their decisions are based on studying probabilities that a 20-year-old will live to be more than 70 years old and that 50 years worth of premiums will be collected before a death benefit is claimed. Have students explore how these probability predictions are tied to the insurance company's profit, the policy holder's personal finances, and investments made by the company.

APPLICATION

A survey of a city's high school students showed that 65% liked the sports programs, 40% liked the general activities programs, and 25% liked both. What is the probability that a person selected at random from this high school likes the general activities programs if it is known that the person likes the sports programs?

Let E_1 = {likes sports programs}, and
E_2 = {likes general activities programs}.

Find $P(E_2|E_1)$.

$$P(E_2|E_1) = \frac{P(E_1 \text{ and } E_2)}{P(E_1)}$$
$$= \frac{0.25}{0.65} \approx 0.38$$

Thus, the probability is approximately 38% that a student chosen at random from this high school likes the general activities programs if it is known that he or she likes the sports programs. ❖

CRITICAL
Thinking

Find $P(E_1|E_2)$ in the Application. How does the order of the events E_1 and E_2 affect the conditional probability?

EXERCISES & PROBLEMS

Conditional Probability

$P(E_2 | E_1)$

If E_1 happens, then the probability of E_2 is...

Communicate

1. Explain what it means for two events to be dependent.

2. How can you determine whether two events are dependent or independent?

3. What does the notation $P(E_2|E_1)$ mean in words?

4. Explain the difference between $P(E_1 \text{ and } E_2)$ and $P(E_2|E_1)$.

5. How is the formula for conditional probability affected if E_1 and E_2 are mutually exclusive events?

CRITICAL
Thinking

$\frac{0.25}{0.40} \approx 0.625$. The conditional probability has a different value. Although the numerator will be the same, the denominator will be different.

ASSESS

Selected Answers

Odd-Numbered Exercises 7–17 and 21–25

Assignment Guide

Core 1–26

Core-Plus 1–27

Technology

In calculating conditional probability, suggest that students compute the probability of the first event and store it in memory where it can be readily recalled for later division.

Error Analysis

Figuring these probabilities by hand can cause division errors because students are dividing a decimal by a decimal. Remind students how to divide by a decimal.

INCLUSION **strategies** **Using Visual Strategies** For the visual learner, have a piece of poster board with 3 circles drawn on it. Have 8 discs of colored paper cut out, 4 red and 4 blue. If red indicates a boy and blue a girl, see whether the students can model the situation of the probability that all 3 circles can be filled by a girl.

RETEACHING **the lesson** **Using Cognitive Strategies** Using highway safety statistics, demonstrate the concept of conditional probability by comparing the probabilities of injury with and without a seat belt, assuming there is an accident. Show how $P(E_1 \text{ and } E_2)$ is increased when divided by $P(E_1)$, which is less than 1, in the formula for $P(E_2 | E_1)$.

Practice & Apply

6. One card is drawn from a standard deck of 52 cards. What is the probability that if you know it is a face card, it will also be a king? $\frac{1}{3}$

7. Find the probability of drawing 3 aces from a standard deck of 52 cards if each card is replaced before the next card is drawn. $\frac{1}{2197}$

8. Suppose that $P(A) = 0.25$ and $P(B|A) = 0.55$. Find $P(A \text{ and } B)$. 0.1375

Airline Safety Chelsi flies from Nashville to San Francisco, with a stop in Dallas. She takes Alpha airlines from Nashville to Dallas, and Beta airlines from Dallas to San Francisco. The probability that an Alpha plane will land safely is 0.998, and the probability that a Beta plane will land safely is 0.994. Assume that safety on one airline is independent from the safety on the other.

9. What is the probability that Chelsi will land safely in Dallas *and* San Francisco? ≈ 0.992

10. What is the probability that Chelsi will land safely in Dallas *and* will not land safely in San Francisco? 0.006

11. Medical Tests A test is being developed for the detection of a particular disease. The researchers want to find the probability that the test will show a false positive. (If a test shows a positive result when the disease is not present, the result is called a false positive.) Preliminary results show that out of 48 people without the disease, 18 show a positive result, and that out of 70 people tested, 30 have the disease. Find the probability that the test will show a false positive. $\frac{3}{8}$

Vital Statistics Suppose the probability of living to be 100 years old is found to be $\frac{1009}{100,000}$.

12. Find the probability that two friends will both live to be 100 years old, assuming the events are independent. 0.0001

13. Why might these events *not* be independent?

Automobile Safety Suppose a survey of 1500 traffic accidents produces the data in the table. The table shows the cause of the accident and the gender of the driver of the car.

	Mechanical failure	Intoxication	Poor judgment
Male	415	168	255
Female	360	74	228

14. Based on these data, if an accident is due to poor judgment, what is the probability that the driver is a male? 0.528

15. Based on these data, if the driver of a car involved in an accident is female, what is the probability that the cause is mechanical failure? 0.544

13. Answers may vary. One possible answer is: Significant friendships seem to help people live longer. So, the fact that an important friend does not die could impact the fact the other friend does not die, i.e. lives to be over 100 years old.

16. Suppose that $P(A \text{ and } B) = 0.25$, $P(A) = 0.4$, and $P(B) = 0.6$. Find $P(B|A)$. 0.625

17. Stock Market The probability that the stock market will go up on Wednesday is 0.25. If it goes up on Wednesday, the probability that it will go up on Thursday is 0.6. Find the probability that the stock market will go up on both Wednesday *and* Thursday. 0.15

18. Mass Communications A message will be transmitted from location X to location Y if the relays R_1, R_2, and R_3 are all closed. The probabilities that these relays will be closed are 0.5, 0.3, and 0.45, respectively. Assuming that the relays function independently, what is the probability that a message will get through? 0.0675

19.  **Portfolio Activity** Refer to the Portfolio Activity on page 595. Choose the number of CDs you wish to have in your CD changer, and choose the number of CDs from each category of music you wish to include. Use a mixture of at least three different categories. Assume you have your CD player set on random play. Find the probability that a particular category of music will be played next. Find the probability that a particular song from a particular CD will be played next. Do you need to know how many tracks each CD has? Does it matter what type of music was played last? Does it matter which song was played last? How does conditional probability affect this situation? Explain all of your responses.
Answers may vary.

Look Back

Simplify. **[Lesson 3.5]**

20. $\lceil -2.8 \rceil$ -2

21. $\lfloor 8.07 \rfloor + \lceil 2.91 \rceil$ 11

22. $\lfloor 1.72 \rfloor - \lceil 0.07 \rceil$ 0

23. Find the zeros of the polynomial function $f(x) = x^2 - 8x + 15$.
[Lesson 6.2] 3 and 5

24. Identify the equation $16x^2 + 4y^2 - 96x + 8y + 84 = 0$ as that of an ellipse or hyperbola. Then find the vertices. **[Lessons 10.3, 10.4]**

25. What is the difference between combinations and permutations?
[Lesson 11.4]

Look Beyond

26. If the probability that a specific event will occur is 0.35, what is the probability that it will *not* occur? Explain.
HINT: What is the total of all the probabilities of all possible outcomes in any experiment? 0.65

27. Seven friends jog at various times. Today they are all jogging. If one person jogs every day, a second person jogs every other day, a third person jogs every three days, and so on, in how many days from today will all seven friends be jogging on the same day again? 420 days

24. ellipse; vertices: $(3, 3)$, $(3, -5)$; co-vertices: $(1, -1)$, $(5, -1)$

25. In a combination, the order of the arrangement does not matter; in a permutation, the order does matter.

Look Beyond

Exercise 26 considers the complement of an event occurring, and reminds students that the total of all probabilities of any event is 1. Exercise 27 is a fun teaser-type problem. Students should enjoy trying to find the solution to this exercise.

Objectives

- Use a random number simulation to approximate the probability of an event.

This lesson may be considered optional in the core course. See page 594A.

Assessing Prior Knowledge

A student is generating random numbers between 1 and 6 by rolling a number cube 10 times. How many different sets of 10 numbers are possible? [**60,466,176**]

TEACH

Why Sometimes situations contain random or chance factors that make analytical predictions difficult, if not impossible. To conduct an experiment of this kind is prohibitively costly in both time and resources. In these situations, a simulation technique, known as the random number simulation method is used.

LESSON 11.7 Simulation Methods

Why Simulation techniques can be used to imitate many types of situations involving chance, such as traffic flow on a proposed state highway, human behavior during times of stress, or the number of passengers likely to arrive for an airline flight. In situations such as these, direct experimentation may be very expensive, or even impossible.

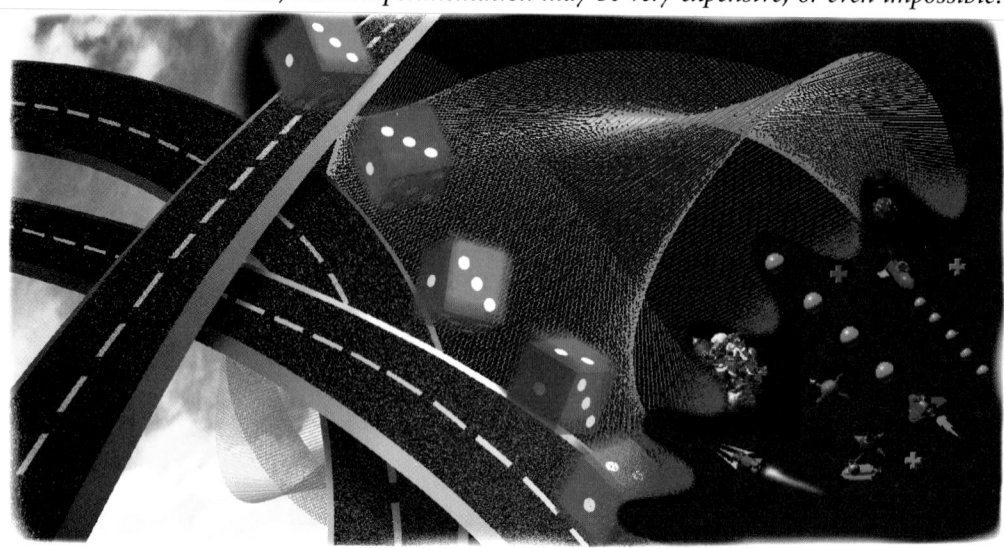

Random Number Simulations

Sometimes it is necessary to approximate probabilities by constructing a procedure that uses random numbers to simulate an actual experiment with an underlying random component. Simulation methods of this type are known as **random number simulations**. Most calculators and computers have a random number generator and the ability to select and fix the number of digits in a computation.

Graphics Calculator

Calculators or computers can be used to simulate tossing a number cube. For example, you can use your calculator to compute 6 random numbers between 1 and 6 inclusive.

Using the command **int**(6 * **rand** + 1), the random number 0.3694 produces a 3, as follows.

$$\text{int}(6 * \text{rand} + 1) = \text{int}(6 * 0.3694 + 1)$$
$$= \text{int}(2.2164 + 1)$$
$$= \text{int}(3.2164)$$
$$= 3$$

```
int (6*rand+1)
                4
                2
                1
                6
                2
                3
```

Use the command **int**(6 * **rand** + 1).

Press [ENTER] 6 times to execute this command 6 times to generate 6 random numbers. ❖

teaching strategy

Technology If the software package **Mathematica** is available, its RandomWalk[m] function generates and plots a two-dimensional random walk. This could be used to show repeated examples of random number simulations to experimentally study random probabilities. For example, a destination rectangle can be drawn on the screen with grease pencil and lengths of time to get there recorded for many trials.

EXAMPLE 1

Suppose experience has shown that the number of customers arriving at an airline ticket counter at any particular minute between 11:00 A.M. and 1:00 P.M. has the following probability distribution. Let X represent the number of customers arriving at the counter. Use random numbers to simulate customers arriving at a ticket counter during 10 one-minute periods.

X	0	1	2	3	4	5	6	7
P(X)	0.004	0.026	0.058	0.092	0.123	0.164	0.152	0.138

X	8	9	10	11	12	13	14	15
P(X)	0.087	0.058	0.035	0.031	0.021	0.007	0.003	0.001

Solution▶

Assign random numbers to each probability so that the number of random numbers is the same as the value of the probability. Notice that, for example, the probability of zero people arriving at the counter is determined to be 0.004, and there are 4 out of the 1000 random numbers assigned to that event. Use cumulative probabilities to make sure that each random number is used only once. Use 3-digit random numbers because each probability is given to the nearest thousandth.

X	P(X)	Cumulative probability	Random numbers
0	0.004	0.004	001–004
1	0.026	0.030	005–030
2	0.058	0.088	031–088
3	0.092	0.180	089–180
4	0.123	0.303	181–303
5	0.164	0.467	304–467
6	0.152	0.619	468–619
7	0.138	0.757	620–757
8	0.087	0.844	758–844
9	0.058	0.902	845–902
10	0.035	0.937	903–937
11	0.031	0.968	938–968
12	0.021	0.989	969–989
13	0.007	0.996	990–996
14	0.003	0.999	997–999
15	0.001	1.000	1000

Graphics Calculator

To simulate the arrival of customers at a counter during 10 one-minute periods, generate ten 3-digit random numbers between 001 and 1000, inclusive. Use the command **int**(1000 * **rand** + 1) with your graphics calculator. For the random numbers 358, 714, 276, 108, 552, 597, 201, 846, 068, and 688, it appears that 5, 7, 4, 3, 6, 6, 4, 9, 2, and 7 customers, respectively, arrived at a ticket counter during 10 one-minute periods between 11:00 A.M. and 1:00 P.M. ❖

Use your own 10 random numbers. How many people arrived at a counter during 10 one-minute periods using your random numbers?

interdisciplinary

CONNECTION

Physics Many problems in physics are too complex to model directly in mathematics, but they can be modeled with random number simulation methods. An example is the interaction of neutrons with the shielding in a nuclear power station. These methods can produce useful results and possible confirmation of assumptions.

Alternate Example 1

Collect random numbers by rolling a number cube 60 times and recording the results. Compare these numbers with the numbers your calculator gives you with the calculator command **int**(6 * **rand** +1) repeated 60 times. Use either set of random numbers to create 30 sets of ordered pairs. The first number in the ordered pairs indicates the number of points. The second number determines a win or a loss, with 4, 5, or 6 being a win and 1, 2, and 3 a loss. Each ordered pair is assigned alternately to each of two students. Each student then calculates the score for his or her set of ordered pairs. The student with the greater score wins.

A ongoing SSESSMENT

Answers may vary.

CRITICAL
Thinking

Explain why the total number of the cumulative probabilities has to be 1. Would changing the distribution of the random numbers in the table in Example 1 affect the simulation or the total number of the cumulative probabilities? Explain.

Simulations for Problems Related to Area

Random number simulations can be used to solve problems that are unrelated to probability.

EXAMPLE 2

Approximate the value of π using random number simulations.

GEOMETRY
Connection

Solution▶

Generate a sequence of random numbers, r, using the **rand** feature on your graphics calculator, so that $0 < r \leq 1$. Using these random numbers, form the ordered pairs (r_1, r_2), (r_3, r_4), (r_5, r_6), (r_7, r_8), and so on. The ordered pairs that satisfy the inequality $x^2 + y^2 < 1$ lie inside the quarter circle. The ordered pairs that satisfy the inequality $x^2 + y^2 \geq 1$ lie outside the quarter circle.

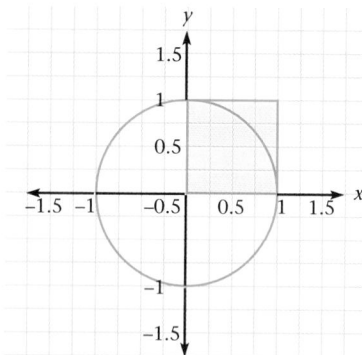

The area of the square shown is 1, and the area of the quarter circle is $\frac{1}{4}\pi(1)^2$, or $\frac{\pi}{4}$. Use proportional reasoning to compare the number of ordered pairs, or points, in the quarter circle and the number of points in the entire square with the area of the quarter circle and the area of the square.

$$\frac{\text{Number of points in the quarter circle}}{\text{Number of points in the square}} \approx \frac{\text{Area of the quarter circle}}{\text{Area of the square}} = \frac{\frac{\pi}{4}}{1}, \text{ or } \frac{\pi}{4}.$$

Multiply both sides of the fraction by 4 to isolate π.

$$4\left(\frac{\text{Number of points in the quarter circle}}{\text{Number of points in the square}}\right) = \pi$$

The table shows some approximations for various numbers of random numbers found.

Total number of random numbers	Random number approximation to π
20	3.45487
100	3.18509
200	3.16612
400	3.16022
600	3.14691
800	3.14506
1000	3.14318
1200	3.14302
1400	3.14223
1600	3.14192
1800	3.14176
2000	3.14163

Graphics Calculator

You can see that as the total number of random numbers increases, the approximation of π, approaches 3.1416. Since π = 3.14159... , this approximation appears to be quite good.

Use your graphics calculator to generate 20 three-decimal place random numbers between 0 and 1 inclusive. Use your numbers to approximate π. For the random numbers 0.952, 0.221, 0.369, 0.008, 0.935, 0.108, 0.006, 0.549, 0.856, 0.977, 0.278, 0.275, 0.122, 0.053, 0.722, 0.013, 0.421, 0.308, 0.972, and 0.029, only 1 ordered pair is outside the quarter circle, (0.856, 0.977).

Proportional reasoning gives

$$4 \left(\frac{9}{10} \right) = 3.6,$$

which is very near the value given for 20 random numbers in the table. ❖

Try This Use your graphics calculator and the command **rand** to generate 100 random numbers. Estimate the value of π using these numbers as ordered pairs, as in Example 2. Compare your estimate with the value of π and with the estimate shown in the table in Example 2.

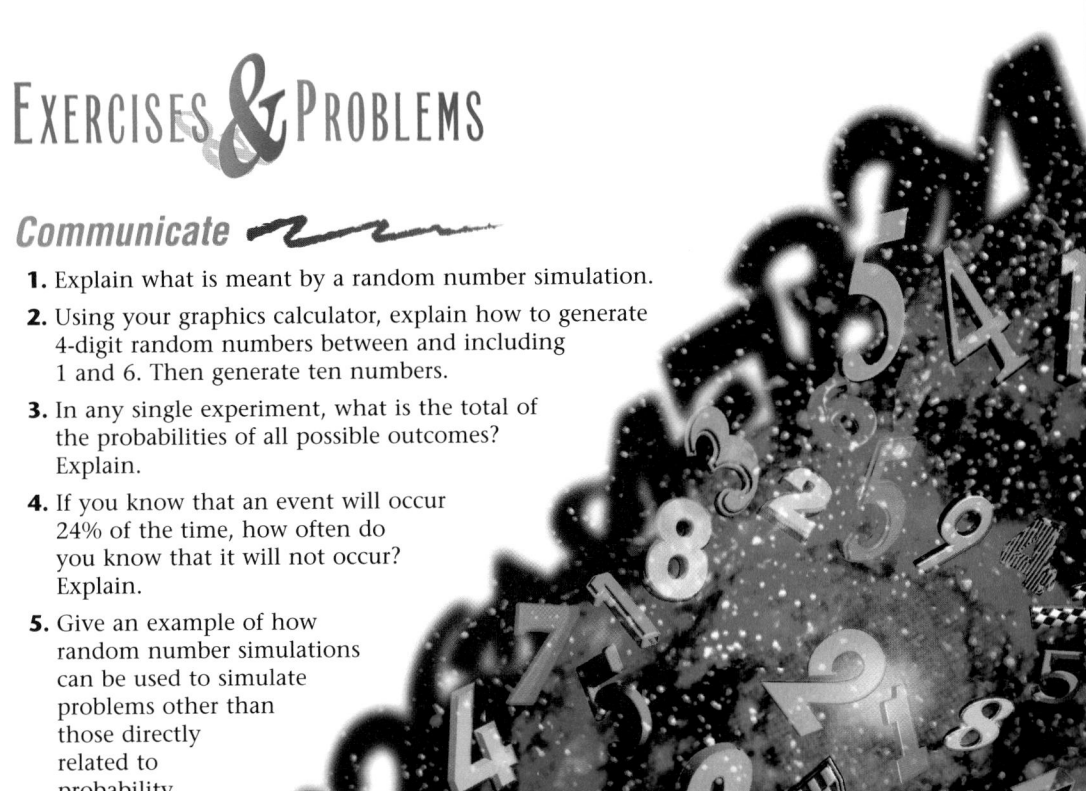

EXERCISES & PROBLEMS

Communicate

1. Explain what is meant by a random number simulation.

2. Using your graphics calculator, explain how to generate 4-digit random numbers between and including 1 and 6. Then generate ten numbers.

3. In any single experiment, what is the total of the probabilities of all possible outcomes? Explain.

4. If you know that an event will occur 24% of the time, how often do you know that it will not occur? Explain.

5. Give an example of how random number simulations can be used to simulate problems other than those directly related to probability.

RETEACHING the lesson

Using Example 2, but without the table filled in, show how the first few entries were determined. Then have the students complete it for themselves and compare it with the correctly completed table. Then show how another sequence of random numbers created a sequence of customer arrivals. Then have the students repeat the process for a third sequence of random numbers.

Practice & Apply

For Exercises 6–12, find the theoretical probability. Then use random number simulations to approximate each probability.

6. In 4 tosses of a number cube, the number 6 will appear exactly once. 0.0031

7. In 3 tosses of a number cube, the number 1 will appear all 3 times. 0.0046

8. In 2 tosses of a number cube, the number 1 will not appear at all. 0.694

9. In 5 tosses of a coin, heads will appear exactly 4 times. 0.156

10. In 2 tosses of a coin, heads will appear exactly 2 times. 0.25

11. In 2 draws, with replacement, from a full deck of 52 cards, a king will appear exactly 1 time. 0.142

12. In 2 draws, without replacement, from a full deck of 52 cards, a king will appear exactly 1 time. 0.145

13. Explain how to use a sequence of random numbers between 0 and 1 to simulate the process of tossing a single number cube.

14. Explain how to use a sequence of random numbers between 0 and 1 to simulate the process of generating random successes and failures in a quality-control situation if the probability of a success is 0.35.

15. **Coordinate Geometry** Example 2 demonstrates how to use a random number simulation to approximate the value of π. Use this procedure to approximate the area of the circle $x^2 + y^2 = 1$.

Real Estate The probabilities that a real-estate agent will sell 0 to 5 properties in one week are given in the Properties-Sold table.

16. Distribute 2-digit random integers from 01 to 100 among these values so that the corresponding random integers can be used to simulate the number of houses the real-estate agent sells per week.

17. Use the distribution you constructed in Exercise 16 to simulate the agent's weekly sales for 26 consecutive weeks. Answers may vary.

18. How do the probabilities from your simulation compare with the probabilities given? Is the sum of your probabilities 1.0? Yes Answers may vary.

Properties Sold

Number	Probability
0	0.12
1	0.21
2	0.26
3	0.19
4	0.12
5	0.10
Total	1.00

Customer Service The probabilities that a large retail store in a mall receives 0 to 10 complaints in a week are given in the Complaints-Received table.

19. Distribute 2-digit random integers from 01 to 100 among these 11 values so that the corresponding random integers can be used to simulate the number of complaints received per week.

20. Use the distribution you constructed in Exercise 19 to simulate the number of complaints for 10 consecutive weeks. Answers may vary.

21. How do the probabilities from your simulation compare with the probabilities given? Is the sum of your probabilities 1.0?

Complaints Received

Number	Probability
0	0.06
1	0.08
2	0.14
3	0.19
4	0.15
5	0.11
6	0.08
7	0.07
8	0.06
9	0.04
10	0.02
Total	1.00

16. Answers may vary for a simulation.

Number	Random Numbers
0	01–12
1	13–33
2	34–59
3	60–78
4	79–90
5	91–100

The answers to Exercises 19 and 21 can be found in Additional Answers beginning on page 842.

Baseball A baseball player's batting statistics are indicated in the Batting-Statistics table.

22. Generate a 3-digit random number, r, such that $0 < r \leq 1$, and match it to an outcome.

23. Simulate this player's activity at the plate, and record your results.

24. How do your probabilities compare with those given?

Batting Statistics

Outcome	Probability
Single	0.198
Double	0.061
Triple	0.020
Home run	0.032
Walk	0.118
Ground out	0.264
Fly out	0.187
Strike out	0.120
Total	**1.000**

The answers to Exercises 23–30 can be found in Additional Answers beginning on page 842.

Ecology The probabilities that there are 0 to 4 significant leaks of toxins into a large bay on any one day are given in the Toxins table.

25. Distribute random integers from 0001 to 10,000 among these 5 values so that the corresponding random integers can be used to simulate the number of significant toxic spills into the bay each day.

26. Use the distribution you constructed in Exercise 25 to simulate the number of spills into the bay for 30 consecutive days.

27. How do the probabilities from your simulation compare with the probabilities given?

Toxins

Number of leaks	Probability
0	0.2418
1	0.3356
2	0.2814
3	0.1181
4	0.0231
Total	**1.0000**

28. **Coordinate Geometry** Use a random number simulation to approximate the area of the circle $x^2 + y^2 = 2$.

Construction The buyer for a construction company must arrange for roofing tiles to be ordered in such a way that they are sent as efficiently as possible to job sites. Suppose that during any one week, the number of batches of roofing tiles that the buyer orders is a random variable with the probability distribution given in the Batches-of-Tiles table. Let X represent the number of batches of tiles and $P(X)$ the corresponding probability.

29. Simulate the number of batches of roofing tiles the buyer must order for a one-week period, and record your results.

30. How do the probabilities from your simulation compare with the probabilities given?

Batches of Tiles

X	P(X)
10	0.11
20	0.18
30	0.26
40	0.23
50	0.16
60	0.06
Total	**1.00**

22. int (rand *1000)/1000; answers may vary for a simulation.

23.

Outcome	Random Numbers
Single	0.001–0.198
Double	0.199–0.259
Triple	0.260–0.279
Home Run	0.280–0.311
Walk	0.312–0.429
Ground Out	0.430–0.693
Fly Out	0.694–0.880
Strike Out	0.881–1.000

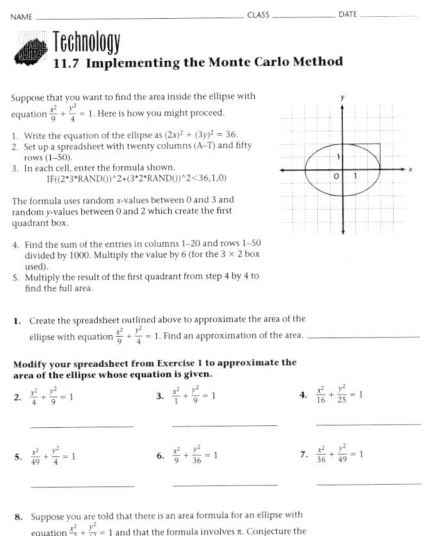

NAME _____ CLASS _____ DATE _____

Technology
11.7 Implementing the Monte Carlo Method

Suppose that you want to find the area inside the ellipse with equation $\frac{x^2}{9} + \frac{y^2}{4} = 1$. Here is how you might proceed.

1. Write the equation of the ellipse as $(2x)^2 + (3y)^2 = 36$.
2. Set up a spreadsheet with twenty columns (A–T) and fifty rows (1–50).
3. In each cell, enter the formula shown.
   IF($(2*3*RAND())^2+(3*2*RAND())^2 < 36$,1,0)

The formula uses random x-values between 0 and 3 and random y-values between 0 and 2 which create the first quadrant box.

4. Find the sum of the entries in columns 1–20 and rows 1–50 divided by 1000. Multiply the value by 6 (for the 3 × 2 box used).
5. Multiply the result of the first quadrant from step 4 by 4 to find the full area.

1. Create the spreadsheet outlined above to approximate the area of the ellipse with equation $\frac{x^2}{9} + \frac{y^2}{4} = 1$. Find an approximation of the area. _____

Modify your spreadsheet from Exercise 1 to approximate the area of the ellipse whose equation is given.

2. $\frac{x^2}{4} + \frac{y^2}{9} = 1$ **3.** $\frac{x^2}{1} + \frac{y^2}{9} = 1$ **4.** $\frac{x^2}{16} + \frac{y^2}{25} = 1$

5. $\frac{x^2}{49} + \frac{y^2}{4} = 1$ **6.** $\frac{x^2}{9} + \frac{y^2}{36} = 1$ **7.** $\frac{x^2}{36} + \frac{y^2}{49} = 1$

8. Suppose you are told that there is an area formula for an ellipse with equation $\frac{x^2}{a^2} + \frac{y^2}{b^2} = 1$ and that the formula involves π. Conjecture the formula by examining your results from Exercises 1–7.

9. Approximate the area between $y = x^2 + 1$ and the x-axis by using the Monte Carlo Method.

HRW Advanced Algebra   Technology   21

31. Answers may vary for the simulation

X	Random Numbers
0	01–04
1	05–11
2	12–20
3	21–37
4	38–59
5	60–75
6	76–84
7	85–89
8	90–94
9	95–98
10	99–100

Medicine Let the number of people admitted to a city's hospital on any summer day when the temperature is greater than 100°F be a random variable with the probability distribution given in the Hospital-Admissions table. Let X represent the number of people admitted and $P(X)$ represent the corresponding probability.

31. Simulate the number of people admitted to this hospital during a summer day when the temperature is greater than 100°F. Record your results.

32. How do the probabilities from your simulation compare with the probabilities given?
Answers may vary, but probabilities should be similar.

Hospital Admissions

X	P(X)
0	0.04
1	0.07
2	0.09
3	0.17
4	0.22
5	0.16
6	0.09
7	0.05
8	0.05
9	0.04
10	0.02

Look Back

33. What is the inverse of $f(x) = \log x$?   **[Lesson 7.5]** $f^{-1}(x) = 10^x$

Use your calculator in radian mode to graph the function $f(x) = 5 - \cos\left(\frac{\pi x}{4}\right)$.   [Lesson 8.7]

34. **Maximum/Minimum** Find the maximum and minimum values of f. max:6; min:4

35. **Maximum/Minimum** Find the period of f. 8

36. In a certain state, all domestic animals (dogs and cats) must have a coded identification tag around their neck. The first place of each identification tag must be an 8, the second and third places must be letters, and the fourth and fifth places can be any numbers, with repetition allowed. How many different identification tags can be made?   **[Lesson 11.2]** 67,600

37. **Geometry** How many lines are determined by 8 points, no 3 of which are collinear? How many triangles are determined by the same points if no 4 are coplanar?   **[Lesson 11.4]** 28; 56

Look Beyond

38. Starting 3 miles away, a girl walks home at an average of 3 miles per hour. Her dog, starting with her, runs home at an average of 5 miles per hour. The dog then immediately turns around and runs back to the girl, and then back home, and then back to the girl, and so on until the girl reaches home. How far does the dog run? 5 miles

Look Beyond

The next chapter studies number patterns called sequences and develops students mathematics skill for working with these sequences and obtaining the sums of related series.

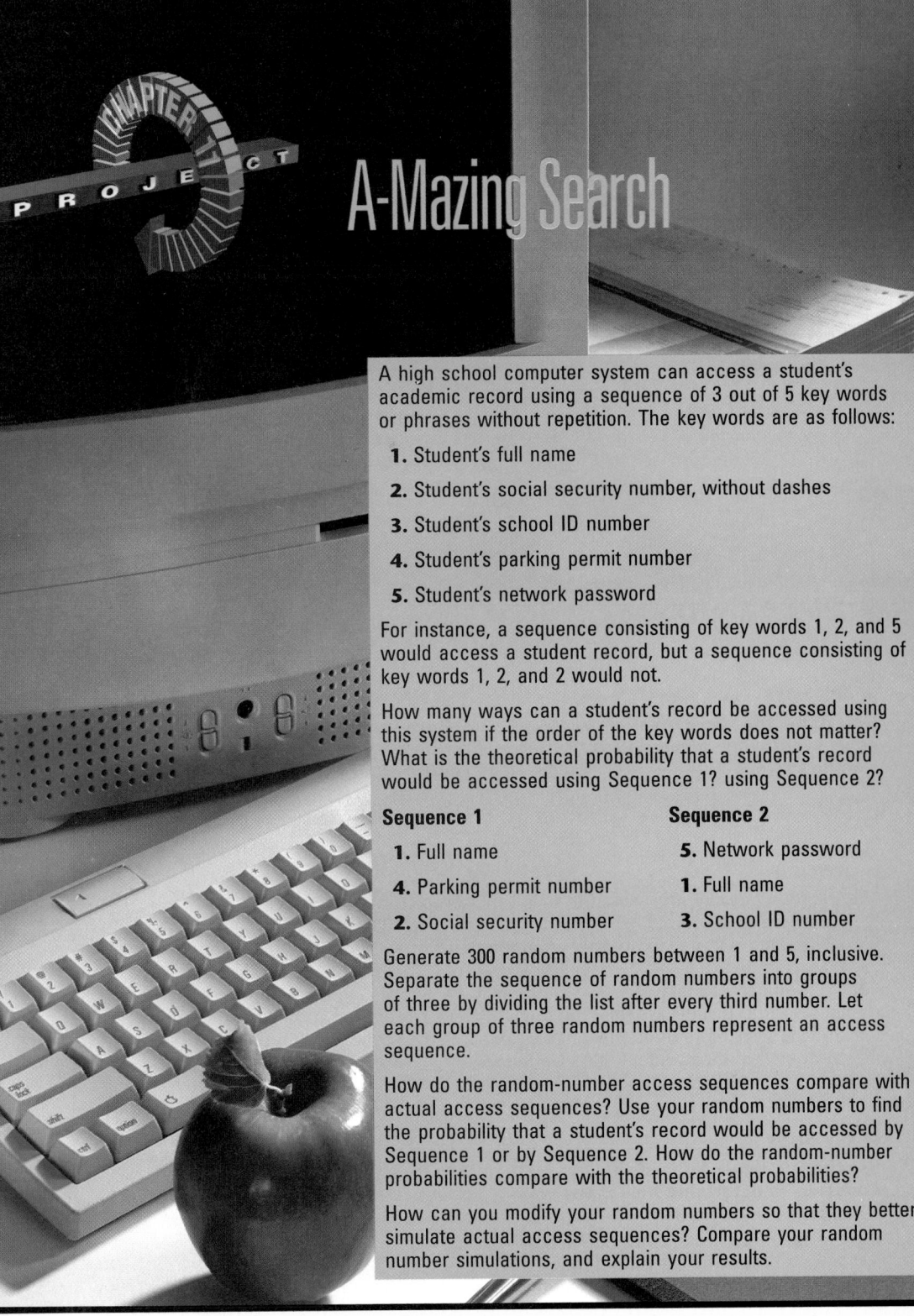

A-Mazing Search

A high school computer system can access a student's academic record using a sequence of 3 out of 5 key words or phrases without repetition. The key words are as follows:

1. Student's full name
2. Student's social security number, without dashes
3. Student's school ID number
4. Student's parking permit number
5. Student's network password

For instance, a sequence consisting of key words 1, 2, and 5 would access a student record, but a sequence consisting of key words 1, 2, and 2 would not.

How many ways can a student's record be accessed using this system if the order of the key words does not matter? What is the theoretical probability that a student's record would be accessed using Sequence 1? using Sequence 2?

Sequence 1
1. Full name
4. Parking permit number
2. Social security number

Sequence 2
5. Network password
1. Full name
3. School ID number

Generate 300 random numbers between 1 and 5, inclusive. Separate the sequence of random numbers into groups of three by dividing the list after every third number. Let each group of three random numbers represent an access sequence.

How do the random-number access sequences compare with actual access sequences? Use your random numbers to find the probability that a student's record would be accessed by Sequence 1 or by Sequence 2. How do the random-number probabilities compare with the theoretical probabilities?

How can you modify your random numbers so that they better simulate actual access sequences? Compare your random number simulations, and explain your results.

FOCUS

In this project, students will see how the concepts of permutation groups could apply to them directly.

MOTIVATE

In today's information age, the control of access to personal information is increasingly important. The security of information systems hinges upon the combinations and permutations of the keys selected.

Cooperative Learning

In groups of three, have the students discuss what the problem is asking and how they might find the answer. Then hold a class-wide discussion to clarify the project. Students return to their groups to work the problem, and then reconvene to compare and discuss answers.

DISCUSS

What were the information security issues raised in this project? What changes, if any, ought to be made to the system as described?

Chapter 11 Review

Vocabulary

Key Skills & Exercises

Lesson 11.1

➤ Key Skills

Determine the theoretical probability of an event when given an appropriate sample space.

$$P(E) = \frac{\text{number of elements in an event}}{\text{number of elements in sample space}}$$

Determine the experimental probability of an event when given the results of an experiment.

$$P(E) = \frac{\text{number of times the event occurred}}{\text{number of times experiment was conducted}}$$

➤ Exercises

1. A compact disc of music used in major motion pictures has 5 tracks from each of 4 decades—the 1960s, the 1970s, the 1980s, and the 1990s. Find the probability that a random selector will first select a track from before 1980. $\frac{1}{2}$

2. Of 28 students, 7 take French, 11 take Spanish, 5 take German, and 3 take Latin. If these are the only foreign languages offered and no student takes more than one language, what is the probability that one randomly selected student takes German? $\frac{5}{28}$

Lesson 11.2

➤ Key Skills

Use the Fundamental Principle of Counting to determine how many ways a decision can be made.

If there are m ways of making Decision 1 and n ways of making Decision 2, then there are $m \times n$ ways to make the two decisions.

Determine the number of permutations of n distinct objects taken r at a time.

$$_nP_r = \frac{n!}{(n-r)!}$$

➤ Exercises

3. A new car is available in 3 choices of interior fabric, 5 choices of interior color, and 7 choices of exterior color. Assuming that all combinations of fabric, interior color, and exterior color are available, how many different cars could be ordered? 105

4. In how many ways can 8 trophies be arranged side by side on a shelf? 40,320

Lesson 11.3

➤ Key Skills

Find the number of distinct permutations of n objects of which r_1 objects are alike and r_2 objects are alike.

$$\frac{n!}{r_1! \, r_2!}$$

Find the number of circular permutations of n objects.

$$(n-1)!$$

➤ Exercises

5. If a CD changer can hold 6 discs in a column arrangement, in how many ways could 6 discs be arranged in this changer? What if the changer were circular? 720; 120

6. In how many ways can the expression a^3b be written if exponents are not used? 4

Lesson 11.4

➤ Key Skills

Determine the number of combinations of n objects taken r at a time.

$$_nC_r = \binom{n}{r} = \frac{n!}{r! \, (n-r)!}$$

Use the method for counting combinations with the Fundamental Principle of Counting to determine how many ways a decision can be made.

If Decision 1 can be made in $_4C_2$ ways and Decision 2 can be made in $_5C_1$ ways, the number of ways to make both decisions is $(_4C_2)(_5C_1)$.

➤ Exercises

The Spanish club has 24 girls and 20 boys.

7. In how many ways can a president and a vice-president be chosen? 1892

8. If 4 representatives are randomly chosen, what is the probability that exactly 2 girls and 2 boys will be chosen?

$$\frac{_{24}C_2 \cdot _{20}C_2}{_{44}C_4} \approx 0.386, \text{ or } 38.6\%$$

Lesson 11.5

➤ Key Skills

Find the probability that two independent events A and B will both occur.

$$P(A \text{ and } B) = P(A) \cdot P(B)$$

Find the probability that either event A or event B will occur.

If A and B are any events, then

$$P(A \text{ or } B) = P(A) + P(B) - P(A \text{ and } B).$$

If A and B are mutually exclusive events, then

$$P(A \text{ or } B) = P(A) + P(B).$$

➤ **Exercises**

On his drive to school, Josh must go through two traffic lights. Because of the way the lights are timed, he has a 0.5 chance of being stopped by the first light and a 0.3 chance of being stopped by the second light. These two signals operate independently of each other.

9. What is the probability that Josh will have to stop for both lights?

10. What is the probability that Josh will have to stop for only one of the lights?

11. What is the probability that Josh will not have to stop at either light?

Lesson 11.6

➤ **Key Skills**

Determine the probability of an event occurring that depends on the probability of another event occurring.

The conditional probability that event E_2 will occur, given that event E_1 occurs is

$$P(E_2|E_1) = \frac{P(E_1 \text{ and } E_2)}{P(E_1)}, \text{ where } P(E_1) \neq 0.$$

➤ **Exercises**

Four coins are flipped. Find the probability that they will all land heads up given that the following event occurs.

12. The first coin lands heads up. $\frac{1}{8}$ **13.** The first two coins land heads up. $\frac{1}{4}$

Lesson 11.7

➤ **Key Skills**

Use a random number simulation to approximate the probability of an event.

To simulate on a calculator the random results when there are n number of equally likely possible outcomes, use the command **int(n * rand** + 1).

➤ **Exercises**

14. Use a graphics calculator to simulate tossing 3 coins 50 times to approximate the probability of getting 3 heads on any single toss. ≈ 0.125

Applications

15. Health A test of 100 adults who kept track of their daily sodium intake showed that 40 of them averaged more than the recommended maximum intake of 2400 milligrams of sodium per day. Of these with high sodium intake, 50% had higher-than-normal blood pressure. Of the other 60 adults, who kept their sodium intake at or below 2400 milligrams per day, only 15% had high blood pressure. Using this data collection as a sample of the general population, find the probability that a person with high blood pressure has a daily intake of over 2400 milligrams of sodium per day. 0.69

Public Service According to a recent survey conducted by the Peace Corps, 93% of the volunteers in the Peace Corps are single and 74% are in their 20s. Assume that being single and being in your 20s are independent events.

16. What is the probability that a person in the Peace Corps is both single *and* in their 20s? 0.6882

17. What is the probability that a person in the Peace Corps is either single or in their 20s or both? 0.9818

Chapter 11 Assessment

Suppose you roll two numbered octahedrons (8 faces) with the faces numbered 1 through 8. Find the probability that you will roll the following sum.
1. Less than 5 $\frac{3}{32}$
2. Exactly 15 $\frac{1}{32}$

Student Poll Of the 2000 students at Ridgeview High, 100 were asked whether they spent more time watching television, listening to the radio, or listening to their own tapes and compact discs. Twenty spent more time watching TV, 15 spent more time listening to the radio, 50 spent more time listening to their own tapes and compact discs, and the remainder had no opinion. What is the theoretical probability that a student chosen at random from the total student body does the following? **Complete Items 3 and 4.**

3. Spends more time either watching TV or listening to the radio 35%

4. Gives no opinion as a response 15%

5. In the student council elections, 4 students ran for president, 5 for vice-president, 3 for secretary, and only 1 for treasurer. How many different teams of these 4 officers can possibly be elected? 60

6. How many different "tunes" can be played using the 8 notes in one octave of the C-major scale if each tune is composed of 3 different quarter notes? 336

7. In how many ways could the 20 antique books be arranged on a single display shelf? 20!

8. How many different arrangements can be formed from the letters *VOODOO*? 30

9. In how many ways can 10 students form a ring around a flagpole? 362,880

10. How many different 3-topping pizzas can be made with 10 toppings? 120

11. Consider the number of ways to choose 3 representatives out of a class of 20 as compared with the number of ways to leave 17 students unchosen. Then explain why $_{20}C_3 = {_{20}}C_{17}$.

12. In how many ways can the coach place 9 of his 20 baseball players on the field if he has 4 players that can only be pitchers and 2 that can only be catchers?

13 Use a graphics calculator to simulate the probability of getting a total of 7 when two 6-sided number cubes are rolled. $\frac{1}{6}$

Geometry A diagonal is a line segment that connects any two nonconsecutive vertices of a polygon.

14. Draw a hexagon with all of its diagonals, and show how the total number of diagonals is related to $_6C_2$. (Remember that sides are not diagonals.) 9

15. Use what you learned about the number of diagonals in a hexagon to determine the number of diagonals in a dodecagon (12-sided polygon). 54

---

11. There are 1140 ways to choose 3 students out of 20; therefore, there are 1140 ways to leave 17 students not chosen.

12. 27,456

14. $_6C_2$ would give the number of ways 6 vertices could be connected 2 at a time; diagonals connect non-adjacent vertices, so the number of sides must be subtracted from this number; number of diagonals $= {_6}C_2 - 6 = 9$.

15. $_{12}C_2 - 12 = 54$

12 Discrete Mathematics: Series and Patterns

Meeting Individual Needs

12.1 Exploring Sequences

Core	Core Plus
Inclusion Strategies, p. 656	Practice Master 12.1
Reteaching the Lesson, p. 657	Enrichment, p. 656
Practice Master 12.1	Technology Master 12.1
Enrichment Master 12.1	Lesson Activity Master 12.1
Technology Master 12.1	
Lesson Activity Master 12.1	
Interdisciplinary Connection, p. 655	
[1 day]	**[1 day]**

12.2 Arithmetic Series

Core	Core Plus
Inclusion Strategies, p. 663	Practice Master 12.2
Reteaching the Lesson, p. 664	Enrichment, p. 663
Practice Master 12.2	Technology Master 12.2
Enrichment Master 12.2	Lesson Activity Master 12.2
Technology Master 12.2	
Lesson Activity Master 12.2	
Interdisciplinary Connection, p. 662	
[2 days]	**[1 day]**

12.3 Geometric Series

Core	Core Plus
Inclusion Strategies, p. 673	Practice Master 12.3
Reteaching the Lesson, p. 674	Enrichment, p. 673
Practice Master 12.3	Technology Master 12.3
Enrichment Master 12.3	Lesson Activity Master 12.3
Technology Master 12.3	Mid-Chapter Assessment Master
Lesson Activity Master 12.3	
Interdisciplinary Connection, p. 672	
Mid-Chapter Assessment Master	
[2 days]	**[1 day]**

12.4 Infinite Geometric Series

Core	Core Plus
Inclusion Strategies, p. 682	Practice Master 12.4
Reteaching the Lesson, p. 683	Enrichment, p. 682
Practice Master 12.4	Technology Master 12.4
Enrichment Master 12.4	Lesson Activity Master 12.4
Technology Master 12.4	Interdisciplinary Connection, p. 681
Lesson Activity Master 12.4	
[2 days]	**[2 days]**

12.5 Pascal's Triangle and Probability

Core	Core Plus
Inclusion Strategies, p. 690	Practice Master 12.5
Reteaching the Lesson, p. 691	Enrichment, p. 690
Practice Master 12.5	Technology Master 12.5
Enrichment Master 12.5	Lesson Activity Master 12.5
Technology Master 12.5	Interdisciplinary Connection, p. 689
Lesson Activity Master 12.5	
[2 days]	**[1 day]**

12.6 The Binomial Theorem

Core	Core Plus
Inclusion Strategies, p. 697	Practice Master 12.6
Reteaching the Lesson, p. 698	Enrichment, p. 696
Practice Master 12.6	Technology Master 12.6
Enrichment Master 12.6	Lesson Activity Master 12.6
Technology Master 12.6	
Lesson Activity Master 12.6	
[2 days]	**[1 day]**

Chapter Summary

Core	Core Plus
Eyewitness Math, pp. 686–687	Eyewitness Math, pp. 686–687
Chapter 12 Project, p. 701	Chapter 12 Project, p. 701
Lab Activity	Lab Activity
Long-Term Project	Long-Term Project
Chapter Review, pp. 702–704	Chapter Review, pp. 702–704
Chapter Assessment, p. 705	Chapter Assessment, p. 705
Chapter Assessment, A/B	Chapter Assessment, A/B
Alternative Assessment	Alternative Assessment
Cumulative ssessment, pp. 706–707	Cumulative Assessment, pp. 706–707
[3 days]	**[3 days]**

Reading Strategies

After students study the lesson and work the exercises, they should be able to associate the type of problems in the exercises with the methods and concepts that the lesson emphasizes. In this way, students can identify a general method they can use for solving the type of problems in this chapter. Each student might want to consider in detail one application problem whose solution summarizes the mathematics in the lesson. They can add notes about the solution of the problem and have a good example for their portfolio.

Visual Strategies

Direct the students to look at the pictures and associated graphs in the chapter opener. Encourage them to speculate about how mathematics might be useful in studying such phenomena. Have students look carefully at how the pictures illustrate patterns that can be modeled by numerical sequences. See if they can connect the visual pattern with abstract representations.

Hands-on Strategies

This chapter gives students a number of opportunities to become actively involved with mathematics in the physical world. The handshaking project gives students a tangible understanding of patterns.

Cooperative Learning

GROUP ACTIVITIES	
Infinite geometric series	Lesson 12.4, Example 2
Random sequences	Eyewitness Math: Is It Fake or is it Random?
A triangle pattern	Lesson 12.5, Exploration 1
Heads and tails	Lesson 12.6, Exploration
Geometric models	Chapter Project

You may wish to have students work with partners for some of the above activities. Additional suggestions for cooperative group activities are noted in the teacher's notes in each lesson.

Multicultural

The cultural connections in this chapter include references to Europe and Asia.

CULTURAL CONNECTIONS	
Europe: Bode's Sequence	Lesson 12.1
Asia: The Arithmetic Triangle	Lesson 12.5, Introduction
Circle game	Lesson 12.5, Exercise 44

Portfolio Assessment

Below are portfolio activities for the chapter. They are listed under seven activity domains that are appropriate for portfolio development.

1. **Investigation/Exploration** Differences between terms in a sequence is the focus of Exploration 1 in Lesson 12.1; ratios between terms is the objective of Exploration 2 in Lesson 12.1. Another exploration you may wish to include is the randomness in sequences in the Eyewitness Math feature.

2. **Applications** Recommended are any of the following: health, Lesson 12.1, Exercise 30; investments, Lesson 12.3, Exercises 30–31; transportation, Lesson 12.5, Exercise 22–25.

3. **Non-Routine Problems** The Is it Fake or is it Random? problem from Eyewitness Math presents the students with non-traditional but interesting ways to look at sequences and randomness.

4. **Project Handshakes and Polygons** Students are asked to extend a geometric model to model the number of handshakes when a certain number of people meet.

5. **Interdisciplinary Topics** Students may choose from the following: statistics, Lesson 12.2, Exercises 47–50; statistics, Lesson 12.3, Exercises 32–34; geometry Lesson 12.3, Exercises 35–38; physics, Lesson 12.4, Exercise 18.

6. **Writing** Communicate exercises of the type in which a student is asked to describe basic concepts of sequences, series, summation, convergence, partial and infinite sums, Pascal's triangle and the Binomial Theorem offer excellent writing selections for the portfolio.

7. **Tools** Chapter 12 uses graphics calculators to compare the explicit and recursive forms of arithmetic and geometric sequences and series. To measure the individual student's proficiency, it is recommended that he or she complete selected worksheets from the Technology Masters.

Technology

The primary role of calculators in this chapter is to speed up the repetitive processes of finding and confirming the patterns in sequences, generating more terms, and adding up the terms in the resulting series. The factorial and combination functions are also useful. In many cases, the scientific calculator is easier and quicker because these operations are on hard keys, not on a pull-down menu that requires many key strokes to access.

For example, to generate terms in the arithmetic sequence 5, 11, 17, 23, ... this pattern of key strokes is very efficient: `5` `ENTER` `+` `6` `ENTER` `ENTER`...

The same pattern can be used for geometric sequences if the multiplication key is used instead of the addition key.

If the calculator is programmable, this sequence of steps can be used as the core of a short program. The use of the pause command will allow the data to be recorded as each term is generated and added.

Calculators frequently use the concepts of this chapter to calculate approximations to the transcendental functions. The natural exponential function (base e) is an example. Encourage students to calculate an approximation of e using the first 10 terms of the series.

$$e^x = 1 + x + \frac{x^2}{2!} + \frac{x^3}{3!} + \frac{x^4}{4!} + \dots$$

Have them compare their answer with that of the `eˣ` key. Doing this exercise is good practice in repetitive operations on the calculator.

While the formulas of the chapter (especially those involving $n!$ and $_nC_r$) appear to be shortcuts, especially with a calculator, the danger is that the calculator can conceal much of the underlying mathematics. It is important that these tools be used as a means to the end (a specific numerical result), and do not overshadow the terminology, skills, and basic concepts necessary to learn the basic mathematics.

Integrated Software

f(g) Scholar™ is an integrated computer-based mathematics productivity tool that combines calculator, spreadsheet, and graphics capabilities to provide a dynamic and interactive environment for explorations in mathematics. Use *f(g) Scholar*™ for any lesson needing a spreadsheet, calculator, graphics calculator, or any combination of the three.

Discrete Mathematics: Series and Patterns

ABOUT THE CHAPTER

Background Information

Mathematics is the discovery and representation of patterns. Some patterns are finite and some are infinite. Some are geometric and some are arithmetic. In this chapter, students will explore infinite numerical patterns, called *sequences* and *series*. These patterns can be found in nature or can be invented by mathematicians.

CHAPTER RESOURCES

- Practice Masters
- Enrichment Masters
- Technology Masters
- Lesson Activity Masters
- Lab Activity Masters
- Long-Term Project Masters
- Assessments Masters
    Chapter Assessments, A/B
    Mid-Chapter Assessment
    Alternative Assessment, A/B
- Teaching Transparencies
- Cumulative Assessment
- Spanish Resources

CHAPTER OBJECTIVES

- Determine whether a given sequence is an arithmetic sequence.
- Determine whether a given sequence is a geometric sequence.
- Find the *n*th term of an arithmetic sequence.
- Find the sum of the first *n* terms of an arithmetic series.
- Find the *n*th term of a geometric sequence.
- Find the sum of the first *n* terms of a geometric series.

CHAPTER 12

Discrete Mathematics: Series and Patterns

LESSONS

Mathematics can be described as the science of patterns. Sequences of numbers, such as 1, 4, 9, 16, ... , often occur on aptitude tests, where they are used to measure the ability to discover, extend, and record patterns. Finding the patterns in sequences of numbers is often used in mathematics, science, economics, and even for entertainment. You will investigate various types of sequences in this chapter.

ABOUT THE PHOTOS

More complex living things are made up of smaller elements (cells, seeds, bones, leaves, scales, etc.). As these grow and develop, sequential patterns are formed. The seeds of a sunflower, the segments of a pineapple, and the spikes of a pine cone all contain natural sequential patterns. It is possible to model the sequential patterns in nature by abstract patterns of shapes in geometry and numbers in algebra.

- Find the sum of an infinite geometric series.
- Use Pascal's Triangle to find combinations.
- Use the Binomial Theorem to expand $(a + b)^n$.
- Use the Binomial Theorem to find a particular term in the expansion of $(a + b)^n$.

PORTFOLIO ACTIVITY

In this activity, students can see how sequences naturally arise in everyday situations. They will find that short, simple sequences can be easily listed directly and studied by using simple arithmetic. They can also determine that many sequences and series are too complicated to be analyzed directly.

Organize students into groups of eight. If there are students remaining, appoint each one as a recorder for a group. Have each group act out the problem and calculate the answer by addition, including a running subtotal of the number of handshakes after each person joins the meeting. Discuss with students how they might solve the problem if it is extended to 100 people. Also, have them form a sequence whose terms are the successive subtotals and find the pattern in it. This provides an example of a recursive sequence that is neither arithmetic nor geometric.

PORTFOLIO ACTIVITY

The Handshake Problem

Eight people are attending a meeting. If each person shakes hands with every other person, how many handshakes will occur?

Imagine the 8 people entering a room one at a time. New handshakes occur as each new person enters the room. For example, when the third person enters, that person must shake hands with the two people already there.

You will be asked to solve this problem on page 669. You may wish to include your work in your portfolio.

ABOUT THE CHAPTER PROJECT

In the Chapter Project on page 701, students should recognize the connection between the sequence from the portfolio activity and a geometric sequence. They should see how the underlying mathematics is the same. Additionally, students should have an opportunity to see how different representations can be used to model the same situation.

Prepare

Objectives

- Determine whether a given sequence is an arithmetic sequence.
- Determine whether a given sequence is a geometric sequence.

Resources

- Practice Master 12.1
- Enrichment Master 12.1
- Technology Master 12.1
- Lesson Activity Master 12.1
- Quiz 12.1
- Spanish Resources 12.1

Assessing Prior Knowledge

1. Find the ratio of 12 to 3. **[4:1]**
2. Two numbers differ by 5. Express the other number if the smaller one is n. **[$n + 5$]**

Teach

why For years humans have tried to see order and patterns in the Earth, in the sky, and even in the cells of living things. Often, the numbers associated with measurements taken or intervals counted will produce patterns that increase or decrease in a regular progression or that repeat time after time. These number patterns are called sequences.

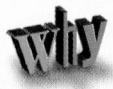

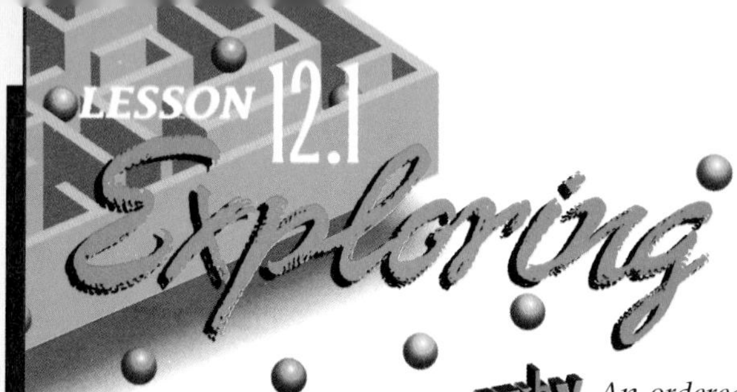

LESSON 12.1 Exploring Sequences

why *An ordered list of numbers can often be used to describe a particular situation. For instance, the numbers 2, 4, 6, 8, 10, 12, … describe the positive even integers. General descriptions of ordered lists of numbers, or sequences of numbers, are very useful in business, science, and art.*

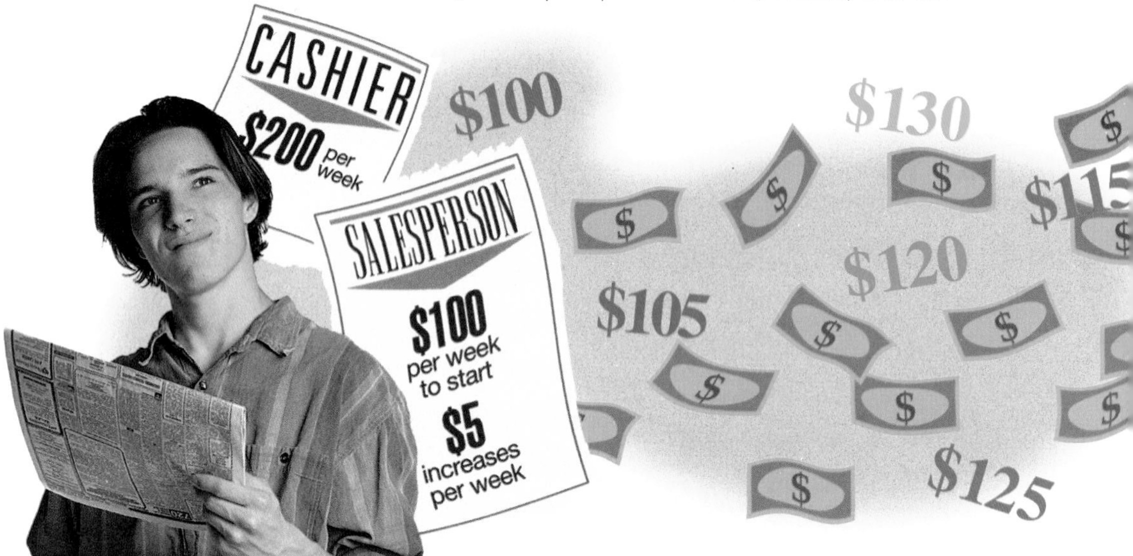

How do the wages for each job compare?

The weekly salaries for the salesperson—$100, $105, $110, and so on—make an ordered list, or **sequence**, of numbers. The numbers in a sequence are the **terms of the sequence**, represented by the letter t.

$$\$100, \ \$105, \ \$110, \ldots, ?, \ ?$$

$$t_1 \quad t_2 \quad t_3 \quad t_{n-1} \quad t_n$$

The third term, t_3, is $110, the second term, t_2, is $105. The difference between the third term and the second term, $t_3 - t_2$, is $5. The difference, d, between *any* two consecutive terms in a sequence is written $t_n - t_{n-1}$.

$$\begin{aligned} d &= t_n - t_{n-1} \\ &= t_3 - t_2 \\ &= \$110 - \$105 \\ &= \$5 \end{aligned}$$

ALTERNATIVE teaching strategy

Hands-On Strategies Put the number 3 on the chalkboard. Pick a row of students and have the first student add 5 to the 3. Record the 8 on the chalkboard, and then have each successive student add a 5 to the last recorded number (3, 8, 13, 18 …). When that row has finished, begin with another 3 and the next row of students. Have each person, in succession, multiply the previously recorded number by 5 (3, 15, 75, 375 …). Compare the two lists. This will serve to introduce, compare, and contrast *arithmetic* and *geometric sequences.*

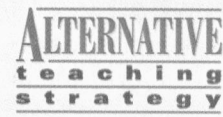

Exploration 1 Differences Between Terms

Let $S = 100, 105, 110, \ldots$ be the sequence of wages for the salesperson.

1 Find the first 10 terms, and complete the table.

n	1	2	3	4	5	6	7	8	9	10
t_n	t_1 100	t_2 105	t_3 110							

2 Find the difference between each consecutive pair of terms in Step 1. How do these differences compare?

3 An **arithmetic** (pronounced ar-ith-MET-ik) **sequence** is a sequence with a constant difference, d, between any two consecutive terms, t_n and t_{n-1}. Explain why the salary sequence is an arithmetic sequence. What is d for the salary sequence?

4 Complete the table using the value of d for this arithmetic sequence. Use the pattern to write a general formula for t_n in terms of t_1 and d.

n	1	2	3	4	5	6	7	8	9	10	$\ldots$	n
t_n	t_1	t_2	t_3	t_4	t_5	t_6	t_7	t_8	t_9	t_{10}	$\ldots$	t_n
	100	105	110								$\ldots$	—
	$100 + 0$	$100 + 5$	$100 + 10$								$\ldots$	—
	$t_1 + 0d$	$t_1 + 1d$	$t_1 + 2d$								$\ldots$	

5 Use the formula for t_n that you wrote in Step 4 to find the salary for the fifteenth week, t_{15}. Check your answer by extending the sequence.

Which sequences below are arithmetic? If the sequence is *not* arithmetic, explain why not.

6 $S_1 = 63, 44, 25, 6, -13, -32, \ldots$

7 $S_2 = 64, 32, 16, 8, 4, 2, 1, \ldots$

8 $S_3 = 1, -1, 1, -1, 1, -1, \ldots$

9 $S_4 = 2, 4, 6, 8, 10, 12, \ldots$ ❖

Examine the following sequence of *powers* of 2.

$$1, 2, 4, 8, 16, 32, \ldots , t_{n-1}, t_n$$

The value of d is different for each pair of consecutive terms.
However, the *ratio* between consecutive terms, $r = \dfrac{t_n}{t_{n-1}}$, is constant.

$$r = \frac{t_2}{t_1} = \frac{2}{1} = 2 \qquad r = \frac{t_3}{t_2} = \frac{4}{2} = 2 \qquad r = \frac{t_4}{t_3} = \frac{8}{4} = 2 \qquad r = \frac{t_5}{t_4} = \frac{16}{8} = 2$$

Write each term in the powers-of-2 sequence as 2 raised to some power. How would you describe the nth term of this sequence?

Exploration 1 Notes

Students should learn that an arithmetic sequence is generated by adding a common difference to each term. If they subtract two successive terms they will find this common difference.

A ongoing SSESSMENT

3. The difference between each pair of consecutive terms is constant, $5; $d = 5$.

4. $t_1 + (n - 1)d = t_n$

A ongoing SSESSMENT

$2^0, 2^1, 2^2, 2^3, 2^4, 2^5, \ldots$; the nth term is 2^{n-1}

---

interdisciplinary

CONNECTION

Business Present students with the following scenario, and have them solve the following problem.

A certain company wants to ensure quality, yet expects each new employee to quickly reach a production level of 100 parts per day. The policy of the company calls for the new employee to produce 34 parts the first day, and then increase production by 3 parts each day. Decide how many parts must be produced each successive day and how many days it will take for each new employee to reach the production goal. [**34, 37, 40, ... , 100; 23**]

Students should learn that a geometric sequence is generated by multiplying a term by a specific number to produce the next term. This specific number is called the common ratio.

A **ongoing**
SSESSMENT

3. the ratio between consecutive terms is the same; $r = 2$

4. $t_n = t_1 r^{n-1}$

CRITICAL
Thinking

Yes, if all the terms are the same; for example, 3, 3, 3, ... where $d = 0$ and $r = 1$

Use Transparency ▶ 47

•Exploration 2 Ratios Between Terms

Let $S = 5, 10, 20, 40, ...$

1 Complete the following table.

n	1	2	3	4	5	6	7	8	9	10
t_n	5	10	20	40						

2 Find the ratio, r, between each consecutive pair of terms in Step 1. How do these ratios compare?

3 A **geometric sequence** is a sequence with a constant ratio between consecutive terms. Explain why S is a geometric sequence. What is r for this geometric sequence?

4 Complete the table using the value of r for this geometric sequence. Use the pattern to write in the last column a general formula for t_n in terms of t_1 and r.

n	1	2	3	4	5	6	7	...	n
t_n	t_1	t_2	t_3	t_4	t_5	t_6	t_7	...	t_n
	5	10	20					...	—
	$5 \cdot 1$	$5 \cdot 2$	$5 \cdot 4$					...	—
	$t_1 \cdot r^0$	$t_1 \cdot r^1$	$t_1 \cdot r^2$					...	

5 Use the formula for t_n that you wrote in Step 4 to find t_{15}. Check your answer by extending the sequence.

Which sequences below are geometric? If the sequence is *not* geometric, explain why not.

6 $S_1 = 63, 44, 25, 6, -13, -32, ...$

7 $S_2 = 64, 32, 16, 8, 4, 2, 1, ...$

8 $S_3 = 1, -1, 1, -1, 1, -1, ...$

9 $S_4 = 2, 4, 6, 8, 10, 12, ...$ ❖

CRITICAL
Thinking

Is it possible to have a sequence that is both arithmetic and geometric? Explain.

SEQUENCES

Arithmetic Sequence	**Geometric Sequence**
A constant *difference* is *added*.	A constant *ratio* is *multiplied*.

Arithmetic Sequence

$21, 18, 15, 12, ...$

$\frac{1}{4}, \frac{1}{2}, \frac{3}{4}, 1, ...$

$-3, -0.5, 2.0, 4.5, ...$

Geometric Sequence

$\frac{1}{2}, 2, 8, 32, ...$

$\frac{2}{3}, 2, 6, 18, ...$

$-3, 1, -\frac{1}{3}, \frac{1}{9}, ...$

ENRICHMENT Show students that the terms of a geometric sequence are exponential in nature. Show this by picking a first term and a common ratio of 3, for example. To get the fifth term, perform four multiplications by 3. Comparing the fifth term to the first term will show that the fifth term is 81 times the value of the 1st term, and $81 = 3^4$. The exponent 4 is the difference between 5 (for the fifth term) and 1 (for the first term).

INCLUSION **strategies** **Using Algorithms** Give the students the function $f(10) = ar^9$, and then have them choose a first term, a, and a common ratio, r. Have them build the first 10 terms of the geometric sequence, and then check the 10th term by evaluating the function above. This should serve to reinforce the formula for the nth term of a geometric sequence.

The sequence 1, 1, 2, 3, 5, 8, 13, 21, 34, ... is known as the Fibonacci sequence. Each term is the sum of the preceding two terms. A sequence of Fibonacci fractions can be obtained by dividing each term by the preceding term.

$$1 \quad \frac{1}{1} \quad \frac{2}{1} \quad \frac{3}{2} \quad \frac{5}{3} \quad \frac{8}{5} \quad \frac{13}{8} \quad \frac{21}{13} \quad \frac{34}{21}$$

GEOMETRY
Connection

Each quotient from the Fibonacci sequence gets closer to a number called the **Golden Ratio**, usually represented by the Greek letter phi (ϕ).

$$\phi = \frac{1 + \sqrt{5}}{2} \approx 1.618$$

In the figure at the right, rectangle *ABCD* has sides *AB* and *BC* such that $\frac{AB}{BC} = \phi$. Therefore, *ABCD* is called a **Golden Rectangle**.

When the square *AEFD* is cut from *ABCD*, the remaining rectangle, *EBCF*, is also a Golden Rectangle.

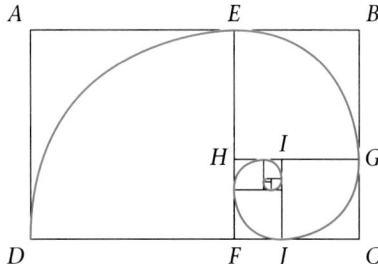

Likewise, when the square *EBGH* is removed from *EBCF*, the remaining rectangle, *HGCF*, is another Golden Rectangle.

In this manner, a sequence of Golden Rectangles can be produced.

Leonardo Fibonacci, also known as Leonardo de Pisa, 1175 – 1250

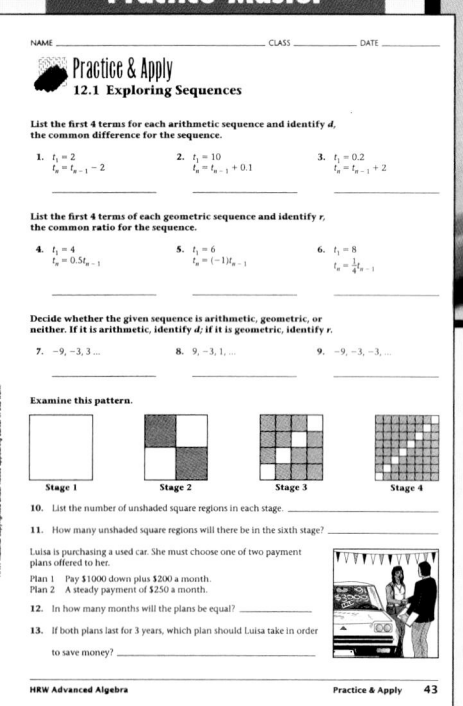

Aongoing ASSESSMENT

For planets, except for Jupiter, the distances calculated using Bode's sequence are slightly greater than the actual distances; for Jupiter the distances are the same.

Astronomy   **Cultural Connection: Europe** In 1866, the German astronomers Johann Daniel Titius and Johann Elert Bode (BO-day) developed a sequence that described the distance of each planet from the Sun.

Begin with the sequence 0, 3, 6, 12, 24, 48, ... , in which each term (except 0 and 3) is double the preceding term.

To get Bode's sequence, add 4 to each term, and divide by 10.

$$0.4, 0.7, 1.0, 1.6, 2.8, 5.2, ...$$

Each number in Bode's sequence is a **Bode number**, and **Bode's Law** describes the relationship between the numbers in Bode's sequence and the distance of each planet from the Sun.

In Bode's Law, the distance from the Earth to the Sun is 1 unit. All the numbers in Bode's sequence give the distance from the corresponding planet to the Sun in units relative to this unit.

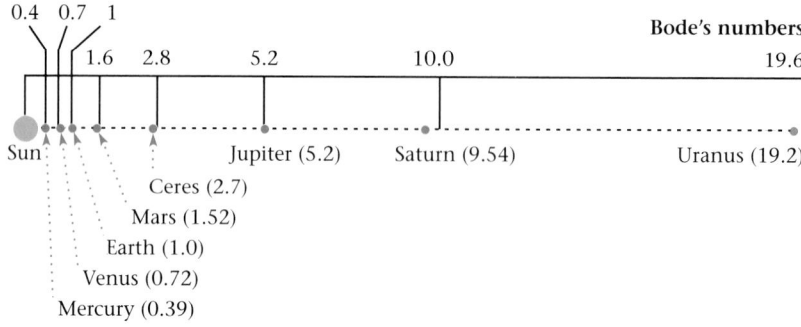

The first Bode number is 0.4, and the first planet from the Sun is Mercury. Thus, according to Bode's Law, the first planet, Mercury, is $\frac{4}{10}$, or 0.4, of Earth's distance from the Sun; the second planet, Venus, is 0.7 of Earth's distance from the Sun; and so on. Notice from the diagram that the actual proportions are very close to Bode's numbers.

When Bode and Titius published their work, the only known planets were Mercury, Venus, Earth, Mars, Jupiter, and Saturn. The fifth planet, Jupiter, corresponded to the *sixth* Bode number, not the fifth. Based on this discrepancy, astronomers began to look for another planet between Mars and Jupiter.

In 1801, Giuseppi Piazzi discovered and named Ceres, whose distance from the Sun corresponded to the fifth Bode number, 2.8. Within the next several years, however, many additional "planets" were discovered at essentially the same distance from the Sun. What Piazzi had assumed to be a new planet was actually the largest of the asteroids in the asteroid belt located between Mars and Jupiter.

Generate the first 10 terms of Bode's sequence. Assuming the distance from the Earth to the Sun is 1.495×10^8 kilometers, use the sequence and the diagram to determine the distances of the eight other planets and Ceres from the Sun. How do these distances compare with the actual distances of each planet from the Sun?

EXERCISES & PROBLEMS

Communicate

1. Name and describe two different kinds of sequences.

2. Create an arithmetic sequence. What rule did you use to define your sequence? What is d for your sequence?

3. Create a geometric sequence. What rule did you use to define your sequence? What is r for your sequence?

4. Create a sequence that is neither geometric nor arithmetic. What rule did you use to define your sequence?

5. Consider the exponential function $S(n) = 10^n$, where $n = 1, 2, 3, \ldots$ Is this an arithmetic or geometric sequence? Explain.

Practice & Apply

Determine whether the given sequence is arithmetic, geometric, or neither. If it is arithmetic, identify d; if it is geometric, identify r.

6. 6, 30, 150, ... $r = 5$ **7.** 4, –12, 36, ... $r = -3$ **8.** 10, 14, 18, ... $d = 4$

9. $1, \frac{1}{2}, \frac{1}{3}, \frac{1}{4}, \ldots$ **10.** $\frac{1}{2}, \frac{1}{4}, \frac{1}{8}, \ldots$ $r = \frac{1}{2}$ **11.** 6, –6, 6, ... $r = -1$
   neither

Determine whether the given sequence is arithmetic. If it is, identify the common difference, d.

12. 6, 10, 14, ... $d = 4$ **13.** 5, 10, 20, ... no   **14.** 5, 6, 8, 11, ... no

15. 8, 5, 2, ... $d = -3$ **16.** –20, –17, –14, ...   **17.** 13, 13, 13, ... $d = 0$
                                      $d = 3$

List the first 4 terms for each arithmetic sequence, and identify d, the common difference for the sequence.

18. $t_1 = 5$ **19.** $t_1 = 18$ **20.** $t_1 = 0$
    $t_n = t_{n-1} + 2$ $t_n = t_{n-1} - 3$ $t_n = t_{n-1} + 0.1$

5, 7, 9, 11; $d = 2$ 18, 15, 12, 9; $d = -3$ 0, 0.1, 0.2, 0.3; $d = 0.1$

List the first 4 terms of each geometric sequence, and identify r, the common ratio for the sequence.

21. $t_1 = 3$ **22.** $t_1 = 2$ **23.** $t_1 = 5$ **24.** $t_1 = 12$
   $t_n = 2t_{n-1}$ $t_n = 3t_{n-1}$ $t_n = (-2)t_{n-1}$ $t_n = \frac{1}{2}t_{n-1}$
                 2, 6, 18, 54; $r = 3$ 12, 6, 3, 1.5; $r = \frac{1}{2}$

25. Economics The starting salary for a teacher in one district is $23,000. If the salary increases by $800 each year, what will the salary be in the tenth year? $30,200

21. 3, 6, 12, 24; $r = 2$

23. 5, –10, 20, –40; $r = -2$

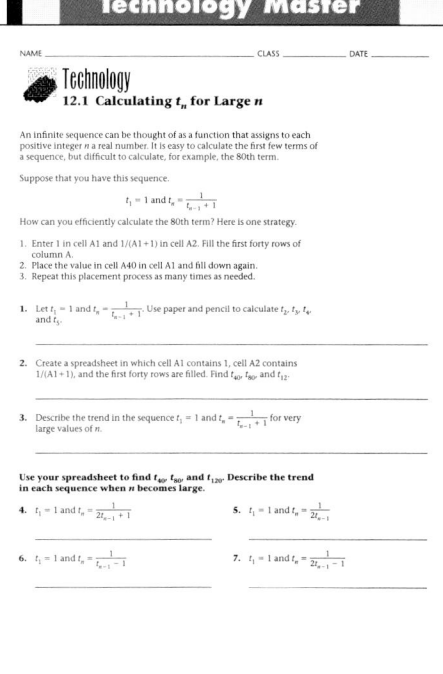
ASSESS

Selected Answers
Odd-Numbered Exercises 7–37

Assignment Guide

Core 1–10, 12–15, 18–19, 21–23, 25–37

Core-Plus 1–5, 8–11, 14–17, 19–20, 22–38

Technology

Invite students to explore the use of the memory keys on their calculators to streamline the successive addition or multiplication process in computing the values of successive terms in sequences.

Error Analysis

A careless error often causes a mistake in determining the common difference or ratio. To avoid this, stress the need for students to check the sequence in more than one place. If the results are the same, chances are the calculated difference or ratio is correct.

Technology Master

Authentic Assessment

Give the students a few days to come across a sequence in another class. This might occur during the study of cell division in biology, increasing velocity in physics, reduction in loan balances in accounting, election years in history, or the like. Have them write a short report or essay about this sequence and its application.

26. Examine the pattern of spheres below. How many spheres will there be in the fourteenth square? 27

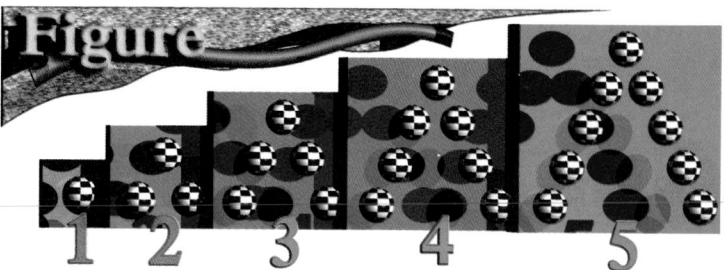

27. In a parking lot, each new row has 3 more cars than the previous row. If the first row has 20 cars, how many cars will the fifteenth row have? 62

Employment Franco has two part-time job offers. His salary options are given below.

Company *A*: $50 the first week with an increase of $5 each week

Company *B*: A steady salary of $80 per week

28. During what week will the salaries be equal? 7th

29. If both jobs last for 13 weeks, which job allows Franco to make more money? Both will pay the same, $1040

30. Health Amanda is beginning a fitness program. During the first week, she will do 25 sit-ups each day. Each week she will increase the number of daily sit-ups by 3. How many sit-ups will Amanda do in the 20th week? 82

Look Back

Solve and check each system of nonlinear equations. [Lesson 10.5]

31. $\begin{cases} 2x + y = 1 \\ 3x^2 + 12x + y = -6 \end{cases}$ **32.** $\begin{cases} 2y^2 = x \ (18, 3); \\ x - 2y = 12 \ (8, -2) \end{cases}$ **33.** $\begin{cases} y^2 + 2x = 17 \\ x + 4y = -8 \end{cases}$ $(4, -3); (-52, 11)$

Representing male with *M* and female with *F*, specify a sample space for the number of male children and female children in a family with 2 children. Then determine the event(s) that correspond(s) to each of the following outcomes. **[Lesson 11.1]** {MM, MF, FM, FF}

34. A family has exactly two girls. FF **35.** A family has at least one girl. MF, FM, FF

36. A family has at most one girl. MM, MF, FM

37. In how many different ways can you choose a combination of 6 objects from a collection of 30 objects? **[Lesson 11.4]** $_{30}C_6 = 593,775$

Look Beyond

Look Beyond

This exercise allows students to work with sequential notation. Students also have the opportunity to explore the sums of a certain number of terms in a sequence. This will lead into series, which they will study in the next lesson.

38. Find the sum of the first 8 terms of the given arithmetic sequence. 96

$$\begin{cases} t_1 = 5 \\ t_n = t_{n-1} + 2 \end{cases}$$

31. $(-1, 3)$ and $\left(-\frac{7}{3}, \frac{17}{3}\right)$

In Lesson 12.1, you explored a sequence formed by increasing a salary by $5 per week. There is a method for finding the sum at the end of n weeks.

$100 + $105 + $110 + $115 + ... + ? =
WEEK 1   WEEK 2   WEEK 3   WEEK 4   ?

SALESPERSON
$100
per week to start
• • •
$5 increases per week.

The salary sequence can be written as a set of ordered pairs,

$$(1, 100), (2, 105), (3, 110), \dots , (n, t_n),$$

defined by the function $t(n) = 100 + 5(n - 1)$, where n is a positive integer.

The Salary Sequence

n	t_n
1	100
2	105
3	110
4	115

What is the domain of the salary sequence function? Notice that the graph of a sequence function is a graph of discrete, or disconnected, points.

A sequence can be considered a function, where n is considered the independent variable and t_n the dependent variable.

nth Term of an Arithmetic Sequence

A **recursive formula** for a sequence defines any term of the sequence using the previous term or terms. A recursive formula for the salary sequence is given below.

$$t_1 = 100$$
$$t_n = t_{n-1} + 5 \qquad n = 2, 3, 4, \ldots$$

To find the tenth term of the salary sequence, substitute 10 for n.

$$t_n = t_{n-1} + 5$$
$$t_{10} = t_9 + \mathbf{5}$$
$$= (t_8 + \mathbf{5}) + \mathbf{5}$$
$$= ((t_7 + \mathbf{5}) + \mathbf{5}) + \mathbf{5}$$
$$= (((t_6 + \mathbf{5}) + \mathbf{5}) + \mathbf{5}) + \mathbf{5}$$

Notice that the recursive formula is not very helpful for finding the tenth term unless the preceding terms are known.

$$\vdots$$
$$t_{10} = t_1 + (\mathbf{10} - \mathbf{1})(\mathbf{5})$$
$$= 100 + \mathbf{9} \cdot \mathbf{5}$$
$$= 145$$

You can use the recursive formula to write an **explicit formula**, a formula for the sequence in terms of n.

$$t_{10} = t_1 + (\mathbf{10} - \mathbf{1})(\mathbf{5})$$
$$t_n = t_1 + (\mathbf{n} - \mathbf{1})(\mathbf{d})$$

$t_1 \quad t_2 \quad t_3 \quad t_n$

The recursive formula for a sequence is not very useful when finding the nth term of the sequence. In contrast, an explicit formula can be used to find *any* term of the sequence.

nth TERM OF AN ARITHMETIC SEQUENCE

If $t_1, t_2, \ldots, t_n$ represents an arithmetic sequence with constant difference d, then the nth term is given by
$$t_n = t_1 + (n - 1)d.$$

Architecture Many modern structures contain sloping, slanted, or angled surfaces and edges. The lengths of beams, columns, rods, and many other structural elements form sequences. Mathematics from this section is often used to calculate these lengths and to add them to determine the total amount of materials needed. Ask the students to find an example of sequential relationships in the school buildings or facilities.

EXAMPLE 1

Find the eighth term of the sequence 37, 34, 31, ...

Solution►

The sequence is arithmetic, and d is -3. To find the nth term of an arithmetic sequence, use $t_n = t_1 + (n-1)d$.

Substitute 8 for n, 37 for t_1, and -3 for d.

$$t_n = t_1 + (n-1)d$$
$$t_8 = 37 + (8-1)(-3)$$
$$= 37 + (-21)$$
$$= 16$$

Check►

Extend the sequence until you reach the eighth term.

n	1	2	3	4	5	6	7	8
t_n	37	34	31	28	25	22	19	16

Thus, the eighth term of the sequence is 16. ❖

Arithmetic Series

The *sum* of the terms of an arithmetic sequence is an **arithmetic series**. The sum of the first n terms of a sequence is indicated by S_n. How can you find S_{100}, the sum of the first 100 terms? You could write out every term and use a calculator to find the sum. However, there is an easier way.

Consider S_{10} for the salary sequence.

$$S_{10} = t_1 + t_2 + \ldots + t_9 + t_{10}$$
$$= 100 + 105 + \ldots + 140 + 145$$

S_{10} can also be found by writing the terms of the sequence in reverse order.

$$S_{10} = t_{10} + t_9 + \ldots + t_2 + t_1$$
$$= 145 + 140 + \ldots + 105 + 100$$

When you add the two forms, all of the terms are the same.

$$2S_{10} = (t_1 + t_{10}) + (t_2 + t_9) + \ldots + (t_9 + t_2) + (t_{10} + t_1)$$
$$2S_{10} = 245 + 245 + \ldots + 245 + 245$$
$$2S_{10} = 10(245)$$

Solving for S_{10} reveals a pattern.

$$S_{10} = 10\left(\frac{245}{2}\right)$$
$$= 10\left(\frac{t_1 + t_{10}}{2}\right)$$

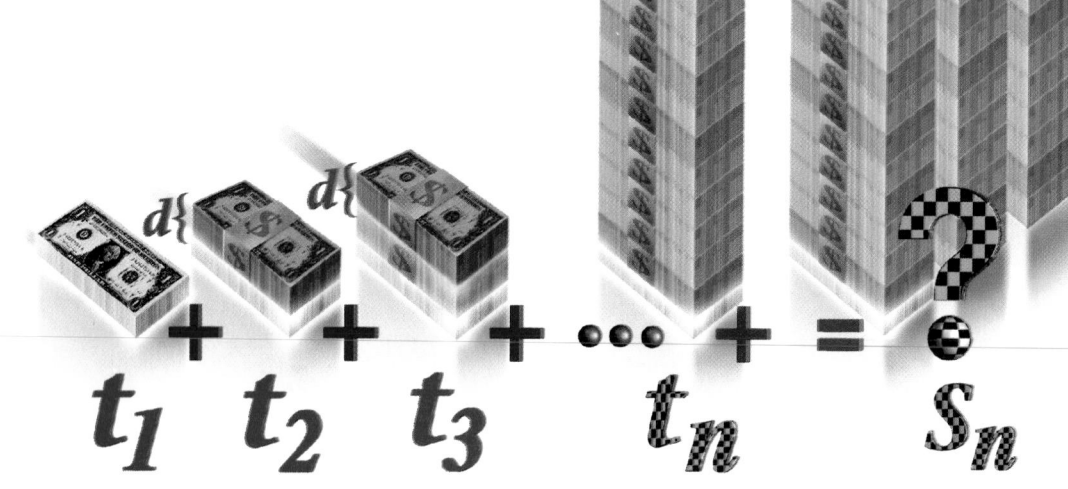

$$t_1 \quad + \quad t_2 \quad + \quad t_3 \quad + \cdots + \quad t_n \quad = \quad S_n$$

SUM OF THE FIRST n TERMS OF AN ARITHMETIC SEQUENCE

If $t_1, t_2, t_3, t_4, \ldots$ is an arithmetic sequence with constant difference d, the sum, S_n, of the first n terms is given by

$$S_n = n\left(\frac{t_1 + t_n}{2}\right).$$

EXAMPLE 2

Graphics Calculator

An auditorium has 20 seats in the first row, and each row has 2 more seats than the row immediately in front of it. There are 26 rows in the auditorium.

Ⓐ How many seats are there in the last row?

Ⓑ What is the seating capacity of the auditorium?

Solution▸

Ⓐ There are 26 rows in the auditorium. To find the number of seats in the last row, find t_{26}.

Substitute 26 for n, 20 for t_1, and 2 for d.

$$t_n = t_1 + (n-1)d$$
$$t_{26} = 20 + (26-1)2$$
$$= 70$$

Graph of $t(n) = 20 + (n-1)2$

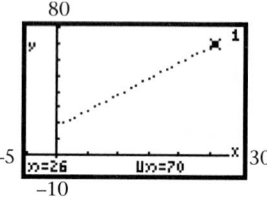

Ⓑ To find the seating capacity, find S_{26}.

$$S_n = n\frac{t_1 + t_n}{2}$$

Substitute 26 for n, 20 for t_1, and 70 for t_{26}.

$$S_{26} = 26 \cdot \frac{20 + 70}{2} = 1170$$

Thus, the seating capacity is 1170. ❖

The formula for S_n can be altered so that the sum can be found without first finding t_n. By substituting $\boldsymbol{t_n = t_1 + (n-1)d}$ in $S_n = n \cdot \frac{t_1 + t_n}{2}$, the formula for S_n is given in terms of n, t_1, and d.

$$S_n = n \cdot \frac{t_1 + t_1 + (n-1)d}{2}$$
$$S_n = n \cdot \frac{2t_1 + (n-1)d}{2}$$

RETEACHING
the
l e s s o n

Using Manipulatives Present the following demonstration to show the concept of an arithmetic series and its sum. Use a large number of identical, stackable objects to make a *trapezoidal* stack in which each row is smaller than the one below it by a constant difference. Count the number of objects. Count the number of objects in the middle of the stack if the number of rows is odd. Otherwise, average the top and bottom rows. Show how multiplying this number times the number of rows gives the sum from the count of the objects. Next, restack the objects into an equivalent *rectangular* pile, whose width will be the middle term. Relate the S_n formula to the area of the rectangle.

Try This A triangular display of cans has 15 rows. The top row has one can, and each row has 3 more cans than the row above it. Find the total number of cans in the display. The shopkeeper wants to add another row of cans to the display. Find the number of cans to be added.

The Greek letter sigma (Σ) is used to represent the sum of a sequence. For example,

$$1 + 2 + 3 + 4 + 5 = \sum_{n=1}^{5} n \qquad 10 + 20 + 30 + 40 + 50 = \sum_{n=1}^{5} 10n$$

$\sum_{n=1}^{5} 10n$ is read *the sum of the terms 10n as n goes from 1 to 5*.

$\sum$ is the *summation* symbol, and *n* is the *index of summation*.

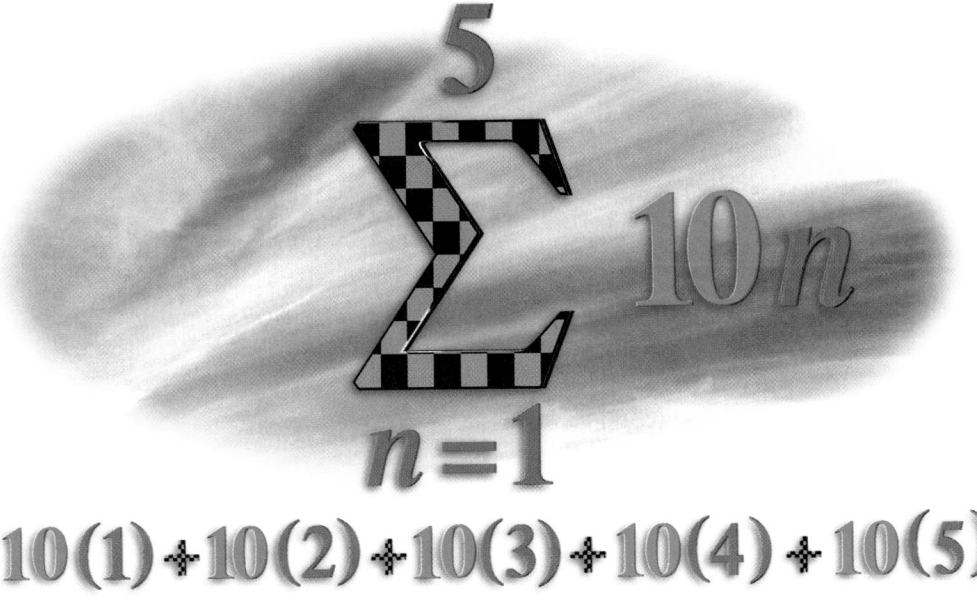

The first 10 terms of the salary sequence can be written in this notation.

$$100 + 105 + 110 + \cdots + 140 + 145 = \sum_{n=1}^{10}(100 + 5(n-1))$$

Try This

330 cans; 46 cans

TEACHING *tip*

Remind the students that multiplication is just repeated addition. In this situation, we have a repeated addition of terms whose values have a special pattern and we are exploiting that pattern to turn the repeated addition into a multiplication.

Cooperative Learning

Divide the class in teams of two. Have the first students choose an arithmetic sequence pattern and read off 10 terms while the second student adds them. Then they work together to write the sum using the Σ notation and verify the sum using the S_n formula. Team members exchange roles and repeat the process.

Use Transparency ▶ **49**

Alternate Example 3

Evaluate each summation.

A. $\sum_{k=2}^{9}(3k-1)$ **[124]**

B. $\sum_{n=1}^{50}\left(\frac{n}{2}+3\right)$ **[787.5]**

CRITICAL
Thinking

See proof below.

ASSESS

Selected Answers

Odd-Numbered Exercises 9–65

Assignment Guide

Core 1–13, 17–19, 21–23, 25–65

Core-Plus 1–10, 12–16, 18–20, 22–68

Technology

The memory feature on a calculator can be used to find the sum of a sequence. Enter the first term into memory, and then add enter each successive term to the memory. When all entries are completed, the memory-recall function will display the sum.

Error Analysis

One of the most common errors is the failure to discriminate between a sequence and its associated series. Emphasize that a sequence is a list of terms and a series is the sum of the terms in the list.

EXAMPLE 3

Find each summation.

Ⓐ $\displaystyle\sum_{k=5}^{8}(2k+1)$

Ⓑ $\displaystyle\sum_{n=1}^{100}(n+1)$

Solution▶

Ⓐ To find the sum $\displaystyle\sum_{k=5}^{8}(2k+1)$, replace k with 5, 6, 7, and 8.

$$\sum_{k=5}^{8}(2\mathbf{k}+1) = [2(\mathbf{5})+1] + [2(\mathbf{6})+1] + [2(\mathbf{7})+1] + [2(\mathbf{8})+1]$$
$$= 11 + 13 + 15 + 17 = 56$$

Ⓑ To find the sum $\displaystyle\sum_{n=1}^{100}(n+1)$, you can replace n with each number from 1 to 100. However, the series is arithmetic, and $\displaystyle\sum_{n=1}^{100}(n+1)$ is the same as S_{100}. Thus, you can find the average value of all the terms using the average of the first and last terms, $\frac{2+101}{2} = 51.5$. Then, multiply by the number of terms, 100.

$$S_{100} = 100(51.5) = 5150 \;\; ❖$$

CRITICAL
Thinking

Show that for *any* arithmetic sequence, the sum of the first n terms is $S_n = n \cdot \frac{t_1 + t_n}{2}$, where n is some positive integer, t_1 is the first term of the sequence, and t_n is the nth term.

EXERCISES & PROBLEMS

Communicate

1. Describe how to find the general term of any arithmetic sequence.

2. Describe how to find the sum of a finite number of terms of an arithmetic sequence.

3. Explain what would happen to the general term of an arithmetic sequence if the difference, d, were doubled.

4. Explain what would happen to the general term of an arithmetic sequence if the first term, t_1, were doubled.

5. Describe the effects of changing t_1 and changing d on the graph of the sequence.

6. Explain what happens to the sum of the first 10 terms of an arithmetic series if each term is doubled.

7. Explain what happens to the sum of the first 10 terms of an arithmetic series if each term is increased by 2.

CRITICAL
Thinking

$$S_n= \quad t_1 \qquad + t_1 + 1d + t_1 + 2d + \ldots + t_n - 2d + t_n - 1d + \quad t_n$$
$$S_n= \quad t_n \qquad + t_n - 1d + t_n - 2d + \ldots + t_1 + 2d + t_1 + 1d + \quad t_1$$
$$\overline{2S_n= \quad t_1 + t_n \quad + t_1 + t_n \;\; + \; t_1 + t_n \; + \ldots + \; t_1 + t_n \; + \; t_1 + t_n \; + t_1 + t_n}$$
$$2S_n= n(t_1 + t_n)$$
$$S_n= \frac{n}{2}(t_1 + t_n)$$

Practice & Apply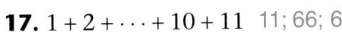

List the first 4 terms for each arithmetic sequence, and identify *d*, the common difference for the sequence.

8. $t_n = 3n + 4$
7, 10, 13, 16; $d = 3$

9. $t_n = -2n + 4$
2, 0, -2, -4, $d = -2$

10. $t_n = \frac{1}{2}n + 4$
$4\frac{1}{2}, 5, 5\frac{1}{2}, 6; \ d = \frac{1}{2}$

For Exercises 11–16, write an explicit formula for the *n*th term of each arithmetic sequence.

11.

n	t_n
1	6
2	8
3	10

$t_n = 2n + 4$

12.

n	t_n
1	11
2	15
3	19

$t_n = 4n + 7$

13.

n	t_n
1	1
2	-6
3	-13

$t_n = -7n + 8$

14.

n	t_n
1	19
2	14
3	9

$t_n = -5n + 24$

15. 7, 10, 13, ...
$t_n = 3n + 4$

16. 10, 9, 8, ...
$t_n = -n + 11$

Find the number of terms and the sum of each series. Then find the average value of the terms in each series.

17. $1 + 2 + \cdots + 10 + 11$ 11; 66; 6

18. $20 + 24 + 28 + \cdots + 60$ 11; 440; 40

19. $5 - 4 - 13 - \cdots - 76$ 10; -355; -35.5

20. $8 + 6 + 4 + \cdots - 4$ 7; 14; 2

For each arithmetic sequence, find S_1, S_2, S_3, and S_4.

21. 3, 7, 11, 15, ...   3; 10; 21; 36

22. 25, 24, 23, 22, ...   25; 49; 72; 94

23. 4, 14, 24, ...   4; 18; 42; 76

24. 6, 2, -2, ...   6; 8; 6; 0

Architecture A new football stadium is to be built in the shape of the Roman Colosseum. The bottom row will have 250 seats, and each row above the bottom row will have 40 seats more than the row in front of it. The design calls for 30 rows.

25. How many seats will there be in the 30th row?   1410

26. What will be the seating capacity of the new stadium?   24,900

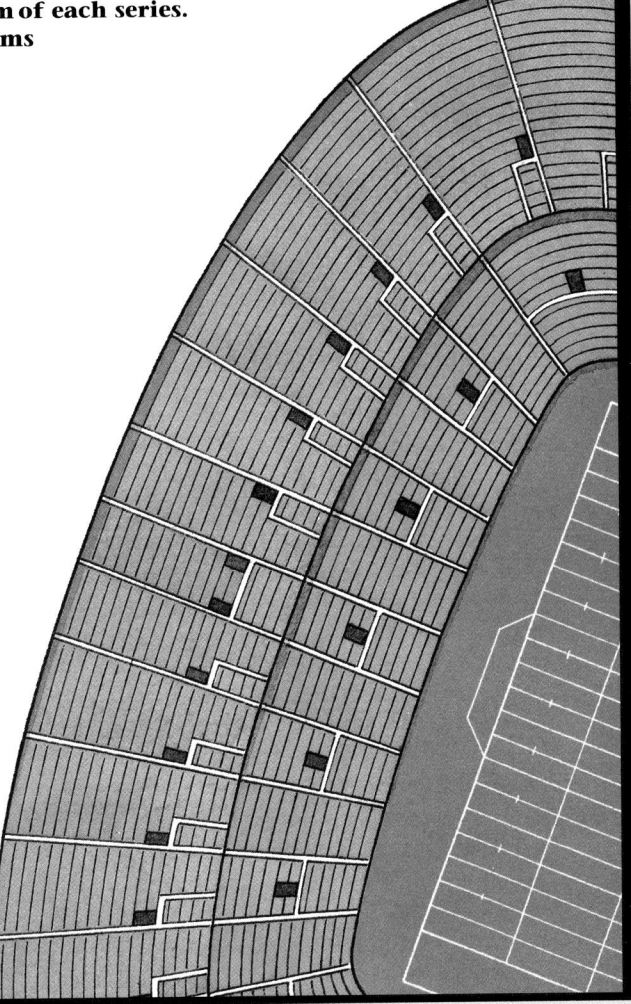

27. In an arithmetic sequence, t_1 is 6, d is 3, and n is 40. Find t_n. 123

28. In an arithmetic sequence, t_1 is 65, d is –3, and n is 20. Find t_n. 8

29. Find the 30th term of the sequence 9, 15, 21, ... 183

30. Find the 25th term of the sequence 15, 13, 11, ... –33

31. In an arithmetic sequence, t_1 is 7, and t_{10} is 52. Find S_{10}. 295

For Exercises 32 and 33, use the sequence 17, 21, 25, ...

32. Find t_{30}. 133   **33.** Find S_{30}. 2250

For Exercises 34 and 35, use the sequence 14, 10, 6, ...

34. Find t_{20}. –62   **35.** Find S_{20}. –480

36. Find S_{12} for the sequence 8, 13, 18, ... 426

37. Find S_{18} for the sequence 8, 3, –2, ... –621

Identify t_1, n, t_n, and d. Then evaluate each sum.

38. $\displaystyle\sum_{k=1}^{25} 3k$

39. $\displaystyle\sum_{k=1}^{17} -2k$

40. $\displaystyle\sum_{k=1}^{16} (3k-1)$

41. $\displaystyle\sum_{k=1}^{10} (9-2k)$

Architecture A mason and his crew have been hired to tile the floor of a shopping mall. A tile pattern like the one shown appears in several locations and requires a special color tile. This tile pattern will appear in three different sizes:

small (base has 6 tiles, as shown),
medium (base has 12 tiles), and
large (base has 24 tiles).

42. Does a medium pattern require twice as many tiles as a small pattern? Does a large pattern require twice as many tiles as a medium pattern? Explain.
no; no; small: 21; med:78; lg:300

43. If the mall has 40 small, 20 medium, and 4 large patterns, how many of the special color tiles must be ordered to complete the job, assuming that the mason adds 10% to the actual tile total to allow for breakage. 3960

44. Does this tile pattern exist with a total of 120 tiles? Explain. yes; base of 15

45. Does this tile pattern exist with a total of 150 tiles? Explain. no

38. $t_1 = 3, n = 25, t_n = 75, d = 3, \Sigma = 975$

39. $t_1 = -2, n = 17, t_n = -34, d = -2, \Sigma = -306$

40. $t_1 = 2, n = 16, t_n = 47, d = 3, \Sigma = 392$

41. $t_1 = 7, n = 10, t_n = -11, d = -2, \Sigma = -20$

45. no; a tile pattern with base 16 tiles has 136 tiles and a pattern with base 17 tiles has 153 tiles.

46. **Portfolio Activity** To complete the Portfolio Activity on page 653, complete the following table for $n = 10$.

Column 1 Number of people	Column 2 Number of new handshakes	Column 3 Sum of handshakes
1	0	0
2 = 1 + 1 new person	1	1
3 = 2 + 1 new person	2	3
4 = 3 + 1 new person	3	6
⋮	⋮	⋮
n	t_n	S_n

- What type of sequence appears in Column 2? What is the eighth term in this sequence? Describe a general term, t_n, from this sequence using n. arithmetic; 7; $t_n = n - 1$
- What type of sequence appears in Column 3? What is S_8 for the sequence in Column 3? Describe the sum of the first n terms, S_n, using t_n and n. sequence of partial sums; 28; $S_n = \frac{n}{2} \cdot t_n$

Statistics In statistics, the *mean* of two numbers, a and b, is $\frac{a+b}{2}$, the average of the numbers. The mean can also be thought of as the term between two numbers in an arithmetic sequence. This idea of mean can be extended to *arithmetic means* in an arithmetic sequence. For example, to find 3 arithmetic means between 8 and 44, first draw a diagram.

$$8, \quad \underset{+d,}{\_\_}, \quad \underset{+d,}{\_\_}, \quad \underset{+d,}{\_\_}, \quad \underset{+d}{44}$$

There are 36 integers between 8 and 44, including 44, and 4 values of d to be added. Since $36 \div 4 = 9$, add 9 to generate the three missing terms. Thus, the 3 arithmetic means are 17, 26, and 35.

Find the following arithmetic means.

47. Insert 1 arithmetic mean between 6 and 20. 13

48. Insert 2 arithmetic means between 9 and 30. 16, 23

49. Insert 3 arithmetic means between 40 and 16. 34, 28, 22

50. Insert 4 arithmetic means between −8 and 22. −2, 4, 10, 16

Find the sum of the first n counting numbers for the given value of n.

51. 10 55 **52.** 20 210 **53.** 50 1275 **54.** 100 5050

55. Create a formula to find the sum of the first n counting numbers, where n is *any* positive integer. $S_n = \frac{n}{2}(n + 1)$

46.

Number of people	Number of new handshakes	Sum of handshakes
5 = 4 + 1	4	10
6 = 5 + 1	5	15
7 = 6 + 1	6	21
8 = 7 + 1	7	28

56. A grandfather clock chimes once at 1 o'clock, twice at 2 o'clock, and so on. In addition, it chimes once at the quarter hour, twice at the half hour, and three times at the third-quarter hour. Write an arithmetic series to determine how many times the clock will chime between 12:01 A.M. and 12:01 P.M. $7 + 8 + 9 + \cdots + 18 = 150$

57. Sports In a softball league, there are 16 teams. Each season, each team plays every other team exactly one time. How many games must be scheduled each season? 120

Income Amelia has just graduated from college and is starting a new job. Her salary is based solely on commission. She hopes to earn $100 during her first week and then increase her salary by $20 each week.

58. Amelia's friend Marty earns a fixed salary of $600 per week. During which week does Amelia hope to earn $600? 26th

59. What is Amelia's projected annual salary, assuming that she works all 52 weeks? $31,720

60. What is Amelia's average weekly salary for the year? $610

61. Marty's job allows a 2-week paid vacation during the first year. Which job would you rather have, Amelia's or Marty's? Explain. Answers may vary.

Look Back

Write an equation to represent each variation. Use k as the constant of variation. [Lesson 9.1]

62. y varies directly as x and z, and y varies inversely as the square of m. $y = \frac{kxz}{m^2}$

63. y varies directly as x^2 and inversely as z^3. $y = \frac{kx^2}{z^3}$

64. Two cards are drawn in succession and without replacement from a full deck of 52 cards. Find the probability that the first card is a 9 of any suit and the second card is a 7 of any suit. **[Lesson 11.5]** $\frac{4}{663} \approx 0.6\%$

65. Two number cubes, one red and one blue, are tossed. The sum of the numbers appearing on the top faces is recorded. What is the probability that the number rolled on one cube is 4 if the sum of the numbers rolled on both cubes is 6? **[Lesson 11.6]** $\frac{2}{5}$

Look Beyond

Find the sum of the first 8 terms of each geometric sequence.

66. 2, 4, 8, 16, ... 510

67. 1, $\frac{1}{3}$, $\frac{1}{9}$, $\frac{1}{27}$, ... $\frac{3280}{2187} \approx 1\frac{1}{2}$

68. Look at the pattern of terms in the sequence in Exercise 67. Consider the 100th term and the 1000th term of this sequence. What value are the terms approaching? 0

LESSON 12.3 Geometric Series

why In Lesson 12.2, you developed formulas for the general term and the sum of n terms of an arithmetic sequence. Formulas can also be derived for the general term and the sum of n terms of a geometric sequence.

WEEK 1 WEEK 2 WEEK 3 WEEK 4 WEEK 10

$\$2.50 + \$5 + \$10 + \$20 + \dots ? =$

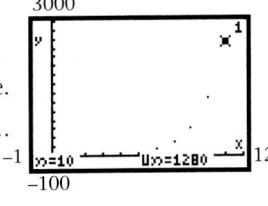

BICYCLE MECHANIC
★
$200
per week

BICYCLE MECHANIC
$$$ **$2.50** $$$
for the first week
and <u>doubled</u>
each week

In order to entice students to work through the entire summer, one employer pays $2.50 for the first week and doubles the salary each week. Which job pays more over the 10 weeks of the summer?

This situation involves the following sequence.

$S = 2.5, 5, 10, 20, 40, \dots$

A graphics calculator in disconnected sequence mode can be used to find the sum of the first n terms of a sequence.

PREPARE

Objectives
- Find the nth term of a geometric sequence.
- Find the sum of the first n terms of an geometric series.

RESOURCES

- Practice Master   12.3
- Enrichment Master   12.3
- Technology Master   12.3
- Lesson Activity Master   12.3
- Quiz   12.3
- Spanish Resources   12.3

Assessing Prior Knowledge

Evaluate.
A. 3^7 [2187]
B. $4\left[\dfrac{(1+5^3)}{(1-5)}\right]$ [−126]

TEACH

why This lesson extends the process of finding a general term and the sum of terms of a sequence to geometric sequences.

Cooperative Learning

In groups of two, have the students describe geometric sequences to each other in either recursive or explicit form. Then convert the description to the other form.

ALTERNATIVE teaching strategy

Using Algorithms List the first five terms of a geometric sequence with a common ratio of 3; for example 2, 6, 18, 54, 162. Write the factors of the terms. Rewrite the 6 in the second term as $2 \cdot 3$ (the product of the previous term and the common ratio). Then rewrite the next term, 18, as $6 \cdot 3 = 2 \cdot (3 \cdot 3)$; the 54 as $18 \cdot 3 = 2 \cdot (3 \cdot 3 \cdot 3)$ and the 162 as $54 \cdot 3 = 2 \cdot (3 \cdot 3 \cdot 3 \cdot 3)$. Call students attention to the repeated factors of 3, and write the terms as powers of 3. Then summarize the nth term as $t_n = t_1 r^{n-1}$.

nth Term of a Geometric Sequence

The doubling salary sequence, S, is a geometric sequence because the ratio, r, between consecutive terms is constant. The first term, t_1, is 2.5, and the common ratio, r, is 2. A recursive formula for the sequence is given.

$$t_1 = 2.5$$
$$t_n = r \cdot t_{n-1} = 2 \cdot t_{n-1}$$

To see how an explicit formula for this geometric sequence is written, examine the following table.

n	t_n	t_n
1	2.5	$2.5 \cdot 2^0$
2	$2 \cdot 2.5 = 5$	$2.5 \cdot 2^1 = 5$
3	$2 \cdot 5 = 10$	$2.5 \cdot 2^2 = 10$
4	$2 \cdot 10 = 20$	$2.5 \cdot 2^3 = 20$
5	$2 \cdot 20 = 40$	$2.5 \cdot 2^4 = 40$
6	$2 \cdot 40 = 80$	$2.5 \cdot 2^5 = 80$
⋮	⋮	⋮

In the explicit formula, notice that the exponent of r is 1 less than the value of n. This pattern can be generalized for the nth term of any geometric sequence.

nth TERM OF A GEOMETRIC SEQUENCE

If $t_1, t_2, \ldots, t_n, \ldots$ represents a geometric sequence with constant ratio r, then for any positive integer n, the nth term is given by
$$t_n = t_1 \cdot r^{n-1}.$$

EXAMPLE 1

A geometric sequence is given by the recursive formula $t_1 = 8$ and $t_n = 3t_{n-1}$. Write an explicit formula for this geometric sequence.

Solution

Substitute 8 for t_1 and 3 for r.
$$t_n = t_1 \cdot r^{n-1}$$
$$t_n = 8 \cdot 3^{n-1} \quad ❖$$

Try This   A geometric sequence is given by the recursive formula $t_1 = 3$ and $t_n = 4t_{n-1}$. Write an explicit formula for this geometric sequence.

interdisciplinary
CONNECTION

Biology In biology, cells multiply by dividing in half, which means that each successive generation has twice as many cells as the previous one. The cell population growth forms a geometric sequence with a common ratio equal to two.

CRITICAL Thinking

Describe two differences between the recursive and the explicit formulas for a geometric sequence. Which formula is easier to use to find many consecutive terms? Which formula is easier to use to find many nonconsecutive terms? Explain.

EXAMPLE 2

Write a recursive formula and an explicit formula for the geometric sequence 8, 4, 2, 1, ...

Solution▸

To find r, form the ratio of any two consecutive terms: $r = \frac{4}{8}$, or $\frac{1}{2}$. The first term, t_1, is 8.

Recursive formula

$$t_1 = 8$$
$$t_n = r \cdot t_{n-1}$$
$$= \frac{1}{2} \cdot t_{n-1}$$

Explicit formula

$$t_n = t_1 \cdot r^{n-1}$$
$$= 8 \cdot \left(\frac{1}{2}\right)^{n-1}$$

Check▸

Use each formula to examine some terms in the sequence. Check these with the original sequence: 8, 4, 2, 1, ...

n	**Recursive formula** $t_1 = 8$ and $t_n = \frac{1}{2} \cdot t_{n-1}$	**Explicit formula** $t_n = 8 \cdot \left(\frac{1}{2}\right)^{n-1}$
1	$t_1 = 8$	$t_1 = 8 \cdot \left(\frac{1}{2}\right)^{0} = 8$
2	$t_2 = \frac{1}{2} \cdot 8 = 4$	$t_2 = 8 \cdot \left(\frac{1}{2}\right)^{1} = 4$
3	$t_3 = \frac{1}{2} \cdot 4 = 2$	$t_3 = 8 \cdot \left(\frac{1}{2}\right)^{2} = 2$
4	$t_4 = \frac{1}{2} \cdot 2 = 1$	$t_4 = 8 \cdot \left(\frac{1}{2}\right)^{3} = 1$
5	$t_5 = \frac{1}{2} \cdot 1 = \frac{1}{2}$	$t_5 = 8 \cdot \left(\frac{1}{2}\right)^{4} = \frac{1}{2}$

Try This Find two forms for the geometric sequence 1, 8, 64, 512, ...

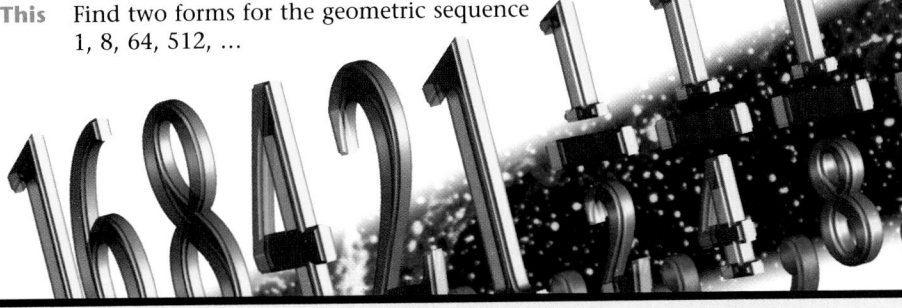

CRITICAL Thinking

The *recursive form* $t_n = r \cdot t_{n-1}$ is useful for extending a sequence from any one term to the next; you don't need to know the first term. The *explicit form* $t_n = t_1 \cdot r_{n-1}$ is different in that it involves the first term, but not the previous term. The recursive form is easier to use to find many consecutive terms. The explicit form is easier to use to find many non-consecutive terms.

Alternate Example 2

Write a recursive formula and explicit formula for the geometric sequence, $270, 90, 30, 10, \frac{10}{3}$, ...

[Recursive form:
$t_1 = 270, t_n = \frac{1}{3} \cdot t_{n-1}$
Explicit form:
$t_n = 270 \cdot \left(\frac{1}{3}\right)^{n-1}$]

Aongoing ASSESSMENT

Try This

recursive: $t_1 = 1$; $t_n = 8 \cdot t_{n-1}$
explicit: $t_n = 1 \cdot 8^{n-1}$

ENRICHMENT Have students explore the convergence of an infinite geometric series by using the series $48 + 24 + 12 + 6 + 3 + \frac{3}{2} + ...$ Add the terms of each series. Students should see that insignificant terms are eventually added to the series and that the series is converging to some finite value. In this case, it is 96.

INCLUSION strategies

Using Algorithms For students who have trouble with the concept of the recursive form, show that it can be obtained from the explicit form

$$t_n = t_1 r^{n-1} = t_1 r r^{n-2} = r t_1 r^{n-2}$$
$$= r(t_1 r^{n-2}) = r t_{n-1}$$

Note that $t_1 r^{n-2} = t_{n-1}$ from the explicit form.

EXAMPLE 3

Teaching tip

Remind the students that the *r* used in Example 3 is the correlation coefficient of the regression, *not* the common ratio.

Alternate Example 3

Find a best fit model for the sequence $t_n = 6 \cdot 2^{n-1}$, where $n = 1, 2, 3, ... [y = 3 \cdot 2^x]$

CRITICAL Thinking

$y = (1.25) \cdot 2^x$ is the same as the explicit form $t_n = 2.5 \cdot 2^{n-1}$. Note that $2.5 \cdot 2^{n-1} = \frac{2.5}{2} \cdot 2 \cdot 2^{n-1} = \frac{2.5}{2} \cdot 2^n$.

Graphics Calculator

STATISTICS *Connection*

CRITICAL *Thinking*

Refer to the bicycle mechanic salaries on page 671. Find a best-fit model for the doubling salary sequence $t_n = 2.5 \cdot 2^{n-1}$, where $n = 1, 2, 3, ...$

Solution▸

Make a scatter plot of the first five terms of the sequence using (n, t_n) as the coordinates.

n (x)	1	2	3	4	5
t_n (y)	2.5	5	10	20	40

With your graphics calculator in connected function mode, use the regression feature to find the model with the correlation coefficient, r, closest to 1, and plot this curve.

The exponential regression model has an r-value of 1. The equation is $y = (1.25) \cdot 2^x$. Graph this equation in connected function mode on your calculator. ❖

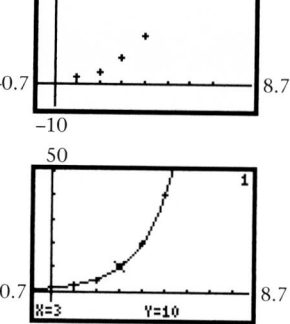

How does the function found in Example 3 compare with the explicit form of the doubling salary sequence?

Sum of a Geometric Series

A **geometric series** is an indicated sum of terms of a geometric sequence.

To find the sum, S_5, of the first 5 terms in the bicycle mechanic doubling salary sequence, you can add the terms.

$$S_5 = 2.5 + 5 + 10 + 20 + 40 = 77.5$$

However, there is another method.

Write another sequence, $2S_5$, where each term is multiplied by 2, the value of r. The first term of S_5 is eliminated in $2S_5$, and each term of the sequence appears to move to the left by one term.

$$S_5 = 2.5 + 5 + 10 + 20 + 40$$
$$2S_5 = 5 + 10 + 20 + 40 + 80$$

Find the difference between S_5 and $2S_5$.

$$\begin{aligned} S_5 &= 2.5 + 5 + 10 + 20 + 40 \\ -(2S_5) &= -(5 + 10 + 20 + 40 + 80) \\ \hline S_5 - 2S_5 &= 2.5 \qquad\qquad -80 \end{aligned}$$

Factor S_5 from the left side.
Notice that 80 is t_6. Substitute $2.5 \cdot 2^5$ for 80.
Factor 2.5 from the right side.
Solve for S_5.

$$S_5(1 - 2) = 2.5 - \mathbf{80}$$
$$S_5(1 - 2) = 2.5 - \mathbf{2.5 \cdot 2^5}$$
$$S_5(1 - 2) = 2.5(1 - 2^5)$$
$$S_5 = 2.5 \left(\frac{1 - 2^5}{1 - 2} \right) = 77.5$$

Using Formulas Show students the following proof of the sum of a geometric series. Have them explain each step.

$$\begin{aligned} S_n &= a + ar + ar^2 + \ldots + ar^{n-2} + ar^{n-1} \\ r \cdot S_n &= ar + ar^2 + \ldots + ar^{n-2} + ar^{n-1} + ar^n \\ S_n - rS_n &= a - ar^n \\ S_n(1 - r) &= a(1 - r^n) \\ S_n &= a\left(\frac{1 - r^n}{1 - r} \right) = t_1\left(\frac{1 - r^n}{1 - r} \right) \end{aligned}$$

This pattern leads to the general formula for the sum of the first n terms of a geometric sequence.

SUM OF THE FIRST n TERMS OF A GEOMETRIC SEQUENCE

Let t_1, t_2, t_3, t_4, ... represent a geometric sequence with constant ratio r. Then S_n, the sum of the first n terms, is given by

$$S_n = t_1 \left(\frac{1 - r^n}{1 - r} \right), \text{ where } r \neq 1.$$

CRITICAL
Thinking

Use a process similar to that used to find the sum of the first 5 terms of the salary sequence to show why the formula for the sum of the first n terms is correct for *any* geometric sequence.

EXAMPLE 4

Again, refer to the bicycle mechanic salaries on page 671. Find the sum of the first 10 terms in the bicycle mechanic doubling salary sequence.

Solution

In the bicycle mechanic doubling salary sequence, $t_1 = 2.5$, $r = 2$, and $n = 10$. Substitute these values into the formula for the sum of n terms of a geometric sequence.

$$S_n = t_1 \left(\frac{1 - r^n}{1 - r} \right)$$
$$S_{10} = 2.5 \left(\frac{1 - 2^{10}}{1 - 2} \right) = 2557.5 \qquad ❖$$

Which of the two bicycle mechanic salaries would you prefer? Explain.

EXAMPLE 5

Find the sum $\displaystyle\sum_{p=1}^{6} 2(3)^{p-1}$ using the formula for the sum of n terms of a geometric sequence.

Solution

n is 6, r is 3, and t_1 is $2 \cdot 3^0$, or 2.

$$S_n = t_1 \left(\frac{1 - r^n}{1 - 4} \right)$$
$$S_6 = 2 \left(\frac{1 - 3^6}{1 - 3} \right)$$
$$= 728 \qquad ❖$$

Try This   Find the sum $\displaystyle\sum_{t=1}^{10} 3(-1)^{t+2}$ using the formula for the sum of n terms of a geometric sequence.

CRITICAL
Thinking

$$S_n = t_1 + rt_1 + r^2 t_1 + \ldots + r^{n-1} t_1$$
$$rS_n = rt_1 + r^2 t_1 + \ldots + r^{n-1} t_1 + r^n t_1$$
$$S_n(1 - r) = t_1 - r^n t_1$$
$$S_n = \frac{t_1(1 - r^n)}{1 - r}$$

Alternate Example 4

Find the sum of the first 7 terms in the bicycle mechanic salary sequence. [$S_7 = 317.50$]

Aongoing
SSESSMENT

For the first 10 weeks, it is better to have the $200 a week. After 10 weeks the salary that doubles is much better.

Alternate Example 5

Find $\sum_{k=1}^{5} 3(4)^{k-1}$. [**1023**]

Aongoing
SSESSMENT

Try This

0

Assignment Guide

Core 1–13, 16–19, 27–53

Core-Plus 1–5, 11–54

Technology

Graphics calculators are needed for Exercises 47–48 and 51–52.

Error Analysis

The most common errors in the sum formula will occur in the order of operations. It is essential that the first computation be the exponents, followed by the simplification of numerators and denominators, and finally the division.

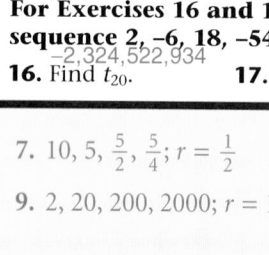

EXERCISES & PROBLEMS

Communicate

1. Describe how to find the general term of any geometric sequence.

2. Describe how to find the sum of a given number of terms of a geometric sequence.

3. Explain what would happen to the general term of a geometric sequence if *r* were doubled.

4. Explain what would happen to the general term of a geometric sequence if t_1 were doubled.

5. Compare your response to Exercise 3 with your response to Exercise 4. Describe the relationship between t_1 and *r* based on your comparison.

Practice & Apply

List the first 4 terms of each geometric sequence, and identify *r*, the common ratio for the sequence.

6. $t_n = 4(2)^n$
8, 16, 32, 64; $r = 2$

7. $t_n = 20\left(\frac{1}{2}\right)^n$

8. $t_n = 4(-3)^n$
−12, 36, −108, 324; $r = -3$

9. $t_n = \frac{1}{5}(10)^n$

10. In a geometric sequence, $t_1 = 6$, $r = 4$, and $n = 7$. Find t_n. 24,576

11. In a geometric sequence, $t_1 = 5$, $r = -2$, and $n = 7$. Find t_n. 320

12. In a geometric sequence, $t_1 = 3$, $r = \frac{1}{10}$, and $n = 20$. Find t_n. 3×10^{-19}

13. In a geometric sequence, $t_1 = 3$, $r = -\frac{1}{10}$, and $n = 20$. Find t_n. -3×10^{-19}

14. Find the 20th term of the sequence 10, 30, 90, ... 11,622,614,670

15. Find the 25th term of the sequence 40, 20, 10, ... 2.384×10^{-6}

For Exercises 16 and 17, use the sequence 2, −6, 18, −54, ...
−2,324,522,934

16. Find t_{20}.   17. Find S_{20}. −1,743,392,200

7. $10, 5, \frac{5}{2}, \frac{5}{4}; r = \frac{1}{2}$

9. $2, 20, 200, 2000; r = 10$

For Exercises 18–20, use the sequence 12, 3, $\frac{3}{4}$, ...

18. Find t_{10}. 4.578×10^{-5} **19.** Find S_{10}. ≈ 15.999 **20.** Find S_{20}. ≈ 16

For Exercises 21 and 22, use the geometric sequence $t_1 = 5$, $r = -5$, and $t_n = 3125$.

21. Find n. 5 **22.** Find the sum of the first n terms. 2605

23. 162 is which term of the geometric sequence 2, –6, 18, ... ? 5th

24. Find the sum $2 - 6 + 18 - 54 + \cdots + 162$. 122

25. The 7th term of a geometric sequence is 256, and the 1st term is 4. What is the 5th term? 64

26. Find the 1st term in a geometric sequence for which $r = 2$ and $t_6 = 96$. 3

On a number line, plot the terms of the following sequence.
$$1, -\frac{1}{2}, \frac{1}{4}, -\frac{1}{8}, \frac{1}{16}, -\frac{1}{32}, \dots$$

27. Is the sequence arithmetic, geometric, or neither? geometric

28. Describe the behavior of the points that you plot.

29. As n gets larger, what happens to t_n? It approaches zero.

Investments Suppose a grandfather decides to give his granddaughter $1 on her first birthday and to double the gift each following year.

30. How much will the granddaughter receive on her 21st birthday? $1,048,576

31. What is the total amount of money received by the granddaughter after her 21st birthday? $2,097,151

Statistics A *geometric mean* of two numbers, a and b, can be considered the term between a and b in a geometric sequence. For example, to find two geometric means between 3 and 192, first draw a diagram.

$$3, \underset{\times r}{\underline{\quad}}, \underset{\times r}{\underline{\quad}}, \underset{\times r}{192}$$

The terms in this geometric sequence are 3, $3r$, $3r^2$, and $3r^3$, with $3r^3 = 192$. So, r is 4, and the 2 geometric means between 3 and 192 are 12 and 48.

Find the following geometric means.

32. Insert 1 geometric mean between 5 and 320. 40

33. Insert 3 geometric means between 12 and 7500. 60, 300, 1500

34. Insert 4 geometric means between 4 and 972. 12, 36, 108, 324

28.

The points get closer to 0 from the left and the right alternately.

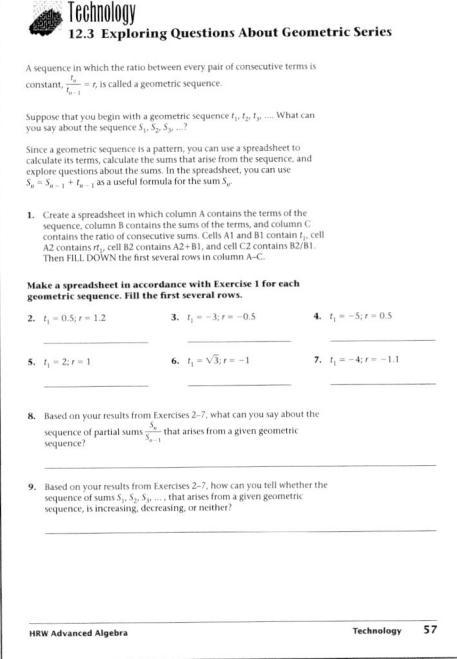

Geometry A side of a square is 8 centimeters long. A second square is inscribed in it by joining the midpoints of the sides of the first square. The process is continued as shown in the diagram.

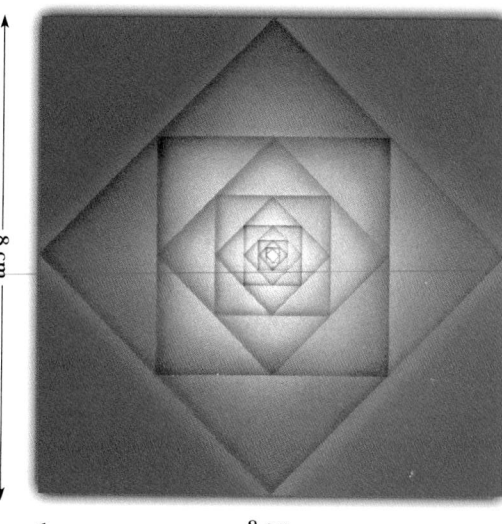

8 cm

8 cm

35. Write the sequence of side-lengths for each of the first 7 squares.

36. Write an explicit formula for the square side-length sequence. $t_n = 8 \cdot \left(\frac{1}{\sqrt{2}}\right)^{n-1}$

37. Find the sum of the side-lengths for the first 7 squares. $15 + 7\sqrt{2} \approx 24.9$ cm

38. Develop a formula for S_n, the sum of the side-lengths for the first n squares, in terms of t_1 and n.

Identify t_1, r, and t_n, and evaluate each sum.

39. $\sum_{k=1}^{10} 5^k$ $t_1 = 5$; $r = 5$; $t_{10} = 9{,}765{,}625$; $s_{10} = 12{,}207{,}030$

40. $\sum_{k=1}^{12} 3 \cdot 2^k$ $t_1 = 6$; $r = 2$; $t_{12} = 12{,}288$; $s_{12} = 24{,}570$

Games In a lottery, the first ticket drawn pays $10,000. Each succeeding ticket pays half as much as the preceding one.

41. If 5 tickets are drawn, how much does the last ticket pay? $625

42. What would be the total amount of prize money given away? $19,375

43. A tortoise travels 1 yard in 1 minute. In the next minute it travels $\frac{3}{4}$ of a yard. Similarly, in each succeeding minute, the tortoise travels $\frac{3}{4}$ as far as in the preceding minute. If the tortoise continues in this way, what is the total distance that it would travel in 15 minutes? approx 3.95 yd

35. $8, 4\sqrt{2}, 4, 2\sqrt{2}, 2, \sqrt{2}, 1$

38. $S_n = 8 \left(\dfrac{1 - \left(\frac{1}{\sqrt{2}} \right)^n}{1 - \left(\frac{1}{\sqrt{2}} \right)} \right)$, where $r = 8$ and $t_1 = \dfrac{1}{\sqrt{2}}$

44. Physics A ball is dropped from a height of 8 feet. It rebounds $\frac{1}{2}$ the distance and again falls to the floor. If the ball keeps rebounding in this manner, what is the total distance, to the nearest tenth, that the ball travels after 10 rebounds? approx 24.0 ft

45. Annuities At age 21, Keri began receiving annual payments from a trust fund. On each birthday thereafter, she received 10% more than she had received the preceding year. If she receives a total of $30,380 by age 25, how much did she receive at age 21? $4976.17

Look Back

Geometry A piece of cardboard measures 6 feet by 3 feet. **[Lesson 6.1]**

46. Write a function, in terms of x, for the volume of a box created by cutting squares, with sides of length x, from the cardboard piece. $f(x) = x(3 - 2x)(6 - 2x)$

47 What is the maximum volume of this box? approx 5.2 ft³

48 What are the real-world domain and range for this function? $0 < x < 1.5$, approx $0 < f(x) < 5.2$

Accounting Personal computers often depreciate rapidly in value due to advances in technology. Suppose a system that originally cost $3800 loses 10% of its value every 2 months.   **[Lesson 7.1]**

49. What is the multiplier for this exponential decay function? 0.9

50. Write a formula for the value of the system, $V(t)$, after t 2-month periods. $V(t) = 3800(0.9)^t$

51 What is the value of the system after a year? approx 53%; approx $2019.48

52 Graph $V(t)$, and trace to estimate the value after 15 months. approx $1724

53. If all the letters are used, how many different permutations are there of the letters in *ROOMMATE*? **[Lesson 11.3]** 10,080

Look Beyond

54.   Coordinate Geometry
An ordered triple lists the coordinates of a point in a three-dimensional coordinate space. Sketch a three-dimensional coordinate space, and plot the points (1, 2, 3), (–1, –2, 3), and (1, 2, –3).

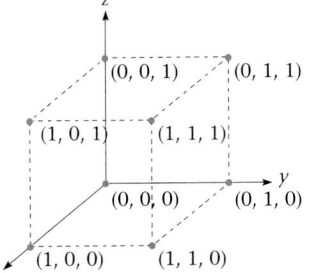

54.

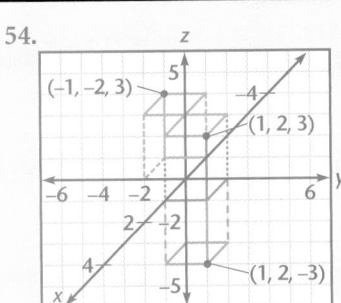

Objectives

• Find the sum of an infinite geometric series.

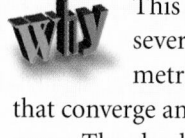

Assessing Prior Knowledge

1. Evaluate $\sum_{n=1}^{8} 3 \cdot (3)^{n+2}$.
   [265,680]

2. Define what is meant by a *geometric sequence*. [a set of numbers in a specific order in which each term after the first term is the product of the common ratio and the preceding term]

TEACH

This lesson looks at several infinite geometric series, some that converge and some that diverge. The absolute value of the common ratio determines convergence or divergence.

There is an old story, called Zeno's paradox, that says Achilles can never catch the tortoise in a race because each time Achilles gets closer to the tortoise, the tortoise also advances. For Example, if Achilles starts at point *A* and the tortoise starts at point *B*, then when Achilles reaches *B*, the tortoise is at *C*, and when Achilles reaches *C*, the tortoise is at *D*, and so on.

Why You have seen how arithmetic and geometric series can be used to model some real-life situations. An infinite *geometric series* can also be used as a model.

Theoretically, this pattern can go on forever without Achilles ever reaching the tortoise. This is an example of a paradox that can be modeled with an infinite sequence. In reality, we know that if Achilles runs faster than the tortoise, he will win the race.

ALTERNATIVE teaching strategy

Using Models Use a string, a weight, a meter stick, and a thumbtack to make a pendulum. Ask students to name what forces, such as friction, decrease the distance traveled during each successive swing. Give students several theoretical distances for successive swings, such as 30 cm, 27 cm, 24.3 cm, … . Ask them to determine *r*. [0.9] Emphasize that theoretically the pendulum would swing an infinite number of times with the distance traveled becoming so small as to be unmeasurable. Next, add the first 40 terms of this sequence (use a calculator). Record that sum and then add the 41st term. Note that the 41st term does not change the sum very much. This will illustrate that the sum is approaching a limit.

The following sequence is an example of an infinite geometric sequence.

$$\frac{1}{2}, \frac{1}{4}, \frac{1}{8}, \frac{1}{16}, \cdots, \frac{1}{2^n}, \cdots$$

This sequence can be graphed in disconnected sequence mode using the explicit formula $t_n = \frac{1}{2^n}$.

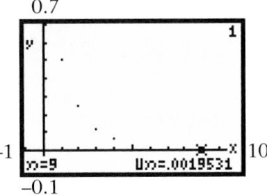

Examine the following *partial sums* of this sequence.

$$S_1 = \frac{1}{2}$$
$$S_2 = \frac{1}{2} + \frac{1}{4} = \frac{3}{4}$$
$$S_3 = \frac{1}{2} + \frac{1}{4} + \frac{1}{8} = \frac{7}{8}$$
$$\vdots$$
$$S_n = \frac{1}{2} + \frac{1}{4} + \frac{1}{8} + \cdots + \frac{1}{2^n}$$

The number S_n is called the **nth partial sum** of the sequence, indicated by $\sum_{n=1}^{n} \frac{1}{2^n}$.

Sigma notation can also be used to indicate an *infinite* geometric series.

$$S_\infty = \frac{1}{2} + \frac{1}{4} + \frac{1}{8} + \cdots + \frac{1}{2^n} + \cdots = \sum_{n=1}^{\infty} \frac{1}{2^n}$$

The symbol ∞ is used for infinity.

S_∞ will be called the infinite sum of a series and will be written S.

It is impossible to add an infinite number of terms. However, by examining the partial sums of the infinite geometric series, S, you can find the number that the sum approaches.

$$S_1 = \frac{1}{2}, \ S_2 = \frac{3}{4}, \ S_3 = \frac{7}{8}, \ \ldots, \ S_n = \frac{2^n - 1}{2^n}$$

Notice that in each partial sum, the denominator doubles *and* the numerator is always 1 less than the denominator. As values of n get larger, the quotient approaches, or *converges*, to 1. Since the infinite series $\sum_{n=1}^{\infty} \frac{1}{2^n}$ converges to 1, its sum is defined to be $\sum_{n=1}^{\infty} \frac{1}{2^n} = 1$.

Give an example of an infinite geometric series that does *not* converge.

SUM OF AN INFINITE GEOMETRIC SERIES

The sum, S, of an infinite geometric series, where $|r| < 1$, is
$$S = \frac{t_1}{1 - r}.$$

An infinite geometric series converges only when $-1 < r < 1$.

interdisciplinary

CONNECTION

Physics Although mathematics is the only field of study that includes the concept of infinity, certain physical processes occur so quickly that a near-infinite number of steps can happen in a reasonably short length of time. For example, damped oscillatory vibration represents a near-infinite geometric sequence with r less than 1 and the amplitude of the vibration quickly converges to zero.

A. Find $\sum_{n=1}^{\infty} \frac{1}{2^{n+1}}$ $\left[\frac{1}{2}\right]$

B. Find the sum $0.2 + 0.02 + 0.002 + 0.0002 + \ldots$ $\left[\frac{2}{9}\right]$

EXAMPLE 1

Graphics Calculator

Ⓐ Find $\sum_{n=1}^{\infty} \frac{1}{3^{n+1}}$.

Ⓑ Find the sum: $0.1 + 0.01 + 0.001 + 0.0001 + \cdots$

Solution▸

Ⓐ First substitute $n = 1$ and $n = 2$ to find t_1 and r.

$$t_n = \frac{1}{3^{n+1}}$$
$$t_1 = \frac{1}{3^{1+1}} = \frac{1}{9}$$
$$t_2 = \frac{1}{3^{2+1}} = \frac{1}{27}$$

So, $t_1 = \frac{1}{9}$ and $r = \frac{t_2}{t_1} = \frac{1}{3}$.

Then use the formula for the sum of an infinite geometric series.

$$\sum_{n=1}^{\infty} \frac{1}{3^{n+1}} = \frac{t_1}{1-r}$$

$$= \frac{\frac{1}{9}}{1-\frac{1}{3}}$$

$$= \frac{\frac{1}{9}}{\frac{2}{3}} = \frac{1}{9} \cdot \frac{3}{2} = \frac{1}{6}$$

Use the formula for the sum of the first n terms of a geometric series,

$$S_n = t_1 \left(\frac{1-r^n}{1-r}\right) = \frac{1}{9}\left(\frac{1-\left(\frac{1}{3}\right)^n}{1-\left(\frac{1}{3}\right)}\right),$$

with your graphics calculator in disconnected sequence mode to graph this geometric series.

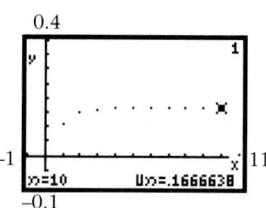

The graph of the series shows that terms approach $\frac{1}{6}$, or $0.1666\ldots$ as x increases.

Ⓑ The geometric series $0.1 + 0.01 + 0.001 + 0.0001 + \cdots$ is the repeating decimal $0.11111\ldots$, or $0.\overline{1}$. The ratio, r, of the series is 0.1, and t_1 is 0.1.

$$S = \frac{t_1}{1-r} = \frac{0.1}{1-0.1} = \frac{1}{9}$$

Thus, the sum of the infinite series is $\frac{1}{9}$, or $0.\overline{1}$.

From the graph of the geometric series,

$$S_n = t_1\left(\frac{1-r^n}{1-r}\right) = 0.1\left(\frac{1-0.1^n}{1-0.1}\right),$$

you can see that the series converges to $0.\overline{1}$.

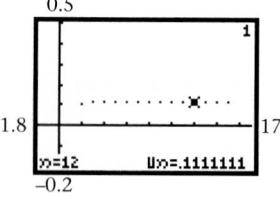

ENRICHMENT Have students compare the terms of Alternate Example 1, Part A, to the exponential function $y = 2^{-x}$ and to the function $-y = \log_2 x$. The functions are continuous, but the points on the example graph refer to a sequence, so they are limited to the points, $(1, \frac{1}{4}), (2, \frac{1}{8}), (3, \frac{1}{16}),\ldots$

INCLUSION **strategies** **Hands-On Strategies** Have students stand at one side of the classroom. Have them measure 48 feet. From the end of the measure, have them return and mark half that distance. They then turn again and measure half the distance to the mark. They should repeat the process several times. Students should determine that the total distance that they can measure will be 96 feet.

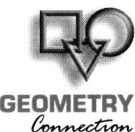

GEOMETRY
Connection

EXAMPLE 2

A side of an equilateral triangle is 20 inches long.
A second equilateral triangle is inscribed in it by joining the midpoints of the sides of the first triangle. The process is continued as shown in the diagram. A path is constructed using a side of each triangle as shown.

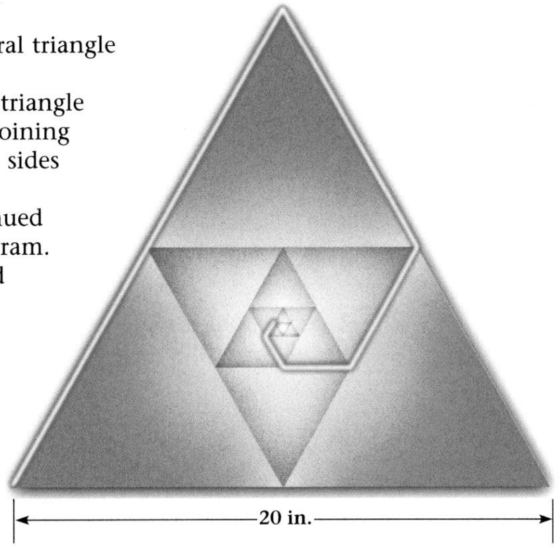

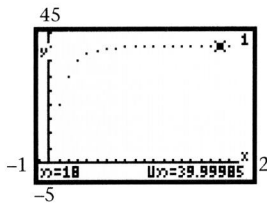

|← —————————— 20 in. —————————— →|

A Find the length of the path in the first 4 equilateral triangles.

B Use your graphics calculator to show that this infinite series converges. What is its sum?

Solution►

The sides of the equilateral triangles form a geometric sequence,
$S = 20, 10, 5, 2.5, \ldots$, where r is 0.5 and t_1 is 20.

A The length of the path in the first 4 triangles is the sum of the first 4 terms.

$$S_4 = 20 + 10 + 5 + 2.5 = 37.5.$$

So the length of the path of the first 4 triangles is 37.5 inches.

B Write the sequence related to the sum of the first n terms for the infinite series, and graph it.

$$S_n = t_1 \left(\frac{1 - r^n}{1 - r} \right)$$
$$= 20 \left(\frac{1 - 0.5^n}{1 - 0.5} \right)$$

The graph shows that the series approaches 40. Thus, the infinite series converges.

$$S_n = \frac{t_1}{1 - r} = \frac{20}{1 - 0.5} = 40 \qquad ❖$$

```
45

          · · · · · · · · · · · · ·•· ¹
       ·
      ·
     ·
    ·
   ·
  ·
 ·
-1 ┤x=18        y=39.99985      21
  -5
```

Try This Find the sum of the perimeters of all the equilateral triangles in Example 2. How does this sum compare with the length of the path?

Alternate Example 3

Use a calculator to find $\sum_{n=1}^{7} 3\left(\frac{1}{2}\right)^n$ [**2.9765625**]
Compare that result with the infinite sum. [3]

CRITICAL
Thinking

The graph plots partial sums of the sequence for successive values of n, so the horizontal asymptote is the y-value that the sum approaches. If the sum is a, then the asymptote has equation $y = a$.

ASSESS

Selected Answers

Odd-Numbered Exercises 5–25

Assignment Guide

Core 1–25

Core-Plus 1–26

Technology

Graphics calculators are needed for Exercises 16–17 and 20.

Practice Master

EXAMPLE 3

Graphics Calculator

Find $\sum_{n=1}^{14} 2\left(\frac{1}{3}\right)^n$. Then compare your result with the infinite series $\sum_{n=1}^{\infty} 2\left(\frac{1}{3}\right)^n$.

Solution▶

Some graphics calculators will find the sum of a finite sequence.

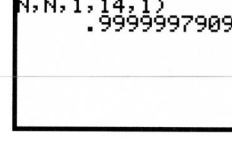

Thus, $\sum_{n=1}^{14} 2\left(\frac{1}{3}\right)^n \approx 1$.

The graph of the sequence related to the sum of the first n terms,

$$S_n = \frac{2}{3}\left(\frac{1-\left(\frac{1}{3}\right)^n}{1-\left(\frac{1}{3}\right)}\right),$$

also shows that the series converges to 1.

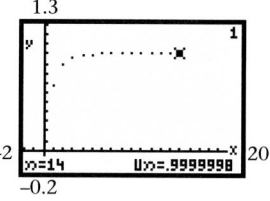

Since $|r|$ is $\frac{1}{3}$, and $\frac{1}{3} < 1$, the sum of the infinite geometric series is found using $S = \frac{t_1}{1-r}$.

$$S = \sum_{n=1}^{\infty} 2\left(\frac{1}{3}\right)^n = \frac{t_1}{1-r} = \frac{\frac{2}{3}}{1-\frac{1}{3}} = 1 \;❖$$

CRITICAL
Thinking

Explain how the asymptote of the graph of an infinite geometric series with $|r| < 1$ illustrates its sum.

EXERCISES & PROBLEMS

Communicate

1. Explain why the absolute value of r must be less than 1 for the sum of an infinite geometric series to converge.

2. Describe the relationship between the sum of the first 20 terms and the sum of *all* the terms of $t_n = 4\left(\frac{1}{6}\right)^n$.

3. Explain what would change in Exercise 2 if r were 0.3.

4. What is the difference between the graph of an infinite series that converges and the graph of an infinite series that does not converge?

Practice & Apply

Find the sum of the infinite geometric series if it exists.

5. $t_1 = 6$
$r = 0.4$ 10

6. $t_1 = 5$
$r = -2$
does not exist

7. $t_1 = 3$
$r = \frac{1}{10}$ $\frac{10}{3}$

8. $t_1 = 3$
$r = -\frac{1}{10}$ $\frac{30}{11}$

Given the sequence $S = 10, 5, 2.5, \ldots$, complete the following.

9. Write the general term of S.
$t_n = 10(0.5)^{n-1}$

10. Find the sum of S. 20

Given the sequence $S = 40, 20, 10, \ldots$, complete the following.

11. Find the 25th term of S.
0.00000238

12. Find the sum of S. 80

Given the sequence $S = 12, 3, \frac{3}{4}, \ldots$, complete the following.

13. Find t_{10}.
0.0000458

14. Find S_{10}.
15.99998474

15. Find $\displaystyle\sum_{n=1}^{\infty} S_n$. 16

Find each sum. Verify your result by graphing the related sequence.

16 $\displaystyle\sum_{k=1}^{\infty} 0.5^k$ 1

17 $\displaystyle\sum_{k=1}^{\infty} 3 \cdot (0.2)^k$ 0.75

18. Physics A golf ball dropped from a height of 81 yards rebounds on each bounce $\frac{2}{3}$ of the distance from which it falls. How far does it travel before coming to rest? 405 yards

19.  **Geometry** The midpoints of the sides of a square are joined to create a new square. This process is performed on each new square. Find the sum of the areas of an infinite sequence of such squares if one side of the first square is 10 feet long. 200 sq ft

Look Back

20 Write the matrix equation that represents the following system. Use your graphics calculator to solve it, and check your answer. **[Lesson 4.6]**

$$\begin{cases} -2x + y + 6z = 18 \\ 5x + 8z = -16 \\ 3x + 2y - 10z = -3 \end{cases}$$

Solve and check. [Lesson 7.7]

21. $9^{x-2} = 27$ $x = 3.5$ **22.** $2^{-x} = 16$ $x = -4$ **23.** $8^x = 0.5$ $x = -\frac{1}{3}$ **24.** $2^x = \frac{1}{2}$ $x = -1$

25. Construction Management The manager of a construction company currently is managing 3 projects. The probability that each will be completed on schedule is 0.85, 0.72, and 0.94. If the completion times are independent events, what is the probability, to the nearest hundredth, that all 3 projects will be completed on schedule? **[Lesson 11.5]**   0.58

Look Beyond

26. Using an 11-minute sand timer and a 7-minute sand timer, what is the easiest way to time the boiling of an egg for 15 minutes?

20. $\begin{bmatrix} -2 & 1 & 6 \\ 5 & 0 & 8 \\ 3 & 2 & -10 \end{bmatrix} \begin{bmatrix} x \\ y \\ z \end{bmatrix} = \begin{bmatrix} 18 \\ -16 \\ -3 \end{bmatrix}$;

$x = -4$
$y = 7$
$z = 0.5$

26. Start both timers. When the 7-min timer finishes turn it over. When the 11-min timer finishes, 4-min will have fallen through the 7-min timer. Turn the 7-min timer over again to time the last 4-min.

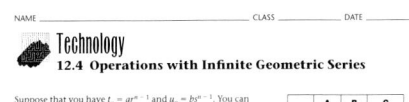

FOCUS

A test to determine whether a sequence of numbers is truly random prepares the way for exploring such topics as the meaning of randomness, the way we test for randomness, and the reason our perception of random events differs from actual occurrences. A geometric model helps students understand and apply one test.

MOTIVATE

Before students read the article, they should discuss the term *random*. They should establish in the discussion an acceptable working definition for random numbers. This definition will allow students to see the difficulty in creating a computer program for generating random numbers. They can then be presented with the problem of generating a set of random numbers on their own and determining whether the numbers are actually random.

Is it FAKE or is it RANDOM?

The Quest for True Randomness Finally Appears Successful

One of the strangest quests of modern computer science seems to be reaching its goal; mathematicians believe they have found a process for making perfectly random strings of numbers.

Sequences of truly patternless, truly unpredictable digits have become a perversely valuable commodity, in demand for a wide variety of applications in science and industry. Randomness is a tool for insuring fairness in statistical studies or jury selection, for designing safe cryptographic schemes and for helping scientists simulate complex behavior.

Yet random numbers—as unbiased and disorganized as the result of millions of imaginary coin tosses—have long proved extremely hard to make, either with electronic computers or mechanical devices. Consumers of randomness have had to settle for numbers that fall short, always hiding some subtle pattern.

Random number generators are sold for every kind of computer. Every generator now in use has some kind of flaw, though often the flaw can be hard to detect. Furthermore, in a way, the idea of using a predictable electronic machine to create true randomness is nonsense. No string of numbers is really random if it can be produced by a simple computer process. But in a more practical sense, a string is random if there is no way to distinguish it from a string of coin flips.

Several theorists presented details of the apparent breakthrough in random-number generation. The technique will now be subjected to batteries of statistical tests meant to see whether it performs as well as the theorists believe it will. The

way people perceive randomness in the world around them differs sharply from the way mathematicians understand it and test for it.

The need for randomness in human institutions seems to begin at whatever age "eeny-meeny-miny-moe" becomes a practical decision-making procedure: randomness is meant to insure fairness. Like "eeny-meeny-miny-moe," most such procedures prove far from random. Even the most carefully designed mechanical randomness makers break down under scrutiny.

One such failure, on a dramatic scale, struck the national draft lottery in 1969, its first year. Military officials wrote all the possible birthdays on 366 pieces of paper and put them into 366 capsules. Then they poured the January capsules into a box and mixed them. Then they added the February capsules and mixed again—and so on.

At a public ceremony, the capsules were drawn from the box by hand. Only later did statisticians establish that the procedure had been far from random: people born toward the end of the year had a far greater chance of being drafted than people born in the early months.

An expert in exposing the flaws in pseudo random-number generators, George Marsaglia of Florida State University, has begun to test the new technique. Dr. Marsaglia judges sequences not just by uniformity—a good distribution of numbers in a sequence—but also by "independence." No number or string of numbers should change the probability of the number or numbers that follow, any more than flipping a coin and getting 10 straight tails changes the likelihood of getting heads on the 11th flip.

How hard can it be to write a bunch of random numbers? Do you think you can write a sequence of 0s and 1s that looks like it came from coin flips? Do you think you could fool a psychologist or mathematician?

To find out, start by writing down a sequence of 120 zeros and ones in an order that you think looks random.

---

2. Answers may vary. Sample:
Sequence A is a random sequence.
Sequence B is a sequence made by guessing.
Sequence C is a fake sequence.

Triple	Frequency		
	Sequence A	Sequence B	Sequence C
(0, 0, 0)	5	2	3
(1, 1, 1)	6	2	4

3. The probability of all heads is $\frac{1}{8}$ and of all tails is also $\frac{1}{8}$. The events are independent and the basic probability of heads is $\frac{1}{2}$, so $\frac{1}{2} \cdot \frac{1}{2} \cdot \frac{1}{2} = \frac{1}{8}$. The same is true for tails.

Before you put your made-up sequence to the test, you need to find out how to check for randomness. One way is by breaking the sequence into smaller sequences, such as sets of 3 digits. Then you can see whether some sets occur too often or not often enough.

You can visualize these sets of digits by plotting them as points. Think of the first 3 digits in your sequence as an ordered triple, the x, y, and z coordinates of a point in space. Using the digits 0 and 1 to model heads and tails, respectively, there are 8 possible ordered triples, as shown.

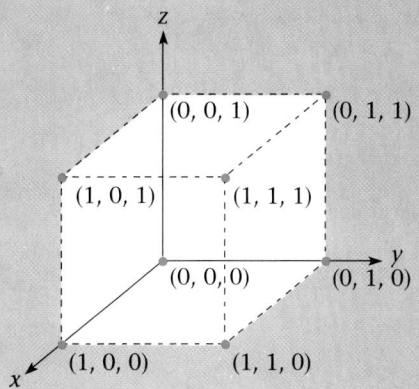

Cooperative Learning

Now you can see whether your classmates are able to distinguish your made-up sequence from an actual sequence.

1. Generate a random sequence of 120 zeros and ones by tossing a coin (heads = 0, tails = 1) or by using a random-number table (even numbers = 0, odd numbers = 1).

   Give this actual sequence and your made-up sequence to another group. (First, mark the sequences so that only you know which one is made up.)

2. Test each sequence you are given by following these steps.
   a. Separate each sequence into 40 ordered triples, as shown here.

   1 0 1/1 1 0/1 0 0/1 1

   b. Count the number of times the triples 000 and 111 appear. Record your results in a chart.

	Sequence A	Sequence B	Sequence C
0 0 0			
1 1 1			

3. If you flip a coin 3 times, what is the probability of getting 3 heads? 3 tails? Explain.

4. In each new sequence you test, about how many of the 40 triples would you expect to be 000? 111?

5. How might you use your answer to Step 4 to distinguish real random sequences from fake ones?

6. Use your answer to Step 5 to tell which sequence is the fake one. Then find out whether you are correct.

7. Do you think the test would work better if you used sequences of 1200 zeros and ones instead of 120? Explain.

8. Write a definition for *random numbers* based on what you think the term means.

---

4. Out of 40 ordered triples, there should be about 5.

5. In actual sequences, the triples 000 and 111 will probabily appear about 5 times each. In sequences people make up, these triples are more likely to appear just once or twice each or even not at all.

7. The longer the actual sequences, the more likely the results will resemble expected results.

Cooperative Learning

Have students try out the test on friends and family members, who are not familiar with problem. Some students can also create their own ways to test randomness. For example, after a head, about what fraction of the times should the next flip be a head? a tail? How can that be used to test for randomness? Try it on the sequences used in the lesson.

DISCUSS

Discuss the importance of random number generators for various fields of study. For example, have students consider their use in the

• physical sciences (weather, chemistry, physics, biology) or

• medical research (surveys, testing results for bias, choosing samples)

Have students discuss how graphing can be used to determine randomness.

Tests of randomness can be very sophisticated, but as the level of sophistication increases, so does the cost. Students can discuss how to determine the degree of sophistication in randomization, and how to fit that degree to the purpose for using the random numbers.

- Use Pascal's Triangle to find combinations.

RESOURCES

- Practice Master   12.5
- Enrichment Master   12.5
- Technology Master   12.5
- Lesson Activity Master   12.5
- Quiz   12.5
- Spanish Resources   12.5

Assessing Prior Knowledge

1. If you flip a coin 10 times and get 8 heads, what is the probability that your 11th toss will be a head? $\left[\frac{1}{2}\right]$

2. What is the probability that a die will come up with an odd number 6 times in a row? $\left[\frac{1}{64}\right]$

TEACH

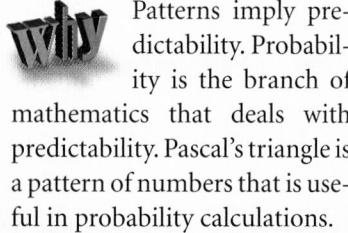

Patterns imply predictability. Probability is the branch of mathematics that deals with predictability. Pascal's triangle is a pattern of numbers that is useful in probability calculations.

Cooperative Learning

Divide the class into teams of 5 members and give each team a 10-row Pascal's Triangle. Have each team find and document as many patterns and/or sequences as possible in the numbers of the triangle. Expect at least seven.

LESSON 12.5 Pascal's Triangle and Probability

Cultural Connection: Asia Education was at a low point in Europe during the Middle Ages. But in China, rapid advances in mathematical discoveries were taking place. Many of the Chinese achievements predated those made in Europe during the Renaissance. One of these achievements was the presentation of the arithmetic triangle. This triangle first appeared in the following form.

Patterns in arithmetic and geometric sequences lead to formulas for finding special sums. A famous pattern in the form of an arithmetic triangle connects combination formulas to probability.

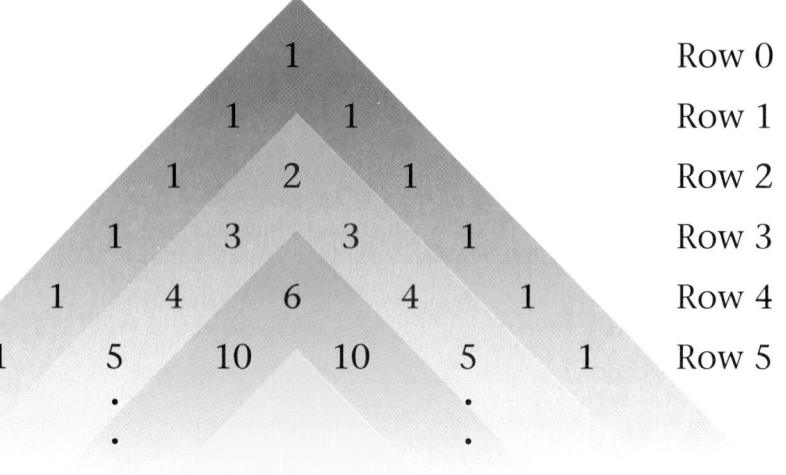

	Row 0
1	Row 0
1 1	Row 1
1 2 1	Row 2
1 3 3 1	Row 3
1 4 6 4 1	Row 4
1 5 10 10 5 1	Row 5

Blaise Pascal (1623 – 1662)

In 1665, the French mathematician Blaise Pascal published a math text containing another form of the arithmetic triangle.

Row 0	1	1	1	1	1	1
Row 1	1	2	3	4	5
Row 2	1	3	6	10
Row 3	1	4	10
Row 4	1	5
Row 5	1

In his text, Pascal developed many patterns. He found one pattern particularly useful in the study of probability. Because Pascal discovered so many patterns and because he was one of the first Europeans to write about them, the arithmetic triangle became known in the western world as **Pascal's Triangle**.

ALTERNATIVE
teaching
strategy

Using Hands-On Strategies Give each student a penny. Each student should toss the penny 10 times and carefully record the outcome, including the number of heads and tails. Then poll the class for the number of times anyone got zero heads. Write a 1 at the lower left of the board to represent that there is one way that zero heads could have happened. Continue polling the class for 1 head, 2 heads, etc.,

and record the results across the bottom of the chalkboard. Ask the students what they would expect the results to be if they each flipped the coin one time. Enter those two numbers (1, 1) at the center top of the chalkboard. Now ask them to flip the coin twice and record their results (1, 2, 1). From this model of Pascal's Triangle, generate the actual pattern and discuss various interpretations of the numbers in the triangle.

Exploration 1 · A Triangle Pattern

1 Examine both forms of Pascal's Triangle for patterns. Find and explain at least two different sequences in the triangles. Find and describe at least two other patterns.

2 Choose any number in the top triangle other than 1. Look at the two numbers closest to it in the preceding row. What is the relationship between the numbers?

3 Repeat Step 2 using a different value.

4 Use the relationship you found in Steps 2 and 3 to construct rows 6 and 7 of the first triangle. Use this information to add the equivalent entries in the second triangle. ❖

EXAMPLE 1

Suppose a coin is flipped 4 times.

A How many different outcomes are possible?

B List all the possible outcomes for each of the following events.

0 heads, 1 heads, 2 heads, 3 heads, and 4 heads

In how many ways can each event occur?

Solution▶

A Each coin flip has 2 possible outcomes: H (heads) or T (tails).

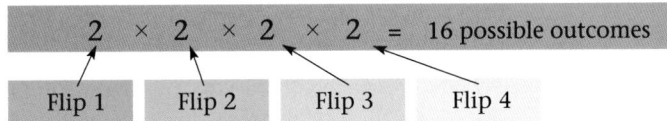

$2 \times 2 \times 2 \times 2 = 16$ possible outcomes

Flip 1   Flip 2   Flip 3   Flip 4

Thus, there are 16 possible outcomes.

B

0 heads	1 head	2 heads	3 heads	4 heads
TTTT	HTTT	HHTT	HHHT	HHHH
	THTT	HTHT	HHTH	
	TTHT	HTTH	HTHH	
	TTTH	THHT	THHH	
		THTH		
		TTHH		
1 way	4 ways	6 ways	4 ways	1 way

❖

How do the values of the frequencies of the events in Example 1 relate to Pascal's Triangle?

The frequency of each event is a combination. For example, the column with 2 heads shows the number of ways you can get 2 heads from tossing 4 coins, or $_4C_2$.

$$_nC_r = \frac{n!}{r!\,(n-r)!}$$

$$_4C_2 = \frac{4!}{2!\,(4-2)!} = 6$$

interdisciplinary

CONNECTION

Genetics In genetics, offspring will have various combinations of the genetically determined characteristics of the parents. Combinatorial mathematics allows predictions of the overall distribution of characteristics among future generations. Have students research the connection between binomial powers and genetics.

Exploration 1 Notes

Students should find the fundamental pattern of Pascal's Triangle — that an entry is the sum of those numbers above it to the immediate left and right.

A ongoing SSESSMENT

2. The chosen number is equal to the sum of the two numbers closest to it in the row above.

Alternate Example 1

A basketball player attempts 3 free throws. How many different shot patterns are possible? Hint: List all the possible ways she might make 3 baskets, 2 baskets, 1 basket, and zero baskets. Each throw has two possible results, B (basket) or M (miss). That's $2 \cdot 2 \cdot 2 = 8$ possible patterns. **[8]**

TEACHING tip

Remind students, factorials in fractions do not divide like factors. In other words, $\frac{8!}{4!}$ is not $2!$.

A ongoing SSESSMENT

The number of possible outcomes are distributed like the 4th row of Pascal's Triangle.

Exploration 2 Notes

Students should find the less obvious patterns of the triangle, including the ways that n and r are related in successive values of $_nC_r$.

Aongoing ASSESSMENT

3. It is possible to predict from the triangular clusters of even numbers, where any even entry will be .

4. It is possible to predict from the triangular arrangement of odd numbers, where any odd entry will be .

5. The pattern repeats every four rows. If n is a whole number, $4n$ rows have odd numbers in the first and last entries; $4n + 1$ rows have odd numbers in first, second, second from last, and last entries; $4n + 2$ rows have odd numbers in the first, third, third from last, and last entries; $4n + 3$ rows have odd numbers in the first four and last four entries.

Alternate Example 2

Students are given a test with 8 questions. If they are to answer only 5 of the questions, how many possibilities are there for questions to answer? [56]

Use Transparency ▶ 53

The following form of Pascal's Triangle is useful for finding combinations.

	Taken 0 at a time	Taken 1 at a time	Taken 2 at a time	Taken 3 at a time	Taken 4 at a time	Taken 5 at a time
0 items	1					
1 item	1	1				
2 items	1	2	1			
3 items	1	3	3	1		
4 items	1	4	6	4	1	
5 items	1	5	10	10	5	1

Exploration 2 Patterns and Probability

1 Describe Pascal's Triangle using $_nC_r$ notation.

$$_0C_0$$
$$_1C_0 \qquad _1C_1$$
$$_?C_? \qquad _?C_? \qquad _?C_?$$
$$_?C_? \qquad _?C_? \qquad _?C_? \qquad _?C_?$$

2 Using a large piece of paper, write the first 15 rows (20 rows if you have a spreadsheet program) of the Arithmetic Triangle.

Make several copies of your triangle. (You may wish to keep a copy handy as a reference.)

3 On one copy, shade all even numbers. If you are using a spreadsheet, shade each cell containing an even number. Describe the pattern this creates in your triangle. How can this pattern help you to predict more entries in the triangle?

4 On another copy, shade all odd numbers. How can the resulting pattern help you predict probabilities using the triangle? How does the pattern of odd numbers compare with the pattern of even numbers? Is one pattern more useful than the other? Explain.

5 Use $_nC_r$ notation to fill in all the entries of one copy of your triangle. Describe the patterns you see in the triangle. ❖

EXAMPLE 2

Use Pascal's Triangle to answer each question.

A If a coin is flipped 5 times, in how many ways can 2 heads occur?

B If a coin is flipped 5 times, what is the probability of getting exactly 2 heads?

ENRICHMENT If a sequence is made by listing the sums of the terms in the rows of Pascal's Triangle, there is an obvious pattern. Less obvious is why there should be a pattern. Show how the pattern *sum of two numbers from above* essentially adds each row to itself. For example,

```
  1  3  3  1
+    1  3  3  1
-------------
  1  4  6  4  1
```

INCLUSION
strategies

Hands-On Strategy Using the row-and-column form of Pascal's Triangle from page 688, identify and number the diagonals and columns. Then ask students to experiment, using a calculator, to determine the relationship between the diagonal and row number of a number entry, and $_nC_r$.

Solution

Ⓐ To find probabilities for 5 flips of a coin, use Row 5 of Pascal's Triangle. Row 5 of Pascal's Triangle has the following entries.

$_5C_0 = 1$	$_5C_1 = 5$	$_5C_2 = 10$	$_5C_3 = 10$	$_5C_4 = 5$	$_5C_5 = 1$
0 heads	1 heads	2 heads	3 heads	4 heads	5 heads

The entries are the number of ways that 0, 1, 2, 3, 4, and 5 heads, respectively, can occur. Thus, when flipping a coin 5 times, 2 heads can occur in 10 ways.

Ⓑ The total number of possible outcomes when a coin is flipped 5 times is $2^5 = 32$. Thus, the probability of getting 2 heads when a coin is flipped 5 times is $\frac{10}{32} = 0.3125$, or about 31%. ❖

EXERCISES & PROBLEMS

Communicate

1. Describe two patterns in Pascal's Triangle that you can see when the triangle is written using $_nC_r$ notation.

2. Describe two patterns in Pascal's Triangle that you can see when the triangle is written with integer terms.

3. Explain how Pascal's Triangle can be used to find the probability that 4 tails will result if 6 coins are tossed.

4. What values of n and r should be used with $_nC_r$ to complete Exercise 3?

5. Explain how Pascal's Triangle can be used to determine the probability that 20 out of 43 coin tosses will land heads up.

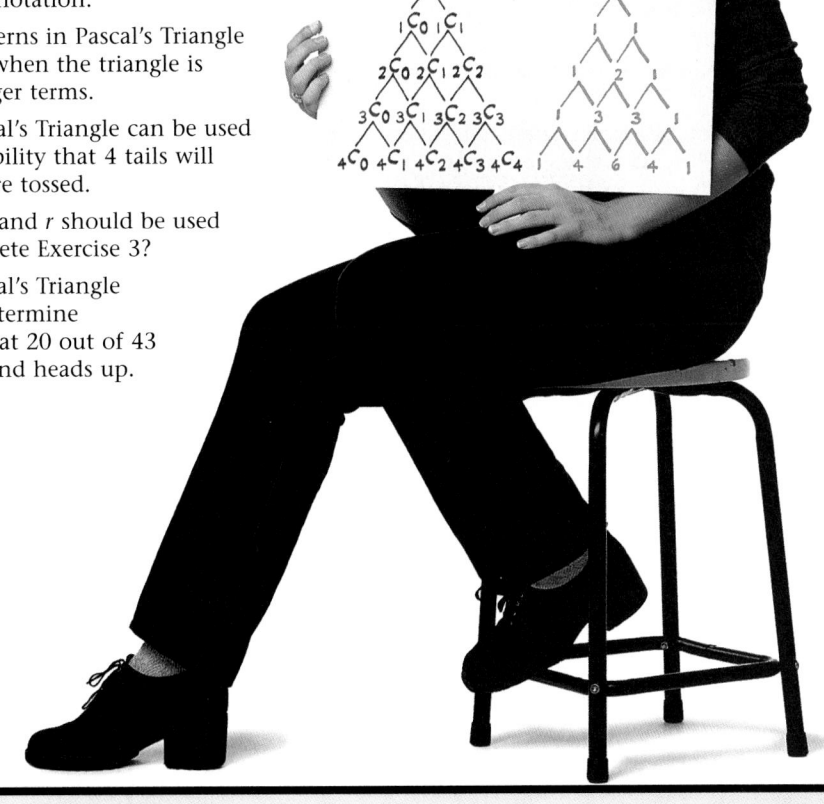

Technology Ask the students to use their calculator's $_nC_r$ procedure to find the value of $_nC_r$ for several values of n and r beginning with $n = 1$ and $r = 0$. Write the values in the pattern shown on page 688, second figure. Explain the connection of these values with the number of ways r things can be selected from a group of n things.

Practice & Apply

6. Which row of Pascal's Triangle contains $_7C_3$?   7th

7. Add the numbers in each row of the triangle. Look for a pattern. Describe the pattern you found.   1, 2, 4, 8, 16,...

8. Use the pattern you found in Exercise 7 to guess the sum of the elements in row 20. $2^{20} = 1,048,576$

9. The sum of the elements of a row of Pascal's Triangle is 4096. Which row is it?   12th

10. A row of Pascal's Triangle contains 17 numbers. Which row is it?   16th

11. How could you answer Exercise 10 without using your copy of the triangle? What pattern is there that connects the number of entries in a row to the row number?   number of entries is one more than number of rows

12. The second entry in a row of Pascal's Triangle is 23. Which row is it? What pattern did you use to answer this question?   23rd; 2nd entry=row number

13. The next-to-last entry in a row of Pascal's Triangle is 23. Which row is it? What pattern did you use to answer this question?   23rd

14. Use one of your copies of Pascal's Triangle to shade all multiples of 3. What pattern do you notice?

15. Use one of your copies of Pascal's Triangle to shade all multiples of 5. What pattern do you notice?

16. Examine each row of Pascal's Triangle where Column 2 contains a prime number. What pattern do you notice about each of these rows?

17. A portion of Pascal's Triangle is given. Entries are adjacent to each other, either in the same row or in consecutive rows. Replace each question mark with the correct number.

a. 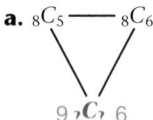   **b.** 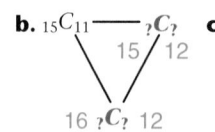   **c.** 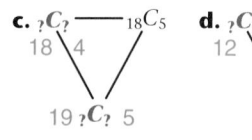   **d.**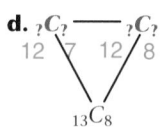

18. Use the results of Exercise 17 to generalize the pattern. Replace each question mark with an expression in terms of n and r.

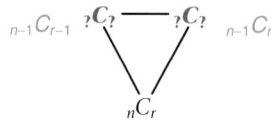

19. Describe the relationship between the values of the three entries in Exercise 18. $_{n-1}C_{r-1} + {}_{n-1}C_r = {}_nC_r$

20. The number 66 appears in Pascal's Triangle. In which row does it occur as the third entry from the right?   12th

21. Give three additional locations for the number 66. What pattern is there between the locations?

14. The pattern forms inverted triangular groups of three, starting in rows divisible by three.

The answers to Exercises 15, 16, and 21 can be found in Additional Answers beginning on page 842.

Transportation Diana drives a taxicab in a large Midwestern city whose streets are laid out as in a coordinate plane. In an effort to beat the competition, Diana decides to familiarize herself with alternate routes to avoid high-use intersections in the city.

22. For each diagram, count the number of routes from point P to point Q, assuming that only right and down directions are legal (to avoid backtracking). HINT: To keep track of alternate routes, describe each route with a sequence of Rs (rights) and Ds (downs). The route indicated in the 3×3 grid can be described as *DDRRDR*.

2

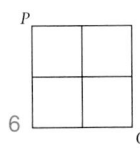
6

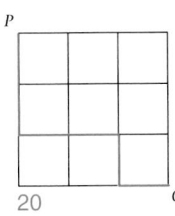
20

23. Predict the number of paths for a 4×4 grid, and write your answer using $_nC_r$ notation. 70; $_8C_4$

24. How many Ds (downs) appear in each route for a 3×3 grid? a 4×4 grid? 3; 4

25. Explain how to find the number of paths for a 5×5 grid. Describe the number of paths using $_nC_r$ notation, and give its numerical value.

A family has 4 children. Find the probability of the family having each of the following numbers of girls.

26. 4 girls $\frac{1}{16}$ **27.** 3 girls $\frac{1}{4}$ **28.** 2 girls $\frac{3}{8}$ **29.** 1 girl $\frac{1}{4}$ **30.** 0 girls $\frac{1}{16}$

31. What is the sum of the probabilities in Exercises 26–30? 1

32. Rewrite each probability in Exercises 26–30 in terms of the number of boys. What is the sum of these probabilities? Explain.

25. use the 5th entry (since dimensions of grid are 5 by 5) of 10th row of Pascal's Triangle (because there are 10 blocks in the grid in the path from P to Q); $_{10}C_5 = 252$

32. $\frac{1}{16}, \frac{1}{4}, \frac{3}{8}, \frac{1}{4}, \frac{1}{16}$; 1; sum of all probabilities is 1 corresponding to 100% probability that one of the possibilities will happen

34. $_8C_4$, $_{10}C_5$, $_{12}C_6$, $_{14}C_7$, $_{16}C_8$

35. $n = 2r$, this is the middle number of a row.

Transformations Reflect rows 0 through 7 of Pascal's Triangle, using $_nC_r$ form, about a vertical line that passes through $_0C_0$ and $_2C_1$.

33. How does this new triangle compare with the original triangle? identical

34. List 5 additional $_nC_r$ entries that appear on the axis of symmetry.

35. What is the relationship between n and r for $_nC_r$ entries on the axis of symmetry? Explain.

Look Back

Determine whether each is a permutation or a combination, and solve. [Lesson 11.2]

36. The number of ways to select the 3 CDs you can afford to buy out of the 8 you like   combination; 56

37. The number of ways to choose a 1st prize, 2nd prize, and 3rd prize out of 10 floats entered in a homecoming parade   permutation; 720

38. The number of ways of selecting a committee of 5 senators from a group of 100   combination; 75,287,520

39. The number of 13-card bridge hands possible in a 52-card deck
combination; 635,013,559,600
40. The number of ways to line up 5 students in a class of 20 students
permutation; 1,860,480
41. The number of ways to place 20 students into groups of 4 students each   combination; 4845

For each sequence, find t_{10} and S_{10}. [Lessons 12.2, 12.3]

42. 6, 12, 18, ...
$t_{10} = 60, S_{10} = 330$

43. 6, 12, 24, ...
$t_{10} = 3072, S_{10} = 6138$

Look Beyond

Look Beyond

The next lesson will show how the entries in Pascal's Triangle are also the coefficients in the binomial expansion.

44. Cultural Connection: Asia In 1270 C.E., Yang Hui introduced an early representation of the Arithmetic Triangle in his book *Arithmetic in Nine Sections*. In the same text, Yang Hui solved the following problem. See if you can solve it too.

Arrange the numbers 1 to 33 so that every circle has the same total and every diameter has the same total.

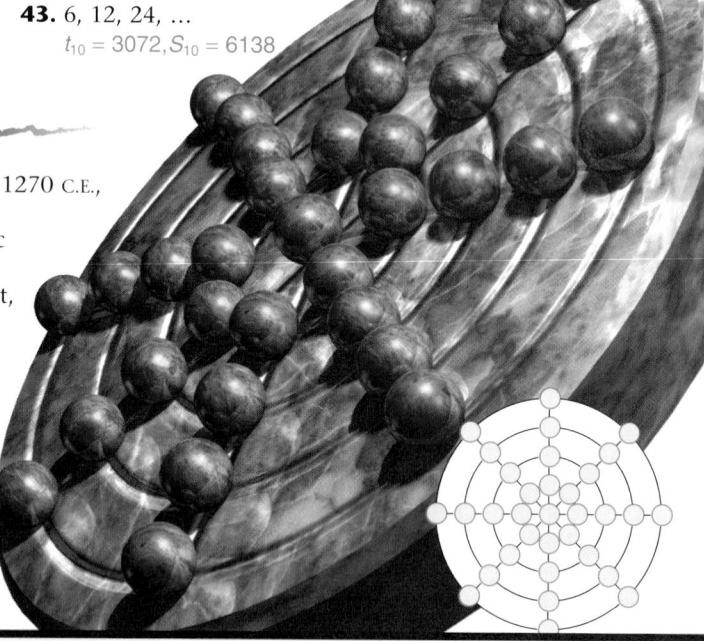

44. Answers may vary. One sample answer is shown: diameters total 153 and circumferences total 136

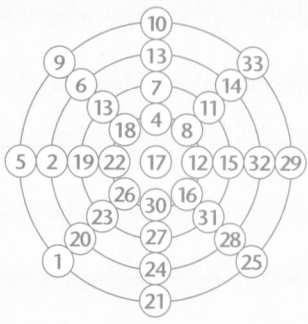

You know that $(a + b)^2 = a^2 + 2ab + b^2$. When any binomial $(a + b)^n$, where n is a positive integer, is written in expanded form, the result is known as a **binomial expansion**.

Consider the case when n is 3. Use the Distributive Property to find this binomial expansion.

$$(a + b)^3 = (a + b)^2(a + b)$$
$$= (a^2 + 2ab + b^2)(a + b)$$
$$= (a^2 + 2ab + b^2)a + (a^2 + 2ab + b^2)b$$
$$= a^3 + 2a^2b + ab^2 + a^2b + 2ab^2 + b^3$$
$$(a + b)^3 = a^3 + 3a^2b + 3ab^2 + b^3$$

In Lesson 6.5, you expanded the expressions $(a + b)^2$ and $(a + b)^3$. If you examine the coefficients of each term, you will discover a relationship found in Pascal's Triangle.

EXAMPLE 1

Use a tree diagram to expand $(a + b)^4$.

Solution

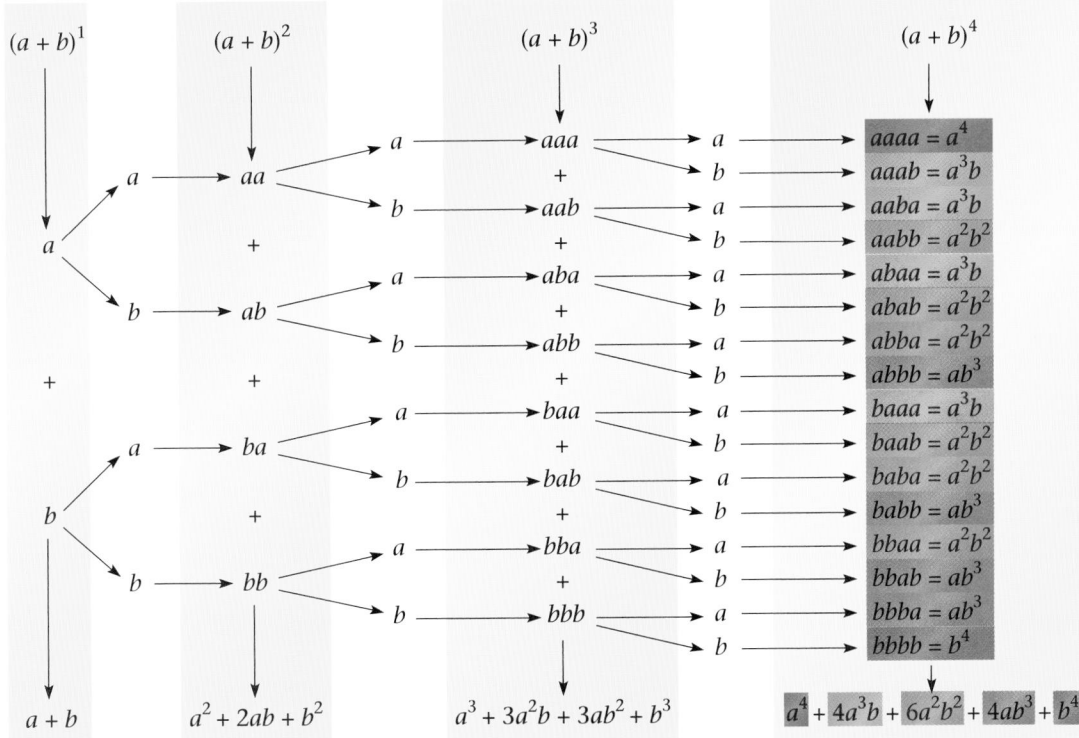

Thus, $(a + b)^4 = a^4 + 4a^3b + 6a^2b^2 + 4ab^3 + b^4$. ❖

ALTERNATIVE teaching strategy

Review the term-by-term pattern used to create Pascal's Triangle. Ask the students to generate the first six rows of Pascal's triangle. Then, make a table of the numbers from the triangle and include the powers of the binomial variables, a and b. The first variable should be in descending powers, the second in ascending powers. Then expand $(x - 2y)^5$, by making a similar table. Begin with the coefficients taken from the sixth row of Pascal's triangle. Then replace a, the first binomial variable, with x and calculate the descending powers. Replace b, the second variable, with $-2y$ and calculate the ascending powers. Finally, combine the three components for each term, simplify the results, and rewrite the terms.

PREPARE

Objectives
- Use the Binomial Theorem to expand $(a + b)^n$
- Use the Binomial Theorem to find a particular term in the expansion of $(a + b)^n$

RESOURCES

- Practice Master   12.6
- Enrichment Master   12.6
- Technology Master   12.6
- Lesson Activity Master 12.6
- Quiz   12.6
- Spanish Resources   12.6

Assessing Prior Knowledge

1. Expand.

   a. $(2x - y)^2$
   $[4x^2 - 4xy + y^2]$
   b. $(x + 2y)^3$
   $[x^3 + 6x^2y + 12xy^2 + 8y^3]$

TEACH

The use of Pascal's Triangle is an efficient model for raising any binomial to any power. It reduces the process of finding the coefficients to that of following simple patterns.

Alternate Example 1

Use a tree diagram to expand $(a + b)^5$. $[a^5 + 5a^4b + 10a^3b^2 + 10a^2b^3 + 5ab^4 + b^5]$

Exploration Notes

This exploration offers a model for determining possible combinations from the coefficients of the Pascal Triangle.

3. Powers of *a* decrease by 1, while powers of *b* increase by 1.

6. The expansion gives the number of ways to get each combination of Hs and Ts.

Alternate Example 2

Find the fourth term in the expansion of $(a + b)^8$. [$56a^5b^3$]

•Exploration• *Heads and Tails*

In Example 1, a pattern is used to expand $(a + b)^n$, when *n* is 1, 2, 3, or 4. There is a way to apply this pattern to real-world experiments.

1 How many terms are there in the expansion of $(a + b)^4$? How does this relate to the exponent 4?

2 How many terms will there be in the expansion of $(a + b)^{10}$? of $(a + b)^n$?

3 Examine the pattern of powers as you proceed from left to right through the terms of the expansion of $(a + b)^4$. What happens to the powers of *a*? What happens to the powers of *b*?

4 What is the sum of the exponents for each term in the expansion of $(a + b)^4$?

5 The coefficients of the terms in $(a + b)^4$ expanded are 1 4 6 4 1. Where do these numbers appear in Pascal's Triangle?

6 Explain how the expansion of $(H + T)^4$, where *H* is heads and *T* is tails, is related to flipping a coin 4 times. ❖

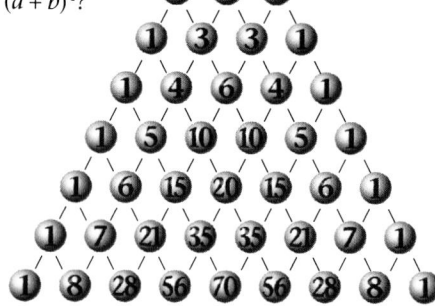

EXAMPLE 2

Graphics Calculator

A Expand $(a + b)^7$.

B Find the coefficient of a^6b^4 in the expansion of $(a + b)^{10}$.

Solution

A First write the variables, *a* and *b*, in the expansion.

$$\_a^7b^0 + \_a^6b^1 + \_a^5b^2 + \_a^4b^3 + \_a^3b^4 + \_a^2b^5 + \_a^1b^6 + \_a^0b^7$$

Next, insert the coefficients.

$$_7C_0 \qquad _7C_1 \qquad _7C_2 \qquad _7C_3 \qquad _7C_4 \qquad _7C_5 \qquad _7C_6 \qquad _7C_7$$

$$\_a^7b^0 + \_a^6b^1 + \_a^5b^2 + \_a^4b^3 + \_a^3b^4 + \_a^2b^3 + \_a^1b^4 + \_a^0b^7$$
$$1\,a^7b^0 + 7\,a^6b^1 + 21a^5b^2 + 35a^4b^3 + 35a^3b^4 + 21a^2b^3 + 7a^1b^4 + 1a^0b^7$$

B The term with a^6b^4 has 6 factors *a* out of the 10 variable factors that are in the term.

To find the coefficient for this term, find the number of possible combinations of 10 items taken 6 at a time.

$$_{10}C_6 = \frac{10!}{6!\,4!} = 210$$

Thus, the coefficient that belongs with a^6b^4 in the expansion of $(a + b)^{10}$ is 210. ❖

ENRICHMENT Often in applied mathematics an approximation over a restricted interval of *x*-values provides a number that is sufficient for the needs of a task. By expressing the binomial $(a + bx)^n$ in the form $a^n \left(1 + \left(\frac{b}{a}\right) x\right)^n$, when $\frac{b}{a}$ is a small number less than 1, it is sufficient to use only 2 or 3 terms of the expression for an adequate approximation. Compare the actual value of $(a + bx)^4$ with the value of the first 2 terms, $a^4 \left(1 + 4 \left(\frac{b}{a}\right) x\right)$, of the approximation when $\frac{b}{a}$ is 0.1 and $0 < x \le 1$. Explore this further by varying the values of *n* and varying *a* and *b* so that $\frac{b}{a} = 0.01$ and 0.001.

Try This **Ⓐ** Expand $(a + b)^6$.

Ⓑ Find the coefficient of a^8b in the expansion of $(a + b)^9$.

The explorations and examples in this lesson suggest a theorem for expanding any binomial.

BINOMIAL THEOREM

$(a + b)^n = {}_nC_0a^nb^0 + {}_nC_1a^{n-1}b^1 + {}_nC_2a^{n-2}b^2 + \cdots + {}_nC_{n-1}a^1b^{n-1} + {}_nC_na^0b^n$

$= \displaystyle\sum_{k=0}^{n} {}_nC_ka^{n-k}b^k$, where $n \geq 1$

$$(a + b)^8 = {}_8C_0\ a^8\ b^0 +$$

$$ {}_8C_1\ a^7\ b^1 + {}_8C_2\ a^6\ b^2 + \ldots$$

$$+ {}_8C_7\ a^1\ b^7 + {}_8C_8\ a^0\ b^8$$

When you use the Binomial Theorem, you usually know the value of n. The expanded form of $(a + b)^n$ has $n + 1$ terms.

EXAMPLE 3

Expand $(2x - 3y)^4$.

Solution➤

Think of $(2x - 3y)^4$ as $[(2x) + (-3y)]^4$, and use the Binomial Theorem.

Treat **2x** as the *a*-term and **−3y** as the *b*-term.

$[(2x) + (-3y)]^4$
$= {}_4C_0(2x)^4(-3y)^0 + {}_4C_1(2x)^3(-3y)^1 + {}_4C_2(2x)^2(-3y)^2 + {}_4C_3(2x)^1(-3y)^3 + {}_4C_4(2x)^0(-3y)^4$
$= 1(16x^4)(1) + 4(8x^3)(-3y) + 6(4x^2)(9y^2) + 4(2x)(-27y^3) + 1(1)(81y^4)$
$= 16x^4 - 96x^3y + 216x^2y^2 - 216xy^3 + 81y^4$

Thus, $(2x - 3y)^4 = 16x^4 - 96x^3y + 216x^2y^2 - 216xy^3 + 81y^4$. ❖

CRITICAL *Thinking*

Explain why the terms in the expansion of $(2x - 3y)^4$ have alternating positive and negative signs.

INCLUSION **strategies**
Using Visual Models
Demonstrate with several examples how the visual pattern in the tree diagram on page 695 represents the pattern in the exponents of the binomial theorem.

Aongoing
ASSESSMENT

Try This

A. $a^6 + 6a^5b + 15a^4b^2 + 20a^3b^3 + 15a^2b^4 + 6ab^5 + b^6$

B. 9

Alternative Example 3
Expand $(4x - y)^3$.
$[64x^3 - 48x^2y + 12xy^2 - y^3]$

CRITICAL *Thinking*

Even powers of $(-3y)^n$ are positive; odd powers are negative.

EXAMPLE 4

Find the third term in the expansion of $(a + b)^9$.

Solution►

The sum of the exponents for each term is 9. The first term is a^9b^0. In each successive term the exponent for a decreases by 1, and the exponent for b increases by 1. Thus, the variables in the third term are a^7b^2.

$$\underline{\quad ?\quad} a^7b^2$$

The coefficient is $_9C_3$.

$$_9C_3 = \frac{9!}{3!\, 6!} = 84$$

Thus, the third term is $84a^7b^2$. ❖

CRITICAL
Thinking

Find the sixth term in the expansion of $(a + b)^9$. Explain why the coefficient of the sixth term is the same as the coefficient of the fifth term.

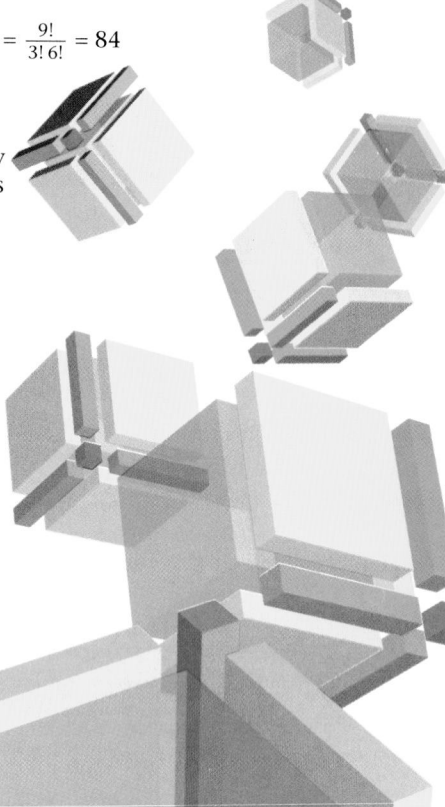

EXERCISES & PROBLEMS

Communicate

1. Describe the connection between the Binomial Theorem and Pascal's Triangle.

2. Explain how to use the Binomial Theorem to expand $(2a - 3b)^5$.

3. Explain how to use the Binomial Theorem to find the theoretical probability that when 4 coins are tossed, exactly 3 will be heads.

4. Explain the meaning of the summation notation $\sum_{n=0}^{6} {_6C_n} a^{6-n} b^n$.

5. Write the binomial expression that is equivalent to the notation in Exercise 4.

Practice & Apply

Use the Binomial Theorem to expand each expression.

6. $(a + b)^7$ **7.** $(a - b)^7$ **8.** $(p + q)^6$

9. $(p - q)^6$ **10.** $(4a + 3b)^5$ **11.** $(a - 2b)^4$

12. Write $\sum_{k=0}^{4} {_4C_k} a^{4-k} b^k$ in expanded form. $a^4 + 4a^3b + 6a^2b^2 + 4ab^3 + b^4$

13. Write your answer to Exercise 12 as a binomial raised to a power. $(a + b)^4$

14. Write $\sum_{k=0}^{6} {_6C_k} a^{6-k} b^k$ in expanded form. $a^6 + 6a^5b + 15a^4b^2 + 20a^3b^3 + 15a^2b^4 + 6ab^5 + b^6$

15. Write your answer to Exercise 14 as a binomial raised to a power. $(a + b)^6$

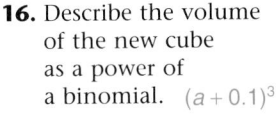 **Geometry** A cube has dimensions $a \times a \times a$. Its length, width, and height are all increased by 0.1.

16. Describe the volume of the new cube as a power of a binomial. $(a + 0.1)^3$

17. Expand the binomial you wrote for Exercise 16. Describe the connection between the dimensions of the cube and the terms of the expanded binomial.

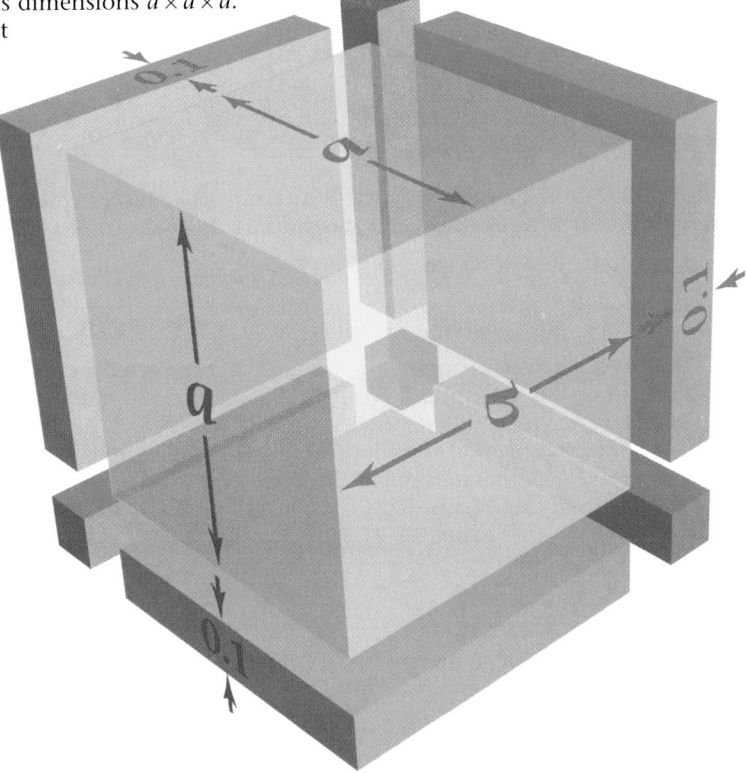

18. How many terms are there in the expansion of $(x + y)^{18}$?   19

19. In the expansion of $(a + b)^{10}$, a certain term contains b^3. What is the exponent of a in this term? Which term is it?   7; 4th

20. In the expansion of $(x + y)^{15}$, a certain term contains x^{10}. What is the exponent of y in this term? Which term is it?   5; 6th

21. The term $36a^7b^2$ appears in the expansion of $(a + b)^n$. What is n?   9

6. $a^7 + 7a^6b + 21a^5b^2 + 35a^4b^3 + 35a^3b^4 + 21a^2b^5 + 7ab^6 + b^7$

7. $a^7 - 7a^6b + 21a^5b^2 - 35a^4b^3 + 35a^3b^4 - 21a^2b^5 + 7ab^6 - b^7$

8. $p^6 + 6p^5q + 15p^4q^2 + 20p^3q^3 + 15p^2q^4 + 6pq^5 + q^6$

9. $p^6 - 6p^5q + 15p^4q^2 - 20p^3q^3 + 15p^2q^4 - 6pq^5 + q^6$

10. $1024a^5 + 3840a^4b + 5760a^3b^2 + 4320a^2b^3 + 1620ab^4 + 243b^5$

11. $a^4 - 8a^3b + 24a^2b^2 - 32ab^3 + 16b^4$

22. The numerical value of $_{10}C_7$ appears as a coefficient of two different terms in the expansion of $(a+b)^{10}$. What are the two terms? $120a^7b^3$, $120a^3b^7$

23. Find the eighth term in the expansion of $(p+q)^{14}$. $3432\,p^7q^7$

24. The sum of the coefficients in the expansion of $(a+b)^n$ is 64. What is n? 6

25. The second term in an expansion of $(a+b)^n$ has a coefficient of 13. Find the value of n, and write the complete second term. 13; $13a^{12}b$

26. The next-to-last term in an expansion of $(a+b)^n$ has a coefficient of 8. Find the value of n, and write the complete term. 8; $8ab^7$

27. The third term from the right in an expansion of $(a+b)^n$ has a coefficient of 36. Find the value of n, and write the complete term. 9; $36a^7b^2$

28. The third term of an expansion of $(a+b)^n$ has a coefficient of 28. Find the value of n, and write the complete term. 8; $28a^6b^2$

29. Find $(2+i)^2$ and $(2+i)^3$ using the Binomial Theorem. $3+4i$; $2+11i$

30. Use your results to Exercise 29 to show that $2+i$ is a solution to $x^3 - x^2 - 7x + 15 = 0$. $2+11i-(3+4i)-7(2+i)+15=0$

Look Back

Graph and identify each equation as that of a circle, an ellipse, a parabola, or a hyperbola.   [Lessons 10.1, 10.2, 10.3, 10.4]

31 $4x^2 + 9y^2 = 25$ ellipse   **32** $\frac{x^2}{100} + \frac{y^2}{36} = 1$ ellipse   **33** $x = y^2 - 6y + 5$ parabola

34 $y = -x^2 + 5x + 6$ parabola   **35** $(x+2)^2 + (y-5)^2 = 10$ circle   **36** $3x^2 - 2y^2 = 30$ hyperbola

37. Passenger Arrivals Suppose that the number of passengers arriving at an airline ticket counter between 7:00 A.M. and 7:30 A.M. has the following probability distribution. Let X represent the number of arriving passengers at the counter, and $P(X)$ the corresponding probability. Use random number simulation to predict the number of passengers arriving at this ticket counter in one day between 7:00 A.M. and 7:30 A.M.   [Lesson 11.7]

X	0	1	2	3	4	5
P(X)	0.0026	0.0138	0.0643	0.0902	0.1968	0.2113

X	6	7	8	9	10	11
P(X)	0.1641	0.0830	0.0545	0.0156	0.0120	0.0118

X	12	13	14	15	16	17
P(X)	0.0107	0.0105	0.0099	0.0092	0.0089	0.0088

X	18	19	20	21
P(X)	0.0076	0.0071	0.0070	0.0003

Look Beyond

The exercise challenges students to find a way to solve a factoring problem.

Look Beyond

38. Find a positive integer, n, such that $n(n-1)(n-2)(n-3)(n-4) = 95{,}040$. 12

37. Construct a cumulative probability table and assign consecutive 4-digit random numbers to the range of each cumulative probability. Then simulate passengers arriving by generating 4-digit random numbers and comparing with the table to see to which counter they go.

PROJECT 9 CHAPTER 12

Handshakes and Polygons

Refer to the portfolio activity described on pages 653 and 669.

Assume 5 people meet at a party. If each person shakes hands with every other person, how many handshakes will occur?

Create a geometric model to illustrate this situation. For example, the handshakes can be drawn using a pentagon and its diagonals.

Extend the geometric model to illustrate how many handshakes will occur when 6, 7, or 8 people meet.

How does the number of diagonals compare with the number of sides of the polygons?

How many diagonals will an *n*-gon have?

Use this information to determine how many handshakes will occur when *n* people meet.

FOCUS

In this project, students will apply their knowledge and skills with sequences to a study of geometry. They will discover that the same patterns occur and that the general formulas and expressions apply.

MOTIVATE

Patterns appeal to people because they are predictable. The patterns in the sequences of this project lead to predictions of future patterns that can be confirmed by experiments. This is the essence of inductive reasoning and scientific method.

Cooperative Learning

Working in pairs, one student should draw regular polygons and all the diagonals. The other student records the number of sides, number of diagonals, and total line segments drawn. Be sure both students have the opportunity to do both tasks. They should do this for 4-, 5-, 6-, 7-, 8-, and 12-sided figures.

DISCUSS

Are there other sequences in the geometry of regular polygons? What sequence describes the number of line segments that meet at each vertex? What about the total number of line segments drawn?

The number of diagonals $= \frac{n(n-3)}{2}$, where n is the number of sides.

The number of handshakes when n people meet $= \frac{n(n-1)}{2}$, which is $\frac{n(n-3)}{2} + n$.

Chapter 12 Review

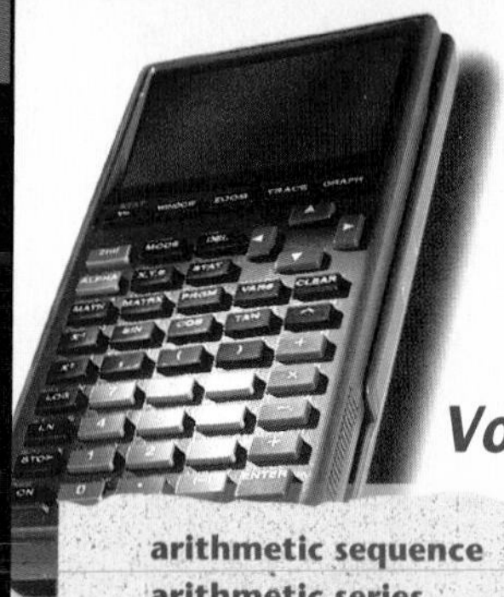

Vocabulary

Key Skills & Exercises

Lesson 12.1

➤ **Key Skills**

Determine whether a given sequence is an arithmetic sequence.

A sequence is arithmetic if for every pair of consecutive terms, $t_n - t_{n-1} = d$, a common difference.

Determine whether a given sequence is a geometric sequence.

A sequence is geometric if for every pair of consecutive terms, $t_n \div t_{n-1} = r$, a common ratio.

➤ **Key Skills**

List the first 5 terms of each sequence, and determine whether it is arithmetic, geometric, or neither. If it is arithmetic, identify d. If it is geometric, identify r.

1. $t_1 = 3$
$t_n = t_{n-1} + 8$
3, 11, 19, 27, 35; arith; $d = 8$

2. $t_1 = 1,\ t_2 = 1$
$t_n = t_{n-1} + t_{n-2}$
1, 1, 2, 3, 5; neither

3. $t_1 = 9$
$t_n = \frac{2}{3} t_{n-1}$
9, 6, 4, $\frac{8}{3}$, $\frac{16}{9}$; geo; $r = \frac{2}{3}$

Lesson 12.2

➤ **Key Skills**

Find the nth term of an arithmetic sequence.

If the first term is t_1 and the common difference is d, then $t_n = t_1 + (n-1)d$.

Find the sum of the first n terms of an arithmetic series.

$$S_n = n \cdot \frac{t_1 + t_n}{2} \quad \text{or} \quad S_n = n \cdot \frac{2t_1 + (n-1)d}{2}$$

➤ **Exercises**

Find the 50th term in each arithmetic sequence.

4. 5, 11, 17, ... 299

5. 20, 13, 6, ... −323

Identify t_1, n, t_n, and d. Then evaluate each sum.

6. $\displaystyle\sum_{k=1}^{30} 5k$ 5; 30; 150; 5; 2325

7. $\displaystyle\sum_{k=1}^{15} (11 - 3k)$ 8; 15; –34; –3; –195

Lesson 12.3

> ### Key Skills

Find the nth term of a geometric sequence.

If the first term is t_1 and the common ratio is r, then $t_n = t_1 \cdot r^{n-1}$.

Find the sum of the first n terms of a geometric series.

$S_n = t_1 \left(\dfrac{1-r^n}{1-r} \right)$, where $r \neq 1$

> ### Exercises

Find the 15th term in each geometric sequence.

8. 3, 6, 12, ... 49,152

9. 100, –50, 25, ... Approx 0.0061

Identify t_1, n, t_n, and r. Then evaluate each sum.

10. $\displaystyle\sum_{k=1}^{10} 0.5^k$ 0.5; 10; ≈ 0.000977; 0.5; ≈ 0.9990

11. $\displaystyle\sum_{k=1}^{6} 3(1.5)^k$ 4.5; 6; ≈ 34.172; 1.5; ≈ 93.516

Lesson 12.4

> ### Key Skills

Find the sum of an infinite geometric series.

$$S = \frac{t_1}{1-r}, \text{ where } |r| < 1$$

> ### Exercises

Find each sum.

12. $\displaystyle\sum_{k=1}^{\infty} 0.8^k$ 4

13. $0.3 + 0.03 + 0.003 + 0.0003 + \cdots$ $\dfrac{1}{3}$

Lesson 12.5

> ### Key Skills

Use Pascal's Triangle to find combinations.

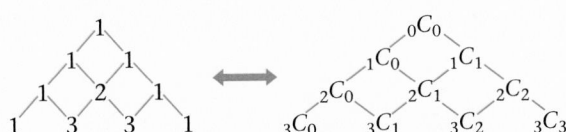

> ### Exercises

14. Write the tenth row of Pascal's Triangle using $_nC_r$ notation.

15. If you flip a coin 10 times, in how many ways can 3 heads occur? 120

16. If you flip a coin 10 times, in how many ways can 7 heads occur? 120

14. $_{10}C_0$; $_{10}C_1$; $_{10}C_2$; $_{10}C_3$; $_{10}C_4$; $_{10}C_5$; $_{10}C_6$; $_{10}C_7$; $_{10}C_8$; $_{10}C_9$; $_{10}C_{10}$

Lesson 12.6

➤ *Key Skills*

Use the Binomial Theorem to expand $(a + b)^n$.

$(a + b)^n = {}_nC_0a^nb^0 + {}_nC_1a^{n-1}b^1 + {}_nC_2a^{n-2}b^2 + \cdots + {}_nC_na^0b^n$

$$= \sum_{k=0}^{n} {}_nC_ka^{n-k}b^k, \text{ where } n \geq 1$$

Use the Binomial Theorem to find a particular term in the expansion of $(a + b)^n$.

To find the fifth term of $(3x - 2)^7$, find ${}_7C_4(3x)^3(-2)^4$.

$${}_7C_4(3x)^3(-2)^4 = 35(27x^3)(16)$$
$$= 15{,}120x^3$$

➤ *Exercises*

Use the Binomial Theorem to expand each of the following.

17. $(a - b)^5$
$a^5 - 5a^4b + 10a^3b^2 - 10a^2b^3 + 5ab^4 - b^5$

18. $\sum_{k=0}^{4} {}_4C_ka^{4-k}b^k$
$a^4 + 4a^3b + 6a^2b^2 + 4ab^3 + b^4$

Find the third term in the expansion of each of the following.

19. $(x + y)^8$ $28x^6y^2$

20. $(3a - 5)^5$ $6750a^3$

Applications

Medicine At 1:00 P.M., Mr. Valdez took a 100-milligram pill of his medication. At the end of each hour, the concentration of the medicine is 75% of the amount present at the beginning of the hour.

21. How many milligrams, to the nearest whole number, remain at the end of 3 hours? 42 mg

22. At approximately what time will there be only 10 milligrams of the medication left in his body? 9:00 P.M.

Physics A ball is dropped from a height of 12 meters. Each time it bounces on the ground, it rebounds to $\frac{2}{3}$ the distance it has fallen.

23. How many meters will it rebound after the third bounce? 3.55 m

24. Theoretically, how far will the ball travel before coming to rest? 60 m

Chapter 12 Assessment

List the first 4 terms of each sequence, and determine whether each is arithmetic or geometric. If it is arithmetic, identify d. If it is geometric, identify r.

1. $t_1 = 12$, $t_n = -2t_{n-1}$ **2.** $t_1 = 8$, $t_n = t_{n-1} + 2.5$

12, -24, 48, -96; geometric; $r = -2$ 8, 10.5, 13, 15.5; arithmetic; $d = 2.5$

Fitness Sam and David have decided that they need to do more push-ups.

3. If Sam starts with 50 push-ups on the first day and adds 5 each day, how many push-ups will he do on the seventh day? 80

4. If David starts slowly with 3 push-ups and doubles this amount each day, how many will he do on the seventh day? 192

Find the 100th term of each arithmetic sequence.

5. -11, -5, 1, ... 583   **6.** 1.05, 1.10, 1.15, ... 6

Find the tenth term of each geometric sequence.

7. 2, 4, 8, ... 1024   **8.** -1000, 500, -250, ... 1.953125

Identify t_1, n, t_n (if possible), and either d or r. Evaluate each of the following.

9. $\displaystyle\sum_{k=1}^{60}(3k+2)$ **10.** $\displaystyle\sum_{k=1}^{8} 3^k$ **11.** $\displaystyle\sum_{k=1}^{10} 5(0.9)^k$ **12.** $\displaystyle\sum_{k=1}^{\infty} 0.5^k$

13. Investments Lydia is investing \$400 every 3 months in a retirement account that pays an annual interest rate of 12% compounded quarterly. How much is in Lydia's account after 2 years, before she makes the ninth deposit? \$3663.64

14. Find the sum of $0.12 + 0.0012 + 0.000012 + \cdots$ $\frac{4}{33}$

15. Write the ninth row of Pascal's Triangle using ${}_nC_r$ notation.

16. What is the coefficient of the 4th term in the expansion of $(a+b)^9$? 84

17. Write $\displaystyle\sum_{k=0}^{5} {}_5C_k a^{5-k}b^k$ in expanded form. $a^5 + 5a^4b + 10a^3b^2 + 10a^2b^3 + 5ab^4 + b^5$

18. Write the term in the expansion of $(x+y)^8$ that contains y^3. $56x^5y^3$

Biology A certain type of bacteria doubles every 2 hours. A laboratory culture contains a count of 120 of these bacteria.

19. How many bacteria will there be at the end of 24 hours? 491,520

20. How many hours will it take for the number of bacteria to surpass 1 billion? 46 hr

9. $t_1 = 5$; $n = 60$; $t_n = 182$; $d = 3$, $S_{60} = 5610$

10. $t_1 = 3$; $n = 8$; $t_n = 6561$; $d = 3$, $S_8 = 9840$

11. $t_1 = 4.5$; $n = 10$; $t_n \approx 1.743$; $r = 0.9$; $S_{10} \approx 29.31$

12. $t_1 = 0.5$; $n =$ the natural numbers; $r = 0.5$; $S = 1$

15. ${}_9C_0$; ${}_9C_1$; ${}_9C_2$; ${}_9C_3$; ${}_9C_4$; ${}_9C_5$; ${}_9C_6$; ${}_9C_7$; ${}_9C_8$; ${}_9C_9$

Chapters 1–12 Cumulative Assessment

Multiple-Choice and Quantitative-Comparison Samples

The first half of the Cumulative Assessment contains two types of items found on standardized tests — multiple-choice questions and quantitative-comparison questions. Quantitative-comparison items emphasize the concepts of equality, inequality, and estimation.

Free-Response Grid Samples

The second half of the Cumulative Assessment is a free-response section. A portion of this part of the Cumulative Assessment consists of student-produced response items commonly found on college entrance exams. These questions require the use of machine-scored answer grids. You may wish to have students practice answering these items in preparation for standardized tests.

Sample answer-grid masters are available in the *Chapter Teaching Resources Booklets*.

College Entrance Exam Practice

Quantitative Comparison For Items 1–5, write
A if the quantity in Column A is greater than the quantity in Column B;
B if the quantity in Column B is greater than the quantity in Column A;
C if the quantities are equal; or
D if the relationship cannot be determined from the given information.

Column A	Column B	Answers
1. A The minimum value in the range of the function $f(x) = |x - 2|$	The minimum value in the range of the function $f(x) = |x| - 2$	Ⓐ Ⓑ Ⓒ Ⓓ [Lesson 3.4]
2. B $(5 - 2i)(5 + 2i)$, where $i = \sqrt{-1}$	$6i^2 \cdot 7i^2$, where $i = \sqrt{-1}$	Ⓐ Ⓑ Ⓒ Ⓓ [Lesson 5.7]
3. A The number of distinct zeros of f $f(x) = (x - 2)^2 (x + 5)(x - 1)$ $f(x) = (x - 4)(x + 1)^3$		Ⓐ Ⓑ Ⓒ Ⓓ [Lesson 6.4]
4. C The x-coordinate of the center of the graph of $x^2 + (y - 3)^2 = 4$	The x-coordinate of the vertex of the graph of $x = 2(y - 1)^2$	Ⓐ Ⓑ Ⓒ Ⓓ [Lessons 10.1, 10.2]
5. C $_{10}C_2$	$_{10}C_8$	Ⓐ Ⓑ Ⓒ Ⓓ [Lesson 11.4]

6. What is the slope of the line that contains the points (6, –8) and
c (–2, –4)? **[Lesson 1.2]**

   a. 2   **b.** –2   **c.** $-\frac{1}{2}$ **d.** $\frac{1}{2}$

7. Which of the following terms describes the
b system of equations? **[Lesson 4.3]** $\begin{cases} 2x - y = 7 \\ 2y - 4x = -14 \end{cases}$

   a. inconsistent   **b.** dependent   **c.** independent   **d.** incompatible

8. Which of the following is a solution for $3^{x-1} = \frac{1}{81}$? **[Lesson 7.6]**

d   **a.** 3   **b.** $\frac{1}{3}$ **c.** $-\frac{1}{3}$ **d.** –3

9. In the triangle at the right, what is the value of $c \sin A$? **[Lesson 8.1]**

d **a.** $\frac{1}{2}$ **b.** $4\sqrt{3}$ **c.** $\frac{\sqrt{3}}{2}$ **d.** 4

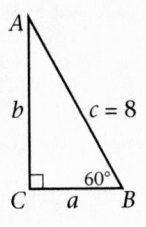

10. What is the number of solutions for the system? $\begin{cases} y = x^2 + 3 \\ x^2 + y^2 = 9 \end{cases}$
b **[Lesson 10.5]**

a. 0 **b.** 1 **c.** 2 **d.** 3

11. If $f(x) = x^2 - 2x$, find $f(x+a) - f(x)$. **[Lesson 2.4]** $2ax + a^2 - 2a$

12. Find the matrix product. **[Lesson 4.2]** $\begin{bmatrix} 3 & -6 \\ 5 & 4 \end{bmatrix}\begin{bmatrix} 1 & 0 \\ 0 & 1 \end{bmatrix}$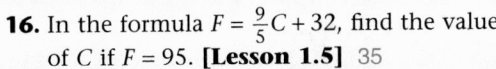

13. Factor to find the zeros of $f(x) = x^3 + 5x^2 - 6x$. **[Lesson 6.2]** $0, 1, -6$

14. Simplify $\frac{5}{x^2 - 3x} + \frac{x}{x^2 - 9}$. **[Lesson 9.5]** $\frac{x^2 + 5x + 15}{x(x+3)(x-3)}$

15. In how many different ways can a committee of 3 students be chosen from a class of 15? **[Lesson 11.4]** 455

Free-Response Grid The following questions may be answered using a free-response grid commonly used by standardized-test services.

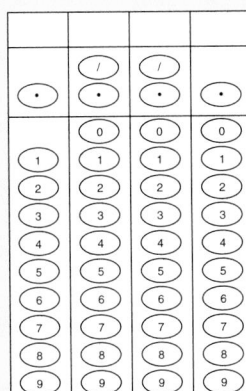

16. In the formula $F = \frac{9}{5}C + 32$, find the value of C if $F = 95$. **[Lesson 1.5]** 35

17. What is the value of $\log_2 \frac{1}{8}$? **[Lesson 7.3]** -3

18. What is the minimum value of $f(x) = 3 + 2\cos\left(x + \frac{\pi}{4}\right)$? **[Lesson 8.4]** 1

19. What is the constant of variation if y varies inversely as the square of x, and y is 2 when x is 4? **[Lesson 9.1]** 32

20. What is the 30th term of the arithmetic sequence 50, 46, 42, ...? **[Lesson 12.2]** -66

13 CHAPTER Discrete Mathematics: Statistics and Probability

Meeting Individual Needs

13.1 Averages and Frequency Tables

Core	Core Plus
Inclusion Strategies, p. 711	Practice Master 13.1
Reteaching the Lesson, p. 712	Enrichment, p. 711
Practice Master 13.1	Technology Master 13.1
Enrichment Master 13.1	Lesson Activity Master
Technology Master 13.1	13.1
Lesson Activity Master 13.1	
[1 day]	**[2 days]**

13.4 Measures of Dispersion

Core	Core Plus
Inclusion Strategies, p. 742	Practice Master 13.4
Reteaching the Lesson, p. 743	Enrichment, p. 742
Practice Master 13.4	Technology Master 13.4
Enrichment Master 13.4	Lesson Activity Master
Technology Master 13.4	13.4
Lesson Activity Master 13.4	Interdisciplinary
	Connection, p. 741
[1 day]	**[1 day]**

13.2 Histograms and Stem-and-Leaf Displays

Core	Core Plus
Inclusion Strategies, p. 723	Practice Master 13.2
Reteaching the Lesson, p. 724	Enrichment, p. 723
Practice Master 13.2	Technology Master 13.2
Enrichment Master 13.2	Lesson Activity Master
Technology Master 13.2	13.2
Lesson Activity Master 13.2	
[1 day]	**[1 day]**

13.5 Exploring Binomial Distributions

Core	Core Plus
Inclusion Strategies, p. 748	Practice Master 13.5
Reteaching the Lesson, p. 749	Enrichment, p. 747
Practice Master 13.5	Technology Master 13.5
Enrichment Master 13.5	Lesson Activity Master
Technology Master 13.5	13.5
Lesson Activity Master 13.5	
[2 days]	**[2 days]**

13.3 Percentiles and Box-and-Whisker Plots

Core	Core Plus
Inclusion Strategies, p. 731	Practice Master 13.3
Reteaching the Lesson, p. 732	Enrichment, p. 731
Practice Master 13.3	Technology Master 13.3
Enrichment Master 13.3	Lesson Activity Master
Technology Master 13.3	13.3
Lesson Activity Master 13.3	Mid-Chapter Assessment
Mid-Chapter Assessment Master	Master
[1 day]	**[1 day]**

13.6 The Normal Distribution

Core	Core Plus
Inclusion Strategies, p. 756	Practice Master 13.6
Reteaching the Lesson, p. 757	Enrichment, p. 755
Practice Master 13.6	Technology Master 13.6
Enrichment Master 13.6	Lesson Activity Master
Technology Master 13.6	13.6
Lesson Activity Master 13.6	Interdisciplinary
	Connection, p. 754
[Optional*]	**[2 days]**

Chapter Summary

Core	Core Plus
Eyewitness Math, pp. 738–739	Eyewitness Math, pp. 738–739
Chapter 13 Project, pp. 762–763	Chapter 13 Project, pp. 762–763
Lab Activity	Lab Activity
Long-Term Project	Long-Term Project
Chapter Review, pp. 764–766	Chapter Review, pp. 764–766
Chapter Assessment, p. 767	Chapter Assessment, p. 767
Chapter Assessment, A/B	Chapter Assessment, A/B
Alternative Assessment	Alternative Assessment
[4 days]	**[3 days]**

Reading Strategies

Before reading each Lesson, students should skim the word problems in the exercises. This will give them an idea of the sort of applications that justify the study of this material. As they read each Lesson, students can try to characterize the type of problem covered. Each student might want to pick one application that they find personally interesting and use this as the anchor-point for the Lesson. They can add these to their portfolio to develop a portfolio that reflects their interests.

Visual Strategies

Direct students to look at the pictures and associated graphs in the chapter opener. Encourage them to speculate about how mathematics might be useful in studying such phenomena. Also have them look at the graphs in the entire chapter and discuss how they differ from graphs in earlier chapters. What aspect of these graphs makes them different from graphs in earlier chapters? Can students imagine how such functions could model the phenomena in the photographs?

Hands-On Strategies

This chapter presents students with a number of opportunities to become actively involved with mathematics in the physical world. A variety of survey projects gives students a chance to see the relevance of statistics to society.

Cooperative Learning

GROUP ACTIVITIES	
Measures of central tendency	Lesson 13.1, Example 1
Fuzzy logic	Eyewitness Math: What's So Fuzzy?
Binomial distributions and multiple choice	Lesson 13.5, Exploration 1
Normal distribution	Lesson 13.6, Example 1
Surveys	Chapter Project

You may wish to have students work with partners for some of the above activities. Additional suggestions for cooperative group activities are noted in the teacher's notes in each lesson.

Multicultural

The cultural connections in this chapter include references to Asia.

CULTURAL CONNECTIONS	
Asia: Statistics Table	Lesson 13.1

Portfolio Assessment

Below are portfolio activities for the chapter. They are listed under seven activity domains that are appropriate for portfolio development.

1. **Investigation/Exploration** Determining the probability of guessing the correct choice in a multiple-choice test is the objective of Exploration 1 in Lesson 13.5; binomial distribution is the focus of Exploration 2 in Lesson 13.5. Another exploration you may wish to include is the study of fuzzy sets and logic in the Eyewitness Math feature.

2. **Applications** Recommended to be included are any of the following: business, Lesson 13.1, Exercises 20–23; marketing, Lesson 13.2, Exercises 15–20; nutrition, Lesson 13.4, Exercise 15–19; automotive mechanics, Lesson 13.6, Exercises 19–21.

3. **Non-Routine Problems** The What's So Fuzzy? problem from Eyewitness Math presents the students with non-traditional but interesting ways to look at statistics and probability.

4. **Project** That's Not Fair! is found on pages 762–763.

Students are asked to look at whether or not a sample in a poll should represent the group or the population.

5. **Interdisciplinary Topics** Students may choose from the following: technology, Lesson 13.2, Exercises 10–14; science, Lesson 13.5, Exercises 18–21.

6. **Writing** Communicate exercises of the type in which a student is asked to describe basic concepts of measures of central tendency, stem-and-leaf and box-and-whisker plots, quartiles, measures of dispersion, binomial distributions, the normal curve, and z-scores offer excellent writing selections for the portfolio.

7. **Tools** Chapter 13 uses graphics calculators to analyze and model data to help solve problems in statistics and probability. To measure the individual student's proficiency, it is recommended that he or she complete selected worksheets from the Technology Masters.

Technology

In this chapter, technology becomes a significant ally. For calculations that involve large lists of numbers upon which statistics and probability depend, calculators and computer spreadsheets make the manipulation and analysis of data easy, efficient, and relatively error-free. Calculators, spreadsheets, and statistical software allow experimentation that would otherwise be prohibitively time-consuming and cumbersome. Graphs can be made and compared quickly to determine the most effective way to display analytical results. Technology is a powerful tool that makes the study of statistics and probability more interesting and workable. Refer to your *HRW Technology Handbook* or the calculator manuals for specific instructions on entering, computing, and graphing statistical data on specific calculators.

One graphics calculator, for example, places data onto images of the traditional index cards and allows the user to leaf through them, correct them, and sort them. When the user is satisfied that the data is ready, a few keystrokes produce a complete statistical analysis of the data set.

Invite students to make a list of 50 single-digit data values on their calculators using the random number command **int(10*rand)**. If a computer spreadsheet is used, the command is INT(10*RAND()).

Then have students graph a box-and-whisker plot and a histogram of the data. For example, one list of 50 random single-digit numbers produced the following mean, median, and mode.

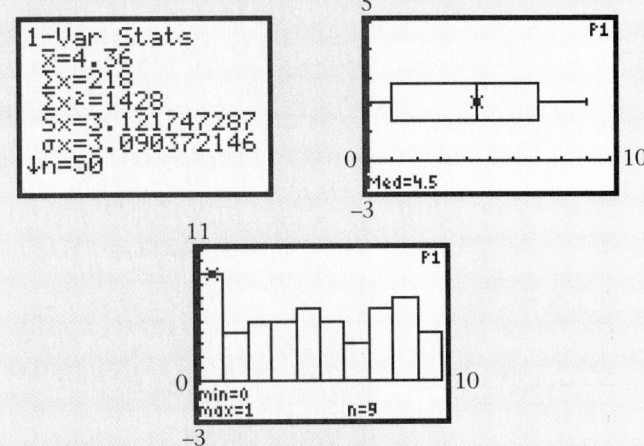

The mean, $\bar{x}$, is 4.36, the median, Med, is 4.5, and the mode is 0.

This kind of personal statistical exploration, made possible by graphics calculators and computers, makes work with statistics more enjoyable for students. In the past, the amount of work discouraged many students from experimenting with statistics. Encourage students to extend their experimenting beyond the classroom, using data from sports, games, and consumer situations.

The concept of a normally distributed data set is easily understood in graphs. The plotting of graphs with a graphics calculator or computer statistics software makes the visual demonstration of the fundamental ideas in statistics easy.

Integrated Software

f(g) Scholar™ is an integrated computer-based mathematics productivity tool that combines calculator, spreadsheet, and graphics capabilities to provide a dynamic and interactive environment for explorations in mathematics. Use *f(g) Scholar*™ for any lesson needing a spreadsheet, calculator, graphics calculator, or any combination of the three.

13 Discrete Mathematics: Statistics and Probability

ABOUT THE CHAPTER

Background Information

Statistics as a formal field of study began with the compiling of mortality data by insurance companies. Today, statistics are used in business, in the physical and life sciences, and in the social sciences.

CHAPTER RESOURCES

- Practice Masters
- Enrichment Masters
- Technology Masters
- Lesson Activity Masters
- Lab Activity Masters
- Long-Term Project Masters
- Assessments Masters
  Chapter Assessments, A/B
  Mid-Chapter Assessment
  Alternative Assessment, A/B
- Teaching Transparencies
- Spanish Resources

CHAPTER OBJECTIVES

- Find the measures of central tendency, mean, median, and mode of a given set of data.
- Construct a frequency table by dividing the data into classes, and find the class mean for the data.
- Construct a histogram for a given set of data.
- Construct a stem-and-leaf plot to represent a given set of data.
- Find the quartiles and interquartile range for a given set of data.
- Construct a box-and-whisker plot to display a given set of data.
- Find the range of a given set of data.

CHAPTER 13

Discrete Mathematics: Statistics and Probability

Statistics is the science of collecting, analyzing, describing, and interpreting data. Types of statistics include descriptive and inferential. Descriptive statistics uses tables, graphs, and summary measures to describe data. Inferential statistics consists of analyzing and interpreting data to make predictions.

Gray whales, which live in the North Pacific Ocean, were recently taken off the endangered species list.

Gray whales have no teeth. They scoop up small fish and plankton from the ocean floor, and strain these through fingernail-like plates that hang from the roof of their mouth.

ABOUT THE PHOTOS

In a controversy like that surrounding the gray whales, each side proclaims to know the "facts" of an issue. Often, both sides are wrong. By using statistics, it is possible to get a more accurate picture of the true state of affairs. In a statistical study, people collect and organize data. The data can then be analyzed to provide a summary as well as a basis for inferences about what to expect and what to do.

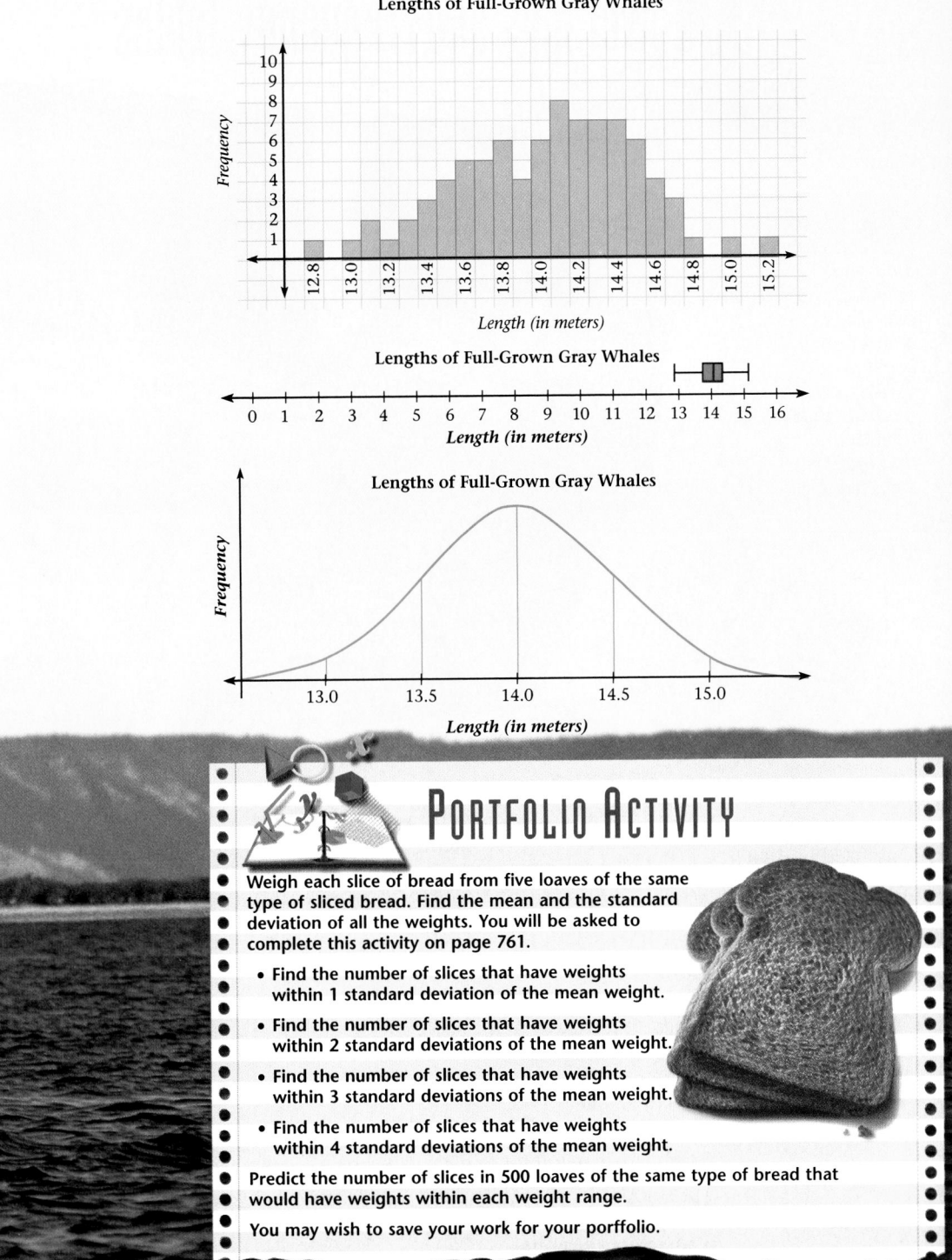

Lengths of Full-Grown Gray Whales

Frequency

Length (in meters)

Lengths of Full-Grown Gray Whales

0 1 2 3 4 5 6 7 8 9 10 11 12 13 14 15 16

Length (in meters)

Lengths of Full-Grown Gray Whales

Frequency

13.0   13.5   14.0   14.5   15.0

Length (in meters)

PORTFOLIO ACTIVITY

Weigh each slice of bread from five loaves of the same type of sliced bread. Find the mean and the standard deviation of all the weights. You will be asked to complete this activity on page 761.

• Find the number of slices that have weights within 1 standard deviation of the mean weight.

• Find the number of slices that have weights within 2 standard deviations of the mean weight.

• Find the number of slices that have weights within 3 standard deviations of the mean weight.

• Find the number of slices that have weights within 4 standard deviations of the mean weight.

Predict the number of slices in 500 loaves of the same type of bread that would have weights within each weight range.

You may wish to save your work for your porffolio.

• Determine the mean deviation and the standard deviation for a given set of data.
• Find the probability of r successes in n trials of a binomial experiment.
• Given a set of data that is normally distributed, find the probability of an event if the mean and standard deviation are known.

PORTFOLIO ACTIVITY

This activity will introduce the realities of experimental measurement of data, the use of statistics to analyze the results, and the formation of predictions about how larger populations will behave.

Divide the class into groups of three, and have students take turns weighing, recording, and analyzing the recorded data.

Ask them to identify and discuss sources of error that are not controlled in this experiment, such as the moisture content, which changes with time. Have them separately weigh a fresh slice and a stale slice to see how big a change is possible.

ABOUT THE CHAPTER PROJECT

In the Chapter Project on pages 762 and 763, students will study and discuss surveys and polls to see how important the sample selection method can be to the validity of the results.

Objectives

- Find the measures of central tendency, mean, median, and mode of a given set of data.
- Construct a frequency table by dividing the data into classes, and find the class mean for the data.

RESOURCES

- Practice Master   **13.1**
- Enrichment Master   **13.1**
- Technology Master   **13.1**
- Lesson Activity Master   **13.1**
- Quiz   **13.1**
- Spanish Resources   **13.1**

Assessing Prior Knowledge

Ask students to explain the term "grade point average," or GPA. How is it computed? [**GPA is the numerical representation of a student's average grade; (number of A's × 4.0 + number of B's × 3.0 + number of C's × 2.0 + number of D's × 1.0)/total number of grades**]

TEACH

Collected data can involve many numbers that are difficult to analyze and interpret. The simplest way to summarize a large quantity of numbers is to reduce them to a few special numbers. These are the mean, the median, and the mode.

ongoing ASSESSMENT

Answers may vary. A student spends approximately 3.0 hours reading.

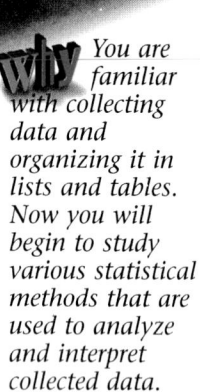

LESSON 13.1 Averages and Frequency Tables

Why *You are familiar with collecting data and organizing it in lists and tables. Now you will begin to study various statistical methods that are used to analyze and interpret collected data.*

The following list represents the approximate number of weekend hours that 36 high-school juniors spent reading material not related to schoolwork.

Weekend Hours Spent Reading

| 1 | 3 | 5 | 5 | 2 | 3 | 0 | 4 | 3 | 5 | 0 | 1 | 2 | 4 | 3 | 2 | 4 | 1 |
| 4 | 2 | 3 | 4 | 1 | 2 | 1 | 2 | 7 | 3 | 4 | 4 | 4 | 2 | 0 | 3 | 1 | 2 |

What is your estimate of the average time a student spends reading on a weekend?

Graphics Calculator

There are 36 data values, one for each student interviewed in the survey. The students are represented on the horizontal axis (Student 1, Student 2, Student 3, and so on).

The numbers of hours spent reading, 0 through 7, are on the vertical axis.

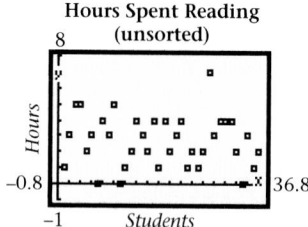
Hours Spent Reading (unsorted)

The scatter plot shows the *distribution* of the data. However, this is not very useful when you wish to find the *average* time that a student spends reading on a weekend.

ALTERNATIVE teaching strategy

Using Discussion Have each student pick 5 numbers between 1 and 30 and write them down. Record all responses on the chalkboard and discuss how they might be analyzed, using the ideas of mean, median, mode, and frequency. The goal might be to answer the question, "What is the favorite number of the class?" Which analysis would be best for that question? What is the average of all the numbers? Does 15 turn out to be the middle? Are any numbers avoided entirely? Why? Is this explanation significant?

Central Tendency: Mean, Median, and Mode

There are three measures of **central tendency**, commonly called averages, used in statistics. They are the *mean*, the *median*, and the *mode*.

The most appropriate measure to use when finding central tendency often depends on the information that you need from the data. However, the most common measure is the arithmetic mean.

MEAN

The **mean**, $\bar{x}$, is found by using the formula

$$\bar{x} = \frac{\sum\limits_{i=1}^{n} x_i}{n}$$

$\left\{\begin{array}{l} \text{where } \sum\limits_{i=1}^{n} x_i \text{ is the sum of the values in the data set,} \\ \text{and } n \text{ is the number of data values.} \end{array}\right.$

There are 36 data values in the list of weekend hours spent reading, so n is 36. The sum of all 36 values is 97.

Therefore, the mean is $\bar{x} = \dfrac{\sum\limits_{i=1}^{n} x_i}{n} = \dfrac{97}{36} \approx 2.7$ hours.

When *all* of the data values are to be added, the summation notation $\sum\limits_{i=1}^{n} x_i$ is often abbreviated $\sum x$.

Graphics Calculator

Once the data values are entered in a calculator, many statistical measures can be computed.

This calculator screen shows the one-variable statistical measures for the hours-spent-reading data.

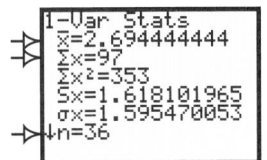

Within this list of statistical measures, you can find the mean, $\bar{x}$, the sum of all data values, $\sum x$, and the number of data values, n.

In Lesson 13.3, you will learn the meaning of other statistical measures provided by the calculator.

When the data values are arranged in numerical order, the **median** is the *middle* data value.

Ordered List of Hours Spent Reading

0 0 0 1 1 1 1 1 1 2 2 2
2 2 2 2 2 **3 3** 3 3 3 3 3
4 4 4 4 4 4 4 4 5 5 5 7

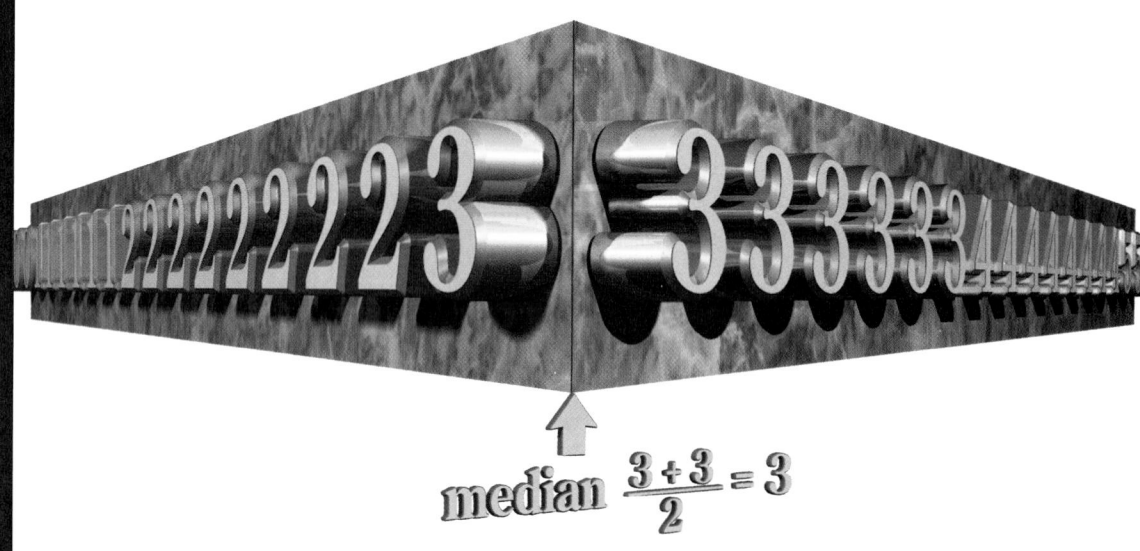

$$\text{median } \frac{3+3}{2} = 3$$

There are *two* middle values in the ordered list. When this occurs, the median of the data set is the mean of the two values. Since the middle two values are both equal to 3, their mean is 3, and the median value is 3.

Graphics Calculator

The **sort** command in the statistics menu of your calculator orders any data list.

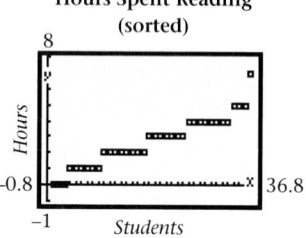

Using the trace feature, you can find that the middle values (the 18th and 19th data values) of the ordered list are both 3.

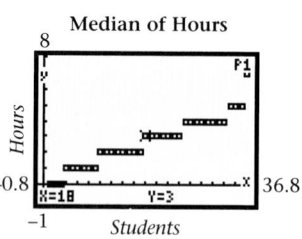

RETEACHING the lesson

Using Visual Models

Sometimes it is easier to visualize recorded data by using a pictograph. A pictograph is a graph in which, for example, the number of new houses built in an area are "graphed" using diagrams of houses, where one diagram might represent 100 houses built. Completed pictographs resemble bar graphs, but the bars are stacks or rows of diagrams. Most graphics software packages include pictures of dogs, houses, trees, etc. Generate a pictograph of this type that can be shown on a transparency to emphasize some of the concepts of this section. Point out that a bar graph and histogram are similar. These graphs allow for the visualization of central tendencies.

The **mode** is the value that appears most often. Since there are eight 2s and eight 4s, the data list has two modes, 2 and 4. It is possible to have more than one mode, but it is not possible to have more than one median.

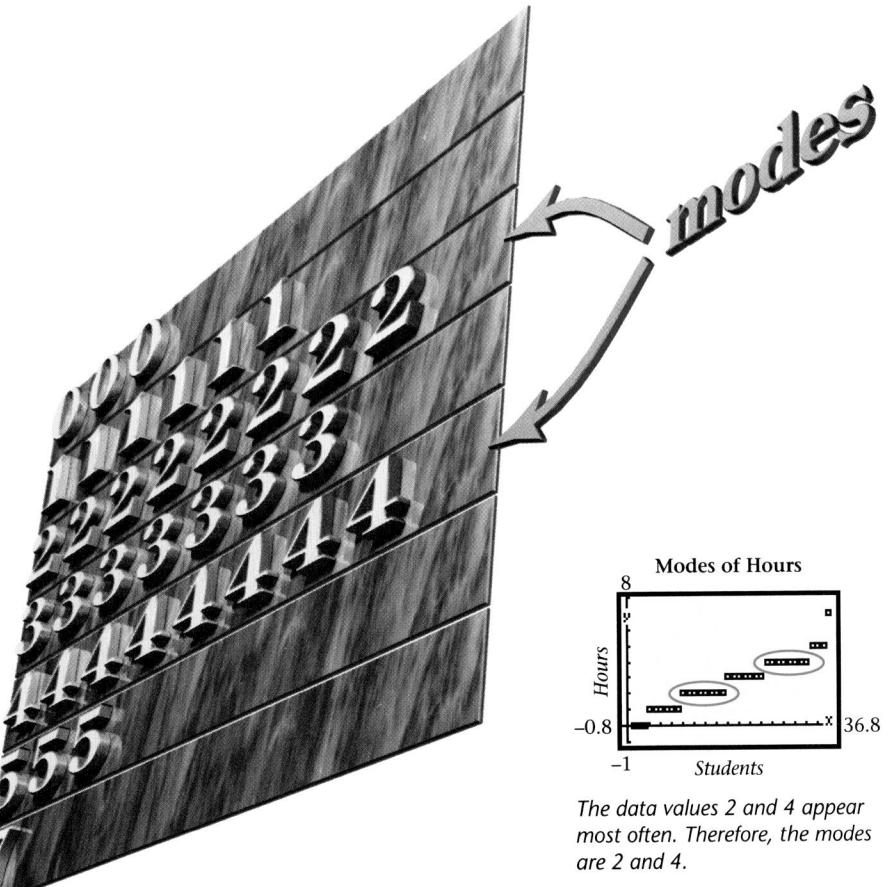

Modes of Hours

The data values 2 and 4 appear most often. Therefore, the modes are 2 and 4.

The following table summarizes how the measures of central tendency are determined.

Measures of Central Tendency	
Mean, $\bar{x}$	Add all the data, and then divide by the number of data values. The mean may also be called the *arithmetic mean*.
Median	Find the middle value of a collection of data values arranged in increasing or decreasing order. If there are two middle numbers, the median is the mean of the two middle numbers.
Mode	Find the value that occurs most often in a collection of data values. If two values occur most often, there are two modes and the set of data is *bimodal*. If more than two values occur most often, the mode is not reported.

EXAMPLE 1

Radio Broadcasting

The following table lists the number of minutes that two radio stations, W°op and KRAD, devote to music each day between the hours of 4:00 P.M. and 6:00 P.M. Find and compare the mean, median, and mode of music-broadcasting times for W°op and KRAD.

W°op	75	52	78	56	52	60	48	59	76	65
KRAD	60	56	62	65	67	60	60	58	68	60

Solution➤

Mean

The number of data values, n, for each station is 10.

$$\text{W°op} \quad \frac{\sum x}{n} = \frac{621}{10} = 62.1 \qquad \text{KRAD} \quad \frac{\sum x}{n} = \frac{616}{10} = 61.6$$

The 62.1-minute mean for W°op is slightly higher than the 61.6-minute mean for KRAD.

Median

Place all of the values in increasing or decreasing order.

W°op   48   52   52   56   **59**   **60**   65   75   76   78

Since n is even, there are two middle numbers, 59 and 60.

$$\frac{59 + 60}{2} = 59.5$$

The median for W°op is 59.5 minutes.

KRAD   56   58   60   60   **60**   **60**   62   65   67   68

Since n is even, there are two middle numbers, 60 and 60.

$$\frac{60 + 60}{2} = 60$$

The median for KRAD is 60 minutes. Thus, the median for KRAD is slightly higher than the median for W°op.

Mode

The mode, or most frequently occurring data value, for W°op is 52 minutes, and the mode for KRAD is 60 minutes. ❖

Try This   Find the mean, median, and mode for the following test scores.

92, 84, 78, 81, 90, 77, 93, 89, 92

CRITICAL Thinking

A company advertises a job opening. The "average" annual salary of all employees is advertised as $45,000. Which average measure do you think the company is probably using? How might a different average measure be misleading?

Cultural Connection: Asia Statistics may have been one of the first subjects studied as *applied* mathematics. The word *statistics*, derived from the word *state*, initially referred to data collected by the state. Organizing data in tables can be traced to almost every early civilization that kept official records. During the Chou dynasty, around 2000 B.C.E., the Chinese kept tables of revenues and expenditures and listed various members of their population.

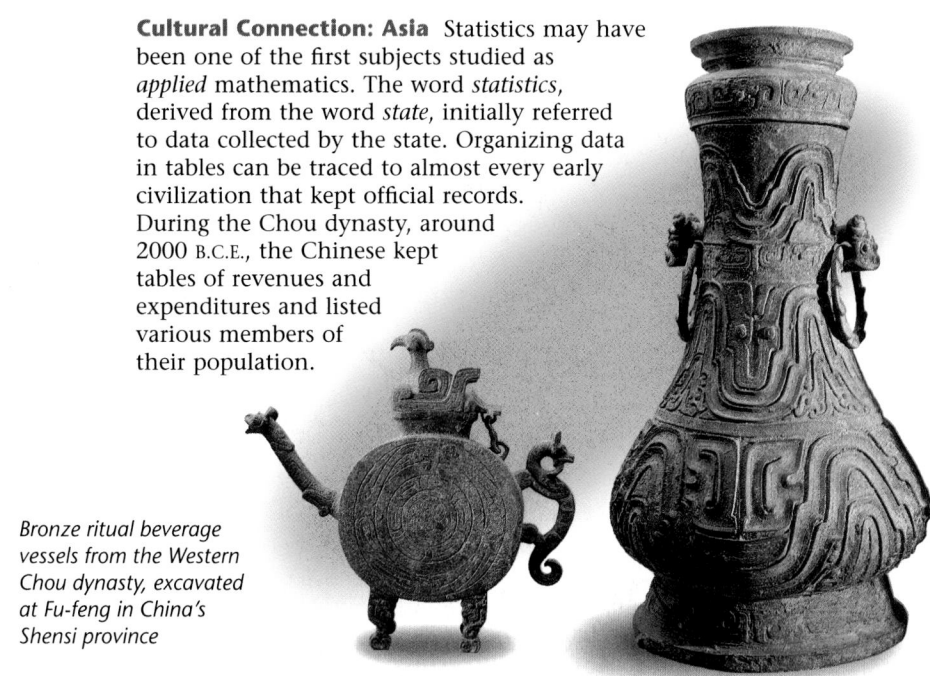

Bronze ritual beverage vessels from the Western Chou dynasty, excavated at Fu-feng in China's Shensi province

Use Transparency ▶ **54**

Frequency Tables

Statistical tables can illustrate how a data set is distributed. A table that records the number of times, or the *frequency*, that each data value occurs is a **frequency table**.

There are 36 data items in the list of weekend hours spent reading, and there are 7 distinct data values: 0, 1, 2, 3, 4, 5, and 7.

To construct a frequency table for this data set, divide the data into 8 groups or classes: 0, 1, 2, 3, 4, 5, 6, and 7. There are no data values for class 6, but it is still necessary to include this class for comparison.

In the following frequency table, the classes are in the first row, and the frequencies for each class are in the second row. The products of each class value, c, and its frequency, f, are in the third row.

Frequency Table of Hours Spent Reading

Class (c)	0	1	2	3	4	5	6	7	Totals
Frequency (f)	3	6	8	7	8	3	0	1	36
$c \times f$	0	6	16	21	32	15	0	7	97

The mean is $\bar{x} = \frac{\sum(c \times f)}{\sum f} = \frac{97}{36} \approx 2.7$.

Explain why the sum of the products, $c \times f$, divided by the sum of the frequencies is the mean.

Lesson 13.1 **715**

TEACHING *tip*

Point out to students that a mean that is computed from data classes and frequencies is sometimes called a *weighted average*. Weighted averages can be used to calculate grade point averages and scores of sporting events, for example, 2 touchdowns at 6 points each, 3 field goals at 3 points each, etc.

Aongoing **SSESSMENT**

The sum of the products represents the total number of hours spent reading, and the sum of the frequencies represents the total number of students. The quotient of these two values is the mean.

Teaching *tip*

If there are large amounts of data, grouping often makes it easier to handle. The size of the groups is often based upon nothing more than a convenient factor of the total number of data values. For example, if there are 65 values, convenient choices are 13 groups of 5 or 5 groups of 13.

When there are many different data values, a *grouped* frequency table is useful. In a **grouped frequency table**, the data values are grouped by a *range* of values rather than a single value.

Suppose 80 local musicians are asked to estimate the number of hours they spend each week rehearsing.

Weekly Rehearsal Times (hours per week)

2	31	9	34	24	15	44	6	27	28
29	15	24	35	23	5	15	29	24	2
5	4	10	25	3	25	47	21	24	19
6	42	17	27	6	36	20	24	34	20
26	35	0	21	14	6	3	8	15	23
48	16	23	25	0	35	34	20	25	2
9	20	14	2	24	27	43	31	20	21
16	36	28	1	51	22	24	39	15	3

To decide how many classes are needed, find the range of the data values.

Range: 0 to 51 = 52 values

For manageability, divide the data set into between 5 and 20 classes. Each class should contain the same number of data values, if possible.

Since the range, 52, is divisible by 4, it is convenient to form classes containing 4 data values. This will create 13 classes. Group the values that fall into each class, and list the frequency, or number of items, in each class. For example, the first class consists of class boundaries, or limits, from 0 to 3. There are 10 values in this class. Thus, the frequency for the first class is 10.

The **relative frequency** of a class is the ratio of its frequency to the total number of data values, n. Since 10 out of the total 80 values are in class 1, the relative frequency for class 1 is $\frac{10}{80}$, or 0.125.

Class number	Class (c)	Frequency (f)	Relative frequency
1	0–3 hr	10	0.125
2	4–7 hr	7	0.0875
3	8–11 hr	4	0.05
4	12–15 hr	7	0.0875
5	16–19 hr	4	0.05
6	20–23 hr	12	0.15
7	24–27 hr	15	0.1875
8	28–31 hr	6	0.075
9	32–35 hr	6	0.075
10	36–39 hr	3	0.0375
11	40–43 hr	2	0.025
12	44–47 hr	2	0.025
13	48–51 hr	2	0.025
	Totals	80	1.000

Aongoing
SSESSMENT

Each relative frequency represents a certain percentage of the total data values; therefore, the total of all the data values must equal 100%, or 1.

The relative frequency rescales each class frequency to a number from 0 to 1.

A grouped frequency table provides various information. For example:

- There are $12 + 15 = 27$ local musicians in classes 6 and 7. Thus, 27 musicians estimate that they practice between 20 and 27 hours each week. This is the largest frequency for any two consecutive classes.

- The sum of the frequencies in classes 1 through 4 is

$$10 + 7 + 4 + 7 = 28.$$

So there are 28 local musicians who practice 15 hours or less each week.

- Probability statements can be made using the relative frequencies. The experimental probabilities are the same as the relative frequencies.
  Thus, the experimental probability that a local musician in this population practices 3 hours or less is $\frac{10}{80}$, or 0.125.
  The experimental probability that a local musician in this population practices 40 hours or more (classes 11 through 13) is $\frac{6}{80}$, or 0.075.

Why must the sum of relative frequencies be 1?

CLASS MEAN

A **class mean** can be found using the following steps.

1. Find the midpoint of each class using the upper and lower values of the class. For example:
$$\text{class 1 midpoint} = \frac{0+3}{2} = 1.5$$

2. Multiply the midpoint of each class by the frequency of that class.
$$\text{For class 1: } 1.5 \times 10 = 15$$

3. Add these products for each class, and divide by the total number of data values, n.

EXAMPLE 2

Find the class mean of the weekly rehearsal times.

Solution▸

First, find the midpoint of each class.

Class number	Class	Class midpoint	Frequency	Class midpoint × frequency
1	0–3	1.5	10	15.0
2	4–7	5.5	7	38.5
3	8–11	9.5	4	38.0
4	12–15	13.5	7	94.5
5	16–19	17.5	4	70.0
6	20–23	21.5	12	258.0
7	24–27	25.5	15	382.5
8	28–31	29.5	6	177.0
9	32–35	33.5	6	201.0
10	36–39	37.5	3	112.5
11	40–43	41.5	2	83.0
12	44–47	45.5	2	91.0
13	48–51	49.5	2	99.0
			Total	1660.0

CRITICAL
Thinking

Since n is 80, the class mean is $\frac{1660}{80} = 20.75$ hours. ❖

Is the *class* mean always equal to the *arithmetic* mean? Explain.

EXERCISES & PROBLEMS

Communicate

1. Name three measures of central tendency, and explain how they are determined.

2. Explain how measures of central tendency can be misleading.

3. Describe the difference between a grouped and an ungrouped frequency table.

4. What is wrong with the following class setup for a grouped frequency table? Classes: 15–20, 20–25, 25–30, 30–35, 35–40, 40–45, 45–50, 50–55, and 55–60

5. Explain how the class mean for a grouped frequency table may be defined.

6. Explain how to construct a list of 5 numbers with a mean of 7, a median of 6, and a mode of 3.

Practice & Apply

For Exercises 7–10, find the mean, median, and mode (if it exists) for each set of data.

7. The number of traffic tickets issued over a 5-day period by a police officer who patrols near a high school is 7, 9, 3, 5, and 5. 5.8; 5; 5

8. The number of students advised by a teacher each day over a normal 20-day registration period is 7, 12, 10, 8, 7, 12, 15, 10, 10, 7, 2, 9, 12, 14, 6, 10, 12, 9, 8, and 8. 9.4; 9.5; 12 and 10

9. The numbers of student immigrants (in thousands) admitted into the United States for selected years from 1970 to 1990 are given in the following table.

Year	1970	1975	1985	1986	1987	1988	1989	1990
Students	107	118	286	288	289	338	361	355

10. The sodium content (in milligrams) of 30 poultry hot dogs is listed. 431.5; 419; 396, 419

420	370	396	385	376	396	422	359	412	517
387	345	392	458	527	534	461	508	397	419
428	402	463	529	399	488	419	437	384	515

11. The mean of 6 pieces of data is $7\frac{2}{3}$. Five of the data values are 9, 4, 8, 8, and 7. What is the sixth data value? 10

12. The mean of 5 pieces of data is 0.21. Four of the data values are 0.12, 0.36, 0.15, and 0.17. What is the fifth data value? 0.25

Education The table at the right lists 5 exam scores of two algebra students.

13. Find the mean score for each student. both are 84

14. Which student do you think has the most consistent scores? Explain how you determined this. Tricia

15. Which student do you think has the best chance of scoring above 80 on the next exam? Why? Tricia

Tricia	Morgan
81	98
84	68
88	99
82	59
85	96

For Exercises 16–19, refer to the frequency table at the right.

16. Find the mean, median, and mode. 13.26; 13; 10

17. What percentage of the data values are less than 12? 43.9%

18. What percentage of the data values are greater than 15?

19. Describe the pattern you see in the frequency table.

c	f
10	20
11	16
12	4
13	6
14	3
15	2
16	14
17	17

9. mean: 267.75; median: 288.5; mode: none

14. Tricia; all scores are close to the mean.

15. Tricia; all of her previous scores are above 80.

18. 37.8%

19. most values are at the extreme ends of the data set

20.

Class	Frequency
7.5	3
7.6	10
7.7	0
7.8	15
7.9	17
8.0	17
8.1	14
8.2	9
8.3	3

21. mean: 7.93; median: 7.9; mode: 7.9 and 8.0

22. 7.9; the mean, median, and mode are all 7.9

23. contradict; all measures of central tendency are below 8.0

Business A vending company claims that the beverage machine it has placed on a school campus dispenses 8 fluid ounces in each cup. The student government wants to verify the validity of this claim. Over several weeks, student volunteers test 88 randomly selected cups of beverage dispensed by the machine.

20. Construct an ungrouped frequency table for the beverage-machine data.

21. Find the mean, median, and mode.

22. Based on the frequency table, predict the number of fluid ounces that a randomly selected cup is likely to contain. Explain why you chose that number.

23. Based on the frequency table, does this set of data tend to confirm or contradict the beverage company's claim? Explain.

Education Suppose 55 high-school sophomores from several randomly chosen schools in a school district are asked how many papers they were assigned to write in one semester of English composition class.

English Composition Papers

6	8	6	4	8	9	10	8	9	8	9
8	10	8	9	8	5	9	9	6	8	8
5	8	9	8	9	6	4	8	10	8	9
8	4	6	9	7	9	10	5	8	9	7
6	10	9	6	10	8	9	9	7	5	8

24. Construct an ungrouped frequency table for the composition-papers data.

25. Find the mean, median, and mode.

26. Based on the data, how many English composition papers were assigned to a randomly selected sophomore student in this district?

27. If you met a student from this district who claimed that he or she was expected to write only 2 papers in one semester of sophomore English composition, would you tend to believe him or her? Explain.

7.8	7.9	7.6	8.0
7.8	8.0	7.9	7.6
8.0	8.0	7.5	7.9
7.8	7.9	7.8	8.0
8.2	7.9	7.6	8.0
7.6	8.1	8.1	7.9
7.5	8.0	7.8	7.8
8.0	8.2	7.9	7.6
8.2	7.8	8.0	8.3
8.0	8.1	7.8	7.6
7.9	8.0	8.3	7.6
8.2	7.8	7.9	8.1
8.0	8.0	8.1	7.8
8.1	8.1	7.6	7.8
7.9	8.1	7.9	8.0
8.2	8.0	8.1	8.0
7.9	8.0	7.5	7.9
8.1	8.1	7.8	8.1
8.2	8.1	7.8	7.8
7.9	8.1	7.8	8.2
8.2	7.9	7.9	7.6
8.2	7.6	8.3	7.9

24.

Class	Frequency
4	3
5	4
6	7
7	3
8	17
9	15
10	6

25. mean: 7.75; median: 8; mode: 8

26. 8

27. No; this is a value less than any value in the data set.

Entertainment Suppose 40 randomly selected CDs are examined for their playing times in minutes.

Compact Disc Playing Times

38.18	43.22	46.22	34.36	43.19
34.87	48.44	32.57	54.51	44.03
40.54	35.20	47.37	41.55	37.54
30.17	38.52	39.46	30.55	42.62
40.13	33.61	31.51	45.57	34.09
29.52	46.70	40.08	43.38	39.73
42.32	37.05	31.80	28.88	33.25
48.55	42.81	35.62	33.51	29.21

28. Construct a grouped frequency table for this set of data.

29. Name at least three pieces of information that the table tells you about the CD playing times.

30. Can you tell from the grouped frequency table how many CDs have less than 35 minutes of playing time? If so, how many are there? If not, estimate the number of CDs that have less than 35 minutes of playing time.

31. Expand your frequency table to include the relative frequency for each class.

32. Estimate the probability that a CD has a playing time of more than 45 minutes.

Look Back

33. Given that $f(x) = x^3$ and $g(x) = x^3 + 2x - 3$, find $(f + g)(x)$ and $(f \cdot g)(x)$. **[Lesson 3.3]** $(f + g)(x) = 2x^3 + 2x - 3$; $(f \cdot g)(x) = x^6 + 2x^4 - 3x^3$

34. Solve the system of linear equations by elimination. **[Lesson 4.3]** $\begin{cases} 2x - 7y = 15 \\ 5x - 4y = 10 \end{cases}$ $\left(\frac{10}{27}, -\frac{55}{27} \right)$

35. Describe the behavior of the graph of a function near a zero of even multiplicity. **[Lesson 6.3]** turning point; does not cross x-axis

Write equivalent logarithmic equations. **[Lesson 7.3]**

36. $16^{\frac{1}{2}} = 4$ **37.** $5^4 = 625$ **38.** $\left(\frac{1}{3} \right)^2 = \frac{1}{9}$ **39.** $2^{-3} = \frac{1}{8}$

$\log_{16} 4 = \frac{1}{2}$ $\log_5 625 = 4$ $\log_{\frac{1}{3}} \frac{1}{9} = 2$ $\log_2 \frac{1}{8} = -3$

Write equivalent exponential equations. [Lesson 7.3]

40. $\log_5 125 = 3$ **41.** $4 = \log_5 625$ **42.** $-4 = \log_3 \frac{1}{81}$ **43.** $\log_{10} 0.001 = -3$

$5^3 = 125$ $5^4 = 625$ $3^{-4} = \frac{1}{81}$ $10^{-3} = 0.001$

Look Beyond

44. Use a graphics calculator to graph $f(x) = \frac{1}{\sqrt{2\pi}} e^{\frac{-x^2}{2}}$. Use the trace feature to obtain several ordered pairs, and then use these pairs to transfer the graph to graph paper. Estimate the area, in square grid units, between the curve and the x-axis by counting squares on your graph paper.

28–31. Answers may vary.

28. A sample grouped frequency table is:

Class	Frequency
25.00–29.99	3
30.00–34.99	11
35.00–39.99	8
40.00–44.99	11
45.00–49.99	6
50.00–54.99	1

29. Answers may vary. One sample answer is: For the table of data in Exercise 28, 100% of all values are between 25 and 55 minutes; 97.5% of all values are below 50 minutes; the mean is 38.76, the median is 38.99, 55% of the CDs last less than 40 minutes.

30. Answer depends on grouped frequency table; for the grouped frequency table in Exercises 28: yes; 14.

31. A sample grouped frequency table is shown.

Class	Frequency	Relative Frequency
25.00–29.99	3	0.075
30.00–34.99	11	0.275
35.00–39.99	8	0.200
40.00–44.99	11	0.275
45.00–49.99	6	0.150
50.00–54.99	1	0.025

32. Answer depends on grouped frequency table; for the grouped frequency table in Exercises 28, the result is $\frac{7}{40}$, or 17.5%.

44. Answers may vary. One example is: approximately $(-1, 0.2)$ and $(0, 0.4)$; area ≈ 1

Look Beyond

This exercise gives students an opportunity to explore the normal curve using their graphics calculators and graph paper.

LESSON 13.2 Histograms and Stem-and-Leaf Plots

 The study of statistics is sometimes made easier by visual presentations of statistical data.

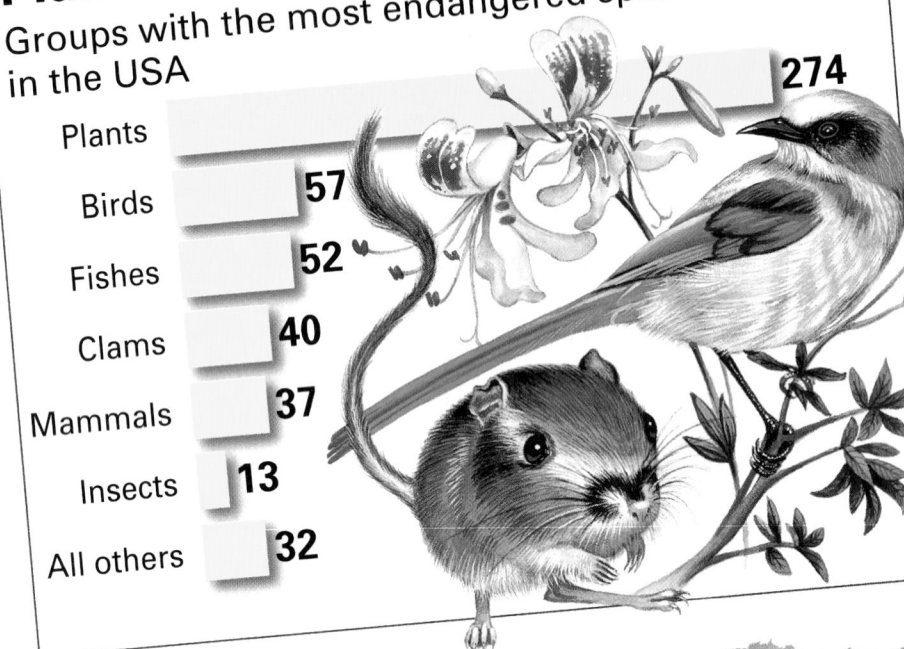

USA Facts

How statistics shape our lives

Plants are the most endangered

Groups with the most endangered species in the USA

Plants	274
Birds	57
Fishes	52
Clams	40
Mammals	37
Insects	13
All others	32

When you scan through a newspaper, you will very likely see information presented in forms such as line graphs, bar graphs, or pie charts. Displaying information in this way enables the reader to see patterns in data.

What information may be read from the bar graph shown?

Histograms

The endangered-species bar graph is one type of a *histogram*. A **histogram** displays the frequency of data values.

Refer to the Frequency Table of Hours Spent Reading on page 715. You can display this information in a histogram.

- Construct the horizontal and vertical axes. Place the frequencies on the vertical axis. Place the data values, hours, on the horizontal axis.

- Construct the rectangular bars with each base centered over the corresponding data values or class number. The height of the rectangular bar for each class represents its frequency.

Hours Spent Reading

Class (*c*)	Frequency (*f*)
0	3
1	6
2	8
3	7
4	8
5	3
6	0
7	1

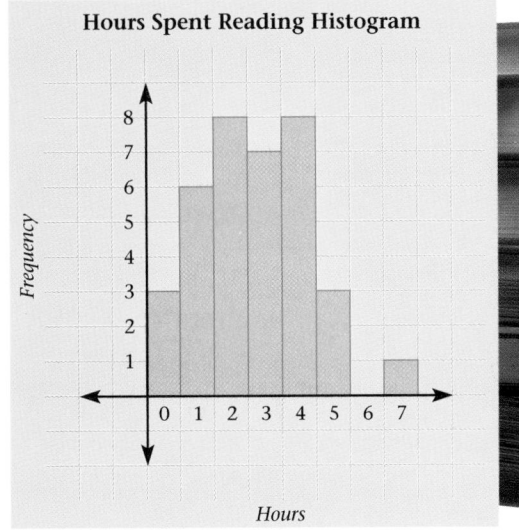

Hours Spent Reading Histogram

A histogram displays the following information.

- The range of all data values

- The value(s) about which the data values tend to cluster

- Whether the data values are skewed, or are clustered to either side of the mean or median

- Gaps or low points in the data distribution

- Outliers from the general distribution of values

ENRICHMENT Put a histogram on the chalkboard. Then have students calculate and draw a dashed line at the mean and another dashed line at the median. Circle the mode. Calculate the mean value between adjacent columns and place a solid dot at those values. Sketch a line of best-fit through the points.

INCLUSION strategies **Using Tables** Ask students to select a past season of a school sports team and make histograms of the points scored by the team and by the opponents. Encourage students to propose hypotheses based on the displayed data.

Alternate Example 1

Poll students about how many hours they work per week. Determine the number of hours students spend in school during the week and convert the given data into a table of percentages of work versus school. Make a histogram and discuss the results it shows. [**Answers may vary.**]

CRITICAL *Thinking*

30–34 hours; yes; each class to the right of the mode is about the same size as the corresponding class to the left.

Use Transparency ▶ 56

EXAMPLE 1

Employment   The grouped frequency table lists the number of part-time hours worked by a group of 261 randomly selected high-school students during the last two weeks of May. Construct a histogram to display the data.

Summer Employment

Class number	Hours	Frequency
1	15–19	25
2	20–24	31
3	25–29	44
4	30–34	60
5	35–39	42
6	40–44	30
7	45–49	27
8	50–54	0
9	55–59	0
10	60–64	0
11	65–69	2

Solution►

Label the vertical axis to include values from 0 to 60.

Label the horizontal axis with hours.

Then, for each class, draw a rectangular bar to represent the frequency of that class.

The histogram shows one peak at 30–34 hours. The data set is nearly symmetric about this class. There are two outliers at 65–69 hours. ❖

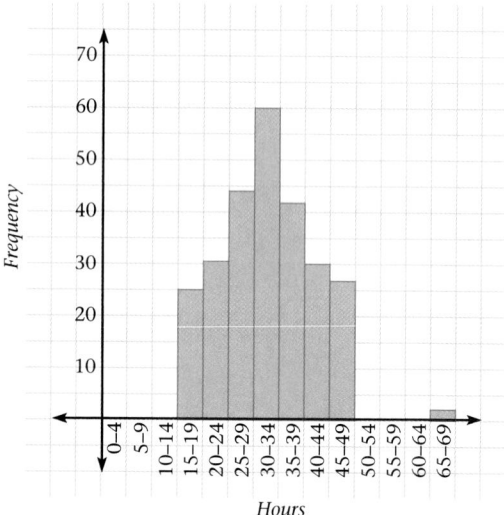

Summer Employment Histogram

CRITICAL *Thinking*

Which class represents the mode of the distribution in Example 1? Is the distribution also nearly symmetric about the mode? Explain.

RETEACHING **the lesson**

Guided Research Ask students to collect data from particular personal interests such as sports, cars, or music. Then have them report the data to you in histogram and stem-and-leaf form.

724 Lesson 13.2

Stem-and-Leaf Plots

In a histogram for grouped data, it may be difficult to examine specific values. A **stem-and-leaf plot** allows you to view patterns within grouped data. Each piece of data is grouped together on a specific row and is arranged in two columns. The left column is called the *stem*, and the right column is called the *leaf*.

EXAMPLE 2

A high-school sociology class conducts a study of the relationship between academic performance and the number of hours spent watching television. To study this relationship, class members find the average number of hours that students spend watching television each week. Construct a stem-and-leaf plot to display the data.

Hours Spent Watching Television

10	43	5	27	25	14	26	42
18	8	7	10	16	30	35	22
52	18	30	11	27	22	60	31

Solution➤

Split the numbers into two parts, a stem and a leaf. Some of the numbers are single-digit numbers, and some are two-digit numbers. Let the stem be the tens digit and the leaves be the units digits.

It is often helpful to arrange the leaves in increasing order.

Stems	Leaves
0	5, 7, 8
1	0, 0, 1, 4, 6, 8, 8
2	2, 2, 5, 6, 7, 7
3	0, 0, 1, 5
4	2, 3
5	2
6	0

Try This The following data represent the average number of dollars that 44 families spent each year on the dental care of one child. Make a stem-and-leaf plot to represent the data. Use the first two digits as the stem and the third digit as the leaf.

Dollars Spent on Dental Care

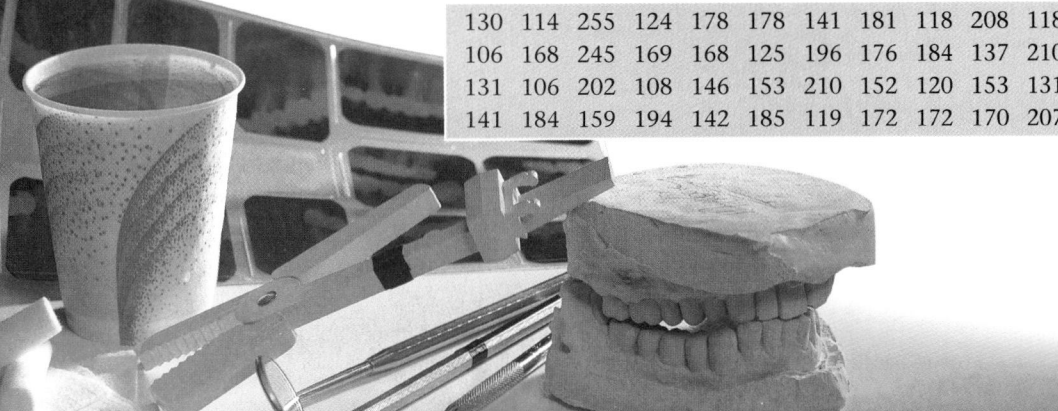

130	114	255	124	178	178	141	181	118	208	118
106	168	245	169	168	125	196	176	184	137	210
131	106	202	108	146	153	210	152	120	153	131
141	184	159	194	142	185	119	172	172	170	207

Alternate Example 2

Determine the number of hours students spend in school per week and the number of hours per week they watch TV. Have them convert TV hours to a percentage of hours in school. Have them make a stem-and-leaf plot of the percentages.

TEACHING *tip*

Discuss how stems (the most significant digits) of the data can have many leaves, which are the least significant digits of the data.

Ongoing ASSESSMENT

Try This

10	6, 6, 8
11	4, 8, 8, 9
12	0, 4, 5
13	0, 1, 1, 7
14	1, 1, 2, 6
15	2, 3, 3, 9
16	8, 8, 9
17	0, 2, 2, 6, 8, 8
18	1, 4, 4, 5
19	4, 6
20	2, 7, 8
21	0, 0
22	
23	
24	5
25	5

A ssess

Selected Answers

Odd-Numbered Exercises 5–47

Assignment Guide

Core 1–49

Core Plus 1–49

Technology

Graphics calculators are needed for Exercise 43. Some graphics calculators have the ability to make histograms. Ask students to use the data from any of the earlier examples or exercises to construct a histogram on their calculators. Compare the various displays and responses, and critique them.

Error Analysis

A statistical graph of data is *skewed* in the direction of the opening tail. The direction of the skew will often be confusing to students.

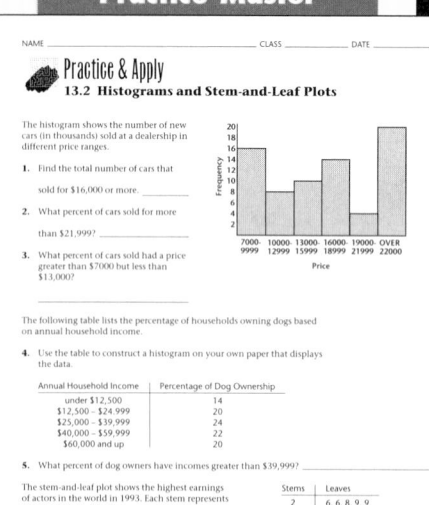

Practice Master

NAME _____ CLASS _____ DATE _____

Practice & Apply
13.2 Histograms and Stem-and-Leaf Plots

The histogram shows the number of new cars (in thousands) sold at a dealership in different price ranges.

1. Find the total number of cars that sold for $16,000 or more. _____

2. What percent of cars sold for more than $21,999? _____

3. What percent of cars sold had a price greater than $7000 but less than $13,000? _____

The following table lists the percentage of households owning dogs based on annual household income.

4. Use the table to construct a histogram on your own paper that displays the data.

Annual Household Income	Percentage of Dog Ownership
under $12,500	14
$12,500 – $24,999	20
$25,000 – $39,999	24
$40,000 – $59,999	22
$60,000 and up	20

5. What percent of dog owners have incomes greater than $39,999? _____

The stem-and-leaf plot shows the highest earnings of actors in the world in 1993. Each stem represents $10,000,000. Each leaf represents $1,000,000.

Stems	Leaves
2	6, 6, 8, 9, 9
3	0, 2, 7
4	9, 8
6	6

6. The highest earning actor was William H. Cosby, Jr. How much money did he earn? _____

7. Arnold Schwarzenegger and Kevin Costner earned the same amount. If they were the second highest earners, how much did each make? _____

8. How many actors made at least $30 million in 1993? _____

HRW Advanced Algebra   Practice & Apply   81

Exercises & Problems

Communicate

1. Explain how to construct a histogram.
2. Explain how to construct a stem-and-leaf plot.
3. What statistical information is easy to determine from a stem-and-leaf plot, but difficult to determine from a histogram?
4. Explain how to find the mode of a data set using a stem-and-leaf plot.

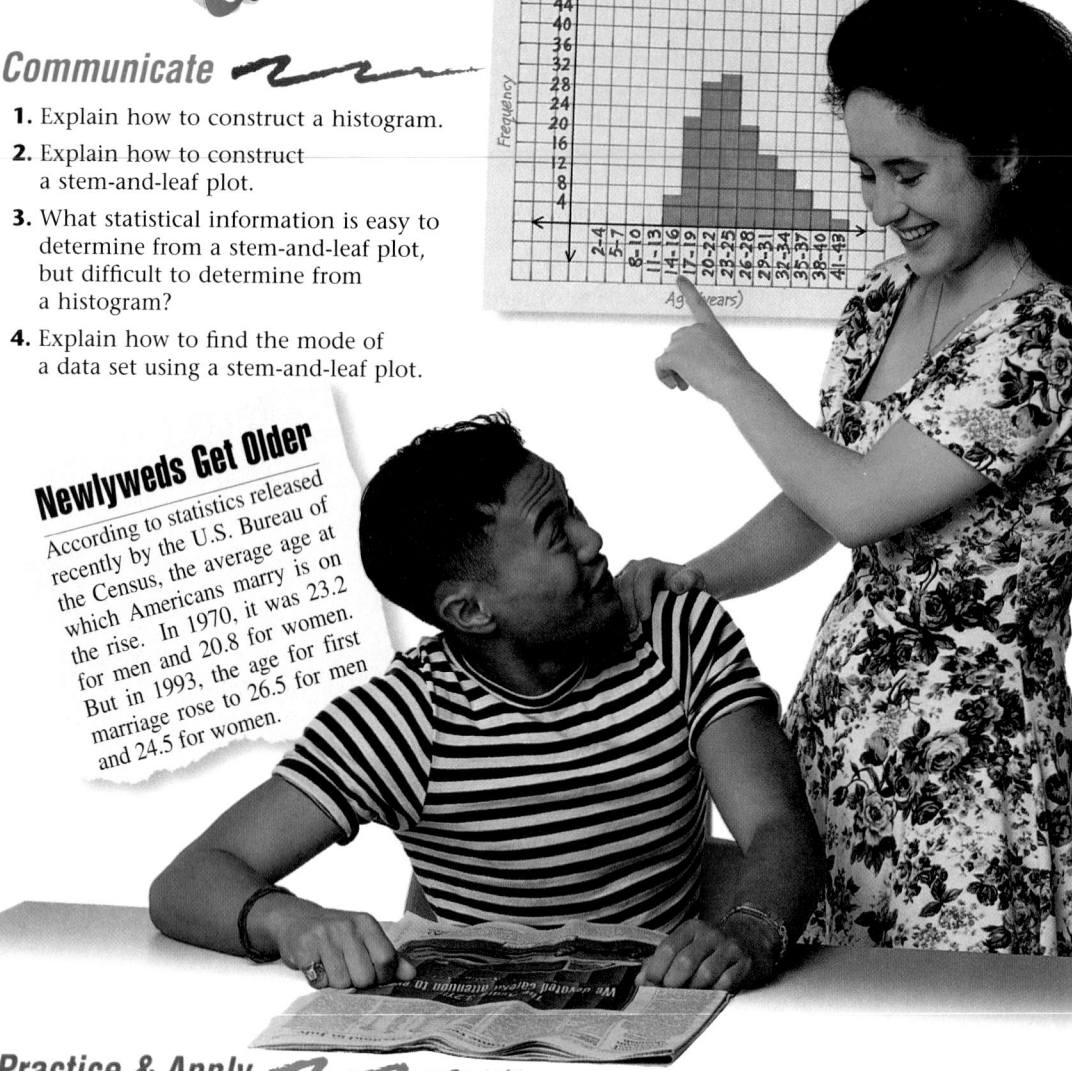

Age at which 150 women in an American city married for the first time

Newlyweds Get Older

According to statistics released recently by the U.S. Bureau of the Census, the average age at which Americans marry is on the rise. In 1970, it was 23.2 for men and 20.8 for women. But in 1993, the age for first marriage rose to 26.5 for men and 24.5 for women.

Practice & Apply

5. Construct a stem-and-leaf plot for the age data given.
6. Describe the distribution pattern of this data set.
7. Does the data set appear to be symmetric about some number? If so, what is that number?
8. Does the data set appear skewed in either direction? If so, describe the skewness.
9. Are there any outliers in this data set? If so, what are they?

Ages of 35 People at a Family Reunion

91	89	58	47	68	97	44
55	80	77	37	48	43	70
89	78	12	30	35	79	46
45	53	72	41	20	34	37
57	66	54	24	91	13	52

5.

1	2, 3
2	0, 4
3	0, 4, 5, 7, 7
4	1, 3, 4, 5, 6, 7, 8
5	2, 3, 4, 5, 7, 8
6	6, 8
7	0, 2, 7, 8, 9
8	0, 9, 9
9	1, 1, 7

6. most of the values are clustered in the 30–50 range

7. no

8. Answers may vary. One sample answer is: yes; data values are slightly skewed towards the upper ages.

9. no

Technology The computer screen shows the percentage of students with personal computers at some universities.

10. Construct a histogram to display the percentages.

11. Describe any trends that are visible in the distribution.

12. Find the mean percentage of students with PCs at these universities. 64.17%

13. Does the distribution appear to be symmetric about the mean percentage? yes

14. Is the distribution skewed about the mean percentage? no

University	Students With PCs
Northwestern	70%
Cornell	75%
Michigan	50%
Penn State	60%
Stanford	65%
U.C.L.A.	65%

Marketing The number of vending-machine purchases made by 21 high-school seniors over a 3-month period is given.

Vending-Machine Purchases

10	43	5	27	25	14	42
18	8	63	10	16	30	22
52	18	30	11	27	22	31

15. Construct a stem-and-leaf plot for this set of data.

16. What is the mean number of vending-machine purchases? 24.95

17. Does the data set appear to be symmetric about the mean? no

18. Does the data set appear skewed in either direction? If so, describe the skewness. yes; toward the higher values

19. Are there any outliers? If so, what are they? yes; 52 and 63

20. How might the answers to Exercises 15–19 affect how often the vending machines are stocked? the greater the mean, the more often it should be stocked

Demographics The number (in millions) of households that were headed by a person between the ages of 15 and 74 in the year 1991 is given.

21. Construct a grouped frequency histogram that displays the head-of-household data.

22. Describe any trends that are visible in the distribution.

23. Find the class mean of the distribution. 44.8 yr

24. Does the distribution appear to be nearly symmetric about the class mean? no

25. Is there any skewness about the class mean? If so, how does this relate to trends in heads of households?

Heads of Households

Age	Number
15–24	4.9
25–34	20.3
35–44	21.3
45–54	14.8
55–64	12.5
65–74	12.0

Answers to Exercises 10, 21, 22 and 25 can be found in Additional Answers beginning on page 842.

11. At least 50% of the students surveyed have PCs; Cornell has the highest percentage (75%) of students with PCs.

15.

0	5, 8
1	0, 0, 1, 4, 6, 8, 8
2	2, 2, 5, 7, 7
3	0, 0, 1
4	2, 3
5	2
6	3

26.

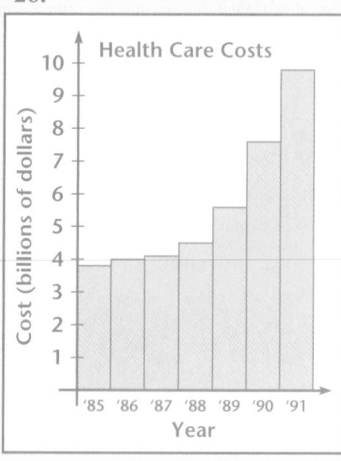

Health Care Costs

Cost (billions of dollars) vs Year ('85 '86 '87 '88 '89 '90 '91)

27. The cost of home health care is increasing over time

30. Yes; skewed to the left which indicates a rise in amount spent over time for home health care

31. The trend is the same. Costs are still rising, but the trend can be seen over a longer period of time.

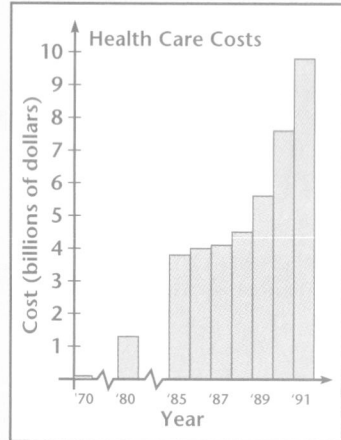

Health Care Costs

Cost (billions of dollars) vs Year ('70 '80 '85 '87 '89 '91)

Health Care Home health care has become an option for many United States citizens. The amount, in billions of dollars, spent in the United States on home health care during the years 1985–1991 is given.

26. Construct a histogram to display the health-care data.

27. Describe any trends that are visible in the distribution.

28. Find the mean amount spent on home health care from 1985 to 1991. $5.63 billion

29. Does the distribution appear to be symmetric about the mean? no

30. Is the distribution skewed about the mean? If so, how does this relate to trends in home health care?

31. In 1970, $0.1 billion was spent on home health care; in 1980, $1.3 billion was spent on home health care. Incorporate these data values with the data aready presented, and make a new histogram. How does the new histogram compare with the first histogram? Are the trends the same? Explain.

United States Home Health Care

Year	Amount ($ in billions)
1985	3.8
1986	4.0
1987	4.1
1988	4.5
1989	5.6
1990	7.6
1991	9.8

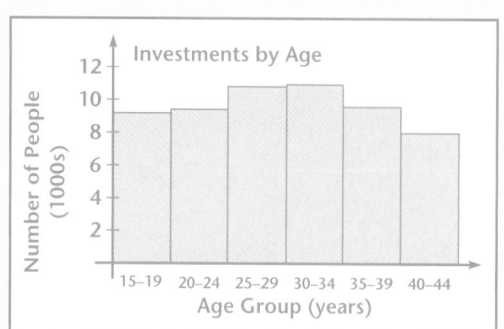

Investment The table lists the number of people aged 15 to 44 who began investments for their future in 1990.

32. Construct a grouped frequency histogram to display the data.

33. Describe any trends that are visible in the ages of beginning investors.

34. Find the class mean of the distribution. 29.3 yr

35. Does the distribution appear to be nearly symmetric about the class mean? yes

36. Is there any skewness about the class mean? If so, how does this relate to trends in the ages of beginning investors?

Beginning Investors

Age	Number
15–19	9179
20–24	9413
25–29	10,796
30–34	10,930
35–39	9583
40–44	7999

32.

Investments by Age

Number of People (1000s) vs Age Group (years) (15–19, 20–24, 25–29, 30–34, 35–39, 40–44)

33. The distribution is relatively flat. There are slightly more beginning investors in the 25–34 age group.

36. No; the data are not skewed.

Health Care The table lists the number of surgeries performed in hospitals with 500 or more beds for the years 1983–1990.

Hospital Surgeries

Year	Number
1983	1167
1984	1225
1985	1465
1986	1725
1987	1942
1988	1963
1989	2049
1990	2162

37. Construct a histogram to display the data.

38. Describe any trends that are visible in the distribution.

39. Find the mean number of hospital surgeries between 1983 and 1990. 1712.25

40. Does the distribution appear to be symmetric about the mean? no

41. Is the distribution skewed about the mean? If so, how does this relate to trends in the number of hospital surgeries?
yes, toward the left;
increasing number of surgeries over time

Look Back

42. Suppose that the value of k depends on the value of j in such a way that the value of k is always 5 more than 4 times the value of j. Use function notation to describe this relationship. **[Lesson 2.4]** $k(j) = 4j + 5$

43. Use your graphics calculator to find a pair of inverse 2×2 matrices. Then show that they are inverse matrices using matrix multiplication. **[Lesson 4.5]**

44. **Transformations** Explain how the graph of $g(x) = (x + 4)^2 + 3$ is related to the graph of $f(x) = x^2$. **[Lesson 5.3]**

Three coins are tossed. Use a diagram to find the probability that 2 coins land tails up, given that the following events occur. [Lesson 11.6]

45. The first coin lands tails up.

46. The first coin lands heads up.

47. The first two coins land heads up. 0

Look Beyond

48. Three partners in an appliance store appear before a bankruptcy judge. Each partner makes four statements. Mr. Houlihan says, "Washington owes me $5000. Steinberg owes me $10,000. Everything Steinberg says is true. Nothing Washington says is true." Mr. Washington replies, "Everything Houlihan says is wrong. I don't owe him anything. Steinberg owes me $5000. I should know because I was the bookkeeper." Finally, Mr. Steinberg says, "Washington was not the bookkeeper. And only two of his statements are true. Furthermore, I don't owe anyone anything. And I always tell the truth." Only one of the partners is telling the complete truth. Which one is it? Washington

38. There is a steady increase in the number of surgeries performed over time.

43. Answers may vary. An example is shown.
$$\begin{bmatrix} 1 & 2 \\ -2 & 4 \end{bmatrix} \cdot \begin{bmatrix} 0.5 & -0.25 \\ 0.25 & 0.125 \end{bmatrix} = \begin{bmatrix} 1 & 0 \\ 0 & 1 \end{bmatrix}$$

44. The graph is the same as the graph of f, but translated 4 units to the left and 3 units up.

45. $\frac{1}{2}$

46. $\frac{1}{4}$

Look Beyond

This exercise allows students to have fun with a math "teaser."

37.

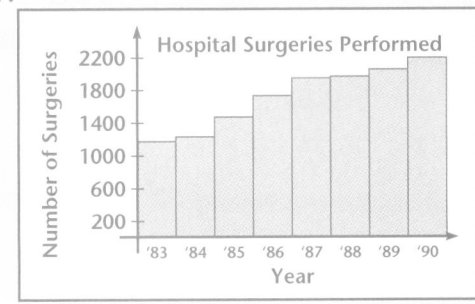

- Find the quartiles and interquartile range for a given set of data.
- Construct a box-and-whisker plot to display a given set of data.

RESOURCES

- Practice Master — 13.3
- Enrichment Master — 13.3
- Technology Master — 13.3
- Lesson Activity Master — 13.3
- Quiz — 13.3
- Spanish Resources — 13.3

Assessing Prior Knowledge

1. A complete stereo system is advertised on sale for 15% off its regular price of $799. What is the sale price? [**$679.15**]

2. Matilda scored 42 points out of a possible 58 on a history quiz. What is her percentage score? [**72.4%**]

TEACH

why Frequently it is of greater interest to know about where a single data value lies with respect to others. Knowing whether it is above or below the mean, median, or mode may not be enough. In this lesson, we extend the median concept (50% below, 50% above) to smaller subdivisions — quartiles and percentiles.

LESSON 13.3 **Percentiles and Box-and-Whisker Plots**

why *Many standardized test scores are expressed in terms of a percentile ranking. Percentile and related rankings can be used to compare a large group of data values.*

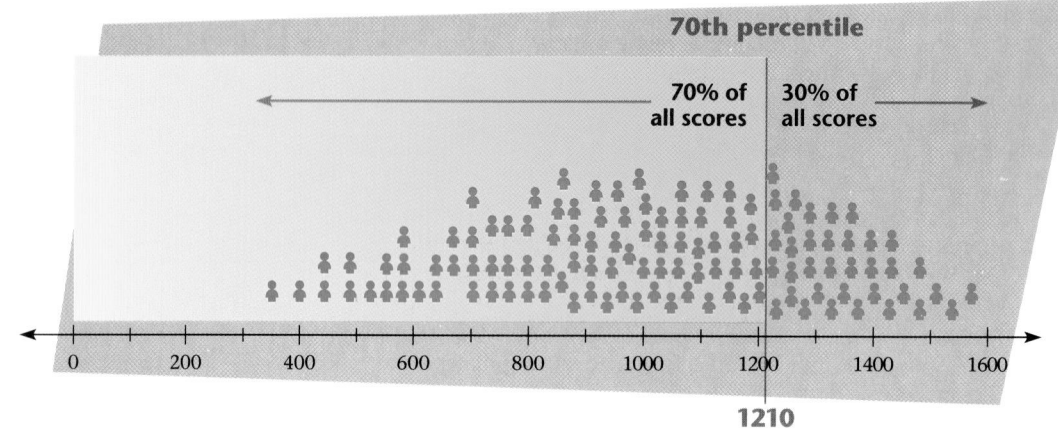

A student who scores at the 70th percentile scores above 70% of the other students.

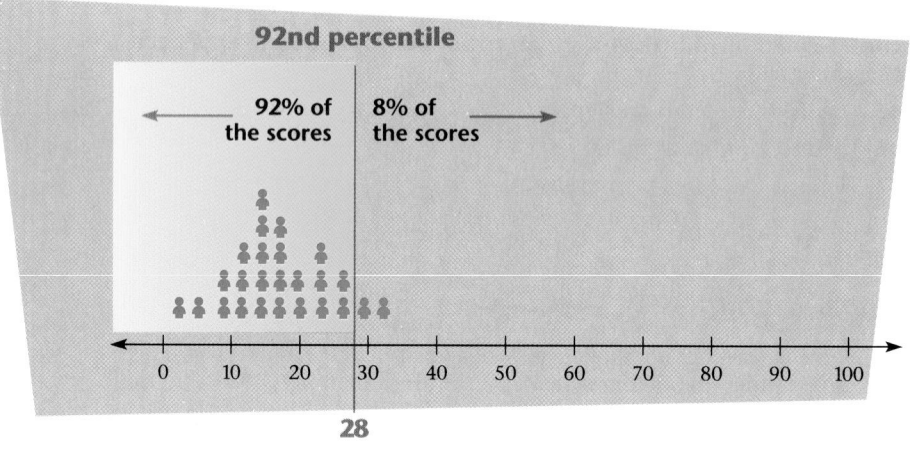

A data value in the 92nd percentile is greater than 92% of the other data values.

Percentiles are found by arranging the data in order. A value is in the *n*th **percentile** if *n*% of the data values are equal to or less than this value.

ALTERNATIVE teaching strategy

Hands-On Strategies Begin with a 10-unit number line. Ask each member of the class to write a number between zero and 10 on a piece of paper. Have each student, in turn, read aloud his or her number. As a number is read, record it with a dot on the number line, stacking repeated number dots vertically. When done, ask the students to divide the class size by 4 and then start counting dots from the left, marking off the quartile boundaries. Then draw the box-and-whisker plot immediately above the number line and dots.

Quartiles

In statistics, the 25th, 50th, and 75th percentiles are frequently used to divide a data set into **quartiles**, or fourths. The 25th percentile, Q_1, ends the 1st quartile; the 50th percentile, Q_2, ends the 2nd quartile; and the 75th percentile, Q_3, ends the 3rd quartile.

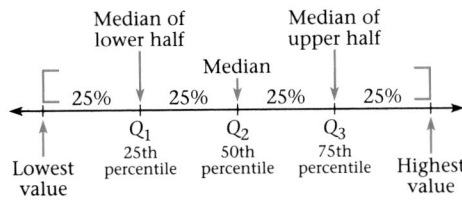

The difference between the high and low quartiles, $Q_3 - Q_1$, is called the **interquartile range**.

To find the quartiles of a data set:

1. Find the median of the entire data set. This is Q_2.

2. Find the median of the data below Q_2. This is the low quartile, Q_1.

3. Find the median of the data above Q_2. This is the high quartile, Q_3.

EXAMPLE 1

On the first test of the year, an algebra class of 21 students made the following scores. Find the quartiles of this data set.

75	90	53	85	75	83	73
80	46	89	91	93	85	95
68	88	97	70	96	93	86

Solution

First, write the 21 data values in increasing order.

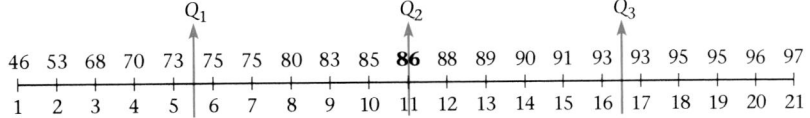

The median of all data values is the 11th score, 86. Thus, $Q_2 = 86$.

The median of the 10 scores below Q_2 is $\frac{73 + 75}{2} = 74$. Thus, $Q_1 = 74$.

The median of the 10 scores above Q_2 is $\frac{93 + 93}{2} = 93$. Thus, $Q_3 = 93$. ❖

CRITICAL Thinking
Would the quartiles be as meaningful for comparison if all of the scores were 90 or above? Explain your answer.

Take the NBA percentage stand-
ings from Alternate Example 1
and make a box-and-whisker
plot of the data.

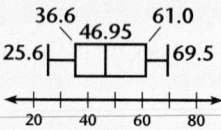

Box-and-Whisker Plots

A **box-and-whisker plot** illustrates how the quartiles are distributed within the data set. Examine the box-and-whisker plot for the scores from Example 1.

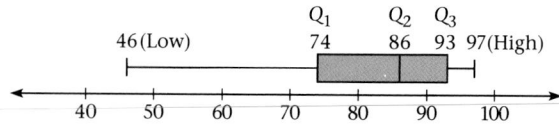

EXAMPLE 2

The ages of the members of the Reuter Family Reunion are given. Make a box-and-whisker plot for the data, and find the interquartile range.

4	7	9	31	34	2	11	33	36	2	8	13
35	37	24	34	31	50	52	57	60	69	78	83

Solution➤

First, write the 24 data values in increasing order.

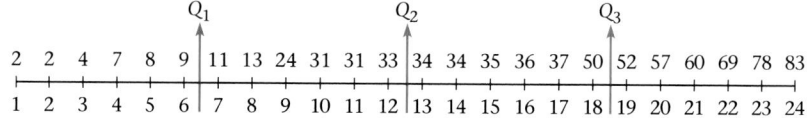

The median of all 24 scores is the mean of the 12th and 13th scores.

$$Q_2 = \frac{33 + 34}{2} = 33.5$$

The median of the 10 scores below Q_2 is $\frac{9 + 11}{2} = 10$. Thus, $Q_1 = 10$.

The median of the 10 scores above Q_2 is $\frac{50 + 52}{2} = 51$. Thus, $Q_3 = 51$.

Locate the quartiles on a number line. Draw a rectangle with the ends of the rectangle at the first and third quartiles. Mark Q_2 in the interquartile range, and draw the whiskers to the lowest and highest data values.

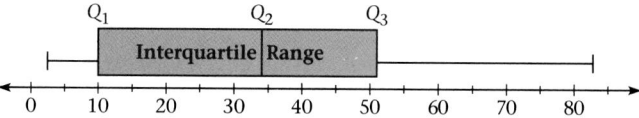

The interquartile range is the data spread from Q_1 to Q_3.

$$\text{Interquartile Range} = Q_3 - Q_1$$
$$= 51 - 10 = 41 \ \diamond$$

Try This Construct a box-and-whisker plot for the data below, and find the interquartile range.

82, 65, 11, 31, 50, 95, 33, 88, 79, 10, 15, 45, 51, 66, 53

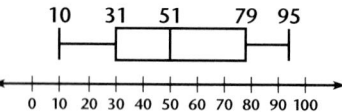
RETEACHING the lesson

Using Algorithms Help the students make a list of the steps used to make a quartile chart, using the data in Alternate Example 1 as a model. Then include steps to plot the quartiles on a number line and draw the box-and-whisker plot above it. After reviewing the students' lists, have them use their own instructions to do Exercises 29–31 on page 735.

EXAMPLE 3

Technology
Graphics
Calculator

In 1995, the earthquake in Kobe, Japan, spurred researchers to compare the intensity of earthquakes in Japan with the intensity of earthquakes all over the world. The intensity of an earthquake is measured using the Richter scale, an exponential function. The following tables contain recent earthquake data for Japan and the world. Make a box-and-whisker plot for each set of data.

Earthquake Intensities for Japan (1970–1995)

6.8	7.7	7.5	6.9	7.0
8.1	7.8	7.8	7.2	7.4

Earthquake Intensities for the World (1993–1995)

7.2	6.8	8.2	6.8	6.8	7.0	6.5
7.2	7.0	7.3	6.9	7.1	6.4	7.0
6.6	6.8	7.2	6.5	7.7	6.5
5.8	7.2	7.1	8.0	6.7	7.1

Solution▶

Using the statistics menu, find the quartile divisions. Recall that the second quartile, Q_2, is the median for the entire data set.

Data for Japan

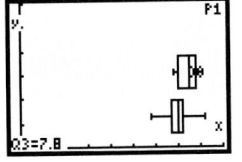

Data for the World

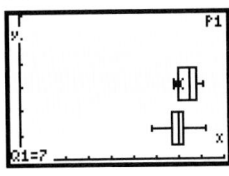

Using the statistics graphing features, graph the box-and-whisker plots for both sets of data, and find their interquartile ranges.

Data for Japan

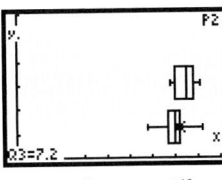

Data for the World

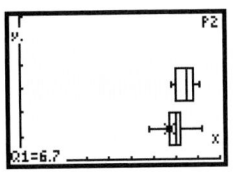

Interquartile range for Japan:
$Q_3 - Q_1 = 7.8 - 7 = 0.8$

Interquartile range for the World:
$Q_3 - Q_1 = 7.2 - 6.7 = 0.5$

CRITICAL Thinking

How do you think the intensities of Japan's earthquakes compare with the intensities of other earthquakes around the world? Explain.

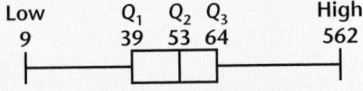

Assignment Guide

Core 1–11, 13–15, 17–19, 23–28, 32–56

Core Plus 1–6, 9–12, 14–16, 20–25, 29–57

Technology

Encourage students to use the **SORT** feature on their graphics calculators to help them solve these exercises. Also, determine how to graph quartiles on each calculator.

Error Analysis

Students will want to simplify the concepts of this lesson. Some may confuse Q_1 with 25%, Q_2 with 50% and Q_3 with 75%. Make sure students understand the definitions of quartile boundaries.

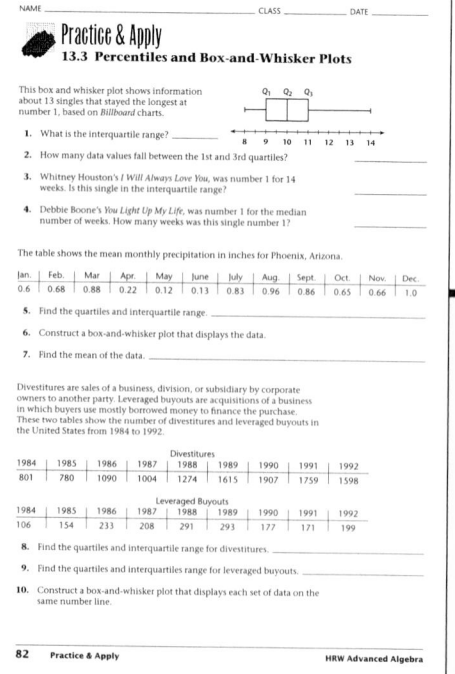

EXERCISES & PROBLEMS

Communicate

1. What percentile contains all but 10% of the data values? Explain.

2. What percentiles correspond to the 1st, 2nd, and 3rd quartiles?

3. Explain how the quartiles of a data set are determined.

4. What is the interquartile range, and what does it measure? How much of the data set does the interquartile range include?

5. Explain what information a box-and-whisker plot displays.

6. Describe how to construct a box-and-whisker plot.

Practice & Apply

Given 1650 data values, find the number of data values equal to or less than the given percentile.

7. 23rd percentile 379   **8.** 55th percentile 907   **9.** 3rd percentile 49

10. 93rd percentile 1534 **11.** 68th percentile 1122 **12.** 95th percentile 1567

Given 930 data values, find the number of data values *between* the given percentiles.

13. 12th percentile and 25th percentile 120

14. 70th percentile and 95th percentile 232

15. 30th percentile and 60th percentile 279

16. 90th percentile and 99th percentile 83

The number of students advised by a teacher over a 20-day registration period is shown.

| 7 | 12 | 10 | 8 | 7 | 12 | 15 | 10 | 10 | 7 | 2 | 9 | 12 | 14 | 6 | 10 | 12 | 9 | 8 | 8 |

17. Find the low quartile. 7.5

18. Find the high quartile. 12

19. What is the interquartile range? 4.5

For Exercises 20–22, use the following SAT scores.

420	370	396	385	376	396	422	359	412	517
387	345	392	458	527	534	461	508	397	419
428	402	463	529	399	488	419	437	384	515

20. Find the low quartile. 392

21. Find the high quartile. 463

22. What is the interquartile range? 71

Ecology The lengths of 24 American burying beetles are given. Use the data for Exercises 23–25.

Lengths of American Burying Beetles (cm)

2.5	2.8	3.1	3.6	3.4	3.8	3.0	2.8
3.5	3.3	2.6	3.0	2.9	2.7	3.4	3.2
3.7	2.5	3.1	2.9	2.5	3.1	3.8	2.9

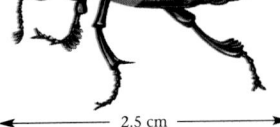

The American burying beetle is endangered. There are fewer than 1000 individual burying beetles in the United States.

23. Find the quartiles. $Q_1 = 2.8$, $Q_2 = 3.05$, $Q_3 = 3.4$

24. Find the interquartile range. 0.6

25. Construct a box-and-whisker plot to display the data.

For Exercises 26–28, use the following scores on a 10-point quiz.

Quiz Scores

6	8	6	4	8	9	10	8	9	8	9
8	10	8	9	8	5	9	9	6	8	8
5	8	9	8	9	6	4	8	10	8	9
8	4	6	9	7	9	10	5	8	9	7
6	10	9	6	10	8	9	9	7	5	8

26. Find the quartiles. $Q_1 = 6$, $Q_2 = 8$, $Q_3 = 9$

27. Find the interquartile range. 3

28. Construct a box-and-whisker plot to display the data.

For Exercises 29–31, use the following high temperatures for various cities on December 15.

High Temperatures

91	89	58	47	68	97	44
55	80	77	37	48	43	70
89	78	12	30	35	79	46
45	53	72	41	20	34	37
57	66	54	24	91	13	52

29. Find the quartiles. $Q_1 = 37$, $Q_2 = 53$, $Q_3 = 77$

30. Find the interquartile range. 40

31. Construct a box-and-whisker plot to display the data.

25.

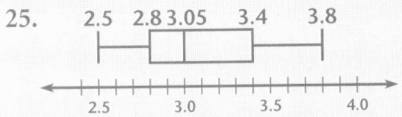

28.

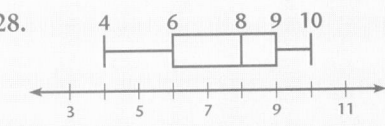

31.

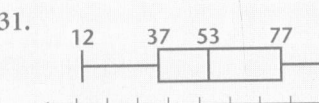

Beetle length labels: 2.5 cm, 2.8 cm, 3.1 cm, 3.6 cm, 3.4 cm

Portfolio Assessment

Have each student select a data set of at least 25 items from any source. Students should prepare a complete report on the data they chose that includes a scatter plot, such as the one on page 730, a quartile plot, such as the one in Example 1 on page 731, and a box-and-whisker plot, such as the one on page 732. Each student should prepare a written one-page report of their results. Have interested students discuss with the class the relative merits of each presentation.

alternative ASSESSMENT

Technology Master

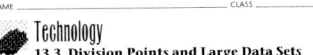

NAME _____ CLASS _____ DATE _____

Technology
13.3 Division Points and Large Data Sets

The median of a set of numerical data is a way to divide the data into two equal groups. When you compute the lower and upper quartiles along with the median, you divide the data into four equal groups. If the data set is large enough, such as this data set, you can subdivide the data set into even more equal parts.

32	86.08	67.52	62.08	40.64	54.72	90.88	43.84	69.76
83.52	33.28	45.44	67.84	59.2	49.6	67.2	37.12	68.16
64.64	32.64	39.68	87.36	53.44	60.16	68.8	51.84	47.68
80.96	58.88	76.48	82.24	91.84	73.6	69.44	35.52	52.8
73.28	38.08	45.44	67.84	93.44	40	92.8	52.8	64.64
45.44	67.84	45.12	33.6	53.12	84.16	72.96	51.52	88.96
90.56	87.68	66.88	80.64	77.44	45.12	35.52	60.48	38.72
87.36	43.84	37.44	74.56	64	48.32	37.12	56.64	82.24
79.68	51.52	80.96	50.24	65.28	56	33.28	73.92	51.52
67.84	32.96	94.08	56.32	88	46.72	54.08	68.48	83.2
86.08	64.64	86.4	58.88	59.84	62.4	52.48	74.24	61.44
44.16	47.36	35.52	45.12	66.88	82.56	73.6	63.04	73.28
37.76	88.96	49.92	53.76	83.52	63.68	67.2	86.72	82.24
67.2	32.32	63.04	80.32	33.92	35.52	65.6	79.04	46.4
72.64	73.92	57.28	91.52	82.24	83.2	87.04	43.2	47.36
80.32	59.52	92.16	91.84	96	38.72	60.48	54.4	75.52
65.6	51.2	59.52	56.32	39.68	88.64	80.64	69.12	69.44
42.24	62.4	90.88	34.56	58.56	49.6	66.88	36.48	81.92
74.56	34.24	90.88	55.68	33.28	63.68	86.4	37.76	69.12
93.12	65.92	72.64	63.68	72	66.56	88.64	62.08	61.12

Use a spreadsheet for Exercises 1–6.

1. Enter the data into a spreadsheet column. SORT it from least to greatest.

2. Find the median.   **3.** Find the lower quartile.   **4.** Find the upper quartile.

5. Find the division points of the data set when it is divided into 8 equal parts.

6. Find the division points of the data set when it is divided into 16 equal parts.

94   Technology   HRW Advanced Algebra

34.

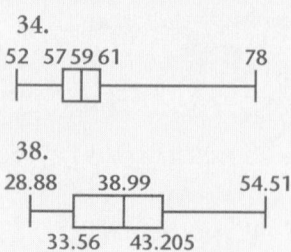

52 57 59 61   78

38.

28.88   38.99   54.51

    33.56 43.205

For Exercises 32–35, use the following radar-measured speeds of cars in a 60-mile-per-hour zone.

Highway Speeds

56	59	60	57	64	78	61	55
58	63	59	62	57	59	58	58
58	58	59	57	62	62	55	62
60	55	59	57	60	59	61	52

32. Find the quartiles. $Q_1 = 57$, $Q_2 = 59$, $Q_3 = 61$

33. Find the interquartile range. 4

34. Construct a box-and-whisker plot to display the data.

35. What percentage of the data lies at or below the 3rd quartile? 78.125%

For Exercises 36–38, use the following data set.

38.18	43.22	46.22	34.36	43.19	34.87	48.44	32.57	54.51	44.03
40.54	35.20	47.37	41.55	37.54	30.17	38.52	39.46	30.55	42.62
40.13	33.61	31.51	45.57	34.09	29.52	46.70	40.08	43.38	39.73
42.32	37.05	31.80	28.88	33.25	48.55	42.81	35.62	33.51	29.21

36. Find the quartiles. $Q_1 = 33.56$, $Q_2 = 38.99$, $Q_3 = 43.205$

37. Find the interquartile range. 9.645

38. Construct a box-and-whisker plot to display the data.

For Exercises 39–41, use the following data set.

7.8	7.9	7.6	8.0	7.8	8.0	7.9	7.6	8.0	8.0	7.5
7.9	7.8	7.9	7.8	8.0	8.2	7.9	7.6	8.0	7.6	8.1
8.1	7.9	7.5	8.0	7.8	7.8	8.0	8.2	7.9	7.6	8.2
7.8	8.0	8.3	8.0	8.1	7.8	7.6	7.9	8.0	8.3	7.6
8.2	7.8	7.9	8.1	8.0	8.0	8.1	7.8	8.1	8.1	7.6
7.8	7.9	8.1	7.9	8.0	8.2	8.0	8.1	8.0	7.9	8.0
7.5	7.9	8.1	8.1	7.8	8.1	8.2	8.1	7.8	7.8	7.9
8.1	7.8	8.2	8.2	7.9	7.9	7.6	8.2	7.6	8.3	7.9

39. Find the quartiles.

40. Find the interquartile range. 0.3

41. Construct a box-and-whisker plot to display the data.

Ecology The wingspans of 16 California condors are given. Use the data for Exercises 42–44.

42. Find the quartiles.

43. Find the interquartile range. 0.3

44. Construct a box-and-whisker plot to display the data.

Captive breeding programs were established to increase the number of California condors in the United States, and remove them from the endangered species list.

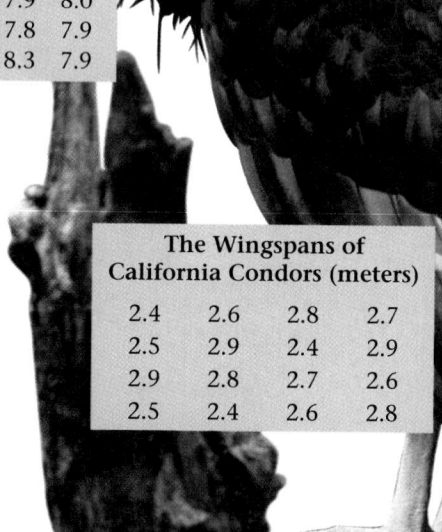

California condors are the largest flying land bird in North America.

The Wingspans of California Condors (meters)			
2.4	2.6	2.8	2.7
2.5	2.9	2.4	2.9
2.9	2.8	2.7	2.6
2.5	2.4	2.6	2.8

39. $Q_1 = 7.8$, $Q_2 = 7.9$, $Q_3 = 8.1$

41.

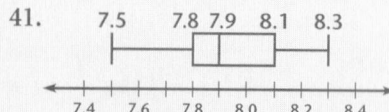

42. $Q_1 = 2.5$, $Q_2 = 2.65$, $Q_3 = 2.8$

44.

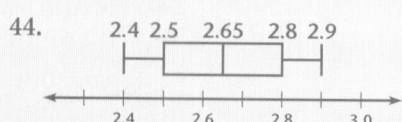

Ecology The lengths of 24 Houston toads are given.

Lengths of Houston Toads (cm)

5.4	5.9	6.8	6.2	6.5	5.8
6.0	5.8	6.3	6.4	5.3	5.5
6.2	5.7	6.0	6.7	6.4	6.2
5.7	5.4	5.6	5.9	6.1	5.6

45. Find the quartiles.

46. Find the interquartile range. 0.60

47. Construct a box-and-whisker plot to display the data.

The population of Houston toads is endangered. There are fewer than 10,000 Houston toads living in the wild today.

Look Back

48. Describe the behavior of the polynomial function $f(x) = (x + r)(x - r)^2$ as values of x approach r.   **[Lesson 6.4]**

For each exponential function, give the y-intercept and indicate whether the function represents exponential growth or decay. [Lesson 7.2]

49. $f(x) = 0.7^x$
(0, 1); decay

50. $f(x) = 0.7^{-x}$
(0, 1); growth

51. $f(x) = 7^x$
(0, 1); growth

52. $f(x) = 7^{-x}$
(0, 1); decay

Find the constant of variation, and write the equation of variation.   [Lesson 9.1]

53. y varies inversely as x, and y is 20 when x is 3. $k = 60$; $y = \frac{60}{x}$

54. y varies inversely as the square of x, and y is 12 when x is 3.

55. y varies inversely as the cube of x, and y is 2 when x is 3.

56. Solve the system of nonlinear equations.   **[Lesson 10.5]**
no real solution
$$\begin{cases} 5x^2 + 4y^2 = 12 \\ 2x + 2y^2 = -15 \end{cases}$$

Look Beyond

57. Make a scatter plot of the movie-attendance data, and find the mean, median, and mode. Determine the overall spread of the data values. What can you determine about the overall trend in this data? What further information would assist you in making a more complete analysis?

Percentage of the U.S. Population That Went to the Movies Every Week

Year	Percentage
1945	56%
1950	40%
1955	23%
1960	12%
1965	9%
1970	7%
1975	10%
1980	10%
1985	9%
1990	9%
1995	10%

45. $Q_1 = 5.65$, $Q_2 = 5.95$, $Q_3 = 6.25$

47.

54. $k = 108$; $y = \frac{108}{x^2}$ or $x^2y = 108$

55. $k = 54$; $y = \frac{54}{x^3}$ or $x^3y = 54$

The answer to Exercise 57 can be found in Additional Answers beginning on page 842.

48. $x = r$ is a turning point for the function; as x approaches r, the graph of f approaches the x-axis; at $x = r$, the graph of f touches the x-axis.

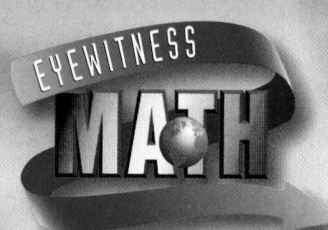

What's So Fuzzy?

Time for Some Fuzzy Thinking

BY PHILIP ELMER-DEWITT

Fuzzy-logic controllers create a smoother ride for passengers and use less energy than human conductors or automated systems that are based on traditional logic.

In the pages of *Books in Print*, listed among works like *Fuzzy Bear* and *Fuzzy Wuzzy Puppy*, are some strange sounding titles: *Fuzzy Systems, Fuzzy Set Theory* and *Fuzzy Reasoning & Its Applications*. The bedtime reading of scientists gone soft in the head? No, these academic tomes are the collected output of 25 years of mostly American research in fuzzy logic, a branch of mathematics designed to help computers simulate the various kinds of vagueness and uncertainty found in everyday life. Despite a distinguished corps of devoted followers, however, fuzzy logic has been largely relegated to the back shelves of computer science—at least in the U.S.

But not, it turns out, in Japan. Suddenly the term *fuzzy* and products based on principles of fuzzy logic seem to be everywhere in Japan: in television documentaries, in corporate magazine ads and in novel electronic gadgets ranging from computer-controlled air conditioners to golf-swing analyzers.

What is fuzzy logic? The original concept, developed in the mid-'60s by Lofti Zadeh, a Russian-born professor of computer science at the University of California, Berkeley, is that things in the real world do not fall into neat, crisp categories defined by traditional set theory like the set of even numbers or the set of left-handed baseball players.

But this on-or-off, black-or-white, 0-or-1 approach falls apart when applied to many everyday classifications, like the set of beautiful women, or the set of tall men or the set of very cold days.

This mathematics turns out to be surprisingly useful for controlling robots, machine tools and various electronic systems. A conventional air conditioner, for example, recognizes only two basic states: too hot or too cold. When geared for thermostat control, the cooling system either operates at full blast or shuts off completely. A fuzzy air conditioner, by contrast, would recognize that some room temperatures are closer to the human comfort zone than others. Its cooling system would begin to slow down gradually as the room temperature approached the desired setting. Result: a more comfortable room and a smaller electric bill.

Fuzzy logic began to find applications in industry in the early '70s, when it was teamed with another form of advanced computer science called the expert system. A product of research into artificial intelligence, expert systems solve complex problems somewhat like humans experts do—by applying rules of thumb. (Example: when the oven gets very hot, turn the gas down a bit.)

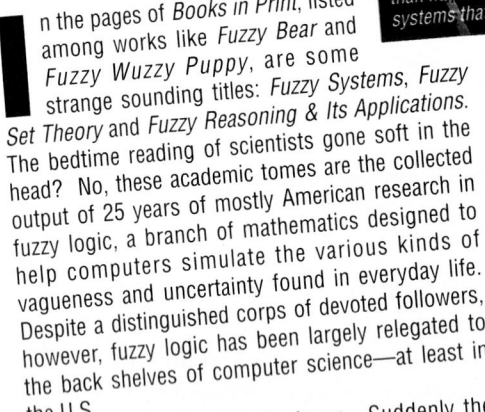

Answers may vary for A1 and A3, as well as for B1 and C.

A2. The middle heights 5'6", 5'7", and 5'8" are hardest to place because students can easily fit in either category.

A4. There is no precise height above which someone is considered tall and below which someone is considered not tall. Naturally, people's opinions will differ more dramatically when the clarity of the boundaries between sets becomes more obscure.

What is fuzzy logic? How can it make things work more efficiently? Does fuzzy logic resemble the way we think more than traditional logic does?

To answer questions like these, you need to explore the basis of fuzzy logic — something called *fuzzy sets*.

You may be familiar with how sets work in traditional logic. If someone belongs to the set of *people who are tall*, then they cannot belong to the set of *people who are not tall*. With fuzzy sets, the distinction is not so sharp, as you will see.

"It is dark in here."

"He lives in a big house."

"He drives so slowly."

"The meal was good."

Cooperative Learning

For activities A and B, use the data in Table 1 or use the heights of 12 students from your class.

A. Before looking at fuzzy sets, see how you might use sets in traditional logic to categorize a group of students by height.

1. Complete Table 2. In the left column, list the heights of the students you think are tall. List the rest in the right column.

2. Which heights were hardest to place? Why?

3. At what height did you draw the line between *tall* and *not tall*? Compare your response with those of your classmates. Describe the disagreements.

4. Why would you expect there to be disagreements?

B. Now try using fuzzy sets.

1. For each name on the list, assign a value from 0 to 1 to indicate the extent to which that person belongs to the set *tall*.

   For example:

0	for someone who definitely does not belong
0.5	for someone who *may be* tall
0.8	for someone who is *fairly* tall
1	for someone who definitely belongs on the *tall* list

2. On a graph, plot the 12 values you assigned to the names on the list. Draw a smooth curve through the points.

3. How does the graph show what you think the word *tall* means?

C. What sort of difficulties can you run into when you apply traditional logic to real situations? How can fuzzy sets help resolve those difficulties?

Table 1	
Student	Height
A	5' 11"
B	5' 3"
C	6' 1"
D	4' 11"
E	5' 4"
F	5' 10"
G	5' 1"
H	5' 8"
I	5' 5"
J	5' 7"
K	5' 6"
L	5' 9"

Table 2	
Tall	Not Tall

"We live close to each other."

"That is a pretty painting."

"It is warm outside."

Cooperative Learning

Have students conduct a survey to see what schoolmates mean by terms like *often*, *sometimes*, and *rarely*. Make the context specific. For example, ask: *If someone watches TV often, how many hours of TV do you think they watch per day?* Students can then explore fuzzy subsets. For example, have them describe how the graph of the fuzzy set for *very tall* would compare with the graph for *tall*.

DISCUSS

Do not try to resolve disagreements, rather encourage students to use their disagreements to refine their definition of the degree of the characteristic. This is the nature of fuzzy sets and fuzzy logic. The lesson will also work better if there are substantial disagreements. Disagreements will occur most commonly in the categorization of middle ranges where difficulty in categorization increases.

B3. The graph shows that students with heights between about 5'1" and 5'10" can be considered tall or not tall to different extents. The taller you are in that range, the more you belong to the set *tall*.

B2. Student	Height	Value
A	5'11"	1.0
B	5'3"	0.05
C	6'1"	1.0
D	4'11"	0.0
E	5'4"	0.1
F	5'10"	0.95
G	5'1"	0.0
H	5'8"	0.8
I	5'5"	0.3
J	5'7"	0.65
K	5'6"	0.5
L	5'9"	0.9

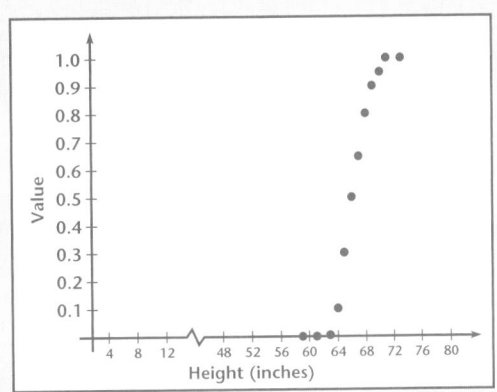

PREPARE

Objectives

- Find the range of a given set of data.
- Determine the mean deviation and the standard deviation for a given set of data.

Assessing Prior Knowledge

1. What is the range of $f(x) = \sqrt{x-1}$? $[y \geq 0]$

2. What is the range of $f(x) = \frac{1}{x-1}$? $[R, \text{ where } y \neq 0]$

TEACH

This section will use a single value, such as the mean deviation, variance, or the standard deviation, to measure how data are distributed over the range.

A ongoing ASSESSMENT

The range of Mary Ann's score is very small compared with the range of Kenneth's scores. Therefore, Mary Ann's scores are more consistent.

LESSON 13.4 Measures of Dispersion

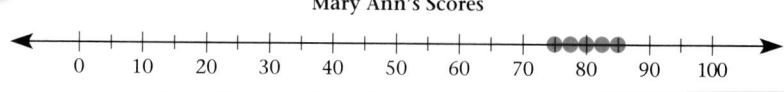

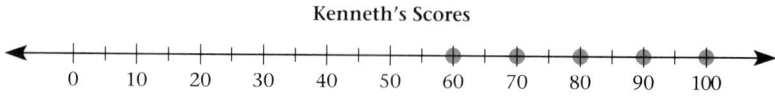

Why *Quartiles and the interquartile range describe how a collection of data is distributed, or dispersed. The dispersion of data can also be described with single-number measures.*

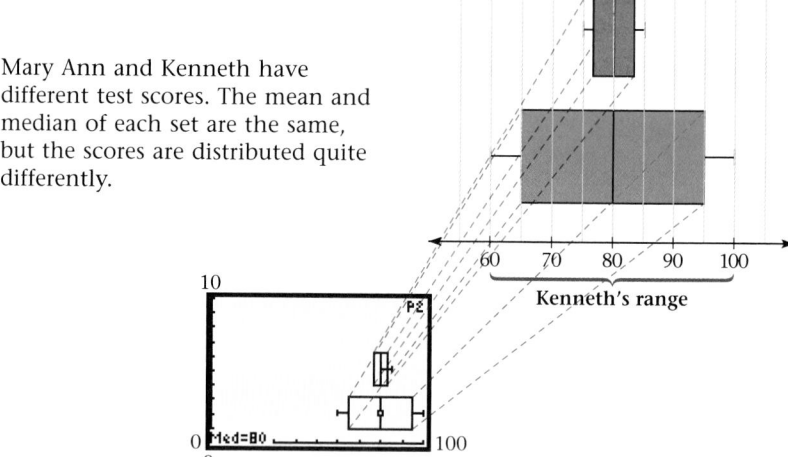

Mary Ann and Kenneth have different test scores. The mean and median of each set are the same, but the scores are distributed quite differently.

The *range* is defined mathematically in terms of absolute value.

RANGE

The **range** of a set of data is the absolute value of the difference between the largest value and the smallest value.

Range = |largest value − smallest value|

The range of Mary Ann's scores: $|85 - 75| = 10$
The range of Kenneth's scores: $|100 - 60| = 40$

Explain how the range can be used to compare Mary Ann's scores with Kenneth's scores?

The range is one *measure of dispersion*. A **measure of dispersion** is a numerical value that indicates the extent to which data values are dispersed, or spread around, a center number.

Using Discussion Begin with a realistic, but hypothetical, set of scores on a 100-point test. Show the range and quartiles. Demonstrate how the mean deviation is calculated and illustrate what the distribution of grades would be if the mean were a C and the other letter grades were defined in terms of intervals with width equal to the mean deviation. Then add the squares of the individual deviations from the mean, divide the total by *n*, and take the square root. Use the resulting standard deviation to determine the letter grades. Ask the students to compare and contrast these two possible grading systems with the more traditional decimal method.

Mean Deviation

The range gives you information about the spread of values in a set of data, but it does not give you any idea about how the data set is distributed or the central tendency of its values.

The mean gives you information about the central tendency of the values in a set of data, but it does not give you information about the spread of values.

The *mean deviation* combines the two measures, range and mean, to describe an average measure of how the values differ, or deviate, from the mean.

To find the mean deviation:

1. Find the absolute value of the difference between each value, x_n, and the mean, $\bar{x}$: $|x_n - \bar{x}|$.
2. Find the sum of these absolute values.
3. Divide the sum by the number of values in the data set, n.

MEAN DEVIATION

$$\frac{\sum |x_n - \bar{x}|}{n}$$

EXAMPLE 1

Find the mean deviation for Mary Ann's scores.

75 78 80 82 85

Solution

First, find the mean score. $\bar{x} = \frac{75 + 78 + 80 + 82 + 85}{5} = 80$

Then find the absolute value of the sum of differences from the mean. It is usually easier to organize your information in a table.

n	x_n	$\bar{x}$	$\lvert x_n - \bar{x} \rvert$
1	75	80	5
2	78	80	2
3	80	80	0
4	82	80	2
5	85	80	5
		Total	14

$$\sum |x_n - \bar{x}| = 14$$

Divide the sum by 5, the value of n. The mean deviation is

$$\frac{\sum |x_n - \bar{x}|}{n} = \frac{14}{5} = 2.8.$$

Thus, each score differs from the arithmetic mean by an average (mean) of 2.8 points. ❖

Try This Find the mean deviation for Kenneth's scores. 60 70 80 90 100

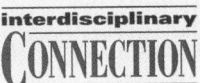

Cooperative Learning

Divide the class into groups of two. Request that one student calculate the mean deviation and the other calculate the standard deviation for one of the data sets in the exercises from Lesson 13.3. When each group has finished, have them report their opinions concerning which method produced a more realistic measure of dispersion.

Alternate Example 1

Find the mean deviation for the data set 12, 19, 17, 13, 20, 37. **[5.89]**

ongoing ASSESSMENT

Try This

12

interdisciplinary CONNECTION

Quality Control Quality control is the science of using mathematics to measure and control the quality of a product. This field utilizes the measures of this lesson to monitor a process and to tell whether or not it is in control or needs to be improved. Have the class choose an appropriate company in the local area (possibly from a list you have compiled). Have the class select a committee of four to contact the company president and the quality control representative. If possible, have the student committee setup a presentation for the class by either the president or the quality control representative. If this is not possible, have the student team interview these two people and report to the class what they learned about real business and its use of mathematics, especially the mathematics of this lesson.

The *standard deviation*, usually denoted by the lowercase Greek letter sigma, σ, is the measure of dispersion most often used by statisticians.

By taking the absolute value of each deviation from the mean, you avoid a negative value for any deviation. For the same reason, calculating the *standard deviation* involves *squaring* each deviation from the mean.

STANDARD DEVIATION

$$\sigma = \sqrt{\frac{\sum (x_n - \bar{x})^2}{n}}$$

CRITICAL
Thinking

What is the difference in the information provided by the standard deviation and the information provided by the mean deviation?

EXAMPLE 2

Find the standard deviation for Mary Ann's test scores: 75, 78, 80, 82, 85.

Solution

Make a table of values.

n	x_n	$\bar{x}$	$x_n - \bar{x}$	$(x_n - \bar{x})^2$
1	75	80	−5	25
2	78	80	−2	4
3	80	80	0	0
4	82	80	2	4
5	85	80	5	25
			Total	58

$$\frac{\sum (x_n - \bar{x})^2}{n} = \frac{58}{5} = 11.6 \text{ and } \sigma = \sqrt{\frac{\sum (x_n - \bar{x})^2}{n}} = \sqrt{11.6} \approx 3.4$$

The standard deviation for Mary Ann's scores is approximately 3.4. ❖

Try This Find the standard deviation for Kenneth's test scores: 60, 70, 80, 90, 100.

The standard deviation is the square root of $\frac{\sum (x_n - \bar{x})^2}{n}$, or the square root of the *variance*.

VARIANCE

$$\sigma^2 = \frac{\sum (x_n - \bar{x})^2}{n}$$

Exercises & Problems

Communicate

1. Explain what a measure of dispersion is.

2. Describe how the range is a measure of dispersion.

3. Explain how to find the mean deviation and the standard deviation for a set of data.

4. Compare and contrast the standard deviation and the mean deviation. Which do you think gives a better measure of dispersion?

5. Give an example of two sets of data that have the same mean and median, but have different measures of dispersion. Explain which set of data is more predictable.

6. For the set of data given below, which measure gives the best information about the distribution: the mean, median, mean deviation, or standard deviation? Explain.

   100 93 81 75 40 66 61 82 71 88 29

Practice & Apply

7. Find the range of the following data values. 42

   13 46 34 16 15 21 55 32 26 23 20
   25 21 23 24 27 26 21 22 25 22 26

8. Find the range of the following data values. 24

   −84 −104 −97 −101 −108 −99

Sports The winning times given (in minutes : seconds . hundredths of a second) for the men's 5000-meter speed-skating competition for several Olympics are shown.

1976	1980	1984	1988	1992	1994
7:24.48	7:02.29	7:12.28	6:44.63	6:59.97	6:34.96

9. Find the mean, median, and mode for the winning times. 6:59.77; 7:01.13; none

10. Find the range, mean deviation, and standard deviation for the winning times.

11. Which measure(s) in Exercises 9 and 10 give(s) the best indication of the trend in men's 5000-meter speed-skating times? Explain.

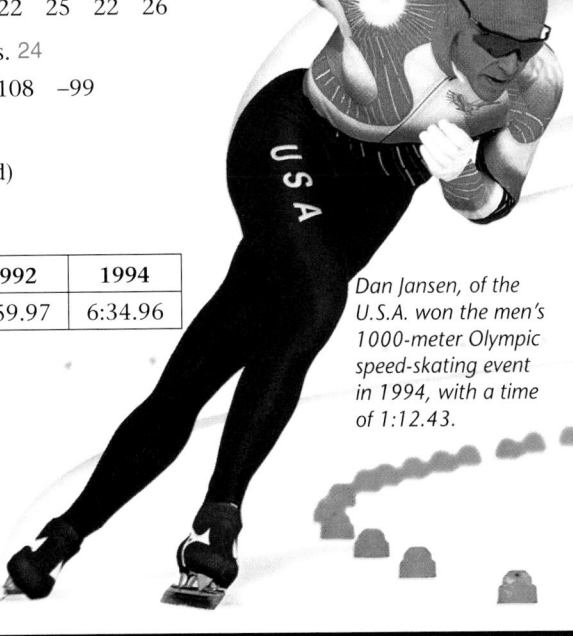

Dan Jansen, of the U.S.A. won the men's 1000-meter Olympic speed-skating event in 1994, with a time of 1:12.43.

RETEACHING the lesson

Hands-On Strategies Ask students to create a data set that represents a sequence of 20 measurements of the same thing, with random errors. For example, assume the data set contains measurements of the length of boards represented as being 48 inches long . Have students toss a single number cube to determine the size of the error of the first measurement. Then have them toss a coin to indicate whether the measurement is too long or too short (heads indicates too long; tails, too short). For example, a roll of 3 on the number cube and a coin flip of tails indicates a board that actually measures 45 inches long. Have students repeat this process for 19 other boards. Have students use the ideas of this lesson to investigate and report on the use of statistics in dealing with modeling errors in real-world measurements.

alternative ASSESSMENT

Portfolio Assessment

A statistics book once stated, "Statistics is *not* mathematics." Have the students research these two topics, mathematics and statistics, and report on why this statement might have been written.

14. Answers may vary and will depend on the year.

16. A deviation from the mean by 2.47 grams of fat per serving

17.

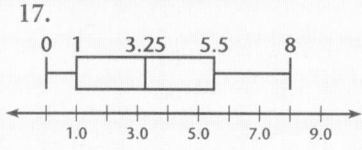

0 1   3.25   5.5   8

1.0   3.0   5.0   7.0   9.0

Sociology The percentage of women who work seems to grow every year. The approximate percentages of women in the work force in the years 1988–1993 are given.

12. Find the range, mean, and median for the data. 3; 55.67; 56

13. Find the standard deviation for the data. 0.94

14. Based on trends in the data, predict the percentage of women in the work force this year. How does this amount compare with the actual percentage? Explain.

1988	1989	1990	1991	1992	1993
54	56	56	55	56	57

Nutrition The fat content, in grams per serving, of 20 different popular types of crackers is given.

15. Find the mean, the range, and the standard deviation for the amount of fat in popular crackers. 3.27; 8; 2.47

16. Explain the meaning of the standard deviation in this case.

17. Make a box-and-whisker plot for this data.

18. What does the interquartile range tell you about the fat content of popular crackers?

19. Which piece of information, the standard deviation or the interquartile range, gives the most information about the data? Explain.

Fat Content of Crackers

8	6	0	6	1
1.5	0	4	0	3
4.5	2	5	3	1
0	7	3.5	6	4

Sociology The table lists average life expectancies for several countries.

20. Find the mean, median, range, and standard deviation for the data. 72.6; 73.6; 18.3; 5.6

21. Make a box-and-whisker plot of this data.

22. What does the standard deviation tell you about life expectancies? Explain.

23. What does the interquartile range tell you about life expectancies? Explain.

Life Expectancies

Japan	78.7
Israel	76.5
U.S.	75.9
Argentina	71.3
Mexico	70.3
India	60.4
Russia	70.0
Canada	77.4

18. The interquartile range indicates that half of the popular brands of crackers have a fat content between 1.0 and 5.5 grams per serving.

19. Answers may vary. In this case, the interquartile range seems to be a better indicator. The standard deviation gives an idea of the average distance from the points to the mean. However, it can be affected by extreme values or outliers.

21.

60.4   73.6   78.7

70.15   76.95

60 62 64 66 68 70 72 74 76 78 80

22. The standard deviation shows how the life expectancies are distributed about the mean. On the average the life expectancy of one particular country differs by 5.6 years from the mean life expectancy of all the countries, 72.6 years.

Sports The winning times (minutes : seconds . hundredths of a second) for the women's 100-meter breast-stroke for several recent Olympics are given below.

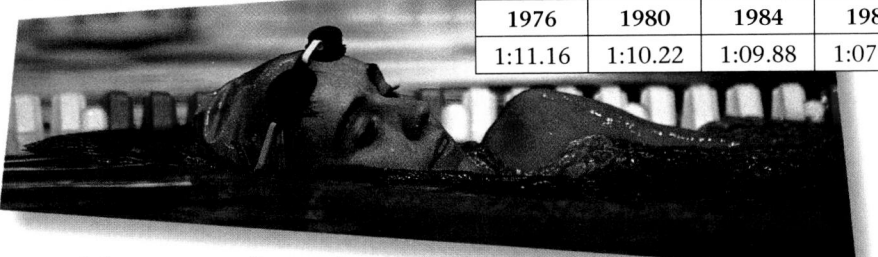

1976	1980	1984	1988	1992
1:11.16	1:10.22	1:09.88	1:07.95	1:08.00

24. Find the mean, median, and mode for these winning times. 1:09.44; 1:09.88; none
25. Find the range, mean deviation, and standard deviation for this data.
26. Which measure from Exercises 24 and 25 gives the best indication of the trend in winning times for 100-meter breaststroke? Explain.
27. Use the trends in these swimming times to predict the time of the 1996 women's breast-stroke winner. Compare this time with the actual time. Explain how the times compare and how you could have improved your prediction.

Look Back

Sports Trying to take a shortcut by going over a tree and a stream, a golfer hits a drive at an angle, θ, of 40° above the horizontal and with an initial velocity of 145 feet per second. The tree is 70 feet tall and stands 190 yards (570 feet) away from the golfer, directly in the path of the ball. The far bank of the 10-yard-wide stream is 205 yards (615 feet) away from the golfer. The parametric equations describing the position of the ball in feet, where t is in seconds, are $x(t) = 145t \cos 40°$ and $y(t) = 145t \sin 40° - 16t^2$. Graph the path of the drive on your graphics calculator. **[Lesson 3.6]**

28. Disregarding the tree, is the ball hit hard enough to clear the stream? How far does it travel (to the nearest yard) before landing? yes; 216 yd

29. Will the shot clear the tree? If so, by how much? If not, how much too low is it?

30. **Maximum/Minimum** By varying the angle, θ, in the parametric equations, try to find the minimum angle needed with an initial velocity of 145 feet per second to clear both the tree and the stream.

Look Beyond

31. A mule and a donkey were carrying bundles of wheat to a market. The mule said, "If you give me one bundle of wheat, then I will have twice as many bundles as you do. However, if I give you one bundle of wheat, then we will each be carrying the same number." How many bundles of wheat was each carrying? mule: 7; donkey: 5

23. The interquartile range indicates that half of the countries have a life expectancy ranging from 70.2 to 76.9 years.

25. range: 0:03.21; mean deviation: 0:01.17; standard deviation: 0:01.27

26. Since all of the times are very close, the range may be the best indicator of the trend since it shows the dispersion in times from the shortest to the longest.

27. 1:06.87; Answers for how the 1996 time affects the comparison may vary.

29. no; about 13 feet too low

30. 44°

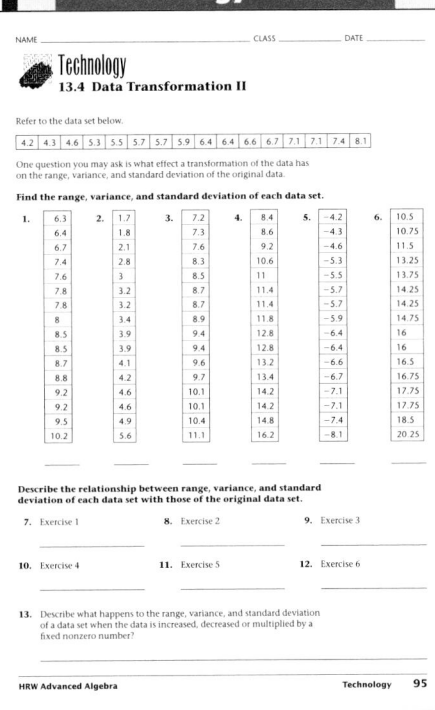

Exploring
Binomial Distributions

PREPARE

Objectives

• Find the probability of r successes in n trials of a binomial experiment.

RESOURCES

• Practice Master	13.5
• Enrichment Master	13.5
• Technology Master	13.5
• Lesson Activity Master	13.5
• Quiz	13.5
• Spanish Resources	13.5

Assessing Prior Knowledge

1. Find 5! [120]

2. Find $_9C_4$. [126]

TEACH

Weather forecasters give us the probability of rain tomorrow, yet it will either rain or it won't. The prefix *bi*, meaning "two" in this context, is used to help statisticians predict the probability of events that either occur or do not occur. In other words, there are only *two* possibilities.

Use Transparency ▶ 57

There are many events for which the outcome is either a success or a failure. For example, each response on a multiple choice question is either correct or incorrect. Statistics is often used to analyze situations for which there are only two possible outcomes.

Suppose you guess all three answers to a true-false test. What is the probability that all three guesses will be correct? The probability of a correct answer for each question is $p = \frac{1}{2}$, and the probability of an incorrect answer for each question is $q = \frac{1}{2}$.

By the Fundamental Principle of Counting, there are 2^3, or 8, distinct ways to answer the 3 questions.

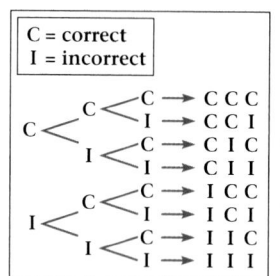

Getting all 3 questions correct can happen in only 1 of 8 ways, so $P(\text{exactly 3 correct}) = \frac{1}{8}$.

Likewise, $P(\text{exactly 2 correct}) = \frac{3}{8}$.

The sum of the outcome probabilities for any experiment must be 1. Thus, for any 2-outcome experiment, where p is the probability of getting an answer correct and q is the probability of getting an answer incorrect, $p + q = 1$.

Examine the expansion of the binomial in $(p + q)^3 = 1$. In this expansion, p and q represent the probabilities of the two possible choices. Since the answer to each question is a *guess*, the probability of being correct, p, and the probability of being incorrect, q, are both $\frac{1}{2}$. The exponent 3 represents the total number of test items.

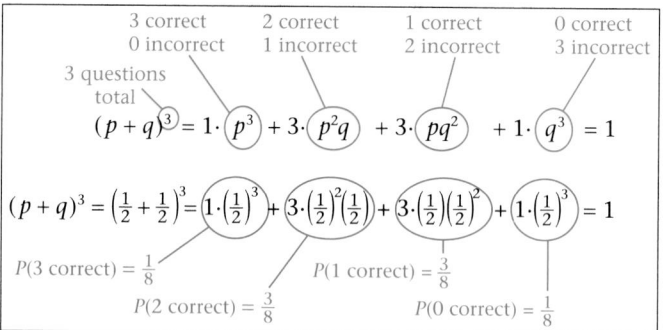

Using Discussion Discuss the following scenarios with students. In throwing dice, the outcome will be either even or odd. When you draw a card from a deck of cards, the card will either be red or black. A newborn child will either be female or male. The result on a multiple-choice question will either be a correct or incorrect. In any of these situations, one outcome is called

a *success* and the other is called a *failure*. If the probability of a success is p and the probability of a failure is q, show students that $p + q = 1$. Explain how the binomial expansion $(p + q)^n = 1$ is really just expanding an identity $(1 = 1)$ to find the details of the left side. Relate this expansion, term by term, back to Pascal's triangle for the coefficients, which are the number of times a given combination *might* happen and the $p^r q^{n-r}$ factors that the combination *will* happen.

Explain how the probability of each event is related to the coefficients in the expansion of $(p + q)^3$. How do these coefficients compare with a row of Pascal's Triangle, given in Lesson 12.5?

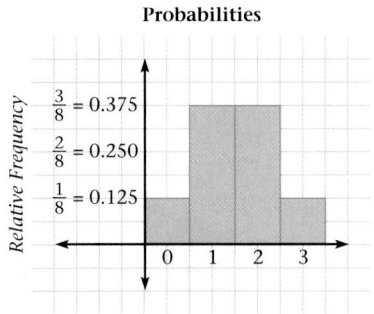

Number of Correct Responses

A list of probabilities, such as $\frac{1}{8}$, $\frac{3}{8}$, $\frac{3}{8}$, $\frac{1}{8}$ (the probabilities for guessing correct answers on the 3-question true-false test), is called **a probability distribution.**

An experiment with exactly 2 choices, such as correct or incorrect, is called a **binomial experiment**. It is also called a Bernoulli experiment, after Jakob Bernoulli (1655–1705), who studied probability theory in the late seventeenth century.

The probability of being correct on each true-false question is $\frac{1}{2}$, or 50%, because each answer is a *guess* in this binomial experiment.

Would studying for the true-false test affect the probability percentage of being correct? Explain.

·Exploration 1 *Multiple Choice*

Suppose a multiple-choice test has 3 questions with 4 choices each. Only 1 of the 4 choices is correct for each question, so the probability of *guessing* a correct choice is 25%, or $p = 0.25$, and the probability of guessing an incorrect answer is 75%, or $q = 0.75$.

 Use the binomial theorem to expand $(p + q)^3$ for a 3-question test (3 trials).

Each term in the expansion represents an outcome probability. For example, $P(3 \text{ correct})$ is represented by the first term in the expansion of $(p + q)^3$: $1 \cdot p^3 = 1 \cdot (0.25)^3 = 0.015625$. Use the binomial expansion to complete the table.

Number correct	Probabilities
3	$P(3 \text{ correct}) = 0.015625$
2	
1	
0	

 Find the probability of getting 2 successes in 3 trials if the probability of being correct is 80%. ❖

ENRICHMENT Ask students to explain the significance of the coefficients in the binomial expansion. Do the *a* and *b* represent the two outcomes? Could the "+" be the logical "or"? Does the exponent, which represents repeated multiplication, also represent repeated trials? Encourage students to study these questions and experimentally test their conclusions.

CRITICAL Thinking

$(p + q)^3 = p^3 + 3p^2q + 3pq^2 + q^3$. The coefficient of each term denotes the number of outcomes in which a specific number of "successes" will occur out of a given number of trials. The coefficient is then multiplied by the probability that the particular event will occur. The exponent of $(p + q)^3$ denotes the total number of trials, also the maximum number of successes. The coefficients of the expansion are identical to the third line in Pascal's Triangle.

Aongoing ASSESSMENT

Yes, because the binomial experiment or test is based on guessing (where each response is equally likely to occur). If you study, then you will reduce the likelihood of needing to guess. Therefore, you will increase the probability of getting a correct response.

Exploration 1 Notes

In Exploration 1, students should learn that they can use the binomial theorem to find the probability that a certain number of successes will occur out of a specified number of trials.

Aongoing ASSESSMENT

2. The probability of 2 successes in three trials is 0.384, or ≈ 38%.

Lesson 13.5 **747**

Cooperative Learning

Divide the class into groups of three. One student should ask a question relating to a binomial probability outcome. The other two group members should answer the question, one using Pascal's triangle, the other using the equation $P(r) = {_nC_r}p^rq^{n-r}$. Have students rotate roles until each student understands how to perform all three tasks.

TEACHING *tip*

Point out the strong connection between the coefficient of a particular term in the binomial expansion and the probability that an event will or will not occur. Remind students that $_nC_r$ refers to the number of ways of getting r successes in n trials.

Exploration 2 Notes

This exploration is like the first one, only the probability of success has been changed in each event from 0.25 to 0.4, and there are 4 trials instead of 3. It shows that the method used in Exploration 1 can become difficult to work with and that the binomial probability distribution formula is more efficient.

ongoing ASSESSMENT

2. **0.1536, or ≈ 15%**

•Exploration 2 Outcomes With Four Trials

1 Use the model in Exploration 1 to write the binomial distribution for any number of successes in 4 trials when the probability of success is 40%.

2 What is the probability of getting 2 successes in 4 trials if the probability of being correct is 80%? ❖

In the explorations, you built models for the probability of 1, 2, 3, or 4 successes in three or four trials. A model can be built for more trials, but the computation becomes tiresome as the number of trials, n, becomes larger. Fortunately, there is a formula that gives the binomial probability distribution in one step.

BINOMIAL PROBABILITY DISTRIBUTION

The **binomial probability distribution** for r successes in n trials is

$$P(r) = {_nC_r}p^rq^{n-r},$$

where p is the probability of success, and $q = 1 - p$ is the probability of failure.

INCLUSION strategies

Using Cognitive Strategies Have one student select a sequence of four letters using a, b, c, and d. For example, *dacb* might be chosen. The student should tell no one what the sequence is. Have each member of the class choose a sequence of four of these letters, as though they were guessing the answers on a four-question multiple choice quiz where there are four choices to each question. The first student's sequence contains the "correct" answers. Have the other students determine the probability that all four answers from any particular student will be correct. Then poll the class to see whether the experiment confirmed the prediction. Discuss why it did or did not.

The probability of getting 2 successes in 3 trials

$$P(2) = {}_3C_2 \cdot p^2 \cdot q^{3-2}$$

if $p = 30\%$ and $q = 70\%$

$$P(2) = {}_3C_2 \cdot (0.3)^2 \cdot (0.7)^{3-2}$$

When you have a 30% chance of succeeding

CRITICAL *Thinking*

Refer to the formula for the binomial probability distribution. Explain why the formula for r successes is given by the term of the binomial expansion of $(p + q)^n$ in which the exponent of p is r.

APPLICATION

About 85% of students at a local high school favor school-sponsored extra curricular activities. If 20 randomly selected students are asked whether they support school-sponsored extra-curricular activities, what is the probability that 19 will say yes?

Since the question can only be answered yes or no, this is a binomial experiment.
The number of trials, n, is 20. The number of successes, r, is 19. The probability of a yes result for each trial, p, is 85%, or 0.85. The probability of a no result, q, is $1 - 0.85$, or 0.15.

The binomial probability distribution for 19 successes in 20 trials is

$$P(r) = {}_nC_r p^r q^{n-r}$$
$$P(19) = {}_{20}C_{19}(0.85)^{19}(0.15)^{20-19}$$
$$= {}_{20}C_{19}(0.85)^{19}(0.15)^1$$
$$\approx 0.1368$$

The probability that 19 out of 20 students will say yes is approximately 13.7%. In other words, 19 out of 20 will say yes about 13.7% of the time. ❖

Technology

Graphics calculators are needed for Exercises 44–46 and 50. Remind students that the binomial probability distribution on page 748 is easily checked using the combination feature on their calculator.

Error Analysis

Students must be careful to discriminate between the probability of a successful trial and the overall probability in a sequence of trials. These two ideas need to be carefully discussed and separated every time a problem is to be solved.

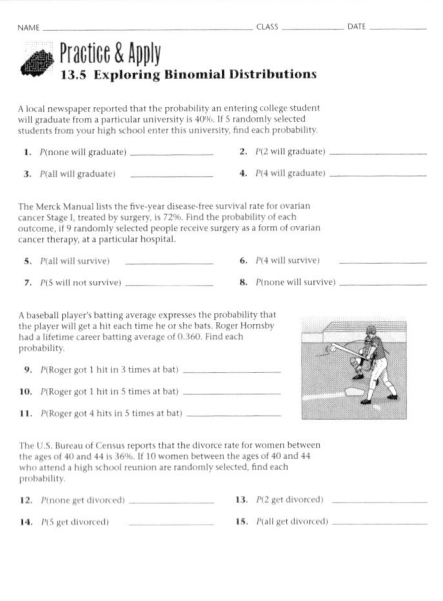

Practice Master

NAME _____ CLASS _____ DATE _____

Practice & Apply
13.5 Exploring Binomial Distributions

A local newspaper reported that the probability an entering college student will graduate from a particular university is 40%. If 5 randomly selected students from your high school enter this university, find each probability.

1. P(none will graduate) _____   2. P(2 will graduate) _____
3. P(all will graduate) _____   4. P(4 will graduate) _____

The Merck Manual lists the five-year disease-free survival rate for ovarian cancer Stage I, treated by surgery, is 72%. Find the probability of each outcome, if 9 randomly selected people receive surgery as a form of ovarian cancer therapy, at a particular hospital.

5. P(all will survive) _____   6. P(4 will survive) _____
7. P(5 will not survive) _____   8. P(none will survive) _____

A baseball player's batting average expresses the probability that the player will get a hit each time he or she bats. Roger Hornsby had a lifetime career batting average of 0.360. Find each probability.

9. P(Roger got 1 hit in 3 times at bat) _____
10. P(Roger got 1 hit in 5 times at bat) _____
11. P(Roger got 4 hits in 5 times at bat) _____

The U.S. Bureau of Census reports that the divorce rate for women between the ages of 40 and 44 is 36%. If 10 women between the ages of 40 and 44 who attend a high school reunion are randomly selected, find each probability.

12. P(none get divorced) _____   13. P(2 get divorced) _____
14. P(5 get divorced) _____   15. P(all get divorced) _____

84   Practice & Apply   HRW Advanced Algebra

EXERCISES & PROBLEMS

Communicate

1. Describe two characteristics of a binomial experiment.

2. How are probabilities related to the coefficients of the binomial theorem?

3. Explain how the number of trials is related to the power of a binomial.

4. Explain how to find the probability of getting 3 out of 5 true-false questions correct if you guess on each response.

5. Each question on a multiple-choice test has 4 choices, only one of which is correct. Explain how to find the probability of getting 3 out of 5 questions correct if you guess on each response.

Practice & Apply

A true-false test has 8 questions. Assuming the answers are guesses, find each probability.

6. P(4 answers are correct) 0.273   7. P(5 answers are correct) 0.219

8. P(7 answers are correct) 0.031   9. P(8 answers are correct) 0.004

A multiple-choice exam has 8 questions. Each question has 4 choices. Assuming the answers are guesses, find each probability.

10. P(4 answers are correct) 0.087   11. P(5 answers are correct) 0.023

12. P(7 answers are correct) 0.00037   13. P(8 answers are correct) 0.000015

To simulate 100 coin tosses, a group of students places 100 pennies in a shoe box and shakes it. They count the number of heads that turn up (successes) and the number of tails that turn up (failures). Find the probability of each outcome assuming there is a 50% probability that any coin will land heads up.

14. P(30 heads) 0.000023   15. P(55 heads) 0.048

16. P(70 heads) 0.000023   17. P(95 heads) 5.9×10^{-23}

Science 28% of the students in Grades 3–12 say that, other than their parents and teachers, *Star Trek* television programs most strongly influence their interests in science. If 18 randomly selected students in Grades 3–12 are asked whether *Star Trek* programs most influence their interests in science (yes or no), find each probability.

18. P(4 students say yes) 0.189   19. P(5 students say yes) 0.206

20. P(6 students say yes) 0.174   21. P(7 students say yes) 0.116

Biology One-fourth of all mammals in the world are bats. If 80 mammals are gathered randomly, find each probability.

22. $P(18 \text{ are bats})$

23. $P(21 \text{ are bats})$

24. $P(23 \text{ are bats})$

25. $P(25 \text{ are bats})$

26. $P(62 \text{ are } not \text{ bats})$

27. $P(none \text{ are bats})$

Technology 74% of people aged 18–29 say they are comfortable programming a VCR. If 12 randomly selected people aged 18–29 are asked whether they are comfortable programming a VCR, find each probability.

28. $P(7 \text{ will say yes})$ 0.114   **29.** $P(8 \text{ will say yes})$ 0.203

30. $P(9 \text{ will say yes})$ 0.257   **31.** $P(10 \text{ will say yes})$ 0.220

32. If 16 instead of 12 people are polled, how will your responses to Exercises 28–31 change? 0.008; 0.024; 0.061; 0.122; respectively

Education A recent Harris survey reported that 30% of high-school graduates get information from newspapers, whereas 42% of college graduates get information from newspapers.
Complete Exercises 33 and 34.

33. If 15 high-school graduates are polled, what is the probability that 10 of them get information from newspapers? 0.003

34. If 15 college graduates are polled, what is the probability that 10 of them get information from newspapers? 0.034

35. A recent survey of randomly selected adults found that 66% would rather have more money than more time. What is the probability that 8 out of 10 randomly selected adults would rather have more money than more time? 0.187

36. A recent survey found that 59% of all books sold are fiction. Find the probability that 16 of the next 30 books sold will be fiction. 0.119

37. Suppose 51% of all adventure travelers are between the ages of 25 and 44. If 13 randomly selected adventure travelers are polled, what is the probability that 10 will *not* be between the ages of 25 and 44? 0.030

38. Suppose 30% of all people eat dessert in the kitchen. If 8 randomly selected people are polled, what is the probability that 3 of them eat dessert in the kitchen? 0.254

22. 0.093

23. 0.098

24. 0.073

25. 0.043

26. 0.093

27. 1.01×10^{-10}

In the United States, 57% of cat owners and 32% of dog owners say their pet usually sleeps on the bed with them.

39. Out of 20 randomly selected cat owners, what is the probability that 15 will say their cat usually sleeps on the bed with them? 0.050

40. Out of 20 randomly selected dog owners, what is the probability that 10 will say their dog usually sleeps on the bed with them? 0.044

41. Technology Approximately 42% of all computers are located in private homes. Out of 20 computers, what is the probability that half of them are located in private homes? 0.136

Look Back

42. What is an invertible function? **[Lesson 3.2]** An invertible function is a function whose inverse is also a function.

43. Find two functions, f and g, such that $(f \circ g)(x)$ is the same as $(g \circ f)(x)$. **[Lesson 3.3]**

Physics A miniature rocket is fired vertically into the air from a height of 16 feet and with an initial velocity of 160 feet per second. The height of the rocket is given by $h(t) = -16t^2 + 160t + 16$.
Complete Exercises 44–46. **[Lesson 5.3]**

44 **Maximum/Minimum** What is the maximum height achieved by the projectile? 416 ft

45 **Maximum/Minimum** At what time does the projectile reach its maximum height? 5 sec

46 At what time does the projectile return to ground level? 10.1 sec

47. Construct an arithmetic sequence of numbers with a common difference of 6. **[Lesson 12.2]**

48. Construct a geometric sequence of numbers with a common ratio of 6. **[Lesson 12.3]**

49. The mean for 8 scores is 94. Seven of the scores are 99, 98, 90, 95, 98, 92, and 96. Find the missing score. 84 **[Lesson 13.1]**

Look Beyond

This exercise allows students to explore the reciprocal of several trigonometric functions using their calculator. This foreshadows Chapter 14 where the secant, cosecant, and cotangent functions are introduced.

Look Beyond

50 What keystrokes on a calculator can you use to find $\frac{1}{\tan 28°}$, $\frac{1}{\cos 28°}$, and $\frac{1}{\sin 28°}$?

43. Answers may vary. One example is $f(x) = 3x + 2$ and $g(x) = \frac{x-2}{3}$.

47. Answers may vary. An example is $-3, 3, 9, 15, \ldots$

48. Answers may vary. An example is 1, 6, 36, 216, ...

50. Answers may vary depending on the student's calculator. Calculators should be set for degrees. For example:

| TAN | 28 | x^{-1} | ;
| COS | 28 | x^{-1} | ;
| SIN | 28 | x^{-1} | .

LESSON 13.6 The Normal Distribution

Objectives

- Given a set of data that is normally distributed, find the probability of an event if the mean and standard deviation are known.

RESOURCES

- Practice Master   13.6
- Enrichment Master   13.6
- Technology Master   13.6
- Lesson Activity Master   13.6
- Quiz   13.6
- Spanish Resources   13.6

 *In the binomial experiments of Lesson 13.5, you used a finite number of trials. When the number of trials is considered to be infinite, the curve generated is the normal curve. The normal curve is perhaps the most famous of all those in statistics and probability.*

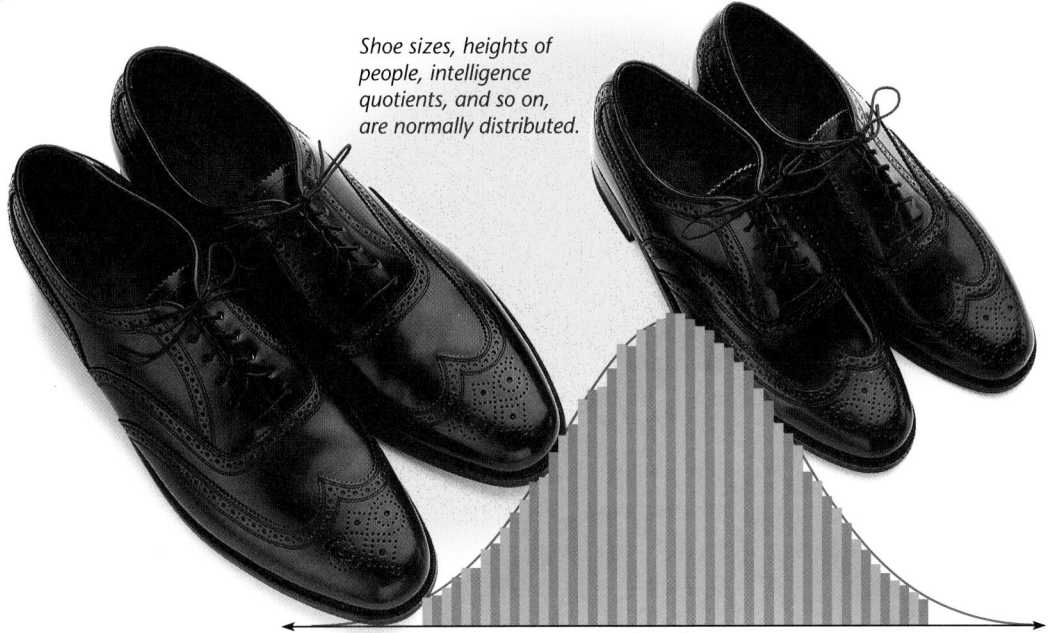

Shoe sizes, heights of people, intelligence quotients, and so on, are normally distributed.

Shoe sizes

The Normal Curve

This graph shows the theoretical probabilities of a coin landing heads up when it is tossed 10 times.

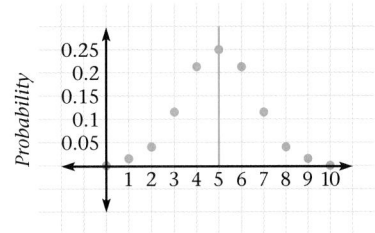

Number of heads

Notice that the graph is symmetric about a vertical line in the center of the graph. As the number of trials increases, a bell-shaped curve, called the **normal curve**, emerges.

The normal curve describes the theoretical outcomes of an infinite number of trials in a binomial experiment. The probability distribution modeled by the normal curve is called the **normal distribution**.

Guided Research
Have students measure the height of each member of the class and graph the results. Then have them bring to class the heights of their family members. Continue to gather data on heights of people. Show, by the graph of the data, how the accumulation of data shapes a normal curve as the number of measurements increases. When a reasonably good curve has developed, have students determine the mean and standard deviation. Then have students use the information to find several *z*-scores based on questions about the data.

This lesson may be considered optional in the core course. See page 708A.

Assessing Prior Knowledge

1. Find the mean of the data set 3, 5, 1, 2, 1, 1, 4, 2. $\left[\frac{19}{8}\right]$

2. Find the standard deviation, σ, for the data set 3, 5, 1, 2, 1, 1, 4, 2. **[1.41]**

3. Find the median of the data set 3, 5, 1, 2, 1, 1, 4, 2. **[2]**

TEACH

Large populations under study seem to have a *normal* distribution of their values about the mean. A normal distribution is one in which 68% of the population falls within 1 standard deviation of the mean, and 95% falls within 2 standard deviations. Ages of all the people in any large city, scores on a national standardized exam, and the number of bank accounts across the nation are some quantities that are large enough to appear infinite. These data fall into a normal distribution.

The **standard normal curve** is a normal curve that is used as a tool for approximating the probabilities of many normally distributed data. Therefore, it is important to know some properties of the standard normal curve.

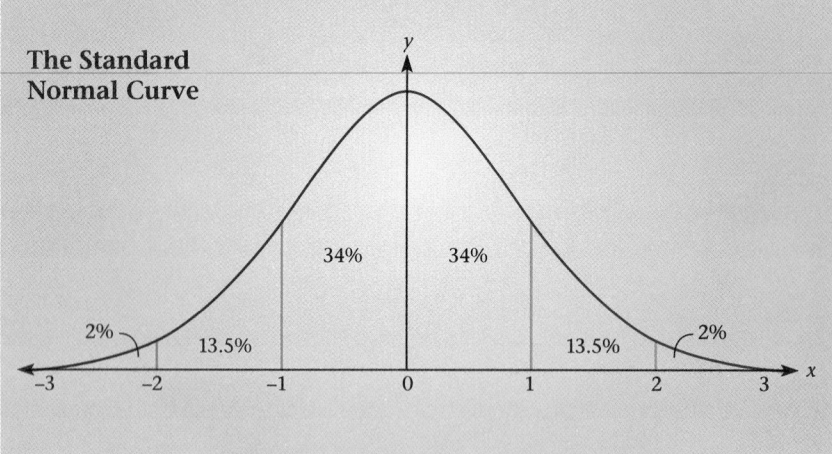

The Standard Normal Curve

Recall that $\bar{x}$ is the mean, and the standard deviation, σ, is $\sigma = \sqrt{\dfrac{\sum(x_n - \bar{x})^2}{n}}$. The standard normal curve has a mean of 0 and a standard deviation of 1. The area under the entire standard normal curve is 1.

Graphics Calculator

Some calculators have built-in features for finding the area under the standard normal curve directly. However, you can use any graphics calculator to graph the equation for the standard normal curve,

$$f(x) = \frac{1}{\sqrt{2\pi}} e^{-\frac{x^2}{2}}.$$

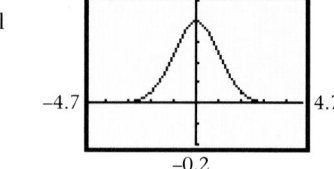

On some graphics calculators, you can use the $\int f(x)\,dx$ feature to approximate areas under the standard normal curve between two x-values.

First choose a lower x-value, or lower limit. A lower limit of −1 is shown here.

Then choose an upper x-value, or upper limit. An upper limit of 1 is shown here.

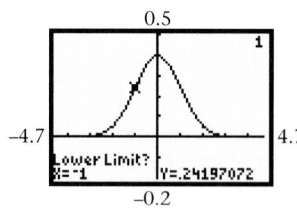

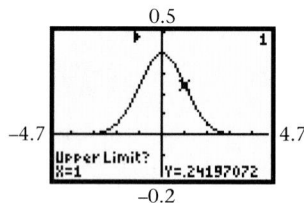

The $\int f(x)dx$ feature gives the area under the curve between the chosen x-values.

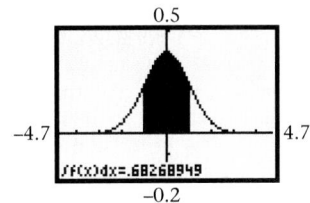

Since the lower x-value is −1 and the upper x-value is 1, the area within 1 standard deviation of the mean is approximately 68%.

A Normal Distribution

A normal distribution has the following properties.

- The normal curve has a vertical line of symmetry at the mean value. For the standard normal curve, this line of symmetry is the y-axis.

- About 68% of the area is within 1 standard deviation of the mean.

- About 95% of the area is within 2 standard deviations of the mean.

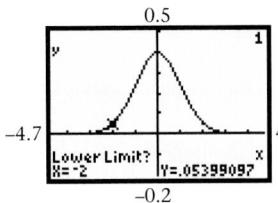 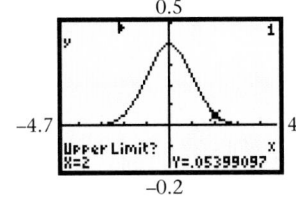 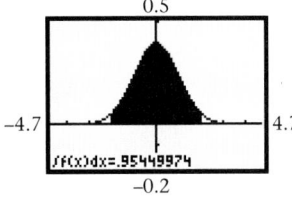

- Over 99% of the area is within 3 standard deviations of the mean.

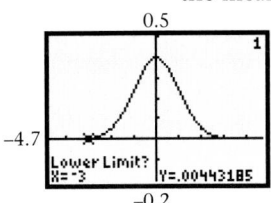 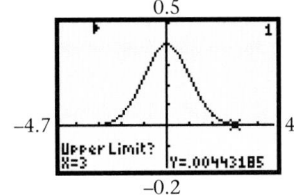 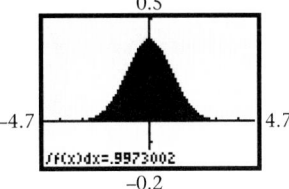

The area under the standard normal curve between two data values, or limits, is directly related to probability. When an event has a normal distribution, the area under the curve between two limits is equal to the probability that a data value in that distribution falls between these two limits.

Since real-world events do not usually have a mean of 0 and a standard deviation of 1, a method for using the standard normal curve for real-world problems is needed.

A television set manufacturer offers a 6-year warranty on the picture tube. The mean lifetime of the tubes has been tested to be 10 years, with a standard deviation of 2 years. Out of 250,000 sets sold, how many can be expected to be returned due to faulty picture tubes within the warranty period? [**Six years is 2 standard deviations below the mean, or about 2.5% of the picture tubes. 0.025 × 250,000 is about 6250 television sets**]

EXAMPLE 1

Automotive Maintenance

An automobile tire manufacturer tested a large number of tires. The life span of the manufacturer's tire approximates a normal distribution. The mean life of a tire is 35,000 miles, with a standard deviation of 4000 miles.

Ⓐ How many tires out of 1000 should last between 31,000 and 39,000 miles?

Ⓑ How many tires out of 500 can be expected to last between 27,000 and 43,000 miles?

Ⓒ William has a tire dealership. He buys 1400 tires. He plans to sell the tires with a warranty stating that the tire will be replaced if it wears out before 31,000 miles. Approximately how many tires will be replaced?

Solution➤

First, find the values for this situation that mark the mean and standard deviations of the normal curve. The mean, $\bar{x}$, is 35,000 miles, and the standard deviation, σ, is 4000 miles.

Find the life expectancies of the tire that are within 1 standard deviation.

$$\bar{x} - \sigma = 31,000 \quad \text{and} \quad \bar{x} + \sigma = 39,000$$

Find the life expectancies of the tire that are within 2 standard deviations.

$$\bar{x} - 2\sigma = 27,000 \quad \text{and} \quad \bar{x} + 2\sigma = 43,000$$

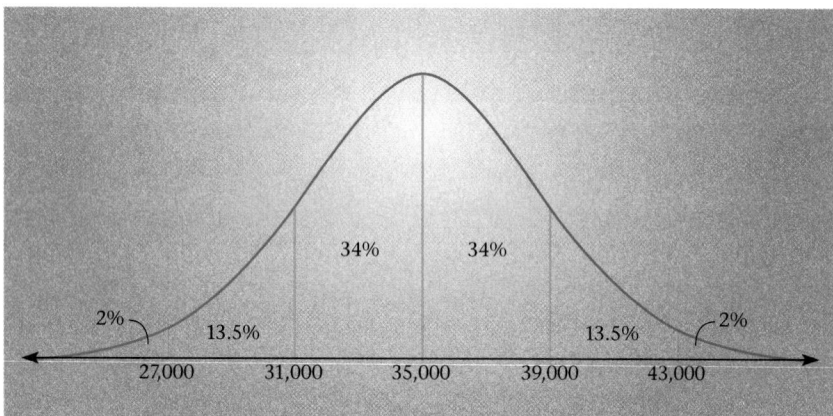

Ⓐ The interval from 31,000 to 39,000 miles is within 1 standard deviation of the mean. Thus, about 68% of 1000 tires will last between 31,000 and 39,000 miles.

$$0.68 \times 1000 = 680$$

Thus, about 680 out of 1000 tires can be expected to last between 31,000 and 39,000 miles.

 **Using Visual Models**
Have students carefully graph the normal curve from −3 to 3 on graph paper. Then, students should count the number of squares under the 1, the 2, and the 3σ intervals. Each of these values should be expressed as a ratio to the total number of squares under the curve.

B The interval between 27,000 and 43,000 is within 2 standard deviations of the mean. Thus, about 95% of the 500 tires can be expected to last between 27,000 and 43,000 miles.

$$0.95 \times 500 = 475$$

Thus, about 475 out of 500 tires can be expected to last between 27,000 and 43,000 miles.

C 31,000 miles is 1 standard deviation below the mean. Approximately 16% fall below 1 standard deviation.

$$1400(0.16) = 224$$

16%

31,000

Thus, about 224 of the 1400 tires purchased will need to be replaced. ❖

Try This The scores on a college entrance exam for graduates of a particular high school are normally distributed with a mean of 580 and a standard deviation of 120. Find the probability that a randomly selected student from this high school scored between 460 and 700 on this exam.

The probabilities for Example 1 are easy to find because the tire-life limits are constant multiples of the standard deviation. What do you think happens when the limits are *not* represented by constant multiples?

RETEACHING
t h e
l e s s o n

Using Visual Models
Give each student a very accurate graph of the normal curve. Explain that the areas under the graph represent the probability that a sample data value, taken from a population that varies normally about its mean, will be in that particular interval. Show students the function that describes the curve, and point out the percentage probability intervals bordered by multiples of the standard deviation. Ask questions of students concerning the probability that an exact value will occur, the probability that a data value in a very small interval will occur, and the probability that a data value in intervals 1σ wide will occur. Ensure that the students understand how to answer these questions graphically, using the curve.

Alternate Example 2

Paula leaves for work at 7:35. It takes her a mean driving time of 20 minutes, with a standard deviation of 2 minutes, to arrive at work. What is the probability that Paula will arrive at work at 7:52? [≈ **6.7%**]

CRITICAL *Thinking*

If the z-score for x is a whole number, then $x - \bar{x}$ must be a multiple of the standard deviation.

---

z-scores allow you to use the standard normal curve to find the probabilities associated with any normal distribution. To find the upper or lower x-limit for a particular value, use the following formula to obtain a z-score.

z-SCORE

$$z = \frac{x - \bar{x}}{\sigma}, \text{ where } \begin{cases} x \text{ is the data value} \\ \bar{x} \text{ is the mean for the data} \\ \sigma \text{ is the standard deviation} \end{cases}$$

EXAMPLE 2

Time Management

Delmer finds that the amount of time it takes him to drive to work has a mean of 35 minutes and a standard deviation of 7 minutes. Assume the times are normally distributed. What is the probability that Delmer will arrive at work by 8:00 A.M. if he leaves home at 7:35 A.M.?

Solution➤

If Delmer leaves home at 7:35, he will need to arrive at work within 25 minutes. Find a z-score for 25.

$$z = \frac{x - \bar{x}}{\sigma} = \frac{25 - 35}{7} \approx -1.43$$

A z-score of –1.43 is located to the left of the mean.

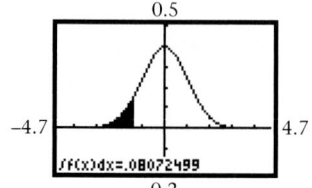

Since 99.99% of the area is within 4 standard deviations of the mean, a lower limit of –4 is sufficient.

Use your graphics calculator to find the area from $x = -4$ to $x = -1.4$.

The area under the curve is about 0.08. Thus, Delmer will get to work within 25 minutes only about 8% of the time. ❖

CRITICAL *Thinking*

What must be true about the difference $x - \bar{x}$ if the z-score for x is a whole number? Explain.

When it is necessary to find areas under a normal curve without a graphics calculator, you can use the Standard Normal Distribution Table.

EXERCISES & PROBLEMS

Communicate

1. Give two examples of real-world data that are normally distributed.
2. Describe four characteristics of the normal curve.
3. Explain how area under a normal curve is related to the standard deviation and probability of a normal distribution.
4. Describe how to use a *z*-score to find an area under the normal curve.
5. Explain how to find the *z*-score for an *x*-value of 27 if $\bar{x}$ is 25 and σ is 3.

Practice & Apply

Determine the indicated areas under the normal curve.

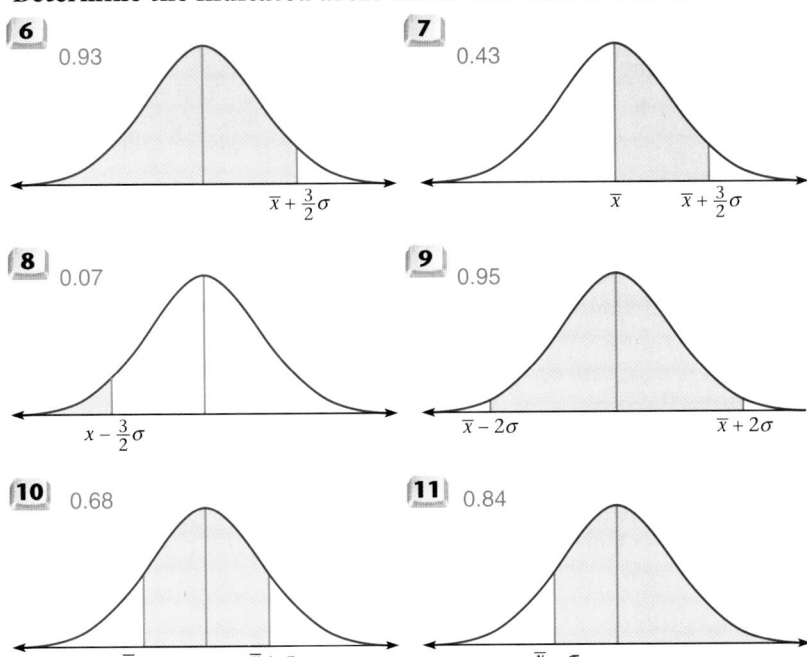

6 0.93 $\bar{x} + \frac{3}{2}\sigma$

7 0.43 $\bar{x}$ $\bar{x} + \frac{3}{2}\sigma$

8 0.07 $x - \frac{3}{2}\sigma$

9 0.95 $\bar{x} - 2\sigma$ $\bar{x} + 2\sigma$

10 0.68 $\bar{x} - \sigma$ $\bar{x} + \sigma$

11 0.84 $\bar{x} - \sigma$

Academics The scores on a particular college entrance exam for graduates of a particular high school are normally distributed with a mean of 520 and a standard deviation of 100. Find the probability that a randomly selected student from this high school scored within each of the following ranges on the exam.

12 Lower than 400 · 0.12 **13** Higher than 600 · 0.21 **14** Between 450 and 550 · 0.38

Assignment Guide

Core 1–10, 12–38

Core Plus 1–5, 8–39

Technology

Graphics calculators are needed for Exercises 6–17, 19–25, 29–31, and 39.

Error Analysis

Encourage students to make a sketch of the problem conditions on the normal curve to check their results from the calculator for reasonableness. This will help students find errors due to inaccurate data entry. Also, student errors will frequently arise when students do not correctly translate the words of a problem into its mathematical form. Have students illustrate, with a sketch, what information the problem has given and what information is requested.

alternative ASSESSMENT

Authentic Assessment

Give the students a typical problem from this lesson and a well-made normal curve of the population in question, with large amounts of blank space around it. Require students to create an illustrated solution, using colored pencils. They should include explanatory notes concerning each graphically depicted answer.

Error Analysis

In computing answers for this text, z-scores and probabilities are measured between -3 and 3. Therefore, if students are using tables or automatic statistics features on their calculators, there may be a small difference between student answers and the answers provided.

Grocery Sales The mean shelf life of a particular dairy product is approximately 11 days with a standard deviation of 3 days. If one of these dairy products is randomly selected, find the probability that it will last the following lengths of time.

15 Between 10 and 15 days 0.52

16 Less than 7 days 0.10

17 More than 14 days 0.16

18. Quality Control A machine is supposed to fill containers with 4 fluid ounces of perfume. When the machine is adjusted properly, it fills containers with 4 fluid ounces, with a standard deviation of 0.1 fluid ounces. If 400 containers are tested, how many containers with less than 3.8 fluid ounces of perfume will indicate that the machine needs to be adjusted? About 9

Automotive Mechanics An auto parts manufacturer finds that the useful life of a spark plug in a car is approximately normally distributed with a mean of 9 months and a standard deviation of 2.3 months.

19 What is the probability that a randomly selected spark plug will last more than 1 year? 0.10

20 What is the probability that a randomly selected spark plug will last less than 7 months? 0.19

21 An auto parts salesperson offers a guarantee that any spark plug that lasts less than 6 months will be replaced. What percentage of spark plugs sold can the salesperson expect to replace? 9.7%

Find the z-score for the indicated data value, x, in a normal distribution. Then find the probability that a value less than or equal to x occurs in a random selection.

22 $\bar{x} = 78$, $\sigma = 5.2$, $x = 65$ **23** $\bar{x} = 112$, $\sigma = 3.2$, $x = 117$

24 $\bar{x} = 7$, $\sigma = 2.1$, $x = 11$ **25** $\bar{x} = -54$, $\sigma = 3.6$, $x = -47$

26. A normally distributed data set of 140 values has a mean of 265 and a standard deviation of 14. About how many of the values are between 237 and 293, inclusive? 133

27. A normally distributed data set of 585 values has a mean of 3.2 and a standard deviation of 0.6. About how many of the values are between 1.4 and 5, inclusive? 580

28. A normally distributed data set of 2006 values has a mean of 120 and a standard deviation of 26. About how many of the values are between 94 and 146, inclusive? 1364

22. z-score: -2.50; probability: 0.006

23. z-score: 1.56; probability: 0.941

24. z-score: 1.91; probability: 0.971

25. z-score: 1.94; probability: 0.973

Mortgage Mortgage statistics collected by a bank indicate that a number of years the average new homeowner will occupy the house before moving or selling is normally distributed with a mean of 6.3 years and a standard deviation of 2.31 years.

29 If a homeowner is selected at random, what is the probability that the homeowner will sell the house or move within 3 years? 0.08

30 What is the probability that a randomly selected homeowner will keep the house more than 10 years? 0.05

31 If a homeowner is selected at random, what is the probability that the homeowner will keep the house from 6 to 8 years? 0.30

32.  **Portfolio Activity** Complete the Portfolio Activity described on page 709.

Look Back

A box contains 50 graphics calculators, some of which are broken. Two of the calculators are randomly selected from the box. Using *B* for broken and *N* for not broken, specify a sample space for this experiment. Then determine the event that corresponds to each of the following outcomes. **[Lesson 11.1]** {BB,BN,NB,NN}

33. At most, one calculator is broken. {BN,NB,NN}

34. No calculator is broken. {NN}

35. Both calculators are broken. {BB}

36. At least one calculator is broken. {BB,BN,NB}

Suppose that two committees are each composed of 3 men and 3 women. One member from each committee is randomly chosen to serve on a joint subcommittee.   **[Lesson 11.5]**

A = {the person chosen from the first committee is a man}

B = {the person chosen from the second committee is a woman}

C = {both people chosen are of the same gender}

37. Show that each pair of events, *AB*, *AC*, and *BC*, are independent.

38. Show that the three events *A*, *B*, and *C*, are not independent.

Look Beyond

39 With your graphics calculator in degree mode, graph the functions $y = \sin(x + 32°)$ and $y = \sin x + \sin(32°)$. Do you think the following statement is true? Explain.

$$\sin(A + B) \stackrel{?}{=} \sin A + \sin B$$

32. Answers may vary.

37. *AB*: the random selection for the first committee comes from a pool of 3 men and 3 women; the random selection for the second committee also comes from a pool of 3 men and 3 women; therefore the first event does not affect the second event.

38. Answers may vary; One sample answer is: They are dependent because if events *A* and *B* occur, then event *C* cannot occur.

39. no; $\sin(x + 32°) \neq \sin x + \sin 32°$,   so $\sin(A + B) \neq \sin A + \sin B$.

That's Not Fair!

You have probably seen and heard the results of many different opinion surveys given with percentages. For instance, a recent Gallup Poll reported that about 48% of the people surveyed felt that the next generation (your generation) will enjoy *less* personal freedom than the *current* generation (your parents). How can you determine whether the survey was conducted fairly? How can you tell whether the results that are reported have any meaning?

In general, it is very difficult to tell, from the results of a poll, whether the poll was conducted and reported fairly. However, there are things you can look for. If a poll reports that 25% of all adults favor automobiles with no sunroof and only people who did not own cars were polled, the results may not be credible.

Consider each of the following surveys. Determine whether the results of the survey would be considered credible. Explain your answers.

1. To determine how athletic interest among female students compares with athletic interest among male students, a survey team questions students in the school orchestra.

2. To determine the opinions on the animal rights movement, an interviewer surveys pet owners attending a dog show.

3. To determine the proportion of students at a high school that favor on-campus parking priveleges, an interviewer surveys the first three students that get off each school bus one morning.

4. To determine opinions on agricultural and rural issues, an interviewer questions the students of an inner-city high school.

1–4. Answers to the project questions may vary.

Chapter 13 Review

Vocabulary

Key Skills & Exercises

Lesson 13.1

➤ Key Skills

Find the measures of central tendency, mean, median, and mode of a given set of data.

The mean is found by using the formula $\bar{x} = \frac{\sum x}{n}$, which is the sum of the values in the data set divided by the number of data values. The median is the middle value when the data set is arranged in numerical order. The mode is the value that occurs the most often.

Construct a frequency table by dividing the data into classes, and find the class mean for the data.

Separate the range of the data into several classes, and determine the frequency (number of data values) for each class. Find the midpoint of each class and multiply by the frequency of that class. Add all of these products and divide by the total number of data values to find the class mean.

➤ Exercises

For Exercises 1–3, use the following heights (in inches) of basketball players: 72, 75, 77, 75, 70, 75, 76, 73, 78, 75, 71, 77, 76, 73, 80, 77.

1. Find the mean. 75   **2.** Find the median. 75   **3.** Find the mode. 75

The following values represent the high temperature on each Monday in one year at a local airport.

15, 7, 11, 21, 38, 29, 45, 40, 52, 48, 56, 68, 52, 65, 71, 66, 72, 78, 73, 73, 81, 78, 85, 92, 89, 77, 85, 93, 95, 86, 90, 104, 91, 90, 82, 86, 78, 75, 79, 65, 73, 68, 62, 71, 55, 58, 49, 55, 61, 55, 40, 18

4. Separate the data into 11 classes, with each class having a range of 10, and construct a frequency table.

5. Find the class mean for these temperatures.

Lesson 13.2

➤ Key Skills

Construct a histogram for a given set of data.

Draw and label the horizontal axis with the data values or class ranges of data values; label the vertical axis with frequencies. Construct vertical bars to represent the frequency of each data value.

Construct a stem-and-leaf plot to represent a given set of data.

Split each data value into two parts. Often you may use the units digit of each number as the leaves and the other digits as the stems.

➤ Exercises

6. Use the heights of the basketball players given for Exercises 1–3 to construct a histogram.

7. Does the distribution appear to be symmetric about the mean or skewed in a particular direction? slightly skewed to the left

8. Construct a stem-and-leaf plot using the temperature data given for Exercises 4–5.

9. What is the median for the temperature data? 69.5

Lesson 13.3

➤ Key Skills

Find the quartiles and interquartile range for a given set of data.

When the data set is arranged in order, find the median of the entire data set, Q_2. The median of the data below Q_2 is labeled Q_1, and the median of the data above Q_2 is Q_3. The interquartile range is the difference $Q_3 - Q_1$.

Construct a box-and-whisker plot to display a given set of data.

Draw a rectangle with the ends at the first and third quartiles. Then draw whiskers that extend from the lowest to the highest data values.

➤ Exercises

The total amount of music on each of 25 compact discs, rounded to the nearest minute, is given.

48, 57, 42, 38, 56, 56, 65, 51, 48, 44, 47, 43, 55, 59, 62, 78, 45, 56, 60, 61, 52, 65, 51, 48, 54

10. Find the quartiles for this data set.

11. Find the interquartile range. 12

12. Construct a box-and-whisker plot to display the data.

Lesson 13.4

➤ Key Skills

Find the range of a given set of data.

The range is the absolute value of the difference between the largest and smallest values.

Determine the mean deviation and the standard deviation for a given set of data.

For n values with a mean of $\bar{x}$, the mean deviation is $\dfrac{\sum |x_n - \bar{x}|}{n}$, and the standard deviation is $\sigma = \sqrt{\dfrac{\sum (x_n - \bar{x})^2}{n}}$.

8.

Stem	Leaf
0	7
1	1, 5, 8
2	1, 9
3	8
4	0, 0, 5, 8, 9
5	2, 2, 5, 5, 5, 6, 8
6	1, 2, 5, 5, 6, 8, 8
7	1, 1, 2, 3, 3, 3, 5, 7, 8, 8, 8, 9
8	1, 2, 5, 5, 6, 6, 9
9	0, 0, 1, 2, 3, 5
10	4

10. $Q_1 = 47.5$, $Q_2 = 54$, $Q_3 = 59.5$

12.

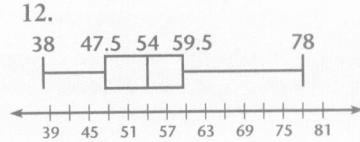

4.

Class	Frequency
0–9	1
10–19	3
20–29	2
30–39	1
40–49	5
50–59	7
60–69	7
70–79	12
80–89	7
90–99	6
100–109	1

5. class mean: 64.31

6.

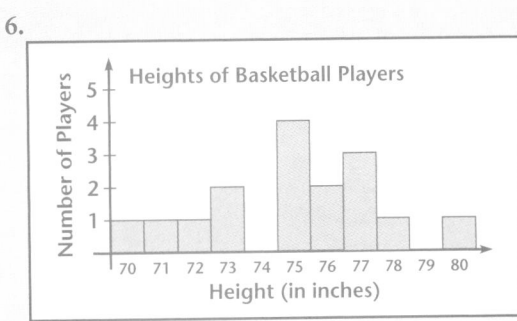

➤ **Exercises**

Tom walks 4 miles each morning. Depending on the weather and how many friends go with him, his total time for the walk varies. For 5 days last week his times (in minutes) were: 61, 58, 66, 56, 62.

13. Find the median and range for these times. 61; 10
14. Calculate the mean deviation. 2.88
15. Calculate the standard deviation. 3.44

Lesson 13.5

➤ *Key Skills*

Find the probability of *r* successes in *n* trials of a binomial experiment.

The binomial probability distribution is $P(r) = {}_nC_r p^r q^{n-r}$, where p is the probability of success, and $q = 1 - p$ is the probability of failure.

➤ *Exercises*

16. If 16 of 25 students in a class are girls, what is the probability that 3 students chosen at random will all be girls? 0.262

17. Genetics The Wylies know from their family history and genetics that there is a 25% chance of one of their children being born with red hair and a 75% chance of one being born with brown hair. If the Wylies plan to have 3 children, what is the probability that exactly one of them will have red hair? 0.422

Lesson 13.6

➤ *Key Skills*

Given a set of data that is normally distributed, find the probability of an event if the mean and standard deviation are known.

If test scores are normally distributed around a mean of 500 with a standard deviation of 80, the probability of earning a score over 600 is:

$$z = \frac{600 - 500}{80} = 1.25$$

Use a *z*-score table or the $\int f(x)dx$ feature on a graphics calculator to find that the probability for a *z*-score of 1.25 or more is approximately 0.106. The probability of earning a score of 600 or more is approximately 10.6%.

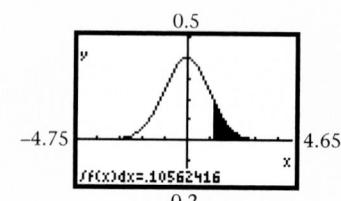

➤ *Exercises*

18. On the test described above, what is the probability of getting a score of 420 or less? about 0.16

19. What percentage of the students that take the test earn a grade of 460 or more?

Applications

Sports Sarah has achieved a free-throw percentage of 0.800 this season.

20. What is the probability that she will make 3 of her next 4 free throws? 0.410
21. What is the probability that she will make at least 3 of her next 4 free throws? 0.819
22. What is the probability that she will make exactly 4 of her next 5 free throws? 0.410

19. 69%

Chapter 13 Assessment

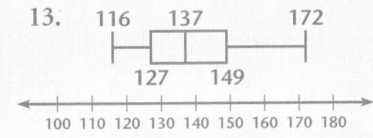

Employment Ten employees at a local shop make the following hourly wages.

$5.75 $5.15 $7.25 $6.15 $5.15
$5.85 $9.15 $7.15 $5.75 $5.75

1. Find the mean for these hourly wages. $6.31
2. Find the median hourly wage. $5.80
3. What is the mode of these wages? $5.75
4. If the minimum hourly wage is raised from $5.15 to $5.50, how might this affect the three measures of central tendency?
mean increases to $6.38; median and mode remain the same

Health The following blood pressures (systolic number) were recorded for 20 people enrolled in a night course at a local college.

116 120 134 154 140 140 151 128 136 120
156 128 130 147 124 145 172 152 138 126

5. Construct a histogram with class ranges of 10 for the data.
6. Is the histogram symmetric, or is it skewed in a particular direction?
7. What is the class mean for these blood-pressure readings? 138.5
8. Construct a stem-and-leaf plot for the blood-pressure readings.
9. What is the median for this set of data? 137
10. What are the outliers for this set of data? 116 and 172
11. Find the quartiles for the blood-pressure readings. $Q_1 = 127$, $Q_2 = 137$, $Q_3 = 149$
12. What is the interquartile range for this data set? 22
13. Construct a box-and-whisker plot for the blood-pressure data.

Entertainment For one weekend, the top 10 movies grossed the following amounts of money (in millions of dollars).

14.8 10.6 6.1 5.6 4.2 4.0 3.7 3.0 2.9 2.5

14. Find the range for this set of data. 12.3
15. Determine the mean deviation for this set of data. 2.86
16. Determine the standard deviation for this set of data. 3.76

At a particular high school, 25% of the students drive themselves to school.

17. What is the probability that out of 8 students chosen at random, exactly 2 of them drive to school? 0.311
18. What is the probability that out of 10 students chosen at random, none of them drive to school? 0.056

Education The teacher of a calculus class wants to raise the grades of his students on a test that proved to be particularly difficult. He decides to grade the test so that any grade of one standard deviation above the mean ($\bar{x} + 1\sigma$) or higher will be recorded as an A; from $\bar{x}$ to $\bar{x} + 1\sigma$ as a B; from $\bar{x}$ to $\bar{x} - 1\sigma$ as a C; from $\bar{x} - 1\sigma$ to $\bar{x} - 2\sigma$ as a D; and below $\bar{x} - 2\sigma$ as an F.

19. If the grades are normally distributed, what percentage of the students will receive each grade? A: about 16%; B: 34%; C: 34%; D: about 14%; F: 2%
20. If the mean score is 65 and the standard deviation is 10 points, what is the lowest score that will qualify for a passing grade of D? 45

13.

116   137   172
127   149

100 110 120 130 140 150 160 170 180

5.

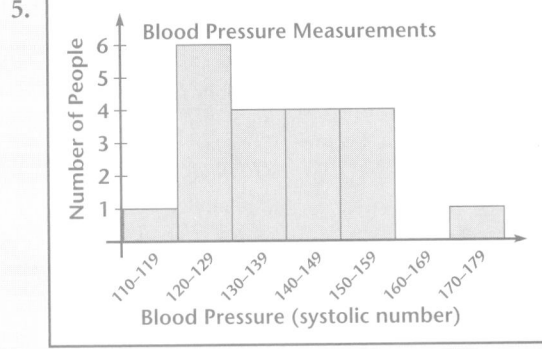

Blood Pressure Measurements

6. skewed to the right

8.
11	6
12	0, 0, 4, 6, 8, 8
13	0, 4, 6, 8
14	0, 0, 5, 7
15	1, 2, 4, 6
16	
17	2

14 CHAPTER

Further Topics in Trigonometry

Meeting Individual Needs

14.1 The Law of Sines

Core

Inclusion Strategies, p. 771
Reteaching the Lesson, p. 772
Practice Master 14.1
Enrichment Master 14.1
Technology Master 14.1
Lesson Activity Master 14.1

[Optional*]

Core Plus

Practice Master 14.1
Enrichment, p. 771
Technology Master 14.1
Lesson Activity Master
14.1

[1 day]

14.2 The Law of Cosines

Core

Inclusion Strategies, p. 778
Reteaching the Lesson, p. 779
Practice Master 14.2
Enrichment Master 14.2
Technology Master 14.2
Lesson Activity Master 14.2

[Optional*]

Core Plus

Practice Master 14.2
Enrichment, p. 778
Technology Master 14.2
Lesson Activity Master
14.2

[1 day]

14.3 Circular Function Relationships

Core

Inclusion Strategies, p. 786
Reteaching the Lesson, p. 787
Practice Master 14.3
Enrichment Master 14.3
Technology Master 14.3
Lesson Activity Master 14.3
Mid-Chapter Assessment Master

[Optional*]

Core Plus

Practice Master 14.3
Enrichment, p. 785
Technology Master 14.3
Lesson Activity Master
14.3
Mid-Chapter Assessment
Master

[1 day]

14.4 Exploring Sum and Difference Identities

Core

Inclusion Strategies, p. 792
Reteaching the Lesson, p. 793
Practice Master 14.4
Enrichment Master 14.4
Technology Master 14.4
Lesson Activity Master 14.4

[Optional*]

Core Plus

Practice Master 14.4
Enrichment, p. 791
Technology Master 14.4
Lesson Activity Master
14.4

[1 day]

14.5 Solving Trigonometric Equations

Core

Inclusion Strategies, p. 797
Reteaching the Lesson, p. 798
Practice Master 14.5
Enrichment Master 14.5
Technology Master 14.5
Lesson Activity Master 14.5

[Optional*]

Core Plus

Practice Master 14.5
Enrichment, p. 797
Technology Master 14.5
Lesson Activity Master
14.5

[1 day]

Chapter Summary

Core

Chapter 14 Project,
   pp. 802–803
Lab Activity
Long-Term Project
Chapter Review, pp. 804–806
Chapter Assessment, p. 807
Chapter Assessment, A/B
Alternative Assessment
Cumulative Assessment,
   pp. 808–809

[Optional*]

Core Plus

Chapter 14 Project,
   pp. 802–803
Lab Activity
Long-Term Project
Chapter Review,
   pp. 804–806
Chapter Assessment,
   p. 807
Chapter Assessment, A/B
Alternative Assessment
Cumulative Assessment,
   pp. 808–809

[3 days]

Reading Strategies

Students should first read the Chapter opener and review the photographs in the opener. Students should then read the WHY section on the first page of each Lesson. This will provide an overview of the entire chapter and motivation for studying it. The Chapter Review provides a vocabulary list that the students should also look at before reading the chapter. Suggest that they keep the list in mind as they read and compile a list of definitions for these terms as they read. They can then skim the Portfolio Activity and the Chapter Project to get a feel for the practical applications of the material in the chapter. Before reading each Lesson, students should skim the word problems in the exercises. This will give them a clear idea of the sort of applications that justify the study of this Lesson. Each student might want to pick one application problem that they find personally interesting and use this as the anchor-point for the Lesson. They can add these to their portfolio and develop a portfolio that reflects their interests.

Visual Strategies

Direct the students to look at the pictures in the chapter opener. Encourage them to speculate about how mathematics might be useful in studying such phenomena. Also have them look at the graphs in the chapter and discuss how they differ from graphs in earlier chapters. What aspect of these graphs makes them different from graphs in earlier chapters? Can they imagine how such functions could model the phenomena in the photographs?

Hands-on Strategies

This chapter presents students with a number of opportunities to become actively involved with mathematics in the physical world. Students will enjoy a variety of surveying projects and similar indirect measurement activities.

Cooperative Learning

GROUP ACTIVITIES	
Sines and triangles	Lesson 14.1, Exploration
Angle sums, differences, and double angles	Lesson 14.4, Explorations 1–3

You may wish to have students work with partners for some of the above activities. Additional suggestions for cooperative group activities are noted in the teacher's notes in each lesson.

Multicultural

The cultural connections in this chapter include references to Asia and Africa.

CULTURAL CONNECTIONS	
Asia: Law of Sines	Lesson 14.1
Africa: Sum and Double Angle Formulas	Lesson 14.4

Portfolio Assessment

Below are portfolio activities for the chapter. They are listed under seven activity domains that are appropriate for portfolio development.

1. **Investigation/Exploration** The relationship among the ratios in a triangle is the focus of the exploration in Lesson 14.1; the sum and difference of angles is the objective of Explorations 1–3 in Lesson 14.4.

2. **Applications Recommended** To be included are any of the following: surveying, Lesson 14.1, Exercises 33–35; navigation, Lesson 14.2, Exercises 20 and 21; electricity, Lesson 14.5, Exercises 34–36.

3. **Non-Routine Problems** The Surveying by Triangulation problem from Chapter Project presents the students with non-traditional but interesting ways to look at the applications of trigonometry.

4. **Project** Survey by Triangulation See pages 802–803. Students are asked to determine the usable area of a piece of land.

5. **Interdisciplinary Topics** Students may choose from the following: coordinate geometry, Lesson 14.2, Exercises 15 and 16; Lesson 14.3, Exercises 20–25.

6. **Writing** Communicate exercises of the type in which a students is asked to describe the Law of Sines, the Law of Cosines, sum and difference formulas, and solving trigonometric equations offer excellent writing selections for the portfolio.

7. **Tools** Chapter 14 uses graphics calculators to help students solve equations involving trigonometric functions. To measure the individual student's proficiency, it is recommended that he or she complete selected worksheets from the Technology Masters.

Technology

Graphics Calculators

The graphics calculator is very helpful in allowing students to view the reciprocal trigonometric functions. The confusion about why cotangent, secant, and cosecant are undefined at particular values of θ will be avoided when students can actually see the patterns of these functions. Also, students will be able to observe that the graphs of these functions have very different shapes from the sine and cosine graphs. Still, they are cyclic functions and will therefore model naturally cyclic phenomena from the real world.

The graphics calculator gives students an additional tool to explore, visualize, and understand concepts like angle sum, angle difference, and double angle identities developed in Lesson 14.4. One of the advantages of the graphics calculator is its capabilities for solving trigonometric equations. The mystery of why there are infinitely many solutions to a single equation becomes clear when the student sees a horizontal line intersecting a cyclic function at many points within a given interval. For example, to find all the values of x for which $\cos 3x$ is 0.5, where $0 \leq x \leq 2\pi$, simply graph the functions $y = \cos 3x$ and $y = 0.5$ on the same coordinate plane. Make certain that the viewing window includes the desired domain. The following image uses the TI-82 in radian mode. Since 2π is about 6.3 radians, the friendly viewing window shown includes the required domain for x.

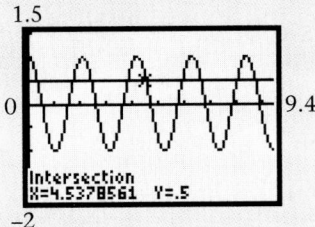

Notice that, even though the intersection feature was used, this x-value does not represent an exact radian angle measure. This angle value divided by π is approximately 1.5. However, if these functions had been plotted in degree mode, the exact degree angle measure, if one existed, could be easily found. This value could then be converted to radians. Have students review the concepts of graphing in degree mode as discussed in the Chapter 8 interleaf. Remind students that if they graph in the degree mode, they must convert all the values, including the solution restrictions, to degrees. They should set their calculators for degrees and enter the minimum and maximum degree values for x. For example, using the functions $y = \cos 3x$ and $y = 0.5$, and using the converted domain restrictions $0° \leq x \leq 360°$, the trace feature gives the exact degree value.

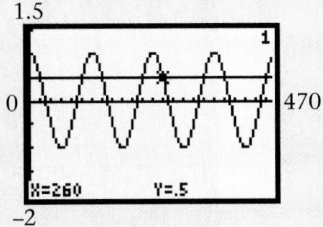

Students can see not only that there are several solutions, but that they can also find exact degree values for these solutions. They can then convert these degree solutions to obtain exact radian solutions. For instance, $260°$ is exactly $\frac{13\pi}{9}$, so one solution to the original question is $\frac{13\pi}{9}$. Student can use trace with a friendly window or a solve feature to find that the remaining solutions within the restricted domain are $20°$, $100°$, $140°$, $220°$, $260°$, and $340°$, or $\frac{\pi}{9}$, $\frac{5\pi}{9}$, $\frac{7\pi}{9}$, $\frac{11\pi}{9}$, $\frac{13\pi}{9}$, and $\frac{17\pi}{9}$.

Integrated Software

f(g) Scholar™ is an integrated computer-based mathematics productivity tool that combines calculator, spreadsheet, and graphics capabilities to provide a dynamic and interactive environment for explorations in mathematics. Use *f(g) Scholar*™ for any lesson needing a spreadsheet, calculator, graphics calculator, or any combination of the three.

ABOUT THE CHAPTER

Background Information

In Chapter 8, the lessons on triangle trigonometry dealt primarily with right triangles. In many applications, however, the triangles that occur are not right triangles. The trigonometric properties and concepts that are developed in this chapter allow students to determine unknown angles, sides, and areas that apply to acute and obtuse triangles as well as right triangles.

CHAPTER RESOURCES

- Practice Masters
- Enrichment Masters
- Technology Masters
- Lesson Activity Masters
- Lab Activity Masters
- Long-Term Project Masters
- Assessment Masters
  Chapter Assessments, A/B
  Mid-Chapter Assessment
  Alternative Assessments, A/B
- Teaching Transparencies
- Cumulative Assessment
- Spanish Resources

CHAPTER OBJECTIVES

- Find all the parts of a triangle when given the measure of two angles and the length of the included side.
- Find all the parts of a triangle when given the lengths of two sides and the measure of the included angle.

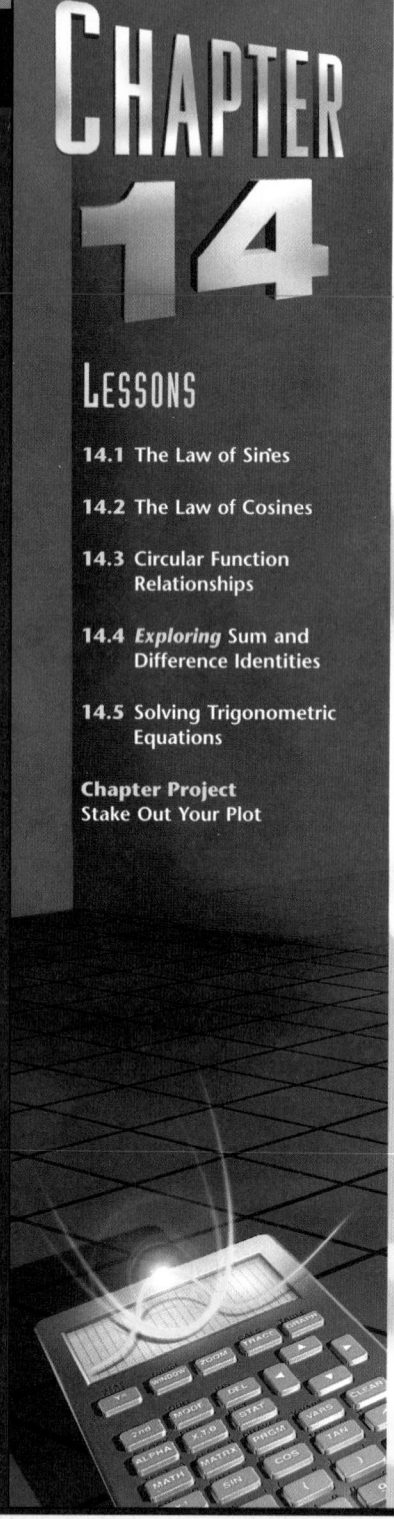

CHAPTER 14

LESSONS

14.1 The Law of Sines

14.2 The Law of Cosines

14.3 Circular Function Relationships

14.4 *Exploring* Sum and Difference Identities

14.5 Solving Trigonometric Equations

Chapter Project
Stake Out Your Plot

Further Topics in Trigonometry

Trigonometry is used to measure angles, sides, and areas of triangles. Trigonometric functions model rotational motion, such as in an engine, and in vibrating or oscillating systems, such as a plucked guitar string, a revolving door, an earthquake, or planetary motion. In this chapter, you will study triangle measurement, oscillating systems, and several new circular functions and the relationships between them.

ABOUT THE PHOTOS

A common pattern of motion that occurs naturally is the cycle. A cycle is caused by such events as plucking a guitar string or the motions of the gears of a working machine. In both of these cases, a particular quantity increases to a maximum, decreases to a minimum, and returns again to the maximum where the cycle begins again. When the paths of two objects converge at a particular location in space, such as the space shuttle and the orbit of the Hubble telescope, their paths do not always converge at right angles. The techniques of this chapter permit the modeling of these paths and their intersections.

The vibrations of this guitar string can be modeled by a trigometric function.

- Find all the parts of a triangle when given the lengths of all three sides.
- Find the value of any of the other five trigonometric functions when given the value of one function and the quadrant in which the angle lies.
- Find the exact values of the trigonometric functions by using the angle sum and difference identities, and the double angle identities.
- Solve equations containing trigonometric functions.

PORTFOLIO ACTIVITY

In this activity, students will be asked to find the distance between two airplanes 90 minutes after they leave the same airport heading in different directions.

Have students work in groups of two. One student can draw the diagram as accurately as possible. The other student can attempt to solve the problem using the Law of Cosines.

Discuss with students the reasons for using the Law of Cosines to solve this problem. Information will be given on page 783 of Lesson 14.2 that will be used to solve this problem.

PORTFOLIO ACTIVITY

Two airplanes take off at the same time from the same airport. The planes fly at the same altitude with different cruising speeds and different headings. On page 783, you will be asked how trigonometry can be used to find out how far apart the planes are at a later time, assuming both planes fly along a straight route.

You may wish to save your work for your portfolio.

ABOUT THE CHAPTER PROJECT

In the Chapter Project on pages 802–803, students will attempt a realistic surveying problem. They will be asked to determine the area of a triangular plot of land with a river running through it. The activity will draw on many trigonometric relations from several lessons in this chapter.

Objectives

- Find all the parts of a triangle when given the measure of two angles and the length of the included side.

RESOURCES

- Practice Master   **14.1**
- Enrichment Master   **14.1**
- Technology Master   **14.1**
- Lesson Activity Master   **14.1**
- Quiz   **14.1**
- Spanish Resources   **14.1**

This lesson may be considered optional in the core course. See page 768A.

Assessing Prior Knowledge

Find *x* in each of these proportions.

1. $\frac{x}{6} = \frac{3}{21}$ [0.857]

2. $\frac{51}{90} = \frac{30}{x}$ [52.94]

TEACH

 In trigonometry, the goal is typically to find one side or one angle of a triangle. In Chapter 8, students learned how to solve this situation for right triangles. Now, students will be able to handle this situation for triangles with no right angles.

LESSON 14.1 The Law of Sines

Surveyors often must compute a distance between two points that is difficult or impossible to measure directly. You have used right triangles to solve this type of problem. However, the unknown distance may not be the side of a right triangle. The Law of Sines can be used to calculate the unknown distance in a triangle without a right angle.

A surveying crew needs to find the distance between two points, *A* and *B*, but a boulder blocks the line of sight. The surveyors are able to find some measurements using a third point, *C*. With this information, the surveying crew is able to find the distance, *AB*, using a formula called the Law of Sines.

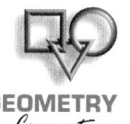

Exploration Sines and Triangles

In this exploration, you will need a protractor, a ruler, and graph paper.

GEOMETRY
Connection

1. Draw any large non-right triangle on a piece of graph paper. Label the sides and angles as shown.

2. Measure each side and each angle of the triangle.

3. Use your measurements to complete the table on the following page.

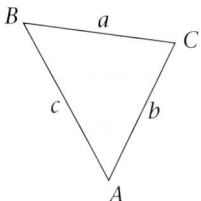

m∠A = ?	sin∠A = ?	a = ?	$\frac{\sin A}{a}$ = ?
m∠B = ?	sin∠B = ?	b = ?	$\frac{\sin B}{b}$ = ?
m∠C = ?	sin∠C = ?	c = ?	$\frac{\sin C}{c}$ = ?

4 Describe the ratios in the far-right column of your table. How do all the values in this column compare? How do your results compare with the results of other students in the class?

5 Write an equation that gives the relationship among the ratios. ❖

The Law of Sines

The relationship you found in Step 5 is called the *Law of Sines*. In triangle *ABC*, the sides opposite ∠A, ∠B, and ∠C are *a*, *b*, and *c*, respectively.

THE LAW OF SINES

The **Law of Sines** relates the three angle measures of a triangle to the lengths of the corresponding three sides.

$$\frac{\sin A}{a} = \frac{\sin B}{b} = \frac{\sin C}{c}$$

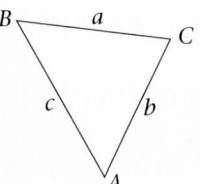

The Law of Sines can be used to determine all the parts of a triangle if the measures of two angles and their included side are known.

EXAMPLE 1

GEOMETRY
Connection

Find *b* of △ABC.

Solution▶

First find m∠C. Recall from geometry that the sum of the angle measures of any triangle is 180°.

$$m\angle A + m\angle B + m\angle C = 180°$$
$$62° + 53° + m\angle C = 180°$$
$$m\angle C = 65°$$

Now you can find *b* using the Law of Sines.

$$\frac{\sin B}{b} = \frac{\sin C}{c}$$

Substitute 53° for *B*, 65° for *C*, 5 for *c*, and solve for *b*.

$$\frac{\sin 53°}{b} = \frac{\sin 65°}{5}$$
$$b \sin 65° = 5 \sin 53°$$
$$b = \frac{5 \sin 53°}{\sin 65°} \approx 4.4$$

Thus, *b* is approximately 4.4 units long. ❖

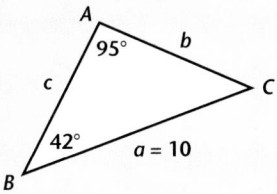

Alternate Example 2

A state highway department is surveying a bridge for a proposed roadway. The engineers have found two anchor points for the bridge at *A* and *C* on each side of the river. To design the bridge, they need to determine the distance between the two anchor positions. The diagram below indicates the measurements they made at *A* and *B* and between *A* and *B*. Find the distance between the anchor points of the proposed bridge to the nearest foot.

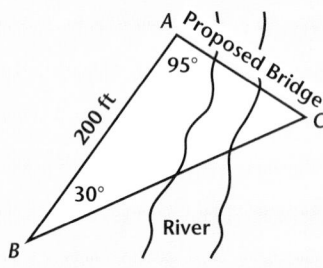

[**122 feet**]

EXAMPLE 2

Surveying   A state highway department is surveying a bridge site for a proposed roadway. The engineers have found two anchor positions for the bridge at *A* and *C* on each side of the river. To design the bridge, they need to determine the distance between the two anchor positions. The diagram below indicates the measurements they made from points *A* and *B*. Find the length of the bridge, to the nearest foot, in the proposed roadway project.

Solution▸

The measures of two angles, *A* and *B*, and their included side, *c*, are known. Therefore, to find *b* use the Law of Sines.

$$\frac{\sin B}{b} = \frac{\sin C}{c}$$

The measures for ∠*B* and ∠*A* are given. Find m∠*C*.

$$m\angle C = 180° - (18° + 138°)$$
$$= 24°$$

Substitute 24° for *C*, 18° for *B*, and 317 for *c*.

$$\frac{\sin 18°}{b} = \frac{\sin 24°}{317}$$
$$b = \frac{317 \sin 18°}{\sin 24°} \approx 240.84$$

Thus, the bridge should be about 241 feet long. ❖

Try This   Refer to the surveying problem on page 770. Use the Law of Sines to find the distance, to the nearest foot, between points *A* and *B*.

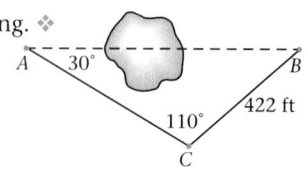

---

ongoing
ASSESSMENT

Try This

AB ≈ 793 feet

RETEACHING
the
lesson

Using Visual Models
Have students work in groups for this activity. Have each group draw two triangles on the chalkboard. For each triangle, have students use a ruler and a protractor to measure the lengths of each side and angle. Then, for each triangle, have students setup the proportions in the Law of Sines. Have each group write a short explanation of how the Law of Sines applies to their triangles.

Proof of the Law of Sines

Consider the acute triangle ABC. When you draw a line segment from point C to point D that is perpendicular to $\overline{AB}$, you form two right triangles, $\triangle ADC$ and $\triangle BDC$.

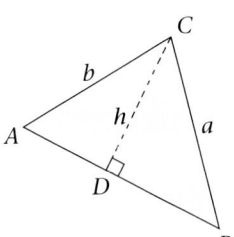

Find $\sin A$ and $\sin B$ in terms of a, b, and h.

$$\sin A = \frac{h}{b} \qquad \sin B = \frac{h}{a}$$

Solve both equations for h.

$$b \sin A = h \qquad a \sin B = h$$

Using substitution, set the two values for h equal to each other.

$$b \sin A = a \sin B$$

Divide both sides by ab, and simplify.

$$\frac{b \sin A}{ab} = \frac{a \sin B}{ab}$$

$$\frac{\sin A}{a} = \frac{\sin B}{b}$$

By the transitive property of equality, it follows that

$$\frac{\sin A}{a} = \frac{\sin B}{b} = \frac{\sin C}{c}.$$

CRITICAL Thinking

Suppose you know the measures of two sides of a triangle and their included angle. Can the Law of Sines be used to find any additional information about the triangle? Explain.

Cultural Connection: Asia The Law of Sines was probably first proved by the central Asian mathematician Nasir-al-Din (1201–1274). After Baghdad was pillaged by the Mongols in the 1250s, a new observatory was built there by the Mongol ruler Hulagu. Nasir-al-Din was the director of this institute in Baghdad, where African and Asian learning was pooled with the learning of Greece. Nasir and his colleagues extended and refined early Indian trigonometry and were among the first to treat trigonometry as separate from astronomy. Together, they developed many of the trigonometric relations that you will study in this chapter.

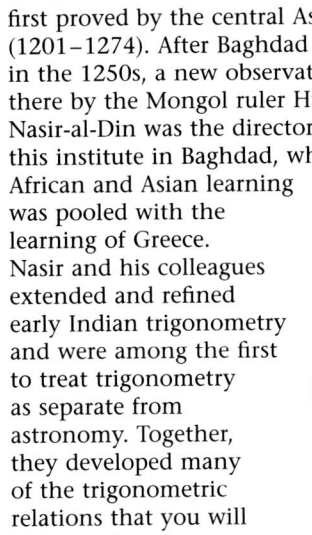

Astrolabe, dated 1216

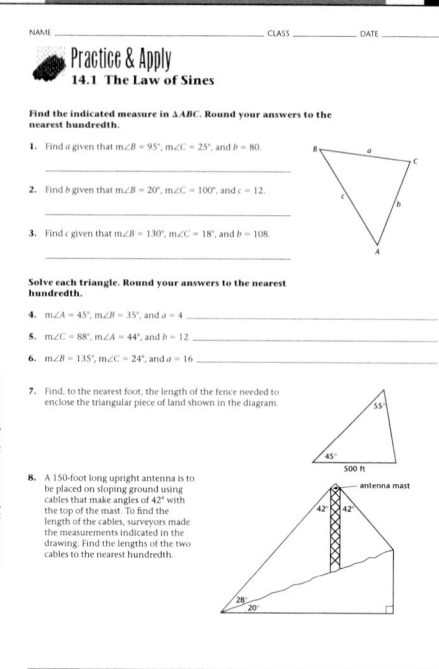
EXERCISES & PROBLEMS

Communicate

1. Explain how the Law of Sines relates the angles of a triangle to the sides of a triangle.

2. Discuss why the length of a side and the angle opposite it must be known in order to apply the Law of Sines.

3. What are two sets of triangle measures that enable you to apply the Law of Sines?

4. Do both triangle measures of Exercise 3 uniquely determine the size and shape of the triangle? Explain.

5. Explain how right-triangle trigonometry is used in the proof of the Law of Sines.

Practice & Apply

For Exercises 6–13, find the indicated measure of △*ABC*, and show your work. Round your answers to the nearest hundredth.

6. Find *b* given that m∠*B* = 37°, m∠*C* = 58°, and *a* = 14.30. 8.64

7. Find *c* given that m∠*B* = 42°, m∠*A* = 65°, and *b* = 212. 302.99

8. Find *b* given that m∠*C* = 49°, m∠*A* = 17°, and *c* = 3.24. 3.92

9. Find *a* given that m∠*A* = 53°, m∠*B* = 22°, and *c* = 8. 6.61

10. Find m∠*A* given that *a* = 1.17, *b* = 2.83, and m∠*B* = 97°. 24.23°

11. Find *c* in Exercise 10. 2.44

12. Find m∠*C*, given that *c* = 582, *b* = 736, and m∠*B* = 67°. 46.7°

13. Find *a* in Exercise 12. 732.13

14. Does the Law of Sines hold for right triangles? Explain. Yes

15. To find the distance from point *A* to point *B* across a river, a base line, $\overline{AC}$, is established. Find the distance, to the nearest meter, from *A* to *B*. Approx 690 m

16. A ship at sea is sighted from two observation posts, *A* and *B*, on shore. Find the distance, to the nearest tenth of a kilometer, from observation post *A* to the ship. 14.5 km

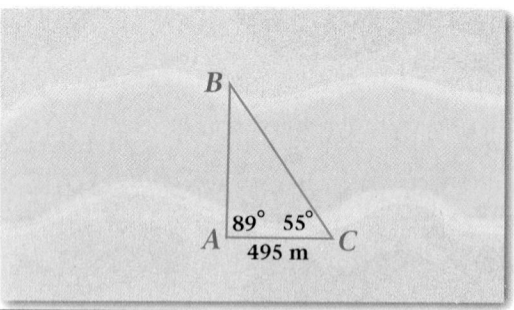

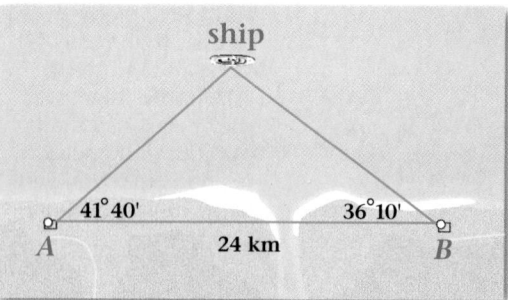

14. Yes, all three ratios will be equivalent to $\frac{1}{c}$ where sin *C* = sin 90° = 1 and *c* is the hypotenuse.

17. Find, to the nearest meter, the distance AB across the lake. 156 m

18. Find, to the nearest meter, the distance BC across the lake. 184 m

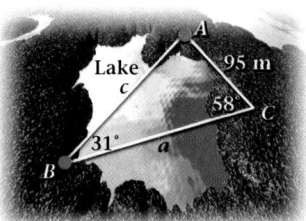

Solve each triangle for the missing angle measures and lengths of sides. Round your answers to the nearest hundredth.

19. $a = 13$, m$\angle A = 41°$, m$\angle B = 75°$

20. m$\angle A = 71°$, $a = 20$, m$\angle C = 62°$

21. m$\angle A = 71°$, m$\angle B = 42°$, $c = 15$

22. $a = 12$, m$\angle B = 110°$, m$\angle C = 35°$

23. $b = 503$, m$\angle A = 15°$, m$\angle B = 105°$

24. m$\angle B = 125°$, m$\angle A = 28°$, $b = 14$

25. $c = 16.5$, m$\angle A = 38°$, m$\angle C = 54°$

26. $b = 14.4$, m$\angle A = 72°$, m$\angle C = 19°$

27. $b = 224$, m$\angle A = 21°10'$, m$\angle B = 84°40'$

28. $c = 916$, m$\angle A = 15°40'$, m$\angle B = 60°30'$

29. m$\angle A = 101°$, m$\angle C = 37°10'$, $a = 23$

30. m$\angle B = 152°$, $b = 95$, m$\angle C = 12°10'$

31. $a = 1.50$, m$\angle B = 32°30'$, m$\angle C = 54°50'$

32. $a = 75.36$, m$\angle A = 18°25'$, m$\angle C = 32°05'$

33. Surveying A bridge will be built across a chasm between 2 cliffs at points A and B. Find the length of the bridge, to the nearest meter, using the information given in the diagram at right. 1073 m

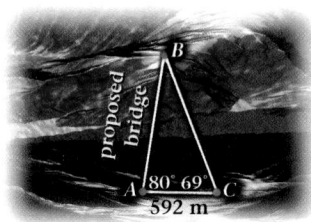

34. Two points, M and N, are separated by a swamp. A base line, $\overline{MK}$, is established on one side of the swamp. $\overline{MK}$ is 180 meters in length. The measures of angles NMK and MKN are 44° and 62°, respectively. Find the distance, to the nearest meter, between M and N. 165 m

35. Surveying Suppose you wish to estimate the height of a mountain 8817 feet that you can see as you drive toward it. At a rest stop, you use a clinometer, which is a pocket-sized device that can measure angles of elevation. You measure the angle of inclination from your eye to the top of the mountain to be 11.5°. You then set your car's mileage indicator to zero and drive toward the mountain for 5 miles. You stop there and measure the angle of inclination to be 27.5°. How high (to the nearest foot) is the mountain?

11.5°   27.5°

5 mi

Clinometer

19. $b = 19.14$, $c = 17.81$, m$\angle C = 64°$

20. $b = 15.47$, $c = 18.68$, m$\angle B = 47°$

21. $a = 15.41$, $b = 10.90$, m$\angle C = 67°$

22. $b = 19.66$, $c = 12$, m$\angle A = 35°$

23. $a = 134.78$, $c = 450.98$, m$\angle C = 60°$

24. $a = 8.02$, $c = 7.76$, m$\angle C = 27°$

25. $a = 12.56$, $b = 20.38$, m$\angle B = 88°$

26. $a = 13.70$, $c = 4.69$, m$\angle B = 89°$

27. $a = 81.23$, $c = 216.44$, m$\angle C = 74°10'$

28. $a = 254.75$, $b = 821.06$, m$\angle C = 103°50'$

29. $b = 15.63$, $c = 14.16$, m$\angle B = 41°50'$

30. $a = 55.21$, $c = 42.65$, m$\angle A = 15°50'$

31. $b = 0.81$, $c = 1.23$, m$\angle A = 92°40'$

32. $b = 184.06$, $c = 126.70$, m$\angle B = 129°30'$

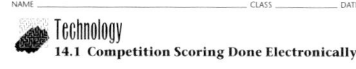

36. Sports In a hot-air-balloon competition, a prize is awarded to the balloon that is highest as it passes over a straight road not far from the takeoff point. To measure the height of the balloons as they pass over the road, two judges position themselves about 500 yards apart on the road so that the balloons will pass over the road between them. As the first balloon passes over the road, one judge measures the angle of elevation to be 44.8°, and the other measures the angle of elevation to be 54.4°. Determine the height of the balloon using these measurements. ≈ 290 yd

37. Two ranger stations, *A* and *B*, located 10 kilometers apart receive a distress call from a camper. Electronic equipment allows them to determine that the camper is at an angle of 71° from Station *A* and an angle of 100° from Station *B*. The line segment connecting the stations serves as one side of each of these angles. Which of the stations is closer to the camper? How far away is it from the camper?
Station B; ≈ 60.4 km

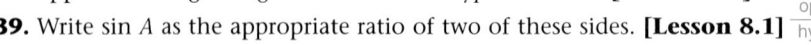

Look Back

38. Copy the right triangle shown, and label the sides adjacent and opposite angle *A* with the words *adjacent* and *opposite*. Label the side opposite the right angle with the word *hypotenuse*. **[Lesson 8.1]**

39. Write sin *A* as the appropriate ratio of two of these sides. **[Lesson 8.1]** $\frac{\text{opp}}{\text{hyp}}$

40. Write cos *A* as the appropriate ratio of two of these sides. **[Lesson 8.1]** $\frac{\text{adj}}{\text{hyp}}$

41. Write tan *A* as the appropriate ratio of two of these sides. **[Lesson 8.1]**

$$\tan A = \frac{\text{opposite}}{\text{adjacent}}$$

Look Beyond

42. **Geometry** The Law of Sines can be used to investigate triangles when the measures of two sides and an angle opposite one of the sides are known. It is possible that more than one triangle may be constructed having these same measures.

Use the Law of Sines to prove that side-side-angle (SSA) alone is not sufficient to determine the unique size and shape of a triangle.

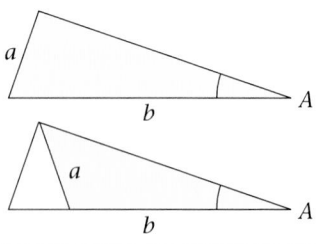

42. Since $\sin(180° - A) = \sin A$, if $\frac{\sin A}{a} = \frac{\sin B}{b}$,

then $\frac{\sin(180° - A)}{a} = \frac{\sin B}{b}$.

LESSON 14.2 The Law of Cosines

why *You know how to determine measures of a triangle when you know the measures of two angles and the length of the included side. You can also find the measures of a triangle when two sides of a triangle and their included angle are known.*

Using the Law of Cosines

Navigation Two fishing boats leave the same dock at the same time and travel along straight paths that together make an angle of 115° with each other. One boat moves at 24 nautical miles per hour while the other moves at 20 nautical miles per hour. After three hours, the slower boat sends a distress signal to the other boat. What is the distance between the two boats?

In this problem, the measures of two sides and the included angle are known. To find the rescue distance, estimated time of arrival, and direction, a new formula is needed.

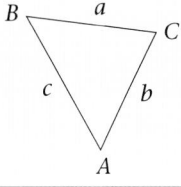

Slower boat

115° Dock

Faster boat

LAW OF COSINES

Given two sides, a and b, of a triangle and the measure of angle C included between them, the **Law of Cosines** states that
$$c^2 = a^2 + b^2 - 2ab \cos C,$$
where c is the length of the side opposite angle C.

CRITICAL Thinking The Law of Cosines can be applied to any triangle when two sides and an included angle are known. Write the Law of Cosines when a, c, and m∠B are known, or when b, c, and m∠A are known.

ALTERNATIVE teaching strategy

Hands-On Strategies Have two students stand in a corner of the classroom. Have one student walk a given distance along a wall of the room. Have the second student walk a given distance at a given angle with respect to the first student's path. These distances and angles should be mea- sured using a meter stick and a large protractor. Have each student in the class use the Law of Cosines to determine the distance between the two students. Then have a third student mea- sure this distance to confirm that the Law of Cosines works. Repeat this exercise with a dif- ferent groups of students.

PREPARE

Objectives

- Find all the parts of a triangle when given the lengths of two sides and the measure of the included angle.
- Find all the parts of a triangle when given the lengths of all three sides.

RESOURCES

- Practice Master   14.2
- Enrichment Master   14.2
- Technology Master   14.2
- Lesson Activity Master   14.2
- Quiz   14.2
- Spanish Resources   14.2

This lesson may be considered optional in the core course. See page 768A.

Assessing Prior Knowledge

In each of the following prob- lems, find an acute angle that satisfies theta, θ. Round θ to the nearest degree.

1. $\theta = \sin^{-1}(0.39)$ [23°]

2. $\theta = \sin^{-1}(0.96)$ [74°]

TEACH

why Students will learn that it is possible to find all the measures of any triangle when they are given either two sides and the included angle or all three sides.

CRITICAL Thinking
$$b^2 = a^2 + c^2 - 2ac \cos B$$
$$a^2 = b^2 + c^2 - 2bc \cos A$$

Use Transparency ▶ 60

Cooperative Learning

Have students work in groups of three to solve problems where three sides and no angles are given. Each student can solve for one of the angles. Then have them total their angles to confirm that they add up to 180°.

The Law of Cosines can be used to solve the navigation problem. Assume the faster boat moves at the same speed with which it left the dock. First find the travel distance and the time needed to reach the boat in distress. Then find the angle through which the rescue boat travels to reach the boat in distress.

Find the distance of each boat from the dock.

Slower boat: $d = rt$
$d = (20)(3) = 60$ nautical miles

Faster boat: $d = rt$
$d = (24)(3) = 72$ nautical miles

Sketch a triangle to represent the problem.

To find the side opposite the dock in the triangle, which is the distance the rescue boat must travel, use the Law of Cosines.

$c^2 = a^2 + b^2 - 2ab \cos(C)$
$c^2 = 60^2 + 72^2 - 2(60)(72) \cos 115°$
$c^2 \approx 12{,}435$
$c \approx \pm 111.5$

Thus, the distance the rescue vessel must travel is approximately 111.5 nautical miles.

Assuming the faster boat continues to travel at 24 nautical miles per hour, it will take $\frac{112}{24}$ hours, or about 4 hours and 40 minutes, to reach the disabled boat.

To determine the angle through which the rescue boat must turn, use the Law of Sines to find the measure of $\angle B$.

$\frac{\sin B}{60} = \frac{\sin 115°}{112}$

$\sin B = \frac{60 \sin 115°}{112}$

$B = \sin^{-1}\left(\frac{60 \sin 115°}{112}\right)$

$B \approx 29°$ or $B \approx 151°$

Since one angle in the triangle is already obtuse, the only choice is the acute angle, so $B \approx 29°$.

Thus, the navigator reports to the captain that it will take approximately 4 hours and 40 minutes to reach the disabled boat, moving in a direction that makes a 29° angle with respect to the initial direction line.

Slower boat

A

60 nautical miles

Doc

115°

c

C

72 nautical miles

B

Faster boat

ENRICHMENT The problem of finding all angles, when given all the sides, requires a different formulation of the Law of Cosines. This form is derived by basic algebra. It is expressed by the formula, $C = \cos^{-1}\left(\frac{a^2 + b^2 - c^2}{2ab}\right)$.

INCLUSION strategies **Hands-On Strategies** Have each student draw two sides of a triangle. Have the student measure the length of the sides and the size of the included angle and then compute the length of the missing side using the Law of Cosines. Then have the student measure the length of the third side and compare the results of the computation with the measured lengths.

Why are there two possible answers for the measure of $\angle B$ on page 778? Explain how you can tell which is the correct answer.

Sometimes it may not be obvious which angle to choose. When this is the case, it is better to determine the unknown angle by applying the Law of Cosines again.

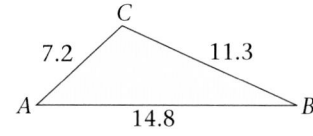

EXAMPLE 1

Find the measure of all angles, given that $a = 11.3$, $b = 7.2$, and $c = 14.8$.

Solution▶

Sketch a diagram roughly to scale.

First use the Law of Cosines to find $\angle B$.

$$b^2 = a^2 + c^2 - 2ac \cos B$$
$$7.2^2 = 11.3^2 + 14.8^2 - 2(11.3)(14.8) \cos B$$
$$\frac{7.2^2 - 11.3^2 - 14.8^2}{-2(11.3)(14.8)} = \cos B$$
$$\cos^{-1}\left(\frac{7.2^2 - 11.3^2 - 14.8^2}{-2(11.3)(14.8)}\right) = B$$
$$28.16° \approx B$$

Then use the Law of Sines to find $\angle A$.

$$\frac{\sin A}{a} = \frac{\sin B}{b}$$
$$\sin^{-1}\left(\frac{a \sin B}{b}\right) = A$$
$$\sin^{-1}\left(\frac{11.3 \sin 28.16}{7.2}\right) \approx A$$
$$48° \approx A \text{ or } 132° \approx A$$

Since $\angle C$ is opposite the longest side, its measure is the largest angle measure. Thus, $m\angle A \approx 48°$.

You can check the measure of $\angle A$ using the Law of Cosines.

$$a^2 = b^2 + c^2 - 2bc \cos A$$
$$11.3^2 = 7.2^2 + 14.8^2 - 2(7.2)(14.8) \cos A$$
$$\cos^{-1}\left[\frac{11.3^2 - 7.2^2 - 14.8^2}{-2(7.2)(14.8)}\right] = A$$
$$48° \approx A$$

To find $m\angle C$, use $m\angle A$ and $m\angle B$.

$$m\angle C \approx 180° - 48° - 28°, \text{ or } 104° \quad ❖$$

Try This Find the measure of each angle of a triangle, given that the lengths of the sides are 2, 1.64, and 1.82.

Explain why it is possible to get two solutions when using the Law of Sines to find an angle measure, but only one angle measure results when using the Law of Cosines (see Example 1). HINT: Consider the domains of the inverse sine and cosine functions.

RETEACHING the lesson

Technology Have students in groups use computerized geometry software to draw a triangle and measure the sides and angles. Students drag one vertex of the triangle in order to examine many different triangles. For each triangle, have students confirm the computerized data by using the Law of Cosines to find

1. the measures of the angles, given the lengths of the sides and

2. the remaining parts of the triangle, given the measures of two sides and their included angle.

Some geometry software has the ability to compute. If this is possible, have students plug in the appropriate parts of the Law of Cosines using all three forms. As the vertex of the triangle is dragged, have students notice that the Law of Cosines continues to give correct values for all angles and lengths of sides.

$\cos C = \dfrac{x\text{-coordinate of } A}{b}$, so the x-coordinate of A is $b \cos C$.

$\sin C = \dfrac{y\text{-coordinate of } A}{b}$, so the y-coordinate of A is $b \sin C$.

CRITICAL
Thinking

Yes. When $C = 180°$, $\cos C = -1$, so the Law of Cosines becomes $c^2 = a^2 + b^2 + 2ab$. By factoring, $c^2 = (a+b)^2$. So $c = a + b$. In this case, ABC is not a triangle but a straight line.

Proof of the Law of Cosines

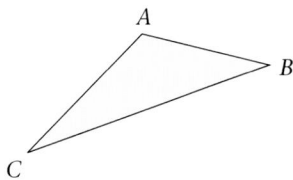

A

B

C

COORDINATE
GEOMETRY
Connection

To prove the Law of Cosines, consider the triangle shown.

Place the triangle in the coordinate plane with point C at the origin and the line CB on the x-axis.

A line is drawn from vertex A to side CB, so that the line is perpendicular to $\overline{CB}$. The sides opposite A, B, and C are labeled a, b, and c, respectively.

The coordinates of the vertices C and B are then $C(0, 0)$ and $B(a, 0)$.

Using trigonometric ratios for $\angle C$, the coordinates of A are $(b \cos C, b \sin C)$.

y

$A(b\cos C, b\sin C)$

b

c

$C\,(0, 0)$

a

$B\,(a, 0)$

x

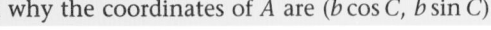

Explain why the coordinates of A are $(b \cos C, b \sin C)$.

Use the distance formula to find the distance, c, between A and B.

$$c = \sqrt{(b \cos C - a)^2 + (b \sin C - 0)^2}$$
$$c^2 = b^2 \cos^2 C - 2ab \cos C + a^2 + b^2 \sin^2 C$$
$$c^2 = b^2 \cos^2 C + b^2 \sin^2 C - 2ab \cos C + a^2$$
$$c^2 = b^2 \left(\cos^2 C + \sin^2 C\right) - 2ab \cos C + a^2$$

Recall from the unit circle in Chapter 8 that the coordinates of a point on the unit circle can be expressed $(\cos \theta, \sin \theta)$, where θ is in standard position. Therefore, from the "Pythagorean" Right-Triangle Theorem, you know that $\cos^2 C + \sin^2 C = 1$. Substitute 1 for $\cos^2 C + \sin^2 C$ in the equation above.

$$c^2 = b^2 \cdot 1 - 2ab \cos C + a^2$$
$$c^2 = a^2 + b^2 - 2ab \cos C$$

Suppose the angle included between sides a and b is a right angle. Since $\cos 90° = 0$, the Law of Cosines reduces to the "Pythagorean" Right-Triangle Theorem.

$$c^2 = a^2 + b^2 - 2ab \cos 90°$$
$$c^2 = a^2 + b^2$$

Notice that any side may be considered to lie along the x-axis. By considering vertices A or B to be at the origin, two equivalent forms of the Law of Cosines may be derived.

CRITICAL
Thinking

In the standard form of the Law of Cosines, $c^2 = a^2 + b^2 - 2ab \cos C$, does the formula hold when $C = 180°$? If so, what is the corresponding length c?

EXERCISES & PROBLEMS

Communicate

1. Explain how to use the Law of Cosines.

2. If $m\angle C = 90°$, what does the Law of Cosines equation become? What is another name for this relation?

3. If $m\angle C < 90°$, is the expression $a^2 + b^2 - 2ab \cos C$ for c^2 less than or greater than the square of the hypotenuse of a right triangle with legs a and b? Explain.

4. If $m\angle C > 90°$, is the expression $a^2 + b^2 - 2ab \cos C$ for c^2 less than or greater than the square of the hypotenuse of a right triangle with legs a and b? Explain.

5. Describe the two forms of information necessary to use the Law of Cosines. How does each set of information compare with the information necessary to use the Law of Sines? Explain.

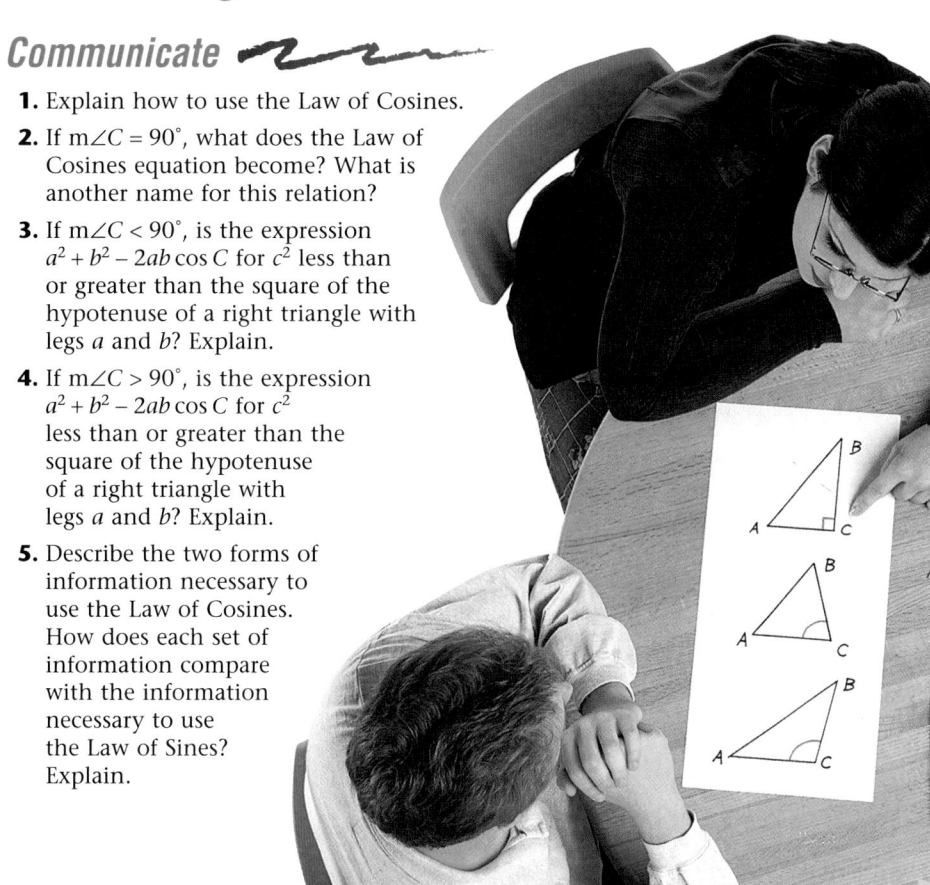

ASSESS

Selected Answers
Odd-Numbered Exercises 7–23 and 27–33

Assignment Guide
Core 1–9, 12–34

Core-Plus 1–5, 8–36

Technology
Graphics calculators are needed for Exercise 36. If students use calculators for any other exercise in this lesson, have them remember that their calculators should be in degree mode.

Error Analysis
Students must be careful to substitute the values for a, b, and c into the correct position in the formula.

Practice & Apply

Find the indicated measure for $\triangle ABC$, and show your work.

6. Find side b given $a = 123$, $c = 97$, and $m\angle B = 22°$. ≈ 49.13

7. Find $m\angle C$ in Exercise 6. $\approx 47.7°$

8. Find side a given $b = 82$, $c = 63.2$, and $m\angle A = 114°$. ≈ 122.20

9. Find $m\angle B$ in Exercise 8. $\approx 37.8°$

10. Find $m\angle A$ given $a = 2.47$, $b = 3.80$, and $c = 4.24$. $\approx 35.2°$

11. Find $m\angle B$ and $m\angle C$ in Exercise 10. $\approx 62.6°$; $\approx 82.2°$

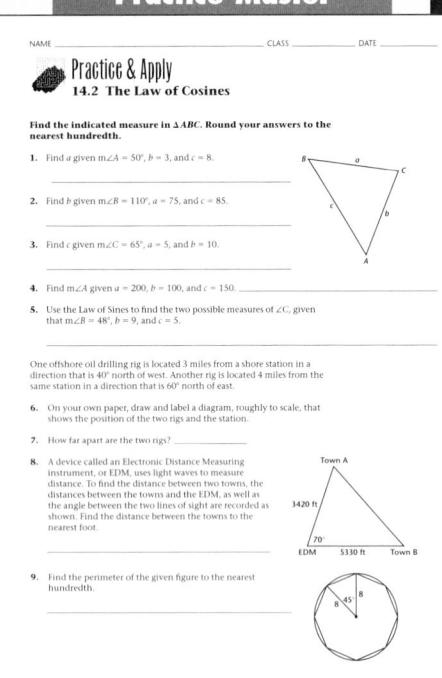

Alternative
ASSESSMENT

Portfolio Assessment

Have each student make up three problems using the Law of Cosines. Students should explain why the Law of Cosines is used to solve their problems. Have students pair up, switch problems, and solve their partner's problem. Then each pair of students can evaluate their work and write a short summary.

Use Transparency ▶ 61

12. All three angles measure 60°; this is an equilateral triangle

15. The diagram differs in that the angle at the center, $\angle A$ in this case, is obtuse. In the proof given on page 780, the angle at the center, $\angle C$, is acute.

16. No, the distance formula is valid between any two points on the plane.

---

12. Using the Law of Cosines, find all the angle measures of a triangle whose sides each measure 1 unit. Explain your answers. 60°

13. Find all the angle measures of an isosceles triangle whose base is $\frac{1}{3}$ as long as its legs. 80.4°; 80.4°; 19.2°

14. Find all the angle measures of a triangle whose sides measure 6, 3, and 5 units. 93.8°; 56.3°; 29.9°

Coordinate Geometry In $\triangle ABC$ shown, the altitude from vertex B is drawn.

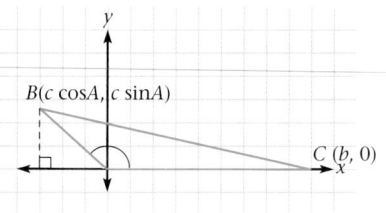

15. How does this diagram differ from the diagram shown in the proof of the Law of Cosines given on page 780?

16. Does the difference in diagrams make a change necessary in the proof of the Law of Cosines given in the lesson? Why or why not?

For Exercises 17–19, use the form of the Law of Cosines
$$c^2 = a^2 + b^2 - 2ab \cos C.$$

17. If $\angle C$ is close to 0°, show that $c \approx |a - b|$.

18. If $\angle C$ is close to 180°, show that $c \approx |a + b|$.

19. Use the results of Exercises 17 and 18 to prove the *triangle inequality*
$$|a - b| < c < |a + b|.$$

Navigation A tanker leaves port at 9:00 A.M. in a direction that is 37° east of North. Its speed is 10 nautical miles per hour (knots). A pleasure boat leaves the same port at 10:00 A.M. in a direction that is 61° east of North with a speed of 15 knots.

20. Draw and label a diagram roughly to scale that shows each boat's position at 1:00 P.M.

21. How far apart will they be at 1:00 P.M.? ≈ 18.34 nautical miles

Maximum/Minimum Suppose you need to install a TV antenna on your roof. From the manufacturer's instructions, you read that the angle the two guy wires make with the antenna pole should be no less than 30°.

22. What is the minimum length of guy wire that you should buy? ≈ 45.8 ft

23. What is the height of the longest antenna pole that can fit on this roof? What amount of guy wire will be required? ≈ 22.7 ft; ≈ 64 ft

24. Is it possible for the guy wire to make a 40° angle with the 16-foot antenna pole? No

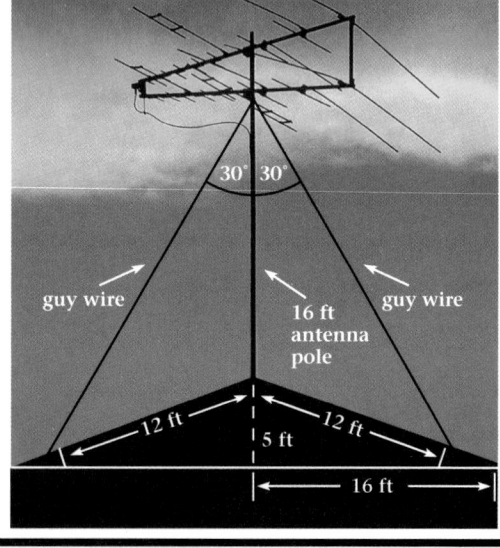

The pitch of this roof is 5 in 16. This means that for every 16 feet of run, there is a rise of 5 feet.

---

17. If $\angle C = 0$, then $\cos C$ is almost 1; so $c^2 \approx a^2 + b^2 - 2ab(1)$, or $c^2 \approx (a - b)^2$ so $c \approx \pm(a - b)$ or $c \approx |a - b|$.

18. If $\angle C = 180°$ then $\cos C = -1$; so $c^2 = a^2 + b^2 + 2ab = (a + b)^2$ so $c = \pm(a + b)$ or $c = |a + b|$.

19. If $0° < C < 180°$ then $|a - b| < c < |a + b|$

20. North

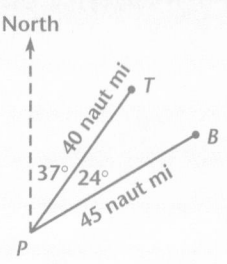

Portfolio Activity Refer to the Portfolio Activity on page 769. Two airplanes take off from the same airport at the same time. One flies at a cruising speed of 480 miles per hour headed in a direction that is 33° east of North. The other is flying at 540 miles per hour in a direction that is 48° west of North.

25. Draw and label a diagram roughly to scale that shows each plane's position after $1\frac{1}{2}$ hours.

26. How far apart are the planes after $1\frac{1}{2}$ hours? 996 miles

Look Back

27. Solve the inequality $3x^2 + 5x - 9 > 0$. **[Lesson 5.9]**
$x < \frac{-5 - \sqrt{133}}{6}$ or $x > \frac{-5 + \sqrt{133}}{6}$

Write each polynomial in factored form. **[Lesson 6.2]**

28. $f(x) = 3x^3 - 12x$
$f(x) = 3x(x + 2)(x - 2)$

29. $g(x) = 2x^4 - 12x^3 + 18x^2$ $g(x) = 2x^2(x - 3)^2$

30. Physics If you have 12 grams of a certain radioactive substance that has a half-life of 47 years, how much will remain after 76 years? **[Lesson 7.4]** 3.91 gm

31. Solve the equation $\log x + \log(2x - 3) = 3$.   **[Lesson 7.5]** $x = \frac{3 + \sqrt{8009}}{4}$

32. Find the horizontal asymptotes of the function $f(x) = \frac{2x^2 + 10x}{x^2 + 2x - 15}$.
[Lesson 9.4] $y = 2$

33. Find the remaining parts of $\triangle ABC$ given $m\angle B = 63.8°$, $m\angle C = 44.3°$, and $b = 0.825$.   **[Lesson 14.1]** $m\angle A = 71.9°$; $a = 0.874$; $c = 0.642$

34. Find the remaining parts of $\triangle ABC$ given $a = 7.15$, $b = 4.78$, and $m\angle B = 33.5°$. Is there only one triangle satisfying these conditions? Explain.   **[Lesson 14.1]** no $c = 3.26$; $m\angle A = 124.4°$; $m\angle C = 22.1°$
or $c = 8.66$; $m\angle A = 55.6°$; $m\angle C = 90.9°$

Look Beyond

35. Consider the equation $\sin^2 t + 2 \sin t - 3 = 0$. Make the substitution $w = \sin t$, and solve the resulting equation for w. 1; −3

36. Use your graphics calculator in radian mode to graph $g(x) = \sin\left(x + \frac{\pi}{6}\right)$.
Find the roots of $\sin\left(x + \frac{\pi}{6}\right) = 0$ for $-\pi < x < 2\pi$. −0.524; 2.618; 5.760

25.

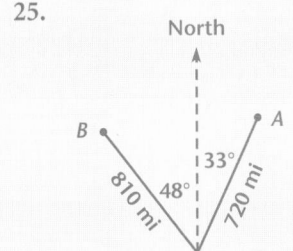

Objectives

• Find the value of any of the other five trigonometric functions when given the value of one function and the quadrant in which the angle lies.

RESOURCES

• Practice Master   14.3
• Enrichment Master   14.3
• Technology Master   14.3
• Lesson Activity Master   14.3
• Quiz   14.3
• Spanish Resources   14.3

This lesson may be considered optional in the core course. See page 768A.

Assessing Prior Knowledge

Use your calculator in degree mode to compute the following.

1. $\tan^{-1}(0.587)$ [**30°**]

2. $\sin^{-1}(0.743)$ [**48°**]

3. $\cos^{-1}(0.242)$ [**76°**]

TEACH

There are six basic trigonometric functions. All six functions can all be defined in terms of the sine and the cosine.

LESSON 14.3 Circular Function Relationships

why *The basic circular functions, sine and cosine, can be defined as the coordinates of a point on the unit circle. Other circular functions, which also appear in physics, engineering, astronomy, and surveying, can be constructed from the two basic functions.*

EXAMPLE 1

Home Improvement

Suppose you are installing a 25-foot-tall shortwave radio antenna in your backyard. The antenna manufacturer recommends that three guy wires be attached to the top of the antenna and that they make angles of 55° with the ground. How far from the base of the antenna should you anchor each guy wire?

Solution

Draw a diagram. One way to solve for x is to use the tangent ratio in the form $\tan \theta = \frac{\sin \theta}{\cos \theta}$.

$$\frac{25}{x} = \tan 55°, \text{ or } \frac{25}{x} = \frac{\sin 55°}{\cos 55°}$$

$$x = 25 \left(\frac{\cos 55°}{\sin 55°} \right) \approx 17.5$$

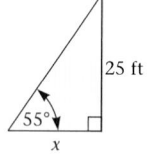

The wires should be anchored about 17.5 feet from the antenna base. ❖

Using Visual Models

Draw the graph of the sine function on the chalkboard. By taking reciprocals of selected values of θ, draw the graph of cosecant on the board. Repeat the process for cosine and secant and for tangent and cotangent.

In Example 1, the reciprocal of the tangent ratio is used to solve the equation $x = 25\left(\frac{\cos 55°}{\sin 55°}\right)$, or $x = 25\left(\frac{1}{\tan 55°}\right)$. The reciprocal of the tangent ratio is called the *cotangent* ratio. All three trigonometric functions, sine, cosine, and tangent, have reciprocal functions.

THE RECIPROCAL TRIGONOMETRIC FUNCTIONS

The **cotangent** function, $y = \cot \theta$, is the reciprocal of the tangent function. $\qquad \cot \theta = \frac{\cos \theta}{\sin \theta}$

The **cosecant** function, $y = \csc \theta$, is the reciprocal of the sine function. $\qquad \csc \theta = \frac{1}{\sin \theta}$

The **secant** function, $y = \sec \theta$, is the reciprocal of the cosine function. $\qquad \sec \theta = \frac{1}{\cos \theta}$

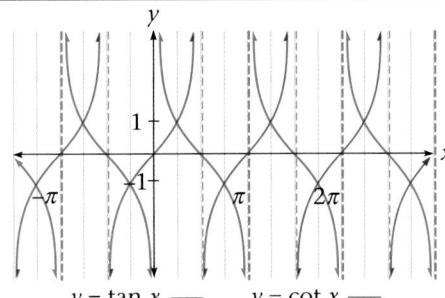

$y = \tan x$ —— $y = \cot x$ ——

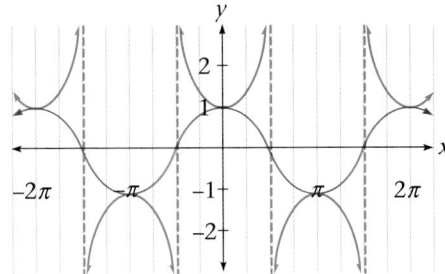

$y = \cos x$ —— $y = \sec x$ ——

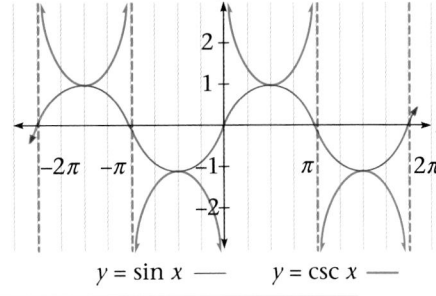

$y = \sin x$ —— $y = \csc x$ ——

Describe the relationship between the graph of each trigonometric function and the graph of its reciprocal function.

CRITICAL *Thinking*

GEOMETRY *Connection*

Let θ be an acute angle. The diagram is based on the unit circle. Explain the connection between the definition of each trigonometric function and its geometric representation.

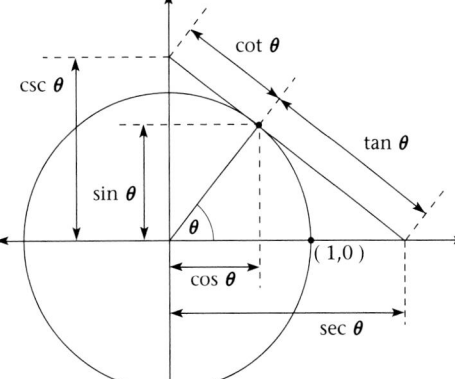

ENRICHMENT Because sine and cosine sometimes equal zero, some of the other functions will be undefined for certain values. For example, if $\sin \theta = 0$, then $\csc \theta$ is undefined for that value of θ. Explore with students the values for which each function is undefined.

EXAMPLE 2

If $\sin \theta = \frac{3}{5}$ and $\cos \theta = \frac{4}{5}$, find $\cot \theta$, $\csc \theta$, and $\sec \theta$.

Solution▸

Since both sine and cosine values are positive, θ must be in Quadrant I. So all trigonometric functions of θ will be positive.

$$\cot \theta = \frac{\cos \theta}{\sin \theta} = \frac{\frac{4}{5}}{\frac{3}{5}} = \frac{4}{3}$$

$$\csc \theta = \frac{1}{\sin \theta} = \frac{1}{\frac{3}{5}} = \frac{5}{3}$$

$$\sec \theta = \frac{1}{\cos \theta} = \frac{1}{\frac{4}{5}} = \frac{5}{4}$$

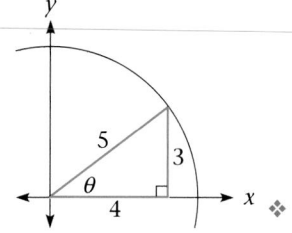

Try This If $\tan x = -1$ and $\sin x = \frac{\sqrt{2}}{2}$, find $\cos x$, $\cot x$, $\csc x$, and $\sec x$.

Pythagorean Identities

The Pythagorean identity $\cos^2 \theta + \sin^2 \theta = 1$ is a direct consequence of the "Pythagorean" Right-Triangle Theorem. There are other forms of this Pythagorean identity that use trigonometric notation. For example, divide both sides by $\cos^2 \theta$.

$$\cos^2 \theta + \sin^2 \theta = 1$$

$$\frac{\cos^2 \theta}{\cos^2 \theta} + \frac{\sin^2 \theta}{\cos^2 \theta} = \frac{1}{\cos^2 \theta}$$

$$1 + \tan^2 \theta = \sec^2 \theta$$

The three Pythagorean identities are given below.

THE PYTHAGOREAN IDENTITIES

$$\sin^2 \theta + \cos^2 \theta = 1$$
$$1 + \tan^2 \theta = \sec^2 \theta$$
$$1 + \cot^2 \theta = \csc^2 \theta$$

EXAMPLE 3

If $\tan t = \frac{5}{12}$ and t is in Quadrant III, use a Pythagorean identity to find $\cos t$ and $\sin t$.

Solution►

Draw a diagram. In Quadrant III, $\sin t$ and $\cos t$ are negative.

Substitute $\frac{5}{12}$ for $\tan t$ in the identity $1 + \tan^2 t = \sec^2 t$.

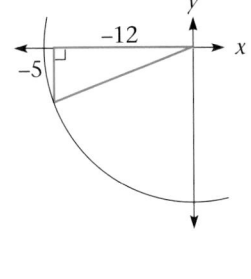

$$1 + \left(\frac{5}{12}\right)^2 = \sec^2 t$$
$$\frac{169}{144} = \sec^2 t$$
$$\pm\frac{13}{12} = \sec t$$

Cosine is the reciprocal of secant, and cosine is negative in the third quadrant. Therefore, $\cos t = -\frac{12}{13}$.

To find $\sin t$, use $\tan t = \frac{\sin t}{\cos t}$.

$$\tan t = \frac{\sin t}{\cos t}$$
$$(\cos t)(\tan t) = \sin t$$
$$\left(-\frac{12}{13}\right)\left(\frac{5}{12}\right) = \sin t$$
$$-\frac{5}{13} = \sin t$$

Thus, $\cos t = -\frac{12}{13}$ and $\sin t = -\frac{5}{13}$. ❖

EXERCISES & PROBLEMS

Communicate

For Exercises 1–4, explain how each function can be expressed in terms of sine, cosine, or both.

1. $\tan t$ **2.** $\cot t$

3. $\sec t$ **4.** $\csc t$

5. If the sine of an angle is negative and the cosine of that angle is positive, explain how to determine the quadrant in which the angle lies.

6. Explain how the three trigonometric Pythagorean identities are related to the "Pythagorean" Right-Triangle Theorem, $a^2 + b^2 = c^2$.

RETEACHING **the lesson**

Using Tables The objective of this lesson can be accomplished by demonstrating the mathematics encountered when completing the table for $\sin \theta = \frac{1}{2}$. Other values for the sine can be used in addition to values for cosine until students understand the relationship between the six trigonometric functions and their signs in the various quadrants.

	Quadrant in which θ lies			
	I	II	III	IV
$\cos \theta$	$\frac{\sqrt{3}}{2}$	$\frac{-\sqrt{3}}{2}$		
$\tan \theta$	$\frac{\sqrt{3}}{3}$	$\frac{-\sqrt{3}}{3}$		
$\cot \theta$	$\sqrt{3}$	$-\sqrt{3}$		
$\sec \theta$	$\frac{2\sqrt{3}}{2}$	$\frac{-2\sqrt{3}}{2}$		
$\csc \theta$	2	2		

Alternate Example 3

If $\tan t = -0.75$ and t is in Quadrant IV, use a Pythagorean identity to find $\cos t$ and $\sin t$. [$\cos t = 0.8$, $\sin t = -0.6$]

ASSESS

Selected Answers

Odd-Numbered Exercises 7–27

Assignment Guide

Core 1–8, 11–12, 15–17, 20–31

Core-Plus 1–6, 9–10, 13–15, 18–31

Technology

Graphics calculators are needed for Exercises 11–14, 16–19, and 29–31.

Most of the exercises in this lesson require students to use the radian mode of their calculators. However, some exercises require the degree mode. Make sure that students are aware of which mode to use for each problem.

Also, most calculators have only three keys for trigonometric functions. Students will have to use the reciprocal key when they use their calculators to find the remaining three functions.

Error Analysis

Most errors result when students give the incorrect sign for the requested trigonometric function. Students can avoid this error by drawing a triangle in the appropriate quadrant on a coordinate plane. They can then use the coordinates of the vertex and the trigonometric ratios to verify the sign.

ALTERNATIVE ASSESSMENT

Performance Assessment

Have students describe the step-by-step method they would use to determine the value of one trigonometric function of an angle when they are given the quadrant in which it lies and the value of some other trigonometric function of that angle.

 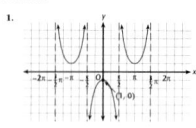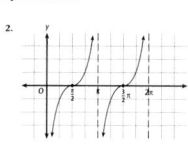
788 Lesson 14.3

Practice & Apply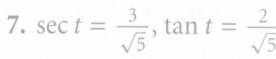

7. Given that $\sin t = \frac{2}{3}$ and $0 \le t \le \frac{\pi}{2}$, find *exact* values of $\sec t$ and $\tan t$.

8. Repeat Exercise 7 for $\frac{\pi}{2} \le t \le \pi$.

9. Given that $\cos t = \frac{1}{5}$ and that $0 \le t \le \frac{\pi}{2}$, find *exact* values of $\csc t$ and $\cot t$.

10. Repeat Exercise 9 for $\frac{3\pi}{2} \le t \le 2\pi$.

11 Use the inverse trigonometric function keys on your calculator to find solutions accurate to the nearest thousandth for $\sin t$ and $\cot t$, given that $\cos t = \frac{2}{3}$ and $0 \le t \le \frac{\pi}{2}$. $\sin t \approx 0.745$ $\cot t \approx 0.894$

12 Repeat Exercise 11 for $\frac{3\pi}{2} \le t \le 2\pi$. $\sin t \approx -0.745$ $\cot t \approx -0.894$

13 Use the inverse trigonometric function keys on your calculator to find solutions accurate to the nearest thousandth for $\cos t$ and $\tan t$, given that $\sin t = \frac{1}{5}$ and $0 \le t \le \frac{\pi}{2}$. $\cos t \approx 0.980$ $\tan t \approx 0.204$

14 Repeat Exercise 13 for $\frac{\pi}{2} \le t \le \pi$. $\cos t \approx -0.980$ $\tan t \approx -0.204$

15. Prove the Pythagorean identity $1 + \cot^2 x = \csc^2 x$ using the Pythagorean identity $\sin^2 x + \cos^2 x = 1$.

For Exercises 16–19, use the inverse trigonometric keys on your calculator to solve each problem to the nearest thousandth.

16 Given $\cos t = \frac{3}{7}$ and $\sin t < 0$, find $\sin t$ and $\cot t$. $\sin t \approx -0.904$ $\cot t \approx -0.474$

17 Given $\sin t = \frac{3}{5}$ and $\tan t < 0$, find $\cos t$ and $\tan t$. $\cos t \approx -0.800$ $\tan t \approx -0.750$

18 Given $\sec t = -\frac{5}{3}$ and t is in Quadrant III, find $\sin t$ and $\tan t$.

19 Given $\cot t = \frac{7}{4}$ and t is in Quadrant III, find $\cos t$ and $\csc t$.

---

7. $\sec t = \frac{3}{\sqrt{5}}$, $\tan t = \frac{2}{\sqrt{5}}$

8. $\sec t = \frac{-3}{\sqrt{5}}$, $\tan t = \frac{-2}{\sqrt{5}}$

9. $\csc t = \frac{5}{\sqrt{24}}$, $\cot t = \frac{1}{\sqrt{24}}$

10. $\csc t = -\frac{5}{\sqrt{24}}$, $\cot t = -\frac{1}{\sqrt{24}}$

15. $\sin^2 x + \cos^2 x = 1$; $\frac{\sin^2 x}{\sin^2 x} + \frac{\cos^2 x}{\sin^2 x} = \frac{1}{\sin^2 x}$; $1 + \cot^2 x = \csc^2 x$

18. $\sin t \approx -0.800$, $\tan t \approx 1.333$

19. $\cos t \approx -0.868$, $\csc t \approx -2.016$

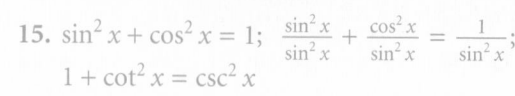

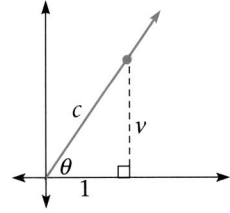

Coordinate Geometry Use the triangle at the right for Exercises 20–22.

20. Write the relationship between c and v.

21. Show that $c = \sec \theta$ and $v = \tan \theta$.

22. Use your results from Exercises 20 and 21 to prove that $1 + \tan^2 \theta = \sec^2 \theta$.

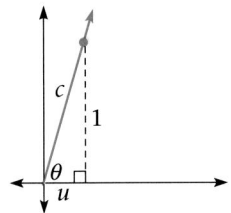

Coordinate Geometry Use the triangle at the right for Exercises 23–25.

23. Write the relationship between c and u.

24. Show that $c = \csc \theta$ and $u = \cot \theta$.

25. Use the results of Exercises 23 and 24 to prove that $1 + \cot^2 \theta = \csc^2 \theta$.

Look Back

26. Graph $y = 2(x - 3)^2 + 5$. Identify the shape of the graph, and indicate the vertex, focus, and axis of symmetry.   **[Lesson 10.1]**

27. Graph $\frac{x^2}{4} + \frac{y^2}{9} = 1$. Identify the shape of the graph, and specify the vertices, co-vertices, and foci.   **[Lesson 10.3]**

28. Given triangle ABC with sides $b = 1.73$, $c = 2.18$, and $m\angle A = 52.5°$, find $m\angle B$.   **[Lesson 14.2]** 50.6°

Look Beyond

29 Use your graphics calculator to graph $f(x) = \sin\left(x + \frac{\pi}{6}\right)$ and $g(x) = \sin x \cos \frac{\pi}{6} + \cos x \sin \frac{\pi}{6}$ on the same screen. What appears to be true? same graphs

30 Use your graphics calculator to graph $f(x) = \cos\left(x - \frac{\pi}{8}\right)$ and $g(x) = \cos x \cos \frac{\pi}{8} + \sin x \sin \frac{\pi}{8}$ on the same coordinate axes. Describe the relationship between the two graphs. same graphs

31 Use your graphics calculator to graph $f(x) = \cos(2x)$ and $g(x) = \cos^2 x - \sin^2 x$ on the same coordinate axes. Describe the relationship between the two graphs. same graphs

20. By the "Pythagorean" Right-Triangle Theorem, $c^2 = 1^2 + v^2$.

21. $\sec \theta = \frac{c}{1}$, so $c = \sec \theta$; $\tan \theta = \frac{v}{1}$ so $v = \tan \theta$

22. $c^2 = 1 + v^2$, so $\sec^2 \theta = 1 + \tan^2 \theta$

23. By the "Pythagorean" Right-Triangle Theorem, $c^2 = 1^2 + u^2$.

24. $\csc \theta = \frac{c}{1}$ so $c = \csc \theta$; $\cot \theta = \frac{u}{1}$ so $u = \cot \theta$

25. $c^2 = 1 + u^2$, so $\csc^2 \theta = 1 + \cot^2 \theta$

26.

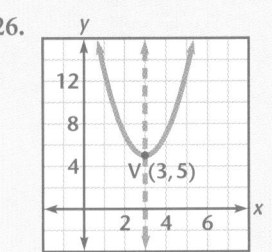

parabola; $V(3, 5)$; $F(3, \frac{41}{8})$; axis of symmetry: $x = 3$

27.

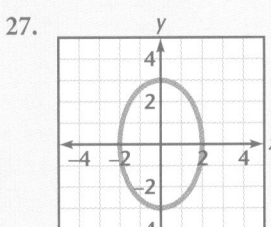

ellipse; vertices $(0, 3)$ and $(0, -3)$; co-vertices $(2, 0)$ and $(-2, 0)$; foci $(0, \sqrt{5})$ and $(0, -\sqrt{5})$

Look Beyond

Exercises 29–31 allow students to explore the angle sum and difference identities and the double angle identities using their graphics calculator.

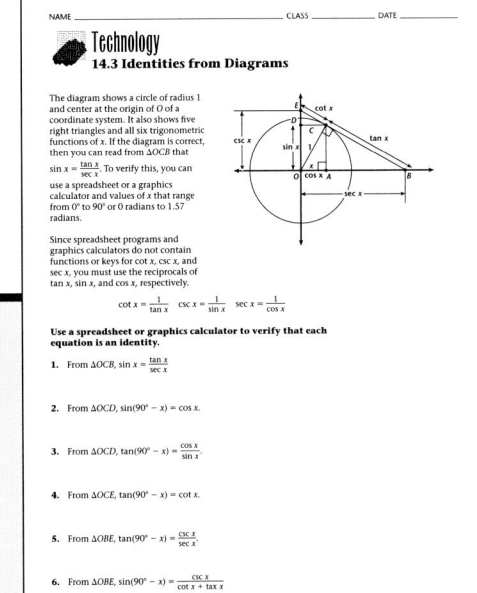

Objectives

• Find the exact values of the trigonometric functions by using the angle sum and difference identities, and the double angle identities.

RESOURCES

• Practice Master	14.4
• Enrichment Master	14.4
• Technology Master	14.4
• Lesson Activity Master	14.4
• Quiz	14.4
• Spanish Resources	14.4

This lesson may be considered optional in the core course. See page 768A.

Assessing Prior Knowledge

Identify each function as even or odd.

1. $f(x) = x^2$ [**even**]

2. $f(x) = -3x^2$ [**even**]

3. $f(x) = \frac{-2}{x^3}$ [**odd**]

TEACH

Students may not readily know the exact value of sin 105° or cos 105°. The sum and difference identities permit students to compute these values exactly, because 105° = 60° + 45°, which are special angles for values of trigonometric functions.

LESSON 14.4

Exploring

Sum and Difference Identities

why *Trigonometric functions are used in modeling real-world events, such as alternating current circuits, the motion of planets, the motion of gears in engines, and the release of energy created by earthquakes. The models of such events usually contain trigonometric expressions, which can be simplified by using trigonometric identities.*

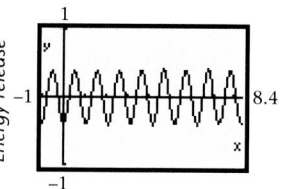

Time (seconds)
The graph models the energy release of an earthquake at 2.5 kilometers from the epicenter

ALTERNATIVE teaching strategy

Using Visual Models

Each identity can be illustrated using a graphics calculator. For each possible identity, graph the left side and the right side on the same coordinate plane. Students will see that, if the identity is true, the graphs of the two sides will be identical.

Exploration 1 Angle Sums

Graphics Calculator

Use your calculator to determine which of the following equations are true.

Substitute real numbers for A and B on the left side of the equation, and evaluate. Then substitute the same real numbers for A and B on the right side of the equation, and evaluate. Repeat this with each equation, using different numbers for A and B, until you are convinced that the equation is true or false.

1 $\cos(A + B) = \cos A + \cos B$

2 $\cos(A + B) = \cos A - \cos B$

3 $\cos(A + B) = \cos A \cos B + \sin A \sin B$

4 $\cos(A + B) = \cos A \cos B - \sin A \sin B$

5 $\sin(A + B) = \sin A + \sin B$

6 $\sin(A + B) = \sin A - \sin B$

7 $\sin(A + B) = \sin A \cos B + \cos A \sin B$

8 $\sin(A + B) = \sin A \cos B - \cos A \sin B$

9 Use a graph to verify the statements that you found to be true. Describe the graphing method that you used.

10 List the correct sum identities for sine and cosine. ❖

Exploration 2 Angle Differences

Graphics Calculator

Recall the definitions of even and odd functions from Chapter 6.

Even function: $f(-x) = f(x)$ *Odd function:* $f(-x) = -f(x)$

1 Graph $y = -\cos x$ and $y = \cos(-x)$.
Is $y = \cos x$ an even or an odd function? Explain.

2 Graph $y = -\sin x$ and $y = \sin(-x)$.
Is $y = \sin x$ an even or an odd function? Explain.

3 Substitute $-B$ for B in the identity you found for $\cos(A + B)$ in Exploration 1. Use the even or odd function properties of sine and cosine to write an identity for $\cos(A - B)$. Use your calculator to test the identity as you did in Exploration 1.

4 Substitute $-B$ for B in the identity you found for $\sin(A + B)$ in Exploration 1. Use the even or odd function properties of sine and cosine to write an identity for $\sin(A - B)$. Use your calculator to test the identity as you did in Exploration 1.

5 List the correct difference identities for cosine and sine. ❖

Exploration 3 Notes

Students are led to derive the double angle identities from the sum identities. Encourage students to compare the analytical, the calculator, and the graphics calculator approaches.

Aongoing SSESSMENT

4. $\cos(2A) = \cos^2 A - \sin^2 A$,
$\sin(2A) = 2 \sin A \cos A$

•Exploration 3 *Double Angles*

Graphics Calculator

1 Substitute A for B in the sum identity for cosine from Exploration 1, and simplify to obtain an identity for $\cos(2A)$.

2 Substitute A for B in the sum identity for sine from Exploration 1, and simplify to obtain an identity for $\sin(2A)$.

3 Test the double angle identities as you did in Exploration 1.

4 List the correct double angle identities for sine and cosine. ❖

EXTENSION

To find the exact values of the expressions $\cos\left(\frac{\pi}{6} + \frac{\pi}{4}\right)$ and $\sin\left(\frac{2\pi}{3} + \frac{\pi}{4}\right)$, use the angle sum and difference identities.

$$\cos\left(\frac{\pi}{6} + \frac{\pi}{4}\right)$$
$$= \left(\cos\frac{\pi}{6}\right)\left(\cos\frac{\pi}{4}\right) - \left(\sin\frac{\pi}{6}\right)\left(\sin\frac{\pi}{4}\right)$$
$$= \left(\frac{\sqrt{3}}{2}\right)\left(\frac{\sqrt{2}}{2}\right) - \left(\frac{1}{2}\right)\left(\frac{\sqrt{2}}{2}\right)$$
$$= \frac{\sqrt{6} - \sqrt{2}}{4}$$

$$\sin\left(\frac{2\pi}{3} - \frac{\pi}{4}\right)$$
$$= \left(\sin\frac{2\pi}{3}\right)\left(\cos\frac{\pi}{4}\right) - \left(\cos\frac{2\pi}{3}\right)\left(\sin\frac{\pi}{4}\right)$$
$$= \left(\frac{\sqrt{3}}{2}\right)\left(\frac{\sqrt{2}}{2}\right) - \left(-\frac{1}{2}\right)\left(\frac{\sqrt{2}}{2}\right)$$
$$= \frac{\sqrt{6} + \sqrt{2}}{4}$$

Suppose that $\sin t = \frac{1}{3}$ and t is in the second quadrant.

You can use the double angle identities to find the exact values of $\cos 2t$ and $\sin 2t$.

First find $\cos t$.

$$\cos^2 t = 1 - \sin^2 t$$
$$\cos^2 t = 1 - \left(\frac{1}{3}\right)^2$$
$$\cos t = \pm\sqrt{\frac{8}{9}}, \text{ or } \pm\frac{\sqrt{8}}{3}$$

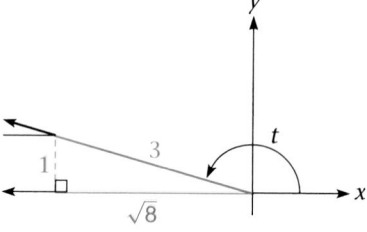

Since t is in the second quadrant, $\cos t$ is $-\frac{\sqrt{8}}{3}$. Substitute $-\frac{\sqrt{8}}{3}$ for $\cos t$ and $\frac{1}{3}$ for $\sin t$ in the double angle identities.

$$\cos 2t = \cos^2 t - \sin^2 t$$
$$= \left(-\frac{\sqrt{8}}{3}\right)^2 - \left(\frac{1}{3}\right)^2 \quad \text{and}$$
$$= \frac{7}{9}$$

$$\sin 2t = 2 \sin t \cos t$$
$$= 2\left(\frac{1}{3}\right)\left(-\frac{\sqrt{8}}{3}\right)$$
$$= -\frac{2\sqrt{8}}{9}$$

Thus, given that $\sin t = \frac{1}{3}$ and t is in the second quadrant, $\cos 2t = \frac{7}{9}$ and $\sin 2t = -\frac{2\sqrt{8}}{9}$. ❖

 **Using Cognitive Strategies** Students can improve their understanding of proofs by rewriting the derivations of the identities given on page 793 as a step-by-step procedure. Students should give a reason for each step.

ANGLE SUM AND DIFFERENCE IDENTITIES

$$\cos(A + B) = \cos A \cos B - \sin A \sin B$$
$$\sin(A + B) = \sin A \cos B + \cos A \sin B$$
$$\cos(A - B) = \cos A \cos B + \sin A \sin B$$
$$\sin(A - B) = \sin A \cos B - \cos A \sin B$$

DOUBLE ANGLE IDENTITIES

$$\cos 2A = \cos^2 A - \sin^2 A$$
$$\sin 2A = 2 \sin A \cos A$$

Claudius Ptolemy, around 85–165 C.E.

Cultural Connection: Africa Ptolemy was an astronomer who lived and worked in the North African city of Alexandria during the second century C.E. The most authoritative work on trigonometry at that time was the *Almagest*, which was written by Ptolemy. Included in the *Almagest* are the sum and double angle formulas for the sine of an angle, which Ptolemy discovered.

EXERCISES & PROBLEMS

Communicate

1. Give a counterexample to explain why $\sin(A + B) \neq \sin A + \sin B$.

2. Give a counterexample to explain why $\cos(A + B) \neq \cos A + \cos B$.

3. How are the angle difference identities obtained from the sum identities for the sine and cosine?

4. How are the double angle identities obtained from the sum identities for the sine and cosine?

5. How could the identity for $\cos\left(\frac{1}{2}x\right)$ be derived from the identity for $\cos 2x$?

6. Explain how you could find the identity for $\sin\left(\frac{1}{2}x\right)$ if you know the identity for $\cos\left(\frac{1}{2}x\right)$.

Using Cognitive Strategies Have students verify each identity. Have them compute both sides of a identity for given values of *A* and *B*. Show students that if they rewrite 75° as 50° + 25°, the identities do not help them compute sin 75° exactly. However, if students write 75° as 30° + 45°, the correct identity will give exact values for each of the trigonometric functions. For each identity, show students a sample decomposition that does not give exact values and one which does. Discuss techniques for finding decompositions that work well.

ASSESS

Selected Answers
Odd-Numbered Exercises 7–47

Assignment Guide
Core 1–12, 15–17, 19–21, 23–28, 30–32, 34–36, 38–47

Core-Plus 1–6, 11–14, 16–18, 20–25, 27–29, 31–33, 35–48

Technology
Graphics calculators are needed for Exercises 45 and 48. Review instructions for obtaining exact values, when they exist, for intercepts, intersections, maximums, and minimums with each type of calculator. Explain to students that graphing in degree mode will often give exact degree values that can then be converted to radians.

Error Analysis
Most errors result when students confuse the signs of one identity with the signs of another.

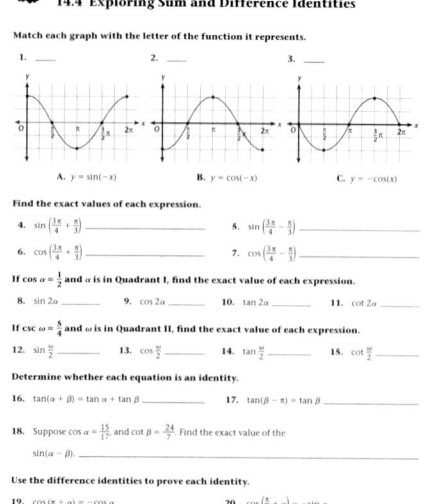

Performance Assessment

Encourage students to try to discover identities for sin 3A and cos 3A. Remind students how the double angle identities are derived from the sum identities.

7. $\dfrac{\sqrt{2}+\sqrt{6}}{4}$ **8.** $\dfrac{\sqrt{2}-\sqrt{6}}{4}$

9. $\dfrac{\sqrt{6}-\sqrt{2}}{4}$

19. $\sin\left(\dfrac{\pi}{2}-\theta\right)=$
$\sin\dfrac{\pi}{2}\cos\theta-\cos\dfrac{\pi}{2}\sin\theta=$
$1\cdot\cos\theta-0\cdot\sin\theta=\cos\theta$

20. $\cos\left(\dfrac{\pi}{2}-\theta\right)=$
$\cos\dfrac{\pi}{2}\cos\theta+\sin\dfrac{\pi}{2}\sin\theta=$
$0\cdot\cos\theta+1\cdot\sin\theta=\sin\theta$

21. $\sin(\pi-\theta)=$
$\sin\pi\cos\theta-\cos\pi\sin\theta=$
$0\cdot\cos\theta-(-1)\sin\theta=$
$\sin\theta$

22. $\cos(\pi-\theta)=$
$\cos\pi\cos\theta+\sin\pi\sin\theta=$
$-1\cos\theta+0\cdot\sin\theta=$
$-\cos\theta$

Practice Master

Practice & Apply

For Exercises 7–14, find the exact value of the expression.

7. $\sin\left(\dfrac{\pi}{6}+\dfrac{\pi}{4}\right)$ **8.** $\sin\left(\dfrac{\pi}{6}-\dfrac{\pi}{4}\right)$ **9.** $\cos\left(\dfrac{\pi}{6}+\dfrac{\pi}{4}\right)$ **10.** $\cos\left(\dfrac{\pi}{6}-\dfrac{\pi}{4}\right)$ $\dfrac{\sqrt{6}+\sqrt{2}}{4}$

11. $\sin\left(\dfrac{2\pi}{3}+\dfrac{\pi}{4}\right)$ **12.** $\sin\left(\dfrac{2\pi}{3}-\dfrac{\pi}{4}\right)$ **13.** $\cos\left(\dfrac{2\pi}{3}+\dfrac{\pi}{4}\right)$ **14.** $\cos\left(\dfrac{2\pi}{3}-\dfrac{\pi}{4}\right)$

$\dfrac{\sqrt{6}-\sqrt{2}}{4}$ $\dfrac{\sqrt{6}+\sqrt{2}}{4}$ $\dfrac{-\sqrt{2}-\sqrt{6}}{4}$ $\dfrac{-\sqrt{2}+\sqrt{6}}{4}$

For Exercises 15–18, $\cos x=\dfrac{1}{3}$, and x is in the fourth quadrant. Find the exact value of each expression.

15. $\cos 2x$ $-\dfrac{7}{9}$ **16.** $\sin 2x$ $-\dfrac{2\sqrt{8}}{9}$ **17.** $\tan 2x$ $\dfrac{2\sqrt{8}}{7}$ **18.** $\cot 2x$ $\dfrac{7}{2\sqrt{8}}$

For Exercises 19–22, use the difference identities to show that each equation is true.

19. $\sin\left(\dfrac{\pi}{2}-\theta\right)=\cos\theta$ **20.** $\cos\left(\dfrac{\pi}{2}-\theta\right)=\sin\theta$

21. $\sin(\pi-\theta)=\sin\theta$ **22.** $\cos(\pi-\theta)=-\cos\theta$

23. Show how the cosine double angle formula, $\cos 2\theta=\cos^2\theta-\sin^2\theta$, can be expressed in terms of only sine or only cosine. HINT: Use the Pythagorean identities.

24. Use the double angle identity for cosine and the result of Exercise 23 to derive the formula for $\cos\left(\dfrac{1}{2}x\right)$ in terms of $\cos x$.

25. Use the formula for $\cos 2x$ and the result of Exercise 23 to derive the formula for $\sin\left(\dfrac{1}{2}x\right)$ in terms of $\cos x$.

For Exercises 26–29, find the exact value of each expression using the angle sum and differnce identities.

26. $\cos 15°$ $\dfrac{\sqrt{6}+\sqrt{2}}{4}$ **27.** $\sin 15°$ $\dfrac{\sqrt{6}-\sqrt{2}}{4}$ **28.** $\cos 105°$ $\dfrac{\sqrt{2}-\sqrt{6}}{4}$ **29.** $\sin 75°$ $\dfrac{\sqrt{6}+\sqrt{2}}{4}$

For Exercises 30–33, find the exact value for each expression using the sum and differnce identities.

30. $\tan 15°$ $\dfrac{\sqrt{6}-\sqrt{2}}{\sqrt{6}+\sqrt{2}}$ **31.** $\cot 15°$ $\dfrac{\sqrt{6}+\sqrt{2}}{\sqrt{6}-\sqrt{2}}$ **32.** $\sec 15°$ $\dfrac{4}{\sqrt{6}+\sqrt{2}}$ **33.** $\csc 15°$ $\dfrac{4}{\sqrt{6}-\sqrt{2}}$

For Exercises 34–37, $\sec x=\dfrac{4}{3}$ and x is in the fourth quadrant. Find the exact value of each expression.

34. $\cos\left(\dfrac{1}{2}x\right)$ $-\sqrt{\dfrac{7}{8}}$ **35.** $\sin\left(\dfrac{1}{2}x\right)$ $\dfrac{1}{\sqrt{8}}$ **36.** $\tan\left(\dfrac{1}{2}x\right)$ $\dfrac{-1}{\sqrt{7}}$ **37.** $\cot\left(\dfrac{1}{2}x\right)$ $-\sqrt{7}$

38. Use the sum and difference for sine to show that the following equation is true.

$$\sin(A+B)+\sin(A-B)=2\sin A\cos B$$

39. Use your results from Exercise 38 to show that if $a=A+B$ and $b=A-B$, then the following equation is true.

$$\sin a+\sin b=2\sin\left(\dfrac{a+b}{2}\right)\cos\left(\dfrac{a-b}{2}\right)$$

---

23. $\cos 2\theta=\cos^2\theta-\sin^2\theta=\cos^2\theta-$
$(1-\cos^2\theta)=2\cos^2\theta-1$
$\cos 2\theta=\cos^2\theta-\sin^2\theta=(1-\sin^2\theta)-$
$\sin^2\theta=1-2\sin^2\theta$

24. $\cos 2\theta=2\cos^2\theta-1$
$2\cos^2\theta=\cos 2\theta+1$
$\cos\theta=\pm\sqrt{\dfrac{\cos 2\theta+1}{2}}$
So, $\cos\left(\dfrac{1}{2}x\right)=\pm\sqrt{\dfrac{\cos x+1}{2}}$

25. $\cos 2\theta=1-2\sin^2\theta$
$2\sin^2\theta=1-\cos 2\theta$
$\sin\theta=\pm\sqrt{\dfrac{1-\cos 2\theta}{2}}$
So, $\sin\left(\dfrac{1}{2}x\right)=\pm\sqrt{\dfrac{1-\cos x}{2}}$

The answers to Exercises 38 and 39 can be found in Additional Answers beginning on page 842.

Sports The range of a golf ball that is struck with an initial velocity of v_0, in feet per second, and that leaves the ground at an angle x is given by the function $h(x) = \dfrac{(v_0)^2 \sin x \cos x}{16}$.

40. Write the function $h(x)$ in terms of the double angle, $2x$. $\quad h(x) = \dfrac{(v_0)^2 \sin(2x)}{32}$

41. Explain how the range of a golf ball is determined by the initial velocity with which it is struck.

42.  **Maximum/Minimum** What is the angle at which a golf ball must be hit to achieve the maximum possible range for a given initial velocity? Explain.
$45°$; maximum value of $h(x)$ over the domain $0 < x < \pi$ occurs when $x = \frac{\pi}{4}$.

Look Back

43. Given the function $f(x) = \sqrt{5-x} + 3$, find the inverse of $f(x)$, and find the domain and range of the inverse. **[Lesson 3.2]**
$f^{-1}(x) = -x^2 + 6x - 4;\ R;\ y \le 5$

44. Given the function $f(x) = \sqrt{5-x} + 3$, find $\dfrac{f(x+h) - f(x)}{h}$, and simplify. **[Lesson 3.3]**
$\dfrac{\sqrt{5-(x+h)} - \sqrt{5-x}}{h}$

45 Solve the following system of equations using matrix algebra. **[Lesson 4.6]**
$$2x + y - 6z = -12$$
$$-x + y + z = -7 \qquad \text{Approx } (-2.7,\ -9.3,\ -0.5)$$
$$5x - 3y + 7z = 11$$

46. Two ranger stations, A and B, located 8 miles apart receive a distress call from a camper. Electronic equipment allows them to determine that the camper is at an angle of $46°$ from Station A and at an angle of $31°$ from Station B. The line segment connecting the stations serves as one side of each of these angles. Which station is closer to the camper? How far away is it from the camper? **[Lesson 14.1]** Station A; approx 4.2 mi

47. Find the remaining parts of triangle ABC given that the lengths of the sides are $a = 2.96$, $b = 3.78$, and $c = 4.54$. **[Lesson 14.2]**
$m\angle A = 40.4°;\ m\angle B = 55.9°;\ m\angle C = 83.7°$

Look Beyond

48 Find one solution to the equation $\cos^2 t + \cos t - 1 = 0$ by graphing $y = \cos^2 t + \cos t - 1$ in radian mode and finding the smallest positive zero of the function. $\quad 0.905$ radians

41. The height varies directly as the square of the initial velocity (the greater the velocity, the greater the height). The angle also plays a role in the height.

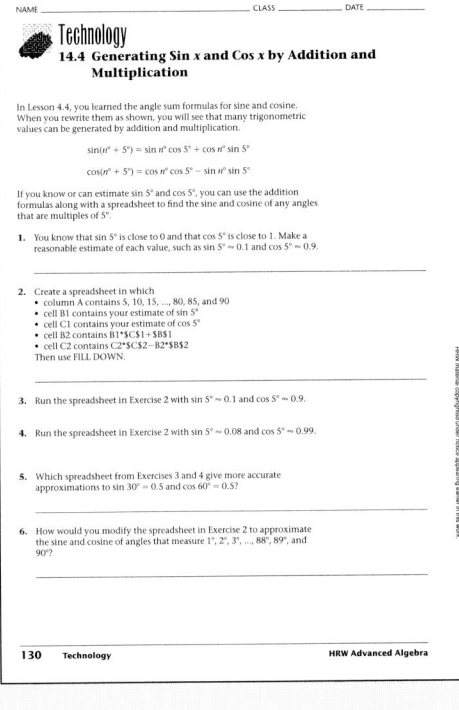

Objectives

• Solve equations containing trigonometric functions.

• Practice Master   14.5
• Enrichment Master   14.5
• Technology Master   14.5
• Lesson Activity Master   14.5
• Quiz   14.5
• Spanish Resources   14.5

This lesson may be considered optional in the core course. See page 768A.

Assessing Prior Knowledge

1. Factor $x^2 - 2x - 8$
   $[(x + 2)(x - 4)]$

2. In the function $f(x) = 4 + 3 \cos 2(x + \pi)$, identify the vertical shift, horizontal phase shift, amplitude and period.
   [4, none, 3, π]

TEACH

Natural phenomena that are cyclic can be modeled by trigonometric functions. The use of a graphics calculator provides a valuable tool for obtaining detailed numerical information about these cyclic phenomena.

Use Transparency ▶ 64

LESSON 14.5 Solving Trigonometric Equations

Why *Many real-world problems are modeled by trigonometric equations. Solving these equations often involves using trigonometric identities and graphing.*

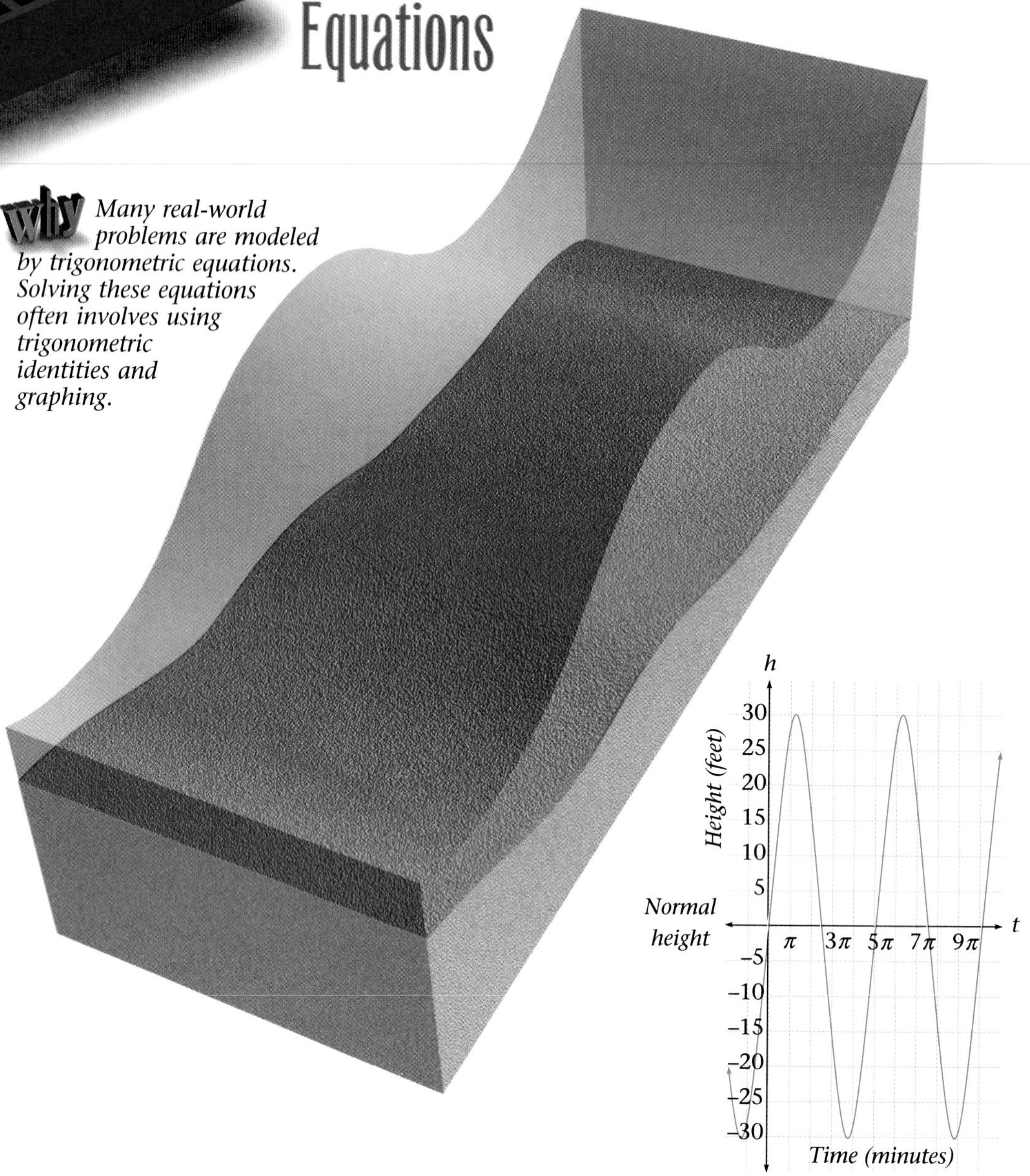

Seismology The height in feet, relative to normal height, of a point on a tidal wave can be modeled by the function $h(t) = 30 \sin\left(\frac{\pi}{8}t\right)$, where t is in minutes and where $t = 0$ is the time of the earthquake that caused the tidal wave. How long after the earthquake occurred will the point on the tidal wave first reach a height of 20 feet above normal?

ALTERNATIVE teaching strategy

Technology Use a graphics calculator to graph a function like $f(x) = 2 + 3 \sin 2(x - \pi)$. Illustrate that the amplitude is 3 and the period is π. By varying these values it is possible to generate graphs of related functions with amplitudes and periods that match the desired natural cyclic event.

Use your graphics calculator to graph $y = 30 \sin\left(\frac{\pi}{8}x\right)$ and $y = 20$. Find the first positive x-value for the intersection of these two functions. About 1.86 minutes after an earthquake occurs, the height of a tidal wave is about 20 feet above normal.

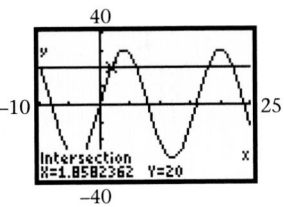

EXAMPLE 1

Graphics
Calculator

Find all the solutions of the trigonometric equation $\sin 3x = \frac{\sqrt{3}}{2}$ for $0 \le x \le 2\pi$.

Solution▸

Use your calculator in radian mode to graph $y = \sin 3x$ and $y = \frac{\sqrt{3}}{2}$, and find all the intersections for $0 \le x < 2\pi$.

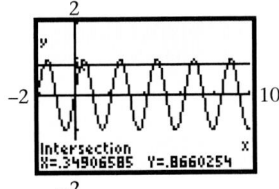

Since the result is not an exact value in x, try graphing the function in degree mode instead of radian mode.

For $0° \le x < 360°$, the graph of $y = \sin 3x$ intersects the graph of $y = \frac{\sqrt{3}}{2}$ when x is $20°$, $40°$, $140°$, $160°$, $260°$, and $280°$.

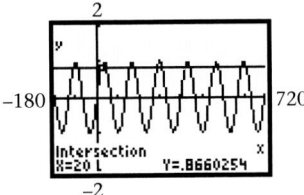

Converting to radian measures, the graphs intersect when x is $\frac{\pi}{9}$, $\frac{2\pi}{9}$, $\frac{7\pi}{9}$, $\frac{8\pi}{9}$, $\frac{13\pi}{9}$, and $\frac{14\pi}{9}$. ❖

Try This Find all the solutions of $\cos 3t = -\frac{\sqrt{3}}{2}$ for $0 \le t < 2\pi$.

EXAMPLE 2

Solve $\sin^2 t - 2\sin t - 3 = 0$ for $0 \le t < 2\pi$.

Solution▸

Since the equation is quadratic, you can solve by factoring or by using the quadratic formula.

$$\sin^2 t - 2\sin t - 3 = 0$$
$$(\sin t + 1)(\sin t - 3) = 0$$

$\sin t + 1 = 0$ | $\sin t - 3 = 0$
$\sin t = -1$ | $\sin t = 3$
$t = \frac{3\pi}{2}$ | no solution

There is only one solution, $t = \frac{3\pi}{2}$, for $0 \le t < 2\pi$. ❖

Alternate Example 1

Find all solutions of the equation $\sin 2t = \frac{1}{2}$ for $0 \le t < 2\pi$.
$$\left[\frac{\pi}{12}, \frac{5\pi}{12}, \frac{13\pi}{12}, \frac{17\pi}{12}\right]$$

Aongoing ASSESSMENT

Try This

$$\frac{5\pi}{18}, \frac{7\pi}{18}, \frac{17\pi}{18}, \frac{19\pi}{18}, \frac{29\pi}{18}, \frac{31\pi}{18}$$

Alternate Example 2

Solve $\sin^2 t + \sin t - 2 = 0$ for $0 \le t < 2\pi$. [t is $\frac{\pi}{2}$]

ENRICHMENT In each example in this lesson, one can use another trigonometric function in place of $\sin x$ or $\cos x$. For example, $\sec x$ can be used in place of $\sin x$. Ask students to explain whether this changes the solution or the method of solution. [**It does change the solution but not the method.**]

INCLUSION strategies **Using Visual Models** Have students make posters representing the graphs of each of the equations from the exercises. Each function should be in a different color and the intersection should be clearly labeled. Graphics calculators with a table feature can be used to assist students in plotting the graphs.

Graphics Calculator

You can also solve $\sin^2 t - 2\sin t - 3 = 0$, the equation in Example 2, by using your graphics calculator. Graph $y = (\sin x)^2 - 2\sin x - 3$, and find the zeros between 0 and 2π.

Notice that over the interval from 0 to 2π, there is only one zero.

Using the calculation feature, the resulting value is very close to the exact value $t = \frac{3\pi}{2} \approx 4.71238898$ radians.

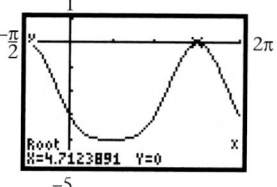

You can often use trigonometric identities to simplify a trigonometric equation before solving it.

EXAMPLE 3

Graphics Calculator

Solve $4\sin^2 x + 3\cos x - 2 = 0$ for $0 \le x < 2\pi$.

Solution

Use the Pythagorean identity, $\cos^2 x + \sin^2 x = 1$, or $\sin^2 x = 1 - \cos^2 x$, to write the equation in terms of only $\cos x$.

$$4\sin^2 x + 3\cos x - 2 = 0$$
$$4(1 - \cos^2 x) + 3\cos x - 2 = 0$$
$$4 - 4\cos^2 x + 3\cos x - 2 = 0$$
$$4\cos^2 x - 3\cos x - 2 = 0$$

Since the resulting quadratic expression, $4\cos^2 x - 3\cos x - 2$, in the equation is not factorable, use the quadratic formula to solve for $\cos x$.

$$\cos x = \frac{3 \pm \sqrt{9 - 4(4)(-2)}}{2 \cdot 4}$$
$$= \frac{3 \pm \sqrt{41}}{8}$$

$$\cos x = \frac{3 + \sqrt{41}}{8} \qquad \cos x = \frac{3 - \sqrt{41}}{8}$$

$$\cos x \approx 1.1754 \qquad \cos x \approx -0.4254$$

Since $\cos x$ cannot be greater than 1, then $\cos x \approx -0.4254$.

$$x \approx \cos^{-1}(-0.4254)$$
$$\approx 2.01 \text{ radians}$$

2.01 radians is in the second quadrant. To find $\cos^{-1}(-0.4254)$ in the third quadrant, subtract 2.01 from 2π.

$$x \approx 2\pi - 2.01$$
$$\approx 4.27 \text{ radians}$$

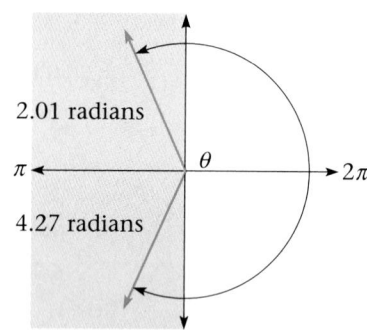

Check►

To check your solutions, graph the function $y = 4\sin^2 x + 3\cos x - 2$, and find the zeros in the interval $0 \le x < 2\pi$.

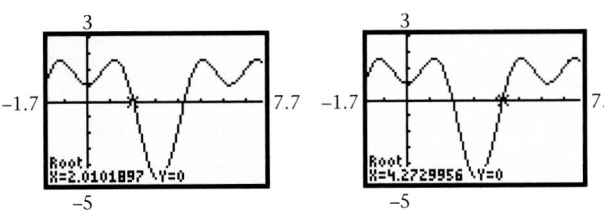

Since the roots are approximately 2.01 and 4.27, the solutions are about 2.01 radians and 4.27 radians. ❖

Try This Solve $1 + \tan^2 x + \sec x = 0$ for $0 \le x < 2\pi$.

CRITICAL
Thinking

Explain how to use your graphics calculator to solve the trigonometric equation in Example 3 *without* using a Pythagorean identity.

EXERCISES & PROBLEMS

Communicate

1. Explain how to use your calculator to find solutions for t from the equation $a\cos(bt) = c$, if a, b, and c are given.

2. Suppose the cosine of an angle is negative, and you have found one solution between 0 and 2π. Explain how to find the second solution.

3. Describe two different methods of solving the equation $\tan^3 t + 2\tan^2 t + \tan t = 0$.

4. Give an example of an equation that is quadratic in the variable $\cos t$. Explain how you would try to solve your equation for t.

5. How are trigonometric identities sometimes useful in solving trigonometric equations?

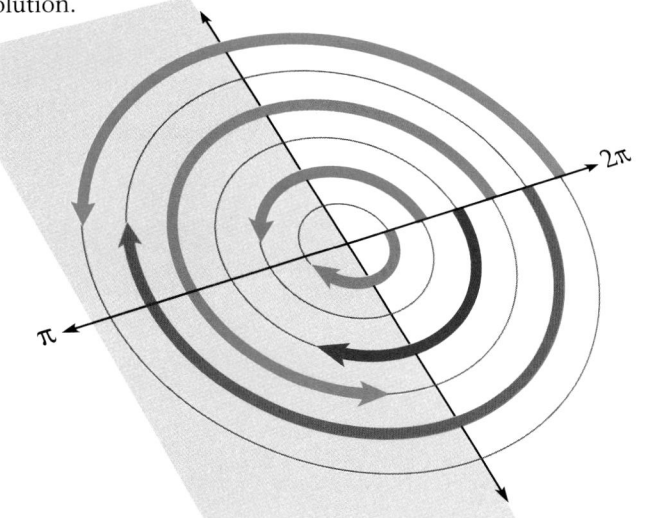

Practice & Apply

For Exercises 6–8, find the smallest value of $t > 0$ that solves each equation.

6 $3\pi \sin\left[\pi\left(t - \frac{1}{27}\right)\right] = 4$ 0.1766   **7** $6\cos\left[2\left(t + \frac{1}{36}\right)\right] = 5$ 0.265   **8** $2\tan\left[3\left(t - \frac{\pi}{8}\right)\right] = 3$ 0.720

For Exercises 9–17, find the *exact* solutions for $0 \le x < 2\pi$.

9. $2\sin x - 1 = 0$ $\frac{\pi}{6}$, $\frac{5\pi}{6}$ 10. $2\cos x + 1 = 0$ $\frac{2\pi}{3}$, $\frac{4\pi}{3}$ 11. $4\sin x + 2\sqrt{3} = 0$ $\frac{4\pi}{3}$, $\frac{5\pi}{3}$

12. $\tan x - \sqrt{3} = 0$ $\frac{\pi}{3}$, $\frac{4\pi}{3}$ 13. $\cot x + 1 = 0$ $\frac{3\pi}{4}$, $\frac{7\pi}{4}$ 14. $4\cos 3x - 2 = 0$ $\frac{\pi}{9}$, $\frac{5\pi}{9}$, $\frac{7\pi}{9}$, $\frac{11\pi}{9}$, $\frac{13\pi}{9}$, $\frac{17\pi}{9}$

15. $\tan^2 2x - 1 = 0$ 16. $2\sec 2x + 4 = 0$ $\frac{\pi}{3}$, $\frac{2\pi}{3}$, $\frac{4\pi}{3}$, $\frac{5\pi}{3}$ 17. $2\sin 3x - \sqrt{3} = 0$ $\frac{\pi}{9}$, $\frac{2\pi}{9}$, $\frac{7\pi}{9}$, $\frac{8\pi}{9}$, $\frac{13\pi}{9}$, $\frac{14\pi}{9}$

For Exercises 18–21, find the approximate solutions accurate to the nearest thousandth for $0 \le x < 2\pi$.

18 $\sin x - \frac{2}{3} = 0$ 0.730; 2.412   **19** $\sec x + \frac{3}{2} = 0$ 2.301; 3.982

20 $\cot x - 4 = 0$ 0.245; 3.387   **21** $2\cot x + 1 = 0$ 2.034; 5.176

22 Astronomy The number of hours of daylight in a given region varies with the time of year. The number of hours of daylight in a 365-day year can be modeled with the function $D(t) = 14 + 3\sin\left(\frac{2\pi}{365}(t - 5)\right)$, where t is the day of the year, and t is 1 on January 1. Find two days of the year when the region gets 15 hours of daylight.
Jan 24 and June 17

For Exercises 23–28, find approximate solutions accurate to the nearest tenth for $0 \le x < 2\pi$.

23 $2\cos^2 x - \cos x - 1 = 0$ **24** $2\sin^2 x + 3\sin x - 2 = 0$

25 $2\cos^2 x - \sin x - 1 = 0$ **26** $\tan 2x + \cot 2x + 2 = 0$

27 $\cos 2x - \sin x = 0$ **28** $\sin 2x + \sin x = 0$

29 $\cos^2 x + 5\cos x + 2 = 0$ **30** $4\sin^2 x - 3\sin x - 2 = 0$

31 $\tan^2 x + \tan x - 1 = 0$ **32** $3\sin 2x + \cos x = 0$

33 Business A consultant tells the owner of a new business that annual profits, in \$10,000s, of successful similar businesses can be modeled by the function $P(t) = 1.08t - 2.3\sin\left(\frac{2\pi}{5.2}(t + 3)\right)$, where t is in years, and t is 0 during the year that the business is started. Use your graphics calculator to graph the function, and estimate how long it will take for the business to reach a yearly profit of \$100,000. Approx 9yr, $9\frac{1}{2}$ mos

23. 0; 2.1; 4.2   28. 0; 2.1; 3.1; 4.2

24. 0.5; 2.6   29. 2.0; 4.3

25. 0.5; 2.6; 4.7   30. 3.6; 5.8

26. 1.2; 2.7; 4.3; 5.9   31. 0.5; 2.1; 3.7; 5.3

27. 0.5; 2.6; 4.7   32. 1.6; 3.3; 4.7; 6.1

Electricity The voltage V in the power cord to a toaster is given by $V(t) = 120\cos(120\pi t)$.

34 Find all the times at which the voltage is 0 for $t > 0$.

35 Find all the times at which the voltage is -120 for $t > 0$.

36 **Maximum/Minimum** Find all the times for which the voltage is maximum. What is the maximum voltage?

Look Back

Acoustics If i is the intensity of sound measured in watts per meter per meter, then $D(i) = 10\log\left(\frac{i}{i_0}\right)$ gives the intensity of sound in decibels (after Alexander Graham Bell), where i_0 is the minimum intensity that can be heard by a human ear (about 10^{-16} watts per meter per meter).
Complete Exercises 37 and 38. **[Lesson 7.5]**

37 Normal conversation has an intensity of 10^{-6} watts per meter per meter. Find the decibel level of conversation. 100 decibels

38 If Deanna shouts loudly enough to be heard over the Niagara Falls (which have an intensity of 10^{-3} watts per meter per meter), what is the minimum decibel level of her voice? 130 decibels

39. **Transformations** Given $f(t) = 3\sin\left(2t - \frac{\pi}{3}\right)$, specify the amplitude, period, and phase shift. Then graph the function. **[Lesson 8.4]**

40. Given triangle ABC with $a = 12.7$, $c = 10.4$, and $\angle B = 26.5°$, find the remaining parts of the triangle. **[Lesson 14.2]**
$b \approx 5.7$; $m\angle A \approx 99.7°$; $m\angle C \approx 53.8°$

Look Beyond

41. **Geography** How is the radius of the Earth measured? One way is to climb to the top of a mountain whose height above sea level is known, measure the angle between the vertical and the horizon (assumed to be at sea level), and then do some calculating. The height of Mount Shasta (California) is 14,162 feet. From the top of Mount Shasta, the horizon (on the Pacific Ocean) is at an angle of $87°53'$ with the vertical. Use these data to estimate the radius of the Earth in miles. 3928 miles

34. $\approx 0.0042 + 0.0166n$ and $0.0125 + 0.0166n$

35. $0.0083 + 0.0166n$

36. $0 + 0.0166n$; 120

39. amplitude 3, period π, phase shift $\frac{\pi}{6}$ to the right

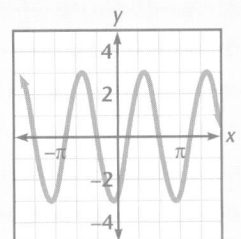

Stake Out Your Plot

CHAPTER 14 PROJECT

Sally is surveying a triangular plot of land with a river running through it. Two of the surveying stakes are 500 meters apart on the corners of the plot, and they are on the same side of the river. The other corner of the plot is marked by a 120-year-old live-oak tree. She uses a transit to measure the angles at the stakes and finds them to be 37° and 48°, as shown in the diagram. The river that crosses the plot of land is approximately 40 meters wide and 100 meters from the tree.

1. Explain how Sally can determine the usable area of the plot.

2. What is the area of the plot?

3. If the land sells for $2.00 per square meter, how much is the land worth?

1. The area of large triangle minus the area of the trapezoid.

2. 45,800 square meters

3. $91,600

Discuss

Activity

Find out how surveyor instruments that are used today incorporate many of the trigonometric concepts that you have learned. Compare the methods of surveying with modern surveyor instruments with the methods of surveying without modern instruments.

Chapter 14 Review

Vocabulary

Key Skills & Exercises

Lesson 14.1

➤ **Key Skills**

Find all the parts of a triangle when given the measure of two angles and the length of the included side.

The measure of the third angle is 180° minus the sum of the other two angle measures. The two missing side lengths can be found by using the Law of Sines.

$$\frac{\sin A}{a} = \frac{\sin B}{b} = \frac{\sin C}{c}$$

➤ **Exercises**

Solve each triangle for the missing angle measures and length.

1. m∠A = 23°, c = 15, m∠B = 70°
2. m∠B = 40°, m∠C = 30°, c = 9 $a \approx 16.9$; $b \approx 11.6$; m∠A = 110°
3. m∠A = 30°, m∠C = 30°, c = 50
$a = 50$; $b \approx 86.6$; m∠B = 120°
4. m∠B = 25°, m∠C = 89°, b = 24 $a \approx 51.9$; $c \approx 56.8$; m∠A = 66°

Lesson 14.2

➤ **Key Skills**

Find all the parts of a triangle when given the lengths of two sides and the measure of the included angle.

Use the Law of Cosines.

$$c^2 = a^2 + b^2 - 2ab \cos C$$
$$b^2 = a^2 + c^2 - 2ac \cos B$$
$$a^2 = b^2 + c^2 - 2bc \cos A$$

Find all the parts of a triangle when given the lengths of all three sides.

Use the Law of Cosines and inverse cosine. If $a = 2$, $b = 3$, and $c = 4$, then to find m∠B, use $b^2 = a^2 + c^2 - 2ac \cos B$.

➤ **Exercises**

Solve each triangle for the missing angle measures and lengths.

5. a = 55, b = 25, m∠C = 115°
6. a = 6, b = 8, c = 10 m∠A ≈ 36.9°; m∠B ≈ 53.1°; m∠C = 90°
7. a = 4, b = 4, m∠C = 90°
8. a = 12, c = 14, m∠B = 140°

1. $a \approx 5.9$, $b \approx 14.1$, m∠C = 87°

5. $c \approx 69.4$, m∠A ≈ 45.9°, m∠B ≈ 19.1°

7. m∠A = 45°, m∠B = 45°, $c \approx 5.7$

8. $b = 24.4$, m∠A = 18.4°, m∠C = 21.6°

Lesson 14.3

➤ *Key Skills*

Find the value of any of the other five trigonometric functions when given the value of one function and the quadrant in which the angle lies.

Use the definitions $\cot \theta = \frac{1}{\tan \theta}$, $\sec \theta = \frac{1}{\cos \theta}$, and $\csc \theta = \frac{1}{\sin \theta}$; the Pythagorean identities, $\cos^2 \theta + \sin^2 \theta = 1$, $1 + \tan^2 \theta = \sec^2 \theta$, and $1 + \cot^2 \theta = \csc^2 \theta$; and the fact that $\sin \theta$ is positive in the first and second quadrants, while $\cos \theta$ is positive in the first and fourth quadrants.

$$\tan t = -\tfrac{4}{3}; \ \cot t = -\tfrac{3}{4}; \ \sec t = \tfrac{5}{3}; \ \csc t = -\tfrac{5}{4}$$

➤ *Exercises*

9. If $\sin t = -\frac{4}{5}$ and $\cos t = \frac{3}{5}$, find $\tan t$, $\cot t$, $\sec t$, and $\csc t$.

10 If $\sin t = \frac{2}{3}$ and $\tan t < 0$, use a calculator to find $\cos t$, $\tan t$, $\cot t$, $\sec t$, and $\csc t$. $\cos t \approx -0.745$; $\tan t \approx -0.894$; $\cot t \approx -1.118$; $\sec t \approx -1.342$; $\csc t \approx 1.5$

Lesson 14.4

➤ *Key Skills*

Find exact values of the trigonometric functions by using the angle sum and difference identities, and the double angle identities.

$$\cos(A + B) = \cos A \cos B - \sin A \sin B \qquad \sin(A + B) = \sin A \cos B + \cos A \sin B$$
$$\cos(A - B) = \cos A \cos B + \sin A \sin B \qquad \sin(A - B) = \sin A \cos B - \cos A \sin B$$

$$\cos 2A = \cos^2 A - \sin^2 A$$
$$\sin 2A = 2 \sin A \cos A$$

To find $\tan 2\theta$, given that $\cos \theta = -\frac{1}{2}$ and θ is in the third quadrant, use $\cos^2 \theta + \sin^2 \theta = 1$ to find $\sin \theta = -\frac{\sqrt{3}}{2}$. Then use the double angle identity.

$$\tan 2\theta = \frac{\sin 2\theta}{\cos 2\theta}$$
$$= \frac{2 \sin \theta \cos \theta}{\cos^2 \theta - \sin^2 \theta}$$
$$= \frac{2 \left(-\frac{\sqrt{3}}{2}\right)\left(-\frac{1}{2}\right)}{\left(-\frac{1}{2}\right)^2 - \left(-\frac{\sqrt{3}}{2}\right)^2}$$
$$= -\sqrt{3}$$

➤ *Exercises*

Given that angles A and B are in the first quadrant and that $\sin A = \frac{\sqrt{3}}{2}$ and $\sin B = \frac{\sqrt{2}}{2}$, evaluate each expression.

11. $\sin(A + B)$ **12.** $\cos(A - B)$

13. $\sin 2A$ **14.** $\cot 2B$

9. $\tan t = -\frac{4}{3}$, $\cot t = -\frac{3}{4}$, $\sec t = \frac{5}{3}$, $\csc t = -\frac{5}{4}$

11. $\frac{\sqrt{6} + \sqrt{2}}{4}$

12. $\frac{\sqrt{2} + \sqrt{6}}{4}$

13. $\frac{\sqrt{3}}{2}$

14. 0

Lesson 14.5

➤ *Key Skills*

Solve equations containing trigonometric functions.

For approximate solutions, use the graphics calculator to find the solutions over a given interval for the equation. For exact solutions, use trigonometric identities and algebraic techniques, such as factoring and the quadratic formula.

➤ *Exercises*

Find exact solutions for $0 \le x < 2\pi$.

15. $2\cos x - \sqrt{3} = 0$ $\frac{\pi}{6}, \frac{11\pi}{6}$

16. $2\sin^2 x + \sin x = 0$ $0; \pi; \frac{7\pi}{6}, \frac{11\pi}{6}$

Use a graphics calculator to find approximate solutions for $0 \le x < 2\pi$.

17. $\cos x + \frac{1}{3} = 0$ 1.91 radians; 4.37radians

18. $3\cos^2 x - 2\sin x + 1 = 0$ 1.05 radians; 2.09 radians

Applications

Recreation At a distance of 3000 yards from the base of a mountain, the angle of elevation to the top is 20°. 1000 yards from the base of the mountain, a ski lift goes to the top of the mountain at an inclination of 35°.

19. What is the length of the ski lift, to the nearest yard? 2643 yd

20. What is the height of the mountain, to the nearest yard? 1516 yd

Sports A baseball player hits the ball 3 feet above ground level with an initial velocity of 100 feet per second and at an angle of 30° with the horizontal. The height of the ball after t seconds is given by $y(t) = -16t^2 + (100\sin 30°)t + 3$, and the horizontal distance it travels is given by $x(t) = (100\cos 30°)t$.

21. Find the height of the ball and the horizontal distance it has traveled after 2 seconds. height: 39 ft; horizontal distance: 173.2 ft

22. If the angle were 60°, would the height double? Would the horizontal distance be half as much? Explain.

Construction A 10-meter utility pole is on a hill. The pole is held in place by several guy wires. The longest wire is anchored downhill from the pole.

23. If the ground forms an angle of 102° with the pole and an angle of 40° with the wire in the triangle with the longest wire, find the length of the longest wire. Approx 15.2 m

24. If it were necessary for the utility pole to be 15 meters long, how long would the guy wires be? Approx 22.8 m

Road Construction State Road 25 from Triton to Tall Oak turns at a 160° angle after 15 miles. The state wants to extend the road so that it goes straight and misses Tall Oak. The citizens of Tall Oak want a new road that goes directly from their town to the extension.

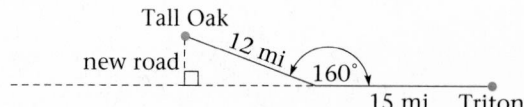

25. If it is 12 miles from the bend in the road to Tall Oak, how long will the new road be? Approx 4.1 mi

22. No; No; $\sin 60° \ne 2(\sin 30°)$ and $\cos 60° \ne 0.5\cos 30°$.

Chapter 14 Assessment

Solve each triangle.

1. m∠A = 37°, c = 26, m∠B = 56° a ≈ 15.7; b ≈ 21.6; m∠C = 87°

2. m∠A = 20°, m∠B = 103°, c = 58 a ≈ 23.7; b ≈ 67.4; m∠C = 57°

3. m∠B = m∠C = 45°, a = 17 b ≈ 12.0; c ≈ 12.0; m∠A = 90°

4. a = 115, b = 83, m∠C = 135° c ≈ 183.3; m∠A = 26.3°; m∠B = 18.7°

5. a = 12, b = 12, c = 10 m∠A ≈ 65.4°; m∠B ≈ 65.4°; m∠C ≈ 49.2°

6. $b = 5\sqrt{3}$, c = 54, m∠A = 60°

7. Sports Jules wants to determine the swimming distance from Point A to Point B across the inlet of a lake. He measures the distance on land and the angle shown. What is the distance AB? Approx 178 yd

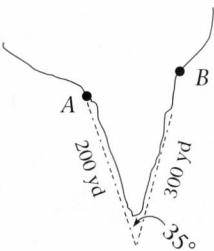

Given that $\sin A = -\frac{\sqrt{2}}{2}$, $\cos A = \frac{\sqrt{2}}{2}$, $\sin B = \frac{\sqrt{3}}{2}$, **and B is in the second quadrant, find exact values for each expression.**

8. tan A   −1

9. cot A   −1

10. cos B $-\frac{1}{2}$

11. sec B   −2

12. tan B $-\sqrt{3}$

13. sin(A − B) $\frac{\sqrt{2}-\sqrt{6}}{4}$

14. cos(A + B) $\frac{\sqrt{6}-\sqrt{2}}{4}$

15. sin 2A   −1

16. tan 2B $\sqrt{3}$

17. Find the exact solution for $\sin^2 x + 2\sin x - 3 = 0$ in the interval $0 \le x < 2\pi$. $\frac{\pi}{2}$

Use a graphics calculator to find approximate solutions in the interval $0 \le x < 2\pi$.

18. $3\tan 2x + \sqrt{3} = 0$ 1.3, 2.9, 4.6, 6.0

19. $\sec^2 x - \sec x - 3 = 0$ 1.12; 2.45; 3.84; 5.16

20. **Maximum/Minimum** As Kelli and Nell drive along the highway from Town A to Town B, they want to listen to their favorite radio station, which broadcasts from a tower located as shown. The station sends a signal that can be received clearly at a maximum distance of 50 miles. Between what two distances from Town B will they be able to hear this station clearly?

19.1 mi and 89.8 mi

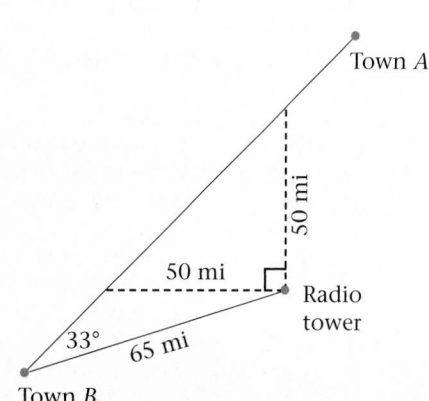

6. $a ≈ 50.2, m∠B ≈ 8.6°, m∠C ≈ 111.4°$

Chapters 1–14
Cumulative Assessment

College Entrance-Exam Practice

Multiple-Choice and Quantitative-Comparison Samples

The first half of the Cumulative Assessment contains two types of items found on standardized tests—multiple-choice questions and quantitative-comparison questions. Quantitative-comparison items emphasize the concepts of equality, inequality, and estimation.

Free-Response Grid Samples

The second half of the Cumulative Assessment is a free-response section. A portion of this part of the Cumulative Assessment consists of student-produced response items commonly found on college entrance exams. These questions require the use of machine-scored answer grids. You may wish to have students practice answering these items in preparation for standardized tests.

Sample answer-grid masters are available in the *Chapter Teaching Resources Booklets*.

College Entrance Exam Practice

Quantitative comparison For items 1–5, write
- **A** if the quantity in Column A is greater than the quantity in column B;
- **B** if the quantity in Column B is greater than the quantity in column A;
- **C** if the two quantities are equal;
- **D** if the relationship cannot be determined from the given information.

	Column A	Column B	Answers
1. A	The value of x in the solution of the system $\begin{cases} 2x + 3y = 21 \\ x - 2y = -7 \end{cases}$	$\begin{cases} -2x + y = 8 \\ 5x - 3y = -21 \end{cases}$	(A) (B) (C) (D) [Lesson 4.3]
2. B	The value of the greatest real zero of f $f(x) = (x - 3)^2(x + 2)(x + 5)$	$f(x) = (2x - 7)(x + 4)^3$	(A) (B) (C) (D) [Lesson 6.2]
3. A	The y-coordinate of the y-intercept of $f(x) = 2\left(\frac{2}{3}\right)^x$	The x-coordinate of the x-intercept of $f(x) = \log_{1.5}x$	(A) (B) (C) (D) [Lessons 7.2, 7.3]
4. B	The period of the graph of f $f(x) = 1 - 2\sin 3x$	$f(x) = 2.3\sin\left(\frac{\pi}{2}x\right)$	(A) (B) (C) (D) [Lesson 8.4]
5. B	The mean of 76, 78, 71, 78, 72, 75	The median of 76, 78, 71, 78, 72, 75	(A) (B) (C) (D) [Lesson 13.1]

6. Which of the following expressions is *not* equivalent to the others? b
[Lesson 2.2]
- **a.** $\left(\dfrac{x^3}{x^{-3}}\right)^2$
- **b.** $\left(\dfrac{x^2}{x^{-6}}\right)^3$
- **c.** $\left(x^3\right)^4$
- **d.** $\dfrac{(xy)^{-3}}{(x^{-5}y^{-1})^3}$

7. What is the equation for the axis of symmetry in the graph of b $f(x) = -2(x + 3)^2 - 1$?   **[Lesson 5.3]**
- **a.** $y = -2$
- **b.** $x = -3$
- **c.** $y = -1$
- **d.** $x = 3$

8. Which of the following is a vertical asymptote for the graph of d $f(x) = \dfrac{x^2 - 1}{x^2 + 4x + 3}$?   **[Lesson 9.4]**
- **a.** $x = 1$
- **b.** $x = -1$
- **c.** $x = 3$
- **d.** $x = -3$

9. 🔲 **Probability**   If three coins are flipped, what is the probability ᶜ
that they will all land heads up, if the first coin lands heads up?
[Lesson 11.6]

    a. $\frac{1}{8}$ **b.** $\frac{1}{2}$ **c.** $\frac{1}{4}$ **d.** $\frac{3}{8}$

10. Which of the following is an identity for $\cos(x+y)$?   **[Lesson 14.4]** ᵇ
    a. $\cos x \sin y + \sin x \cos y$ **b.** $\cos x \cos y - \sin x \sin y$
    c. $\cos x \cos y + \sin x \sin y$ **d.** $\cos x + \cos y$

11 🔲 **Transformations**   The matrix $\begin{bmatrix} 2 & -1 & 0 \\ 5 & 3 & -2 \end{bmatrix}$ represents a

triangle with vertices (2, 5), (−1, 3), and (0, −2). Find the vertices of $(-5, 2), (-3, -1), (2, 0)$
the triangle after it has been rotated 90° counterclockwise. **[Lesson 4.7]**

12. Sports   The path of a football after kickoff is in the shape of a
parabola. Assume a football is kicked from the point (0, 0) in the
coordinate plane and that it reaches a height of 5 yards. The football $V: (20, 5)$
is not caught and lands 40 yards down field. Find the coordinates of
the vertex of this parabola, and use the coordinates of the vertex to $\quad y = -\frac{1}{80}(x-20)^2 + 5$
write an equation for this parabola. **[Lesson 5.2]**

13. Geology   A seismograph shows that the epicenter of an earthquake
is 80 miles from station A. Station B, which is 50 miles due south of $\qquad x^2 + y^2 = 80^2$
station A, determines that the epicenter is 40 miles away. Place station
A at (0, 0), and write two equations for the possible locations of the $\qquad x^2 + (y+50)^2 = 40^2$
epicenter as determined by each station. **[Lesson 10.2]**

14. 🔲 **Geometry**
The Golden Ratio,   small part   big part
mentioned in Lesson 12.1,
is defined as
$\dfrac{\text{small part}}{\text{big part}} = \dfrac{\text{big part}}{\text{whole}}$.   whole

Let 1 be the small part, and let x be the big part. Substitute into $\dfrac{1+\sqrt{5}}{2} \approx 1.618$
the proportion and solve for x. What value do you get for x?
[Lesson 12.1]

15. 🔲 **Probability**   If a coin is tossed 10 times, what is the probability
that it will land heads up exactly 4 times? **[Lesson 13.5]** $\frac{_{10}C_4}{2^{10}} \approx 20.5\%$

Free-Response Grid The following questions may be answered using a
free-response grid commonly used by standardized test services.

16. If the inverse function for
$f(x) = \frac{2}{3}x - 7$ is written in the
form $g(x) = ax + b$, what is the value of a?
[Lesson 3.2] 1.5

17. Find the number of degrees in $\frac{3\pi}{4}$ radians.
[Lesson 8.5] 135°

18. Find the length of the radius of the circle
defined by the equation $x^2 + y^2 - 4x + 6y + 4 = 0$.
[Lesson 10.2] 3

19. How many different 4-letter "words" can
be formed from the letters of *TEEN*?
[Lesson 11.3] 12

20. In triangle ABC, $a = 20$, $b = 14$, and
$m\angle C = 115°$. Find c to the nearest tenth.
[Lesson 14.2] 28.9

INFO BANK

Functions and Their Graphs

Throughout the text a variety of functions were studied. The simplest form of any function is called the parent function. Each parent function has a distinctive graph. This page summarizes the basic graphs.

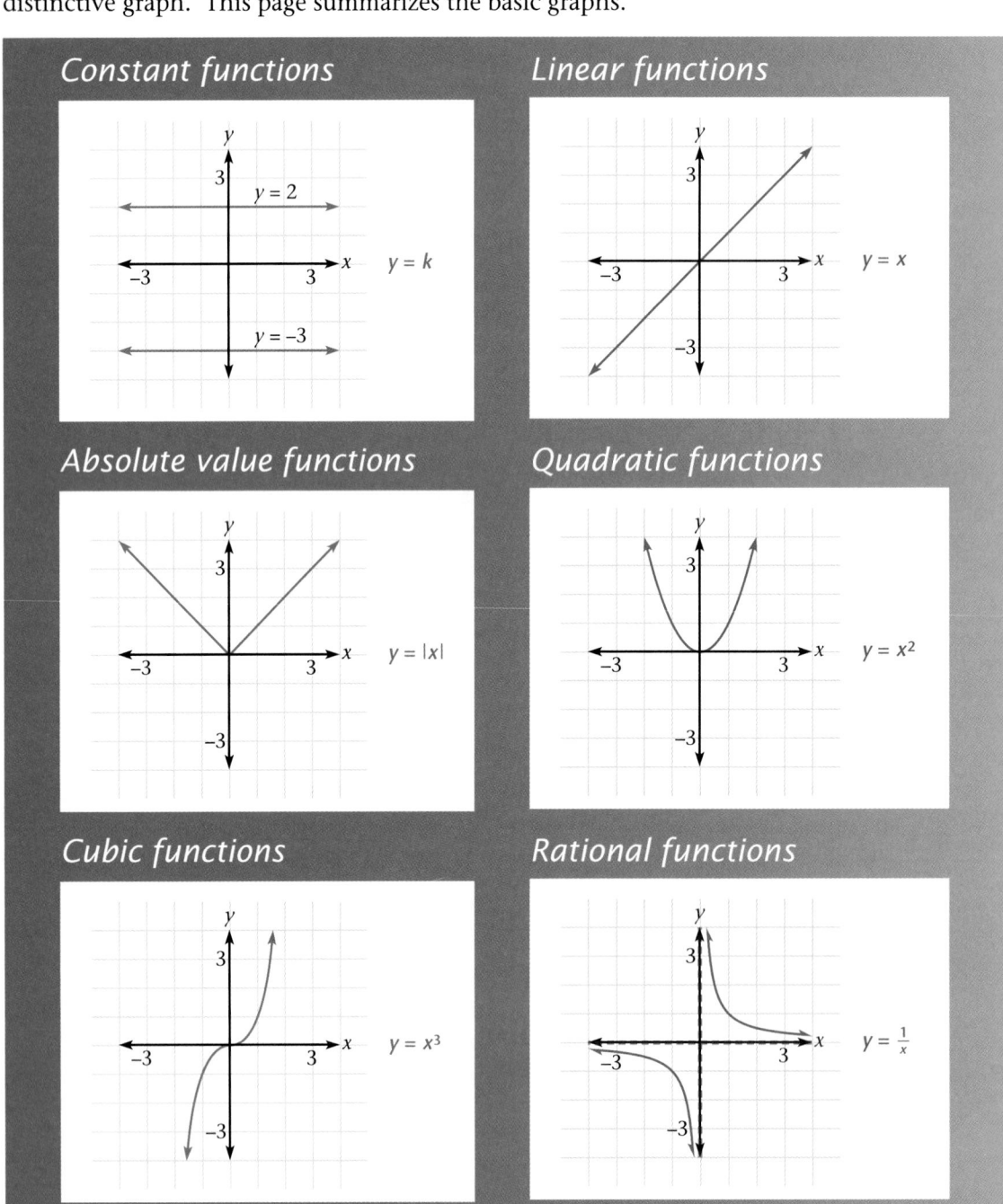

Constant functions

$y = 2$

$y = k$

$y = -3$

Linear functions

$y = x$

Absolute value functions

$y = |x|$

Quadratic functions

$y = x^2$

Cubic functions

$y = x^3$

Rational functions

$y = \frac{1}{x}$

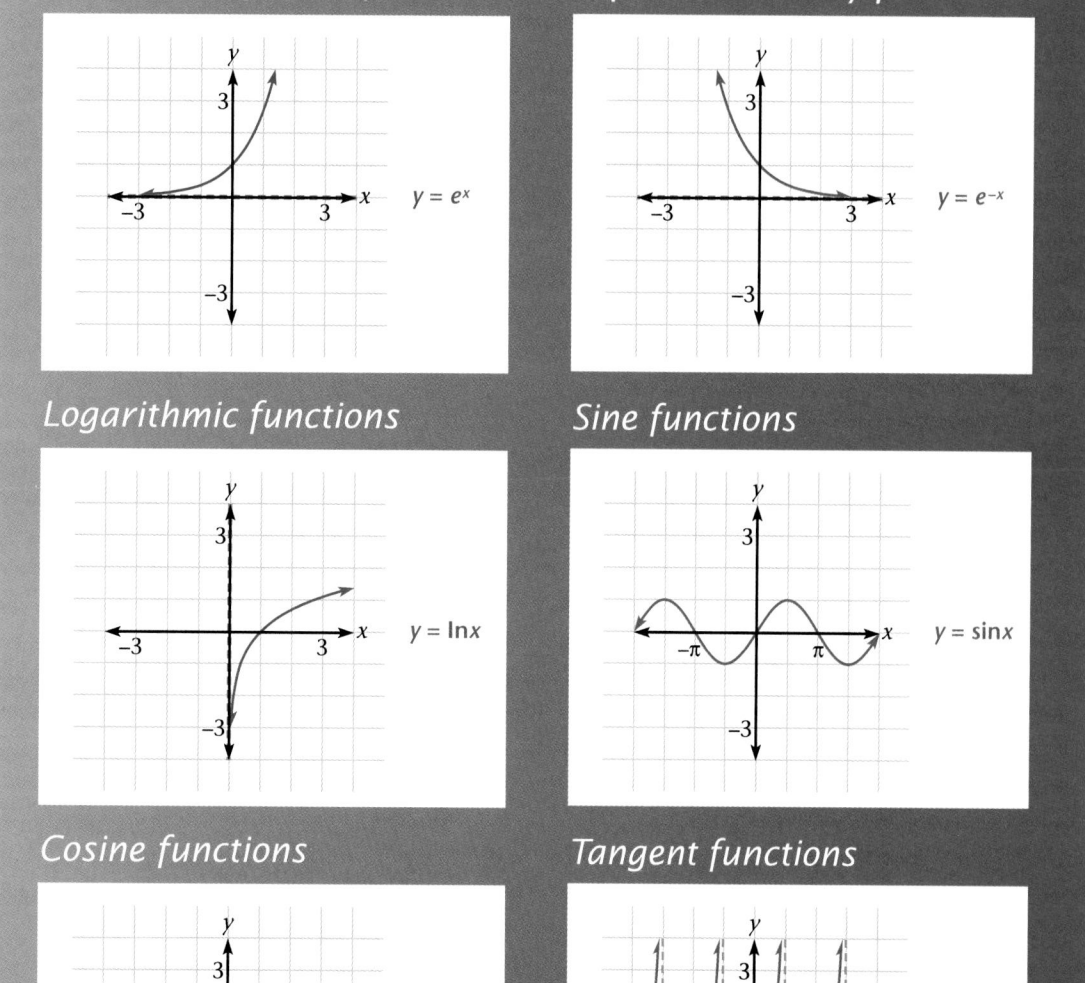

Exponential growth functions

$y = e^x$

Exponential decay functions

$y = e^{-x}$

Logarithmic functions

$y = \ln x$

Sine functions

$y = \sin x$

Cosine functions

$y = \cos x$

Tangent functions

$y = \tan x$

Table **813**

Table of Random Digits

Line\Col	(1)	(2)	(3)	(4)	(5)	(6)	(7)	(8)	(9)	(10)	(11)	(12)	(13)	(14)
1	10480	15011	01536	02011	81647	91646	69179	14194	62590	36207	20969	99570	91291	90700
2	22368	46573	25595	85393	30995	89198	27982	53402	93965	34095	52666	19174	39615	99505
3	24130	48360	22527	97265	76393	64809	15179	24830	49340	32081	30680	19655	63348	58629
4	42167	93093	06243	61680	07856	16376	39440	53537	71341	57004	00849	74917	97758	16379
5	37570	39975	81837	16656	06121	91872	60468	81305	49684	60672	14110	06927	01263	54613
6	77921	06907	11008	42751	27756	53498	18602	70659	90655	15053	21916	81825	44394	42880
7	99562	72905	56420	69994	98872	31016	71194	18738	44013	48840	63213	21069	10634	12952
8	96301	91977	05463	07972	18876	20922	94595	56869	69014	60045	18425	84903	42508	32307
9	89579	14342	63661	10281	17453	18103	57740	84378	25331	12566	58678	44947	05585	56941
10	85475	36857	53342	53988	53060	59533	38867	62300	08158	17983	16439	11458	18593	64952
11	28918	69578	88231	33276	70997	79936	56859	05859	90106	31595	01547	85590	91610	78088
12	63553	40961	48235	03427	49626	69445	18663	72695	52180	20847	12234	90511	33703	90322
13	09429	93969	52636	92737	88974	33488	36320	17617	30015	08272	84115	27156	30613	74952
14	10365	61129	87529	85689	48237	52267	67689	93394	01511	26358	85104	20285	29975	89868
15	07119	97336	71048	08178	77233	13916	47564	81056	97735	85977	29372	74461	28551	90707
16	51085	12765	51821	51259	77452	16308	60756	92144	49442	53900	70960	63990	75601	40719
17	02368	21382	52404	60268	89368	19885	55322	44819	01188	65255	64835	44919	05944	55157
18	01011	54092	33362	94904	31273	04146	18594	29852	71585	85030	51132	01915	92747	64951
19	52162	53916	46369	58586	23216	14513	83149	98736	23495	64350	94738	17752	35156	35749
20	07056	97628	33787	09998	42698	06691	76988	13602	51851	46104	88916	19509	25625	58104
21	48663	91245	85828	14346	09172	30168	90229	04734	59193	22178	30421	61666	99904	32812
22	54164	58492	22421	74103	47070	25306	76468	26384	58151	06646	21524	15227	96909	44592
23	32639	32363	05597	24200	13363	38005	94342	28728	35806	06912	17012	64161	18296	22851
24	29334	27001	87637	87308	58731	00256	45834	15398	46557	41135	10367	07684	36188	18510
25	02488	33062	28834	07351	19731	92420	60952	61280	50001	67658	32586	86679	50720	94953
26	81525	72295	04839	96423	24878	82651	66566	14778	76797	14780	13300	87074	79666	95725
27	29676	20591	68086	26432	46901	20849	89768	81536	86645	12659	92259	57102	80428	25280
28	00742	57392	39064	66432	84673	40027	32832	61362	98947	96067	64760	64584	96096	98253
29	05366	04213	25669	26422	44407	44048	37937	63904	45766	66134	75470	66520	34693	90449
30	91921	26418	64117	94305	26766	25940	39972	22209	71500	64568	91402	42416	07844	69618
31	00582	04711	87917	77341	42206	35126	74087	99547	81817	42607	43808	76655	62028	76630
32	00725	69884	62797	56170	86324	88072	76222	36086	84637	93161	76038	65855	77919	88006
33	69011	65795	95876	55293	18988	27354	26575	08625	40801	59920	29841	80150	12777	48501
34	25976	57948	29888	88604	67917	48708	18912	82271	65424	69774	33611	54262	85963	03547
35	09763	83473	73577	12908	30883	18317	28290	35797	05998	41688	34952	37888	38917	88050
36	91567	42595	27958	30134	04024	86385	29880	99730	55536	84855	29080	09250	79656	73211
37	17955	56349	90999	49127	20044	59931	06115	20542	18059	02008	73708	83517	36103	42791
38	46503	18584	18845	49618	02304	51038	20655	58727	28168	15475	56942	53389	20562	87338
39	92157	89634	94824	78171	84610	82834	09922	25417	44137	48413	25555	21246	35509	20468
40	14577	62765	35605	81263	39667	47358	56873	56307	61607	49518	89656	20103	77490	18062
41	98427	07523	33362	64270	01638	92477	66969	98420	04880	45585	46565	04102	46880	45709
42	34914	63976	88720	82765	34476	17032	87589	40836	32427	70002	70663	88863	77775	69348
43	70060	28277	39475	46473	23219	53416	94970	25832	69975	94884	19661	72828	00102	66794
44	53976	54914	06990	67245	68350	82948	11398	42878	80287	88267	47363	46634	06541	97809
45	76072	29515	40980	07391	58745	25774	22987	80059	39911	96189	41151	14222	60697	59583
46	90725	52210	83974	29992	65831	38857	50490	83765	55657	14361	31720	57375	56228	41546
47	64364	67412	33339	31926	14883	24413	59744	92351	97473	89286	35931	04110	23726	51900
48	08962	00358	31662	25388	61642	34072	81249	35648	56891	69352	48373	45578	78547	81788
49	95012	68379	93526	70765	10592	04542	76463	54328	02349	17247	28865	14777	62730	92277
50	15664	10493	20492	38391	91132	21999	59516	81652	27195	48223	46751	22923	32261	85653

Standard Normal Distribution Table

This table gives the area under the standard normal curve to the left of a given positive number *a*.

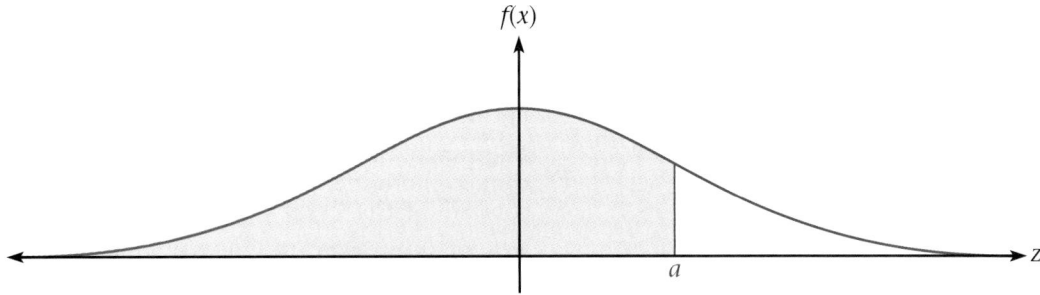

					P(z < a) for a ≥ 0					
a	0	1	2	3	4	5	6	7	8	9
0.0	.5000	.5040	.5080	.5120	.5160	.5199	.5239	.5279	.5319	.5359
0.1	.5398	.5438	.5478	.5517	.5557	.5596	.5636	.5675	.5714	.5753
0.2	.5793	.5832	.5871	.5910	.5948	.5987	.6026	.6064	.6103	.6141
0.3	.6179	.6217	.6255	.6293	.6331	.6368	.6406	.6443	.6480	.6517
0.4	.6554	.6591	.6628	.6664	.6700	.6736	.6772	.6808	.6844	.6879
0.5	.6915	.6950	.6985	.7019	.7054	.7088	.7123	.7157	.7190	.7224
0.6	.7257	.7291	.7324	.7357	.7389	.7422	.7454	.7486	.7517	.7549
0.7	.7580	.7611	.7642	.7673	.7704	.7734	.7764	.7794	.7823	.7852
0.8	.7881	.7910	.7939	.7967	.7995	.8023	.8051	.8078	.8106	.8133
0.9	.8159	.8186	.8212	.8238	.8264	.8289	.8315	.8340	.8365	.8389
1.0	.8413	.8438	.8461	.8485	.8508	.8531	.8554	.8577	.8599	.8621
1.1	.8643	.8665	.8686	.8708	.8729	.8749	.8770	.8790	.8810	.8830
1.2	.8849	.8869	.8888	.8907	.8925	.8944	.8962	.8980	.8997	.9015
1.3	.9032	.9049	.9066	.9082	.9099	9115	.9131	.9147	.9162	.9177
1.4	.9192	.9207	.9222	.9236	.9251	.9265	.9279	.9292	.9306	.9319
1.5	.9332	.9345	.9357	.9370	.9382	.9394	.9406	.9418	.9429	.9441
1.6	.9452	.9463	.9474	.9484	.9495	.9505	.9515	.9525	.9535	.9545
1.7	.9554	.9564	.9573	.9582	.9591	.9599	.9608	.9616	.9625	.9633
1.8	.9641	.9649	.9656	.9664	.9671	.9678	.9686	.9693	.9699	.9706
1.9	.9713	.9719	.9726	.9732	.9738	.9744	.9750	.9756	.9761	.9767
2.0	.9772	.9778	.9783	.9788	.9793	.9798	.9803	.9808	.9812	.9817
2.1	.9821	.9826	.9830	.9834	.9838	.9842	.9846	.9850	.9854	.9857
2.2	.9861	.9864	.9868	.9871	.9875	.9878	.9881	.9884	.9887	.9890
2.3	.9893	.9896	.9898	.9901	.9904	.9906	.9909	.9911	.9913	.9916
2.4	.9918	.9920	.9922	.9925	.9927	.9929	.9931	.9932	.9934	.9936
2.5	.9938	.9940	.9941	.9943	.9945	.9946	.9948	.9949	.9951	.9952
2.6	.9953	.9955	.9956	.9957	.9959	.9960	.9961	.9962	.9963	.9964
2.7	.9965	.9966	.9967	.9968	.9969	.9970	.9971	.9971	.9973	.9974
2.8	.9974	.9975	.9976	.9977	.9977	.9978	.9979	.9979	.9980	.9981
2.9	.9981	.9982	.9982	.9983	.9984	.9984	.9985	.9985	.9986	.9986
3.0	.9987	.9987	.9987	.9988	.9988	.9989	.9989	.9989	.9990	.9990

Table **815**

absolute value function   The distance from the origin to a point x units from the orgin is called the absolute value of x. The absolute value function is defined as:
$$f(x) = |x| = \begin{cases} x, & \text{if } x \geq 0 \\ -x, & \text{if } x < 0 \end{cases}$$ (135)

addition property of equality   Let a, b, and c represent any real numbers. If $a = b$, then $a + c = b + c$. (38)

addition property of inequality   Let a, b, and c represent any real numbers. If $a < b$, then $a + c < b + c$. If $a > b$, then $a + c > b + c$. (46)

amplitude   If $y = b \sin \theta$ or $y = b \cos \theta$, $|b|$ is the amplitude. (454)

angle difference identities   Trigomometric identities that involve differences of angles. They are $\cos(A - B) = \cos A \cos B + \sin A \sin B$ and $\sin(A - B) = \sin A \cos B - \cos A \sin B$. (793)

angle sum identities   Trigomometric identities that involve sums of angles. They are $\cos(A + B) = \cos A \cos B - \sin A \sin B$ and $\sin(A + B) = \sin A \cos B + \cos A \sin B$. (793)

arc length   The length, d, of $\overset{\frown}{AB}$ intercepted by central angle θ, measured in radians, in a circle with radius r is $d = r\theta$. (468)

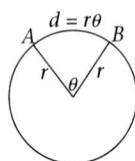

area of a sector   The area of sector AOB with central angle θ, measured in radians, in a circle with radius r is $A = \frac{1}{2}\theta r^2$. (471)

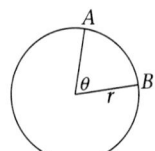

arithmetic sequence   A sequence with a constant difference, d, between any two consecutive terms, t_n and t_{n-1}. (655)

associative property of real numbers   For any real numbers a, b, and c, $(a + b) + c = a + (b + c)$ and $(a \cdot b) \cdot c = a \cdot (b \cdot c)$. (65)

asymptotes of a hyperbola   The dashed lines through the center of a hyperbola to which the hyperbola gets closer and closer to but never touches. (562)

augmented matrix   A single matrix that contains both the coefficient matrix and the constant matrix. (190)

axis of symmetry   A line over which an image mirrors itself. (114)

axis of symmetry of a parabola   The line that goes through the vertex of a parabola and divides the parabola in half. (255)

back substitution   A method of solving a system of equations which involves solving for one variable first, and working backward to solve for the other variables. (189)

base (of an exponential function)   The number b in the exponential function $f(x) = ab^x$. (384)

binomial experiment   An experiment with exactly 2 choices, such as correct or incorrect. (747)

binomial probability distribution   The binomial probability distribution for r successes in n trials is $P(r) = {}_nC_r p^r q^{n-r}$, where p is the probability of successes and $q = 1 - p$ is probability of failure. (748)

Bode number   Each number in the sequence 0.4, 0.7, 1.0, 1.6, 2.8, 5.2, ... , which is called Bode's sequence. (658)

Bode's Law   Describes the relationship between the numbers in Bode's sequence and the distance of each planet from the Sun. (658)

Bode's sequence The sequence 0.4, 0.7, 1.0, 1.6, 2.8, 5.2, ... , which is derived from adding 4 to each term of the sequence 0, 3, 6, 12, 24, 48, ... and then dividing each sum by 10. (658)

boundary line A line for a linear inequality that divides the coordinate plane into two half-planes. (48)

bounded region A region that contains the solutions to a system of inequalities and that does not extend beyond its boundary lines. (223)

box-and-whisker plot A graph that illustrates how quartiles are distributed within a data set. (732)

center of a circle The fixed point in a plane from which all points of a circle are the same distance. (544)

central angle An angle with its vertex at the center of a circle. (460)

central tendency The three averages used in statistics. They are the mean, the median, and the mode. (711)

circle A set of points in a plane that are the same distance from a fixed point in the plane, called the center. (544)

circular functions Trigonometric functions that can be defined in terms of a point $(x, y) = (\cos \theta, \sin \theta)$ on the unit circle. (784)

$\cos \theta = x$ $\qquad$ $\tan \theta = \frac{y}{x}$ $\qquad$ $\sec \theta = \frac{1}{x}$

$\sin \theta = y$ $\qquad$ $\cot \theta = \frac{x}{y}$ $\qquad$ $\csc \theta = \frac{1}{y}$

circular permutation The number of circular permutations of n objects is $(n - 1)!$. (614)

class mean A measure of central tendency found using the data in a grouped frequency table. The class mean can be calculated by finding the midpoint of each class, multiplying the midpoint by the frequency of the corresponding class, adding the products for each class, and dividing the sum by the total number of data values. (717)

closure property of real numbers For any real numbers a and b, $a + b$ and $a \cdot b$ are real numbers. (65)

combinations Different selections from a set of objects without regard to order. The number of combinations of n distinct objects taken r at a time is

$C(n,r) = {}_nC_r = \binom{n}{r} = \frac{n!}{r!\,(n-r)!}$. (620)

common logarithmic function (log) The logarithmic function with base 10, written $f(x) = \log_{10} x$. (403)

commutative property of real numbers For any real numbers a and b, $a + b = b + a$ and $a \cdot b = b \cdot a$. (65)

complex conjugates Complex numbers of the form $a + bi$ and $a - bi$. The complex conjugate of $a + bi$ is denoted by $\overline{a + bi}$. (298)

complex number A number that can be written in the form $a + bi$, where a and b are real numbers and i is the imaginary unit. (294)

complex plane The plane on which complex numbers are graphed. The horizontal axis is called the real axis and the vertical axis is called the imaginary axis. (298)

composition of functions The composition of functions f and g, $(f \circ g)(x)$, is defined as $f(g(x))$. The domain of f must include the range of g. (130)

compound interest The interest on the principal *plus* the interest on interest already earned. In general, if r is the annual percentage rate, n is the number of times per year that the interest is compounded, then the amount after t years is

$A(t) = P\left(1 + \frac{r}{n}\right)^{nt}$. (386)

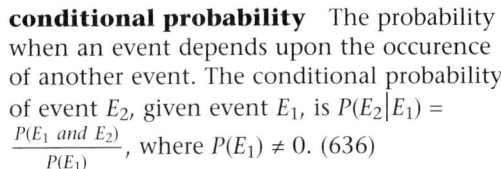

conditional probability   The probability when an event depends upon the occurence of another event. The conditional probability of event E_2, given event E_1, is $P(E_2|E_1) =$ $\frac{P(E_1 \text{ and } E_2)}{P(E_1)}$, where $P(E_1) \neq 0$. (636)

conic section   The graphs of any figure, such as circles, parabolas, ellipses, and hyperbolas, that can be represented by passing a plane through a hollow double cone. The general equation of a conic section can be written in the form $Ax^2 + Bxy + Cy^2 + Dx + Ey + F = 0$, where A, B, C, D, E, and F are not all zero. (569)

conjugate axis   The axis of a hyperbola that does *not* connect the vertices. It is determined by the rectangle that contains the transverse axis and the asymptotes as diagonals. (562)

consistant system   A system of linear equations that has at least one solution. (182)

constant function   A constant function is a linear function with slope zero. (91)

constant of variation (direct variation)   In a direct variation represented by the equation $y = kx$, where $k \neq 0$, k is called the constant of variation. (31)

constant of variation (indirect variation)   In an inverse variation represented by the equation $xy = k$ or $y = \frac{k}{x}$, where $k \neq 0$ and $x \neq 0$, k is called the constant of variation. (494)

correlation coefficient   A number represented by the variable r, where $-1 \leq r \leq 1$, that describes how closely points in a scatter plot cluster around the line of best fit. (25)

cosecant   The reciprocal function of the sine function, given by the ratio $\frac{1}{\sin \theta}$ and abbreviated csc θ. (785)

cosine of an angle   In a right traingle, the cosine of an acute angle A is given by the ratio $\frac{\text{length of side adjacent } \angle A}{\text{length of hypotenuse}}$, and is abbreviated $\cos A = \frac{\text{adj}}{\text{hyp}}$. (434)

cotangent   The reciprocal function of the tangent function, given by the ratio $\frac{\cos \theta}{\sin \theta}$ and abbreviated cot θ. (785)

coterminal angles   Angles that share the same terminal side. (440)

crossing point   The point at which the value of a polynomial function changes signs. (351)

decreasing function   For any real numbers a and b where $a < b$, a linear function f is decreasing if $f(a) > f(b)$. (94)

degree of a polynomial   The highest power of x with a nonzero coefficient in a polynomial written in expanded form. (329)

dependent events   When the occurrence of one event has an effect on the occurrence of a following event, the events are said to be dependent. (627)

dependent system   A system of linear equations that has an infinite number of solutions. (182)

dependent variable   The output of a function. For $y = f(x)$, $f(x)$ is the dependent variable. (85)

difference of functions   The difference found by subtracting the dependent variables of the functions. (99)

difference of two cubes   A polynomial of the form $a^3 - b^3$ that can be written as the product of two factors, $x^3 - b^3 = (x - b)(x^2 + bx + b^2)$. (342)

difference of two squares   A polynomial of the form $x^2 - a^2$ that can be written as the product of two factors, $x^2 - a^2 = (x - a)(x + a)$. (333)

direct variation   The equation $y = kx$ describes a direct variation where y varies directly as x, k is the constant of variation, and $k \neq 0$. (31)

directrix The fixed line from which the set of points that form a parabola are all the same distance as they are from the fixed point called the focus. (537)

discriminant Let $ax^2 + bx + c = 0$ be any quadratic equation, where a, b, and c are real numbers and $a \neq 0$. The value of $b^2 - 4ac$ is the discriminant and can be used to determine the number of real solutions of a quadratic equation. (288)

distance formula The distance, d, between the points (x_1, y_1) and (x_2, y_2) is $d = \sqrt{(x_2 - x_1)^2 + (y_2 - y_1)^2}$. (266)

distinct permutations of indistinguishable objects The number of distinct permutations of n objects, of which r objects are alike, is $\frac{n!}{r!}$, where $r \leq n$. The number of distinct permutation of n objects, of which r_1 are alike, r_2 are alike, r_3 are alike, ..., and r_k are alike is $\frac{n!}{r_1! \, r_2! \, r_3! \ldots r_k!}$. (613)

distributive property of real numbers For any real numbers a, b, and c, $a(b + c) = ab + ac$ and $(b + c)a = ba + ca$. (39)

division algorithm If the polynomial P is divided by the polynomial D, then $P = DQ + R$, where Q is the quotient polynomial, and R is the remainder polynomial of degree less than D. If R is 0, then D divides P exactly. (342)

division property of equality Let a, b, and c represent any real numbers, where $c \neq 0$. If $a = b$, then $\frac{a}{c} = \frac{b}{c}$. (38)

division property of inequality Let a, b, c, and d represent any real numbers, where $c > 0$ and $d < 0$. If $a < b$, then $\frac{a}{c} < \frac{b}{c}$ and $\frac{a}{d} > \frac{b}{d}$. If $a > b$, then $\frac{a}{c} > \frac{b}{c}$ and $\frac{a}{d} < \frac{b}{d}$. (46)

domain The set of possible values for the first coordinates of a relation. (79)

double angle identities Trigonometric identities that involve double angles. They are $\cos^2 A = \cos^2 A - \sin^2 A$ and $\sin 2A = 2 \sin A \cos A$. (793)

elementary row operations Operations performed on a matrix that result in an equivalent matrix. The elementary row operations are interchanging two rows, multiplying all elements of a row by a nonzero constant, and adding a constant multiple of the elements of one row to the corresponding elements of another row. (192)

ellipse A set of points in a plane such that the sum of the distances from two fixed points in the plance, called foci, to any point in the set is a constant. (552)

equivalent To have the same value. (39)

equivalent equations Equations that have the same solutions. (39)

even function A function with the property $f(-x) = f(x)$. (510)

event A set that includes elements chosen from a sample space. (597)

expanded form of a polynomial function A polynomial function written in the form $f(x) = a_n x^n + a_{n-1} x^{n-1} + \cdots + a_2 x^2 + a_1 x + a_0$, where the subscripted a-terms are the coefficients. (329)

experimental probability The experimental probability for the occurence of a particular event, E, is
$$P(E) = \frac{\text{number of times the event occurred}}{\text{number of times the experiment was conducted}}.$$
(597)

explicit definition of a sequence A definition in which the nth term of sequence is defined in terms of n and either d, the constant difference, or r, the constant ratio. (662)

exponential decay The exponential function $f(x) = ab^x$ is a function of exponential decay when $a > 0$ and $0 < b < 1$. (384)

exponential function The function $f(x) = ab^x$, where a and b are real numbers such that $a \neq 0$, $b > 0$, and $b \neq 1$, and x is any real number. (384)

exponential growth   The exponential function $f(x) = ab^x$ is a function of exponential growth when $a > 0$ and $b > 1$. (384)

exponential-log inverse properties   For $x > 0$, $b^{\log_b x} = x$. For all x, $\log_b b^x = x$. (393)

extraneous roots   False roots of a quadratic equation that may be obtained when both sides of the equation are squared. (264)

Factor Theorem   If f is any polynomial function, $x - a$ is a factor of f if and only if a is a zero of f. Also, a is a zero of f if $x - a$ divides f exactly. (336)

factor over the integers   A polynomial is factored over the integers when it is expressed as a product of two or more factors having integral coefficients. (335)

factored form of a polynomial function   A polynomial function written in the form $f(x) = C(x - r_1)(x - r_2) \ldots (x - r_{n-1})(x - r_n)$, where the subscripted r-values, $r_1, r_2, \ldots, r_n$, are the roots, or zeros, of the function and C is constant. (328)

feasible region   A region that contains points that are the possible, or feasible, solutions to a system of linear inequalities. (229)

Fibonacci sequence   The sequence 1, 1, 2, 3, 5, 8, 13, 21, 34, ... , in which each term is obtained by adding the preceding two terms. (657)

foci of a hyperbola   Two fixed points in a plane such that the difference of the distances from these points to any point on a hyperbola is a constant. (560)

foci of an ellipse   Two fixed points in a plane such that the sum of the distances from these points to any point on an ellipse is a constant. (552)

focus of a parabola   The fixed point from which the set of points that form a parabola are all the same distance as they are from the fixed line called the directrix. (537)

frequency table   A table that shows the number of times, or frequency, that each data value in a data set occurs. (715)

function   A relation in which, for each first coordinate, there is exactly *one* corresponding second coordinate. (77)

function notation   A function is usually defined in terms of x and y, where $y = f(x)$; x is the independent variable, and $f(x)$ is the dependent variable. (85)

Fundamental Principle of Counting   If there are m ways of making Decision 1 and n ways of making Decision 2, then there are $m \cdot n$ ways of making both decisions. (606)

Fundamental Theorem of Algebra   Every polynomial function of degree $n > 0$ has at least one complex zero. (343)

geometric sequence   A sequence with a constant ratio between consecutive terms. (656)

Golden Ratio   The ratio $\frac{1 + \sqrt{5}}{2} \approx 1.618$, usually represented by the Greek letter phi (ϕ). (657)

Golden Rectangle   A rectangle with consecutive sides whose lengths form the Golden Ratio. (657)

greatest-integer function   The function denoted $f(x) = [x]$ that converts a real number, x, into the largest integer that is less than or equal to x. (143)

grouped frequency table   A frequency table in which the data values are grouped by a range of values. (716)

histogram A bar graph in which the length of the bars shows the frequency of data values. (723)

hole in a graph For any real number a, when $x - a$ is a factor of both the numerator and the denominator of a function, the graph of the function has a hole in the graph at $x = a$. (519)

horizontal-line test If a horizontal line crosses the graph of a function in more than one point, the inverse of the function is not a function. (123)

hyperbola A set of points in a plane such that the difference of the distances from two fixed points in the plane, called foci, to any point on the hyperbola is a constant. (560)

identity matrix (I_n) An $n \times n$ matrix with ones along the main diagonal and zeros elsewhere. (198)

identity property of real numbers For any real number a, there is a number 0 such that $a + 0 = a = 0 + a$, and a number 1 such that $a \cdot 1 = a = 1 \cdot a$. (65)

image point The result of reflecting a point over a given line. (115)

imaginary number For any positive real number a, $\sqrt{-a}$ is an imaginary number, where $\sqrt{-a} = i\sqrt{a}$ and $(i\sqrt{a})^2 = -a$. (290)

inconsistent system A system of equations that has no solution. (182)

increasing function For any real numbers a and b, where $a < b$, a linear function f is increasing if $f(a) < f(b)$. (94)

independent events When the occurrence of one event has no effect on the occurrence of the second event, the events are said to be independent. (627)

independent system A system of equations that has a unique solution. (182)

independent variable The input of a function. For $y = f(x)$, x is the independent variable. (85)

inequality A mathematical sentence that contains >, <, ≥, ≤, or ≠. (45)

initial side (of an angle) The side of an angle in standard position in the coordinate plane that is on the positive x-axis. (438)

inner dimensions When two matrices are to be multiplied, the number of columns of the first matrix and the number of rows of the second matrix are called the inner dimensions. To perform the multiplication, the inner dimensions of the two matrices must be the same. (172)

integers (I) The set of whole numbers combined with their opposites. (62)

interquartile range The difference between the third and the first quartiles for a data set. (731)

inverse cosine relation (cos^{-1}) The relation used to find the measure of an angle when the cosine of that angle is known. (447)

inverse of a function The inverse of a function f can be obtained by reversing the order of the coordinates in each ordered pair of f. The function, f, will have an inverse function, or be invertible, when each second coordinate of f corresponds to exactly one first coordinate. (121)

inverse matrix If A is an $n \times n$ matrix with an inverse, then A^{-1} is its inverse matrix, and $A \cdot A^{-1} = I_n = A^{-1} \cdot A$. (199)

inverse property of real numbers For any real number a, there is a number $-a$ such that $a + (-a) = 0 = (-a) + a$. For any nonzero real number a, there is a number $\frac{1}{a}$ such that $a \cdot \frac{1}{a} = 1 = \frac{1}{a} \cdot a$. (65)

inverse sine relation (sin⁻¹) The relation used to find the measure of an angle when the sine of that angle is known. (447)

inverse tangent relation (tan⁻¹) The relation used to find the measure of an angle when the tangent of that angle is known. (447)

inverse variation The equation $xy = k$, or $y = \frac{k}{x}$, describes an inverse variation, where y varies inversely as x, k is the constant of variation, and $k \neq 0$ and $x \neq 0$. (494)

invertible A function is invertible when it has an inverse that is a function. (122)

irrational numbers The set of numbers that cannot be written as the quotient of two integers. (63)

Law of Cosines Given two sides, a and b, of a triangle and the measure of angle C included between them, the Law of Cosines states that $c^2 = a^2 + b^2 - 2ab \cos C$, where c is the length of the side opposite angle C. (777)

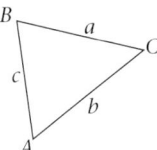

Law of Sines The Law of Sines relates the three angle measures of a triangle to the lengths of the corresponding three sides.
$\frac{\sin A}{a} = \frac{\sin B}{b} = \frac{\sin C}{c}$ (771)

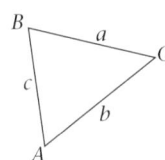

leading coefficient The coefficient of the highest power of x, a_n, in the expanded form of a polynomial function. (329)

line of best fit The straight line that fits closest to all of the data points in a scatter plot. (25)

linear equation An equation whose graph is a line. (10)

linear factor Any factor of the form $x - r$, where r can be a real or complex number. (328)

linear function A function that can be written in the form $f(x) = mx + b$, where m is the slope and b is the y-intercept. (91)

linear inequality An inequality with a boundary line that can be expressed in the form $y = mx + b$. The solution to a linear inequality is the set of all ordered pairs that make the inequality true. (49)

linear programming A method of finding a maximum or a minimum value that satisfies all of the given conditions of a particular situation. (230)

linear relationship A relationship in which a constant difference in consecutive x-values results in a constant difference in consecutive y-values. (9)

logarithmic function The logarithmic function $y = \log_b x$ is the inverse of the exponential function $y = b^x$, where $b \neq 1$ and $b > 0$; and $y = \log_b x$ if and only if $x = b^y$. (392)

main diagonal The diagonal containing elements of the coefficient matrix from the upper left hand corner to the lower right hand corner. (190)

major axis The longer axis of an ellipse which connects the vertices. (553)

matrix Any rectangular arrangement of elements in rows and columns enclosed by brackets. (165)

matrix algebra A method for solving a matrix equation of the form $AX = B$ for X. The method involves multiplying both sides of the matrix equation by A^{-1}. (204)

matrix dimensions The dimensions of a matrix are given by the number of rows and the number of columns. (166)

matrix multiplication If A_{mn} is an $m \times n$ matrix and B_{np} is an $n \times p$ matrix, then matrix multiplication AB is defined because the inner dimensions, n, are the same. The product matrix, AB, found by pairing rows of A with columns of B, has dimenstions $m \times p$. (173)

matrix notation For a matrix named M, m_{ab} represents the element in row a and column b. (165)

mean The mean, $\bar{x}$, is found using the formula $\bar{x} = \dfrac{\sum\limits_{i=1}^{n} x_i}{n}$, where $\sum\limits_{i=1}^{n} x_i$ is the sum of the values in the data set, and n is the number of data values. May also be called the arithmetic mean. (711)

mean deviation Combines the range and mean to describe an average measure of how the values differ, or deviate, from the mean. The formula for the mean deviation is $\dfrac{\sum |x_n - \bar{x}|}{n}$, where x_n is the nth value in the data set, $\bar{x}$ is the mean of the data set, and n is the number of values in the data set. (741)

measure of dispersion A numerical value that indicates the extent to which data values are dispersed, or spread around, a center number. (740)

median The middle data value in an ordered data set. When there are two middle values, the median of the data set is the mean of these two values. (712)

minor axis The shorter axis of an ellipse which connects the co-vertices. (553)

mode The value that appears most often in a data set. (713)

multiple zeros Multiple roots produced by repeated linear factors of a polynomial function. (336)

multiplication property of equality Let a, b, and c represent any real numbers. If $a = b$, then $ac = bc$. (38)

multiplication property of inequality Let a, b, c, and d represent any real numbers, where $c > 0$ and $d < 0$. If $a < b$, then $ac < bc$ and $ad > bd$. If $a > b$, then $ac > bc$ and $ad < bd$. (46)

multiplier In exponential growth or decay, the number multiplied by each value to obtain the next value. (377)

mutually exclusive events When two events cannot occur at the same time, the events are said to be mutually exclusive. (630)

natural exponential function The function $f(x) = e^x$ is called the natural exponential function. (408)

natural logarithmic function (ln x) The inverse of the natural exponential function $y = e^x$ is the natural logarithmic function, $y = \log_e x$, and is usually written $y = \ln x$. (409)

natural number e In the exponential function $f(x) = e^x$, the base is the natural number e. (408)

negative measure The measure of an angle if the rotation from the initial side is measured in the clockwise direction. (440)

normal curve A bell-shaped curve that describes the theoretical outcomes of an infinite number of trials in a binomial experiment. (753)

normal distribution The probability distribution of the theoretical outcomes of an inifinite number of trials in a binomial experiment. (753)

nth partial sum The sum of the first n terms of a geometric series, indicated by $\sum\limits_{n=1}^{n}$. (681)

nth term of a geometric sequence If t_1, t_2, ..., t_n represents a geometric sequence with constant ratio, r, then for any positive integer n, the nth term is given by $t_n = t_1 \cdot r^{n-1}$. (672)

nth term of an arithmetic sequence If $t_1, t_2, \ldots, t_n$ represents an arithmetic sequence with constant difference, d, then the nth term is given by $t_n = t_1 + (n - 1)d$. (662)

number of polynomial roots Every polynomial $P_n(x)$ of degree n can be written as a factored product of exactly n linear factors, corresponding to exactly n zeros (counting multiplicities). $P_n(x) = C(x - r_1)(x - r_2) \ldots \ldots (x - r_{n-1})(x - r_n)$ (343)

objective function The function in linear programming that states the object of a situation, to minimize or to maximize a particular quantity. (230)

odd function A function with the property $f(-x) = -f(x)$. (510)

outer dimensions When two matrices are multiplied, the number of rows of the first matrix and the number of columns of the second matrix are called the outer dimensions. The outer dimensions become the dimensions of the resulting product matrix. (172)

outlier A point on a scatter plot that appears to stand out from the rest of the points. (24)

parabola The graph of a quadratic function. (255) This graph is the set of points in a plane that are the same distance from the directrix as they are from the focus. (537)

parameter The variable t in a system of parametric equations. (147)

parametric equations A system of equations that examines two quantities in terms of a third quantity. (147)

percentile A data value in the nth percentile is greater than $n\%$ of the other data values in a data set. (730)

perfect-square trinomial A trinomial of the form $x^2 + 2xb + b^2$ or $x^2 - 2xb + b^2$. (276)

period of a function The length along the x-axis of one cycle of a periodic function. (453)

permutation An arrangement of a set of distinct objects into a specific order. (607)

phase shift For $y = \sin(\theta - d)$ and $y = \cos(\theta - d)$, d is called the phase shift of the function. The phase shift is a horizontal shift of its graph. (455)

polynomial function Any function that can be expressed as a product of linear factors multiplied by a constant. (328)

population An entire group for which a survey is conducted. (596)

positive measure The measure of an angle if the rotation from the initial side is measured in the counterclockwise direction. (440)

power property of logarithms For any positive number m and any real number p, the power property states that $\log_b m^p = p \log_b m$. (399)

pre-image point A name for the point before it has been reflected over a given line. (115)

principal square root The positive square root of a number. It is indicated by the square root sign. (262)

principal The original amount of money invested. (386)

principal values The numbers in the restricted range of an inverse trigonometric function. The principal values for the inverse cosine function are $0° \leq \cos^{-1} x \leq 180°$, the inverse sine are $-90° \leq \sin^{-1} x \leq 90°$, and the inverse tangent are $-90° < \tan^{-1} x < 90°$. (448)

probability A numerical value that measures the likeliness of an event occurring. (596)

probability distribution A list of probabilities for a given event. (747)

probability of independent events
If A and B are independent events, then
$P(A \text{ and } B) = P(A) \cdot P(B)$. (628)

product of functions   The product found
by multiplying the dependent variables of the
functions: $(f \cdot g)(x) = f(x) \cdot g(x)$. (100)

product property of logarithms   For any
two positive numbers m and n, the product
property states that $\log_b(mn) = \log_b m + \log_b n$.
(397)

proportion   An equation that states that
two ratios are equal. (31)

Pythagorean identities   Three identities
derived from the "Pythagorean" Right-Triangle
Theorem. They are $\sin^2 \theta + \cos^2 \theta = 1$, $1 +
\tan^2 \theta = \sec^2 \theta$, and $1 + \cot^2 \theta = \csc^2 \theta$. (786)

"Pythagorean" Right-Triangle Theorem
In a right triangle, the length of the
hypotenuse, c, is related to the lengths of the
legs, a and b, by the equation $c^2 = a^2 + b^2$.
(265)

quadratic formula   If $ax^2 + bx + c = 0$,
where $a \neq 0$, then $x = \frac{-b \pm \sqrt{b^2 - 4ac}}{2a}$. (282)

quadratic function   A function that can be
written in the form $f(x) = ax^2 + bx + c$, where a,
b, and c are real numbers and $a \neq 0$. The graph
of a quadratic function is a parabola. (255)

quadratic inequality   An inequality that
contains one or more quadratic expressions.
(308)

quartile   The values used to divide a data set
into fourths. The 25th percentile, Q_1, ends the
1st quartile; the 50th percentile, Q_2, ends the
2nd quartile; and the 75th percentile, Q_3, ends
the 3rd quartile. (731)

quotient of functions   The quotient found
by dividing the dependent variables of the
functions: $(f \div g)(x) = (f(x)) \div (g(x))$, where
$g(x) \neq 0$. (100)

quotient property of logarithms   For any
two positive numbers m and n, the quotient
property states that $\log_b \frac{m}{n} = \log_b m - \log_b n$.
(398)

radian measure   The measure of a central
angle with radii and intercepted arc the same
length is 1 radian. (460)

radius   The segment from the center of a
circle to any point on the circle is called the
radius of the circle. (544)

random number simulation   A procedure
that uses random numbers to simulate an
actual experiment with an underlying random
component. (640)

range   The set of possible values for the
second coordinates in a relation. (79) The
absolute value of the difference between the
largest value and the smallest value of a data
set. (740)

rational function   The function
$f(x) = \frac{p(x)}{q(x)}$, where both the numerator,
$p(x)$, and the denominator, $q(x)$, are
polynomials. The domain of a rational
function is all real numbers, x, such that the
denominator $q(x) \neq 0$. (502)

rational numbers (Q)   The set of all
numbers that can be written as the quotient
of two integers, where the denominator is not
equal to zero. (62)

real numbers (R)   The combination of all
rational and all irrational numbers. (63)

realistic domain   Meaningful x-values of
a function that models a real-world situation.
(327)

recursive definition of a sequence   A
definition in which any term is defined using
the previous term or terms. (662)

reflection   A transformation that flips a
figure over a given line. (137)

relation   Any set of ordered pairs. (77)

relative frequency The ratio of the frequency of a data value or class to the total number of data values. (716)

roots of an equation The *x*-values of an equation that correspond to *y* = 0. (258)

rounding-up function The function, denoted $f(x) = \lceil x \rceil$, that converts a real number, *x*, into the smallest integer greater than or equal to *x*. (143)

row reduction method The process of performing elementary row operations on an augmented matrix to solve a system of equations. (193)

rules of exponents Let *a* represent any real number, where $a \ne 0$, and let *m* and *n* be integers, where $n \ne 0$. Then $a^0 = 1$, $a^m \cdot a^n = a^{(m+n)}$, $a^{-n} = \frac{1}{a^n}$, $a^m \div a^n = a^{(m-n)}$, $(a^m)^n = a^{m \cdot n}$, and $a^{\frac{m}{n}} = \sqrt[n]{a^m}$. (71)

sample space A set, *S*, of outcomes of an experiment where any outcome of the experiment corresponds to exactly one element of *S*. (597)

scatter plot A graph of ordered pairs that displays the relationship between two sets of data. (23)

secant The reciprocal function of the cosine function, given by the ratio $\frac{1}{\cos \theta}$ and abbreviated sec θ. (785)

sector of a circle The area of a circle enclosed by a central angle and the arc intercepted by the central angle. (470)

sequence An ordered list of numbers. (654)

series The indicated sum of terms of a sequence. (663, 674)

simple radical form The form of a radical in which the number under the square root sign does not contain any perfect-square factors. (263)

simplex method A common method of performing linear programming with a large number of constraints and variables. (238)

sine of an angle In a right traingle, the sine of an acute angle *A* is given by the ratio $\frac{\text{length of side opposite } \angle A}{\text{length of hypotenuse}}$ and is abbreviated $\sin A = \frac{\text{opp}}{\text{hyp}}$. (434)

slope The ratio of the change in vertical direction to the change in horizontal direction: slope $= \frac{\text{change in } y}{\text{change in } x} = \frac{y_2 - y_1}{x_2 - x_1}$ (17)

slope formula Let *f* represent any linear function, and let $(x_1, f(x_1))$ and $(x_2, f(x_2))$ be two points on the graph of *f*. The slope of the graph of *f* is $m = \frac{f(x_2) - f(x_1)}{x_2 - x_1}$. (91)

slope-intercept form A linear equation in the form $y = mx + b$, where *m* represents the slope and *b* represents the *y*-intercept. (19)

square of a binomial The factored form of a perfect-square trinomial, either $(x + b)^2$ or $(x - b)^2$. (276)

square roots Let *a* and *b* represent real numbers, with $b \ge 0$. If $a^2 = b$, then $a = \sqrt{b}$ or $a = -\sqrt{b}$, also written $a = \pm\sqrt{b}$. (263)

standard deviation A measure of dispersion for a data set, given by the formula $\sigma = \sqrt{\frac{\sum (x_n - \bar{x})^2}{n}}$, where x_n is the *n*th value in the data set, $\bar{x}$ is the mean of the data set, and *n* is the number of values in the data set. (742)

standard equation of a circle The standard equation of a circle in the coordinate plane is $(x - h)^2 + (y - k)^2 = r^2$, where (h, k) is the center and $r > 0$ is the radius. (546)

standard equation of a hyperbola The standard equation of a hyperbola centered at (h, k) is $\frac{(x - h)^2}{a^2} - \frac{(y - k)^2}{b^2} = 1$, where the transverse axis is horizontal through the center, or $\frac{(y - k)^2}{a^2} - \frac{(x - h)^2}{b^2} = 1$, where the transverse axis is vertical through the center. (562)

standard equation of a parabola The standard equation of a parabola with vertex (h, k) is $y = a(x - h)^2 + k$, where $x = h$ is the axis of symmetry, $\left(h, k + \frac{1}{4a}\right)$ is the focus, and $y = k - \frac{1}{4a}$ is the directrix, or $x = a(y - k)^2 + h$, where $y = k$ is the axis of symmetry, $\left(h + \frac{1}{4a}, k\right)$ is the focus, and $x = h - \frac{1}{4a}$ is the directrix. (540)

standard equation of an ellipse The standard equation of an ellipse centered at (h, k) is $\frac{(x - h)^2}{a^2} + \frac{(y - k)^2}{b^2} = 1$, where the major aixs is horizontal, or $\frac{(x - h)^2}{b^2} + \frac{(y - k)^2}{a^2} = 1$, where the major axis is vertical. (553)

standard normal curve A normal curve that is used as a tool for approximating the probabilities of normally distributed data. (754)

standard position (of an angle) An angle in the coordinate plane with its vertex at the origin and its initial side on the positive x-axis. (438)

statistics The science of collecting, analyzing, describing, and interpreting data. (708)

stem-and-leaf plot A display of a set of data in which each piece of data is grouped together on a specific row, and arranged in two columns. The left column is called the stem, and the right column is called the leaf. (725)

step function A function whose graph looks like a series of steps. (143)

substitution property If two expressions are equivalent, you can replace one with the other. (39)

subtraction property of equality Let a, b, and c represent any real numbers. If $a = b$, then $a - c = b - c$. (38)

subtraction property of inequality Let a, b, and c represent any real numbers. If $a < b$, then $a - c < b - c$. If $a > b$, then $a - c > b - c$. (46)

sum of an infinite geometric series The sum, S, of an infinite geometric series, where $|r| < 1$, is $S = \frac{t_1}{1 - r}$. (681)

sum of functions The sum found by adding the dependent variables of the functions: $(f + g)(x) = f(x) + g(x)$ (99)

sum of the first n terms of an arithmetic sequence If $t_1, t_2, t_3, t_4, \ldots$ is an arithmetic sequence with constant difference d, then the sum of the first n terms is given by $S_n = n\left(\frac{t_1 + t_n}{2}\right)$. (664)

sum of two cubes A polynomial of the form $x^3 + b^3$ that can be written as the product of two factors, $x^3 + b^3 = (x + b)(x^2 - bx + b^2)$. (342)

symmetry with respect to the line $y = x$ Two points have symmetry with respect to the line $y = x$ if the x- and y-coordinates of the pre-image point are switched to form its image point. (116)

symmetry with respect to the x-axis Two points have symmetry with respect to the x-axis if the x-coordinates are the same, but the y-coordinates are opposites. (115)

symmetry with respect to the y-axis Two points have symmetry with respect to the y-axis if the y-coordinates are the same, but the x-coordinates are opposites. (115)

synthetic division A compact form of division that can be used to divide polynomials without writing out all of the steps of long division. (346)

tangent of an angle In a right traingle, the tangent of an acute angle A is given by the ratio $\frac{\text{length of side opposite } \angle A}{\text{length of side adjacent } \angle A}$ and is abbreviated $\tan A = \frac{\text{opp}}{\text{adj}}$. (434)

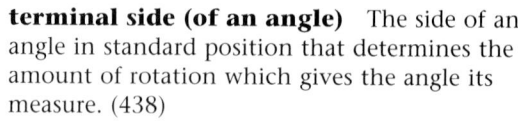

terminal side (of an angle) The side of an angle in standard position that determines the amount of rotation which gives the angle its measure. (438)

terms of a sequence The numbers in a sequence, represented by the letter t. (654)

theoretical probability When all the elements of a sample space are equally likely, the theoretical probability that an event, E, will occur is given by

$$P(E) = \frac{\text{number of elements in the event}}{\text{number of elements in the sample space}}. \quad (598)$$

transformation A translation, reflection, rotation, and/or stretch of an image. (210)

translation A transformation that slides a figure. (136)

transverse axis The axis of a hyperbola that connects the vertices. (562)

tree diagram A visual display of a sequence of choices. (606)

turning point A point on the graph of a polynomial function where there is a peak or a valley. (351)

unbounded region A region that contains the solutions to a system of inequalities and that is not limited by a boundary. (223)

unit circle A circle with a radius of 1 unit centered at the origin of the coordinate plane. (438)

variance The standard deviation of a data set is the square root of the variance, which is given by the formula $\sigma^2 = \dfrac{\sum(x_n - \bar{x})^2}{n}$, where x_n is the nth value in the data set, $\bar{x}$ is the mean of the data set, and n is the number of values in the data set. (742)

vertex The turning point of a parabola. (255)

vertical-line test If a vertical line crosses the graph of a relation in more than one point, the relation is not a function. (78)

whole numbers (W) The set of numbers $\{0, 1, 2, 3, 4, \ldots\}$. (62)

x-intercept The point where the graph of a line crosses the x-axis. (258)

y-intercept The point where the graph of a line crosses the y-axis. (18)

zero product property If a and b are real numbers and $ab = 0$, then either $a = 0$ or $b = 0$ must be true. (258)

zeros of a function The zeros of a function, f, are the values of x that make $f(x)$ equal to zero. (258)

INDEX

Definitions of bold face entries can be found in the glossary.

Back substitution, 189
Banneker, Benjamin, 74–75
Base (of an exponential function), 384
Bernoulli, Jakob, 747
Binomial distributions, 746–749
   probability, 748
Binomial expansion
   using combinations, 696
   using Pascal's Triangle, 696
   using tree diagram, 695
Binomial experiment, 747
Binomial probability distribution, 748
Binomial Theorem, 695–698
   and Pascal's Triangle, 696
Bode number, 658
Bode's Law, 658
Bode's sequence, 658
Bode, Johann Elert, 658
Boundary line, 48–51
   dashed, 49
   solid, 49
Bounded region, 223
Box-and-whisker plots,
   732–733
   interquartile range, 732
   quartiles, 732
Boyle's law, 492
Buffon, Comte du, 603

Cardano, Girolamo, 290
Center
   of a hyperbola, 561–564
   of an ellipse, 552–553
Central angle, 460
Central tendency, 711
   average, 711
   mean, 711, 713
   median, 711–712
   mode, 711, 713
Charles, Jacques, 97
Chinese counting board, 191
Chou dynasty, 715

Circles, 544–548, 569
   applications of, 547
   center, 544
   equation of, 545
   radius, 544
   standard equation of, 546
Circular functions, 784–787
Circular permutation, 620
Class(es), 715–718
   frequency, 717–718
   mean, 717
   midpoint of, 717–718
Closure property, 65
Combinations, 595, 620–624
   and Pascal's Triangle, 689–691
   and the Fundamental Principle
      of Counting, 622–624
   number of, 621
Common logarithmic function, 403
Communicate (*See Applications, language arts.*)
Commutative property, 65
Completing the square, 276–279
   solving quadratic equations by,
      278
Complex conjugates, 298
Complex numbers, 294–299
   operating with, 295
Complex plane, 298–299
   complex axis, 298
   real axis, 298
Composition of functions,
   128–131
   commutative property, 130
   inverse property, 131
Compound interest, 386–387
Conditional probability,
   635–637
Conic sections, 569
   circle, 569
   ellipse, 569
   general equation of, 569
   hyperbola, 569
   parabola, 569
   parametric representations of,
      578–582
Conjugate axis of a hyperbola,
   562
Connections (*See Applications, Cultural Connections, Math Connections.*)

Consistant system, 182
Constant difference, 9–10
Constant function, 91
Constant of variation , 31,
   494–495
Converge, 681
Correlation, 54–55
   coefficient, 25–25
   negative, 24–26
   positive, 24–26
   zero, 24
Cosecant, 785–786
Cosine function
   graph of, 453–455
   horizontal shift of, 455, 464
   horizontal stretch of, 454, 455
   period of, 453, 464
   reflection of, 454
   vertical shift of, 453, 455, 464
   vertical stretch of, 454, 455, 464
Cosine, 434
Cotangent, 785–786
Coterminal angles, 440–441
Counting, 605–608
   combinations, 620–624
   Fundamental Principle of, 606
Critical Thinking, 10, 16, 17, 24,
   26, 32, 37, 50, 51, 63, 70, 80,
   85, 93, 99, 117, 124, 131, 138,
   143, 167, 175, 181, 191, 194,
   200, 206, 213, 223, 230, 235,
   264, 266, 272, 279, 284, 291,
   295, 298, 304, 311, 327, 328,
   338, 342, 351, 359, 361, 379,
   384, 387, 392, 393, 399, 410,
   419, 434, 441, 442, 446, 449,
   460, 464, 470, 497, 498, 503,
   510, 513, 519, 525, 537, 538,
   541, 547, 552, 553, 562, 565,
   570, 571, 579, 599, 608, 614,
   622, 631, 637, 642, 656, 666,
   672, 674, 675, 681, 684, 697,
   698, 714, 718, 724, 731, 733,
   742, 747, 749, 772, 773, 777,
   779, 785, 799
Cross multiplication, 31–32
Crossing point, 351
Cubit, 29
Cultural Connections
   Africa, 134, 279, 793
      Egypt, 3, 29, 41, 43, 92,
      428, 447

feasible, 228–230, 234–238
  unbounded, 223
Regression models, 379
Relation, 77
Relative frequency, 716–718
Richter scale, 733
Roots of an equation, 258
  extraneous, 264
Rounding-up function, 143
Row reduction method, 193
Rules of exponents, 71

Sample space, 597–599
Scalar transformation, 138
Scatter plot, 7, 23–26
Scientific notation, 68–69, 71
Secant, 785–787
Sector of a circle, 470
Seked, 447
Sequences, 652–654
  arithmetic, 655, 656, 662–663
  Bode's, 658
  Fibonacci, 657
  geometric, 656, 671–674
  recursive definition of, 662
Series, 663–666
  arithmetic, 661–666
  geometric, 671–675
  infinite geometric, 680–684
Signum function, 146
Simple radical form, 263
Simplex method, 238
Sine function
  graph of, 453–455
  horizontal shift of, 455, 463
  horizontal stretch of, 454, 455
  period of, 453, 463
  reflection of, 454
  vertical shift of, 453, 455, 463
  vertical stretch of, 454, 455, 463
Sine, 434
Slope formula, 91
Slope, 15–19
  correlation and, 24
Slope-intercept form, 18–19, 90, 91
Spreadsheets, 6, 11, 383

Square of a binomial, 276
Square roots, 262
  principal, 262
Standard deviation, 742
Standard equation of a circle, 546
Standard equation of a hyperbola, 562
Standard equation of a parabola, 540
Standard equation of an ellipse, 562
Standard normal curve, 754–758
  z-score, 758
Standard position (of an angle), 438, 439
Statistics, 594, 708
  descriptive, 708
  inferential, 708
Stem-and-leaf plots, 725
  leaf, 725
  stem, 725
Step functions, 142–144
Substitution property, 39
Subtraction property of equality, 38–40
Subtraction property of inequality, 46
Sum of an infinite geometric series, 681
Sum of functions, 99
Sum of the first n terms of an arithmetic sequence, 664
Sum of two cubes, 342
Summation symbol, 665
Symmetry, 114
  axis of, 114
  functional, 154–155
  with respect to the line $y = x$, 116
  with respect to the x-axis, 115
  with respect to the y-axis, 115
Synthetic division, 346
System of parametric equations, 147
Systems of linear equations, 180–184
  consistent, 182
    independent, 182

dependent, 182
  inconsistent, 182
  representing with augmented matrix, 190
  solving
    by back substitution, 189
    by elimimation, 183–184
    using Chinese counting board, 191, 195
    using matrix algebra, 204–206
    using row reduction method, 193–194
Systems of linear inequalities, 220
  applying, 223–224
  graphing, 221–223
Systems of nonlinear equations
  applications of, 575
  first- and second-degree, 570–571
  two second-degree, 571–574

Tangent, 6, 434
Terminal side (of an angle), 438
Terms of a sequence, 654
Theoretical probability, 596, 598–600
Titius, Johann Daniel, 658
Transformation, 210
Transformation matrices, 210
  using to enlarge and reduce objects, 212
  using to reflect objects, 242
  using to reverse transformations, 213
  using to rotate objects, 211–212, 243
Translation, 136–137
Transverse axis of a hyperbola, 562
Tree diagram, 606
  using to expand $(a + b)^n$, 695
Trigonometric equations, 796–799
  graphs of, 796–799
  solving, 797–799
Trigonometric functions, 428

CREDITS

PHOTOS

Abbreviations used: (t) top, (c) center, (b) bottom, (l) left, (r) right, (bckgd) background.
FRONT COVER: (l) Dick Frank/The Stock Market; (ct) Dr. E.R. Degginger; (cb) Sam Dudgeon/HRW Photo; (r) Ric Ergenbright.
TABLE OF CONTENTS: Page v(t), Rob Waymen Photography; v(c), The Granger Collection; v(b), Ron Tanaka; vi(t), Courtesy of the National Museum of the American Indian Smithsonian Institution, S2379; vi(b), Rob Waymen Photography; vii(t), The Bettmann Archive; vii(c), Rob Waymen Photography, vii(b) Pronk&Associates; viii(t), Rob Waymen Photography; viii(b), Telegraph Colour Library/Masterfile; ix(t), Thomas Kitchin/First Light; ix(c), Stuart Westmoreland/Tony Stone Images; ix(b), Jeff Kaufman/FPG/Masterfile; x(t), Rob Waymen Photography; x(b), Ron Tanaka; xi(t), ©1996 R. Andrew Odum/Peter Arnold, Inc.; xi(b), Giraudon/Art Resource, NY. **CHAPTER ONE:** Page 2(tr), Ron Tanaka; 2(br),3(tr), Jan Becker; 3(br),©Richard T. Nowitz; 4(cl), NASA/Image Bank; 5(tl), Charles Thatcher/Tony Stone Images; 5(cr), Roger Tully/Tony Stone Images; 5(bc), Don Smetzer/Tony Stone Images; 7(cr), Paul Chesley/Tony Stone Images; 7(br), Tim Brown/Tony Stone Images; 7(cl), Andy Sacks/Tony Stone Images; 8(t,bckgd), Aldo Torelli/Tony Stone Images; 11(l), Ron Tanaka; 12(tr), Rob Waymen Photography; 13(br), Chad Slattery/Tony Stone Images; 14(t), Ron Tanaka; 15(c), Jan Becker; 16(c),18(c), Ron Tanaka; 20(br), Ron Tanaka; 21(b), Jan Becker; 23(tr), Rob Waymen Photography; 24(b),27(br), Ron Tanaka; 29(tl),(bckgd), Copyright British Museum; 30(tr), Tony Stone Images; 30(bl), Rob Waymen Photography; 32(bckgd), Ron Tanaka; 32(tr), U.S. Geological Survey/Photo Researchers; 32(tl), Earth Imaging; 33(tr),34(r), Ron Tanaka; 34(tl), Kevin Aitken/First Light; 34(tc), Al Grotell; 35(r),36(tr),(b), Ron Tanaka; 40(b), Hugh Sitton/Tony Stone Images; 41(tr),(l), Copyright British Museum; 42(b), Jan Becker; 45(tl), Rob Waymen Photography; 47(l), Rob Waymen Photography-photo manipulation by Jun Park; 50(br), Gary Hood/Inner Visions; 51(br), Rob Waymen Photography; 53(tr), Ron Tanaka; 54(r),55(l), Rob Waymen Photography; 55(c), Ron Tanaka. **CHAPTER TWO:** 60(cl),(bc),61(tl), Alexander Marshack; 62(t) Rob Waymen Photography; 62(c) Ron Tanaka; 64(br), Rob Waymen Photography; 67(tr), The Granger Collection; 68(tr), Ronald Royier, SPL/Photo Researchers; 68(b), Jan Becker; 70(tr), Rob Waymen Photography; 70(bl), Ms. Glaser 40: Orientabteilung, Staatsbibliothek zu Berlin - Preußischer Kulturbesitz; 72(tr), Rob Waymen Photography; 73(r), Ron Tanaka; 76(c), Adventure Photo/Masterfile; 80(br), Rob Waymen Photography; 83(tr), Jan Becker; 86(br),89(cr),90(tr), Ron Tanaka; 90(cr), Jan Becker; 91(bl),95(tr), Rob Waymen Photography; 97(tr), Jan Becker; 98(t),(c), Ron Tanaka; 101(c), Jan Becker; 102(b), courtesy of Muriel Porter Weaver; 105(b), Rob Waymen Photography. **CHAPTER THREE:** 112(b), Rob Waymen Photography; 113(tl), The Granger Collection; 113(tr), The Bettmann Archive; 113(b), Rob Waymen Photography; 114(tr), Courtesy of the National Museum of the American Indian, Smithsonian Institution, S2379; 114(bc), Courtesy of the National Museum of the American Indian, Smithsonian Institution, S2438; 116(b),(cl), Rob Waymen Photography; 117(br), Ross Hamilton/Tony Stone Images-photo manipulation by Jun Park; 118(bl), Jim Zuckerman/First Light; 118(bc), Paul Berger/Tony Stone Images; 118(br), Charly Franklin/FPG/Masterfile; 120(c) Rob Waymen Photography; 121(tl), Ron Tanaka; 122(b), Rob Waymen Photography; 123(br),127(cr), Ron Tanaka; 128(r), Rob Waymen Photography; 128(l), Jan Becker; 133(tl), The Bettmann Archive; 141(tc),(tr), Ron Tanaka; 141(cr), Courtesy of the Oriental Institute of the University of Chicago; 142(tr), Jan Becker; 144(cr), Jan Becker-photo manipulation by Jun Park; 145(br), Ron Tanaka; 146(tr), Lucille Roybal-Allard, U.S. House of Representatives, California 33rd District; 146(cl), Assemblymember Martha Escutia, California State Legislature; 146(cr), Assemblymember Joe Baca, California State Legislature; 147(tc), Assemblymember Grace Napolitano, California State Legislature; 147(tc), U.S. Department of Defense/Cpl. Haynes/(DM-SC-89-05560); 147(cr), Charles Thatcher/Tony Stone Images; 148(cl), Jan Becker; 150(l), Ron Tanaka; 151(br), Lisa Valder/Tony Stone Images; 152(bl), Rob Waymen Photography; 152(br), Ron Tanaka-photo manipulation by Jun Park; 153(r), Rob Waymen Photography; 154-155(bckgd), Ron Tanaka. **CHAPTER FOUR:** 162(br), Jan Becker; 162(tl), Grant Black/First Light; 162(tr), Pronk&Associates/David Montle; 162(br), Ron Tanaka; 164(t), Rob Waymen Photography; 164-165(c), Ron Tanaka; 165(t), Rob Waymen Photography; 166(tl), Jan Becker; 168(br), Ron Tanaka; 169(br), Jan Becker; 170(tr), Ron Tanaka; 171(t), Jonathon Nowrok/Tony Stone Images; 171(cl), Lewis Portnoy/Spectra-Action, Inc.; 174(l), Ron Tanaka; 176(b), Rob Waymen Photography; 177(b),178(tr), Ron Tanaka; 179(cr), Jan Becker; 180(t),181(l),182-183(c),185(l),(br),186(b),191(c),195(b),196(r),197(t),198(c), Ron Tanaka; 199(cl), The University of Chicago; 199(cr), Pronk&Associates/David Montle; 213(b), Rob Waymen Photography; 215(c), Jan Becker; 220(b), Craddock/First Light; 223(c), Ron Tanaka; 225(tl), Mark Tomalty/Masterfile; 226(t),(br),227(tr), Ron Tanaka; 228(t), Rob Waymen Photography; 231(l), Ron Tanaka; 232(cl), Jan Becker; 234(t), Rob Waymen Photography-photo manipulation by Jun Park; 236(b), Ron Tanaka-photo manipulation by Jun Park; 239(tr), Rob Waymen Photography; 241(tr), Jan Becker. **CHAPTER FIVE:** 252-253(bckgd),(c), Rob Waymen Photography; 254(c), Jan Becker; 256(b),258(l), Rob Waymen Photography; 259(r), Ron Tanaka; 260(br), David de Lossy/Image Bank; 261(tr), Ron Tanaka; 265(tl), The Bettmann Archive; 265(tr), G.A. Plimpton Collection, Rare

Book & Manuscript Library, Columbia University; 267(tr), Jan Becker; 273(tr), Sherman Hines/Masterfile; 273(br), D.S. Henderson/Image Bank; 274(t),(r),275(tr), Rob Waymen Photography; 276-277(t), Sam Dudgeon/HRW Photo; 279(tl),(tc),(tr), Freer Gallery of Art/Smithsonian Institution; 280(tr), Sam Dudgeon/HRW Photo; 281(tr), Ron Tanaka; 285(b), Rob Waymen Photography; 286(tr), Ron Tanaka; 286(cr), Jan Becker; 287(tr), Matt Lambert/Tony Stone; 290(b), Rob Waymen Photography; 290(inset left), The Bettmann Archive; 292(br),297(t), Rob Waymen Photography; 299(b), Jan Becker; 301(cr), Ron Tanaka; 302(bckgd), T. Tracy/FPG/Masterfile; 302(bl), Jan Becker; 308(b),310(b), Rob Waymen Photography; 312(r), Ron Tanaka-photo manipulation by Jun Park; 313(t),314(r), Ron Tanaka; 315(tr), NASA/Science Source/Photo Researchers; 316-317(bckgd), Image Tech West/First Light; 316(bl),317(t), Ron Tanaka. **CHAPTER SIX:** 324-325(bckgd), Romilly Lockyer/Image Bank; 324(c), Fulrio Boiter/Image Bank; 324(br), The Granger Collection; 325(br), Ron Tanaka; 326(b), Jan Becker; 328(cr), Ron Tanaka-photo manipulation by Jun Park; 332(c), Jan Becker; 339(r), Jan Kopec/Tony Stone Images; 341(cr), Ron Tanaka-photo manipulation by Jun Park; 343(tl), The Bettmann Archive; 347(b), Rob Waymen Photography; 348(b), Larry Ulrich/Tony Stone Images; 350(c), Ron Tanaka; 352(br), Rob Waymen Photography; 356(tl),(br), Cedar Point Photos by Dan Feicht; 358(c), Ron Tanaka; 361(bl), The Bettmann Archive; 361(br), Ron Tanaka; 363(br),364(br), Rob Waymen Photography; 366-367(bckgd), Ron Tanaka. **CHAPTER SEVEN:** 374(bc), Howard Sochurek/First Light; 374(br), Dennis Coleman/Peter Arnold Inc.; 375(tl), Reuters/Bettmann; 375(cr), UPI/Bettmann; 376(t), Paolo Negri/Tony Stone Images; 376-377(b), David Scharf/Peter Arnold Inc.; 376(inset), Lloyd Sutton/Masterfile; 379(b), Rob Waymen Photography-photo manipulation by Jun Park; 380(r), Telegraph Colour Library/Masterfile; 381(inset), Paul Chesley/Tony Stone Images; 382(left), Ken Straiton/First Light; 382(br), ZEFA/Masterfile; 383(t), Rob Waymen Photography; 389(b), Jan Becker; 395(r), Ron Tanaka; 398(t),400(t), Rob Waymen Photography; 401(cr), Science VU-©WHO1/Visuals Unlimited; 402(bckgd), Tom Carroll/FPG/Masterfile; 402(bl), The Bettmann Archive; 404(br), Ron Tanaka; 405(tr), The Bettmann Archive; 410(l), ©Rich Treptow/Visuals Unlimited; 411(r), Rob Waymen Photography; 413(r), Jerry Kobalenko/First Light; 414-415(bckgd), Peter Pearson/Tony Stone Images; 414(t), Ron Tanaka, 414(b), Jefferson National Expansion Memorial/National Park Service; 416-417(t), Rob Waymen Photography; 419(c), Jan Becker; 422(b), Rob Waymen Photography; 423(l),(b), Rob Waymen Photography-photo manipulation by Jun Park. **CHAPTER EIGHT:** 429(cl), Telegraph Colour Library/Masterfile; 429(bl), Stephen Simpson/FPG/Masterfile; 430(r),(inset), Jan Becker; 432(b), Rob Waymen Photography; 443(br), Boden/Ledingham/Masterfile; 445(t), Ron Tanaka; 446(tr), The Granger Collection; 450(br), Rob Waymen Photography; 451(tr), Stock Market/Masterfile; 451(c), Tony Stone Images; 452(tr),453(tr), Ron Tanaka; 455(t), Rob Waymen Photography; 456(br), J.A.Kraulis/Masterfile; 465(tr), Rob Waymen Photography; 467(r), Telegraph Colour Library/Masterfile; 467(tc), Rob Waymen Photography; 470(tr), Ron Tanaka; 471(b), Rob Waymen Photography; 473(tr), Ron Tanaka; 474(tr), Ed Gifford/Masterfile; 476(tl), Telegraph Colour Library/Masterfile; 477(tl),(tr), Jan Becker; 478(cl), Jurgen Vogt/Image Bank; 479(r), David W. Hamilton/Image Bank; 480(br), Ron Tanaka; 482-483(bckgd), Tony Stone Images; 482(bl), Mary Ellen McQuay/First Light; 483(tr), The Granger Collection; 483(cr), Rob Waymen Photography; 483(bl), Hans Blohm/Masterfile. **CHAPTER NINE:** 492-493(bckgd),492(bl),493(br), Rob Waymen Photography; 494(t), Mike Dobel/Masterfile; 495(bckgd), Larry Keenan Assoc./Image Bank; 495(br),(bl), The Bettmann Archive; 498-499(b), Rob Waymen Photography; 500(r), Ron Tanaka; 501(tr), Tom Raymond/Tony Stone Images; 506(b), Rob Waymen Photography; 507(r), Ron Tanaka; 513(bl), Thomas Kitchin/First Light; 515(b), Frank Whitney/Image Bank; 516(t), Ron Tanaka; 517(tr), Jeff Kaufman/FPG/Masterfile; 517(bc),(bl), Sam Dudgeon/HRW Photo; 522(tc), ©Thinking Machines Corporation 1995/Steve Grohe; 523(br), Imperial War Museum; 524(tl), Allsport/Masterfile; 524(tc), Bruce Wodder/Image Bank; 524(tr), William Sallaz/Image Bank; 526(tr), Rob Waymen Photography; 528-529(bckgd), Ron Tanaka; 529(tr), Jon Ortner/Tony Stone Images. **CHAPTER TEN:** 534(cr), Bill Brooks/Masterfile; 541(br), Bill Brooks/Masterfile; 542(r), Rob Waymen Photography; 543(tr), Cosmo Condina/Tony Stone Images; 544(t), David Ximeno Tejada/Tony Stone Images; 544(inset), Larry MacDougal/First Light; 548(b), Ron Tanaka; 550(tr), William E. Ferguson; 550(inset), Reuters/Bettmann; 551(t), David Jeffrey/Image Bank; 551(cl),(b), Ron Tanaka; 556(b), U.S. Geological Survey/Photo Researchers; 557(tr),(br), Telegraph Colour Library/Masterfile; 558(tr), Architect of the Capitol; 559(tr), Yoav Levy/First Light; 564-565(b), Stuart Westmoreland/Tony Stone Images; 567(t), John Foster/Masterfile; 574(cr), The Bettmann Archive; 575(tr), Ron Tanaka-photo manipulation by Jun Park; 577(tr),578(br), Ron Tanaka; 581(b), Rob Waymen Photography-photo manipulation by Jun Park; 583(tr), Rob Waymen Photography; 584(t),585(br), Ron Tanaka; 586-587(bckgd), John Lawlor/Tony Stone Images; 587(tr), J.A.Kraulis/Masterfile; 587(cl), ©Bruno J. Zehnder/Peter Arnold, Inc. **CHAPTER ELEVEN:** 594-595(bckgd),595(br),597(tr),(l),598(br), Ron Tanaka; 600(bl), Rob Waymen Photography; 601(tr), Ron Tanaka; 603(tc), Rob Waymen Photography; 605(t),(br), Jan Becker; 606(b), Ron Tanaka; 609(r), Rob Waymen Photography; 612(t),614(c), Ron Tanaka; 615(c), Jan Becker;

616(cl), National Archives (127-GR-137-57875); 616(cr), National Archives (127-N-69889-B); 617(tr), Darwin R. Wiggett/First Light; 617(tl), Charles Michael Murray/First Light; 618(tl),619(br), ABC Productions; 619(cr),620(b), Ron Tanaka; 624(br), Rob Waymen Photography; 625(b), Ron Tanaka; 626(tr),(c), The Bettmann Archive; 627(b),628(l),(c),629(cr),630(br), Ron Tanaka; 631(bl), Jan Becker; 632(tr), Ron Tanaka; 633(tr),(cr), Jan Becker; 637(b), Rob Waymen Photography; 638(br), Leverett Bradley/Tony Stone Images; 644(br), Ron Tanaka; 645(tr), Tony Stone Images; 647(bckgd), Ron Tanaka. **CHAPTER TWELVE:** 652-653(bckgd), Steve Taylor/Tony Stone Images; 652-653(bc), Robert Young Pelton/First Light; 652(cl), Jean-Paul Manceau/Tony Stone Images; 653(tl), Dallas & John Heaton/First Light; 653(tr), Lewis Kemper/Tony Stone Images; 653(br), Sobel/Klonsky/Image Bank; 654(bl), Rob Waymen Photography-photo manipulation by Jun Park; 657(br), Murray Alcosser/Image Bank; 658(b), Sightseeing Archives/Image Bank; 661(bl),663(bl), Rob Waymen Photography-photo manipulation by Jun Park; 665(tr),668(br), Ron Tanaka; 669(tr),(cr), Rob Waymen Photography; 671(c), Rob Waymen Photography; 676(r), Rob Waymen Photography; 677(c), Ron Tanaka; 679(r), Ron Tanaka-photo manipulation by Jun Park; 684(br), Rob Waymen Photography; 686(cr), Ron Tanaka-photo manipulation by Jun Park; 692(bl), Ron Tanaka; 693(tl), Sherman Hines/Masterfile-photo manipulation by Jun Park; 693(tr), Phil Habib/Tony Stone Images. **CHAPTER THIRTEEN:** 708-709(bckgd), Vince Streano/Tony Stone Images; 709(br), Ron Tanaka; 710(t), Rob Waymen Photography; 710(cl), Ron Tanaka; 715(tr) Erich Lessing/Art Resource, NY; 716(c),717(t), Rob Waymen Photography-photo

manipulation by Jun Park; 720(br),721(tr),723(cr)725(b), Ron Tanaka; 726(t), Rob Waymen Photography; 727(t), Ron Tanaka; 728(tr), Pelaez/First Light; 733(tr), Reuters/Bettmann; 736(br), Animals Animals/Mark Chappell; 737(tr), ©1996 R. Andrew Odum/Peter Arnold, Inc.; 738(tr), Brett Froomer/Image Bank; 738(bckgd), Ron Tanaka; 743(br), William Sallaz/Image Bank; 744(tr), Dawn Goss/First Light; 744(cl), Robert W. Allen/First Light; 745(t), Allsport USA/Tim De Frisco; 746(tr), Ron Tanaka; 748(t), Rob Waymen Photography; 749(cr), Frank Siteman/Tony Stone Images; 749(br), Brian Milne/First Light; 753(c), Ron Tanaka; 757(t), Jan Becker; 760(t), Ron Tanaka; 761(tr), LD Gordon/Image Bank; 762(bl),(br),763(tl), Rob Waymen Photography; 763(tr), Jan Becker. **CHAPTER FOURTEEN:** 768(cr), Roger Ressmeyer-©1996 NASA/Corbis; 768(bc), F. Whitney/Image Bank; 769(tr), Ron Tanaka; 770(cr),772(cr), Jan Becker; 773(br), Giraudon/Art Resource, NY; 775(b), Grant V. Faint/Image Bank; 775(br), Ron Tanaka; 776(tr), Dan Ham/Tony Stone Images; 777(tr), Grant V. Faint/Image Bank; 781(r), Rob Waymen Photography; 782(bl), Walter Bibikow/Image Bank; 783(tc), Charles Thatcher/Tony Stone Images; 784(t), Jan Becker; 787(br),791(r), Rob Waymen Photography; 793(cl), Erich Lessing/Art Resource, NY; 793(br), Rob Waymen Photography; 795(tr), Ron Tanaka-photo manipulation by Jun Park; 800(cr), Stephen Studd/Tony Stone Images; 801(tr), Ron Tanaka; 801(cr), Cosmo Condina/Tony Stone Images; 802-803(bckgd), Telegraph Colour Library/Masterfile-photo manipulation by Jun Park; 802-803(bckgd),(b), World Perspectives/Tony Stone Images; 802(bc), Peter Millar/Image Bank; 803(tr), Telegraph Colour Library/Masterfile.

ILLUSTRATIONS

Abbreviations used: (t) top, (c) center, (b) bottom, (r) right, (l) left.

Bardell, Graham page 680

Diggle, Justin pages 4-5 (bckgd), 7 (bckgd), 217, 428 (bc), 428-429 (bckgd), 429 (t), 534-535 (bckgd)

Elsom, Vicki page 350 (c)

Ghiglione, Kevin pages 60-61 (bckgd), 132-133 (bckgd), 203, 269 (bckgd), 374-375, 396-397, 406-407, 520-521, 640

Herman, Michael pages 50 (cr), 82, 119, 120 (br), 126, 187, 216, 254 (br), 288, 336, 384-386, 392-394, 437, 438, 463, 469, 503-505, 538, 539, 567 (br), 584 (tr), 610, 636, 643, 660, 662, 664, 665 (b), 672, 673, 678 (t), 683, 711, 749 (t), 774, 777 (cr), 778 (cr), 790

Lau, Bernadette pages 96 (br), 457, 611, 678 (b), 722, 735, 750-751

Lightfall Art & Design/Paul Rivoche pages 304-306

McMaster, Jack pages 83 (bckgd), 86 (br), 125 (tr), 188, 343 (c), 436, 444, 449, 667

Nasmith, Ted pages 39 (l), 81, 475

Newbigging, Martha pages 240, 390

Park, Jun pages 10 (cr), 13 (cr), 15 (bl), 21(cr), 28 (br), 32 (cr,cl), 71, 104, 135-138, 201, 208, 210, 267 (cr), 344-345, 481, 508, 527, 536, 543 (br), 639(tr), 657 (t), 688 (t), 695, 696

Phillips, Ian pages 8 (tc), 19, 22, 30 (c), 43, 44, 84, 88, 96 (t), 98 (b), 102 (c), 140, 381 (b), 382 (tr), 497, 547, 613, 634, 635, 714, 719, 769 (br)

Pronk&Associates pages 9, 10 (cl), 37, 38, 39 (cr), 87, 130, 189, 209, 262-263, 271, 282, 294-295, 330-331 (bckgd), 354, 365, 459, 462, 569, 572, 602, 694, 698, 699, 712, 730, 788, 796

Puckett, David page 468

Stahl, Margo pages 29 (tr, cr), 92, 232 (b), 375 (b), 428 (cr), 447, 549, 560, 604, 641

Smith Kent pages 421, 623, 701, 772 (cl), 775 (t, c),

Visual Sense Illustration/Margo Davies Leclair pages 28 (tl), 207, 225 (tr), 415 (tr)

PERMISSIONS

For permission to reprint copyrighted material, grateful acknowledgment is made to the following sources:

Kentucky Department of Education: "Kentucky Mathematics Portfolio Holistic Scoring Guide" (Retitled: "Portfolio Holistic scoring Guide") from *Kentucky Department of Education*, 1994–1995. Copyright ©1994 by Kentucky Department of Education.

The New York Times: From "The Quest for True Randomness Finally Appears Successful" by James Gleick from *The New York Times*, April 19, 1988. Copyright ©1988 by The New York Times Company. From "Biggest Division a Giant Leap in Math" by Gina Kolata and "Factoring a 155-Digit Number: The Problem Solved" from *The New York Times*, June 20, 1990. Copyright ©1990 by The New York Times Company.

Omni Publications International, Ltd.: From "Traditional 3-D glasses won't help you find the message concealed in these random dots" (Retitled: "3-D glasses won't help") by Scot Morris from *Omni*, November 1991, p. 128. Copyright ©1991 by Omni Publications International, Ltd.

Marilyn vos Savant and Parade: Excerpt (Retitled: "Eyewitness Math: Let's Make a Deal") from the column "Ask Marilyn™" by Marilyn vos Savant from *PARADE*, September 9, 1990, p. 15. Copyright ©1990 by Parade.

Scientific American: Adapted illustration (Parts 1 & 3 only) of two blood pressure graphs by Eric Mose and Laszlo Kubinyi from page 97 of "The Physiology of the Giraffe" by James V. Warren from

Scientific American, November 1974. Copyright ©1974 by Scientific American, Inc. All rights reserved.

Time, Inc.: From "Time for Some Fuzzy Thinking" by Philip Elmer-Dewitt from *Time*, September 25, 1989, p. 79. Copyright ©1989 by Time, Inc.

Page 60(tc,cr), Tempus Books of Microsoft Press.
Page 61(tr), by permission of Oxford University Press, from Chinese Mathematics, A Concise History, 1987, by Li Yan and Du Shiran, translated by John N. Crossley and Anthony W.-C Lun
Page 61(c), drawing by Persis B. Clarkson after ceramic vessel in the National Museum of the American Indian
Page 74(c),(tl),(tr), Maryland Historical Society, Baltimore
218(c), from "The Magic Eye" ©N.E. Thing, Inc. reprinted with permission of Andrews & McMeel. All rights reserved.
Page 356-357(bckgd), courtesy of Arrow Dynamics, Clearfield, Utah
Page 574(b), Abaris Books/Norwalk, CT.
Page 596(tl,cl), copyright ©1993 USA Today. Reprinted with Permission. (manipulation by Brian Hughes)
Page 596(bl), Copyright ©1994 USA Today. Reprinted with Permission. (manipulation by Brian Hughes)

ADDITIONAL ANSWERS

Lesson 1.1, Pages 8–14

Exploration

1. $x = 6$, $y = 120$

2. The y-value is 20 times the x-value.

3.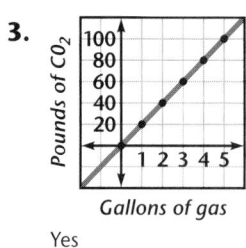

Gallons of gas

Yes

4. All the tables have a constant difference between consecutive x-values. Tables 1, 3, and 5 have constant differences between consecutive y-values.

5. Graph of Table 1

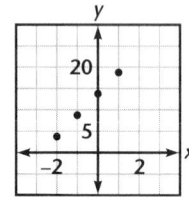

straight line

Graph of Table 2

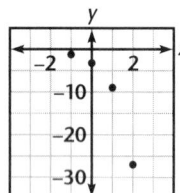

not a straight line

Graph of Table 3

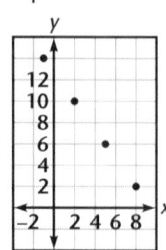

straight line

Graph of Table 4

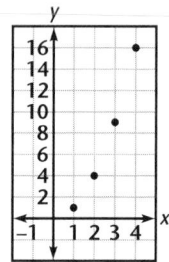

not a straight line

Graph of Table 5

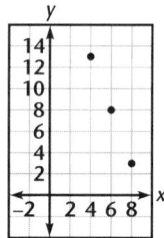

straight line

Graph of Table 6

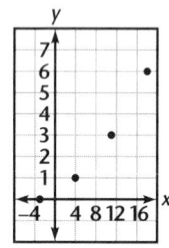

not a straight line

Tables 1, 3, and 5 represent points that fit on a straight line. These are also the tables with constant differences in both consecutive x- and y-values in Step 4.

6. If the differences between consecutive x-values and consecutive y-values are constant, then the points will all fit on one line.

Communicate

1. Make a table of values for the equation. If consecutive x- and y-values both have constant differences, then the equation is linear.

2.

p	6	8	10	12	14
t	0.42	0.56	0.70	0.84	0.98

The variables are linearly related, since both consecutive sets of values have constant differences. The prices, p, have a constant difference of \$2, and the tax amounts, t, have a constant difference of \$0.14.

3. The tax is 7% of the price. To calculate 7% of a number, you can multiply it by 0.07. So the equation is $t = 0.07p$.

4. The graph will be steadily decreasing if, in the table of values, the constant differences between consecutive x- and y-values have opposite signs.

5. The graph will be steadily increasing if, in the table of values, the constant differences between consecutive x- and y-values have the same sign.

Practice & Apply

24. linear; the constant difference in consecutive x-values is 2, and the constant difference in consecutive y-values is 4.

25. not linear; consecutive x-values have a constant difference of 1, but consecutive y-values do not have a constant difference.

26. linear; consecutive values of x have a constant difference of -3, and consecutive vaules of y have a constant difference of 12.

27. Let c represent the total cost and r the number of rides. With a green ticket, $c = 6 + 0.25r$. With a red ticket, $c = 3 + 0.75r$.

28. yes

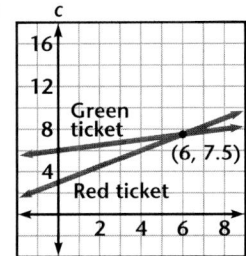

29. The best option depends on the planned number of rides. For fewer than 6 rides, buying a red ticket and paying $3 initially is cheaper. For more than 6 rides, buying a green ticket and paying $6 initially is cheaper. If 6 rides are taken, both plans cost the same.

Lesson 1.2, Pages 15–22

Exploration 1

1.

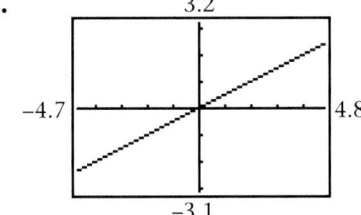

Samples answers given for Steps 2–4.

2. $P_1(0, 0)$ and $P_2(2, 1)$

3. For P_1 and P_2: $1 - 0 = 1$

4. For P_1 and P_2: $2 - 0 = 2$

5. The quotient is $\frac{1}{2}$.

6. All quotients should be $\frac{1}{2}$.

7. The quotient is the same as the x-coefficient in the linear equation.

8. The quotient and the x-coefficient are both -3.

9. The quotient is 5.

10. The larger the absolute value of the x-coefficient, the steeper the graph.

Exploration 2

1.

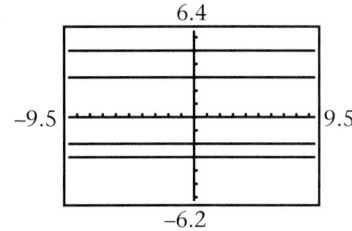

2. They are all horizontal lines.

3. The slopes all equal 0.

4. Slope is 0.

5. Horizontal lines have a slope of 0.

6.

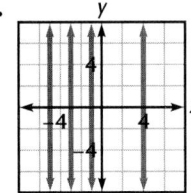

7. They are all vertical lines.

8. The slopes are all undefined.

9. The slope is undefined.

10. Vertical lines have an undefined slope.

Exploration 3

1.

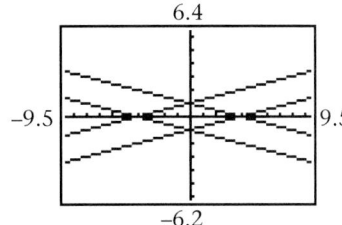

2. Answers will vary for the first two columns. For the third column containing the quotients, the answers from top to bottom are $\frac{1}{4}$, $-\frac{1}{4}$, $\frac{1}{4}$, and $-\frac{1}{4}$.

3. & 4. The graphs of $y = \frac{1}{4}x + 1$ and $y = \frac{1}{4}x - 1$ have equal slopes. The graphs of $y = -\frac{1}{4}x + 1$ and $y = -\frac{1}{4}x - 1$ have equal slopes. These pairs of lines are parallel, but they have different y-intercepts.

5. The graphs of linear equations with equal slopes are parallel.

6. The lines $y = \frac{1}{4}x + 1$ and $y = -\frac{1}{4}x + 1$ both cross the y-axis at $(0, 1)$. Both equations have constant term $+1$.

7. The lines $y = \frac{1}{4}x - 1$ and $y = -\frac{1}{4}x - 1$ cross the y-axis at $(0, -1)$. Both equations have constant term -1.

8.

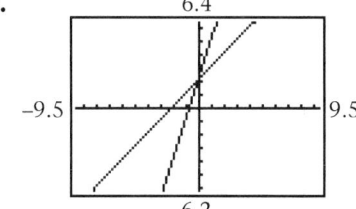

Both graphs cross the y-axis at $(0, 2)$. Both equations have constant term $+2$.

9. $y = \frac{1}{3}x + 4$ crosses at $(0, 4)$; $y = -\frac{1}{3}x - 1$ crosses at $(0, -1)$; $y = \frac{3}{4}x + 8$ crosses at $(0, 8)$; $y = x - 5$ crosses at $(0, -5)$.

10. The constant term in the equation is the y-coordinate of the point where the graph crosses the y-axis.

Communicate

1. Any two points on a line may be used to find its slope. The slope ratio describes slant or steepness, which is the same for any segment of the line. The slope of the line passing through the points $(-1, 7)$, $(0, 4)$, and $(2, -2)$ is -3.

2. The graph of a line that has negative slope slants downward

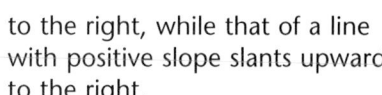

to the right, while that of a line with positive slope slants upward to the right.

3. A horizontal line has a slope of 0. A vertical line has an undefined slope.

4. Substitute the values for the slope and y-intercept into the slope-intercept form of a line, $y = mx + b$. Since the slope, m, is -3 and the y-intercept, b, is 7, the equation is $y = -3x + 7$.

5. First, find the slope by calculating the change in y divided by the change in x for the given points, $\frac{y_2 - y_1}{x_2 - x_1}$. This gives $\frac{5 - (-2)}{4 - 3} = 7$ for the slope. Next, substitute the point (4, 5) and the slope, $m = 7$, into the slope-intercept form, $y = mx + b$. This gives the equation $5 = 7(4) + b$, which can be solved for b, so $b = -23$. Now that the slope and the y-intercept are known, the equation of the line can be written in the slope-intercept form: $y = 7x - 23$.

Lesson 1.3, Pages 23–29

Exploration

1.

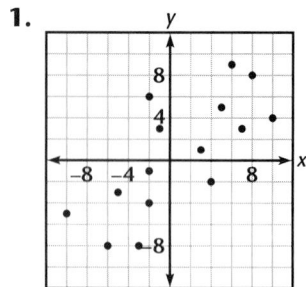

2. Answers may vary but the line should pass through the "center" of the data points, with a similar number of data points on either side of it.

3. Probably not, though the lines should be similar.

4. A line of best fit is a line that is closest to most of the points in a scatter plot.

Communicate

1. Answers may vary. Examples of high positive correlation are feeding and growth, drug dosage and effects, and study time and test scores. Example of high negative correlations are cost and demand, pesticide use and the number of infesting pests, and the weight of a motor vehicle and its gas mileage.

2. The points in the scatter plot are widely dispersed, with no apparent line of best fit.

3. The points in the scatter plot lie roughly in a line that slopes downward to the right. However, since r is not close to -1, there will be many outliers.

4. The points in the scatter plot lie close to a line that slopes upward to the right. Since r is very close to 1, most of the points will be close to the line of best fit.

5. Negative correlation: As the number of hours of watching television increases, grades will probably decrease.

6. Positive correlation: The faster a car is going, the longer the brakes must work to stop it, and the farther the car will travel before stopping.

Lesson 1.4, Pages 30–35

Communicate

1. Find the value of k in the direct variation equation, $y = kx$, by substituting 9 for x and 18 for y. This gives $k = 2$. Write the equation of the direct variation using $k = 2$: $y = 2x$.

2. All linear equations are not direct variations. For example, $y = x + 1$ is linear, but is not a direct variation because it cannot be written in the form $y = kx$. All direct variations are linear equations, since $y = kx$ is a linear equation of the form $y = mx + b$, with $m = k$ and $b = 0$.

3. The formula $C = 2\pi r$ expresses how the circumference varies with the radius. The constant of variation is 2π.

4. Since "y varies directly as x" is written $y = kx$, then "p varies directly as q" is written $p = kq$, where k is the constant of variation.

5. The statement "a is directly proportional to b" means that "a varies directly as b," which is written $a = kb$, where k is the constant of variation. You can also write "a is directly proportional to b" as $\frac{a}{b} = k$, where k is the constant of variation.

Lesson 1.5, Pages 36–44

Communicate

1. To solve $x - 2 = -5x + 30$ by graphing, graph the lines $y = x - 2$ and $y = -5x + 30$ on the same coordinate plane and find the x-coordinate of the intersection point. The x-coordinate of the intersection point is 5.333..., or $5\frac{1}{3}$.

2. Choose two variables to represent the two numbers: $x + y = 45$. Let x represent the smaller number. Then the greater number, y, is $45 - x$. Write an equation for the second sentence: The greater number, $45 - x$, is, =, 1 more than 3 times the smaller number, $3x + 1$. So the equation is $45 - x = 3x + 1$.

3. Use the addition and the subtraction properties of equality to regroup the x-terms and the integer terms: $45 - 1 = 3x + x$. Simplify: $44 = 4x$. Use the division property of equality to find x: $\frac{44}{4} = \frac{4x}{4}$, or $11 = x$.

4. 11 is not the complete answer because it is only one of the two numbers (the smaller number). According to Exercise 2, the object is to find both numbers. The larger number is 34.

Lesson 1.6, Pages 45–53

Exploration 1

1.
$$-3 < 5$$
$$-3 + (-4) < 5 + (-4)$$
$$-7 < 1 \quad \text{True}$$

2. Answers may vary but should be similar to
$$5 > 2$$
$$5 - (-7) > 2 - (-7)$$
$$12 > 9 \quad \text{True}$$

3. Adding or subtracting any number to both sides of an inequality produces an equivalent inequality.

Exploration 2

1.
$$-3 < 5$$
$$(-3)(-2) < (5)(-2)$$
$$6 < -10 \quad \text{False}$$

2. Answers may vary but should be similar to
$$5 > 2$$
$$(-4)5 > (-4)2$$
$$-20 > -8 \quad \text{False}$$

3. Multiplying both sides of an inequality by a *negative* number produces an equivalent inequality provided *the order of the inequality is reversed*. Multiplying both sides of $-3 < 5$ by -2 gives $6 > -10$, which is true. Multiplying both sides

of an inequality by a positive number directly produces an equivalent inequality. For example, multiplying both sides of $5 > 2$ by 2 gives $10 > 4$, which is true.

Communicate

1. The method of solving an inequality is the same as solving an equation except, when multiplying or dividing both sides of an inequality by a negative number, the inequality symbol must be reversed to produce an equivalent inequality.

2. "Nonnegative" means "greater than or equal to zero," which is expressed as $x \geq 0$.

3. First, solve the inequality for x, which leads to $x > 1$. Draw a broken vertical line passing through (1, 0) to show $x = 1$. Shade to the right of this vertical line to show where the inequality, $x > 1$, is true.

4. By the Multiplication Property of Inequalities, when you multiply both sides of $x < y$ by -1, you get the equivalent inequality $-x > -y$. This is not the same inequality as $-x < -y$, since the inequality symbol has been reversed.

Practice & Apply

11. Solve for y to obtain $y \geq -\frac{5}{2}x - 4$. for $Y1$, enter $(-5/2)X - 4$. Define the viewing window similar to $X\min = -5$, $X\max = 5$, $Y\min = -5$, $Y\max = 5$. Use the test point (0, 0) to find that the solution set is all the points above the line, including the line.

15.
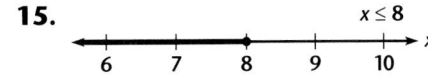

16.
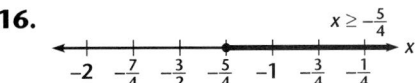

17.
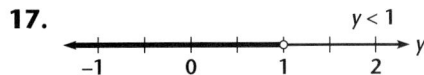

18.
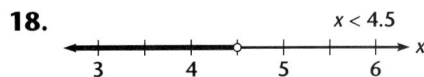

19.
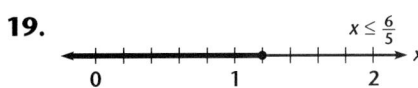

20.
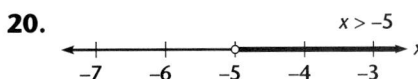

21.
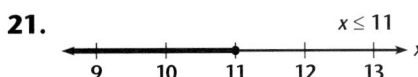

22.
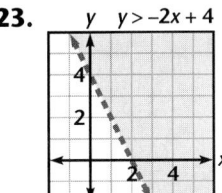

23.
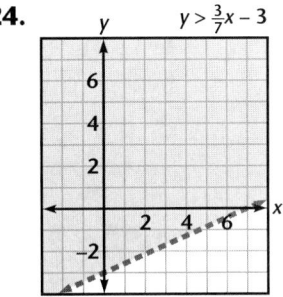

24.

25. $y \leq -\frac{1}{5}x - 3$
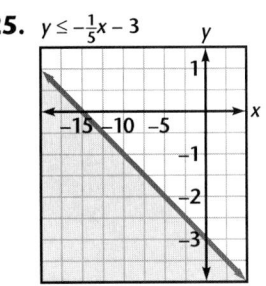

26.

$$x \le -\frac{5}{2}$$

27.

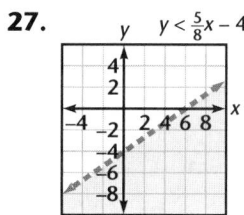

$$y < \frac{5}{8}x - 4$$

28.

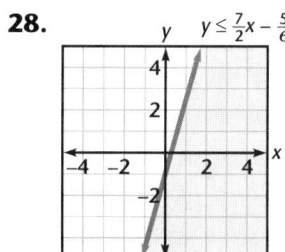

$$y \le \frac{7}{2}x - \frac{5}{6}$$

29. $y > \frac{4}{3}x + 8$

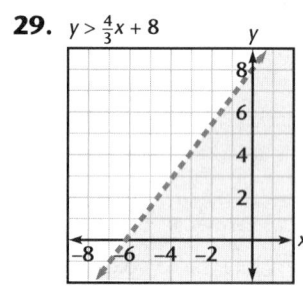

30.

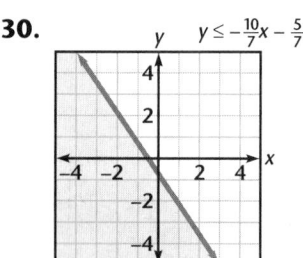

$$y \le -\frac{10}{7}x - \frac{5}{7}$$

Lesson 2.1, Pages 62–67

Exploration 1

1. $-2 = -2$; true

2. $-\frac{1}{4} = \frac{1}{4}$; false

3. $-\frac{13}{4} = -\frac{5}{4}$; false

4. $-\frac{31}{6} = -\frac{31}{6}$; true

5. $-1 = -1$; true

6. $\frac{17}{4} = -\frac{23}{4}$; false

7. $-\frac{7}{6} = -\frac{7}{6}$; true

8. $\frac{13}{6} = \frac{1}{6}$; false

9. $-4 = 6$; false

10. $\frac{1}{12} = \frac{1}{12}$; true

11. Answers may vary. Students should discover that the true expressions (as listed in the next answer) are true no matter which rational numbers are chosen.

12. The statements in 1, 4, 5, 7, and 10 seem to be true for all rational numbers.

Exploration 2

1. $-15 = -15$; true

2. $\frac{2}{3} = \frac{3}{2}$; false

3. $-\frac{1}{6} = -\frac{3}{8}$; false

4. $-\frac{4}{3} = -\frac{4}{3}$; true

5. $\frac{1}{5} = \frac{1}{5}$; true

6. $-\frac{40}{3} = -\frac{40}{3}$; true

7. $-\frac{3}{20} = -\frac{15}{4}$; false

8. $-\frac{1}{2} = -\frac{1}{2}$; true

9. $\frac{5}{3} = \frac{5}{3}$; true

10. $-\frac{1}{4} = -\frac{9}{4}$; false

11. Answers may vary. Students should discover that the true expressions (as listed in the next answer) are true no matter which rational numbers are chosen.

12. The statements in 1, 4, 5, 6, 8, and 9 seem to be true for all rational numbers. However, division by zero is not allowed.

Exploration 3

1. $-\frac{4}{17} = \frac{17}{15}$; false

2. $-5 = -5$; true

3. $-\frac{11}{3} = -\frac{7}{3}$; false

4. $\frac{33}{8} = \frac{33}{8}$; true

5. $-\frac{5}{3} = \frac{7}{3}$; false

6. $\frac{1}{12} = \frac{1}{12}$; true

7. $-\frac{35}{8} = -\frac{35}{8}$; true

8. $-12 = 3$; false

9. Answers may vary. Students should discover that the true expressions (as listed in the next answer) seem to be true no matter which rational numbers are chosen. However in statement 1, c and d cannot equal zero.

10. The statements in 2, 4, 6, and 7 seem to be true for all rational numbers.

Communicate

1. Answers may vary.

2. To add a positive integer x, move x units to the right of the original integer on a number line. To add a negative integer x, move $|x|$ units to the left of the original integer on a number line.

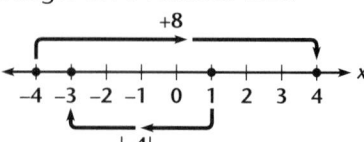

3. The commutative property means that the order in which you add or multiply two real numbers does not change the sum or product. The word *commutative* has the root *commute*, which means something moving from one place to another.

4. The associative property means that the way you group three

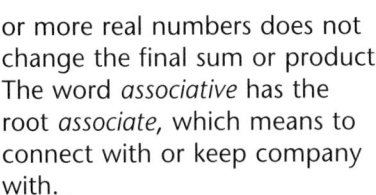

or more real numbers does not change the final sum or product. The word *associative* has the root *associate*, which means to connect with or keep company with.

5. Answers may vary. Examples are: the need to keep track of people, seasons, crops and animals; the development of trade would expand the need for numbers; numbers have developed from simple tallies to systems that can be manipulated quickly and efficiently to meet the needs of modern business, science, and technology.

Look Beyond

56. The graphs of $y = x^{-1}$ and $y = \frac{1}{x}$ are the same. The graph of $y = -\frac{1}{x}$ is the mirror image of the first two graphs across either the x- or y-axis, and the graph of $y = -x$ is a straight line through the origin with a slope of –1.

x	–2	–1	0	1	2
$y = x^{-1}$	$-\frac{1}{2}$	–1	undefined	1	$-\frac{1}{2}$
$y = \frac{1}{x}$	$-\frac{1}{2}$	–1	undefined	1	$\frac{1}{2}$
$y = -\frac{1}{x}$	$\frac{1}{2}$	1	undefined	–1	$-\frac{1}{2}$
$y = -x$	2	1	0	–1	–2

The y-values are the same for $y = x^{-1}$ and $y = \frac{1}{x}$.

Lesson 2.2, Pages 68–73

Exploration 1

1.

Column 1	Column 2	Column 3
3125	3125	15,625
117,649	117,649	5,764,801
243	243	81

2. Columns 1 and 2 have the same values. The exponent in Column 2 is the sum of the exponents in Column 1.

---

3. $8^{15} \cdot 8^9 = 8^{(15+9)} = 8^{24}$

Exploration 2

1.

Column 1	Column 2	Column 3
32,768	32,768	256
64	64	11.3137...
1	1	3

Exploration 3

1.

Column 1	Column 2	Column 3	Column 3
0.0625	16	0.0625	–0.0625
0.5	–2	0.5	–0.5
0.004115...	–243	0.004115...	–0.004115...

2. Columns 1 and 3 have the same values. The exponent of the denominator in Column 3 has the opposite sign of the exponent in Column 1.

3. $6^{-9} = \frac{1}{6^9}$

Exploration 4

1.

x	x^2
2	4
3	9
4	16
5	25

x	$x^{\frac{1}{2}}$
4	2
9	3
16	4
25	5

x	x^3
2	8
3	27
4	64
5	125

x	$x^{\frac{1}{3}}$
8	2
27	3
64	4
125	5

2. Tables 1 and 2 represent the inverse operations "square of x" and "square root of x." Tables 3 and 4 represent the inverse operations "cube of x" and "cube root of x." The terms $x^{\frac{1}{n}}$ and x^n represent the inverse operations "x to the nth" and "nth root of x," where x is a real number. In other words, $(x^{\frac{1}{n}})^n = x$ and $(x^n)^{\frac{1}{n}} = x$ for $x > 0$.

3. $64^{\frac{1}{5}} = \sqrt[5]{64} = 2$

---

2. Columns 1 and 2 have the same values. The exponent in Column 2 is the difference of the exponents in Column 1.

3. $\frac{7^{212}}{7^{210}} = 7^{(212-210)} = 7^2$

Communicate

1. Add the exponents and keep the same base: the result is x^{a+b}.

2. Subtract the exponent of the denominator from the exponent of the numerator and keep the same base: the result is x^{a-b}.

3. A number with a negative exponent is the same as 1 divided by the number with a positive exponent. For example, 3^{-4} is the same as $\frac{1}{3^4}$.

4. A number with a zero exponent is equal to 1. For example, $2^0 = 1$. This agrees with the rule for dividing powers. For example, $\frac{2^5}{2^5} = 1$ and $2^{5-5} = 2^0 = 1$.

5. To find a power with a fractional exponent, $x^{\frac{a}{b}}$, first find the bth root of x. Then find the ath power of the result. For example, to find $4^{\frac{3}{2}}$, first find the square root of 4, which is 2. Then find the third power of 2, or 2^3. So,

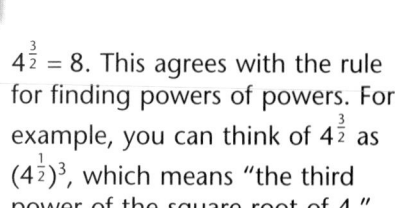

$4^{\frac{3}{2}} = 8$. This agrees with the rule for finding powers of powers. For example, you can think of $4^{\frac{3}{2}}$ as $(4^{\frac{1}{2}})^3$, which means "the third power of the square root of 4."

Practice & Apply

30.

x	−2	−1	0	1	2
$y_1 = (x + 3)^2$	1	4	9	16	25
$y_2 = x^2 + 6x + 9$	1	4	9	16	25
$y_3 = x^2 + 3^2$	13	10	9	10	13

31. The graphs and tables for $y_1 = (x + 3)^2$ and $y_2 = x^2 + 6x + 9$ are the same. The graph of $y_3 = x^2 + 3^2$ also seems to be congruent to the others but in a different position. All three graphs pass through the point (0, 9).

32. The graph of $y_3 = x^2 + 3^2$ has its minimum point at (0, 9), while $y_1 = (x + 3)^2$ and $y_2 = x^2 + 6x + 9$ both have their minimum at (−3, 0). The table shows that the points for y_3 are different than the other two functions for every point except for (0, 9).

33. For any real number a, $(x + a)^2 = x^2 + 2ax + a^2$.

Lesson 2.3, Pages 76–82

Communicate

1. A *relation* is any set of ordered pairs. A *function* is a relation in which, for each choice for the first variable, there is only one possible value for the second variable.

2. You can represent a function as a set of ordered pairs, using a table showing corresponding values of the two variables, or using a graph.

3. The domain of a set of ordered pairs is the set of all first

coordinates of the ordered pairs. The range is the set of all second coordinates of the ordered pairs. Examples may vary but should be equivalent to: for the set of ordered pairs {(a, b), (c, d), (e, f)} the domain is a, c, d and the range is b, d, f.

4. To find the domain from a graph, you need to observe what set of x-values have corresponding y-values. If there appears to be a minimum or maximum x-value, you can use the *trace* feature to find its value. The range is the set of values that y takes. The *trace* feature can be used to find any minimum or maximum values of the range.

5. Answers may vary but should be such that at least one vertical line will intersect the graph at more than one point. An example is shown.

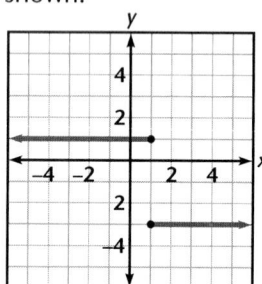

6. The vertical-line test is a visual method for determining whether a relation is a function. If at least one vertical line intersects the graph in more than one point, then the relation is not a function.

Lesson 2.4, Pages 83–89

Communicate

1. $f(x) = 3x$ means that for each choice for the independent variable x, multiply by 3 to obtain the corresponding value of the dependent variable $f(x)$. Ordered

pairs for the function are of the form $(x, 3x)$.

2. The function $a(r) = \pi r^2$, relates the area a of a circle to its radius r. For each choice of the independent variable r, multiply the square of the r-value by π to obtain the corresponding value of the dependent variable. Ordered pairs for the function are of the form $(r, \pi r^2)$.

3. The independent variable is x and the dependent variable is $g(x)$.

4. Answers may vary. Two sample answers are shown.

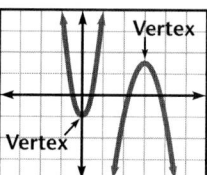

5. The perimeter of a square is the sum of the lengths of the four equal sides. In function notation, $p(s) = 4s$, where s represents the length of a side.

Lesson 2.5, Pages 90–97

Communicate

1. Linear functions, including functions of direct variation and constant functions.

2. Any function that can be written in the form $f(x) = mx + b$, where m represents the slope and b represents the y-intercept, is a linear function. m and/or b can be any real numbers.

3. The slope formula for functions, $m = \dfrac{f(x_2) - f(x_1)}{x_2 - x_1}$, is essentially the same as the slope formula for linear equations, $m = \dfrac{y_2 - y_1}{x_2 - x_1}$. Any point on the line (x_1, y_1) is expressed using function notation as $(x_1, f(x_1))$.

4. Constant functions are linear functions that have slope zero. An example is $f(x) = c$, where c is any real number.

5. Functions of direct variation are linear functions that have a y-intercept of zero. An example is $f(x) = mx$, where m is any real number.

6. First use the two given points to find the slope using the formula $m = \dfrac{f(x_2) - f(x_1)}{x_2 - x_1}$. Next substitute the calculated value for m in the slope-intercept form a linear function, $f(x) = mx + b$. Then substitute the x-value and corresponding $f(x)$-value of one of the given points into the equation and solve for b. Finally write the linear function with the calculated values of m and b.

Lesson 2.6, Pages 98–103

Exploration 1

	x	Y_1	Y_2	Y_3
1.	–2	–5	–1	–6
2.	–1	–4	0	–4
3.	0	–3	1	–2
4.	1	–2	2	0
5.	2	–1	3	2

6. The domain and range of all three functions are all real numbers.

7. For each value of x, Y_3 is the sum of the values of Y_1 and Y_2.

Exploration 2

	x	Y_1	Y_2	Y_3
1.	–2	–5	–1	5
2.	–1	–4	0	0
3.	0	–3	1	–3
4.	1	–2	2	–4
5.	2	–1	3	–3

6. The domain of all three functions is all real numbers. The range of Y_1 and Y_2 is all real numbers. The range of Y_3 is all real numbers such that $y \geq -4$.

7. For each value of x, Y_3 is the product of Y_1 and Y_2.

Communicate

1. When two functions are added, the y-values for each function for each x-value are added. The domain of the sum function is the set of numbers that are in the domain of *both* of the original functions. The range of the sum function is the set of numbers that is the sum of the corresponding y-values of each of the original functions, for each x-value that is in the domain of both.

2. No. As a result of the Commutative Property of Addition for real numbers, $(f + g)(x) = (g + f)(x)$.

3. No. As a result of the Commutative Property of Multiplication for real numbers, $(f \cdot g)(x) = (g \cdot f)(x)$.

4. When two functions are multiplied, the y-values for each function for each x-value are multiplied. The domain of the product function includes all numbers that are in the domain of *both* of the original functions. The range of the product function is the set of numbers that is obtained by multiplying the corresponding y-values of the two original functions for each x-value that is in the domain of both.

5. When two functions are divided, the domain of the quotient function cannot include any x-values that make the denominator of the quotient function equal to zero. In other words, the domain

of $\dfrac{f(x)}{g(x)}$ excludes any x-values where $g(x) = 0$.

Look Back

31.

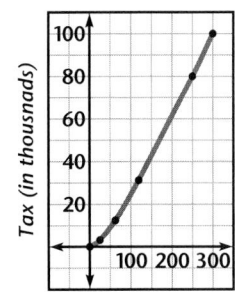

Income (in thousands)

The endpoints of each of the graphs connect to form one graph that gets steeper at each change of tax brackets. The graphs do not intersect because their domains do not overlap.

Lesson 3.1, Pages 114–119

Exploration 1

1. & 2.

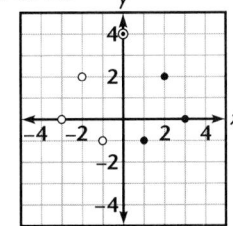

3. $A'(-1, -1)$, $B'(-2, 2)$, $C'(-3, 0)$, $D'(0, 4)$

4. The y-coordinate of each image point is the same as the y-coordinate of its corresponding pre-image point. The x-coordinate of each image point has the opposite sign of the x-coordinate of its corresponding pre-image point.

5. The line segment joining each pre-image point to its image point is perpendicular to the y-axis. These line segments have the same length on each side of the y-axis.

6. The image of $E(a, b)$ with respect to the y-axis is $E'(-a, b)$. The pre-image and image points share the same y-coordinate, but have opposite x-coordinates.

Exploration 2

1. & 2.

3. $A'(1, 1)$, $B'(2, -2)$, $C'(3, 0)$, $D'(0, -4)$

4. The x-coordinate of each image point is the same as the x-coordinate of its corresponding pre-image point. The y-coordinate of each image point has the opposite sign of the y-coordinate of its corresponding pre-image point.

5. The line segment joining each pre-image point to its image point is perpendicular to the x-axis. These line segments have the same length on each side of the x-axis.

6. The image of $E(a, b)$ with respect to the x-axis is $E'(a, -b)$. The pre-image and image points share the same x-coordinate, but have opposite y-coordinates.

Exploration 3

1–3.

4. $A'(-1, 1)$, $B'(2, 2)$, $C'(0, 3)$, $D'(4, 0)$

5. The x-coordinate of each image point is the same as the y-coordinate of its corresponding pre-image point. The y-coordinate of each image point is the same as the x-coordinate of its corresponding pre-image point.

6. The line segment joining each pre-image point to its image point is perpendicular to the line $y = x$. These line segments have the same length on each side of the line $y = x$.

7. & 8.

9. Answers may vary. Students should find that their image points are (3, 2), (6, 4), (0, 1), and (−4, −3) and that their classmates will have chosen the same image points.

10. The image of $E(a, b)$ with respect to the line $y = x$ is $E'(b, a)$.

Communicate

1. Symmetry means that a part of a graph or drawing mirrors another part of the graph or drawing. Points or figures might be mirror images of other points or figures across a line (meaning that they would match up if you folded the plane on that line), or a part of a figure may mirror another part of the same figure, as in a circle, which is symmetric about any diameter.

2. An axis of symmetry is a line that serves as a "mirror" for a symmetric graph or figure. It is the line along which you would fold the figure to see the two sides of the graph or figure match up.

3. An image point is a point found by reflecting the pre-image point across an axis of symmetry.

4. A pre-image point is a point that is reflected across an axis of symmetry to find a new symmetric point (the image point).

5. Two points that are symmetric to each other with respect to the y-axis share the same y-coordinate. Their x-coordinates, however, are opposites of each other. Two points are symmetric with respect to the x-axis when their x-coordinates are the same and their y-coordinates have opposite signs.

6. Two points that are symmetric to each other with respect to the line $y = x$ have x- and y-coordinates that are switched. The x-coordinate of the image point is the y-coordinate of the pre-image point, and vice versa.

7. The pre-image points A, B, C, and D on the butterfly are symmetric with respect to the line $y = x$ to the image points A', B', C', and D' respectively.

Practice & Apply

15. image points:
$\{(2, -1), (4, -2), (6, -3), (8, -4)\}$

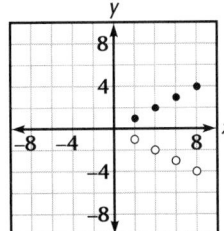

16. image points:
$\{(-3, 6), (-1, 2), (1, -2), (3, -6)\}$

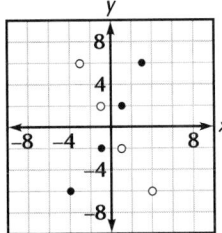

17. image points:
$\{(5, -2), (4, -3), (3, -4), (2, -5)\}$

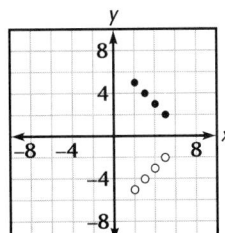

18. image points:
$\{(9, 2), (3, 1), (1, 0),$
$(3, -1), (9, -2)\}$

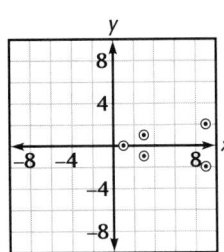

22. image points:
$\{(1, 2), (2, 4), (3, 6), (4, 8)\}$

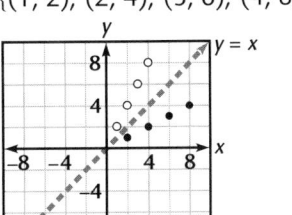

23. image points:
$\{(-6, -3), (-2, -1), (2, 1), (6, 3)\}$

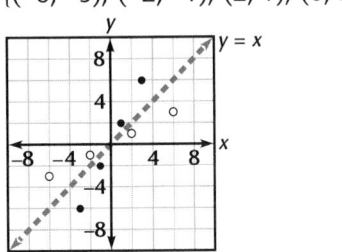

24. image points:
$\{(2, 5), (3, 4), (4, 3), (5, 2)\}$

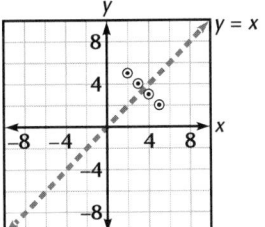

25. image points: $\{(-2, 9), (-1, 3),$
$(0, 1), (1, 3), (2, 9)\}$

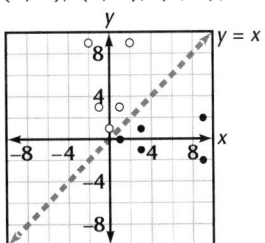

26.

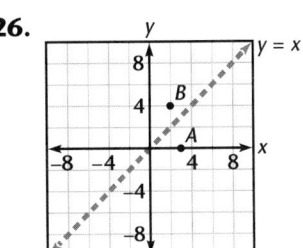

27. $y = -4x + 12$

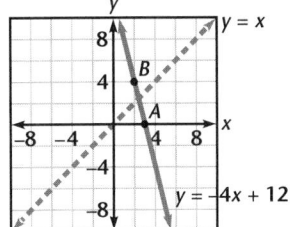

29. $y = -\frac{1}{4}x + 3$

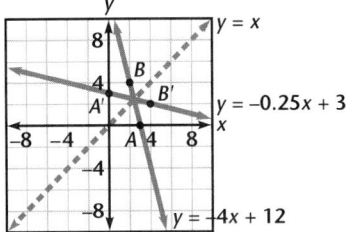

34. Answers may vary. Sample
answers: $x = 1$ and $y = 49$, $x = 12$
and $y = 38$, $x = 22$ and $y = 28$.

Lesson 3.2, Pages 120–127

Exploration

1–3.

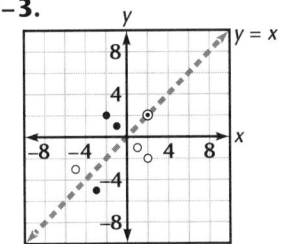

The inverse of f is
$\{(2, -2), (1, -1), (2, 2), (-5, -3)\}$.

4. The inverse of f is not a function
because the first coordinate
2 corresponds to two second
coordinates, -2 and 2.

5. For f, the domain is $\{-3, -2,$
$-1, 2\}$ and the range is
$\{-5, 1, 2\}$. For the inverse of f,
the domain is $\{-5, 1, 2\}$ and
the range is $\{-3, -2, -1, 2\}$.
The domain of f is the same as
the range of its inverse, and the
range of f is the same as the
domain of its inverse.

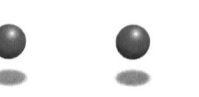

Communicate

1. Because finding the inverse of a function is the same as reflecting each point across the line $y = x$, you need only reverse the order of coordinates of each ordered pair of the function to find its inverse. If an invertible function is written in function notation, let $y = f(x)$, interchange x and y, and solve for y. Then you can replace y with $f^{-1}(x)$.

2. A horizontal line intersects the graph of $y = x^2$ at two points. This means that for the inverse relation, a vertical line will intersect the graph at two points. Any graph that is intersected by a vertical line at more than one point does not represent a function.

3. The horizontal-line test states that if a function has an inverse, then any horizontal line that intersects the graph of the function will do so at only one point. The vertical-line test ensures that a function has only one second coordinate for any first coordinate, while the horizontal-line test ensures that a function has only one first coordinate for any second coordinate. Because the first and second coordinates of a function and its inverse are reversed, the horizontal-line test is performing the same function as the vertical-line test applied to the inverse.

4. The *inverse of a function* is found by reversing the order of the coordinates of each ordered pair of the function. It may or may not be a function itself. If the inverse is itself a function, then it is an *inverse function*.

5. If the graphs of two functions are mirror images across the line $y = x$, then the functions are inverses. One way to help spot

this is to note that if a function has y-intercept $(0, a)$ and x-intercept $(b, 0)$, the inverse will have x-intercept $(a, 0)$ and y-intercept $(0, b)$. Also, if one graph contains any points on the line $y = x$, then the other graph will also contain those points.

Practice & Apply

20. The inverse function exists and is the same function.

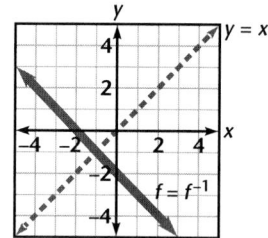

22. The inverse function exists and is shown sketched in bold in the graph below.

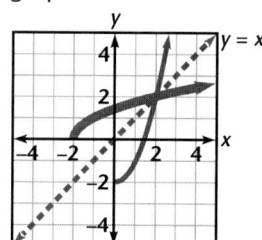

24. $f^{-1}(x) = 5 - x$
The inverse is the same as the original function.

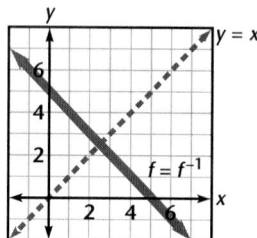

25. $f^{-1}(x) = 8x + 3$
The inverse is shown in bold in

the graph below.

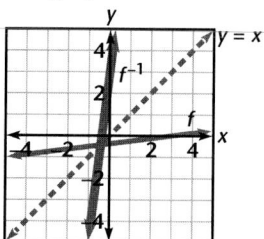

26. $f^{-1}(x) = -\dfrac{1}{2}x + \dfrac{1}{2}$

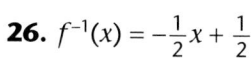
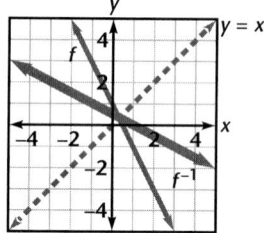

27. $f^{-1}(x) = -7x + 5$

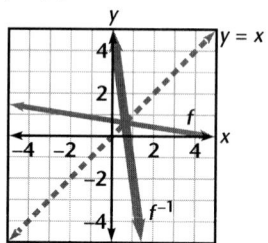

28. $f^{-1}(x) = \dfrac{2}{x + 7}$

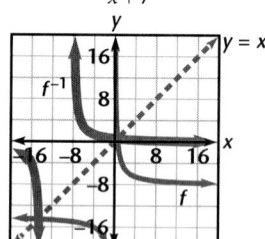

29. $f^{-1}(x) = \dfrac{1}{x}$. The inverse is the same as f.

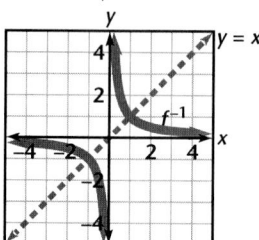

Exploration 1

1. $f \circ h = \frac{x-1}{2}$; $h \circ f = \frac{1+x}{2}$.
Answers may vary. Students may notice that the functions are not exactly the same but that they both have the same denominator.

2. $g \circ h = \frac{1 - 2x + x^2}{8}$; $h \circ g = \frac{1 - 0.5x^2}{2}$
Answers may vary. Students should notice that the two functions are different.

3. Answers may vary; most compositions $y \circ h$ and $h \circ y$ will not be equivalent.

4. The composition of functions is not always commutative. Student answers may vary. An example is: For most functions, the result will be different depending on the order in which the functions are composed.

Exploration 2

1. $f^{-1}(x) = x - 3$;
$h^{-1}(x) = -2x + 1$; $I^{-1}(x) = x$

2.

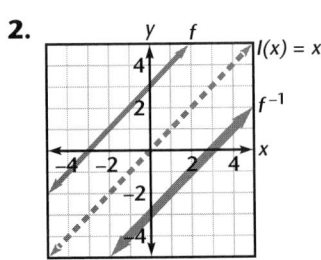

f^{-1} is shown in bold in the graph. The axis of symmetry of the graphs of f and f^{-1} is $y = x$.
$(f \circ f^{-1})(x) = x$ and $(f^{-1} \circ f)(x) = x$; The compositions of f and f^{-1} and of f^{-1} and f are the same as the equation of their axis of symmetry, $y = x$.

3.

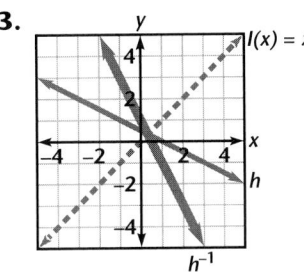

h^{-1} is shown in bold in the graph. The axis of symmetry of the graphs of h and h^{-1} is $y = x$.
$(h \circ h^{-1})(x) = x$ and $(h^{-1} \circ h)(x) = x$; The compositions of h and h^{-1} and of h^{-1} and h is the same as the equation of their axis of symmetry, $y = x$.

4.

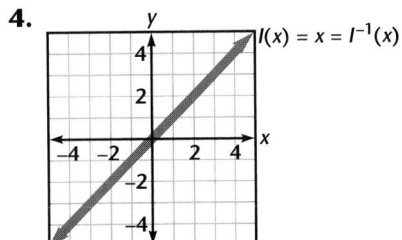

The graph of I is the same as the graph of the axis of symmetry of the functions f and h and their inverses.

5. The composition of any function and its inverse is the identity function, $y = x$.

Communicate

1. Composing two functions consists of using the dependent variable of one function as the independent variable of another function and simplifying the result.

2. The domain of f and g is the set of real numbers. The range of f is all $y \geq 0$, and the range of g is all real numbers.

3. $(f \circ g)(x) = x^2 + 4x + 4$;
$(g \circ f)(x) = x^2 + 2$

4. The domain of $f \circ g$ and of $g \circ f$ is all real numbers; the range of $f \circ g$ is all nonnegative real numbers; the range of $g \circ f$ is all real numbers greater than or equal to 2.

5. If the range of g is not in the domain of f, then f will be undefined for some values of $g(x)$. So, the domain of f must always include all the values $g(x)$ (the range of g).

6. The identity function, $I(x) = x$.

7. Composition in general is not a commutative operation. For, example, for the functions $f(x) = x + 1$ and $g(x) = x^2 + 1$, $f \circ g \neq g \circ f$. However, when f and g are inverse functions, then $f \circ g = g \circ f$.

Lesson 3.4, Pages 135–141

Communicate

1. The absolute value function results in a nonnegative value regardless of the sign of the independent variable. If a quantity is already positive, then applying the absolute value function to it does not change its sign. If a quantity is negative, then applying the absolute value function makes the quantity positive.

2. The function $g(x) = |x + c|$ has the same graph shape, domain (the real numbers), and range (the nonnegative real numbers) as the function $f(x) = |x|$. The difference lies in the axis of symmetry. The axis of symmetry of g is the line $x = -c$. The graph of g is the graph of f translated c units to the left for positive c and $|c|$ units to the right for negative c.

3. The function $h(x) = |x| + d$ has the same graph shape, domain (the real numbers), and axis of symmetry ($x = 0$) as the function $f(x) = |x|$. The difference lies in the range. The range of h is $y \geq d$. The graph of h is the graph of f translated d units upward for positive d and $|d|$ units downward for negative d.

4. The function $k(x) = a|x|$ (for positive a) has the same domain (the real numbers), range (the nonnegative real numbers), and axis of symmetry, $x = 0$, as the function $f(x) = |x|$. When $a = 1$, k and f are identical. For $a < 1$, the graph becomes increasingly wider than the graph of f as a approaches 0. For $a > 1$, the graph becomes increasingly narrower than the graph of f as a increases.

5. The function $j(x) = -a|x|$ (for positive a) has the real numbers for its domain, and the nonpositive real numbers for its range. It shares with $f(x) = |x|$ its axis of symmetry, $x = 0$. When $a = 1$, the graph of j is a simple reflection of the graph of f across the x-axis. For $a < 1$, the downward-opening graph becomes increasingly wider than the graph of f as a approaches 0. For $a > 1$, the downward-opening graph becomes increasingly narrower than the graph of f as a increases.

6. Absolute value graphs form rays that meet at right angles when the graphs are translations or reflections of $f(x) = |x|$. Graphs of $g(x) = a|x - h| + k$ have rays that meet at right angles only when $|a| = 1$. For example, the graphs of $10|x|$ and $0.1|x|$ have rays that meet respectively at angles much less than and greater than $90°$.

7. To solve an absolute value equation by graphing, graph the equation and then trace to find the value or values of x that satisfy the equation. To solve an equation using the definition

of absolute value, isolate the absolute value symbol on one side of the equation, replace the absolute value symbol with $\pm()$, and then solve two equations, one with the $+$ sign, and the other with the $-$ sign, separately.

Lesson 3.5. Pages 142–146

Communicate

1. A step function is a function whose graph looks like a series of disconnected steps. The graph is horizontal along each step, but jumps suddenly to the next step. The steps contain no points that fall directly above or below any other step, so that the graph is that of a function.

2. The greatest-integer function is a step function that returns the value of the greatest integer that is less than or equal to a given quantity. When you apply the greatest-integer function to an integer, you get back the integer. When you apply the greatest-integer function to a non-integer, you get back the next integer that is smaller than that quantity.

3. The rounding-up or least-integer function is a step function that returns the value of the smallest integer that is greater than or equal to a given quantity. When you apply the rounding-up function to an integer, you get back the integer. When you apply the rounding-up function to a non-integer, you get back the next integer that is larger than that quantity.

4. Two similarities of the greatest-integer and rounding-up functions are that they are both step functions and that neither changes the value of an integer.

5. The greatest-integer function and the rounding-up function differ in that the greatest-integer function returns smaller values when applied to a non-integer, while the rounding-up function returns larger values when applied to a non-integer. Also, the graph of the greatest-integer function contains the left endpoint of each step, but not the right, while the graph of the rounding-up function contains the right endpoint of each step, but not the left.

Practice & Apply

26. The graph of f:

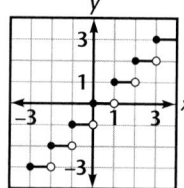

The graph of g:

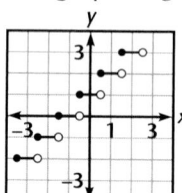

The graph of h:

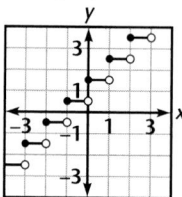

29.

time (hours)	charge ($)
$0 < t \le 1$	45
$1 < t \le 1.5$	65
$1.5 < t \le 2$	85
$2 < t \le 2.5$	105
$2.5 < t \le 3$	125
$3 < t \le 3.5$	145
$3.5 < t \le 4$	165
$4 < t \le 4.5$	185
$4.5 < t \le 5$	205
$5 < t \le 5.5$	225
$5.5 < t \le 6$	245
$6 < t \le 6.5$	265
$6.5 < t \le 7$	285
$7 < t \le 7.5$	305
$7.5 < t \le 8$	325
$8 < t \le 8.5$	345
$8.5 < t \le 9$	365
$9 < t \le 9.5$	385
$9.5 < t \le 10$	405

30.

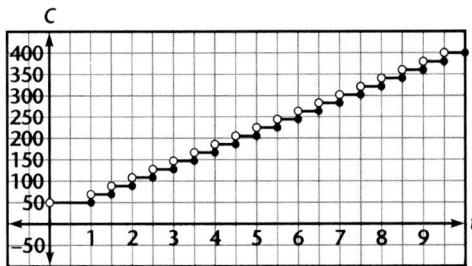

34.

time (minutes)	charge ($)
$0 < t \le 1$	0.29
$1 < t \le 2$	0.59
$2 < t \le 3$	0.88
$3 < t \le 4$	1.18
$4 < t \le 5$	1.47
$5 < t \le 6$	1.77
$6 < t \le 7$	2.06
$7 < t \le 8$	2.35
$8 < t \le 9$	2.65
$9 < t \le 10$	2.94

Look Beyond

47. domain: R; range: $\{-1, 0, 1\}$

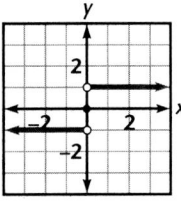

Lesson 3.6, Pages 147–153

Communicate

1. Parametric representation of functions allows you to see how one variable affects two variables which depend on it. Also, parametric representation lets you simulate many physical events involving one independent and two dependent variables.

2. Parametric representation is based upon a variable, t, which is called the parameter. In many real-world situations, the parameter, t, is time. The other two variables, $x(t)$ and $y(t)$, describe equations that are dependent on t. For example, $x(t)$ might describe horizontal position as a function of time and $y(t)$ might describe vertical position as a function of time.

3. The parametric representation of a function shows the dependence of x and y on a third variable, t. In the function itself, you see the relationship of x and y, but any dependence on the parameter t is hidden. For example, in the parametric representation of a baseball's flight, you see the relationship between the height and the horizontal position, and you also see the time for each position point (x, y). In the function itself, you see only the relationship between the height and the horizontal position.

4. To convert the parametric equations for a line into the function they represent, first find two ordered pairs (x, y) corresponding to two values of t in the parametric equations. Use these points to find the slope, m, and then substitute m and the x- and y-values from one of the ordered pairs into the point-slope formula of a line to solve for the y-intercept, b.

5. As noted in Exercise 3, combining two parametric equations into one function hides any dependence that the two function variables have on the parameter. For example, in an equation describing the path of an object in motion, you would no longer know the time that corresponds to any given position.

Lesson 4.1, Pages 164–170

Exploration

1. & 2. Dimensions of A: 3×4; dimensions of B: 3×4; matrices A and B have the same dimensions, 3×4.

3. $A + B = \begin{bmatrix} 12 & -7 & 9 & 4 \\ 2 & 5 & -1 & -2 \\ -8 & 8 & -3 & 0 \end{bmatrix}$

The calculator found the sum of each element in matrix A and its corresponding element in matrix B. Matrices A and B must have the same dimensions so that there is a corresponding element in matrix B for each element in matrix A. The dimensions of the sum matrix are the same because each of the elements of A and B are added, producing a sum matrix with exactly the same number of corresponding elements.

4. $B - A = \begin{bmatrix} 4 & -1 & -5 & 2 \\ -2 & 5 & -11 & 2 \\ 4 & 6 & 5 & -6 \end{bmatrix}$

Each element in A was subtracted from its corresponding element in B. Matrices A and B must have the same dimensions so that there is a corresponding element in matrix A for each element in matrix B. The dimensions of $B - A$ are the same as A and B because each of the elements of B and A are subtracted, producing a difference matrix with exactly the same number of elements.

5. $A - A = \begin{bmatrix} 0 & 0 & 0 & 0 \\ 0 & 0 & 0 & 0 \\ 0 & 0 & 0 & 0 \end{bmatrix} = C$

6. $C + B = \begin{bmatrix} 8 & -4 & 2 & 3 \\ 0 & 5 & -6 & 0 \\ -2 & 7 & 1 & -3 \end{bmatrix}$

Matrix C is the *zero* matrix, or the *additive identity* matrix.

7. $-1 \cdot B = \begin{bmatrix} -8 & 4 & -2 & -3 \\ 0 & -5 & 6 & 0 \\ 2 & -7 & -1 & 3 \end{bmatrix}$

Each element in the resulting matrix is the opposite of its corresponding element in matrix B.

Communicate

1. Answers may vary. For example, Doctor's office: social security number, date of birth, date of last physical, vaccination record, date of last visit, cost, payment.

2. The dimensions of a matrix are given by the number of rows by the number of columns.

3. m_{52} is the element found at the intersection of the 5th row and the 2nd column.

4. b_{21}, b_{22}, b_{23}, and b_{24} are all located consecutively from left to right in row 2.

5. a_{11}, a_{22}, and a_{33} are located on the diagonal of matrix A.

6. Answers may vary. One example is

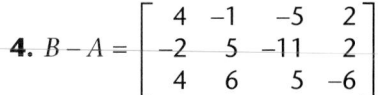

Practice & Apply

30. $W = \begin{bmatrix} \text{plastic} & 4-\text{top} & \text{picnic} \\ \text{wood} & 6-\text{top} & \text{dinner} \\ \text{marble} & 2-\text{top} & \text{round} \end{bmatrix}$

31. A sample answer is

	Squash	Tomatoes	Peppers	Melons	
Farmer	27	31	24	18	$= P$
Store	48	72	61	25	

32. Yes; a 4×2 matrix could be created with the produce listed in the rows and the sellers listed in the columns.

33. Answers may vary. Using the sample answer given for Exercise 31, the number of peppers is located in p_{13}.

34. Answers may vary. Using the sample answer above, the location p_{21} contains the information that 48 squash were sold at the produce section of a grocery store.

39. Answers may vary. An example is:

	Country	Jazz	Rock	Blues	Classical	
Records	10	15	5	8	3	$= M$
Tapes	5	9	1	13	6	
CDs	2	1	9	5	10	

49. Answers may vary, but the diagram should resemble the pattern used in matrix multiplication.

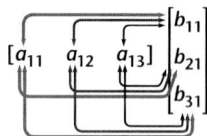

Exploration

1. $AB = \begin{bmatrix} 2 & 3 \\ 7 & 11 \end{bmatrix}$; $BA = \begin{bmatrix} 25 & -43 \\ 7 & -12 \end{bmatrix}$; no

2. $AC = \begin{bmatrix} 1 & 0 \\ 0 & 1 \end{bmatrix}$; $CA = \begin{bmatrix} 1 & 0 \\ 0 & 1 \end{bmatrix}$; yes, $AC = CA$

3.

	Col 1	Col 2	Col 3
	$AI = A$	$BI = B$	$CI = C$
	$IA = A$	$IB = B$	$IC = C$

	Col 4	Col 5	Col 6
	$(AB)C =$ $\begin{bmatrix} -19 & 7 \\ -68 & 25 \end{bmatrix}$	$(CB)A =$ $\begin{bmatrix} -111 & 191 \\ -68 & 117 \end{bmatrix}$	$BC =$ $\begin{bmatrix} -41 & 15 \\ -11 & 4 \end{bmatrix}$
	$A(BC) =$ $\begin{bmatrix} -19 & 7 \\ -68 & 25 \end{bmatrix}$	$C(BA) =$ $\begin{bmatrix} -111 & 191 \\ -68 & 117 \end{bmatrix}$	$CB =$ $\begin{bmatrix} -18 & -31 \\ -11 & -19 \end{bmatrix}$

4. yes, multiplication with the identity matrix is commutative.

5. The parentheses determine the order of matrix multiplication and, therefore, determine the product; yes, matrix multiplication is associative.

6. No

7. Multiplication with the identity matrix *always* results in the original matrix.

Communicate

1. To perform matrix multiplication, the inner dimensions of the two matrices must be the same. The outer dimensions of the two matrices become the dimensions of the product matrix.

2. Answers may vary. One answer is: Multiply $\begin{bmatrix} 3 & -2 \end{bmatrix} \begin{bmatrix} -3 \\ -1 \end{bmatrix}$ to get the element in row 1, column 2: $3(-3) + (-2)(-1) = -7$.

3. yes; no; all the elements will be positive because the product of

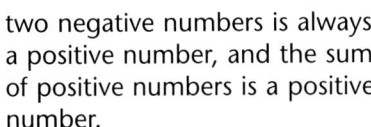

two negative numbers is always a positive number, and the sum of positive numbers is a positive number.

4. Answers may vary depending on the calculator used. Refer to your *HRW Technology Handbook* or the calculator manual to determine how to multiply matrices on a particular calculator. On the TI-82, use the following directions: Use the **MATRX** EDIT menu to enter the elements for each matrix. Clear the screen using **QUIT**, then multiply the matrices using the **MATRX** NAMES menu.

5. An error message is given.

6. The row labels of the first matrix become the row labels of the product matrix. The column labels of the second matrix become the column labels of the product matrix. The labels indicate the type of data and its arrangement. Also, if the inner labels do not match, the product matrix has no real-world meaning.

7. The identity matrix is a square matrix with 1s along the diagonal and 0s for all of the other elements. The identity matrix is the multiplicative identity for square matrices. Multiplying any square matrix by an identity matrix of the same dimensions yields the original matrix.

Lesson 4.3, pages 180–188

Exploration

1. Each pair of lines intersect at one point. The point of intersection is different for each pair.

2. Each pair represents the same line. Therefore, each pair of lines intersect at all the points on

the line. Each pair represents a different line.

3. No pairs of lines intersect. Each pair of lines has the same slope but a different y-intercept. Each pair is a different set of parallel lines.

4. Type 1 systems of equations intersect at one point. Type 2 systems of equations intersect at an infinite number of points. Type 3 systems of equations do not intersect.

Communicate

1. Answers may vary. An example is: Independent and dependent systems are consistent since they are systems which have solutions. In other words, it is possible to solve and get a usable result for these systems.

2. The graph of an independent system is the graph of two lines which intersect in exactly one point. The graph of a dependent system is the graph of one line with an infinite number of solutions. The graph of an inconsistent system is the graph of a pair of parallel lines; there is no solution.

3. Answers may vary. Graphs of systems should match the description in the answer for Exercise 2.

4. Add like terms of the two equations, eliminate the y-term, and solve for x by dividing by 3. Substitute the value of x, 3, in one of the original equations, and solve for y. The solution is $(3, -1)$. Check your solution in *both* original equations.

5. Multiply each side of the second equation by 4. Add like terms of resultant equations to eliminate the y-terms. Solve for x by

dividing by 11. Substitute $-\frac{17}{11}$, the value of x, into one of the original equations, and solve for y. The solution is $(-\frac{17}{11}, -\frac{21}{11})$. Check your solution in *both* original equations.

6. An inconsistent system yields a false numerical statement (i.e. $5 = 4$); a dependent system yields a numerical statement that is always true and contains no variables (i.e. $4 = 4$).

7. To check your solution, substitute its x- and y-values into each original equation of the system. If your solution is correct, you will produce true numerical statements.

Lesson 4.4, Pages 189–197

Communicate

1. Solve the third equation for z by dividing by -1. Then substitute the value for z, -3, into the second equation, and solve for y by simplifying and dividing by 5. Finally, substitute the values for z and y, -3 and $-\frac{6}{5}$, into the first equation, and solve for x by simplifying and dividing by 2. The result is $\left(\frac{57}{10}, \frac{-6}{5}, -3 \right)$.

2. Coefficient matrix: $\begin{bmatrix} 3 & -2 & 1 \\ 1 & 1 & 0 \\ 4 & 2 & 0 \end{bmatrix}$;

   Constant matrix: $\begin{bmatrix} 6 \\ 3 \\ 0 \end{bmatrix}$;

   Augmented matrix:
   $\begin{bmatrix} 3 & -2 & 1 & | & 6 \\ 1 & 1 & 0 & | & 3 \\ 4 & 2 & 0 & | & 0 \end{bmatrix}$

3. Coefficient matrix: $\begin{bmatrix} 1 & -4 & 7 \\ 2 & 1 & -1 \\ 1 & 0 & 4 \end{bmatrix}$;

Constant matrix: $\begin{bmatrix} 17 \\ -5 \\ 13 \end{bmatrix}$;

Augmented matrix:
$\begin{bmatrix} 1 & -4 & 7 & | & 17 \\ 2 & 1 & -1 & | & -5 \\ 1 & 0 & 4 & | & 13 \end{bmatrix}$

4. Main diagonal: –3, 2, 1;
System of equations:
$\begin{cases} -3x - 4y = 2 \\ 4x + 2y - 3z = 6 \\ -2x + z = -6 \end{cases}$

5. Represent the system in an augmented matrix. Use elementary row operations on the augmented matrix to obtain a triangle of zeros under the main diagonal. Identify the system of equations that the augmented matrix represents, and use back substitution to solve for the unknowns.

6. Multiply all elements of row 1 by –5, add the resulting values to row 3, and replace row 3 with these new values.

7. Interchange rows 1 and 3.

8. Multiply all the elements of row 1 by 2. Replace row 1 with the result.

Look Beyond

46. $\frac{1}{5}$; If $5x = 10$, then both sides of the equation can be multiplied by the multiplicative inverse of 5, $\frac{1}{5}$, to solve for x:
$\frac{1}{5}(5x) = \frac{1}{5}(10) \rightarrow \frac{5}{5}x = \frac{10}{5} \rightarrow$
$x = 2$.

47. 4; If $\frac{1}{4}y = -11$, then both sides of the equation can be multiplied by the multiplicative inverse of $\frac{1}{4}$, 4, to solve for y:
$4\left(\frac{1}{4}y\right) = 4(-11) \rightarrow \frac{4}{4}y = -44 \rightarrow$
$y = -44$.

48. –5; If $-\frac{1}{5}z + \frac{4}{5} = 6$, then both sides of the equation can be

multiplied by the multiplicative inverse of $-\frac{1}{5}$, –5, to solve for z:

$-5\left(-\frac{1}{5}z + \frac{4}{5}\right) = -5(6)$
$-\frac{5}{5}z - \frac{5 \cdot 4}{5} = -30$
$z - 4 = -30$
$z = -26$

Lesson 4.5, Pages 198–202

Communicate

1. The product is the original left matrix $\begin{bmatrix} 1 & 3 \\ 2 & -1 \end{bmatrix}$.

2. The product is the original right matrix $\begin{bmatrix} -2 & 1 \\ 5 & 3 \end{bmatrix}$.

3. The inner dimensions are not the same.

4. $I_4 = \begin{bmatrix} 1 & 0 & 0 & 0 \\ 0 & 1 & 0 & 0 \\ 0 & 0 & 1 & 0 \\ 0 & 0 & 0 & 1 \end{bmatrix}$

5. Enter the matrix; use the 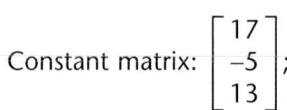 key to find its inverse.

Look Back

39. Division Property of Equality;
$x = -\frac{1}{4}$

40. Division Property of Equality;
$x = -11\frac{3}{11}$

5. Write the matrix equation, $AX = B$: $\begin{bmatrix} 2 & 3 & -2 \\ 3 & -3 & 2 \\ 6 & -2 & 8 \end{bmatrix}\begin{bmatrix} x \\ y \\ z \end{bmatrix} = \begin{bmatrix} 4 \\ 16 \\ 10 \end{bmatrix}$.

Determine A^{-1}: $\begin{bmatrix} 0.2 & 0.2 & 0 \\ 0.12 & -0.28 & 0.1 \\ -0.12 & -0.22 & 0.15 \end{bmatrix}$. Multiply both sides of the matrix

equation by A^{-1} on the left side:

$\begin{bmatrix} 0.2 & 0.2 & 0 \\ 0.12 & -0.28 & 0.1 \\ -0.12 & -0.22 & 0.15 \end{bmatrix}\begin{bmatrix} 2 & 3 & -2 \\ 3 & -3 & 2 \\ 6 & -2 & 8 \end{bmatrix}\begin{bmatrix} x \\ y \\ z \end{bmatrix} = \begin{bmatrix} 0.2 & 0.2 & 0 \\ 0.12 & -0.28 & 0.1 \\ -0.12 & -0.22 & 0.15 \end{bmatrix}\begin{bmatrix} 4 \\ 16 \\ 10 \end{bmatrix}$

Look Beyond

42. $A^{-1} = \begin{bmatrix} 5 & -7 \\ -2 & 3 \end{bmatrix}$; $\begin{bmatrix} 1 & 0 \\ 0 & 1 \end{bmatrix}\begin{bmatrix} x \\ y \end{bmatrix} = \begin{bmatrix} 13 \\ -5 \end{bmatrix}$; $\begin{bmatrix} x \\ y \end{bmatrix} = \begin{bmatrix} 13 \\ -5 \end{bmatrix}$

Lesson 4.6, pages 203–209

Communicate

1. Coefficient matrix, A, times a variable matrix, X, equals the constant matrix, B, or $AX = B$.

2. $\begin{cases} 2x - y + 3z = 4 \\ -3x - z = 1 \\ x - 3y + z = 5 \end{cases}$

3. Represent the system of equations by the matrix equation, $AX = B$. Find the inverse of the coefficient matrix, A. Multiply both sides of the matrix equation $AX = B$ by A^{-1} which results in the equation $X = A^{-1}B$. X will contain the values of each variable.

4. Matrix multiplication is not commutative. The matrix, $A^{-1}B$, is not equivalent to the matrix BA^{-1}. Therefore, since it is necessary to multiply the left side of the matrix equation, $AX = B$, by A^{-1} *on the left* in order to isolate X on one side of the equation, $A^{-1}AX = (A^{-1}A)X = IX = X$, it is also necessary to multiply matrix B on the left.

The result is the solution: $\begin{bmatrix} 1 & 0 & 0 \\ 0 & 1 & 0 \\ 0 & 0 & 1 \end{bmatrix} \begin{bmatrix} x \\ y \\ z \end{bmatrix} = \begin{bmatrix} x \\ y \\ z \end{bmatrix} = \begin{bmatrix} 4 \\ -3 \\ -2.5 \end{bmatrix}$,

or $x = 4$, $y = -3$, and $z = -2.5$.

6. Write the matrix equation, $AX = B$: $\begin{bmatrix} 1 & 2 & 0 \\ 0 & 1 & 2 \\ 2 & 0 & 1 \end{bmatrix} \begin{bmatrix} x \\ y \\ z \end{bmatrix} = \begin{bmatrix} -6 \\ 11 \\ 16 \end{bmatrix}$.

Determine A^{-1}: $\begin{bmatrix} 0.11\ldots & -0.22\ldots & 0.44\ldots \\ 0.44\ldots & 0.11\ldots & -0.22\ldots \\ -0.22\ldots & 0.44\ldots & 0.11\ldots \end{bmatrix}$ Multiply both sides of

the matrix equation by A^{-1} on the left side:

$\begin{bmatrix} 0.11\ldots & -0.22\ldots & 0.44\ldots \\ 0.44\ldots & 0.11\ldots & -0.22\ldots \\ -0.22\ldots & 0.44\ldots & 0.11\ldots \end{bmatrix} \begin{bmatrix} 1 & 2 & 0 \\ 0 & 1 & 2 \\ 2 & 0 & 1 \end{bmatrix} \begin{bmatrix} x \\ y \\ z \end{bmatrix} =$

$\begin{bmatrix} 0.11\ldots & -0.22\ldots & 0.44\ldots \\ 0.44\ldots & 0.11\ldots & -0.22\ldots \\ -0.22\ldots & 0.44\ldots & 0.11\ldots \end{bmatrix} \begin{bmatrix} -6 \\ 11 \\ 16 \end{bmatrix}$ The result is the solution:

$\begin{bmatrix} x \\ y \\ z \end{bmatrix} = \begin{bmatrix} 4 \\ -5 \\ 8 \end{bmatrix}$, or $x = 4$, $y = -5$,

and $z = 8$.

7. dependent and inconsistent systems

Practice & Apply

18. $x + y + z = 180$
$30 + 60 + 90 = 180$
            $180 = 180$ True

$z = 3x$ $y = \frac{1}{2}(x + z)$
$90 = 3(30)$ $60 = \frac{1}{2}(30 + 90)$
$90 = 90$ True $60 = 60$ True

19. $\begin{bmatrix} 1 & 1 & 1 & 1 \\ 2 & -1 & 1 & -3 \\ 3 & 1 & -1 & -1 \\ 2 & -3 & 1 & -1 \end{bmatrix} \begin{bmatrix} x \\ y \\ z \\ w \end{bmatrix} =$
$\begin{bmatrix} 10 \\ -9 \\ -2 \\ -5 \end{bmatrix} \rightarrow \begin{bmatrix} x \\ y \\ z \\ w \end{bmatrix} = \begin{bmatrix} 1 \\ 2 \\ 3 \\ 4 \end{bmatrix}$

20. $\begin{bmatrix} 1 & 2 & -6 & 1 \\ -2 & -3 & 9 & 1 \\ 1 & 2 & -5 & 2 \\ 2 & 4 & -12 & 3 \end{bmatrix} \begin{bmatrix} x \\ y \\ z \\ w \end{bmatrix} =$
$\begin{bmatrix} 12 \\ -19 \\ 15 \\ 24 \end{bmatrix} \rightarrow \begin{bmatrix} x \\ y \\ z \\ w \end{bmatrix} = \begin{bmatrix} 2 \\ 14 \\ 3 \\ 0 \end{bmatrix}$

21. $\begin{bmatrix} 1 & -1 & 0 \\ 0 & 1 & -1 \\ 2 & 0 & -1 \end{bmatrix} \begin{bmatrix} x \\ y \\ z \end{bmatrix} = \begin{bmatrix} 5 \\ -6 \\ 2 \end{bmatrix} \rightarrow$
$\begin{bmatrix} x \\ y \\ z \end{bmatrix} = \begin{bmatrix} 3 \\ -2 \\ 4 \end{bmatrix}$

22. Let x represent amount invested in Plan A,
    y represent amount invested in Plan B, and
    z represent amount invested in Plan C.

Then $\begin{cases} 0.80x + 0.20y + 0.50z = 16{,}110.00 \\ 0.15x + 0.70y + 0.10z = 9016.75 \\ 0.05x + 0.10y + 0.40z = 5698.75 \end{cases} \rightarrow$

$\begin{bmatrix} 0.80 & 0.20 & 0.50 \\ 0.15 & 0.70 & 0.10 \\ 0.05 & 0.10 & 0.40 \end{bmatrix} \begin{bmatrix} x \\ y \\ z \end{bmatrix} = \begin{bmatrix} 16{,}110.00 \\ 9016.75 \\ 5698.75 \end{bmatrix}$

23. $\begin{bmatrix} 11275 \\ 8950 \\ 10600 \end{bmatrix}$ $11,275$ was invested in Plan A, $8950 was invested

in Plan B, and $10,600 was invested in Plan C.

24. Check: $0.80(11{,}275) + 0.20(8950) + 0.5(10{,}600) \overset{?}{=} 16{,}110.00$
$16{,}110.00 \overset{?}{=} 16{,}110.00$ True

$0.15(11{,}275) + 0.70(8950) + 0.10(10{,}600) \overset{?}{=} 9016.75$
$9016.75 \overset{?}{=} 9016.75$ True

$0.50(11{,}275) + 0.10(8950) + 0.40(10{,}600) \overset{?}{=} 5698.75$
$5698.75 \overset{?}{=} 5698.75$ True

$0.50(12{,}000) + 0.10(8000) + 0.40(9500) \overset{?}{=} 10{,}600$
$10{,}600 \overset{?}{=} 10{,}600$ True

Look Back

29. $\begin{bmatrix} 5 & -4 \\ -3 & 3 \end{bmatrix} \begin{bmatrix} -5 & -2 \\ 8 & 3 \end{bmatrix} =$
$\begin{bmatrix} (5)(-5) + (-4)(8) & (5)(-2) + (-4)(3) \\ (-3)(-5) + (3)(8) & (-3)(-2) + (3)(3) \end{bmatrix}$
$= \begin{bmatrix} -57 & -22 \\ 39 & 15 \end{bmatrix}$

Lesson 4.7, pages 210–217

Exploration 1

1. $BA = \begin{bmatrix} -5 & 5 & 5 & -5 \\ 3 & 6 & -6 & -3 \end{bmatrix}$

2. 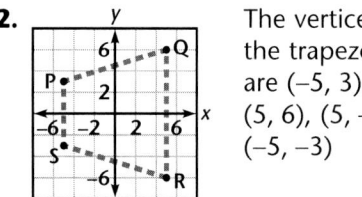 The vertices of the trapezoid are $(-5, 3)$, $(5, 6)$, $(5, -6)$, $(-5, -3)$

3. Both figures are trapezoids of the same shape and size, centered at the origin.

4. The figure from matrix BA is the same as the figure represented by matrix A but is rotated $90°$ counterclockwise.

        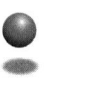

Exploration 2

1.

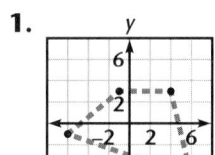

2. $BA = \begin{bmatrix} -3 & 1 & 5 & -3 \\ -1 & -6 & 6 & 4 \end{bmatrix}$

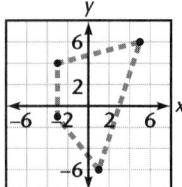

The quadrilateral represented by BA is congruent to the figure represented by A but is rotated 90° counterclockwise.

3. $B(BA) = \begin{bmatrix} 1 & 6 & -6 & -4 \\ -3 & 1 & 5 & -3 \end{bmatrix}$

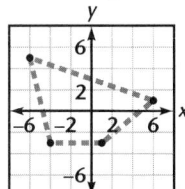

The quadrilateral represented by $B(BA)$ is congruent to the quadrilaterals represented by A and BA. The quadrilateral represented by $B(BA)$ is rotated 180° counterclockwise from the quadrilateral represented by A, and rotated 90° counterclockwise from that represented by BA.

4. The quadrilateral represented by A can be rotated 270° counterclockwise by multiplying A by BBB on the left.

5. $BBB = \begin{bmatrix} 0 & 1 \\ -1 & 0 \end{bmatrix}$

$(BBB)A = \begin{bmatrix} 3 & -1 & -5 & 3 \\ 1 & 6 & -6 & -4 \end{bmatrix}$

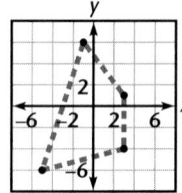

The quadrilateral represented by $(BBB)A$ is congruent to the quadrilateral represented by A but is rotated 270° counterclockwise.

6. For a 90° counterclockwise rotation, multiply by $\begin{bmatrix} 0 & -1 \\ 1 & 0 \end{bmatrix}$. For a 180° rotation, multiply by $\begin{bmatrix} -1 & 0 \\ 0 & -1 \end{bmatrix}$. For a 270° counterclockwise rotation, multiply by $\begin{bmatrix} 0 & 1 \\ -1 & 0 \end{bmatrix}$.

Exploration 3

Matrix A

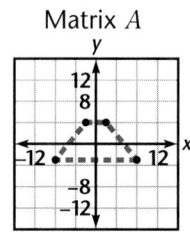

Isosceles trapezoid

1. $BA = \begin{bmatrix} 4 & -4 & -16 & 16 \\ 8 & 8 & -6 & -6 \end{bmatrix}$

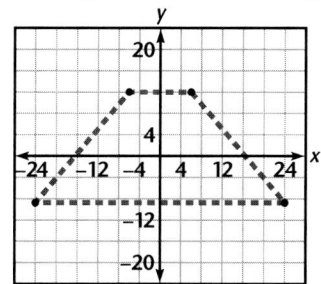

---

2. $CA = \begin{bmatrix} 6 & -6 & -24 & 24 \\ 12 & 12 & -9 & -9 \end{bmatrix}$

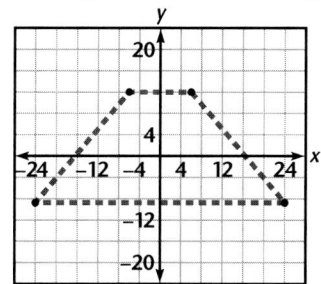

3. Matrix B increased the size of the object by 2. Matrix C increased the size of the object by 3.

4. $D(BA) = \begin{bmatrix} 2 & -2 & -8 & 8 \\ 4 & 4 & -3 & -3 \end{bmatrix}$

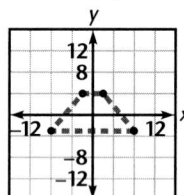

5. Matrix D reduced the size of the object by $\frac{1}{2}$.

6. $X = \begin{bmatrix} \frac{1}{3} & 0 \\ 0 & \frac{1}{3} \end{bmatrix}$

7. To enlarge or reduce an object represented by a matrix, multiply the matrix by a 2 by 2 matrix of the form $\begin{bmatrix} a & 0 \\ 0 & a \end{bmatrix}$, where a can be any positive real number, greater than 1 for an enlargement or less than 1 for a reduction.

Exploration 4

1. Matrix A

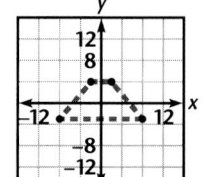

Isosceles trapezoid

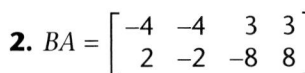

2. $BA = \begin{bmatrix} -4 & -4 & 3 & 3 \\ 2 & -2 & -8 & 8 \end{bmatrix}$

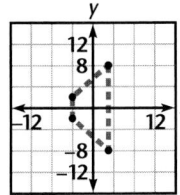

The trapezoid represented by matrix A is rotated counterclockwise by 90°.

3. $B^{-1}A = \begin{bmatrix} 4 & 4 & -3 & -3 \\ -2 & 2 & 8 & -8 \end{bmatrix}$

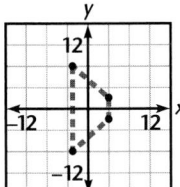

The trapezoid represented by matrix A is rotated clockwise by 90°.

4. $BB^{-1}A = \begin{bmatrix} 2 & -2 & -8 & 8 \\ 4 & 4 & -3 & -3 \end{bmatrix}$

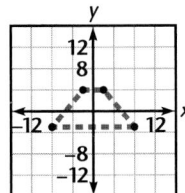

The trapezoid represented by $BB^{-1}A$ is the same as the trapezoid represented by matrix A.

5. The objects are congruent. Multiplication by matrix B or matrix B^{-1} rotates the object 90°. Matrix B rotates *counterclockwise*, while matrix B^{-1} rotates *clockwise*.

6. $CA = \begin{bmatrix} 4 & -4 & -16 & 16 \\ 8 & 8 & -6 & -6 \end{bmatrix}$

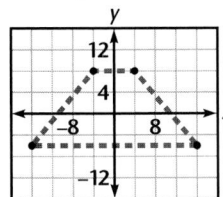

The transformation enlarged the trapezoid by 2.

7. $C^{-1}A = \begin{bmatrix} 1 & -1 & -4 & 4 \\ 2 & 2 & -1.5 & -1.5 \end{bmatrix}$

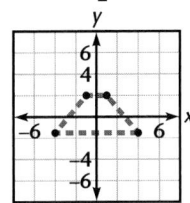

The transformation reduced the trapezoid by $\frac{1}{2}$.

8. $CC^{-1}A = \begin{bmatrix} 2 & -2 & -8 & 8 \\ 4 & 4 & -3 & -3 \end{bmatrix}$

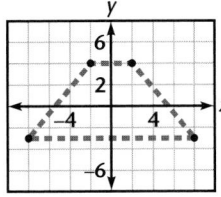

The trapezoid represented by CC^{-1} is the same as the trapezoid represented by A.

Communicate

1. $\begin{bmatrix} 0 & -1 \\ 1 & 0 \end{bmatrix}$

2. $\begin{bmatrix} 0 & -1 \\ 1 & 0 \end{bmatrix}\begin{bmatrix} 0 & -1 \\ 1 & 0 \end{bmatrix} = \begin{bmatrix} -1 & 0 \\ 0 & -1 \end{bmatrix}$

3. $\begin{bmatrix} 0 & -1 \\ 1 & 0 \end{bmatrix}\begin{bmatrix} 0 & -1 \\ 1 & 0 \end{bmatrix}\begin{bmatrix} 0 & -1 \\ 1 & 0 \end{bmatrix}$
$= \begin{bmatrix} 0 & 1 \\ -1 & 0 \end{bmatrix}$

4. Multiply on the left by the inverse of the transformation matrix.

5. $\begin{bmatrix} -1 & 0 \\ 0 & -1 \end{bmatrix}^{-1} = \begin{bmatrix} -1 & 0 \\ 0 & -1 \end{bmatrix}$

6. $\begin{bmatrix} 5 & 0 \\ 0 & 5 \end{bmatrix}$

Practice & Apply

10. $\begin{bmatrix} 0 & 1 \\ -1 & 0 \end{bmatrix}^{-1}\begin{bmatrix} 8 & 2 & -5 & 1 \\ 6 & -7 & -7 & 6 \end{bmatrix} =$
$\begin{bmatrix} -6 & 7 & 7 & -6 \\ 8 & 2 & -5 & 1 \end{bmatrix}$

11. $\begin{bmatrix} 5 & 0 \\ 0 & 5 \end{bmatrix}\begin{bmatrix} 8 & 2 & -5 & 1 \\ 6 & -7 & -7 & 6 \end{bmatrix} =$
$\begin{bmatrix} 40 & 10 & -25 & 5 \\ 30 & -35 & -35 & 30 \end{bmatrix}$

12. $\begin{bmatrix} \frac{2}{3} & 0 \\ 0 & \frac{2}{3} \end{bmatrix}\begin{bmatrix} 0 & -1 \\ 1 & 0 \end{bmatrix}\begin{bmatrix} 8 & 2 & -5 & 1 \\ 6 & -7 & -7 & 6 \end{bmatrix}$
$= \begin{bmatrix} -4 & 4.67 & 4.67 & -4 \\ 5.33 & 1.33 & -3.33 & 0.67 \end{bmatrix}$

13. $\begin{bmatrix} -1 & 0 \\ 0 & -1 \end{bmatrix}\begin{bmatrix} 3 & 0 \\ 0 & 3 \end{bmatrix}\begin{bmatrix} 8 & 2 & -5 & 1 \\ 6 & -7 & -7 & 6 \end{bmatrix}$
$= \begin{bmatrix} -24 & -6 & 15 & -3 \\ -18 & 21 & 21 & -18 \end{bmatrix}$

14. $\begin{bmatrix} 4 & 0 \\ 0 & 4 \end{bmatrix}\begin{bmatrix} -1 & 0 \\ 0 & -1 \end{bmatrix}\begin{bmatrix} 3 & 5 & -5 & -3 \\ 6 & -4 & -4 & 6 \end{bmatrix}$
$= \begin{bmatrix} -12 & -20 & 20 & 12 \\ -24 & 16 & 16 & -24 \end{bmatrix}$

15. $\begin{bmatrix} \frac{1}{2} & 0 \\ 0 & \frac{1}{2} \end{bmatrix}\begin{bmatrix} 7 & 0 \\ 0 & 7 \end{bmatrix}\begin{bmatrix} 3 & 5 & -5 & -3 \\ 6 & -4 & -4 & 6 \end{bmatrix}$
$= \begin{bmatrix} 10.5 & 17.5 & -17.5 & -10.5 \\ 21 & -14 & -14 & 21 \end{bmatrix}$

16. $\begin{bmatrix} -1 & 0 \\ 0 & -1 \end{bmatrix}\begin{bmatrix} 0 & -1 \\ 1 & 0 \end{bmatrix}^{-1}\begin{bmatrix} 3 & 5 & -5 & -3 \\ 6 & -4 & -4 & 6 \end{bmatrix}$
$= \begin{bmatrix} -6 & 4 & 4 & -6 \\ 3 & 5 & -5 & -3 \end{bmatrix}$

17. $\begin{bmatrix} 3 & 0 \\ 0 & 3 \end{bmatrix}\begin{bmatrix} -1 & 0 \\ 0 & -1 \end{bmatrix}\begin{bmatrix} 2 & 8 & -3 \\ 5 & -7 & 2 \end{bmatrix}$
$= \begin{bmatrix} -6 & -24 & 9 \\ -15 & 21 & -6 \end{bmatrix}$

18. $\begin{bmatrix} 0 & -1 \\ 1 & 0 \end{bmatrix}^{-1}\begin{bmatrix} \frac{3}{4} & 0 \\ 0 & \frac{3}{4} \end{bmatrix}\begin{bmatrix} 2 & 8 & -3 \\ 5 & -7 & 2 \end{bmatrix}$
$= \begin{bmatrix} -3.75 & 5.25 & -1.5 \\ 1.5 & 6 & -2.25 \end{bmatrix}$

26. $\begin{bmatrix} 5.22 & 0 & 3.22 \\ -3 & 0 & -6.48 \end{bmatrix}$

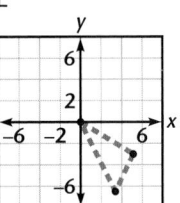

31. Both are rotations in the counterclockwise direction, but by different degrees.

32. Both are rotations in the clockwise direction, but by different degrees.

33. Find the area of the floor plan and change the units, square inches, to square feet.

34. Find the area of the building and change the units, square feet, to square inches.

35. $\begin{bmatrix} 7 & -7 & -7 & 7 \\ 2 & 2 & -2 & -2 \end{bmatrix}$

36. $\begin{bmatrix} 0 & 1 \\ -1 & 0 \end{bmatrix} \begin{bmatrix} 7 & -7 & -7 & 7 \\ 2 & 2 & -2 & -2 \end{bmatrix} = \begin{bmatrix} 2 & 2 & -2 & -2 \\ -7 & 7 & 7 & -7 \end{bmatrix}$

37. If the grid unit is $\frac{1}{a}$, where a is a positive real number, and then the matrix is $\begin{bmatrix} a & 0 \\ 0 & a \end{bmatrix}$.
The multiplication equation is
$\begin{bmatrix} a & 0 \\ 0 & a \end{bmatrix} \begin{bmatrix} -7 & 7 & 7 & -7 \\ 7 & 7 & -7 & -7 \end{bmatrix}$.

Lesson 4.8, Pages 220–227

Exploration 1

1.
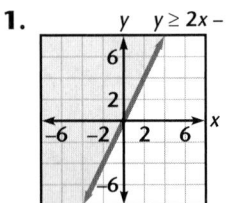
$y \geq 2x - 1$

2.
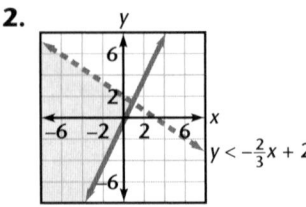
$y < -\frac{2}{3}x + 2$

3. One boundary line, $y = 2x - 1$, is solid since its inequality has a $\geq$ symbol. The other boundary line, $y = -\frac{2}{3}x + 2$, is dashed since its inequality has a $<$ symbol. The region with both colors (overlapping area) indicates the

intersection of the two shaded regions. It is not possible to view all the points in this region of the graph since it extends infinitely in the negative x and negative y directions.

4. Answers may vary, but must be in the region shaded with both colors. Examples are

Point	$y \geq 2x - 1$	$y = -\frac{2}{3}x + 2$
(0, 0)	$0 \geq -1$ True	$0 < 2$ True
(0, 1)	$1 \geq -1$ True	$1 < 2$ True
(−1, 0)	$0 \geq -3$ True	$0 < \frac{8}{3}$ True

5. Answers may vary, but none of the ordered pairs chosen may be in the region shaded with *both* colors. Examples are

Point	$y \geq 2x - 1$	$y = -\frac{2}{3}x + 2$
(4, 0)	$0 \geq 7$ False	$0 < -\frac{10}{3}$ False
(0, 3)	$3 \geq -1$ True	$3 < 0$ False
(2, 0)	$0 \geq 3$ False	$0 < \frac{2}{3}$ True

6. None of the ordered pairs chosen in Step 4 should make any of the inequalities false. Any points substituted from the region shaded with *both* colors (the solution region) will make *both* inequalities true.

7. None of the ordered pairs chosen in Step 5 should make both of the inequalities true. Points substituted from a region that is *not* shaded with *both* colors will *not* make *both* inequalities true.

8. Shade the area which makes both inequalities true. This shaded area contains all the solution points.

Exploration 2

1.
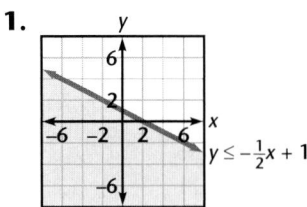
$y \leq -\frac{1}{2}x + 1$

2.
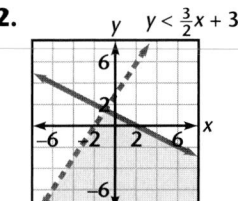
$y < \frac{3}{2}x + 3$

3.
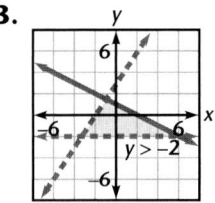
$y > -2$

4. The intersection of all three shaded regions is a triangle with vertices (−1, 1.5), (6, −2), and $(-\frac{10}{3}, -2)$. The entire solution region can be viewed since it is bounded on all sides.

5. Answers may vary, but must be in the region shaded with all three colors. Examples are

Point	$y \leq -\frac{1}{2}x + 1$	$y < \frac{3}{2}x + 3$	$y > -2$
(0, 0)	$0 \leq 1$ True	$0 < 3$ True	$0 > -2$ True
(0, 1)	$1 \leq 1$ True	$1 < 3$ True	$1 > -2$ True
(0, −1)	$-1 \leq 1$ True	$-1 < 3$ True	$-1 > -2$ True

6. Answers may vary, but none of the ordered pairs chosen may be in the region shaded with *all three* colors. Examples are

Point	$y \leq -\frac{1}{2}x + 1$	$y < \frac{3}{2}x + 3$	$y > -2$
(0, 2)	$2 \leq 1$ False	$2 < 3$ True	$2 > -2$ True
(0, −3)	$-3 \leq 1$ True	$-3 < 3$ True	$-3 > -2$ False
(1, 1)	$1 \leq \frac{1}{2}$ False	$1 < \frac{9}{2}$ True	$1 > -2$ True

7. None of the ordered pairs chosen in Step 5 should make any of the inequalities false. Any points substituted from the region shaded with *all three* colors (the solution region) will make *all three* inequalities true.

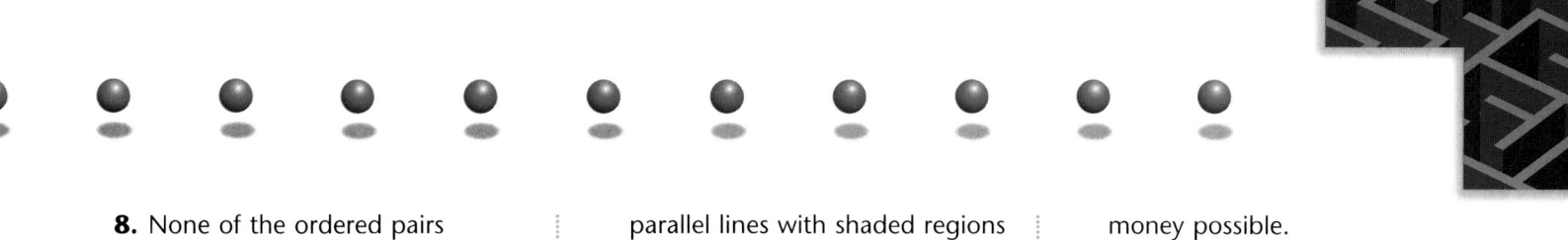

8. None of the ordered pairs chosen in Step 6 should make *all three* inequalities true. Points substituted from a region that is *not* shaded with *all three* colors will *not* make *all three* inequalities true.

9. Shade the area that makes each inequality true. The intersection of all the shaded areas is the solution region.

Exploration 3

1.

	Number of scoops	Amount of fat (g)	Amount of salt (mg)
Meat	x	10x	9x
Rice	y	2y	6y
Meat & Rice	$x + y$	10x + 2y	9x + 6y

2. a. $10x + 2y \leq 20$
b. $9x + 6y \leq 30$

3. $x > 0,\ y > 0$

4.

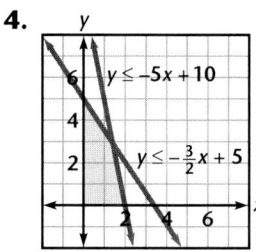

5. 1, 2, or 3

6. 1

7. 2; 0; no, a meat-and-rice dish implies both items are included in the dish

8. 5; 0; no, a meat-and-rice dish implies both items are included in the dish

9. 0; 0; no, neither meat nor rice are used and a meat-and-rice dish implies both items are included in the dish.

Communicate

1. Yes; the two inequalities could be parallel lines with shaded regions in opposite directions which do not intersect.

2. The shaded region where all of the individual solution regions intersect is completely surrounded by boundary lines.

3. An unbounded solution region has individual solution regions which all extend infinitely in at least one direction.

4. Answers may vary. An example is

	Number hours	Amount earned, ($)
Programming	x	20x
Tutoring	y	10y
Total for week	$x + y$	20x + 10y

The total amount earned, $20x + 10y$, must be at least $500. The total hours worked, $x + y$, can be at most 40.

5. $x > 0$ and $y > 0$

6. $\begin{cases} x + y \leq 40 \\ 20x + 10y \geq 500 \end{cases}$;

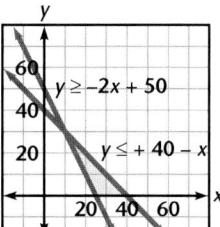

bounded

7. Answers may vary. An example of a point in the solution region is (22, 11) where 22 represents 22 hours of programming and 11 represents 11 hours of tutoring. Angela will make $22(20) + 11(10) = \$550$ and she will work $22 + 11 = 33$ hours.

8. Answers may vary. Examples follow:
Angela should program 40 hours and do no tutoring. This way, she makes $40(20) = \$800$, the most money possible.
Angela should program for 25 hours. This is the least amount of time she can spend working and still make $25(20) = \$500$.
Angela should spend 20 hours programming and 20 hours tutoring for variety.

Practice & Apply

14.

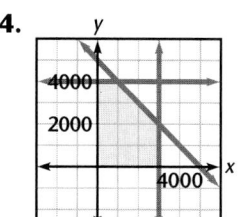

15. The inequality $x + y \leq 5000$ changes to $x + y \leq 5500$. The total capacity for both types of seats is increased by 500, which expands the region of possible values. Now, the auditorium can sell 3000 reserved tickets and 2500 regular admission tickets for the most profit.

16.

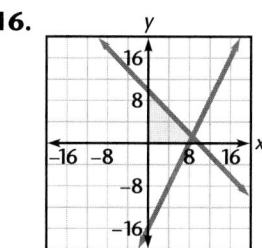

17.

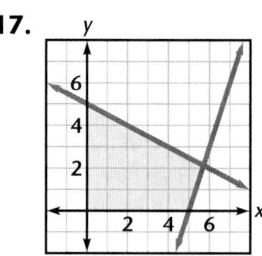

18.

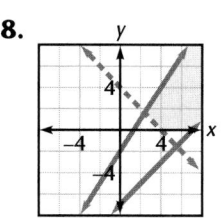

19.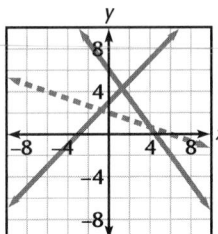

There is no region where all three inequalities overlap, so there is no solution to this system.

22. The possible combinations of jackets are:

Leather	Imitation
15	0
14	0 – 4
13	0 – 8
12	0 – 12
11	0 – 16
10	0 – 20
9	0 – 24
8	0 – 28
7	0 – 32
6	0 – 36
5	0 – 40
4	0 – 44
3	0 – 48
2	0 – 52
1	0 – 56
0	0 – 60

Look Back

36. $x = \frac{3}{5}y + 3$; $y = \frac{5}{3}x - 5$

37. $x = \frac{-2y + 13}{7}$; $y = \frac{-7x + 13}{2}$

38. $x > y + 9$; $y < x - 9$

Look Beyond

45. Answers may vary. An example is *feasible* means "capable of being done or carried out;" a synonym for *feasible* is *possible*.

Lesson 4.9, pages 228–233

Exploration

1.

Profit	Profit equation	Slope-intercept form
400	400 = 15r + 8s	$s = -\frac{15}{8}r + 50$
600	600 = 15r + 8s	$s = -\frac{15}{8}r + 75$
800	800 = 15r + 8s	$s = -\frac{15}{8}r + 100$
1000	1000 = 15r + 8s	$s = -\frac{15}{8}r + 125$
1200	1200 = 15r + 8s	$s = -\frac{15}{8}r + 150$

2. The slopes are the same whereas the y-intercepts are different for each profit equation.

3.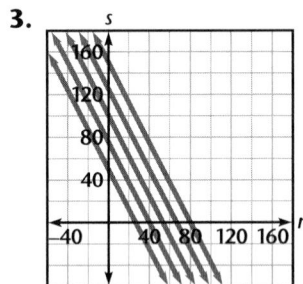

All five profit equations have the same slope, therefore, the lines of the equations are parallel. Each line has a unique x- and y-intercept.

4. Answers may vary. Using the point (40, 20) from the feasible region of Example 1 on page 228; $P = 760$.

5. Answers may vary. Using the sample answer from Step 4, the profit equation is $s = -\frac{15}{8}r + 95$.

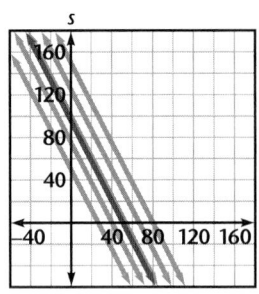

6. Yes, because the coordinates (r, s) are taken from the feasible region.

Communicate

1. The feasible region would be shifted to the left and would have vertices (20, 10), (20, 30), (50, 30), (60, 20), and (60, 10).

2. The profit function would change to $P = 30r + 50s$, or $s = -\frac{3}{5}r + \frac{P}{50}$. The profit lines would still be parallel, but they would have shallower slope of −0.6.

Practice & Apply

6.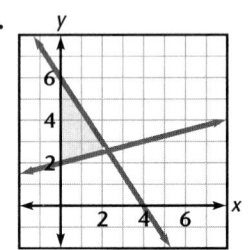

Four possible points are (0, 3), (1, 3), (1, 4), and (0, 5).

7. Answers may vary. With $P = 5x + 3y$, examples are:
- Using the point (1, 4) from the feasible region of Exercise 3: $P = 17$
- Using the point (2, 3) from the feasible region of Exercise 4: $P = 19$
- Using the point (2, 7) from the feasible region of Exercise 5: $P = 31$
- Using the point (1, 3) from the feasible region of Exercise 6: $P = 14$

8. Answers may vary. With $P = 3x + 2y$, examples are:
- Using the point (1, 2) from the feasible region of Exercise 3: $P = 7$
- Using the point (1, 3) from the feasible region of Exercise 4: $P = 9$
- Using the point (2, 4) from the feasible region of Exercise 5: $P = 14$
- Using the point (1, 4) from the

feasible region of Exercise 6:
$P = 11$

9. Let d represent the number of downhill skis. Let c represent the number of cross-country skis.

	d $(d \geq 0)$	c $(c \geq 0)$	d and c
Hours to fabricate	$6d$	$4c$	$6d + 4c \leq 108$
Hours to finish	$1d$	$1c$	$1d + 1c \leq 24$
Profit	$40d$	$30c$	$P = 40d + 30c$

10. $\begin{cases} 6d + 4c \leq 108 \\ d + c \leq 24 \\ d \geq 0 \\ c \geq 0 \end{cases}$

11. $P = 40d + 30c$

12. Answers may vary. Examples, where ordered pairs represent (d, c), are
(2, 15): $P = 530$;
(15, 2): $P = 660$;
(18, 5): $P = 870$;
(5, 18): $P = 740$.

13. Let n represent the number of nonstop flights. Let d represent the number of direct flights.

	n	d	d and c
Number of planes	$n \geq 0$	$d \geq 0$	$n + d \leq 10$
Number of pasengers	$150n$	$100d$	$150n + 100d \leq 1200$
Cost	$1200(150)n$	$900(100)d$	$C = 1200(150)n + 900(100)d$

17. Let m represent the number of acres of millet. Let a represent the number of acres of alfalfa.

	m	a	m and a
# of acres	$m \geq 0$	$a \geq 0$	$m + a \leq 90$
Seed costs	$4m$	$6a$	$4m + 6a \leq 480$
Labor costs	$20m$	$10a$	$20m + 10a \leq 1400$
Income	$110m$	$150a$	$I = 110m + 150a$

21. Let v represent the number of vans. Let b represent the number of buses.

	v	b	v and b
# of vehicles	$v \geq 0$	$b \geq 0$	$v + b \leq 10$
Vehicle cost	$10{,}000v$	$20{,}000b$	$10{,}000v + 20{,}000b \leq 100{,}000$
Maintenance cost	$100v$	$75b$	$100v + 75b \leq 500$
Total riders	$15v$	$25b$	$T = 15v + 25b$

Lesson 4.10, Pages 234–241

Exploration

1.

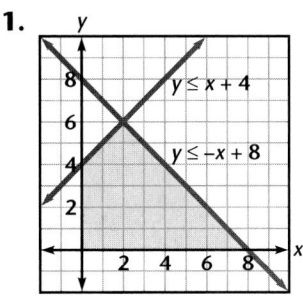

2.

P	Objective equation	Slope-intercept form
6	$6 = 2x + 3y$	$y = -\frac{2}{3}x + 2$
12	$12 = 2x + 3y$	$y = -\frac{2}{3}x + 4$
18	$18 = 2x + 3y$	$y = -\frac{2}{3}x + 6$
21	$21 = 2x + 3y$	$y = -\frac{2}{3}x + 7$
24	$24 = 2x + 3y$	$y = -\frac{2}{3}x + 8$
30	$30 = 2x + 3y$	$y = -\frac{2}{3}x + 10$

3.

Intersects: $6 = 2x + 3y$, $12 = 2x + 3y$, $18 = 2x + 3y$, and $21 = 2x + 3y$
Does not intersect: $24 = 2x + 3y$, and $30 = 2x + 3y$

4. Feasible: 6, 12, 18, 21; Not feasible: 24, 30; approx. $0 \leq P \leq 22$

5. Maximum $P = 22$; the maximum value of P intersects the feasible region at (2, 6) which is a vertex of the feasible region. Minimum $P = 0$; the minimum value of P intersects the feasible region at (0, 0) which is a vertex of the feasible region.

6. At one of the vertices of the feasible region.

Communicate

1. At a vertex of the feasible region

2. Determine the vertices of the feasible region. Find the value of the objective function at each vertex. The greatest and the least values are the maximum and the minimum values, respectively.

3. Use the trace or solve feature to find the coordinates of the

 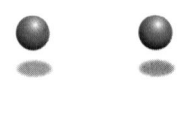 

vertices of the feasible region. Find the value of the objective function at each vertex. The greatest and the least values are the maximum and the minimum values, respectively.

Look Back

29.

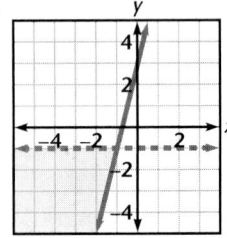

30.

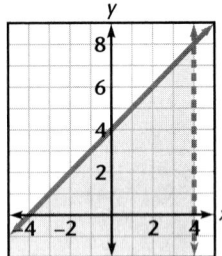

31.

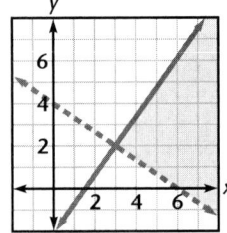

Look Beyond

37. (3, 0) and (–2, 0)

38. (–4, 0) and (9, 0)

39. (0, 0) and (–12, 0)

37–39. For each function of the form $f(x) = (x - a_1)(x - a_2)$, the graph crosses the x-axis at the values of a_1 and a_2.

Lesson 5.1, pages 254–261

Exploration 1

1. Domain for $w(x)$ in the real world: $x > 0$

2. Domain for $l(x)$ in the real world: $0 < x \le 10$

3. $A(x)$ is not a linear function. Domain for $A(x)$ in the real world: $0 < x \le 10$

4. The maximum possible area is $A = 50$ square yards; $w = 5$ yards, length = 10 yards.

5. The x-intercepts are:
$w(x)$: (0, 0)
$l(x)$: (10, 0)
$A(x)$: (0, 0) and (10, 0)

6. The x-intercepts of the area function are the x-intercepts of $w(x)$ and $l(x)$.

Exploration 2

1.

x	Ticket Price	Attendance	Revenues
–2	3	340	$1020
–1	4	320	$1280
0	5	300	$1500
1	6	280	$1680
2	7	260	$1820
3	8	240	$1920
4	9	220	$1980
x	$5 + x$	$300 - 20x$	$(5 + x)(300 - 20x)$

2. $t(x) = 5 + x$, a linear function

3. $a(x) = 300 - 20x$, a linear function

---

4. $r(x) = (5 + x)(300 - 20x) = 10(-2x^2 + 20x + 150)$; a quadratic function; y-intercept is (0, 1500); the y-intercept indicates that the revenue is $1500 when the increase in price is 0.

5. The coordinates of the maximum point on the graph are (5, 2000). The student council should increase the ticket price by $5. The new price will be $10, and the maximum revenue will be $2000.

6. x-intercept of $t(x)$ is (–5, 0); x-intercept of $a(x)$ is (15, 0); x-intercepts of $r(x)$ are (–5, 0) and (15, 0). The x-intercepts of the revenue function, r, are the combined x-intercepts of the price and attendance functions, t and a.

Exploration 3

1. & 2. The graph of the revenue function is increasing for $x < 5$ and decreasing for $x > 5$.

3. r changes from increasing to decreasing at $x = 5$. Any parabola changes from increasing to decreasing or from decreasing to increasing at its vertex.

Exploration 4

1.

	x-intercepts	midpoint of x-coordinates of x-intercepts	vertex	x-coordinate of vertex
$y = (x - 4)(x + 2)$	(4, 0), (–2, 0)	1	(1, –9)	1
$y = (x + 1)(x + 6)$	(–1, 0), (–6, 0)	–3.5	(–3.5, –6.25)	–3.5
$y = (x - 3)(x - 5)$	(3, 0), (5, 0)	4	(4, –1)	4

2. The midpoint of the x-coordinates of the x-intercepts is the same as the x-coordinate of the vertex. Thus, the vertex is located on a vertical line that passes midway between the x-intercepts.

3. x-intercepts: (–1, 0), (9, 0); x-coordinate of the vertex: 4; answers may vary, but students should find that their predictions were correct.

Communicate

1. If width is represented by $w(x) = x$, then the length is represented by $l(x) = \frac{60 - 2x}{2}$. The area = $wl = x(30 - x)$. The maximum area occurs at the vertex of the area function, (15, 225). Thus, the maximum area occurs when the width *and* length are 15 yards long, a square pen. The maximum area is 225 square yards.

2. If the function can be written in the form $f(x) = ax^2 + bx + c$ and $a \neq 0$, the function is quadratic.

3. Let x = the number of 50¢ increases in price. Then, sales = $s(x) = 30 - 2x$, selling price = $p(x) = 5 + (0.50)x$, and revenue = $r(x) = s(x) \cdot p(x) = (30 - 2x)(5 + (0.50)x)$. The linear components are $30 - 2x$, and $5 + (0.50)x$.

4. To maximize $r(x)$, find the vertex of its graph, (2.5, 156.25). So, Julia should charge $5 + (0.50)(2.5)$, or \$6.25 per dozen. Her revenue will be maximized at \$156.25 per day.

5. If a function f is increasing, when x increases, $f(x)$ also increases. If a function f decreases when x increases, $f(x)$ is decreasing.

6. To find the x-intercepts of a quadratic function f, find the zeros of the function. Find the midpoint of the x-intercepts to find the x-coordinate of the vertex. Name this value x_0. Then the y-coordinate of the vertex is $f(x_0)$.

Lesson 5.2, Pages 262–268

Communicate

1. Since $x^2 = a$, where $a > 0$, is satisfied by $x = \sqrt{a}$ and $x = -\sqrt{a}$, both roots must be included.

2. When both sides of an equation are squared, extraneous roots may be introduced.

3. Possible roots should be checked by substitution into the original equation. When the substitution yields a false statement, the root substituted is extraneous and is therefore not a correct response.

4. If the variable represents a quantity that is known to be positive, such as length or price, then only the principal root is used.

5. Use the "Pythagorean" Right-Triangle Theorem: $c^2 = 3^2 + 4^2$, so the hypotenuse, $c = \sqrt{25}$, or 5.

6. Use the distance formula, $d = \sqrt{(x_1 - x_2)^2 + (y_1 - y_2)^2}$, and the coordinates of the two points, where the points are represented in the formula as (x_1, y_1) and (x_2, y_2), to find the distance between the points.

Practice & Apply

41. $R(-4, 1)$, $S(5, 4)$, $T(2, -2)$
$RS = \sqrt{(5 + 4)^2 + (4 - 1)^2} = \sqrt{90}$,
$ST = \sqrt{(5 - 2)^2 + (4 + 2)^2} = \sqrt{45}$,
$RT = \sqrt{(2 + 4)^2 + (-2 - 1)^2} = \sqrt{45}$.
Since $ST = RT$, the triangle is isosceles.

42. $A(-4, 1)$, $B(5, 4)$, $C(2, -2)$
$AB = \sqrt{(5 + 4)^2 + (4 - 1)^2} = \sqrt{90}$,
$BC = \sqrt{(5 - 2)^2 + (4 + 2)^2} = \sqrt{45}$,
$AC = \sqrt{(2 + 4)^2 + (-2 - 1)^2} = \sqrt{45}$.
$BC^2 + AC^2 = 45 + 45 = 90$, and $AB^2 = 90$.
Therefore, since $BC^2 + AC^2 = AB^2$, the triangle is a right triangle.

Lesson 5.3, Pages 269–275

Exploration 1

1.

	$f(x) = \frac{1}{2}x^2$	$f(x) = x^2$	$f(x) = 2x^2$	$f(x) = 5x^2$
Vertex	(0, 0)	(0, 0)	(0, 0)	(0, 0)
Axis of symmetry	y-axis	y-axis	y-axis	y-axis

As the coefficient of x^2 increases, the parabola becomes narrower.

2.

	$f(x) = -\frac{1}{2}x^2$	$f(x) = -x^2$	$f(x) = -2x^2$	$f(x) = -5x^2$
Vertex	(0, 0)	(0, 0)	(0, 0)	(0, 0)
Axis of symmetry	y-axis	y-axis	y-axis	y-axis

A negative coefficient of x^2 causes the parabola to open downward.

3.

	$f(x) = x^2 + 4$	$f(x) = x^2 + 1$	$f(x) = x^2 - 2$	$f(x) = x^2 - 5$
Vertex	(0, 4)	(0, 1)	(0, -2)	(0, -5)
Axis of symmetry	y-axis	y-axis	y-axis	y-axis

For $f(x) = x^2 + 3$, the vertex is (0, 3); the axis of symmetry is the y-axis.

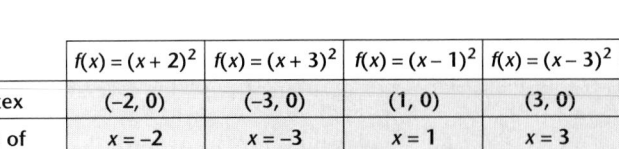

4.

	$f(x) = (x + 2)^2$	$f(x) = (x + 3)^2$	$f(x) = (x - 1)^2$	$f(x) = (x - 3)^2$
Vertex	(–2, 0)	(–3, 0)	(1, 0)	(3, 0)
Axis of symmetry	$x = -2$	$x = -3$	$x = 1$	$x = 3$

For $f(x) = (x - 6)^2$, the vertex is (6, 0), and the axis of symmetry is $x = 6$.

5.

	$f(x) = (x - 1)^2 + 3$	$f(x) = (x - 1)^2 - 3$	$f(x) = (x + 1)^2 + 3$	$f(x) = (x + 1)^2 - 3$
Vertex	(1, 3)	(1, –3)	(–1, 3)	(–1, –3)
Axis of symmetry	$x = 1$	$x = 1$	$x = -1$	$x = -1$

For $f(x) = (x + 4)^2 - 2$, the vertex is (–4, –2), and the axis of symmetry is $x = -4$.

6.

	$f(x) = 3(x - 1)^2 - 2$	$f(x) = -2(x + 3)^2 - 1$	$f(x) = -\frac{1}{2}(x + 2)^2 + 3$	$f(x) = -\frac{1}{3}(x - 4)^2$
Vertex	(1, –2)	(–3, –1)	(–2, 3)	(4, 0)
Axis of symmetry	$x = 1$	$x = -3$	$x = -2$	$x = 4$

For $f(x) = -6(x - 4)^2 - 5$, the vertex is (4, –5), and the axis of symmetry is $x = 4$.

7. If $f(x) = a(x - h)^2 + k$, the vertex is at (h, k), and the axis of symmetry is $x = h$. If $a > 0$, the parabola opens upward, if $a < 0$, the parabola opens downward. As $|a|$ increases, the parabola becomes narrower.

Communicate

1. Use $y = a(x - h)^2 + k$ where $h = -3$, $k = 2$, and $a > 0$: $y = a(x + 3)^2 + 2$, where $a > 0$.

2. For any quadratic function written in the form $y = a(x - h)^2 + k$, the coordinates of the vertex are (h, k), and the axis of symmetry is $x = h$. So for $f(x) = -5(x + 3)^2 - 2$, the vertex is (–3, –2), and the axis of symmetry is $x = -3$.

3. For any quadratic function written in the form $f(x) = a(x - h)^2 + k$, the graph opens upward if $a > 0$ and downward if $a < 0$.

4. Let $f(x) = 0$ and solve the resulting quadratic equation. If $-3(x + 1)^2 - 7 = 0$, then $(x + 1)^2 = -\frac{7}{3}$. This equation has no solution, so the graph has no x-intercepts.

5. The x-intercepts represent the times when the object will reach the ground. Generally, only positive values of time are realistic solutions. The maximum height of the object occurs at the vertex of the graph of the function. Since $h(t) = -16t^2 + 5t + 12$, use your graphics calculator to find the positive x-intercept, (1.04, 0), and the vertex, (0.16, 12.39). The object will strike the ground after approximately 1 second, and the maximum height of the object is approximately 12.4 feet.

6. Use the quadratic model for the height of a falling object affected by gravity, $h(t) = -\frac{1}{2}gt^2 + v_0t + h_0$. The initial velocity, v_0, of a dropped object is 0, and the initial height, h_0, is given as 10 feet. So the equation that models the height of the object in feet is $h(t) = -16t^2 + 10$.

Lesson 5.4, Pages 276–281

Communicate

1. Take twelve x-tiles and line half of them horizontally beside one side of an x^2-tile and the other half

vertically below the x^2-tile. There are now six x-tiles on two sides of the x^2-tile, so add 6^2, or 36, unit tiles to complete the square. $x^2 + 12x + 36$ is a perfect-square trinomial.

2. Leave the variable terms on the left of the equal sign, and move the constant term, –13, to the left side of the equal sign: $x^2 + 4x = 13$. Add the square of half the coefficient of x, 4, to both sides of the equation: $x^2 + 4x + 4 = 13 + 4$. The trinomial on the left is now a perfect square. Write it as the square of a binomial: $(x + 2)^2 = 17$. Take the square root of both sides of the equation, and solve the resulting linear equations for x: $x + 2 = \pm\sqrt{17}$, so $x = \sqrt{17} - 2$ or $x = -\sqrt{17} - 2$.

3. Divide both sides of the equation by the coefficient of x^2, 2: $x^2 + 2x = 7.5$. Add the square of half the coefficient of x, 1, to both sides of the equation: $x^2 + 2x + 1 = 7.5 + 1$. The trinomial on the left is now a perfect square. Write it as the square of a binomial: $(x + 1)^2 = 8.5$. Take the square root of both sides of the equation and solve the resulting linear equations for x: $x + 1 = \pm\sqrt{8.5}$, so $x = \sqrt{8.5} - 1$ or $x = -\sqrt{8.5} - 1$.

4. Divide 20 x-tiles into four groups of five, and arrange them on the 4 sides an x^2-tile. The area of this shape is 108. To complete a square that will have side length of $5 + x + 5$, add 25 unit tiles in all four corners. The area of this square shape is $108 + 4(25)$, or 208. In terms of the x-tiles, the area of the square is $(x + 10)^2$, so by solving the equation $(x + 10)^2 = 208$, you can find the solutions for x: $x = \sqrt{208} - 10$ or $x = -\sqrt{208} - 10$.

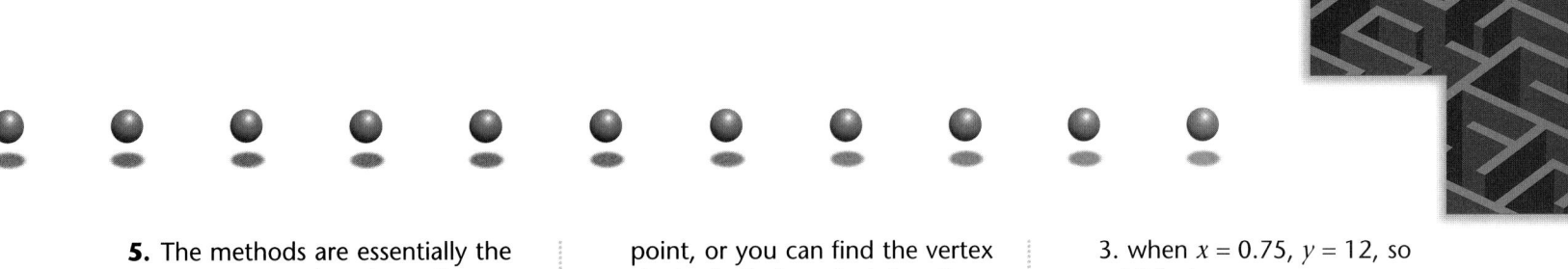

5. The methods are essentially the same except that the x-tiles are divided into four groups, placed on the four sides of an x^2-tile, in the al-Khowarizmi method and into two groups, placed on two adjacent sides of an x^2-tile, in the modern method. In both cases, the new figure is filled out with unit tiles to form a square that has side length $x + \frac{b}{2}$, where b is the coefficient of x in the original equation.

Lesson 5.5, Pages 282–287

Communicate

1. Move all terms to the same side of the equal sign: $-2x^2 + 3x + 8 = 0$. Now substitute $a = -2$, $b = 3$, and $c = 8$ into the quadratic formula, and simplify to find the roots of the equation:
$$x = \frac{-3 \pm \sqrt{73}}{-4}$$

2. Answers may vary but must include three of the following:
- You can draw the graph of the function and find the point(s) at which the parabola intersects the x-axis;
- You can model the equation using algebra tiles with the modern method;
- You can model the equation using algebra tiles with the al-Khowarizmi method;
- You can find the roots of $f(x) = 0$ algebraically by factoring;
- You can find the roots of $f(x) = 0$ algebraically by completing the square; or
- You can find the roots of $f(x) = 0$ algebraically using the quadratic formula.

3. You can draw the graph of the function and find the coordinates of the maximum or minimum point, or you can find the vertex algebraically by calculating the value of $x = -\frac{b}{2a}$ and substituting this value into f to determine the corresponding y-value for the coordinates of the vertex.

4. The axis of symmetry formula can be used to determine the x-coordinate of the vertex. Then substitute this x-value into the function to get the corresponding y-value, which is the minimum or maximum value.

5. Answers may vary, but most students will choose the quadratic formula because it seems to be less work and lends itself to the use of the calculator. Many students will appreciate that the quadratic formula is just the end result of completing the square on a general quadratic equation. Using the quadratic formula saves steps.

Practice & Apply

42. Since the maximum height occurs after 6 seconds, the pyrotechnician would want the fireworks to begin firing after $6 - 2.5$, or 3.5 seconds. If the sparks are to occur equally on either side of the maximum, then they should start firing after $6 - 1.25$, or 4.75 seconds.

43. To find the equation of the parabolic curve shown on page 253, let $h(t) = at^2 + bt + c$ and choose three points from which to calculate the constants a, b, and c. Using $(0, 3)$, $(0.7, 11.96)$, and $(1.2, 8.76)$, $h(t) = -16t^2 + 24t + 3$. Answers to questions from page 253:
1. when $y = 10$, x is about 1.104, so $\approx$ 1.1 sec
2. when $y = 0$, x is about 1.616, so $\approx$ 1.6 sec
3. when $x = 0.75$, $y = 12$, so $\approx$ 12 feet.
Answers may vary.

Lesson 5.6, Pages 288–293

Exploration

1. Even powers of i are real, odd powers of i are imaginary numbers. Numbers with exponents that are multiples of 4 have value 1, other numbers with even exponents have value -1.

2.

$i^9 = i$	$i^{13} = i$	$i^{17} = i$
$i^{10} = -1$	$i^{14} = -1$	$i^{18} = -1$
$i^{11} = -i$	$i^{15} = -i$	$i^9 = -i$
$i^{12} = 1$	$i^{16} = 1$	$i^{17} = 1$

3. When the powers of i become greater than 4, the cycle of i, i^2, i^3, and i^4 is repeated. $i^{4a} = 1$ and $i^{4a-2} = -1$, where a is an integer. i^n is a positive real number if n is divisible by 4, and it is a negative real number if n is even but not divisible by 4.

Communicate

1. Calculate the value of the discriminant, $D = b^2 - 4ac$. If $D > 0$, there are two real-number solutions; if $D = 0$, there is one real-number solution; and if $D < 0$, there are no real-number solutions.

2. The graph of a quadratic equation with a negative discriminant lies completely above or below the x-axis; the graph of a quadratic equation with a discriminant equal to zero touches the x-axis at one point. Both graphs are parabolas.

3. $\sqrt{-a} = \sqrt{a(-1)} = i\sqrt{a}$, where a is a real number and $a > 0$.

4. No; $\sqrt{-2} \cdot \sqrt{-3} = i\sqrt{2} \cdot i\sqrt{3} = -\sqrt{6}$, which is not equal to $\sqrt{6}$.

5. If n is divisible by 4, then $i^n = 1$.

6. Isolate the x^2 term: $x^2 = -5$. Then take the square root of both sides: $x = \pm\sqrt{-5}$. The right side is the square root of a negative number which may be expressed as an imaginary number using i to replace $\sqrt{-1}$: $\sqrt{-5} = \sqrt{-1}\sqrt{5} = i\sqrt{5}$. Now, solve for x: $x = i\sqrt{5}$ or $x = -i\sqrt{5}$.

Lesson 5.7, Pages 294–301

Communicate

1. The set of complex numbers consists of numbers of the form $a + bi$, where a and b are real numbers. If $b = 0$, then the number is real. If $a = 0$, then the number is imaginary.

2. Equate the real parts to find the solution for x and the imaginary parts to find the solution for y: $2x = 9$, or $x = 4.5$; $5i = 20yi \rightarrow 5 = 20y$, or $y = 0.25$.

3. $(a + bi) \pm (c + di) = (a \pm c) + (b \pm d)i$. Add, or subtract, the real and imaginary parts separately. $(a + bi)(c + di) = ac + (bc + ad)i + bdi^2 = (ac - bd) + (bc + ad)i$. Find the product in the same way as multiplying two binomials, then collect the real and imaginary parts.

4. If the graph of $f(x) = ax^2 + bx + c$ has no x-intercepts, then the solutions of $ax^2 + bx + c = 0$ are complex conjugates. The real parts are the same; the imaginary parts have opposite signs.

5. The conjugate of $a + bi$ is $a - bi$. Complex conjugates will have the same real parts and opposite imaginary parts.

6. The horizontal axis is used to represent the real part and the vertical axis is used to represent

the imaginary part. To graph the complex number $a + bi$, plot the point (a, b) on the coordinate system described.

Practice & Apply

18. Real part: -5; imaginary part: $6i$

19. Real part: 0; imaginary part: $6i$

20. Real part: 8; imaginary part: $0i$

21. Real part: 0; imaginary part: $4i$

26.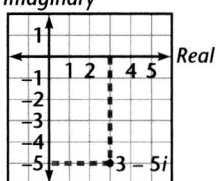

27.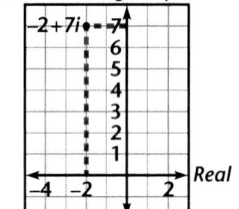

28.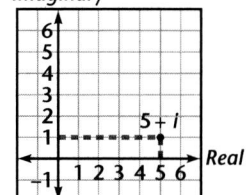

29.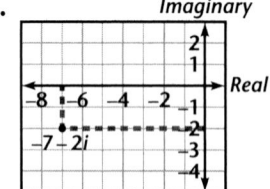

30.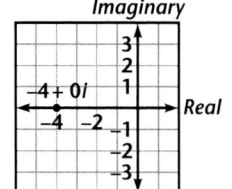

31.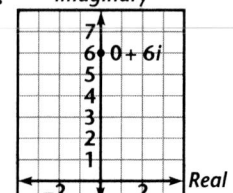

32.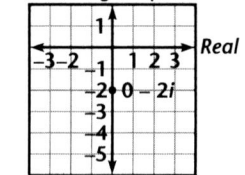

33.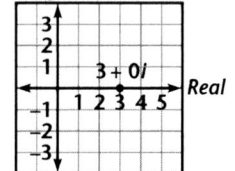

34.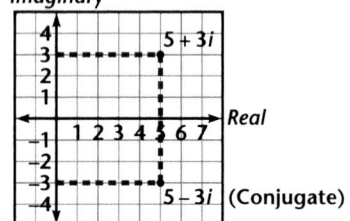

35.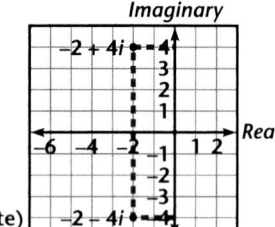

36.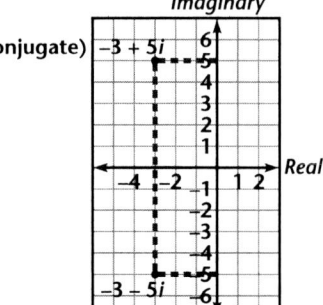

37. *Imaginary*

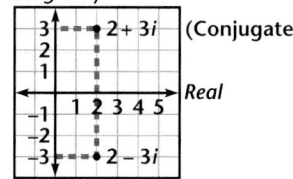

38. *Imaginary*

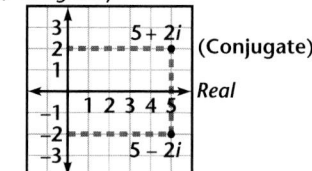

39. *Imaginary*

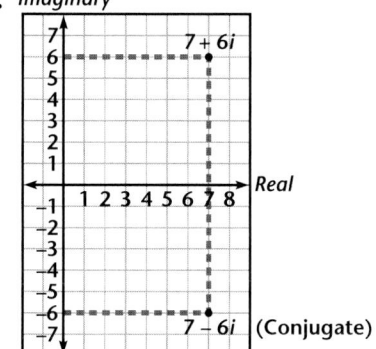

40.

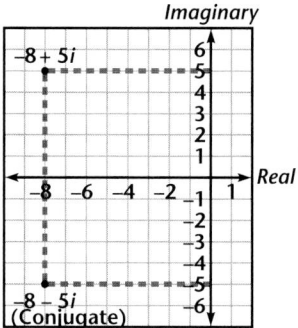

41.

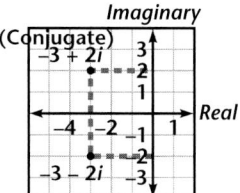

46. $-2 - 4i$; $3 + i$; $-3 + 3i$. A complex number and its conjugate are symmetric with respect to the real axis.

47. $(-2 + 4i)(-2 - 4i) = 20$; $(3 - i)(3 + i) = 10$; $(-3 - 3i)(-3 + 3i) = 18$.

The product is on the real axis and is the square of the distance of the point (or its conjugate) from the origin: e.g., for the point $3 - i$, the distance from the origin is $\sqrt{3^2 + 1^2}$, or $\sqrt{10}$.

Look Beyond

61.

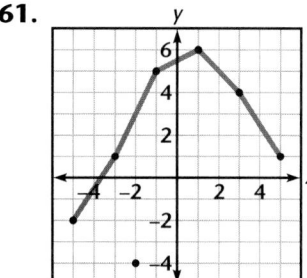

The graph appears to be approximately parabolic. At $x = 8$, y would be about -7.

62. Answers may vary. A sample answer is: The predicted values are different because a linear function is a poor approximation of the data. As Exercise 61 reveals, the data are much better approximated by a quadratic function.

Lesson 5.8, Pages 302–307

Communicate

1. Use the general equation for a parabola, $y = ax^2 + bx + c$, and substitute the x- and y-values from the given data, one point at a time, to obtain three equations in three unknowns, a, b, and c: $a + b + c = 6$; $9a - 3b + c = 8$; $25a + 5b + c = -2$

2. Solve the three equations for a, b, and c, then substitute these values into the general equation for a parabola to get $y = -0.1875x^2 - 0.875x + 7.0625$.

3. Check without a calculator by substituting the given (x, y) pairs into the equation to see if the

equation holds true for *all* pairs. With a graphics calculator, graph the function and check that the given values lie on the curve.

Lesson 5.9, Pages 308–315

Exploration 1

1. The zeroes of $f(x) = x^2 - 2x - 3$ are $x = -1$ and $x = 3$.

2. Answers may vary. Examples are given.

Region 1		Region 2		Region 3	
x	$x^2 - 2x - 3$	x	$x^2 - 2x - 3$	x	$x^2 - 2x - 3$
-4	21	0	-3	4	5
-3	12	1	-4	5	12
-2	5	2	-3	6	21

3. Region 2, where $-1 < x < 3$, contains values that result in $x^2 - 2x - 3 < 0$. Region 1, where $x < -1$, and Region 3, where $x > 3$, both contain values that result in $x^2 - 2x - 3 > 0$.

4. $f(x) > 0$: $x < -1$ or $x > 3$
$f(x) < 0$: $-1 < x < 3$
These are the same values for x as in Step 3.

5. $x^2 - 2x - 3 > 0$: With a graphics calculator, use the trace or solve feature to find the x-intercepts and, from the graph, determine values of x for which the graph is above the x-axis.
Without a graphics calculator, find the zeros of the function, and test x-values on either side of the zeros to determine in which region(s) the function is greater than zero.
$x^2 - 2x - 3 < 0$: The methods are similar except, on the graphics calculator, trace to determine values of x for which the graph is below the x-axis. Using the algebraic method, test values of x to determine where the function is less than zero.

6. The graph of the solutions to $x^2 - 2x - 3 < 0$ does not include the zeros of $x^2 - 2x - 3 = 0$; whereas, the graph of the solutions to $x^2 - 2x - 3 \le 0$ does include the zeros. Similarly, the graph of the solutions to $x^2 - 2x - 3 > 0$ does not include the zeros, while the graph of the solutions to $x^2 - 2x - 3 \ge 0$ does.

Exploration 2

1. & 2. The revenue line is above the cost parabola for $10 < x < 40$. Kate needs to sell between 10 and 40 T-shirts per week for the revenue to exceed the cost.

3. $x = 10$ or $x = 40$.

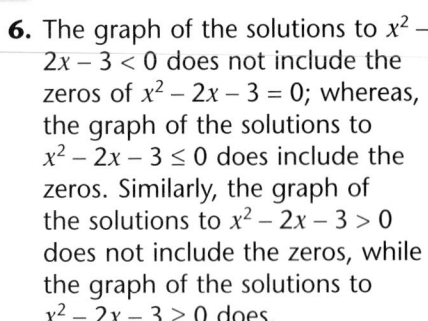

Region 1 contains points less than 10, Region 2 contains points in the interval $10 < x < 40$, and Region 3 contains points greater than 40.

4.

	x	Revenue $R(x) = 10x$	Cost $C(x) = 0.1x^2 + 5x + 40$	True or False $R(x) > C(x)$?
Region 1 $x < 10$	0	0	40	False
Region 2 $10 < x < 40$	20	200	180	True
	30	300	280	True
Region 3 $x > 40$	50	500	540	False

Region 2, where $10 < x < 40$, contains function values where $R(x) > C(x)$.

5. $R(x) > C(x)$: With a graphics calculator, graph the two functions $Y_1 = R(x)$ and $Y_2 = C(x)$. Use the trace or solve feature to find their points of intersection. By observation, record the region(s), on either side of or between the points of intersection for which the graph of $R(x)$ lies above the graph of $C(x)$. Without a graphics calculator, find the points of intersection by solving the equation $R(x) = C(x)$. Then test x-values in each region to determine where $R(x) > C(x)$. The values of x for which $R(x) < C(x)$ are found in a similar way, except that, with a graphics calculator, you can observe where $R(x)$ is below $C(x)$.

Communicate

1. Graph $Y_1 = x^2 - 2x - 8$ and trace or solve to find the zeros. The zeros are the endpoints of the regions where the graph is above the x-axis, i.e. where $x^2 - 2x - 8 > 0$. The solutions are $x < -2$ or $x > 4$.

2. Solve the equation $x^2 - 2x - 8 = 0$. Use the two zeros, -2 and 4, to divide the number line into three regions. Test an x-value from each of the three regions to determine in which region(s) $x^2 - 2x - 8 > 0$. The solutions are $x < -2$ or $x > 4$.

3. First solve the equation $x^2 - 2x - 8 = 4x + 3$. Use the two solutions to divide the real number line into three regions. Test an x-value from each region to determine where $x^2 - 2x - 8 < 4x + 3$. The solutions are $-1.5 < x < 7.5$.

4. The graph is the region bounded by the parabola $f(x) = x^2 - 2x - 8$ and the linear function $g(x) = 4x + 3$, and contains all the points for which $f(x) < g(x)$: $-1.5 < x < 7.5$.

5. The zeros of f are -1 and 4. If $f(x) > 0$ when $-1 < x < 4$ then f must be above the x-axis in this region, i.e. it must have a maximum between the two zeros. Thus $f(x) = a(x + 1)(x - 4) > 0$ or $a(x^2 - 3x - 4) > 0$ where $a < 0$.

6. $R(x) = 12x$;
$P(x) = -0.1x^2 + 7x - 40$

Lesson 6.1, pages 326–331

Exploration 1

1. The zero of the width function is 0; since $w > 0$, the realistic domain is $x > 0$.

2. The zero of the length function is 9; since $l > 0$, the realistic domain is $0 < x < 9$.

3. The zero of the height function is -1; since $h > 0$, the realistic domain is $x > -1$.

4. V is neither linear (does not have a constant slope) nor quadratic (V has more than two zeros); $V(x) = 0$ if x is 0, 9, or -1.

5. Realistic domain of V: $0 < x < 9$

6. The maximum of V, to the nearest tenth, is 126.2 ft^3. The corresponding dimensions are width: 5.8 feet, length: 3.2 feet, and height: 6.8 feet. These dimensions are not reasonable for most people because the compost bin would be too tall.

Exploration 2

• $f(x) = \frac{1}{5}(x)(x - 1)(x + 2)(x - 3)$

1. There are four linear factors. The constant is $\frac{1}{5}$.

2. The zeros of f are 0, 1, -2, and 3. The zeros are the r-values.

3. $f(x) = \frac{1}{5}x^4 - \frac{2}{5}x^3 - x^2 + \frac{6}{5}x$; the highest power of x is 4, which is

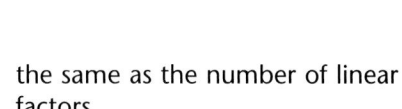

the same as the number of linear factors.

- $g(x) = -(x + 1)(x - 2)(x + 3)$

  1. There are three linear factors. The constant is –1.

  2. The zeros of g are –1, 2, –3. The zeros are the r-values.

  3. $g(x) = -x^3 - 2x^2 + 5x + 6$; the highest power of x is 3, which is the same as the number of linear factors.

- $h(x) = 2(x + 3)(x - 1)$

  1. There are two linear factors. The constant is 2.

  2. The zeros of h are –3, 1. The zeros are the r-values.

  3. $h(x) = 2x^2 + 4x - 6$; the highest power of x is 2, which is the same as the number of linear factors.

- $j(x) = -\frac{3}{2}(x + 1)$

  1. The is one linear factor. The constant is $-\frac{3}{2}$.

  2. The zero of j is –1. The zero is the r-value.

  3. $j(x) = -\frac{3}{2}x - \frac{3}{2}$; the highest power of x is 1, which is the same as the number of linear factors.

- $k(x) = 3$

  1. There are no linear factors. The constant is 3.

  2. There are no zeros of k. There are no r-values.

  3. $k(x) = 3$; the highest power of x is 0, which is the same as the number of linear factors.

- $l(x) = (x - 3)(x + 1)$

  1. There are two linear factors. The constant is 1.

  2. The zeros of l are 3, and –1. The zeros are the r-values

3. $l(x) = x^2 - 2x - 3$; the highest power of x is 2, which is the same as the number of linear factors.

Communicate

1. A polynomial function is any function that can be expressed as a product of linear factors multiplied by a constant.

2. The r-values of the linear factors of a polynomial function are the roots or zeros of the function.

3. The factored form of a polynomial function is $f(x) = C(x - r_1)(x - r_2) \ldots (x - r_{n-1})(x - r_n)$ where C is a constant and the subscripted r-values, $r_1, r_2, \ldots r_n$, are the roots, or zeros, of the function. The degree of the polynomial in factored form is equal to the number of factors.

4. The expanded form of a polynomial function is $f(x) = a_n x^n + a_{n-1} x^{n-1} + \ldots + a_2 x^2 + a_1 x + a_0$, where the subscripted a-terms are the coefficients. The coefficient a_n is called the leading coefficient of the polynomial and is the same as the value of the constant C in factored form. a_0 is a constant coefficient. It is also the y-intercept of the function. The degree of a polynomial function in expanded form is the highest power of the variable with a nonzero coefficient.

5. A quadratic function can be expressed in the form $a_2 x^2 + a_1 x + a_0$, where $a_2 \neq 0$. Therefore, a quadratic function is a polynomial function.

6. A constant linear function has the form $f(x) = a_0$. A nonconstant linear function has the form $f(x) = a_1 x + a_0$, where $a_1 \neq 0$. Therefore, all linear functions are polynomial functions.

3. $l(x) = x^2 - 2x - 3$; the highest power of x is 2, which is the same as the number of linear factors.

Look Beyond

37. In each pair of linear factors, the form is $(x + r)(x - r)$. The values of r differ.

42. a. $x^2 - 16 = (x + 4)(x - 4)$
b. $x^2 - 36 = (x + 6)(x - 6)$
c. $x^2 - 49 = (x + 7)(x - 7)$
d. $x^2 - 256 = (x + 16)(x - 16)$

Lesson 6.2, Pages 332–340

Exploration 1

1. $(x + 1)(x - 1) = x^2 - 1$
$(x + 4)(x - 4) = x^2 - 16$
$(5x + 2)(5x - 2) = 25x^2 - 4$
The product is the square of the first term minus the square of the second term. $(a + b)(a - b) = a^2 - b^2$.

2. The zeros of f are ± 1, the zeros of g are ± 4, and the zeros of h are ± 0.4. The zeros of each function are the same as the r-values in the linear factors from Step 1.

3. $f(x) = x^2 - a^2 = (x + a)(x - a)$. The zeros of f are $\pm a$.

Exploration 2

1. zeros of f are –0.5 and 3

2. $r_1 = -\frac{1}{2}$, $r_2 = 3$; they are the same.

3. $(x - r_1)(x - r_2) = \left(x + \frac{1}{2}\right)(x - 3) = x^2 - \frac{5}{2}x - \frac{3}{2}$. The product is f divided by 2. If C is 2, then $2\left(x^2 - \frac{5}{2}x - \frac{3}{2}\right) = f(x)$. C is the same as the leading coefficient of f.

4. $x^2 - \frac{5}{2}x - \frac{3}{2} = \frac{f(x)}{2}$. The roots are $x = 3$ and $-\frac{1}{2}$, the same as found in Step 2.

5. Use the quadratic formula to find the roots of g: $x = \frac{1}{5}, -\frac{5}{2}$. Now, $\left(x - \frac{1}{5}\right)\left(x + \frac{5}{2}\right)$ is the factored

form of $\frac{1}{10}g$. The factored form of g is $g(x) = 10\left(x - \frac{1}{5}\right)\left(x + \frac{5}{2}\right)$. Clear fractions to get g factored over the integers, $g(x) = (5x - 1)(2x + 5)$

Communicate

1. The roots of a polynomial can be easily determined from the factored form of the polynomial.

2. The zeros of the linear factors of a polynomial are the zeros of the polynomial function.

3. A quadratic polynomial of the form $x^2 + 2kx + k^2 = (x + k)^2$ with $k \neq 0$ is a perfect square trinomial. A quadratic polynomial of the form, $x^2 - a^2 = (x - a)(x + a)$ is a special product called the difference of two squares.

4. yes; by multiplying the constant, C, by the common denominator in the linear factors

5. no; if the discriminant is not a perfect square, it is an irrational number. Therefore, no common denominator could be found for the linear factors.

Look Beyond

47.
$$
\begin{array}{r}
412 \\
28\overline{)11536} \\
\underline{112} \\
33 \\
\underline{28} \\
56 \\
\underline{56} \\
0
\end{array}
$$

48. The divisor is subtracted from the dividend, beginning at the left. Thus 4 times 28 is subtracted from 115, leaving a remainder of 3. The next digit of the quotient is brought down, and 1 times 28 is subtracted from 33, leaving 5. Finally 6 is brought down from the quotient, giving 56. 2 times 28 is subtracted from 56, leaving a remainder of 0.

The repeated steps are that, one digit at a time, the quotient is divided by the divisor, and the remainder at each of these steps is combined with the next digit of the quotient until there are no digits left in the quotient.

Lesson 6.3, Pages 341–349

Communicate

1. The graph of a polynomial function indicates real zeros of the function at each x-intercept of the graph of the function.

2. Locate the zeros and write the linear factors using these zeros. For example, if 2 and –1 are zeros of a function f, then $f(x) = (x - 2)(x + 1)$.

3. As a result of the Fundamental Theorem of Algebra, a polynomial of degree n has exactly n zeros, including multiplicities. Therefore, since each zero corresponds to a linear factor, a polynomial of degree n will have n linear factors.

4. A polynomial of degree n has a maximum of n complex zeros if n is an even positive integer. If n is 0, there will be no real zeros and no complex zeros. If n is 2, there can be 2 real zeros (including multiplicities) and no complex zeros, or no real zeros and 2 complex zeros. If n is 4, there can be 4 real zeros and no complex zeros, or 2 real zeros and 2 complex zeros, or no real zeros and 4 complex zeros. If n is 6, there can be 6 real zeros and no complex zeros, 4 real zeros and 2 complex zeros, 2 real zeros and 4 complex zeros, or no real zeros and 6 complex zeros. In general, if n is a positive even integer, then there will be an even number of both the

real and the complex zeros. Because complex zeros come in pairs of complex conjugates, the number of complex zeros in any polynomial will always be even.

5. A polynomial of degree n, where n is a positive odd integer, has a maximum of $n - 1$ complex zeros and at least one real zero. If n is 1, there will be 1 real zero and no complex zeros. If n is 3, there can be 3 real zeros and no complex zeros, or 1 real zero and 2 complex zeros. If n is 5, there can be 5 real zeros and no complex zeros, 3 real zeros and 2 complex zeros, or 1 real zero and 4 complex zeros. In general, if n is a positive odd integer, for every $n > 1$, there will be 1 real zero and at most $n - 1$ complex zeros, where the number of complex zeros is always even.

Lesson 6.4, Pages 350–355

Exploration 1

• Linear functions
  $f(x) = x$:

1. Increasing everywhere, decreasing nowhere, no turning points.

2. One zero at $x = 0$; a crossing point.

3. The leading coefficient is positive; the far left of the graph has negative values and the far right has positive values.

  $g(x) = x - 2$:

1. Increasing everywhere, decreasing nowhere, no turning points.

2. One zero at $x = 2$; a crossing point.

3. The leading coefficient is positive; the far left of the graph has negative values and the far right has positive values.

$h(x) = -x$:

1. Decreasing everywhere, increasing nowhere, no turning points.

2. One zero at $x = 0$; a crossing point.

3. The leading coefficient is negative; the far left of the graph has positive values and the far right has negative values.

• Quadratic functions
$f(x) = x^2$:

1. Increasing for $x > 0$, decreasing for $x < 0$, turning point at $x = 0$.

2. A zero of multiplicity 2 at $x = 0$; a turning point.

3. The leading coefficient is positive; the graph has positive values at both the far left and the far right ends of the graph.

$g(x) = x(x - 2)$:

1. Increasing for $x > 1$, decreasing for $x < 1$, turning point at $x = 1$.

2. Zeros at $x = 0$ and at $x = 2$; both are crossing points.

3. The leading coefficient is positive; the graph has positive values at both the far left and the far right ends of the graph.

$h(x) = -x^2$

1. Increasing for $x < 0$, decreasing for $x > 0$, turning point at $x = 0$.

2. A zero of multiplicity 2 at $x = 0$; a turning point.

3. The leading coefficient is negative; the graph has negative values at both the far left and the far right ends of the graph.

• Cubic functions
$f(x) = x^3$:

1. Increasing everywhere, decreasing nowhere, no turning points.

2. A zero of multiplicity 3 at $x = 0$; a crossing point.

3. The leading coefficient is positive; the graph has negative values at the far left end of the graph and positive values at the far right end of the graph.

$g(x) = x^2(x - 2)$

1. Increasing for $x < 0$ and $x > 1.33$, decreasing for $0 < x < 1.33$; 2 turning points at $x = 0$ and $x = 1.33$.

2. A zero of multiplicity 2 at $x = 0$, a turning point, and a zero at $x = 2$, a crossing point.

3. The leading coefficient is positive; the graph has negative values at the far left end of the graph and positive values at the far right end of the graph.

$h(x) = x(x - 2)(x + 4)$

1. Increasing for $x < -2.4$ and $x > 1.1$, decreasing for $-2.4 < x < 1.1$; 2 turning points at $x = -2.4$ and at $x = 1.1$.

2. 3 zeros at $x = 0$, $x = 2$, and $x = -4$; all 3 zeros are crossing points.

3. The leading coefficient is positive; the graph has negative values at the far left end of the graph and positive values at the far right end of the graph.

$k(x) = -x^3$:

1. Decreasing everywhere, increasing nowhere, no turning points.

2. A zero of multiplicity 3 at $x = 0$; a crossing point.

3. The leading coefficient is negative; the graph has positive values at the far left end of the graph and negative values at the far right end of the graph.

• Quartic functions

$f(x) = x^4$:

1. Increasing for $x > 0$, decreasing for $x < 0$, turning point at $x = 0$.

2. A quadruple zero at $x = 0$, a turning point.

3. The leading coefficient is positive; the graph has positive values at both the far left end and the far right end of the graph.

$g(x) = x^3(x - 2)$:

1. Increasing for $x > 1.5$, decreasing for $x < 1.5$.

2. A zero of multiplicity 3 at $x = 0$, a crossing point; a zero at $x = 2$, a crossing point.

3. The leading coefficient is positive; the graph has positive values at both the far left end and the far right end of the graph.

$h(x) = x^2(x - 2)(x + 4)$

1. Increasing for $x > 1.39$, $-2.89 < x < 0$; decreasing for $0 < x < 1.39$ and $x < -2.89$; f has 1 turning point.

2. A zero of multiplicity 2 at $x = 0$, and zeros at $x = 2$, and $x = -4$; the zero at $x = 0$ is a turning point; the zeros at $x = 2$ and $x = -4$ are crossing points.

3. The leading coefficient is positive; the graph has positive values at both the far left end and the far right end of the graph.

$k(x) = x(x - 2)(x + 4)(x - 6)$

1. Increasing for $-2.6 < x < 1$ and $x > 4.6$, decreasing for $x < -2.6$ and $1 < x < 4.6$; three turning points at $x = -2.6$, 1, and 4.6.

2. Zeros at $x = 0$, 2, -4, and 6; all are crossing points.

3. The leading coefficient is positive; the graph has positive values at both the far left end and the far right end of the graph.

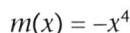

$m(x) = -x^4$

1. Increasing for $x < 0$, decreasing for $x > 0$, turning point at $x = 0$.

2. A zero of multiplicity 4 at $x = 0$; one turning point at $x = 0$.

3. The leading coefficient is negative; the graph has negative values at both the far left end and the far right end of the graph.

	Degree 1: Linear	Degree 2: Quadratic	Degree 3: Cubic	Degree 4: Quartic
Domain	all Reals	all Reals	all Reals	all Reals
Range	all Reals	≥ min for $a > 0$; ≤ max for $a < 0$	all Reals	≥ y-value of lowest turning point for $a > 0$, ≤ y-value of highest turning point for $a < 0$
Num. of distinct zeros	1	0, 1, or 2	1, 2, or 3	0, 1, 2, 3, or 4
Num. of turning points	0	1	0 or 2	1 or 3
Num. of crossing points	1	0 or 2	1 or 3	0, 2, or 4
Far left and far right behaviour	$a > 0$: upward to the right $a < 0$: downward to the right	$a > 0$: upward at the left and the right $a < 0$: downward at the left and the right	$a > 0$: downward at the left, upward at the right $a < 0$: upward at the left, downward at the right	$a > 0$: upward at the left and the right $a < 0$: downward at the left and the right

4. The quadratic and quartic functions have an odd number of turning points and an even number of crossing points. With a positive leading coefficient, both open up. All polynomials of even degree have these characteristics.

5. Linear and cubic functions have an even number of turning points (where 0 is considered an even number) and an odd number of crossing points. With a positive leading coefficient, the graphs are negative at the far left and positive at the far right. All polynomials of odd degree have these characteristics.

Exploration 2

1. & 2. f_1 has one turning point zero at $x=3$ and no crossing

points. f_2 has one turning point at $x=1.5$ and crossing points at $x=1$ and $x=3$. f_3 has turning points at $x=1$, $x=2$, and $x=3$ and no crossing points. f_4 has turning points at $x=1$, $x=-1.15$, and $x=2.40$ and crossing points at $x=-2$ and $x=3$.

3. A zero of odd multiplicity is a crossing point, and a zero of even multiplicity is a turning point.

4. Answers may vary. An example is $f(x) = (x-1)(x+2)(x-4)^2$. In this example, there are 3 turning points, one of which is a zero. For any example, a quartic function with exactly 2 crossing points will have one turning point zero.

5. Answers may vary. An example is $f(x) = (x+3)^2(x-2)^2$. All students should find that their function

has one additional turning point. For this example, it is located at $(-0.5, 39.1)$.

Communicate

1. Zeros at $x = 2, -4$ and 5; degree is 3.

2. No; the function is a cubic; it has 3 crossing points and 2 turning points. It must be both increasing and decreasing for some values of x.

3. far left: downward far right: upward

4. The graph changes from increasing to decreasing or from decreasing to increasing at a turning point. f does not have any zeros that are turning points.

5. At a crossing point, the value of the function changes signs; i.e. the graph crosses the x-axis. f has 3 crossing points.

6. Yes; if the zeros have even multiplicity. For example: $y = x^4$, or $y = (x-2)^2(x+2)^2$. These functions have turning point zeros, but no crossing points.

Lesson 6.5, Pages 358–365

Communicate

1. $V(x) = (6 - 2x)(4 - 2x)x$

2. From the graph of a real-world function, the realistic domain occurs when the function has positive values *and* when the domain gives each linear factor of the function positive, or realistic, values.

3. Up to, but not including, 6 inches

4. Graph and trace the volume function. For each value of x (the length of the side of the cut-out square), the y-value gives the volume of the box with x as the

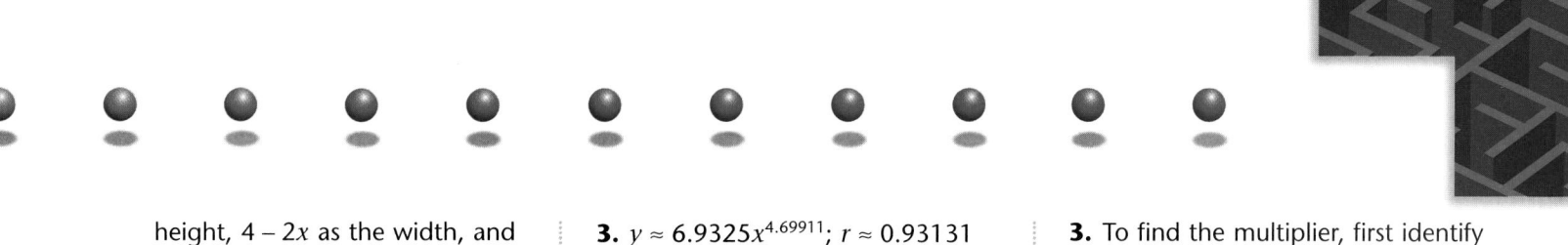

height, $4 - 2x$ as the width, and $6 - 2x$ as the length.

5. The volume function increases for $x < \approx 0.8$ inches and for $x > \approx 2.55$ inches and decreases for about 0.8 inches $< x < 2.55$ inches.

6. $\$400(1.05)^3 = \$463.05$

Lesson 7.1, Pages 376–382

Exploration 1

1–4. answers given in the form (hour, population): (3, 800), (4, 1600), (5, 3200), (6, 6400), (7, 12,800), (8, 25,600), (9, 51,200), (10, 102,400), (11, 204,800), (12, 409,600), (13, 819,200), (14, 1,638,400), (15, 3,276,800), (16, 6,553,600), (17, 13,107,200), (18, 26,214,400), (19, 52,428,800), (20, 104,857,600)

5. 102,400; 104,857,600

6. During the 14th hour; during the 17th hour; during the 20th hour

7. 2

Exploration 2

1–3. (1992, 490,149), (1993, 502,893), (1994, 515,968), (1995, 529,383), (1996, 543,147), (1997, 557,269), (1998, 571,758), (1999, 586,624), (2000, 601,876)

4. 601,876

5. During the year 2020

6. 1.026

Exploration 3

1. & 2. $y \approx 2680573x - 17,660,273$; $r \approx 0.61924$

3. $y \approx 6.9325x^{4.69911}$; $r \approx 0.93131$

4. $y = 100(2^x)$; $r = 1$

5. Modeling the bacteria population growth with an exponential function gives a correlation coefficient of 1, which means that this model is a perfect fit to the data.

6. Let the year 1992 be represented by 2 and begin with the data point (1, 477728). Then, the linear regression: $y \approx 13,786x + 461858$; $r \approx 0.99936$; power regression: $y \approx 460,427x^{0.10089}$; $r \approx 0.95166$; exponential regression; $y \approx 465,622(1.026^x)$; $r \approx 0.9999999999$. The exponential function best models the growth, with a correlation coefficient that is almost exactly 1. (The coefficient would have been exactly 1 if the population values had not been rounded to the nearest whole number.)

Communicate

1. To model population growth using a calculator, first enter the original population. Then multiply by the percentage the population grows in the given time interval to find the new population. Continue to calculate successive populations by repeatedly pressing ENTER, since doing so multiplies each previous population by the percentage, or multiplier.

2. After entering population data on a calculator's statistical registers, choose various regression models (linear, power, exponential, etc.) for the data. For each type of regression, the calculator presents a correlation coefficient. The function that best models the population growth is the one that gives a correlation coefficient closest to 1.

3. To find the multiplier, first identify the original population and the new population after the given time period. Determine the multiplier by dividing the new amount by the original amount. For example, in Exploration 1, the original amount of bacteria is 100. After 1 hour, the bacteria doubled, so the new amount is 200. 200 divided by 100 is 2, the growth multiplier for the number of bacteria. In Exploration 2, the original population of the city is 465,622 people. After 1 year, the new population is 477,728 people. 477728 divided by 465,622 is 1.026, the growth multiplier for the population of the city.

4. The multiplier of 2 in the bacteria growth experiment gives an exponential curve that very rapidly increases in steepness. The multiplier of 1.026 in the city growth model also gives an exponential curve that increases in steepness, but much more gradually than the bacteria growth curve.

5. The population is decreasing.

Lesson 7.2, Pages 383–390

Communicate

1. An exponential function has a constant base raised to a variable power. The base must be positive and not equal to one. The domain is the set of real numbers. For $a > 0$ in $f(x) = ab^x$, the range is the set of positive real numbers. The y-intercept is a.

2. For $a > 0$ in $f(x) = ab^x$, f is an exponential decay function if $0 < b < 1$, and an exponential growth function if $b > 1$. Though both have the set of all positive reals as their range, the decay

function continually decreases as x increases, while the growth function continually increases as x increases.

3. An asymptote is a line to which a graph gets closer and closer without ever touching.

4. The domain of an exponential function is the set of real numbers. The range of $f(x) = ab^x$ depends on a. For $a > 0$, the range is the set of positive real numbers. For $a < 0$, the range is the set of negative real numbers.

5. The terms of a polynomial function are constant powers of a variable, while an exponential function contains variable powers of a constant.

6. Using the definition of negative exponents, $\frac{1}{4} = 4^{-1}$. Therefore, $\left(\frac{1}{4}\right)^x = (4^{-1})^x = 4^{-x}$.

Look Back

50.

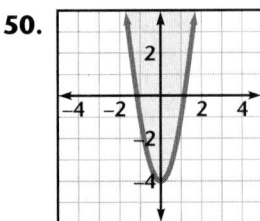

51.

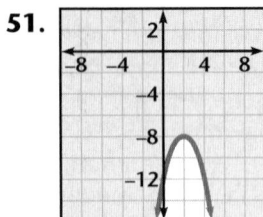

52.

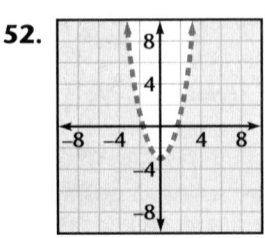

Look Beyond

53. The domain of $x = 4^y$ is the set of all positive real numbers, and the range is the set of all real numbers. It is a function since it is always increasing, and therefore passes the vertical line test.

Lesson 7.3, Pages 391–395

Exploration

1. Table 1: 8, 16, 32, 64; Table 2: 1, 2, 3, 4, 5, 6

2. The tables contain the same row values, but the x-values in Table 1 are the y-values in Table 2, and the y-values in Table 1 are the x-values in Table 2.

Communicate

1. For any base b, where $b > 0$ and $b \neq 1$, the exponential and logarithmic functions, $f(x) = b^x$ and $g(x) = \log_b x$, are inverse functions.

2. The expression $\log_b x$ represents the power of b that gives x. So, the logarithm is an exponent.

3. The domain of a logarithmic function is the set of positive real numbers, and the range is the set of real numbers.

4. For $b > 0$, $\log_b 1$ is 0, since 0 is the power for any base b that gives 1.

5. A value in the domain of an exponential function is an exponent, and the corresponding range element is the value of that power of a given base. The logarithmic function is the inverse of an exponential function, so its range values are the domain values of the exponential function. Thus, the logarithm values are exponents.

Lesson 7.4, Pages 396–401

Exploration 1

1. 4, 5, 6, 7

2. a. Each expression equals 3.
 b. Each expression equals 4.
 c. Each expression equals 5.

3. The examples in Step 2 indicate that $\log_2(ab) = \log_2 a + \log_2 b$.

4. Answers may vary. One possible answer is that since $512 = 8(64)$, then
$$\log_2 512 = \log_2[8(64)]$$
$$= \log_2 8 + \log_2 64$$
$$= 3 + 6$$
$$= 9.$$

Exploration 2

1. a. Each expression equals 3.
 b. Each expression equals 1.
 c. Each expression equals 2.

2. The examples in Step 1 indicate that $\log_2\left(\frac{a}{b}\right) = \log_2 a - \log_2 b$. The logarithm of a quotient is the difference of the logarithms of the numerator and the denominator of that quotient.

3. $\log_3\left(\frac{243}{9}\right) = \log_3 243 - \log_3 9$
$$= 5 - 2$$
$$= 3$$

Exploration 3

1. a. $\log_b 3^2 = \log_b(3 \cdot 3)$
$$= \log_b 3 + \log_b 3$$
$$= 2\log_b 3;$$
 b. $\log_b 5^3 = \log_b(5 \cdot 5 \cdot 5)$
$$= \log_b 5 + \log_b 5 + \log_b 5$$
$$= 3\log_b 5;$$
 c. $\log_b 7^4 = \log_b(7 \cdot 7 \cdot 7 \cdot 7)$
$$= \log_b 7 + \log_b 7 + \log_b 7 + \log_b 7$$
$$= 4\log_b 7.$$

2. $p \log_b m$

3. The logarithm of a number raised to a power is that power times

the logarithm of the number.

4. $\log_{10} 100^4 = 4 \log_{10} 100$
$= 4(2) = 8$

Communicate

1. The logarithm of a product of two positive numbers is the sum of the logarithms of the two numbers.

2. The logarithm of the quotient of two positive numbers is the difference of the logarithms of the numerator and the denominator of that quotient.

3. The logarithm of a number raised to a power is the power times the logarithm of the number.

4. By the associative property, $\log_5 uvw = \log_5 u(vw)$. Then, by the product property, $\log_5 u(vw) = \log_5 u + \log_5 vw$. Applying the product property to the second term gives $\log_5 vw = \log_5 v + \log_5 w$, so $\log_5 uvw = \log_5 u + \log_5 v + \log_5 w$.

5. First, apply the product property:
$\log_3 x(x + 3)^2 = \log_3 x + \log_3(x + 3)^2$
Now apply the power property to the second term:
$\log_3 x(x + 3)^2 = \log_3 x + 2\log_3(x + 3)$

6. Apply the reverse of the quotient property:
$\log_4(x + 3)^2 - \log_4 3(x + 3)$
$= \log_4 \left(\frac{(x + 3)^2}{3(x + 3)} \right)$.
Factor out $x + 3$ from the numerator and denominator:
$\log_4(x + 3)^2 - \log_4 3(x + 3)$
$= \log_4 \left(\frac{x + 3}{3} \right)$, where $x \ne -3$.

Lesson 7.5, Pages 401–406

Communicate

1. Napier's rods allowed the user to find the products, quotients, and powers of large numbers

by adding, thus reducing the number and difficulty of the calculations.

2. The common logarithmic function has base 10.

3. The inverse function of $f(x) = \log x$ is $f^{-1}(x) = 10^x$.

4. The domain of $y = \log x$ is the set of all positive real numbers, and the range is the set of all real numbers. The domain of $f^{-1}(x) = 10^x$ is all real numbers, and the range is the positive real numbers. The domains and ranges of the two functions are reversed.

5. To find common logs using a calculator, use the ⌐LOG⌐ key.

Lesson 7.6, Pages 407–413

Exploration

1. $f(1) = 2$; $g(1) = 2$

2. $f(10) = 1.1$; $g(10) \approx 2.5937$

3. $f(100) = 1.01$; $g(100) \approx 2.7048$

4. $f(1000) = 1.001$;
$g(1000) \approx 2.7169$

5. $f(10,000) = 1.0001$;
$g(10,000) \approx 2.7181$

6. As n becomes larger, g seems to approach e.

Communicate

1. The natural exponential base is the number e, which is about 2.718.

2. Continuous compounding of interest is an example of exponential growth with e as a base. Radioactive decay is an example of exponential decay with e as a base.

3. The natural logarithmic function and the natural exponential

function are inverse functions.

4. The function $f(x) = ce^{ax}$, where c and a are positive constants, describes continuous exponential growth. When c represents the initial amount, a the interest rate, and x the time period, f gives the formula for the amount of money resulting from continuous compounding of interest.

5. The function $f(x) = ce^{ax}$, where c is a positive constant and a is a negative constant, describes continuous exponential decay. When c represents the original amount of a radioactive isotope, a equals $-k$, the opposite of the decay constant for the isotope, and x is the time period, f gives the formula for the remaining amount of an isotope undergoing radioactive decay after a certain time period.

6. The half-life of a radioactive substance is the time it takes the substance to break down until only half the mass of the original isotope is remaining.

Lesson 7.7, Pages 416–421

Communicate

1. To solve an exponential equation by graphing, put the equation in the form of an exponential expression equal to a constant. Then graph the expression, and trace or solve to find the value of the variable when the expression equals the constant.

2. To solve an exponential equation without graphing, it helps to simplify the exponential expression. You can use the definition of an exponential function, and the properties of exponential functions. It will often help to take the logarithm

of both sides and use the exponential-log inverse property to solve.

3. To solve a logarithmic equation by graphing, put the equation in the form of a logarithmic expression equal to a constant. Then graph the expression, and trace or solve to find the value of the variable when the expression equals the constant.

4. To solve a logarithmic equation without graphing, it helps to use the product, quotient, and power properties of logarithms, and the definition of the logarithmic function, so that the equation contains only one logarithmic expression. Then, use the exponential-log inverse property to solve.

5. Often, you encounter logarithms of an expression containing a variable with exponents containing variable expressions. The inverse relation between exponential and logarithmic functions allows you to simplify these equations so that the expressions with variables are isolated, and thus are no longer part of a logarithmic or exponential expression.

Lesson 8.1, Pages 430–437

Exploration 1

1–3. $m\angle B = 90°$, $m\angle D = 90°$, $m\angle F = 90°$; $m\angle C = 45°$, $m\angle E = 45°$, $m\angle G = 45°$

4. Answers may vary. For each triangle, the length of the vertical leg and the horizontal leg should be the same, the hypotenuse should be about 1.414 times the length of either leg, and the ratio of either leg to the hypotenuse should be about 0.707.

5. The ratios are the same (about 0.707) for all three triangles.

6. 45°; because the legs of the right triangle are the same length, the angles opposite them must have the same measure, as is true of the triangles in Steps 3 and 4. So, $m\angle X = 0.5(180° - 90°) = 45°$.

7. hypotenuse: $\sqrt{2}$; $\frac{XY}{YZ} = \frac{1}{\sqrt{2}} \approx$ 0.707; The ratios should be the same.

8. $RT = a\sqrt{2}$; $ST = a$

Exploration 2

1–3. $m\angle B = 90°$, $m\angle D = 90°$, $m\angle F = 90°$; $m\angle C = 60°$, $m\angle E = 60°$, $m\angle G = 60°$

4. Answers may vary. For each triangle, the length of the horizontal leg should be about 1.732 times the length of the vertical leg, the hypotenuse should be about twice as long as the vertical leg, the ratio of the length of the vertical leg to the length of the hypotenuse should be about one-half, and the ratio of the length of the horizontal leg to the length of the hypotenuse should be about 0.866.

5. For each triangle, the ratios of the length of the vertical leg to the length of the hypotenuse (0.5) should be the same, and the ratios of the length of the horizontal leg to the length of the hypotenuse (about 0.866) should be the same.

6. 30°; the ratio of the length of the vertical leg to the length of the hypotenuse is the same as that of the triangles in Steps 3 and 4, so the angles opposite them must be the same. In all of them, the side opposite the vertical (shorter) leg has a measure of 30°.

7. $XY = \sqrt{3}$; $\frac{XY}{XZ} = \frac{\sqrt{3}}{2} \approx 0.866$; The ratios should be the same.

8. $RS = a\sqrt{3}$; $RT = 2a$

Communicate

1. Using the 45-45-90 triangle relationships, the length of each leg is represented by a, so in this case, $a = 5$. The length of the hypotenuse is represented by $a\sqrt{2}$, so the hypotenuse has length $5\sqrt{2}$, or about 7.07.

2. Using the 30-60-90 triangle relationships, the length of the shorter leg is represented by a, so in this case, $a = 4$. The length of the longer leg is represented by $a\sqrt{3}$, so the longer leg has length $4\sqrt{3}$, or about 6.93. The length of the hypotenuse is represented by $2a$, so the hypotenuse has length $2 \cdot 4$, or 8.

3. In a 45-45-90 triangle, the length of each leg opposite a 45° angle has length a, and the hypotenuse has length $a\sqrt{2}$. Since the sine ratio is given by $\frac{\text{opp}}{\text{hyp}}$, $\sin 45° = \frac{a}{a\sqrt{2}} = \frac{1}{\sqrt{2}}$, or $\frac{\sqrt{2}}{2}$, which is about 0.707.

4. In a 30-60-90 triangle, the length of the leg adjacent to the 30° angle has length $a\sqrt{3}$, and the hypotenuse has length $2a$. Since the cosine ratio is given by $\frac{\text{adj}}{\text{hyp}}$, $\cos 30° = \frac{a\sqrt{3}}{2a} = \frac{\sqrt{3}}{2} \approx 0.866$.

5. The ratio of any two sides of a 30-60-90 triangle is the same as the ratio of the corresponding two sides of any other 30-60-90 triangle. Thus, you can set up a proportion to show how one ratio equals the corresponding ratio. This allows you to solve for a missing side in one of the triangles.

Lesson 8.2, Pages 438–445

Communicate

1. For an angle θ in standard position with terminal side intersecting the unit circle at (x, y), the hypotenuse determined by the terminal side of the angle always has a length of one unit. The side adjacent to θ has a length of x, and the side opposite to θ has a length of y. So, $\cos\theta = \frac{x}{1} = x$ and $\sin\theta = \frac{y}{1} = y$, or $(x, y) = (\cos\theta, \sin\theta)$.

2. The coordinates are given by $(x, y) = (\cos 60°, \sin 60°)$.

3. Because $\tan\theta = \frac{\sin\theta}{\cos\theta}$, $\tan 90° = \frac{\sin 90°}{\cos 90°}$. But $\cos 90° = 0$, and division by 0 is not defined, so $\tan 90°$ is undefined.

4. Because 360° is coterminal with 0°, you can use the trigonometric ratios of 0°. Because an angle of 0° intersects the unit circle at $(1, 0)$, and $(x, y) = (\cos\theta, \sin\theta)$, $\cos 360° = 1$, $\sin 360° = 0$, and $\tan 360° = \frac{\sin 360°}{\cos 360°} = \frac{0}{1} = 0$.

Practice & Apply

8. 260°

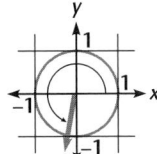

9. 132°

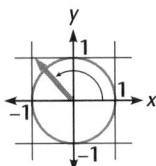

10. −125°

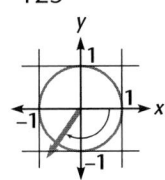

11. 318°

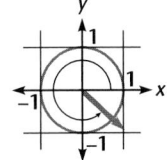

12. −245°

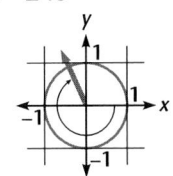

13. 54°

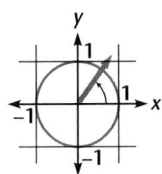

14. 94°

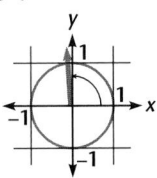

15. −315°

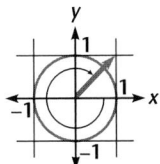

16. −12°

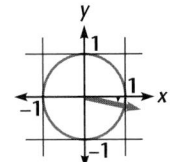

17. 890°

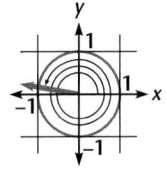

18. −490°

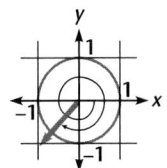

19. −756°

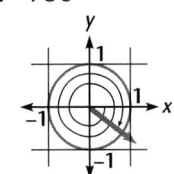

20. 45°

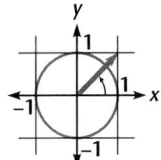

21. 60°

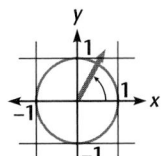

22. 90°

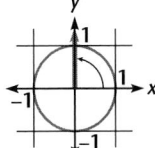

23. −30°

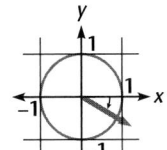

24. 135°

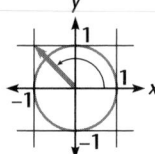

25. 210°

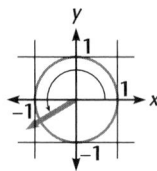

26. 300°

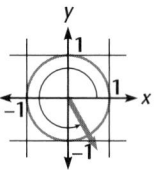

27. 152°

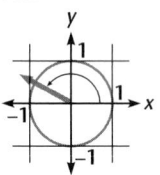

28. −340°

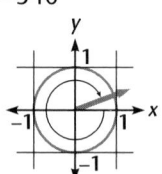

29. −114°

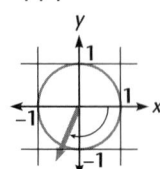

30. 745°

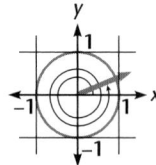

31. −400°

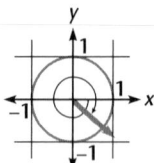

Lesson 8.3, Pages 446–452

Communicate

1. If you know the measure of an acute angle and the length of one of the sides of a right triangle, you can find the lengths of the other sides by using the sine, cosine, or tangent ratios. To find the desired side, write an equation for a trigonometric ratio of the acute angle that involves the given side and the desired side. Then solve for the length of the desired side.

2. When you know the lengths of the sides of a right triangle, you can find the measures of the acute angles by using the inverse trigonometric ratios. Write a ratio of two of the sides as a trigonometric equation involving the desired angle. Then use a calculator to find the inverse trigonometric function value ($\cos^{-1} x$, $\sin^{-1} x$, or $\tan^{-1} x$) for the trigonometric ratio. This value is the desired angle.

3. Because (−0.88, 0.47) = ($\cos x$, $\sin x$), use your calculator to find the angle whose cosine is −0.88, $\cos^{-1}(-0.88)$, or the angle whose sine is 0.47, $\sin^{-1}(0.47)$. Because the terminal side of the angle is in the second quadrant, however, use $\cos^{-1}(-0.88)$. The principal value for an angle with a cosine of −0.88 has terminal side in the second quadrant, but the principal value for an angle with a sine of 0.47 has terminal value in the first quadrant. The

desired angle has a measure of about 152°.

4. Without principal values, the inverse trigonometric relations would not be functions, since there would be infinitely many angles that have the given trigonometric ratio. For example, all coterminal angles have the same trigonometric ratios, so you cannot give a unique angle with that ratio unless you restrict the domain.

5. An angle of 58° has terminal side in the first quadrant. Since the cosine corresponds to the *x*-value on the unit circle, the reflection of the angle over the *x*-axis will have the same cosine. This angle has terminal side in the fourth quadrant, and a measure of −58°. To find the value for an angle coterminal with this angle between 0° and 360°, 360° − 58° = 302°.

6. The functions $f(x) = \cos^{-1} x$ and $f(x) = \sin^{-1} x$ have domains $1 \leq x \leq 1$. The function $f(x) = \tan^{-1} x$ has the set of all real numbers for its domain.

Lesson 8.4, Pages 453–458

Exploration 1

1. domain: all real numbers; range: $1 \leq y \leq 3$; period: 360°

2. domain: all real numbers; range: $-4 \leq y \leq -2$; period: 360°

3. domain: all real numbers; range: $-0.5 \leq y \leq 1.5$; period: 360°

4. For $a > 0$, the graph of $y = \sin \theta$ is translated upward a units. For $a < 0$, the graph of $y = \sin \theta$ is translated downward $|a|$ units.

5. (1) domain: all real numbers; range: $1 \leq y \leq 3$; period: 360° (2) domain: all real numbers; range:

−4 ≤ y ≤ −2; period: 360° (3) domain: all real numbers; range: −0.5 ≤ y ≤ 1.5; period: 360° (4) For a > 0, the graph of y = cos θ is translated upward a units. For a < 0, the graph of y = cos θ is translated downward |a| units.

Exploration 2

1. domain: all real numbers; range: −0.5 ≤ y ≤ 0.5; period: 360°

2. domain: all real numbers; range: −3 ≤ y ≤ 3; period: 360°

3. domain: all real numbers; range: −5 ≤ y ≤ 5; period: 360°

4. For b > 1, the graph of y = b sin θ is the graph of y = sin θ vertically stretched b units. For 0 < b < 1, the effect is a vertical compression of the graph.

5. (1) domain: all real numbers; range: −0.5 ≤ y ≤ 0.5; period: 360° (2) domain: all real numbers; range: −3 ≤ y ≤ 3; period: 360° (3) domain: all real numbers; range: −5 ≤ y ≤ 5; period: 360° (4) For b > 1, the graph of y = b cos θ is the graph of y = cos θ vertically stretched b units. For 0 < b < 1, the effect is a vertical compression of the graph.

Exploration 3

1. The graphs have the same shape, domain, range, and period. The graphs are reflections over the x-axis.

2. The graphs have the same shape, domain, range, and period. The graphs are reflections over the x-axis.

Exploration 4

1. domain: all real numbers; range: −1 ≤ y ≤ 1; period: 720°

2. domain: all real numbers; range: −1 ≤ y ≤ 1; period: 120°

3. domain: all real numbers; range: −1 ≤ y ≤ 1; period: 1440°

4. For c < 1, the graph of y = sin cθ increases in period (stretches horizontally) from the period of y = sin θ, 360°, to $\frac{360°}{c}$. For c > 1, the graph of y = sin cθ decreases in period (compresses horizontally) from a period of 360° to $\frac{360°}{c}$.

5. (1) domain: all real numbers; range: −1 ≤ y ≤ 1; period: 720° (2) domain: all real numbers; range: −1 ≤ y ≤ 1; period: 120° (3) domain: all real numbers; range: −1 ≤ y ≤ 1; period: 1440° (4) For c < 1, the graph of y = cos cθ increases in period (stretches horizontally) from 360° to $\frac{360°}{c}$. For c > 1, the graph of y = cos cθ decreases in period (compresses horizontally) from 360° to $\frac{360°}{c}$.

6. The graphs of y = sin(−θ) and y = − sin θ are the same. The graph of y = − cos θ is a reflection of the graph of y = cos(−θ) with respect to the x-axis.

Exploration 5

1. For d > 0, the graph of y = sin(θ − d) is the graph of y = sin θ translated d units to the right. For d < 0, the graph of y = sin(θ − d) is the graph of y = sin θ translated |d| units to the left.

2. For d > 0, the graph of y = cos(θ − d) is the graph of y = cos θ translated d units to the right. For d < 0, the graph of y = cos(θ − d) is the graph of y = cos θ translated |d| units to the left.

Communicate

1. The period of sin θ is 360°, so the sine of any angle of the form 30° + 360n, for integer n in degrees is $\frac{1}{2}$. For example, sin 390° and sin(−330°) will have the same value as sin 30°, $\frac{1}{2}$.

2. The graph of g(θ) = −3 cos θ is the graph of f(θ) = cos θ stretched vertically by a factor of 3 and reflected over the x-axis.

3. The graph of g(θ) = cos $\frac{1}{2}$ θ is the graph of f(θ) = cos θ stretched horizontally so that the period increases from 360° to 720°.

4. The graph of g(θ) = 5 + cos θ is the graph of f(θ) = cos θ shifted 5 units upward.

5. The graph of g(θ) = − cos $\frac{1}{2}$ θ is the graph of f(θ) = cos θ stretched horizontally so that the period increases from 360° to 720°, and reflected over the x-axis.

8. vertical shift: none; amplitude: 4; period: 360°; phase shift: none

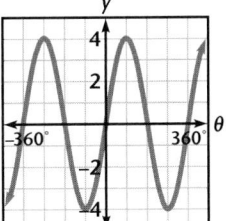

9. vertical shift: none; amplitude: 4 with a reflection over the x-axis; period: 360°; phase shift: none

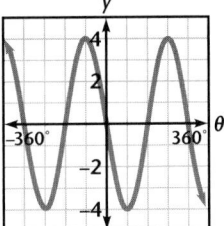

10. vertical shift: 2 units upward; amplitude: 1; period: 360°; phase shift: none

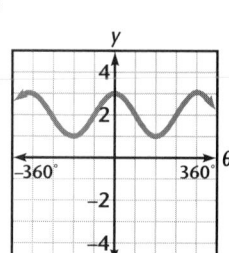

11. vertical shift: none; amplitude: 1; period: 360°; phase shift: 4° to the left

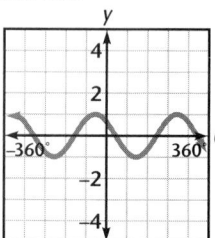

12. vertical shift: none; amplitude: 1; period: 180°; phase shift: none

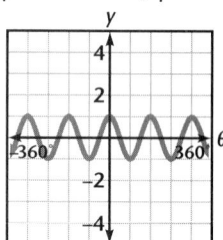

13. vertical shift: 3.5 units upward; amplitude: 1; period: 360°; phase shift: 120° to the left

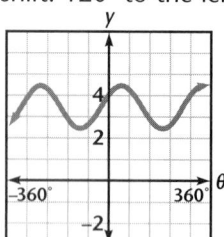

14. vertical shift: none; amplitude: 3; period: 360°; phase shift: 30° to the right

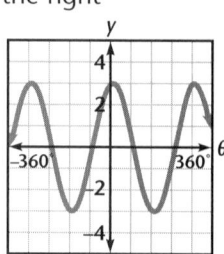

15. vertical shift: none; amplitude: 5.1; period: 360°; phase shift: none

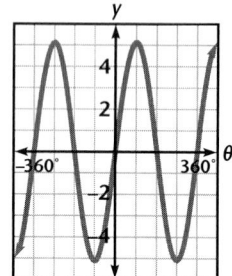

16. vertical shift: 4 units upward; amplitude: 1 with a reflection over the x-axis; period: 720°; phase shift: 270° to the right

19. Answers may vary. A sample answer is that the period of the lying graph is greater than the period of the standing graph, possibly because the giraffe is more relaxed when lying down and his heart beats more slowly. The vertical shift of the lying graph is upward about 60 units compared with the standing graph, possibly because the heart is working against a lower force of gravity at the giraffe's head when the giraffe is standing. When the giraffe is lying down, the force of gravity at the head is higher, so the blood pressure at the head is higher.

Lesson 8.5, Pages 459–467

Exploration

1. missing values given counterclockwise from 0 radians:
$\frac{\pi}{3}, \frac{2\pi}{3}, \frac{5\pi}{6}, \frac{4\pi}{3}, \frac{3\pi}{2}, \frac{5\pi}{3}, \frac{11\pi}{6}$

2. missing values given counterclockwise from 0 radians:
$\frac{3\pi}{4}, \frac{5\pi}{4}, \frac{3\pi}{2}, \frac{7\pi}{4}$

3. $60°; \frac{\pi}{3}$

4. $\frac{1}{12}$ of a rotation is $\frac{2\pi}{12} = \frac{\pi}{6}$ radians, $\frac{5}{12}$ of a rotation is $\frac{5 \cdot 2\pi}{12} = \frac{5\pi}{6}$ radians, and $\frac{n}{12}$ of a rotation is $\frac{n \cdot 2\pi}{12} = \frac{n\pi}{6}$ radians.

5. missing values given counterclockwise from 0 radians: 0.52, 0.79, 1.05, 2.09, 2.36, 2.62, 3.67, 3.93, 4.19, 4.71, 5.24, 5.50, 5.76

Communicate

1. Radian measure reflects the distance along the circumference of the unit circle that is intercepted by a central angle. Thus, a complete rotation represents 2π radians. Degree measure is based on arbitrarily assigning a measure of 360° to a full rotation. You can change radian measure to degree measure, and vice versa. Degree measures are used most frequently for measures of geometric figures or directions, while radian measure is used most frequently for physical circular functions and problems.

2. Because a full rotation is 2π radians, one-third of a rotation is $\frac{2\pi}{3}$ radians.

3. 360° is equivalent to 2π radians, so 1° is equivalent to $\frac{2\pi}{360}$ or $\frac{\pi}{180}$ radians. To convert degrees to radians, multiply by $\frac{\pi}{180}$. To convert radians to degrees, multiply by $\frac{180}{\pi}$.

4. The phase shift from $f(x) = \cos x$ to $g(x) = \sin x$ is $\frac{\pi}{2}$ radians to the right. That is, $\sin x = \cos \left(x - \frac{\pi}{2} \right)$.

Practice & Apply

24. $-\frac{3\pi}{4}$ radians $= -135°$

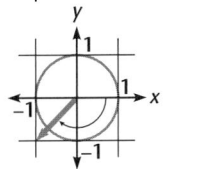

25. $450° = \frac{5\pi}{2}$ radians

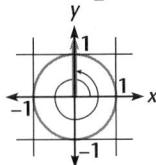

26. 6.28 radians $\approx 360°$

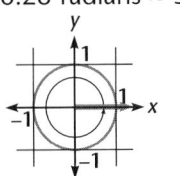

27. $158° \approx 2.75$ radians

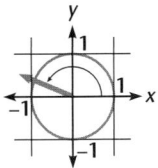

28. 1 radian $\approx 57°$

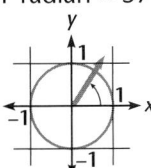

29. $-85° = -\frac{17\pi}{36}$ radians

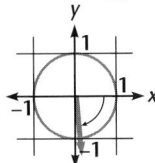

30. -10 radians $\approx -573°$

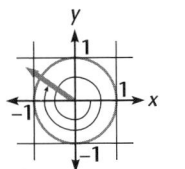

31. $-1440° = -8\pi$

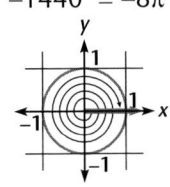

Lesson 8.6, Pages 468–474

Communicate

1. The area of a sector is the same fraction of the area of the circle as the central angle of the sector, θ, is of the central angle of the circle. The area of a circle is πr^2, and its central angle is 2π, so this gives $\frac{\text{sector area}}{\pi r^2} = \frac{\theta}{2\pi}$. So sector area $= \frac{\pi r^2 \theta}{2\pi}$, or $\frac{1}{2}\theta r^2$.

2. To find the length of an arc on a circle with a radius of 13, multiply the central angle that intercepts the arc (in radians) by the radius, 13.

3. First, find the distance that the object travels by multiplying the angle through which it travels (in radians) by the radius of the circle. Then you can find the velocity by dividing this distance by the time taken to travel the distance.

4. To find the area of a sector of a circle, multiply the product of the central angle of the sector (in radians) and the square of the radius of the circle by one half.

Look Beyond

45.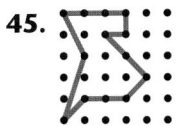

46.

Lesson 8.7, Pages 475–481

Communicate

1. It is best to use radian measure in functions that model changes over time because radian measure lets you use real numbers instead of degrees. In this case, the variable represents time, so the use of radians avoids confusing variable meanings.

2. As a increases for $a > 0$, the graph of $f(\theta) = a + \sin\theta$ is translated upward. As a decreases, the graph of $f(\theta) = a + \sin\theta$ is translated downward.

3. As $|b|$ increases, the graph of $f(\theta) = b\sin\theta$ increases in amplitude, that is, it is stretched vertically. As $|b|$ decreases, the graph of $f(\theta) = b\sin\theta$ decreases in amplitude, that is, it is compressed vertically.

4. As $|c|$ increases, the graph of $f(\theta) = \sin(c\theta)$ decreases in period, that is, it is compressed horizontally. As $|c|$ decreases, the graph of $f(\theta) = \sin(c\theta)$ increases in period, that is, it is stretched horizontally.

Practice & Apply

39. 1. intellectual: $f(x) = \sin\left(\frac{2\pi}{33}x\right)$;
emotional: $f(x) = \sin\left(\frac{2\pi}{28}x\right)$
2. The y-values must at least include those from $y = -1$ to $y = 1$.
3. They never simultaneously reach a maximum or a minimum, but they come closest to reaching a simultaneous maximum every $23 \cdot 28 \cdot 33 = 21{,}252$ days, or about once every 58 years. The same period holds for the closest approach to a simultaneous minimum.
4. Answers may vary.
5. Answers may vary.

Look Beyond

47. The periods are the same. The range of the reciprocal function is $y \geq 1$ and $y \leq -1$ instead of $1 \leq y \leq -1$. The domain of the reciprocal function excludes all values of x such that $x = \frac{\pi}{2} + n\pi$, for integer n. The reciprocal function has asymptotes at these values instead. The reciprocal function is not continuous like the cosine function. It increases where the cosine function decreases, and decreases where the cosine function increases. When the cosine function has a maximum, the reciprocal has a minimum, and when the reciprocal function has a maximum, cosine has a minimum.

Lesson 9.1, Pages 494–501

Communicate

1. $y = \frac{10}{x}$

2. $y = \frac{x}{k}$; it is equivalent to $y = \left(\frac{1}{k}\right)x$ which is a *direct* variation with constant of variation $\frac{1}{k}$.

3. Since $xy = 12$ for every pair of values (e.g. $3 \cdot 4 = 12$, $6 \cdot 2 = 12$), the constant of variation, k, is 12.

4. After entering the data, check to see that $r = -1$ when the data is fitted to a power regression model.

5. Yes; y varies inversely as the square root of x. The constant of variation is k.

Practice & Apply

14. $l = \frac{24}{w}$; answers on the table may vary. A sample is shown.

l	24	16	12	10	8	6	4	2	1
w	1	1.5	2	2.4	3	4	6	12	24

Lesson 9.2, Pages 502–508

Communicate

1. vertical asymptote is $x = -2$; horizontal asymptote is $y = 0$

2. vertical asymptote is $x = 0$; horizontal asymptote is $y = -4$

3. vertical asymptote is $x = 5$; horizontal asymptote is $y = 2$

4. vertical asymptote is $x = -1$; horizontal asymptote is $y = -6$

5. $f(x) = \frac{1}{x-3} + 2$

6. vertical asymptote is $x = 3$; horizontal asymptote is $y = 2$

Practice & Apply

26. Graph of C has a vertical asymptote at $x = -25$, and a horizontal asymptote at $y = 0$.

Look Back

28.

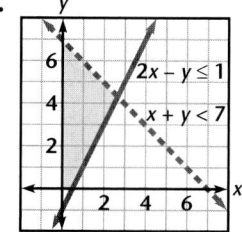

Answers may vary, but the points should be on the boundary of, or inside of, the quadrilateral whose vertices are $(0, 0)$, $(0.5, 0)$, $(2\frac{2}{3}, 4\frac{1}{3})$, and $(0, 7)$.

29.

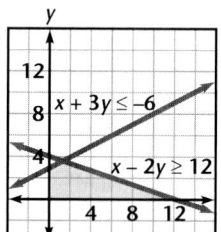

Answers may vary but the points should be on the boundary of, or inside of, the region whose vertices are $(0, 0)$, $(0, 3)$, $(1.2, 3.6)$, and $(12, 0)$.

Lesson 9.3, Pages 509–516

Exploration 1

	$f(-3)$	$f(-2)$	$f(-1)$	$f(0)$	$f(1)$	$f(2)$	$f(3)$
$f(x) = x^2$	9	4	1	0	1	4	9
$f(x) = x^3$	−27	−8	−1	0	1	8	27
$f(x) = x^4$	81	16	1	0	1	16	81
$f(x) = x^5$	−243	−32	−1	0	1	32	243
$f(x) = x^4 + x^2$	90	20	2	0	2	20	90
$f(x) = 3x^5 + x$	−732	−98	−4	0	4	98	732
$f(x) = 5x^4 - 3x^2$	378	68	2	0	2	68	378
$f(x) = x^4 - 2$	79	14	−1	−2	−1	14	79

1. $f(x) = x^2$, $f(x) = x^4$, $f(x) = x^4 + x^2$, $f(x) = 5x^4 - 3x^2$, and $f(x) = x^4 - 2$. These functions contain only even powers of x.

2. $f(x) = x^3$, $f(x) = x^5$ and $f(x) = 3x^5 + x$. These functions contain only odd powers of x.

3. If all the powers of x are even, then $f(-x) = f(x)$.

4. If all the powers of x are odd, then $f(-x) = -f(x)$.

Exploration 2

		Zeros of f	x-coordinate of the vertex
1.	$f(x) = (x + 4)(x - 3)$	–4 and 3	–0.5
2.	$f(x) = (x - 2)(x + 1)$	2 and –1	0.5
3.	$f(x) = (x + 3)(x + 3)$	–3	–3
4.	$f(x) = (x - 4)(x - 4)$	4	4

		Vertical asymptotes of f	Maximum point of f between the asymptotes
5.	$f(x) = \frac{1}{(x + 4)(x - 3)}$	$x = -4$, $x = 3$	(–0.5, –0.082)
6.	$f(x) = \frac{1}{(x - 2)(x + 1)}$	$x = 2$, $x = -1$	(0.5, –0.444)
7.	$f(x) = \frac{1}{(x + 3)(x + 3)}$	$x = -3$	none
8.	$f(x) = \frac{1}{(x - 4)(x - 4)}$	$x = 4$	none

9. The vertical asymptotes intercept the zeros.

10. The reciprocal function will have a local maximum if the quadratic function has a local minimum. The x-coordinate of the local maximum point of the reciprocal function is the same as the x-coordinate of the vertex (local minimum point) of the quadratic function.

11. The vertical asymptotes intercept the zeros.

12. No. A quadratic function with a zero of multiplicity 2 has the form $f(x) = a(x - b)^2$, where a and b are any real numbers and $a \neq 0$. The vertex of this function is at $(b, 0)$. The asymptote of the reciprocal function is the vertical line $x = b$, and since the graph of the reciprocal function increases in value on either side of its asymptote, there is no maximum.

Communicate

1. Factor the denominator of the f. The zeros of the denominator are the x-values of the asymptotes of f; $x^2 - 6x - 7 = (x - 7)(x + 1)$, so the vertical asymptotes of f are $x = 7$ and $x = -1$.

2. Answers may vary but should be equivalent to $f(x) = \frac{a}{b(x + c)^2}$

where a, b, and c are real numbers and a, $b \neq 0$.

3. All real numbers except 1 and –1.

4. Reflect the graph of $f(x) = \frac{1}{x^2}$ with respect to the x-axis.

5. even: $f(-x) = f(x)$; symmetric about the y-axis. Examples may vary but should include functions with only even powers of x; e.g. $f(x) = x^4 - 3x^2$ or $g(x) = x^6 + 8$. odd: $f(-x) = -f(x)$; symmetric about the origin. Examples may vary but should include functions with only odd powers of x.; e.g. $f(x) = \frac{1}{x^3}$ or $g(x) = x^9 - \frac{1}{x}$.

6. Answers may vary but should be of the form $f(x) = \frac{a}{b(x - 1)^p(x - 3)^q(x + 4)^r}$, where a and b are non-zero real numbers and p, q, and r are non-zero whole numbers.

Practice & Apply

22.

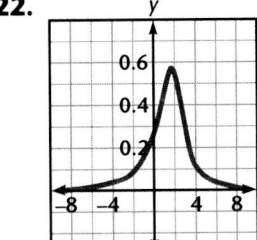

23.

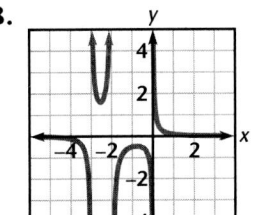

24. f and g are even, h and i are odd; the inverse of f is the relation $y = \pm\sqrt{x}$; the inverse of g is the relation $y = \pm\frac{1}{\sqrt{x}}$, where $x > 0$; $h^{-1}(x) = \sqrt[3]{x}$; $i^{-1}(x) = \frac{1}{\sqrt[3]{x}}$, $x \neq 0$.

25. h and i have inverses that are functions; both of these functions are odd.

30. 1. $i = \frac{k}{d^2}$, where i is the intensity of the light, d is the distance, and k is the constant of variation.
If k is 4, the graph would look like the following:

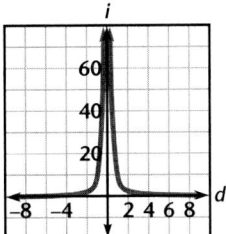

2. $i = \frac{600}{d^2}$

3. Answers may vary but the three points should be of the form $(d, \frac{600}{d^2})$.

4. The measure of intensity is never zero. As the distance becomes larger, the intensity becomes smaller. It approaches, but never reaches, zero. This

is the real-world meaning of the horizontal axis being an asymptote for the function $i = \frac{600}{d^2}$.

Lesson 9.4, Pages 517–521

Communicate

1. $x = 6$, $x = 1$

2. $x = -5$, $x = 5$

3. The degree of the denominator is equal to the degree of the numerator.

4. Answers may vary but should be equivalent to $f(x) = \frac{(x+2)p(x)}{(x+2)q(x)}$, where $p(x)$ and $q(x)$ are polynomials and neither has $x + 2$ as a factor. For example, $f(x) = \frac{(x+2)(x)}{(x+2)(x-2)}$.

5. The point $(-4, -0.125)$ is a hole in the graph of f.

6. A hole is a single point in the domain for which the function is not defined. When the function is factored and simplified, the curve becomes smooth at that point. A vertical asymptote is a line which the function approaches but never crosses; it cannot be removed by simplifying the function.

7. Divide the numerator by the denominator. The quotient is the value of the horizontal asymptote. In this example, $f(x) = \frac{x^2-1}{x^2-9} = 1 + \frac{8}{x^2-9}$, so the horizontal asymptote is $y = 1$.

Practice & Apply

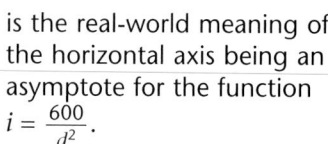

	Holes	Vertical asymptotes	Horizontal asyptotes	Domain
14.	none	$x = \frac{1}{4}$, $x = -2$	$y = 0.25$	$x \neq \frac{1}{4}$, -2
15.	none	$x = -1$	$y = 0$	$x \neq -1$
16.	none	$x = 1$, $x = -1$	$y = 0$	$x \neq \pm 1$
17.	none	$x = -1$, $x = -2$	$y = 0$	$x \neq -1$, -2
18.	$x = 3$	$x = -\frac{1}{2}$	$y = 0$	$x \neq -\frac{1}{2}$, 3
19.	$x = -5$	$x = 0$, $x = 3$	$y = 0$	$x \neq 0$, $3, -5$
20.	none	$x = 0$, $x = 2$	$y = 0$	$x \neq 0$, 2
21.	none	$x = -1$, $x = -2$, $x = -3$	$y = 1$	$x \neq -1$, $-2, -3$
22.	$x = 1$	$x = -3$	$y = 1$	$x \neq 1$, -3

23. No; the only value of x for which f is undefined is 1, but since the function is still undefined at $x = 1$ after the common factor, $x - 1$, has been divided out, $x = 1$ is just a vertical asymptote of f.

Lesson 9.5, Pages 524–527

Communicate

1. A rational equation is an equation that contains one or more rational expressions.

2. Rational expressions can be combined by finding the least common denominator (LCD) of the denominators of every expression.

3. Some rational equations can be solved by multiplying both sides of the equation by the LCD to eliminate the denominators of all the rational expressions in the equation. The resulting polynomial equation can then be solved by factoring in combination with the quadratic formula. All the resulting roots

must be checked in the original equation to ensure that no extraneous roots have been introduced.

4. An extraneous root is any root found in the solving process that does not make the original equation true when substituted into the original equation.

5. The possible roots may not be in the domain of the function or they may be extraneous roots.

Lesson 10.1, Pages 536–543

Exploration

1. No; it is half of a parabola since $y > 0$.

2. A parabola opening to the left with vertex at $(0, 0)$ and axis of symmetry $y = 0$.

3. $x = ay^2$ is not a function. For each value of x there are two values for y. Solve for y and, for simplicity, let $a = 1$. Now, $y = \pm\sqrt{x}$, where $x \geq 0$. To graph $x = y^2$ on the graphics calculator, enter two functions: $y = \sqrt{x}$ and $y = -\sqrt{x}$.

Communicate

1. vertex is at (h, k); focus is at $\left(h, k + \frac{1}{4a}\right)$; directrix is $y = k - \frac{1}{4a}$

2. vertex is at (h, k); focus is at $\left(h + \frac{1}{4a}, k\right)$; directrix is $x = h - \frac{1}{4a}$

3. Complete the square on y and solve for x: $x = \frac{1}{5}(y + 6)^2 - 3$. Since $h = -3$ and $k = -6$, the vertex is $(-3, -6)$, and the focus and directrix can be found using the expressions given in Exercise 2.

4. The parabola will open in the direction that the focus is from the directrix.

Practice & Apply

11. $V(0, 0)$; $F(2, 0)$; directrix: $x = -2$
$y^2 = 8x$

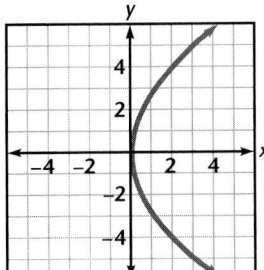

12. $V(0, 0)$; $F\left(-\frac{3}{4}, 0\right)$; directrix:
$x = \frac{3}{4}$
$y^2 = -3x$

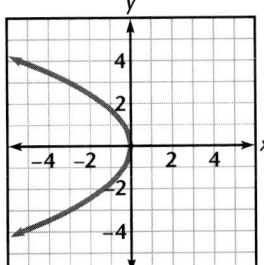

13. $V(0, 0)$; $F\left(0, -\frac{5}{4}\right)$; directrix:
$y = \frac{5}{4}$
$x^2 = -5y$

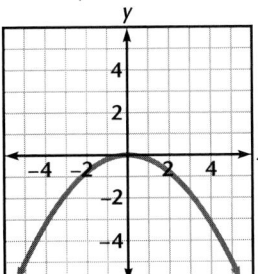

14. $V(0, 0)$; $F(0, 3)$; directrix: $y = -3$
$x^2 = 12y$

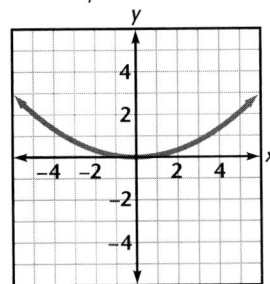

15. $V(-2, 1)$; $F\left(-2, \frac{5}{2}\right)$; directrix:
$y = -\frac{1}{2}$
$y = \frac{1}{6}(x + 2)^2 + 1$

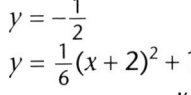 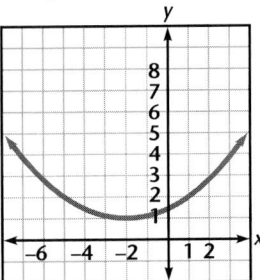

16. $V(4, 4)$; $F\left(4, \frac{17}{4}\right)$; directrix:
$y = \frac{15}{4}$
$y = (x - 4)^2 + 4$

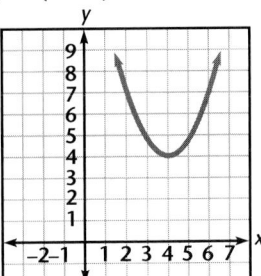

17. $V(-2, 2)$; $F\left(-\frac{1}{4}, 2\right)$; dir. $x = -\frac{15}{4}$
$x = \frac{1}{7}(y - 2)^2 - 2$

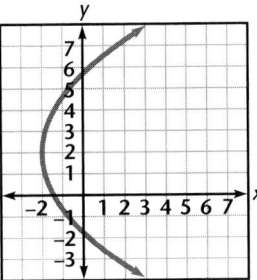

18. $V\left(-\frac{11}{16}, \frac{-3}{2}\right)$; $F\left(-\frac{27}{16}, \frac{-3}{2}\right)$;
directrix: $x = \frac{5}{16}$
$x = -\frac{1}{4}\left(y + \frac{3}{2}\right)^2 - \frac{11}{16}$

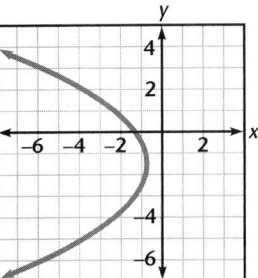

20. $y = \frac{-1}{46,875}(x - 7500)^2 + 1200$

Lesson 10.2, Pages 544–550

Communicate

1. $(x - h)^2 + (y - k)^2 = r^2$ where (h, k) are the coordinates of the center and r is the length of the radius.

2. Locate the center at (h, k); determine the radius and write the equation in the form $(x - h)^2 + (y - k)^2 = r^2$.

3. Complete the square for both x and y to write the equation in the form $(x - h)^2 + (y - k)^2 = r^2$. h and k are the center of the circle.

4. Solve for y:
$y = -4 \pm \sqrt{16 - (x - 1)^2}$
and graph two half circles.

Lesson 10.3, Pages 551–559

Communicate

1. Substitute the coordinates of the foci, F_1 and F_2, the constant sum d, and a point $P(x, y)$ on the ellipse into the equation $F_1P + F_2P = d$. Simplify the resulting equation to express it in the standard form for an ellipse.

2. $\frac{(x - h)^2}{a^2} + \frac{(y - k)^2}{b^2} = 1$, where a and b are the lengths of the major and minor axes, respectively.

3. The major axis is horizontal when the x-term has the larger

denominator. The major axis is vertical when the y-term has the larger denominator.

4. Place the foci at (3, 0) and (−3, 0) and use the distance formula as described in Exercise 1.

Practice & Apply

33. No; if the center of the highway is the origin, when x is 24, then y is 9 which is not high enough.

34. Answers may vary depending on the source used, and the accuracy of the data in the particular source. Have students provide their source and justify the accuracy of the data used. Information on the comet Kohoutek will be hardest to find. Have students refer to journals on astronomy for this information.

Look Back

39. $f \circ (g \circ h) = f(1 + \frac{2}{4 - x^2})$
$= -2 - \frac{4}{4 - x^2}$, and
$(f \circ g) \circ h = (-2 - \frac{4}{x}) \circ h$
$= -2 - \frac{4}{4 - x^2}$.

40. $f \circ (g \circ h) = f(3(7 - x) - 4)$
$= (3(7 - x) - 4)^2 + 3$
$= 9x^2 - 102x + 292$, and
$(f \circ g) \circ h = ((3x - 4)^2 + 3) \circ h$
$= (3(7 - x) - 4)^2 + 3$
$= 9x^2 - 102x + 292$

Look Beyond

48. Both graphs have two parts— they are both hyperbolas; the graph from Exercise 46 crosses the x-axis at 1 and −1, and the graph from Exercise 47 crosses the y-axis at 1 and −1.

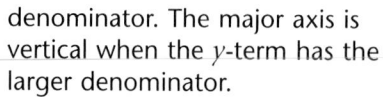
Lesson 10.4, Pages 560–568

Communicate

1. If the x^2 term is negative the transverse axis is vertical; if the y^2 term is negative, the transverse axis is horizontal.

2. The vertices are on the transverse axis at a distance a from the center (h, k). The foci are at a distance c from the center and the relationship $c^2 = a^2 + b^2$ can be used to find b. Then $\frac{(x - h)^2}{a^2} - \frac{(y - k)^2}{b^2} = 1$ or $\frac{(y - h)^2}{a^2} - \frac{(x - k)^2}{b^2} = 1$.

3. Use the coordinates of the foci to find the center. The distance from the center to either focus is c. Half of the constant difference is a. Use the relation $b^2 = a^2 - c^2$ to find b. Then the equation of the hyperbola is $\frac{(x - h)^2}{a^2} - \frac{(y - k)^2}{b^2} = 1$ or $\frac{(y - h)^2}{a^2} - \frac{(x - k)^2}{b^2} = 1$ where the center is (h, k).

4. The distance from center to foci is c, the distance from center to a vertex is a, and the distance from center to the end of the conjugate axis is b where $c^2 = a^2 + b^2$.

5. The asymptotes are lines through the diagonals of the rectangle formed by the points at the ends of the transverse and conjugate axes. They approach, but never touch, the graph of the hyperbola.

6. Since the sound takes 1 second longer to reach the second microphone, that microphone is 5000 feet farther from the whale. The microphones are 25,000 feet apart so, working in thousands, place the foci at (−12.5, 0) and (12.5, 0), and let the constant difference d be 5. Following the steps described in

Exercise 3, the possible locations of the whale are on the hyperbola
$\frac{4x^2}{25} - \frac{y^2}{150} = 1$.

Practice & Apply

21. $\frac{(x + 3)^2}{4} - \frac{(y - 2)^2}{2} = 1$; $C(-3, 2)$; axes 4 and $2\sqrt{2}$; foci: $(-3 \pm \sqrt{6}, 2)$

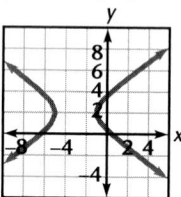

22. $\frac{(x - 3)^2}{4} - \frac{(y + 2)^2}{25} = 1$; $C(3, -2)$; axes 4 and 10; foci: $(3 \pm \sqrt{29}, -2)$

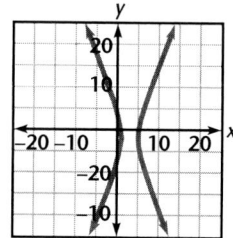

23. $(x - 1)^2 - (y + 2)^2 = 1$; $C(1, -2)$; axes are both 2; foci: $(1 \pm \sqrt{2}, -2)$

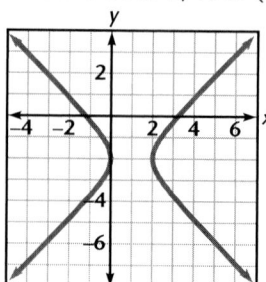

24. $\frac{(x - 2)^2}{9} - \frac{(y + 1)^2}{9} = 1$; $C(2, -1)$; axes are both 6; foci: $(2 \pm \sqrt{18}, -1)$

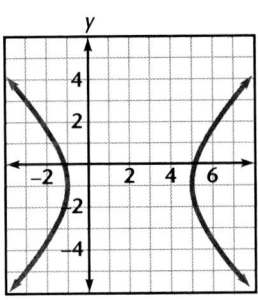

25. $\frac{(y-6)^2}{4} - \frac{(x+1)^2}{4} = 1$; $C(-1, 6)$;
axes are both 4; foci: $(-1, 6 \pm \sqrt{8})$

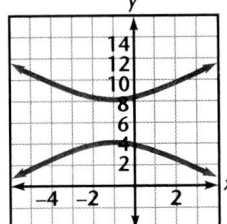

26. $\frac{(y+8)^2}{49} - \frac{3(x-2)^2}{49} = 1$, or
$\frac{(y+8)^2}{49} - \frac{(x-2)^2}{\frac{49}{3}} = 1$; $C(2, -8)$;
axes are 14 and $\frac{14}{\sqrt{3}}$; foci:
$\left(2, -8 \pm \sqrt{49 + \frac{49}{\sqrt{3}}}\right)$

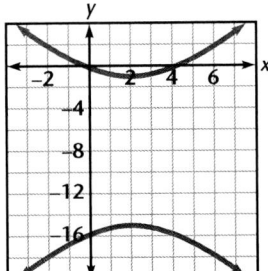

27. $\frac{(y+2)^2}{1} - \frac{(x-\frac{3}{2})^2}{\frac{1}{2}} = 1$, or $(y+2)^2 -$
$2(x-\frac{3}{2})^2 = 1$; $C\left(\frac{3}{2}, -2\right)$; axes are
2 and $\frac{2}{\sqrt{2}}$; foci: $\left(\frac{3}{2}, -2 \pm \sqrt{\frac{3}{2}}\right)$

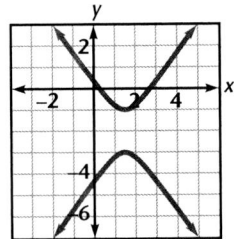

28. $\frac{(x-\frac{5}{2})^2}{\frac{3}{8}} - \frac{(y-\frac{3}{2})^2}{\frac{3}{20}} = 1$, or
$\frac{8(x-\frac{5}{2})^2}{3} - \frac{20(y-\frac{3}{2})^2}{3} = 1$;
$C\left(\frac{5}{2}, \frac{3}{2}\right)$; axes are $2\sqrt{\frac{2}{8}}$ and
$2\sqrt{\frac{3}{20}}$; foci: $\left(-\frac{5}{2} \pm \sqrt{\frac{21}{40}}, \frac{3}{2}\right)$

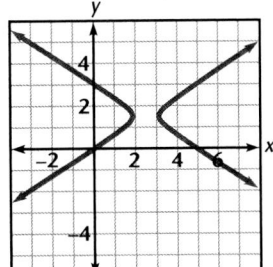

29. $\frac{(x-\frac{1}{2})^2}{\frac{1}{3}} - \frac{(y+\frac{1}{2})^2}{\frac{1}{3}} = 1$, or
$3\left(x-\frac{1}{2}\right)^2 - 3\left(y+\frac{1}{2}\right)^2 = 1$;
$C\left(\frac{1}{2}, -\frac{1}{2}\right)$; axes are both $\frac{2}{\sqrt{3}}$;
foci: $\left(\frac{1}{2} \pm \sqrt{\frac{2}{3}}, -\frac{1}{2}\right)$

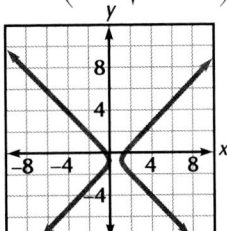

Lesson 10.5, Pages 569–577

Communicate

1. 0, 1, or 2 intersections

2. 0, 1, 2, 3, or 4 intersections

3. yes; a circle or an ellipse may touch one branch of a hyperbola and intersect the other branch in two points, an ellipse may be inside a circle just touching it at one vertex but intersecting it at two points on the other side of the circle.

4. none; an asymptote is a line that approaches, but never touches, a graph.

5. The graphs do not intersect.

Lesson 10.6, Pages 578–585

Exploration 1

1. ellipse

2. major axis length 6 units, minor axis 4 units

3. (0, 0)

4. $\frac{x^2}{4} + \frac{y^2}{9} = 1$

	length of major axis	length of minor axis	coordinates of center	equation in rectangular form
5.	6	6	(0, 0)	$\frac{x^2}{9} + \frac{y^2}{9} = 1$
6.	7	6	(0, 0)	$\frac{x^2}{9} + \frac{y^2}{12.25} = 1$
7.	7	6	(0, 0)	$\frac{x^2}{12.25} + \frac{y^2}{9} = 1$
8.	4	4	(0, 0)	$x^2 + y^2 = 16$

9. All have center at (0, 0) and are symmetrical about the *x*-axis and the *y*-axis. The graph in Steps 5 and 8 are circles; all the others are ellipses.

10. The major axis will be horizontal of length 8; the minor axis will have length 6. Student predictions should be correct: the length of the major axis is double the constant coefficient of cos *t* and the length of the minor axis is double the constant coefficient of sin *t*.

11. $x(t) = 2.5 \cos t$, $y(t) = 1.5 \sin t$

12. $x(t) = 6 \cos t$, $y(t) = 6 \sin t$

Exploration 2

1. parabola

2. $x(t)$ same; $y(t)$ different; both parabolas; the parabola in Step 1 opens downward and the parabola in Step 2 opens upward.

3. The equations for $x(t)$ and $y(t)$ have been switched; both are parabolas; the parabola in Step

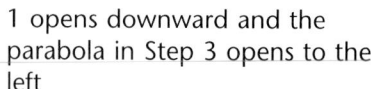

1 opens downward and the parabola in Step 3 opens to the left

4. $x(t) = 20t + 16t^2$, $y(t) = 20t$

5. B

6. A

7. D

8. C

Exploration 3

	starting point	direction
1.	(10, 0)	counterclockwise
2.	(–10, 0)	counterclockwise
3.	(0, 10)	clockwise
4.	(0, –10)	clockwise

5. The function, sine or cosine, and whether the coefficient is negative or positive affects the starting point of the graph. The direction of the graph is controlled by whether $x(t)$ is in terms of $\cos t$ or $\sin t$.

Exploration 4

1. The radius of the circle is determined by the constant coefficient a in the system $x(t) = a \cos t$ and $y(t) = a \sin t$.

2. (4, –9)

3. (–10, 8)

4. (–8, –3)

5. The x-coordinate is the constant term in $x(t)$; the y-coordinate is the constant term in $y(t)$.

Communicate

1. $x(t) = a \cos t + h$, $y(t) = b \sin t + k$

2. $x(t) = r \cos t + h$, $y(t) = r \sin t + k$

3. Ellipse: $\cos t = \frac{x(t)}{a}$, $\sin t = \frac{y(t)}{b}$; $\frac{x^2(t)}{a^2} + \frac{y^2(t)}{b^2} = 1$

Circle: $\cos t = \frac{x(t)}{r}$, $\sin t = \frac{y(t)}{r}$; $x^2(t) + y^2(t) = r^2$

Parabola: $x(t) = at$, $y(t) = at - bt^2$; $y(t) = x(t) - bt^2$.

4. If $y(t)$ contains t^2, the parabola opens upward; if $y(t)$ contains $-t^2$, the parabola opens downward; if $x(t)$ contains t^2, the parabola opens to the right; if $x(t)$ contains $-t^2$, the parabola opens to the left. If $x(t)$ has a constant term $+k$, translate the parabola horizontally by k units; if $y(t)$ has a constant term h, translate the parabola vertically by h units.

5. The function, sin or cos, and whether the coefficient is negative or positive; whether $x(t)$ is in terms of $\cos t$ or $\sin t$

Lesson 11.1, Pages 596–604

Exploration 1

1. & 2. Answers may vary.

3. $\frac{\text{number of heads}}{\text{number of tosses}}$; answers may vary.

4. 25; answers may vary; it is very likely that the experimental results would get closer to the estimate as the number of coin tosses increases.

Exploration 2

1. Each element in S corresponds to exactly one possible outcome; yes, the events are equally likely; a penny and a dime are each equally likely to land either heads or tails and the result of one coin does not affect the other.

2. a. *HH*
   b. *HH, HT, TH*
   c. *HH, HT, TH*
   d. *HH, HT, TH, TT*
   e. *HT, TH*
   f. *TT*

g. *TT*
h. Not possible—you cannot get 3 heads with the toss of 2 coins.

3. Yes; f and g; the events in each pair are the same.

4. a. $\frac{1}{4}$; **b.** $\frac{1}{2}$; **c.** $\frac{3}{4}$; **d.** 1; **e.** $\frac{1}{2}$; **f.** $\frac{1}{4}$; **g.** $\frac{1}{4}$; **h.** 0

5. $S = \{HH, HT, TH, TT\}$; The answers for repeating Steps 2–4 with two tosses of a penny are exactly the same as those for the original Steps 2–4. The sample spaces and theoretical probabilities for tossing any two coins and for tossing one coin twice are exactly the same.

Exploration 3

1. 81π, or approximately 254.47 in.^{2}

2. $\frac{\pi}{16}$, or approximately 0.196 in.^{2}

3. ≈ 0.0008; The probability of hitting a bull's-eye is theoretical.

4. $\approx 8.5 \times 10^{-6}$, or ≈ 0.0000085

Communicate

1. The term *experiment* refers to a test performed by carrying out an event over a repeated number of trials through a poll or through compiling research data. Examples include counting and recording the number of cars that complete a stop at a stop sign during a specific time period, or finding the height of each student in a particular grade at a particular school.

2. The term *probability of an event* refers to the likelihood that an event will occur.

3. Theoretically, the event is likely to occur 3 out of 4 times. The probability that the event will not occur is 1 out of 4 times. The sum of the events that will

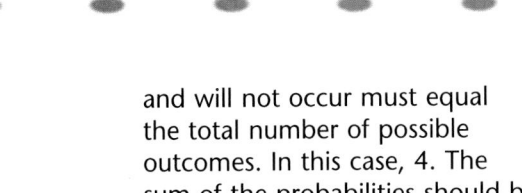

and will not occur must equal the total number of possible outcomes. In this case, 4. The sum of the probabilities should be 1.0.

4. Experimental probability is based on collected data; theoretical probability is a forecast about what should occur with equally likely events.

5. Experimental, since the prediction is based on the player's previous performance.

6. Johanna is incorrect, since her chances of winning are based on the number of tickets sold or printed. The greater the number of tickets sold, the lower the chances of winning.

7. Outcomes must be equally likely because the events must be considered independent, that is, not affected by any external factors.

Lesson 11.2, Pages 605–611

Communicate

1. A tree diagram gives a visual display of all of the possible choices.

2. The number of ways to make all 5 decisions is $a \cdot b \cdot c \cdot d \cdot e$.

3. A permutation is an arrangement of a set of n distinct objects in a specific order.

4. Answers may vary. One sample answer is ABC, CDA, ACE, CEB.

5. n is either greater than or equal to r. If n is equal to r, then the formulas are the same, but the variables are not technically being reversed. If n is greater than r, $r - n$ would be a negative number. The factorial of a negative number is not defined.

Practice & Apply

19. There are 8 arrangements.

p	q	r
T	T	T
T	T	F
T	F	T
T	F	F
F	T	T
F	T	F
F	F	T
F	F	F

20. There are 16 arrangements.

p	q	r	s	p	q	r	s
T	T	T	T	F	T	T	T
T	T	T	F	F	T	F	F
T	T	F	T	F	T	F	T
T	T	F	F	F	T	T	F
T	F	T	T	F	F	T	T
T	F	T	F	F	F	T	F
T	F	F	T	F	F	F	T
T	F	F	F	F	F	F	F

Lesson 11.3, Pages 612–617

Communicate

1. A permutation is the arrangement of a set of distinct objects in a specific order.

2. Yes, when $r_1! = r_2! = r_3! = \ldots = r_k! = 1!$, or 1.

3. No, if 8 people were seated with 2 people on each side of a square table, a shift of one person would create different pairs of people on each side of the table.

4. $r_1 + r_2 + \ldots \leq n$

Lesson 11.4, Pages 620–626

Communicate

1. In a permutation, order matters; in a combination, order does not matter.

2. $C(n, r) = \frac{P(n, r)}{r!}$. Both find the number of arrangements that are possible. Combinations reduce the total number of arrangements by $r!$ because order does not matter, whereas with permutations, order does matter.

3. Answers may vary. One possible answer is ABC, BCD, CDE, ACE.

4. $r \leq n$; therefore, $r - n < 0$ and $(r - n)!$ would be undefined, or $r = n$ and the formula has not really been changed.

5. $_nC_n = \frac{n!}{n!(n-n)!}$
$= \frac{n!}{n!0!}$
$= \frac{n!}{n!}$
$= 1$

6. $\frac{8!}{5!(8-5)!} = \frac{8!}{5!3!}$ and $\frac{8!}{3!(8-3)!} = \frac{8!}{3!5!}$;
$\frac{9!}{2!(9-2)!} = \frac{9!}{2!7!}$ and $\frac{9!}{7!(9-7)!} = \frac{9!}{7!2!}$;
$\frac{12!}{8!(12-8)!} = \frac{12!}{8!4!}$ and $\frac{12!}{4!(12-4)!} = \frac{12!}{4!8!}$

Lesson 11.5, Pages 627–634

Communicate

1. Two events are mutually exclusive if they cannot both occur at the same time.

2. Two events are independent if the occurrence of one event does not affect the occurrence of the other event.

3. Two mutually exclusive events are affected by each other in that if one occurs, the other cannot occur at the same time. Dependent events are events such that the occurrence of one has an affect on the occurrence of the other, but one *must* occur for the other to occur.

4. Answers may vary. For example, roll the blue cube and then roll the blue cube again.

5. Answers may vary. For example, rolling a 5 on the blue cube and rolling a 3 on the red cube.

6. Answers may vary. For example, rolling a 5 on the blue cube and rolling a 5 on the red cube when the experiment was to find the number of pairs of five's.

7. If A and B are mutually exclusive events, then if one event occurs, the other cannot occur at the same time, so $P(A \text{ and } B) = 0$.

Lesson 11.6, Pages 635–639

Exploration

1.

g_1g_2	g_2g_1	g_3g_1	g_4g_1	b_1g_1	b_2g_1
g_1g_3	g_2g_3	g_3g_2	g_4g_2	b_1g_2	b_2g_2
g_1g_4	g_2g_4	g_3g_4	g_4g_3	b_1g_3	b_2g_3
g_1b_1	g_2b_1	g_3b_1	g_4b_1	b_1g_4	b_2g_4
g_1b_2	g_2b_2	g_3b_2	g_4b_2	b_1b_2	b_2b_1

2. 30, 40%

3.

g_1g_2	g_2g_1	g_3g_1	g_4g_1
g_1g_3	g_2g_3	g_3g_2	g_4g_2
g_1g_4	g_2g_4	g_3g_4	g_4g_3
g_1b_1	g_2b_1	g_3b_1	g_4b_1
g_1b_2	g_2b_2	g_3b_2	g_4b_2

4. 20, 60%

5. There is a greater probability of selecting two good plugs if you know that the first plug is good.

Communicate

1. The outcome of one event affects the outcome of the other event.

2. They are dependent if the outcome of one event affects the outcome of the other; they are independent if the outcome of one event does not affect the outcome of the other.

3. The probability of event E_2 occurring given that event E_1 has occurred

4. $P(E_2|E_1)$ assumes that E_1 has occurred. $P(E_1 \text{ and } E_2)$ does not assume that E_1 has occurred.

5. If E_1 and E_2 are mutually exclusive, $P(E_2|E_1) = 0$.

Lesson 11.7, Pages 640–646

Communicate

1. A random number simulation involves using random numbers to model probabilities and then determining possible outcomes using these numbers.

2. int(10,000(6*rand)+1))/10,000

3. 1; the total probabilities of all outcomes is always 100%.

4. 65%; the probability of an event occurring plus the probability of an event not occurring is 100%.

5. It can be used to find the areas of circles, sectors, polygons, or any enclosed regions.

Practice & Apply

19. Answers may vary for a simulation.

Number	Random Numbers
0	01-06
1	07-14
2	15-28
3	29-47
4	48-62
5	63-73
6	74-81
7	82-88
8	89-94
9	95-98
10	99-100

21. Answers may vary, but the sum of the probabilities should be 1.00.

23. Answers may vary.

24. Answers may vary, but probabilities should be similar.

25. Answers may vary for the simulation

Number	Random Numbers	Cumulative Probability
0	0001-2418	0.2418
1	2419-5774	0.5774
2	5775-8588	0.8588
3	8589-9769	0.9769
4	9770-10,000	1.0000

26. Answers may vary.

27. Answers may vary but the probabilities from the simulation and the given probabilities should be similar.

28. Answers may vary. The area should be approximately 12.57, or about 4π, square units.

29. Answers may vary for the simulation

X	Random Numbers	Cumulative Probability
10	01-11	0.11
20	12-29	0.29
30	30-55	0.55
40	56-78	0.78
50	79-94	0.94
60	95-100	1.00

30. Answers may vary, but the probabilities should be similar.

Lesson 12.1, Pages 652–660

Exploration 1

1. $t_4 = 115$, $t_5 = 120$, $t_6 = 125$, $t_7 = 130$, $t_8 = 135$, $t_9 = 140$, $t_{10} = 145$

2. Each difference is 5.

3. The difference between each pair of consecutive terms is constant, $\$5$; $d = 5$.

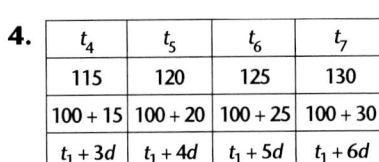

4.

t_4	t_5	t_6	t_7
115	120	125	130
$100 + 15$	$100 + 20$	$100 + 25$	$100 + 30$
$t_1 + 3d$	$t_1 + 4d$	$t_1 + 5d$	$t_1 + 6d$

t_8	t_9	t_{10}	...	t_n
135	140	145	...	
$100 + 35$	$100 + 40$	$100 + 45$	...	$100 + (n-1)5$
$t_1 + 7d$	$t_1 + 8d$	$t_1 + 9d$	...	$t_1 + (n-1)d$

$t_n = t_1 + (n-1)d$

5. $t_{15} = t_1 + (n-1)d = 100 + (15-1)5$
$\qquad\qquad\qquad = 100 + 70$
$\qquad\qquad\qquad = 170;$
The salary for the fifteenth week is $170.
Check: $t_{11} = 150$, $t_{12} = 155$, $t_{13} = 160$, $t_{14} = 165$, $t_{15} = 170$

6. Arithmetic; common difference of -19.

7. Not arithmetic; the differences are not constant.

8. Not arithmetic; the differences are not constant.

9. Arithmetic; common difference of $+2$.

Exploration 2

1. $t_5 = 80$, $t_6 = 160$, $t_7 = 320$, $t_8 = 640$, $t_9 = 1280$, $t_{10} = 2560$

2. Each ratio is 2.

3. The ratio between consecutive terms is the same; $r = 2$.

4.

4	5	6	7		n
t_4	t_5	t_6	t_7	...	t_n
40	80	160	320	...	$5 \cdot 2^{n-1}$
$5 \cdot 8$	$5 \cdot 16$	$5 \cdot 32$	$5 \cdot 64$	...	$5 \cdot 2^{n-1}$
$t_1 r^3$	$t_1 r^4$	$t_1 r^5$	$t_1 r^6$	...	$t_1 r^{n-1}$

$t_n = t_1 r^{n-1}$

5. $t_{15} = t_1 r^{n-1} = (5)(2)^{15-1} =$
$(5)(16,384) = 81,920;$
check: $t_{11} = 5120$, $t_{12} = 10,240$, $t_{13} = 20,480$, $t_{14} = 40,960$, $t_{15} = 81,920$

6. Not geometric; the constant difference is -19 so S_1 is arithmetic.

7. Geometric; common ratio is 0.5.

8. Geometric; common ratio is -1.

9. Not geometric; the common difference is a constant $+2$, so S_4 is arithmetic.

Communicate

1. Arithmetic sequences: a constant difference between each pair of consecutive terms;
Geometric sequences: a constant ratio between each pair of consecutive terms.

2. Answers may vary but should be of the form a, $a + d$, $a + 2d$, ..., $a + (n-1)d$, where a is the first term, d is the common difference, and n is the number of terms.

3. Answers may vary but should be of the form a, ar, ar^2, ar^3, ..., ar^{n-1}, where a is the first term, r is the common ratio, and n is the number of terms.

4. Answers will vary but they should *not* have either of the patterns described in Exercises 2 and 3. For example: 6, 9, 14, 21, 30,... . $t_n = n^2 + 5$ where $n \geq 1$

5. 10, 100, 1000, ... is a geometric sequence with $r = 10$ because each term is 10 times the previous term.

Lesson 12.2, Pages 661–670

Communicate

1. For the general, or nth term, add $(n-1)$ times the common difference, d, to the first term, t_1. In other words, $t_n = t_1 + d(n-1)$.

2. Multiply the average of the terms in the arithmetic sequence by the number of terms in the sequence to find the sum of the sequence.

3. The nth term would be $t_1 + 2(n-1)d$, where d is the first sequence's difference. This is $(n-1)d$ larger than the original nth term.

4. The nth term would be $t_1 + t_1 + (n-1)d$. This is t_1 more than the nth term of the original sequence.

5. In the graph of the sequence, changing t_1 results in a vertical shift of the graph; and changing d results in a change in the slope of the line that goes through the points in the sequence.

6. The sum is doubled.

7. The sum is increased by 20.

Lesson 12.3, Pages 671–679

Communicate

1. Multiply the first term by the common ratio $n-1$ times, to find the general, or nth, term.

2. Multiply the first term by 1 minus the common ratio raised to the nth power and then divide by 1 minus the common ratio.

3. The nth term would be 2^{n-1} times as large.

4. The nth term would be twice as large.

5. t_1 is a multiplication factor; r is an exponential factor.

Lesson 12.4, Pages 680–685

Communicate

1. Successive terms must get smaller in absolute value or the series will not converge to any one value but will keep getting larger in absolute value.

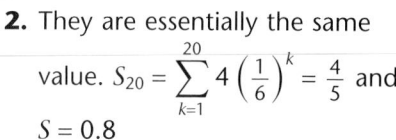

2. They are essentially the same value. $S_{20} = \sum_{k=1}^{20} 4\left(\frac{1}{6}\right)^k = \frac{4}{5}$ and $S = 0.8$

3. The $\frac{1}{6}$ would be multiplied by 1.8 to get 0.3, so the sum would be increased exponentially to ≈ 1.71.

4. For series that converge, the graph has a horizontal asymptote; for series that do not converge, the graph will keep increasing or decreasing as n gets larger.

Lesson 12.5, Pages 688–694

Exploration 1

1. Students' answers may vary. For example: 1, 2, 3, 4, 5, … is an arithmetic sequence, 1, 1, 1, 1, 1, 1 is a geometric sequence; 1, 5, 10, 10, 5, 1 is symmetric; row n has $n + 1$ terms.

2. & 3. The chosen number is equal to the sum of the two numbers closest to it in the row above.

4. Row 6: 1, 6, 15, 20, 15, 6, 1; Row 7: 1, 7, 21, 35, 35, 21, 7, 1

Exploration 2

1. $_2C_0$ $_2C_1$ $_2C_2$
 $_3C_0$ $_3C_1$ $_3C_2$ $_3C_3$

2. Row 8: 1, 8, 28, 56, 70, 56, 28, 8, 1
Row 9: 1, 9, 36, 84, 126, 126, 84, 36, 9, 1
Row 10: 1, 10, 45, 120, 210, 252, 210, 120, 45, 10, 1
Row 11: 1, 11, 55, 165, 330, 462, 462, 330, 165, 55, 11, 1
Row 12: 1, 12, 66, 220, 495, 792, 924, 792, 495, 220, 66, 12, 1
Row 13: 1, 13, 78, 286, 715, 1287, 1716, 1716, 1287, 715, 286, 78, 13, 1
Row 14: 1, 14, 91, 364, 1001,

2002, 3003, 3432, 3003, 2002, 1001, 364, 91, 14, 1
Row 15: 1, 15, 105, 455, 1365, 3003, 5005, 6435, 6435, 5005, 3003, 1375, 475, 115, 15, 1

3. It is possible to predict, from the triangular clusters of even numbers, where any even entry will be.

4. It is possible to predict, from the triangular arrangement of odd numbers, where any odd entry will be. Answers may vary. Sample answers: since the even numbers appear in clusters, it is easier to use the even-coded triangle.

5. For example Row 6:
  $_6C_0$ $_6C_1$ $_6C_2$ $_6C_3$ $_6C_4$
 $_6C_5$ $_6C_6$
For each row, in $_nC_r$, n is the row number and r increases from 0 to n. The pattern repeats every four rows. If n is a whole number, $4n$ rows have odd numbers in the first and last entries; $4n + 1$ rows have odd numbers in the first, second, second from last, and last entries; $4n + 2$ rows have odd numbers in the first, third, third

from last, and last entries; $4n + 3$ rows have odd numbers in the first four and last four entries.

Communicate

1. Answers may vary. A sample answer is: the subscript n is the row number, and the subscripts r are the whole numbers from 0 to n, increasing from left to right.

2. Answers may vary but should include that the first and last entries are always 1, the second and second last entries are always n.

3. 6th row, fifth entry = 15; total of row entries = 64; probability = $\frac{15}{64}$ = 23.4% chance that four out of six coins tossed will be tails. (Use the fifth entry because the first entry is the number of zero tails.)

4. $n = 6$, $r = 4$

5. 43rd row, 20th entry divided by the sum of the entries in the 43rd row will give the probability that 20 out of 43 coins will land heads up.

Practice & Apply

14. The patterns form inverted triangular groups of three starting on rows divisible by three.

```
                              1
                           1     1
                        1     2     1
                     1     3     3     1
                  1     4     6     4     1
               1     5    10    10     5     1
            1     6    15    20    15     6     1
         1     7    21    35    35    21     7     1
      1     8    28    56    70    56    28     8     1
   1     9    36    84   126   126    84    36     9     1
 1    10    45   120   210   252   210   120    45    10     1
1    11    55   165   330   462   462   330   165    55    11     1
1  12   66  220  495  792  924  792  495  220   66   12   1
1  13  78  286  715 1287 1716 1716 1287 715  286  78  13   1
1  14  91  364 1001 2002 3003 3432 3003 2002 1001 364  91  14   1
```

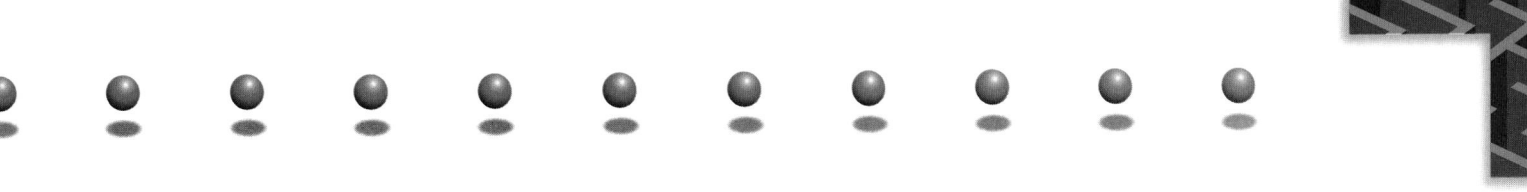

15. The patterns form inverted triangular groups of ten starting in rows divisible by 5.

```
                           1
                        1     1
                     1     2     1
                  1     3     3     1
               1     4     6     4     1
            1     5    10    10     5     1
         1     6    15    20    15     6     1
      1     7    21    35    35    21     7     1
   1     8    28    56    70    56    28     8     1
1     9    36    84   126   126    84    36     9     1
   1    10    45   120   210   252   210   120    45    10     1
   1    11    55   165   330   462   462   330   165    55    11     1
 1    12    66   220   495   792   924   792   495   220    66    12     1
1    13    78   286   715  1287  1716  1716  1287   715   286    78    13     1
1    14    91   364  1001  2002  3003  3432  3003  2002  1001   364    91    14     1
```

16. The second and second from last entries in prime rows are prime.

```
                           1
                        1     1
                     1     2     1
                  1     3     3     1
               1     4     6     4     1
            1     5    10    10     5     1
         1     6    15    20    15     6     1
      1     7    21    35    35    21     7     1
   1     8    28    56    70    56    28     8     1
1     9    36    84   126   126    84    36     9     1
   1    10    45   120   210   252   210   120    45    10     1
   1    11    55   165   330   462   462   330   165    55    11     1
 1    12    66   220   495   792   924   792   495   220    66    12     1
1    13    78   286   715  1287  1716  1716  1287   715   286    78    13     1
1    14    91   364  1001  2002  3003  3432  3003  2002  1001   364    91    14     1
```

21. 3rd from the last entry in 12th row; second and second from the last entries in 66th row

Lesson 12.6, Pages 695–700

Exploration

1. 5; one more than 4

2. 11; $n + 1$

3. Powers of a decrease by 1; while powers of b increase by 1.

4. 4

5. Fourth row of Pascal's Triangle

6. The expansion gives the number of ways to get each combination of Hs and Ts.

Communicate

1. The nth row of Pascal's Triangle gives the coefficients of the terms of $(a + b)^n$.

2. Write combinations of $(2a)$ and $(-3b)$ starting with $_5C_0(2a)^5(-3b)^0$, using the coefficients in the fifth row of Pascal's triangle, with powers of $(2a)$ decreasing from 5 to 0 and powers of $(-3b)$ increasing from 0 to 5, and ending with $_5C_5(2a)^0(-3b)^5$.

3. The coefficient of H^3T in $(H + T)^4$

is $_4C_3$, or 4; $2^4 = 16$, so the probability is $\frac{4}{16}$ or $\frac{1}{4}$.

4. This gives a series when values are substituted for n from $n = 0$ to $n = 6$ and is equivalent to the binomial expansion $(a + b)^6$.

5. $(a + b)^6$

Lesson 13.1, Pages 710–721

Communicate

1. • Mean: the sum of all the data values divided by the number of data values
   • Median: the middle value of a data set when the data are arranged in numerical order
   • Mode: the data value which occurs most often

2. If some values are extremely small or large when compared to other values, the mean may not give the best representation of the central tendency.

3. Ungrouped: the frequency of each value is recorded. Grouped: only the frequencies of distinct classes of values are recorded.

4. The values 20, 25, 30, 35, 40, 45, 50, and 55 are each in two different classes.

5. Separate the range of the data into several grouped classes and determine the frequency of each class. Find the midpoint of each class and multiply by the frequency of that class. Add all of these products and divide by the total number of data values to determine the grouped class mean.

6. List two 3s for the mode. Since the median is 6, then 6 is the central number. Now we have 3, 3, 6, $\underline{a}$, $\underline{b}$. The mean of all of these numbers is 7, so $\frac{3+3+6+a+b}{5} = \frac{12+a+b}{5}$, or $7 \cdot 5 = 12 + a + b$. Thus, the sum of a and b must be 23. Thus, a can be 10 and b can be 13. Now we have a list of 5 numbers, 3, 3, 6, 10, and 13, whose mode is 3, whose median is 6, and whose mean is 7.

Lesson 13.2, Pages 722–729

Communicate

1. Construct the horizontal and vertical axes. Label the vertical axis with the frequencies, and label the horizontal axis with the data values or data classes. Construct vertical bars with each base centered over the corresponding data value or data class and so that each bar has the height of the frequency of its data value or class.

2. Split the numbers into a stem and a leaf. Let the stems represent the tens digits and the leaves represent the units digits. Draw a vertical line, and on the left of this line, list the stems (or tens digits) in numerical order from top (lowest) to bottom (highest). Next to each stem, on the right side of the vertical line, list the leaves (the ones digits) that correspond to that stem. The leaves can be listed in numerical order from largest to smallest, but this is not necessary.

3. Which specific data values appear most often. In general, any information pertaining to individual data values.

4. Determine which leaf on a particular stem occurs the most often; this stem and leaf represent the mode.

Practice & Apply

10.

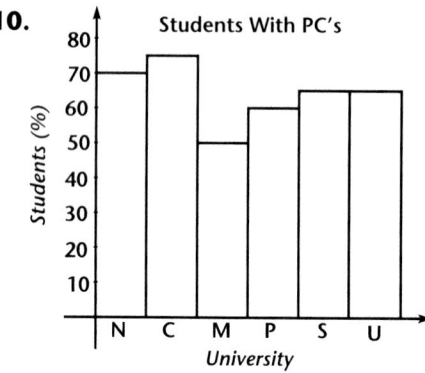

21.

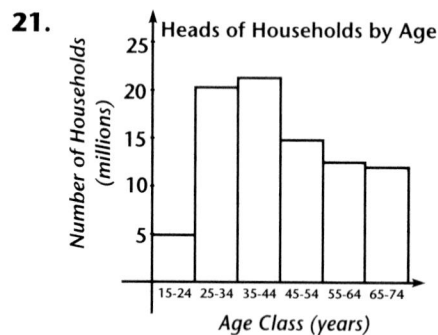

22. Answers may vary. Examples are: most heads of households are older than 24 years of age, the largest number of household heads are in the 25–44 years-of-age class.

25. Yes, the skew is slightly towards the right; the number of heads of households in an age group decline as the ages increase.

Lesson 13.3, Pages 730–737

Communicate

1. The 90th percentile indicates that 90% of the data are at or below this number; therefore, 10% are above this number.

2. 25%, 50%, and 75%, respectively

3. Arrange the data in order. Find the median of the data set; this is the 2nd quartile, Q_2. Determine the median of the data below Q_2;

this the 1st quartile, Q_1. Then find the median of the data above Q_2; this is the 3rd quartile, Q_3.

4. The interquartile range is the difference, $Q_3 - Q_1$; it measures the data range from 25% to 75% of the data set; the interquartile range contains 50% of the data.

5. The box displays the interquartile range; the vertical line in the box indicates the median of the data; and the whiskers extend below the box to the least data value and above the box to the greatest data value.

6. Draw a rectangle with the ends at Q_1 and Q_3. Draw whiskers that extend from Q_1 to the lowest data value and from Q_3 to the highest data value. Include the median of the data set by drawing a vertical line at this value inside the rectangular box.

Look Beyond

57.

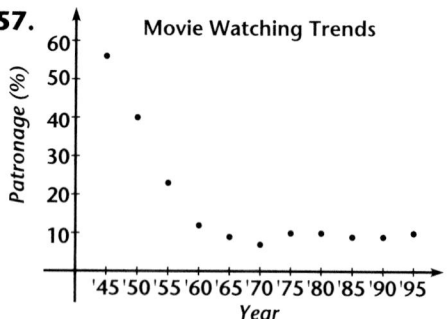

mean = 17.7%, median = 10%, mode = 9% and 10%, spread = 49%; from 1945 to 1975: decrease in movie watching activity; from 1975: the activity leveled off; Answers may vary. Examples are: How many movies are produced each year?, How many theaters have opened?, How many televisions are in each home?

Lesson 13.4, Pages 740–745

Communicate

1. A measure of dispersion is a numerical value that indicates how much spread the data values have from a central number, usually the mean.

2. The range is a measure of dispersion because it gives the absolute value of the difference between the largest and smallest value in a data set.

3. To find the mean deviation, take the sum of the absolute values of the differences between each value and the mean and divide by the number of values in the data set. To determine the standard deviation of a set of data, find the sum of squared differences of each value and the mean. Divide the sum by the number of values in the data set, then take the square root of the quotient.

4. The standard deviation squares the differences from the mean to make the differences positive. The mean deviation takes the absolute value of the differences from the mean to make the differences positive. Both statistics yield positive results and are measures of dispersion or variation around the arithmetic mean. The standard deviation is considered to be the better measure of dispersion because the sum of squared deviations about the mean is always less than the value found with any other measure of central tendency. Furthermore, the standard deviation can be used for other statistical calculations.

5. Answers may vary. An example is: 10, 50, 90 and 45, 50, 55. Both sets have a mean of 50 and a median of 50. However, their standard deviations are 40 and 5, respectively. This indicates that the second set is more consistent in data values.

6. The mean is 71.45, the median is 75, the mean deviation is 16.36, and the standard deviation is 20.67. The standard deviation is the best measure because, to determine its value, the mean must be calculated and the number of values counted. Also, the differences of each value from the mean is squared, then the square root is taken, thus giving the best value for the amount of clustering about the mean.

Lesson 13.5, Pages 746–752

Exploration 1

1.

Number Correct	Probabilities
3	$P(3 \text{ correct}) = 0.015625$
2	$P(2 \text{ correct}) = 3p^2q = 3(0.25)^2(0.75) = 0.140625$
1	$P(1 \text{ correct}) = 3pq^2 = 3(0.25)(0.75)^2 = 0.421875$
0	$P(0 \text{ correct}) = 1q^3 = (0.75)^3 = 0.421875$

2. In binomial expansion $(p + q)^3$, 2 successes in 3 trials corresponds to the term $3p^2q$, where p is the probability of a success, and q is the probability of a failure. Substitute 0.80 and 0.20 for p and q, respectively, and solve. The probability of 2 successes out of 3 is ≈ 0.384, or $\approx 38\%$.

Exploration 2

1.

Number of successes	Probabilities
4	$P(4 \text{ successes}) = p^4 = (0.4)^4 = 0.0256$
3	$P(3 \text{ successes}) = 4p^3q = 4(0.4)^3(0.6) = 0.1536$
2	$P(2 \text{ successes}) = 6p^2q^2 = 6(0.4)^2(0.6)^2 = 0.3456$
1	$P(1 \text{ success}) = 4pq^3 = 4(0.4)(0.6)^3 = 0.3456$
0	$P(0 \text{ successes}) = q^4 = (0.6)^4 = 0.1296$

2. The probability of getting 2 successes in 4 trials is ≈ 0.1536, or $\approx 15\%$.

Communicate

1. There are only two possible outcomes; these outcomes may or may not have an equal probability of occurring.

2. Each probability has a coefficient that is the same as its corresponding coefficient in the binomial theorem.

3. They are the same number.

4. The probability of giving a correct response on a true-false question is 0.5 if you guess on each response. This is also the probability of giving an incorrect response. To determine the probability of getting 3 out of 5 questions correct, find the number of combinations of getting 3 correct questions out of 5, $_5C_3$, or 10. Multiply this value by the probability of 3 correct

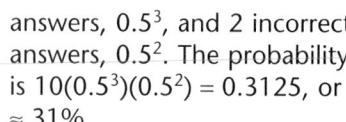

answers, 0.5^3, and 2 incorrect answers, 0.5^2. The probability is $10(0.5^3)(0.5^2) = 0.3125$, or $\approx 31\%$.

5. The probability of giving a correct answer is 0.25 when you are guessing on each question. The probability of giving an incorrect response is 0.75. To determine the probability of getting 3 out of 5 multiple choice questions correct, find the number of combinations of getting 3 correct questions out of 5, $_5C_3$, or 10. Multiply this value by the probability of 3 correct questions, $(0.25)^3$, and by the probability of 2 incorrect questions, $(0.75)^2$. The probability is $10(0.25)^3(0.75)^2 \approx 0.088$, or $\approx 9\%$.

Lesson 13.6, Pages 753–761

Communicate

1. Answers may vary. Examples are weights of new born babies and hat sizes.

2. Answers may vary but may include the following:
- The distribution is bell-shaped.
- It is symmetrical about a vertical line going through the mean value of the distribution.
- It describes the theoretical outcome of an infinite number of trials in a binomial experiment.
- Approximately 99% of the data values lie within three standard deviations of the mean.

3. The area within one standard deviation of the mean represents a 68% probability that a value will fall within this range. The area within two standard deviations of the mean represents a 95% probability that a value will fall within this range. The

area within three standard deviations represents a 99% probability that a value will fall within this range.

4. In general, use the $\int f(x)dx$ function on your calculator to find the area under the curve from $x = z$ to four standard deviations above the mean, or from $x = z$ to four standard deviations below the mean, depending on the probability desired.

5. Find the difference between x and $\bar{x}$ and divide by σ. This is the z-score.

Lesson 14.1, Pages 770–776

Exploration

1–3. Answers may vary.

4. The values should be the same.

5. $\dfrac{\sin A}{a} = \dfrac{\sin B}{b} = \dfrac{\sin C}{c}$

Communicate

1. In a triangle, the ratio of the sine of any angle to the length of the side opposite that angle is constant.

2. To be able to solve the proportion, at least one angle and the length of the side opposite that angle must be known so that the constant ratio $\dfrac{\sin A}{a}$ can be calculated.

3. ASA (Angle-Side-Angle) and ASS (Angle-Side-Side)

4. If ASA is given, then the size and shape of the triangle is unique. If ASS is given, then sometimes two possible triangles exist, one with one obtuse angle and one with three acute angles.

5. The original non-right triangle is split into two right triangles. Then

the basic sine ratio is used in each right triangle. These sine ratios are equated to form the Law of Sines.

Lesson 14.2, Pages 777–783

Communicate

1. When two sides and the included angle are given, the Law of Cosines is used to find the measure of the side opposite the given angle. When three sides are given, the Law of Cosines is used to find the measures of each of the angles.

2. Since $\cos 90° = 0$, the Law of Cosines becomes $a^2 = b^2 + c^2$. This is the "Pythagorean" Right-Triangle Theorem.

3. If $m\angle C < 90°$ then $0 < \cos C < 1$, and therefore $2ab \cos C$ is a positive quantity. So $a^2 + b^2 - 2ab \cos C < a^2 + b^2$.

4. If $m\angle C > 90°$, then $-1 < \cos C < 0$, and therefore $2ab \cos C$ is a negative quantity. So $a^2 + b^2 - 2ab \cos C > a^2 + b^2$

5. The Law of Cosines is needed to find an angle measure when SSS (Side-Side-Side) is given, or to find the third side when SAS (Side-Angle-Side) is given. The Law of Sines can be used to find a second angle when SSA is given or to find a second side when ASA is given. If three sides are given, the Law of Cosines must be used first. If two sides and one angle are given, and if the given angle is not contained by the two given sides, then the Law of Cosines must be used.

Lesson 14.3, Pages 784–789

Communicate

1. $\tan t = \frac{\sin t}{\cos t}$

2. $\cot t = \frac{\cos t}{\sin t}$

3. $\sec t = \frac{1}{\cos t}$

4. $\csc t = \frac{1}{\sin t}$

5. If $\sin x$ is negative, then x is in Quadrant 3 or 4. If $\cos x$ is positive, then x is in Quadrant 1 or 4. If both conditions apply, the angle must lie in Quadrant 4.

6. For a point (x, y) on the unit circle, $x = \cos\theta$, $y = \sin\theta$, and $c = 1$. Relating the sides of the right triangle : $\sin^2\theta + \cos^2\theta = 1$. Dividing each term by $\sin^2\theta$ gives the Pythagorean identity $1 + \cot^2\theta = \csc^2\theta$. Dividing each term of $\sin^2\theta + \cos^2\theta = 1$ by $\cos^2\theta$ gives the Pythagorean identity $\tan^2\theta + 1 = \sec^2\theta$.

Lesson 14.4, Pages 790–795

Exploration 1

1. F

2. F

3. F

4. T

5. F

6. F

7. T

8. F

9. Graph the two functions, $y =$ left-side of the equation, $y =$ right-side of the equation. Since the same graph is drawn, the two functions must be equal.

10. $\sin(A + B) = \sin A \cos B + \cos A \sin B$, $\cos(A + B) =$

$\cos A \cos B - \sin A \sin B$

Exploration 2

1. Even; graph of $y = \cos(-x)$ is the same as the graph of $y = \cos x$

2. Odd; graph of $y = -\sin x$ is the same as the graph of $y = \sin(-x)$

3. $\cos(A - B) = \cos(A + (-B)) =$
$\cos A \cos(-B) - \sin A \sin(-B) =$
$\cos A \cos B - (-\sin A \sin B) =$
$\cos A \cos B + \sin A \sin B$

4. $\sin(A - B) = \sin(A + (-B)) =$
$\sin A \cos(-B) + \cos A \sin(-B) =$
$\sin A \cos B + (-\cos A \sin B) =$
$\sin A \cos B - \cos A \sin B$

5. $\sin(A - B) = \sin A \cos B - \cos A \sin B$, $\cos(A - B) = \cos A \cos B + \sin A \sin B$

Exploration 3

1. $\cos(2A) = \cos(A + A) =$
$\cos A \cos A - \sin A \sin A =$
$\cos^2 A - \sin^2 A$

2. $\sin(2A) = \sin(A + A) = \sin A \cos A + \cos A \sin A = 2\sin A \cos A$

3. $\cos(2A) = \cos^2 A - \sin^2 A$, $\sin(2A) = 2\sin A \cos A$

Communicate

1. Answers may vary. A sample answer is: $\sin\left(\frac{\pi}{6} + \frac{\pi}{3}\right) =$ $\sin\left(\frac{\pi}{2}\right) = 1$, but $\sin\frac{\pi}{6} + \sin\frac{\pi}{3} =$ $\frac{1}{2} + \frac{\sqrt{3}}{2} = \frac{1 + \sqrt{3}}{2} \neq 1$.

2. Answers may vary. A sample answer is: $\cos\left(\frac{\pi}{6} + \frac{\pi}{3}\right) =$ $\cos\left(\frac{\pi}{2}\right) = 0$, but $\cos\frac{\pi}{6} + \cos\frac{\pi}{3} =$ $\frac{\sqrt{3}}{2} + \frac{1}{2} = \frac{\sqrt{3} + 1}{2} \neq 0$.

3. In $A + B$, replace B with $-B$ and remember that $\sin(-B) = -\sin B$ and $\cos(-B) = \cos B$

4. $\sin 2A = \sin(A + A) = \sin A \cos A + \cos A \sin A = 2\sin A \cos A$;

$\cos 2A = \cos(A + A) =$
$\cos A \cos A - \sin A \sin A = \cos^2 A - \sin^2 A$

5. Let x be $\frac{x}{2}$ in the double angle identity $\cos 2x = \cos^2 x - \sin^2 x$. So, $\cos 2\frac{x}{2} = \cos x = \cos^2 \frac{x}{2} - \sin^2 \frac{x}{2}$. Solve for $\cos \frac{x}{2}$. Students will need to use a Pythagorean identity to eliminate the sine term.

6. Since $\left(\sin\frac{x}{2}\right)^2 + \left(\cos\frac{x}{2}\right)^2 = 1$, if you knew the identity for $\cos\frac{x}{2}$, it could be substituted into the first equation which could then be solved for $\sin\frac{x}{2}$.

Practice & Apply

38. $\sin(A + B) + \sin(A - B) =$
$(\sin A \cos B + \cos A \sin B) +$
$(\sin A \cos B - \cos A \sin B) =$
$2\sin A \cos B$

39. Given $a = A + B$ and $b = A - B$, by addition $A = \frac{a + b}{2}$ and by subtraction $B = \frac{a - b}{2}$. Using the result from Exercise 38, $\sin a + \sin b = \sin(A + B) + \sin(A - B) = 2\sin A \cos B = 2\sin\left(\frac{a + b}{2}\right)\cos\left(\frac{a - b}{2}\right)$.

Lesson 14.5, Pages 796–801

Communicate

1. Graph $y = a\cos(bt)$ and $y = c$ and find the points of intersection.

2. Usually the first solution found will be in the second quadrant. Determine $2\pi - x$ to find the solution that lies in the third quadrant.

3. Graph the function $y = \tan^3 t + 2\tan^2 t + \tan t - 1$, and find its zeros; or solve the equation algebraically, letting $x = \tan t$.

4. Answers may vary but should be of the form $a\cos^2 t + b\cos t + c = 0$, where a, b, and c are real numbers and $a \neq 0$. Graph the equation and find its zeros to find the values of t that solve the equation.

5. Trigonometric identities can enable you to express all the terms of an equation in terms of a single trigonometric function.